Peptides

Chemistry, Structure and Biology

Peptides

Chemistry, Structure and Biology

Proceedings of the Eleventh American Peptide Symposium
July 9-14, 1989, La Jolla, California, U.S.A.

Edited by

Jean E. Rivier
The Salk Institute for Biological Studies
La Jolla, California 92037, U.S.A.

and

Garland R. Marshall
Department of Pharmacology
Washington University School of Medicine
St. Louis, Missouri 63110, U.S.A.

ESCOM ▪ Leiden ▪ 1990

CIP-Data Koninklijke Bibliotheek, Den Haag

Peptides

Peptides : Chemistry, Structure and Biology : Proceedings of the Eleventh American Peptide Symposium, July 9-14, 1989, La Jolla, CA, USA / ed. by Jean E. Rivier and Garland R. Marshall. – Leiden : ESCOM. – Ill.
With index, ref.
SISO 546 UDC 547.96(063)
Subject heading: Proteins.

ISBN-13:978-94-010-9062-9   e-ISBN-13:978-94-010-9060-5
DOI: 10.1007/978-94-010-9060-5

Published by:

ESCOM Science Publishers B.V.
P.O. Box 214
2300 AE Leiden
The Netherlands

Preface

The Eleventh American Peptide Symposium was held on the San Diego campus of the University of California in picturesque La Jolla on July 9-14, 1989. More than 1100 participants from around the world came to attend a scientific program comprised of an assortment of oral presentations, posters and exhibits. Most rewarding to us was the regular attendance at all the scientific sessions by our registered guests despite the beautiful scenery offered by La Jolla and San Diego county. Indeed for all of us, the purpose of the biennial American Peptide Symposium is to try to keep up with a fast evolving and fascinating field. The goal of the proceedings of these meetings is to provide the scientific community unable to attend some or all of the Eleventh American Peptide Symposium with a synopsis of current results in peptide research. Of 65 oral presentations and approximately 390 poster displays, the Program Committee selected 393 articles for publication. A few manuscripts by distinguished invited speakers were unfortunately not received in time to appear in these proceedings. All manuscripts herein were selected on the basis of their originality at the time of presentation and their significance in the development of peptide-related research.

As in the past, this Eleventh meeting brought together scientists with a wide range of expertise, which was harnessed to address and solve basic as well as applied scientific problems in broad areas.

As expected, synthetic aspects of peptide chemistry, together with structural biology, gathered the largest number of contributions. Most important were the recognition of the usefulness of a series of new coupling reagents and, possibly for the first time, credible evidence of the power of solid phase peptide synthesis for the synthesis of small proteins that can ultimately be purified. Persistence in the study of enzymes, including lipases, for peptide bond formation yielded unique and fascinating results albeit with limited applicability to date.

The development of more potent antagonists of such peptides as bombesin, gonadotropin releasing hormone, parathyroid hormone or vasopressin and oxytocin, should be recognized for its intellectual value and potential impact on human welfare. Peptide agonists that release growth hormone, whether related to growth hormone releasing factor or derived from enkephalins, may also in the near future become an integral part of the armamentarium of our medical and veterinary community. Approaches to the design of more potent peptides often take advantage of the introduction of unnatural amino acids and peptide bond mimetics that may render specific bonds resistant to enzymic degradation, or of conformational constraints brought about by substitutions on the peptide backbone or through side-chain–side-chain covalent bonds (GnRH analogs, CCK analogs, cyclosporin A analogs and others). More than ever, most analog design was, in some ways, directed through the use of molecular modeling and/or spectroscopic analyses.

Major developments in analytical chemistry, as applied to peptides and small proteins, were illustrated by original contributions in the field of mass spectroscopy and capillary zone electrophoresis.

Design of enzyme inhibitors with unique specificity and high affinity, used as tools to modulate pathological conditions, remain an art despite a better knowledge of a given enzyme's structures and mechanisms of action. Tremendous progress in this area – including the search for inhibitors of the HIV protease – was reported.

New peptide structures and peptide or protein precursors isolated from different tissues, including those from mammals, marine organisms and insects, were described; some of these peptides are still in need of a function, yet development of antibodies and localization studies can often guide the physiologists in their search.

As in the recent past, considerable attention was dedicated to new approaches for structural analysis and their impact on the determination of peptide and protein folding: among the structures studied, models for those of the leucine zippers and zinc fingers were reported.

A few pioneering reports on the synthesis of peptide mimetics (peptide azoles or carbohydrate-derived peptidomimetics) point in a direction that most peptide chemists interested in developing new drugs may have to take in order to confer oral activity to their target peptides.

Major developments in immunology resulted from the use of new immunogens, such as those generated on resins or those generated on poly-lysine scaffolds.

Protein/DNA interactions were the subject of several presentations and addressed the targeting of DNA sites with simple metal complexes, structural considerations for the hydrolysis of oligoribonucleotides by basic peptides, the design of ion-carriers and that of "sticky fingers" for peptides, among others.

Finally, part of a morning was dedicated to the peptide chemist's contribution to the study of AIDS. Major progress in understanding the role of HIV protease and in design of synthetic inhibitors of this enzyme was made possible by the total synthesis and structural analysis (by two independent groups of investigators) of this 99-amino acid aspartic acid protease. Other contributions were concerned with the development of new sensitive tests for the determination of HIV infection, as well as the design and synthesis of peptides for the development of neutralizing antibodies.

Overall, we recognize that peptide synthesis *per se* is still a major challenge, with the need for new strategies, coupling reagents and protecting groups, and that custom-designed and synthesized peptides (including peptide mimetics) are in fact the tools that have permitted such disciplines as structural biology (both theoretical and applied), physiology, molecular biology and medicinal chemistry to progress so rapidly.

It was indeed appropriate that this year's Pierce Award was bestowed on Professor Murray Goodman, a scientist who, with enduring success and through exacting efforts at integrating synthesis, spectroscopic analysis, molecular modeling and bioassay, de facto, epitomizes the renaissance peptide chemist. As

Chairman of the Fifth American Peptide Symposium, the Editor of Biopolymers and Mentor of numerous young scientists, Professor Goodman has exemplified outstanding leadership and service in peptide chemistry. His memorable lecture gave a synopsis of his research as well as illustrated the breadth of his interests.

A scientific session was dedicated to one of our cherished and outstanding colleagues, Professor Theodore Wieland, in recognition of his many contributions to peptide chemistry and structure activity relationships. Professor Wieland unfortunately could not attend that most remarkable session, which was introduced and chaired by two of his respected friends, Drs. Ralph Hirschmann and Christian Birr.

It is a pleasure for the Chairman (JER) to acknowledge the contributions of those groups, individuals and firms who contributed intellectually, physically, and financially to the successful meeting we all enjoyed. Both the Planning Committee and the Program Committee were instrumental in suggestions and recruitment for the program. The local committee helped in solving the logistic challenges associated with an overwhelming representation.

The Salk Institute for Biological Studies, presided over by Dr. Renato Dulbecco and with the help of many members of its administrative staff, was supportive in many substantive ways. Mr. Jamie Simon contributed his creative talents to the design of the logo that merged two remarkable architectural landmarks: the courtyard of the Salk Institute for Biological Studies and the library of the University of California, San Diego. Members of the Salk photo-laboratory, headed by Mr. Jim Cox, are to be commended for their efficiency in duplicating hundreds of slides in order to allow for high quality presentations in the overflow auditorium.

The support from the University of California, San Diego, offered by Drs. Richard Atkinson (Chancellor) and Murray Goodman, was decisive in the warm welcome and efficiency of services provided on campus. Ms. Lene Hartman, Conference Coordinator for the University, was of great assistance at many stages of the meeting.

The professional assistance of Meeting Management was outstanding in the organizational details of the meeting, the welcome reception at the Salk Institute and the banquet at Sea World. Our sincere thanks go to Ms. Nomi Feldman, Ms. Shirley Kolkey, Ms. Gail Reed and Ms. Mildred Robinson for consistently acting upon all administrative matters of the meeting with equanimity.

Also for the first time, travel grants were made available to young American and established foreign investigators, thanks to the enlightened generosity of one of our sponsors.

Then, there were those at the meeting who wore the deep blue T-shirts displaying the symposium logo. These individuals are members of the research groups of Dr. Murray Goodman and the Chairman. They were the courteous people who never said "no" and made the Eleventh American Peptide Symposium a success by attending to every situation, managing audiovisual equipment, and monitoring access to the meeting. To the following individuals go our heartfelt thanks for their efforts: Dr. Jerry Boublik, Dr. Alastair Douglas, Dr. Carl Hoeger, Dr.

Yuji Nishiuchi, Dr. Alexander Polinsky, Dr. Albert Probstl, Dr. Mark Spicer, Dr. Joseph Taulane, Dr. Paula Theobald, Dr. Toshimasa Yamazaki, Mr. John Andrews, Ms. Karen Braun, Mr. Robert Feinstein, Mr. Ziwei Huang, Mr. Ron Kaiser, Mr. David Karr, Mr. Dean Kirby, Ms. Charleen Miller, Mr. Darryl Palmer, Mr. Duane Pantoja, Mr. John Porter, Mr. Seonggu Ro, Ms. Odile Said-Nejad, Mr. Richard Schumann, Mr. Jeff Spencer, Ms. Anna Toy and Mr. Xavier Vidal.

Publication of the proceedings could not have been accomplished without the help of several individuals. Two editorial reviewers, Ms. Neely Swanson and Ms. Phyllis Minick, first reviewed all manuscripts and brought order to chaos, while Dr. Carl Hoeger and Mr. John Porter searched the literature to update references. Dr. Elizabeth Schram, of ESCOM Science Publishers, contributed her understanding and effort to bring this volume to press in a timely fashion with the high publishing standards that it exhibits; a book whose size is considerably expanded over that published at the occasion of any previous American Peptide Symposium.

Finally, there is Becky, who has truly been responsible for the success of this meeting. Ms. Rebecca Hensley, who worked as the special project assistant for this meeting, has earned the gratitude and respect of all those with whom she has dealt, for her dedication to the organization of this meeting and her ability to integrate communications between the University of California, San Diego Campus, Meeting Management, the authors of the proceedings, the Publisher and the Chairman's office.

Ultimately, the major contributors were all the symposium participants who we thank for their presentations: be they in private or in public, in the form of an oral communication or that of a written manuscript. We feel that the quality of the proceedings volume reflects the diversity of our field and the commitment of all authors to outstanding scientific integrity and dedication.

Jean E. Rivier
Garland R. Marshall

Eleventh American Peptide Symposium
The Salk Institute and University of California, San Diego
July 9-14, 1989

Chairman

Jean E. Rivier, *The Salk Institute for Biological Studies*

Planning Committee

Irwin M. Chaiken, *Smith Kline & French Laboratories*
Charles M. Deber, *The Hospital for Sick Children, Toronto*
Bruce W. Erickson, *University of North Carolina, Chapel Hill*
Arthur M. Felix, *Hoffmann-La Roche Inc.*
Lila M. Gierasch, *University of Texas Southwestern Medical Center*
Ralph F. Hirschmann, *University of Pennsylvania*
Victor J. Hruby, *University of Arizona*
Maurice Manning, *Medical College of Ohio*
Garland R. Marshall, *Washington University School of Medicine*
Jean E. Rivier, *The Salk Institute for Biological Studies*
Clark Smith, *The Upjohn Company*
John A. Smith, *Massachusetts General Hospital*
Arno F. Spatola, *University of Louisville*
John M. Stewart, *University of Colorado Health Sciences Center*

Program Committee

Ralph A. Bradshaw, *University of California, Irvine*
Arthur M. Felix, *Hoffmann-La Roche Inc.*
Lila M. Gierasch, *University of Texas Southwestern Medical Center*
Richard G. Hiskey, *University of North Carolina*
Daniel S. Kemp, *Massachusetts Institute of Technology*
Hugh D. Niall, *Genentech Inc.*
Jean E. Rivier, *The Salk Institute for Biological Studies*
John A. Smith, *Massachusetts General Hospital*
Alfonso Tramontano, *Scripps Clinic and Research Foundation*
Wylie W. Vale, *The Salk Institute for Biological Studies*

Local Committee

Murray Goodman, *University of California, San Diego*
Richard A. Houghten, *Scripps Clinic and Research Foundation*
Thomas Lobl, *Immunetech Pharmaceuticals*

The Eleventh American Peptide Symposium greatly appreciates the support and generous financial assistance of the following organizations:

Sponsors

Applied Biosystems Inc.
Bachem Feinchemikalien AG
Bachem Laboratories Inc.
CALBIOCHEM Corporation
CIBA-GEIGY, Pharmaceuticals Division
Hoffmann-La Roche Inc.
MilliGen/Biosearch, Division of Millipore
Pierce Chemical Company
PROPEPTIDE
Sigma Chemical Company
The Upjohn Company
VYDAC, A Divison of The SEP/A/RA/TIONS Group

Donors

Advanced ChemTech Inc.
Biomeasure Inc.
Carlsberg Biotechnology Ltd.
Chemical Dynamics Corporation
E.I. du Pont de Nemours & Co.
Ferring Pharmaceuticals AB
Glaxo Research Laboratories
Hoechst-Roussel Pharmaceuticals Inc.
Merck Sharp & Dohme Research Laboratories
Monsanto Company
Multiple Peptide Systems
Organon International B.V.
ORPEGEN
Peninsula Laboratories Inc.
Peptide Institute Inc.
Peptides International Inc.
PEPTISYNTHA
Pharmacia Inc.
SANDOZ Research Institute
Smith Kline & French Laboratories
Tanabe Seiyaku Co. Ltd.
UCB-Bioproducts S.A.

Contributors

Abbott Laboratories
AMGEN Inc.
American Cyanamid Co.

Boehringer Ingelheim Ltd.
Bristol-Myers Company
Burroughs Wellcome Company
Immunobiology Research Institute
Johnson & Johnson
Lilly Research Laboratories
Parke-Davis, Pharmaceutical Division, Warner-Lambert Company
The Peptide Laboratory
Peptide Technology Limited
Pfizer Central Research
Squibb Institute for Medical Research
Sterling-Winthrop Research Institute
Syntex Research
Triton Biosciences Inc.
Wyeth-Ayerst Research

Abbreviations

Abbreviations used in the proceedings volume are defined below:

AA, aa	amino acids	APP	avian pancreatic polypeptide
AAA	α-alkyl-α-amino acids; amino acid analysis	APY	anglerfish peptide YG
		AR	adrenergic receptor
Aab	3-aminomethyl-4-amino-butanoic acid	ARC	AIDS related complex
		Asu	aminosuberic acid
ABTS	2,2′-azinobis(3-ethylbenz-thiazoline sulfonic acid)	ASW	artificial sea water
		ATR	attenuated total internal reflection
AC	adenylate cyclase		
Ac$_5$c	aminocyclopentane carboxylic acid	AVP	arginine-8-vasopressin
Ac$_6$c	aminocyclohexane carboxylic acid	Bab	3,5-bis(2-aminoethyl)benzoic acid
ACE	angiotensin-converting enzyme	Bal	β-alanine
		BBB	blood brain barrier
ACh	acetylcholine	BHAR	benzhydrylamine resin
ACHPA	4-amino-3-hydroxy-5-cyclo-hexylpentanoic acid	BHK	baby hamster kidney
		BK	bradykinin
Acm	acetamidomethyl	BME	β-mercaptoethanol
ACP	acyl carrier protein	BN	bombesin
ACSA	adenylate cyclase-stimulating activity	BnPeOH	2,2-[bis(4-nitrophenyl)]-ethanol
AcT	N^α-acetyltransferase	Boc	tert-butyloxycarbonyl
ACTH	corticotropin	Boc-ON	2-tert-butyloxycarbonyl-amino-2-phenylacetonitrile
Ada	adamantyl		
ADR	adriamycin	BOI	2-(benzotriazol-1-yl)-oxy-1,3-dimethyl-imidazolinium
AEC	3-amino-9-ethylcarbazol		
AFP	antifreeze polypeptides	Bom	benzyloxymethyl
Ahx	aminohexyl	BOP	benzotriazolyl-oxy-tris-(dimethylamino)-phospho-nium hexafluorophosphate
Aib	aminoisobutyric acid		
AIBN	azoisobutyronitrile		
AIDS	acquired immune deficiency syndrome	BOP-Cl	bis(2-oxo-3-oxazolidinyl)-phosphinic chloride
AMD	actinomycin D	BPA	benzylphenoxyacetamido-methyl
AMP	aminomethylpiperidine		
ANF	atrial natriuretic factor	Bpoc	biphenylylpropyloxycarbonyl
Ang,	angiotensin;	BPTI	bovine trypsin inhibitor
ANG	angiotensinogen	Br$_2$Dmb	3,5-bis(bromomethyl)-benzoate
Ang II	angiotensin II		
ANP	atrial natriuretic peptide	BroP	bromo tris(dimethylamino)-phosphonium hexafluoro-phosphate
AP	aminopeptidase		
APC	antigen presenting cell		
APG	azidophenyl glyoxal	BSA	bovine serum albumin
APM	aminopeptidase M	BTD	bicyclic β-turn dipeptide
Apo	apolipoprotein	Bum	tert-butyloxymethyl

Bzl	benzyl	DEPC	diethylphosphorocyanidate
		DEPE	dielaidoylphosphatidyl-ethanolamine
CAM	computer assisted manufacturing	Dha	dehydroalanine
cAMP	cyclic adenosyl mono-phosphate	Dhb	dehydrobutyrene
		DHP	dihydroxypropyl
CAT	chloramphenicol acyl-transferase	DIBAL	diisobutyl aluminium hydride
		DIC	diisopropylcarbodiimide
Cbz, Z	carbobenzoxy: benzyloxycarbonyl	DIEA	diisopropylethylamine
		DIP	4,7-diphenyl phenanthroline
CCD	countercurrent distribution	DKP	diketopiperazine
CCK	cholecystokinin	DLPS	dilauroylphosphatidylserine
CD	circular dichroism	DMA	dimethylacetamide
CDD	cardiodilatin	DMAP	dimethylaminopyridine
CE	carbetocin	DMBHA	dimethoxybenzhydryl amine
CHA	cyclohexylamine	DMF	dimethylformamide
cHex	cyclohexyl	Dmp	dimethylphosphinyl
CHO	chinese hamster ovary; aldehyde	DMPC	dimyristoylphosphatidyl-choline
ChTX	charybdotoxin	DMPG	dimyristoyl phosphatidyl glycerol
CID	chemically ionized desorption		
CLA	cyclolinopeptide	DMS	dimethyl sulfide
CM	chloromethyl	DMSO	dimethyl sulfoxide
CNS	central nervous system	Dmt-OH	2,2-dimethyl-L-thiazolidine-4-carboxylic acid
Cpa	4-chlorophenylalanine		
CPD	carboxypeptidase	DNA	deoxyribonucleic acid
CPG	controlled pore glass	DNP	dinitrophenyl
CPMAS	cross-polarization/magic angle spinning	Dns	dansyl
		DOPC	dioleoyl-*sn*-glycerophospho-choline
CPZ	chlorpromazine		
CRF	corticotropin releasing factor	DOPE	dioleoylphosphatidylethanol-amine
CsA	cyclosporin A		
CT	calcitonin; chymotrypsin	Dpa	diphenylalanine
		DPBT	diphenylphosphorylbenz-oxazolethione
CTMS	chlorotrimethylsilane		
CVS	cardiovascular system	DPCDI	diisopropylcarbodiimide
CZE	capillary zone electrophoresis	DPP	dipeptidyl peptidase
		DPPA	diphenylphosphorylazide
Dab	diaminobutyric acid	DPPC	dipalmitoylphosphatidyl-choline
DABITC	4-dimethylaminophenyl-4'-isothiocyanate		
		DPPG	dipalmitoyl phosphatidyl glycerol
DBU	1,8-diazobicyclo[5.4.0]-undec-7-ene		
		Dpr	1,3-diaminopropionic acid
DCCI, DCC	dicyclohexylcarbodiimide	DSP	dimethylsulfonium methyl sulfate
DCHA	dicyclohexylamine	Dtc	5,5-dimethylthiazolidine-4-carboxylic acid
DCM	dichloromethane		
Dcp	dichlorophenyl	DTNB	dithiobis(2-nitrobenzoic acid)
DCU	dicyclohexylurea	Dts	dithiasuccinoyl
DDQ	dichlorodicyanoquinone	DTT	dithiothreitol
DEAE	diethylamino ethanol	Dyn	dynorphin
DEDTC	*N,N*-diethyldithiocarbamate		
Deg	diethylglycine	EA	ergotamine

EDAC	1-(3-dimethylaminopropyl)-3-ethylcarbodiimide hydrochloride
EDRF	endothelium-derived relaxing factor
EDTA	ethylenediaminetetraacetic acid
EGF	epidermal growth factor
ELISA	enzyme-linked immunosorbent assay
Enk	enkephalin
EP	endorphin
EPR	electron paramagnetic resonance
ER	endoplasmic reticulum
ESR	electron spin resonance
ET	endothelin
EtA	α-ethylalanine
Et_3N	triethylamine
FABMS	fast atom bombardment mass spectrometry
Fg	fibrinogen
FGF	fibroblast growth factor
FID	flame ionization detector
Flg	fluorenylglycine
Fm, fm	fluorenylmethyl
FMDV	foot-and-mouth disease virus
Fmoc	fluorenylmethoxycarbonyl
Fpa	4-fluorophenylalanine
FPLC	fast protein liquid chromatography
FRET	fluorescence resonance energy transfer
FSH	follicle stimulating hormone; follitropin
FTIR	Fourier transform infrared
GA	gramicidin A
GABA	gamma aminobutyric acid
GAL	galanin
GC	gas chromatography
GDA	glutaraldehyde
GEMSA	guanidino-ethylmercaptosuccinic acid
GH	growth hormone
GHRH, GRF	growth hormone releasing hormone
GHRP	growth hormone releasing peptide
GITC	2,3,4,6-tetra-*O*-Ac-β-D-glucopyranosyl isothiocyanate
Gla	D-galactopyranosyl

Glc	D-glycopyranosyl
GLP	glucagonlike peptide
Gn	guanidine
GnRH	gonadotropin releasing hormone
GPI	guinea pig ileum
GRF, GHRH	growth hormone releasing factor
GRP	gastrin releasing peptide
GTP	guanosine triphosphate
hBP	human serum binding protein
HBV	hepatitis B virus
hCG	human chorionic gonadotropin
hCgA	human chromogranin A
hCGRP	human calcitonin gene-related peptide
hCys, Hcys	homocysteine
HDL	high density lipoprotein
HF	hydrogen fluoride
HFBA	heptafluorobutyric acid
hGH	human growth hormone
Hip	hydroxyisovalerylpropionic acid
HIV	human immunodeficiency virus
HLE	human leukocyte elastase
HMP	hydroxymethylphenoxyacetic acid; hydroxymercaptopropionic acid
HMPA	hexamethylphosphoric triamide
hNP	human neutrophil peptide
HOBt	hydroxybenzotriazole
HODhbt	hydroxyoxodihydrobenzotriazine
HOHAHA	homonuclear Hartman Hahn spectroscopy
HONp	nitrophenol
HOOBt	hydroxyoxodihydrobenzotriazine
HOSu	*N*-hydroxysuccinimide
HPLC	high pressure liquid chromatography
Hpp	3-(4-hydroxyphenyl)-propionyl
Hse	homoserine
HSV	herpes simplex virus
HTLV	human T-cell leukemia virus
HVE	high voltage paper electrophoresis

Hyp	hydroxyproline		Mbh	methoxybenzhydryl
Hz	hertz		MBHA	methylbenzhydrylamine
			MBHAR	methylbenzhydrylamine resin
Ia	class II major histo-compatibility complex		MBP	myelin basic protein
			MBS	*m*-maleimidobenzoyl-*N*-hydroxysuccinimide ester
IB	inhibin			
IC	inhibitory concentration		MCH	melanin concentrating hormone
ICDH	isocitrate dehydrogenase			
i.c.v.	intra-cerebro-ventricular		MCPBA	*m*-chloroperbenzoic acid
IEC	ion-exchange chromato-graphy		MD	molecular dynamics
			MHC	major histocompatibility complex
IEF	isoelectric focusing			
IFNα	interferon α		MIC	minimal inhibitory concentration
IGF	insulin-like growth factor			
IgG	immunoglobulin		MIR	main immunogenic region
IL	interleukin		MNE	magnetic nonequivalence
i.m.	intramuscular; imidazole		MPS	multiple peptide synthesis
			MS	mass spectrometry
i.n.	intranasal		MSH	melanocyte stimulating hormone; melanotropin
INEPT	insensitive nuclei enhance-ment by pulse transfer			
Ing	indenylglycine		Msob	methylsulfinylbenzyl
IP	inositol phosphate		MT	metallothionein
IR	infrared; insulin receptor		MTC	medullary thyroid carcinoma
			Mtr	methoxytrimethylphenyl-sulphonyl
IRMA	immunoradiometric assay			
i.v.	intravenous		Mts	mesitylene sulfonyl
			MuLV	murine leukemia virus
KL	kallidin		MVD	mouse vas deferens
KLH	keyhole limpet hemocyanin			
			Nal	2-naphthylalanine
Lan	lanthionine		Nbb	*o*-nitrobenzamidobenzyl
LCAT	lecithin cholesterol acyl transferase		NBS	*N*-bromosuccinimide
			NE	norepinephrine
LDA	lithium diisopropylamide		NEI	neuropeptide Glutamic-Iso-leucine
LDH	lactate dehydrogenase			
LEC	ligand-exchange chromato-graphy		NEM	*N*-ethylmaleimide
			NGE	neuropeptide Glycine-Glutamic
LH	luteinizing hormone; lutropin			
			NIS	*N*-iodosuccinimide
LPH	lipotropin		NK	neurokinin
LSIMS	liquid secondary ion mass-spectrometry		NM	neuromedin
			NMM	*N*-methylmorpholine
LVP	lysine-8-vasopressin		NMP	*N*-methylpyrrolidinone
			NMR	nuclear magnetic resonance
MAb	monoclonal antibody		NMT	*N*-myristoyl transferase
MAG	magainin		NOE	nuclear Overhauser effect
MAP, MAp	membrane anchored protein; multiple antigen peptide; mean arterial pressure		NP	neutrophil peptide; neurophysin
			NPY	neuropeptide Y
MAPs	macromolecule associated proteins		NT	neurotensin
MAS	magic angle spinning		ONb	*o*-nitrobenzyl

OPA	o-phthaldialdehyde	Pqt	3-(1'-methyl-4,4'-bipyridi-nium-1-yl) propyl
OPA-ME	o-phthaldialdehyde-2-mercapto ethanol	PRL	prolactine
O-Su	O-succinimide ester	PS	pancreastatin
OT	oxytocin	PSA	preformed symmetrical anhydride
OTf	O-triflate		
OVLT	organum vasculosum laminate terminalis	PT	pertussis toxin
		PTH	phenylthiohydantoin; parathyroid hormone
PA	palmitic acid	PTK	protein tyrosine kinase
PAB	p-alkoxybenzyl	Ptz	3-(10-phenothiazinyl) propanol
Pac	phenacyl		
PAF, Paf	p-aminophenylalanine	PTZ	phenothiazine
PAGE	polyacrylamide gel electro-phoresis	PVDF	polyvinylidene fluoride
		PVN	paraventricular nuclei
Pal	3-pyridylalanine	PyBOP	(benzotriazolyl)-N-oxy-pyrro-lidinium phosphonium hexafluorophosphate
PAM	phenylacetamidomethyl		
Pas	6,6-pentamethylene-2-aminosuberic acid		
		PYY	peptide Tyrosine-Tyrosine
pBNP	porcine brain natriuretic peptide		
		RIA	radioimmunoassay
PBS	phosphate buffered saline	RMSD, rmsd	root mean square distance
PCP	phencyclidine		
Pen	penicillamine	RNase	ribonuclease
Pfp	pentafluorophenyl ester	ROE	rotating frame nuclear Overhauser effect
PFP	pentafluoropropionyl		
PG	proteoglycan	RPHPLC	reversed-phase high pressure liquid chromatography
PGE_2	prostaglandin E_2		
PGF	proteoglycan growth factor	RR	ribonucleotide reductase
PGPR	plant growth-promoting rhizobacteria	RRS	rat renin substrate
		RSV	rous sarcoma virus
PhA	phenoxyacetal		
PHA	phytohemagglutinin	SA	symmetrical anhydrides
PHBT	polymeric hydroxybenzo-triazole	SAP	saporin
		Sar	sarcosyl; sarcosine
Phi	4-iodophenylalanine		
pI	isoelectric point	SAR	structure-activity relations
Pip	piperidine	s.c.	subcutaneous
Piv	pivaloyl mixed anhydride	SCLC	small cell lung carcinoma
PK	protein kinase	Scm	methoxycarbonyldithia
PLP	poly-L-proline	SCP	small cardioactive peptide
PMA	phorbol myristate acetate	SDC	sample displacement chromatography
Pmc	2,2,5,7,8-pentamethyl-chroman-6-sulfonyl		
		SDS	sodium dodecyl sulfate; single dose suppression
Pmp	3,3-pentamethylene-3-mercaptopropionic acid		
		SEB	staphylococcal enterotoxin B
PP	pancreatic polypeptide	SEC	size exclusion chromato-graphy
PPA	n-propylphosphoric anhydride		
		SEM	standard error of the mean
PPE	porcine pancreatic elastase	SFGF	Shope fibroma growth factor
PPK	porcine pancreatic kallikrein	SGPA	*Streptomyces griseus* protease A
PPL	porcine pancreatic lipase		
Ppt	diphenylphosphinothionyl	SH	sulfhydryl

SLE	systemic lupus erythematosus		Tmob	trimethoxy benzyl
SMPS	simultaneous multiple peptide synthesis		TMP	3,4,7,8-tetramethylphenanthroline
SP	substance P		TMS	trimethylsilyl
SPPS	solid phase peptide synthesis		TMSOTf	trimethylsilyl trifluoromethanesulfonate
SPS	solid phase synthesis		TNF	tumor necrosis factor
SRIF	somatostatin		TNP	trinitrophenyl
SRP	signal recognition peptide		TR	time resolved
Sta	statin		TRH	thyrotropin releasing hormone
Su	succinimide			
SV	simian virus		Tris	tris(hydroxymethyl)-aminomethane
SWM	sperm whale myoglobin			
			TRNOE	transferred nuclear Overhauser effect
Tacm	*S*-trimethylacetamidomethyl			
TASP	template-assembled synthetic proteins		TSH	thyroid stimulating hormone
			TT	tetanus toxoid
TBDMSCl	*tert*-butyldimethylsilyl chloride			
			UV	ultraviolet
TEA	triethylamine			
TEAP	triethylamine phosphate		VCD	vibrational circular dichroism
TFA	trifluoroacetic acid		VIP	vasoactive intestinal peptide
TFE	trifluoroethanol		VN	vitronectin
TFMSA	trifluoromethanesulfonic acid		VNA	virus neutralizing antibody
TGF	transforming growth factor		VSMC	vascular smooth muscle cells
THF	tetrahydrofuran		VT	vasotocin
Thi	*see* Dtc			
THTP	tetrahydrothiophene		WSCI	water soluble carbodiimide
Tiq	1,2,3,4-tetrahydroquinoleic-3-carboxylic acid			
			Z, Cbz	carbobenzoxy; benzyloxycarbonyl
TLC	thin layer chromatography			
TM	transmembrane			

Contents

Seventh Alan E. Pierce Award Lecture

Session I: Structural biology

Contents

Contents

Contents

Contents

Contents

Session II: Enzymology

Session III: Analytical methodologies

Session IV: Recent bioactive peptides and biology

Contents

Session V: Peptide/protein folding

Contents

Contents

Contents

Session VI: Immunology

Contents

Session VII: Protein/DNA interactions

Session VIII: HIV and related areas

Session IX: Synthetic methodologies and peptide bond mimetics

Contents

Contents

Contents

l-

Contents

Seventh Alan E. Pierce Award Lecture

Professor Murray Goodman

Introduced by: Jean E. Rivier
The Salk Institute
La Jolla, California, U.S.A.

and

Charles M. Deber
The Hospital for Sick Children
Toronto, Ontario, Canada

Professor Murray Goodman

Recipient of the Seventh Alan E. Pierce Award

Synthesis, spectroscopy and computer simulations in peptide research

Murray Goodman

Department of Chemistry, University of California, San Diego, CA 92093, U.S.A.

Let me begin by expressing my gratitude to the Awards Committee who selected me as a recipient of the Seventh Alan E. Pierce Award. I am honored to receive this award at a meeting chaired by Jean Rivier and to deliver the lecture in San Diego where my co-workers can participate. They have been instrumental in creating much of the science that will be discussed in my presentation.

I owe a great debt to my mentors. Melvin Calvin, my Ph.D professor who in addition to introducing me to science, instilled in me the excitement of interdisciplinary research. His insight and scientific perceptions remain benchmarks for me. John Sheehan taught me peptide chemistry. I learned synthesis from him and recognized that he has made major contributions to modern peptide synthetic methodologies. George Kenner and I had the temerity to attempt to write an all-encompassing review [1] of peptide chemistry in the 1950s. From him I learned to integrate the chemistry and biology of peptides. Last, but not least, Herman Mark gave me my first opportunity to be an independent researcher. He taught me many things, but above all that science is international and that there are no national boundaries to knowledge. In fact, there is a synergism when scientists from different countries work together.

In the time allotted, it is not possible to cover all our published work. However, I do want to show where we have come from, where we are and, hopefully, where we are going. To do this I must select specific topics to discuss. I cannot list all of my co-workers and collaborators, but there is an implicit recognition by me of each and every one of them. This lecture is a tribute to the associates, students, postdoctoral researchers, visiting professors and collaborators who have worked with us over the years.

Let us go back to the early 1950s, to the heady days of the discovery of the α-helix by Linus Pauling [2,3]. Professor Pauling gave a lecture at the University of California at Berkeley where I was a graduate student. The elegance of the structural elucidation stimulated me as a chemist to think about characterizing certain features of the α-helix. Several years later, we commenced our independent research with an attempt to establish the critical length that peptide chains require to form helices.

Oligopeptides

By stepwise synthetic techniques, Schmitt prepared a series of glutamate oligomers [4,5]. He employed reactions developed by Steuben [6-8] in our laboratories. The chiro-optical properties of these novel structures were studied to establish some important physical/chemical features of their structure. In those days there were no CD instruments available. Schmitt [9,10], Listowsky [9–12], Boardman [10–13], Rosen [14] and Langsam [14] measured optical rotations as a function of chain length and optical rotatory dispersions based upon measurements in the visible and near ultraviolet spectral regions.

We were able to determine the optical rotation of oligopeptides as a function of the number of residues in the chain and in various solvents (Fig. 1). In dichloroacetic acid, this dependence established rotations for disordered structures in solution. In dimethylformamide, the oligoglutamates were shown to exhibit rotations that deviated from values of disordered structures at and above the heptamer, whereas in dioxane there was found a discontinuity in the values of the optical rotation between the tetra- and pentapeptides. Such complex behavior arises from the onset of secondary structure.

We also employed the Moffitt equation [15,16] in which the coefficient of the higher term b_0 relating optical rotation to wavelength was used as an indication

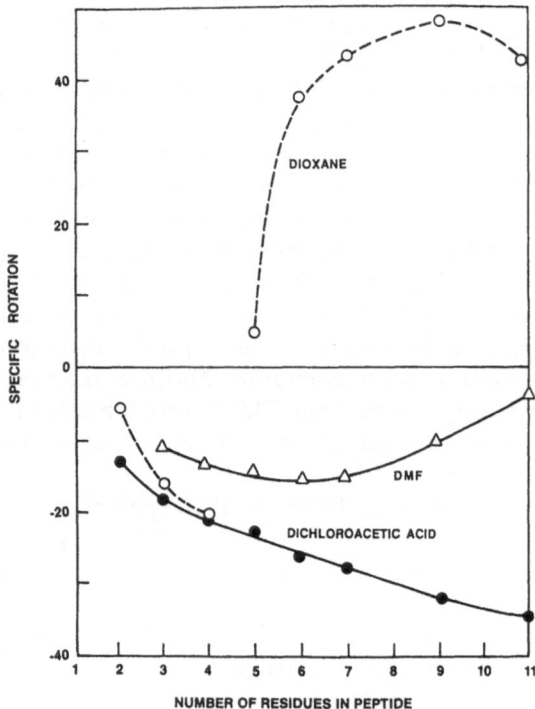

Fig. 1. *Specific rotations (measured at 589 nm) of oligomeric peptides derived from γ-methyl-L-glutamate as a function of solvent and number of residues. Rotations were measured in dioxane (open circles) and dichloroacetic acid (filled circles) at 2% concentration except the hepta- and nonapeptides in dioxane solution. These rotations were measured in a 1.43% and 0.22% solution, respectively. Rotations measured in dimethylformamide (open triangles) were at a concentration of 1%. Rotations in all solvents were determined at 25°C.*

of secondary structure. The characteristics of highly ordered polypeptides were established in the seminal work of Blout [17–20], Doty [20–23] and Katchalski [24–26]. The values for b_0 in helix supporting solvents are low for the dimer through hexamer indicating a lack of secondary structure for these molecules. With the heptamer, the values of b_0 increase sharply so that L-oligoglutamates with about 20 residues exhibit essentially total helicity, i.e., showing values of b_0 of about –600 (Fig. 2).

In the 1960s, CD became available [27]. We examined the CD for the same series of oligopeptides through the visible and ultraviolet regions of the spectrum [10–13,28,29]. It was shown that highly helical polypeptides exhibited enhanced $n \rightarrow \pi^*$ transitions at 220 nm and a split $\pi \rightarrow \pi^*$ set of transitions at 206 nm and 194 nm [30]. Toniolo and Verdini [31] observed the onset of helicity in helix supporting solvents. In solvents such as trifluoroethanol and trimethyl-phosphate helicity begins with a heptaglutamate (Fig. 3). Naider and Rupp [32, 33] found that a very similar relationship is observed for alanine oligopeptides with benzyloxycarbonyl groups at the N-termini and morpholino groups at the C-termini as shown in Fig. 4. Since that time, our laboratory and other researchers have examined many linear oligopeptide and co-oligopeptide series [34]. Much information on conformations has been gathered using optical rotation, optical rotatory dispersion, CD and other techniques [34].

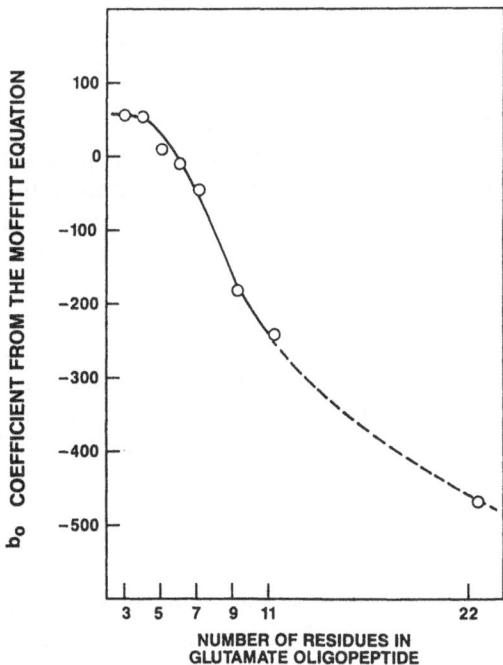

Fig. 2.   The b_0 coefficients of the Moffitt equation [15, 16] of the oligomeric peptides derived from γ-methyl-L-glutamate in dimethylformamide at 25° C. The b_0 coefficient for an oligomeric mixture with chain lengths centering about 22 is also included in this plot.

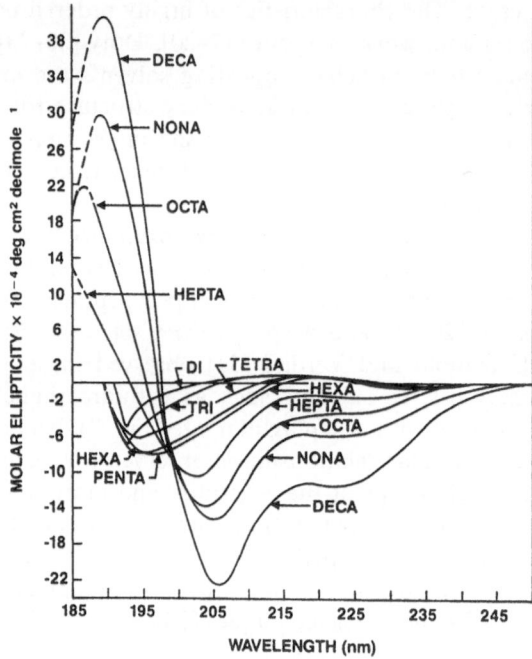

Fig. 3. CD spectra of N-benzyloxycarbonyl-γ-ethyl-L-glutamate *oligomers in trimethylphosphate as a solvent. The curves for the molar ellipticity of each oligomer are included.*

In 1958, we commenced our love affair with nuclear magnetic resonance (NMR) spectroscopy. Following the pioneering work of Bovey [35,36] and his co-workers on polypeptides and stereoregular polymers, we attempted to measure the NMR spectrum of an L-glutamate heptamer in TFE. As can be seen from Fig. 5A, no useful resonances could be discerned over the NH and aryl proton spectral region. Masuda was able to obtain some important information on the helix to coil transition by NMR [37]. By 1968 technology had advanced so that we could make measurements at 220 MHz. As in Fig. 5B, Toniolo and Verdini working with Bovey and Phillips showed that all of the NHs of an L-glutamate hexamer could be discerned in TFE [31]. By 1978, we had available to us a 360-MHz instrument and examined the L-glutamate heptamer, as can be seen in Fig. 5C [38]. All the NHs and most of the envelopes of the α-CH resonances could be resolved. By techniques that I will shortly describe, we could actually assign each of the resonances to specific residues in the chain. By 1988, we had progressed to a 500-MHz NMR instrument that is particularly useful for complex peptides in the 2D mode of measurement. I will return to these types of measurements in the section of this presentation dealing with bioactive peptides.

We were able to extend our work on oligopeptides by taking advantage of the liquid phase synthetic approach of Bayer and Mutter [39]. A series of oligopeptides of L-glutamates tethered to polyoxyethylene (MW 5000) were prepared by us and studied by NMR in a variety of solvents [40–43]. Mutter

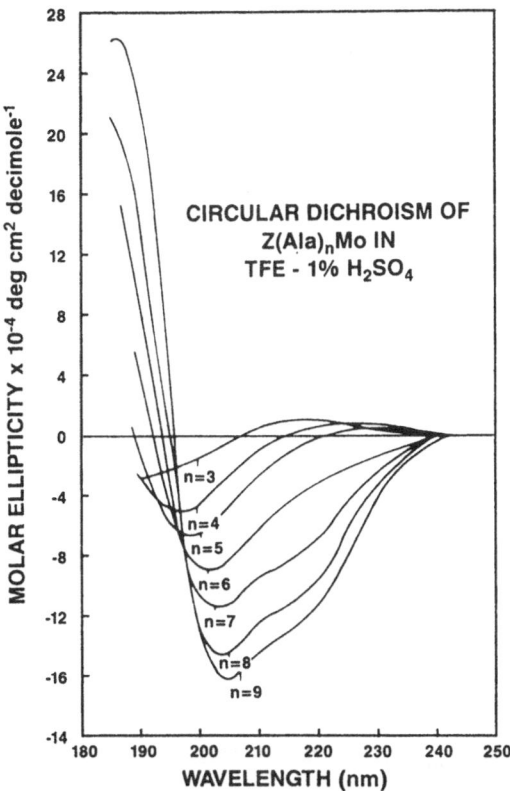

Fig. 4. CD spectra of benzyloxycarbonyl L-alanyl-N-methylmorpholinoamide oligomers measured in a trifluoroethanol 1% sulfuric acid solvent mixture. The curves for the molar ellipticity of each oligomer are included.

showed earlier that the polyoxyethylene chain does not affect the preferred conformational characteristics of the oligopeptide chain [44]. By selective placement of α-deuterated L-glutamates, Saltman and Ribeiro were able to assign all the NHs and most of the α-CHs to specific residues in the oligopeptide chain [40–43]. The effects of solvents and end groups on preferred conformations of these peptide–polyoxyethylene conjugates were carefully examined by NMR [40–43]. For example, chloroform favors C-7 conformations, whereas, TFE induces α-helical preferences for the heptamer oligoglutamates [41,42].

Implicit in our study of oligopeptides was the fact that optically pure peptides could by synthesized. To ensure such peptide bond formation, we undertook to study racemization during coupling reactions via oxazolinone intermediates. In our laboratories, Levine [45] and McGahren [46–49] synthesized the first crystalline and optically active amino acid and peptide oxazolinones. The reactions of these oxazolinones with L-phenylalanine methyl ester under typical conditions for peptide synthesis showed that racemization is an order of magnitude more rapid than ring opening reactions (Table 1). Glaser established that rates of

7

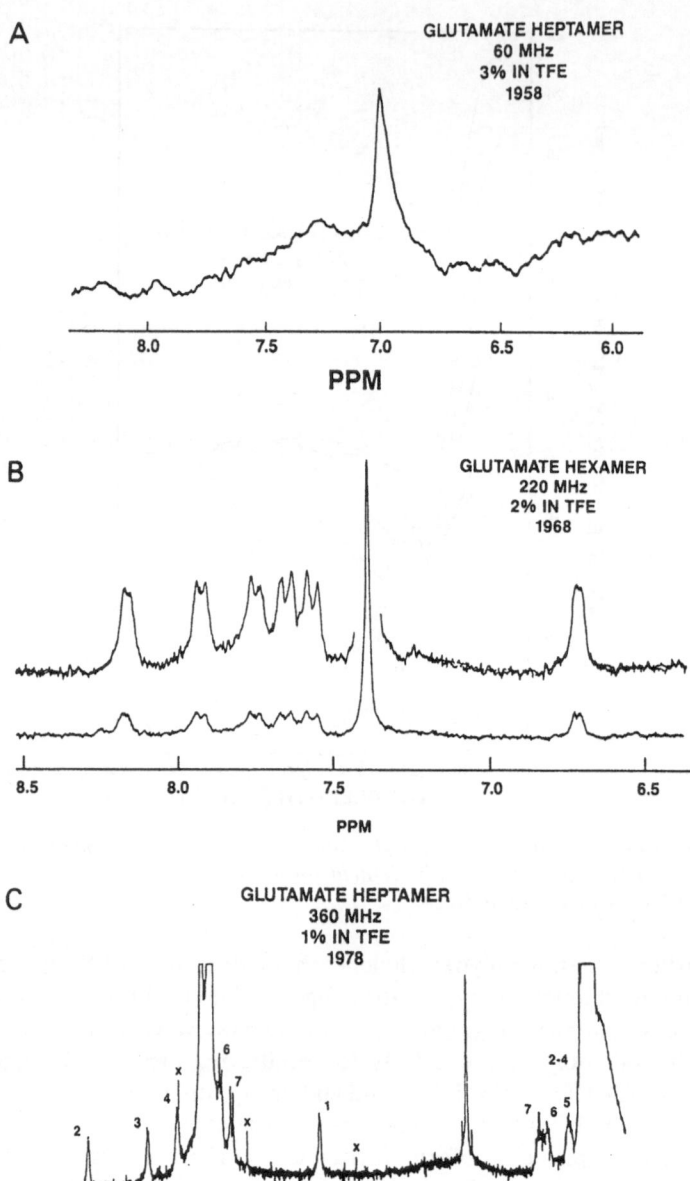

Fig. 5. *Partial NMR spectra of urethane-protected γ-ethyl-L-glutamate heptamer at (A) 60 MHz and (C) 360 MHz and the hexamer (B) at 220 MHz.*

Table 1 *Amino acid and peptide oxazolinones*

	2-Phenyl-L-4 benzyloxazolinone	2-(1′-Benzyloxycarbonylamino-1′-methyl)-ethyl-L-4-benzyloxazolinone
m.p.	88–89°C	97.4–98.8°C
$[\alpha]_D^{25}$	71.2° (C = 0.5, dioxane)	131.2° (C = 1, dioxane)
k_{rac}(l/mol·min)[a]	3.69	0.750
k_{ro} (l/mol·min)[a]	0.019	0.065

[a] Second order rate constants for racemization (k_{rac}) and ring opening (k_{ro}) of oxazolinones by phenylalanine methyl ester in dioxane at 25°C.

racemization can be substantially slowed by α-dinucleophiles such as azide ion, *N*-hydroxysuccinimide, hydroxybenzotriazole and others, whereas peptide bond formation is facilitated [50–52].

Polypeptides and Polydepsipeptides

Let us turn our attention to another important area of our research. Early on, we became interested in high molecular weight polypeptides. Such polymers can be prepared from α-amino acid *N*-carboxyanhydrides (NCAs) [53–56]. Our studies were commenced by Arnon [53,54] and Hutchison [55,56], who examined the strong base initiated polymerizations of NCAs including γ-benzyl L-glutamate and other NCAs. Peggion extended the mechanistic studies initially with Scoffone [57] and then in our laboratories [58]. We followed the kinetics for the initiation and propagation reactions that produce the polymers. Our results were best explained by an 'active monomer' mechanism. In 1977, Peggion, Szwarc, Bamford and I showed that sodium hydride initiates polymerization of γ-benzyl-L-glutamate and ϵ-benzyloxycarbonyl-L-lysine NCAs [59]. The kinetics of these polymerizations are consistent with an 'active monomer' mechanism (Fig. 6) [59].

We have prepared many different types of polypeptides in order to understand the forces governing conformational preferences of these biopolymers. In this presentation, I have selected polymers that involve unusual monomeric residues designed to illustrate specific side-chain and main-chain effects on secondary structure. To this end, Felix made poly-δ-hydroxy-L-α-aminovaleric acid by the polymerization of *O*-acetyl-δ-hydroxy-L-α-aminovaleric acid *N*-carboxyanhydride [60]. By use of chiro-optical properties, he was able to show that this neutral water-soluble polypeptide is a highly helical structure that can be denatured upon addition of a low concentration of lithium bromide [60]. Other researchers established that poly β-benzyl-L-aspartate assumes a left-handed α-

9

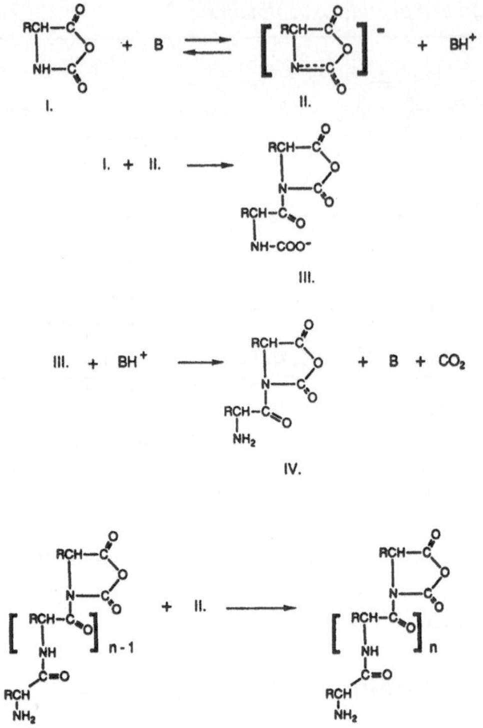

Fig. 6. A representation of the 'active monomer' mechanism for the polymerization of an α-amino acid N-carboxyanhydride monomer.

helix in solution [18,61]. Under the same conditions, poly β-*p*-nitrobenzyl-L-aspartate forms right-handed α-helices [60]. Felix and Deber prepared and studied copolymers of *p*-nitrobenzyl-L-aspartate with the benzyl-L-aspartate [60]. A transition from left- to right-handed helix was observed as the fraction of *p*-nitrobenzyl-L-aspartate content was increased (Fig. 7). In a related system, poly L-*p*-aminophenylalanine was synthesized in our laboratories by Peggion and Toniolo [62,63]. This analog of poly-L-tyrosine was shown to go through a helix to coil transition in aqueous organic solvents as the *p*-amino groups were protonated. Kossoy, Falxa and Benedetti synthesized derivatives of L-*p*-amino-phenylalanine, which were converted to L-arylazophenylalanine [64–68]. From these amino acids, we were able to make appropriate NCA monomers that were polymerized and copolymerized. Irradiation of solutions of azo-polymers in the visible spectral region, where the azo-chromophore absorbs, shows a transformation of the Cotton effect from a positive band to a negative absorption (Fig. 8). Azo-aromatic chromophores provided a basis to relate optical rotatory dispersion to CD and the structure of the polypeptides. Following our initial studies, work on azo-aromatic containing polypeptides has been continued and

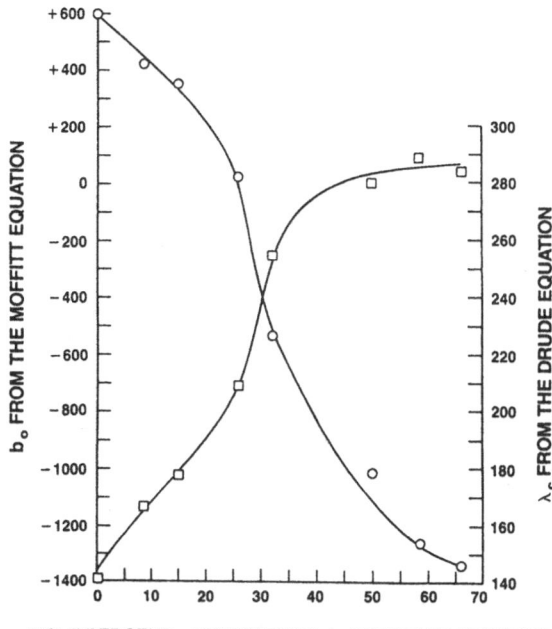

Fig. 7. Values of b_o (o) from the Moffitt equation [15, 16] and λ_c (□) from the Drude equation in chloroform at 25.0°C for copolymers of p-nitrobenzyl-L-aspartate with benzyl-L-aspartate as a function of mole percent of p-nitrobenzyl-L-aspartate residues.

greatly expanded in the laboratories of Ciardelli and Pieroni [69,70], Irie [71], Ueno [72,73] and Yamamoto [74,75].

During the 1960s, we turned our attention to semi-empirical calculations to determine preferred conformations for polypeptides [76–78]. These theoretical approaches were based on rigid geometries, torsional modes and Lennard–Jones potentials [79,80]. Organic, physical and polymer chemists also utilized similar semi-empirical relationships [81–87]. Peptide chemists constructed the same type of contour maps and called them Ramachandran plots [88,89] after one of the pioneers in this field. I want to illustrate the use of calculations for a residue of poly N-methyl-L-alanine. It shows the variations of ϕ and ψ while the peptide bond ω is maintained in a planar form, *trans* or *cis*, throughout the calculations (Fig. 9). Using this approach, Mark predicted four minima for poly N-methyl-L-alanine as seen in Fig. 10 [90,91]. Fried [92] and Chen [93] synthesized the polymer and model compounds and examined the CD and NMR spectra [92,93]. The nature of the Cotton effects and the chemical shifts are consistent with the structure predicted to be the most stable by Mark's calculations, that is, a right-handed 3-fold helix with the amide bonds all *trans*. The next most favored conformation is a left-hand α-helix, which is preferred by the calculations of Liquori and DeSantis [94].

An ester group is nearly isosteric with respect to a peptide bond. D'Alagni

11

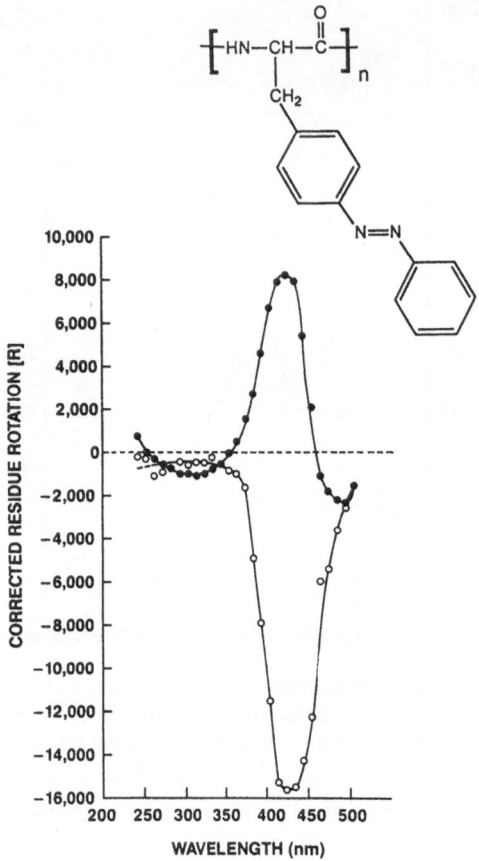

Fig. 8. Optical rotatory dispersion of poly-ʟ-p-(phenylazo)-phenylalanine before (●) and after (○) irradiation at 425 nm in trifluoroacetic acid.

examined the NMR spectra of poly-S-lactic acid and its model compound [95]. Gilon and Nissen [96] developed a general method to prepare polydepsipeptides using sublimation of the monomeric pentachlorophenyl esters. This polymerization route was further refined by Mathias [97]. In addition, an approach to polymerization in solution was developed by Katakai [98]. Using rigid geometry semi-empirical calculations, Ingwall created Ramachandran plots of many different series of polydepsipeptides and predicted their preferred conformations [99,100]. All of the insights gained in our laboratories were extended by Becktel who investigated the CD and helix-coil transitions of polydepsipeptides [101–103]. From such studies, we have been able to present key thermodynamic properties of polydepsipeptides, as shown in Table 2. These results are important for understanding peptide and protein structure.

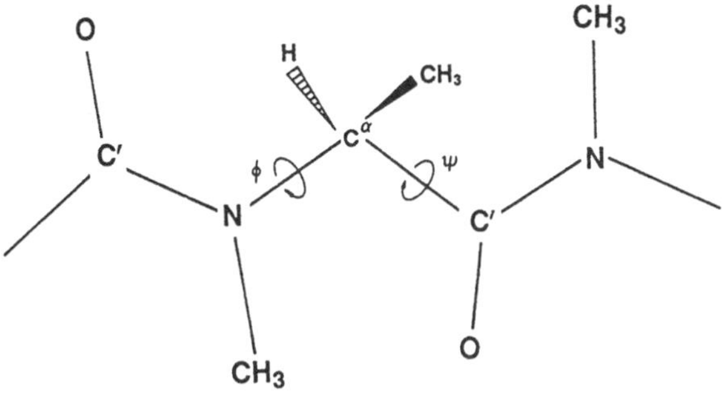

Fig. 9. Schematic representation of a segment of the poly-N-methyl-L-alanine chain denoting the internal rotation angles ϕ (N-C$_\alpha$) and ψ(C$_\alpha$-C').

Table 2 *Thermodynamic parameters of a series of polydepsipeptides[a]*

Compound	Solvent	σ	ΔH	ΔH°	ΔS	T$_c$
Poly(Ala-Lac)	CHCl$_3$	0.001	338	627	1.33	−36.7
Poly(Ala-Lac)	THF	0.007	344	687	1.47	−40
Poly(Val-Lac)	THF	0.003	272	544	1.04	−10.5
Poly[(Ala)$_2$-Lac]	CHCl$_3$	0.003	579	867	1.81	47.5
Poly[(Ala)$_2$-Lac]	THF	0.005	513	770	1.67	34.5
Poly[(Ala)$_3$-Lac]	TFE	0.006	163	217	0.67	−30
Poly[(Leu)$_2$-Lac]	TFE	0.002	136	205	0.55	−37
Poly[Glu(OCH$_3$)$_2$-Lac]	TFE	0.0009	969	1435	3.13	36.2
Poly[Glu(OCH$_3$)$_2$-Lac]	THF	0.0006	1156	1735	3.73	36.8
Poly[Ala-Glu(OBzl)-Lac]	THF	0.0006	1119	1678	3.63	35.6
Poly[Lys(Z)-Glu(OCH$_3$)-Lac]	THF	0.0006	1117	1766	3.83	34.6

[a] σ is the nucleation parameter from the Zimm–Bragg equations, ΔH is the average enthalpy/residue for the polydepsipeptide, ΔH° is the average enthalpy for the amides, ΔS is the average entropy/residue in the polydepsipeptide and T$_c$ is the melting temperature for the helix to coil transition for the polydepsipeptides.

Bioactive Peptides

When our research group moved to San Diego in 1970, we launched a major program to study bioactive peptides and peptidomimetics. Our approach combined synthesis, spectroscopy and computer simulations together with bioassays. From such studies, we hoped to develop structure–biological activity relationships for a wide variety of peptidic molecules. In this presentation I wish to select representative systems to discuss including opioids, somatostatin, neuropeptide Y and nisin. Our early work developed from the pioneering discoveries by Guillemin and his associates with respect to hypothalamic hormone releasing hormones [104,105]. Gilon, Donzel, Sakarellos and Blagdon studied analogs of TRH and GnRH by CD, NMR and molecular modeling [106–109]. They were able to propose conformational preferences for the molecules under study.

Before considering other systems, I want to call your attention to the retro-

13

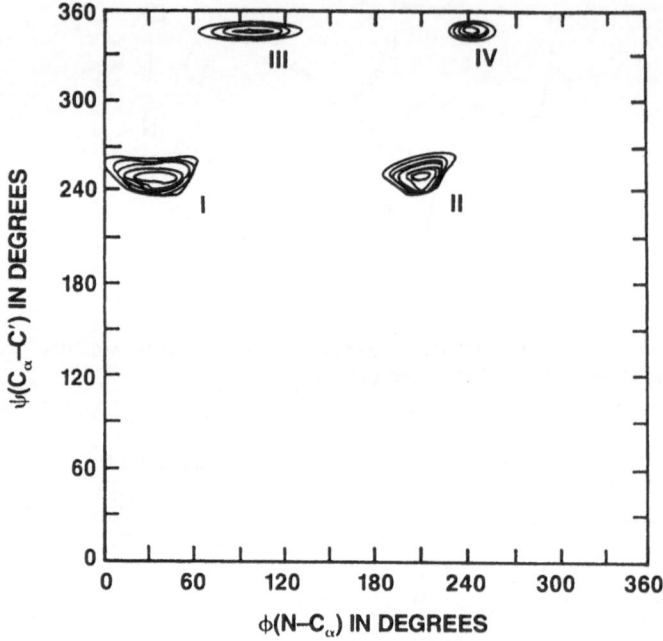

Fig. 10. *Low energy conformations of the poly-N-methyl-ʟ-alanine chain. Intervals between contours represent 1 kcal/mol of peptide units. The corresponding energies are tabulated in the chart below the contour map.*

Helix	ϕ, deg	ψ, deg	E, kcal mole^{-1}
I	30	250	−0.85
II	210	250	−0.33
III	80	345	2.54
IV	240	345	1.50

inverso modifications of peptide composition that Chorev and Pallai developed over a decade ago (Fig. 11) [110–113]. To reverse a peptide bond, Chorev carefully investigated the use of the Curtius rearrangement [114,115]. Pallai utilized the benzene iodonium bistrifluoracetate method published by Loudon for the generation of amines from amides [112,116]. Chorev, Pallai and Richman firmly

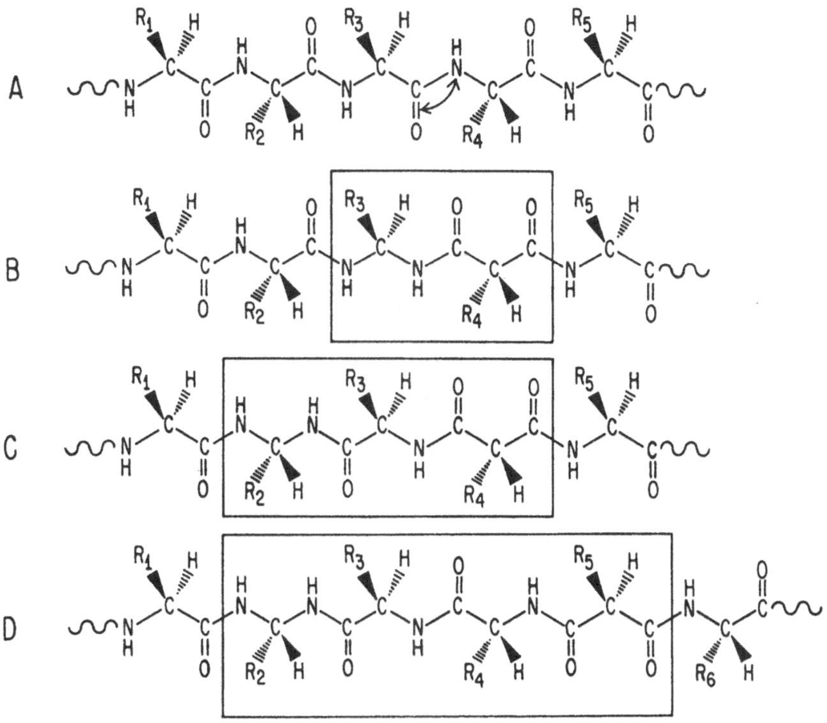

Fig. 11.   Schematic representation of partially retro-inverso modifications of a parent peptide (A). The modification (B) involves two adjacent amino acid residues containing side chains R_3 and R_4, by which the gemdiaminoalkyl and malonyl peptidomimetic structure is incorporated. In (C) the altered residues containing side chains R_2 and R_4 are moved apart so that an amino acid with reversed direction and chirality separates the diaminoalkyl and malonyl residues. In (D) a reversed peptide is inserted between the peptidomimetic residues.

established the stereochemistry for gemdiamino alkyl and alkylated malonyl-diamido residues [113,116].

In the area of opioids, Schiller and DiMaio [117] synthesized cyclic enkephalin analogs containing a 14-membered ring. This structural feature created constraints in the molecule that led to conformational preferences. Work in our laboratory commenced with this parent constrained molecule and included peptidomimetic residues such as those based upon retro-inverso modifications. Berman [118], Richman [119] and Lucietto [120] synthesized most of these molecules. The analogs are shown in Fig. 12. Schiller measured the biological activity and selectivity of our analogs [121] (Table 3). Hassan, Mammi and Mierke studied the molecular modeling and the NMR spectra [120–124]. We integrated the constraints as seen in the NMR (NOE values) [125] with computer modeling based on the important research of Hagler, Dauber and Osguthorpe, who developed the system DISCOVER [126]. The molecular dynamics, with and without NOE constraints, of these enkephalin analogs led us to propose a model

15

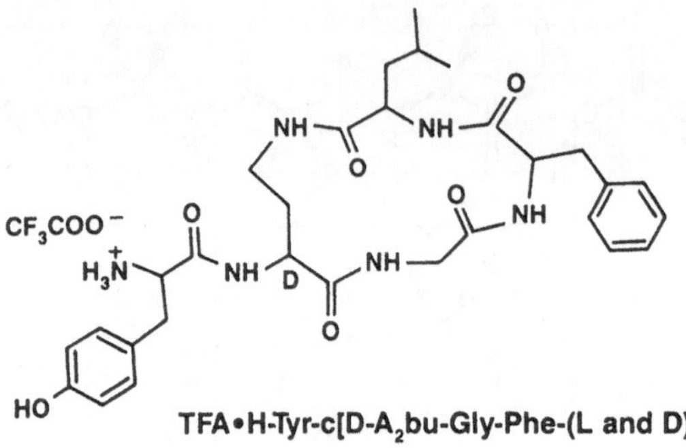

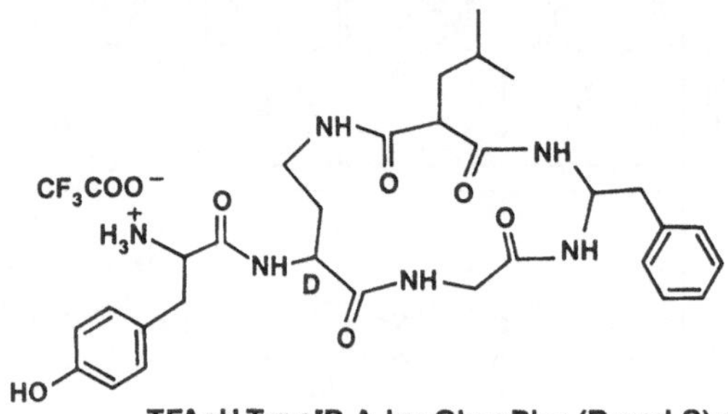

TFA•H-Tyr-c[D-A₂bu-Gly-Phe-(L and D)-Leu]

TFA•H-Tyr-c[D-A₂bu-Gly-gPhe-(R and S)-mLeu]

Fig. 12. Structure of the cyclic enkephalin derivatives Tyr-c[D-A₂bu-Gly-Phe-L and D-Leu], and their isomeric partial retro-inverso analogs, Tyr-c[D-A₂bu-Gly-gPhe-R and S-mLeu].

Table 3   *Guinea pig ileum (GPI) and mouse vas deferens (MVD) assay of various enkephalin and dermorphin analogs[a]*

Compound	GPI IC$_{50}$ (nM)	MVD IC$_{50}$ (nM)	MVD/GPI IC$_{50}$ ratio
Tyr-c[D-A₂bu-Phe-Phe-Leu]	1.07 ± 0.09	4.70 ± 0.92	4.39
Tyr-c[D-A₂bu-Phe-Phe-D-Leu]	1.14 ± 0.12	13.9 ± 0.9	12.2
Tyr-c[D-A₂bu-Phe-gPhe-S-mLeu]	0.518 ± 0.099	6.30 ± 1.11	12.2
Tyr-c[D-A₂bu-Phe-gPhe-R-mLeu]	17.6 ± 2.8	117 ± 31	10.1
Tyr-c[D-Glu-Phe-gPhe-L-retroLeu]	39.0 ± 9.3	3.64 ± 0.10	0.09
Tyr-c[D-Glu-Phe-gPhe-D-retroLeu]	2.75 ± 0.63	49.1 ± 9.9	17.9
Tyr-c[D-A₂bu-Gly-Phe-Leu]	14.1 ± 2.9	81.4 ± 5.8	5.77
Tyr-c[D-A₂bu-Gly-Phe-D-Leu]	66.1 ± 8.4	27.1 ± 1.6	0.409
Tyr-c[D-A₂bu-Gly-gPhe-S-mLeu]	1.51 ± 0.19	7.76 ± 3.17	5.14
Tyr-c[D-A₂bu-Gly-gPhe-R-mLeu]	25.5 ± 2.0	14.9 ± 5.0	0.584
Tyr-c[D-Glu-Gly-gPhe-D-retroLeu]	19.4 ± 3.4	313 ± 102	16.1
[Leu⁵]enkephalin	246 ± 39	11.4 ± 1.1	0.0463

[a] These determinations were carried out in the laboratories of Dr. Peter Schiller and his co-workers at the Clinical Institute of Montreal.
Mean of 3 determinations ± SEM.

16

for selectivity. In Fig. 13, the superpositions of 40 ps of molecular dynamics are shown for two pairs of peptides and peptidomimetic opioids. The parent peptide Tyr-c[D-A₂bu-Gly-Phe-Leu] is highly constrained and is similar in its molecular topology to Tyr-c[D-A₂bu-Gly-gPhe-S-mLeu] (μ-selective opioid). The peptide Tyr-c[D-A₂bu-Gly-Phe-D-Leu] is much more flexible. Its molecular topology is very similar to that found for Tyr-c[D-A₂bu-Gly-Phe-R-mLeu] (nonselective opioid). The peptide and peptidomimetics with L and S configurations at the Leu residue maintain an extended structure with maximal distance

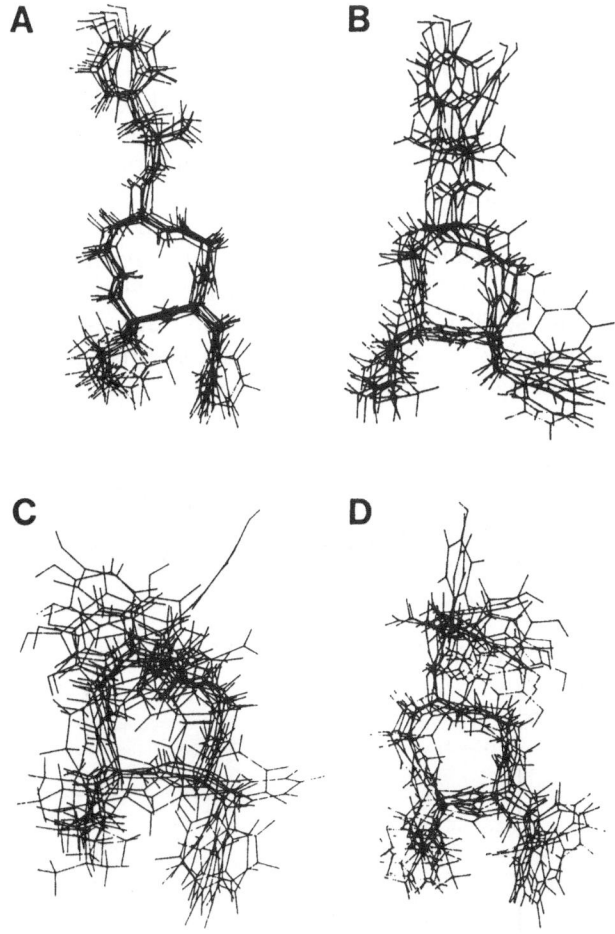

Fig. 13. A superposition of molecular dynamics simulations in vacuo over 20 ps using the DISCOVER program of the cyclic enkephalin analogs: Tyr-c[D-A₂bu-Gly-Phe-L and D-Leu]and Tyr-c[D-A₂bu-Gly-gPhe-S and R-mLeu]. It should be noted that the parent peptide: Tyr-c[D-A₂bu-Gly-Phe-Leu] (A) exhibits the same overall restricted conformation as the peptidomimetic: Tyr-c[D-A₂bu-Gly-gPhe-S-mLeu] (B), whereas the peptide: Tyr-c[D-A₂bu-Gly-Phe-D-Leu] (C) is much more flexible and is similar in overall topology to Tyr-c[D-A₂bu-Gly-gPhe-R-mLeu](D).

17

between the tyrosyl and phenylalanyl aryl side chains, whereas the molecules with the D and R configurations at the Leu residue alternate between an extended structure and folded arrays in which these aryl groups are much closer together. We believe that the extended structure is necessary for recognition at the guinea pig ileum (μ) receptor and the folded form is required for activity at the mouse vas deferens (δ) receptor. The alteration of configurations noted above leads to nonselective opioids. Hruby and Karplus carried out conformational analysis of a cyclic 14-membered ring enkephalin containing two D-penicillamine residues at positions 2 and 5, Tyr-c[D-Pen-Gly-Phe-D-Pen], which is highly δ-selective. In their studies the two aromatics are in very close proximity to each other. These results support our model for selective activity of opioids [127].

We have extended our integrated approach to consider cyclic dermorphin analogs in which the glycine of the Schiller parent cyclic enkephalin analog is replaced by L-phenylalanine. Specifically, the compound Tyr-c[D-A$_2$bu-Phe-gPhe-R-mLeu] was synthesized, and preferred conformations were studied by molecular dynamics simulations and NMR spectroscopy. Figure 14 shows the distance of the aryl group of the tyrosine residue to the aryl groups at positions 3 and 4 for the entire NOE-constrained molecular dynamics simulation. The distance of the tyrosine to the aryl group at the g-Phe[4] residue is similar to that found for the Schiller parent enkephalin and our μ-selective enkephalins containing peptidomimetics. It is, therefore, not surprising that Schiller found the dermorphin analog Tyr-c[D-A$_2$bu-Phe-gPhe-R-mLeu] to be quite potent and

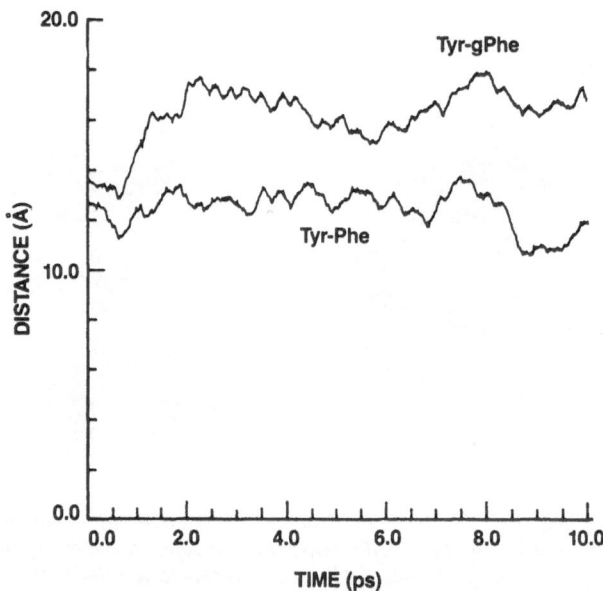

Fig. 14. Time variations of the distances between Tyr[1]-Phe[3] and Tyr[1]-gPhe[4] aromatic ring centers during the molecular dynamics carried out for Tyr-c[D-A$_2$bu-Phe-gPhe-R-mLeu] at 300 K using the DISCOVER program.

μ-selective (Table 3). At this stage, we are examining steric and electronic contributions arising from the phenylalanine at position 3. This and other dermorphin-like molecules have been prepared and examined by molecular modeling and NMR spectroscopy in our laboratories by Said-Nejad, Felder, Mierke and Yamazaki [128]. We hope to relate the structures of these compounds to our model for opioid bioactivity and selectivity.

In a related series of experiments, we have examined cyclic hexapeptides related to somatostatin. Veber and his associates at Merck synthesized a large number of cyclic hexapeptides incorporating residues 7–10 of the somatostatin sequence [129]. Some of these proved to be superactive in in vitro tests on the inhibition of growth hormone release. We have chosen one of these, c[Phe-D-Trp-Lys-Thr-Phe-Pro], as the parent structure into which peptidomimetics would be incorporated. Figure 15 shows the Merck compound and two of the analogs we have prepared. In this presentation it is impossible to discuss the synthesis and molecular modeling for each of the compounds. Therefore, I want to summarize the work on the analog incorporating the gSar-mPhe peptidomimetic into the cyclohexamer structure [130].

Pattaroni synthesized the compounds c[R and S-mPhe-D-Trp-Lys-Thr-Phe-gSar] as shown in Fig. 16 [130]. Noteworthy in this stepwise synthetic scheme

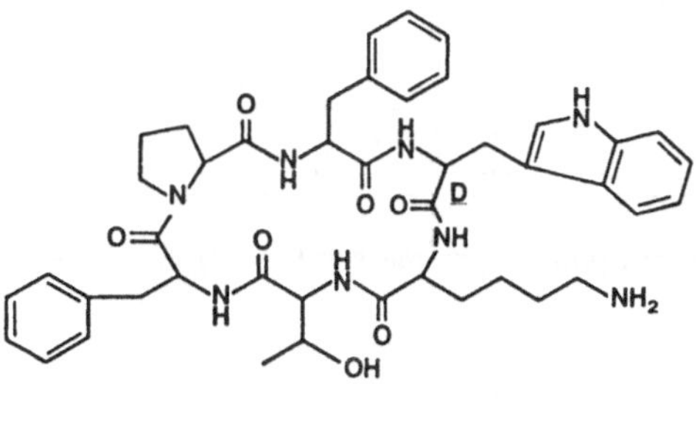

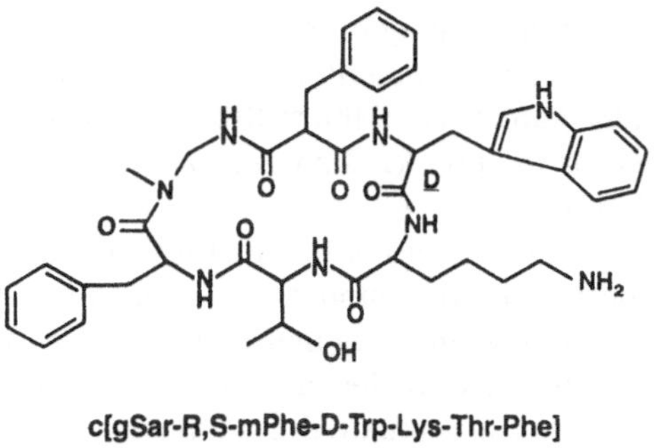

c[Pro-Phe-D-Trp-Lys-Thr-Phe]

c[gSar-R,S-mPhe-D-Trp-Lys-Thr-Phe]

Fig. 15. *Structures of the parent cyclic hexapeptide c[Pro-Phe-D-Trp-Lys-Thr-Phe] and the retro-inverso modified molecules c[gSar-R and S-mPhe-D-Trp-Lys-Thr] related to somatostatin.*

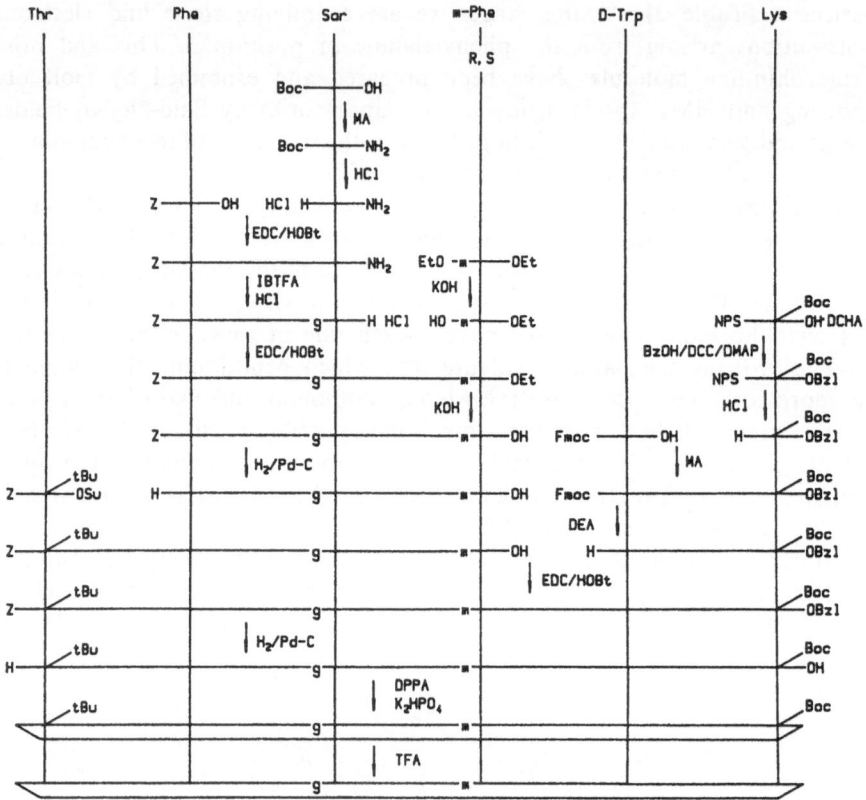

Fig. 16. *Schematic representation for the synthesis of c[gSar-R and S-mPhe-D-Trp-Lys-Thr-Phe].*

are the use of benzene iodonium bistrifluoroacetate for the Hofmann rearrangement and the separation of the mixture of diastereomeric final products by HPLC. Pattaroni and Mierke carried out extensive NMR analyses on these diastereomers and were able to assign the configuration at the malonyl residues for each of the diastereomers unambiguously. By use of NOEs, distance constraints were obtained and used in the molecular dynamics simulations. From such studies, a model for the preferred conformations of the two diastereomers has been developed [130].

It is important to note that c[S-mPhe-D-Trp-Lys-Thr-Phe-gSar] (the S analog) is 50% as active as the Veber compound as determined by Vale and Yamamoto in the growth hormone inhibition assay and that c[R-mPhe-D-Trp-Lys-Thr-Phe-gSar] (the R analog) is inactive in the same assay. Preliminary in vivo studies on inhibition of pentagastrin-stimulated acid release in dogs carried out by Beglinger and Gyr show that the S analog is superactive, but the R analog is inactive [131]. Figure 17 shows the distance of the mPhe, D-Trp and Lys side chains from each other as a function of the simulation times of both the

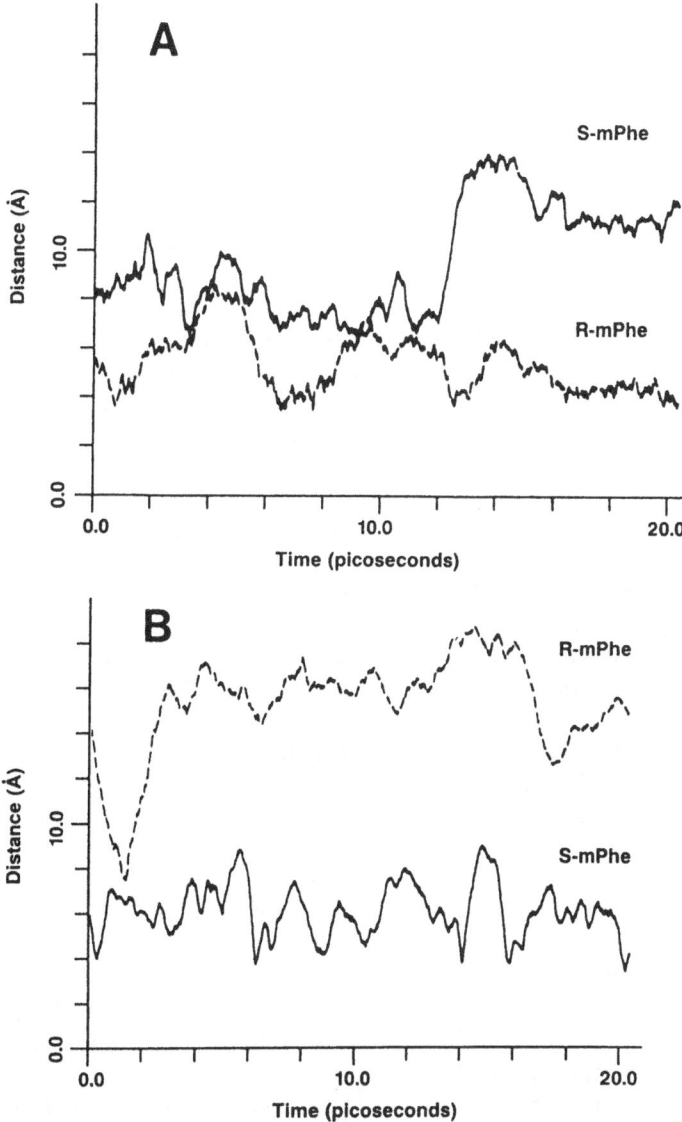

Fig. 17.   *Distance vs. time for specific residues in c[gSar-R and S-mPhe-D-Trp-Lys-Thr-Phe]. (A) Distance between D-Trp and Lys side chains. (B) Distance between D-Trp and Phe side chains.*

S and R isomers. The side chains of the mPhe, the D-Trp and the Lys residues in the S isomer are close to each other for much of the time, whereas in the R isomer the side chain of the mPhe residue is far from the other two throughout the simulation. We propose that the proximity of the side chains of the three residues, as seen in the S isomer, is necessary for somatostatin-like bioactivity.

Neuropeptide Y

Elucidation of the preferred conformations of the 36-amino acid neuropeptide Y (NPY) and biologically active fragments of NPY in solution is underway using CD and NMR techniques. This work represents a collaboration with Rivier and Boublik of the Salk Institute and Brown of the UCSD School of Medicine [132]. Taulane examined the CD spectra of NPY and showed that the neuropeptide is highly helical in TFE and H_2O. The fragment NPY(18–36), which causes a relatively long-term hypotension in conscious rats when administered intra-arterially, is highly helical in TFE and TFE/H_2O mixtures with H_2O content below 20% [132]. In pure H_2O, this fragment is only partially helical (~25%). The NPY(1–19) is not helical in TFE or H_2O [132]. Feinstein has fully assigned the protons of the fragment NPY(18–36) by NMR in dimethylformamide-d_7 (DMF-d_7). More than 50 interresidue NOEs have been identified in DMF-d_7 [133]. The assignment of NPY(18–36) in water is underway. In addition, numerous C-terminal NPY fragments with different biological activity are being studied. These include N-MeNPY(18–36), N-AcNPY(18–36), [D-Ala^{18}]NPY(18–36) and the NPY series (16–36), (19–36), (20–36) and (21–36).

Molecular dynamics simulations are underway on NPY(18–36) using the NOEs obtained in DMF-d_7. A protocol has been developed to explore the conformational space consistent with the observed NOE data. This methodology should result in a family of preferred conformations of NPY(18–36) in solution. The NOE data obtained in water will be treated similarly. In addition, as the aforementioned analogs are investigated, the postulation of structure–activity relationships should become possible.

Nisin

Wakamiya, Shiba and their co-workers have carried out a remarkable synthetic program on nisin, a naturally occurring peptide antibiotic [134]. Through a collaborative study with them, we have made considerable progress in elucidating the structure of nisin [135]. This molecule was a major focus of the research efforts of my colleague and friend, the late Erhard Gross, who made important contributions to this field [136,137].

Jung called our attention to the pronounced homology of nisin, subtilin, epidermin, pep 5 and gallidermin. These structures include unsaturated amino acids and various lanthionine bridges [138]. These bridges define conformationally constrained segments. In a molecular *aufbau* approach, we have characterized these individual segments and are assembling the resulting conformations to aid in determining the structure of the whole molecule.

Palmer, Pattaroni and Mierke began with NMR studies of these fragments using HOHAHA, ROESY and temperature coefficient measurements [135]. These molecules posed new computational challenges. The novel residues required new force constants and parameters that we are developing though calculations, IR spectroscopy, and X-ray diffraction studies [139]. Our approach to molecular

22

dynamics has also been expanded. We now begin with high temperature 750–1000 K) dynamics in an attempt to search thoroughly the conformational space of each molecule. At regular intervals conformations are extracted and energy-minimized. Some of these minimizations involve NOE constraints. However, we are not trying to fit experimental data but to use these constraints as a short-cut in our computational search. Starting from the lowest energy conformations, we proceed with molecular dynamics at 300 K. Again, at regular intervals, we carry out unconstrained minimizations to generate families of low energy conformations.

In such a manner, we have characterized each of the rings of nisin. Time-averaged properties from these low temperature dynamics are in better agreement with our NMR results than individual minimum energy conformations [139]. One example is the observed NOE between the histidine ring and each of the lanthionine bridges in rings D and E. The dynamics clearly show the highly mobile side chain passing near one lanthionine and then the next. We are carrying out additional spectroscopic measurements and calculations on the entire nisin molecule, and several large, active fragments [139]. By our approach we hope to elaborate on the channel-forming properties described by Jung [138].

Urethane-Protected α-Amino Acid *N*-Carboxyanhydrides

This presentation began with the step-wise synthesis of oligopeptides. I want to end with an exciting, novel approach to peptide synthesis. Naider and I working together with Fuller and his co-workers have developed a general method for the preparation of urethane-protected α-amino acid *N*-carboxyanhydrides (NCAs):

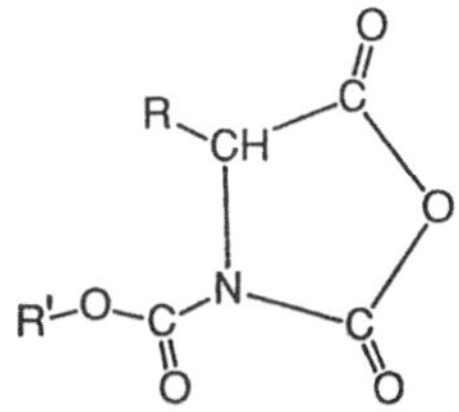

In the past, it has been claimed that such molecules were inherently unstable and could not be prepared [140–145]. We have made many derivatives in which the urethane portion of the NCA is composed of 9-Fmoc, Z and Boc [146]. The urethane-protected NCAs are crystalline, stable compounds that react rapidly with nucleophiles such as amines to form amides or peptide bonds. The general synthesis of urethane-protected NCAs was accomplished by allowing the NCAs to react with the appropriate chloroformate in the presence of *N*-methylmorpholine (a base that does not promote polymerization of NCAs) [146].

We have undertaken some preliminary peptide syntheses with flow system solid phase methodology [146]. The resin was derivatized using an appropriate

urethane-protected NCA and 4-alkoxy-2′, 4′-dimethoxybenzhydrol on a polystyrene resin. The excess hydroxyl groups were capped and the protecting groups
removed. Sequentially, the urethane-protected NCAs were added after removal
of protecting groups and washing steps to build up the desired peptide chain.
In this manner Fuller, Cohen and Shabankareh have prepared [D-Ala²]-leucine
enkephalin, the acyl carrier decapeptide and thymosin α-1 (16–28). Independently
of this work, Felix and Danho have used Fmoc-NCA chemistry in comparison
with a water-soluble carbodiimide, hydroxybenzotriazole approach to prepare
the heptapeptide CCK. Both syntheses involved batchwise methods. Figure 18
shows the HPLC for both schemes. The results are essentially equivalent, although
in the carbodiimide reactions four steps involved double coupling whereas in
the Fmoc-NCA route only one step required double coupling (i.e., the methionine
to aspartic acid linkage). We are actively examining the urethane-protected NCA
methodology as a general approach to peptide synthesis.

Concluding Remarks

I have attempted to cover more than thirty years of our research and develop
a story indicating the great challenges that remain for the future. In listening
to the talks at this Eleventh American Peptide Symposium and perusing the
posters, I see peptide chemistry and biology as a vibrant field. With the support

Ac-Tyr¹-Met²-Gly³-Trp⁴-Met⁵-Asp⁶-Phe⁷-NH₂

FMOC - NCA Couplings   DIC / HOBt Couplings

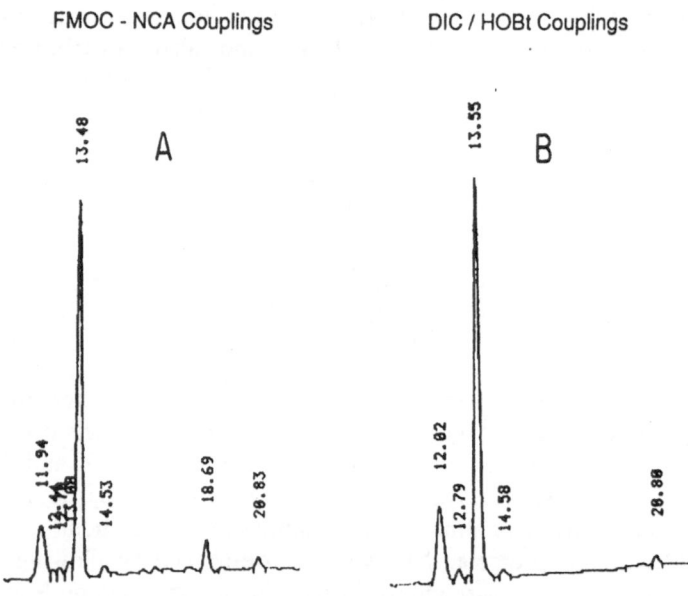

*Fig. 18. HPLC traces of the CCK-7 products from the syntheses (A) utilizing the urethane-
protected NCA method and (B) using the diisopropylcarbodiimide-hydroxybenzotriazole
approach.*

of people such as Dr. Pierce, peptide scientists will continue to make remarkable discoveries. Once again, I wish to thank my co-workers, colleagues and friends. I am honored to accept the Seventh Alan E. Pierce Award.

Acknowledgements

I want to thank Joseph Taulane for helping me design the presentation for this Pierce Award lecture. His tireless efforts, together with my research group, ensured that the presentation would be an accurate and meaningful summary of our research. I also want to indicate my gratitude to the National Institutes of Health and the National Science Foundation for their years of support, which enabled us to carry out the work that led to this recognition.

References

1. Goodman, M. and Kenner, G.W., Adv. Prot. Chem., 12 (1957) 465.
2. Pauling, L. and Corey, R.B., Proc. Natl. Acad. Sci. U.S.A., 37 (1951) 729.
3. Pauling, L. and Corey, R.B., Proc. Natl. Acad. Sci. U.S.A., 37 (1951) 235.
4. Goodman, M. and Schmitt, E.E., J. Am. Chem. Soc., 81 (1959) 5507.
5. Goodman, M. and Schmitt, E.E., Ann. N.Y. Acad. Sci., 88 (1960) 669.
6. Goodman, M. and Steuben, K.C., J. Am. Chem. Soc., 24 (1959) 3980.
7. Goodman, M. and Steuben, K.C., J. Am. Chem. Soc. 84 (1962) 1279.
8. Goodman, M. and Steuben, K.C., J. Org. Chem., 27 (1962) 3409.
9. Goodman, M., Listowsky, I. and Schmitt, E.E., J. Am. Chem. Soc., 84 (1962) 1296.
10. Goodman, M., Schmitt, E.E., Listowsky, I., Boardman, F., Rosen, I.G. and Stake, M.A., Polyamino Acids, Polypeptides and Proteins, 1963, p. 195.
11. Goodman, M., Listowsky, I., Masuda, Y. and Boardman, F., Biopolymers, 1 (1963) 33.
12. Goodman, M., Boardman, F. and Listowsky, I., J. Am. Chem. Soc., 85 (1963) 2491.
13. Goodman, M. and Boardman, F., J. Am. Chem. Soc., 85 (1963) 2483.
14. Goodman, M., Langsam, M. and Rosen, I.G., Biopolymers, 4 (1966) 305.
15. Moffitt, W., J. Am. Phys., 25 (1956) 467.
16. Moffitt, W. and Yang, J.T., Proc. Natl. Acad. Sci. U.S.A., 43 (1956) 723.
17. Blout, E.R. and Karlson, R.H., J. Am. Chem. Soc., 80 (1958) 1259.
18. Karlson, R.H., Norland, K.S., Fasman, G.D. and Blout, E.R., J. Am. Chem. Soc., 82 (1960) 2268.
19. Blout, E.R. and Asadourian, A., J. Am. Chem. Soc., 78 (1956) 955.
20. Blout, E.R., Holtzer, A.M., Bradbury, J.H. and Doty, P., J. Am. Chem. Soc., 76 (1954) 4493.
21. Doty, P. and Yang, J.T., J. Am. Chem. Soc., 78 (1956) 498.
22. Doty, P. and Lundberg, R.D., J. Am. Chem. Soc., 78 (1956) 4810.
23. Rosenheck, K. and Doty, P., Proc. Natl. Acad. Sci. U.S.A., 47 (1961) 1775.
24. Berger, A., Kurtz, J. and Katchalski, E., J. Am. Chem. Soc., 76 (1954) 5552.
25. Steinberg, I.Z., Harrington, W.F., Berger, A., Sela, M. and Katchalski, E., J. Am. Chem. Soc., 82 (1960) 5263.
26. Katchalski, E. and Sela, M., Advances in Protein Chemistry, Academic Press, New York, 1958, p. 243.
27. Woody, R.W., Macromol. Rev., 12 (1977) 181.
28. Goodman, M. and Rosen, I.G., Biopolymers, 2 (1964) 519.
29. Goodman, M. and Rosen, I.G., Biopolymers, 2 (1964) 537.

30. Holzwarth, G. and Doty, P., J. Am. Chem. Soc., 87 (1965) 218.
31. Goodman, M., Verdini, A.S., Toniolo, C., Phillips, W.D. and Bovey, F.A., Proc. Natl. Acad. Sci. U.S.A., 64 (1969) 444.
32. Goodman, M., Rupp, R. and Naider, F., Bioorg. Chem., 1 (1971) 294.
33. Goodman, M., Rupp, R. and Naider, F., Bioorg. Chem., 1 (1971) 310.
34. Goodman, M., Verdini, A.S., Choi, N.S. and Masuda, Y., Topics in Stereochemistry, Vol. 5, John Wiley and Sons, New York, 1970, p. 69.
35. Bovey, F.A., Tiers, G.V.P. and Filipovich, G., J. Polym. Sci., 38 (1959) 73.
36. Johnson, Jr., C.E. and Bovey, F.A., J. Chem. Phys., 29 (1958) 1012.
37. Goodman, M. and Masuda, Y., Biopolymers, 2 (1964) 107.
38. Goodman, M. and Saltman, R.P., Biopolymers, 20 (1981) 1929.
39. Bayer, E. and Mutter, M., Nature, 237 (1972) 512.
40. Ribeiro, A.A., Saltman, R.P. and Goodman, M., Biopolymers, 24 (1985) 2431.
41. Ribeiro, A.A., Saltman, R.P. and Goodman, M., Biopolymers, 24 (1985) 2449.
42. Ribeiro, A.A., Saltman, R.P. and Goodman, M., Biopolymers, 24 (1985) 2469.
43. Ribeiro, A.A., Saltman, R.P. and Goodman, M., Biopolymers, 24 (1985) 2495.
44. Pillai, V.N.R. and Mutter, M., Acc. Chem. Res., 14 (1981) 122.
45. Goodman, M. and Levine, L., J. Am. Chem. Soc., 86 (1964) 2918.
46. Goodman, M. and McGahren, W.J., J. Am. Chem. Soc., 87 (1965) 3028.
47. Goodman, M. and McGahren, W.J., J. Am. Chem. Soc., 88 (1966) 3887.
48. Goodman, M. and McGahren, W.J., Tetrahedron, 23 (1967) 2017.
49. Goodman, M. and McGahren, W.J., Tetrahedron, 23 (1967) 2031.
50. Goodman, M. and Glaser, C.B., Tetrahedron Lett., 40 (1969) 3473.
51. Goodman, M. and Glaser, C.B., In Lande, S. (Ed.) Proceedings of the 1st American Peptide Symposium, Yale University, Gordon and Breach, New York, 1968, p. 267.
52. Goodman, M. and Glaser, C.B., J. Org. Chem., 35 (1970) 1954.
53. Goodman, M. and Arnon, U., Biopolymers, 1 (1963) 500.
54. Goodman, M. and Arnon, U., J. Am. Chem. Soc., 86 (1964) 3384.
55. Goodman, M. and Hutchison, J., J. Am. Chem. Soc., 87 (1965) 3524.
56. Goodman, M. and Hutchison, J., J. Am. Chem. Soc., 88 (1966) 3627.
57. Peggion, E., Scoffone, E., Cosani, A. and Portolan, A., Biopolymers, 4 (1966) 695.
58. Goodman, M. and Peggion, E., Pure Appl. Chem., 3 (1981) 699.
59. Peggion, E., Goodman, M., Szwarc, M. and Bamford, C.H., Macromolecules, 10 (1977) 1299.
60. Goodman, M. and Felix, A.M., Biochemistry, 3 (1964) 1529.
61. Bradbury, E.M., Downey, A.R., Elliott, A. and Hanby, W.E., Proc. Roy. Soc. Sect. A, 259 (1960) 110.
62. Goodman, M. and Peggion, E., Biochemistry, 6 (1967) 1533.
63. Goodman, M., Toniolo, C. and Peggion, E., Biopolymers, 6 (1968) 1691.
64. Goodman, M. and Kossoy, A., J. Am. Chem. Soc., 88 (1966) 5010.
65. Davis, G.W., Goodman, M. and Benedetti, E., Acc. Chem. Res., 1 (1968) 275.
66. Goodman, M. and Benedetti, E., Biochemistry, 7 (1968) 4226.
67. Benedetti, E., Kossoy, A., Falxa, M.L. and Goodman, M., Biochemistry, 7 (1968) 4234.
68. Benedetti, E. and Goodman, M., Biochemistry, 7 (1968) 4242.
69. Fissi, A., Pieroni, O. and Ciardelli, F., Biopolymers, 26 (1987) 1993.
70. Pieroni, O., Fabbri, D., Fissi, A. and Ciardelli, F., Makromol. Chem., Rapid Commun., 9 (1988) 637.
71. Irie, M., In Winnik, M.A. (Ed.) Photophysical and Photochemical Tools in Polymer Science, Reidel, Dordrecht, 1986, p. 269.
72. Ueno, A., Morikawa, Y., Anzai, J. and Osa, T., Makromol. Chem., Rapid Commun., 5 (1984) 639.
73. Ueno, A., Takahashi, K., Anzai, J. and Osa, T., J. Am. Chem. Soc., 103 (1981) 6410.
74. Yamamoto, H. and Nishida, A., J. Photochem. Photobiol., A. Chem., 42 (1988) 149.

75. Yamamoto, H. and Nishida, A., Bull. Chem. Soc. Jpn., 61 (1988) 2201.
76. Westheimer, F.A., In Newman, M.S. (Ed.) Steric Effects in Organic Chemistry, John Wiley and Sons, New York, 1956, p. 523.
77. Poland, D. and Scheraga, H.A., In Fasman, G.D. (Ed.) Poly-α-amino Acids, Marcel Dekker, New York, 1967, p. 391.
78. Nemethy, G. and Scheraga, H.A., Biopolymers, 3 (1965) 155.
79. Pitzer, K.S., Science, 101 (1945) 672.
80. Pitzer, K.S., Disc. Faraday Soc., 10 (1951) 66.
81. Eliel, E.L., Allinger, N.L., Angyal, S.J. and Morrison, G.A. (Eds.) Conformational Analysis, Interscience Publishers, New York, 1965.
82. Lifson, S., J. Chem. Phys., 30 (1959) 964.
83. Lifson, S., J. Chem. Phys., 40 (1964) 3705.
84. Liquori, A.M., Acta Crystallogr., 8 (1955) 345.
85. DeSantis, P., Giglio, E., Liquori, A.M. and Ripamonti, A., J. Polymer Sci. A., 1 (1963) 1383.
86. Flory, P.J., Statistical Mechanics of Chain Molecules, John Wiley and Sons, New York, 1969.
87. Brant, D.A. and Flory, P.J., J. Am. Chem. Soc., 87 (1965) 2791.
88. Ramakrishnan, C. and Ramachandran, G.N., Biophys. J., 5 (1965) 909.
89. Ramachandran, G.N., Ramakrishnan, C. and Sasiskharan, V., J. Mol. Biol., 7 (1963) 95.
90. Mark, J.E. and Goodman, M., J. Am. Chem. Soc., 89 (1967) 1267.
91. Mark, J.E. and Goodman, M., Biopolymers, 5 (1967) 809.
92. Goodman, M. and Fried, M., J. Am. Chem. Soc., 89 (1967) 1264.
93. Chen, F., Prince, F.R. and Goodman, M., Biopolymers, 12 (1973) 2549.
94. Liquori, A.M. and DeSantis, P., Biopolymers, 5 (1967) 815.
95. Goodman, M. and D'Alagni, M., J. Polymer Sci. (Polymer Lett.), 5 (1967) 515.
96. Nissen, D., Gilon, C. and Goodman, M., Makromol. Chem., 1 (1975) 25.
97. Mathias, L.J., Fuller, W.D., Nissen, D. and Goodman, M., Macromolecules, 11 (1978) 534.
98. Katakai, R. and Goodman, M., Macromolecules, 15 (1982) 25.
99. Ingwall, R.T. and Goodman, M., Macromolecules, 7 (1974) 598.
100. Ingwall, R.T., Gilon, C. and Goodman, M., Macromolecules, 9 (1976) 802.
101. Becktel, W., Goodman, M., Katakai, R. and Wouters, G., Makromol. Chem. Suppl., 4 (1981) 100.
102. Wouters, G., Katakai, R., Becktel, W.J. and Goodman, M., Macromolecules, 15 (1982) 31.
103. Becktel, W.J., Wouters, G., Simmons, D.M. and Goodman, M., Macromolecules, 18 (1985) 630.
104. Burgus, R., Butcher, M., Amoss, M., Ling, N., Monahan, M., Rivier, J., Fellows, R., Blackwell, R., Vale, W. and Guillemin, R., Proc. Natl. Acad. Sci. U.S.A., 69 (1972) 278.
105. Burgus, R. and Guillemin, R., Ann. Rev. Biochem., 39 (1970) 499.
106. Blagdon, D.E., Rivier, J. and Goodman, M., Proc. Natl. Acad. Sci. U.S.A., 70 (1973) 1166.
107. Donzel, B., Gilon, C., Blagdon, D., Erisman, M., Burnier, J., Goodman, M., Rivier, J. and Monahan, M., In Walter, R. and Meienhofer, J. (Eds.) Proceedings of the 4th American Peptide Symposium, Ann Arbor Science Publishers, Ann Arbor, MI, 1975, p. 165.
108. Donzel, B., Goodman, M., Rivier, J., Ling, N. and Vale, W., Nature, 256 (1975) 750.
109. Sakarellos, C., Donzel, B. and Goodman, M., Biopolymers, 15 (1976) 1835.
110. Goodman, M. and Chorev, M., Acc. Chem. Res., 12 (1979) 1.
111. Chorev, M., Willson, G.C. and Goodman, M., J. Am. Chem. Soc., 99 (1977) 8075.

112. Pallai, P. and Goodman, M., J. Chem. Soc., Chem. Commun., (1982) 280.
113. Goodman, M. and Chorev, M., In Eberle, A., Geiger, R. and Wieland, T. (Eds.) Perspectives in Peptide Chemistry, Karger, Basel, 1981, p. 283.
114. Chorev, M. and Goodman, M., Int. J. Pept. Prot. Res., 21 (1983) 258.
115. Chorev, M., MacDonald, S.A. and Goodman, M., J. Org. Chem., 49 (1984) 821.
116. Pallai, P., Richman, S., Struthers, R.S. and Goodman, M., Int. J. Pept. Prot. Res., 21 (1983) 84.
117. DiMaio, J. and Schiller, P.W., Proc. Natl. Acad. Sci. U.S.A., 77 (1980) 7162.
118. Berman, J. and Goodman, M., Int. J. Pept. Prot. Res., 23 (1984) 610.
119. Richman, S., Goodman, M., Nguyen, T. and Schiller, P.W., Int. J. Pept. Prot. Res., 25 (1985) 648.
120. Mierke, D.F., Lucietto, P., Schiller, P.W. and Goodman, M., Biopolymers, 26 (1987) 1573.
121. Berman, J., Goodman, M., Nguyen, T. and Schiller, P.W., Biochem. Biophys. Res. Commun., 115 (1983) 864.
122. Mammi, N.J. and Goodman, M., Biochemistry, 25 (1986) 7607.
123. Hassan, M. and Goodman, M., Biochemistry, 25 (1986) 7596.
124. Mammi, N., Hassan, M. and Goodman, M., J. Am. Chem. Soc., 107 (1985) 4008.
125. Wüthrich, K., NMR of Proteins and Nucleic Acids, John Wiley and Sons, New York, 1986, p. 176.
126. Dauber, P., Goodman, M., Hagler, A.T., Osguthorpe, D., Sharen, R. and Stern, P., In Lykos, P. and Shavitt, I. (Eds.) Proceedings of ACS Symposium on Supercomputers in Chemistry, ACS, Washington, DC, 1981, p. 161.
127. Hruby, V.J., Kao, L.-F., Pettitt, B.M. and Karplus, M., J. Am. Chem. Soc., 110 (1988) 3351.
128. Mierke, D.F., Said-Nejad, O.E., Yamazaki, T., Felder, E., Schiller, P.W. and Goodman, M., In Rivier, J.E. and Marshall, G.R. (Eds.) Peptides: Chemistry, Structure and Biology (Proceedings of the 11th American Peptide Symposium), ESCOM, Leiden, 1990, p. 348.
129. Veber, D.F., Freidinger, R.M., Perlow, D.S., Palaveda, Jr., W.J., Holly, F.W., Strachan, R.G., Nutt, R.F., Arison, B.J., Homnick, C., Randall, W.C., Glitzer, M.S., Saperstein, R. and Hirschmann, R., Nature, 292 (1981) 55.
130. Mierke, D.F. and Pattaroni, C., (1989) manuscript in preparation.
131. Beglinger, L. and Gyr, K., personal communication.
132. Boublik, J., Scott, N., Taulane, J., Goodman, M., Brown, M. and Rivier, J., Int. J. Pept. Prot. Res., 33 (1989) 11.
133. Boublik, J.H., Scott, N.A., Feinstein, R.D., Goodman, M., Brown, M.R. and Rivier, J.E., In Rivier, J.E. and Marshall, G.R. (Eds.) Peptides: Chemistry, Structure and Biology (Proceedings of the 11th American Peptide Symposium), ESCOM, Leiden, 1990, p. 317.
134. Fukase, K., Kitazawa, M., Sano, A., Shimbo, K., Fujita, H., Horimoto, S., Wakamiya, T. and Shiba, T., Tetrahedron Lett., 29 (1988) 795.
135. Palmer, D.E., Mierke, D.F., Pattaroni, C., Goodman, M., Wakamiya, T., Fukase, K., Kitazawa, M., Fujita, H. and Shiba, T., Biopolymers, 28 (1989) 397.
136. Gross, E., In Meienhofer, J. (Ed.) Proceedings of the 3rd American Peptide Symposium, Ann Arbor Science Publishers, Ann Arbor, MI, 1972, p. 671.
137. Gross, E., In Walter, R. and Meienhofer, J. (Eds.) Proceedings of the 4th American Peptide Symposium, Ann Arbor Science Publishers, Ann Arbor, MI, 1975, p. 31.
138. Jung, G., In Rivier, J.E. and Marshall, G.R. (Eds.) Peptides: Chemistry, Structure and Biology (Proceedings of the 11th American Peptide Symposium), ESCOM, Leiden, 1990, p. 865.
139. Palmer, D.E., Mierke, D.F., Pattaroni, C., Nunami, K., Ro, S., Huang, Z., Wakamiya, T., Fukase, K., Kitazawa, M. Fujita, H., Shiba, T. and Goodman, M., In Rivier,

J.E. and Marshall, G.R. (Eds.) Peptides: Chemistry, Structure and Biology (Proceedings of the 11th American Peptide Symposium), ESCOM, Leiden, 1990, p. 616.

140. Kricheldorf, H.R., α-Amino Acid-*N*-Carboxy-Anhydrides and Related Heterocycles, Springer-Verlag, New York, 1987, p. 59.

141. Bailey, J.L., J. Chem. Soc., (1950) 3461.

142. Denkewalter, R.G., Schwam, H., Strachan, R.G., Beesley, T.E., Veber, D.F., Schoenewaldt, E.F., Barkemeyer, H., Paleveda, Jr., W.J., Jacob, T.A. and Hirschmann, R., J. Am. Chem. Soc., 88 (1966) 3163.

143. Halstrom, J.B. and Kovacs, K.G., U.S. Patent 4267344, 1981 (*N*-benzyl, *N*-benzhydryl, *N*-xanthyl, *N*-pentamethoxybenzene).

144. Block, H. and Cox, M.E., In Young, G.T. (Ed.) Peptides (Proceedings of the 5th European Symposium) , MacMillan, New York, 1963, p. 83.

145. Kricheldorf, H.R., Makromol. Chem., 178 (1977) 905.

146. Fuller, W.D., Cohen, M., Shabankareh, M., Naider, F. and Goodman, M., J. Am. Chem. Soc., (1989) submitted.

Session I
Structural biology

Chairs: Ralph F. Hirschmann
University of Pennsylvania
Philadelphia, Pennsylvania, U.S.A.

Kenneth D. Kopple
Smith Kline & French Laboratories
Swedeland, Pennsylvania, U.S.A.

Christian Birr
ORPEGEN
Heidelberg, F.R.G.

and

Antonio C.M. Paiva
Escola Paulista de Medicina
São Paulo, Brazil

Bicyclic gonadotropin releasing hormone (GnRH) antagonists

J.E. Rivier[a], C. Rivier[a], W. Vale[a], S. Koerber[a], A. Corrigan[a], J. Porter[a], L.M. Gierasch[b] and A.T. Hagler[a]

[a]*The Clayton Foundation Laboratories for Peptide Biology, The Salk Institute, 10010 North Torrey Pines Road, La Jolla, CA 92037, U.S.A.*
[b]*Department of Pharmacology, University of Texas, Southwestern Medical Center, 5323 Harry Hines Blvd., Dallas, TX 75235-9041, U.S.A.*

Introduction

The secretion of luteinizing hormone (LH) and follicle-stimulating hormone (FSH) is under the stimulatory control of the decapeptide amide, gonadotropin releasing hormone (pGlu-His-Trp-Ser-Tyr-Gly-Leu-Arg-Pro-Gly-NH$_2$). Because mammalian GnRH assumes mostly a random coil conformation in aqueous solution, and because the GnRH pituitary receptor has not been chemically or structurally characterized, the nature of their physico-chemical interaction cannot be studied spectroscopically. In order to define the 'bioactive conformation' of GnRH, we embarked on a multidisciplinary study whose aim was to obtain a fully rigid and maximally potent analog for spectroscopic analysis. Because we had found that bridging GnRH sequences at positions 1 and 10 or 4 and 10 led to partial agonists or antagonists repectively, and because we have more latitude in the selection of amino acid substitutions in the design of antagonists, we designed, synthesized and tested cyclic GnRH antagonists [1–4]. Here, we show that some antagonists of GnRH which are bridged at positions 4 and 10 and are equipotent to the corresponding and most potent linear analogs [2], can be further constrained by the introduction of a second bridge between residues 5 and 8 with retention of high potency. This major success could be achieved through the development and implementation of a concerted stragegy involving the use of peptide synthesis and biological testing, computer simulations and NMR spectroscopy [2,5,6].

Methods

Peptide synthesis

Peptides were synthesized by SPPS using the Boc strategy on a MBHAR and the general techniques previously reported by this laboratory [2] including HF treatment. The 5/8 side-chain amide bond of **3** and **4** was generated first on the resin using the method of Felix et al. [7] followed by the azide mediated closure of the 4/10 ring also used for **1** and **2**. Final preparative purification was carried out with HPLC [8]. Peptides were judged to be greater than 95%

pure using HPLC. AAA were consistent with expected values. Calculated values for protonated molecular ions were in agreement with those obtained using FABMS.

Bioassays

For the purpose of this limited study, analogs were tested for their ability to inhibit ovulation in the cycling rat [9] and to inhibit LH release by the pituitary in an in vitro cell culture assay [10].

Computer simulations

Computer simulations at the San Diego Supercomputer Center Cray X-MP/ 48, were conducted using methods previously described [11].

Results and Discussion

Our approach to obtaining extremely potent monocyclic antagonists of GnRH has been described elsewhere [2,5]. Combined molecular dynamics and NMR studies on the constrained cyclo (4–10)[Ac-Δ^3Pro[1],DFpa[2],D-Trp[3],Asp[4],D-Nal[6], Dpr[10]]GnRH antagonists have provided structural information [12,13].

Because of the constrained nature of this analog, this analysis will help refine our model of the binding conformation. NMR results (J. Rizo et al., in preparation) indicate the presence of a stable β-turn, around the D-Nal[6]-Leu[7] positions, a feature that has been postulated to be necessary for biological activity. Temperature dependence of NH resonances are consistent with three transannular hydrogen bonds involving the NHs of Asp, Tyr and Arg. By NMR, the N-terminal tripeptide of this analog displays some conformational preference, although it is less rigid than the cyclic portion. Interestingly, molecular dynamics calculations (S. Koerber et al., in preparation) yield two conformational families for this peptide with the N-terminal tripeptide either below or above the ring. NMR studies are underway to analyze whether one of these two conformational families is preferred, or alternatively, if there is an intermediate, more dynamic conformational distribution for the N-terminal tripeptide. However, careful examination of the N-terminal tripeptide-down conformer indicates that the following side chains are in close proximity, 2 and 7, 3 and 9, 5 and 6, and 5 and 8 suggesting that they may be replaced by bridging elements.

We have synthesized a large number of cyclo 2/7 antagonists of GnRH that were at best marginally potent [3]. On the other hand the bridging of residues 5 to 8 seemed very promising. Indeed, we reported a year ago on the synthesis and biological activity of bicyclo(4/10,5/8)[Ac-D-Nal[1],D-Cpa[2], D-Pal[3],Asp[4], Cys[5],D-Arg[6],Cys[8],Dpr[10]]GnRH (analog A) [3]. While this work was being carried out, Dutta et al. [14] described the synthesis and biological potencies of several monocyclic (5–8)GnRH antagonists. These analogs in which the side chains of different amino acids at positions 5 and 8 were bridged using either amide bonds or disulfide bridges in a systematic manner similar to that reported by our group [2] are less potent (approximately 5- to 10-fold) than those having similar bridges at positions 4 and 10. Whether the low potency of the cyclo (5–8) analogs reported by Dutta et al. results from the use by these authors of nonoptimized substitutions at positions 1 through 3 or from unfavorable constraints brought about by the bridging elements, is yet to be determined.

Table 1   *In vitro and in vivo potencies and receptor affinities of cyclic GnRH antagonists*

Compounds	AOA[a]		In vitro potencies[b]
(Cyclo 4/10)[Ac-Δ^3Pro^1,D-Fpa^2,D-Trp^3, Asp^4,D-Nal^6,Dpr^{10}]GnRH (**1**)	10	(2/10)	2.0 (1.5–2.6)
(Cyclo 4/10)[Ac-D-Nal^1,D-Fpa^2,D-Trp^3, Asp^4,D-Arg^6,Dpr^{10}]GnRH (**2**)	2.5	(2/10)	0.3 (0.2–0.4)
bicyclo(4/10,5/8)[Ac-D-Nal^1,D-Cpa^2,D-Pal^3, Asp^4,Glu^5,D-Nal^6,Lys^8,Dpr^{10}]GnRH (**3**)	50	(2/ 8)	0.6 (0.4–0.9)
bicyclo(4/10,5/8)[Ac-D-Nal^1,D-Cpa^2,D-Trp^3, Asp^4,Glu^5,D-Arg^6,Lys^8,Dpr^{10}]GnRH (**4**)	5	(2/10)	

[a] AOA-antiovulatory assay: dosage in micrograms (rats ovulating/total).
[b] [Ac-Δ^3Pro^1,D-Fpa^2,D-Trp^{3,6}]GnRH $= 1$.

Since we had reported that analog A inhibited ovulation by 86% at 200 μg and since the Cys (5 to 8) ring was not the optimized ring size according to Dutta et al. [14], we synthesized 3 and 4 (Table 1) which, to our knowledge, encompassed some of the most favored substitutions. These analogs were tested for biological activity and their structures analyzed in order to compare them with those of analog A, and 1 and 2 resp. We found 3 and 4 to inhibit ovulation by 75% at 50 μg and 80% at 5 μg per rat, respectively. While 4 is the most potent bicyclic GnRH analog ever reported, 3 is the first analog of GnRH that does not carry a positive charge but that has significant affinity for the GnRH receptor in an in vitro assay. A molecular mode of an homolog of 4 is shown in Fig. 1 and provides the rationale for the introduction of additional constraints.

Conclusions

Using a model derived from theoretical studies of a 4–10 bridged monocyclic GnRH analog (1), we observed that several pairs of amino acid side chains were close enough in space to be bridged. We synthesized a series of bicyclic analogs with either disulfide or amide bonds. Compounds 3 and 4 were found to be the most potent bicyclic analogs tested so far. Preliminary theoretical studies suggest that the conformation of the 4–10 cycle is conserved in these analogs despite the introduction of the second ring. We realize that, because the N-terminal tripeptide can be located either above or below (Fig. 1) the ring formed by the 4–10 cycle, additional constraints will have to be introduced in order to obtain a fully rigid analog. It is believed that full structural charac-terization of such an analog could help us understand the mechanism of action of GnRH and the molecular basis of activation of its receptor.

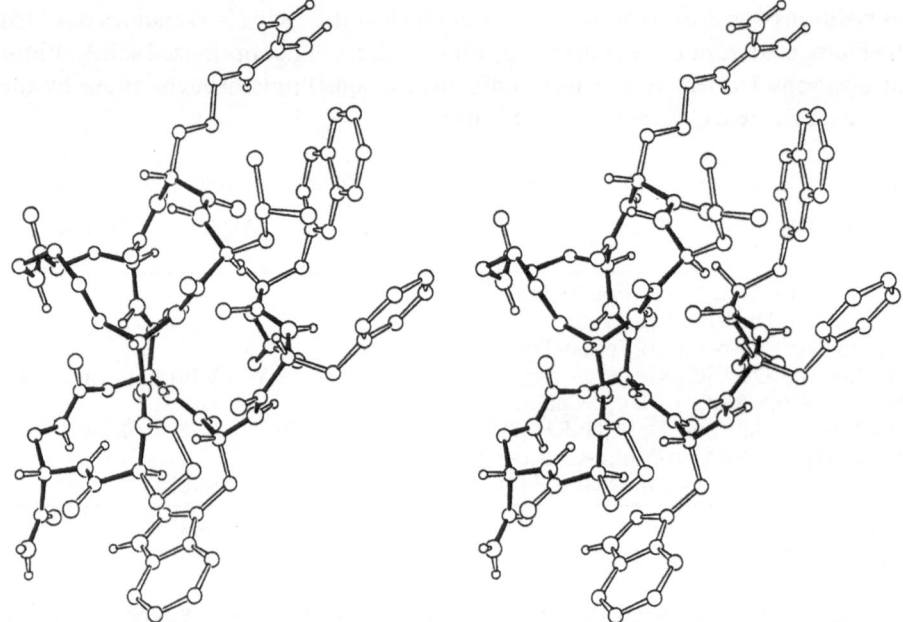

Fig. 1.   ORTEP drawing of (bicyclo 4/10,5/8) Ac-D-Nal-D-Phe-D-Trp-Asp-Glu-D-Arg-Leu-Lys-Pro-Dpr-NH₂ analogous to 4.

Acknowledgements

Research supported by NIH contract NO1-HD-4-2833. Research conducted in part by The Clayton Foundtion for Research, California Division. CR and WV are Clayton Foundation Investigators. Computer simulations were carried out in part on the Cray X-MP/48 at the San Diego Supercomputer Center. The authors thank Ron Kaiser, John Dykert, Duane Pantoja, Rosalia Chavarin and Leatrice Gandara for excellent technical assistance, and Rebecca Hensley for manuscript preparation.

References

1. Rivier, J., Rivier, C., Perrin, M., Porter, J. and Vale, W., In Zatuchni, G.I., Shelton, J.D. and Sciarra, J.J. (Eds.) LHRH Peptides as Female and Male Contraceptives, Harper and Row, Philadelphia, PA, 1981, p. 13.
2. Rivier, J., Kupryszewski, G., Varga, J., Porter, J., Rivier, C., Perrin, M., Hagler, A., Struthers, S., Corrigan, A. and Vale, W., J. Med. Chem., 31 (1988) 677.
3. Rivier, J., Koerber, S., Rivier, C., Hagler, A., Perrin, M., Gierasch, L., Corrigan, A., Porter, J. and Vale W., In Chen, H.-C. (Ed.) International Symposium on Frontiers in Reproduction Research: The Role of Growth Factors, Oncogenes, Receptors and Gonadal Polypeptides, Beijing, China, 1988, in press.
4. Rivier, J., Koerber, S., Rivier, C., Porter, J. and Hagler, A., In Villafranca, J. (Ed.)

Current Research in Protein Chemistry, Academic Press, Orlando, FL, 1989, in press.

5. Struthers, R.S., Hagler, A. and Rivier, J., In Vida, J.A. and Gorden, M. (Eds.) Conformationally Directed Drug Design, American Chemical Society, Washington, D.C., Series No. 251, 1984, p. 239.

6. Struthers, R.S., Rivier, J. and Hagler, A.T., In Venkataraghavan, B. and Feldman, R.J. (Eds.) Macromolecular Structure and Specificity: Computer Assisted Modeling and Applications, Ann. New York Acad. Sci., 459 (1985) 81.

7. Felix, A.M., Wang, C.-T., Heimer, E., Fournier, A., Bolin, D.R., Ahmad, M., Lambros, T., Mowles, T. and Miller, L., In Marshall, G.R. (Ed.) Peptides: Chemistry and Biology (Proceedings of the 10th American Peptide Symposium), ESCOM, Leiden, 1988, p. 465.

8. Hoeger, C., Galyean, R., Boublik, J., McClintock, R., Rivier, J., Biochromatography, 2 (1987) 134.

9. Corbin, A. and Beattie, C.W., Endocr. Res. Commun., 2 (1975) 1.

10. Vale, W. and Grant, G. In O'Malley, B.W. and Hardman, J.G. (Eds.) In vitro pituitary hormone secretion for hypophysiotropic substances, Methods in Enzymology: Hormones and Cyclic Nucleotides. Academic Press, New York, NY, 1974, p. 82.

11. Hagler, A.T., In Hruby, V.J. (Ed.) The Peptides: Analysis, Synthesis, Biology – Conformation in Biology and Drug Design, Vol. 7, Academic Press, Orlando, FL, 1985, p. 213.

12. Baniak II, E.L., Rivier, J.E., Struthers, R.S., Hagler, A. and Gierasch, L.M., Biochemistry, 26 (1987) 2642.

13. Baniak II, E.L., Gierasch, L.M., Rivier, J.E. and Hagler, A.T., In Marshall, G.R. (Ed.) Peptides: Chemistry and Biology (Proceedings of the 10th American Peptide Symposium), ESCOM, Leiden, 1988, p. 457.

14. Dutta, A.S., Gormley, J.J., McLachlan, P.F. and Woodburn, J.R., Biochem. Biophys. Res. Commun., 159 (1989) 1114.

Physiologically important peptidases and proteases: Ideal targets for the design of new therapeutics

James C. Powers, Chih-Min Kam, Józef Oleksyszyn and Toshihisa Ueda
School of Chemistry, Georgia Institute of Technology, Atlanta, GA 30332, U.S.A.

Introduction

Peptidases and proteases are involved in a wide variety of important physiological processes. Serine proteases are important for blood coagulation, the immune defense, and the destruction of invading organisms. Metallopeptidases and metalloproteases are involved in the processing of most hormones, including angiotensin and enkephalins. Aspartic proteases are important in viral infections and the control of blood pressure. Cysteine proteases process hormones, and are involved in tumors. Protease inhibitors thus have great potential for the treatment of many disease states. Important target enzymes for the design of new inhibitors included human leukocyte elastase (emphysema, inflammation), chymases and tryptases (inflammation and blistering), blood coagulation enzymes (intravascular clotting), renin and angiotensin-converting enzyme (hypertension), and the HIV protease (AIDS). The tertiary structures of several important proteases, including human leukocyte elastase and the HIV protease, have recently been determined and should be invaluable in the design of new classes of inhibitors.

Results and Discussion

Arginine fluoroketone transition state inhibitors

Arginine fluoroalkyl ketones are potential transition-state inhibitors for trypsin-like enzymes due to their ability to form tetrahedral adducts with the active site serine residue. Seven inhibitors have been synthesized using a modified Dakin-West procedure (Table 1). The structure of $Bz-Arg-CF_2CF_3$ was analyzed by ^{19}F NMR and MS, and was shown to exist in solution primarily as a hydrate or cyclic carbinolamine. Arginine fluoroalkyl ketones are good inhibitors of blood coagulation serine proteases and were found to be slow binding inhibitors for bovine trypsin with K_I values of 0.2–56 μM. $Bz-Arg-CF_2CF_3$ was the best inhibitor for bovine thrombin and human factor XIa, and inhibited thrombin and factor XIa competitively with K_I values of 13 and 62 μM, respectively. $Bz-Arg-CF_3$ and $Bz-Arg-CF_2CF_3$ inhibited human plasma kallikrein competitivelywith K_I values of 50 μM. None of the seven arginine fluoroalkyl ketones were good inhibitors of human factor Xa and factor XIIa.

Arginine fluoroketones showed some selectivities toward various coagulation

Table 1 *Reversible inhibition of trypsin-like serine proteases by arginine fluoroketones[a]*

Inhibitors	K_I (μM)			
	Trypsin	Thrombin	Kallikrein	XIa
Bz-Arg-CF$_3$	0.6	>60	50	>410
Bz-Arg-CF$_2$CF$_3$	7.4	13	52	62
Bz-Arg-CF$_2$CF$_2$CF$_3$	6.3	>90	>170	>1130
p-Toluoyl-Arg-CF$_3$	56	>210	>300	>1130
Admantanoyl-Arg-CF$_3$	1.9	>45	>330	>1130
1-Naphthoyl-Arg-CF$_3$	0.2	>40	>60	>500
Bz-Arg-CF$_2$Cl	3.8	>56	>80	180

[a] Inhibition constants were measured at 0.1 M Hepes, 0.01 M CaCl$_2$, pH 7.5, 9% Me$_2$SO and 25°C. All inhibitors show slow-binding type of inhibition with trypsin and the rates were measured after incubation of enzyme with inhibitor for 5 min.

enzymes. Bz-Arg-CF$_2$CF$_3$ is the best inhibitor for thrombin, plasma kallikrein and factor XIa; however, it didn't inhibit pancreatic kallikrein, human factor Xa or human factor XIIa. Bz-Arg-CF$_3$ inhibited pancreatic and plasma kallikrein effectively, but not other coagulation enzymes. Although these arginine fluoroketones contain only one amino acid, their K_I values toward trypsin are comparable to that of the tripeptide fluoromethyl ketone Lys-Ala-Lys-CH$_2$F ($K_I = 1$ μM, [1]). The tripeptide inhibitor D-Phe-Pro-Arg-CH$_2$F, which contains a thrombin specific sequence, was a 50-fold better inhibitor than Bz-Arg-CF$_2$CF$_3$ [2].

The arginine fluoroalkyl ketones were tested in the prothrombin time (PT) and activated partial thromboplastin time (APTT) coagulant assays. Two fluoroketones, Bz-Arg-CF$_2$CF$_3$ and 1-naphthoyl-Arg-CF$_3$, had significant anticoagulant activity. Bz-Arg-CF$_2$CF$_3$ was found to prolong PT 1.8-fold at 120 μM and to prolong APTT 2.4-fold at 90 μM, while 1-naphthoyl-Arg-CF$_3$ only prolonged APTT 1.7-fold at 100 μM.

α-Aminoalkylphosphonates diphenyl esters

Peptidyl derivatives of α-aminoalkylphosphonate diphenyl ester [3] are effective and specific inhibitors of serine proteases at low concentrations. Cbz-Phe^P(OPh)$_2$ (8.2 μM) reacts with chymotrypsin (k_{obsd}/[I] = 1200 M^{-1} s^{-1}) and does not react with elastases. Cbz-Val^P(OPh)$_2$ (4.1 μM) reacts with HLE (280 M^{-1} s^{-1}) and does not react with chymotrypsin at this concentration. The data suggest that good interactions with the S$_1$ pocket of the target serine protease are necessary before nucleophilic substitution on the phosphorus atom occurs to give a stable phosphonyl derivative (Fig. 1).

Longer peptides with a C-terminal phosphonate related to phenylalanine are good inhibitors for chymotrypsin-like enzymes (Table 2, RMCP II = rat mast cell protease II), but not for elastases. The best inhibitor is Suc-Val-Pro-Phe^P(OPh)$_2$, corresponding to the sequence of an excellent 4-nitroanilide substrate for these enzymes. Only one of the two steroisomers reacts with chymotrypsin (146 000 M^{-1} s^{-1}). The ^{31}P NMR of chymotrypsin inhibited by this peptide shows

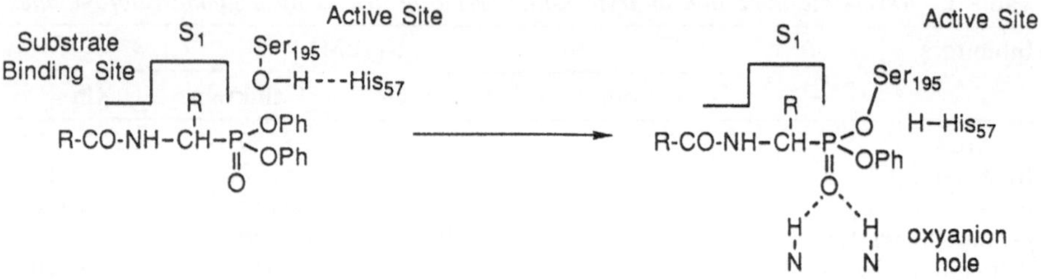

Fig. 1. Schematic drawing of the reaction of a peptidyl α-aminoalkylphosphonate diphenyl ester with the active site of a serine protease.

one broad signal at 25.98 ppm, corresponding to the Ser^{195} phosphonate ester. The best inhibitor for elastases was Boc-Val-Pro-Val^P(OPh)$_2$ with k_{obsd}/[I] value of 27 000 M^{-1} s^{-1} for human leukocyte elastase (HLE), and 11 000 M^{-1} s^{-1} for porcine pancreatic elastase (PPE). Again, this sequence corresponds to a good HLE substrate sequence, and at low concentrations this peptide did not react with chymotrypsin.

Phosphonate diphenyl ester inhibitors are chemically stable, relatively easy to synthesize, do not react with acetylcholinesterase, form very stable derivatives, possibly due to their resemblance to the tetrahedral intermediate involved in peptide bond hydrolysis, and have considerable potential utility as therapeutic agents.

Isocoumarin anticoagulants

7-Amino-4-chloro-3-(3-isothiureidopropoxy)isocoumarin (ACITIC, Fig. 2), and other isocoumarins with basic substituents have been synthesized recently as inhibitors of trypsin-like enzymes [4]. ACITIC is a potent inhibitor of several coagulation enzymes such as thrombin, factor Xa, factor XIa, factor XIIa, factor VIIa, kallikrein, plasmin and plasminogen activator. ACITIC inhibited trypsin, porcine pancreatic kallikrein, and human factor XIa most effectively with k_{obs}/[I] values of $(2.2–4.1) \times 10^4$ M^{-1} s^{-1}, and inhibited human factor XIIa, plasmin, and plasminogen activator less potently with inhibition constants in the range of 10^3 M^{-1} s^{-1}.

Inactivated trypsin by ACITIC was quite stable and regained only 4% activity

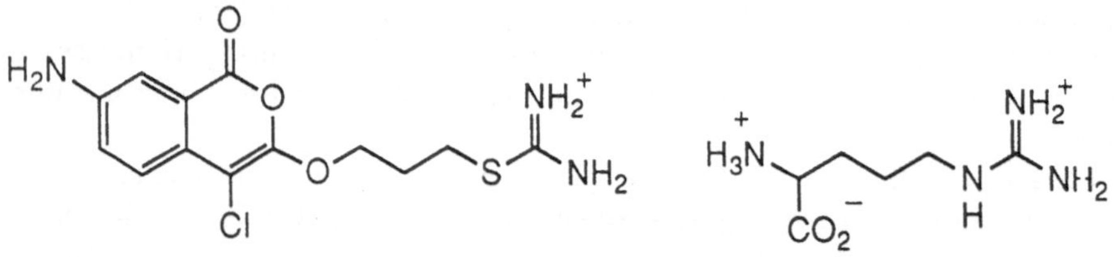

Fig. 2. Drawing of 7-amino-4-chloro-3-(3-isothiureidopropoxy)isocoumarin (ACITIC) showing its relationship to arginine.

Table 2 *Irreversible inactivation of serine proteases by peptidyl α-aminoalkylphosphonate diphenyl esters*[a]

Inhibitor	$k_{obsd}/[I]$ $(M^{-1} s^{-1})$					
	chymotrypsin	Cathepsin G	RMCP II	PPE	HLE
MeO-Suc-Ala-Ala-Pro-Phe^P(OPh)$_2$	11 000	4 400	–	NI	NI
Z-Phe-Pro-Phe^P(OPh)$_2$	17 000	5 100	32	NI	NI
Suc-Val-Pro-Phe^P(OPh)$_2$	41 000	36 000	15 000	NI	NI
MeO-Suc-Ala-Ala-Pro-Val^P(OPh)$_2$	NI	NI	NI	7 100	7 100
Boc-Val-Pro-Val^P(OPh)$_2$	NI	NI	NI	11 000	27 000

[a] Inactivation rate constants were measured in 0.1 M Hepes, 0.5 M NaCl, pH 7.5 buffer containing 9% Me$_2$SO at 25°C. Inhibitor concentrations were 4–10 μM. NI = less than 5% inhibition after 1 h.

41

upon standing at 25°C for one day. Even with the addition of hydroxylamine (0.29 M), the inhibited trypsin regained only 45% of the original enzymatic activity. ACITIC is stable in plasma, and the half-lives for spontaneous hydrolysis in human and rabbit plasma are 165 and 140 min, respectively. The stability of ACITIC in plasma indicates that this compound should be useful as an anticoagulant in vitro and in vivo. ACITIC has excellent anticoagulant activity in human and rabbit plasma using the prothrombin time (PT) and the activated partial thromboplastin time (APTT) assays. ACITIC prolongs PT ca. 1.1–2-fold and prolongs APTT 3.6–4.5-fold in human or rabbit plasma at the concentration of 20–30 μM. ACITIC was also administered into rabbits with continuous infusion, and prolonged APTT two-fold over the control value at the dose of 0.4 mg/ml; whereas, at the highest dose of 1.2 mg/ml, APTT was prolonged 3–4 times the control value.

Conclusion

It is clear that small molecular weight peptide and heterocyclic molecules can be very effective inhibitors for serine proteases and have considerable therapeutic potential. While some progress is being made in the construction of specific heterocyclic inhibitors for serine proteases, introduction of specificity into an inhibitor structure is still more easily accomplished using a peptide backbone. Design of specific inhibitors for individual trypsin-like enzymes still remains a major challenge.

Acknowledgements

The authors wish to thank Dr. James Travis and his research group at the University of Georgia for the neutrophil enzymes used in this research and Dr. Kazuo Fujikawa for the blood coagulation enzymes. The research was supported by grants from NIH (HL 29307 and HL 34035).

References

1. McMurray, J.S. and Dyckes, D.F., Biochemistry, 25 (1986) 2298.
2. Angliker, H., Wikstrom, P., Rauber, P. and Shaw, E., Biochem. J., 241 (1987) 871.
3. Oleksyszyn, J.and Powers, J.C., Biochem. Biophys. Res. Commun., 161 (1989) 143.
4. Kam, C.-M., Fujikawa, K. and Powers, J.C., Biochemistry, 27 (1988) 2547.

Approaches to renin inhibitors with increased duration of action

Peter D. Williams, Linda S. Payne, Debra S. Perlow, G.F. Lundell,
Norman P. Gould, Peter K.S. Siegl, Terry W. Schorn, Robert J. Lynch,
John J. Doyle, John F. Strouse, Charles S. Sweet, Phyllis T. Arbegast,
Inez I. Stabilito, George P. Vlasuk, Roger M. Freidinger and Daniel F. Veber
Merck Sharp and Dohme Research Laboratories, West Point, PA 19486, U.S.A.

Introduction

A major factor limiting the duration of action of metabolically stable peptide derivatives is hepatic uptake and biliary excretion, a problem that has been identified with short-acting renin inhibitors [1]. Thus, we have been seeking ways to overcome this limitation by reducing the molecular size and peptide character of the inhibitor structure and by changing physical properties, such as lipophilicity and ionic state. A general solution to this problem would clearly be of great value, not only in the renin inhibitors area, but also in the many other therapeutic areas that involve peptides as drug targets.

Results and Discussion

Our first concern was to reduce the molecular size and peptide character by modifying the ACHPA design, so as to include a structural element that could mimic the hydrophobic P_1' side chain, and thereby obviate the necessity for inclusion of amino acids at the C-terminus that correspond to the P_2' and P_3' positions. Previous attempts at enhancing potency by introducing a P_1' side chain

Table 1 *Conformationally constrained 'ACHPA-lactam' analogs*

	R_1	R_2	IC_{50} (nM)[a]
1a	H	H	10
1b	H	$CH=CH_2$	8.0
1c	$-CH=CH_2$	H	2.5
1d	$-CH_3$	CH_3	1.3

[a]Human plasma renin assay.

43

P.D. Williams et al.

at the 2-position of statine met with limited success [2]. Our approach utilizes conformationally constrained 'ACHPA-lactam' analogs as shown in Table 1. The lactam ring can be utilized as a template upon which to place substituents to map the enzyme active site in a spatially defined manner. The ACHPA-lactams were synthesized by aldol addition of the appropriate lactam enolate to Boc-L-cyclohexylalaninal. Initial studies showed that the 2S, 3S, 4S ACHPA-lactam diastereomer was greatly preferred, and that a 5-membered lactam ring size was optimal. Substitution on the 5-membered lactam ring with small hydrophobic groups, as shown in Table 1, improved binding, with geminal substitution bringing potency to the 1 nM range (e.g., **1d**). Also a variety of

Table 2 *Structures of ACHPA diamine amides and quaternary ammonium derivatives*

	R	IC$_{50}$ (nM)^a	t$_{1/2}$ (h)^b
2a		30	<0.5
2b		22	2.5
2c		13	<0.5
2d		34	2.2
2e		0.97	1.0

^{a} Human plasma renin assay.
^{b} Plasma half-life of drug after bolus i.v. administration to conscious, sodium-deficient rhesus monkeys.

44

substituents on the lactam nitrogen varying in size and polarity were found to be well-tolerated (data not shown). Although potent compounds with a reduction in molecular weight and peptide character were realized, the ACHPA-actam variation was not sufficient in and of itself to improve the i.v. duration of action, nor the oral bioavailability in the rat or rhesus monkey.

Another approach was pursued in which ACHPA diamine amides and their quaternary ammonium derivatives were studied (Table 2). Thus, the amide of racemic 3-aminoquinuclidine **2a** and the quaternized derivative **2b** derived from the more potent 3S enantiomer were administered intravenously to conscious, sodium-deficient rhesus monkeys. The time course inhibition of plasma renin activity indicated that the quaternized amine possesses a much greater duration of action. The rate of biliary excretion in the rat was demonstrably slower for **2b** than for **2a**, with a total of 35% of the administered dose of **2b** and 40% of the administered dose of **2a** appearing in the bile after 2 h. In the rhesus monkey, the plasma half-life of drug, as measured by bioassay after i.v. administration, was 2.5 h for **2b**, whereas the half-life of the unquaternized analog fell within the distribution phase, and thus was too short to be measured (<0.5 h). Similarly, a comparison of the monocyclic amine **2c** and its quaternized derivative **2d** showed a substantially increased half-life for **2d**, indicating that this phenomenon is not restricted to quaternary ammonium salts of quinuclidine. The more potent tetrapeptide quinuclidinium analog **2e**, which has a higher molecular weight and an additional amide bond, exhibited a plasma half life of 1.0 h after i.v. administration. This contrasts with previous findings where tetrapeptide inhibitors containing polar end groups at the C-terminus were found to be uniformly short-acting [3]. These findings suggest that a quaternary ammonium group may be of general utility in reducing the clearance rate of small peptides.

References

1a. Boger, J., Bennett, C.D., Payne, L.S., Ulm, E.H., Blaine, E.H., Homnick, C.F., Schorn, T.W., LaMont, B.I. and Veber, D.F., Regulatory Peptides, Supplement 4 (1985) 8.

 b. Plattner, J.J., Marcotte, P.A., Kleinert, H.D., Stein, H.H., Greer, J., Bolis, G., Fung, A.K.L., Bopp, B.A., Luly, J.R., Sham, H.L., Kempf, D.J., Rosenberg, S.H., Dellaria, J.F., De, B., Merits, I. and Perun, T.J., J. Med. Chem., 31 (1988) 2277.

2. Veber, D.F., Bock, M.G., Brady, S.F., Ulm, E.H., Cochran, D.W., Smith, G.M., LaMont, B.I., DiPardo, R.M., Poe, M., Freidinger, R.M., Evans, B.E. and Boger, J., Biochem. Soc. Trans., 12 (1984) 956.

3. Bock, M.G., DiPardo, R.M., Evans, B.E., Freidinger, R.M., Rittle, K.E., Payne, L.S., Boger, J., Whitter, W.L., LaMont, B.I., Ulm, E.H., Blaine, E.H., Schorn, T.W. and Veber, D.F., J. Med. Chem., 31 (1988) 1918.

Highly potent ψ[CH$_2$NH]-modified pseudopeptidyl inhibitors of renin: Molecular modeling and aspartyl protease selectivity studies

Tomi K. Sawyer[a],*, Linda L. Maggiora[a], Li Liu[a], Douglas J. Staples[a], V. Susan Bradford[a], Boryeu Mao[b], Donald T. Pals[c], Ben M. Dunn[d], Roger A. Poorman[a], Jessica Hinzmann[a], Anne E. DeVaux[a], Joseph A. Affholter[a] and C.W. Smith[a]

[a]Biopolymer Chemistry, [b]Computational Chemistry and [c]Cardiovascular Diseases Research Units, The Upjohn Company, Kalamazoo, MI 49001, U.S.A.
[d]Department of Biochemistry, University of Florida, Gainesville, FL 32610, U.S.A.

Introduction

The development of renin inhibitors has significantly advanced based on the discoveries of synthetic pseudodipeptides that mimic the proposed P$_1$–P$_1'$ Leuψ[C(OH)$_2$NH]Val tetrahedral intermediate of substrate (angiotensinogen, ANG) hydrolysis catalyzed by this aspartyl protease. The major conceptual approach to design such nonhydrolyzable P$_1$–P$_1'$ mimetics of this proposed aminol intermediate has been related to Xaaψ[CH(OH)CH$_2$]Yaa analogs, including those that conserve the CH(OH) functionality but possess different C-terminal modifications [for a review, see Ref. 1]. In contrast, there exists a paucity of literature related to synthetic nonhydrolyzable P$_1$–P$_1'$ Xaaψ[CH$_2$NH]Yaa analogs [1–4]. In this study, we investigated the comparative SAR of a series of ANG-based pseudopeptides having P$_1$–P$_1'$ Xaaψ[CH$_2$NH]Yaa modifications against human renin, pepsin, cathepsin D, cathepsin E and gastricsin. Molecular modeling of inhibitor–renin interactions was performed as previously reported [4] using CHARMM and a 3D computer graphics model of human renin, CKH-RENIN.

Results and Discussion

The octapeptide H-D-His-Pro-Phe-His-Pheψ[CH$_2$NH]Phe-Val-Tyr-OH(U-70531E) is the only P$_1$–P$_1'$ Pheψ[CH$_2$NH]Phe-modified, ANG-based aspartyl protease inhibitor for which a crystallographic structure of the ligand, bound to the active site of a structurally homologous enzyme, rhizopuspepsin, has been determined [5]. The renin inhibitory activities of U-70531E derivatives are summarized in Table 1. Noteworthy was the observation that both the P$_5$ and P$_2'$–P$_3'$ residues of U-70530E (IC$_{50}$ = 4.2 × 10^{-7} M) could be eliminated to yield a more potent pentapeptide, U-71909E (IC$_{50}$ = 1.3 × 10^{-8} M). In contrast, the contribution of P$_4$ Pro and P$_3$ Phe were very critical to the renin inhibitory potency of inhibitory activity per se of these P$_1$–P$_1'$ Pheψ[CH$_2$NH]Phe-modified

---

*To whom correspondence should be addressed.

Table 1   *Structure-activity properties of $\psi[CH_2NH]$-modified pseudopeptides*

Entry	Compound[a]	Renin IC$_{50}$ (M)[b]
1	H-D-His-Pro-Phe-His-Pheψ[CH$_2$NH]Phe-Val-Tyr-OH	4.2×10^{-7}
2	Ac-Ftr-Pro-Phe-His-Pheψ[CH$_2$NH]Phe-Val-Tyr-NH$_2$	3.0×10^{-9}
3	Ac-Ftr-Pro-Phe-His-Pheψ[CH$_2$NH]Phe-NH$_2$ (U-71908E)	5.6×10^{-10}
4	Ac-His-Pro-Phe-His-Pheψ[CH$_2$NH]Phe-NH$_2$	1.0×10^{-8}
5	Ac-Pro-Phe-His-Pheψ[CH$_2$NH]Phe-NH$_2$ (U-71909E)	1.3×10^{-8}
6	Ac-Phe-His-Pheψ[CH$_2$NH]Phe-NH$_2$	4.5×10^{-7}
7	Ac-His-Pheψ[CH$_2$NH]Phe-NH$_2$	$\gg 1.0 \times 10^{-5}$
8	Ac-Pro-Phe-His-Pheψ[CH$_2$NH]Phe-N(Me)$_2$	2.6×10^{-6}
9	Ac-Pro-Phe-His-Pheψ[CH$_2$NH]Phe-OH	3.7×10^{-6}
10	Ac-Pro-Phe-His-Pheψ[CH$_2$NH]Phe-ol	2.2×10^{-6}
11	Ac-Pro-Phe-His-Pheψ[CH$_2$NH]Pgl-NH$_2$	$\gg 1.0 \times 10^{-5}$
12	Ac-Pro-Phe-His-Pheψ[CH$_2$NH]Hph-NH$_2$	1.5×10^{-7}
13	Ac-Pro-Phe-His-Pheψ[CH$_2$NH]Tyr-NH$_2$	2.5×10^{-8}
14	Ac-Pro-Phe-His-Pheψ[CH$_2$NH]Phe(p-Cl)-NH$_2$	3.8×10^{-8}
15	Ac-Pro-Phe-His-Pheψ[CH$_2$NH]Phe(p-NO$_2$)-NH$_2$	3.4×10^{-8}
16	Ac-Pro-Phe-His-Chaψ[CH$_2$NH]Cha-NH$_2$	3.1×10^{-7}
17	Ac-Pro-Phe-His-Chaψ[CH$_2$NH]Val-NH$_2$	ca. 1 $\times 10^{-7}$
18	Ac-Pro-Phe-His-Chaψ[CH$_2$NH]β-Me-Phe-NH$_2$	$\gg 1.0 \times 10^{-6}$
19	Ac-Pro-Phe-His-Chaψ[CH$_2$NH]Phe-NH$_2$ (U-79465E)	1.7×10^{-9}
20	Ac-Pro-Phe-His-Chaψ[CH$_2$NH]Pea	6.0×10^{-6}
21	Ac-Pro-Phe-His-Chaψ[CH$_2$NMe]Phe-NH$_2$	$\gg 1.0 \times 10^{-6}$
22	Ac-Pro-Phe-*N*-Me-His-Chaψ[CH$_2$NH]Phe-NH$_2$	1.0×10^{-8}
23	Ac-Pro-Hph-*N*-Me-His-Chaψ[CH$_2$NH]Phe-NH$_2$	9.8×10^{-10}

[a] Ftr = N^{in}-formyl-Trp; Pgl = phenylglycine; Hph = homophenylalanine;
Cha = cyclohexylalanine; Pea = phenylethylamine.
[b] Human angiotensinogen, pH 6.0.

pseudopeptides. The conformational, topographical and enzyme active site intermolecular binding properties of U-71909E were explored by molecular modeling studies using CKH-RENIN.

In brief, an extended conformation of the inhibitor was determined subsequent to molecular dynamics and energy minimization of the ligand–enzyme complex. The P$_1'$ Phe side chain was observed to occupy a binding pocket of the CKH-RENIN that was different relative to that of P$_1'$ Val-substituted derivatives having either ψ[CH(OH)CH$_2$] or ψ[CH$_2$NH] modification. Also noteworthy was the discovery that the C-terminal CONH$_2$ moiety was proximate and H-bonded to the catalytic Asp residues of the CKH-RENIN active site. To test these molecular modeling observations of U-71909E, we explored the effects of both P$_1$–P$_1'$ Xaaψ[CH$_2$NH]Yaa and C-amide modification of the pseudopeptide by SAR (see Table 1). Overall, we determined that the C-terminal P$_1'$ Phe-NH$_2$ moiety was indeed important to the inhibitory activity of U-71909E. Substitution of the P$_1$ Phe by Cha in U-71909E resulted in significantly increased potency (U-79465E, IC$_{50}$ = 1.7×10^{-9} M). Based on enzyme degradation studies (data not shown) performed on U-79465E, we observed that both elastase and chymotrypsin cleaved the P$_4$–P$_2$ sequence (i.e., Pro-Phe and Phe-His, respectively); however, the C-terminal CONH$_2$ moeity was stable to carboxypeptidase-Y. Structure–

47

activity and kinetic analysis of U-79465E ($K_i = 1.3 \times 10^{-10}$ M, recombinant human renin) and selected derivatives having P_4–P_2 modifications resulted in the identification of a highly potent and enzymatically-stable derivative, Ac-Pro-Hph-*N*-Me-His-Chaψ[CH$_2$NH]Phe-NH$_2$ ($K_i = 1.6 \times 10^{-10}$ M). Finally, we further determined the comparative aspartyl protease inhibitory SAR of three different P_1–P_1' Xaaψ[CH$_2$NH]Yaa-modified pseudopeptides against human pepsin, cathepsin-D, cathepsin-E or gastricsin. In summary, U-71909E, compound **17** and U-79465E were all essentially inactive ($K_i \gg 1.0 \times 10^{-6}$ M) to inhibit these aspartyl proteases. Therefore, our results suggest that such P_1–P_1' Xaaψ[CH$_2$NH]Yaa-modified ANG derivatives may be designed to exhibit high potency, metabolic stability and aspartyl protease selectivity, and provide new lead compounds in the development of renin-inhibitory therapeutic agents useful in the treatment of hypertension.

References

1. Greenlee, W.J., Pharm. Res., 4 (1987) 364.
2. Szelke, M., Leckie, B., Hallett, A., Jones, D.M., Sueiras, J., Atrash, B. and Lever, A.F., Nature, 299 (1982) 555.
3. Plattner, J.J., Greer, J., Fung, A.K.L., Stein, H., Kleinert, H.D., Sham, H.L., Smital, J.R. and Perun, T.J., Biochem. Biophys. Res. Commun., 139 (1986) 982.
4. Sawyer, T.K., Pals, D.T., Mao, B., Maggiora, L.L., Staples, D.J., DeVaux, A.E., Schostarez, H.J., Kinner, J.H. and Smith, C.W., Tetrahedron, 44 (1988) 661.
5. Suguna, K., Padlan, E.A. Smith, C.W., Carlson, W.D. and Davies, D.R., Proc. Natl. Acad. Sci. U.S.A., 84 (1987) 7009.

Synthesis of a 1-position analog of cyclosporin with selective biological activities

Daniel H. Rich[a],*, Chong-Qing Sun[a], Kevin J. Plzak[a], Johannes D. Aebi[a],
Brian E. Dunlap[a], P.L. Durette[b], F. Dumont[b] and M.J. Staruch[b]

[a]*School of Pharmacy, University of Wisconsin-Madison, 425 N. Charter St., Madison,
WI 53706, U.S.A.*
[b]*Merck Sharp & Dohme Research Laboratories, Rahway, NJ 07065, U.S.A.*

Introduction

The unique amino acid, (4R)-4-[(E)-butenyl]-4,*N*-dimethyl-L-threonine [MeBmt (**1**)] found in the 1-position of the immunosuppressive drug, cyclosporin A (CsA), is critical for maximal immunosuppressive activity[1,2]. In order to determine the effects that structural changes in this residue have on immuno-suppressive activity, we have synthesized a variety of 1-position analogs of CsA (Table 1) and determined their immunosuppressive activities and binding to cyclophilin, the CsA binding protein [3] recently shown to be identical to peptidyl prolyl isomerase (PPIase)[4,5].

Results and Discussion

Three methods were used to synthesize the needed β-hydroxy-α-*N*-methyl amino acids. MeBm$_2$t (**2**) was synthesized (Scheme 1) by the reaction of Pmz-Sar-O^{t}Bu (**4**) with LDA at −78°C followed by condensation with 2,2-dimethyl-4-hexenal (**5**) as described by Aebi et al. [6] for the synthesis of MeBmt (**1**). Residual starting material could be separated from oxazolidinone by reaction

Scheme I

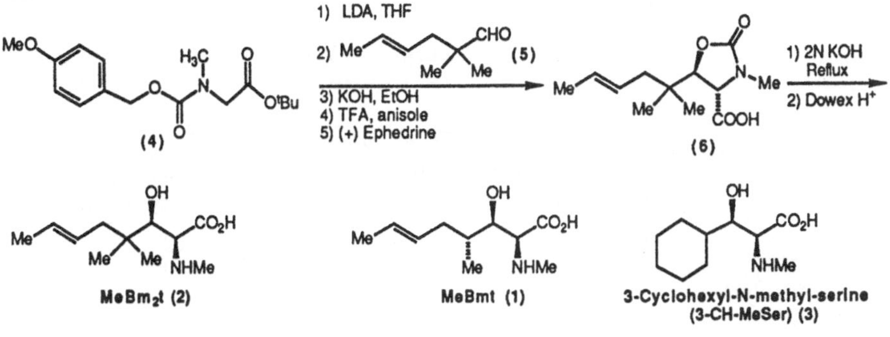

*To whom correspondence should be addressed.

with TFA/anisole to give, after aqueous workup, a mixture of oxazolidinone acids, which were resolved by fractional crystallization of the (+)-ephedrine salt. The enantiomerically pure oxazolidinone (6) was hydrolyzed in refluxing KOH and chromatographed over Dowex H[+] cationic exchange resin to give the amino acid MeBm$_2$t (2) in 21% yield. (2S,3R)-3-Cyclohexyl-N-methyl-serine (3) was prepared in the same way from cyclohexyl-carboxaldehyde in 20% yield.

The novel 3-substituted MeBmt analog, (2S,3R)-3,7-dimethyl-MeBth (7), was prepared from nerol (Scheme 2) [7]. Sharpless asymmetric epoxidation of nerol gave the chiral epoxide (9), which was treated with methyl isocyanate followed by base catalyzed rearrangement to give the oxazolidinone (10). Oxidation of the hydroxyl with K_2RuO_4 afforded the oxazolidinone acid (11), which was hydrolyzed in refluxing KOH. Ion exchange chromatography over Dowex H[+] gave the pure amino acid (7) in 44% overall yield. In a similar fashion, (2S,3S)-3,7-dimethyl-MeBth (8) was synthesized from geraniol in 41% yield. (4S)-MeBmt was synthesized in 49% yield by using the asymmetric glycine enolate aldol reaction of Evans and Weber [8].

Biological activities of several analogs were determined in three assay systems by using the previously described methods [2,9]. The results are shown in Table 1. The analog prepared from 7 retained little activity in these assays. With one exception, low immunosuppressive activity parallels low affinity for cyclophilin.

CsA's relatively specific immunosuppressive activity against defined target cells, the existence of analogs with a range of activity, and its reasonably low K_d suggest a receptor-mediated mechanism of action. In an attempt to isolate the CsA receptor, Handschumacher et al. [3] discovered an 18 kDa protein, called cyclophilin. While fulfilling many of the requirements for a CsA receptor, cyclophilin has subsequently been shown to exist in all human tissues tested,

Table 1   *Biological properties of some cyclosporin analogs*

```
MeLeu — MeVal — MeBmt¹ — Abu — Sar
  |                                |
MeLeu                              |
  |                                |
D-Ala —— Ala —— MeLeu⁶ —— Val — MeLeu
```

Schematic structure of CsA

Compound	Inhibition Con A stimulated thymocytes	Inhibition IL-2 secretion	Inhibition binding to cyclophilin
CsA	100	100	100[a]
Dihydro-CsA	25–50	48	40[a]
(3-CH-MeSer)¹CsA	7–8	16	7–8
(MeBm₂t)¹CsA	20–30	25	<1
(MeAla)⁶CsA	<1	<1	40[a]

[a]Data taken from [9].

Scheme II

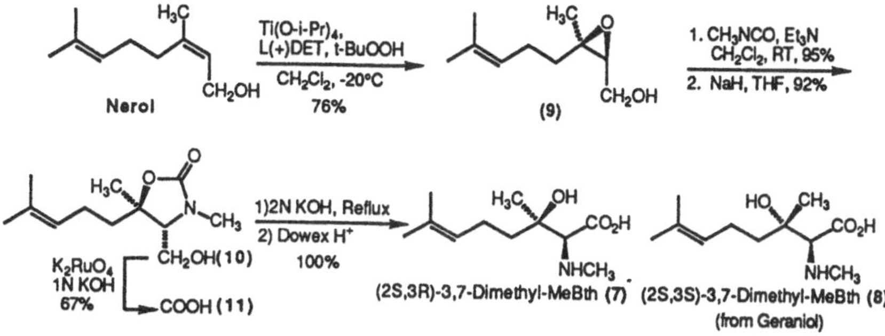

in all eukaryotic cells examined, and in relatively high concentration. Recently, cyclophilin has been shown to be identical with peptidyl prolyl isomerase, (PPI-ase), an enzyme that catalyzes the cis to trans isomerization of acyl-prolyl amide bonds [4,5,10].

If inhibition of PPI-ase is solely responsible for the immunosuppressive activity of CsA, then a one to one correlation of binding to cyclophilin and immu-nosuppressive activity should be found. However, Durette et al. [9] found that the CsA analog, MeAla⁶-CsA, lacked immunosuppressive activity even though it bound well to cyclophilin. Our results, described here, are even more difficult to reconcile with immunosuppressive activity being caused solely by inhibition of cyclophilin binding, because MeBm₂t¹-CsA retains more immunosuppressive activity than predicted by the binding to cyclophilin. Harding and Handschu-macher [11] have suggested that multiple isoforms of cyclophilin exist. If so, it is possible that other, as yet undetected, minor forms of PPI-ase exist, and that these are more important for selective immunosuppression than the form isolated to date. In that event, we suggest that MeBm₂t¹-CsA may be binding efficiently in vivo to a minor component of cyclophilin or PPI-ase to produce the observed immunosuppressive response. It is clear that the synthetic methods described here have led to the discovery of a highly selective CsA analog that can prove useful in the characterization of CsA receptors.

Acknowledgements

Financial support from the National Institutes of Health (AR32007) is greatly acknowledged. CQS thanks Applied Biosystems Inc. for a travel award to attend the 11th American Peptide Symposium.

References

1. Wenger, R.M., Prog. Allergy, 38 (1986) 46.
2. Rich, D.H., Dhaon, M.K., Dunlap, B.E. and Miller, S.P.F., J. Med. Chem., 29 (1986) 978.

3. Handschumacher, R.E., Harding, M.W., Rice, J., Drugge, R.J. and Speicher, D.W., Science, 226 (1984) 544.
4. Takahashi, N., Hayano, T. and Suzuki, M., Nature, 337 (1989) 473.
5. Fischer, G., Wittmann-Liebold, B., Lang, K., Kiethaber, T. and Schmid, F.X., Nature, 337 (1989) 476.
6. Aebi, J.D., Dhaon, M.K. and Rich, D.H., J. Org. Chem., 52 (1987) 2881.
7. Sun, C.Q. and Rich, D.H., Tetrahedron Lett., 29 (1988) 5205.
8. Evans, D.A. and Weber, A.E., J. Am. Chem. Soc., 108 (1986) 6757.
9. Durette, P.L., Boger, J., Dumont, F., Firestone, R., Frankshun, R.A., Koprak, S.L., Lin, C.S., Melino, M.R., Pessolano, A.A., Pisano, J., Schmidt, J.A., Sigel, N.H., Staruch, M.J. and Witzel, B.E., Transplant. Proc., 20 (Suppl. 2) (1988) 51.
10. Fischer, G., Bang, H. and Mech, C., Biomed. Biochim. Acta, 43 (1984) 1101.
11. Harding, M.W. and Handschumacher, R.E., Transplantation, 46 (Suppl. 2) (1988) 29.

Cholecystokinin analogs with high affinity and selectivity for brain versus peripheral membrane receptors

Victor J. Hruby, Sunan Fang, Richard Knapp, Wieslaw Kazmierski, G.K. Lui and Henry I. Yamamura

Departments of Chemistry and Pharmacology, University of Arizona, Tucson, AZ 85721, U.S.A.

Introduction

Cholecystokinin-8 (CCK-8, CCK[26-33]) H-Asp-Tyr(SO_3^-)-Met-Gly-Trp-Met-Asp-Phe-NH$_2$ is the C-terminal octapeptide of a larger 33 amino acid peptide which has multiple biological activities as a hormone in the periphery, and as a neurotransmitter in the CNS. Efforts to determine the specific biological activities associated with the different sites of action will be greatly aided by the design of agonist and antagonist analogs, and mimetics with high potency and selectivity for these CNS and peripheral receptors. Indeed, recent studies in several laboratories [e.g., 1, 2] have demonstrated that there are at least 2 classes of receptors for CCK, those primarily in the central nervous system (B type), and those primarily in peripheral organs (A type). However, the role(s) of each class of these receptors in the numerous putative CCK-induced biological effects is still largely unknown. To help elucidate the role of these receptor systems, we have sought to obtain highly receptor-selective CCK analogs. We report here some of our studies which have led to the development of highly potent and selective analogs for the CCK brain receptor.

Results and Discussion

In previous investigations, we [3] and others have demonstrated that replacement of both the Met[28] and Met[31] residues by the pseudoisosteric amino acid norleucine (Nle) provides CCK analogs with virtually identical biological activities to those with Met residues. This led us to the design of CCK analogs with N-methylnorleucine (N-MeNle) residues. The use of N-methyl residues in peptide design has several potential advantages related to the conformational and physical-chemical consequences of such substitutions, including: (1) ready accessibility of the *cis* conformation of the peptide bond; (2) steric constraint, including a bias of the χ_1 angle to *trans* (180°); (3) no proton donating ability of the amide bond; (4) increased basicity of the amide carbonyl; and (5) stability of the peptide bond to enzyme degradation.

The desired CCK[26-33] and CCK[29-33] analogs were prepared by SPPS on a MBHAR. Coupling of amino acid residues such as Trp and Tyr to the growing peptide chain when the N-terminal residue was N-MeNle proved troublesome using standard coupling protocols, such as DIC/HOBt or symmetrical anhydrides. However, we found that use of the BOP reagent at room temperature generally gave satisfactory results. Work-up and purification followed standard RPHPLC procedures. Sulfation was performed using pyridine/SO_3 followed by RPHPLC purification. Proof of structure included AAA, analytical TLC and RPHPLC, and FABMS. Binding-displacement assays were performed on membrane preparations from the brain and pancreas of rats and guinea pigs using [^{125}I]CCK[26-33] as the radioligand.

The affect of N-MeNle[31]-substitution in CCK[29-33] (3 and 4, Table 1) was most promising in that, not only was substantial selectivity for central vs. peripheral receptors observed, but potency at the central receptor was increased almost 50-fold. When the same substitution was made in CCK[26-33], however, a 200-fold loss in potency at the central receptor was observed (compare 4 and 5, Table 1), and the analogs appeared to bind better to peripheral receptors. Interestingly, however, when N-MeNle was substituted into position 28 of CCK[26-33] a large increase in potency occurred at the central receptor, and the analog (6, Table 1) was modestly selective for the central vs. peripheral receptor. Most interestingly, when N-MeNle was placed in both the 28 and 31 position of CCK[26-33] (Table 1), a highly potent ($IC_{50} = 0.13$ nM) and selective analog for the B type vs. the A type receptor was seen. To the best of our knowledge, this is the most selective analog for the B class CCK receptor reported thus far. We have prepared a highly pure ^{3}H-labeled derivative of [N-MeNle[28,31]]CCK[26-33], and preliminary binding studies in guinea pig whole brain membranes indicate the presence of two distinct populations of binding sites with K_Ds, of about 0.4 nM and 6 nM, respectively. This may suggest the presence of two different CCK B class receptors in the brain.

Table 1 *Binding of CCK[26-33] analogs vs. [^{125}I]CCK[26-33] in rats*

	Analogs	Cortex IC_{50}(nM)	Pancreas IC_{50}(nM)	Ratio Panc./Cortex
1	CCK[26-33] SO_3	0.32	0.13	0.40
2	[Nle[28,31]]CCK[26-33] Bis SO_3	0.70	0.30	0.41
3	[Nle[31]]CCK[29-33]	230	9 700	42
4	[N-MeNle[31]]CCK[29-33]	4.7	> 10 000	> 2 000
5	[Nle[28],N-MeNle[31]]CCK[26-33]	980	170	0.17
6	[N-MeNle[28],Nle[31]]CCK[26-33]	1.45	47	32
7	[N-MeNle[28,31]]CCK[26-33]	0.13	1 030	7 900

Sulfation (either mono or bis) led to analogs of CCK[26-33] (Table 2) with greatly increased binding to peripheral receptors and variable (though small) changes in binding to central receptors relative to the non-sulfated derivatives. Most of these analogs showed only modest selectivity for A or B type receptors. One striking exception was [N-MeNle[28],Nle[31]]CCK[26-30] Bis SO_3^- that appears to be highly selective for the brain receptor (about 9000-fold). This unexpected result is under further investigation.

Table 2 *Binding of CCK$^{26-33}$ analogs vs. [^{125}I]CCK$^{26-33}$ in rats*

	Analogs	Cortex IC$_{50}$(nM)	Pancreas IC$_{50}$(nM)	Ratio Panc./Cortex
1	CCK$^{26-33}$ SO$_3$	0.32	0.13	0.40
8	[Nle^{28},N-MeNle^{31}]CCK$^{26-33}$-Mono SO$_3^-$	0.82	2.80	3.4
9	[N-MeNle^{28},Nle^{31}]CCK$^{26-33}$-Mono SO$_3^-$	1.85	0.42	0.23
10	[N-MeNle^{28,31}]CCK$^{26-33}$-Mono SO$_3^-$	0.090	1.4	15
11	[Nle^{28},N-MeNle^{31}]CCK$^{26-33}$-Bis SO$_3^-$	0.50	1.4	2.8
12	[N-MeNle^{28,31}]CCK$^{26-33}$-Bis SO$_3^-$	1.40	0.41	0.29

Finally, we have begun to examine the conformational features of our analogs which may be responsible for the observed selectivities. ^{1}H and ^{13}C NMR studies clearly demonstrate the presence of two conformations involving *cis-trans* isomerism about the Trp^{30}-N-MeNle^{31} amide bond in 4 with the *trans* isomer predominating. The corresponding [N-MeNle^{28,31}] derivative 7 shows 4 distinct isomers (2 major, 2 minor). It is suggested that the *cis* isomer is preferred for the 'biologically active' conformation in the brain system. Very recently, Carpentier et al. [4] reported on a cyclic lactam analog of CCK$^{26-33}$, Boc-γ-D-
-Glu-Tyr(SO$_3$H)-Nle-D-Lys-Trp-Nle-Asp-Phe-NH$_2$ which is highly selective for the B type CCK receptor. It will be interesting to determine the conformational similarities between these two classes of compounds.

Acknowledgement

This work was supported by U.S. Public Health Service Grant DK 36289.

References

1. Innis, R.B. and Snyder, S.H., Proc. Natl. Acad. Sci. U.S.A., 77 (1980) 6917.
2. Sakamoto, C., William, J.A. and Goldfine, I.D., Biochem. Biophys. Res Commun., 124 (1984) 497.
3. Sugg, E.E., Shook, J.E., Serra, M., Korc, M., Yamamura, H.I., Burks, T.F. and Hruby, V.J., Life Sci., 39 (1986) 1623.
4. Carpentier, B., Dor, A., Roy, P., England, P., Pham, H., Durieux, C. and Roques, B.P., J. Med. Chem., 32 (1989) 1184.

Design and evaluation of novel gastrin-releasing peptide antagonists for the treatment of small cell lung cancer

David C. Heimbrook[a], Walfred S. Saari[b], Nancy L. Balishin[a],
Thorsten W. Fisher[b], Arthur Friedman[a], David M. Kiefer[a], Nicola S. Rotberg[a],
John W. Wallen[a] and Allen Oliff[a]

[a]Department of Cancer Research and [b]Department of Medicinal Chemistry,
Merck Sharp and Dohme Research Laboratories, West Point, PA 19486, U.S.A.

Introduction

Gastrin-releasing peptide (GRP) is a peptide hormone containing 27 amino acids that is structurally analogous to the amphibian peptide bombesin [1]. GRP stimulates a wide variety of biological responses in different tissues and cell lines, including mitogenesis in 3T3 mouse fibroblasts [2,3]. GRP has also been proposed to play a central role in the pathophysiology of small cell lung cancer (SCLC) via an autocrine growth mechanism [4,5]. These observations suggest that GRP antagonists may have clinical utility as inhibitors of the pathophysiological response to GRP in human diseases.

Results and Discussion

The observation that GRP derivatives containing C-terminal ester linkages are potent GRP antagonists in vitro and in vivo [6] suggests that other C-terminal modifications might yield even more potent antagonists. A series of GRP derivatives containing C-terminal ether (Table 1) and alkyl amide (Table 2) moieties were prepared and tested for GRP antagonist activity in competitive binding inhibition and mitogenic inhibition assays in Swiss 3T3 fibroblasts [6]. Several trends emerge from the data obtained on these compounds. All of the ether and alkyl amide derivatives tested were mitogenic antagonists in Swiss 3T3 fibroblasts. In general, the alkyl ether derivatives displayed activity comparable to that of the alkyl ester derivatives described previously [6], while the corresponding alkyl amide derivatives exhibit improved potency (e.g., compare compounds 2, 5, and 14). In addition, representative ether (5) and alkyl amide (16) derivatives were shown to competitively block GRP-dependent elevation of intracellular calcium concentration in H345 SCLC cells in vitro [6]. This result demonstrates that these compounds are GRP antagonists in human cancer cells, and suggests that the GRP receptors on mouse fibroblasts and human SCLC cells bind GRP in a similar manner.

As has been previously suggested, substitution of Ala for Gly[24] of N-acetyl-GRP 20-27 yields a peptide with reduced activity, while the D-Ala[24] derivative

Table 1   *Activity of alkyl ether GRP derivatives in Swiss 3T3 fibroblasts*

#	Structures	Binding Inhibition (nM)	Mitogenic Stimulation (nM)	Mitogenic Inhibition (nM)
1	Ac-His-Trp-Ala-Val-Gly-His-Leu-Met-NH$_2$ (positions 20 21 22 23 24 25 26 27)	1.8	0.20	—
2		3.9	—	20
3		81	—	300
4		8.3	—	91
5		6.2	—	31
6		3.2	—	18
7		11	—	32
8	Ac-His-Trp-Ala-Val-Ala-His-Leu-Met-NH$_2$	180	20.0	—
9	D-Ala -Leu-Met-NH$_2$	3.2	0.20	—
10	Ala	410	—	> 300
11	D-Ala	5.0	—	35

retains full potency (**1, 8,** and **9**) [2]. Similar results were obtained when these substitutions were made in the ethyl ether antagonist (**5, 10,** and **11**). These observations suggest that a β-turn might be important for high-affinity binding of both GRP and GRP antagonists to the GRP receptor [7].

The primary motivation for the development of GRP antagonists was their

Table 2 *Activity of alkyl amide GRP derivatives in Swiss 3T3 fibroblasts*

#	Structures	Binding Inhibition	Mitogenic Stimulation	Mitogenic Inhibition
12	Ac-His-Trp-Ala-Val-Gly-His- N(R) (positions 20 21 22 23 24 25)	2.6	—	10
13	-N(S)	4.9	—	35
14	-N(R)	1.4	—	3
15	-N(R,S)	2.4	—	6
16	-N(R)	2.0	—	4
17	-N(S)	11	—	22
18	-N(R)	2.8	—	7

anticipated utility for inhibiting the growth of small cell lung cancer cells. Effects of GRP and GRP antagonists on the growth rate of SCLC cells in vitro were monitored in both suspension culture and in anchorage-independent growth in semi-solid media. SCLC cell lines H345, H69, H417, H209, and N592 were maintained in SIT medium [6,8]. Cells were transferred to insulin-free media approximately 96 h prior to use in mitogenesis assays.

Growth of SCLC cells in suspension culture was monitored by viable cell staining or by [³H]thymidine uptake. Under the conditions tested, N-acetyl-GRP (20-27) did not stimulate growth of these cells at any concentration between 1 nM and 50 µM. Neither **5** nor **16** inhibited thymidine uptake compared to untreated cells at concentrations up to 10 µM. Growth stimulation was observed when H345, N592, or H209 cells were treated with 25 nM epidermal growth factor, or when H345 or N592 cells were treated with 10 nM IGF-1. As expected, neither of the GRP antagonists inhibited growth stimulation by EGF or IGF-1.

SCLC cell lines H345, N592 and H69 formed colonies in soft agarose when plated as small clumps, but not when plated as single cells [9]. No stimulation of colony formation by N-acetyl-GRP (20-27) was observed at concentrations from 1 nM to 10 μM. Compound **16** did not inhibit colony formation at concentrations up to 50 μM compared to untreated cells. Assay responsiveness was demonstrated by the addition of 5% FBS, which caused a 100-fold increase in colony formation. However, **16** did not inhibit FBS stimulation of colony formation.

We have described the development of GRP antagonists that are equipotent to native GRP in competitive binding inhibition assays, are stable in human serum in vitro, and block GRP-stimulated effects in human SCLC cells in vitro and in murine tissues in vivo. Unfortunately, in our hands these GRP antagonists were ineffective at inhibiting the growth of SCLC cells in vitro. Additional in vivo studies will be required to demonstrate whether the GRP antagonists described above might have utility as therapeutic agents for the treatment of SCLC.

Acknowledgements

The authors are grateful to L. Wassel for determination of amino acid compositions, E. Sausville (G. Washington University, Washington, DC) for the SCLC cell lines, and M. Riemen and R. Freidinger for helpful discussions.

References

1. Anastasi, A., Erspamer, V. and Bucci, M., Experientia, 27 (1971) 166.
2. Rivier, J.E. and Brown, M.R., Biochemistry, 17 (1978) 1766.
3. Rozengurt, E. and Sinnett-Smith, J., Proc. Natl. Acad. Sci. U.S.A., 80 (1983) 2936.
4. Cuttitta, F. Carney, D.N., Mulshine, J., Moody, T.W., Fedorko, J., Fischler, A. and Minna, J.D., Nature, 316 (1985) 823.
5. Minna, J.D., Cuttitta, F., Battey, J.F., Mulshine, J.L., Linnoila, I., Gazdar, A.F., Trepel, J., and Sausville, E.A., In DeVita, V.T., Hellman, S. and Rosenberg, S.A. (Eds.) Important Advances in Oncology, Vol. 4, J.B. Lippincott, Philadephia, PA, 1988, p. 55.
6. Heimbrook, D.C., Saari, W.S., Balishin, N.L., Friedman, A., Moore, K.S., Riemen, M.W., Kiefer, D.M., Rotberg, N.S., Wallen, J.W. and Oliff, A., J. Biol. Chem., 264 (1989) 11258.
7. Rose, G.D., Gierasch, L.M. and Smith, J.A., Adv. Prot. Chem., 37 (1985) 1.
8. Carney, D.N., Bunn, P.A., Gazdar, A.F., Pagan, J.A. and Minna, J.D., Proc. Natl. Acad. Sci. U.S.A., 78 (1981) 3185.
9. Carney, D.N., Cuttitta, F., Moody, T.W. and Minna, J.D., Cancer Res., 47 (1987) 821.

Lanthionine peptide nisin: Study of a structure-activity relationship

Tateaki Wakamiya[a,*], Koichi Fukase[a], Manabu Kitazawa[a], Hiroshi Fujita[a], Akira Kubo[a], Yasushi Maeshiro[a] and Tetsuo Shiba[b]

[a]Department of Chemistry, Faculty of Science, Osaka University, Toyonaka, Osaka 560, Japan
[b]Peptide Institute, Protein Research Foundation, Minoh, Osaka 562, Japan

Introduction

The antibiotic nisin was the first lanthionine and dehydroalanine-containing peptide found in nature [1,2], and its unique structure made up of five cyclic sulfide parts, termed rings A, B, C, D and E, was proposed by E. Gross et al. in 1971 [3] (Fig. 1). This compound is now widely used as a food preservative in Europe particularly because of its activity against *Clostridium botulinum*. Recently, we successfully synthesized nisin and confirmed the proposed structure synthetically [4,5]. Through the synthetic study, versatile preparative methods for dehydroalanine and lanthionine peptides were established that were applicable to syntheses of other natural peptides containing these amino acids.

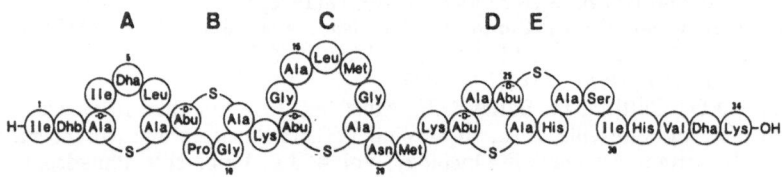

Fig. 1. The structure of nisin. Abu = α-aminobutyric acid; Dha = dehydroalanine; Dhb = dehydrobutyrine; Ala—S—Ala = lanthionine (meso form); Abu—S—Ala = methyllanthionine (threo form).

Although some interesting biological activities, such as antimalarial activity, in vitro release of lysosomal enzymes, induction of fetal resorption, as well as antibacterial activity, were mentioned for nisin [6], relationships between these activities and the structure of nisin had not yet been studied in detail. In the present study, we aimed to elucidate the structure and antibacterial activity relation of this compound. For this purpose, many peptide fragments were obtained, not only by chemical degration or enzymatic digestion of natural compound, but also synthetically.

---

*To whom correspondence should be addressed.

60

Results and Discussion

We first isolated a degradated peptide **1** of nisin that was included in a commercially available nisin as a by-product. This compound seemed to be formed by a decomposition of Dha[33] residue in nisin. Cyanogen bromide degradation of nisin gave peptide fragments **2** and **3** that were separated by preparative HPLC. When the reaction was carried out by use of a limited amount of the reagent, a fragment **4** was obtained, as well as the fragments **2** and **3**.

A tryptic digestion of nisin produced four peptide fragments **5** to **8**. The peptide bonds between not only Asn[20] and Met[21], but also His[31] and Val[32] were abnormally cleaved during the digestion by an unknown mechanism. These fragments were also isolated by preparative HPLC (Fig. 2).

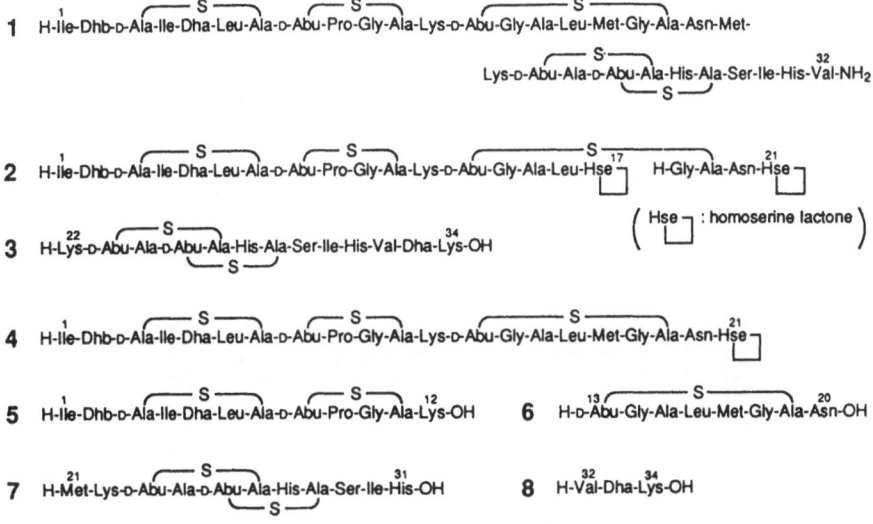

Fig. 2. Compounds 1–8.

Measurement of antibacterial activity of the peptides obtained above suggested that the N-terminal segment containing three disulfide ring moieties might be important for the exhibition of the activity (Table 1). It was especially noteworthy that deletion of ring C caused an extreme or rather almost complete loss of the activity, though ring C itself did not show the activity at all. Additionally, it was concluded that the C-terminal region in the nisin molecule, including ring D–E and one Dha residue, was not essential for the exhibition of the activity, but necessary for enhancement of the activity.

Since it was not easy to obtain any more fragments from natural compound, we next carried out a synthetic work to prepare several peptide fragments for a systematic elucidation of the SAR of nisin. On the basis of the strategy for total synthesis of nisin, three cyclic peptide segments were first prepared by

Table 1 *Minimum inhibitory concentrations (MIC) of nisin fragments*

Test organisms	MIC (µg/ml) of the peptides[a]					
	Nisin	1	2	4	5	13
Staphylococcus aureus ATCC 6538P	6.3	12.5	>100	25	>100	NT[b]
Staphylococcus aureus MS 353	3.1	12.5	>100	12.5	100	12.5
Staphylococcus aureus MS 353 C36	3.1	12.5	>100	25	>100	50
Staphylococcus aureus MS 353 AO	6.3	12.5	>100	25	>100	50
Staphylococcus aureus 0116	12.5	50	>100	25	>100	NT
Staphylococcus aureus 0119	6.3	50	>100	25	>100	25
Staphylococcus aureus 0126	50	>100	>100	50	>100	25
Staphylococcus aureus 0127	12.5	25	>100	50	>100	50
Staphylococcus epidermidis sp-al-1	3.1	6.3	>100	3.1	100	50
Streptococcus pyogenes N.Y.5	<0.2	0.8	100	3.1	6.3	12.5
Streptococcus pyogenes 1022	0.8	3.1	100	3.1	25	12.5
Streptococcus faecalis 1501	50	>100	>100	>100	>100	>100
Streptococcus agalactiae 1020	3.1	6.3	>100	12.5	>100	>100
Sarcina lutea ATCC 9341	3.1	3.1	>100	25	>100	50
Micrococcus flavus ATCC 10240	<0.2	0.8	6.3	0.4	25	NT
Corynebacterium diphteriae P.W.8	0.8	3.1	100	6.3	25	12.5
Bacillus subtilis ATCC 6633	6.3	25	>100	25	>100	25

[a] MIC values of all other peptides which were not listed in the Table were of >100.
[b] NT = not tested.

desulfurization of cystine peptides [7–9] and subjected to condensation to yield peptides **13** to **17**, as summarized in Fig. 3. Of these synthetic peptides, only peptide **13** was active against test organisms (Table 1).

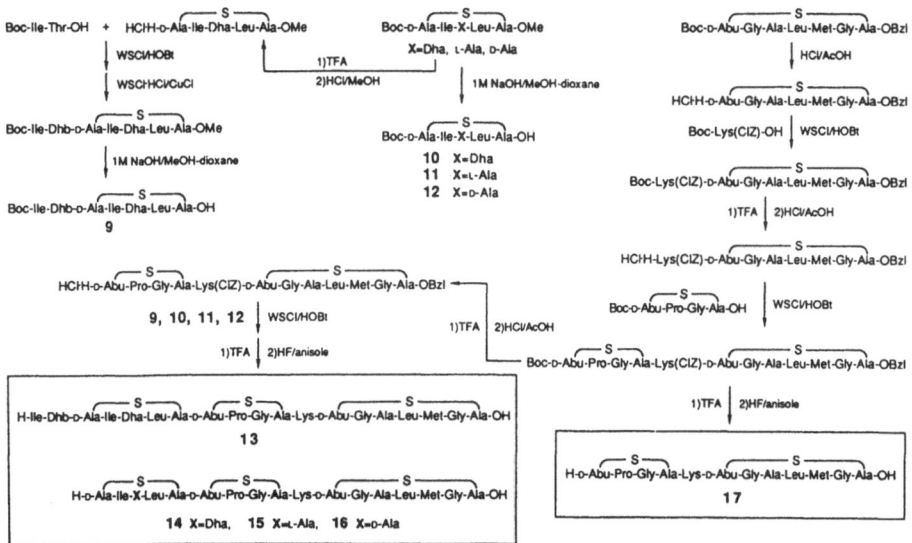

Fig. 3. Synthetic scheme of peptide fragments related to N-terminal region in nisin.

Consequently, the N-terminal 19-amino acid segment of the nisin might be a minimal requisite for the exhibition of antibacterial activity of this unique lanthionine peptide. Our study is currently focused on syntheses of the following analogs related to N-terminal (1–19)-peptide, i.e., [L-Abu²]-, [D-Abu²]-, [L-Ala⁵]-, and [D-Ala⁵]-(1–19), to clarify the role of two dehydroamino acid residues in the active sequence.

Acknowledgements

We would like to thank the research laboratories, Toyo Jozo Co., Ltd., for the measurement of antibacterial activity. This work was supported in part by Grants-in-Aid for Scientific Research No. 60540352 and No. 62790153 from the Ministry of Education, Science and Culture.

References

1. Rogers, L.A. and Whittier, E.O., J. Bacteriol., 16 (1928) 211, 321.
2. Mattick, A.T.R. and Hirsch, A., Lancet, ii (1947) 5.
3. Gross, E. and Morell, J.L., J. Am. Chem. Soc., 93 (1971) 4634.

4. Fukase, K., Kitazawa, M., Sano, A., Shimbo, K., Fujita, H., Horimoto, S., Wakamiya, T. and Shiba, T., In Shiba, T. and Sakakibara, S. (Eds.) Peptide Chemistry 1987, Protein Research Foundation, Osaka, 1988, p. 337.
5. Fukase, K., Kitazawa, M., Sano, A., Shimbo, K., Fujita, H., Horimoto, S., Wakamiya, T. and Shiba, T., Tetrahedron Lett., 29 (1988) 795.
6. Gross, E. and Morell, J.L., J. Am. Chem. Soc., 89 (1967) 2791.
7. Wakamiya, T., Shimbo, K., Sano, A., Fukase, K., Yasuda, H. and Shiba, T., In Sakakibara, S. (Ed.) Peptide Chemistry 1982, Protein Research Foundation, Osaka, 1983, p. 149.
8. Wakamiya, T., Shimbo, K., Sano, A., Fukase, K. and Shiba, T., Bull. Chem. Soc. Jpn., 56 (1983) 2044.
9. Fukase, K., Wakamiya, T. and Shiba, T. Bull., Chem. Soc. Jpn., 59 (1986) 2505.

Short-chain bombesin receptor antagonists with IC_{50}s for cellular secretion and growth approaching the picomolar region

David H. Coy[a], John E. Taylor[b], Ning-Yi Jiang[a], Sun Hyuk Kim[b], Lu-Hua Wang[c], Shih Che Huang[c], Jacques-Pierre Moreau[b] and Robert T. Jensen[c]

[a]*Peptide Research Laboratories, Department of Medicine, Tulane University Medical Center, New Orleans, LA 70112, U.S.A.*
[b]*Biomeasure Inc., Hopkinton, MA 01748, U.S.A.*
[c]*Digestive Diseases Branch, NIH, Bethesda, MD 20892, U.S.A.*

Introduction

Although the first reported [1] really potent bombesin (BN) antagonist, (Leu[13]ψ[CH$_2$NH]Leu[14])-BN (A), has proved to be an effective and useful inhibitor of BN actions in several in vitro systems (see Table 1), including growth of human small cell lung carcinoma cells and inhibition of in vivo rat gastric acid release [2–5], its 30-fold lower binding affinity than BN itself suggested that further improvements to K_d and IC_{50} values could be made. Indeed, in the reduced peptide bond pseudopeptide series, it has been possible to substantially increase binding affinity and antagonist potency, while shortening the chain length from 14 to 9 amino acids [6] (see Table 1). This was accomplished using D-amino acid substitutions in position 6, accompanied by the removal of the first 5 amino acids. A recent report by Heimbrook et al. [7], in which a weak antagonist was created by removal of the C-terminal Met residue in GRP(20–27), prompted us to examine a similar design strategy in a number of des-Met[14]-BN peptides with position 6 modifications, N-terminal deletions, and alkylamide and ester substituents at the C-terminus.

Methods

Peptides were synthesized by normal SPPS and cleaved from Merrifield resins with an appropriate amine, hydrazine/MeOH or MeOH or EtOH containing 10% Et$_3$N, in the case of the peptide esters, and, additionally, with heating at 60°C for the latter. Antagonists were examined for their ability to displace [^{125}I-Tyr^4]-BN binding to guinea pig acini cells and murine Swiss 3T3 cells, and to inhibit BN-stimulated amylase release from the former, and growth of the latter as described [1,6].

Results and Discussion

In the pseudopeptide series [6], there was a substantial loss of activity when

Table 1 *Ability of BN analogs to bind to guinea pig acini and murine Swiss 3T3 cells and inhibit BN-stimulated amylase release from the former and growth of the latter*

BN analog		Acini		3T3	
		K_d	IC_{50} (nM)	K_d	IC_{50} (nM)
A.	$[\psi Leu^{13,14}]$	60	31	65	18
B.	$[\psi Leu^{13,14}](6-14)$	327	150	–	–
C.	$[D\text{-}Phe^6,\psi Leu^{13,14}](6-14)$	14	5	7	9
D.	$[(1-13)]NH_2$	216	33	18	1303
E.	$[(6-13)]NH_2$	239	41	1216	1627
F.	$[D\text{-}Phe^6](6-13)NH_2$	96	24	23	29
G.	$[D\text{-}Nal^6](6-13)NH_2$	95	34	28	55
H.	$[D\text{-}Phe^6](6-13)OH$	7931	1026	–	–
I.	$(6-13)$ethylamide	95	54	–	–
J.	Ac-$(7-13)$ethylamide	50	29	18	262
K.	$[D\text{-}Phe^6](6-13)$ethylamide	16	7	3	0.7
L.	$[D\text{-}Phe^6](6-13)$propylamide	4	1.6	1.7	0.8
M.	$[D\text{-}Phe^6](6-13)$butylamide	p.a.[a]	43	5	0.1
O.	$[D\text{-}Phe^6](6-13)$phenethylamide	12	6	18	2
P.	$[D\text{-}Phe^6](6-13)$methyl ester	7	2	1	0.16
Q.	$[D\text{-}Phe^6](6-13)$ethyl ester	5	1.5	1.6	0.19
R.	Ac-GRP$(20-26)$ethyl ester	10	4	2	1.4
S.	$[D\text{-}Phe^6](6-13)$ NHNH$_2$	5	4	2	–

[a]p.a. = partial agonist.

amino acids were removed from the N-terminus of the BN antagonists. Thus, $[\psi Leu^{13,14}]$-BN(6–14) (B) was about 5 times less potent than the full sequence analog. However, in the present des-Met^{14}-analogs, little loss of potency was found in going from 13 to 8 amino acids (D to E in Table 1). The presence of D-Phe^6 in both sets of shortened analogs dramatically increased binding affinities and inhibitory potencies (C and F). The size of the aromatic side chain appeared to be of marginal importance, as evidenced by the retention of high potency with a D-β-Nal analog (G in Table 1) and a variety of other aromatic replacements (not shown).

Inhibitory potency was also greatly improved by alkyl substituents on the C-terminal amide, with [D-Phe^6]-BN(6–13) ethylamide (K) being 5–20 times more potent than the amide. The propylamide analog (L) was even more potent, but potency began to decrease in the acini system with the butylamide analog (M). This peptide, however, still displayed high affinity for 3T3 cells and was surprisingly effective in blocking their growth with an IC_{50} of only 100 pM. This particular analog also displayed partial agonist activity in the amylase release assay and may, thus, be revealing some important differences between BN receptors on guinea pig pancreas and mouse 3T3 fibroblasts. Work is presently under way to examine this phenomenon using other cell preparations from different animal species.

Another highly effective C-terminal modification strategy was that of esterification. Both a methyl and an ethyl ester (P and Q) were uniformly potent antagonists in all systems examined, and were in the picomolar range for inhibition of cell growth. The presence of an acetyl group in place of D-Phe in position 6 was also quite effective (R), but the acetylated analogs were usually 2–10 times less potent than their D-Phe counterparts, depending on the assay system.

It was also found that a hydrazide group at the C-terminus also produced a highly effective antagonist (analog S). This is a significant result, from a structural viewpoint, since it seems to indicate that it is the electron-donating capacities of the amide substituents, rather than their hydrophobic/hydrophilic characteristics, that are primarily responsible for enhanced binding affinity.

Finally, [D-Phe^6]-BN(6–13) propylamide has been·tested in vivo in the rat for inhibition of BN-stimulated pancreatic amylase release, and was very potent and relatively long-acting. A bolus injection of 100 nM/kg produced inhibition lasting for about 60 min, whereas the same dose of the original pseudopeptide A was without significant effect. In summary, a number of highly potent BN antagonists now exist which should be valuable aids in elucidating the various GI and growth-promoting effects of endogenous BN, and which may soon lead to therapeutically useful compounds.

Acknowledgements

This work was supported in part by NIH Grant CA-45153.

References

1. Coy, D.H., Heinz-Erian, P., Jiang, N.-Y., Sasaki, Y., Taylor, J., Moreau, J.-P., Wolfrey, W.T., Gardner, J.D. and Jensen R.T., J. Biol. Chem., 263 (1988) 5056.
2. Woll, P.J., Coy, D.H. and Rozengurt, E., Biochem. Biophys. Res. Commun., 155 (1988) 359.
3. Trepel, J.B., Moyer, J.D., Cutitta, F., Fruchi, F., Coy, D.H., Natale, R.B., Mulshine, J.L., Jensen, R.T. and Sausville, E.A., Biochem. Biophys. Res. Commun., 156 (1988) 1383.
4. Mahmoud, S., Palaszynski, E., Fiskum, G., Coy, D.H. and Moody, T.W., Life Sci., 44 (1988) 367.
5. Rossowski, W.J., Murphy, W.A., Jiang, N.-Y., Yeginsu, O., Ertan, A. and Coy, D.H., Scand. J. Gastroenterol., 24 (1989) 121.
6. Coy, D.H., Taylor, J., Jiang, N.-Y., Kim, S.H., Wang, L.-H., Huang, S.C., Moreau, J.-P., Gardner, J.D. and Jensen, R.T., J. Biol. Chem., 25 (1989) 14691.
7. Heimbrook, D.C., Boyer, M.E., Garsky, V.M., Balishin, N.L., Kiefer, D.M., Oliff, A. and Riemen, M.W., In Tam, J. and Kaiser, E.T. (Eds.) Synthetic Peptides: Approach to Biological Problems (Proceedings of the UCLA Symposium on Molecular and Cellular Biology), New Series, 86 (1989) 295.

Synthesis of some endothelin analogs and big endothelin: Structure-activity relationships

K. Nakajima, S. Kubo, S. Kumagaye, H. Kuroda, H. Nishio, M. Tsunemi, T. Inui, N. Chino, T.X. Watanabe, T. Kimura and S. Sakakibara

Peptide Institute Inc., Protein Research Foundation, Minoh-shi, Osaka 562, Japan

Introduction

Endothelin (ET), a potent vasoconstrictor peptide isolated from the culture media of porcine aortic endothelial cells, is a 21 amino acid peptide having two intramolecular disulfide bonds [1]. The structures of human and rat ET have also been deduced from cDNA sequences; human ET was identical with porcine ET, but distinct from rat ET [2,3]. Recently, the structures of three types (ET-1, ET-2 and ET-3) of ET were deduced from human genomic DNA, in which ET-1 and ET-3 were found to be identical with porcine (human) and rat ET, respectively [4]. ET-2 differs from ET-1 at positions 6 and 7, whereas ET-3 had six different amino acid residues at positions 2, 4, 5, 6, 7 and 14 from ET-1 as shown in Fig. 1. In previous papers, we reported the synthesis and disulfide structure determination of ET-1, ET-3, and sarafotoxin S6b, which is a similar vasoconstrictor peptide isolated from snake venom [5], and found that two disulfide bonds in these molecules are located at positions 1–15 and 3–11 [6,7]. In the present study, we synthesized various ET related peptides, including ET-2 and big ET, ET-1(1–39), and measured their vasoconstrictor activities in order to elucidate their SAR.

Results and Discussion

ET-1, ET-2, ET-3, sarafotoxin S6b, [Asu[1,15], Ala[3,11]]-ET-1 and ET-1(1–39) were synthesized by the segment condensation procedure in solution applying our maximum protection strategy. Figure 2 shows a typical synthetic route for ET-1 (1–39), in which two sets (1 and 15, 3 and 11) of Cys residues were protected

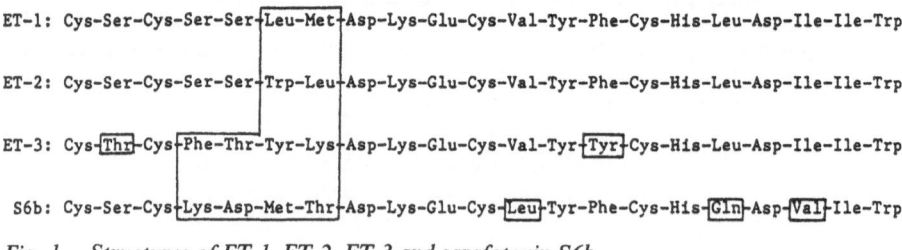

ET-1: Cys-Ser-Cys-Ser-Ser-Leu-Met-Asp-Lys-Glu-Cys-Val-Tyr-Phe-Cys-His-Leu-Asp-Ile-Ile-Trp

ET-2: Cys-Ser-Cys-Ser-Ser-Trp-Leu-Asp-Lys-Glu-Cys-Val-Tyr-Phe-Cys-His-Leu-Asp-Ile-Ile-Trp

ET-3: Cys-Thr-Cys-Phe-Thr-Tyr-Lys-Asp-Lys-Glu-Cys-Val-Tyr-Tyr-Cys-His-Leu-Asp-Ile-Ile-Trp

S6b: Cys-Ser-Cys-Lys-Asp-Met-Thr-Asp-Lys-Glu-Cys-Leu-Tyr-Phe-Cys-His-Gln-Asp-Val-Ile-Trp

Fig. 1. Structures of ET-1, ET-2, ET-3 and sarafotoxin S6b.

68

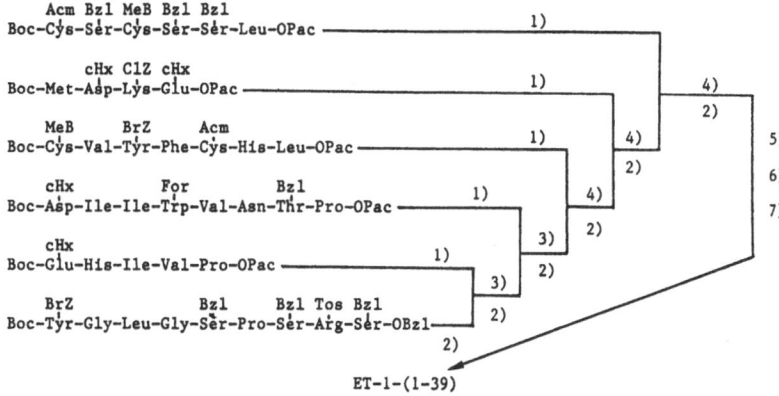

Fig. 2. Synthesis of ET-1(1–39).
1) Zn/AcOH, 2) TFA, 3) WSCI/HOBt, 4) WSCI/HOOBt, 5) HF/p-cresol (85/15),
6) K₃Fe(CN)₆, 7) I₂/MeOH.

by acetamidomethyl (Acm) and 4-methylbenzyl (MeBzl) groups, respectively, and crosslinked the disulfide bonds specifically. Conditions for assembling the molecule, deprotection, and crosslinking were almost the same as reported in our previous paper [6]. The crude product was purified by DEAE cellulose chromatography and RPHPLC, successively. Other ET-1 related peptides were synthesized by SPPS, applying the same procedure as described in [6]. The homogeneities of the synthetic peptides were confirmed by analytical HPLC and amino acid analysis. The proper formation of the disulfide structure in big ET was confirmed by comparing the enzyme mapping on HPLC using our ET-1 as the standard. The vasoconstrictor activities of these synthetic peptides were determined using rat pulmonary artery ring preparations; the conditions are the same as reported previously [6,7]. The results are summarized in Table 1. The potencies of ET-2, ET-3 and sarafotoxin S6b were one-half, one-sixtieth and one-third that of ET-1, respectively. Such differences in biological potencies should primarily arise from sequence heterogeneity at the amino terminal portion, especially at position 4 to 7. The vasoconstrictor potencies of ET-1 related peptides are also summarized in Table 2. All of the monocyclic and linear peptides showed

Table 1   Vasoconstrictor activity of ET-1, ET-2, ET-3 and sarafotoxin S6b using rat pulmonary artery ring preparations

Peptide	Potency ratio	
ET-1	100[a]	($n=7$)
ET-2	49.0	($n=6$)
ET-3	1.9	($n=6$)
Sarafotoxin S6b	35.1	($n=6$)

The biological potency of each peptide was expressed as % of the standard ET-1.
[a] $ED_{50}=0.65\pm0.05$ nM.

Table 2 *Vasoconstrictor activity of ET-1 related peptides using rat pulmonary artery ring preparations*

Peptide	Potency ratio	
[Cys[1-15], Cys[3-11]]-ET:(ET-1)	100	(n = 7)
[Cys[1-11], Cys[3-15]]-ET	0.8	(n = 8)
[Cys[1-15], Cys(Acm)[3,11]]-ET	0	(n = 6)[a]
[Cys[3-11], Cys(Acm)[1,15]]-ET	0	(n = 8)[b]
[Cys(Acm)[1,3,11,15]]-ET	0	(n = 5)
[Cys[1-15], Ala[3,11]]-ET	29.6	(n = 6)
[Ala[1,15], Cys[3-11]]-ET	0.4	(n = 6)
[Asu[1,15], Ala[3,11]]-ET	0	(n = 4)
ET(1–15)-NH$_2$	0	(n = 6)
ET(16–21)	0	(n = 6)
Ac-ET	0.5	(n = 8)
Lys-Arg-ET	0.2	(n = 5)
Des-Trp[21]-ET	0	(n = 5)
ET-NH$_2$	5.6	(n = 6)
ET-1(1–39)	1.8	(n = 8)
[Ala[4]]-ET	35.6	(n = 2)
[Ala[5]]-ET	24.2	(n = 4)
[Gly[6]]-ET	78.1	(n = 6)
[Met(O)[7]]-ET	69.1	(n = 6)
[Asn[8]]-ET	0.8	(n = 7)
[Leu[9]]-ET	54.8	(n = 5)
[Gln[10]]-ET	0	(n = 4)
[Phe[13]]-ET	63.3	(n = 2)
[Ala[14]]-ET	0[b]	(n = 4)
[Tyr[21]]-ET	32.8	(n = 5)
[Phe[21]]-ET	18.4	(n = 5)

[a]$ED_{25} = 300$ nM; [b]$ED_{25} = 1000$ nM.

greatly reduced or complete lack of activity; however, among monocyclic analogs, [Cys[1-15], Ala[3,11]]-ET showed one-third that of ET-1, indicating that formation of the inner disulfide bond is not necessary for expressing its activity. Nevertheless, its deamino dicarba analog, [Asu[1,15], Ala[3,11]]-ET-1 was almost completely inactive. These results indicate the importance of the terminal NH$_2$ group for its biological activity. This assumption was also supported by the fact that the amino terminal blocked or extended peptides, such as Ac-ET or Lys-Arg-ET, greatly reduced the activity. A truncated peptide from the C-terminal, (1–15)-NH$_2$, was also completely inactive even though two disulfide bonds were formed properly in the molecule. Although removal of the C-terminal Trp residue resulted in complete loss of the activity, replacement of Trp residue by Tyr or Phe still retained the biological activity. However, carboxyl terminal blocked or extended peptides such as ET-NH$_2$ and ET-1(1–39) greatly decreased the potency, indicating that the C-terminal COOH group, rather than the Trp side chain, is important for the biological activity. Analogs [Ala[4]], [Ala[5]], [Gly[6]], [Met(O)[7]], [Leu[9]] and

[Phe[13]]-ET-1, substituted at the positions of Ser[4], Ser[5], Leu[6], Met[7], Lys[9] and Tyr[13], respectively, still retained one-fifth to four-fifths of the potency. On the other hand, replacement of Asp[8], Glu[10] and Phe[14] by Asn[8], Gln[10] and Ala[14], respectively, resulted in greatly decreased potency.

Conclusion

Although two sets of disulfide bonds in the ET molecule are not always necessary for expression of activity, terminal NH_2 and COOH groups, ω-COOH groups of Asp[8] and Glu[10], and the aromatic moiety of Phe[14] are important for binding of the ET molecule to its receptor.

References

1. Yanagisawa, M., Kurihara, H., Kimura, S., Tomobe, Y., Kobayashi, M., Mitsui, Y., Yazaki, Y., Goto, K. and Masaki, T., Nature, 332 (1988) 411.
2. Ito, Y., Yanagisawa, M., Ohkubo, S., Kimura, C., Kosaka, T., Inoue, A., Ishida, N., Mitsui, Y., Onda, H., Fujino, M. and Masaki, T., FEBS Lett, 231 (1988) 440.
3. Yanagisawa, M., Inoue, A., Ishikawa, T., Kasuya, Y., Kimura, S., Kumagaye, S., Nakajima, K., Watanabe, T.X., Sakakibara, S., Goto, K. and Masaki, T., Proc. Natl. Acad. Sci. U.S.A., 85 (1988) 6964.
4. Inoue, A., Yanagisawa, M., Kimura, S., Kasuya, Y., Miyauchi, T., Goto, K. and Masaki, T., Proc. Natl. Acad. Sci. U.S.A., 86 (1989) 2863.
5. Takasaki, C., Tamiya, N., Bdolah, A., Wolleberg, Z. and Kochva, E., Toxicon, 26 (1988) 543.
6. Kumagaye, S., Kuroda, H., Nakajima, K. Watanabe, T.X., Kimura, T., Masaki, T. and Sakakibara, S., Int. J. Pept. Prot. Res., 32 (1988) 519.
7. Nakajima, K., Kumagaye, S., Nishio, H., Kuroda, H., Watanabe, T.X., Kobayashi, Y., Yamaoki, H., Kimura, T. and Sakakibara, S.J., Cardiovasc. Pharmacol., 13 (Suppl.5) (1989) S8.

Novel cyclic hexapeptide oxytocin antagonists derived from *Streptomyces silvensis*

Roger M. Freidinger[a], George F. Lundell[a], Mark G. Bock[a], Robert M. DiPardo[a],
Carl F. Homnick[a], Paul S. Anderson[a], Douglas J. Pettibone[a],
Bradley V. Clineschmidt[a], Lawrence R. Koupal[b], Cheryl D. Schwartz[b],
Joanne M. Williamson[b], Michael A. Goetz[b], Otto D. Hensens[b], Jerrold M. Liesch[b]
and James P. Springer[b]

[a]*Merck Sharp & Dohme Research Laboratories, West Point, PA 19486, U.S.A.*
[b]*Merck Sharp & Dohme Research Laboratories, Rahway, NJ 07065, U.S.A.*

Introduction

The neurohypophyseal hormone, oxytocin (OT), functions in parturition to contract the uterine myometrium during labor, and the mammary myoepithelium postpartum to elicit milk letdown [1]. During the last 20 years, structural modifications of OT have led to the development of potent antagonists [2].

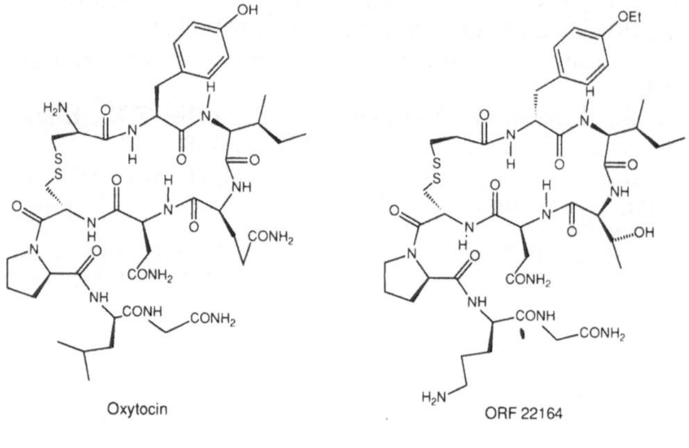

Fig. 1.   *Oxytocin and a typical antagonist, ORF22164.*

One such antagonist, ORF 22164 (Fig. 1) is currently undergoing human clinical trials for use in the prevention of premature birth [3].
   We describe here a novel class of cyclic hexapeptide oxytocin antagonist.

Results and Discussion

Receptor-based screening resulted in the discovery of an extract of a fermentation broth from *Streptomyces silvensis* which inhibited binding of ^{3}H-OT to

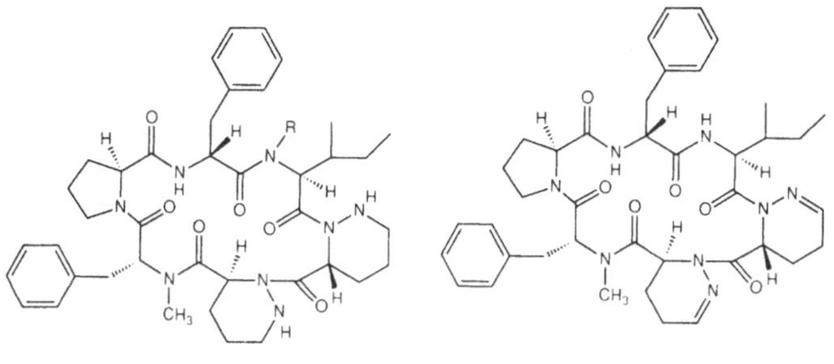

L·156.373 · R=OH
L·364.918 · R=H

L·365.209

Fig. 2.   Structures of cyclic hexapeptide oxytocin antagonists.

rat uterine receptors [4]. Isolation and structural characterization of the active components of this extract produced L-156,373, a structurally unique cyclic hexapeptide containing 5 unusual amino acids (Fig. 2). Features of the structure include 3 D-residues, 2 piperazic acids, an N-OH isoleucine, and an N-Me group. The antibacterial cyclic depsipeptide monamycins have some related structural features such as alternating D- and L-amino acids and the piperazic acid residues [5]. L-156,373 has moderate affinity for the oxytocin receptor ($K_i = 150$ nM).

In order to improve the potency and selectivity of L-156,373, direct chemical modifications of the structure were explored. Treatment with $TiCl_3$ in aqueous MeOH buffered with NaOAc resulted in selective dehydroxylation of the N-OH-Ile in about 90% yield to produce the Ile-containing analog (L-364,918). This compound has 5-fold higher affinity for oxytocin receptors.

On exposure to air, L-364,918 undergoes slow oxidation to a mixture of components that are suggested to contain 5,6-dehydropiperazic acid based on MS and NMR analysis. Oxidation under defined conditions with tert-butyl hypochlorite in pyridine gives in 96% yield L-365,209 in which both piperazic acids are oxidized. L-365,209 exhibits further increased oxytocin receptor affinity ($K_i = 7$ nM) and 50–100-fold selectivity over inhibition of binding of [3H]-arginine vasopressin (AVP) to V_1 and V_2 receptors (rat liver and kidney, respectively). It is comparable in these properties to ORF 22164 (see Table 1).

Table 1   *Affinities of cyclic hexapeptide oxytocin antagonists for rat OT and AVP receptors*

	K_i(nM)[a]		
		[3H]AVP	
Compound	[3H]OT(uterus)	V_1(liver)	V_2(kidney)
L-156,373	150 ±23	2 200 ± 260	3 400 ± 420
L-364,918	30 ± 3.5	1 300 ± 120	2 400 ± 510
L-365,209	7.3± 1.1	370 ± 9.8	820 ± 160
ORF22164	3.1± 0.52	220 ± 19	79 ± 4

[a] Group mean ±SE for 3–6 determinations; K_i's were determined from IC_{50} values generated using 5–8 concentrations of compound in triplicate.

73

L-365,209 has been further characterized as an OT antagonist [4]. In the isolated rat uterus, L-365,209 was a potent competitive OT antagonist ($K_B = 1.7$ nM), but at concentrations of 100–150 nM, it did not block the contractile effects of bradykinin or $PGF_{2\alpha}$. The effect of AVP in vitro were also antagonized by L-365,209, although its potency was weaker in accord with the binding data. In vivo L-365,209 antagonized the contractile action of exogenous oxytocin on the rat uterus with a relatively long duration of action. At a submaximal dose of 1 mg/kg i.v., the antagonist effect of L-365,209 was still observed after 110 min. In none of the OT or AVP assays did L-365,209 exhibit agonist activity.

Previously described OT antagonists are analogs of the native hormone and are heterodetic cyclic peptides [2,6]. Comparison of L-365,209 with antagonists such as ORF 22164 shows that the former compound represents a significant structural departure from known antagonists. L-365,209 is smaller in molecular size, with 6 amino acids compared to the usual 9. It is also a homodetic cyclic peptide without the cystine disulfide of OT. Within the macrocycle of L-365,209 are D-amino acids and cyclic amino acids not found in other OT antagonists. The cyclic amino acids are especially unique. Molecular modeling comparisons suggest that the one region of correspondence between L-365,209 and ORF 22164 may be the dipeptides D-Phe-Ile and D-Tyr(OEt)-Ile. SAR studies are currently probing this possibility.

The potency, selectivity, and novel structural features of L-365,209 make it an important new lead toward the development of OT antagonists. This, and related, structures will also be valuable for further defining the structural and conformational requirements of the oxytocin receptor.

Acknowledgement

We thank Mrs. Vera Finley for typing the manuscript.

References

1. Pritchard, J.A., MacDonald, P.C. and Gant, N.F., Williams Obstetrics, 17th ed., Appleton-Century-Crofts, Norwalk, 1985, p. 295.
2. Sawyer, W.H. and Manning, M., In Amico, J.A. and Robinson, A.G. (Eds.) Oxytocin: Clinical and Laboratory Studies, Elsevier, Amsterdam, 1985, p. 423.
3. Akerlund, M., Stromberg, P., Hauksson, A.K., Andersen, L.F., Lyndrup, J., Trojnar, J. and Melin, P., Br. J. Obstet. Gynaecol., 94 (1987) 1040.
4. Pettibone, D.J., Clineschmidt, B.V., Anderson, P.S., Freidinger, R.M., Lundell, G.F., Koupal, L.R., Schwartz, C.D., Williamson, J.M., Goetz, M.A., Hensens, O.D., Liesch, J.M. and Springer, J.P., Endocrinology, 125 (1989) 217.
5. Hassall, C.H., Morton, R.B., Ogihara, Y. and Phillips, D.A.S., J. Chem. Soc. (C), (1971) 526.
6. Melin, P., Trojnar, J., Johansson, B., Vilhardt, H. and Akerlund, M., J. Endocrinol., 111 (1986) 125.

Systematic approach to study the structure-activity of transforming growth factor α

James P. Tam, Yao-Zhong Lin, Cui-Rong Wu, Zhi-Yi Shen, Mauro Galantino, Wen Liu and Xiao-Hong Ke

The Rockefeller University, 1230 York Ave., New York, NY 10021, U.S.A.

Introduction

The proliferation of normal cells is a well-controlled and highly conserved process that requires the regulation by hormone-like growth factors. One of the best characterized family of growth factors is the transforming growth factor type α (TGFα; Fig. 1) and epidermal growth factor (EGF) family, which are small mitogenic proteins containing three disulfide linkages [1]. Homologous members of the TGFα/EGF family now encompass a diverse group of proteins that include domains of blood coagulation factors, extracellular matrix, DNA-tumor virus early gene products, and invertebrate embryonic proteins [2].

V¹ V S H F N D C P D S H T Q F C F H G T C R F L V Q E D K P A C V C H S G Y V G A R C E H A D L L A⁵⁰

Fig. 1. Structure of human transforming growth factor-α.

To understand the relationships between the structures and functions of the 50-residue TGFα, we used a systematic approach to scan two complementary series of point-substituted analogs. In the first series (D-amino acid scan), each amino acid was replaced singly by its corresponding D-amino acid in 41 point-substituted analogs mainly to examine the structural effect of D-amino acid substitution on biological activities. Nine positions consisting of 6 cysteines and 3 glycines were not replaced in the D-amino acid scan. In the second series (Ala scan), each amino acid was replaced individually with Ala (Ala scan) to examine the role of the side chain in its receptor-contact roles. The D-amino acid scan of the whole TGFα sequence has been completed, but the Ala scan was only partially completed. By examining the ratio of biological activity due to the D-amino acid and the Ala point-substituted analogs (DA factor analysis), it is possible to determine residues important for the receptor-contact and structural integrity of TGFα.

Results and Discussion

The point-substituted analogs were synthesized by SPPS [3,4], cleaved from the resin supports by the low-high HF method [5], refolded to the oxidized forms by an improved and efficient refolding method [3], and purified to homogeneity by RPHPLC. The integrity of each analog was established by fission ion MS. The refolding strategy made use of the thermodynamic nature of the EGF/TGFα structure and allowed them to refold to their most thermodynamic stable form under denaturing and constant disulfide reshuffling conditions. Essentially, the refolding method consisted of two sequential gradients: a decreasing gradient of denaturants for refolding and an increasing gradient of oxidants for disulfide formation. Both processes were conducted in a single reaction vessel to minimize loss due to handling. Crude peptides were reduced, after HF cleavage, by 0.2 M dithiothreitol in 8 M urea, 0.1 M Tris-HCl, pH 8.4. Refolding was conducted in a gradual gradient of reducing concentrations of urea under the dialysis condition to remove dithiothreitol and other small molecules $M_r < 1000$ (18–36 h). At 2 M urea concentration, the peptide solution was diluted with Tris-HCl buffer and 1 mM each of oxidized and reduced glutathione were added to initiate the second gradient refolding method with an increasing concentration of oxidized glutathione. The disulfide formation monitored by HPLC showed that the peak corresponding to the refolded peptide usually eluted 3–5 min earlier than the reduced form. This simplified refolding procedure produced 70 TGFα analogs in this study.

In the D-amino acid scan, large variations of more than 100 000-fold of biological activities of the point-substituted analogs were observed in the competitive EGF receptor and mitogenic assays. The largest decrease of potencies (more than 500-fold) was found in those residues at the middle loop with the β-sheet structure (Thr[20]-Leu[24]) and those at the COOH domain (Val[33], Tyr[38], Arg[42] to Asp[47]), while the smallest decrease (2–10-fold) was seen with those residues at the NH$_2$ domain (residues 1–14). Decrease of activity (100–500-fold) was observed with analogs substituted at three other positions (Pro[9], Phe[15] and Pro[30]). When compared with the known solution NMR structure [5], it became evident that the D-amino acid substitution was tolerated at the structural flexible regions such as NH$_2$ and COOH termini, reverse turns, or helix region (residues 10–17), but not tolerated at the β-sheet region such as the middle loop (Thr[20]-Leu[24],Arg[42]-Asp[47]).

In contrast to the D-amino acid scan, the decrease in biological potency was small to moderate in the Ala-substituted analogs we synthesized. Most analogs produced less than 10-fold decrease in potency. Large decrease in potencies (40–250-fold) was observed with Ala[15], Ala[22], Ala[23], Ala[38], Ala[42], Ala[44], Ala[47] and Ala[48]. To distinguish whether the lost potency is due to the structural or receptor-contact factor, we propose the use of the 'DA Factor' analysis, which computes the relative biological potency of each analog due to substitutions by the corresponding D-amino acid and by Ala (Activity$_{D-amino\ acid}$/Activity$_{Ala}$ = DA factor). High DA factor values (>25) indicate that the residues are structural

Table 1 *DA factor analysis for structural/receptor-contact residues of TGFα*

Structural residues	DA factor	Receptor-contact residues	DA factor
His[18](32/1)[a]	32	Phe[15](108/136)	1
Phe[23](21 000/260)	81	Tyr[38](800/78)	10
Arg[22](100 000/133)	750	Arg[42](4 667/1 300)	4
Val[33](100 000/20)	5 000	Glu[44](2 000/189)	11
		Asp[47](125/45)	3
		Leu[48](10/40)	0.3

[a] The receptor-binding activities due to D-amino acid and Ala substitutions are in parenthesis and expressed relative to human TGFα (TGFα = 1).

rather than receptor-contact (Table 1). These include His[12], Arg[22], Leu[24] and Val[33]. Low DA factor values (<25) indicate that the residues are nonstructural and are responsible for receptor-contacting roles and include Phe[15], Tyr[38], Arg[42], Glu[44], Asp[47] and Leu[48]. Interestingly, those residues with low DA factor (0.5–10) are either invariant or highly conserved in the EGF/TGFα family and are predicted to be centrally involved in the receptor-binding functions [2]. Since the DA values of Tyr[38] and Asp[47] are still high, it is possible that Tyr[38] and Asp[47] also play structural roles. Thus, based on the complementary scan approach and DA Factor analysis, the receptor binding region is located at the carboxy two-thirds of the molecule (residues 15–50) and located at the same face of the molecule contributed by parts of the first disulfide (residues 15–17) and third disulfide loops (residues 38–48).

Acknowledgements

This work was supported by NIH Grant CA36544.

References

1. Marquardt, H., Hunkapiller, M.W., Hood, L.E., Twardzik, D.R., DeLarco, J.E., Stephenson, J.R. and Todaro, G.J., Proc. Natl. Acad. Sci. U.S.A., 80 (1983) 4684.
2. Campbell, I.D., Cooke, R.M., Baron, M., Harvey, T.S. and Tappin, M.J., Progr. Growth Factor Res., 1 (1989) 13.
3. Merrifield, R.B., J. Am. Chem. Soc., 85 (1963) 2149.
4. Tam, J.P., Sheikh, M.A., Salmon, D.S. and Osowski, L., Proc. Natl. Acad. Sci. U.S.A., 83 (1986) 8082.
5. Tam, J.P., Heath, W.F. and Merrifield, R.B., J. Am. Chem. Soc., 105 (1983) 6442.

Transition-state analog inhibitors of human renin: A comparison of six different types

Michael Szelke[a], D. Michael Jones[a], Philippa A. Dudman[a], Butrus Atrash[a], Javier Sueiras-Diaz[a] and Brenda J. Leckie[b]

[a]Ferring Research Institute, Southampton University Research Centre, Chilworth, Southampton S01 7NP, U.K.
[b]MRC Blood Pressure Unit, Western Infirmary, Glasgow G11 6NT, U.K.

Introduction

Renin inhibitors block the production of angiotensins and thus represent a possible alternative to angiotensin-converting enzyme inhibitors in the treatment of hypertension. Our laboratory was the first one to initiate a systematic development of stable isosteric transition-state analogs as inhibitors of renin. These exploit the much greater binding affinity of the enzyme for the tetrahedral transition state **B** than for the ground state **A** of its substrate [1] (Fig. 1).

Results and Discussion

The first successful transition-state mimic developed by us [2,3] was the methylene-amino or reduced isostere $-CH_2-NH-$ shown in structure **C** (Fig. 1). When incorporated in place of the scissile bond into the minimum equine substrate sequence **1** comprising residues (6–13) (Table 1), it produced a 10^4-fold increase in potency in vitro and a dose-related hypotensive activity in vivo [4,5] (cf. **2** in Table 1).

The S-hydroxy-ethylene analog **E** (Fig. 1) was found to be an even more efficient mimic of the transition state [6]: when introduced into **3** it gave a 10^5-fold increase in potency (cf. **4** in Table 1) and the corresponding Boc-derivative **5** was a million times more potent than its peptide parent **3**. Using such a tight-binding isostere at P_1-P_1', it is possible to omit secondary binding sites (e.g. P_5 and P_4) and still retain nanomolar potency (cf. **6** in Table 1). We have found the Boc-protected (8–13)-hexapeptide sequence a convenient 'test-bed' for comparing the different transition-state analogs. Recently we have synthesized the closely related S-amino-ethylene isostere **J** [7] (Fig. 1), which provides an inhibitor of comparable potency but greater selectivity than the corresponding hydroxy-ethylene analog (cf. **7** in Table 1).

Keto-methylene isosteres (**F** in Fig. 1) were first introduced by us in 1976 [8] and first used in a renin inhibitor in 1980 [3]. Although the keto-methylene moiety **F** itself is not tetrahedral, it is capable of undergoing hydration to give the geminal diol **G**, which is a good mimic of the transition state **B**. In our 'test-bed', the keto isostere produced an inhibitor of moderate potency (**8** in

78

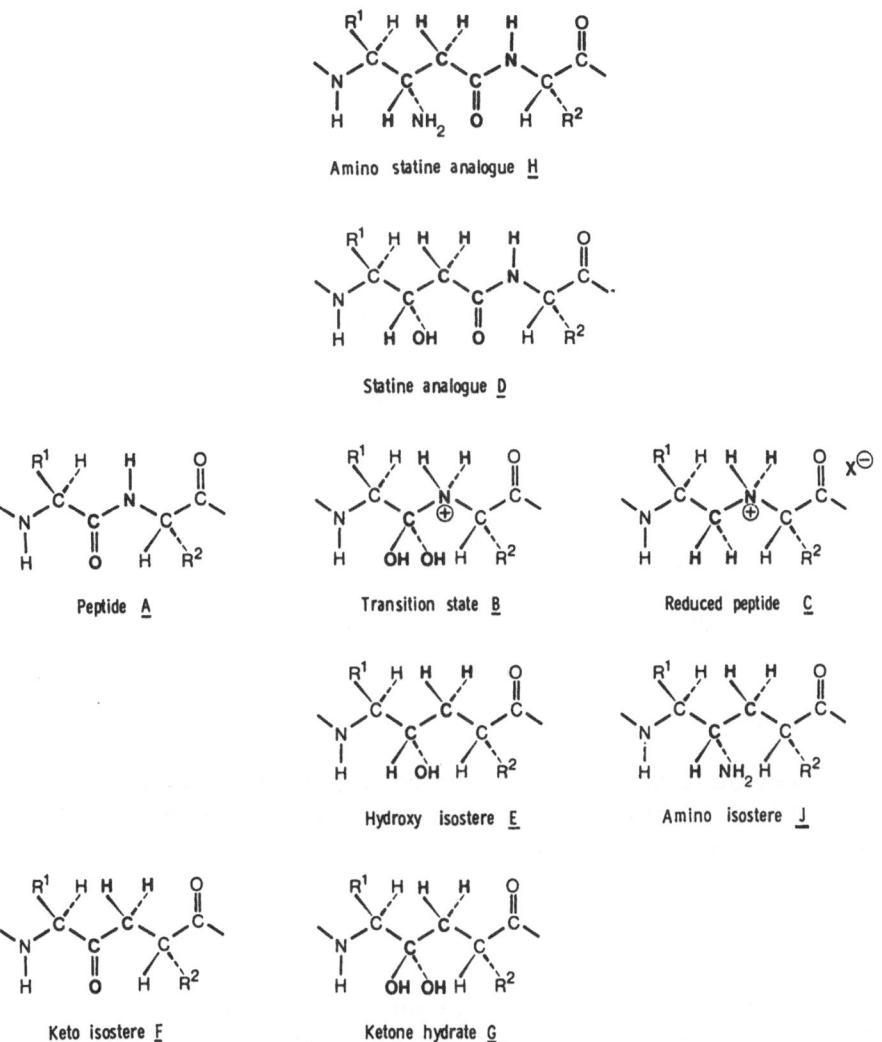

Fig. 1. *Stereochemical relationship of the peptide bond A and its transition state B to the non-hydrolyzable mimics C – J.*

Table 1), indicating that the equilibrium is in favor of the non-hydrated form **F**.

Statine (Sta) has been incorporated into renin substrate sequences both by us [9] and by others, either as a substitute for the P_1 residue alone or to replace the P_1-P_1' dipeptide [10]. Although in neither case does it provide an isosteric mimic of the transition state [6], inhibitors containing Sta still exhibit good potency provided that the surrounding molecular framework contains sufficient secondary binding sites. Clearly, the human (8–13)-hexapeptide is inadequate to support Sta, as shown by the low potency of the Sta analog 11 (Table 1).

Table 1 *In vitro potencies of renin inhibitors*

No.	Code no.	6 P$_5$ – 7 P$_4$ – 8 P$_3$ – 9 P$_2$ – 10 P$_1$ $\;\;$ 11 P$_1'$ – 12 P$_2'$ – 13 P$_3'$	IC$_{50}$ (μM)[a]
1	H-114	H-His-Pro-Phe-His-Leu——Leu-Val-Tyr-OH	200[b]
2	H-77	H-DHis-Pro-Phe-His-Leu—$\overset{R}{}$—Leu-Val-Tyr-OH	0.02[b]
3	H-112	H-His-Pro-Phe-His-Leu——Val- Ile- His-OH	313
4	H-194	H-His-Pro-Phe-His-Leu$\overset{OH}{}$Val- Ile- His-OH	0.0030
5	H-261	Boc-His-Pro-Phe-His-Leu$\overset{OH}{}$Val- Ile- His-OH	0.0002
6	H-269	Boc-Phe-His-Leu$\overset{OH}{}$Val- Ile- His-OH	0.0024
7	H-301	Boc-Phe-His-Leu—$\overset{A}{}$—Val- Ile- His-OH	0.0065
8	H-289	Boc-Phe-His-Leu—$\overset{K}{}$—Val- Ile- His-OH	0.095
9	H-294	Boc-Phe-His-Leu—$\overset{R}{}$—Val- Ile- His-OH	0.30
10	H-293	Boc-Phe-His-Ads————Ile- His-OH	0.44
11	H-312	Boc-Phe-His-Sta ————Ile- His-OH	>1.3

[a] vs. human plasma renin
  OH:-CH(OH)–CH$_2$- in place of -CO–NH-
  A: -CH(NH$_2$)–CH$_2$- in place of -CO–NH-
  K: -CO–CH$_2$- in place of -CO–NH-
[b] vs. dog plasma renin
  R: -CH$_2$–NH- in place of -CO–NH-
  Ads:3-(S)-amino-deoxystatine
  Sta: 3-(S)-statine

In 1985 we introduced 3(S)-amino-deoxystatine [10] (Ads) in order to provide additional ionic binding for Sta at the negatively charged active site of renin. It produces a more potent inhibitor than Sta (cf. **10** vs. **11** in Table 1), particularly against rat renin.

Conclusion

We have shown that all five new transition-state mimics introduced by us will provide active inhibitors of renin when incorporated into the human substrate (8–13)-hexapeptide, their rank order of potencies vis-à-vis statine being: hydroxy ≧ amino ≫ keto > reduced ≧ Ads > Sta. Out of this series, the hydroxy and amino isosteres promise to be the most efficient in producing potent inhibitors of relatively low molecular weight, such as those now being sought as orally effective antihypertensive agents.

Acknowledgements

We acknowledge financial support by Ferring AB, Ferring Pharmaceuticals Ltd., AB Hässle and the Medical Research Council.

References

1. Wolfenden, R., Nature, 223 (1969) 704.
2. Parry, M.J., Russell, A.B. and Szelke, M., In Meienhofer, J. (Ed.) Chemistry and Biology of Peptides, Ann Arbor Science Publishers, Ann Arbor, 1972, p. 541.
3. Szelke, M., Jones, D.M. and Hallett, A., European Patent 45665 (1980).
4. Szelke, M., Leckie, B.J., Tree, M., Brown, A., Grant, J., Hallett, A., Hughes, M., Jones, D.M. and Lever, A.F., Hypertension, Suppl. II, 4 (1982) 59.
5. Szelke, M., Leckie, B., Hallett, A., Jones, D.M., Sueiras-Diaz, J., Atrash, B. and Lever, A.F., Nature, 229 (1982) 555.
6. Szelke, M., Jones, D.M., Atrash, B., Hallett, A. and Leckie, B.J., In Hruby, V.J. and Rich, D.H. (Eds.) Peptides: Structure and Function, Pierce Chemical Co., Rockford, IL, 1983, p. 579.
7. Jones, D.M., Leckie, B.J., Svensson, L. and Szelke, M., In Rivier, J.E. and Marshall, G.R. (Eds.) Peptides: Chemistry, Structure and Biology (Proceedings of the 11th American Peptide Symposium), ESCOM, Leiden, 1990, p. 971.
8. Sharpe, R. and Szelke, M., British Patent 1, 587, 809 (1976).
9. Tree, M., Atrash, B., Donovan, B., Gamble, J., Hallett, A., Hughes, M., Jones, D.M., Leckie, B., Lever, A.F., Morton, J.J. and Szelke, M., J. Hypertension, 1 (1983) 399.
10. Jones, D.M., Sueiras-Diaz, J., Szelke, M., Leckie, B. and Beattie, S., In Deber, C.M., Hruby, V.J. and Kopple, K.D. (Eds.) Peptides: Structure and Function, Pierce Chemical Co., Rockford, IL, 1985, p. 759.

Synthetic peptide with antithrombotic activity

M.H. Charon[a], A. Poggi[b], M.B. Donati[b] and G. Marguerie[a]

[a]DRF/LBio/Hématologie, Unité INSERM 217, CENG 85X, F-38041 Grenoble Cedex,
France
[b]Consorzio Mario Negri Sud, via Nazionale, I-66030 Santa Maria Imbaro, Italy

Introduction

Platelet GPIIbIIIa is an adhesive receptor for fibrinogen, fibronectin, and von Willebrand factor. It has been shown [1] that its interaction with fibrinogen mediates blood platelet aggregation. Since RGD-containing peptides seemed to be implicated in adhesive reactions, we have realized a structure–function relationship study of these peptides in the platelet system. It has led us to a powerful in vitro antiaggregant, and this effect has been confirmed in an in vivo antithrombotic assay in mouse.

Results and Discussion

First, a series of RGDX peptides has been synthesized and tested in vitro on human platelets to evaluate their capacity to inhibit fibrinogen binding and platelet aggregation after stimulation with ADP (Table 1). We observed that the most important inhibitions in both systems were obtained by RGDX peptides,

Table 1 *Inhibition of fibrinogen binding and platelet aggregation after stimulation with ADP (5 μM)*

RGDX	Hydrophilicity of X (Hopp and Woods)	Fibrinogen binding (IC$_{50}$ μM)	Platelet aggregation (IC$_{50}$ μM)
RGDS	+0.3	55	170
RGDQ	+0.2	45	207
RGDC	−1.0	35	175
RGDV	−1.5	18	150
RGDL	−1.8	7	48
RGDY	−2.3	4	25
RGDF	−2.5	9	37
RGDW	−3.4	4	14

in which the residue X was strongly hydrophobic. Thus, we obtained the greatest antiaggregant effect with the peptide RGDW. This result has been confirmed with other stimulating platelet agents (collagen, adrenalin, thrombin) and compared with a control peptide RGGW that was inactive. We have also shown, in vitro, that RGDW had no effect on human endothelial cell adhesion on fibronectin.

82

The peptides were subsequently tested in vivo in a mouse antithrombotic assay [2]. After simultaneous intravenous injection of collagen, adrenalin and peptide, we observed (Fig. 1a) that, at the concentration of 1.5 mg/mouse, RGDW protected more than 90% of the mice from death or paralysis. The negative peptide RGGW did not show any effect. For comparison, Ticlopidin, which is used as a therapeutic drug against thrombosis, was injected per os at 2.5 mg/mouse 24 h and 1 h before testing. It gave a protective effect clearly weaker than that of RGDW. The effect of RGDW was dose-dependent (Fig. 1b) and the IC_{50} appeared to be about 0.3 mg/mouse.

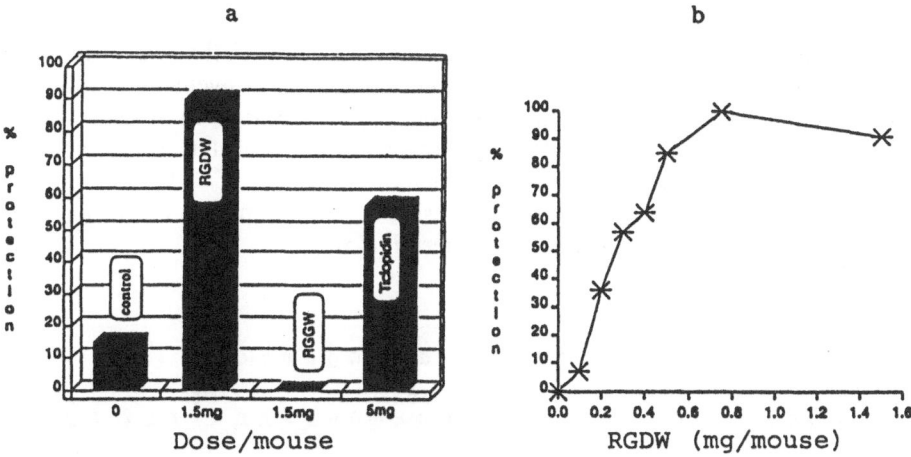

Fig. 1. *Mouse antithrombotic assay. Protection against embolism induced by a collagen/adrenalin mixture.*

Finally, the effectiveness of RGDW, RGGW and Ticlopidin on bleeding time in mice was measured. As RGGW, RGDW did not prolong bleeding at the doses where it had a potent antithrombotic effect. On the other hand, Ticlopidin showed a significant prolongation of bleeding time.

The results obtained with RGDW suggest new promising avenues for the design of antithrombotic drugs.

References

1.   Marguerie, G.A., Plow, E.F. and Edgington, T.S., J. Biol. Chem., 254(1979)5357.
2.   Di Minno, G. and Silver, M., J. Pharmacol. Exp. Ther., 255(1983)57.

The effect of modifications to the N-terminal region of hirudin on thrombin inhibition

Andrew Wallace, Stanley Dennis, Jan Hofsteenge and Stuart R. Stone*
Friedrich Miescher-Institut, P.O. Box 2543, CH-4002 Basel, Switzerland

Introduction

Hirudin is a polypeptide of 65 or 66 amino acids (Fig. 1) that was originally isolated from the leech *Hirudo medicinalis* but is now more conveniently produced by recombinant methods. It reacts rapidly with thrombin to form a tight, noncovalent complex ($K_d = 20$ fM; [1]) and shows an absolute specificity for thrombin [2]. Its 3D structure has been determined using 2D NMR studies [3, 4]. These studies indicate that hirudin consists of three domains: a central core, a protruding 'finger' and a disordered C-terminal region.

```
                         10                            20                            30
NH2-V V Y T D C T E S G Q N L C L C E G S N V C G Q G N K C I L

                         40                            50                            60          65
     G S D G E K N Q C V T G E G T P K P Q S H N D G D F E E I P E E Y* L Q-COOH
```

Fig. 1. Sequence of hirudin. The sequence given is that found in [6]; Y indicates a sulphated tyrosyl residue.*

Results and Discussion

Importance of the α-amino group

Addition of a single methionine to the N-terminus of hirudin led to a decrease in binding energy of 26 kJ mol^{-1} (Fig. 2). Removal of this methionyl residue by CNBr resulted in the recovery of full inhibitory activity. The amount by which the binding energy decreased was dependent on the nature of residue used to extend the N-terminus. An additional glycyl residue resulted in a smaller (20 kJ mol^{-1}) decrease in binding energy (Fig. 2).

A large part of the decrease in binding energy caused by the additional amino acid was due to the displacement of the positively charged α-amino group, as shown by experiments in which this group was specifically modified. In order to conduct these experiments, a mutant was constructed in which all the other primary amino groups were replaced by site-directed mutagenesis. Removal of the positive charge of the α-amino group by acetylation of the mutant resulted in a reduction of binding energy by 27 kJ mol^{-1}. Acetamidination of the

*To whom correspondence should be addressed.

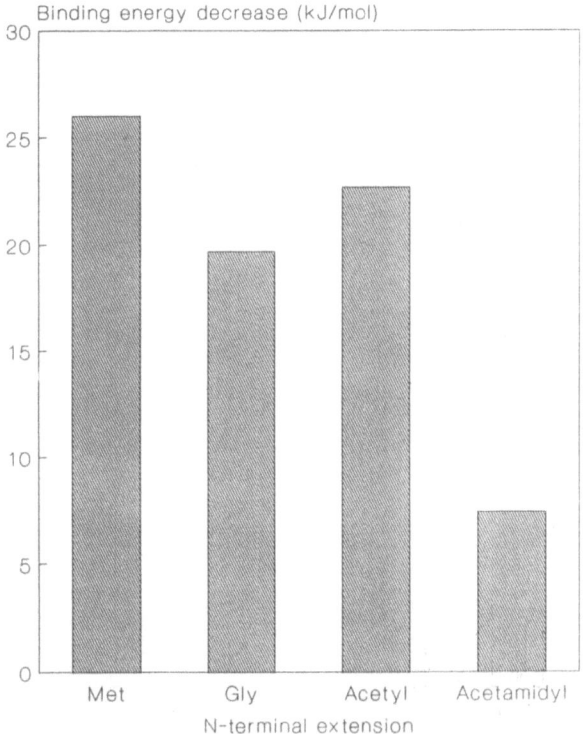

Fig. 2.   Effect of N-terminal additions on the binding energy of hirudin. Hirudin was modified by either site-directed mutagenesis to produce the Met and Gly extensions or chemically to produce acetylated and acetamidinated extensions.

α-amino group, which adds a group of similar size to the acetyl group but retains a positive charge, caused only a 12 kJ mol^{-1} decrease in binding energy (Fig. 2). Thus, it appears that a positive charge immediately adjacent to the N-terminal valyl residue is required for optimal binding to thrombin. The positively charged α-amino group is presumably involved in an ionic interaction with a carboxylate of thrombin. The results obtained by Chang [5] also indicate that the α-amino group is involved in an interaction with thrombin. This group is readily modified in free hirudin but is protected from chemical modification in the complex with thrombin.

Importance of the hydrophobic N-terminal residues
    The importance of the hydrophobic nature of the N-terminus to the strength of the hirudin–thrombin interaction was demonstrated by the observation that replacement of the two N-terminal valyl residues by polar amino acids caused a marked decrease in the binding energy (Fig. 3). In contrast, conservative replacements by other hydrophobic amino acids resulted in only moderate changes in affinity (Fig. 3). By far the largest decrease in binding energy was observed

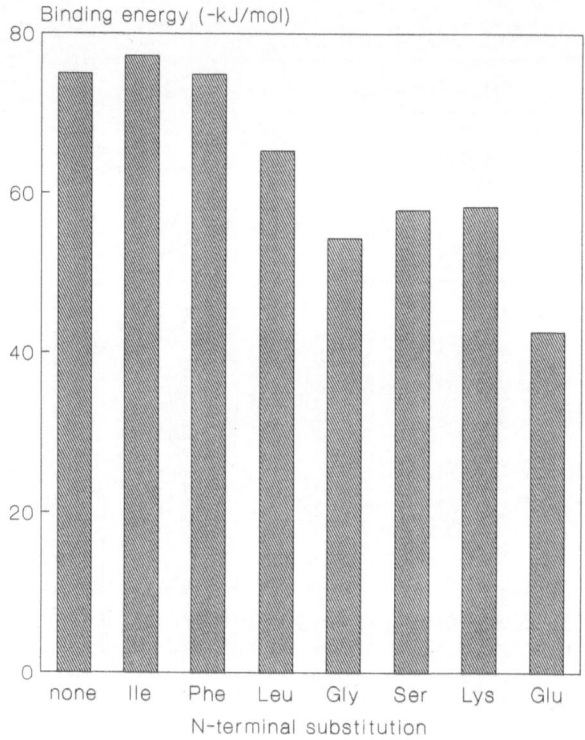

Fig. 3.   Effect of N-terminal substitutions on the binding energy of hirudin. The paired N-terminal valyl residues of recombinant hirudin were replaced by the amino acids indicated.

when Val[1] and Val[2] were replaced by glutamate. In contrast, the effect of substitution by lysine was much smaller (Fig. 3). The differential effect observed between the glutamyl and lysyl mutants is presumably due to the effect of the glutamate replacement on the proposed ionic interaction made by the α-amino group. The presence of a negatively charged residue in the first and second positions would disrupt such an interaction. On the other hand, the positively charged lysyl residue would also be able to participate in this ionic interaction, while the presence of an ionized carboxyl would make it energetically less unfavourable to bind the positively charged α-amino group in a hydrophobic pocket.

References

1.   Stone, S.R. and Hofsteenge, J., Biochemistry, 25 (1986) 4622.
2.   Walsmann, P. and Markwardt, F., Pharmazie, 36 (1981) 633.
3.   Folkers, P.J.M., Clore, G.M., Driscoll, P.C., Dodt, J., Köhler, S. and Gronenborn, A.M., Biochemistry, 28 (1989) 2601.
4.   Haruyama, H. and Wüthrich, K., Biochemistry, 28 (1989) 4301.
5.   Chang, Y.-J., J. Biol. Chem., 264 (1989) 7141.
6.   Dodt, J., Müller, H., Seemüller, U. and Chang, Y.-J., FEBS Lett., 165 (1984) 180.

Interaction of the fibrin α- and β-chain creates a polymerization site

Kun-Hwa Hsieh

Department of VCAPP, Washington State University, Pullman, WA 99164-6520, U.S.A.

Introduction

Fibrinogen is a soluble plasma protein (M_r 340 000), consisting of two identical units, each containing three different peptide chains. The six chains are held together by 29 disulfide bonds in the structure of $(A\alpha, B\beta, \gamma)_2$. Following injury, thrombin is generated locally, and its limited proteolysis of fibrinogen leading to clot formation is central to normal hemostasis [1]. To account for the sequential cleavage of two pairs of fibrinopeptides (FPA and FPB), accompanied by the transformation of an initially loose $(\alpha, \beta, \gamma)_2$ fibrin to the compact and branched clot, fibrin polymerization has been proposed to involve a two-step mechanism: binding of the newly exposed NH_2 termini (E domain) with the COOH termini (D domain) of an adjacent unit results in an end-to-end aggregate of the type I fibrin; subsequent reinforcement by lateral association generates the more compact type II fibrin [2].

This proposed mechanism is consistent with the observation that peptide analogs corresponding to the fibrin α-chain NH_2-terminal region (GPRP/V, Aα 17–20) could bind to fibrinogen and inhibit fibrin polymerization [3,4]. However, further extension of the peptide chain to Aα 17–22 or Aα 17–26 resulted in homologs inactive in the repolymerization assay, and extensive studies showed that the NH_2-terminal homolog of fibrin β-chain (GHRP, Bβ 15–18) could bind to fibrinogen, but dit not inhibit reaggregation [3,5]. Since promotion of the lateral assembly of fibrin at the expense of linear association has been observed, during which a reduced protofibril mass : length ratio was accompanied by a decreased fibrin gel turbidity but an unchanged fibril density [6], the turbidity repolymerization assay [2,3] may be inherently insensitive to lateral association. In addition, the sequential cleavage of fibrinopeptides exposes two α-chains and two β-chains. Thus, D : E binding may involve simple ionic and H-bonding interaction [7], whereas fibrin aggregation is conformation-dependent [8] and may require dimeric and tetrameric polymerization sites.

Methods

Turbidity assay

Fibrin monomer was prepared by clotting fibrinogen with thrombin. Subsequent dispersion of the clot in bromide buffer followed by centrifugation (2000 rpm, 10 min) removed the undispersed material. Linear aggregation of fibrin was monitored at 350 nm, following dilution of fibrin monomer solution in low-ionic strength phosphate buffer according to reported procedures [3,4].

Modified assay

Fibrin aggregation was initiated as above. Reaggregated fibrin was removed by centrifugation (2000 rpm, 10 min), and % unpolymerized fibrin in the supernatant was estimated using the absorbency constant of A_{280} (1%) = 15.5. The extent of polymerization was: 100% minus % unpolymerized fibrin.

This assay utilizes the principle of fibrin clottability [9] and should, therefore, be sensitive to both linear and lateral associations. A comparison of the turbidity and modified assay showed that GPRP inhibited fibrin polymerization to comparable levels in both procedures (data not shown).

Results and Discussion

In corroboration of the reported studies, we observed that neither fibrin Aα 17–29 (Gly-Pro-Arg-Val-Val-Glu-Arg-His-Gln-Ser-Ala-Cys-Lys(TFA)), nor Bβ 15–24 (Gly-His-Arg-Pro-Leu-Asp-Lys-Lys-Arg-Glu), alone inhibited fibrin re-polymerization in the turbidity assay. In the modified assay (Fig. 1), although neither peptide alone was active, a mixture of Aα 17–29 and Bβ 15–24 was at least as effective as Gly-Pro-Arg-Pro. This finding strongly suggests that interaction of the α- and β-chains can create a polymerization site. Because the native sequences of Bβ 23–25 (Arg-Glu-Glu) and Aα 22–24 (Glu-Arg-His), and of Bβ 20–23 (Asp-Lys-Lys) and Aα 29–32 (Lys-Asp-Ser-Asp), are mutually complementary and can form multiple ionic bonds, it appears likely that ionic

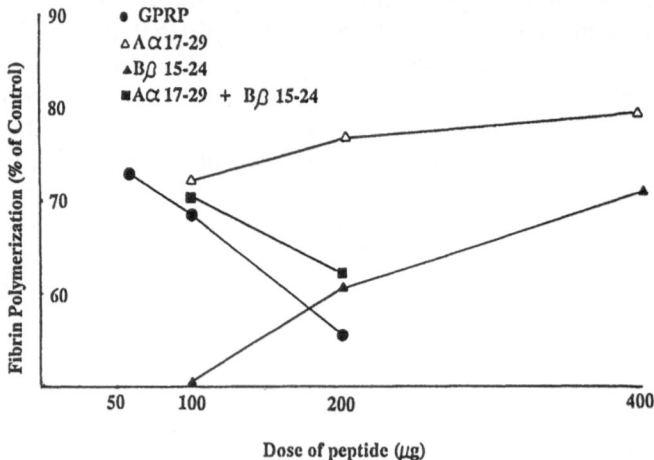

Fig. 1. Inhibition of fibrin polymerization in the modified clottability assay. Symbols are: ● *GPRP;* Δ *Aα 17–29;* ▲ *Bβ 15–24;* ■ *Aα 17–29 + Bβ 15–24.*

interaction of fibrin α- and β-chains may create a dimeric binding site. Whether this site is involved in the linear or lateral association of fibrin remains to be determined.

Acknowledgements

Support of this work by the American Heart Association, Washington Affiliate, is gratefully acknowledged.

References

1.   Shafer, J.A. and Higgins, D.L., CRC Crit. Rev. Clin. Lab. Sci., 26 (1988) 1.
2.   Blomback, B., Hessel, B., Hogg, D. and Therkildsen, L., Nature, 275 (1978) 501.
3.   Laudano, A.P. and Doolittle, R.F., Proc. Natl. Acad. Sci. U.S.A., 78 (1978) 3085.
4.   Hsieh, K.H., Mudd, M.S. and Wilner, G.D., J. Med. Chem., 24 (1981) 322.
5.   Laudano, A.P. and Doolittle, R.F., Biochemistry, 19 (1980) 1013.
6.   Galanakis, D.K., Lane, B.P. and Simon, S.R., Biochemistry, 26 (1987) 2389.
7.   Wiesel, J.W., Biophys. J., 50 (1986) 1079.
8.   Cierniewski, C.S., Kloczewiak, M. and Budzynski, A.Z., J. Biol. Chem., 261 (1986) 9116.
9.   Mosesson, M.W. and Sherry, S., Biochemistry, 5 (1966) 2829.

RGD analog antagonists of GP IIb/IIIa stabilized against proteolysis are active platelet aggregation inhibitors in vivo

Steven P. Adams[a],*, Foe S. Tjoeng[a], Kam F. Fok[a], Mark E. Zupec[a],
Mark H. Williams[a], Larry P. Feigen[b], Susan G. Knodle[b], Nancy S. Nicholson[b],
Beatrice B. Taite[b], Masateru Miyano[b] and Richard J. Gorczynski[b]

*[a]Biological Sciences Department, Monsanto Company, and [b]Cardiovascular Research,
G.D. Searle Company, 700 Chesterfield Village Pkwy., St. Louis, MO 63198, U.S.A.*

Introduction

Fibrinogen (Fg) is an essential cytoadhesive molecule for the aggregation of platelets and formation of hemostatic plugs. Recent data suggest that the platelet glycoprotein IIb/IIIa complex (gp IIb/IIIa) is the Fg binding site on platelets that mediates the adhesive function required for platelet aggregation [1]. The binding of Fg to gp IIb/IIIa involves determinants on both the α and the γ chains of Fg, and one essential site of interaction is the sequence RGDF (α95–98) [2]. Synthetic peptides containing the RGD sequence effectively antagonize Fg binding to gp IIb/IIIa and prevent in vitro platelet aggregation induced by a variety of platelet activating agents. The ability to inhibit platelet aggregation irrespective of the activating agent by blocking gp IIb/IIIa makes it an attractive antithrombotic approach.

RGD peptides are relatively ineffective agents in vivo. The present study examines the fate of RGDS and several analogs in human plasma and explores approaches to overcome the effects of proteolysis.

Results and Discussion

RGDS and a related peptide, RGDY(Me)-NH$_2$, were incubated in fresh human plasma and their disappearance was monitored by RPHPLC. As shown in Fig. 1, both peptides were rapidly degraded. Products from the degradation of RGDY(Me)-NH$_2$ were resolved by HPLC and identified as GDY(Me)-NH$_2$, DY(Me)-NH$_2$ and Y(Me)-NH$_2$. The time-course for their appearance was consistent with sequential degradation from the N-terminus (Fig. 2). When bestatin, a specific aminopeptidase inhibitor [3,4] was included in the incubation, degradation was completely inhibited. To further confirm the role of aminopeptidase in the degradation, and to identify improved compounds for in vivo studies, two analogs were synthesized in which the α-amino group was acetylated (Ac-RGDY(Me)-NH$_2$) or deleted (des-NH$_2$-RGDY(Me)-NH$_2$). Both analogs were stable in plasma; moreover, they were good inhibitors of ADP-induced platelet

---

*To whom correspondence should be addressed.

aggregation ($IC_{50} = 10\ \mu M$). In collagen-induced thrombocytopenia, an in vivo rat model of platelet aggregation, RGDS was an ineffective inhibitor even at high doses, ~30% inhibition at a dose of 10 mg/kg. On the other hand, the improved analogs inhibited aggregation 65–90% at 1 mg/kg.

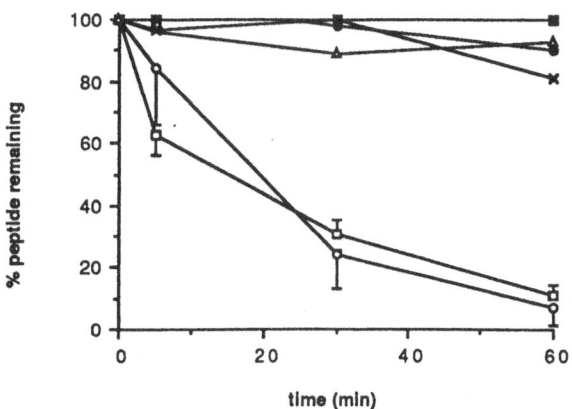

Fig. 1. Degradation of RGD peptides (~150 μM) in fresh human plasma: RGDS (□); RGDY(Me)-NH₂(o); RGDS(M + 100 mM bestatin (■); RGDY(Me)-NH₂ + 100 mM bestatin (●); Ac-RGDY(Me)-NH₂ (x); des-NH₂-RGDY(Me)-NH₂ (Δ).

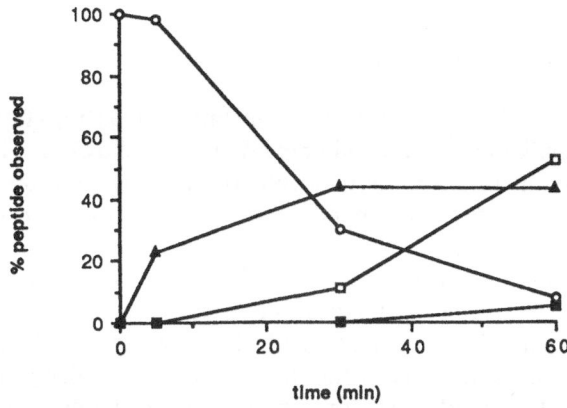

Fig. 2. Products evolving during the degradation of 150 μM RGDY(Me)-NH₂ (o) in fresh human plasma: GDY(Me)-NH₂ (▲), DY(Me)-NH₂ (□), Y(Me)-NH₂ (■).

References

1. Plow, E.F., Marguerie, G. and Ginsberg, M., Biochem. Pharmacol., 36 (1987) 4035.
2. Andrieux, A., Hudry-Clergeon, G., Ryckewaert, J.-J., Chapel, A., Ginsberg, M.H., Plow, E.F. and Marguerie, G., J. Biol. Chem., 264 (1989) 9258.
3. Suda, H., Aoyagi, T., Takeuchi, T. and Umezawa, H., Arch. Biochem. Biophys., 177 (1976) 196.
4. Harbeson, S.L. and Rich, D.H., Biochemistry, 32 (1989) 1378.

MDL 28 050, representative of a new class of anticoagulant: Development, in-vitro and in-vivo actions

John L. Krstenansky, Thomas J. Owen, Marguerite H. Payne,
Robert J. Broersma, Mark T. Yates and Simon J.T. Mao

Merrell Dow Research Institute, 2110 E. Galbraith Road, Cincinnati, OH 45215, U.S.A.

Introduction

MDL 28 050, N$^\alpha$-succinyl-Tyr-Glu-Pro-Ile-Pro-Glu-Glu-Ala-Cha-D-Glu-OH, is a synthetic inhibitor of α-thrombin that acts at a site distinct from the catalytic site of the enzyme. It has an in vitro potency of 29 nM against human α-thrombin, and 150 nM against bovine α-thrombin. It was developed from systematic SAR [1–4] which demonstrated that a minimal functional domain of the leech antithrombin, hirudin, was represented in residues 56–64 of the protein [5]. Hirudin (56–65) itself had a potency of 34 000 nM against bovine α-thrombin. Thus, MDL 28 050 represents a 225-times more potent material than the corresponding native sequence of the same length, and it also incorporates a number of unnatural modifications that provide increased enzymatic stability.

Results and Discussion

The anticoagulant effect observed after the intravenous injection of MDL 28 050 at 1 and 5 mg/kg in normal and anephric rats, is shown in Fig. 1. Doses of 1 or 5 mg/kg result in strong anticoagulation at early time points, and at 30 min the higher dose was able to maintain this level of activity. Removal of the kidneys led to further prolongation of the anticoagulant activity, indicating that the kidneys represent a major route of elimination of the peptide analog.

Total protection towards thromboplastin-induced disseminated intravascular coagulation in mice has been demonstrated. The dose required for complete protection increases with the time of the thromboplastin challenge after dosing with MDL 28 050. Increased doses or removal of the kidneys prolongs the anticoagulant activity, as was observed in the above ex-vivo rat studies. A similar dose-dependent effect is seen in a rat model of stasis-induced venous thrombosis, with anephric animals having prolonged protection.

Conclusion

We found MDL 28 050 and related materials to be effective antithrombin agents with several advantages over heparin or the coumarins. This class of antithrombin agent does not require the presence of cofactors. They have less

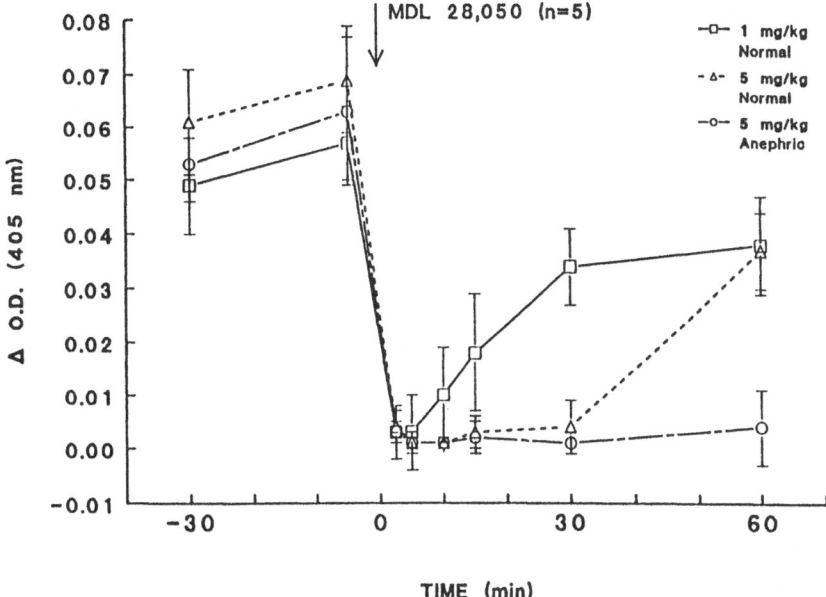

Fig. 1.   Anticoagulant activity of MDL 28 050 after intravenous injection (at time = 0 min) in rats. The level of anticoagulant activity is measured by recording the O.D. 405 nm, which is a measure of the turbidity resulting from fibrin-clot formation, of a diluted sample of rat plasma 12 min after the addition of bovine thrombin. The basal response is established by the readings at the –30 and –5-min time points.

of a tendency to cause bleeding, which is a major undesired result of current anticoagulant therapy. Thus, these agents may be generally applicable when anticoagulation is desired (even in the absence of antithrombin III), and safer to use by having less hemorrhagic potential.

References

1.   Krstenansky, J.L., Mao, S.J.T., Owen, T.J. and Yates, M.T., In Marshall, G.R. (Ed.) Peptides: Chemistry and Biology (Proceedings of the 10th American Peptide Symposium), ESCOM, Leiden, 1988, p. 447.
2.   Krstenansky, J.L., Owen, T.J., Yates, M.T. and Mao, S.J.T., J. Med. Chem., 30 (1987) 1688.
3.   Owen, T.J., Krstenansky, J.L., Yates, M.T. and Mao, S.J.T., J. Med. Chem., 31 (1988) 1009.
4.   Krstenansky, J.L., Owen, T.J., Yates, M.T. and Mao, S.J.T., Throm. Res., 52 (1988) 137.
5.   Mao, S.J.T., Yates, M.T., Owen, T.J. and Krstenansky, J.L., Biochemistry, 27 (1988) 8170.

Structure-activity studies toward the improvement of antiaggregatory activity of Arg-Gly-Asp-Ser (RGDS)

Fadia El-Fehail Ali[a],*, Raul Calvo[a], Todd Romoff[a], James Samanen[b], Andrew Nichols[b] and Barbra Store[b]

[a]Department of Peptide Chemistry and [b]Department of Pharmacology, Smith Kline & French Laboratories, P.O. Box 1539, King of Prussia, PA 19406-0939, U.S.A.

Introduction

The antiaggregatory activity of the $F(ab')_2$ fragment of MAb against the platelet fibrinogen (Fg) receptor GPIIb-IIIa in humans demonstrated the utility of inhibition of Fg/platelet binding as antithrombotic therapy [1]. Demonstration of thrombus formation inhibition in dogs by intracoronary infusion of Ac-RGDS-NH_2 (**1**) established that analogs with high potency and duration could constitute a peptide approach to antithrombotic therapy [2]. We report a series of analogs of **1** synthesized by SPPS that culminated in analogs **21** and **22** displaying high potency (IC_{50} 4.0 μM) and duration of action in plasma (up to 3 h).

Results and Discussion

Side-chain modifications in **1** failed to improve potency dramatically. Modifications at the Ser residue as in **14** and **15** mainly enhanced duration of action (Table 1). Ser could be deleted without dramatic loss of potency as in **16**. Inactive β-Asp (**13**) did not form in samples of **1** stored at RT for 6 months. Unlike RGDV [3], an additional Arg residue as in **8** did not enhance potency. Cyclization of the -RGD-sequence in disulfide analogs **18–25** proved to enhance potency and plasma stability. In contrast to linear *N*-ethylamide (**17**), cyclic *N*-ethylamide (**22**) retained good potency. The increased lipophilicity in (*N,N'*-diethyl[guan])Arg substitution [4] failed to enhance potency in **23**. Pentapeptides typified by **20** displayed superior potency over hexapeptides of the types **18** or **24**. Future reports will describe in vivo efficacy of analogs with even greater potency.

---

*To whom correspondence should be addressed.

Table 1 *Inhibition[a] of ADP(10 µM)-induced platelet aggregation in dog PRP[b] of -RGD-peptides*

No.	Compounds	IC$_{50}$ µM Dog PRP/ADP	%Activity at 3 h[c]
1	Ac-Arg-Gly-Asp-Ser-NH$_2$	91.3 ± 0.1	0.00[d]
2	D-Arg	> 200	–
3	D-Asp	> 200	–
4	D-Ser	138 ± 4	33.3
5	Lys	> 200	–
6	Cit	> 200	–
7	His	> 200	–
8	Ac-Arg-Arg	113 ± 14	5.9
9	D-Ala	> 200	–
10	Pro	> 200	–
11	Asn	> 200	–
12	Glu	> 200	–
13	β-Asp	> 200	–
14	Val	56 ± 14	92
15	Tyr	102 ± 14	98
16	---	138 ± 16	54
17	NHEt	> 200	–
18	Ac-Cys-Arg-Gly-Asp-Ser-Cys-NH$_2$	32.7 ± 11	100
19[e]	Mpr-Arg-Gly-Asp-Cys-NH$_2$	10.9 ± 2.3	100
20	Ac-Cys-Arg-Gly-Asp-Cys-NH$_2$	16.2 ± 5.9	55[f]
21	Ac-Cys-Arg-Gly-Asp-Pen-NH$_2$	4.1 ± 0.6	100
22	Ac-Acy-Arg-Gly-Asp-Pen-NHEt	9.5 ± 2	52[f]
23	Ac-Cys-Arg(N,N'-Et$_2$guan)-Gly-Asp-Pen-NH$_2$	82 ± 5	5[f]
24	Ac-Cys-Gly-Arg-Gly-Asp-Cys-NH$_2$	52 ± 18	100
25	Ac-Cys-Gly-Arg-Gly-Asp-Pen-NH$_2$	11.4 ± 2	100

[a] Measured by light transmittance in Chronolog aggregometer [Zucker, M., Methods in Enzymology, 169 (1988) 117].
[b] PRP = platelet-rich plasma.
[c] Activity measured after incubation in PRP for 3 h at ambient temperature, final peptide concentration 200 µM.
[d] t$_{1/2}$ = 90 min
[e] Mpr = mercaptopropionyl, Pen = penicillamine.
[f] Incubation of IC$_{50}$ concentration peptide, 3 h at 37°C.

References

1. Coller, B.S., Scudder, L.E., Berger, H.J. and Iuliucci, J.D., Ann. Int. Med., (1988) 635.

2. Samanen, J.M., Ali, F.E., Berry, D.E., Bennett, D.B., Romoff, T. and Storer, B.L., FASEB J., 3 (1989) A744; b. Thromb. Haemost, 61 (1989) 183.
3. Ruggeri, Z.M., Houghten, R.A., Russell, S.R. and Zimmerman, T.S., Proc. Natl. Acad. Sci. U.S.A., 83 (1986) 5708.
4. Nestor, J.J., Tahilramani, R., Ho, T.L., McRae, G.I. and Vikery, B.H., J. Med. Chem., 31 (1988) 65.

Biological activities of synthetic EGF-like domains in blood coagulation factor IX

Linda H. Huang[a], William Sweeney[b] and James P. Tam[a]

[a]The Rockefeller University, 1230 York Ave., New York, NY 10021, U.S.A.
[b]Hunter College, CUNY, 695 Park Ave., New York, NY 10021, U.S.A.

Introduction

A sequential repeat of epidermal growth factor (EGF)-like domains is a common feature among blood clotting proteins. Factor IX, Factor X, protein S, and protein Z each have two EGF-like domains, and protein C has four [1]. To determine the structure and function of these domains, we synthesized the first (peptide A, residues 45–87) and second (peptide B, residues 84–128) EGF-like domains, and a peptide containing both domains (peptide C, residues 45–128) (Fig. 1).

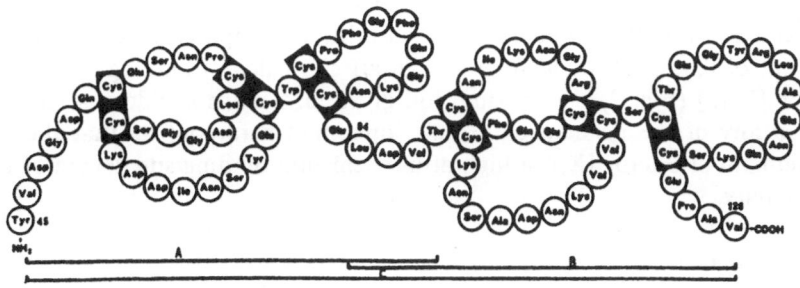

Fig. 1.   Primary structure of EGF.

Results and Discussion

These syntheses were performed using the Boc-benzyl protecting group approach in SPPS. The crude and deprotected peptides, after HF cleavage from the resin, were folded by the mixed disulfide exchange procedure [2]. All peptides were purified using C_{18} RPHPLC. The AAA of all synthesized peptides were consistent with their expected compositions. The molecular weights of peptides A and B were confirmed by MS. The CD spectrum of peptide A is consistent with the presence of β-sheet structure. The NMR spectrum of peptide A also suggests this, as a number of slowly exchangeable resonances are seen downfield of 8 ppm, implying a hydrogen bond network, and a number of resonances suggestive of β-sheet structure between 4.8 and 5.8 ppm [3] are observed.

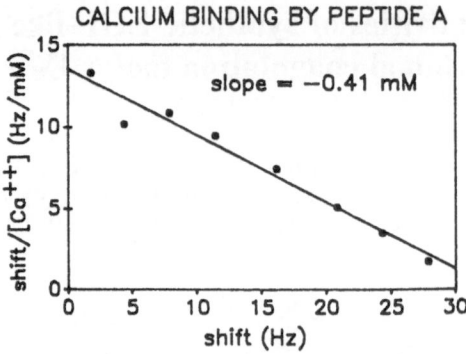

Fig. 2.   *A plot of shift/[Ca⁺⁺] vs. shift; the slope is equal to minus the calcium binding constant.*

None of the peptides showed an ability to compete effectively with mouse [¹²⁵I]-EGF for EGF receptors in A431 cells. Similarly, none of the peptides exhibited significant mitogenicity in normal rat kidney cells clone 49F in culture. However, calcium binding activity was observed. On addition of calcium ion, a resonance in the NMR spectrum at 5.4 ppm moves upfield, with a maximum shift at high calcium concentration of 33 Hz. A plot of shift/[Ca^{2+}] vs. shift [4] yielded a dissociation constant of 0.4 mM, as compared to the dissociation constant of 0.08 mM found for the high affinity calcium binding site of native Factor IX [5] (Fig. 2). Magnesium ion at 23 mM did not induce a significant shift in any of the resonances. Results presented here confirm that the EGF-like domain of Factor IX is a high affinity calcium binding site devoid of EGF-like activity.

Acknowledgements

This work was supported in part by PHS grants CA36544 and LH41935, and a travel grant from Applied Biosystems Inc.

References

1.   Rees, D.J.G., Jones, I.M., Handford, P.A., Walter, S.J., Esnouf, M.P., Smith, K.J. and Brownlee, G.G., EMBO J., 7 (1988) 2053.
2.   Stenflo, J., Holme, E., Lindstedt, S., Chandramouli, N., Huang, L.H.T., Tam, J. P. and Merrifield, R.B., Proc. Natl. Acad. Sci. U.S.A., 86 (1989) 444.
3.   Wagner, G., Pardi, A. and Wüthrich, K., J. Am. Chem. Soc., 105 (1984) 5948.
4.   Connors, K.A., Binding Constants: The Measurement of Molecular Complex Stability, John Wiley & Sons, New York, NY, 1987, p. 194.
5.   Morita, T., Issacs, B.S., Esmon, C.T. and Johnson, A.E., J. Biol. Chem., 259 (1984) 5698.

Biosynthesis and characterization of insulin-like growth factor I (IGF-I)

R.D. DiMarchi, H. Long, J. Epp, B. Schoner and R. Belagaje

Lilly Research Laboratories, Indianapolis, IN 46285, U.S.A.

Introduction

The genetic manipulation of *E. coli* to code for expression of an appropriate IGF-I fusion protein has been previously described by this laboratory [1]. A precursor bearing a single tryptophan, immediately adjacent to the natural IGF-I sequence, was expressed in the form of cytoplasmic inclusion bodies. Physical isolation of these inclusion bodies through differential centrifugation yielded a trp Ie'-IGF-I fusion protein of approximately 60% purity. Four previously described methods of tryptophan cleavage were assessed with this partially pure fusion-protein (Table 1). Two principal improvements in current methodology were required to produce IGF-I with reasonable efficiency. The first being elimination of cysteic acid formation without an adverse effect on cleavage yield. The second improvement required a highly efficient and cleavage-compatible method of methionine regeneration.

Results and Discussion

Treatment of the fusion protein with DMSO/HCl/TFA (1/1/98) formed a nearly equal mixture of IGF-I and its methionine sulfoxide analog in approximately 40% yield. AAA revealed cysteic acid formation to have occurred at less than 1% of the initial cysteine/cystine content. Reduction of the sulfoxide analog of IGF-I, immediately following cleavage, was achieved in 95% yield through DMS/HCl/TFA (10/1/89) treatment. The IGF-I was purified as its *S*-sulfonate derivative by cation-exchange chromatography.

Through analogous conditions developed for proinsulin formation, natural IGF-I was obtained from its *S*-sulfonate in nearly 50% yield. Unlike proinsulin, a major single impurity was formed in this conversion at one-half the yield of IGF-I. The impurity and IGF-I were respectively purified to near homogeneity. Each peptide was treated with *S. aureus* V8 to generate a series of fragments that were chromatographically purified and characterized by AAA and MS. The difference between the impurity and the IGF-I was identified to exist in the heterodimeric fragment consisting of B5-B10 and A6-A12. The impurity was presumed to be a monomeric disulfide isomer of IGF-I.

The heterodimeric disulfide peptides of B5-B10 and A6-A12, where B7 is linked to A6 or A7, were separately synthesized by unambiguous routes. This entailed

Table 1 *Tryptophan cleavage of IGF-I fusion protein*

Amino acid[a]	Untreated	BNPS-skatole[2]	NCS[3]	CNBr[4]	DMSO-HCl[5]	DMSO-HBr[5]
Cysteic acid	<0.1	0.6	0.7	0.3	0.1	0.8
Homoserine	<0.1	<0.1	<0.1	<0.1	<0.1	<0.1
Methionine	2.0	<0.1	<0.1	<0.1	<0.1	<0.1
Meth. sulfoxide	<0.1	1.7	1.9	1.6	1.7	1.9
Meth. sulfone	<0.1	<0.1	<0.1	<0.1	<0.1	0.1
Tyrosine	3.0	2.8	2.8	2.3	2.9	2.8
Cleavage	–	33%	25%	33%	1%	30%

[a] AAA was performed following hydrolysis in 4N-methanesulfonic acid. The results are expressed in amino acid equivalents in the IGF-I fusion protein. The cysteine content in untreated fusion protein is six.

the differential protection of the respective disulfides with Acm and 4-MeBzl. In each synthesis, the A-B disulfide was formed by B-chain thiol-mediated exchange with the A-chain S-sulfonate fragment. The second disulfide was subsequently generated by iodide-mediated formation under acidic conditions. The two heterodimeric isomers were chromatographically distinct by RPHPLC analysis. Comparison of the IGF-I and impurity peptide fragmentation pattern with the synthetic standards revealed the latter to be an A6-B7, A7-A11 disulfide isomer of the naturally occurring A7-B7, A6-A11 structure.

Conclusion

The synthetic methodology described within this report provides a highly efficient alternative to cyanogen bromide cleavage for the biosynthesis of peptides. The efficient regeneration of methionine, following tryptophan cleavage, facilitates synthesis and permits additional fragmentation, if so desired, through the action of cyanogen bromide. For the first time, the natural disulfide bonding pattern of IGF-I has been definitively characterized [6]. In contrast to proinsulin, a specific isomer was formed in high yield and characterized as the monomeric A6-B7, A7-A11 disulfide analog of IGF-I.

References

1. DiMarchi, R., Long, H., Epp, J., Schoner, B. and Belagaje, R., In Kaiser, G. and Tam, J., (Eds.) Synthetic Peptides: Approaches to Biological Problems, 1989, p. 283.
2. Omenn, G.S., Fontana, A. and Anfinsen, C.B., J. Biol. Chem., 245 (1970) 1895.
3. Schechter, Y., Patchornik, A. and Burstein, Y., Biochemistry, 15 (1976) 5071.
4. Huang, H.V., Bond, M.W., Hunkapiller, M.W. and Hood, L.E., Methods Enzymol., 91 (1983) 318.
5. Savige, W.E. and Fontana, A., Methods Enzymol., 47 (1977) 459.
6. Rinderknecht, E. and Humbel, R.E., J. Biol. Chem., 253 (1978) 2769.

Direct identification of disulfide bond linkages in insulin-like growth factor I (IGF-I) by chemical synthesis

Michio Iwai[a], Masakazu Kobayashi[b], Kouichi Tamura[b], Yoshinori Ishii[b], Hisashi Yamada[b] and Mineo Niwa[b]

[a]Marine Technical College, Faculty of Liberal Arts, Nishikura-cho, Ashiya 659, Japan
[b]Research Laboratories, Fujisawa Pharmaceutical Co. Ltd., 2-1-6 Kashima, Yodogawa-ku, Osaka 532, Japan

Introduction

IGF-I, or Somatomedin C, is a 70-residue serum polypeptide with growth-promoting and insulin-like hypoglycemic activities [1]. The primary structure of IGF-I was chemically determined by Rinderknecht and Humbel in 1978 [2], except for three disulfide linkages that were postulated based on the homology with insulin. We found several IGF-I derivatives in a refolding (air oxidation) mixture of IGF-I produced by a recombinant DNA method [3]. Isomers I – IV were isolated from the mixture by RPHPLC. Isomer III was identified as [Glu^{15}]-IGF-I, and isomer IV proved to be a disulfide bond isomer of IGF-I, having Cys^6-Cys^{52}, Cys^{18}-Cys^{61} and Cys^{47}-Cys^{48}. Isomer II was identified as natural human IGF-I with respect to its retention time on HPLC and its biological activities, and Isomer I as another disulfide bond isomer of Isomer II around Cys^6, Cys^{47}, Cys^{48} and Cys^{52}. However, the exact disulfide bond linkage system of IGF-I has never been proved by chemical methods. The fact that Isomer I and IV are 8 and 68% active in [^{3}H]-thymidine uptake stimulation, and 5 and 22.4% active in Radio-receptor Assay, like Isomer II (natural form) [3], respectively, prompted us to determine the disulfide bond linkage system in IGF-I to establish further its SAR.

Results and Discussion

Isomer I and II afforded, by V$_8$-protease digestion, the characteristic peptide fragments (Type I and Type II) with the disulfide linkage formed from Cys^6, Cys^{47}, Cys^{48} and Cys^{52}. Two possible fragments, one with Cys^6-Cys^{47} and Cys^{48}-Cys^{52} (Type I) and the other with Cys^6-Cys^{48} and Cys^{47}-Cys^{52} (Type II), were synthesized by SPPS (Fig. 1). All the blocking groups except the Acm group were removed by HF followed by intramolecular disulfide bond formation with air or K$_3$Fe(CN)$_6$ oxidation to give 48-Acm IGF-I(47–53) and 47-Acm IGF-I(47–53), respectively. 6-Acm IGF-I(4–9) was prepared in a similar manner to that shown in Fig. 1. Coupling of 48-Acm IGF-I(47–53) and 6-Acm IGF-I(4–9) by I$_2$/HCl-MeOH produced a heterodimer peptide, IGF-I(4–9, 47–53) Type II, along with the respective homodimers (Fig. 2). Isolated IGF-I(4–

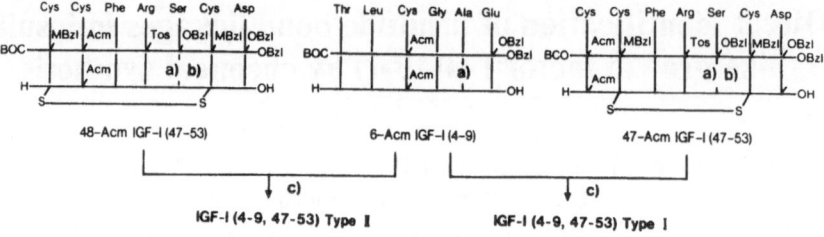

Fig. 1. Synthetic scheme of two IGF-I disulfide containing peptides, Type I and Type II. (a) HF/anisole, (b) air or $K_3 Fe(CN)_6$, (c) I_2/HCl-MeOH.

9, 47–53) Type II was identified by HPLC and AAA as the characteristic peptide obtained by V_8-protease digestion of natural type IGF-I. In a similar manner, IGF-I(4–9, 47–53) Type I was identified as a peptide obtained from Isomer I of IGF-I.

Thus, the disulfide bond linkage system of IGF-I was absolutely determined to be Cys^6-Cys^{48}, Cys^{18}-Cys^{61} and Cys^{47}-Cys^{52}, and that of the less active Isomer I to be Cys^6-Cys^{47}, Cys^{18}-Cys^{61} and Cys^{48}-Cys^{52}.

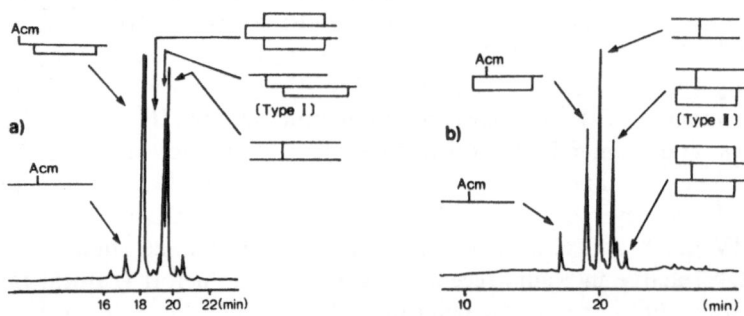

Fig. 2. RPHPLC of the total crude synthetic products of Type I peptide (a) and Type II (b). The peptides were loaded on YMC AP-302 200 Å ODS column (4.6×150 mm) and eluted with 0.01 M TFA v.s. acetonitrile, 0–60% over a period of 0–30 min. Flow rate, 1 ml/min. Detection was at 214 nm.

References

1. Froesch, E.R., Burgi, H., Ramseier, E.B., Bally, P. and Labhart, A., J. Clin. Invest., 42 (1963) 1816.
2. Rinderknecht, E. and Humbel, R.E., J. Biol. Chem., 253 (1978) 2769.
3. Tamura, K., Yamada, H., Ishii, Y., Koyama, S., Asada, T., Horiai, H., Kobayashi, M., Niwa, M., Shibiyama, F. and Iwai, M., 8th International Congress of Endocrinology, July 1988, Kyoto, Japan, Abstr. No. 23-18-40.

Synthesis of biologically active analogs of a lipopeptide mating pheromone from *Saccharomyces cerevisiae*

Ariel Ewenson[a], Chu-Biao Xue[a], Stevan Marcus[b], Guy Caldwell[b],
Jeffrey M. Becker[b] and Fred Naider[a]

[a]*Department of Chemistry, College of Staten Island, CUNY, Staten Island, N.Y. 10301, U.S.A.*
[b]*Department of Microbiology, University of Tennessee, Knoxville, TN 37996, U.S.A.*

Introduction

The recognition of yeast sex pheromones by membrane-bound receptors provides an excellent model system for understanding the mechanism of action of mammalian peptide hormones. Mating in *Saccharomyces cerevisiae* involves the reciprocal action of two peptide pheromones, the α-factor and the *a*-factor. The latter pheromone was recently shown to be a lipopeptide with a structure YIIKGVFWDPAC(S-farnesyl)OCH$_3$ [1], and was synthesized using both solution-phase and solid-phase strategies [2].

Many peptide hormones are biosynthesized as preprohormones and processed by cellular enzymes to the mature biologically active molecule. In contrast to the α-factor, which is bio-synthesized as a long polypeptide precursor containing a signal sequence and glycosylation sites, the *a*-factor appears to be synthesized by a unique pathway [3]. In order to gain information on the mechanism of *a*-factor biosynthesis and excretion, and to learn more about the SAR of this lipopeptide, we decided to prepare analogs in which the cysteine sulfur atom and the carboxyl terminal methyl ester were modified.

Results and Discussion

The following *a*-factor analogs, YIIKGVFWDPAC, YIIKGVFWDPAC(S-ethyl), YIIKGVFWDPAC(SH)OCH$_3$, and YIIKGVFWDPAC(S-farnesyl) were synthesized and tested for bioactivity.

The dodecapeptides were synthesized by both solid-phase and solution phase procedures. For YIIKGVFWDPAC(S-farnesyl), two independent routes were used. In one of these, a protected dodecapeptide with the structure Boc-Y(2BrZ)IIK(2ClZ)GVFW(For)D(Chx)AC(Acm) – was assembled on a PAM resin. All coupling reactions were carried out using three equivalents of Boc amino acid activated with an equal amount of DIC and HOBt in DMF/CH$_2$Cl$_2$. Each residue was routinely double-coupled, and the completion of the coupling reaction was monitored using ninhydrin. After removal of the Boc group, using TFA/CH$_2$Cl$_2$ in the presence of anisole (5%), the peptide was cleaved from the PAM resin using HF. The crude peptide [YIIKGVFWDPAC(S-Acm)] was purified to greater than 97% homogeneity using RPHPLC. The Acm group was cleaved from this precursor using $Hg(CF_3COO)_2$, and the free SH dodecapeptide was liberated by treatment with H_2S to remove the mercury salts. In syntheses where the cysteine sulfur was protected with a 4-methylbenzyl group, difficulty was encountered during HF cleavage, and complete removal of this protecting group could only be achieved at temperatures of 10–25°C.

All attempts to directly farnesylate the sulfur atom of YIIKGVFWDPAC with farnesyl bromide resulted in a complex reaction product. Consequently, we protected the α-amine and the Lys side chain using the methylsulfonyl-ethyloxycarbonyl group [Msc] [4]. The Acm group was then removed from the di-Msc-protected dodecapeptide, as above, and the Msc-protected intermediate was farnesylated on the sulfur atom using farnesyl bromide in the presence of DIEA. The Msc groups were removed by treatment with 0.1 N NaOH for 5 min at room temperature, and the final peptide was isolated using RPHPLC. The farnesylated dodecapeptide free acid was also made by saponifying the authentic a-factor which had been synthesized by an independent strategy [2]. Both products were identical as judged by RPHPLC, and had the expected AAA and FABMS molecular ion. The free SH dodecapeptide methyl ester was prepared by condensing [Ala-Cys(S-)OCH_3]$_2$ with Fmoc-YIIK(Fmoc)GVFW(For-)D(Ofm)P using the BOP reagent. The resulting 24-peptide was deprotected, treated with Zn/acetic acid, and purified using HPLC. All peptides subjected to biological assay were greater than 97% homogeneous.

The a-factor analogs were bioassayed using a strain that is supersensitive to the pheromone (*S. cerevisiae* 757 MATα). Activity is reported as the lowest amount of pheromone that results in discernable growth arrest on a lawn of strain 757. The activities of the pheromones were: YIIKGVFWDPAC(S-farnesyl)OCH_3, 6 pg/disc; YIIKGVFWDPAC(S-farnesyl)COOH, 8 ng/disc; and YIIKGVFWDPAC(SH)OCH_3, 8 ng/disc. All other pheromones exhibited biological activity at concentrations higher than 100 ng/disc.

These results demonstrate that neither the farnesyl group nor the methyl ester are absolutely essential for activity of the a-mating factor of *S. cerevisiae*. Nevertheless, removal of either of these groups reduces activity as judged by a growth-arrest assay by a factor of approximately 1000. Substitution of the farnesyl moiety by an ethyl group also results in a reduction in activity. These findings indicate that a large hydrophobic group at the carboxyl end of the a-factor makes an important contribution to pheromone activity.

Acknowledgements

Support from the National Institutes of Health GM 22086 and GM 22087 is gratefully acknowledged.

References

1. Anderegg, R.J., Betz, R., Carr, S.A., Crabb, J.W. and Duntze, W., J. Biol. Chem., 263 (1988) 18236.
2. Xue, C.-B., Caldwell, G.A., Becker, J.M. and Naider, F., Biochem. Biophys. Res. Commun., in press.
3. Fuller, R.S., Sterne, R.E. and Thorner, J., Annu. Rev. Physiol., 50 (1988) 345.
4. Tesser, G.I. and Balvert-Geers, I.C., Int. J. Pept. Protein Res., 7 (1975) 295.

Solid-phase peptide synthesis of an analog of CCK-8 containing the non-hydrolyzable sulfated tyrosine residue

Rosario Gonzalez-Muniz, Fabrice Cornille, Christiane Durieux,
Florence Bergeron, Joël Pothier and Bernard P. Roques

*Département de Chimie Organique, U266 INSERM, UA498 CNRS, UER des Sciences
Pharmaceutiques et Biologiques, 4 avenue de l'Observatoire, F-75006 Paris, France*

Introduction

The sulfate ester of CCK-8, borne by the tyrosine residue, is a critical determinant for its biological activity. Recently, we have described a peptide, Ac-[L-Phe(CH$_2$SO$_3$Na)^{27}, Nle^{28}, Nle^{31}]CCK$^{27-33}$, in which the OSO$_3$H of the sulfate tyrosine has been replaced by the non-hydrolyzable CH$_2$SO$_3$H group [1]. This was the first described analog of CCK-8 modified on the sulfate ester which recognizes both central and peripheral receptors that retain all the agonist properties of CCK-8 [2]. In order to study whether or not this modified amino acid is suitable for SPPS, Boc-(L,D)-Phe(CH$_2$SO$_3$Na)-OH was prepared, and its chemical stability was verified using conventional conditions associated with the Boc/Bzl strategy. We report here the use of this synthetic amino acid derivative for SPPS of CCK-8 analogs.

Results and Discussion

Asp-(L,D)-Phe(CH$_2$SO$_3$H)-Met-Gly-Trp-Met-Asp-Phe-NH$_2$ was synthesized on MBHAR using N$^\alpha$-Boc protection for all amino acids and BOP as coupling reagent [3]. Side-chain protecting groups were: Asp(OcHex), Phe(CH$_2$SO$_3$Na), Met(O), Trp(CHO). The peptide was cleaved from the resin following the low-high HF procedure and purified by semi-preparative HPLC. Finally, diastereoisomers L (**1**) and D (**2**) were easily separated using semipreparative HPLC and characterized by UV and NMR spectroscopy [4]. The absolute configuration of Phe(CH$_2$SO$_3$H)^{27} residue in compounds **1** and **2** has been unambiguously confirmed by preparation of Boc-L-Phe(CH$_2$SO$_3$Na)-OH, which afforded compound **1** by SPPS.

Table 1 *Biological activities of CCK-8 analogs 1 and 2*

Compound	Binding K$_I$ (M)		Amylase release EC$_{50}$ (M)
	Brain	Pancreas	
CCK-8 (CCK$^{26-33}$)	$0.28 \pm 0.01 \times 10^{-9}$	$0.64 \pm 0.05 \times 10^{-9}$	$2.80 \pm 0.11 \times 10^{-11}$
1 [L-Phe(CH$_2$SO$_3$H)^{27}]CCK$^{26-33}$	$1.00 \pm 0.28 \times 10^{-9}$	$1.04 \pm 0.06 \times 10^{-9}$	$3.98 \pm 0.28 \times 10^{-11}$
2 [D-Phe(CH$_2$SO$_3$H)^{27}]CCK$^{26-33}$	$2.91 \pm 0.31 \times 10^{-9}$	$1.74 \pm 0.03 \times 10^{-8}$	$1.58 \pm 0.19 \times 10^{-10}$

These CCK-8 analogs, which have the same biological properties as the parent compound (Table 1) but with an enhanced chemical stability, could be of great interest to biochemical and, more particularly, to pharmacological studies. Thus, the 'stabilized-CCK-8' (1), easily prepared by SPPS (yield of pure compound = 16%), could replace CCK-8.

Results are the mean ±SEM of three separate experiments, each value in triplicate.

References

1. Marseigne, I. and Roques, B.P., J. Org. Chem., 53 (1988) 3621.
2. Marseigne, I., Roy, P., Dor, A., Durieux, C., Pélaprat, D., Reibaud, M., Blanchard, J.C. and Roques, B.P., J. Med. Chem., 32 (1989) 445.
3. Nguyen, D., Heitz, A. and Castro, B., J. Chem. Soc. Perkin Trans I, (1987) 1915.
4. Fournié-Zaluski, D.M., Lucas-Soroca, E., Devin, J. and Roques, B.P., J. Med. Chem., 29 (1986) 751.

Highly selective cyclic cholecystokinin analogs for central receptors

Marc Rodriguez, Marie-Françoise Lignon, Marie-Christine Galas, Muriel Amblard and Jean Martinez

Centre CNRS-INSERM de Pharmacologie-Endocrinologie, rue de la Cardonille, F-34094 Montpellier Cedex, France

Introduction

It is well known that mammalian pancreatic and brain CCK receptors differ in structure and ligand specificity [1]. Pancreatic CCK-receptor (CCK-A) is selective for sulfated forms of CCK [2], but sulfated and unsulfated CCK-8 (and shorter C-terminal fragments, down to CCK-4) interact about equally with the brain receptor (CCK-B) [3]. Thus, it is of interest to design CCK analogs as specific ligands for each class of receptors, in order to assess the structural requirements for receptor selectivity, and further study the physiological role of CCK. Introduction of conformational constraints, i.e., by peptide cyclization, is a way to achieve receptor selectivity that was successfully carried out in the enkephalins series [4]. Previously described conformational studies of CCK [5] have shown that the backbone of CCK is highly folded, and that the side chains of the methionine residues 28 and 31 are pointing outwards. We thus investigated the synthesis of cyclic CCK analogs in which residues 28 and 31 have been replaced by lysines, whose side chains are linked by a succinyl bridge.

$$\text{Ac-Tyr(SO}_3\text{H)-Lys-Gly-Trp-Lys-Asp-Phe-NH}_2 \quad \text{JMV310}$$
$$\underset{\text{CO-(CH}_2)_2\text{-CO}}{\rule{4em}{0pt}}$$

$$\text{Ac-Tyr-Lys-Gly-Trp-Lys-Asp-Phe-NH}_2 \quad \text{JMV320}$$
$$\underset{\text{CO-(CH}_2)_2\text{-CO}}{\rule{4em}{0pt}}$$

$$\text{H-Lys-Gly-Trp-Lys-Asp-Phe-NH}_2 \quad \text{JMV328}$$
$$\underset{\text{CO-(CH}_2)_2\text{-CO}}{\rule{4em}{0pt}}$$

$$\text{Ac-Lys-Gly-Trp-Lys-Asp-Phe-NH}_2 \quad \text{JMV332}$$
$$\underset{\text{CO-(CH}_2)_2\text{-CO}}{\rule{4em}{0pt}}$$

Fig. 1. Structures of CCK analogs.

Results and Discussion

The cyclic CCK analogs obtained in this work are listed in Fig. 1. The synthesis

108

was carried out in solution by fragment condensation (bridging of the lysine side chains) and cyclization at the Gly-Trp bond by the BOP reagent. A detailed synthetic procedure will be published elsewhere. All final compounds were extensively characterized by ^{1}H NMR and FABMS.

Table 1   *Apparent affinities of cyclic CCK analogs*

	Binding		sf[a]
	Rat pancreatic acini IC$_{50}$ (nM)	Guinea pig brain membranes IC$_{50}$ (nM)	
Boc-[Nle^{28,31}]-CCK-7	2.32 ± 0.48	0.29 ± 0.05	8.54 ± 3.13
Boc-Trp-Leu-Asp-Phe-NH$_2$	4000 ± 851	2.22 ± 0.29	1884 ± 629
JMV310	13000 ± 2345	1.32 ± 0.14	10151 ± 2854
JMV320	21750 ± 2688	2.03 ± 0.26	11065 ± 2741
JMV328	28250 ± 3614	12.0 ± 0.82	2386 ± 463
JMV332	16750 ± 4534	5.0 ± 0.71	3550 ± 1411

[a]sf: IC$_{50}$ (rat pancreatic acini)/IC$_{50}$ (guinea pig brain membranes).

Compounds JMV310, JMV320, JMV328 and JMV332 were tested for their ability to inhibit binding of [^{125}I]-BH-CCK-8 to rat pancreatic acini and to guinea pig brain membranes (Table 1), and were compared to the potent CCK analog Boc-[Nle^{28,31}]-CCK-7 and to the CCK-4 analog Boc-Trp-Leu-Asp-Phe-NH$_2$. The cyclic CCK derivatives were weakly potent in inhibiting binding of labeled CCK-8 to rat pancreatic acini, showing apparent affinities in the 10 μM range. Thus, they are less potent than Boc-Trp-Leu-Asp-Phe-NH$_2$ and Boc-[Nle^{28,31}]-CCK-7, respectively, by one and four orders of magnitude. However, all cyclic CCK analogs were very potent in inhibiting binding of [^{125}I]-BH-CCK-8 to guinea pig brain membranes (IC$_{50} \approx$ 1–10 nM). Compounds JMV310 and JMV 320 (IC$_{50}$ 1.32 and 2.03 nM) were the most potent, comparable to the CCK-4 analog (IC$_{50}$ 2.22 nM). These results clearly show that sulfation of the tyrosine residue of the cyclic analog does not dramatically affect the binding to either the central or peripheral CCK receptors. However, removal of the tyrosine residue, leading to compounds JMV328 and JMV332, mainly affects the apparent affinity for central receptors (IC$_{50}$ 12 and 5 nM, respectively). The selectivity factor, ratio of apparent affinity for pancreatic receptors over apparent affinity for central receptors, of compounds JMV328 and JMV332 is about that of Boc-Trp-Leu-Asp-Phe-NH$_2$ (approximately 2000), whereas, it is equal to about 10 000 for compounds JMV310 and JMV320. Thus, derivatives JMV310 and JMV320 appear to be the most selective CCK analogs for central receptors described to date, and are the first example of CCK analogs modified at the C-terminal tetrapeptide end of the hormone that are able to retain significant affinity for central CCK receptors. They might be an interesting tool for studying the physiological role of the CCK central receptor (CCK-B). Therefore, extensive pharmacological and conformational studies are now in progress in our laboratory.

References

1. Sakamoto, C., Williams, J.A. and Goldfine, I.D., Biochem. Biophys. Res. Commun., 124 (1984) 49.
2. Jensen, R.T., Lemp, G.F. and Gardner, J.D., Proc. Natl. Acad. Sci. U.S.A., 77 (1980) 2079.
3. Innis, R.B. and Snyder, S.H., Proc. Natl. Acad. Sci. U.S.A., 77 (1980) 6917.
4. Di Maio, J. and Schiller, P., Proc. Natl. Acad. Sci. U.S.A., 77 (1980) 7162.
5. Fournié-Zaluski, M.C., Belleney, J., Lux, B., Durieux, C., Gérard, D., Gacel, G., Maigret, B. and Roques, B., Biochemistry, 25 (1986) 3778.

Synthesis and biological activity of a cyclic analog of α-factor

Chu-Biao Xue[a], Effimia Bargiotta[b], Jeffrey M. Becker[b] and Fred Naider[a]

[a]Department of Chemistry, College of Staten Island, CUNY, Staten Island,
NY 10301, U.S.A.
[b]Department of Microbiology, University of Tennessee, Knoxville, TN 37996, U.S.A.

Introduction

The α-mating factor of *Saccharomyces cerevisiae* is a linear tridecapeptide with the sequence of Trp-His-Trp-Leu-Gln-Leu-Lys-Pro-Gly-Gln-Pro-Met-Tyr [1]. Numerous investigations have been carried out on the relationships between the primary structure, the conformation, and the biological activity of the α-factor. In particular, proton NMR studies in solution [2] and in the presence of lipid [3], in combination with the biological activities of analogs that can and cannot form β-turns [4] have provided the evidence that the biologically active conformation of the mating factor contained a Type II β-turn spanning residues 7 and 10. In order to get further insight into the mode of interaction of this yeast-mating pheromone with the receptor, we undertook the synthesis of a cyclic analog of α-factor, cyclo^{7,10}[Nle^{12}]-α-factor, in which the critical region was constrained by a covalent link between the ε-amine of Lys^7 and the γ-carboxyl of the Glu^{10}.

Results and Discussion

The synthesis of the cyclic analog of α-factor used the SPPS on a PAM resin following a strategy recently developed by Felix et al. [5]. Boc group was used for the Nα-protection of all residues. The side-chain protection groups were Trp(For), His(Tos), Lys(Fmoc) and Glu(OFm). Deprotection of the Boc group was accomplished with 45% TFA/2% dimethyl sulfide in CH_2Cl_2, except for Gln where 4 N HCl in dioxane was employed. Neutralization was carried out with a low concentration (5%) of diisopropylethylamine in CH_2Cl_2. After assembly of BocTrp(For)-His(Tos)-Trp(For)-Leu-Gln-Leu-Lys(Fmoc)-Pro-Gly-Glu(OFm) Pro-Nle-Tyr(2-BrZ)-PAM-resin, the Fmoc and OFm groups were deprotected with 20% piperidine in DMF, and cyclization was effected on the resin by using BOP reagent as coupling agent. The cyclized peptide was cleaved from the resin with high HF using anisole as scavenger at 0–2°C for 1.5 h. It was found by analytical HPLC that the crude product contained two major peaks at 12.1 and 17.1 min (Fig. 1A). The latter peak completely disappeared after treatment of the crude product with HOBt in water and acetonitrile for 2 h. The cleaved peptide was incubated with 1 N piperidine in DMF and water to remove the

111

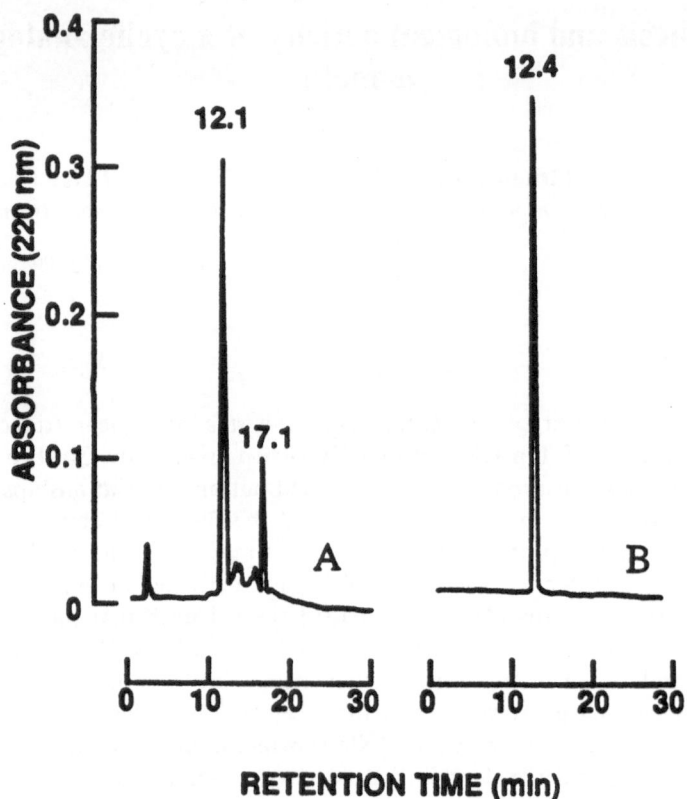

Fig. 1. HPLC of the synthetic cyclic peptide. (a) The crude product from HF cleavage; (b) The purified peptide.

formyl group. However, no change in retention time was observed, indicating that the formyl group had been completely removed during the treatment of the resin with piperidine in DMF prior to cyclization.

The crude cyclic peptide was purified by preparative RPHPLC. The total yield of the purified peptide (>99% homogeneous, Fig. 1B) was 30%, on the basis of the initial amino-acid content on the resin. In comparison with our previous results on the synthesis of α-factor analogs by SPPS [6], significant improvements were achieved in both the homogeneity of the crude product and the total yield of the purified peptide. These improvements are attributed to the utilization of HCl in dioxane for the deprotection of the Boc group on Gln, and longer duration and slightly higher temperature in the HF cleavage. The homogeneity of the peptide was confirmed by TLC on silica gel thin layers using two systems, and the structure was verified using AAA, FABMS, peptide sequencing and 400 MHz NMR.

The biological activity of the synthetic peptide was investigated in the growth arrest assay. In comparison with the linear [Nle[12]]-α-factor, the cyclic analog was 1/4 as active using a wild-type tester, 1/5 as active with *S.cerevisiae* 631, and 1/20 as active against *S.cerevisiae* 629 and *S.cerevisiae* 4202-15-3. These results show that this constrained peptide retains high biological activity and provides further evidence that the biologically active conformation of the α-mating pheromone might possess a bend involving Lys^7-Pro^8-Gly^9-Gln^{10} residues.

Acknowledgements

Support from the National Institutes of Health GM-22086 and GM-22087 is gratefully acknowledged.

References

1. Stotzler, D., Kiltz, H.H. and Duntze, W., Eur. J. Biochem., 69 (1976) 397.
2. Jelicks, L.A., Naider, F., Shenbagamurthi, P., Becker, J.M. and Broido, M.S., Biopolymers, 27 (1988) 431.
3. Naider, F., Jelicks, L.A., Becker, J.M. and Broido, M.S., Biopolymers, 28 (1989) 487.
4. Shenbagamurthi, P., Kundu, B., Raths, S., Becker, J.M. and Naider, F., Biochemistry, 24 (1985) 7070.
5. Felix, A.M., Wang, C.-T., Heimer, E.P. and Fournier, A., Int. J. Pept. Prot. Res., 31 (1988) 231.
6. Tallon, M.A., Shenbagamurthi, P., Marcus, S., Becker, J.M. and Naider, F., Biochemistry, 26 (1987) 7767.

Anti-aspartame type sweeteners, chemically stable peptides

Masahiro Tamura and Hideo Okai

*Department of Fermentation Technology, Faculty of Engineering, Hiroshima University,
Higashi-hiroshima, Hiroshima 724, Japan*

Introduction

Since the discovery of aspartame [1] many studies have been done to establish the molecular requirements for sweet taste. It would be most reasonable, as reported by many workers, that the sweet taste of aspartame is elicited by the trifunctional unit AH-B-X, were AH is an acidic proton, B is an electronegative atom or center, and X is a hydrophobic group, as indicated in Fig. 1.

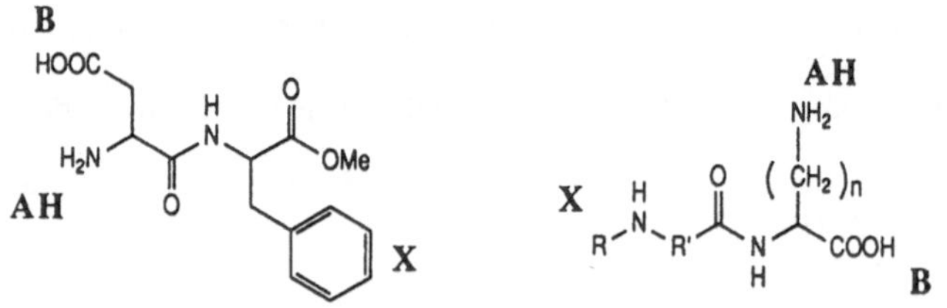

Fig. 1.   Aspartame (1) and a model of anti-aspartame-type sweetener (2).

Recently, during the study of salty peptides, we found that the basic dipeptides, Gly-Lys and γ-Abu-Lys, produced a sweet taste at about the same level as sucrose [2]. In comparing the structure of these basic dipeptides to that of aspartame, it was predicted that the ϵ-NH$_2$ group and the α-COOH group in C-terminal Lys and N-terminal amino acid residue corresponded to AH, B and X in aspartame, respectively (Fig. 1). We named these kinds of dipeptides 'anti-aspartame-type sweeteners', and synthesized a series of such dipeptides to investigate their utility.

Results and Discussion

Results of sensory analyses are shown in Table 1. In the case of Bz-Gly-Lys-OH and its derivatives (3–5), the sweet potencies were increased, but the quality was not good enough because of other co-existing tastes. The Lys derivative, however, was superior in both taste qualities and potencies to Orn and Dab derivatives. Thereafter, the basic amino acid was chosen to be Lys.

Next, the main peptide chain (R′) was elongated (6 and 7). Both compounds

114

Table 1   *Results of sensory analyses of anti-aspartame-type sweeteners*

Compounds		T.V.(mM)	Taste
3	Bz-Gly-Lys-OH	0.67	sweet > sour
4	Bz-Gly-Orn-OH	1.97	sweet > sour
5	Bz-Gly-Dab-OH	0.88	sweet > bitter
6	Bz-β-Ala-Lys-OH	0.24	sweet
7	Bz-γ-Abu-Lys-OH	0.20	sweet
8	Ph-ac-Gly-Lys-OH	0.10	sweet
9	Ph-pr-Gly-Lys-OH	0.34	sweet
10	Ph-ac-β-Ala-Lys-OH	0.53	sweet > bitter
11	Ph-pr-β-Ala-Lys-OH	0.66	sweet > bitter
12	Ac-Phe-Lys-OH	0.22	sweet > sour
13	Ac-D-Phe-Lys-OH	0.32	sweet > sour > bitter

Ph-ac- = phenylacetyl; Ph-pr- = phenylpropionyl; T.V. = threshold value.

had a sweet taste without bitterness, and the potency was about 20 times stronger than that of sucrose. Further, the length of the acyl group (R) was elongated, and phenylacetyl-Gly-Lys-OH (**8**) was found to have a sweet taste about 50 times stronger than that of sucrose. Compounds **10** and **11**, whose peptide chains were further elongated,had a bitter taste in addition to their sweet taste.

It is believed that aspartame produces the strong sweet taste because it has a branched structure at the Phe moiety so that the flexibility of its X functional group (phenyl group) is limited. Therefore, we introduced the branched structure into our anti-aspartame-type sweeteners. Among them, Ac-Phe-Lys-OH (**12**) was found to be useful with a strong sweet taste and high taste quality.

We developed the new type of sweeteners according to the AH-B-X theory and successfully designed the anti-aspartame-type sweeteners. All the synthetic compounds reported in this paper do not contain a methyl ester. This is one of the more remarkable features and is thought to be a great advantage with regard to stability in solution and nutritional problems, in comparison with aspartame.

References

1.   Mazur, R.H., Schlatter, J.M. and Goldkamp, A.H., J. Am. Chem. Soc., 91 (1969) 2684.
2.   Tada, M. Shinoda, I. and Okai, H., J. Agric. Food Chem., 32 (1984) 992.

Agonistic and antagonistic activities of neuromedin U-8 analogs on isolated smooth muscles

T. Hashimoto*, H. Masui, N. Sakura, K. Okimura and Y. Uchida

Faculty of Pharmaceutical Sciences, Hokuriku University, Ho-3, Kanagawa-machi, Kanazawa 920-11, Japan

Introduction

Neuromedin U-8 (NMU-8: Tyr-Phe-Leu-Phe-Arg-Pro-Arg-Asn-NH$_2$) is a novel neuropeptide isolated from porcine spinal cord [1]. The octa- or hepta-peptide amide structure of NMU-8 is conserved in neuromedin U-25 [1] or rat neuromedin U [2] at their C-terminals. The peptides have potent uterus contractile activity, and effects on blood pressure and flow [3,4]. There is little information on the structure requirement for the activities of NMU-8. SAR of NMU-8 and the development of an antagonist against NMU-8 are described.

Methods

Peptide synthesis: NMU-8 and 18 related peptides (Table 1) were synthesized by automatic SPPS using BHAR and were purified by RPHPLC.

Bioassay: Biological properties of the NMU-8 related peptides were examined on isolated guinea pig trachea and chick crop. The method used for experiments were essentially the same as those previously described [5].

Results and Discussion

On guinea pig trachea preparation, **2, 3, 5, 7, 10** and **12** possessed no intrinsic activity, while the activity partially remained in **1, 4, 6, 8** and **9**. None of these analogs showed any antagonistic activity except [D-Phe^2]-NMU-8(**12**), which exerted some antagonistic effect (x = 5.12) against NMU-8.

On isolated chick crop, the replacement of Phe^2, Leu^3, Phe^4, Arg^5, Pro^6, Arg^7 or Asn^8 with Gly, and the truncation of the N-terminal, two residues of NMU-8 brought about a drastic decrease of the contractile activity (Table 1). The substitution of the corresponding D-moiety for Phe^2, Phe^4, Arg^5, Pro^6 or Asn^8 caused a marked decrease of the agonistic activities, while the replacement of Tyr^1 with D-form enhanced the activity. Interestingly, **16** was found to act as an antagonist against NMU-8 when **12, 14, 15** and **18** did not show any antagonistic activity against NMU-8.

The systematic study of the SAR of NMU-8 revealed that (a) the active site

---

*To whom correspondence should be addressed.

116

Table 1 *Agonistic and antagonistic activities of NMU-8 analogs on isolated chick crop*

Analog No.	Peptide	RA[a],*	x[b]
	NMU-8	1	
1	[Gly^1]-NMU-8	0.40 ± 0.10	NT[c]
2	[Gly^2]-NMU-8	<0.01	NT
3	[Gly^3]-NMU-8	0.08 ± 0.02	NT
4	[Gly^4]-NMU-8	<0.01	NT
5	[Gly^5]-NMU-8	<0.01	NT
6	[Gly^6]-NMU-8	<0.01	NT
7	[Gly^7]-NMU-8	NT	NT
8	[Gly^8]-NMU-8	<0.05	NT
9	NMU-8 (2-8)	0.53 ± 0.14	NT
10	NMU-8 (3-8)	<0.01	NT
11	[D-Tyr^1]-NMU-8	2.47 ± 0.17	NT
12	[D-Phe^2]-NMU-8	<0.09	–
13	[D-Leu^3]-NMU-8	0.33 ± 0.02	NT
14	[D-Phe^4]-NMU-8	<0.08	–
15	[D-Arg^5]-NMU-8	<0.11	–
16	[D-Pro^6]-NMU-8	–	5.22 ± 0.12
17	[D-Arg^7]-NMU-8	0.75 ± 0.18	NT
18	[D-Asn^8]-NMU-8	<0.12	–

[a] Relative affinity: antilog of [pD_2 of an analog minus pD_2 of NMU-8] (mean ± SE).
[b] Empirical x-intercept of the line of Scild plot as determined by the linear regression analysis (mean ± SE).
[c] Not tested.
* Numbers of determinations were 6–8.

of NMU-8 may exist between positions 2 and 8; (b) the side chain of each amino acid at positions 2, 3, 5 and 7 seems to be of relative importance for the expression of contractile acitivity; and, (c) the replacement of Phe^2 or Pro^6 with D-form could change the pharmacologic spectrum of NMU-8 from that of an agonist to that of an antagonist. These results indicate important clues in the further development of potent and specific antagonists against NMUs.

References

1. Minamino, N., Kangawa, K. and Matsuo, H., Biochem. Biophys. Res. Commun., 130 (1985) 1078.
2. Minamino, N., Kangawa, K., Honzawa, M. and Matsuo, H., Biochem. Biophys. Res. Commun., 156 (1988) 355.
3. Conlon, J.M., Domin, J., Thim, L., DiMarzo, V., Morris, H.R. and Bloom, S.R., J. Neurochem., 51 (1988) 988.
4. Sumi, S., Inoue, K., Kogire, M., Doi, R., Takaori, K., Suzuki, T., Yajima, H. and Tobe, T., Life Sci, 41 (1987) 1585.
5. Hashimoto, T., Uchida, Y., Naminohira, S. and Sakai, T., Jpn. J. Pharmcol., 45 (1987) 570.

Structure-activity relationship in cytoprotective peptides

Giancarlo Zanotti[a], Filomena Rossi[b], Benedetto Di Blasio[b], Carlo Pedone[b],
Ettore Benedetti[b], Kornelia Ziegler[c] and Teodorico Tancredi[d]

[a]Centro di Studio per la Chimica del Farmaco del CNR, Dipartimento di Studi
Farmaceutici, Università 'La Sapienza', I-00185 Rome, Italy
[b]Dipartimento di Chimica, Via Mezzocannone, 4, I-80134 Napoli, Italy
[c]Institut für Pharmakologie und Toxikologie, Frankfurter Strasse, D-6300 Giessen, F.R.G.
[d]ICMIB del CNR, Arco Felice, Naples, Italy

Introduction

Cyclolinopeptide A [CLA (I), Table 1], a homodetic cyclic nonapeptide from linseed, defends liver cells against several poisonous substances, such as phalloidin, ethanol, cysteamine and DMSO. This strong cytoprotective activity, shared with antamanide and somatostatin, relies on the high affinity of these peptides for the same liver membrane proteins responsible for the uptake of toxic compounds and bile salts [1]. Recently the conformational analysis of CLA, both in the solid state and in solution, has been performed in our laboratories [2]. In particular, a singular conformational state in solution, consistent with the solid structure, has been detected in $CDCl_3$ at 214 K. This result prompted us to develop a research program aiming to investigate the SAR of CLA and related cytoprotective peptides in more detail. To this purpose, a few CLA analogs have been synthesized and their conformational and biological properties investigated.

Table 1 *Chemical structure and cytoprotective activity of CLA, its analogs Ia–Ie and of the cystin peptide II.*

No.	Position no.									CD_{50} µM
	1	2	3	4	5	6	7	8	9	
I(CLA)	Pro—	Pro—	Phe—	Phe—	Leu—	Ile—	Ile—	Leu—	Val	0.84
Ia	Ala									1.2
Ib		Ala								1.3
Ic			Ala							6.2
Id				Ala						3.2
Ie					Aib	Aib		D-Ala		–
II	Boc—	Cys—	Val—	Pro—	Pro—	Phe—	Phe—	Cys—	Ome	1.8

Results and Discussion

Figure 1 shows the structure and biological activity of CLA and its analogs. CD_{50} values stand for peptide concentration required to inhibit the cholate uptake

118

by 50%. In analogs Ia–Id each member of the postulated active sequence of CLA, -Pro[1]-Pro[2]-Phe[3]-Phe[4] has been replaced by an L-Ala residue. The CD_{50} values found for these compounds reveal that all of them retain biological activity. However, the higher CD_{50} value found for analog Ic indicates that the Phe[3] residue plays the most important role. The introduction in analog Ie of constraints, such as Aib and D-Ala residues, is likely to induce an ordered secondary structure diminishing the mobility of the 9-membered ring. This fact is indicated by a 2D NMR study of Ie (CLAIB) in $CDCl_3$ solution. At room temperature, the CLAIB structure is fairly rigid, resembling that of CLA in the same solvent at 214 K (Fig. 1). Besides, CLA and CLAIB are characterized by almost identical conformational features in the solid state. The biological test for this analog is in progress.

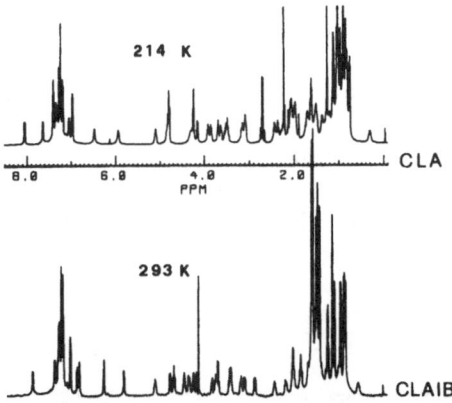

Fig. 1.   Comparison of 500 MHz ^1H NMR spectra of CLA and CLAIB in $CDCl_3$.

Cystin peptide II has been prepared in order to obtain information on the role of the composition and size of the ring in determining the conformation and the biological effect of the active part of the molecule. II retains half of the cytoprotective effect of CLA, thus demonstrating that its 5–8 sequence is not crucial for activity.

References

1.   Kessler, H., Klein, M., Müller, A., Wagner, K., Bats, J.W., Ziegler, K. and Frimmer, M., Angew. Chem. Int. Ed. Engl., 25 (1986) 997.
2.   Tancredi, T., Zanotti, G., Rossi, F., Benedetti, E., Pedone, C. and Temussi, P.A., Biopolymers, 28 (1989) 513.

Effects of cecropin and melittin analogs and hybrids on pro- and eucaryotic cells

D. Wade[a], R.B. Merrifield[a] and H.G. Boman[b]

[a]The Rockefeller University, 1230 York Avenue, New York, NY U.S.A.
[b]Department of Microbiology, University of Stockholm, S-106 91 Stockholm, Sweden

Introduction

The cecropins and melittin are short polypeptides produced by the silk moth, *Hyalophora cecropia*, and the honey bee, *Apis mellifera* [1,2]. Cecropins A, B, and D represent the principal components of the humoral immune system of *H. cecropia*. Melittin is the major protein and toxic component of bee venom. These peptides are 26 (melittin) to 37 (cecropin A) amino acids in length, and each contains a strongly basic region and long stretches of hydrophobic amino acid residues. However, the N-termini of cecropins are basic, whereas in melittin, the C-terminus is basic (Fig. 1). Both the cecropins and melittin will lyse bacteria, but the antibacterial spectrum of melittin is narrower and only melittin will lyse eucaryotic cells.

K-W-K-L-F-K-K-I-E-K-V-G-Q-N-I-R-D-G-I-I-K-A-G-P-A-V-A-V-V-G-Q-A-T-Q-I-A-K-NH$_2$

G-I-G-A-V-L-K-V-L-T-T-G-L-P-A-L-I-S-W-I-K-R-K-R-Q-Q-NH$_2$

Fig. 1.   Amino acid sequences of cecropin A (top) and melittin (bottom).

We are attempting to determine which elements of structure are responsible for the antibiotic and lytic properties of these molecules by constructing analogs and hybrids of cecropins and melittin and examining their effects on procaryotic and eucaryotic cells.

Results and Discussion

Peptides were synthesized by established procedures for SPPS [3], using MBHA resin as the solid support. They were deprotected and cleaved from the resin by the low/high HF method, and purified by gel filtration and RPHPLC. Compositions, purities, and molecular weights were assessed by AAA, analytical HPLC, and MS. Antibacterial activities were determined by inhibition zone assays. Table 1 lists the results obtained with selected peptides.

A small change in cecropin A, deletion of the helix disrupting sequence Gly-Pro, had the effect of enhancing the antibacterial activity 100-fold against *Bacillus*

Table 1 *Lethal concentrations (µM) for cecropin A (CA) and melittin (M) analogs and hybrids*[a]

Compound	E. coli	B. subtilis[b]	S. aureus	S. cerevisiae	SRC[c]
CA(1–37)	0.2	200	>200	62	>200
M(1–26)	0.8	0.3	0.2	13	4–8
Ca(1–22)(25–37)	0.4	2	>300	--	>300
M(16–26)(1–13)	0.7	1	10	25	>200
CA(1–13)M(1–13)	0.5	0.9	2	20	>200

[a] Lethal concentrations were calculated from inhibition zones on thin agarose plates seeded with the cells listed above [1].
[b] B. subtilis with Medium E.
[c] SRC = sheep red cells.

subtilis. Larger changes, such as transposing the N- and C-terminal regions of melittin [M(16–26)(1–13)] or combining the N-termini of cecropin A and melittin [CA(1–13)M(1–13)], greatly reduced red cell lysis with respect to melittin, and enhanced antibacterial activity towards *Bacillus subtilis* and *Staphylococcus aureus* with respect to cecropin A. These results demonstrate that the general approach of synthesis and study of analogs and hybrids of these peptides can lead to an understanding of the structural features that are important for their biological activities, and may lead to improved antibiotics.

Acknowledgements

D.W. gratefully acknowledges Applied Biosystems Inc. for a travel award. This work was supported by USPHS Grant DK-01260.

References

1. Boman, H.G. and Hultmark, D., Annu. Rev. Microbiol., 41 (1987) 103.
2. Habermann, E., Science, 177 (1972) 314.
3. Christensen, B., Fink, J., Merrifield, R.B. and Mauzerall, D., Proc. Natl. Acad. Sci. U.S.A., 85 (1988) 5072.

Magainin analogs: A study of activity as a function of α-helix modification

Hao-Chia Chen[a], Judith H. Brown[a], John L. Morell[a] and Charng-Ming Huang[b]

[a]Endocrinology and Reproduction Research Branch, NICHD, Bethesda, MD 20892, U.S.A.
[b]Clinical Pathology Department, Clinical Center, NIH, Building 6, Room 2A-13,
Bethesda, MD 20892, U.S.A.

Introduction

Magainins (PGS peptides) consist of two analogous peptide sequences of 23 residues [magainin 1 : GIGKFLHSAG(K)KFGKAFVGEIMK(N)S, magainin 2 differs as shown in parentheses] which are the first series of antibiotics reported to be produced by a vertebrate [1–3]. They were isolated from the African clawed frog, and exhibited a wide spectrum of antimicrobial activity and osmotic lysis of protozoa [3]. Substitution of Ala for Gly[13] and Gly[18] resulted in an increase of up to two orders of magnitude in antimicrobial activity, and a greater propensity for amphiphilic α-helical formation in a buffer-TFE mixture [4]. However, change to D-Ala at positions 9, 13 and 18 yielded a peptide with no detectable hemolytic or antimicrobial activity. We report here the synthesis of analogs with all remaining permutations of D- and L-Ala at these positions in order to elucidate the importance of the location of the helix in affecting biological activity.

Methods

Eight magainin 2-amide analogs, designated as F(Ala[13,18]), H(D-Ala[9,13,18]), J(D-Ala[13,18]), K(Ala[13],D-Ala[9,18]), L(Ala[13],D-Ala[18]), M(Ala[18],D-Ala[9,13]), N(Ala[18],D-Ala[13]), P(Ala[13,18],D-Ala[9]) were synthesized by the SPPS method and purified by preparative RPHPLC on a Vydac C_4, 300 Å, 15–20 μ column [4,5]. CD measurements, bacterial and hemolysis assays were similar to those described [4], except that total bacteria growth inhibition for 24 h was the criterion for the determination of the minimal inhibition concentration (MIC).

Results and Discussion

From observations that alternating D,L sequences, such as gramicidin A, exhibit left handed β-helices [6], while all-D sequences, such as RNase S-peptide, form left handed α-helices [7], we infer a strong right handed α-helix breaking tendency for a D-Ala residue in a predominantly L peptide. As shown in Table 1, substitution of D-Ala reduces biological activities, retention time in RPHPLC and α-helical conformation in a buffer-TFE mixture. It could be expected that helix potential in the N-terminal half of the molecule is of principal importance, since the positive

Table 1   *Structure and activity relationship of magainin analogs*

Peptide[a]	E. coli MIC[24 h], μg/ml	% Hemolysis at 100 μg/ml	Retention time (min)	% α-Helix 80% TFE
F	10	6.8	39.4	60
L (D-Ala[18])	50	1.2	31.9	60
N (D-Ala[13])	50	0.75	24.5	45
P (D-Ala[9])	125	0.2	29.7	42
J (D-Ala[13,18])	50	0.2	18.2	47
K (D-Ala[9,18])	250	0.08	17.1	42
M (D-Ala[9,13])	250	0.1	18.3	32
H (D-Ala[9,13,18])	>250	0.06	15.7	35

[a]D-Ala substitution as indicated.

charges which can interact with membrane phospholipids cluster there. This expectation is supported by the trend seen in Table 2. Clearly, modification at position 13 or 18 is less important than at 9 in affecting antimicrobial activity, and the cumulative effect of a modification at 9 and 13 (**M**) is more deleterious than at 13 and 18 (**J**). Thus, molecules with the same apparent helix content can have considerably different activity, as in the cases of **N**, **P**, **J** and **K**, because of the particular location of the helix in the molecule.

Table 2   *Comparison of antimicrobial activity (MIC) reduction resulting from D-Ala substitution*

Position 18	F(L):L(D)	N(L):J(D)	P(L):K(D)	M(L):H(D)
	10 50	50 50	125 250	250 >250
Position 13	F(L):N(D)	L(L):J(D)	P(L):M(D)	K(L):H(D)
	10 50	50 50	125 250	250 >250
Position 9	F(L):P(D)	L(L):K(D)	N(L):M(D)	J(L):H(D)
	10 125	50 250	50 250	50 >250

References

1. Giovannini, M.G., Poulter, L., Gibson, B.W. and Williams, D.H., Biochem. J., 243 (1987) 113.
2. Zasloff, M., Proc. Natl. Acad. Sci. U.S.A., 84 (1987) 5449.
3. Zasloff, M., Martin, B. and Chen, H.C., Proc. Natl. Acad. Sci. U.S.A., 85 (1988) 910.
4. Chen, H.C., Brown, J.H., Morell, J.L. and Huang, C.M., FEBS Lett., 236 (1988) 462.
5. Morell, J.L. and Brown, J.H., Int. J. Pept. Prot. Res., 26 (1985) 49.
6. Urry, D.W., Proc. Natl. Acad. Sci. U.S.A., 68 (1971) 672.
7. Corigliano-Murphy, M.A., Xun, L.A., Ponnamperuma, C., Dalzoppo, D., Fontana, A., Kanmera T. and Chaiken, I.M., Int. J. Pept. Prot. Res., 25 (1985) 225.

Synthesis and antimicrobial activity of magainin alanine substitution analogs

J.H. Cuervo, B. Rodriguez and R.A. Houghten

Scripps Clinic and Research Foundation, Department of Molecular Biology, 10666 North Torrey Pines Road, La Jolla, CA 92037, U.S.A.

Introduction

The observation that the female African-clawed frog, *Xenopus laevis*, rarely experiences infection following surgical removal of its ovaries led to the discovery of a previously unrecognized antimicrobial host-defense system present in this animal [1]. This observation culminated with the discovery of two compounds – magainin 1 and 2 – having significant antimicrobial activity against a number of bacteria species. Analysis of these compounds revealed that each was a peptide having 23 amino acid residues. We have demonstrated that specific single amino acid omissions in magainin 1 or 2 result in analogs having a wide range of increased or decreased antimicrobial activity, depending on the position omitted, and generally correlating with the concept of amphipathicity as a necessary condition for antimicrobial activity [2]. We report here the effects on antimicrobial potency and hemolytic activity of alanine substitution analogs of magainin 2.

Results and Discussion

A complete series of single substitution analogs of magainin 2, in which each amino acid was individually replaced with alanine, was prepared and tested against *P. aeruginosa*, *E. coli* and *S. epidermis*. Since we have demonstrated [2] that the C-terminal amide forms of magainin 1 and 2 are more potent than the C-terminal carboxyl forms as antimicrobial agents, the peptide analogs were prepared as C-terminal amides using simultaneous multiple peptide synthesis (SMPS, [3]).

In this series, alanine an α-helix promoting amino acid [4], was walked through the sequence of magainin 2 to generate a complete set of C-terminal amide substitution analogs. Relative to magainin 2, the analogs [Ala³]-MAG-2-NH$_2$, [Ala⁷]-MAG-2-NH$_2$, [Ala⁸]-MAG-2-NH$_2$, [Ala¹³]-MAG-2-NH$_2$, [Ala¹⁸]-MAG-2-NH$_2$, [Ala¹⁹]-MAG-2-NH$_2$ and [Ala²³]-MAG-2-NH$_2$ showed equal or higher antimicrobial activity against all the bacteria tested. These results provide support for the nonessential nature of the side chains or lack thereof of Gly³, His⁷, Ser⁸, Gly¹³, Gly¹⁸, Glu¹⁹ and Ser²³ on the antimicrobial activity of magainin 2. Recently, Hao-Chia Chen et al. [5] described studies of magainin 1 and 2, in which the simultaneous replacement of Ser⁸, Gly¹³, and Gly¹⁸ with alanine

Table 1   *Antimicrobial activity and cytotoxicity of magainin 2 alanine replacements[a]*

Peptide[b]	Minimum inhibition concentration($\mu g/ml$)			%Hemolysis[c]
	E. coli	*P. aeruginosa*	*S. epidermis*	
MAG-2-NH$_2$	25.0	25.0	50.0	1.0
[Ala^{23}]-MAG-2-NH$_2$	12.5	12.5	50.0	1.0
[Ala^{22}]-MAG-2-NH$_2$	25.0	25.0	100.0	1.0
[Ala^{21}]-MAG-2-NH$_2$	25.0	25.0	100.0	1.0
[Ala^{20}]-MAG-2-NH$_2$	100.0	50.0	> 100.0	0.1
[Ala^{19}]-MAG-2-NH$_2$	12.5	6.25	12.5	1.3
[Ala^{18}]-MAG-2-NH$_2$	25.0	12.5	50.0	1.0
[Ala^{17}]-MAG-2-NH$_2$	50.0	50.0	> 100.0	ND
[Ala^{16}]-MAG-2-NH$_2$	100.0	50.0	> 100.0	0.1
[Ala^{15}]-MAG-2-NH$_2$	> 100.0	50.0	> 100.0	0.1
[Ala^{13}]-MAG-2-NH$_2$	12.5	12.5	25.0	1.5
[Ala^{12}]-MAG-2-NH$_2$	12.5	50.0	100.0	ND
[Ala^{11}]-MAG-2-NH$_2$	100.0	100.0	> 100.0	0.1
[Ala^{10}]-MAG-2-NH$_2$	100.0	50.0	> 100.0	1.0
[Ala^{8}]-MAG-2-NH$_2$	12.5	25.0	12.5	5.0
[Ala^{7}]-MAG-2-NH$_2$	25.0	25.0	50.0	1.0
[Ala^{6}]-MAG-2-NH$_2$	50.0	50.0	100.0	1.0
[Ala^{5}]-MAG-2-NH$_2$	50.0	50.0	100.0	1.0
[Ala^{4}]-MAG-2-NH$_2$	> 100.0	100.0	> 100.0	4.0
[Ala^{3}]-MAG-2-NH$_2$	25.0	25.0	25.0	5.0
[Ala^{2}]-MAG-2-NH$_2$	50.0	50.0	50.0	0.1
[Ala^{1}]-MAG-2-NH$_2$	> 100.0	100.0	50.0	0.1

[a] Approximately 10^5 bacteria were suspended in 225 μl of half strength of trypticase soy broth medium onto tissue culture plates containing a 1:2 serial dilution of peptides.
[b] Magainin 2 amide: GIGKFLHSAKKFGKAFVGEIMNS-NH$_2$.
[c] At 100 μg/ml.
ND = not determinated.

not only increased antimicrobial activity but also hemolysis. We have found that the single replacement of Gly^3 with alanine yielded an analog with equal activity against gram-negative bacteria and increased antimicrobial activity against gram-positive bacteria. Replacement of Ser^8 with alanine increased the antimicrobial activity of magainin 2 against *E. coli* and *S. epidermis*, whereas the activity against *P. aeruginosa* remained unchanged. These two analogs, [Ala^3]-MAG-2-NH$_2$ and [Ala^8]-MAG-2-NH$_2$, while yielding improved antimicrobial activity relative to the parent peptide, also caused substantially increased hemolytic activity (Table 1).

Conclusion

The most striking finding in this study is the five-fold increase in potency upon replacement of Glu^{19} with Ala. The hemolytic potency of this analog, relative to that of the parent peptide, was found to be similar at 100 μg/ml (Table 1). Thus, substitution of alanine analogs of magainin 2 can lead to greater antimicrobial activity than that of the starting peptide, while maintaining the level of cytotoxicity equal to that of the starting peptide.

J.H. Cuervo, B. Rodriguez and R.A. Houghten

References

1. Zasloff, M., Proc. Natl. Acad. Sci. U.S.A., 84 (1987) 5449.
2. Cuervo, J.H., Rodriguez, B. and Houghten, R.A., Pept. Res., 1 (1988) 81.
3. Houghten, R.A., Proc. Natl. Acad. Sci. U.S.A., 82 (1985) 5131.
4. Chou, P.Y. and Fasman, G.D., Biochemistry, 13 (1974) 222.
5. Chen, H-C., Brown, J.H., Morrell, J.L. and Huang, C.M., FEBS Lett., 236 (1988) 462.

Effect of histidine modification in TRH on binding to rat pituitary and brain receptors

Virender M. Labroo[a], Stefan Vonhof[a], Giora Z. Feuerstein[b] and Louis A. Cohen[a]

[a]*Building 8A, Room B1A09, NIH, Bethesda, MD 20892, U.S.A.*
[b]*Department of Neurology, USUHS, 4301 Jones Bridge Road, Bethesda, MD 20814, U.S.A.*

Introduction

TRH (Glp-His-Pro-NH$_2$) is well known to exert a wide variety of effects on both the central nervous (CNS) and cardiovascular systems (CVS), in addition to governing the release of thyrotropin and prolactin. We have shown that His modifications result in differential effects on binding to high affinity TRH receptors and in dissociation of some of these activities [1–3]. To understand the physicochemical parameters affecting binding to these receptors and to investigate if individual biological activities of TRH are mediated through different receptor subtypes, we have prepared additional [4(5)-X-Im]-TRH analogs (X = Cl, Br, I, CN and CO$_2$H, Fig. 1) and determined their affinities to TRH receptors in pituitary and various regions of the rat brain. These data were analyzed by QSAR.

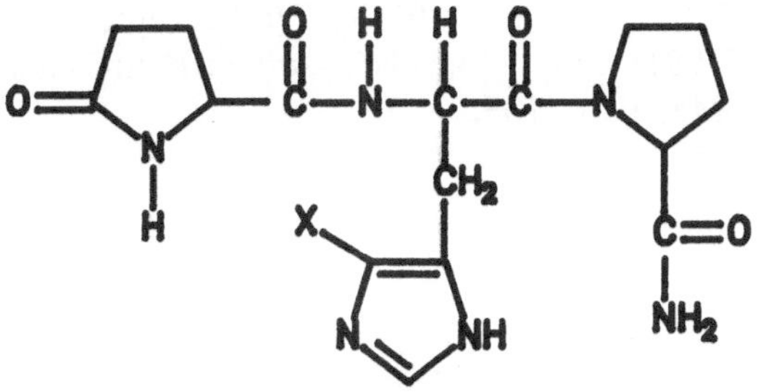

X = H, F, Cl, Br, I, CF$_3$, CN or CO$_2$H

Fig. 1. Structures of TRH and of its 4(5)-imidazole-substituted analogs.

Results and Discussion

[4(5)-X-Im]-TRH (X = Cl, Br or I) were prepared by direct halogenation of TRH with one equivalent of the corresponding halosuccinimide in MeOH. TRH, on photolysis with CF$_3$I in MeOH in the presence of Et$_3$N, furnished a mixture of [4(5)-CF$_3$-Im]-TRH and [2-CF$_3$-Im]-TRH, which were separated by RPHPLC (Beckman Ultrasphere C-18 column). Both [4(5)-CO$_2$H-Im]-TRH and [4(5)-CN-

127

Table 1 *Mean values (μM) for the inhibitor constants (K$_i$) of TRH and of its imidazole-substituted analogs*

Analogs	Pituitary	Hypothalamus	Brain stem	Cortex
TRH	0.019 ± 0.001	0.033 ± 0.004	0.026 ± 0.010	0.048 ± 0.013
[4-F]-TRH	13.5 ± 1.7	8.5 ± 0.8	8.3 ± 0.7	7.5 ± 0.5
[4-Cl]-TRH	39.7 ± 8.7	32.6 ± 5.5	23.1 ± 3.4	26.2 ± 2.9
[4-Br]-TRH	2.5 ± 0.09	1.93 ± 0.15	1.59 ± 0.06	1.81 ± 0.19
[4-I]-TRH	9.4 ± 4.5	15.4 ± 5.3	6.75 ± 2.46	31.1 ± 10.01
[4-CF$_3$]-TRH	569 ± 40	465 ± 148	1010 ± 390	392 ± 51
[4-NO$_2$]-TRH	> 1000.0	> 1000.0	> 1000.0	> 1000.0
[4-CN]-TRH	123 ± 9.0	140 ± 5	150 ± 10	143 ± 4
[4-CO$_2$H]-TRH	13.5 ± 2.5	13.1 ± 1.9	9.1 ± 1.5	11.3 ± 0.7

Im]-TRH were prepared from [4(5)-CF$_3$-Im]-TRH by treatment with 0.5 N NaOH and 4% aqueous NH$_3$, respectively. The products were desalted and purified by RPHPLC. All peptides were characterized by ^{1}H NMR (300 MHz) and MS.

TRH and 4(5)-Im-substituted analogs were evaluated for their abilities to compete with [^{3}H]-[3-Me-His^2]-TRH at high affinity TRH receptor binding sites in the pituitary, hypothalamus, brain stem and cortex. K$_i$ values are given in Table 1. Computerized QSAR analysis (Biosoft) revealed that the strength of receptor binding of 4(5)-Im-substituted analogs of TRH is dependent on a combination of the basicity of the imidazole ring of His and the size of the substituent. The unexpectedly high binding for [4(5)-Br-Im]-TRH indicates that additional factors, such as hydrophobicity, may also be involved in determining the binding ability. Furthermore, the selectivity of an analog such as [4(5)-NO$_2$-Im]-TRH, which possesses *only* the CVS activity of TRH, is explained by its total lack of binding to the high affinity TRH receptors, and implies that the CVS activities of TRH and of [4(5)-NO$_2$-Im]-TRH are mediated through sites not labeled by [3-Me-His^2]-TRH.

References

1. Labroo, V.M., Kirk, K.L., Cohen, L.A., Delbeke, D. and Dannies, P.S., Biochem. Biophys. Res. Commun., 113 (1983) 581.
2. Siren, A.-L., Feuerstein, G., Labroo, V.M., Cohen, L.A. and Lozovsky, D., Neuropeptides, 8 (1986) 63.
3. Vonhof, S., Paakkari, I., Feuerstein, G., Cohen, L.A. and Labroo, V.M., Eur. J. Pharmacol., 164 (1989) 77.

Benzophenone analogs of pepstatin A and cyclosporine A as potential photoaffinity labeling reagents

Petr Kuzmic, Chong-Qing Sun and Daniel H. Rich*

School of Pharmacy, University of Wisconsin, 425 N. Charter St., Madison, WI 53706, U.S.A.

Introduction

Recently, Kauer et al. reported that *p*-benzoyl phenylalanine (**1**, Bpa) is a highly efficient ligand for use in photoaffinity labeling studies [1]. We have extended these results to develop new photoreactive peptides for studying aspartic protease active sites and cyclosporine receptors.

Results and Discussion

Pepstatin A results

Bpa (**1**) (Fig. 1) was synthesized by Kauer's method [1] and converted to the corresponding statine (Sta, **2**) analog (BpaSta, **3**) by using a modification of Castro's procedure [2] (65% yield, >95% e.e.). In the stereoselective reduction of an intermediate tetramic acid to BpaSta-lactam, conditions were optimized so that protection of aromatic carbonyl was not necessary (slow addition of sodium borohydride at −20°C). D-*p*-Benzoyl phenylalanine (D-Bpa, **4**) was synthesized by Schöllkopf's bis-lactim alkylation methodology [3] (37% yield from 4-benzoylbenzyl bromide; 95% e.e.); in the alkylation of bis-lactim ethyl ether derived from L-Val-Gly diketopiperazine, the benzophenone carbonyl group was protected as the dimethylene ketal.

The Boc-protected benzophenone derivatives **1**, **3** and **4** were used to synthesize six analogs of pepstatin A (Iva-Val-Val-Sta-Ala-Sta-OH), **5-10**, by using solution-phase DCC/HOBT coupling [4]; peptides **5-10** were chosen so that the benzophenone side-chain substitution would be placed in the binding subsites P_1, P_2, or P_3. Compounds **5-10** were tested for inhibition of penicillopepsin and porcine pepsin, and the inhibition constants are collected in Table 1.

Cyclosporine A result

An analog of cyclosporine A (CsA) in which the key amino acid (4R)-4-[(E)-butenyl]-4, N-dimethyl-L-threonine (MeBmt,**11**) was replaced by the photolabeling benzophenone analog **12**, was prepared by a multistep synthesis using the procedures developed in this laboratory [5]. In comparison with CsA,

---

*To whom correspondence should be addressed.

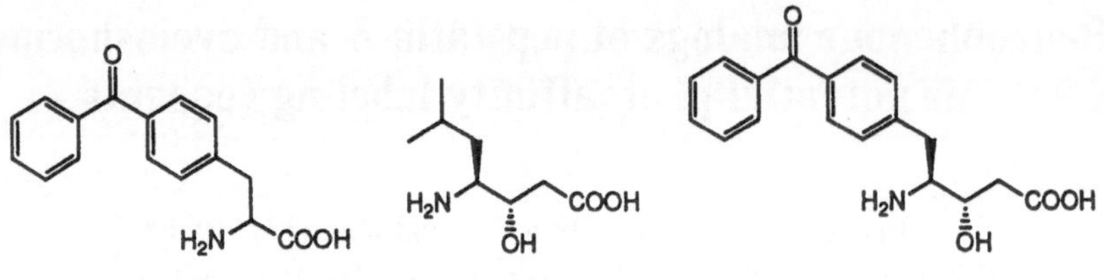

Bpa, 1
D-Bpa, 4

Sta, 2

BpaSta, 3

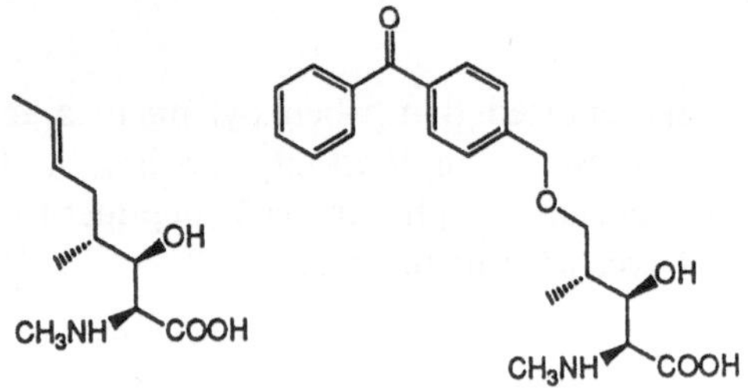

MeBmt, 11

MeBBOmt, 12

Fig. 1.   Structures of unusual amino acids.

[MeBBomt[1]]-CsA showed 5% biological activity in inhibition of Concanavalin A-stimulated murine thymocytes [6].

The use of the photochemical reactivity of the benzophenone group has been shown to be very promising for photoaffinity labeling studies of proteins [1]. The purpose of this work was to determine how incorporation of this group into other amino acid side chains affects the biological activity of the parent peptides, pepstatin and cyclosporine. The biological activity of the new CsA analog was relatively small, but considering the generally high alkylation yields

Table 1 *Inhibition of aspartic proteases by benzophenone-substituted analogs of pepstatin (Boc-X-Y-Z-Ala-Sta-OMe)*

	X	Y	Z	Penicillo-pepsin[a] K_i (nM)	Porcine pepsin[b] K_i (nM)
	Pepstatin A			1.0	0.05
5	Val	Val	Sta	1.1	0.07
6	Val	Val	BpaSta	1.0	0.05
7	Val	Bpa	Sta	3.0	1.1
8	Bpa	Val	Sta	1.5	0.5
9	Val	D-Bpa	Sta	>50	>10
10	D-Bpa	Val	Sta	8.5	0.6

[a]Substrate: Ac-Ala-Ala-Lys-Phe(NO$_2$)-Ala-Ala-NH$_2$, pH 5.5.
[b]Substrate: Phe-Gly-His-Phe(NO$_2$)-Phe-Ala-Phe-OMe, pH 4.0.

130

obtained from aryl ketones in photolabeling studies, the compound still has some potential for use in identification of cyclosporine receptors. Photochemical experiments with CsA will be reported separately.

BpaSta (3) represents a novel analog of statine. When incorporated into the P_1 position of pepstatin, surprisingly good inhibitors of the two aspartic proteinases tested were obtained, in spite of the substantial increase in steric bulk from the added benzophenone moiety. The goal of our preliminary photochemical studies was to determine whether the enzyme becomes permanently inhibited upon irradiation in the presence of a photoreactive inhibitor. Upon UV irradiation (Rayonet photochemical reactor, 350 nm, 30 s, 25°C) in the presence of analog 9 (50 nM), porcine pepsin lost 75% of enzymatic activity. Only 5% loss in activity was observed when photolysis was performed in the absence of inhibitor. Studies to determine the amino acids in the enzyme active sites that are alkylated by 9 and by the 1-position benzophenone (6) are in progress.

Acknowledgements

We gratefully acknowledge support from the NIH (DK20100 and AR32007). C.Q. Sun thanks Applied Biosystems Inc. for a travel award to attend the 11th American Peptide Symposium.

References

1. Kauer, J.C., Erickson-Viitanen, S., Wolfe, H.R. and DeGrado, W.F., J. Biol. Chem., 261 (1986) 10695.
2. Jouin, P., Castro, B. and Nisato, D., J. Chem. Soc. Perkin Trans I, (1987) 1177.
3. Schöllkopf, U., Groth, U. and Deng, C., Angew. Chem. Int. Ed. Engl., 20 (1981) 798.
4. Rich, D.H., Sun, E.T.O. and Ulm, E., J. Med. Chem., 23 (1980) 27.
5. Sun, C.Q., Guillaume, D., Dunlap, B.E. and Rich, D.H., J. Med. Chem., submitted.
6. Dunlap, B.E., Dunlap, S.A. and Rich, D.H., Scand. J. Immunol., 20 (1984) 237.

Analogs of the cyclic depsipeptide antibiotic viscosin

Terrence R. Burke Jr.[a], Bhaskar Chandrasekhar[a] and Michael Gottlieb[b]

[a]Peptide Technologies Corporation, 125 Michigan Avenue N.E., Washington, DC 20017, U.S.A.
[b]The Johns Hopkins University, Baltimore, MD, U.S.A.

Introduction

In an effort to develop new leads in the treament of Chagas' disease, attention was directed to the peptide lactone viscosin **1**, which has antiviral and anti-microbial activity against various mycobacteria. Preliminary in vitro testing against *T. cruzi* produced trypanosomal lysis in the absence of significant hemolysis, which was subsequently supported by promising in vivo activity in mice. A study was undertaken that confirmed the proposed structure of viscosin **1** by total synthesis using solid-phase techniques [1,2]. Methodology developed in that synthesis has been used to prepare peptide analogs of viscosin designed to explore specific features of the molecule as they relate to antitrypanosomal activity. In addition, peptido-mimetics were prepared that replaced the peptide-lactone ring of viscosin with a crown ether.

Results and Discussion

Side-chain analogs consisted of a variety of linear alkyl amides, whereas modifications in the ring junction involved replacing the D-allo-Thr with a more accessible hydroxy amino acid. Synthesis of these analogs relied on SPPS utilizing Fmoc protection with acid-sensitive alkoxybenzylalcohol resin. Linear fragments of the type: R-(OIle)-D-allo-Thr-D-Val-L-Leu-(OBzl)-D-Ser-L-Leu-(OBzl)-D-Ser were prepared. The O-Ile branch point was formed employing a strategy of pentafluorophenyl active ester amidation in the presence of unprotected hydroxyl groups with a coupling order that minimized undesirable N $\Leftrightarrow$ O shifting during esterification. Illustrative of this is the synthesis of viscosin, in which coupling of Fmoc-(OBzl)-D-Glu pentafluorophenyl ester with·unprotected N-terminal D-allo-Thr yielded the corresponding amide, which was then esterified with Boc-L-Ile using DCC/DMAP. Cyclization of finished linear fragments was achieved using BOP·Cl/TEA [1].

A second aspect of this work examined the importance of ion transport in viscosin's anticruzi activity and involved replacement of the peptide lactone ring with a benzo-15-crown-5 polyether. These analogs (**2**) differed from each other in the side chain R which were long-chain alkylamides or simple peptides.

Preliminary antitrypanosomal activity, assayed using *Leishmania donovani*, has been examined by measuring trypanosomal lysis and growth inhibition. Viscosin

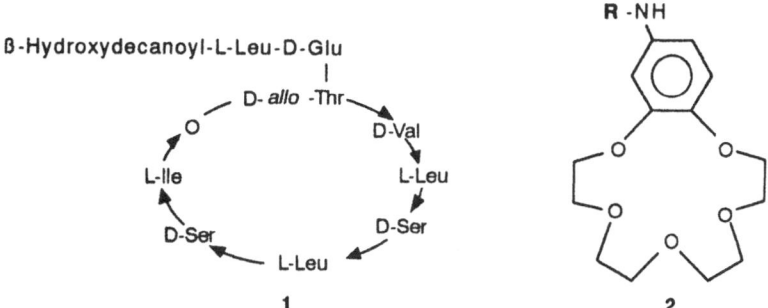

β-Hydroxydecanoyl-L-Leu-D-Glu
|
D-*allo*-Thr
D-Val
L-Ile
L-Leu
D-Ser
D-Ser
L-Leu

1

R -NH

2

*Fig. 1. Viscosin **1** and analogs.*

itself produced both lysis and growth inhibition at 100 μg/ml, but the corresponding linear peptide resulting from ring opening at the lactone bond (viscosic acid) showed no activity at up to 300 μg/ml. In the same fashion, the isolated viscosin side-chain fragment D-β-hydroxydecanoyl-L-Leu-D-Glu-NH$_2$ was also inactive. Crown ether analogs showed a range of potencies, from no lytic effect at 300 μg/ml, to lysis at 30 μg/ml. Growth inhibition in some cases was evidenced at the minimum concentration tested (10 μg/ml). Further testing is presently in progress.

Acknowledgements

Appreciation is expressed to Sumitomo Chemical Company, Osaka, Japan for providing a reference sample of viscosin. Some of the FABMS determinations were carried out at the Middle Atlantic Mass Spectrometry Laboratory, a National Science Foundation Shared Instrumentation Facility. Funding for this work was provided by National Institutes of Health Grant No. AI-23571.

References

1. Burke, Jr., T.R., Knight, M., Chandrasekhar, B. and Ferretti, J., Tetrahedron Lett., 30 (1989) 519.
2. Burke, Jr., T.R. and Chandrasekhar, B., J. Chromatogr., in press.

Synthetic, physico-chemical and pharmacological studies on linear and cyclic analogs of antamanide modified in position 6 and 9

Gianfranco Borin[a], Paolo Ruzza[a], Andrea Calderan[a], Monica Secchieri[a] and Marisa Carrara[b]

[a]Biopolymer Research Center, CNR Department of Organic Chemistry, University of Padua, Via Marzolo 1, I-35131 Padua, Italy
[b]Department of Pharmacology, L.go Meneghetti, University of Padua, I-35131 Padua, Italy

Introduction

Recently Nielsen proposed antamanide, a non-toxic cyclic decapeptide from *Amanita phalloides*, as a cancer chemotherapeutic agent. Our previous studies on F10 metastatic cells of B16 murine melanoma confirmed Nielsen's results, but significantly higher cell-growth inhibition was found using the linear antamanide analog and its des-Phe^5,Phe^6-cyclic octapeptide derivative [1].

Results and Discussion

We have synthesized, by the classical solution methods, the series of linear and cyclic deca- and octapeptide analogs of antamanide reported in Fig. 1.

The CD spectra in acetonitrile of all cyclic decapeptides, linear [Gly^6]- and [Gly^9]-decapeptides and des-Phe^5,Phe^6-Gly^9-analogs are significantly modified by addition of Na$^+$, K$^+$ and Ca^{2+}. On the contrary, we cannot detect any ion-induced spectral change with the linear [Tyr^6]- and [Tyr^9]-derivatives and des-Phe^5,Phe^6-Tyr^9-analogs. The cytotoxic effects of these peptides on B16-F10 transformed cells tested after exposure during 24 and 48 h at different concentrations are reported in Table 1. Among the Gly-analogs, the linear [Gly^6]-decapeptide and [Gly^9]-octapeptide show cytotoxic activity at 10^{-5} M concentration after 24 and 48 h, while the [Gly^9]-antamanide, inactive after 24 h, is cytotoxic after 48 h, both at 10^{-5} and 10^{-6} M. The cyclic [Tyr^9]-octapeptide and the linear [Tyr^9]-decapeptide are also active at these two concentrations, but after both 24 and 48 h exposures.

Conclusions

We observed cell-growth inhibition after treatment also with peptides for which we could not detect any metal ion-induced spectral change in non-polar solvent. Since the Na$^+$-complexation capability of the antamanide and its analogs is

134

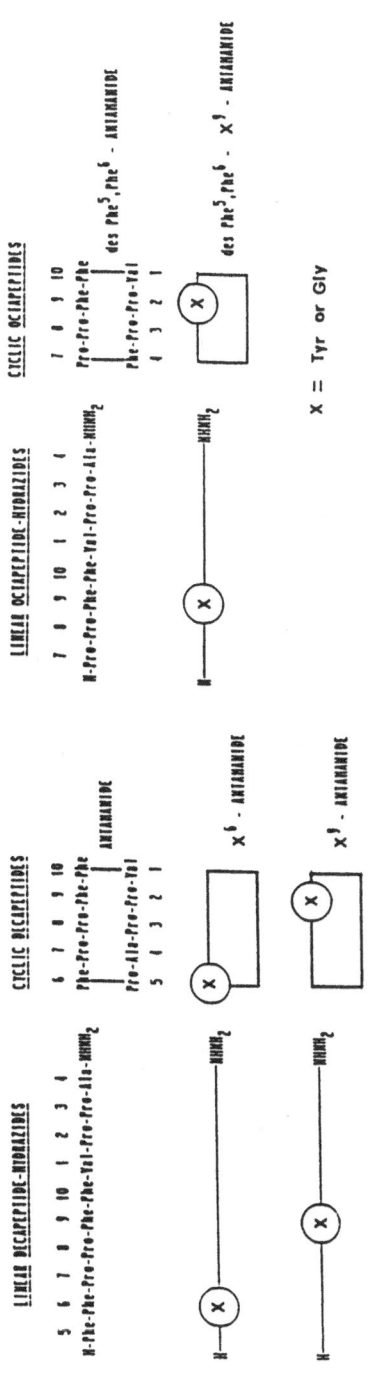

Fig. 1. Linear (L) and cyclic (C) deca- and octapeptide (O) analogs of antamanide.

Table 1 Cell growth of the B16-F10 cellular line after treatment with the synthetic peptides as percent of controls

	ANT	Tyr⁶L	Tyr⁶C	Tyr⁹L	Tyr⁹C	OTyr⁹L	OTyr⁹C	Gly⁶L	Gly⁶C	Gly⁹L	Gly⁹C	OGly⁹L	OGly⁹C
10⁻⁵ M 24 h	59.60**	100.00	100.00	70.40**	95.90	86.70	60.50**	71.80***	76.90*	95.40	100.70	73.10**	80.60*
48 h	53.30**	84.20*	80.90*	80.70*	101.80	73.10**	55.60***	82.10*	85.90	94.30	59.70**	59.30**	86.30
10⁻⁶ M 24 h	84.90*	106.20	107.30	81.50*	106.70	97.90	82.70*	89.70	93.90	107.60	107.10	90.00	100.00
48 h	79.30**	99.30	93.10	91.60	104.60	95.60	82.30*	88.40	98.80	104.00	82.80*	71.10**	96.50
10⁻⁷ M 24 h	93.10	108.60	120.00	90.10	109.30	110.20	90.10	105.10	112.80	110.60	114.30	101.20	105.10
48 h	98.30	99.20	107.30	100.00	106.90	98.70	104.20	105.20	109.20	107.70	101.20	95.50	103.50

*$P > 0.05$; **$P > 0.01$

135

considered a fundamental prerequisite for their biological activity against phalloidin, our results still indicate that a different mechanism for cell-growth inhibition should be contemplated.

References

1. Borin, G., Ruzza, P., Calderan, A. and Carrara, M., In Penke, B. and Torok, A. (Eds.) Peptides: Chemistry, Biology, Interactions with Proteins, De Gruyter, Berlin, New York, 1988, p. 337.

Synthesis and proteolytic cleavage of 3'-N-peptidyl-Adriamycin prodrugs

H. Dalton King, Marcia K. DeVirgilio, John D. Martin, Robert D. Radcliffe and Daniel J. Coughlin

Cytogen Corporation, 201 College Road East, Princeton, NJ 08540, U.S.A.

Introduction

One goal of our antibody-based drug-delivery program has been to develop 'conditionally unstable' cytotoxic drug-monoclonal antibody conjugates for the in vivo treatment of cancer. In this system, the drug ideally would be delivered to its site of action and then released there. This targeted-delivery approach offers the potential advantages of higher efficacy, along with decreased dosage of drug, and lowered side effects.

Depending on the pharmacokinetic requirements of this system, it might be possible to design a linker between the drug and antibody which releases drug upon the initiation of a secondary event after binding to the tumor. One possibility for such an event is proteolysis of the linker by a tumor-endogenous enzyme. Towards this end, we have developed a series of peptide-linked Adriamycin (ADR, Doxorubicin) derivatives which serve as models for ADR prodrugs (Fig. 1). These peptides contain sequences known to be sensitive to proteolysis, and are linked to ADR via the 3'-amino group of the drug [1,2]. Synthesis was carried out by solution-phase methods using an Fmoc-based protecting group strategy. It is known that 3'-acyl derivatives of anthracyclines generally have decreased cytotoxicity compared to the parent [3]; however, degradation of these prodrugs might release free ADR or a fragment which contains ADR.

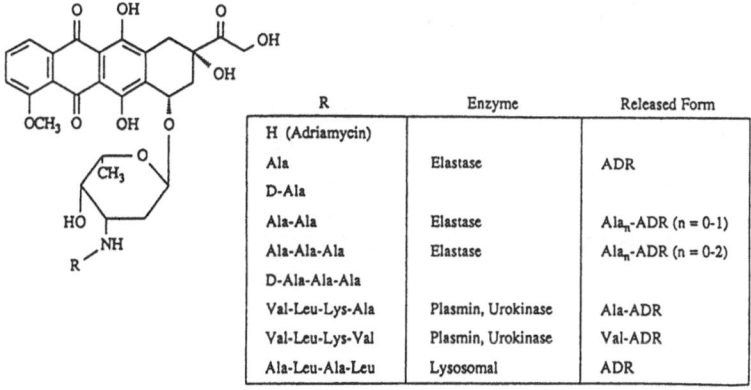

R	Enzyme	Released Form
H (Adriamycin)		
Ala	Elastase	ADR
D-Ala		
Ala-Ala	Elastase	Ala_n-ADR (n = 0-1)
Ala-Ala-Ala	Elastase	Ala_n-ADR (n = 0-2)
D-Ala-Ala-Ala		
Val-Leu-Lys-Ala	Plasmin, Urokinase	Ala-ADR
Val-Leu-Lys-Val	Plasmin, Urokinase	Val-ADR
Ala-Leu-Ala-Leu	Lysosomal	ADR

Fig. 1. ADR prodrug series.

137

In order to ascertain whether these prodrugs are good candidates for linkage to monoclonal antibodies in the drug delivery system described above, we tested them against a panel of proteases which vary widely in specificity (Table 1). Release of drug and peptide fragments was monitored by a modified AAA procedure utilizing 9-fluorenylmethyl-chloroformate as the pre-column derivatizing reagent [4,5]. Use of Fmoc-labeling not only provided an efficient and sensitive assay for all amine components, but also allowed the identification of most amino acid, peptide, and peptidyl-ADR fragments based on comparison to synthetic intermediates. Protease selectivity and specificity of these prodrugs were evaluated as a function of peptide sequence and length.

Table 1 *Qualitative results of enzymatic degradation experiments*

ADR analog	Enzyme[a]								
	Elastase	Chymotrypsin	Trypsin	Plasmin	Urokinase	Pepsin	Papain	Pronase	Subtilisin
Ala-ADR	–	–	–	–	–	–	–	S	–
D-Ala-ADR	–	–	–	–	–	–	–	–	–
Ala-Ala-ADR	–	–	–	–	–	–	–	F	–
Ala-Ala-Ala-ADR	–	–	–	–	–	–	–	F	F
D-(Ala-Ala-Ala)-ADR	–	–	–	–	–	–	–	–	–
Val-Leu-Lys-Ala-ADR	S	S	F	S	–	S	S	F	F
Val-Leu-Lys-Val-ADR	S	S	F	S	–	S	S	F	F
Ala-Leu-Ala-Leu-ADR	S	S	–	–	–	–	–	F	F

[a](–): no cleavage observed; (S): 'slow' cleavage observed; 24 h; (F): 'fast' cleavage observed, 1 h.

Results and Discussion

(1) Ala$_n$-ADR series
In this series, all enzymes were ineffective, with the exception of the bacterial enzymes pronase and subtilisin. Pronase seems sensitive to the proximity of ADR on the peptide chain, since shorter-chain analogs are slower to cleave. Analysis of the degradation products of Ala$_3$-ADR shows a distinct preference for cleavage at both P_2 and P_3 over P_1. This is most likely due to steric bulk associated with the drug.

(2) Val-Leu-Lys-X-ADR
Broad proteolytic activity is demonstrated by the entire panel of enzymes, with the unexpected exception of urokinase. Trypsin shows fast cleavage at the predicted site, next to the positively charged Lys residue, to yield X-ADR and the free peptide Val-Leu-Lys. Elastase cleaves at multiple points along the chain, yielding fragments Lys-X-ADR, Val-Leu, X-ADR, Lys, and free ADR. Elastase activity may be explained by the presence of hydrophobic residues, which are preferred cleavage sites for this enzyme. The failure of urokinase to cleave Val-Leu-Lys-X-ADR was unexpected, since Val-Leu-Lys-AMC is a known substrate for urokinase.

(3) Ala-Leu-Ala-Leu-ADR

Cleavage by elastase and chymotrypsin was observed as expected, as well as pronase and subtilisin. Because of the synthetic route, degradation products could not be positively identified. It can be stated, however, that free ADR was not released from Ala-Leu-Ala-Leu-ADR in any proteolysis.

Conclusion

This experiment demonstrates that ADR can be modified with peptide side chains at the 3′-position to yield prodrugs which are· capable of degrading to ADR-containing species in the presence of some enzymes. These model systems may have applications as monoclonal antibody conjugates. Targeting of a specific tumor-endogenous enzyme is a critical link in this drug delivery system. Once conjugated to a macromolecule, the cleavage rates of these substrates may be different, however, this effect is uncharacterized at this time. Nonetheless, these results, in conjunction with those from other groups [6,7], demonstrate the potential of developing conditionally unstable antibody conjugates of anthracycline drugs.

Acknowledgements

We thank Farmitalia-Carlo Erba for their financial support of this work.

References

1. Chakravarty, P.K., Carl, P.L., Weber, M.J. and Katzenellenbogen, J.A., J. Med. Chem., 26 (1983) 638.
2. Trouet, A., Masquelier, M., Baurain, R. and Deprez-de Campeneere, D., Proc. Natl. Acad. Sci. U.S.A., 79 (1982) 626.
3. Arcamone, F., Doxorubicin: Anticancer Antibiotics, Academic Press, New York, 1981, p. 245.
4. Einarsson, S., Folestad, S., Josefsson, B. and Lagerkvist, S., Anal. Chem., 58 (1986) 1638.
5. Cunico, R., Mayer, A.G., Wehr, C.T. and Sheehan, T.L., Biochromatography, 1 (1986) 6.
6. Kuefner, U., Lohrmann, U., Montejano, Y.D., Vitols, K.S. and Huennekens, F.M., Biochemistry, 28 (1989) 2288.
7. Senter, P.D., Saulnier, M.G., Schreiber, G.J., Hirschberg, D.L., Brown, J.P., Hellstrom, I. and Hellstrom, K.E., Proc. Natl. Acad. Sci. U.S.A., 85 (1988) 4842.

Structure–activity relationship of Ac-CCK-7: Analogs with o-, m- and p-tyrosine sulfate and phosphate

C.F. Stanfield, A.M. Felix and W. Danho

Peptide Research Department, Roche Research Center, Hoffmann-La Roche Inc., Nutley, NJ 07110, U.S.A.

Introduction

Cholecystokinin (CCK), a linear, 33-amino acid polypeptide hormone is located primarily in mammalian digestive tract and central nervous system [1]. Although various forms of CCK have been detected in vivo, the primary circulating form of the hormone is the 26–33 fragment, CCK-8, that was originally proposed by Gibbs et al. [2] as a satiety signal. CCK-8 suppresses feeding in several species including man. An analog of CCK-8, Ac-CCK-7, Ac-Tyr(SO_3H)-Met-Gly-Trp-Met-Asp-Phe-NH_2, possesses the full potency of the intact hormone. Thus, analogs of Ac-CCK-7 may be useful for the reduction of food intake.

Results and Discussion

Previous studies established that the tyrosine sulfate, Tyr(SO_3H), is important for the satiety-inducing activity of CCK-8. The influence of the Tyr(SO_3H) residue was further investigated by varying two key factors: the regiochemistry (distance/proximity to the receptor) and the stereochemistry of this key residue. Analogs of Ac-CCK-7, in which Tyr(SO_3H) was replaced by D-Tyr(SO_3H), m-L-Tyr(SO_3H), m-D-Tyr(SO_3H), o-L-Tyr(SO_3H), o-D-Tyr(SO_3H) and their corresponding phosphate analogs, were synthesized.

SPPS was used for assembling the peptides. Sulfation was carried out in solution using pyridine acetyl sulfate. The direct phosphorylation of Ac-CCK-7 with $POCl_3$/pyridine failed to produce the phosphate analog. An alternative method using N^α-t-Boc-(o-diphenylphospho)-L-tyrosine, Boc-Tyr[PO(OC_6H_5)$_2$], as a monomer for SPPS was developed. This reagent was synthesized in excellent yield by reacting Boc-Tyr-Bzl with diphenylchlorophosphate followed by hydrogenation. The monomer is stable to conditions of SPS; the diphenylphospho group is readily cleaved with tetra(n-butyl)-ammonium fluoride (TBAF). Synthesis of the phosphotyrosine analogs is outlined in the following scheme.

The analogs were characterized by analytical HPLC, MS, UV and IR spectroscopy and bioassayed in vivo by the two-meal feeding assay [3] and in vitro (receptor binding) using bovine striatum and rat pancreas [4] (Table 1). The results can be summarized as follows: (1) Comparison of the binding efficacy relative to food intake indicates poor correlation of pancreatic and striatum

140

Boc-Phe-PAM-Resin

| 1. CF$_3$COOH
| 2. BOP coupling of Boc-Asp(OFm)-OH
| Boc-Met-OH, Boc-Trp(For)-OH, Boc-Gly-OH,
| Boc-Met-OH, Boc-Tyr[PO(OC$_6$H$_5$)$_2$]-OH
| 3. Acetylation with CH$_3$COOH/BOP
↓

Ac-Tyr[PO(OC$_6$H$_5$)$_2$]-Met-Gly-Trp(CHO)-Met-Asp(OFm)-Phe-PAM-Resin

| 1. Piperidine/DMF
| 2. TBAF/THF
| 3. NH$_3$ / methanol
| 4. Preparative HPLC
↓

Ac-Tyr(PO$_3$H$_2$)-Met-Gly-Trp-Met-Asp-Phe-NH$_2$

Table 1   *Comparative activity of Ac-CCK-7 and analogs*

Compound	Receptor binding (IC$_{50}$; nM)		Feeding assay
	Bovine striatum	Rat pancreas	Food Intake (ED$_{50}$; μg/kg i.p.)
Ac-[Tyr(SO$_3$H)]-CCK-7 (reference compound)	0.8	0.2	7
Ac-[D-Tyr(SO$_3$H)]-CCK-7	28	4.6	1
Ac-[m-Tyr(SO$_3$H)]-CCK-7	42	18	15
Ac-[m-D-Tyr(SO$_3$H)]-CCK-7	53	280	32
Ac-[o-Tyr(SO$_3$H)]-CCK-7	66	560	>320 (23%)[a]
Ac-[o-D-Tyr(SO$_3$H)]-CCK-7	72	530	>320 (32%)
Ac-[Tyr(PO$_3$H$_2$)]-CCK-7	85	15	3
Ac-[D-Tyr(PO$_3$H$_2$)]-CCK-7	160	190	>320 (36%)
Ac-[m-Tyr(PO$_3$H$_2$)]-CCK-7	35	130	220
Ac-[m-D-Tyr(PO$_3$H$_2$)]-CCK-7	40	140	>320 (19%)
Ac-[o-Tyr(PO$_3$H$_2$)]-CCK-7	180	500	>320 (6%)
Ac-[o-D-Tyr(PO$_3$H$_2$)]-CCK-7	76	280	>320 (0%)

[a] For compounds with ED$_{50}$ > 320, the number in parenthesis is the percent inhibition of food intake at a dose of 320 μg/kg i.p.

binding with food intake. (2) Ac-[m-Tyr(SO$_3$H)]-CCK-7 retains substantial anorectic activity, whereas the Ac-[o-Tyr(SO$_3$H)]-CCK-7 analog is essentially inactive. (3) Ac-[D-Tyr(SO$_3$H)]-CCK-7 and Ac-[m-D-Tyr(SO$_3$H)]-CCK-7 retain substantial activity. (4) The phosphate moiety is a suitable replacement for sulfate in Ac-CCK-7. (5) Ac-[m-Tyr(PO$_3$H$_2$)]-CCK-7 retains some anorectic activity whereas the Ac-[o-Tyr(PO$_3$H$_2$)]-CCK-7 is essentially inactive. (6) Ac-[D-Tyr(PO$_3$H$_2$)] and Ac-[m-D-Tyr(PO$_3$H$_2$)] are essentially inactive, in contrast to the sulfate series, which may be due to the larger radius of the phosphate (0.15 Å) that forces the tyrosine residue into an unfavorable conformation with respect to the receptor. These results indicate that a negative charge in the proper location is important for anorectic activity, but the chirality at the α-carbon is not important.

References

1. For a review see: Baile, C.A., McLaughlin, C.L. and Della-Fera, M.A., Physiol. Res., 66 (1986) 172.
2. Gibbs, J., Young, R.C. and Smith, G.P., Nature, 245 (1973) 323.
3. Triscari, J., Geiler, V., Nelson, D. and Danho, W., Int. J. Obesity, 11 (1987) 443A.
4. Van Dijk, A., Richards, J.C., Trzeciak, A., Gillessen, D. and Mohler, H., J. Neuroscience, 4 (1984) 1021.

Structure-activity relationships of cholecystokinin: Tyr[SO₃H] substitutions

Tomi K. Sawyer[a,*], Robert T. Jensen[b], Timothy H. Moran[c], Lua-Hua Wang[b],
Da-Hong Yu[b], Douglas J. Staples[a], Anne E. DeVaux[a] and V. Susan Bradford[a]

[a]Peptide Therapeutics and Core Facility, Biopolymer Chemistry, The Upjohn Company,
Kalamazoo, MI 49001, U.S.A.
[b]Cell Biology Section, Digestive Diseases Branch, NIADDKD, National Institutes of Health,
Bethesda, MD 20892, U.S.A.
[c]Department of Psychiatry, Johns Hopkins University School of Medicine, Baltimore,
MD 21205, U.S.A.

Introduction

Cholecystokinin-8 (CCK-8-Asp-Tyr[SO₃H]-Met-Gly-Trp-Met-Asp-Phe-NH₂)
effects a variety of gastrointestinal, endocrine, and CNS activities [1]. Among
the many functions proposed for the endogenous CCK peptides is its anorexigenic
(satiety) effects [2]. A number of CCK target tissues have been identified having
CCK receptors and of potential importance in satiety mediation. Among these
are the pylorus, vagal and brainstem (i.e., area postrema and nucleus tractus
solitarius), CCK receptors described [3] to be of Type-A classification. CCK
agonist binding to these Type-A CCK receptor populations is Tyr[SO₃H]-
dependent in contrast to Type-B CCK/gastrin receptor populations. In some
species, such as the rat, studies [2] suggest that Type-A receptors may mediate
satiety, whereas in other species it is unclear. In this study, we explored the
comparative SAR of CCK-8 using three well-defined Type A CCK receptor
assays (i.e., rat pancreas, pylorus, and brainstem), and an in vivo anorexigenic
assay.

Results and Discussion

The methods for determination of CCK receptor binding using rat pancreatic,
pylorus and brainstem tissues has been previously described [3–5]. Anorexigenic
activities were first determined in nonfasting male Sprague–Dawley rats that
were pre-trained to ingest a daily glucose solution. CCK peptides were administered i.p. (vs. saline-injected control animals 5 min prior to monitoring food
ingestion, which was recorded in 5-min intervals during a 60-min time period.
Relative to a Ac-[Nle³,⁶]-CCK-8 (U-67827E), we first focused our attention on
the Tyr[SO₃H] residue to explore its role in Type-A CCK pancreatic receptor
binding. Analogs of U-67827E included those having Phe[p-NH₂], and Phe[p-

---

*To whom correspondence should be addressed.

143

NHC(NH$_2$) = N] substitutions for Tyr[SO$_3$H]. As shown in Table 1, compounds having *p*-amino or guanidyl modifications of the *p*-sulfate ester in U-67827E showed a 100–1000-fold reduction in potency in vitro. Deletion of the N-terminal Asp residue of U-67827E to yield Ac-[Nle^{2,5}]-CCK-7 (U-86317E) did not significantly alter pancreatic CCK receptor binding affinity of the compound relative to CCK-8. The Phe[*p*-CH$_2$SO$_3$H]-substituted derivative of U-86317E was essentially equipotent to CCK-8 on the rat pancreas CCK receptor assay. A similar result for Ac-[Phe(*p*-CH$_2$SO$_3$H)^1, Nle^{3,6}]-CCK-7 has been very recently reported [6] on guinea pig target tissues. A comparison of CCK-8, U-67827E and [Phe(*p*-NH$_2$)^2,Nle^{3,6}]-CCK-8 to bind brainstem and pylorus CCK receptors was then determined (Table 1). A 10-fold selectivity for [Phe(*p*-NH$_2$)^2,Nle^{3,6}]-CCK-8 to bind brainstem receptors vs. pylorus (or pancreatic) receptors was observed.

Table 1 *Rat pancreas, brainstem and pyloris CCK receptor binding properties of CCK-8/CCK-7 analogs modified at Tyr[SO$_3$H]*

Compound	IC$_{50}$ (M), CCK receptor binding		
	Pancreas	Brainstem	Pylorus
CCK-8	1.1×10^{-9}	6.8×10^{-10}	1.1×10^{-9}
Ac-[Nle^{3,6}]-CCK-8 (U-67827E)	1.4×10^{-9}	5.2×10^{-10}	4.1×10^{-10}
Ac-[Phe(*p*-NH$_2$)^2,Nle^{3,6}]-CCK-8	2.0×10^{-7}	2.9×10^{-8}	7.2×10^{-7}
Ac-[Phe(*p*-NHC(NH$_2$) = N)^2,Nle^{3,6}]-CCK-8	1.1×10^{-6}	ND[a]	ND
Ac-[Nle^{2,5}]-CCK-7 (U-86317E)	1.3×10^{-9}	ND	ND
Ac-[Phe(*p*-CH$_2$SO$_3$H)^1,Nle^{2,5}]-CCK-7	1.2×10^{-9}	ND	ND

[a]ND = not determined.

The [Phe(*p*-NH$_2$)^2,Nle^{3,6}]-CCK-8 was only 50-fold less potent than CCK-8 in brainstem receptor binding. These potencies were essentially identical to that of these compounds on rat pancreatic amylase release and pyloric contraction assays (data not shown). The anorexigenic potency of U-67827E in the rat model was 10–100-fold greater than that of CCK-8, and the analog possessed sustained activity in vivo (Table 2). No in vivo anorexigenic activity was observed for the Phe(*p*-NH$_2$)^2-modified derivative of U-67827E at doses $\geq$ 10–10 000-fold the threshold doses of CCK-8 or U-67827E, respectively. These results extend our previous studies [8] on CCK peptide binding to specific Type-A CCK receptor populations in vitro and the exploration of potentially significant mechanistic pathways of satiety using a rat anorexigenic in vivo assay.

Table 2 *Rat anorexigenic activities of selected CCK-8 analogs modified at the Tyr[SO$_3$H]site*

Compound	Threshold dose[a] (nmol/kg), rat anorexigenic activity	
	30 min	60 min
CCK-8	0.5	5.0
U-67827E	0.05	0.05
Ac-[Phe(*p*-NH$_2$)^2,Nle^{3,6}]-CCK-8	> 100	> 100

[a]Dose of CCK analog that effects approximately a 50% anorexigenic response.

References

1. Beinfeld, M.C., Neuropeptides, 3 (1983) 411.
2. Smith, G.P. and Gibbs, J., In Neurosecretion and Brain Peptides, Raven, New York, NY, 1981, p. 389.
3. Moran, T.H., Robinson, P.H., Goldrich, M.S. and McHugh, P.R., Brain Res., 362 (1986) 175.
4. Smith, G.T., Moran, T.H., Coyle, J.T., Kuhar, M.J., O'Donohue, T.L. and McHugh, P.R., Am. J. Physiol., 246 (1984) R127.
5. Jensen, R.T., Lemp, G.F. and Gardner, J.D., J. Biol. Chem., 257 (1982) 5554.
6. Marseigne, I., Roy, P., Dor, A., Durieux, C., Pelaprat, D., Reibaud, M., Blanchard, J.C. and Roques, B.P., J. Med. Chem., 32 (1989) 445.
7. Sawyer, T.K., Jensen, R.T., Moran, T., Schreur, P.J.K.D., Staples, D.J., DeVaux, A.E. and Hsi, A., In Marshall, G.R. (Ed.) Peptides: Chemistry and Biology, (Proceedings of the 10th American Peptide Symposium), ESCOM, Leiden, 1988, p. 503.

Novel gastrin antagonists and receptor probe

Alastair J. Douglas[a], Akihiro Yasui[b], Brian Walker[a], Colin F. Johnston[a],
Donal F. Magee[b] and Richard F. Murphy[a]

[a]*Departments of Biochemistry and Medicine, The Queen's University of Belfast, Ireland*
[b]*Department of Biomedical Sciences, Creighton University, Omaha, NE 68178, U.S.A.*

Introduction

The C-terminal tetrapeptide of gastrin, Trp-Met-Asp-Phe-NH$_2$, can elicit a full biological response [1]. Analogs devoid of the C-terminal Phe amide are potent competitive antagonists, the smallest being Boc-Trp-Leu-Asp-NH$_2$ [2,3]. We synthesized [4] tripeptide analogs (Table 1) that were modified both N- and C-terminally, and that contained *N*-methylated peptide bonds [5]. The purity of the synthetic sequences was determined by microanalysis, TLC and HPLC.

Table 1 *The inhibition of pentagastrin-stimulated acid release from fundus of dogs with gastric fistulae*

Peptide	Acid release (mmol H$^+$/kg)	% inhibition
Boc-Trp-Leu-β-Ala	617.58	60
IPA-Leu-β-Ala.DCHA	562.22	54
PPA-Leu-β-Ala.DCHA	125.51	12
PPA-NMe-Leu-β-Ala.DCHA	256.77	25
PPA-Leu-NMe-B-Ala.DCHA	185.32	18

IPA = indole-3-propionic acid, PPA = phenyl-3-propionic acid.

Methods

The peptides were infused into dogs with gastric fistulae at a dose of 20 pmol kg^{-1} h^{-1}, in the presence of either pentagastrin (650 pmol kg^{-1} h^{-1}) or histamine (0.13 mmol kg^{-1} h^{-1}). Pentagastrin-stimulated acid release, determined at 10-min intervals, was inhibited by all analogs; Boc-Trp-Leu-β-Ala was the most potent (Table 1). The analogs had no effect on histamine-stimulated acid release.

None of the peptides, up to a concentration of 10^{-6} M, stimulated amylase secretion from isolated pancreatic acini [6]. Neither did they inhibit amylase secretion stimulated by CCK (10^{-10} M). PPA-Leu-β-Ala was a very weak antagonist at non-physiological concentrations (10^{-3} M).

Biotinylated analogs of peptide hormones have been prepared to determine the affinities for receptors [7-9], but only in the case of gonadotropin-releasing hormone [10] and parathyroid hormone [11] have they been used to visualize the receptors on cultured cell monolayers. Because of the apparent specificity

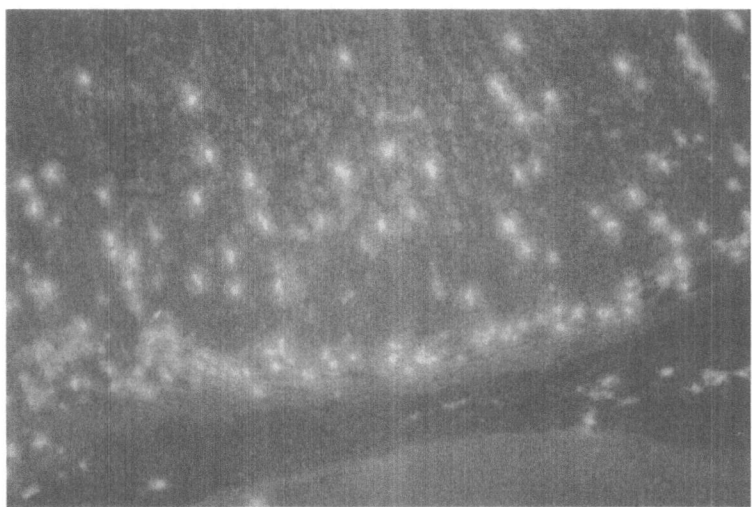

Fig. 1.   *Photomicrograph of rat fundic mucosa after administration of Biotinyl-Trp-Leu-β-Ala.*

of the gastrin antagonists, Biotin-Trp-Leu-β-Ala was synthesized as a possible histological probe for gastrin receptors. Cryostat sections (10 μm) of rat intestinal tissues were pre-fixed for 30 min at room temperature, using 0.04% (w/v) paraformaldehyde in phosphate buffer (0.2 M, NaH_2PO_4-Na_2HPO_4, pH 7.2) containing 0.15 M NaCl. The biotin-labeled peptide was dissolved in phosphate buffer containing ethanol (10%, v/v) . Pentagastrin and gastrin 1-17 solutions were diluted to required concentrations with phosphate buffer. All peptide solutions were mixed with equal volumes of an enzyme inhibitor solution containing bacitracin (1%, w/v), phosphoramidon (10 μM), 1-10 phenanthroline (1 mM) and phenylmethylsulphonyl fluoride (100 μM) in 10 mM Hepes buffer (pH 7.4) containing mannitol (300 mM) and bovine serum albumin (1%, w/v). The peptide-inhibitor solutions (250 μl) were then applied directly onto the tissue sections. Following incubation for 30 min at room temperature, the bathing solutions were removed and the sections were washed thoroughly with phosphate buffer. Streptavidin-fluorescein solution (200 μl stock solution as supplied by Amersham International Inc.) was applied to each tissue section for 20 min at room temperature. Following washing with phosphate buffer, the tissue sections were mounted and viewed under a microscope fitted with a UV light source.

Results and Discussion

When Biotin-Trp-Leu-B-Ala solution (250 μg ml^{-1}) was incubated with sections of fundus from rats that had not been fed for 48 h, submucosal cells, whose tissue location was not consistent with parietal cells, were stained (Fig. 1). Staining was observed with duodenal, ileal and colonic tissues. The probe bound to lung alveolar tissue, apparently near blood vessels. The binding of the probe was not observed in heart, kidney, liver or pancreas, reflecting the apparent lack of interaction of the parent peptides with CCK receptors in pancreatic acini. No non-specific staining was observed with biotin. When the tissue sections were preincubated with excess Boc-Trp-Leu-β-Ala, pentagastrin or gastrin 1–17, fluorescent staining resulting from binding of the probe was completely prevented. This suggests that the cellular binding site recognizes the indole and the β-carboxylic acid structures common to all three peptides. Binding of the probe was also inhibited in tissues taken from animals that had been fed ad libitum, suggesting that endogenous gastrin competes for the binding sites. Thus, the probe appears to be highly specific for a gastrin, rather than CCK receptor.

The histological results are in disagreement with conventional theories of gastrin-stimulated acid release being mediated by a gastrin receptor on parietal cells [2,13], but support data suggesting that gastrin acts at a histamine releasing cell, thereby causing the release of acid from the parietal cells by the action of histamine [14,15]. The binding of the gastrin probe to colon and lung may indicate a growth-promoting or even cancer-supporting role for gastrin. Because of their potency and specificity, the gastrin antagonists prepared here may have therapeutic potential for the control of acid release and to prevent the development of some cancers, while the probe could have diagnostic applications.

References

1. Tracy, H.J. and Gregory, R.A., Nature, 204 (1964) 935.
2. Martinez, J., Magous, R., Lignon, M.F., Laur, J., Castro, B. and Bali, J.-P., J. Med. Chem., 27 (1984) 1597.
3. Lavezzo, A., Bali, J.-P., Magous, R., Lignon, M.F., Nisato, D., Laur, J., Castro, B. and Martinez, J., Regul. Peptides, 15 (1986) 111.
4. Douglas, A.J., PhD Thesis, Queen's University, Belfast, Ireland, 1988.
5. Cheung, S.T. and Benoitin, N.L., Can. J. Chem., 55 (1977) 906.
6. Bruzzone, R., Halban, P.A., Gjinovci, A. and Trimble, E.R., Biochem. J., 226 (1985) 621.
7. Finn, F.M., Titus, G. and Hofmann, K., Biochemistry, 23 (1984) 2554.
8. Brennan, D.P. and Levine, M.A., J. Biol. Chem., 262 (1987) 14795.
9. Hochhaus, G., Gibson, B.W. and Sadee, W., J. Biol. Chem., 263 (1988) 92.
10. Westlund, K.N., Wynn, P.C., Chmielowiec, S., Collins, T.J. and Childs, G.V., Peptides, 5 (1984) 627.
11. Niendorf, A., Dietel, M., Arps, H., Lloyd, J. and Childs, G.V., J. Histochem. Cytochem., 34 (1986) 357.
12. Grossman, M.I., In Sleisenger, M.H. and Fordtran, J.S. (Eds.) Gastrointestinal Disease, Saunders, Philadelphia, PA, 1978, p. 640.
13. Takeuchi, K., Speir, G.R. and Johnson, L.R., Am. J. Physiol., 237 (1979) E284.
14. Irvine, W.T. and Code, C.F., Am. J. Physiol., 195 (1958) 202.
15. Sandvik, A.K., Waldum, H.L., Kleveland, P.M. and Schulze Sognen, B., Scand. J. Gastroenterol., 22 (1987) 803.

CCK antagonists: Investigations of the relationship between benzotript and glutamic acid-based antagonists

James F. Kerwin Jr., Frank Wagenaar, Hana Kopecka, Chun Wel Lin, Thomas Miller, David Witte and Alex M. Nadzan

Neuroscience Research Division, Pharmaceutical Discovery D-47H, Abbott Laboratories, Abbott Park, IL 60064, U.S.A.

Introduction

Our laboratory's research on CCK antagonists has focused on the structural relationships among the various classes of nonpetide CCK antagonists. Recently, we reported on the structural similarities of glutamic acid-based CCK antagonists represented by A-64718 and Merck's benzodiazepine antagonist L-364,718 [1]. From this work, we proposed an overlap model for CCK antagonists, and ensuing SAR studies resulted in the potent CCK antagonist A-65186 [2]. Currently, we have extended our SAR studies to include the weak CCK antagonist benzotript (I) [3], which has resulted in a series of tryptophan-based CCK antagonists (II) whose potencies approach those of our hybrid glutamic acid derivatives.

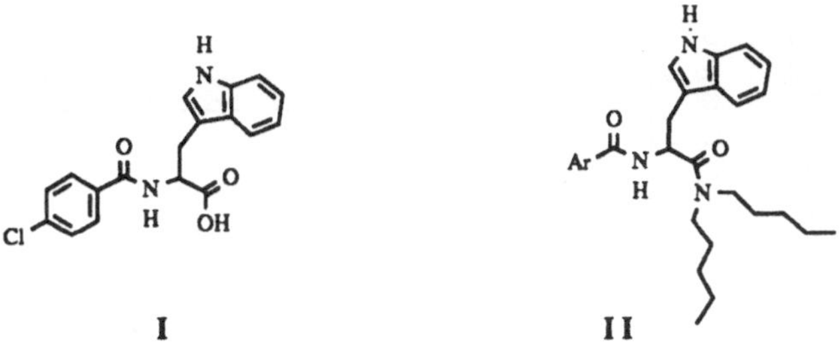

Fig. 1. Structures of benzotript (I) and N-aroyl-Trp-di-n-pentyl amides (II).

Results and Discussion

The series of *N*-aroyl-tryptophan di-*n*-pentyl amides (II), shown in Table 1, was prepared from Boc-tryptophan utilizing a procedure similar to that reported for the glutamic acid series [2]. Compounds **1–17** gave satisfactory ^{1}H NMR, MS, and elemental analyses. Pharmacological assays were conducted as previously described [2,4].

Conversion of **I** to the di-*n*-pentyl amides (**1** and **2**) improved affinity for pancreatic CCK receptors approximately 1000-fold over that of the parent [5], but addition of a second chloro atom to the benzoyl moiety (**3** and **4**, corresponding

Table 1 *SAR of tryptophan-based CCK antagonists (structure II)*

Entry	Ar	Isomer	Pancreas	IC_{50} (nM) Cortex	Inhib. of amylase release
1	4-Chlorophenyl	R	370(2)	> 10 000	10 000 inhib. 81%
2		S	450(2)	≧ 10 000	–
3	3,4-Dichlorophenyl	R	190 ± 70(3)	≧ 10 000	10 000 inhib. 74%
4		S	6 700(2)	> 10 000	–
5	2-Indolyl	R	51 ± 13(5)	8 000 ± 850(3)	670(2)
6		S	520(2)	3 100	30 000 inhib. 69%
7	3-Indolyl	R	430 ± 140(3)	20 000	10 000 inhib. 77%
8	2-Pyrrolyl	R	5 700 ± 820(3)	> 10 000	–
9	3-Quinolinyl	R	23 ± 5.4(6)	15 000	42(2)
10		S	1 100 ± 270(3)	6 300	10 000 inhib. 68%
11	3-Pyridyl	R	15 000 ± 2 200(3)	> 10 000	–
12	2-Quinolinyl	R	120 ± 34(3)	450	3 000 inhib. 79%
13	4-Hydroxy-2-quinolinyl	R	51 ± 14(4)	3 700	110
14	8-Hydroxy-2-quinolinyl	R	390 ± 150(4)	> 10 000	10 000 inhib. 58%
15	4,8-Dihydroxy-2-quinolinyl	R	21 ± 3.4(6)	4 200	10 000 inhib. 92%
16	2-Naphthyl	R	29 ± 3.8(3)	13 000	1 000 inhib. 78%
17	1-Naphthyl	R	500 ± 120(3)	9 700	30 000 inhib. 64%

to (R) and (S)-CR 1409) did not significantly enhance affinity. Further improvements were realized by incorporation of the 2-indolyl and 3-quinolinyl functions. In both cases, the R enantiomers (5 and 9, respectively) proved to be more potent than the S isomers. The importance of a fused-aryl heterocycle fore potency is apparent by comparing 5 with 8 and 9 with 11. Introduction of the 3-indolyl and 2-quinolinyl moieties (7 and 12) afforded weaker binding analogs. The potency of 12, however, could be enhanced by replacement with a 4-hydroxy- (13) or a 4,8-dihydroxy- (15) but not a 8-hydroxy-2-quinolinyl function (14). Finally, introduction of the 2-naphthyl but not the 1-naphthyl, function resulted in an analog with enhanced affinity (16 vs. 17).

The SAR observed for the tryptophan-based antagonist series closely parallels that reported for the glutamic acid series [2]. Both series demonstrate identical preferences for the R-enantiomer, similar selectivities for type A (pancreatic) over type B (cortex) CCK receptors (i.e., 157-fold for 5 and 650-fold for 9) and identical SAR requirements for the *N*-aryl function. Overall, members of the tryptophan series exhibit potencies between one-half and one-fifth those of the corresponding glutamic acid derivatives.

References

1. Nadzan, A.M., Kerwin, Jr., J.F., Kopecka, H., Lin, C.W., Miller, T., Witte, D. and Burt, S., In Wang, R.Y. and Schoenfeld, R. (Eds.) Cholecystokinin Antagonists (Neurology and Neurobiology, Vol. 47), Alan R. Liss Inc., New York, 1988, p. 93.
2. Kerwin, Jr., J.F., Nadzan, A.M., Kopecka, H., Lin, C.W., Miller, T., Witte, D. and Burt, S., J. Med. Chem., 32 (1989) 739.
3. Hahne, W.F., Jensen, R.T., Lemp, G.F. and Gardner, J.D., Proc. Natl. Acad. Sci. U.S.A., 78 (1981) 6304.
4. Lin, C.W., Bianchi, B., Grant, D., Miller, T., Danaher, E.A., Tufano, M.D., Kopecka, H. and Nadzan, A.M., J. Pharmacol. Exp. Ther., 236 (1986) 729.
5. Gardner, J.D. and Jensen, R.T., Am. J. Physiol., 246 (1984) G471.

Use of a novel analog of cholecystokinin for the photoaffinity labeling of the hormone-binding domain of the pancreatic receptor

Ulrich G. Klueppelberg, Delia I. Pinon, Judy Lundy, Herbert Y. Gaisano,
Stephen P. Powers and Laurence J. Miller*

*Gastroenterology Research Unit, Mayo Clinic and Foundation, Rochester, MN 55905,
U.S.A.*

Introduction

Cholecystokinin (CCK) is the major hormonal stimulant of pancreatic enzyme secretion. For the biochemical characterization of the CCK receptor, we have developed a series of photoaffinity-labeling probes based on the native hormone, with sites of cross-linking at the amino-terminus [1], mid-region [2], and carboxyl-terminus [3] of the receptor binding domain. Native hormone, however, has been shown to bind to two sites, though this may represent functional states of the same molecule.

Characterization of the binding site relevant for stimulating secretion may be made possible by a CCK phenylethyl ester [4,5] which is a fully efficacious secretagogue without the supramaximal inhibition of secretion typical of native CCK, and which binds to only a single site on the pancreatic acinar cell [5]. In this work, we have incorporated photolabile residues into the positions of Trp^{30} and Phe^{33}, in the midregion and carboxyl-terminus of phenylethyl ester analogs of CCK to define further the hormone-binding domain of this receptor.

Methods

D-Tyr-Gly-[Nle^{28,31}, 4-NO$_2$-Phe^{33}]CCK(26–33): (4-NO$_2$-Phe^{33}) was prepared as described elsewhere [3]. Phenylethyl ester derivatives were prepared similarly: Fmoc-Asp-phenylethyl ester or Fmoc-Asp-(4-NO$_2$-phenylethyl) ester was coupled to benzhydrylamine resin via a 4-hydroxymethylphenoxyacetic acid spacer, and the Fmoc-octapeptide ester was prepared on the resin. After cleavage from the resin and removal of side-chain protection, the peptide was purified and Tyr^{27} was sulfated using SO$_3$-pyridine complex. N-α-Fmoc protection was removed, and a site for iodination was introduced using N,O-bis-Fmoc-D-Tyr-ONSu or 4-hydroxyphenylpropionic acid-ONSu [3].

Biological activity and binding of the peptides were assessed in dispersed rat-pancreatic acini and enriched pancreatic plasma membranes [5,6]. Following

---

*To whom correspondence should be addressed.

affinity-labeling and resolution on SDS-PAGE, deglycosylation, using endoglycosidase F, and protease peptide-mapping experiments were performed [7,8].

Results and Discussion

Indeed, Hpp-Gly-[Nle^{28,31}]CCK(26–32)-2-(4-nitrophenylethyl)ester (4-NO$_2$-
OPE), stimulated dispersed pancreatic acini to secrete amylase to a maximum,
without displaying supramaximal inhibition. Thus, the dose-response curve of
4-NO$_2$-OPE closely resembled that of the previously reported phenylethyl-ester
analog D-Tyr-Gly[Nle^{28,31}]CCK(26–32)-phenylethyl ester, (OPE), and the analog
also had a similar potency (ED$_{50}$ = 2.5 nM) [5]. Binding of ^{125}I-4-NO$_2$-OPE was
inhibited by both CCK(26–33) (IC$_{50}$ = 0.5 nM) and OPE (IC$_{50}$ = 1 nM), suggesting
that this peptide mediated its biological effects via CCK receptors.

Upon photolysis, both ^{125}I-4-NO$_2$-OPE and ^{125}I-6-NO$_2$-Trp^{30}-OPE specifically
labeled a plasmalemmal protein of apparent M$_r$ = 85 000 – 95 000 (Fig. 1a).
Deglycosylation of the M$_r$ = 85 000 – 95 000 labeled by ^{125}I-4-NO$_2$-OPE, using
endoglycosidase F (endo F), generated a major band of M$_r$ = 42 000 (Fig. 1b),
thus confirming the identity of this species as the N-linked glycoprotein previously
labeled with other CCK receptor probes [1–3, 5–8].

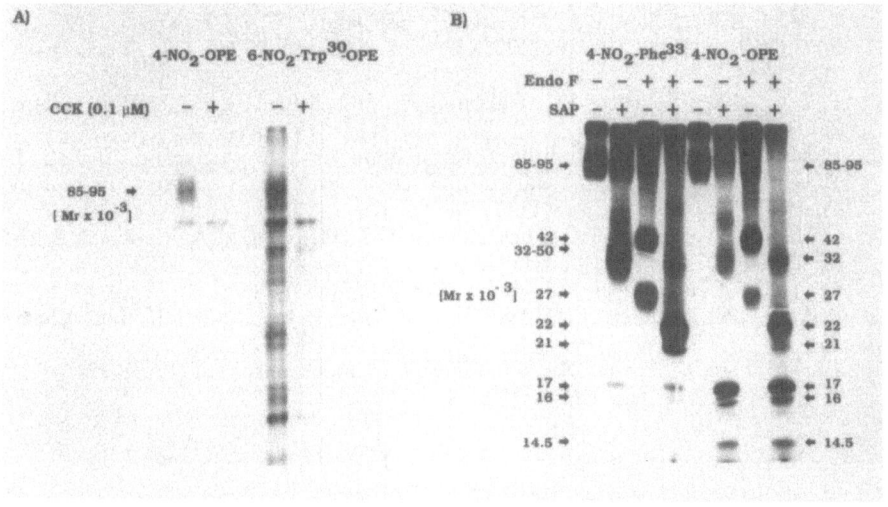

Fig. 1. *(A) 4-NO$_2$-OPE and 6-NO$_2$-Trp^{30}-OPE were used in affinity-labeling studies as
described [2,3]. Both probes specifically labeled a species of apparent M$_r$ = 85 000–95 000.
(B) Following affinity-labeling by 4-NO$_2$-Phe^{33} (left 4 lanes) or 4-NO$_2$-OPE (right 4 lanes)
and resolution on SDS-PAGE, the M$_r$ = 85 000–95 000 was cut from the gels and electroeluted.
The M$_r$ = 85 000 – 95 000 species migrated in the same position on the second gel (lanes
1 and 5). Deglycosylation with endoglycosidase F (Endo F) shifted it to a major M$_r$ = 42 000
(lanes 3 and 7) Staphylococcus aureus V8 protease (SAP) cleavage prior to (lanes 2 and
6) and after deglycosylation (lanes 4 and 8) generated similar fragments after labeling with
both probes.*

In order to elucidate further the labeling generated by ^{125}I-4-NO$_2$-OPE and by ^{125}I-4-NO$_2$-Phe^{33}, we conducted staphylococcus V8 protease peptide-mapping studies prior to and after deglycosylation. Indeed, all fragments labeled by the native CCK analog, ^{125}I-4-NO$_2$-Phe^{33} ($M_r = 32\,000$, $22\,000$, $17\,000$, $16\,000$ and $14\,500$) were also seen after labeling with ^{125}I-4-NO$_2$-OPE (Fig. 2b). Even when the labeled core proteins were digested with trypsin, both probes labeled the same major fragment of $M_r = 8000$ and minor components of $M_r = 14\,000$ and $M_r = 6000$.

These studies provide important evidence that both native CCK and phenylethyl ester analogs mediate their distinctive biological dose-response curves via interaction with the same CCK binding protein and similar domains of this receptor.

Acknowledgements

This work was supported by grants DK32878 and DK34988 from NIH, and by grant KL 607/1-1 from Deutsche Forschungsgemeinschaft (U.G.K.).

References

1. Pearson, R.K., Miller, L.J., Powers, S.P. and Hadac, E.M., Pancreas, 2 (1987) 79.
2. Klueppelberg, U.G., Gaisano, H.Y., Powers, S.P. and Miller, L.J., Biochemistry, 28 (1989) 3463.
3. Powers, S.P., Fourmy, D., Gaisano, H. and Miller, L.J., J. Biol. Chem., 263 (1988) 5295.
4. Lignon, M., Galas, M., Rodriquez, M., Gaillon, G. and Martinez, J., In Bali, J.-P. and Martinez, J. (Eds.) Gastrin and Cholecystokinin: Chemistry, Physiology, Pharmacology, Elsevier, New York, 1987, p. 57.
5. Gaisano, H.Y., Klueppelberg, U.G., Pinon, D.I., Pfenning, M.A., Powers, S.P. and Miller, L.J., J. Clin. Invest., 83 (1989) 321.
6. Pearson, R.K. and Miller, L.J., J. Biol. Chem. 262 (1987) 869.
7. Pearson, R.K., Miller, L.J., Hadac, E.M. and Powers, S.P., J. Biol. Chem., 262 (1987) 13850.
8. Klueppelberg, U., Powers, S.P. and Miller, L.J., Biochemistry, 28 (1989) 7124.

Use of *N*-benzylglycine as a replacement for aromatic amino acid residues. Synthesis of [*N*-benzylglycine⁷]-bradykinin

Janis D. Young[a] and Alexander R. Mitchell[b]

[a]*Department of Biological Chemistry, CHS 33-233, UCLA School of Medicine, Los Angeles, CA 90024-1737, U.S.A.*
[b]*University of California, Lawrence Livermore National Laboratory, Livermore, CA 94550, U.S.A.*

Introduction

N-Benzylglycine (NBzlGly) is an achiral structural isomer of phenylalanine that may be useful as an amino acid replacement in SAR studies. The incorporation of this aromatic secondary amino acid into biologically active peptides has not been reported. The synthesis of Arg-Pro-Pro-Gly-Phe-Ser-NBzlGly-Arg-Phe was undertaken since it is known that replacement of proline⁷ with D-phenylalanine in bradykinin (BK) converts BK agonists into antagonists [1]. We describe the synthesis, characterization, and bioassay of [NBzlGly⁷]-BK.

Results and Discussion

[NBzlGly⁷]-BK was synthesized using SPPS [2] on a Pam-resin [3]. Boc-amino acids were used. Side-chain protection was benzyl for serine and *p*-toluenesulfonyl for arginine. Boc-*N*-benzylglycine was synthesized from *N*-benzylglycine using di-*t*-butyl dicarbonate [4]. The Boc-NBzlGly was incorporated into the peptide using the symmetric anhydride coupling cycle normally used for proline. Double couplings were used throughout the synthesis. Quantitative ninhydrin tests on resin samples taken after each coupling indicated coupling efficiencies of greater than 99%. The peptide-resin was cleaved with anhydrous hydrogen fluoride (0°C, 45 min) containing 10% anisole and 5% dimethylsulfide. The peptide was purified by HPLC using a Waters C-18 Bondapak semipreparative column employing a water and acetonitrile gradient containing 0.1% TFA. The purified peptide was homogeneous by HPLC and gave the expected AAA. FABMS [5] of the peptide verified the expected sequence and molecular weight (1110). The FABMS fragmentation pattern in the Ser⁶-NBzlGly⁷ region of the peptide is given in Fig. 1. [NBzlGly⁷]-BK was bioassayed and is compared with BK, [D-Phe⁷]-BK and [Phe⁷]-BK in Table 1. Bioassays revealed [NBzlGly⁷]-BK to be a potent BK-agonist, especially in the rat blood pressure assay (337% BK potency) following intravenous administration. BK and [NBzlGly⁷]-BK were identical with respect to destruction (98–99%) on passage through the pulmonary circulation.

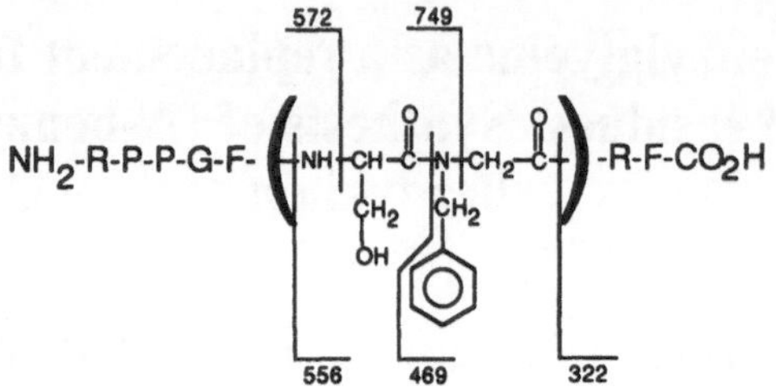

Fig. 1. FABMS fragmentation pattern in the region of Ser⁶-NBzlGly⁷ of [NBzlGly⁷]-BK.

Table 1 *Bioassay of bradykinin and position seven analogs [1]*

Analog	Smooth muscle		Rat blood pressure		
	Rat uterus	GP ileum	IA	IV	Destruction
Bradykinin	100%	100%	100%	100%	99%
[NBzlGly⁷]-Bradykinin	27%	36%	97%	337%	98%
[D-Phe⁷]-Bradykinin	1%	–[a]	2%	4%	36%
[Phe⁷]-Bradykinin[b]	0.6%	0.1%	0	0	ND

[a] Antagonist. $pA_2 = 5.7$.
[b] Unpublished results, R.J. Vavrek.
  Abbreviations: Rat uterus = rat uterus contraction; GP ileum = guinea pig ileum contraction; IA = intraaortic; IV = intravenous; ND = not determined.

Acknowledgements

We thank Terry D. Lee (City of Hope, Duarte, CA) for the FABMS and gratefully acknowledge R.J. Vavrek and J.M. Stewart for the bioassays and informative discussions.

References

1. Vavrek, R.J. and Stewart, J.M., Peptides, 6 (1985) 161.
2. Stewart, J.M. and Young, J.D., Solid Phase Peptide Synthesis, 2nd ed., Pierce Chemical Co., Rockford, IL, 1984.
3. Mitchell, A.R., Erickson, B.W., Ryabtsev, M.N., Hodges, R.S. and Merrifield, R.B., J. Am. Chem. Soc., 98 (1976) 7357.
4. Moroder, L., Hallett, A., Wunsch, E., Keller, O. and Wersin, G., Hoppe-Seyler's Z. Physiol. Chem., 357 (1976) 1651.
5. Lee, T.D., In Shively, J.E. (Ed.) Methods of Protein Microcharacterization, The Humana Press Inc., Clifton, NJ, 1986, p. 403.

Minimum structure bradykinin antagonists

Raymond J. Vavrek and John M. Stewart

Biochemistry Department B126, University of Colorado Medical School,
4200 East 9th Avenue, Denver, CO 80262, U.S.A.

Introduction

The autacoid peptide hormone, bradykinin (BK: Arg-Pro-Pro-Gly-Phe-Ser-Pro-Phe-Arg), plays many vital physiological roles (in cardiovascular control, renal and GI function, reproduction, and wound healing, for example), and is implicated in several significant pathologies (such as pain, inflammation, rhinitis, shock, asthma and allergic responses). The development of specific and competitive BK sequence-related antagonists [1] has provided tools for the study of the physiologically important kallikrein–kinin system. The tissue selectivity of many of these antagonists predicts multiple receptors for BK, each with differing requirements for various segments of the BK molecule. In general, naturally occurring biologically active peptides contain more than the necessary and sufficient information within their structure to elicit a particular physiological or pharmacological response. We report a study of the structure minimization of the BK antagonist sequence (Arg-Pro-Hyp-Gly-Phe-Ser-DPhe-Phe-Arg) [2], using synthetic methods, to define the minimum peptide sequence that can antagonize a classical BK pharmacological action [such as contraction of isolated rat uterus (RUT) or guinea pig ileum (GPI), or lowering of rat blood pressure (RBP)].

Results and Discussion

Removal of single or multiple amino acid residues from either the C- or N-terminal of the antagonist sequence destroys antagonist activity. BK itself suffers a similar loss of biological (agonist) activity with removal of single residues from either end. We demonstrated that replacement of either of the Arg residues at positions 1 or 9 of the antagonist with hydrophobic amino acids [3], or replacement of the Arg at position 1 with hydrophobic acyl groups (i.e., benzoyl-, phenylbutyryl-), with or without single residue deletions at the Pro positions of the antagonist [4], does not eliminate antagonist activity.

Continuing this approach, we acylated antagonist sequences that were truncated from the N-terminal, producing acylated heptapeptide to dipeptide fragments of the antagonist, using acetyl-, phenylacetyl- (PAA), phenylpropionyl- (PPA), phenylbutyryl- (PBA) and adamantylacetyl- (AAA) N-terminal groups. In all of the cases in which antagonist sequence position 7 of the acylated truncated peptide contained a D-Phe residue, that is, with fragments representing positions

7-8-9 (DPhe-Phe-Arg) of the antagonist sequence, BK antagonist activity could be demonstrated. This was especially the case in the GPI assay, in which antagonist activity and potency was relatively constant (pA_2 values of 4.7 – 4.9) in fragments containing from three to seven amino acid residues that were acylated with PBA. Acylated C-terminal dipeptide fragments, corresponding to positions 8 and 9 of both BK and the antagonist (Phe-Arg), had no biological activity.

Preliminary studies indicate that the antagonist specificity of acylated bradykinin antagonist sequences smaller than nine residues may be lost. In addition to antagonizing BK, several hexa- and heptapeptide fragments had the ability to antagonize angiotensin II (AngII) in the RUT assay, and some acylated tetra-, penta-, hexa- and heptapeptide fragments antagonized the action of both AngII and substance P (SP) in the GPI assay. Thus, the full nonapeptide sequence of the BK antagonist is probably required for high antagonist potency and specificity. However, the full antagonist sequence is not required for antagonist activity.

Some acylated fragments, such as AAA-Ser-DPhe-Thi-Arg, PAA-Thi-Ser-DPhe-Thi-Arg, and PBA-Gly-Phe-Ser-DPhe-Phe-Arg (tetra-, penta- and hexa-peptides, respectively) act as antagonists of both BK and SP in the GPI assay, and none antagonize Ang II in either the GPI or RUT assay. Since both BK and SP are involved in peripheral pain signal transmission, this dual antagonist activity may predict therapeutically useful analgesics.

Acknowledgements

Supported by grants from the NHLBI-NIH and from Nova Pharmaceutical Corporation.

References

1. Vavrek, R.J. and Stewart, J.M., Peptides, 6(1985)161.
2. Vavrek, R.J. and Stewart, J.M., In Theodoropoulos, D. (Ed.) Peptides 1986, Walter de Gruyter, Berlin, 1987, p. 655.
3. Vavrek, R.J. and Stewart, J.M., In Bayer, E. and Jung, G. (Eds.) Peptides 1988, Walter de Gruyter, Berlin, 1989, p. 565.
4. Vavrek, R.J. and Stewart, J.M., In Aubry, A., Maurraud, M. and Vitoux, B. (Eds.) Second Forum on Peptides, John Libbey Eurotext, Montrouge, 1989, p. 453.

A selective pseudopeptide analog of bradykinin resistant to carboxypeptidases and angiotensin-converting enzyme

Guy Drapeau, Daniela Jukic, Nour-Eddine Rhaleb, Richard Laprise, Noureddine Rouissi, Stéphane Dion and Domenico Regoli

Department of Pharmacology, Medical School, University of Sherbrooke, Sherbrooke, Quebec, Canada J1H 5N4

Introduction

Bradykinin (BK) and one of its fragments, des-Arg^9-BK, exert potent biological effects in a variety of organs and tissues [1]. Their actions are mediated through two receptor types, B_1 and B_2, that are sometimes present in the same tissue. BK is an agonist of the B_2 receptor, with practically no affinity for the B_1 receptor. However, an interesting feature of the kinin system is that, in several tissues, BK is metabolized by basic carboxypeptidases, like carboxypeptidases N or M into des-Arg^9-BK, a potent and selective B_1 receptor agonist.

Thus, in pharmacological studies, where this type of enzyme may be active, the effects observed with BK may be that of its metabolite or a combined result of the activities of both BK and des-Arg^9-BK on the two receptor types. A specific agonist of the B_2 receptor would, therefore, be helpful in evaluating the role of this receptor in physiological and pathological conditions.

In order to afford such an analog, the bond between Phe^8 and Arg^9 of BK was modified by substituting it with a CH_2-NH pseudopeptide bond to prevent its degradation by carboxypeptidases into des-Arg^9-BK. The analog obtained [Phe$^8\psi$(CH$_2$NH)Arg^9]-BK was tested for biological activity in several pharmacological preparations of both receptor types and was also incubated with enzyme preparations to test its stability.

Results and Discussion

The results summarized in Table 1 on B_2 preparations indicate that the pseudopeptide [Phe$^8\psi$(CH$_2$NH)Arg^9]-BK is more potent than BK in untreated (control) preparations of the rabbit jugular vein (RJV), and the rat blood pressure (RBP) is, however, less potent than BK on the dog carotid artery (DCA) and dog renal artery (DRA). In all tissues treated with captopril, an angiotensin-converting enzyme (ACE) inhibitor, the effect of BK is potentiated significantly, whereas no significant potentiation is observed with the pseudopeptide analog. This is interpreted as indirect evidence that [Phe$^8\psi$(CH$_2$NH)Arg^9]-BK is not degraded by ACE into an inactive fragment. Evaluation on the rabbit aorta (RA), a B_1 receptor preparation, indicates that [Phe$^8\psi$(CH$_2$NH)Arg^9]-BK is

Table 1 *Pharmacological evaluation of BK and [Phe⁸ψ(CH₂NH)Arg⁹]-BK on B_2 and B_1 preparations*

Receptor type	BK				[Phe⁸ψ(CH₂NH)Arg⁹]-BK			
	Control		+ Captopril		Control		+ Captopril	
B_2	pD_2^a	R.A.ª	pD_2	R.A.	pD_2	R.A.	pD_2	R.A.
RJV	7.35	100	8.48	1349*	8.18	676	8.29	870
DCA	8.64	100	9.60	912*	7.48	7	7.46	7
DRA	9.85	100	10.23	240*	9.00	14	8.86	10
RBP		100ᵇ		956*		489		314ᴺˢ
B_1								
RA	6.22	100	6.67	281*	Inactive		Inactive	

ª pD_2 = apparent affinity evaluated from the concentration of agonist that produced 50% of the maximum effect; R.A. = relative affinity in percent of that of BK (BK = 100).
ᵇ Evaluated by i.v. injections and expressed in percent of the hypotensive activity of BK.
* Significant potentiation. RJV = rabbit jugular vein; DCA = dog carotid artery; DRA = dog renal artery; RBP = rat blood pressure; RA = rabbit aorta; NS = not significant.

inactive on this receptor type, while BK is active and its effect is potentiated by captopril. The activity of BK on this tissue has been attributed to the release of its metabolite, des-Arg⁹-BK, by carboxypeptidase N because the effect of BK is blocked by mergetpa, an inhibitor of this enzyme [2].

BK and [Phe⁸ψ(CH₂NH)Arg⁹]-BK were incubated with purified carboxypeptidase B that has an activity similar to that of carboxypeptidase N, and with a lung powder preparation of ACE. Results with carboxypeptidase B (incubation of 1.4×10^{-6} M of BK or [Phe⁸ψ(CH₂NH)Arg⁹]-BK with 0.1 µl carboxypeptidase B (1.1 µg/µl) in 300 µl of Tris(pH 7.6)) show that, in the assay conditions, more than 50% of BK is metabolized in 20 min (complete degradation in 40 min), while [Phe⁸ψ(CH₂NH)Arg⁹]-BK was stable after 5 h of incubation. The ACE preparation degraded BK into BK (1–7) with no marked metabolism of the pseudopeptide analog.

References

1. Regoli, D. and Barabé, J., Pharmacol. Rev., 32 (1980) 1.
2. Regoli, D., Drapeau, G., Rovero, P., Dion, S., Rhaleb, N.E., Barabé, J., D'Orléans-Juste, P. and Ward, P., Eur. J. Pharmacol., 127 (1986) 219.

1,4-Piperazine-derived, partially nonpeptidic analogs of substance P

E. Roubini[a], C. Gilon[b], Z. Selinger[c] and M. Chorev[a]

[a]Department of Pharmaceutical, [b]Department of Organic and [c]Department of Biological Chemistry, The Hebrew University of Jerusalem, Jerusalem 91120, Israel

Introduction

The design of nonpeptidic analogs of bioactive peptides is anticipated to lead to compounds with improved bioavailability, prolonged duration of action and high selectivity. This approach may also lead to the development of highly potent and selective antagonists. We have previously outlined our approach to the design of nonpeptide peptidomimetics and demonstrated it in the synthesis of potent, partially nonpeptidic analogs of substance P (SP), based on 4-hydroxyproline [1]. The 1,4-piperazine system presented in this work represents a larger ring system, as compared to the proline ring, with a greater potential for conformational flexibility. Substitution of 1,4-piperazine by the essential SP pharmacophores resulted in structure I that was examined by CPK models. Only the $(2S,5S)$-1,4-piperazine isomer allowed close reproduction of the topological model obtained for one of the reported low energy conformations of SP [2].

Results and Discussion

The synthetic route developed for the preparation of analog I and II allowed the selection of substituents and chirality at C_2 and C_5 in the 1,4-piperazine ring, and also provided the options either to alkylate or to acylate the heterocyclic nitrogens (Scheme 1). The key intermediate in the synthesis of analogs I and II, 2-keto-3,6-dibenzyl-1,4-piperazine derivative 2, was obtained by cyclization of the pseudodipeptide 1. The biological activities of SP analogs I and II in the guinea pig ileum (GPI) and rat vas deferens (RVD) are summarized in Table 1. The reduced potency of the piperazine analogs, despite their close resemblance to the topological model of SP, may be due to excessive conformational constraints induced by the substituted piperazine ring.

Table 1 *Biological activities of SP analogs in GPI and RVD*

Analog	GPI EC_{50} (μM)	RVD EC_{50} (μM)
I	7.5	>>20.0
II	20.0	>>20.0
[pGlu^6]SP$_{6-11}$	1.2×10^{-3}	20.0

161

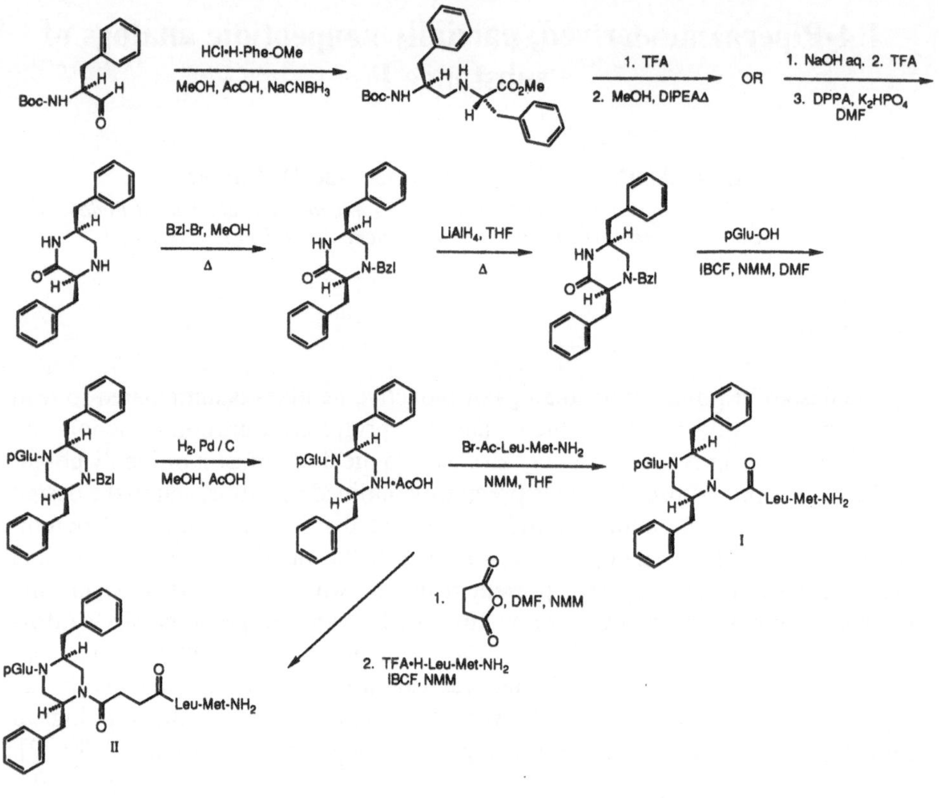

Scheme 1. *Synthesis of 1,4-piperazine analogs of SP.*

Conclusion

Biologically relevant topology can be derived from modeling of low-energy conformations. Small polyfunctional, stereochemically defined heterocyclic rings could be employed for construction of a topological model. The development of these important non-peptidic leads into compounds with higher affinity for the SP receptors is currently being examined in our laboratory.

References

1. Roubini, E., Wormser, U., Levian-Teitelbaum, D., Laufer, R., Gilon, C., Selinger, Z. and Chorev, M., In Hruby, V.H., Kopple, K.D. and Deber, C.M. (Eds.) Peptides: Structure and Function, Pierce Chem. Co., Rockford, IL, 1984, p. 635.
2. Manavalan, P. and Momany, F., Int. J. Pept. Prot. Res., 20 (1982) 3513.

Second generation of potent parathyroid hormone (PTH) antagonists

M. Chorev*, M.E. Goldman, E. Roubini, J.E. Reagan, J.J. Levy, R.F. Nutt, M.P. Caulfield and M. Rosenblatt

Merck Sharp & Dohme Research Laboratories, West Point, PA 19486, U.S.A.

Introduction

Parathyroid hormone (PTH) antagonists are valuable research tools in studying the role of PTH (84 amino acid residues) and parathyroid hormone-related protein (PTHrP, 141 amino acid residues) as calciotrophic hormones, and as potential drugs in treatment of calcium disorders. Sequence homology of both hormones is limited to their amino terminus, in which 8 out of 13 amino acid residues are identical. Recently, we have identified position 12 in both hormones as a site relatively tolerant of structural manipulation [1]. Replacement of Gly[12] with D-Trp led to a 10-fold increase in potency compared to that of the corresponding parent antagonists (see Table 1) [2]. We report here our recent efforts toward the preparation of highly potent PTH antagonists.

Table 1 *Binding and cyclase activity of antagonist analogs derived from PTH and PTHrP sequences using bovine renal cortical membranes[a]*

	Peptide analog	Binding[b] K_b (nM)	Cyclase[c] K_i (nM)
1.	[Nle^{8,18},Tyr^{34}]bPTH(7–34)-NH$_2$	145 ± 10	$1\,600 \pm 300$
2.	[Nle^{8,18},D-Trp^{12},Tyr^{34}]bPTH(7–34)-NH$_2$	15 ± 5	125 ± 10
3.	[Nle^{8},D-Trp^{12,18},Tyr^{34}]bPTH(7–30)-NH$_2$	4 ± 1	30 ± 5
4.	[Nle^{8},D-Trp^{18},Tyr^{34}]bPTH(7–34)-NH$_2$	200 ± 5	700 ± 180
5.	[Nle^{8},L-Trp^{18},Tyr^{34}]bPTH(7–34)-NH$_2$	45 ± 10	420 ± 30
6.	[Nle^{8},D-Ala^{18},Tyr^{34}]bPTH(7–34)-NH$_2$	870 ± 460	640 ± 25
7.	[Nle^{8,18},D-Trp^{12},Lys^{13}(ϵ-X),Tyr^{34}]bPTH(7–34)-NH$_2$[d]	4 ± 1	60 ± 10
8.	[Nle^{8,18},D-Trp^{12},Lys^{13}(ϵ-Y),Tyr^{34}]bPTH(7–34)-NH$_2$[e]	35 ± 5	230 ± 25
9.	[D-Trp^{12}]PTHrP(7–34)-NH$_2$	50 ± 15	390 ± 180
10.	[Nle^{8,18},D-Trp^{12}]bPTH(7–18)PTHrP(19–34)-NH$_2$	190 ± 30	320 ± 60
11.	[D-Trp^{12}]PTHrP(7–18)[Tyr^{34}]bPTH(19–34)NH$_2$	70 ± 15	350 ± 40

[a] Values are mean ($\pm$ SEM) from at least three experiments.
[b] Inhibition of binding of 25 pM of [Nle^{8,18},mono-^{125}I-Tyr^{34}]bPTH(1–34)-NH$_2$.
[c] Antagonizing 3 nM of [Nle^{8,18},Tyr^{34}]bPTH(1–34)-NH$_2$.
[d] $X \equiv CO(CH_2)_2Ph$.
[e] $Y \equiv N,N$-[CH$_2$CH(CH$_3$)$_2$]$_2$.

*Present address: Department of Pharmaceutical Chemistry, School of Pharmacy, The Hebrew University of Jerusalem, Jerusalem 91120, Israel.

Results and Discussion

Peptides were synthesized on a *p*-MBHA resin, purified and characterized as described previously [3]. Modification of ϵ-amino Lys[13] (as in analogs **7** and **8**) was achieved by coupling N^α-Boc-N^ϵ-Fmoc-Lys to the corresponding resin-bound-peptide followed by deprotection with piperidine. Acylation with DCC-mediated preformed hydrocinnamic anhydride and reductive alkylation in the presence of $NaBH_3CN$ and isobutyraldehyde yielded the intermediates, which were then extended to the corresponding analogs **7** and **8**, respectively. Table 1 summarizes the binding affinities and inhibition of PTH-stimulated adenylate cyclase activities, as determined in the bovine renal binding assay. Enhancement of binding affinity was achieved by combining the effect of D-Trp[12] with additional hydrophobic substitutions at positions 13 and 18, sites identified as tolerant of a wide range of structural modifications. The increase in potency due to hydrophobic substitution in position 18 was observed only in combination of D-Trp[12] (cf. **3** and **4–6**). In a similar manner, introduction of a hydrophobic moiety on the ϵ-NH_2 of Lys[13] contributed to the enhanced potency of analogs **7** and **8**. It seems that the *N,N*-diisobutyl substitution may exceed the spacial limitations at this site of interaction. The hybrids of PTH and PTHrP, analogs **10** and **11**, demonstrate that the C-terminal domains of these functionally related analogs of the calciotrophic hormones could be interchanged and still retain potency. This finding may indicate that these structurally different C-terminal domains interact in a similar manner with either one of the N-terminal domains of the hormone, or with the receptor. This concept is currently being explored in our laboratory in studies toward the design of more potent PTH antagonists.

References

1. Chorev, M., Goldman, M.E., Caporale, L.H., Levy, J.J., Reagan, J.E., DeHaven, P., Gay, T., Nutt, R.F. and Rosenblatt, M., In Shiba, T. and Sakakibara, S. (Eds.) Peptide Chemistry 1987, Protein Research Foundation, Osaka, 1988, p. 621.
2. Goldman, M. E., McKee, R. L., Caulfield, M. P., Reagan, J. E., Levy, J. J., DeHaven, P. A., Rosenblatt, M. and Chorev, M., Endocrinology, 123 (1988) 2597.
3. Caporale, L., Nutt, R., Levy, J., Smith, J., Arison, B., Bennett, C., Albers-Schonberg, G., Pitzenberger, S., Rosenblatt, M. and Hirschmann, R., J. Am. Chem. Soc., 54 (1989) 343.

Structure study of human calcitonin

Masamitsu Doi[a], Yuji Kobayashi[a], Yoshimasa Kyogoku[a], Misato Takimoto[b] and Keigo Goda[b]

[a]*Institute for Protein Research, Osaka University, Suita, Osaka 565, Japan*
[b]*International Research Laboratories, CIBA-GEIGY(Japan)Ltd., Takarazuka, Hyogo 565, Japan*

Introduction

Calcitonin (CT) is well known as a key regulatory hormone in calcium metabolism. Since the idea of an amphiphilic helix was proposed, this peptide has been a target of conformational analyses by CD measurements and modeling. Kaiser et al. [1] designed a model of salmon CT to optimize an amphiphilic structure and demonstrated that this model peptide has a helical structure in 50% TFE. Salmon CT also has a helical structure in the presence of phospholipid as Epand et al. [2] found.

We have worked to determine conformations of polypeptides in solution by the combined use of NMR measurements and distance geometry algorithms [3]. Here, we applied our technique to analyzing the structure of human CT (hCT) in a mixture of water and TFE.

Results and Discussion

The structural study of hCT was performed as follows. 2D NMR spectra, such as DQF-COSY, HOHAHA and NOESY, were measured in TFE-d_3/D_2O or H_2O (2:3, v/v). In these spectra, the peaks were assigned by sequential assignment strategy [4].

Then, the NOEs determined from NOESY spectra, were interpreted as distances for use in distance geometry calculations as the constraints [3,5]. The distance geometry algorithm consists of constructing a randomly chosen initial conformation and minimizing the sum of the square root of differences between atomic distance in the calculated structure and the value of the corresponding distances obtained from NMR experiments.

Eight resulting structures with the lowest values of the sum, superimposed to obtain the best fit for minimum root mean square deviation (RMSD) are shown in Fig. 1. These computer drawings show that the center region from Leu[9] to Phe[22] is in a helical conformation. This result confirms that hCT assumes a helix structure as predicted for salmon CT and its analog [1,2].

Figure 2 indicates how this helix differs from a typical α-helix. The arrows show the deviation of the individual residue positions from the ideal positions of the corresponding residues in an α-helix. The deviations are observed in the

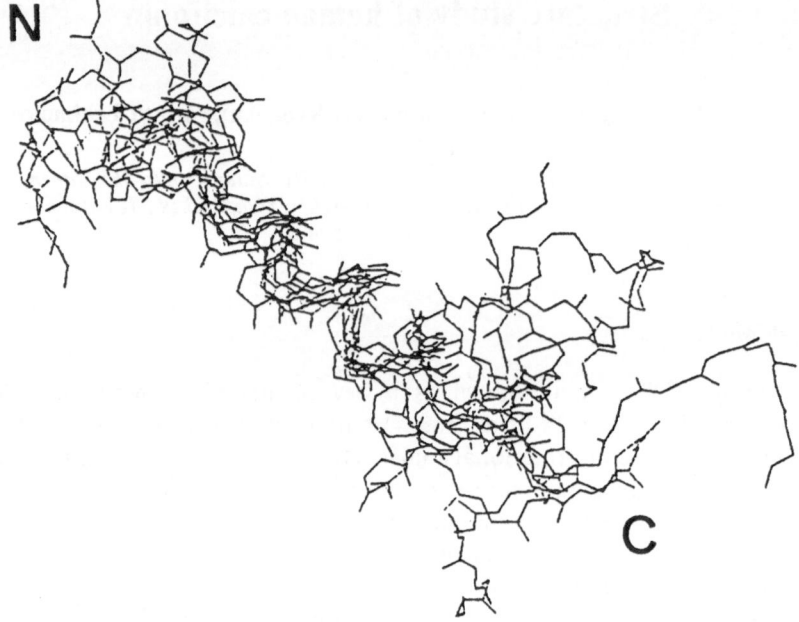

Fig. 1. Computer drawings of eight structures for hCT superimposed to obtain the best fit for minimum RMSD. All covalent bonds between main-chain backbone atoms are shown.

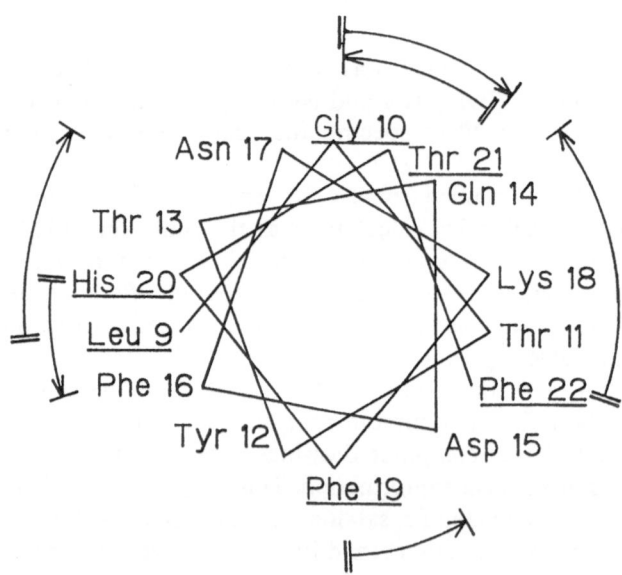

Fig. 2. Helical wheel projection of the Leu^9-Phe^{22} region.

regions from Leu[9] to Gly[10] and from Phe[19] to Phe[22]. Therefore, the region from Thr[11] to Lys[18] is found to be in an α-helix but its neighboring regions at both ends are in an extended helical conformation. In addition, it has been revealed that the region from Leu[9] to Thr[21] has an amphiphilic arrangement of the side chains.

References

1. Moe, G.R., Miller, R.J. and Kaiser, E.T., J. Am. Chem. Soc., 105 (1983) 4100.
2. Epand, R.M. and Epand, R.F., Biochemistry, 25 (1986) 1964.
3. Ohkubo, T., Kobayashi, Y., Shimonishi, Y., Kyogoku, Y., Braun, W. and Gō, N., Biopolymers, 25 (1986) s123.
4. Wüthrich, K., Wilder, G., Wagner, G. and Braun, W., J. Mol. Biol., 155 (1982) 311.
5. Braun, W. and Gō, N., J. Mol. Biol., 186 (1986) 611.

Irreversible ligands for bombesin receptors

Roberto de Castiglione, Luigia Gozzini, Mauro Galantino, Fabio Corradi, Marina Ciomei and Isabella Molinari

Farmitalia Carlo Erba R & D, Erbamont Group, Via dei Gracchi 35, I-20146 Milan, Italy

Introduction

Peptides of the bombesin (BN)/GRP family act as autocrine growth factors for small cell lung carcinoma (SCLC). In the search of BN antagonists as possible therapeutic agents for this kind of tumor, a number of C-terminal BN alkylating analogs have been synthesized and evaluated in Swiss 3T3 fibroblasts for receptor binding affinity and for mitogenic activity, alone and in competition with BN.

Results and Discussion

The Melphalan moiety has been introduced at different positions of the NH_2-terminus of BN analogs. Its location seems to affect the binding affinity of the compounds, position 6 being more favorable than position 3. When the alkylating moiety has been introduced into BN structures showing good receptor affinity and agonistic activity, a slight increase in IC_{50} was observed as compared to the reference compound. Vice versa, the introduction of the same moiety in peptides which display poor receptor binding (and are biologically inactive or weak antagonists), induces at least a 10-fold increase in binding affinity.

Usually, in competition experiments, BN alkylating analogs with intrinsic 'antagonistic' features display comparable antagonism to 25 nM BN, both when given at the same time as or 24 h before the agonist challenge, whereas analogs with intrinsic 'agonistic' features display better antagonism in the sequential than in the contemporaneous treatment.

The number of GRP receptors per cell, 24 h after treatment, is more greatly reduced (up to complete suppression) by these alkylating analogs than by other BN-like peptides, and, in the second challenge, the percent inhibition of thymidine incorporation is inversely related to the number of residual receptors.

The alkylating moiety Melphalan does not bind by itself to GRP receptors in Swiss 3T3 cells, while the alkylating analogs do not bind to other growth factor receptors, such as the EGF receptor in A431 cells. In addition, on a molar basis, the toxicity of these alkylating analogs ($IC_{50} > 50$ μM) is much lower than that of the cytotoxic Melphalan moiety (IC_{50} 3.3 μM).

Peptides containing the alkylating moiety in *meta* position apparently are more cytotoxic than the same peptides carrying the alkylating moiety in position *para*, as in the classical Melphalan.

Probably due to higher chemical and enzymic stability, the N-terminal protected

Table 1 *Effects of selected BN analogs on mouse Swiss 3T3 fibroblasts*

Formula 1 2 3 4 5 6 7 8 9 10 11 12 13 14	A	B	C	D
Glp-Gln-Arg-Leu-Gly-Asn-Gln-Trp-Ala-Val-Gly-His-Leu-Met-NH₂ (bombesin)	12.6 ± 0.6	6.6 ± 2.6*	N.T.	N.T.
Boc-Thr————OH	> 100 000	1.0	N.T.	N.T.
Boc-Thr————	10.7 ± 0.1	6.8**	N.T.	N.T.
Boc-Thr————[box]	1000 ± 100	1.1	53 ± 14	N.T.
Boc-pMel————OH	12 000 ± 400	1.3	64 ± 10	39 ± 7
Boc-pMel————	48 ± 2	6.6	20 ± 3	34 ± 5
H-pMel————	60 ± 14	4.8	17 ± 4	61 ± 9
Boc-pMel————[box]	69 ± 5	1.0	85 ± 2	61 ± 5
H-pMel————[box]	406 ± 49	1.4	44 ± 12	19 ± 11
Boc-pMel————His–Dnp	1170 ± 280	2.3	67 ± 3	68 ± 5
H-pMel————His–Dnp	680 ± 105	1.8	44 ± 12	83 ± 6
Boc-pMel————phe	5240 ± 1809	1.2	37 ± 4	77 ± 11
Boc-mMel-Leu-Gly-Thr————His–Dnp————NH(CH₂)CH₃	731 ± 67	1.0	82 ± 1	90 ± 3

*Value obtained at 25 nM concentration; **Value obtained at 0.5 μM concentration.

A = Inhibition of [^{125}I]GRP binding [IC$_{50}$(nM)]; B = Fold increased of [^{3}H]thymidine incorporation over basal value at 5 μM concentration; C, D = % inhibition of [^{3}H]thymidine incorporation induced by 25 nM BN given at the same time as (C) or 24 h after (D) the challenge with the BN analog.

Identical residues are represented by solid lines, while residues of D-amino acids or deleted amino acids represented by small letters and open boxes, respectively.

Mel = [bis(2-chloroethyl)amino]-L-phenylalanine; N.T. = not tested.

derivatives are more potent antagonists than the unprotected ones. Preliminary results with selected BN alkylating analogs indicate that these compounds are also able to inhibit the growth of SCLC cell lines in soft agar.

Antagonist of gastrin-releasing peptide

Hidehito Mukai[a], Koichi Kawai[b], Seiji Suzuki[b], Kamejiro Yamashita[b] and Eisuke Munekata[a]

[a]Institute of Applied Biochemistry and [b]Institute of Clinical Medicine, University of Tsukuba, Tsukuba, Ibaraki 305, Japan

Introduction

Gastrin-releasing peptide (GRP), its C-terminal decapeptide (GRP-10), and neuromedin B (NMB) are bombesin-like neuropeptides isolated from mammals

GRP-10:   Gly-Asn-His-Trp-Ala-Val-Gly-His-Leu-Met-NH$_2$
NMB: Gly-Asn-Leu-Trp-Ala-Thr-Gly-His-Phe-Met-NH$_2$

and are considered to be involved in the nervous regulation systems as neurotransmitters or neuromodulators, although the actual physiological functions are not yet established. The peptides of the bombesin family have multiple biological activities and are known to stimulate the contraction of smooth muscles, the release of gastropancreatic hormones such as gastrin, insulin and glucagon, the secretion of juice from pancreatic acini and the proliferation of several cell lines. Generally, antagonistic substances play a significant role in the study of the detailed functions of biologically active substances, such as atropine against acetylcholine, and naloxone against opiates.

Recently, we found [1] that GRP-10 analogs, in which the amino acid residue at position 6 is replaced by Ala or Gly, antagonize the insulin secretion induced by GRP-10, based on SAR.

Methods

Synthetic peptides were dissolved in saline solution containing 0.2% bovine serum albumin. In in vivo study, the peptide solution was injected into dogs, and insulin, glucagon and gastrin levels in the plasma were studied as previously described [2]. In in vitro study, the pancreas was isolated from dogs, and perfused by the method of Iversen et al. [3]. Insulin (IRI), glucagon (IRG) and gastrin levels in the plasma and the effluent perfusate were determined by RIA.

Results and Discussion

The prior injection of [Ala^6]-GRP-10 into conscious dogs in doses of 45, 225 and 450 nmol/kg decreased cumulation increase in the plasma insulin level, over a 10-min period following the injection of 4.5 nmol/kg GRP-10, to 67.4,

49.3 and 22.1%, respectively, as shown in Fig. 1. Prior injection at doses of 4.5, 45, 225 and 450 nmol/kg reduced the plasma glucagon level to 94.0, 43.0, 48.3, 47.6 and 58.6%. Plasma gastrin levels were 94.0, 155.4 and 167% following the injection of 45, 225 and 450 nmol/kg as shown in Fig. 2.

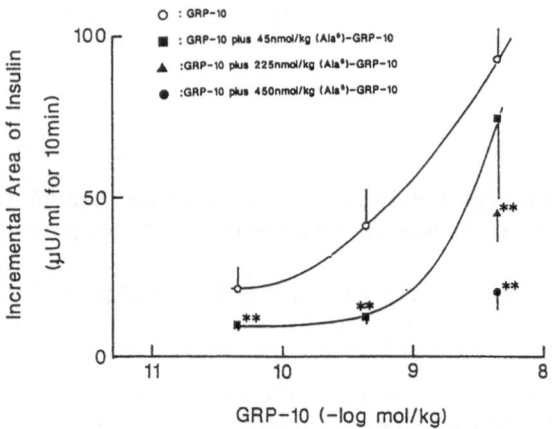

Fig. 1.   Dose–response curves of stimulation of insulin release by GRP-10 and its inhibition by various doses of [Ala⁶]-GRP-10.

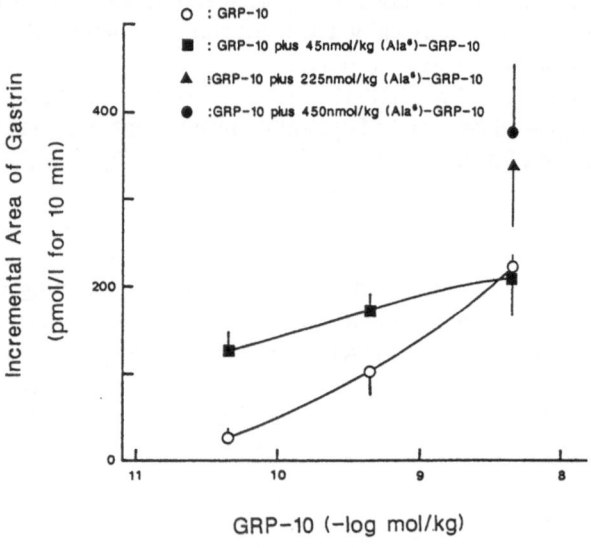

Fig. 2.   Dose–response curves of stimulation of gastrin release by GRP-10 and its enhancement by [Ala⁶]-GRP-10.

In the canine pancreas perfusion experiments, insulin release stimulated by 1 nM of GRP-10 was reduced to 85 and 33% by the infusion of 10 and 100 nM of [Ala^6]-GRP-10, respectively.

[D-Arg^1, D-Pro^2, D-Trp^{7,9}, Leu^{11}]-substance P, a synthetic antagonist against bombesin-like peptides, suppressed insulin release with 1 nM of GRP-10 to 65.9 and 24.1% by the infusion of 100 nM and 1 μM. [D-Phe^8]-GRP-10 is also known as an antagonist of the bombesin family and reduced insulin release to 97.7, and 15.4% by infusion of 1 and 10 μM. Insulin release was almost completely suppressed by infusion of 100 nM of [Gly^6]-GRP-10.

The results are summarized as follows:

(1) [Ala^6]-GRP-10 and [Gly^6]-GRP-10 are proposed to be antagonists for the release of insulin and glucagon stimulated by GRP-10.

(2) Antagonistic activities of both peptides are approximately 10-fold that of [D-Arg^1, D-Pro^2, D-Trp^{7,9}, Leu^{11}]-substance P and 100-fold that of [D-Phe^8]-GRP-10.

(3) The different features of each peptide in biological activity suggest that different receptors are involved in the secretion of insulin and gastrin.

Acknowledgements

This study was supported by a Grant-in-Aid for Scientific Research from the Ministry of Education, Science and Culture, Japan and Ajinomoto Co. Ltd.

References

1. Mukai, H., Kawai, K., Suzuki, S., Yamashita, K. and Munekata, E., Peptide Chemistry 1988, Protein Research Foundation, Osaka, 1989, p. 61.
2. Mukai, H., Kawai, K., Yamashita, K. and Munekata, E., Peptide Chemistry 1987, Protein Research Foundation, Osaka, 1988, p. 463.
3. Iversen, J. and Miles, D.W., Diabetes, 20 (1971) 1.

ICI 216140 and other potent in vivo antagonist analogs of bombesin/gastrin-releasing peptide

R. Camble, R. Cotton, A.S. Dutta, A. Garner, C.F. Hayward, V.E. Moore and
P.B. Scholes

ICI Pharmaceuticals, Alderley Park, Macclesfield, Cheshire SK10 4TG, U.K.

Introduction

Bombesin (BN) and gastrin-releasing peptide (GRP) possess a number of chemical and biological similarities. In addition to various other effects, both BN and GRP have been identified as potent mitogens in Swiss 3T3 cells [1], and GRP has also been proposed as an autocrine growth factor in small cell lung cancer (SCLC) [2]. The antagonists of BN/GRP may, therefore, have therapeutic utility in the treatment of SCLC.

Our attempts to obtain BN/GRP antagonists we based on two approaches: (a) synthesis of truncated and side-chain deletion analogs, and (b), synthesis of analogs based on a substance P antagonist Z-Arg-Pro-Lys(Z)-Pro-Gln-Gln-Phe-Phe-Gly-Leu-OMe [3]. Incorporation of the changes which led to substance P antagonism into the C-terminal decapeptide of GRP gave a potent BN/GRP antagonist (compound 1, Table 1). Further modifications were then attempted to obtain potent in vivo antagonists of BN/GRP with longer duration of action.

Methods

Synthesis
The analogs listed in Table 1 were synthesized by SPPS, purified by RPHPLC, and characterized by AAA and FABMS.

Biological tests
In the in vitro mitogenic test (Table 1), the incorporation of ^{3}H-thymidine into Swiss 3T3 cells was measured over a period of 48 h at 37°C. In the in vivo test (Table 2), the inhibition of bombesin-stimulated amylase secretion was measured in rats after i.v. and s.c. administration of the compounds [3].

Results and Discussion

Modifications of compound 1 by introducing D-amino acid residues in various positions resulted in improved potency (2, 3, 8) in the in vitro test system (Table 1). Although 1–8 were potent antagonists of BN in vitro and in vivo (given i.v.), none of these inhibited bombesin-stimulated amylase secretion in rats when

Table 1 *Chemical structures and in vitro (inhibition of mitogenesis using Swiss 3T3 cells) activities on BN/GRP antagonists*

Compound	Structure	IC_{50} nM
1	Z-Arg-Pro-Lys (Z) -His-Trp-Ala-Val-Gly-His-Leu-OMe	2.3
2	Z-Arg-Pro-Lys (Z) -His-Trp-Ala-Val-D-Ala-His-Leu-OMe	0.7
3	Z-Arg-Pro-Lys (Z) -D-Gln-Trp-Ala-Val-D-Ala-His-Leu-OMe	0.2
4	Boc-D-Deh-Pro-Lys (Z) -His-Trp-Ala-Val-D-Ala-His-Leu-OMe	0.6
5	Ac-Pro-Lys (Z) -His-Trp-Ala-Val-D-Ala-His-Leu-OMe	0.9
6	Ac-Gly-Asn——His-Trp-Ala-Val-D-Ala-His-Leu-OMe	6.4
7	Boc-D-Arg——His-Trp-Ala-Val-D-Ala-His-Leu-OMe	10.0
8	Ac-D-Phe(Cl)-His-Trp-Ala-Val-D-Ala-His-Leu-OMe	0.3
9	Ac-Phe——His-Trp-Ala-Val-D-Ala-His-Leu-OMe	1.0
10	-OCH₂CO——His-Trp-Ala-Val-D-Ala-His-Leu-OMe	1.1
11	-CH₂CO——His-Trp-Ala-Val-D-Ala-His-Leu-OMe	0.9
12	CH₃CO——His-Trp-Ala-Val-D-Ala-His-Leu-OMe	3.0
13	CH₃CH₂CO——His-Trp-Ala-Val-D-Ala-His-Leu-OMe	0.5
14	(CH₃)₂CHCO——His-Trp-Ala-Val-D-Ala-His-Leu-NHMe	3.3
15	CH₃CH₂CO——His-Trp-Ala-Val-D-Ala-His-MeLeu-OMe	3.3

Table 2 *Inhibition of BN-stimulated amylase secretion in rats by the antagonists*

Compound	Dose (mg/kg)	Route of administration	Time of administration (min. before BN)	% inhibition
1	0.5	i.v.	0	86
	20.0	s.c.	60	61
3	0.5	i.v.	0	73
	20.0	s.c.	30	5
4	0.5	i.v.	0	92
	20.0	s.c.	30	17
5	0.5	i.v.	0	91
	20.0	s.c.	30	14
6	0.5	i.v.	0	93
	20.0	s.c.	150	Inactive
7	0.5	i.v.	0	86
	20.0	s.c.	30	13
8	0.5	i.v.	0	80
	20.0	s.c.	150	Inactive
9	0.25	i.v.	0	86
	2.0	s.c.	30	83
10	0.5	i.v.	0	92
	20.0	s.c.	30	26
11	0.25	i.v.	0	95
	0.02	i.v.	0	61
	10.0	s.c.	150	49
12	0.5	i.v.	0	92
	2.0	s.c.	150	6
13	2.0	s.c.	60	96
	2.0	s.c.	150	51
14	2.0	s.c.	150	100
	0.1	s.c.	150	58
15	2.0	s.c.	150	94
	0.02	s.c.	150	45

administered subcutaneously. The heptapeptide derivatives (**9** and **11**) showed modest inhibition, whereas **14** (ICI 216140) and **15** (ICI 216167) were potent inhibitors of amylase secretion when given subcutaneously. Both of these also displayed prolonged duration of action (>3 h). Results on other modifications leading to further improvements in potency and duration of action will be reported elsewhere.

References

1. Zachary, I. and Rozengurt, E., Proc. Natl. Acad. Sci. U.S.A., 82 (1985) 7616.
2. Sporn, M.B. and Roberts, A.B., Nature, 313 (1985) 745.
3. Camble, R., Cotton, R., Dutta, A.S., Garner, A., Hayward, C.F., Moore, V.A. and Scholes, P.B., Life Sci., 45 (1989) 1521.

A structure-activity study of NK-2 selective tachykinin antagonists

Paolo Rovero[a], Vittorio Pestellini[a], Riccardo Patacchini[b], Sandro Giuliani[b], Carlo A. Maggi[b], Alberto Meli[b] and Antonio Giachetti[b]

[a]Peptide Synthesis Laboratory, Chemistry Dept. and [b]Smooth Muscle Division, Pharmacology Dept., A. Menarini Pharmaceuticals, via Sette Santi 3, I-50131 Firenze, Italy

Introduction

Several antagonists of Substance P (SP) containing two or more D-Trp residues in their sequence have been reported [1,2]. Most of these antagonists show limited selectivity for the three tachykinins receptors [3], probably because they were designed as modifications of the SP backbone. In fact, although SP has some preferential activity for the NK-1 receptors, it is active as an agonist at all of the three tachykinin receptors. We attempted to obtain selective NK-2 antagonists by introduction of D-Trp on the backbone of a NK-2 selective agonist, the C-terminal heptapeptide NKA (4–10).

Results and Discussion

The results presented in Table 1 indicate that, by introducing D-Trp in positions 6 and 8 along with pyroglutamic acid pGlu in position 4 or Nle in position 10 of NKA (4–10), selective, although weak, NK-2 tachykinin antagonists are obtained. Similar substitutions, previously reported on the sequence of SP, gave rise to nonselective antagonists, presumably for the limited selectivity of the agonist used as template. Peptides 3 and 4 are interesting examples of selective antagonists developed starting from a selective agonist. Further modifications, like the addition of Phe in position 10 or of a third D-Trp in position 9, gave rise to more potent but less selective antagonists, thus showing that each amino acid substitution can dramatically affect selectivity.

Further studies are necessary for the development of peptides showing enhanced potency along with improved selectivity.

Acknowledgements

This work was supported in part by IMI, Rome (Grant 46287).

Table 1 Biological activity of NKA (4–10) analogs on the guinea pig ileum (GPI), rat vas deferens (RVD) and rat portal vein (RPV)

No.	Sequence	GPI (NK-1)		RVD (NK-2)		RPV (NK-3)	
		pD_2	pA_2	pD_2	pA_2	pD_2	pA_2
1	H-Asp-Ser-Phe-Val-Gly-Leu-Met-NH_2	7.43 ± 0.11		5.98 ± 0.15		6.79 ± 0.05	
2	H-Asp-Ser-DTrp-Val-DTrp-Leu-Met-NH_2	5.06 ± 0.19		n. a.	4.90 ± 0.17	5.58 ± 0.21	
3	pGlu-Ser-DTrp-Val-DTrp-Leu-Met-NH_2	n.a.	n.a.	n. a.	4.83 ± 0.25	<5	
4	H-Asp-Ser-DTrp-Val-DTrp-Leu-Nle-NH_2	n.a.	n.a.	n. a.	4.32 ± 0.19	<5	
5	H-Asp-Ala-DTrp-Val-DTrp-Leu-Met-NH_2	<5		n. a.	4.88 ± 0.20	n.a.	5.56 ± 0.10
6	H-Asp-Ser-DTrp-Val-DTrp-Leu-Phe-NH_2	n.a.		n. a.	5.15 ± 0.23	n.a.	5.20 ± 0.09
7	H-Asp-Ser-DTrp-Val-DTrp-DTrp-Phe-NH_2	n.a.	6.34 ± 0.28	n.a.	5.93 ± 0.29	n.a.	n.a.

n.a. = no activity at 10 nM.
Each value mean ± SE of at least 4 determinations.

References

1. Folkers, K., Horig, J., Rampold, G., Lane, P., Rosell, S. and Bjorkroth, U., Acta Chem. Scand., B 36 (1982) 389.
2. Regoli, D., Escher, E. and Mizrahi, J., Pharmacology, 28 (1984) 301.
3. Buck, S. and Shatzer, S., Life Sci., 42 (1988) 2701.

Synthesis and biological activity of $[\psi(CH_2NH)]$ analogs of neurokinin A_{4-10}

S.L. Harbeson, S.H. Buck, C.F. Hassmann III and S.A. Shatzer

Merrell Dow Research Institute, 2110 E. Galbraith Road, Cincinnati, OH 45215, U.S.A.

Introduction

The tachykinins are a family of peptides that share the C-terminal sequence: Phe-Xxx-Gly-Leu-Met-NH_2. In mammals, Substance P and the neurokinins have been identified in the central and peripheral nervous system and have shown spasmogenic activity in smooth muscle tissue. In view of Coy's discovery that $[\psi(CH_2NH)]$ analogs of bombesin ($<$QQRLGNQWAVGHLM-NH_2) are bombesin antagonists [1], we have synthesized and tested a series of Neurokinin A (NKA, HKTDSFVGLM-NH_2) analogs in which the amide bonds of [Leu[10]]-NKA_{4-10} have been sequentially replaced with the reduced amide $[\psi(CH_2NH)]$ bond.

Results and Discussion

The peptides shown in Table 1 were prepared according to the method of Coy, et al. [1] using SPPS with Boc protection. Compounds I, II, IV and V were prepared by reductive alkylation of resin-bound peptides with N-t-Boc-aminoaldehydes prepared according to the method of Fehrentz and Castro [2]. Compound II was made using N-t-Boc-glycinal synthesized from N-t-Boc-allylamine [3]. Reductive alkylation of the resin-bound peptide with succinic semialdehyde provided compound VI. Compound VI is, therefore, a $[\psi(CH_2NH)]$ analog of succinoyl-NKA_{5-10}. The peptides were deprotected and cleaved from the resin with HF/anisole and purified by semi-prep HPLC on a Vydac C_{18} column (22×250 nm). The final products were characterized by FABMS and AAA. Purity was determined by analytical RPHPLC in two different solvent systems.

The receptor affinity of the analogs was determined by competitive binding experiments in receptor-specific tissues: hamster bladder (NK-2), rat salivary gland (NK-1) and rat cerebral cortex (NK-3) [4]. The in vitro data (Table 1) show that replacement of the amide bonds in [Leu[10]]-NKA_{4-10} with the $[\psi(CH_2NH)]$ moiety results in a pronounced loss in NK-2 receptor affinity in all cases except in I (MDL 28564). MDL 28564 retains significant NK-2 affinity ($K_I = 157$ nM) with greatly reduced NK-1 ($K_I \approx 250$ μM) and NK-3 ($K_I \approx 500$ μM) affinities, making MDL 28564 the most selective NK-2 peptide reported in the literature [5]. This compound showed weak partial agonism in a pI turnover assay ($\approx 10\%$ of NKA), but also showed antagonist activity by its ability to shift the dose–response curve for NKA-induced pI turnover in hamster bladder.

Table 1 *Compounds synthesized, and competitive radioligand binding data*

	Peptide	IC_{50} (nM)		
		NK-1[a]	NK-2[b]	NK-3[c]
I	Asp-Ser-Phe-Val-Gly-Leu[ψ(CH$_2$NH)]Leu-NH$_2$	250×10^3	157 ± 17	250×10^3
II	Asp-Ser-Phe-Val-Gly[ψ(CH$_2$NH)]Leu-Leu-NH$_2$	ND	$> 10 \times 10^3$	ND
III	Asp-Ser-Phe-Val[ψ(CH$_2$NH)]Gly-Leu-Leu-NH$_2$	$> 100 \times 10^3$	$> 5 \times 10^3$	ND
IV	Asp-Ser-Phe[ψ(CH$_2$NH)]Val-Gly-Leu-Leu-NH$_2$	$> 10 \times 10^3$.	$> 5 \times 10^3$	ND
V	Asp-Ser[ψ(CH$_2$NH)]Phe-Val-Gly-Leu-Leu-NH$_2$	$> 40 \times 10^3$	$> 10 \times 10^3$	ND
VI	Suc[ψ(CH$_2$NH)]Ser-Phe-Val-Gly-Leu-Leu-NH$_2$	ND	2×10^3	ND
SP		0.3 ± 0.1	40 ± 11	65 ± 5
NKA		54 ± 0.5	0.7 ± 0.2	35 ± 3
NKB		200 ± 25	18 ± 4	2 ± 0.3

[a] [^{125}I-Bolton-Hunter]-substance P with rat submandibular gland.
[b] [^{125}I-His']^1-neurokinin A with hamster bladder.
[c] [^{125}I-Bolton-Hunter]-eledoisin with rat cerebral cortex.
ND = not determined.

References

1. Coy, D.H., Heinz-Erian, P., Jiang, N.-Y., Sasaki, Y., Taylor, J., Moreau, J.P., Wolfrey, W.T., Garcher, J.D. and Jensen, R.T., J. Biol. Chem., 11 (1988) 5056.
2. Fehrentz, J.-A and Castro, B., Synthesis, (1983) 676.
3. Bischofberger, N., Waldmann, H., Saito, T., Simon, E.S., Lees, W., Bednarski, M.D. and Whitesides, G.M., J. Org. Chem., 53 (1988) 3457.
4. Buck, S.H. and Shatzer, S.A., Life Sci., 42 (1988) 2701.
5. Bristow, D.R., Curtis, N.R., Suman-Chauhan, N., Watling, K.J. and Williams, B., Br. J. Pharmacol., 90 (1987) 211.

C-terminal modified bombesin and litorin analogs with potent in vitro and in vivo antiproliferative activity

S.H. Kim[a], J.E. Taylor[a], R.T. Jensen[b], D.H. Coy[c], A.E. Bogden[a] and
J.-P. Moreau[a]

[a]Biomeasure Inc., Hopkinton, MA 01748, U.S.A.
[b]Digestive Diseases Branch, NIH, Bethesda, MD 20982, U.S.A.
[c]Peptide Research Laboratories, Department of Medicine, Tulane University Medical
Center, New Orleans, LA 70112, U.S.A.

Introduction

Previous studies have shown that C-terminal modification of bombesin and litorin by pseudopeptide bond results in analogs with high receptor antagonist properties [1,2]. In order to study further SAR, a series of peptides with additional C-terminal modifications were synthesized and tested in vitro and in vivo.

Materials and Methods

Synthesis of peptides was carried out by SPPS on MBHAR using commercially available Boc-amino acids. Boc-statine and related amino acids were purchased or prepared by a literature method [3,4]. Cleavage and deprotection of the resin-bound peptide was achieved by treatment with liquid HF containing anisole and dithiothreitol as scavengers. Purification of the crude material by RPHPLC on Vydac C_{18} support yielded a product whose purity was $>95\%$ by analytical HPLC.

In vitro receptor affinities were determined from the competition for [125I]GRP binding to Swiss 3T3 cells, and from the inhibition of BN-stimulated thymidine uptake by 3T3 cells. Some analogs were also tested for their ability to inhibit BN-stimulated amylase release from isolated guinea-pig pancreatic acini.

For in vivo studies, the NCI-H69 human small-cell lung carcinoma line was transplanted from in vitro culture by implanting athymic nude mice with the equivalent of 5 confluent 75-cm^2 tissue-culture flasks, s.c., in the right flank. No attempt was made to disaggregate the cell agglomerates. Tumor size was calculated as the average of two diameters, i.e., (length + width/2) mm. BN antagonists were administered at 50 μg/injection, b.i.d., or s.c., as a perilesional infusion. Care was taken not to inject intralesionally. There were 10 animals in the control and 5 in the test groups.

Table 1   *Structure and in vitro activities*

Analogs	Binding ($K_I = nM$)	Inhibition of thymidine uptake ($IC_{50} = nM$)	Inhibition of amylase release ($IC_{50} = nM$)
[Sta[13]]-des-Met[14] BN	18	agonist	
[Sta[8]]-des-Met[9] Litorin	150	165	150
[D-Phe[1],AHPPA[8]]-des-Met[9] Litorin	233	101	354
[D-Phe[1],Sta[8]]-des-Met[9] Litorin	7.2	37.2	15.1
[D-Phe[1],ACHPA(3R,4S)[8]]-des-Met[9] Litorin	3.9	21.6	28
[D-p-Cl-Phe[1],Sta[8]]-des-Met[9] Litorin	4.5	8.9	26
[D-p-Cl-Phe[1],ACHPA (3R,4S)[8]]-des-Met[9] Litorin	7.9	46.8	30
[D-p-Cl-Phe[1],ACHPA (3S,4S)[8]]-des-Met[9] Litorin	3.8	43.7	36
[D-p-Cl-Phe[1],Sar[6],ACHPA[8]]-des-Met[9] Litorin	1.9	6.7	34
[D-p-Cl-Phe[1],β-Leu[8]]-des-Met[9] Litorin	0.8	5.2	

Statine: (3S,4S)-4-amino-6-methyl-3-hydroxy-heptanoic acid.
AHPPA: (3S,4S)-4-amino-3-hydroxy-5-phenylpentanoic acid.
ACHPA: (3S,4S)-4-amino-5-cyclohexyl-3-hydroxypentanoic acid.
β-Leu: β-homo-L-leucine.
Sar: Sarcosine.
Bombesin: p-Glu-Gln-Arg-Leu-Gly-Asn-Gln-Trp-Ala-Val-Gly-His-Leu-Met-NH$_2$
Litorin: p-Glu-Gln-Trp-Ala-Val-Gly-His-Phe-Met-NH$_2$

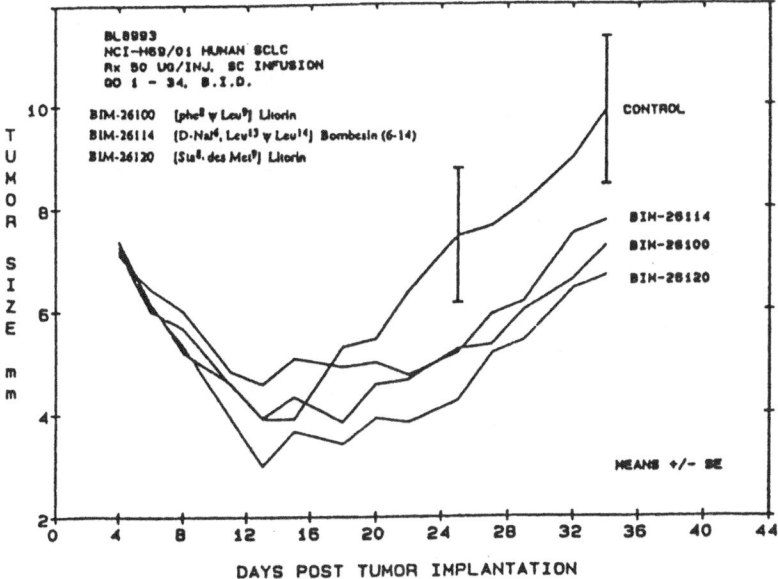

Fig. 1. Evolution of tumor size in athymic nude mice over 34 days of BN and Litorin analogs.

Results and Discussion

Structure and in vitro activities are listed in Table 1; in vivo activities are shown in Fig. 1.

The replacement of the two distal C-terminal amino acids, by statine, in BN, generated potent agonist. Surprisingly, the same kind of modification in the litorin series produced a potent antagonist. Furthermore, replacement at the position 1 by lipophilic D-amino acid generated even more-potent antagonists. Stereochemistry of the hydroxyl group in the statine analogs did not seem to change biological activity. Considering that statine and its related analogs (AHPPA, ACHPA) are γ-amino acid, a β-leucine analog was also prepared, which proved to be a very potent antagonist. Differences between in vitro and in vivo activities of these compounds are under investigation.

References

1. Coy, D.H., Heinz-Erian, P., Jiang, N.-Y., Sasaki, Y., Taylor, J.E., Moreau, J.-P., Wolfey, W.T., Gardner, J.D. and Jensen, R.T., J. Biol. Chem., 263(1988)5056.
2. Coy, D.H., Taylor, J.E., Jiang, N.-Y., Kim, S.H., Wang, H.H., Huang, S.C., Moreau, J.-P., Gardner, J.D. and Jensen, R.T., J. Biol. Chem., 264(1989)14691.
3a. Maibaum, J. and Rich, D., J. Org. Chem., 53(1988)869.
  b. Schuda, P.H., Greenlee, W.J., Chakravarty, P.K. and Eskola, P., ibid., 53(1988)873.
4. Wakamiya, T., Uratani, H., Teshima, T. and Shiba, T., Bull. Chem. Soc. Jpn., 48(1975)2401.

Design of a cyclic bombesin analog

Martha Knight[a], Terrence R. Burke Jr.[a], J. Desiree Pineda[a], Steven L. Cohen[b], Samira Mahmoud[c] and Terry W. Moody[c]

[a]*Peptide Technologies Corporation, 125 Michigan Ave. N.E., Washington, DC 20017, U.S.A.*
[b]*Department of Psychology, Bloomsburg University of Pennsylvania, Bloomsburg, PA 17815, U.S.A.*
[c]*Department of Biochemistry, George Washington University School of Medicine, Washington, DC 20037, U.S.A.*

Introduction

Bombesin (BN) is a neuropeptide that has gastrointestinal effects. Its mammalian congener, gastrin-releasing peptide, is a circulating hormone that functions in digestion and is ectopically produced by small cell lung carcinoma. The peptide is an autocrine growth factor for these cells and related tumors. Analogs were synthesized to determine the smallest biologically active sequence, and other changes were incorporated to increase biological and chemical stability. Met was replaced by Nle to prevent oxidative degradation, and D-Ala was substituted for Gly to stabilize the conformation. Carboxyl-terminal fragment peptides were synthesized, and large amino groups were attached to protect against proteolysis and enhance activity. The partial sequence of BN that possesses activity was further stabilized by forming an amide bond between an ϵ-amino group in position 5 and the carboxyl group of Met at position 14, thus achieving the first active cyclic analog.

Methods

The peptides were synthesized by SPPS on MBHAR using Boc chemistry. Purification was by CCC [1] and preparative RPHPLC. The cyclic peptide was synthesized on Boc-Met chloromethyl resin, purified, cyclized by reaction with BOP-Cl and isolated by RPHPLC. The product had the correct MW as determined by FABMS.

The peptides were analyzed for in vivo and in vitro activity [2]. The reduction in liquid food intake by rats, following intraperitoneal injection, was measured for 30 min. A baseline was established. Each dose of a compound was tested in six animals. The doses were equimolar to 0, 4, and 400 μg/kg of BN. The lower dose of BN lowered food ingestion to 27% of baseline.

The peptides caused a receptor-mediated increase in intracellular Ca^{2+} levels in small cell lung carcinoma cells (NCI-H345) due to increasing phosphatidyl-inositol turnover. The cells were loaded with a fluorescent dye, Fura 2AM at

37°C. When BN or an agonist is added, a rise in the amplitude of fluorescence occurs for 4 min. The effect was measured with varying concentrations to determine the ED_{50}.

Results and Discussion

Concentrations of half-maximal effects by the compounds in both assays are given in Table 1. The agonists showed moderate potency compared to BN in the cell receptor interaction. Not all peptides were tested in both assays. The cyclic BN analog, **137**, that displayed moderately high activity, binds to BN receptors in SCLC, brain membranes, and other cell lines, and stimulates growth of Swiss 3T3 cells. The smallest peptide, **136**, was the most potent in the food inhibition behavior. The bioavailability of i.p.-administered compounds is improved if they are resistant to proteolysis. For this and other reasons, these results suggest that the smaller compounds are more efficacious. The potencies of the octapeptide and the cyclic peptide demonstrate the feasibility of small conformationally stabilized BN receptor-directed analogs as in vivo agents.

Table 1 *Stimulation of intracellular calcium increase in SCLC cells and inhibition of food intake by BN analogs*

Name	Structure	Ca^{2+} levels ED_{50}, μM	Liquid food intake ED_{50}, $\mu g/kg$
	1 2 3 4 5 6 7 8 9		
BN	PCA-Gln-Arg-Leu-Gly-Asn-Gln-Trp-Ala-		
	10 11 12 13 14		
	Val-Gly-His-Leu-Met-NH$_2$	0.005	1.3
ψ	[CH$_2$$^{12/13}$,Leu^{14}]BN [ref. 2]	0.2	n.d.
131	[N-Ac,Nle^{14}]BN(5–14)	0.2	120
137	[cyclo Ac-Leu^4,D-Lys^5,D-Ala^{11},Met^{14}]BN (4–14)	0.3	n.d.
133	[N-β-Naphthoyl,Nle^{14}]BN(5–14)	0.3	60
134	[N-Bz,D-Ala^{11},Nle^{14}]BN(5–14)	0.4	20
136	[N-Ac,D-Ala^{11},Nle^{14}]BN(7–14)	0.5	1.3
132	[N-Bz,Nle^{14}]BN(5–14)	0.5	29
135	[N-β-Naphthoyl,D-Ala^{11},NMeLeu^{13},Nle^{14}]BN(5–14)	n.d.	2 500

n.d. = not determined.

Acknowledgements

Reseach support was provided by grants from the NIH (SBIR CA-44399) and from the NSF (BNS-8815133). FABMS was performed at the National Science Foundation Regional Instrumentation Facility at the Johns Hopkins University School of Medicine.

References

1. Knight, M., Pineda, J.D. and Burke, Jr., T.R., J. Liq. Chromatogr., 11 (1988) 119.
2. Mahmoud, S., Palaszynski, E., Fiskum, G., Coy, D.H. and Moody, T.W., Life Sci., 44 (1989) 367.

NMR and CD conformational studies on bombesin antagonists

C. Di Bello[a], L. Gozzini[b], A. Scatturin[c], M.G. Corradini[d], G.D'Auria[d],
E. Trivellone[d] and L. Paolillo[d]

[a]Institute of Industrial Chemistry, University of Padova, Padova, Italy
[b]Farmitalia – Carlo Erba, R&D Biotechnology, Milan, Italy
[c]Department of Pharmaceutical Chemistry, University of Ferrara, Ferrara, Italy
[d]Department of Chemistry, University of Naples and ICMIB-CNR, Naples, Italy

Introduction

Bombesin (BN) and related peptides are potent mitogens for several cells and function as autocrine growth factors in the development of small cell lung carcinoma (SCLC) [1,2]. This finding has opened the search for BN antagonists able to inhibit the growth of this tumor. Recently, the discovery of [Leu$^{13}\psi$[CH$_2$NH]Leu^{14}]-BN represented a breakthrough in the development of BN antagonists [3]. Since structural information could be of great help both for elucidation of the mode of action and in the formulation and development of more selective and potent antagonists, we have undertaken a systematic study of BN antagonists in solution by using 500 MHz and CD techniques. In particular, [D-Phe^{12}, Leu^{14}]-BN and [Leu$^{13}\psi$[CH$_2$NH]Met^{14}]-BN (6–14) nonapeptide have been examined. NMR and CD measurements in different solvents are discussed together with the data previously collected for the parent BN molecule. A comparative CD investigation on spantide is also reported.

Results and Discussion

NMR measurements

The conformational properties of the two BN antagonists [D-Phe^{12}, Leu^{14}]-BN and [Leu$^{13}\psi$[CH$_2$NH]Met^{14}]-BN (6–14) in solution have been studied in two different systems: the polar proton acceptor DMSO and the polar proton acceptor/donor H$_2$O. Two-dimensional techniques have been used to assign the amino acid resonances.

The chemical shifts of NH and α-CH resonances in DMSO of [D-Phe^{12},Leu^{14}]-BN and [Leu$^{13}\psi$[CH$_2$NH]Met^{14}]-BN (6–14) appear rather similar to each other, and to those of the C-terminal nonapeptide of BN. The NMR spectra of [Leu$^{13}\psi$[CH$_2$NH]Met^{14}]-BN (6–14) show that, in DMSO, H$_2$O and the micellar environment (SDS, 50–100 mM solution), the chemical shift values are rather similar indicating no substantial structural modifications in going from organic to micellar systems. Nevertheless, while in H$_2$O, the NH protons at both ends of the peptide chain exchange rapidly with the solvent molecules; in the presence of SDS, this exchange is somewhat reduced. To gain further information on the structural properties in solution of [Leu¹³ψ[CH₂NH]Met¹⁴]-BN (6–14), the temperature coefficients of amide NHs have been measured. The data obtained show that, in H₂O, the coefficients are rather high and that some get smaller in the presence of SDS micelles. This points to a reduced exposure of some amide protons to interactions with the surrounding environment. Taken together, the results obtained by NMR measurements demonstrate that, as for the parent BN molecule, [D-Phe¹²,Leu¹⁴]-BN and [Leu¹³ψ[CH₂NH]Met¹⁴]-BN (6–14) do not possess a conformational preference in DMSO or H₂O. In the presence of SDS, which in some instances has been shown to modify the conformation of peptides in solution, both chemical shifts and coupling constants measured for [Leu¹³ψ[CH₂NH]Met¹⁴]-BN (6–14), although slightly different, still represent conformation averages. Surprisingly, however, the backbone temperature coefficients decrease in the micellar system. This cannot be taken by itself as an indication of folding, but certainly it implies that the molecule in a membrane-like environment (SDS micelles) is less accessible than in H₂O.

CD measurements

The CD spectrum of [D-Phe¹², Leu¹⁴]-BN in aqueous solution exhibits a negative maximum at about 198 nm. As for BN, the shape of the spectrum is typical of peptides essentially in a random conformation. In TFE and in the micellar environment (SDS 40 mM), both BN antagonists tend to assume a more ordered conformation. In TFE and in SDS the spectrum of [Leu¹³ψ[CH₂NH]Met¹⁴]-BN (6–14), characterized by a negative maximum at 204–205 nm and a shoulder in the 210–220 nm range, is very similar to that observed for BN in the same solvent, thus indicating that this analog also tends to assume a folded conformation (helix, or β-structure) when exposed to a membrane-like environment. Interestingly, spantide, which also possesses some inhibitory activity against BN, seems to assume, even in H₂O, an ordered conformation. The CD spectrum, indeed, shows negative maxima at 222 nm and at about 200 nm. The shape of the curve recalls the classical CD spectrum of an α-helix. In TFE and SDS a red shift of the $\pi-\pi^*$ transition is observed and, most significantly, the intensities $[\theta]_R^{222}/[\theta]_R^{200}$ ratio increases, thus indicating an increase of folded conformations in going from TFE to SDS.

References

1. Zachary, I. and Rozengurt, E., Proc. Natl. Acad. Sci. U.S.A., 85 (1985) 7616.
2. Cuttitta, F., Carney, D.N., Mulshine, J., Moody, T.W., Fedorko, J., Fischller, A. and Minna, J.D., Nature, 316 (1985) 823.
3. Coy, D.H., Heinz-Erian, P., Jiang, N.-Y., Gardner, J.D. and Jensen, R.T., J. Biol. Chem., 263 (1988) 5056.

Cytotoxic metallo-peptide analogs of gonadotropin-releasing hormone

T. Janaky[a], S. Bajusz[a], V. Csernus[a], L. Bokser[a], M. Fekete[a], G. Srkalovic[a],
T.W. Redding[b] and A.V. Schally[a,b]

[a]Department of Medicine, Tulane University School of Medicine, 1601 Perdido Street,
New Orleans, LA 70146, U.S.A.
[b]Endocrine, Polypeptide and Cancer Institute, Veterans Administration Medical Center,
1601 Perdido Street, New Orleans, LA 70146, U.S.A.

Introduction

The treatment of endocrine-dependent or hormone-sensitive tumors, such as breast and prostatic cancers, with GnRH analogs represents a new approach in oncology [1]. The combination of hormonal therapy with chemotherapy may enhance the response. Such combined therapy could be provided by treatment with highly potent analogs of GnRH that contain cytotoxic moieties. Since such a peptide can bind to specific receptors, this can provide some target selectivity for the cytotoxic molecule and make it 'cell specific', decreasing its nonspecific toxicity. In this paper, we report on the synthesis of GnRH analogs containing clinically used antitumor drug Cisplatin (cis-diamminedichloroplatinum) and cytotoxic copper(II) and nickel(II) complexes of hydroxy-oxo compound [2].

Results and Discussion

Metal complexes were introduced into suitable modified analogs of GnRH. The 19 metallopeptide analogs synthesized can be illustrated by the following general formulas:

pGlu-His-Trp-Ser-Tyr-D-Lys[A_2xy(Q)]-Leu-Arg-Pro-Gly-NH_2
pGlu-His-Trp-Ser-Tyr-D-Lys[Ahx-A_2xy(Q)]-Leu-Arg-Pro-Gly-NH_2
Ac-D-Nal(2)-D-Phe(4Cl)-R^3-Ser-Arg-D-Lys[A_2xy(Q)]-Leu-Arg-Pro-D-Ala-NH_2

wherein A_2xy stands for 2,3-diaminopropionic acid (A_2pr) or 2,4-diaminobutyric acid (A_2bu), Ahx is 6-aminohexanoic acid, R^3 is D-Pal [3-(3-pyridyl)alanine] or D-Trp, Q is $PtCl_2$ or B_2Q', wherein B derived from salicyl aldehyde, 5-Cl-salicylaldehyde, pyridoxal or pyridoxal-5'-phosphate and Q' is Cu(II) or Ni(II). As chelating function, an N^ϵ-diaminoacyl-D-lysine or an N^ϵ-[6-(diaminoacyl)-aminohexanoyl]-D-lysine was incorporated into position 6 of GnRH and two of its antagonistic analogs during SPPS, or the D-Lys^6 analog was acylated thereafter with the corresponding diaminoacid. Coordination compounds of $PtCl_2$, in which the platinum is chelated by two amino groups of the diaminoacid side chain (Fig. 1A), could be obtained by 3 days reaction of K_2PtCl_4 with peptides mentioned above. Analogs comprising of a hydroxy-oxo compound

190

and a Cu(II) or Ni(II) complex could be prepared by two methods: reacting a diamino-D-Lys[6] peptide with a preformed aldehydato complex (e.g., bis(sa-licylaldehydato)copper(II)], or first with a hydroxy-oxo compound and then with a metal ion. The diaminoacid forms two Schiff-bases with the hydroxy-oxo compound that provides two $-CH=N<$ groups and two phenolate anions for the coordination of a bivalent metal cation, e.g., Cu^{++}. The result is a six-five-six-membered fused ring system in the reaction with 2,3-diaminopropyl side chain, and a three-fused six-membered chelate ring with 2,4-diaminobutyryl group (Figs. 1B and 1C). Formation of the complexes was followed by RPHPLC (C_{18}, solvents were NH_4OAc, pH = 7.0, and acetonitrile). The reaction of diaminoacyl-D-Lys[6] analog with an aldehydato complex of copper produced a faster moving transition metallopeptide that was transformed into the final product. We could demonstrate the existence of the complexes of the D- and L-forms of A_2pr containing agonists and several times we could separate them by HPLC.

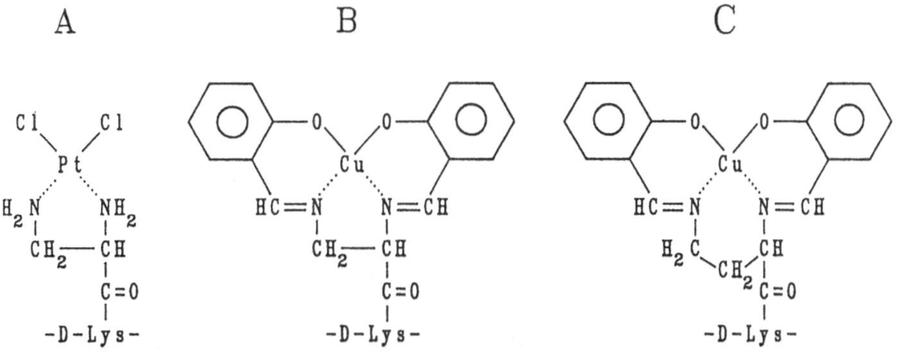

A   B   C

Fig. 1. Structures of metallo-complexes

The agonistic metallopeptides showed higher potency than GnRH, while some of the antagonistic analogs exerted a powerful and long lasting inhibitory effect on LH release in dispersed rat pituitary cell superfusion system. Most of these peptides bind with high affinity to rat pituitary and human breast cancer cell membrane receptors. Some of these GnRH analogs possessed significant cytotoxic activity in cultures of human mammary cancer and prostate cancer cell line.

Acknowledgements

This work was supported by NIH Grants CA 40003 and CA 40004 by the Medical Research Service of the Veterans Administration.

References

1. Schally, A.V., Redding, T. and Comaru-Schally, A.M., Cancer Treat. Rep., 68 (1984) 281.
2. Bajusz, S., Janaky, T., Csernus, V.J., Bokser, L., Fekete, M., Srkalovic, G., Redding, T.W. and Schally, A.V., Proc. Natl. Acad. Sci. U.S.A., 86 (1989) 6313.

Structure-activity relationships, pharmacokinetics, and bioavailability studies of reduced size analogs of gonadotropin-releasing hormone (GnRH)

**Fortuna Haviv, Timothy D. Fitzpatrick, Eugene N. Bush, Gilbert Diaz,
Edwin S. Johnson, Stephen Love and Jonathan Greer**
Pharmaceutical Products Division, Abbott Laboratories, Abbott Park, IL 60064, U.S.A.

Introduction

Recently, we reported [1] the synthesis and the biological activities of the smallest potent agonists and antagonists of GnRH. The structures of these compounds contain the (4–9) portion of GnRH coupled to a carboxylic acid that is designed to mimic Trp^3. As a further step towards the development of orally active analogs of GnRH, we studied the effect of physicochemical properties on the bioavailability of our reduced size analogs.

Methods

Peptides were synthesized using the SPPS on Merrifield resin. The peptides were cleaved from the resin upon treatment with anhydrous ethylamine and subsequently deprotected with anhydrous HF in the presence of anisole. The crude peptides were purified by HPLC using a C_{18} reversed phase column. The purity of the final compound was based on HPLC, FABMS and AAA.

Peptides were tested in vitro in the rat pituitary receptor binding assay [2] and for LH release or suppression using cultured rat pituitary cells [3,4]. The receptor binding affinities are reported as a pK_I. The LH release potencies for agonist are reported as a pD_2, and LH suppression for antagonist as pA_2. To measure bioavailability, the compounds were administered in the rat by bolus iv and id. The serum levels of the compounds were determined by a RIA using an anti-GnRH analog antibody that recognizes the C-terminal residues Leu-Arg-Pro-NHEt. Bioavailability (F) was calculated as the ratio:

$$F = \frac{\text{Area under the curve (id)/dose(id)}}{\text{Area under the curve (iv)/dose(iv)}} \times 100.$$

Results and Discussion

In our previous publication [1] about reduced size GnRH analogs, we reported the SAR studies of the (4–9) GnRH series. We found that the size and the shape of the substituent at position 3 determines agonist or antagonist response. Also, depending on the nature of the substituent at position 6 and 4, the biological response may switch from antagonist to agonist and vice versa. To study the effect of the physicochemical properties on bioavailability (F) and half life ($t_{1/2}$) in the 3-indolepropionyl-Ser^4- (4–9) series, we varied the properties of the substituent at position 6 from highly hydrophobic to highly hydrophilic (Table 1). It appears that a highly hydrophobic amino acid, such as D-2Nal or D-Trp, has a beneficial effect on the binding affinity of the compounds, but tended to give worse pharmacokinetics by shortening the duration of action ($t_{1/2}$) and often increasing the volume of distribution (V_D), perhaps because of more rapid hepatic extraction and tissue uptake, respectively. On the other hand, highly hydrophilic, and especially basic amino acids, such as D-Lys and D-Arg, are less potent, yet significantly prolong the half life of the compounds and have low volumes of distribution giving much better pharmacokinetics (Table 1). No dramatic changes in absorption (F) of compounds were observed for any of the compounds. The best increase of about six-fold was observed with D-Arg and D-Orn derivatives.

Table 1 *The effect of physicochemical properties on pharmacokinetics and bioavailability in the N-[3-indolepropionyl]Ser^4-(4–9) series*

Compd	X^3-Ser-Tyr-Y^6-Leu-Arg-Pro-NHEt							
	X^3	Y^6	pK_I	pD_2	pA_2	$t_{1/2}$ (min)	F (%)	V_D (ml)
1	3-(3-indole)propionyl	D-Leu	6.70	6.93		23.7	0.42	130
2	3-(3-indole)propionyl	D-Trp	8.15	7.60		10	0.08	109
3	3-(3-indole)propionyl	D-2Nal	8.77		7.44	8.6	0.50	240
4	3-(3-indole)propionyl	D-Phe	7.78		6.07			
5	3-(3-indole)propionyl	D-Thi	7.29		6.53			
6	3-(3-indole)propionyl	D-Tyr	7.76		7.06	25.7	0.46	384
7	3-(3-indole)propionyl	D-Cha	7.38		6.44			
8	3-(3-indole)propionyl	D-Ser	5.95		5.83	67	0.17	154
9	3-(3-indole)propionyl	D-Arg	6.70	6.10		163	0.5	136
10	3-(3-indole)propionyl	D-Orn	6.53		5.94	82.5	0.5	116
11	3-(3-indole)propionyl	D-Lys	7.10	5.93		180	0.08	40
12	3-(3-indole)propionyl	D-Lys(Isp)	7.20		5.15	94.6	0.39	89
13	3-(3-indole)propionyl	D-Lys(Nic)	7.30		5.20	56.3	0.13	95
14	3-(3-indole)propionyl	D-Lys(Pic)	6.45		5.95	30	0.37	102
15	3-(1-Naphthyl)propionyl	D-2Nal	9.55		9.00	11.7	0.46	1 878
	Leuprolide		10.37	10.14		26.3	0.08	146
	[Ac-D-Cpa^{1,2},D-Trp^3,D-Arg^6,D-Ala^{10}]GnRH	10.74		9.43		95	0.08	70

D-2Nal = D-3-(2-naphthyl)-alanine; D-Thi = D-3-(2-thienyl)-alanine; D-Cha = D-3- cyclohexyl-alanine; D-Lys(Isp) = D-*N*-epsilon-isopropyl-lysine; D-Lys(Nic) = D-*N*-epsilon-nicotinyl-lysine; D-Lys(Pic) = D-*N*-epsilon-picolinyl-lysine; D-Cpa = D-3-(4-chloro)-phenylalanine).

Conclusion

Changes in the physicochemical properties of the (4–9) GnRH series have a pronounced effect on potency and pharmacokinetics, but only a moderate effect on intestinal absorption.

Acknowledgements

We are indebted to Dr. P. Michael Conn of the University of Iowa for providing the anti-rat-LH and anti-GnRH analog antibodies.

References

1. Haviv, F., Palabrica, C.A., Bush, E.N., Diaz, G., Johnson, E.S., Love, S. and Greer, J., J. Med. Chem., 32 (1989) 2340.
2. Marian, J., Cooper, R.L. and Conn, P.M., Mol. Pharmacol., 19 (1981) 399.
3. Jinnah, H.A. and Conn, P.M., Am. J. Physiol., 249 (1985) E619.
4. Vale, W., Grant, G., Amoss, M., Blackwell, P. and Guillemin, R., Endocrinology, 91 (1974) 562.

Synthesis of a somatostatin analog containing a tetrazole cis-amide bond surrogate

Janusz Zabrocki[a], Urszula Slomczynska[a] and Garland R. Marshall[b],*

[a]Institute of Organic Chemistry, Politechnika, P-90-924 Lodz, Poland
[b]Department of Pharmacology, Washington University School of Medicine, St. Louis, MO 63110, U.S.A.

Introduction

In natural peptides, N-methyl amino acids may play a special role because of their proclivity for cis-trans isomerism of the amide bond. Over 10% of proline residues in protein crystal structures has been found with cis-amide bonds. We have previously suggested [1,2] the 1,5-disubstituted tetrazole ring, $\psi[CN_4]$, as an amide bond surrogate for cis-amide bonds, and reported conversion of dipeptide esters into tetrazole derivatives without racemization. The difficulties in the preparation and use of this dipeptide analog, as reported by Yu and Johnson [3], have been overcome, as witnessed by the successful preparation of bradykinin analogs containing tetrazole dipeptides [4].

We have chosen to use this cis-amide bond surrogate to replace either the proline, or N-methyl alanine, thought to favor a cis-amide bond in cyclic hexapeptide analogs of somatostatin [5]. An analog of the somatostatin compound, cyclo[-D-Trp-Lys-Val-Phe-N-Me-Ala-Tyr-], has been prepared, in which the dipeptide Phe-N-MeAla has been replaced by a dipeptide surrogate, Phe-$\psi[CN_4]$-Ala, with the amide bond converted to a 1,5-disubstituted tetrazole ring [2]. This ring holds the amide bond in the cis-conformer, which is presumed to be the active conformation.

Results and Discussion

The dipeptide Z-Phe-Ala-OBzl (2.3 g, 5 mM) was reacted for 1.5 h at 10°C with PCl_5 (1.05 g) and hydrazoic acid (15 ml) in the presence of quinoline (1.18 ml, 10 mM) to give the tetrazole derivative, Z-Phe-$\psi[CN_4]$-Ala-OBzl, in 62% yield after purification by flash chromatography. Careful treatment of 1.58 g of the dipeptide with 8.2 ml of 30% HBr in acetic acid for 20 min gave HBr·H_2N-Phe-$\psi[CN_4]$-Ala-OBzl in 83% yield, which was extended in solution to give the protected hexapeptide, Boc-D-Trp-Lys(Z)-Val-Phe-$\psi[CN_4]$-Ala-Tyr-OMe, as shown in Fig. 1. The carboxyl protecting group was removed by treatment with chymotrypsin in DMF/water (pH 5.5), due to racemization of the tetrazole dipeptide α-proton on base exposure. For reasons that are not understood,

---

*To whom correspondence should be addressed.

J. Zabrocki, U. Slomczynska and G.R. Marshall

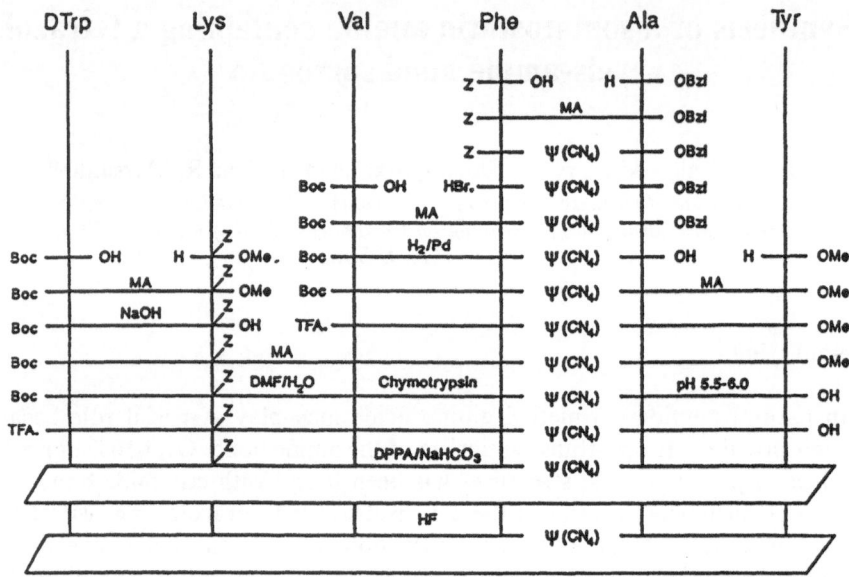

Fig. 1. Synthesis of somatostatin analog containing a tetrazole cis-amide bond surrogate. Abbreviations: $\psi(CN_4) = 1.5$-disubstituted tetrazole ring; MA = mixed anhydride with isobutyl chloroformate; $DPPA$ = diphenylphosphorylazide.

removal of the Boc group by the HCl procedure of Brady et al. [6] was incomplete, and TFA with the associated side reaction of t-butyl alkylation of the indole ring was used instead. The peptide was cyclized for 3 days by use of DPPA in DMF in 85% yield, followed by removal of the Z group on Lys. HPLC purification yielded both the desired product as well as some 7% of side product, presumed to be the t-butyl adduct.

Alternatively, cyclization was attempted enzymatically on hexapeptide sequences containing this amide modification by the use of trypsin. In this case, the linear hexapeptide sequence, Boc-Val-Phe-ψ[CN$_4$]-Ala-Tyr-D-Trp-Lys(Z)-OMe, was prepared in a similar fashion, and the Boc and Z groups removed with TFMSA. The free amine was generated by ion-exchange on Amberlite A-21 in methanol, and enzymatic cyclization attempted under two sets of conditions. Water with 10% methanol at pH 8.5 gave 7.5–9% yield of the desired product, and the remainder was dimer. Reaction in methylene chloride containing 10% water gave 15% of the product with the rest as dimer. These results suggest that enzymatic cyclization is a feasible approach, but that appropriate conditions need further exploration.

While this work was in progress, a report of a cyclic hexapeptide analog of somatostatin by Elseviers et al. [7] appeared, in which N-methyl-α-benzyl-o-aminomethylphenylacetic acid was used to replace the same dipeptide unit. The activity of the active isomer was approximately 1% that of a similar somatostatin analog. This observation is consistent with the cis-amide as the bioactive species. The lower affinity may be due to the choice of analog, in which the Phe-*N*-MeAla dipeptide has been replaced by a surrogate of Gly-Phe, in which the amide has been replaced by a benzene unit. Alternatively, the geometric similarity of this surrogate may not mimic the geometry and conformational effects of the cis-amide bond as well as does the tetrazole unit [2]. The biological activity (yet to be measured) of the tetrazole analog should help to resolve these issues.

Acknowledgements

Supported in part by the National Institutes of Health, Grant GM-24483, as well as the Polish Academy of Sciences, Grant CPBP 01.13.2.5.

References

1. Marshall, G.R., Humblet, C., Van Opdenbosch, N. and Zabrocki, J., In Rich, D.H. and Gross, E. (Eds.) Peptides: Synthesis, Structure, Function (Proceedings of the 7th American Peptide Symposium), Pierce Chemical Co., Rockford, IL, 1981, p. 669.
2. Zabrocki, J., Smith, G.D., Dunbar Jr., J.B., Iijima, H. and Marshall, G.R., J. Am. Chem. Soc., 110(1988) 5875.
3. Yu, K.-L. and Johnson, R.L., J. Org. Chem., 52(1987) 2051.
4. Zabrocki, J., Smith, G.D., Dunbar, Jr., J.B., Marshall, K.W., Toth, M. and Marshall, G.R., In Jung, G. and Bayer, E. (Eds.) Peptides 1988, Walter de Gruyter, New York, 1989, p. 295.
5. Veber, D.F., In Rich, D.H. and Gross, E. (Eds.) Peptides: Synthesis, Structure, Function (Proceedings of the 7th American Peptide Symposium), Pierce Chemical Co., Rockford, IL, 1981, p. 685.
6. Brady, S.F., Freidinger, R.M., Paleveda, W.J., Colton, C.D., Homnick, C.F., Whitter, W.L., Curley, P., Nutt, R.F. and Veber, D.F., J. Org. Chem., 52(1987) 764.
7. Elseviers, M., Van Der Auwera, L., Pepermans, H., Tourwe, D. and Van Binst, G., Biochem. Biophys. Res. Comm., 154(1988) 515.

Synthesis and conformation of new analogs of somatostatin with isosteric replacements

M. Elseviers, H. Jaspers, N. Delaet, S. De Vadder, H. Pepermans*, D. Tourwé and G. van Binst**

Eenheid Organische Chemie, ORGC, Faculteit Wetenschappen, Vrije Universiteit Brussel, Pleinlaan 2, B-1050 Brussels, Belgium

Introduction

One of the most active somatostatin analogs for GH release inhibition has been synthesized by Veber et al. [1] (1 in Table 1). A complete conformational analysis of this peptide has been performed by Kessler et al. [2], who showed that the Phe-Pro peptide bond existed almost exclusively in the cis configuration and that a β-turn type II' structure was present for the core tetrapeptide (Phe-D-Trp-Lys-Thr). In our ongoing approach of mimicking parts of the peptide molecules, we designed spacers that could either fix the cis structure [3], such as the cis-olefinic dipeptide analog (2) and o-aminomethylphenylacetic acid (o-AMPA) (3a) derivatives [4,5], or fix the trans structure with a trans-olefinic (4) dipeptide analog. In addition, we prepared the saturated equivalent (5) and the reduced amide bond (6) in order to determine the consequences on the activity and conformation of the increase of mobility in that part of the molecule [6].

Table 1   *Somatostatin fragments and analogs*

1	cyclo(Pro-Phe-D-Trp-Lys-Thr-Phe)
2	Boc-Pheψ[Z,CH=CH]GlyOH
4	Boc-Pheψ[E,CH=CH]GlyOH
5	Boc-Pheψ[CH$_2$-CH$_2$]GlyOH
6	Boc-Pheψ[CH$_2$-N]ProOH
7a	cyclo(o-AMPA-Phe-D-Trp-Lys-Thr)
7b	cyclo(N-Me-α(R)Bzl-o-AMPA-Phe-D-Try-Lys-Thr)
7c	cyclo(N-Me-α(S)Bzl-o-AMPA-Phe-D-Trp-Lys-Thr)
7d	cyclo(N-Me-o-AMPA-Phe-D-Trp-Lys-Thr)
8a	cyclo(N-Me-α(R)Bzl-o-AMPA-Tyr-D-Trp-Lys-Val)
8b	cyclo(N-Me-α(S)Bzl-o-AMPA-Tyr-D-Trp-Lys-Val)
9	cyclo(Pheψ[E,CH=CH]Gly-Phe-D-Trp-Lys-Thr)
10	cyclo(Pheψ[CH$_2$-CH$_2$]Gly-Phe-D-Trp-Lys-Thr)
11	cyclo(Pheψ[CH$_2$-N]Pro-Phe-D-Trp-Lys-Thr)

---

*Present adress: Unilever Research Laboratories, O. van Noortlaan 120, 3133 AT Vlaardingen, The Netherlands.
**To whom correspondence should be addressed.

Results and Discussion

Cyclo(*o*-AMPA-Phe-D-Trp-Lys-Thr) (**7a**) was devoid of inhibitory activity on GH release and the NMR conformation analysis showed a predominance of a δ-turn conformation stabilized by a strong H-bond between the NH and CO groups of *o*-AMPA. The lack of an equivalent of Phe[11] could also play a negative role.

This situation has been corrected by the synthesis of three new spacers: *N*-methyl-α-benzoyl-*o*-aminomethylphenylacetic acid (**3b**), *N*-methyl-*o*-aminomethylphenylacetic acid (**3c**) and δ-benzyl-*o*-aminomethylphenylacetic acid (**3d**) and their incorporation in the adequate sequence by SPS. This resulted in the new somatostatin analogs (**7b,c,d**).

Until now, we have been unable to prepare suitable amounts of the cis-olefinic dipeptide analog, cyclo(Pheψ[Z,CH = CH]Gly-Phe-D-Trp-Lys-Thr). Incorporation of the trans-olefinic derivative (**4**) yielded analog (**9**), of the saturated derivative (**5**), the analog (**10**) and the reduced amide bond (**6**) the analog (**11**). The results of the activities are summarized in Table 2.

Table 2   *Biological and conformational properties of somatostatin analogs*

Analog	GH release inhibition test	NMR conformation
7a	inactive	δ-turn + β II'-turn
7b	active	β VI-turn + β II'-turn
7c	inactive	β VI-turn + β II'-turn
8a	active	β VI-turn + β II'-turn
8b	inactive	β VI-turn + β II'-turn
7d	inactive	
9	very low	γ-turn + β II'-turn
10	inactive	β VI$_{g+}$-turn + β II'-turn
11	very low	

Conclusions

For GH release inhibitory activity, not only the β-turn type II' is needed in the backbone, but also a β-turn type VI, as a result of a cis peptide bond. These are necessary but not sufficient conditions. The diastereomeric products **7b** vs. **7c** and **8a** vs. **8b** fulfill these conditions but the orientation of the hydrophobic side chain (Phe side chain) plays an important role. The saturated derivatives (**10, 11**) also fulfill the backbone conditions but are inactive, although the ψPhe side chain is equatorial. A further study on the possible orientation of the side chain in these more flexible derivatives is needed.

References

1. Veber, D.F., Freidinger, R.M., Perlow, D.S., Paleveda, W.J., Holly, F.W., Strachan,

R.G., Nutt, R.F., Arison, B.H., Homnick, C., Randall, W.C., Glitzer, M.S., Saperstein, R. and Hirschmann, R., Nature, 292 (1981) 55.

2. Kessler, H., Bernd, M., Kogler, H., Zarboch, J., Sorensen, O.W., Bodenhausen, G. and Ernst, R.R., J. Am. Chem. Soc., 105 (1983) 6944.

3. Yu, K.L. and Johnson, R.L., J. Org. Chem., 52 (1987) 2051.

4. Vander Elst, P., Van den Berg, E., Pepermans, H., Van Der Auwera, L., Zeeuws, R., Tourwé, D. and Van Binst, G., Int. J. Pept. Prot. Res., 29 (1987) 318.

5. Elseviers, M., Van Der Auwera, L., Pepermans, H., Tourwé, D. and Van Binst, G., Biochem. Biophys., Res. Comm., 154 (1988) 515.

6. Jaspers, H., Ph.D. Thesis, Brussels, 1989.

Conformational restriction of glucagon leads to an analog with two distinct binding modes: Further evidence that glycogenolysis follows two signal transduction pathways

Dev Trivedi, Mary Jo Grummel, Beat Gysin, Douglas Sanderson, Klaus Brendel
and Victor J. Hruby

*Department of Chemistry and Department of Pharmacology, University of Arizona,
Tucson, AZ 85721, U.S.A.*

Introduction

Glucagon stimulates glucose production via the second messenger cAMP (GR2 receptor). It has also been postulated that glucagon acts through another second messenger [1]. Houslay et al. [2] showed that both glucagon and its antagonist (THG) are capable of stimulating breakdown of phosphatidyl inositol (GR1 receptor) with an increase in cytosolic inositol triphosphate (IP_3), and Tager et al. [3] have reported a high and low affinity receptor of glucagon in canine and rat liver, though only one protein could be isolated by photoaffinity labeling [4]. We report here that a cyclic, conformationally restricted analog of glucagon, [Asp9, Lys12, Lys17,18, Glu21]glucagon (1), has two distinct binding modes. Only the low affinity state showed partial agonist activity in the adenylate cyclase assay, but glycogenolytic activity paralleled both affinity sites, indicating that two signal transduction mechanisms operate for glucose production.

Results and Discussion

(1) was synthesized, purified and characterized by methods similar to those previously reported [5]. Cyclization was carried out directly on the resin. Receptor binding, adenylate cyclase activities and glycogenolysis were measured as described in [1,5]. (1) was synthesized to examine the relationship of glucagon's 3D structure to its multiple biological effects. This cyclic analog showed a distinct biphasic binding curve with IC_{50}s of 650 nM and 6.5 nM (Fig. 1). Interestingly, only the low affinity binding led to AC activation (10% partial agonist). Furthermore, the cyclic analog displayed two distinct modes of glucose production (not shown) which parallels the binding curve. Glucagon's GR2 receptor-stimulated adenylate cyclase is suggested [6] to undergo desensitization/uncoupling through a cAMP-independent process that involves the stimulation of inositol phospholipid metabolism by glucagon acting through GR1 receptors. In the case of the cyclic analog, glucagon's GR2 receptor apparently undergoes total desensitization/uncoupling from the GR1 receptor, resulting in two separate pathways for glycogenolysis. The high-affinity state may operate for glucagon by mobilizing Ca^{2+}.

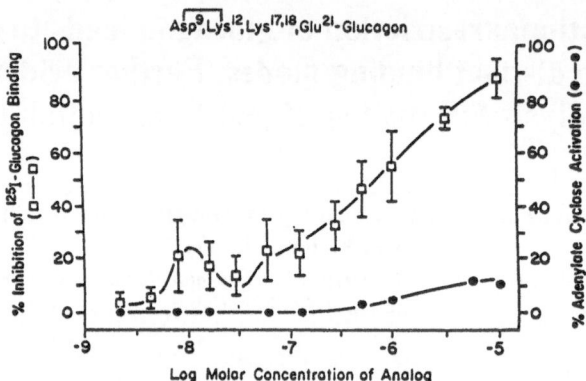

Fig. 1. *Receptor binding and adenylate cyclase activity.*

Conclusion

The conformational restrictions of glucagon via an amide bridge formation, lead to **(1)** with two binding modes, one of which is independent of cAMP. More importantly, the two signal transduction systems in glucagon follow two separate pathways for glycogenolysis that parallel the high- and low-affinity states.

Acknowledgements

Supported by U.S. Public Health Service Grant DK-21085.

References

1. McKee, R.L., Trivedi, D.B., Zechel, C., Johnson, D.G., Brendel, K. and Hruby, V.J., In Marshall, G.R. (Ed.) Peptides: Chemistry and Biology (Proceedings of the 10th American Peptide Symposium), ESCOM, Leiden, 1988, p. 341.
2. Wakelam, M.J.O., Murphy, G.J., Hruby, V.J. and Houslay, M.D., Nature, 323 (1987) 620.
3. Musso, G.F., Assoian, R.K., Kaiser, E.T., Kezdy, F.J. and Tager, H.S., Biochem. Biophys. Res. Comm., 119 (1984) 713.
4. Herberg, J.T., Codina, J., Rich, K.A., Rojas, F.J. and Iyengar, R., J. Biol. Chem., 259 (1984) 9285.
5. Krstenansky, J.L., Trivedi, D.B., Johnson, D.G. and Hruby, V.J., Biochemistry, 25 (1986) 3833.
6. Murphy, G.J., Hruby, V.J., Trivedi, D.B., Wakelam, M.J.O. and Houslay, M.D., Biochem. J., 243 (1987) 39.

Antagonist properties of hydrophobic derivatives of glucagon

Cecilia G. Unson, Keiko Iwasa, Ellen M. Gurzenda, Tracy L. Durrah and
R.B. Merrifield

The Rockefeller University, 1230 York Avenue, New York, NY 10021, U.S.A.

Introduction

The binding of glucagon to its receptor in the plasma membrane initiates a series of reactions leading to the activation of adenylate cyclase and an increase in intracellular levels of cAMP. Structure–activity studies have been directed towards sorting out the chemical features that contribute to receptor recognition and binding, and those that are necessary for the biological response. The ability to make this distinction would allow us to design glucagon analogs that bind but do not activate adenylate cyclase, leading to potentially useful antagonists.

Recently, we reported the synthesis [1] and biological activities [2] of des-His[1] [Glu⁹]glucagon amide, a glucagon antagonist. Synthetic analogs, combining des-His[1][Glu⁹]glucagon amide with structural features designed to stabilize putative secondary structures in a 3D model of glucagon, have resulted in seven other antagonists of glucagon [3]. In this report, we synthesized acylated derivatives of des-His[1][Glu⁹]glucagon amide based on the supposition that increasing the hydrophobicity of the peptide will also increase the binding affinity for its membrane-bound receptor protein.

Results and Discussion

Des-His[1][Glu⁹]glucagon amide was assembled by SPPS, on MBHA resin. After Boc-deprotection, the peptide resin was acylated at its amino terminus (Ser²) with octanoic, myristic, or 2,4-difluorobenzoic acid, by activation with DPPA in DMF containing TEA. The succinylated derivative was prepared by reaction with succinic anhydride in DMF. The N^ϵ-octanoyl Lys¹² and Lys²⁹ glucagon amide analogs were prepared by coupling with N^α-Boc, N^ϵ-Fmoc-lysine, followed by Fmoc–deprotection with 30% piperidine in CH_2Cl_2, and acylation with octanoic acid/DPPA/TEA in DMF. The protected peptide resins were cleaved by the low/high HF procedure, and the crude peptides were purified on octadecyl-silica. Purity was demonstrated by analytical HPLC. MS analysis identified the expected $(M+H)^+$ peaks to be within ± 0.3 Da of theory, and AAA yielded the expected amino acid ratios. The acylated analogs were assayed for their receptor-binding affinity to hepatocyte membranes, as well as the ability to stimulate adenylate cyclase, using methods previously described [1].

Table 1 *Hydrophobic antagonists of glucagon*

No.	Glucagon (-G-) analog	Membrane binding (%)	Relative cAMP activation (%)	Inhibition index $(I/A)_{50}$	pA_2
1	Des-His^1[Glu^9]-G-NH$_2$	41	<0.0001	12	7.2
2	N^α-2,4-Difluorobenzoyl-G-NH$_2$	4	0.1		
3	[N^ϵ-Octanoyl-Lys^{12}],des-His^1[Glu^9]-G-NH$_2$	5.3	0.63		
4	[N^ϵ-Octanoyl-Lys^{29}],des-His^1-G-NH$_2$	8.5	0.14		
5	N^α-Octanoyl, des-His^1-G-COOH	25	0.22		
6	N^α-Octanoyl, des-His^1[Glu^9]-G-NH$_2$	48	0.02	25	7.5
7	N^α-Myristoyl, des-His^1[Glu^9]-G-NH$_2$	12	0.10	47	7.3
8	N^α-Succinyl, des-His^1[Glu^9]-G-NH$_2$	28	<0.004	25	7.4
9	N^α-2,4-Difluorobenzoyl, des-His^1[Glu^9]-G-NH$_2$	75	<0.008	4	8.5

Monoacylation of the amino groups in glucagon analogs produced peptides with binding affinities for the receptor as high as 75% relative to glucagon, but with low adenylate cyclase activity (Table 1). The receptor was sensitive to the position of the fatty acid since octanoylation at the N^α-position of des-His^1[Glu^9]glucagon amide (**6**) bound 48%, while octanoylation at the N^ϵ-Lys^{12} or at N^ϵ-Lys^{29} positions (**3, 4**) reduced receptor binding, suggesting that the bulky hydrocarbon chain does not fit into the putative binding regions near Lys^{12} or the carboxy terminal Lys^{29}. It is already known that a positive charge on Lys^{12} is not necessary for binding and activity, since some N^ϵ-acylated derivatives of native glucagon are reported to be full agonists [4]. The N^α-acylated derivatives in combination with des-His^1[Glu^9] amide substitutions (**6,7,8,9**) were all of very low activity in the adenylate cyclase assay (0.1–0.004% relative to glucagon) and were potent antagonists, with inhibition indices of <50 and good pA_2 values. The best antagonist of this group, and the best reported to date, was N^α-2,4-difluorobenzoyl, des-His^1[Glu^9]glucagon amide, with 75% relative binding, 0.008% relative potency, an inhibition index of 4 and a pA_2 of 8.5.

Acknowledgements

This work was supported by USPHS grant DK-24039.

References

1. Unson, C.G., Andreu, D., Gurzenda, E.M. and Merrifield, R.B., Proc. Natl. Acad. Sci. U.S.A., 84 (1987) 4083.
2. Unson, C.G., Gurzenda, E.M. and Merrifield, R.B., Peptides, 10 (1989) in press.
3. Unson, C.G., Gurzenda, E.M., Iwasa, K. and Merrifield, R.B., J. Biol. Chem., 264 (1989) 789.
4. Carrey, E.A. and Epand, R.M., J. Biol. Chem., 257 (1982) 10624.

Structure-activity relationships of galanin: Importance of the N-terminal sequence for agonist activity on smooth muscle

James W. Aiken*, Farida G.B. Kaddis, Mark A. Connell, Douglas J. Staples, Carol A. Bannow, John H. Kinner and Tomi K. Sawyer

Metabolic Diseases Research and Biopolymer Chemistry Units, The Upjohn Company, Kalamazoo, MI 49001, U.S.A.

Introduction

Galanin (GAL), a 29-amino acid peptide, originally isolated from the upper small intestine of pigs [1], may be an important modulator of a variety of gastrointestinal, endocrine, and CNS activities [2]. Relatively low concentrations contract the isolated stomach fundus strip of the rat [1]. In the present study we have explored the SAR of porcine GAL (H-Gly[1]-Trp-Thr-Leu[4]-Asn-Ser-Ala-Gly[8]-Tyr-Leu-Leu-Gly[12]-Pro-His-Ala-Ile[16]-Asp-Asn-His-Arg[20]-Ser-Phe-His-Asp[24]-Lys-Tyr-Gly-Leu[28]-Ala-NH$_2$) using the isolated stomach fundus strip of the rat as an in vitro bioassay of agonist activity.

Results and Discussion

In the stomach fundus strip of the rat, noteworthy for its sensitivity to serotonin [3], galanin is about 1/10 as potent (ED$_{50}$ for serotonin $= 2 \times 10^{-9}$ M and ED$_{50}$ for galanin $= 2 \times 10^{-8}$ M) and galanin's maximum response is about 55% that of serotonin. Some tachyphylaxis to galanin occurred, but it was not a problem if exposures to agonists were at least 20 min. apart. A concentration of a prostaglandin synthesis inhibitor, indomethacin 3 μM, sufficient to selectively block the response to 0.2 μM arachidonic acid, did not reduce the effects of galanin. Likewise, a concentration of the serotonin antagonist, methysergide 3 μM, that completely blocked the effects of serotonin, had no effect on responses to galanin. Glyburide (3 μM), which produces a functionally opposite effect to galanin on ATP-sensitive K$^+$ channels regulating insulin secretion in rat islet tumor cells [4], failed to alter the contractions produced by galanin in the rat stomach fundus, suggesting that this particular intracellular mechanism is not involved in regulation of the smooth-muscle effects of galanin.

Several GAL fragment derivatives were prepared by SPPS, RPLC purification, and their physicochemical properties determined by AAA and FABMS. Initial studies of synthetic GAL fragment derivatives showed that the carboxy-terminal

---

*To whom correspondence should be addressed.

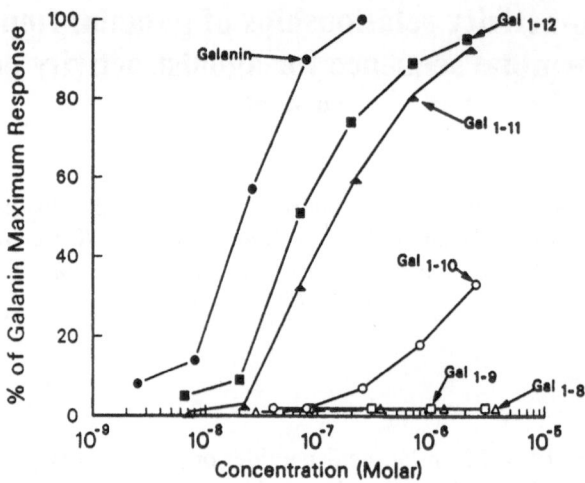

Fig. 1.   Representative experiment in which the effects of GAL ($GAL_{1-29-NH_2}$, $GAL_{1-12-NH_2}$, $GAL_{1-11-NH_2}$, $GAL_{1-10-NH_2}$, GAL_{1-9-NH_2} and GAL_{1-8-NH_2}) were compared. Data are plotted as the percent of the maximum response to GAL (ordinate) produced by various concentrations (Molar) of GAL derivatives (abscissa).

13–29 sequence was not required for full agonist activity. The amino-terminal 1–12 derivative, $GAL_{1-12-NH_2}$, had full efficacy and was nearly as potent (ED_{50} 5.6×10^{-8} M) as GAL, but Ac-$GAL_{13-29-NH_2}$ was completely inactive. The Gly^1 residue did not contribute to agonist properties of GAL, since Ac-$GAL_{2-29-NH_2}$ was equipotent to full-length $GAL_{1-29-NH_2}$. However, the Trp^2 residue was absolutely essential for the agonist properties of GAL since Ac-$GAL_{3-29-NH_2}$ had no agonist (or antagonist) activity at concentrations up to 3×10^{-6} M.

The total lack of agonist activity of GAL_{1-8-NH_2}, compared to the full agonist activity of $GAL_{1-12-NH_2}$, suggested that amino acids in the sequence Tyr^9-Leu^{10}-Leu^{11}-Gly^{12} also contribute to agonist activity. Synthesis and testing of GAL_{1-9-NH_2}, $GAL_{1-10-NH_2}$, and $GAL_{1-11-NH_2}$ revealed that the Leu^{10}-Leu^{11}-amino acids were the key residues involved. Figure 1 illustrates the fall-off in activity that occurs as the N-terminal sequence is shortened from 12 to 8 amino acids in length. Together, these results indicate that the key amino acids for the agonist activity of GAL are Trp^2 and Leu^{10}-Leu^{11}, and suggest that the peptide GAL_{2-11} should have full agonist activity. Indeed, synthesis and testing of Ac-$GAL_{2-11-NH_2}$ showed that it had full efficacy, relative to GAL, and equal potency to $GAL_{1-12-NH_2}$, which is approximately 1/2 to 1/3 the potency of full-length GAL.

Conclusion

In the present study, we used porcine GAL (Bachem) on a rat tissue. However, the only differences between porcine GAL and rat GAL [5] occur near the carboxy terminus, distal to the sites that are important for efficacy. Our results suggest that the amino acid sequence from Gly^{12} to the carboxy terminus plays only a minor role in affinity and no role in efficacy. Further studies are required to determine whether this relationship will hold for receptors in other tissues and species.

References

1. Tatemoto, K., Rökaeus, Å., Jornvall, H., McDonald, T.J. and Mutt, V., FEBS Lett., 164 (1983) 124.
2. Rökaeus, Å., Trends Neurosci., 10 (1987) 158.
3. Vane, J.R., Brit. J. Pharmacol. Chemother., 12 (1957) 344.
4. DeWeille, J., Schmid-Antomarchi, H., Fosset, M. and Lazdunski, M., Proc. Natl. Acad. Sci. U.S.A., 85 (1988) 1312.
5. Kaplin, L.M., Spindel, E.R., Isselbacher, K.J. and Chin, W.W., Proc. Natl. Acad. Sci. U.S.A., 85 (1988) 1065.

N-terminal analogs of vasoactive intestinal peptide: Identification of a binding pharmacophore

David R. Bolin[a], Jeanine M. Cottrell[a], Nancy O'Neill[b], Ralph J. Garippa[b] and Margaret O'Donnell[b]

[a]*Peptide Research Department and* [b]*Department of Pharmacology and Chemotherapy, Roche Research Center, Hoffmann-La Roche, Nutley, NJ 07110, U.S.A.*

Introduction

Vasoactive intestinal peptide (VIP) has been shown to be a potent smooth muscle relaxant and a principal endogenous component in the regulation of airway tone [1]. It is postulated that improper functioning of VIP-associated mechanisms or a reduction of VIP or VIP-containing neurons [2] may be responsible for the bronchoconstriction hyperreactivity, and mucus secretion found in bronchial asthma. An analog with properties superior to those of native VIP may be of significant value as a therapeutic agent for bronchial and allergic asthma.

Results and Discussion

VIP, as well as most other members of the glucagon family of peptide hormones, possesses an N-terminal histidine residue. Several groups have reported that the histidine residue is essential for full biological activity of the individual peptides, as noted for VIP, glucagon and secretin. Our initial structure-activity results for VIP had indicated the His[1] residue to be a potential receptor binding site. We wished to expand these results by defining the structural elements which represent this pharmacophore.

We had previously reported analog **1** (Table 1) as having enhanced potency (3.7-fold) over native VIP ($EC_{50} = 10$ nM) [3]. All indications had pointed to the side-chain imidazole group as being the binding site of His[1]. However analog **2**, in which the imidazole function was replaced by a hydrogen, was found to retain full intrinsic activity and potency, and competed effectively for binding to the VIP receptor. N-methylation of Ala[1], as in compound **3**, yielded only a slight reduction in potency. The unacetylated forms of **2** and **3**, compounds **4** and **5**, respectively, were, however, significantly less potent. These results suggested that the acetyl function retained all features of the imidazole of His[1] necessary for eliciting full biological activity. It is likely that receptor interaction, possibly as hydrogen bonding, occurs through the carbonyl group rather than the amide N-H.

Table 1   *Smooth-muscle relaxant activity of VIP analogs*

Compound	R-[X^1,Lys^{12},Nle^{17},Val^{26},Thr^{28}]-VIP				
	R	X	EC$_{50}$ (nM)[a]	Potency[b]	Binding (nM)[c]
1	CH$_3$CO-	His	2.7	1.00	15
2	CH$_3$CO-	Ala	2.4	1.13	69
3	CH$_3$CO-	N-Me-Ala	7.2	0.37	7.6
4	H-	Ala	250	0.01	35
5	H-	N-Me-Ala	150	0.02	340
6	CH$_3$CO-	D-Ala	3.9	0.69	5.3
7	CH$_3$CO-	Gly	1.7	1.59	0.63
8	CH$_3$CO-	β-Ala	190	0.01	270
9	–	pyro-Glu	45	0.06	40
10	CH$_3$SO$_2$	Ala	150	0.02	56
11	CH$_3$NHCO-	Ala	61	0.04	56
12	CH$_3$OCO-	Ala	23	0.12	17
13	HOOC(CH$_2$)$_2$CO-	–	300	0.01	1800
14	CH$_3$S(CH$_2$)$_2$CO-	–	26	0.10	19
15	CH$_3$SO(CH$_2$)$_2$CO-	–	3.8	0.71	82

[a] EC$_{50}$ for relaxation of guinea pig tracheal rings.
[b] Potency relative to compound 1 ($= 1.00$).
[c] Binding to guinea-pig lung membrane preparations by displacement of [^{125}I-Tyr^{10}]-VIP.

Inversion of the chirality at Ala^1 or elimination of the methyl group, as in compounds 6 and 7, respectively, did not significantly affect activity. Extending the distance of the acyl group from the α-carbon by one methylene unit, as in compound 8, dramatically decreased activity and binding. Compound 9, which conserved the α-carbon-acyl distance and locked the orientation of the acyl group, was also found to be reduced in activity. It appeared that proper distance and spacial arrangement of the acyl group was critical for activity.

The sulfonamide 10, methyl urea 11, and methyl carbamate 12 were found to be very weakly potent, with 12 being tenfold less potent than 2. Three analogs, 13 – 15, were prepared, by design, to be isosteric with N^α-acetyl species. The charged succinate derivative 13 was nearly inactive, whereas the sulfide 14 retained full intrinsic and reasonable binding activity with reduced potency. The methyl sulfoxide 15 was found to be almost equipotent to 2 in relaxant activity. This result may be rationalized by an approximate overlap between the sulfoxide and acetyl carbonyl groups in the two analogs.

These results have illustrated that the acyl and sulfoxide groups are effective bio-isosteric replacements for the imidazole group of the native molecule. Both groups, as in compounds 2 and 15, can be fit so that the oxygens overlap with the ring N^τ of the imidazole in 1. This has led us to propose that either the π or, more likely, the lone pair electrons participate in binding to the receptor and can be considered to be the pharmacophore at this site. Additional investigations, including design of analogs having restricted conformational freedom, will enhance our understanding of the nature of the binding interaction and may lead to more stable peptidomimetics.

References

1. Said, S.I., Kitamura, S., Yoshida, T., Preskitt, J. and Holden, L.D., Ann. N.Y. Acad. Sci., 221 (1974) 103.
2. Ollerenshaw, S., Jarvis, D., Woolcock, A., Sullivan, C. and Scheibner, T., New Engl. J. Med., 320 (1989) 1244.
3. Bolin, D.R., Sytwu, I.-I., Cottrell, J.M., Garippa, R.J., Brooks, C.C. and O'Donnell, M., In Marshall, G.R. (Ed.) Peptides: Chemistry and Biology, (Proceedings of the 10th American Peptide Symposium), ESCOM, Leiden, 1988, p. 441.

Release of acetylcholine from ileum invoked by vasoactive intestinal polypeptides (VIP) correlates with their potency to induce contraction

Peter K. Chiang, R. Richard Gray and Richard K. Gordon

Department of Applied Biochemistry, Walter Reed Army Institute of Research, Washington, DC 20307-5100, U.S.A.

Introduction

The neuropeptide VIP is distributed in the peripheral and central nervous system, and is a potent stimulus of diverse biological actions [1,2]. Previous studies [3–5] have shown that VIP either partially mediates the release of acetylcholine (ACh) from ileum smooth muscle or that there is no effect of VIP on contraction in this tissue.

The present experiments were undertaken to correlate the VIP-induced contraction of guinea pig ileum with the release of ACh from the ileum. Modifications of conventional HPLC-ACh [6] methods were employed to separate ACh from all other contaminants and allow its quantification. The amino acid sequences of the three VIP analogs used in these studies are shown below. The differences in amino acid substitutions are underlined.

(gp)VIP: His-Ser-Asp-Ala-<u>Leu</u>-Phe-Thr-Asp-<u>Thr</u>-Tyr-Thr-Arg-Leu-Arg-Lys-Gln-Met-Ala-<u>Met</u>-Lys-Lys-Tyr-Leu-Asn-Ser-<u>Val</u>-Leu-Asn-NH$_2$

(hpr)VIP: His-Ser-Asp-Ala-<u>Val</u>-Phe-Thr-Asp-<u>Asn</u>-Tyr-Thr-Arg-Leu-Arg-Lys-Gln-Met-Ala-<u>Val</u>-Lys-Lys-Tyr-Leu-Asn-Ser-<u>Ile</u>-Leu-Asn-NH$_2$

Results and Discussion

Two VIP analogs were found to cause a concentration-dependent contraction of guinea pig ileum (Fig. 1). The VIP analog from guinea pig, (gp)VIP, was the most potent inducer of guinea pig ileum contraction, followed by the human-porcine-rat (hpr)VIP, which differs from (gp)VIP by four amino acid substitutions. (hpr)VIP(1–15) was inactive even at 10 μM. However, ileum contraction induced by 1 μM (gp)VIP, the most potent VIP, was only about 50% of the maximum observed with ACh. In the present studies, the contraction of the ileum induced by VIP or ACh was blocked by the specific muscarinic antagonist atropine [7,8] in a dose-dependent manner (not shown). Thus, the contractions observed in these ileum preparations most likely involved only muscarinic receptors and not nicotinic receptors.

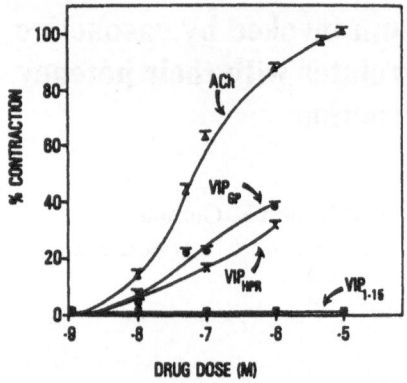

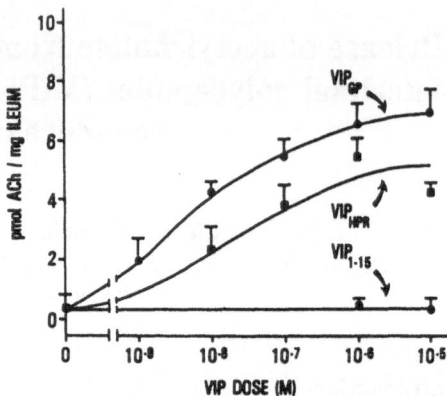

Fig. 1. Contraction of guinea pig ileum by ACh or VIP analogs. Each point is the mean ± SEM of three or more experiments.

Fig. 2. VIP-induced release of ACh from guinea pig ileum. Each point is the mean ± SEM of three or more experiments.

Since the endogenous inducer of muscarinic receptor-mediated ileum contraction is ACh, the mode of action of the VIP analogs was determined by measuring their effects on ACh secretion from the ileum. At 1 μM VIP, (gp)VIP was the most potent inducer of ACh secretion from the ileum (Fig. 2) and caused the largest ileum contraction. (hpr)VIP induced less ACh secretion than (gp)VIP, and was also less potent in inducing ileum contraction. Dose-dependent secretion of ACh from guinea pig ileum was observed for both (gp)VIP and (hpr)VIP, but (hpr)VIP(1–15), even at 10 μM, yielded none.

In both assays for the VIP effects, ileum contraction and ACh secretion (Figs. 1 and 2), the rank order of potency of the VIP analogs was the same: (gp)VIP > (hpr)VIP, whereas (hpr)VIP(1–15) was inactive. These results suggest that the VIP receptor [9] in the guinea pig ileum tolerated the nonpolar amino acid substitutions introduced into (hpr)VIP. Even though the differences in the amino acid content were small between (gp)VIP and (hpr)VIP, the VIP receptors of guinea pig ileum distinguished between these substitutions. However, the VIP receptor did not recognize (hpr)VIP(1–15), thus indicating that the VIP receptor requires a specific peptide length for recognition.

References

1. Itoh, M., Obata, K., Yanaihara, N. and Okamoto, H., Nature 304 (1983) 547.
2. Tatemoto, K. and Mutt, V., Proc. Natl. Acad. Sci. U.S.A., 78 (1981) 6603.
3. Cohen, M.L. and Landry, A.S., Life Sci., 26 (1980) 811.
4. Kusunoki, M., Tsai, L.H., Taniyama, K. and Tanaka, C., Am. J. Physiol., 251 (1986) G51.
5. Okuno, M., Shinomura, Y., Himeno, S., Kashimura, M. and Tarui, S., Biomed. Res., 9 (1988) 119.

6. Eva, C., Hadjiconstantinov, M., Neff, H.F. and Meek, J.L., Anal. Biochem., 143 (1984) 320.
7. Kilbinger, H., Halim, S., Lambrecht, G., Weiler, W. and Wessler, I., Eur. J. Pharmacol., 103 (1984) 313.
8. Gordon, R.K. and Chiang, P.K., J. Pharmacol. Exp. Ther., 236 (1986) 85.
9. Gespach, C., Bawab, W., Chastre, E., Emami, S., Yanaihara, N. and Rosselin, G., Biochem. Biophys. Res. Comm., 151 (1988) 939.

Small, synthetic growth hormone-releasing peptides:
A new class of drugs

T.O. Yellin[a], W.F. Huffman[a], H.W. Alila[b], R.J. Gyurik[b], T.O. Lindsey[b] and B.E. Ilson[c]

[a]Department of Peptide Chemistry, Smith Kline & French Laboratories, Swedeland, PA 19479, U.S.A.
[b]Smith Kline Beckman Animal Health Products, West Chester, PA 19350, U.S.A.
[c]Clinical Research Unit, Smith Kline & French Laboratories, Presbyterian Hospital, Philadelphia, PA 19104, U.S.A.

Introduction

Soon after the discovery of the endogenous opioid pentapeptide [menthionine[5]], and [leucine[5]]-enkephalins [1], it was shown that, like morphine [2,3], these peptides released growth hormone (GH) and prolactin (PRL) in vivo in the rat [4–6]. Metabolically more stable analogs, such as [D-Ala[2]]-methionine enkephalin amide [7], were extraordinarily potent stimulators of GH and PRL release and led to the hypothesis that endogenous opiate peptides are important regulators of neuroendocrine functions [7]. Against this background, Beckman scientists were engaged in an effort to develop enkephalin analogs as medicinal alternatives to opioid alkaloids when, in 1977, Bowers et al. [8] discovered that [D-Trp[2]]-methionine enkephalin amide released GH but not PRL or other anterior pituitary hormones from rat pituitary in vitro. The effect was concentration-dependent and not blocked by naloxone. As a result, the Beckman group determined to develop specific, non-opiate GH-releasing peptides (GHRPs) as growth promoters.

Independently, other early workers at Wyeth [9] and Sandoz [10] also observed that the GH- and PRL-releasing activities of their enkephalin analogs diverged from typical or expected opiate pharmacology, as well as showing differential effects on GH and PRL. These leads and others [7], indicating the involvement of multiple receptors in the neuroendocrine actions of enkephalins, remained in a pregnant background. In any case, the GH-releasing activity of Sandoz FK 33–824, [D-Ala[2], N-MePhe[4], Met(O)-ol]-enkephalin, was dramatically demonstrated in man [10,11].

Trial and error substitutions in the [D-Trp[2]]-pentapeptide, aided by conformational energy calculations, resulted in increasingly more potent analogs. This phase of the work culminated in a peptide, Tyr-D-Trp-Ala-Trp-D-Phe-NH$_2$ (SK&F 110520), that was 1000 times more potent in vitro than the parent enkephalin analog, and led to the hypothesis that two sets of stacked aromatic rings, separated by a spacer (e.g. Ala or Ser), represented the optimal bioactive conformation of GHRPs [12]. The new pentapeptides were not active in vivo [13], but continued empirical modifications resulted in yet more potent analogs with in vivo activity [14].

SK&F 110679, His-D-Trp-Ala-Trp-D-Phe-Lys-NH$_2$, is the outstanding example of this new class of drugs [15]. The hexapeptide, which bears no resemblance to enkephalins or native GHRH [16], is a potent and specific GHRP in chicks, rats, lambs, calves, and rhesus monkeys. Chronic treatment with i.p. SK&F 110679 resulted in significant acceleration of body weight gain in the rat [15].

We have extended previous investigations of SK&F 110679 by studying its effects in the pig and in man.

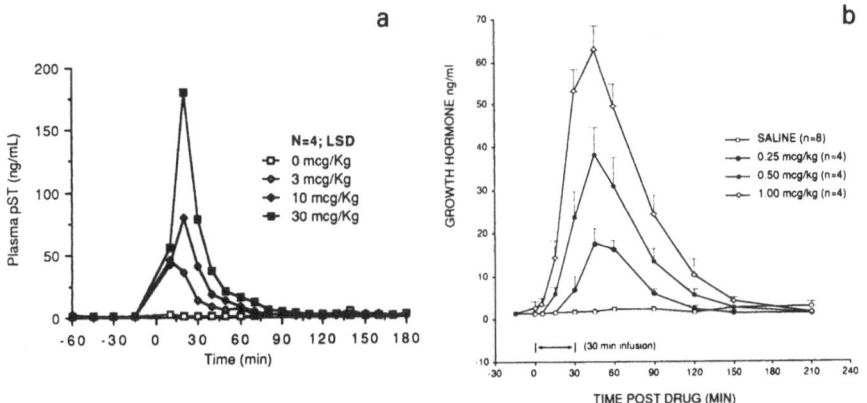

Fig. 1. Effect of SK&F 110679 (i.v. bolus) on porcine plasma somatotropin levels in pigs (a) and on human serum somatotropin levels in male volunteers (b; i.v. infusion).

Results and Discussion

In the pig, baseline GH values are low and steady for extended periods, so it is easy to obtain unequivocal results in a few subjects. The pig also proved to be the most sensitive to GHRPs, among laboratory animals. Bolus i.v. administration of SK&F 110679 (3–30) μg/kg produced sharp, dose-dependent increases in plasma GH levels that peaked in 15 min and returned to baseline after an hour (Fig. 1a). Similar, but more sustained activity was obtained after s.c. administration (25–100 μg/kg). In two of three fasted pigs, there was a marked, prolonged rise in GH after oral administration of SK&F 110679 in solution (0.5 mg/kg). Finally, three of three pigs responded with rapid elevation of GH to peak levels between 45 and 75 ng/ml plasma following intranasal administration of SK&F 110679 (200 μg/kg in 0.4 ml), an effect that subsided during an hour.

Our initial results in male volunteers have encouraged us to believe that SK&F 110679 may be an outstanding clinical candidate for therapeutic use in humans. Intravenous infusions of the peptide (0.25–1.0 μg/kg/30 min) produced dose-dependent elevations in serum GH (Table 1 and Fig. 1b) that peaked 45 min

215

Table 1   *GH response to SK&F 110679*

Dose (µg/kg)	Peak serum GH (ng/ml)		AUC (ng·min/ml)	
Saline (n = 8)	1.9 ± 0.26		488 ± 77	
0.25 (n = 4)	17.8 ± 6.1	(P = .03)	1 178 ± 228	(P = .004)
0.50 (n = 4)	38.3 ± 9.2	(P = .04)	2 460 ± 585	(P = .044)
1.00 (n = 4)	63.0 ± 5.4	(P = .002)	4 289 ± 620	(P = .009)

after the start of the infusion and returned to baseline by the fourth hour. Serum PRL, FSH, LH, TSH and ACTH were not significantly affected by the drug, which was well tolerated. There were no clinically significant changes in cardiovascular parameters, clinical chemistry or hematology. Additional clinical investigations are in progress.

Conclusions

The advantages of a small, synthetic peptide (e.g., cost, formulation, and administration) for present and proposed indications of GH and GHRH are readily apparent. Indeed, compared with results reported in the literature [16], it appears that SK&F 110679 is even more efficacious than GHRH as a GH secretagog in man.

A recent report from Merck [17] is convincing in its arguments that GHRH and SK&F 110679 act on separate and distinct receptors and demonstrates that the two peptides are highly synergistic in combination. But the mechanism of action of this unique drug remains to be discovered, and it is being explored.

We have not excluded the possibility that the action of SK&F 110679 is mediated by an as yet unclassified opiate receptor.

References

1. Hughes, J., Smith, T.W., Kosterlitz, H.W., Fothergill, L.A., Morgan, B.A. and Morris, H.R., Nature, 258 (1975) 577.
2. Kokka, N., Garcia, J.F., George, R. and Elliott, H.W., Endocrinology, 90 (1972) 735.
3. Reed, J.L. and Ghodse, A.H., Br. Med. J., II (1973) 582.
4. Dupont, A., Cusan, L., Garon, M., Labrie, F. and Li, C.H., Proc. Natl. Acad. Sci. U.S.A., 74 (1977) 358.
5. Cocchi, D., Santagostino, A., Gil-Ad, I., Ferri, S. and Muller, E.E., Life Sci., 20 (1977) 2041.
6. Bruni, J.F., Van Vugt, D., Marshall, S. and Meites, J., Life Sci., 21 (1977) 461.
7. Cusan, L., DuPont, A., Kledzik, G.S., Labrie, F., Coy, D.H. and Schally, A.V., Nature, 268 (1977) 544.
8. Bowers, C.Y., Momany, F.A., Chang, D., Hong, A. and Chang, K., Endocrinology, 106 (1980) 663.
9. Lien, E.L., Clark, D.E. and McGregor, W.H., FEBS Lett., 88 (1978) 208.
10. Von Grafferied, B., El Pozo, D., Roubicek, J., Krebs, E., Polding, W., Bumvieister, P. and Kerp, L., Nature, 272 (1978) 729.

11. Stubs, W.A., Jones, A., Edwards, C.R.W., Delitala, G., Jeffcoate, W.J., Ratter, S.J., Besser, G.M., Bloom, S.R. and Alberti, K.G.M.M., Lancet, ii (1978) 1225.
12. Momany, F.A., Bowers, C.Y., Reynolds, G.A., Chang, D., Hong, A. and Newlander, K., Endocrinology, 108 (1980) 31.
13. Bowers, C.Y., Reynolds, G.A., Chang, D., Hong, A. and Momany, F.A., Endocrinology, 108 (1980) 1071.
14. Momany, F.A., Bowers, C.Y., Reynolds, G.A., Hong, A. and Newlander, K., Endocrinology, 114 (1984) 1531.
15. Bowers, C.Y., Momany, F.A., Reynolds, G.A. and Hong, A., Endocrinology, 114 (1984) 1537.
16. Thorner, M.O. and Cronin, M.J., In Muller, E.E., MacLeod, R.M. and Frohman, L.A. (Eds.) Neuroendocrine Perspectives, Vol. 4, Elsevier Science Publishers, Amsterdam, 1985, p. 95.
17. Cheng, K., Chan, W.W.-S., Barrets, Jr., A., Convey, D.M. and Smith, R.G., Endocrinology, 124 (1989) 2791.

Effect of bovine serum albumin (BSA) on the degradation rate of growth hormone-releasing factor (GRF) by dipeptidylpeptidase-IV

Teresa M. Kubiak, Roger A. Martin, Diane L. Cleary and James F. Caputo

The Upjohn Company, Reproduction and Growth Physiology Research, Kalamazoo, MI 49001, U.S.A.

Introduction

A primary site of enzymatic degradation of GRF in vitro [1–3] and in vivo [1] is at the peptide bond between positions 2 and 3. This cleavage, caused by plasma dipeptidylpeptidase-IV (DPP-IV), results in virtually complete loss of hormone activity [1,3].

BSA is the most abundant protein in bovine plasma. Zysk et al. [4] recently reported that GRF shows some affinity towards BSA. In the present study, we examined effects of BSA on the DPP-IV-mediated cleavage kinetics of a bovine GRF analog, [Leu^{27}]-bGRF(1–29)NH$_2$ (1).

Results and Discussion

(1) was made by the solid phase method [3] and served as a model GRF .peptide. The kinetic studies were carried out using either purified porcine kidney-derived DPP-IV or bovine plasma as the DPP-IV source.

Table 1 lists the results of a series of four experiments performed to examine the influence of fatty-acid-free BSA on the cleavage rate of GRF by purified DPP-IV. In each case the parent peptide was enzymatically converted to [Leu^{27}]-bGRF(3–29)NH$_2$. The hydrolysis rate of (1) by DPP-IV in the absence of BSA was taken as the reference point (Expt. 1). (1) pretreatment with 4% BSA in PBS at 37°C for 30 min. prior to the addition of DPP-IV reduced the peptide hydrolysis rate by 29% as compared to reference (Expt. 2). The hydrolysis was slowed by 33% and 57%, respectively, when DPP-IV or both DPP-IV and GRF were preincubated with BSA before initiating the enzymatic reaction (Expts. 3 and 4).

These results can be accounted for, if one assumes that both (1) and DPP-IV bind to BSA in an equilibrium fashion, and that it is the free forms of both that are necessary for the enzymatic reaction to proceed. In support of this, an intermediate inhibition of the (1) hydrolysis rate was observed in the presence of 2% BSA (Expts. 2 and 3), as opposed to the maximum inhibition observed in 4% BSA solution (Expt. 4).

While the present set of experiments did not directly address the BSA interaction

Table 1 *Effects of BSA on the cleavage rate of (1) by DPP-IV**

Exp. No.	Pretreatment with 4% BSA		%BSA in reaction (w/v)	$(k \pm SD) \times 10^{2}**$ (min.$^{-1}$)	$t_{1/2}$ (min.)	Relative stability (%)
	GRF	DPP-IV				
1	–	–	0	3.04 ± 0.06^a	22.8	100
2	+	–	2	2.35 ± 0.08^b	29.5	129
3	–	+	2	2.28 ± 0.19^b	30.4	133
4	+	+	4	1.94 ± 0.04^c	35.7	157

* $(1) = [Leu^{27}]$-bGRF(1–29)NH$_2$; 0.06 mM (1) in PBS at pH 7.6 with or without BSA was mixed 1 : 1 (v/v) with DPP-IV/PBS solution with or without BSA. Reaction mixture was incubated at 37°C. Aliquots of 0.05 ml were taken at 0, 5, 10 ,15, 20, 30, 45 and 60 min, and quenched with 0.05 ml 8M guanidine HCl in 0.5% TFA for HPLC analysis (Vydac C$_{18}$, 0.46 × 5 cm; 0.05% aq.TFA-CH$_3$CN)

** k, Degradation rate of (1) was determined by nonlinear analysis based on first-order kinetics (n = 3); values with different superscripts were statistically different at $p < 0.05$.

with (1) and/or DPP-IV, previous efforts to determine the (1)/BSA binding constant clearly indicated the presence of substantial binding. Due to high nonspecific binding of (1) (data not shown), however, precise numerical estimates were not obtained.

When bovine plasma was used as the DPP-IV source, pretreatment of (1) with BSA had no effect on the peptide hydrolysis rate. The amount of the intact peptide recovered from plasma, by means of a slightly modified solid phase extraction (1), after a 1-h incubation at 37°C was 44.4% ± 4.5% vs. 42.9% ± 1.5% with and without 4% BSA pretreatment, respectively (n = 6). No treatment effect in this case may result from a high endogenous BSA concentration in plasma masking effects of exogenous BSA.

Acknowledgements

We wish to thank Dr. Richard Schowen, the University of Kansas at Lawrence, U.S.A., and Dr. Alfred Barth, Martin Luther University, Halle, D.D.R., for a gift of purified porcine DPP-IV.

References

1. Frohman, L.A., Downs, T.R., Williams, T.C., Heimer, E.P., Pan, Y.C.-E. and Felix, A.M., J. Clin. Invest., 78 (1986) 906.
2. Frohman, L.A., Downs T.R., Heimer E.P., and Felix, A.M., J. Clin. Invest., 83 (1989) 1533.
3. Kubiak, T.M., Kelly, C.R., and Krabill, L.F., Drug Metab. Dispos., 17 (1989) 393.
4. Zysk, J.R., Cronin, M.J., Anderson, J.M. and Thorner, M.O., J. Biol. Chem., 261 (1986) 1671.

Characterization and identification of the major decomposition products of [Leu^{27}]hGRF(1–32)NH$_2$ formed upon incubation in aqueous solution at pH 7.4

A.R. Friedman[a], A.K. Ichhpurani[a], D.M. Brown[a], R.M. Hillman[a], H.A. Zurcher-Neely[b] and D.M. Guido[b]

[a]*Reproduction and Growth Physiology Research and* [b]*Biopolymer Chemistry, The Upjohn Company, Kalamazoo, MI 49001, U.S.A.*

Introduction

Growth hormone releasing factors (GRFs) are a series of closely related 43–44 residue hypothalamic peptides that are involved in the regulation of the synthesis and release of growth hormone [1]. The amino-terminal residues from a number of species are especially conserved. GRF(1–29)NH$_2$ contains the full growth hormone releasing potency of the native 43–44 residue peptides [2]. All reported native GRFs contain Met^{27}. However, this residue can be replaced by Leu with essentially no change in growth hormone-releasing potency [3].

Elevation of growth hormone is associated with increased growth [4] and lactation [5] in a number of animal species. The use of GRF for the regulation of a long term process, such as growth, will require extended treatment. We examined the effect of incubation of [Leu^{27}]hGRF(1–32)NH$_2$ (aqueous buffer, pH 7.4) on its growth hormone-releasing potency and chemical integrity .

Results and Discussion

Upon incubation of a solution of [Leu^{27}]hGRF(1–32)NH$_2$:

Tyr-Ala-Asp-Ala-Ile-Phe-Thr-Asn-Ser-Tyr-Arg-Lys-Val-Leu-Gly-Gln-Leu-Ser-Ala-Arg-Lys-Leu-Leu-Gln-Asp-Ile-Leu-Ser-Arg-Gln-Gln-Gly-NH$_2$

in buffer (200 μM peptide, 20 mM sodium phosphate, 150 mM sodium chloride, 0.02% sodium azide, pH 7.4, 37°C) for 14 days (336 h), the 14 day sample had only 22% of the growth hormone-releasing potency of the 0-day sample from cultured bovine anterior pituitary cells (the '0' day sample and 14 day sample had EC$_{50}$s of 0.14 and 0.65 nanomolar respectively).

The incubation samples (0, 9, 26, 48, 97, 167, and 336 h) were examined by HPLC. At 14 days, 32% of the original [Leu^{27}]hGRF(1–32)NH$_2$ remained. Under these conditions, [Leu^{27}]hGRF(1–32)NH$_2$ had a calculated half-life of 193 h (K$_d$ = 0.0037/h). Two major decomposition products, with appearance constants of 0.0025/h and 0.00093/h, respectively and a minor decomposition product were observed. The predominant decomposition product (46%) was shown to be [isoAsp^8,Leu^{27}]hGRF(1–32)NH$_2$ (AAA, same as parent peptide, Edman sequencing, parent sequence through cycle 7, blocked at cycle 8, FABMS (M + H)$^+$ 3655.1, parent (M + H)$^+$ 3654.4, comparison of HPLC retention times to the synthesized peptide and its Clostripain map, identical). The second major decomposition product (16%) was shown to be [Asp^8,Leu^{27}]hGRF(1–32)NH$_2$ (AAA, same as parent peptide, sequencing parent sequence through cycle 15, FABMS (M + H)$^+$ 3655.7, parent (M + H)$^+$ 3654.4), comparison of HPLC retention times to the synthesized peptide and its Clostripain map, identical). Clostripain selectively hydrolyses peptides containing Arg residues at its carboxyl side, and in the case of [Leu^{27}]hGRF(1–32)NH$_2$ gives fragments 1–11 (C1), 12–20 (C2), 21–29 (C3), and 30–32 NH$_2$ (not observed). Only C1 fragments contain Tyr residues and can be distinguished from C2 and C3 fragments by the presence of UV absorption at 276 nm. The minor decomposition product (5%) was partially characterized and may be [D-isoAsp^8,Leu^{27}]hGRF(1–32)NH$_2$ (AAA, same as parent, sequencing, parent through cycle 7, blocked at cycle 8, Clostripain map, modified C1, intact C2 and C3). During Clostripain mapping of the [Asp^8, Leu^{27}]hGRF(1–32)NH$_2$ fraction from a degraded sample of [Leu^{27}]hGRF (1–32)NH$_2$, a second, minor, unknown C-1 peak was detected. Although not further characterized, this peak may signify the formation of [D-Asp^8, Leu^{27}]hGRF(1–32)NH$_2$ as well.

Deamidation of Asn and Gln residues in peptides has been reported [6]. Generally, Asn residues are more susceptible than Gln residues. Although [Leu^{27}]hGRF(1–32) contains four Gln residues, in addition to the single Asn residue, no Gln deamidation products were detected. Deamidation of Asn residues is reported to involve a succinimide intermediate that gives, upon addition of water, both the isoAsp and Asp peptides. The isoAsp peptide is usually the predominant product [7]. Minor amounts of the epimerized D-isoAsp and D-Asp peptides have also been observed.

When measured in a bovine pituitary cell culture assay for GH release, the synthesized [isoAsp^8,Leu^{27}]hGRF(1–32)NH$_2$ was estimated to be 400–500 times less potent and the [Asp^8,Leu^{27}]hGRF(1–32)NH$_2$ calculated to be 25 times less potent than [Leu^{27}]hGRF(1–32)NH$_2$. The loss of GH releasing potency of the incubated [Leu^{27}]hGRF(1–32)NH$_2$ solution is explained by the diminished content of parent peptide and the low potency of its degradation products.

References

1. Frohman, L.A. and Jansson, J.-O., Endocr. Rev., 7, 3 (1986) 223.
2. Ling, N., Baird, A., Wehrenberg, W.B., Ueno, N., Munegumi, T. and Brazeau, P., Biochem. Biophys. Res. Commun., 123 (1984) 854.
3. Felix, A.M. and Heimer, E.P., U.S. Patent 4, 732, 972, March 22, 1988.
4. Moseley, W.M., Huisman, J. and VanWeerden, E.J., Domestic Anim. Endocrinol., 4, 1 (1987) 51.

5. Enright, W.J., Chapin, L.T., Moseley, W.M. and Tucker, H.A., J. Dairy Sci., 71 (1988) 99.
6. Robinson, A.B. and Rudd, C.J., Curr. Top. Cell. Regul., 8 (1974) 247.
7. Geiger, T. and Clarke, S., J. Biol. Chem., 262, 2 (1987) 785.

A possible β-sheet/β-turn structure in resin-bound segments of a human GRF analog: Conformational origin of a difficult coupling

Charles M. Deber[a], Mary K. Lutek[a], Edgar P. Heimer[b] and Arthur M. Felix[b]

[a]Research Institute, Hospital for Sick Children, Toronto, Canada M5G 1X8 and Department of Biochemistry, University of Toronto, Toronto, Ont., Canada M5S 1A8
[b]Peptide Research Department, Hoffmann-La Roche Inc., Nutley, NJ 07110, U.S.A.

Introduction

'Difficult couplings' represent steps that remain incomplete under conditions in which the majority of couplings in the same overall synthesis proceed readily to essentially 100% reaction, and often require multiple coupling before completion is reached. Such difficult coupling steps have been generally attributed to a sequence-dependent tendency for the growing peptide chain to self-associate, with concomitant aggregation and/or precipitation that can impede the approach of incoming reagents [1].

Human growth hormone releasing factor and a variety of its analogs have been synthesized and assayed for potency (for a review, see Ref. 2). During the solid phase synthesis of one such analog, [Ala15, Leu27, Asn28] – GRF (1-32)-OH, we noted that the incorporation of Boc-Gln16-OH was incomplete even after 3 cycles of coupling [3]. A similar situation had been noted for the corresponding step in the synthesis of GRF (1–27)-OH [4]. No particular feature of primary sequence in the middle of the present analog ([Ala15, Leu27, Asn28] – GRF (14–32) = Leu14-Ala-Gln-Leu-Ser(Bzl)-Ala-Arg(Tos)-Lys(C1Z)-Leu-Leu-Gln -Asp(OcHex)-Ile-Leu-Asn-Arg(Tos)-Gln-Gln-Gly32-PAM resin), nor of the two amino acids specifically involved in the coupling (Gln16 to Leu17) provided an immediate rationale for the observed 'difficult coupling'.

Results and Discussion

We have now obtained proton NMR spectra (500 MHz) of resin-bound GRF analog fragments 19–32 through 14–32 in the region 3.9–5.4 ppm (Fig. 1). In this spectral region, resonances occur for α-protons of most residues, along with benzylic methylene protons (of side-chain protecting groups) near 5.25 ppm. In spectra 18–32 and 17–32, superimposed broader and narrower components are visible; at 16–32, only broad resonance envelopes appear for all protons. Since broadening of line-widths in NMR spectra is fundamentally attributable to a decrease in rates of local molecular motions [5], peptide chains at these stages of the synthesis are interacting with each other, and probably also between polymer beads, to produce particles of higher average molecular weight(s). Upon

223

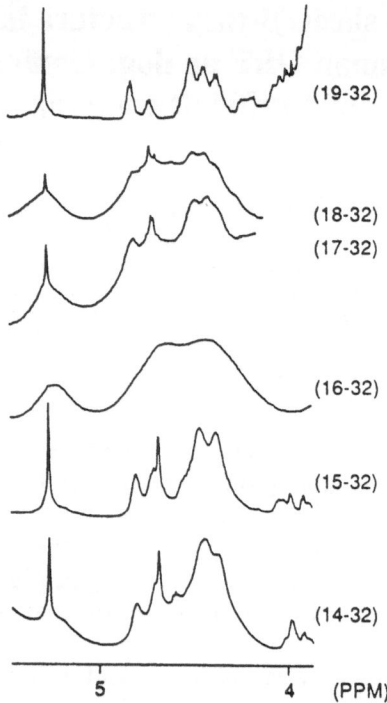

Fig. 1. *^{1}H high-resolution NMR spectra (500 MHz), α-proton region, of GRF analog (19–32) through (14–32) fragment peptides bound to PAM-resin (PAM=p-phenylacetamido-methyl). Concentrations: 30 mg resin-bound peptide/ml dimethylsulfoxide-d_6. Spectra (128 transients) were collected in 16 400 data points; acquisition time=2.98 s; relaxation delay=2.0 s; line-broadening=0.1 Hz. Adapted with permission from Ref. 3.*

addition of Ala^{15}, the spectrum of the 15–32 peptide resolves and once again resembles the spectral situation in fragment 19–32.

We propose that the conformational origin of the difficult coupling of Gln^{16} to Leu^{17} arises from a specific combination of peptide secondary structural features, namely the β-sheet/β-turn structure shown in Fig. 2. Thus, a propensity for β-sheet formation, for example, in the GRF linear sequence 20–27, in conjunction with β-turn formation 'upstream' from the difficult coupling site (at Ala^{19}, Arg^{20} as shown) folds the peptide chain to provide the requisite 'anti-parallel partner' of the sheet structure, thereby converting an incipient β-sheet single chain into the actual two-strand structure. Although Chou–Fasman predictive criteria [6] must be applied with caution to this system (for a discussion, see Ref. 3), we found that the probability of a β-turn involving GRF residues 18–21 (as depicted in Fig. 2) is strikingly higher than that of the surrounding four-residue segments. Finally, since this scheme reflects the dominance of peptide primary sequence in conferring specific secondary structure, addition of Ala^{15} must similarly produce long-range effects on peptide chain conformation, i.e.,

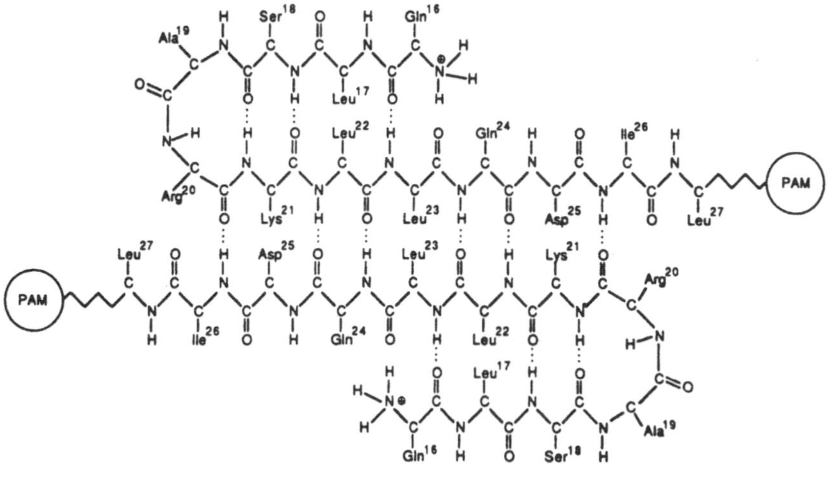

Fig. 2. *Schematic representation of an intermolecular β-sheet/β-turn structure for resin-bound GRF(16–32) proposed to account for a 'difficult coupling' synthesis step at Gln¹⁶ to Leu¹⁷. Residues 16–27 are shown, but the remaining residues (28–32) may also participate in the structure. In this hypothetical scheme, intra-chain anti-parallel β-sheet structure is nucleated by the propensity for β-turn formation around Ala¹⁹, Arg²⁰. Once it is nucleated, intermolecular structure can be propagated indefinitely by inter-chain H-bonding. Adapted with permission from Ref. 3.*

a net preference for intramolecular (helical?) structure in the newly elongated segment (15–27). The β-turn is then eliminated, the aggregates dissociate, and the coupling steps to the growing peptide chain once again proceed normally.

These experimental observations in synthesis and spectroscopy suggest that awareness of potential β-sheet/β-turn sequences can guide analog design and/or facilitate pre-programming of synthesis steps in anticipation of difficult couplings.

Acknowledgements

This work was supported, in part, by a grant to C.M.D. from the Medical Research Council of Canada (MT-5810). M.K.L. held an MRC studentship.

References

1. Kent, S.B.H., Annu. Rev. Biochem., 57 (1988) 957.
2. Felix, A.M., Heimer, E.P. and Mowles, T.F., Annu. Rep. Med. Chem., 20 (1985) 185.
3. Deber, C.M., Lutek, M.K., Heimer, E.P. and Felix, A.M., Pept. Res., 2 (1989) 184.
4. Kent, S.B.H., In Deber, C.M., Hruby, V.J. and Kopple, K.D., (Eds.) Peptides: Structure and Function (Proceedings of the 9th American Peptide Symposium), Pierce Chemical Co., Rockford, IL, 1985, p. 407.
5. Allerhand, A. and Oldfield, E., Biochemistry 12 (1973) 3428.
6. Chou, P. and Fasman, G.D., Annu. Rev. Biochem., 47 (1978) 251.

225

Solid-phase synthesis and biological activity of highly potent cyclic and dicyclic (lactam) analogs of growth hormone-releasing factor

Arthur M. Felix[a], Ching-Tso Wang[a], Edgar P. Heimer[a], Robert M. Campbell[a], Vincent S. Madison[a], David Fry[a], Voldemar Toome[a], Thomas R. Downs[b] and Lawrence A. Frohman[b]

[a]Roche Research Center, Hoffmann-La Roche Inc., Nutley, N.J. 07110, U.S.A.
[b]Division of Endocrinology and Metabolism, University of Cincinnati College of Medicine, Cincinnati, OH 45267, U.S.A.

Introduction

A family of (i, i+4) cyclic GRF analogs was designed to retain α-helical segments. GRF analogs with a lactam ring in the central region ($Asp^8 \rightarrow Lys^{12}$), e.g. cyclo^{8,12}[Asp^8,Ala^{15}]-GRF(1–29)-NH$_2$ have been shown to retain significant biological activity (growth-hormone release) in vitro and in vivo [1,2] (Table 1). The cyclic analogs have also been shown to have improved stability to plasma degradation (Table 1). Although the parent peptide, GRF(1–44)-NH$_2$, has been reported to have a short lifetime in human plasma (92% degraded in 60 min at 37°C) [3], the lifetime of cyclo^{8,12}[Asp^8,Ala^{15}]-GRF(1–29)-NH$_2$ is significantly enhanced [4] (Table 1). Moreover, N-terminal-replacement analogs in the cyclo^{8,12}-GRF series resulted in improved potencies in vitro and in vivo [5]. A series of novel (i, i+4) cyclic GRF analogs, in which the lactam ring is near the COOH-terminus (Lys$^{21} \rightarrow$ Asp^{25}), was designed to evaluate the relative importance of the position of the lactam with respect to potency, duration of activity and conformation.

Table 1 *Relative potencies/stabilities of substituted cyclic analogs of GRF(1–29)-NH$_2$*

GRF analog	Relative potency	Plasma degradation[a]
GRF(1–44)-NH$_2$	1.0	92%
Cyclo^{8,12}-[Asp^8,Ala^{15}]-GRF(1–29)-NH$_2$	0.77	12%
Cyclo^{8,12}-[des-NH$_2$-Tyr^1,D-Ala^{15}]-GRF(1–29)-NH$_2$	2.47	–
Cyclo^{21,25}-[Ala^{15}]-GRF(1–29)-NH$_2$	1.33	23%
Cyclo^{21,25}-[des-NH$_2$-Tyr^1,D-Ala^2,Ala^{15}]-GRF(1–29)-NH$_2$	1.33	–
Cyclo$^{8,12;21,25}$-[Asp^8,Ala^{15}]-GRF(1–29)-NH$_2$	0.69	5%
Cyclo$^{8,12;21,25}$-des-NH$_2$-Tyr^1,D-Ala^2,Asp^8,Ala^{15}]-GRF(1–29)-NH$_2$	3.91	–

[a]% degradation after incubation in human plasma for 60 min at 37°C.

Results and Discussion

Cyclo^{21,25}-[Ala^{15}]-GRF-(1–29)-NH$_2$ was synthesized by the solid-phase strategy, cyclized on the resin, and cleaved by the HF procedure. Although this compound retained high biological activity (potency = 1.33 relative to GRF(1–44)-NH$_2$), N-terminal-replacement analogs did not result in further improvement in potency as was previously observed for the cyclo^{8,12}-GRF series (Table 1). Our recently reported solid-phase strategy [5] using N$^\alpha$-Boc-amino acids together with OFm/Fmoc side-chain protection for Asp and Lys, respectively, was then extended to include two stepwise BOP cyclizations of Lys^{21} to Asp^{25} followed by Asp^8 to Lys^{12} (Fig. 1). The resultant novel dicyclo-GRF analog, dicyclo$^{8,12;21,25}$-[Asp^8,Ala^{15}]-GRF(1–29)-NH$_2$, was equipotent to the two monocyclic peptides. N-terminal replacements in the dicyclo$^{8,12;21,25}$-GRF system resulted in increased potencies. Cyclo$^{8,12;21,25}$-[des-NH$_2$-Tyr^1,D-Ala^2;Asp^8,Ala^{15}]-GRF(1–29)-NH$_2$ was the most active analog in the series with nearly 4 times the in vitro potency of GRF(1–44)-NH$_2$ (Table 1).

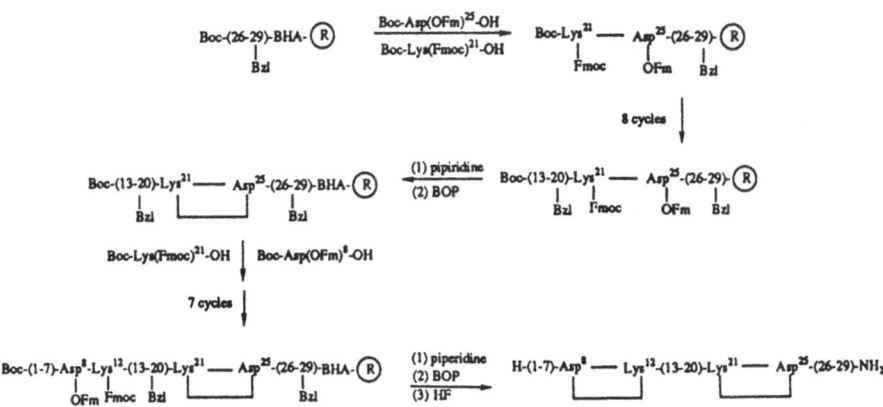

Fig. 1. Synthesis of cyclo$^{8,12;21,25}$-[Ala^{15}]-GRF(1–29)-NH$_2$ and related analogs.

Circular dichroism studies of the monocyclic and dicyclic GRF analogs revealed substantial α-helicity in 75% methanol (>85%). Differences were observed in water at pH 3 where helicity increased in the order [Ala^{15}]-GRF(1–29)-NH$_2$ < cyclo^{21,25}-[Ala^{15}]-GRF-(1–29)-NH$_2$ ~ cyclo^{8,12}-[Asp^8,Ala^{15}-GRF(1–29)-NH$_2$ < dicyclo$^{8,12;21,25}$-[Asp^8, Ala^{15}]-GRF(1–29)-NH$_2$. Similar observations were made from molecular dynamics calculations based on NOE-derived distance constraints. At the receptor, these cyclic peptides should retain helicity in the lactam-containing segments. The high biological potency of these peptides demonstrates that the helical segments are compatible with the bioactive conformer. Stability in human plasma increased in the order [Ala^{15}]-GRF(1–29)-NH$_2$ < cyclo^{21,25}-[Ala^{15}]-GRF-(1–29)-NH$_2$ < cyclo^{8,12}-[Asp^8,Ala^{15}]-GRF(1–29)-NH$_2$ < dicyclo$^{8,12;21,25}$-[Asp^8,Ala^{15}]-GRF-(1–29)-NH$_2$. This enhanced resistance of the dicyclo-GRF analogs to enzymatic degradation (dipeptidylpeptidase Type IV and trypsin-like) may augment the biological activity. In vivo studies with these cyclic-GRF and dicyclic-GRF analogs are in progress.

References

1. Felix, A.M., Heimer, E.P., Wang, C.-T., Mowles, T.F., Toome, V., Fry, D., Madison, V. and Lambros, T.J., In Jung, G. and Bayer, E. (Eds.) Peptides 1988 (Proceedings of the 20th European Peptide Symposium), De Gruyter, Berlin 1989, p. 601.
2. Felix, A.M., Heimer, E.P., Wang, C.-T., Lambros, T.J., Fournier, A., Mowles, T.F., Maines, S., Campbell, R.M., Wegrzynski, B.B., Toome, V., Fry, D. and Madison, V., Int. J. Pept. Prot. Res., 32 (1988) 441.
3. Frohman, L.A., Downs, T.R., Williams, T.C., Heimer, E.P., Pang, Y.-C. and Felix, A.M., J. Clin. Inv., 78 (1986) 906.
4. Frohman, L.A., Downs, T.R., Heimer, E.P. and Felix, A.M., J. Clin. Inv., 83 (1989) 1533.
5. Felix, A.M., Wang, C.-T., Heimer, E.P. and Fournier, A., Int. J. Pept. Prot. Res., 31 (1988) 231.

Chromatography viewed as an integral part of successful solid phase peptide synthesis

R.D. DiMarchi, G.S. Brooke, M. Johnson, H.B. Long and J.P. Mayer

Lilly Research Laboratories, Indianapolis, IN 46285, U.S.A.

Introduction

In the chemical synthesis of long peptides, the individual amino acid coupling efficiencies are readily recognized as contributing sizably to the final yield of product. The particular focus of this synthetic study with growth hormone releasing factor (GRF) was the impurities resulting from incomplete amino acid couplings, and the difficulties that they present in ultimately obtaining a homogeneous product. These synthetic inefficiencies yield a multitude of low-level deletion impurities. Depending on their nature, they can be difficult to readily identify and diminish preparative RPHPLC performance. With increasing peptide length, the requirements for chromatographic resolution generally increase, while the resolving power tends to diminish. Consequently, synthetic methodology that facilitates chromatographic performance becomes of increased importance as target peptides increase in size.

Results and Discussion

Table 1 *Ninhydrin analysis of GRF (1-44)-OCH$_2$-PAM resin[a]*

	Single	Double	Double/capping
Average coupling efficiency	97.8	98.4	99.5
Total projected deletions	62.4	49.3	17.2
Product yield[b]	37.6	50.7	50.7
Product/deletions	0.60	1.03	2.95

[a] Synthesis was performed by automated ABI 430A protocols.
[b] Calculated as the difference between theoretical yield and measured deletion content.

In principle, the effective capping of projected deletion peptides should facilitate chromatography. Ninhydrin analysis of coupling efficiency provides an estimate of the remaining unreacted amine [1,2]. The synthesis of GRF by a double coupling strategy illustrated that, while the projected product yield only increased an absolute amount of 13.1% relative to the single coupled synthesis, the ratio of yield to deletion peptides increased by 71% (Table 1). Analysis of the impact of capping on the double coupled synthesis revealed an absolute decrease in projected deletion peptides of 32.1%. This is more than double the amount of amino group acylation that was achieved by the preceding second coupling.

More importantly, the ratio of yield to deletion peptides increased three-fold.

The chromatographic analysis of the GRFs prepared by three different synthetic methods, but purified identically by preparative RPHPLC, revealed a uniformly high level of apparent purity, with only subtle relative differences. AAA was also incapable of identifying any statistically significant differences. Preview sequence analysis dramatically illuminated the presence of deletion peptides (Table 2). As originally hypothesized, significant improvements in final product purity, while not readily apparent by commonly used techniques, were achieved by an additional coupling, and still further by acetic anhydride capping. In no instance did the purity match that achieved by the rDNA prepared standard.

Table 2   *Preview sequence analysis[a] of purified GRF (1-44)-OH*

	Single cpl. (%)	Double cpl. (%)	Double cpl/capped (%)	rDNA (%)
Ala[1]	3.5	0.2	0.2	<0.1
Ile[4]	8.0	N.D.	N.D.	N.D.
Phe[5]	8.6	N.D.	N.D.	N.D.
Tyr[9]	10.9	N.D.	N.D.	N.D.
Val[12]	10.5	N.D.	N.D.	N.D.
Leu[13]	11.4	N.D.	N.D.	N.D.
Ser[17]	8.2	1.3	0.8	<0.1
Ala[18]	N.D.	1.1	0.2	N.D.
Asp[24]	N.D.	1.6	N.D.	N.D.
Ile[25]	6.9	1.1	0.2	<0.1
Met[26]	7.1	N.D.	N.D.	N.D.
Gly[31]	N.D.	4.1	1.0	N.D.
Gly[38]	25.0	7.5	1.2	<0.1
Ala[39]	22.0	10.0	2.0	<0.1

[a] Percentage shown is that determined for the indicated previewed amino acid at each cycle.
[b] N.D. = not determined.

Conclusion

We have shown, in a series of GRF syntheses, that low-level deletion peptides were not totally removed by conventional preparative RPHPLC. These impurities were observed best by preview sequence analysis of the purified products. Given the unknown biological nature of these impurities, their removal prior to use in human subjects is warranted. A greater awareness of the need to achieve near-quantitative amine acylation with a judicious selection of chromatographic methods, followed by an extensive analysis of product purity, is required to synthesize peptides of the quality currently obtainable by rDNA methods.

References

1.   Sarin, V.K., Kent, S.B.H., Tam, S.P. and Merrifield, R.B., Anal. Biochem., 117 (1981) 147.
2.   Merrifield, R.B., Vizioli, L.D. and Boman, H.G., Biochemistry, 21 (1982) 5020.

Synthesis of GRF analogs with potent in vivo GH-releasing activity

Imre Mezö[a], Balázs Szöke[a], István Teplán[a], Gábor B. Makara[b], György Rappay[b], Magdolna Kovács[c], Judit Horváth[c] and Sándor Vígh[c]

[a] *1st. Institute of Biochemistry, Semmelweis University Medical School, 1444 Budapest 8, P.O. Box 260, Hungary*
[b] *Institute of Experimental Medicine, Hungarian Academy of Science, Budapest, Hungary*
[c] *Institute of Anatomy, University Medical School, Pécs, Hungary*

Introduction

Growth hormone (GH)-releasing hormone (GRF) analogs may have a significant role in both human medicine and veterinary application. Previously, we found that Gaba[30]-substitution is advantageous for the synthesis and the biological activity of such analogs [1]. Presently, we demonstrate that combinations of certain previously known substitutions enhance the in vivo potency of our new Gaba[30]-GRF/1-30/-NH_2 analogs. However, the potencies of these analogs differ, depending on the route of administration.

Results and Discussion

Synthesis

[D-Ala[2],Nle[27],Gaba[30]]-hGRF (1–30)-NH_2 (HB-495)

[D-Ala[2],Leu[15],Nle[27],Gaba[30]]-hGRF (1–30)-NH_2 (HB-515)

[D-Ala[2],D-Arg[11],Leu[15]-Nle[27]-Gaba[30]]-hGRF (1–30)-NH_2 (HB-527)

were synthesized by the usual SPPS methodology. The crude material was purified on Sephadex G-50 column, followed by MPLC methodology (Synchroprep RP-P30μm resin (C-18)) using gradient elution with 0.25 N TEAP solution and CH_3CN. Desalting was performed by MPLC using 10% AcOH and isopropanol. Peptides were characterized by TLC and RPHPLC. Interestingly, the HB-527 analog showed significantly lower k' than that of HB-515. Therefore, we expected a remarkable difference in the conformations of the epimers. From the CD spectra, we conclude that the L-Arg[11]-containing analog (HB-515) exhibits a more ordered conformation in water than its epimer.

In vitro bioassays

Our analogs showed that the potencies were similar to that of the reference hGRF (1–44)-NH_2 preparation.

In vivo bioassays [2]

After intravenous administration of a 2.0 μg/kg dose, HB-495 was twice as active, HB-515 and 527 only 1.5 times more active, than hGRF (1–29)-NH$_2$.

After subcutaneous administration of a 5 μg/kg dose, our analogs caused about the same GH release as hGRF (1–29)-NH$_2$ dose of 250 μg/kg [2]; therefore, our analogs, injected subcutaneously, are about 50 times more potent than hGRF (1–29)-NH$_2$. At the 12.5 μg/kg dose, HB-515 exhibited 1.5 times higher GH-releasing activity than HB-495 at 5 min after injection (see Fig. 1).

Intramuscular administration of 2.0 μg/kg of our analogs elevated the plasma GH level to 2-fold higher than the basal level. In a dose of 5 μg/kg, HB-515 released GH 1.5–2.0 times higher than HB-495 and HB-527, respectively, at 5 min after injection (see Fig. 1).

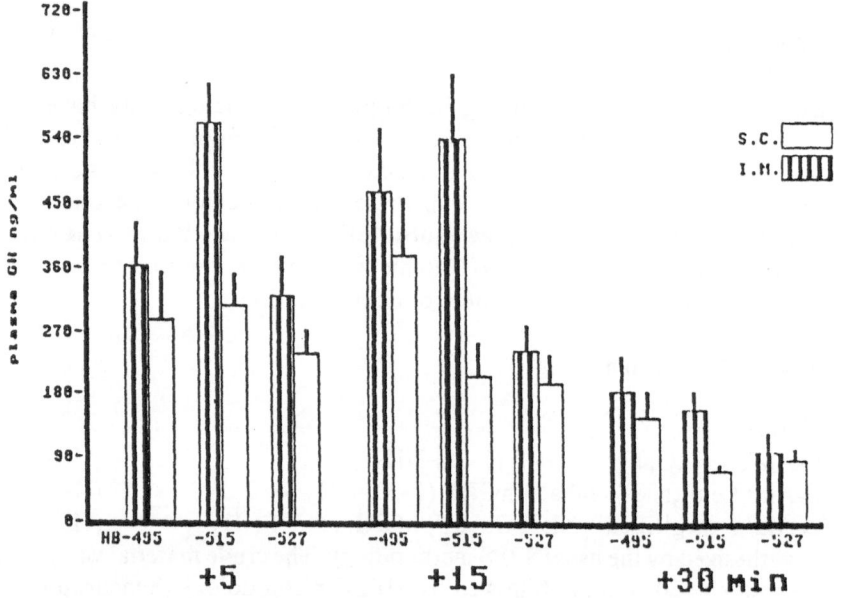

Fig. 1. Comparison between intramuscular (i.m.) and subcutaneous (s.c.) GH-releasing effects of GH-RH analogs at 5 μg/kg doses.

Acknowledgements

Research was supported by the Hungarian Academy of Sciences.

References

1. Mezö, I., Szöke, B., Vadász, Zs., Teplán, I., Makara, B.G., Kovács, M., Horváth, J. and Flerkó, B., In Jung, G. and Bayer, E. (Eds.) Peptides 1988, Walter de Gruyter, Berlin, 1989, p. 604.
2. Kovács, M., Gulyás, J., Bajusz, S. and Schally, A.V., Life Sci. 42 (1988) 27.

Isosteric analogs of growth hormone-releasing factor (1–29)-NH$_2$

Simon J. Hocart, William A. Murphy and David H. Coy

Peptide Research Laboratories, Department of Medicine, Tulane University School of Medicine, 1430 Tulane Avenue, New Orleans, LA 70112, U.S.A.

Introduction

Previous structure–activity investigations with the ψ[CH$_2$NH] peptide bond isostere have produced antagonists when inserted into various peptides. These include bombesin, in which the incorporation of Leu$^{13}\psi$[CH$_2$NH]Leu^{14} produced an antagonist [1], and tetragastrin, in which Boc-Trp-Leuψ[CH$_2$NH]Asp-Phe-NH$_2$ is an antagonist [2]. In this preliminary study, we chose to investigate the effect of the isostere on growth hormone-releasing factor (1–29)-NH$_2$ (GRF).

Results and Discussion

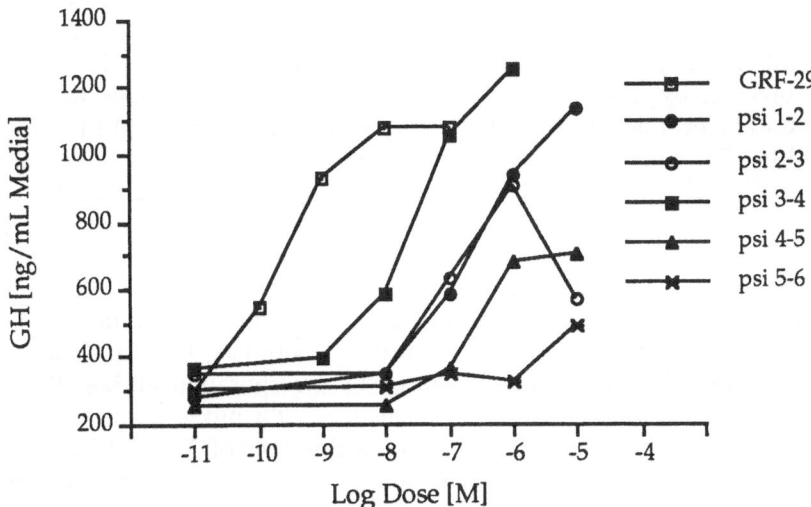

Fig. 1. Effect of ψ[CH$_2$NH]GRF(1–29)NH$_2$ analogs on GH secretion from dispersed rat pituitary cells.

Analogs were prepared by conventional SPPS on MBHAR. The following protocol was used on an Advanced ChemTech ACT 200 synthesizer: deblocking, 33% TFA (1 min, 25 min); DCM wash; PrOH wash; neutralization, 10 % DIEA

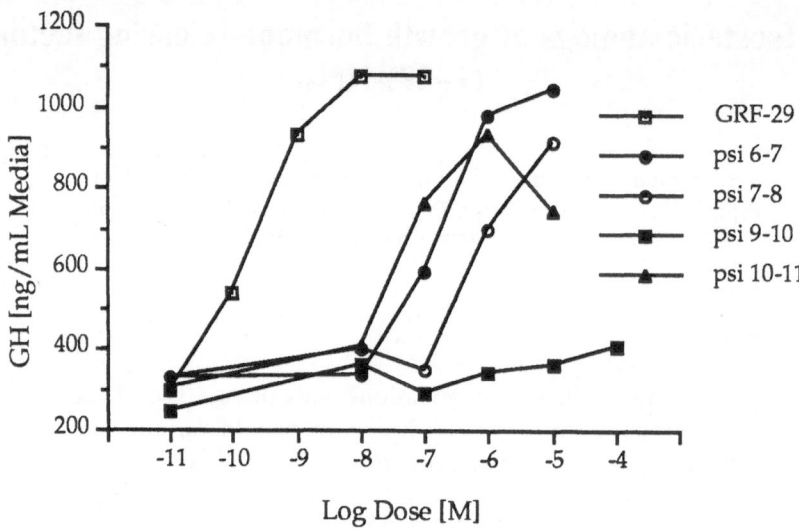

Fig. 2.   *Effect of ψ [CH₂NH]GRF(1–29)NH₂ analogs on GH secretion from dispersed rat pituitary cells.*

(2 washes); DMF wash; coupling of preformed HOBT esters, 45 min in DMF, 15 min DMAP; PrOH Wash; DCM wash. The isosteres were incorporated by racemisation-free reductive alkylation with a preformed protected amino acid aldehyde in the presence of $NaCNBH_3$ in acidified DMF [3,4]. The aldehydes were prepared by reduction of the protected dimethylhydroxamates with $LiAlH_4$ at 0°C. The peptides were cleaved with HF and purified by gel filtration and RPHPLC. The purified analogs were assayed in a 4-day primary culture of male rat anterior pituitary cells for GH release as previously described (See Figs. 1 and 2) [5]. Potential antagonists were retested in the presence of GRF(1–29)NH₂ (1 nM) (see Fig. 3). The following results were obtained: Incorporation of the isostere at position 5–6, gave a very weak agonist with ≪0.1 % activity of the GRF control (see Fig. 1). Incorporation of the isostere in positions 1–2 and 7–8 gave weak agonists with ~0.1% activity (see Fig. 2). Similarly, an agonist with ~1% activity was produced by incorporation at 3–4. Placement at positions 2–3, 6–7 and 10–11 gave analogs with mixed agonist/antagonist activity, dependent upon the dose. The analog [Ser⁹ψ[CH₂NH]Tyr¹⁰]GRF(1–29)NH₂ was found to be inactive in this assay (see Fig. 2) and was retested for antagonist activity in the presence of a stimulating dose of GRF(1–29)NH₂ (see Fig. 3). This analog, [Ser⁹ψ[CH₂NH]Tyr¹⁰]GRF(1–29)NH₂, was found to be an antagonist in the 10 μM range vs. 1 nM GRF and had no agonist activity at doses as high as 0.1 mM (see Fig. 2). The analog [Asn⁸ψ[CH₂NH]Ser⁹]GRF(1–29)NH₂ could not be synthesized and has yet to be investigated.

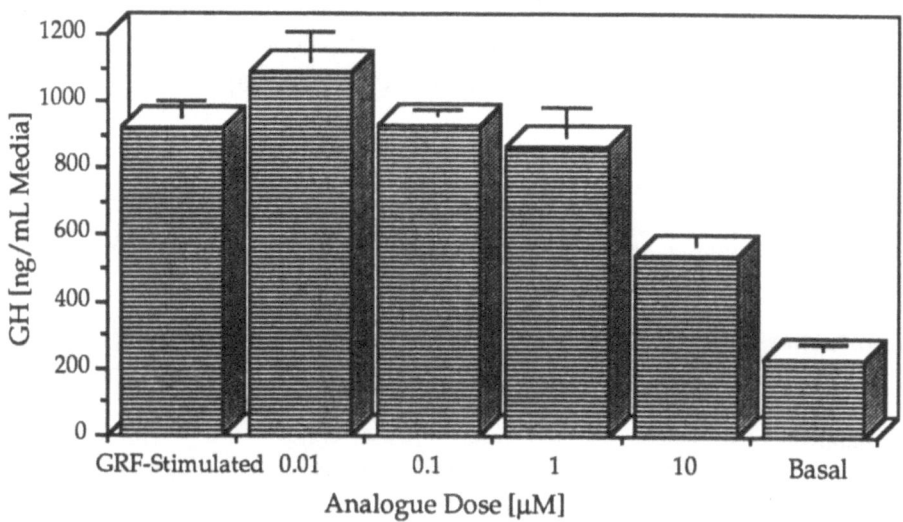

Fig. 3.   Antagonism of GRF-stimulated GH secretion by Ser⁹ψ [CH₂NH]Tyr¹⁰GRF(1–29)NH₂.

Acknowledgements

The authors wish to acknowledge the technical assistance of Etchie Yauger and Vienna Mackay. Supported by NIH grant DK-30167.

References

1.   Coy, D.H., Heinz-Erian, P., Jiang, N-Y., Sasaki, Y., Taylor, J., Moreau, J-P., Wolfrey, W.T., Gardner, J. and Jensen, R., J. Biol. Chem., 283 (1986) 5056.
2.   Martinez, J., Bali, J-P., Rodriguez, M., Castro, B., Magous, R., Laur, J. and Lignon, M-F., J. Med. Chem., 28 (1985) 1874.
3.   Coy, D.H., Hocart, S.J. and Sasaki, Y., Tetrahedron, 44 (1988) 835.
4.   Hocart, S.J., Nekola, M.V. and Coy, D.H., J. Med. Chem., 31 (1988) 1820.
5.   Murphy, W.A. and Coy, D.H., Pept. Res., 1 (1988) 36.

Synthesis and biological activity of growth hormone-releasing factor analogs

J.F. Hernandez, P.G. Theobald, G. Yamamoto, J. Andrews, C. Rivier, W. Vale and J.E. Rivier

The Salk Institute, 10010 N. Torrey Pines Road, La Jolla, CA 92037, U.S.A.

Introduction

New analogs of human (h) GRF(1–29)-NH$_2$ have been developed following two approaches :

(1) It was assumed that the peptide acts at the receptor through an α-helical secondary structure [1]. In order to stabilize this putative bioactive conformation, leucine and alanine residues were substituted with the corresponding α-methyl derivatives, strong helical inductors, that may also provide a higher resistance to biodegradation.

(2) It was postulated that altering the conformation or functionality of the peptide backbone by incorporating the pseudopeptide bond [CH$_2$–NH] might lead to antagonists that retain high receptor affinity [2], and/or to long acting molecules [3].

Results and Discussion

The peptides, synthesized on MBHA resins, were purified by preparative RPHPLC ($>$97% purity) and characterized by optical rotation, AAA and MS. The coupling of sterically hindered Cα-methyl amino acids was accomplished using a BOP/HOBt mixture [4]. The reduced amide bonds were incorporated by solid-phase reductive alkylation [5] with Boc-amino aldehydes in the presence of NaBH$_3$CN [6].

As summarized in Table 1, entries **3–5**, mono Cα-substituted analogs of hGRF(1–29)-NH$_2$ did not differ significantly in potency from the parent compound **2** in an in vitro assay, suggesting that local conformational restrictions imposed by the α-methyl group were tolerated by the receptor. However, **6** and **7** with multiple α-methyl amino acid substitutions were 25–100 times less potent than **2**, indicating that methyl groups may impede some contacts with the receptor or constrain the molecule in a less favorable conformation. The fact that the two latter compounds give a higher helical content, as determined by CD spectroscopy (data not shown), suggests that other parameters in addition to helicity are important for in vitro potency. Interestingly, **6** showed, in an in vivo assay in the rat (data not shown), a higher potency and a longer duration of action than **2** or all of the mono-substituted analogs, suggesting a higher resistance to biodegradation.

Table 1 *Characterization of GRF agonists*

#	Compound	$[\alpha]D^{a}$	RT^{b}	$Potency^{c}$
1	hGRF(1–40)-OH	−64°	7.3	1.00
2	[NMeTyr^1,Ala^{15},Nle^{27},Asn^{28}]-hGRF(1–29)-NH$_2$	−63°	12.5	8.63 (5.34–14.76)
3	[NMeTyr^1,Ala^{15},C$_\alpha$MeLeu^{27},Asn^{28}]-hGRF(1–29)-NH$_2$	−52°	13.9	10.01 (5.22–20.18)
4	[NMeTyr^1,Ala^{15},C$_\alpha$MeLeu^{26},Nle^{27},Asn^{28}]-hGRF(1–29)-NH$_2$	−43°	13.1	6.44 (4.06–10.34)
5	[NMeTyr^1,Ala^{15},C$_\alpha$MeLeu^{22},Nle^{27},Asn^{28}]-hGRF(1–29)-NH$_2$	−44°	14.6	7.18 (4.12–12.79)
6	[NMeTyr^1,C$_\alpha$MeLeu^{5,13,17,22,27},Ala^{15},Aib^{19},Asn^{28}]-hGRF(1–29)-NH$_2$	−26°	14.9	0.28 (0.15– 0.50)
7	[D-Ala^2,C$_\alpha$MeLeu^{5,13,17,22,27},Ala^{15},Aib^{19},Asn^{28}]-hGRF(1–29)-NH$_2$	−18°		0.08 (0.04– 0.17)
8	[NMeTyr^1,Nle^{27},Asn^{28}]-hGRF(1–29)-NH$_2$	−60°	11.46	10.39 (7.4 –14.5)
9	[$^1\psi^2$-CH$_2$NH,Nle^{27},Asn^{28}]-hGRF(1–29)-NH$_2$	−64°	9.15	0.0020 (0.001–0.003)
10	[NMeTyr^1,$^2\psi^3$-CH$_2$NH,Nle^{27},Asn^{28}]-hGRF(1–29)-NH$_2$	−68°	8.53	0.0058 (0.003–0.011)
11	[NMeTyr^1,D-Ala^2,$^2\psi^3$-CH$_2$NH,Nle^{27},Asn^{28}]-hGRF(1–29)-NH$_2$	−56°	8.71	0.0016 (0.001–0.003)
12	[NMeTyr^1,$^4\psi^5$-CH$_2$NH,Nle^{27},Asn^{28}]-hGRF(1–29)-NH$_2$	−56°	8.89	0.0054 (0.004–0.008)
13	[NMeTyr^1,$^5\psi^6$-CH$_2$NH,Nle^{27},Asn^{28}]-hGRF(1–29)-NH$_2$	−56°	7.95	0.0002 (0.000–0.000)
14	[NMeTyr^1,$^6\psi^7$-CH$_2$NH,Nle^{27},Asn^{28}]-hGRF(1–29)-NH$_2$	−48°	8.37	0.0009 (0.000–0.002)

a $c = 1$ in 1% AcOH.

b RT = retention time in minutes on Vydac analytical C$_{18}$ column (0.46×25 cm), 5 μM particle size, detection at 210 nm, A = 0.1% TFA, gradient was 30–39% CH$_3$CN in 15 min, 2.0 ml/min flow rate.

c Potency expressed relative to hGRF(1–40)-OH = 1 with 95% confidence limits. After 3–5 days, cultured rat pituitary cells were incubated with compounds at several doses for 3 h ($n = 3$). GH levels in the assay medium were determined by RIA. Relative potencies were derived from dose–response curves using the BIOPROG program [7].

The potencies of the reduced amide hGRF analogs (Table 1, entries **9-14**) were significantly lowered, relative to that of their parent compound, although nearly full intrinsic activity was retained. These results suggest that hormone-receptor interactions are compromised by backbone modification in the region of residues 1-7.

Acknowledgements

Research was supported by NIH, Grant DK-26741. Our thanks are expressed to R. Chavarin, C. Donaldson, L. Gandara, D. Jolley, R. Kaiser, C. Miller and D. Pantoja for their technical assistance, and to Dr. Terry Lee of the Beckman Institute of Immunology, City of Hope, Duarte, CA, for MS analysis.

References

1. Clore, G.M., Martin, S. and Groneborn, A., J. Mol. Biol., 191 (1986) 553.
2. Martinez, J, Bali, J.P., Rodriguez, M., Castro, B., Magous, R., Laur, J. and Lignon, M.F., J. Med Chem., 28 (1985) 1874.
3. Szelke, M., Leckie, B., Hallet, A., Jones, D.M., Sueras, J., Atrash, B. and Lever, A.F., Nature, 299 (1982) 555.
4. Hudson, D., J. Org. Chem., 53 (1988) 617.
5. Sasaki, Y. and Coy, D., Peptides, 8 (1987) 119.
6. Fehrentz, J.A. and Castro, B., Synthesis, (1983) 676.
7. Vale, W., Vaughan, J., Jolley, D., Yamamoto, G., Bruhn, T., Seifert, H., Perrin, M., Thorner, M. and Rivier, J., Methods Enzymol., 124 (1986) 389.

Characterization of convulsant peptides related to corticotropin-releasing factor (CRF)

Masashi Miyamoto and Motoyoshi Nomizu

Pharmaceutical Laboratory, KIRIN Brewery Co. Ltd., 2-2. Soujamachi 1 Chome, Maebashi Gunma 371, Japan

Introduction

CRF and its related peptides have some conserved primary regions. We have taken an interest in the characteristic conserved sequence of the CRF family, Pro-Pro-Ile-Ser, which is located in the proximal N-terminal region. Biological function of this N-terminal region is not yet known. We synthesized several peptide fragments of human CRF(hCRF) and sauvagine that involved either this entire N-terminal sequence or part of it, and found several peptides caused a transient convulsion in mice after intracerebroventricular (i.c.v.) injection. To determine a further SAR, we examined the activity of analogs of H-Pro-Pro-Ile-OH that seemed to be the active component.

Results and Discussion

Among 17 peptides corresponding to part of the N-terminal region of hCRF and the frog skin peptide, sauvagine, 6 peptides that involved Pro-Pro-Ile at their N-terminal caused a kindling-like convulsion in mice immediately after (0–2 min) i.c.v. injection. H-Pro-Pro-Ile-OH showed the strongest activity and was the shortest active peptide (Table 1). It seemed that this tripeptide was an active center of the convulsant activity of these peptides.

On the other hand, the analogs in which the L-Pro residues of H-Pro-Pro-Ile-OH were substituted with D-Pro did not cause the convulsion. Another type of analog replaced the Ile with other amino acids which decreased the convulsive activity. These results suggest that the convulsions are not caused by non-specific action of H-Pro-Pro-Ile-OH. Furthermore, it was suggested that there was a specific receptor that recognizes the structure, Pro-Pro-X, and Ile was the most suitable amino acid residue for the X position.

When administered either intravenously or intracisternally, H-Pro-Pro-Ile-OH did not stimulate ACTH secretion in rats. Furthermore, H-Pro-Pro-Ile-OH did not inhibit typical radioligands of neurotransmitter receptors which bind to rat brain membranes. These results suggested that this tripeptide had no affinity for CRF receptors causing ACTH secretion nor to other neurotransmitter receptors.

Weiss et al. [1] showed that the ovine CRF itself produced late onset (1–6 h

Table 1 *Potency of synthetic peptide fragments of human CRF and sauvagine causing convulsion in mice*

Peptide	(position)	Number of convulsed mice ($n = 10$)				
		Dose (nmol/head)				
		300	100	30	10	3
ZGPPISIDLS	(Sauvagine 1–10)	0				
ZGPPISIDL	(Sauvagine 1–9)	0				
ZGPPISI	(Sauvagine 1–7)	0				
ZGPPIS	(Sauvagine 1–6)	0				
ZGPPI	(Sauvagine 1–5)	0				
ZGPP	(Sauvagine 1–4)	0				
GPPIS	(Sauvagine 2–6)	0				
PPISI	(Sauvagine 3–7)	9	0			
PPISLDLT	(hCRF 4–11)	10	8	0		
PPISLD	(hCRF 4–9)	9	9	8	0	
PPISL	(hCRF 4–8)	9	0			
PPIS	(hCRF 4–7)	10	10	7	0	
PPI	(hCRF 4–6)	10	10	10	7	1
EPPIS	(hCRF 3–7)	0				
EPPI	(hCRF 3–6)	0				
PIS	(hCRF 5–7)	0				
PP	(hCRF 4–5)	0				

after i.c.v. administration) of kindling-like seizures in rat. Considering their report and our results, we think our synthetic convulsant peptides could be candidates as the ligands for a receptor, which is independent of ACTH secretion in the central nervous system.

Acknowledgements

We thank Ms. M. Ohashi and A. Ohokubo for their skillful technical assistance.

References

1. Weiss, S.R.B., Post, R.M., Gold, P.W., Chrousos, G., Sullivan, T.L., Walker, D. and Pert, A., Brain Res., 372 (1986) 345.

Synthesis and biological activities of rabbit corticostatin (rCS)

Nobutaka Fujii[a], Akira Okamachi[a], Susumu Funakoshi[a], Masataka Kuroda[a], Yoshio Hayashi[a], Haruaki Yajima[a], Junichi Fukada[a], Hiroo Imura[a], Roberto Bessalle[b] and Mati Fridkin[b]

[a]*Faculty of Pharmaceutical Sciences and Faculty of Medicine, Kyoto University, Sakyo-ku, Kyoto 606, Japan*
[b]*Department of Organic Chemistry, The Weizmann Institute of Science, 76100 Rehovot, Israel*

Introduction

In 1988, Zhu et al. [1], isolated rabbit corticostatin (rCS), a 34-residue peptide involving 6 disulfide-linked Cys residues and 9 Arg residues, from rabbit fetal lung. The structure of rCS was coincided with that of one of rabbit neutrophil peptides, NP 3-a, reported by Selsted et al. [2]. In order to evaluate the biological activities of rCS, and examine the usefulness of trimethylsilyl bromide (TMSBr) [3] as a deprotecting procedure for the Fmoc-based SPPS of Arg-rich peptides, and silver trifluoromethanesulfonate (AgOTf) [4] as an S-Acm deprotecting procedure, we undertook the total synthesis of rCS.

Results and Discussion

Fmoc-based SPPS of rCS was carried out manually according to the principle of Sheppard et al.[5] using the following side-chain protected Fmoc amino acids: Arg(Mtr), Cys(Acm), Ser(tBu), Glu(OtBu), and Tyr(tBu). As the starting resin, *p*-alkoxybenzyl-type polystyrene resin (0.70 mequiv./g, 0.2 mmol) was employed. The first C-terminal residue, Fmoc-Arg(Mtr)-OH (5 equiv.), was loaded on the resin by DPCDI (5 equiv.) in the presence of dimethylaminopyridine (1 equiv.). Each residue was introduced by DPCDI + HOBt procedure until the resin became negative to the Kaiser test. In the final step, the Fmoc group was removed by treatment with 20 % piperidine/DMF, and the rest of the protecting groups, except for the Acm group, with 1 M TMSBr-thioanisole/TFA (4°C, 60 min), as shown in Fig. 1. Nine Arg(Mtr) residues were efficiently deprotected by this brief treatment. The resulting Cys(Acm)-peptide was treated with AgOTf (60 equiv.) at 4°C for 60 min to remove Acm groups from the Cys residues. After treatment with DTT, followed by gel-filtration on Sephadex G-15, the reduced peptide was subjected to air oxidation at 4°C for 7 days in 0.1 M ammonium acetate buffer (pH 7.5) to establish the three disulfide bonds. The crude, air-oxidized product was purified to homogeneity by HPLC on TSK-GEL LS 410KG column using a gradient elution with MeCN in 0.1% TFA.

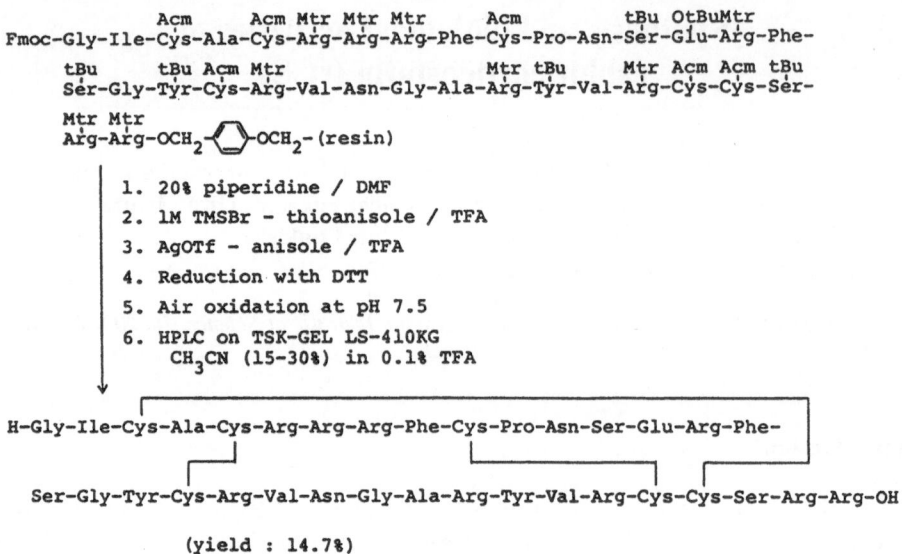

```
          Acm        Acm Mtr Mtr Mtr       Acm              tBu OtBuMtr
Fmoc-Gly-Ile-Cys-Ala-Cys-Arg-Arg-Arg-Phe-Cys-Pro-Asn-Ser-Glu-Arg-Phe-

     tBu        tBu Acm Mtr                    Mtr tBu      Mtr Acm Acm tBu
     Ser-Gly-Tyr-Cys-Arg-Val-Asn-Gly-Ala-Arg-Tyr-Val-Arg-Cys-Cys-Ser-

     Mtr Mtr
     Arg-Arg-OCH₂-⬡-OCH₂-(resin)
```

1. 20% piperidine / DMF
2. 1M TMSBr – thioanisole / TFA
3. AgOTf – anisole / TFA
4. Reduction with DTT
5. Air oxidation at pH 7.5
6. HPLC on TSK-GEL LS-410KG
   CH_3CN (15–30%) in 0.1% TFA

```
H-Gly-Ile-Cys-Ala-Cys-Arg-Arg-Arg-Phe-Cys-Pro-Asn-Ser-Glu-Arg-Phe-

    Ser-Gly-Tyr-Cys-Arg-Val-Asn-Gly-Ala-Arg-Tyr-Val-Arg-Cys-Cys-Ser-Arg-Arg-OH
```

(yield : 14.7%)

Fig. 1. Synthesis of rabbit corticostatin (rCS).

The overall yield calculated from the protected peptide resin was 14.6 %. The purity of synthetic rCS, thus obtained, was ascertained by AAA after 6 N HCl hydrolysis, analytical HPLC, and sequence analysis. In addition, synthetic rCS proved to be a monomer by FABMS.

The disulfide array of synthetic rCS was examined by sequence analysis of proteolytic fragments. Thermolysin treatment of a tryptic fragment of synthetic rCS gave a two-chain fragment cross-linked with Cys5-Cys20, and a three-chain fragment consisting of H-Ile-Cys3-OH, H-Phe-Cys10-Pro-Asn-OH, and H-Cys30-Cys31-OH. The precise disulfide array of the latter three-chain fragment is under investigation. These results suggested that synthetic rCS would have similar disulfide structure to that of human defensin reported by Selsted et al. [6].

Synthetic rCS (0.1–10 μg/ml) dose-dependently inhibited ACTH (100–100 pg/ml)-stimulated corticosterone production in rat adrenal cell suspensions, but showed no cytotoxic activity against several tumor cells (K562, P815, and EL4) at doses less than 100 μg/ml and no bactericidal effect on several strains of *E. coli* at doses even higher than 500 μg/ml.

References

1. Zhu, Q., Hu, J., Mulay, S., Esch, F., Shimasaki, S. and Solomon, S., Proc. Natl. Acad. Sci.U.S.A., 85 (1988) 592.
2. Selsted, M.E., Brown, D.M., DeLange, R.G., Harwig, S.S.L. and Lehrer, R.I., J. Biol. Chem., 260 (1985) 4579.

3. Fujii, N., Otaka, A., Sugiyama, N., Hatano, M. and Yajima, H., Chem. Pharm. Bull., 35 (1987) 3880.
4. Fujii, N., Otaka, A., Watanabe, T., Okamachi, A., Tamamura, H., Yajima, H., Inagaki, Y., Nomizu, M. and Asano, K., J. Chem. Soc., Chem. Comm., (1989) 283.
5. Dryland, A. and Sheppard, R.C., J. Chem. Soc., Perkin Trans I, (1986) 125.
6. Selsted, M.E. and Harwig, S.S.L., J. Biol. Chem., 264 (1989) 4003.

Structure-activity relationships of atrial natriuretic factor (ANF) receptor agonists

T.W. Rockway[a], P.J. Connolly[a], E.M. Devine Jr.[a], G.P. Budzik[a], A.M. Thomas[a], Y. Kiso[b] and W.H. Holleman[a]

[a]Cardiovascular Research Division, Abbott Laboratories, 1 Abbott Park Road, Abbott Park, IL 60064, U.S.A.
[b]Kyoto Pharmaceutical University, Kyoto, Japan

Background

ANF, a 28-amino acid peptide hormone secreted by atrial cardiocytes, is involved in the maintenance and regulation of fluid volume homeostasis. Our group has recently reported [1] the preparation and biological activity of ANF(7–23), a reduced-size analog of ANF(1–28), as well as a study of the solution conformation of the molecule in SDS micelles [2]. ANF(7–23), which constitutes the intact core region of ANF, with the N- and C-terminal tails removed, shows full ANF agonist activity in vitro and in vivo but reduced potency compared to ANF.

Chou–Fasman analysis of this core sequence suggested the possibility of two loops, comprising residues 7–10 and 18–21. In contrast, NMR shows an 18–21 inverse loop as well as a 14–17 loop, yet suggests that the 7–10 region may be extended. We have made a number of substitutions in these regions to examine the possibility of such structural features and to probe the importance of the putative loops.

Cys-Phe-Gly-Gly-Arg-Ile-Asp-Arg-Ile-Gly-Ala-Gln-Ser-Gly-Leu-Gly-Cys
7   15   23

Results and Discussion

We began by exploring the effect of deleting individual glycine residues on the analogs' ability to relax rabbit aortic smooth muscle rings preconstricted with histamine (values reported as pD2, the dose providing half-maximal relaxation; for ANF(7–23), pD2 = 5.70). Consistent with Chou–Fasman predictions, the removal of glycine within the putative loop regions results in significant decreases in activity (des-Gly^9 < 5.15; des-Gly^{20} < 5.15), whereas removal of those outside such regions is better tolerated (des-Gly^{16} = 5.76; des-Gly^{22} = 6.30). Interestingly, several combinations of deletions are also acceptable; most notably, des-Gly^{9,22} is more active in this assay than the parent (pD2 = 6.24).

We also studied the effects of amino acid replacement within these structural regions. In the region comprising residues 7–10, these studies gave mixed results. Replacement of Gly[9] (position i + 2) with (D)-Ala led to a modest improvement (pD2 = 5.81), but replacement with Pro was detrimental (pD2 = 5.22), suggesting the possibility of a Type II turn. On the other hand, conversion of Gly[10] to Pro was beneficial (pD2 = 5.92). Although substitution of the 9,10 dipeptide with either aminopentanoic (= 5.36) or aminohexanoic acid (= 5.89) was tolerated; none of these results is consistent with a turn structure. In any event, the effects of these changes were minimal. Changes in the 18–21 region generally supported the possibility of a loop here; in particular, replacement of Ser[19] (position i + 1) with (D)-Ser gave a dramatic improvement in the vasorelaxant response (pD2 = 6.68), and this residue was also successfully replaced by Pro (= 5.84).

We next looked at the effects of individual side-chain groups on the functional response. It has been reported [3] that the position-8 aromatic group is important for inhibition of aldosterone production. We have observed that this residue is also critical for vasorelaxation, and replacement with cyclohexylalanine (pD2 = 6.13) leads to a more potent analog. Conversion of Arg[11] to lysine gave an equipotent compound (pD2 = 5.95) that showed significantly higher in vivo activity (bolus injection into the rat) than ANF(7–23). Asp[13] seems to be a critical residue; even conservative substitutions, such as Glu (pD2 = 5.26), Asp-OCH$_3$ (< 5.17), or Asn (< 5.0) cause dramatic decreases in activity. Conversely, changes in the side-chains of residues 18 and 19 are tolerated. Gln[18] is readily converted to Asn (pD2 = 5.68) or Thr (= 5.79), and Glu[18] is only slightly less active (pD2 = 5.34). Similarly Ser[19] may be replaced by (D)-Ala (= 5.60) or by Freidinger's lactam [4] (= 5.55) without loss of activity.

Conclusions

The results of our deletion and substitution studies are consistent with prior stuctural analysis and NMR studies on ANF(7–23), suggesting a loop in the region comprising residues 18–21. In the residue 7–10 region, where the previous analyses produced differing conclusions, we could find no clear evidence for either a looped or a linear conformation. Our series of side-chain replacements led us to the conclusion that, while positions 8 and 13 contained functionality critical for the vasorelaxant response, no critical residues were contained within the 18–21 loop.

These studies led us to postulate that we might be able to shrink the size of our ANF analogs further by excising this 18–21 loop entirely; removing, in addition, the adjacent glycine residues at positions 16 and 22, which our deletion studies had shown to be expendable, produced the decamer, Cys[16] ANF(7–16). This compound relaxed rabbit aortic rings with a pD2 (= 5.42) only slightly lower than that recorded for ANF(7–23).

References

1a. Budzik, G.P., Firestone, S.L., Bush, E.N., Devine Jr., E.M., Rockway, T.W., Sarin, V. and Holleman, W.H., Fed. Proc., 46 (1987) 1133.

 b. Budzik, G.P., Firestone, S.L., Bush, E.N., Connolly, P.J., Rockway, T.W., Sarin, V. and Holleman, W.H., Biochem. Biophys. Res. Comm., 144 (1987) 422.

2. Olejniczak, E.T., Gampe Jr., R.T., Rockway, T.W. and Fesik, S.W., Biochemistry, 27 (1988) 7124.

3. Craven, T.G., Kem, D.C. and Schiebinger, R.J., Endocrinology, 122 (1988) 826.

4. Freidinger, R.M., Perlow, D.S. and Veber, D.F., J. Org. Chem., 47 (1982) 104.

UK-69,578: A potent selective inhibitor of E.C.3.4.24.11 which potentiates ANF in vivo

David Brown, Paul L. Barclay, Ian T. Barnish, Simon F. Campbell,
John C. Danilewicz, Peter Ellis, Keith James, Gillian M.R. Samuels,
Nicholas K. Terrett and Martin J. Wythes

Pfizer Central Research, Sandwich, Kent CT13 9NJ, U.K.

Introduction

Atrial Natriuretic Factor (ANF) is a 28-amino-acid peptide secreted by the heart which induces natriuresis and diuresis, lowers blood pressure, and decreases renin and aldosterone in animals and man. However, the potential utility of ANF in hypertension and congestive heart failure is limited by poor oral bioavailability and short plasma half-life. As an alternative approach, we sought a non-peptide agent that would potentiate the physiological actions of ANF via inhibition of the enzyme responsible for ANF degradation in vivo.

Results and Discussion

We have demonstrated that ANF is degraded in vitro by the endopeptidase E.C.3.4.24.11 from human and rat kidney, and that prototype inhibitors protect ANF from degradation, a result which has been subsequently reported by others [1]. Stepwise modification of the endopeptidase inhibitor (1) (Fig. 1) [2] demonstrated that conformational restraint is well tolerated in the P_2' and the P_1' residue (2, 3). Replacement of the backbone aza link by a methylene group gave a substantial increase in potency (4). The cis-4-aminocyclohexane carboxylate system (5) confers similar inhibitory potency to the 3-isomer (4) and removes a chiral center. In contrast to the aza series (1 and 2), the aminocyclohexane carboxylate moiety gives a clear potency advantage (5, 6 and 7) in the glutaramide series.

Compound 8 (UK-69,578) was selected for more detailed evaluation. UK-69,578 blocks degradation of human ANF by human kidney homogenate (complete inhibition at 10^{-6} M); increases the half-life of rat APIII from 0.96 to 2.5 min in nephrectomized rats(3 mg/kg i.v.); and potentiates the natriuretic (250%) and diuretic (300%) action of rat APIII in anaesthetized rat (3.0 mg/kg i.v.), with no significant change in urinary potassium excretion. UK-69,578 does not inhibit angiotensin converting enzyme, trypsin, chymotrypsin, leucine/aminopeptidase, carboxypeptidase A or renin.

Clinical trials in hypertension and congestive heart failure are currently in progress with UK-73,967 (S(+)-enantiomer, 9) and with UK-79,300, an orally active 5-indanyl ester prodrug.

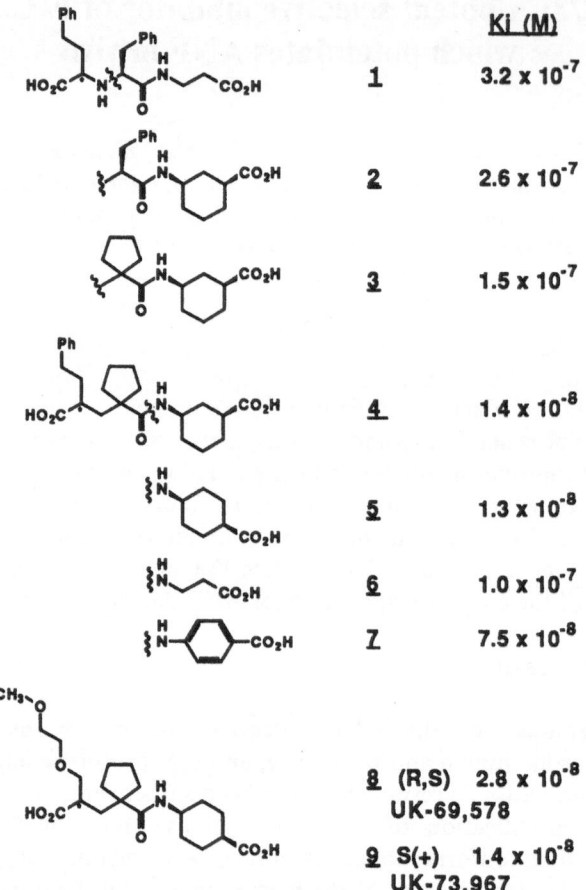

		Ki (M)
	1	3.2×10^{-7}
	2	2.6×10^{-7}
	3	1.5×10^{-7}
	4	1.4×10^{-8}
	5	1.3×10^{-8}
	6	1.0×10^{-7}
	7	7.5×10^{-8}
	8 (R,S) UK-69,578	2.8×10^{-8}
	9 S(+) UK-73,967	1.4×10^{-8}

Fig. 1.   Inhibition of rat kidney endopeptidase 24.11.

Acknowledgements

We thank P.E. Barnes, D.V.J. Batchelor, G. Corless, K.N. Dack, E. Dunn, E. Hawkeswood, H.F. Johnson and A. Takle for technical assistance.

References

1.   Stephenson, S.L. and Kenny, A.J., Biochem. J.,.243 (1987) 183.
2.   Mumford, R.A., et al., Biochem. Biophys. Res. Commun., 109 (1982) 1303.

248

Conformationally constrained peptide analogs of atrial natriuretic factor (ANF)

T.W. Rockway, E.T. Olejniczak, E.M Devine Jr., G.P. Budzik, T.J. Opgenorth and W.H. Holleman

Cardiovascular Research Division, Department 47-V, Abbott Laboratories, Abbott Park, IL 60064, U.S.A.

Introduction

Atrial natriuretic factor (ANF) is a 28 amino acid peptide exhibiting a variety of biological functions, including vasorelaxation, natriuresis and diuresis. Initial investigations from this laboratory of the SAR of analogs of ANF_{1-28} sought to determine the relationship of individual amino acids and observed biological activity. We reported [1] a recent investigation that provided evidence that several amino acid residues and amino acid segments can be deleted from the full size peptide while maintaining agonist activity. The series of small peptide analogs, whose parent is $[Cys^{16}]ANF_{5-16}$ Phe-Arg, exhibited ANF-like agonist activity with reduced potency. In order to define the conformations responsible for observed biological activity, we have designed and synthesized a series of small conformationally constrained ANF peptide analogs. Initial attempts at restricting these small peptide analogs were directed toward replacement of the Cys^7 side chain with the more bulky penicillamine residue. Further restrictions of the peptide analogs were done by preparing bicyclic peptide analogs of $[Cys^{16}]ANF_{5-16}$ Phe-Arg. This report will describe the synthesis, biological evaluation and conformational study of two such bicyclic analogs:

$[Cys^{7,16},des\text{-}Arg^{11},Cys^{13,15a},D\text{-}Arg^{14}]ANF_{5-16}$ Phe-Arg, **1**, and
$[Cys^{7,16},des\text{-}Arg^{11},Cys^{7a,13},D\text{-}Arg^{14}]ANF_{5-16}$ Phe-Arg , **2**.

Results and Discussion

The peptide analogs were prepared by SPPS, followed by treatment of the resin with liquid HF (60 min, 0°C) [2]. The cyclization of the bicyclic peptides was performed utilizing either a one-pot method involving the addition of two equivalents of iodine in ethanol (60 min, 25°C), or a two-step method involving treatment with potassium ferricyanide (60 min, 25°C) followed by iodine in ethanol (60 min, 25°C). Each peptide was purified by preparative HPLC, and the integrity of each peptide was determined by AAA, FABMS, sequence analysis and NMR spectroscopy.

The bicyclic peptide analogs **1** and **2** were evaluated for ANF-like agonist

249

characteristics utilizing rabbit adrenal receptor binding, relaxation of vascular smooth muscle (rabbit aorta) and in vivo natriuresis and diuresis in rats [3]. Bicyclic analog 1 bound to rabbit adrenal receptors (350 nm), relaxed vascular smooth muscle (3 μm) and produced a 10-fold increase in natriuresis, as well as a 10% decrease in blood pressure at an infusion dose of 150 μg/kg/min. Bicyclic analog 2 exhibited 10-fold weaker receptor affinity (3600 nm) and showed no vasorelaxation, hypotension, natriuresis or diuresis at an infusion dose of 300 μg/kg/min.

Since bicyclic analogs 1 and 2 exhibited divergent biological properties, a conformational study of these two analogs was undertaken utilizing 2D NMR techniques. Figure 1 shows the backbone conformation of 1 and Fig. 2 shows the backbone conformation of 2, determined from distance constraints obtained from 2D NOESY experiments. The conformations of 1 and 2 are similar within the disulfide rings; however, the backbone conformation of the amino acid residues outside the disulfide ring differs significantly in the two cases. In bicyclic analog 1, the N-terminal Ser-Ser peptide backbone is found in close proximity

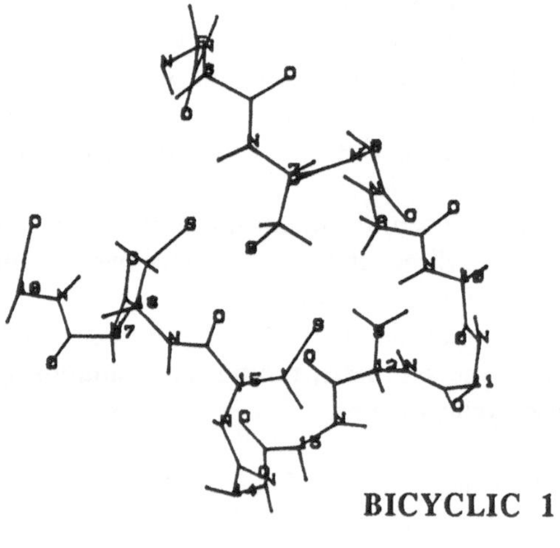

BICYCLIC 1

Fig. 1. Backbone conformation of bicyclic analog 1.

to the C-terminal Phe-Arg, while in bicyclic analog 2, the N-terminal and C-terminal residues are more separated. Although both analogs 1 and 2 bind to ANF receptors, only 1 exhibits additional biological properties similar to that of ANF. The biological results, coupled with the solution conformation observations of the two bicyclic analogs, suggest that the orientation of the amino acid residues outside the disulfide ring have no marked effect on receptor interaction, but are important for biological activity.

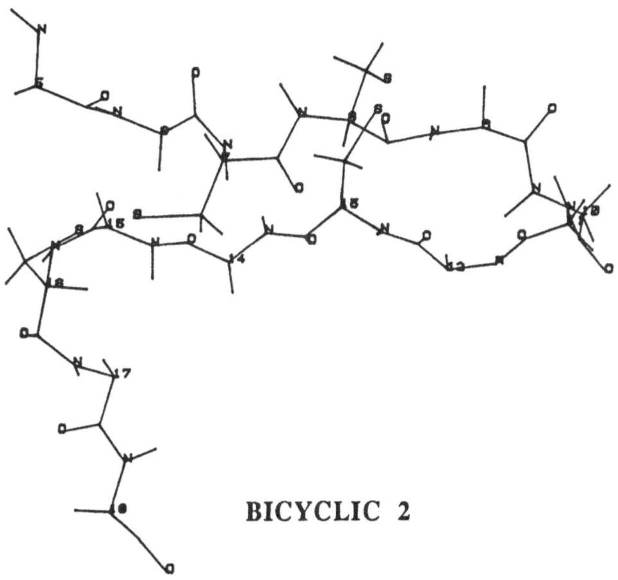

BICYCLIC 2

Fig. 2. Backbone conformation of bicyclic analog 2.

References

1. Rockway, T.W., Oral presentation Medicinal Chemistry Symposium, SUNY, Buffalo, NY, 1989.
2. Stewart, J.M. and Young, J.D., Solid Phase Peptide Synthesis, 2nd ed., Pierce Chemical Company, Rockford, IL, 1984.
3. Budzik, G.P., Firestone, S.L., Bush, E.N., Connolly, P.J., Rockway, T.W., Sarin, V.K. and Holleman, W.H., Biochem. Biophys. Res. Commun., 144 (1987) 422.

Small atrial natriuretic factor (ANF) peptide analogs maintaining agonist activity

T.W. Rockway, G.P. Budzik, E.N. Bush, P.J. Connolly, S.K. Davidsen,
E.D. Devine, S.D. Lucas, T.W. von Geldern and W.H. Holleman

*Cardiovascular Research Division, Department 47V, Abbott Laboratories, Abbott Park,
IL 60064, U.S.A.*

Introduction

ANF is a circulating 28-amino acid peptide hormone that exhibits potent vasodilatory, natriuretic and diuretic activity and inhibits aldosterone biosynthesis and renin release. Our investigations have been directed toward determining the relationships of individual amino acids in the native peptide to observed biological activity. We report the design and synthesis of a series of ANF peptide analogs and provide evidence that several amino acid residues and segments can be deleted from the full size peptide with retention of full agonist activity.

Our initial investigation led to the discovery of a small ANF peptide analog, [Cys[16]]ANF(7–16), which displayed receptor affinity and possessed vasorelaxant activity. However, this peptide exhibited potent pressor activity in the anesthetized rat and led to the discovery that amino acids from both the N-terminus (Ser-Ser) and the C-terminus (Phe-Arg) are required for in vivo agonist activity.

The sequence of rat ANF is shown below. The highlighted amino acids are those essential for full agonist activity. Also shown is the sequence of a reduced size analog that has full agonist activity (compound **2**).

$$
\begin{array}{ccccc}
& 7 & 10 & & 15
\end{array}
$$

Ser-Leu-Arg-Arg-**Ser-Ser-Cys-Phe-Gly**-Gly-**Arg-Ile-Asp-Arg-Ile**-Gly-Ala-

|

-Gln-Ser-Gly-**Leu**-Gly-Cys-Asn-Ser-**Phe-Arg**-Tyr   (1)

$$
\begin{array}{ccc}
21 & 25 & 28
\end{array}
$$

$$
\begin{array}{cccc}
5 & 10 & 15 & 18
\end{array}
$$

Ser-Ser-Cys-Phe-Gly-Gly-Arg-Ile-Asp-Arg-Ile-Cys-Phe-Arg   (2)

Results and Discussion

The peptide analogs were prepared by standard SPS techniques, followed by treatment of the resin with liquid HF (60 min, 0°C). The cyclization of the

252

peptides was achieved by the addition of two equivalents of iodine in ethanol to a dilute solution of the deprotected peptide. Crude peptides were purified by preparative HPLC, and homogeneity was determined by amino acid and sequence analysis, FABMS and NMR spectroscopy. Agonist activities were determined by receptor binding to rabbit adrenal membranes (pK_i), vasorelaxation of histamine-constricted rabbit aortic rings, inhibition of ACTH-induced aldosterone biosynthesis in primary rat adrenal cells, stimulation of cGMP biosynthesis in bovine transformed aortic endothelial cells and natriuretic, diuretic and hypotensive activity in Inactin-anesthetized normotensive rats.

Initial structure–activity studies focused on the role of the individual amino acids in [Cys[16]]ANF(5–16)Phe-Arg, (2), and the relationship between receptor affinity and vasorelaxation. These experiments confirmed our earlier work [1] with ANF(5–28) indicating that Phe[8],Ile[12] and Ile[15] were crucial for receptor binding activity. However, the requirements for agonist activity are more exacting than for receptor binding; e.g., the replacement of Asp[13] by Ser did not alter receptor binding but eliminated functional responses. The replacement of either Ser[5], Ser[6] or C-terminal Phe by 8-aminooctanoic acid yields full agonists and illustrates the importance of a positive charge on the N-terminus and the fixing of the C-terminal Arg at a site most acceptable to the receptor. The importance of a hydrophobic group at position 8 is exemplified by the fact that replacement of Phe[8] by cyclohexylalanine did not alter either receptor affinity or vasorelaxant potency. Systematic replacement of the L-amino acids by the D-stereoisomers resulted in 10–100-fold decrease in activity with the exceptions of Arg[11] and Arg[14] where the D-amino acids were approximately 5-fold and 2-fold more potent, respectively. Compound **2** and the D-Arg[11] analog of (**2**) have the following biological activities: receptor affinity, 0.2 and 0.07 μM; vasorelaxation, 2.0 and 0.24 μM; EC_{50} for aldosterone inhibition, 3 μM and not determined; EC_{50} for cGMP stimulation, 10 and 0.7 μM. A bolus dose (300 μg/kg) of each peptide decreased MAP, 21 and 22 mm Hg, and increased natriuresis 8-fold and 10-fold, respectively.

These results illustrate for the first time that large portions of the ANF(1–28) molecule can be deleted without altering the agonist properties of the molecule, although with a 10–100-fold loss in potency.

References

1. Holleman, W.H., Budzik, G.P., Devine, E.D., Sarin, V., Connolly, P.J., Rockway, T.W. and Bush, E.N., In Brenner, B.M. and Laragh, J.H. (Eds.) Biologically Active Atrial Peptides, Raven Press, New York, 1987, p. 226.

Acyclic 'seco-analogs' of atrial natriuretic peptides have biological activity in vitro: Structure-activity relationships

P.R. Bovy[a], J.M. O'Neal[a], G.M. Olins[a], D.R. Patton[a], E.G. McMahon[a], M.A. Palomo[a], P. Toren[b] and E.W. Kolodziej[a]

[a]*Cardiovascular Research, G.D. Searle & Co. and* [b]*P.S.C., Monsanto Life Sciences Research Center, 700 Chesterfield Village Parkway, St Louis, MO 63198, U.S.A.*

Introduction

The family of recently discovered peptides, atrial natriuretic peptides (ANP), represents a new hormonal system that is integrated positively and negatively with many physiological systems and, in particular, connected with the renin-angiotensin system. Several excellent reviews cover the genesis of this discovery [1–3] and the pharmacology [4–8] of the newly discovered system. The bioactive peptide that is released in the circulation is a 28 amino acid long peptide, cyclic by virtue of a disulfide bond[1–8, 10]. The integrity of the 17-membered cyclic structure has been postulated as a requirement for biological activity [9]. Affinity cross-linking studies have revealed the existence of two distinct types of endogeneous binding proteins specific for ANP. One class of ANP-binding proteins is associated with stimulation of guanylate cyclase activity (cyclase-coupled receptor, CC receptor), while the other is not coupled to any known second messenger (non-coupled or NC receptor) [11–13]. Although the circulating 28-mer bind to both receptor subtypes with similar subnanomolar affinities, various studies have indicated that they have different structural requirements for their ligands [11–13].

Results and Discussion

We now find that some synthetic peptides, obtained by juxtaposition of non-consecutive fragments of the ANP cyclic primary sequence, 'seco-analogs' of ANP, recognize the guanylate cyclase-coupled receptor and are full agonist in smooth muscle vasorelaxant assay (rabbit aorta).

We have designed several seco-peptides to test the hypothesis that the structural elements responsible for recognition and activation of the CC receptor extend over non-adjacent sections of the primary sequence. Thus, we have linked together linear, cysteine-containing fragments of the ANP sequence by a cystine bridge, leading to seco-peptide analogs (i.e., **5**). This approach is restricted to cysteine-containing fragments, but is quite attractive in view of the possibility of connecting various non-adjacent N- and C-terminus fragments of ANP without modifying the primary sequence surrounding the disulfide bridge in the natural compound, a region of the peptide thought to be of importance for activity [8,10].

254

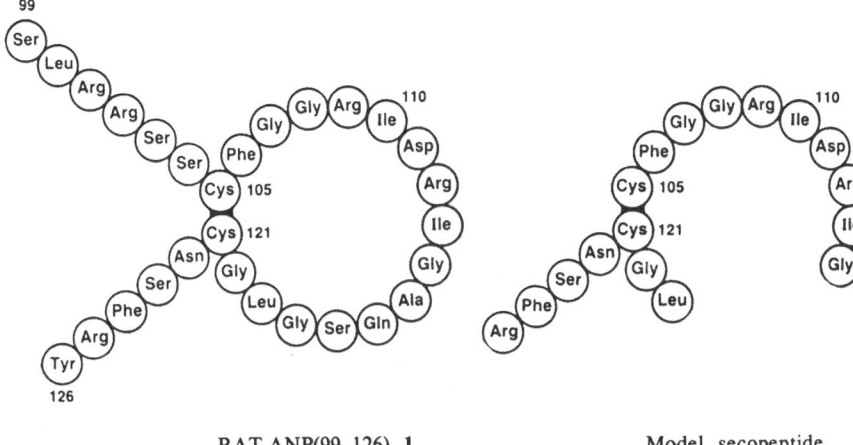

RAT ANP(99–126) **1**   Model secopeptide

Fig. 1.   A schematic representation of ANP(103–126) 1 and of a seco-analog.

The cysteine-containing linear fragments were synthesized by established methodologies of solid phase synthesis as Cys(Acm) derivatives. After deprotection of the cysteines with Hg(OAc)$_2$, the seco-peptides were assembled by oxidative dimerization of various fragments. The desired mixed dimer was obtained as an apparent statistical mixture with the symmetrical dimers. RPHPLC was used for separation and final purification. Peptides were characterized by AAA and FABMS. Biological data (in vitro binding to ANP receptors and in vitro rabbit aorta vasorelaxation), are reported in Table 1. In the binding experiment, 2–5 discriminate guanylate cyclase-coupled ANP receptor (CC) from non-coupled (NC). This observation is well in line with previous reports on structural requirements for affinity to the NC receptor [11]. However, seco-peptides 2 and 3, lacking residues identified as crucial for biological activity, have no measurable affinity for the cyclase-coupled receptor and, accordingly, fail to induce relaxation in rabbit aorta. On the other hand, the seco-peptides 4 and 5, containing all the crucial residues display affinity for the CC receptor, and fully relax precontracted rabbit aorta with a reduced potency. Interestingly, 6, a 'natural' seco-peptide metabolite resulting from the action of endopeptidase on ANP(103–126) [14], has a very weak biological activity, a confirmation that an intact peptide bond between Cys[105]-Phe[106] is an important requirement for activity.

Our data confirm the important role of Leu[117] and Arg[109] and/or Ile[110] in recognition of the CC receptor and indicate that the cyclic structure of ANP, albeit not obligatory for biological activity, participates to stabilize the receptor–hormone complex.

An important feature of the seco-peptide method lies in its versatility in combining two non-consecutive linear fragments from a peptide-hormone primary sequence that both participate in its interaction with a receptor. This method

255

Table 1 Relative vasorelaxant activity (rabbit aorta) and relative binding affinities (rabbit lung membranes) of ANP analogs with ANP(103–126) as the standard

Compounds[a]	Binding rel. affinity[b]		Vasorelaxation rel. potency[b]
	NC	CC	
1 S^{103} S - C - F - G - G - R - I - D - R - I - G ⌐ Y^{126} R - F - S - N - C - G - L - G - S - Q - A - G ⌐	1	1	1
2 C - F D - R - I - G ⌐ R - F - S - N - C - G - L - G - S - Q - A - G ⌐	–	–	<0.001
3 C - F - G - G - R - I - D - R - I - G - NH$_2$ R - F - S - N - C - Ac	2.1	<0.003	<0.001
4 C - F - G - G - R - I - D - R - I - G - NH$_2$ R - F - S - N - C - G - Ac	3.6	0.011	0.01
5 C - F - G - G - R - I - D - R - I - G - NH$_2$ R - F - S - N - C - G - L - G - S - Q - Ac	15	0.12	0.05
6 S - S - C ⌐ F - G - G - R - I - D - R - I - G ⌐ Y R - F - S - N - C - G - L - G - S - Q - A - G	2.2	0.004	<0.01

[a] Some fragments have the N-terminus acylated (Ac) and the C-terminus as an amide (NH$_2$) to avoid the introduction of a charge at a neutral site in the natural sequence; compounds **2–5** have an arginine-amide at the C-terminus; amino acid analysis and FABMS agree with the theoretical value.

[b] See Refs. 11–13 for experimental method; ANP(103–126) **1** has an apparent $K_i = 0.12 \pm 0.16$ nM and, an $EC_{50} = 2.3 \pm 0.6$ nM.

256

could have a general use in helping to identify pharmacophoric pattern(s) in cystine-containing peptides.

Acknowledgements

We acknowledge M.G. Jennings and J.F. Zobel for AAA.

References

1. Currie, M.G., Geller, D.M., Cole, B.R., Siegel, N.R., Fok, K.F., Adams, S.P., Eubanks, S.R., Galluppi, G.R. and Needleman, P., Science, 233 (1983) 67.
2. Cantin M. and Genest, J. Endocr. Rev., 6 (1985) 107.
3. Cantin, M. and Genest, J., Biochem. Invest. Med., 9 (1986) 319.
4 Needleman, P. and Greenwald, J.E., New Engl. J. Med., 314 (1986) 828.
5. Ackermann, U., Clin. Chem., 32 (1986) 24.
6. Lang, R.E., Unger, T. and Ganten, D., J. Hypertension, 5 (1987) 255.
7. Trapani, A., Olins, G.M., Blaine, E.H., In Allen, R. (Ed.) Annual Reports of Medicinal Chemistry, Academic Press, New York, NY, 1988, p. 101.
8. Bovy, P.R., Med. Res. Rev., 10 (1990) in press.
9. Watanabe, T.X., Noda, Y., Chino, N., Nishiuchi, Y., Kimura, T., Sakakibara, S. and Imai, S.M., Eur. J. Pharmacol., 120 (1986) 123.
10. Nutt, R.F. and Veber, D.F., Endocrinol. Metab. Clin. North Am., 16 (1987) 19.
11. Bovy, P.R., O'Neal, J.M. Olins, G.M. and Patton, D.R., J. Med. Chem., 32 (1989) 869.
12. Olins, G.M., Patton, D.R., Bovy, P.R. and Mehta, P.P., J. Biol. Chem., 263 (1988) 10989.
13. Scarborough, R.M., Schenk, D.B., McEnroe, G.A., Arfsten, A., Kang, L., Schwartz, K. and Lewicki, J.A., J. Biol. Chem., 261 (1986) 12960.
14. Olins, G.M., Krieter, P.A., Trapani, A.J., Spear, K.L. and Bovy, P.R., Mol. Cell. Endocrinol., 61 (1989) 201.

Very active ANF analogs with implications for a bioactive conformation

T.M. Williams, R.F. Nutt, S.F. Brady, T.A. Lyle, T.M. Ciccarone, C.D. Colton,
W.J. Palaveda, G.M. Smith, D.F. Veber and R.J. Winquist

Merck, Sharp and Dohme Research Laboratories, West Point, PA 19486, U.S.A.

Introduction

Structure–activity studies in the ring portion of atrial natriuretic factor R^3-R-S-S-C^7-F-G-G-R-I-D-R-I-G-A-Q-S-G-L-G-C^{23}-N-S-F-R-Y^{28} (ANF) were carried out in an effort to define aspects of the bioactive conformation.

Results and Discussion

The glycine residues in ANF were singly substituted by alanine, D-alanine or proline [1]. The best substitutions in terms of activity (inhibition of me-thoxamine-contracted rabbit aorta or rabbit renal artery tissue) were combined to produce composite analogs (Table 1). [D-Ala^9,Ala^{10}]ANF(3–28) was 3 times as active as either of the singly substituted parent analogs, suggesting that a type II' β-turn may be part of the bioactive conformation. Incorporation of a γ-lactam β-turn mimic [2] at position 9–10 resulted in a 10-fold loss in activity. However, [D-Ala^9,Pro^{10}]ANF(3–28), a substitution known to stabilize β-turns [3] retained 75% of the activity of ANF(3–28), in contrast to [Pro^{10}]ANF(5–26)NH$_2$, which is only 5% as active as its parent compound. It is known that Phe^8 is an important residue for activity [1], so conformational and steric changes in this part of the structure may well have a profound effect on bioactivity. A γ-turn at Gly^{10} in the bioactive form of ANF is also consistent with some of the data. Although there is some evidence from SAR of a turn in this region, more conclusive data are required to make a definitive statement.

The achiral α-amino-isobutyric acid (Aib), a stabilizing residue in peptide helical structures [4] was incorporated into the ring of ANF at two positions. [Aib^{13}]ANF(3–28) and [Aib^{16}]ANF(3–28) were respectively 9% and 60% as active as ANF(3–28), and were less potent than D-Asp and D-Ala in these positions [5]. For comparison, [Aib^{13}]ANF(3–28) had 50% of the activity as an uncharged Asp replacement, [Abu^{13}]ANF(3–28) (Abu = 2-(S)-aminobutyric acid).

Computer modeling of the ring portion of the molecule was undertaken to evaluate various conformations on the basis of total energy (energy minimized with AMBER [6], proximity of side chains important for biological activity, and general fit with the observed analog activities. A working model includes a 3_{10} helix in the sequence Ile^{12}-Gln^{18} and γ-turns at Gly^9 and Gly^{20}. Interestingly,

Table 1   *Biological activity of multiple substitutions for glycine in ANF(3–28)*

Substitution analogs	Glycine positions in ANF					Relative potency
	9	10	16	20	22	
1	D-Ala	Ala				3.39
2	D-Ala		Ala			3.76
3	D-Ala		Ala	Pro		1.52
4	D-Ala	Ala	Ala	Pro		1.96
5	D-Ala	Ala	Ala	Pro	D-Ala	3.70
6	D-Ala	Ala	Ala	Ala	D-Ala	1.37

[a]Inhibition of methoxamine-induced contraction of rabbit renal artery tissue.
Relative potency ANF(3–28) = 1.00 (IC$_{50}$ = 0.30 nM).

the γ-turn at Gly[9] preminimization assumed preferentially the characteristics of a type II' β-turn after minimization. The side chains of Phe[8], Ile[12], Ile[15] and Leu[21], known to be important for biological activity [1,5,7], are clustered together in a 10Å×10Å surface, and may define an area for interaction with the high affinity B ANF receptor [8]. It must be stressed that this does not represent a definitive model of the bioactive conformation of ANF, but rather a blueprint for approaching the design and synthesis of conformationally constrained analogs.

References

1. ANF analogs were synthesized as described in Nutt, R.F. and Veber, D.F., Endocrinol. Metabol. Clin. N. Am., 16 (1987) 19.
2. Freidinger, R.M., Veber, D.F., Perlow, D.S., Brooks, J.R. and Saperstein, R., Science, 210 (1980) 656.
3. Rose, G.D., Gierasch, L.M. and Smith, J.A., Adv. Prot. Chem., 37 (1985) 1.
4. Toniolo, C., Bonora, G.M., Bavoso, A., Benedetti, E., Di Blasio, B., Pavone, V. and Pedone, C. Biopolymers, 22 (1983) 205.
5. Nutt, R.F., Brady, S.F., Lyle, T.A., Ciccarone, T.M., Colton, C.D., Paleveda, W.J., Williams, T.M., Smith, G.M., Winquist, R.J. and Veber, D.F., In Tam, J.P. and Kaiser, E.T. (Eds.) Synthetic Peptides: Approaches to Biological Problems, Vol. 86, Proceedings of the UCLA Symposium on Molecular and Cellular Biology, New Series, 1989, p. 267.
6. Weiner, S.J., Kollman, P.A., Case, D.A., Singh, U.C., Caterina, G., Alagona, G., Profeta, Jr., S. and Weiner, P., J. Am. Chem. Soc., 106 (1984) 765.
7. Nutt, R.F., Ciccarone, T.M., Brady, S.F., Colton, C.D., Paleveda, W.J., Lyle, T.A., Williams, T.M., Veber, D.F., Wallace, A. and Winquist, R.J., In Marshall, G.R. (Ed.) Peptides: Chemistry and Biology (Proceedings of the 10th American Peptide Symposium), ESCOM, Leiden, 1988, p. 444.
8. Scarborough, R.M., McEnroe, G.A., Arfsten, A., Kang, L.-L, Schwartz, K. and Lewicki, J.A., J. Biol. Chem., 263 (1988) 16818.

Developing a major pathway for peptide condensation: Application of DPBT method in the synthesis of α-hANP

Gen Li, Tieming Cheng, Chongxi Li, Limin Zhang, Mengshen Cai, Ming Zhao, Jingli Zhang, Chao Wang and Qiuyue Gu

Institute of Peptide Research, Beijing Medical University, Beijing 100083, People's Republic of China

Introduction

In 1978, we observed that tetrahydrothiazole-2-thione (TTT) was a good leaving group in the aminolysis of peptide synthesis [1]. Since then, many heterocyclic compounds structurally related to TTT were studied in our laboratory [2], as well as in others [3]. Some of them are benzoxazolone, benzoxazolethione, 5-substituted oxadiazothione, benzimidazole, etc. The structures of most reactive intermediates related to these acylated heterocyclics are amides, except that of acylated benzothiazolethione. X-ray diffraction shows the acylated benzothiazole-thione was a thiol ester [2]. Most of these heterocyclics are used for the first time in peptide synthesis. The yields are satisfactory and the acylated intermediates can be isolated and purified in crystalline forms.

Results and Discussion

Acylation of these heterocyclics is usually done using DCC as condensing agent. As all peptide chemists know, these reagents cause dehydration of glutamine and asparagine, and other complications with amino acids having hydroxyl groups on the side chains. It is also very tedious to remove the substituted urea formed in the reaction. To avoid using DCC, and to increase further the activities of the heterocyclics as leaving groups, they are allowed to react with diphenyl-chlorophosphate to form the corresponding amides with an active P-N bond in the molecules [4]. We hope that the activity of the P-N bond may be modulated by converting the heterocyclics we studied into phosphoramide that can be prepared easily by the reaction of diphenylchlorophosphate with the heterocyclics. The phosphoamides are fine crystalline compounds and can be kept for a long time. We report here our results on DPBT (diphenylphosphorylbenzoxazolethione).

$$(PhO)_2\text{-}\overset{\overset{\textstyle O}{\|}}{P}\text{-}N$$

Y=O, DPBO

Y=S, DPBT

Using phosphoramides for peptide synthesis, instead of DCC and other methods, the following advantages should be mentioned. In contrast to the DCC method, the acylated intermediate is stable and can be purified before it is aminolyzed. Racemization of the peptides synthesized by our method was found to be less than 6% by Young's and modified Anderson's tests. Dehydration of asparagine and glutamine was not observed. For example, Z-asparagine is treated separately with DCC and DPBT in pyridine. After standing for 2 h, a strong $-C \equiv N$ IR absorbing band is observed only with the DCC-treated sample. Side-chain hydroxyl groups need not be protected. Both DPBO and DPBT can be used for SPPS, as well as for coupling in solution.

Synthesis of α-human atrial natriuretic peptide (α-hANP) [5]

This important peptide has been synthesized by various methods. Because of its strong depression effect on blood pressure, it has been studied extensively to determine the relationship between its structure and physiological effect. It is an octacosanoic peptide with a disulfide linkage. We used DPBT for the synthesis by solution segment condensation procedure and this is outlined in the following figure:

Boc-ANP(17–22)-OH + H-ANP(23–28)-OBzl $\xrightarrow{\text{DPBT}}$ Boc-ANP(17–28)-OBzl

Boc-ANP(11–16)-OH + H-ANP(17–28)-OBzl $\xrightarrow{\text{DPBT}}$ Boc-ANP(11–28)-OBzl

Boc-ANP(1–10)-OH + H-ANP(11–28)-OBzl $\xrightarrow{\text{DPBT}}$ Boc-ANP(1–28)-OBzl

The protected peptide is deprotected by means of hydrofluoric acid and separated by HPLC. The physiological effects of the synthetic product on blood pressure and immunological activity are found to be comparable with the standard. Some segment peptides were also synthesized by the DPBO method. The yields and their optical purity were generally higher than those obtained by the DCC method.

References

1a. Dissertation submitted to the Chemistry Department, Peking University, 1978.
 b. Li, C. et al., Tetrahedron Lett., 22 (1981) 3467.
2a. Li, G., Doctoral Dissertation submitted to the Chemistry Department, Peking University, 1987.
2b. Li, G. et al., International Conference on Biochemistry, Beijing, 1987.
3. Romani, S., Moroder, L., Boverman, G. and Wünsch, E., Synthesis, 738 (1985) 512.
4. Ikota, N., Shioiri, T., Yamada, S., Chem. Pharm. Bull., 28 (1980) 3064.
5. Cai, M., Peng, S., Cheng, T., Dai, D., Wang, S., Li, J., Zhang, J., Cong, Y., Liang, W. et al., Scientia Sinica, B 9 (1987) 969.

Synthesis of human endothelin and its precursors in recombinant cells

Yasuaki Itoh, Takuya Watanabe, Takuo Kosaka, Chiharu Kimura, Kazuhiro Ogi, Kazuki Kubo, Nobuhiro Suzuki, Tadashi Yasuhara, Haruo Onda and Masahiko Fujino

Tsukuba Research Laboratories, Takeda Chemical Industries, Ltd., Wadai 7, Tsukuba, Ibaraki 300-42, Japan

Introduction

Peptides of the endothelin (ET) family act as strong vasoconstrictor/pressor agents in vivo and in vitro [1,2]. We have previously described cloning of human endothelin-1 (ET-1) precursor cDNA [3]. Several proteolytic cleavages of the 212-amino acid residue-prepro-ET-1, including unusual COOH-terminal processing between Trp^{73} and Val^{74} precede the formation of human ET-1. This previously unknown type of peptide processing is thought to be catalyzed by a putative ET converting enzyme. To investigate the manner of ET-1 biosynthesis, we have expressed the human ET-1 cDNA and its derivatives in *E. coli*, yeast and mammalian cells. In this work, we attempted to answer the following questions: (i) Are recombinant *E. coli* and yeast cells practical tools for obtaining ET-1 and its precursors? (ii) Does recombinant ET-1 retain biological activities? (iii) Can mammalian cells that are not ET-producing cells achieve the unique protein processing of ET-1?

Results and Discussion

Amounts of products produced in recombinant cells were measured by sensitive enzyme immunoassays for ET-1 [4] and big-ET-1 [5]. The molecular sizes of the products were determined by SDS–PAGE or HPLC. In *E. coli* N4830 harboring either pTS4007 or pTS4008 (Fig. 1), we detected prepro-ET-1 and COOH-matured ET-1 having the expected molecular weights and ET-1 antigenicity [6]. These *E. coli*-derived peptides will be used as substrates for processing studies of ET-1 and antigens for generating antibodies against the ET-1 precursors.

Saccharomyces cerevisiae 20B-12 harboring either pTS2013 or pTS2014 (Fig. 1) secreted considerable amounts (2–4 mg/l) of fusion proteins (glycosylated mating factor α1-leader/ET-1) into the culture supernatants. To obtain mature ET-1, COOH-matured fusion protein (product of pTS2014) was treated with arginylendopeptidase, which cleaves peptide bonds after arginine residues. The resulting yeast-derived ET-1 was correctly folded in respect to two disulfide bonds and had the expected molecular mass, as shown by HPLC and FABMS,

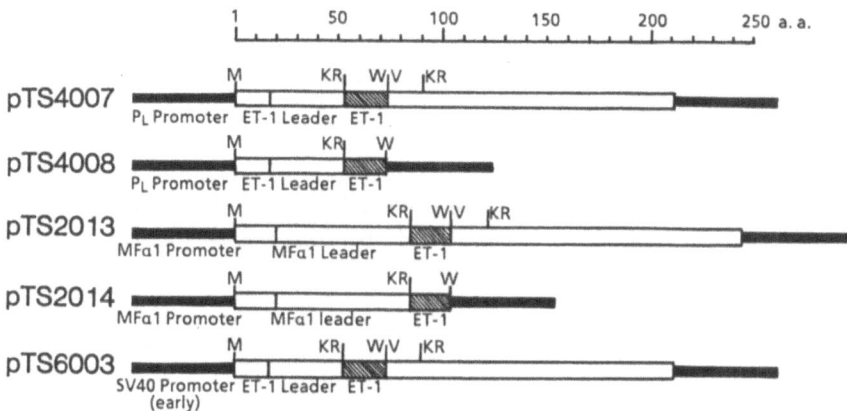

Fig. 1. Schematic representation of coding regions expressed in this work.

respectively. Furthermore, the yeast-derived ET-1 retained a similar pressor effect to that of native ET-1 in a conscious rat.

CHO K1 cells transformed with pTS6003 [7] (Fig. 1) secreted correctly processed big-ET-1 and ET-1 into the culture medium, suggesting that the putative ET converting enzyme exists widely even in cells that are originally not ET-producing cells.

The availability of these recombinant *E. coli*, yeast and CHO cells to obtain ET-1 and its precursors will facilitate studies of this peptide.

References

1. Yanagisawa, M., Kurihara, H., Kimura, S., Tomobe, Y., Kobayashi, M., Mitsui, Y., Yazaki, Y., Goto, K. and Masaki, T., Nature, 332 (1988) 411.
2. Inoue, A., Yanagisawa, M., Kimura, S., Kasuya, Y., Miyauchi, T., Goto, K. and Masaki, T., Proc. Natl. Acad. Sci. U.S.A., 86 (1989) 2863.
3   Itoh, Y., Yanagisawa, M., Ohkubo, S., Kimura, C., Kosaka, T., Inoue, A., Ishida, N., Mitsui, Y., Onda, H. and Fujino, M., FEBS Lett., 231 (1988) 440.
4. Suzuki, N., Matsumoto, H., Kitada, C., Masaki, T. and Fujino M., J. Immunol. Methods, 118 (1989) 245.
5. Suzuki, N., Matsumoto, H., Kitada, C., Kimura, S., Miyauchi, T. and Fujino, M., J. Immunol. Methods, in press.
6. Watanabe, T., Itoh, Y., Ogi, K., Kimura, C., Suzuki, N. and Onda, H., FEBS Lett., 251 (1989) 257.
7. Kosaka, T., Suzuki, N., Matsumoto, H., Itoh, Y., Yasuhara, T., Onda, H. and Fujino, M., FEBS Lett., 294 (1989) 42.

Synthesis and characterization of porcine endothelin and big endothelin

Federico C.A. Gaeta[a],*, Leo B. Slater[a], Brooks R. Sunday[a], Jerry R. Miller[b], Christina L. Ramsaur[b], Lorraine Ghibaudi[b] and Meeta Chatterjee[b]

[a]Department of Medicinal Chemistry and [b]Department of Pharmacology, Schering-Plough Research, 60 Orange Street, Bloomfield, NJ 07003, U.S.A.

Introduction

In 1988, Yanagisawa et al. [1] reported the isolation and sequence of the vasoconstrictor peptide, endothelin, from porcine aortic endothelial cells. They suggested that endothelin was expressed in a pre-pro form and that an unusual Trp-Val cleavage of a 39-residue big endothelin was involved in the processing to a mature peptide. Later [2] they reported three isopeptides, structurally and pharmacologically distinct from the human endothelin family, designated ET-1, ET-2, and ET-3. ET-1 is identical to porcine [1] or 'classical' endothelin. Since endothelin may be involved in a novel cardiovascular control mechanism, we were interested in studying the conversion of big endothelin into endothelin. We report herein the syntheses of both ET-1 and the porcine putative precursor big endothelin (BET) by Fmoc methodology, as well as some of our initial biological results.

CSCSSLMDKECVYFCHLDIIW   CSCSSLMDKECVYFCHLDIIW ↓ VNTPEHIVPYGLGSPSRS
        Human or porcine ET-1   Porcine BET

Results and Discussion

Synthesis. Peptides were assembled on Sasrin™ resins with an ABI 430A synthesizer using Fmoc NMP/HOBt/cap cycles. Cys(Trt) and Arg(Pmc) [3] protecting groups were used. Cleavage with TFA/thioanisole/DMS/EDT (93:3:3:1) at RT was carried out for 1–3 h. Air oxidations were performed as in Yanagisawa et al. [4] and Kumagaye et al. [5] and furnished a 3:1 mixture of types A and B isomers, respectively, which were separated by RPHPLC. The desired isomers (type A) had [1–15,3–11] disulfide links, and had the shortest retention times on HPLC. Peptides were characterized spectroscopically by FABMS, PDMS, and NMR, and pharmacologically by receptor binding and coronary artery contraction assays.

---

*Present address: Cytel Corporation, 11099 N. Torrey Pines Road, La Jolla, CA 92037, U.S.A.

Receptor binding assays. Competition binding assays were performed using a washed membrane fraction from porcine aortic smooth muscle media, as in Hirata et al. [6], with [^{125}I]ET (2000 Ci/mmol, Amersham).

Porcine coronary artery contraction assays. Fresh porcine coronary vessel strips were mounted on SS wires and suspended from isometric force transducers in a muscle bath (37°C) filled with Krebs bicarbonate buffer gassed with 95:5 O_2/CO_2. Tissue viability was assessed via KCl and/or norepinephrine (NE) challenge. The response of each strip to the peptides was expressed as a percent of maximum response to KCl or NE.

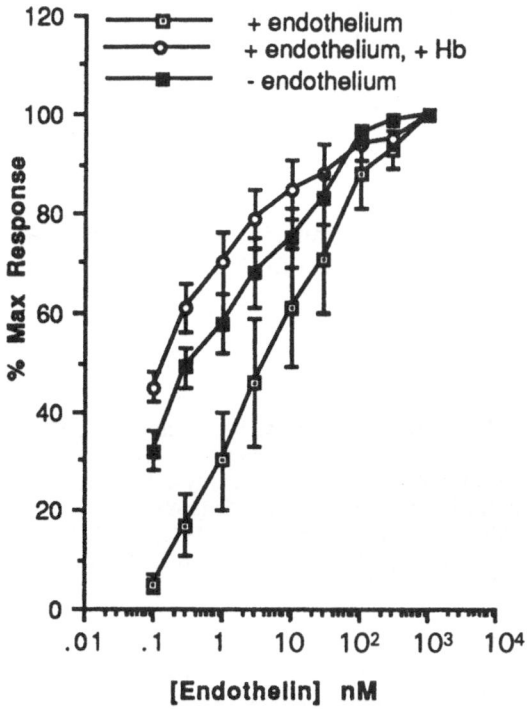

Fig. 1. *Effect of endothelium and hemoglobin (Hb) on the contractile response of coronary arteries to ET-1.*

Type A ET-1 was about two orders of magnitude more potent than type B, in agreement with the literature [5]. ET-1 competed with [^{125}I]ET for specific binding to aortic smooth muscle membranes with IC_{50} of 0.2–0.5 nM, whereas the major isomer (type A) of BET had $IC_{50} = 1$ μM. The minor BET isomer (type B) did not bind at concentrations as high as 0.1 mM and did not elicit smooth muscle contraction. Figure 1 shows dose–response curves for contraction of coronary arteries by ET-1. Responses to ET-1 were attenuated in the presence of endothelial cells. The larger EC_{50} obtained in the presence of endothelium (4.0 nM vs. 0.3 nM) could result from release of endothelium- derived relaxing factor (EDRF) [7], consistent with recent observations [8] that pressor effects of circulating ET-1 are limited, in part, by EDRF. In the presence of endothelium and added hemoglobin (Hb), the EC_{50} of ET-1 was 0.15 nM. Thus EDRF may play a compensatory role in negating the contractile effects of ET.

Interestingly, BET causes a contraction despite low activity in receptor binding assays. Others also have reported [9] weak vasoconstrictor activity of putative precursors in vitro. This result suggests that active ET-1 may be liberated from inactive BET under the experimental conditions. The attenuation of activity in the presence of endothelium is consistent with the results discussed above. The Trp-Val cleavage required to produce ET-1 from BET is an unusual one that could be caused by chymotryptic-like activity. Our results and those of others [10] suggest that if endothelin can be demonstrated to regulate vascular smooth muscle contraction in vivo, novel therapeutic agents that are based on inhibition of 'endothelin-converting' enzymes may evolve.

Acknowledgements

We thank Drs. D. Dalgarno, B. Pramanik, and T. Tsarbopoulos of our Physical and Analytical Chemistry Department for NMR and MS analyses.

References

1. Yanagisawa, M., Kurihara, H., Kimura, S., Tomobe, Y., Kobayashi, M., Mitsui, Y., Yazaki, Y.,Goto, K. and Masaki, T., Nature, 332 (1988) 411.
2. Inoue, A., Yanagisawa, M., Kimura, S., Kasuya, Y., Miyauchi, T., Goto, K. and Masaki, T., Proc. Natl. Acad. Sci. U.S.A., 86 (1989) 2863.
3. Ramage, R. and Green, J., Tetrahedron Lett., 28 (1987) 2287.
4. Yanagisawa, M., Inoue, A., Ishikawa, T., Kasuya, Y., Kimura, S., Kumagaye, S.-I., Nakajima, K., Watanabe, T. X., Sakakibara, S., Goto, K. and Masaki, T., Proc. Natl. Acad. Sci. U.S.A., 85 (1988) 6964.
5. Kumagaye, S.-I., Kuroda, H., Nakajima, K., Watanabe, T.X., Kimura, T., Masaki, T. and Sakakibara, S., Int. J. Pept. Prot. Res., 32 (1988) 519.
6. Hirata, Y., Yoshimi, H., Takata, S., Watanabe, T.X., Kumagai, S., Nakajima, K. and Sakakibara, S., Biochem. Biophys. Res. Commun., 154 (1988) 868 (and references cited therein).
7. Sakuma, I., Stuehr, D.J., Gross, S.S., Nathan, C. and Levi, R., Proc. Natl. Acad. Sci. U.S.A., 85 (1988) 8664 (and references cited therein).
8. De Nucci, G., Thomas, R., D'Orleans-Juste, P., Antunes, E., Walder, C., Warner, T.D. and Vane, J.R., Proc. Natl. Acad. Sci. U.S.A., 85 (1988) 9797.
9. Kashiwabara, T., Inagaki, Y., Ohta, H., Iwamatsu, A., Nomizu, M., Morita, A. and Nishikori, K., FEBS Lett., 247 (1989) 73.
10. McMahon, E.G., Fok, K.F., Moore, W.M., Smith, C.E., Siegel, N.R. and Trapani, A.J., Biochem. Biophys. Res. Commun., 161 (1989) 406.

Synthesis and biological properties of endothelin and endothelin analogs

Carl Hoeger[a], Marvin R. Brown[b] and Jean E. Rivier[a]

[a]Clayton Foundation Laboratories for Peptide Biology, The Salk Institute,
10010 N. Torrey Pines Road, La Jolla, CA 92037, U.S.A.
[b]Department of Medicine, University of California Medical Center, 225 Dickinson Street,
San Diego, CA 92103, U.S.A.

Introduction

The bicyclic 21-amino acid peptide endothelin (human/porcine ET; ET-1; CSCSSLMDKECVYFCHLDIIW; intra-chain disulfide linkages at Cys 1–15 and Cys 3–11) has been isolated and characterized from porcine aortic endothelial cell culture supernatant [1]. ET-1 is one of the most potent vasoconstrictor/ pressors known and may be important in the regulation of the mammalian cardiovascular system (CVS). It is also possible that dysfunction in the control of ET production and/or changes in the sensitivity of vascular smooth muscle cells (VSMC) to this peptide may be a causitive factor in hypertension and spastic disorders of blood vessels [2]. In an effort to elucidate the biological role of this peptide, we have prepared ET-1 and some ET analogs and tested them for their ability to raise blood pressure in the conscious rat.

Results and Discussion

The individual peptides were synthesized on a Boc-L-Trp-polystyrene–1% divinylbenzene cross-linked resin using a gradiative acidolytic (TFA/HF) approach [3]. The crude peptide obtained after HF treatment of the fully assembled and protected peptide resin was either oxidized to the cyclic form directly [4], or converted into its tetra-S-sulfonate for subsequent transformation to the desired ET or ET analog by treatment with DTT [5]. After purification by HPLC techniques, the purified peptides all gave satisfactory amino acid analyses.

The peptides examined in this study are listed in Table 1. Using a conscious rat blood pressure assay, each peptide was tested for its ability to raise mean arterial blood pressure (MAP) [6]. At a dosage of 0.14 nmol/kg (300 ng), ET-1 produces a 30-mm Hg increase in MAP; at 0.42 nmol/kg (1 μg) this increase is 60 mm Hg; [Thr^{2,5},Phe^4,Tyr^{6,14},Lys^7]-endothelin-1 (ET-3; rat ET) is only 60% as potent as ET-1 as previously reported [8]. At the high dose, a time-course of action reveals that ET-1 elicits a rapid increase in MAP, reaching a maximum at 10–15 min, followed by a return to near normal levels by 45 min. The ET analogs examined are less potent (see Table 1). ET-1·4SSO$_3$ is only 1/3 as potent as ET-1 but is able to produce an equivalent elevation of MAP at dosages 10-fold

Table 1 *Biological activities of endothelin and analogs*

Peptide	Relative potency[a]
Endothelin-1 (ET-1; human/porcine ET)	100
[Thr^{2,5},Phe^4,Tyr^{6,14},Lys^7]-endothelin-1 (ET-3; rat ET) [8]	60
[Cys-SSO$_3^{1,3,11,15}$]-endothelin-1 (ET-1·4SSO$_3$)	33
[D-Cys^1]-endothelin-1 ([D-Cys^1]-ET-1)	71
[D-Cys-SSO$_3^1$,Cys-SSO$_3^{3,11,15}$]-endothelin-1 ([D-Cys^1]-ET-1·4SSO$_3$)	16
[Cys(Acm)^{1,3,11,15}]-endothelin-1 ([Cys(Acm)^{1,3,11,15}]-ET-1)	36

[a] Potency as defined by ability of peptide to increase mean arterial pressure (MAP) relatie to ET-1 (n = 3 or 4 animals; p < 0.05).

higher than ET-1 with a prolonged (3-fold) length of duration. [D-Cys^1]-ET-1 is also able to produce substantial increases in MAP at high doses (60 mm Hg increase in MAP at a dose of ca. 5 nmol/kg) with an even longer duration of action than ET-1 (>5 times), yet [D-Cys^1]-ET-1·4SSO$_3$ is essentially inactive. [Cys(Acm)^{1,3,11,15}]-ET-1 has been reported to be inactive in a rat aortic strip assay [7], yet in the conscious rat it does increase MAP at high dosages (ca. 5 nmol/kg), and is approximately as potent as ET-1·4SSO$_3$ without the corresponding prolongation of action. A satisfactory explanation of this observation will require further testing of these and other linear endothelins.

Acknowledgements

Research was supported by NIH Grants HL-43154, HL-27716 (M.B.) and HL-41910 (C.H., J.R.). We are indebted to Duane Pantoja, Dean Kirby, Charleen Miller, Ron Kaiser, Karen Carver, and Drs. Jerry Boublik and Terry Hexum for their expert assistance in the purification and chemical and biological characterization of the peptides.

References

1. Yanagisawa, M., Inoue, A., Ishikawa, T., Kasuya, Y., Kimura, S., Kumagaye, S., Nakajima, K., Watanabe, T.X., Sakakibara, S. and Goto, K., Nature, 332 (1988) 411.
2. Itoh, Y., Yanagisawa, M., Ohkubo, S., Kimura, C., Kosaka, T., Inoue, A., Ishida, N., Mitsui, Y., Onda, H., Fujino, M. and Masaki, T., Febs Lett., 231 (1988) 440.
3. Stewart, J.M. and Young, J.D. Solid Phase Peptide Synthesis, 2nd ed., Pierce Chemical Co., Rockford, IL, 1984.
4. Rivier, J., Galyean, R., Gray, W., Azimi-Zonooz, A., McIntosh, J., Cruz, L. and Olivera, B., J. Biol. Chem., 262 (1987) 1194.
5. Schwartz, G.P. and Katsoyannis, P.G., J. Chem. Soc. Perkin Trans. I, (1973) 2894.
6. Brown, M.R., Tache, Y. and Fisher, D., Endocrinology, 105 (1979) 660.
7. Kumagaye, S., Pept. Chem., 1988, (1989) 215.
8. Yanagisawa, M., Inoue, A., Ishikawa, T., Kasuya, Y., Kimura, S., Kumagaye, S., Nakajima, K., Watanabe, T.X., Sakakibara, S., Goto, K. and Masaki, T., Proc. Natl. Acad. Sci. U.S.A., 85 (1988) 6964.

Endothelin: Solid-phase synthesis and structure-activity studies

Kam F. Fok[a], Marshall L. Michener[a], Steven P. Adams[a], Ellen G. McMahon[b], Maria A. Palomo[b] and Angelo J. Tranpani[b]

[a]Monsanto Company, Biological Sciences, St. Louis, MO 63198, U.S.A.
[b]G.D. Searle Company, Cardiovascular Research, St. Louis, MO 63198, U.S.A.

Introduction

Endothelin is an endothelial cell-derived vasoconstrictor peptide with pressor activity that Yanagisawa et al. originally isolated and sequenced from the culture medium of porcine endothelial cells [1]. Consisting of 21-amino acid residues with two sets of inter-chain disulfide linkages, porcine endothelin is one of the most potent vasoconstrictors known. However, the physiological functions of endothelin are unclear. To explore potential therapeutic targets, structure–function studies on endothelin have been carried out.

Results and Discussion

Endothelin and related peptide fragments (Table 1) were prepared by SPPS. Porcine endothelin contracts the isolated rat aorta with an ED_{50} of ca. 7 nM; rat endothelin was about 1/50 as active as porcine endothelin. The endothelin end ring (Fragment 1) and center ring (Fragment 2) fragments were also prepared; however, neither fragment displayed agonist or antagonist activity when tested up to 1000 nM. The nucleotide sequence of porcine preproendothelin predicted a peptide sequence from Cys^{100} to Ile^{130}, exhibiting significant homology with endothelin. 'Endothelin-like peptide' was synthesized and tested on the rat aorta but was inactive at 50 nM.

Recently, the similarity of a family of snake venom peptides with human/porcine endothelin was reported [2]. The sequence of sarafotoxin S6B, a representative example of these peptides (Table 1), bears a strong resemblance to endothelins suggesting that sarafotoxin S6B may have similar bioactivity. In the rat aorta, synthetic sarafotoxin S6B mediated contraction qualitatively similar to endothelin but was about 1/10 as active. Structure–function studies reported here indicate that the activity of endothelin is structurally dependent and that local changes in charge and hydrophobicity between residues 3 and 7 may affect activity.

Any attempt to dissect the endothelin molecule and discover smaller antagonists based on the endothelin framework, is likely to be a daunting undertaking. Endothelin and related peptides described in this study, are difficult to synthesize

269

K.F. Fok et al.

Table 1 *Sequences and activity of endothelin-related peptides*

Names	Sequences	Activity[a]
Endothelin (human, porcine)[b]	CSCSSLMDKECVYFCHLDIIW	[1.0]
Endothelin (rat)	CTCFTYKDKECVYYCHLDIIW	0.02
Endothelin-like peptide (porcine)[c]	CQCASQKDKKCWSFCQAGKEI	NA[d]
Endothelin, Fragment 1	CSSLMDKEC	NA
Endothelin, Fragment 2	CSCSS/CVYFC heterodimer	NA
Proendothelin (porcine)	CSCSSLMDKECVYFCHLDIIWV-NTPEHIVPYGLGSPSRSR	NA
Sarafotoxin S6b	CSCKDMTDKECLYFCHQDVIW	0.1

[a] Vasoconstricting activity on rat aorta relative to porcine endothelin = 1 (ED_{50} = 77 nM).
[b] Disulfide bridges were determined as Cys^1 to Cys^{15} and Cys^3 to Cys^{11}.
[c] Sequence of preproendothelin Cys^{110} to Ile^{130}; significant homology with endothelin.
[d] Not active.

and exhibit solubility problems. An alternative strategy would be to inhibit a putative 'endothelin converting enzyme'. To this end, proendothelin (40 mer) was also synthesized. It was virtually devoid of vasoconstrictor activity on the rat aorta; however, treatment with chymotrypsin in vitro generated endothelin and vasoconstrictor activity [3]. Therefore, conversion of proendothelin to endothelin may be an important step for expression of activity and a potential target for drug design.

References

1. Yanagisawa, M., Kurihara, H., Kimura, S., Tomobe, Y., Kobayashi, M., Mitsui, Y., Yazaki, Y., Goto, K. and Masaki, T., Nature 332 (1988) 411.
2. Takasaki, C., Tamiya, N., Bdolah, A., Wollberg, Z. and Kochva, E., Toxicon, 26 (1988) 543.
3. McMahon, E.G., Fok, K.F., Moore, W.M., Smith, C.E., Siegel, N.R. and Trapani, A.J., Biochem. Biophys. Res. Commun., 161 (1989) 406.

A novel strategy for the deprotection of S-acetamidomethyl containing peptides: An approach to the efficient synthesis of endothelin

Wen Liu, Gong H. Shiue and James P. Tam

The Rockefeller University, 1230 York Ave., New York, NY 10021, U.S.A.

Introduction

Human endothelin is a long-lasting vasoconstrictor hormone produced by endothelial cells [1]. It consists of 21 amino acid residues and two disulfide linkages. In the SPPS of endothelin, a protecting group such as S-acetamidomethyl (Acm) group [2] would be required. Here, we describe a novel strategy in the deprotection of the Acm group. Conventionally, Cys(Acm) is removed after the deprotection of side-chain protecting groups. Our strategy departs from the conventional approach and removes the Cys(Acm) as a continuing step of the synthetic cycles. Thus, the Cys(Acm) groups are removed in the reaction vessel prior to the removal of other protecting groups and the cleavage of the peptide from the solid support. In this new approach, the cleavage of Cys(Acm) is conducted in organic solvents by a heavy metal salt, $Hg(OAc)_2$, in the milieu of the resins. Subsequent deprotection of the oxygen-linked benzyl side-chain protecting groups by the sulfide-assisted deprotection conditions with trifluoromethanesulfonic acid (TFMSA) in TFA completely avoids the use of HF and allows the deprotection to occur as sequential steps in the reaction vessel.

Results and Discussion

Human endothelin was synthesized on the Pam-resin [3] using the Boc-benzyl strategy by stepwise SPPS. At the completion of the synthesis, Cys(Acm), His(Dnp), Trp(For) and the oxygen-linked protecting groups were sequentially removed while the peptide was still attached on the resins [4]. To remove the Cys(Acm) group, the peptide-resin was treated with 0.06 M $Hg(OAc)_2$ in DMF in the dark at 20°C for 3 h. The Hg^{2+} salt was removed by washings with DMF and DMF-mercaptoethanol (9:1, v/v). His(Dnp) was then removed with 1 M thiophenol in DMF. Finally, the oxygen-linked protecting groups and Trp(For) were removed by the low-TFMSA procedure [5] using a mixture of TFMSA/TFA/DMS(dimethylsulfide)/EDT(ethanedithiol)p-cresol (7.6:57.4:25:5:5, v/v) at 4°C for 1 h. After the removal of the aromatic scavengers by repeating washings, the peptide without the protecting groups was cleaved from the resin by a more acidic mixture of the TFMSA mixture consisting of TFMSA/TFA/thioanisole/EDT (8:80:8:4, v/v) at 4°C for 40

271

min. It is important to point out that the Pam-resin or BHA-resin, which contain a more acid stable linkage, allow this manipulation. The filtrate was collected and the peptide precipitated by cold ether (with a few drops of pyridine). The crude peptide was quickly dialyzed (or through gel permeation) in 4 M urea at pH 8.2, 0.1 M Tris-HCl and refolded by the mixed disulfide method (reduced and oxidized glutathione) in a 2 M urea solution, pH 8.2, 0.1 M Tris-HCl buffer. Purification of the refolded and oxidized endothelin by C_{18} RPHPLC gave a homogeneous product in an 8% overall yield (Fig. 1).

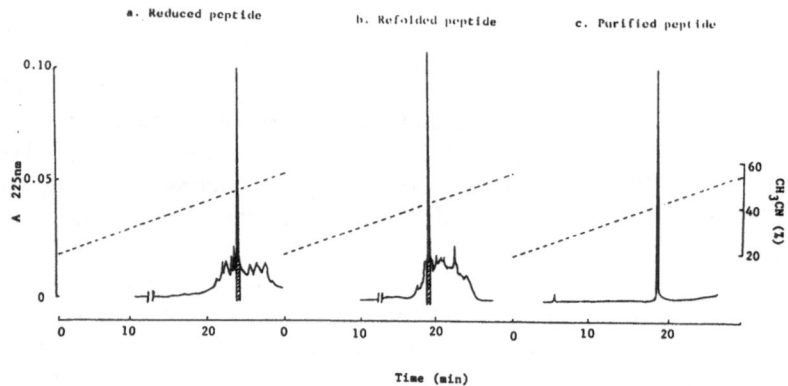

Fig. 1. C_{18} RPHPLC of synthetic endothelin. (a) Crude and reduced; (b) crude and refolded; and (c) purified peptide.

Four advantages were observed from this approach. First, in the sequential deprotection, particularly the removal of the Cys(Acm), a large excess of reagents could be used singly or repeatedly to complete the deprotection reaction. Secondly, reagents and by-products (particularly the heavy metal salts) could be removed from the product by simple washings and filtrations. Thirdly, no loss of material due to precipitation with the heavy metal salt was observed with multiple cysteinyl peptides. This was in strong contrast with the synthesis of endothelin by the conventional approach that resulted in the total loss of material due to precipitation. Finally, selective activation of the deprotected Cys by the mixed disulfide method could be conveniently achieved.

Acknowledgements

This work was supported by USPHS Grant CA 36544 and Bachem U.S.A.

References

1.   Inoue, A., Yanagisawa, M., Kimura, S., Kasuya, Y., Miyauchi, T., Katsutoshi, G. and Masaki, T., Proc. Natl. Acad. Sci. U.S.A., 86 (1989) 2863.

2. Veber, D.F., Milkowshi, J.D., Varga, S.L., Denkewalter R.G. and Hirschmann, R., J. Am. Chem. Soc., 91 (1972) 506.
3. Mitchell, A.R., Erickson, B.W., Ryabtsev, M.N., Hodges, R.S. and Merrifield, R.B., J. Am. Chem. Soc., 98 (1976) 7357.
4. Tam, J.P., J. Org. Chem., 50 (1985) 5291.
5. Tam, J.P., Heath, W.F. and Merrifield, R.B., J. Am. Chem. Soc., 108 (1986) 5242.

Receptor binding and biophysical studies of monocyclic analogs of endothelin

John T. Pelton, Richard Jones, Vladimir Saudek and Robert Miller

Merrell Dow Research Institute, 16 rue d'Ankara, F-67009 Strasbourg, France

Introduction

Endothelin I is a bicyclic, 21-amino acid, vasoconstrictor peptide recently isolated and characterized from cultured porcine aortic endothelial cells [1]. An unusual feature of this relatively short peptide is the presence of two disulfide bonds: one between residues 1–15 and another between residues 3–11. To examine the role of these disulfide bonds in maintaining the conformation of the molecule and their importance to the biological activity of the peptide, we recently synthesized endothelin I analogs in which the cysteine amino acids were replaced by the pseudo-isosteric alanine residue.

Results and Discussion

Peptides were synthesized by standard SPPS using a Boc/Bzl protection scheme on Boc-L-Trp-PAM resin (0.42 mmol/g). The peptides were deprotected and cleaved from the resin with anhydrous liquid HF/anisole (10 : 1, v/v) and cyclized with $K_3Fe(CN)_6$ at high dilution. Purification was accomplished by a combination of gel filtration, ion-exchange chromatography and RPHPLC. The peptides were characterized by AAA, FABMS, TLC and HPLC.

Endothelin I and analogs were assessed for their ability to inhibit the specific binding of [^{125}I]endothelin I to rat cerebellum. Binding constants and Hill slopes are shown in Table 1. Surprisingly, the monocyclic peptides retain considerable binding affinity in this tissue. Furthermore, these analogs are full agonists in the blood-perfused mesenteric artery [2]. More unusual, however, is the acyclic peptide, which has an affinity similar to that of endothelin I in the binding assay, but is a weak partial agonist in the mesenteric bed.

The conformation(s) of endothelin I and analogs was examined by CD and NMR. The CD spectra of endothelin I and [Ala^{3,11}]endothelin I are very similar with a strong negative absorption band around 206 nm. This negative band is red-shifted by about 10 nm in the CD spectra of [Ala^{1,15}]endothelin I and the acyclic analog, the latter showing more beta-like structure with an additional positive absorption band around 193 nm. In the NMR spectra, however, protons in all four analogs show similar chemical shifts and coupling constants. In endothelin I, 19 non-sequential NOEs were observed, whereas in the mono- and acyclic peptides only a few such NOEs were obtained suggesting a less

Table 1 *Binding constants and Hill slopes for inhibition of [^{125}I]endothelin I by endothelin I and alanyl analogs*

	K_i (nM)	Hill slope	Number of experiments
Endothelin I	1.29 ± 0.24	1.1 ± 0.1	6
[Cys(Acm)^{3,11}]Endothelin I	12.6 ± 1.3	0.8 ± 0.1	6
[Ala^{1,15}]Endothelin I	17.5 ± 4.2	1.5 ± 0.1	7
[Ala^{3,11}]Endothelin I	4.43 ± 0.5	1.3 ± 0.1	8
[Ala^{1,3,11,15}]Endothelin I	1.12 ± 0.17	0.9 ± 0.1	7

defined structure. Molecular modeling experiments based on the NMR data for endothelin I yielded a compact structure for this peptide. Similar studies for the moncyclic analogs suggest that these peptides may adopt a conformation similar to that of the native hormone, although the conformation of the acyclic analog is somewhat different.

The results of our present studies indicate that the bicyclic nature of endothelin I is not an absolute requirement for biological activity. Further, the CD studies indicate that endothelin I and [Ala^{3,11}]endothelin I have similar conformations in solution. This is supported by the biological studies in which the affinity of this monocyclic peptide for endothelin I receptors is only slightly less than that of endothelin itself. Furthermore, [Ala^{3,11}]endothelin I is a full agonist in the blood-perfused mesenteric artery. However, conformational changes occur with the other analogs, and although the acyclic peptide has binding affinity for endothelin I receptors similar to the native hormone, it is a weak partial agonist in the mesenteric bed.

References

1. Yanagisawa, M., Kurihara, H., Kimura, S., Tomobe, Y., Kobayashi, M., Mitsui, Y., Yazaki, Y., Goto, K. and Masaki, T., Nature, 322 (1988) 411.
2. Randal, M.D., Douglas, S.A. and Hiley, C.R., Br. J. Pharmacol., (1989) in press.

Application of two-step hard acid deprotection/cleavage procedures to the solid-phase synthesis of the putative precursor of human endothelin

Motoyoshi Nomizu[a], Yoshimasa Inagaki[a], Akihiro Iwamatsu[a],
Tomoko Kashiwabara[a], Hideo Ohta[a], Akihito Morita[a],
Koji Nishikori[a], Akira Otaka[b], Nobutaka Fujii[b] and Haruaki Yajima[b]

[a]Central Laboratories of Key Technology and Pharmaceutical Laboratory,
KIRIN Brewery Co. Ltd., Maebashi Gunma 371, Japan
[b]Faculty of Pharmaceutical Science, Kyoto University, Kyoto 606, Japan

Introduction

A two-step deprotection/cleavage procedure involving a weak hard acid treatment, trimethylsilylbromide (TMSBr) [1], followed by a stronger hard acid for Boc-based SPPS was examined. In the first, deprotection step, protected peptide resin is treated with 1 M TMSBr-thioanisole/TFA. In the second, cleavage step, the peptide resin is treated with a stronger hard acid, such as trimethyl

```
          Acm Bzl Acm Bzl Bzl      O  OBzlClZ OBzlAcm    BrZ
Boc-Cys-Ser-Cys-Ser-Ser-Leu-Met-Asp-Lys-Glu-Cys-Val-Tyr-Phe
          Acm Tos      OBzl       CHO
     -Cys-His-Leu-Asp-Ile-Ile-Trp-PAM resin    (100 mg, 0.188 mmol/g)

   |   1. deprotection procedure
   |        a) 1M TMSBr-thioanisole/TFA, m-cresol, EDT, 0 °C, 1 h
   |         + 1M TMSOTf-thioanisol/TFA, m-cresol, EDT, 0 °C, 1 h
   |        b) HF, m-cresol, 0 °C, 1 h
   |        c) 1M TMSOTf-thioanisol/TFA, m-cresol, EDT, 0 °C, 2 h
   |        d) Low-High HF4); HF (1.3 ml), dimethylsulfide (3.3 ml),
   |              p-cresol (250 µl), p-thiocresol (250 mg),0 °C, 30 min
   |           + HF (4.5 ml),0 °C, 30 min
   ↓   2. Sephadex G-25 (50 % AcOH)

Crude [ET(Cys-Acm)]        Yield; a) 42 mg (78 %)
                                  b) 45 mg (83 %)
   ↓  1. HPLC                     c) 44 mg (81 %)
                                  d) 45 mg (83 %)
Purified [ET(Cys-Acm)]

   |   1 AgOTf/TFA, anisole, 0 °C, 1 h
   |   2. Sephadex G-10 (50 % AcOH)
   |   3. air-oxidation
   ↓   4. HPLC

Purified Synthetic ET-21
[yield; 15.4 % (procedure a), from the Protected Peptide Resin]
```

Fig. 1.   Deprotection procedure, cyclization and purification of ET-21.

silyltrifluoromethanesulfonate (TMSOTf) [2] or HF. The current results demonstrate the usefulness of this procedure, in comparison with other deprotection procedures, for practical SPPS in its application for the synthesis of human endothelin (ET-21) [3]. We applied this procedure to the SPPS of a more complex peptide. The putative precursor of human endothelin (hET-38) was also successfully synthesized using this procedure.

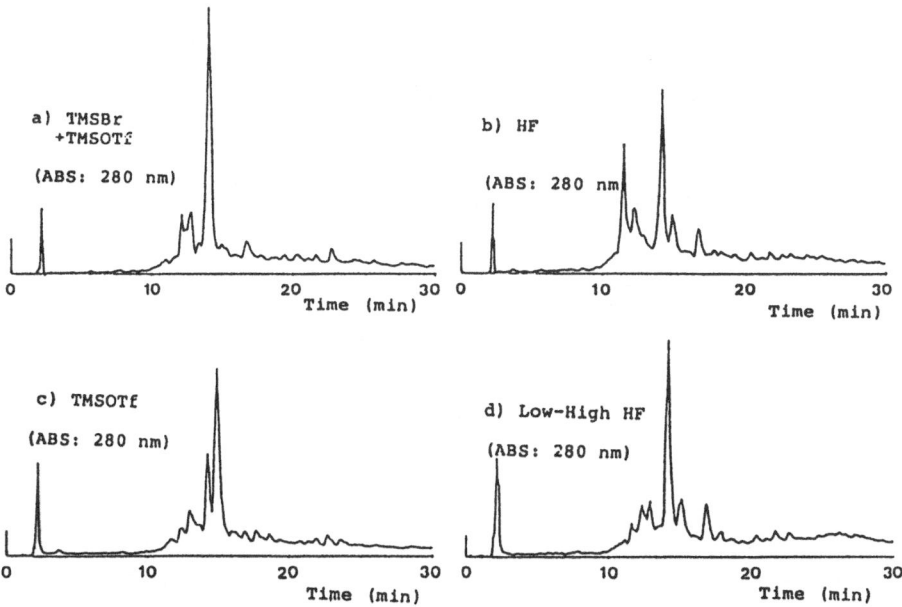

Fig. 2. HPLC pattern of crude ET-21 (Cys-Acm) on Nucleosil 5C$_{18}$ (4.6×150 mm column); gradient elution with CH$_3$CN (25% – 45% in 30 min) in 0.1 % TFA, flow rate 1.0 ml/min.

Results and Discussion

The protected 21-residue peptide resin corresponding to the linear sequence of ET-21 was synthesized using an automated peptide synthesizer and treated with four kinds of deprotection procedures (a)–(d). We examined the purity of these samples using analytical HPLC. As shown in Fig. 2, procedure (a) gave the crude Cys(Acm)-form of ET-21 with the highest purity, thus demonstrating the usefulness of our two-step procedure. Based on the above experimental results, we successfully applied this procedure to the SPPS of hET-38. Bioactivities of these synthetic peptides were tested in vitro by measuring the contraction of rat aorta, and in vivo by recording the arterial pressure in rats. hET-38 had less vasoconstrictor activity in vitro, but a potent pressor effect in vivo [5]. We also found that ET-21 had the strongest vasoconstrictor activity in vitro. The vasoconstrictor activity in vitro of other related peptides that elongated from C-terminal of ET-21 to hET-38, was almost comparable to hET-38. Therefore, ET-21 is the optimum length for expression of vasoconstrictor activity in vitro.

References

1. Fujii, N., Otaka, A., Sugiyama, N., Hatano, M. and Yajima, H., Chem. Pharm. Bull., 35 (1987) 3880.
2. Fujii, N., Otaka, A., Ikemura, O., Akaji, K., Funakoshi, S., Hayashi, Y., Kuroda, Y. and Yajima, H., J. Chem. Soc., Chem. Commun., 1987, 273.
3. Itoh, Y., Yanagisawa, M., Ohkubo, S., Kimura, C., Kosaka, T., Inoue, A., Ishida, N., Mitsui, Y., Onda, H., Fujino, M. and Masaki, T., FEBS Lett., 231 (1988) 440.
4. Tam, J.P., Heath, W.F. and Merrifield, R.B., J. Am. Chem. Soc., 105 (1983) 6442.
5. Kashiwabara, T., Inagaki, Y., Ohta, H., Iwamatsu, A., Nomizu, M., Morita, A. and Nishikori, K., FEBS Lett., 247 (1989) 73.

Energy-based modeling studies of endothelin

J.C. Hempel, W.A. Ghoul, J.-M. Wurtz and A.T. Hagler

Biosym Technologies, 10065 Barnes Canyon Road, San Diego, CA 92121, U.S.A.

Introduction

Endothelin (ET), the most potent endogenous vasoconstrictor known [1], binds to receptors distinct from those of the well-recognized vasoconstrictors [2]. Understanding of the conformational preferences of this 21-amino acid peptide can aid in the search for constrained analogs that bind at the receptor. A new conformational analysis strategy that identifies all regions of the molecule where two or more conformers generated in a modeling study are similar (and, by inference, the regions where they differ) is used to define eight families (sets of conformers similar to within predefined tests) generated in energy-based molecular modeling studies of porcine ET.

Results and Discussion

High-temperature molecular dynamics (MD) was used to search conformation space. The conformer resulting from 5 ps of MD at 900 K (incorporating a bias potential to maintain *trans*-peptide bonds) was stored, the velocities randomized and the procedure repeated. Sixty conformers were generated in this manner, annealed (cooled to 300 K and subjected to 10 ps of MD at 300 K), and minimized. The empirical force field used in this study has been reported previously [3].

Similarity in local backbone conformation at residue i is established when ϕ is equal to within 20 degrees, and ψ is equal to within 20 degrees for two or more conformers (all peptide bonds are *trans*). Similarity in long range conformation is established when a $C\alpha$ to $C\alpha$ distance is equal to within 0.2 Å. Conformational domains are defined as groups of four or more residues that pass all tests for local and long range conformational similarity.

The peptide backbone conformation of the ring defined by residues 3–11 was analyzed. Cys[3] and Cys[11] of ET, as well as Cys[1] and Cys[15], are linked by a disulfide bond [1]. Family I consists of all conformers similar to the lowest energy conformer generated in this study over one or more conformational domains. Succeeding families are defined using the lowest energy conformer not included in the definition of any previous family. Two of four low-energy families defined in this study are illustrated in Fig. 1. Family I is characterized by repeating C-7 motifs (residues 3–5 and 9–11). Family IV is characterized by helical backbone structure (residue 9 through 13).

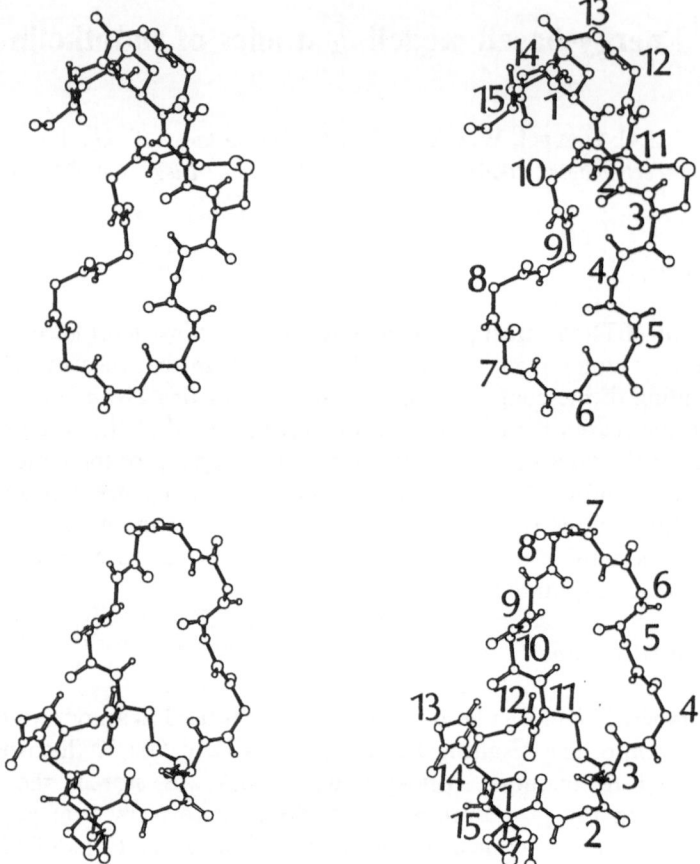

Fig. 1. *Stereo representations of the peptide backbone (residues 1–15) of endothelin for the lowest energy conformer from (top) family I at 366 kcal (note repeating C-7 motifs for residues 3–5 and 9–11) and (bottom) family IV at 380 kcal (note helical structure extending from residue 9 through 13).*

References

1. Yanagisawa, M., Kurihara, H., Kimura, S., Tomobe, Y., Kobayashi, M., Yazaki, Y., Goto, K. and Masaki, T., Nature, 332 (1988) 411.
2. Tomobe, Y., Miyauchi, T., Saito, A., Yanagisawa, M., Kimura, S., Goto, K. and Masaki, T., Eur. J. Pharm., 152 (1988) 373.
3. Dauber-Osguthorpe, P., Roberts, V.A., Osguthorpe, D.J., Wolff, J., Genest, M. and Hagler, A.T., Proteins – Structure and Genetics, 4 (1986) 31.

Analogs of linear and cyclic antagonists of arginine vasopressin: Similarities and some surprising differences

M. Manning[a], M. Kruszynski[a], A.M. Kolodziejczyk[a], W.A. Klis[a], J. Przybylski[a], A. Olma[a], L.L. Cheng[a], A.S. Kolodziejczyk[a], N.C. Wo[b] and W.H. Sawyer[b]

[a]*Department of Biochemistry, Medical College of Ohio, P.O. Box 10008, Toledo, OH 43699-0008, U.S.A.*
[b]*Department of Pharmacology, Columbia University, 630 West 168th Street, New York, NY 10032, U.S.A.*

Introduction

We have previously reported that position 1 in Aaa-D-Tyr(Et)-Phe-Val-Asn-Abu-Pro-Arg-Arg-NH$_2$ (A) (where Aaa = 1-adamantaneacetic acid) tolerates a broad latitude of structural modification [1]. Here, we report a series of analogs of (A) having D-amino acid substitutions at position 2; all directly related to previously reported cyclic AVP antagonists [2]. We also report three series of analogs of (A) with modifications at position 6 all and at positions 1 and 6 combined. In addition, we report 2 related cyclic AVP antagonists that have an Arg-NH$_2$ residue at position 9.

Results and Discussion

The antagonistic potencies for all antagonists are presented in Table 1.

Position 2 modifications (**1–6**: Table 1). By analogy with cyclic AVP antagonists [2,3], aromatic D-amino acids appear to be well-tolerated in the linear antagonists reported here. However, the dramatic losses of anti V$_2$ potency of **4–6**, containing aliphatic D-amino acids at position 2, is in striking contrast to the anti-V$_2$ potency of their counterpart cyclic AVP antagonists [2].

Position 6 modifications (**7–15**: Table 1). Here, the effects of basic amino acids depend very much on the nature of the substituent at position 1. The Nva^6 analogs exhibit surprisingly good V$_2$ antagonism. Remarkably, **13** is 22 times more potent as a V$_2$ antagonist than as a V$_1$ antagonist, and is, in fact, the most selective linear V$_2$ antagonist reported to date.

Position 9 modifications (**16** and **17**: Table 1). Although **16** is virtually equipotent with its parent [3], remarkably, the substitution of Arg-NH$_2$ for Gly-NH$_2$ in a linear antagonist [1] and in a selective V$_2$/V$_1$ cyclic antagonist d(CH$_2$)$_5$[D-Ile^2,Ile^4]AVP [4] resulted, respectively, in the most potent linear V$_2$/V$_1$ antagonist (A) and the most potent cyclic V$_2$/V$_1$ antagonist (**17**) reported to date.

M. Manning et al.

Table 1 *Linear AVP antagonists: Modified at positions 1, 2 and 6 (1–15) and related Arg⁹-NH₂ cyclic AVP antagonists (16,17)*

No.	1 2 3 X-Y-Phe-Val-Asn-Z-Pro-Arg-Arg-NH₂ X^1	4 5 6 7 8 9 Y^2	Z^9	Antiantidiuretic (anti-V_2) pA_2^b	Antivasopressor (anti-V_1) pA_2^b
1	Aaa	D-Tyr	Abu	7.54	7.37
2	Aaa	D-Phe	Abu	7.23	7.33
3	Aaa	D-Ile	Abu	7.04ᵉ	~6.2ᵉ
4	Aaa	D-Ile	Abuᶜ	~5.6	<5.2ᵉ
5	Aaa	D-Val	Abu	~6.2ᵉ	<5.2ᵉ
6	Aaa	D-Leu	Abu	~5.7ᵉ	<5.2ᵉ
7	Aaa	D-Tyr (Et)	Nva	7.72	7.32
8	Aaa	D-Tyr (Et)	Arg	7.98	8.13
9	Aaa	D-Tyr (Et)	Lys	7.70	8.14
10	Phaa	D-Tyr (Et)	Nva	8.06	8.17
11	Phaa	D-Tyr (Et)	Arg	~6.60ᵉ	8.21
12	Phaa	D-Tyr (Et)	Lys	~6.85ᵉ	8.47
13	Pa	D-Tyr (Et)	Nva	8.07	6.73
14	Pa	D-Tyr (Et)	Arg	<5.2ᵉ	7.20
15	Pa	D-Tyr (Et)	Lys	<5.9ᵉ	6.99
16	[d(CH₂)₅Pa¹,D-Tyr²(Et),Val⁴,Arg⁹]AVP¹			7.75	8.31
17	[d(CH₂)₅Pa¹,D-Ile²,Ile⁴Arg⁹]AVP			8.43	7.02

ᵃ Aaa = 1-adamantaneacetyl; Phaa = phenylacetyl; Pa = propionyl.
ᵇ Estimated in vivo pA_2 values represent the negative logarithms of the dose (in nmol/ kg) that reduces the response seen with 2 χ units of agonist to equal the response seen with χ units of agonist administered in the absence of antagonist, divided by the estimated volume of distribution in the rat (67 ml/kg).
ᶜ Contains Gly-NH₂ at position 9.
ᵈ Full chemical name: [1(β-mercapto-β,β-pentamethylene propionic acid) 2-O-ethyl-D-tyrosine, 4-valine, 9-argininamide]arginine vasopressin.
ᵉ These peptides showed some weak agonism in these assays.

Acknowledgements

We thank Ms. Ann Chlebowski for expert help in the preparation of this manuscript, and NIH for grants GM-25280 (MM) and DK-01940 (WHS).

References

1. Manning, M., Klis, W.A., Przybylski, J., Kruszynski, A., Olma, A., Wo, N.C., Pelton, G.H. and Sawyer, W.H., Int. J. Pept. Prot. Res., 32 (1988) 455.
2. Manning, M., Klis, W.A., Olma, A., Seto, J. and Sawyer, W.H., J. Med. Chem., 25 (1982) 414.
3. Manning, M., Olma, A., Klis, W.A., Kolodziejczyk, A.M., Seto, J. and Sawyer, W.H., J. Med. Chem., 25 (1982) 45.
4. Manning, M., Nawrocka, E., Misicka, A., Olma, A., Klis, W.A., Seto, J. and Sawyer, W.H., J. Med. Chem., 27 (1984) 423.

Synthesis of 4-substituted β,β-pentamethylene-β-mercaptopropionic acids

Kenneth A. Newlander[a], Heidemarie G. Bryan[a], James F. Callahan[a],
Drake S. Eggleston[a], Michael L. Moore[a], Nelson C.F. Yim[a], William F. Huffman[a]
and Lloyd M. Jackman[b]

[a]Departments of Peptide and Physical & Structural Chemistry, Smith Kline & French
Laboratories, King of Prussia, PA 19406, U.S.A.
[b]Chemistry Department, The Pennsylvania State University, University Park, PA 16802,
U.S.A.

Introduction

The application of conformationally restricted amino acids has found wide-spread use in peptide chemistry. The utilization of β,β-dialkylcysteines and their desamino derivatives in particular have produced antagonists of oxytocin [1] and vasopressin [2] as well as enkephalin [3] analogs with δ-receptor subtype selectivity. It has recently been found that in vasopressin antagonists, replacement of the β,β-pentamethylene-β-mercaptopropionic acid (Pmp) with cis-4'-MePmp 2 afforded a new analog 5 that displayed substantially reduced partial agonist activity (Scheme 1). The replacement of Pmp with trans-4'MePmp 3, on the other hand, resulted in an antagonist 6 where the effect on partial agonist activity was less demonstrable [4]. Convenient syntheses of 2 and 3 have been accomplished via the 1,4-conjugate addition of 4-methylbenzylmercaptan to 4-methyl-cyclohexylidene acetic acid 1a or its ethyl ester 1b [5]. It was found that by varying the reaction conditions, predominant formation of either the cis or trans isomers could be obtained. A series of experiments was then undertaken to better understand this difference in stereoselectivity.

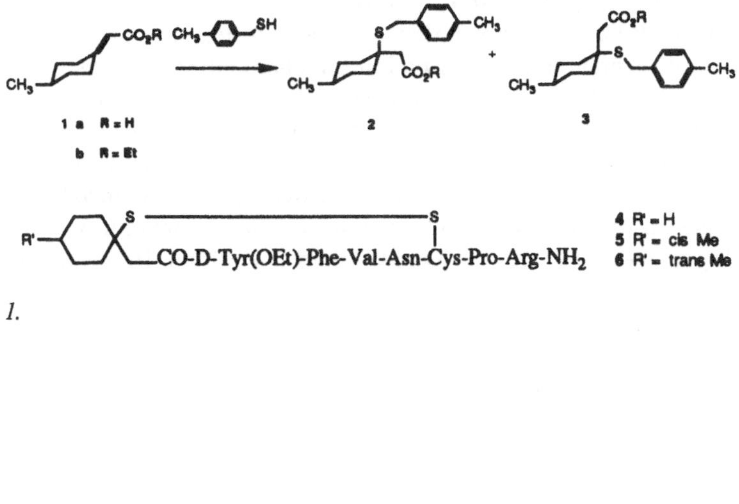

Scheme 1.

Results and Discussion

First, reactions were investigated to test for the possible equilibration of isomeric product. Molecular modeling results [6] showed that there was a 0.96 kcal/mol difference in energy between the lower energy cis isomer **2** and the trans isomer **3**. It was found that purified isomers of **2a,b** and **3a,b** [7] did not interconvert under reaction conditions. It was also found that the change in solvent or starting material from acid **1a** to ester **1b** did not influence the stereochemical preference (see Table 1, entries G,H and J,K). The possibility of a radical-type mechanism was also ruled out, since in the absence of base, addition of a radical initiator such as AIBN produced no detectable product. These results indicate that observed reversal in isomer ratio is caused by the change in base used in the reaction.

A series of reactions were then undertaken to investigate the effect of the base. As seen in Table 1 (entries A–D), a marked increase in the preference for the cis isomer is observed going from potassium to lithium counterions. This strongly suggests that chelation is playing a role in these reactions [8]. This proposed chelation effect was abolished, as expected, by the addition of cryptands [9] (Table 1, entries E and F). Certain amine bases were also found to give a slight preference for the cis isomer, i.e., reactions using DBU and Triton-B (Table 1, entries G–I) both gave ratios comparable to the cryptand results. This implies that these bases are strong enough to abstract the proton from the mercaptan and give reactions similar to the 'free thiolate' reactions in the cryptand experiments. Assuming chelation occurs prior to or simultaneous with thiolate attack, these results suggest that chelation is preferred on the axial face to give more cis product. Inspection of a low energy model of **1** [6] indicates that chelation with the carbonyl oxygen on the equatorial face would be sterically hindered by the axial β-hydrogen. This hindrance would also be more pronounced for strong chelating metals such as lithium, where chelation with solvent and/ or aggregation would increase its effective bulk. The proximity of the carbonyl group to these hydrogens can be seen experimentally from the anisotropic effect in the NMR of **1a** and **1b**. In contrast, amine bases not capable of chelation or of forming thiolate ion (Table 1, entries J,K,O–R) gave reactions that favor trans products. It has been shown that such bases instead form hydrogen bonded complexes with mercaptans [10]. In addition, changing from primary to tertiary amines did not significantly change the isomeric ratio, suggesting that the amine base has no steric effect in discriminating between the two faces of **1** during the reaction. Since chelation in these reactions cannot occur, there is no direct interaction of the base with the axial β-hydrogen in an equatorial attack. The only interaction on the equatorial face comes from a torsional strain between the developing C–S bond and the axial C-2 hydrogen bond. In this model, attack from the top face is disfavored by steric effects between the mercaptan and the axial C-3, C-5 hydrogens of **1** favoring the trans isomer [11]. Although these interactions also exist in the thiolate model, it is evident from the experimental results that the chelation effect predominates.

Table 1   *Ratio of products 2 and 3 using various bases and additives*[a]

Entry	Base	R	Method[b]	%cis 2	%trans 3
A	LiN(SiMe$_3$)$_2$	Et	A	83	17
B	NaH-60 w% dispersion in oil	Et	A	78	22
C	NaN(SiMe$_3$)$_2$	Et	A	76	24
D	KN(SiMe$_3$)$_2$[c]	Et	A	72	28
E	LiN(SiMe$_3$)$_2$ + Kryptofix-211[d]	Et	A	65	35
F	KN(SiMe$_3$)$_2$[c] + Kryptofix-222[d]	Et	A	59	41
G	1,8-diazabicyclo[5.4.0]undex-7-ene	Et	A	65	35
H	1,8-diazabicyclo[5.4.0]undex-7-ene	H	B	65	35
I	Triton B[e]-40 w% in MeOH	Et	A	68	32
J	piperidine[f]	Et	A	38	62
K	piperidine	H	B	38	62
L	piperidine + acetic acid	Et	C	46	54
M	piperidine + pivalic acid	Et	C	47	53
N	piperidine + phenol	Et	C	47	53
O	2,2,6,6-tetramethylpiperidine	H	B	37	63
P	cyclohexylamine	H	B	39	61
Q	triethylamine	H	B	31	69
R	*N*-methylmorpholine	H	B	31	69

[a] Ratios determined using GLC on a HP 530 μ-20 m methylsilicone column (isothermal 190°C, 12 min). The acids (R = H) were first derivatized to their ethyl ester with HCl/EtOH @ 100°C for 1 h.

[b] Method A: 15–20 mol % of base was used. The reaction was run in THF at ambient temperature for 16 h with 1.2 equiv. of mercaptan. Method B: 1.2 equiv. of base was used. The reaction was run in toluene in a sealed tube at 120°C with 1.2 equiv. of mercaptan. Method C: 1.2 equiv. of base was used along with 1 equiv. of the acid. The reaction was run in toluene in a sealed tube at 120°C with 1.21 equiv. of mercaptan.

[c] KN(SiMe$_3$)$_2$ used was 1 N in toluene.

[d] Kryptofix-211 = 4,7,13,18-tetraoxa-1,10-diazabicyclo[8.5.5]hexacosane; Kryptofix-222 = 4,7,13,16,21,24-hexaoxa-1,10-diazabicyclo[8.8.8]hexacosane; 25 mol% of each was added prior to addition of **1**.

[e] Triton-B = benzyltrimethylammonium hydroxide.

[f] Only a trace amount of products was seen after refluxing overnight; no products were seen at room temperature for 16h.

A final observation in these reactions is that amines that give high yields of product with acid **1a** yield little or no product with ester **1b**. This turns out not to be due to the unreactivity of the ester but to the lack of a proton donor in the reaction. For reaction of acid **1a**, no additional acid is needed since **1a** can act both as the reactant and as the proton source. Addition of a Bronsted acid (Table 1, entries L–N) to the reaction with ester **1b** and amine dramatically increased the amount of product obtained without substantially effecting the product ratio.

References

1. Nestor, J.J., Ferger, M.F. and Du Vigneaud, V., J. Med. Chem. 18 (1975) 284.

2. Manning, M., Lammek, B., Kolodziejczyk, A.M., Seto, J. and Sawyer, W.H., J. Med. Chem., 24(1981)701.
3a. Mossberg, H.I., Hurst, R., Hruby, V.J., Galligan, J.J., Burks, T.F., Gee, K. and Yamamura, H.I., Life Sci., 32(1983)2565.
  b. Bryan, W.M., Callahan, J.F., Cod, E.E., Lemieux, C., Moore, M.L., Schiller, P.W., Walker, R.F. and Huffman, W.F., J. Med. Chem., 32(1989)302.
4. Huffman, W.F., Albrightson-Winslow, C., Brickson, B., Bryan, H.G., Caldwell, N., Dytko, G., Eggleston, D.S., Kinter, L.B., Moore, M.L., Newlander, K.A., Schmidt, D.B., Silvestri, J.S., Stassen, F.L. and Yim, N., J. Med. Chem., 32(1989)880.
5. Yim, N. and Huffman, W.F., Int. J. Pept. Prot. Res., 21(1983)568.
6. Molecular modeling was preformed using MacroModel version 2.01 (W.C. Still, Columbia University). Specifically, structures were generated using the multiple conformation mode followed by batch minimization using the MM2 force field.
7. For X-ray structures of **2a** and **3a**, see: Eggleston, D.S., Yim, N., Silvestri, J., Bryan, H. and Huffman, W., Acta Crystallogr., C45(1989)259.
8. For a review on the effects of chelation in stereo- and regio-control, see: Beak, P. and Meyers, A.I., Acc. Chem. Res., 19(1986)356.
9. Parker, D., Adv. Inorg. Radiochem., 27(1983)1.
10. Bicca De Alencastro, R. and Sandorfy, C., Can. J. Chem., 51(1973)985.
11. For a discussion on steric vs. torsional strain interactions in cyclohexanone systems, see: Mukherjee, D., Wu, Y., Fronczek, F.R. and Houk, K.N., J. Am. Chem. Soc., 110(1988)3328 (and references cited therein).

Vasopressin antagonists containing D-cysteine at position 6

William M. Bryan[a], Fadia El-Fehail Ali[a], Michael L. Moore[a],
William F. Huffman[a], Christine Albrightson[b], Bridget Brickson[b], Nancy Caldwell[b],
Lewis B. Kinter[b], Dulcie B. Schmidt[b] and Frans Stassen[b]

[a]Department of Peptide Chemistry and [b]Departments of Pharmacology and Molecular
Pharmacology, Smith Kline & French Laboratories, PO Box 1539, King of Prussia,
PA 19406-0939, U.S.A.

Introduction

Recently we reported that the vasopressin V_2 receptor antagonist **1** displayed unexpected agonist activity in phase-I trials [1]. Subsequent to this finding we developed a sensitive in vivo model which might predict such agonist activity in analogs like **1** [2]. In an attempt to eliminate this intrinsic agonist activity, we studied the effect of D-amino acid substitution at position 6 in several potent antagonists modified at the carboxy-terminal tripeptide [3].

Results and Discussion

The last two columns of Table 1 indicate the relative agonist activities of these analogs. Trained, conscious, water-loaded dogs were given a bolus administration of antagonist 20 min after treatment with indomethacin. With the exception of **11** and **13** the L-Cys antagonists display full agonist activity compared to AVP. The D-Cys-containing analogs in general exhibit far less agonist activity relative to their L-Cys congeners while displaying only slightly reduced in vitro and in vivo antagonist activity. However, with the exception of **4** their affinity for the porcine V_2 receptor is about an order of magnitude less.

While the exact mechanism by which compounds such as **1** display agonist effects is unclear, we have been able to show that this agonism can be demonstrated in vivo. Employing this model we have found a series of analogs containing D-cysteine at position 6 which show that this substitution is compatible with the antagonist pharmacophore at the renal V_2 receptor and with a decrease in intrinsic agonist activity.

Table 1 *Biological activity of D and L cysteine containing vasopressin analogs*

$$\text{CO-D-Tyr(Et)-Phe-Val-Asn-X-R}$$

Cpd.	X	R	Pig K_b (nM)[a]	Dog K_i (nM)[b]	Rat ED_{300} (μg/kg)[c]	Dog* U osm (mOsm/kg)[d]	Vol (mL/min)[e]
1	Cys	Pro-Arg-NH₂	12 ± 1.2	3.9 ± 0.4	8.4 ± 0.7	1625 ± 142	0.06 ± 0.01
2	D-Cys	Pro-Arg-NH₂	220	5.6	73	104 ± 23	1.97 ± 0.63
3	Cys	Arg-D-Arg-NH₂	5.4	1.7	13.5 ± 2.5	469 ± 221	0.6 ± 0.17
4	D-Cys	Arg-D-Arg-NH₂	2.8	1.2	19	150 ± 36	1.05 ± 0.56
5	Cys	Arg-NH₂	9.1	2.5	58	946 ± 370	0.13 ± 0.06
6	D-Cys	Arg-NH₂	82	5.2	>500	81	1.6
7	Cys	Arg-Arg-NH₂	3.4 ± 0.5	17 ± 0.1	7.2	1453	0.18
8	D-Cys	Arg-Arg-NH₂	32	0.835	18	140	1.0
9	Cys	N-MeArg-Arg-NH₂	1.8	0.18	9.3	1713 ± 206	0.1 ± 0.04
10	D-Cys	N-MeArg-Arg-NH₂	48	2.4	9	144	1.3
11	Cys	D-Arg-D-Arg-NH₂	5.2	0.88	11.5 ± 3.3	99	1.26
12	D-Cys	D-Arg-D-Arg-NH₂	42	7.4	42	153	1.5
13	Cys	D-Arg-Arg-NH₂	2.7	1.9	14.6 ± 0.4	177	0.66
14	D-Cys	D-Arg-Arg-NH₂	190	2.9	25	141	0.95

[a] Measured by inhibition of [³H]LVP binding to a renal medullary preparation.
[b] Inhibition of vasopressin-stimulated adenylate cyclase in a renal medullary preparation.
[c] Dose required to lower urine osmolality to 300 mOsm/kg of water.
[d] Maximum urine osmolality at a dose of 100 μg/kg peptide after a pretreatment with indomethacin.
[e] Maximum urine flow at a dose of 100 μg/kg peptide.
* Maximum effect of indomethacin treatment on urine osmolality and flow was 92 ± 12 mOsm/kg and 2.8 ± 0.5 mL/min. Maximum effect of AVP at 3 ng/kg on urine osmolality and flow was 939 ± 76 mOsm/kg and 1.4 ± 0.3 mL/min.

References

1. Dubb, J., Allison, N., Tatoian, D., Blumberg, A., Lee, K. and Stote, R., Kidney Int., 31 (1986) 267.
2. Albrightson, C., Caldwell, N., Brennan, K., DePalma, D. and Kinter, L., Fed. Proc. Fed. Am. Soc. Exp. Biol., 46 (1987) 267.
3. Ali, F.E., Chang, H.-L., Huffman, W.F., Heckman, G., Kinter, L.B., Weidley, E.F., Edwards, R., Schmidt, D., Ashton-Shue, D. and Stassen, F., J. Med. Chem., 30 (1987) 2291.

Quantitative structure-activity relationships (QSAR) of neurohypophyseal hormone analogs acting as oxytocin inhibitors on uterus in vitro

Vladimir Pliška[a] and Marvin Charton[b]

[a]*Institute of Animal Science, ETH Zürich, CH-8092 Zürich, Switzerland*
[b]*Pratt Institute, Chemistry Department, Brooklyn, NY 11205, U.S.A.*

Introduction

Recently, we have applied Free–Wilson analysis [1] to a group of 160 peptides that act as oxytocin inhibitors on the rat uterus in vitro [2]. Results indicate that contributions of side chains in individual backbone positions (SCC) to the pA_2 value are both mutually independent and additive. Since SCCs (like pA_2, EC_{50}, potency, etc.) are specific descriptors of a biological activity, we assume that methods of correlation analysis commonly used in QSAR may detect physicochemical and structural features of side chains that determine biological properties of a peptide molecule. This combined 'Free–Wilson/correlation' analysis [3] may, in the future (i) facilitate the design of peptides by delimiting optimal properties of substituents, (ii) enable analysis of intramolecular forces that influence, in individual chain positions, the biological property in question, and (iii) allow us the derivation (and justification) of a correlation equation that would constitute a substitution in several positions in a generic group of peptides.

Methods

The oxytocin analogs employed, their pA_2 values, SCC values for seven chain positions, and additional abbreviations of amino acids have been listed earlier [2]. The 'intramolecular force' equation (IMF) [4] was used for correlation analysis. Structural descriptors considered were: polarizability (α) [4], localized electrical effect parameter (σ_I) [5], number of hydrogen bond donors in the side chain (n_H), number of full nonbonding orbitals (n_n), indicator variable standing for the presence or absence of an ionized group (i), and the steric parameter (ϑ) defined from van der Waals radii [6]. The analysis was carried out for positions 1 and 2, for which enough SCC values were available. IMF descriptors were computed for the side chain in position 2 (Y), and for two substituent groups of the terminal position 1 (for structural schemes see inserts in Fig. 1): the group to which the SH group (if any) is bonded (X), and the group (other than methyl) attached to the C^α atom (Z). A 'dummy' parameter was further introduced for the presence of an additional methyl on the C^α atom. The values

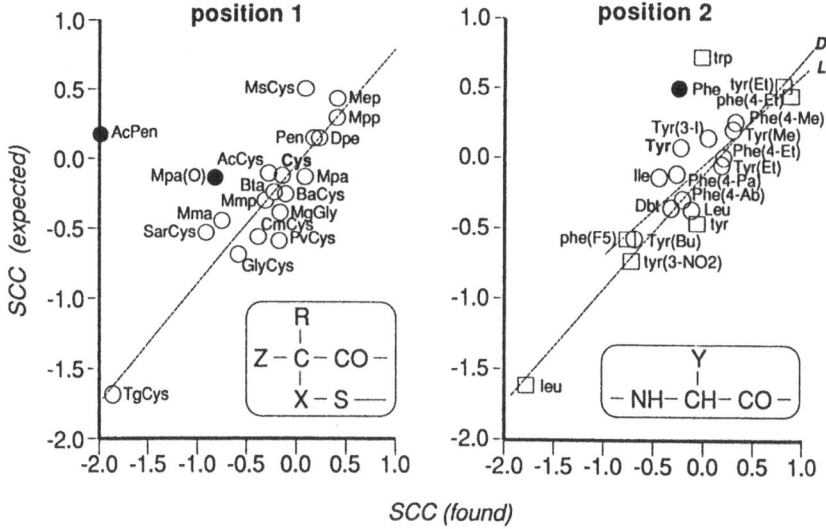

Fig. 1. SCC (OT pos. 1 and 2) found by Free–Wilson analysis (abscissa) and computed as a function of IMF descriptors by regression analysis (ordinate). Circles: open, L-enantiomers; closed, D-enantiomers; shaded, outliers. Significance of all correlations: p < 0.01. Inserts: substitution patterns in positions 1 and 2, respectively. Abbreviations, see Ref. 2.

of the descriptors will be published later. 'Relative importance' of the substituent effect parameters was expressed according to [7] using histidine as a reference.

Results and Discussion

For position 1, the SCCs correlate significantly ($p < 0.001$) with α_X, σ_{IZ}, α_Z, n_{HZ}, n_{nZ} and i_Z (indices X and Z denominate the corresponding group; see above), after the omission of two outliers (cf. Fig. 1). The relative importance of group X is, however, low (18.7%) compared to group Z (81.2%). An additional C^α methylation does not exercise a visible effect. L- and D-enantiomers in position 2 were analyzed separately. For L-forms, the dominating IMF parameter is σ_{IY} (rel. importance 45.2%), besides ϑ_Y (28.9%), i_Y (13.6%), and α_Y (9.3%); one outlier was omitted. α_Y (71.8%) and i_Y (28.2%) are the most relevant parameters for the D-forms. Results of the regression analysis are shown in Fig. 1.

The pA_2 values correlate closely with dissociation constants of the peptide–receptor complex [8] and, thus, the reported analysis permits a conclusion about structural features influencing the binding of neurohypophyseal hormones to the myometrial receptor. This binding is clearly influenced (in a positive sense) by localized electrical effects as well as electrical charge of the N-terminal amino group or its substituents, but less vigorously by the polarizability (tightly correlating with several descriptors of hydrophobicity) of the group, which bonds SH in position 1. Steric effects of any 1-substituent play a rather subordinate role. Opposite effects of the side-chain polarizability and electrical charge were found for L- and D-enantiomers in position 2. Whereas decreasing polarizability and the presence of electrical charge enhance the receptor binding, the binding of D-enantiomers seems to be enhanced by increasing polarizability and in the absence of electrical charge. No systematic effect of D-substitution upon binding could be detected in comparable chains, but its nonspecific 'overall' effects are large.

Acknowledgements

Supported by Swiss National Science Foundation, Grant 3.559-86 (V.P.).

References

1. Free, S.M. and Wilson, J.W., J. Med. Chem., 7 (1964) 395.
2. Pliška, V. and Heiniger, J., Int. J. Pept. Prot. Res., 31 (1988) 520.
3. Pliška, V., Experientia, 34 (1978) 1190.
4. Charton, M., J. Theor. Biol., 91 (1981) 115.
5. Charton, M., Progr. Phys. Org. Chem., 13 (1981) 119.
6. Charton, M., In Roche, E.B. (Ed.) Design of Biopharmaceutical Properties Through Prodrugs and Analogs, American Pharmaceutical Society, Washington, DC, 1977, p. 228.
7. Charton, M. and Charton, B.I., J. Theor. Biol., 99 (1982) 629.
8. Pliška, V., J. Receptor Res., 8 (1988) 245.

Vasotocin prodrug analogs as specific vasopressor agonists

Carl-Johan Aurell[a], Bengt Bengtsson[a], Franciszek Kasprzykowski[b],
Karolina Lawitz[a], Per Melin[a],* and Jerzy Trojnar[a]

[a]*Research Department, Ferring Pharmaceuticals, Box 30561, S-200 62 Malmö, Sweden*
[b]*Institute of Chemistry, University of Gdansk, 80-952 Gdansk, Poland*

Introduction

Prolonged bioactivity of synthetic analogs of vasopressin (VP) has previously been obtained by stabilizing the molecule against enzymatic degradation [1]. One approach to achieve such properties has been the *N*-acylation with one or more amino acid residues in the N-terminus of [Lys^8]VP and [Arg^6]VP [2, 3] or extension of the side chain of the lysine residue at position 8 of [Lys^8]VP [1].

It is generally believed that such analogs, which are low in activity, function as prodrugs, slowly releasing the active principle by enzymatic cleavage of the added amino acid residues. Until now, however, these modifications have produced analogs of low potency and specificity as well as moderate duration of effect. Based on a vasotocin (VT) derivative, [Hmp^1,Phe^2,Orn^8]VT, a vasopressor-specific nonapeptide, we have synthesized and biologically tested some analogs modified by *O*-acylation of hydroxymercaptopropionic acid (Hmp) at position 1 (Fig. 1) as well as *N*-acylation in the side chain of ornithine at position 8.

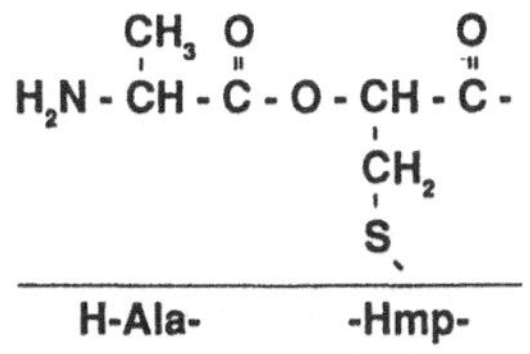

Fig. 1.   H-Alanine (H-Ala) acylation of hydroxymercaptopropionic acid (Hmp).

Results and Discussion

Chemical synthesis. All peptides were synthesized according to SPPS methodology, utilizing the Boc-strategy when acylated at both positions 1 and 8 with the same amino acid, and utilizing Fmoc-strategy when differently acylated or only *O*-acylated. The acylations were carried out by means of activated carbonyl

---

*To whom correspondence should be addressed.

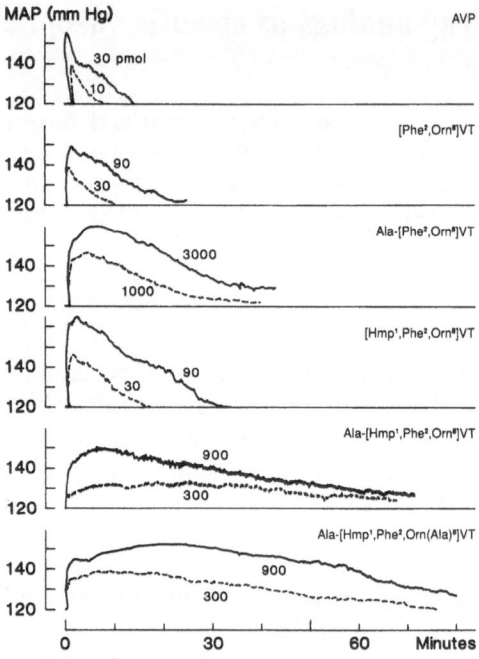

Fig. 2. *Time course of pressor response to vasotocin analogs with and without amino acid extension. Intravenous doses in pmol kg⁻¹ body weight.*

compounds (e.g., Boc-amino acid pentafluorophenylesters) in organic solvents, catalyzed by 4-dimethylamino pyridine.

Bioassays. Vasopressor potency was estimated as maximal mean arterial pressure after i.v. drug injection in anesthetized male rats without premedication; duration of the pressure effect was given as an index of persistance values and was based on the descending part of the pressure curve [4]. Antidiuretic tests were performed in anesthetized hydrated rats, and potency estimation was based on the maximal increase of urine conductivity after i.v. injections [5].

Hmp substitution at position 1 of the pressor specific nonapeptide [Phe², Orn⁸]VT more than doubled pressor potency and slightly elevated the effect's duration compared to the Cys¹ analog (Table 1). When acylated by alanine at Hmp¹ and Cys¹, respectively, the difference in both these pressor parameters became more apparent. The combined Ala acylations at Hmp¹ and at the side chain at position 8 resulted in some further improvement of effect prolongation by a factor of about 12 compared to that of AVP (Fig. 2).

The relatively high potency and long-enduring effect of the present *O*-acetylated analogs likely reflect the slow enzymatic cleavage of the extended amino acid residue leaving Hmp derivatives, which themselves resist further degradation by aminopeptidases.

Table 1 *Biological activities of vasopressor-specific vasotocin hormonogens*[a]

Peptide	Vasopressor (P) (IU/μmol)	Vasopressor Effect duration (I.P)	Antidiuresis (AD) (IU/μmol)	P/AD
Cys- Tyr- Phe-Gln-Asn-Cys-Pro-Arg- Gly-NH$_2$[b]	614±25	1.0	620 ±54	1
– Phe-Ile – – – Orn – – –	172±22	2.8±0.5	2.2± 0.2	78
Ala – Phe-Ile – – – Orn – – –	25± 2	6.0±0.9	0.5± 0.04	50
– Hmp- Phe-Ile – – – Orn – – –	440±17	3.1±0.5	11.3± 2.2	39
Ala-Hmp- Phe-Ile – – – Orn–Ala – –	90±10	9.9±1.8	1.8± 0.3	50
Ala-Hmp- Phe-Ile- – – Orn–Gly –	55± 7	11.6±2.0	6.2± 0.7	9
Ala-Hmp- Phe-Ile – – Orn –	28± 3	7.6±1.8	0.7± 0.02	40

[a] Means ± S.E.M.
[b] AVP.

The present pharmacological data from our extended pressor-specific derivatives suggest that esterification of analogs containing Hmp instead of Cys is advantageous in increasing potency as well as duration of effect by neurohypophyseal hormone analogs.

References

1. Jost, K., In Jost, K., Lebl, M. and Brtnik, F. (Eds.) Handbook of Neurohypophyseal Hormone Analogues, Vol. II, Part 1, CRC Press, Boca Raton, FL, 1987, p. 1.
2. Kyncl, J., Rezabek, K., Kasafirek, E., Pliška, V. and Rudinger, J., Eur. Pharmacol., 28 (1974) 294.
3. Moore, G.J., Kwok, Y.C., Ko, E.M., Severson, D.L. and Rosenior, J.C., Endocrinology, 111 (1982) 1626.
4. Pliška, V., Arzneim.-Forsch., 16 (1966) 886.
5. Larsson, L.-E., Lindberg, G., Melin, P. and Pliška, V., J. Med. Chem., 21 (1978) 352.

Bioavailability of Carbetocin
([2-*O*-methyltyrosine] deamino-1-carba oxytocin) in the rat

Jiřina Slaninová[a], Tomislav Barth[a], Michal Lebl[a], Bohuslav Černý[b] and
Jerzy Trojnar[c]

[a]*Institute of Organic Chemistry and Biochemistry, 16610 Prague 6, Czechoslovakia*
[b]*Institute of Nuclear Biology and Radiochemistry, 14220 Prague 4, Czechoslovakia*
[c]*Ferring AB, Research Laboratories, S-200 62 Malmö, Sweden*

Introduction

The long-acting synthetic analog of oxytocin (OXT), Carbetocin ([2-*O*-methyl-tyrosine]-deamino-1-carba oxytocin) (CE), becomes a drug of choice in veterinary medicine, particularly for reproductive purposes.

Results and Discussion

OXT and CE [1] were synthesized at the Institute of Organic Chemistry and Biochemistry. Tritiated CE was prepared as described earlier [2]. Uterotonic and galactogogic tests were performed as previously described [3]. The half-life was determined as described in Ref. 4.

Comparison of intravenous (i.v.) and intranasal (i.n.) application of OXT and CE is shown in Figs. 1A–D. To obtain a measurable effect after i.n. application a dose of OXT 4 orders of magnitude higher than that used for i.v. administration was required, but with CE, the i.n. dose needed was only 2 orders of magnitude higher than the i.v. dose. The time-course of response to both peptides applied i.n. resembled that of CE after i.v. administration.

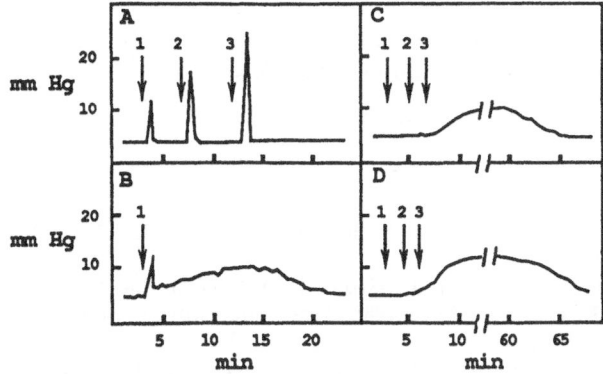

Fig. 1. *Intramammary pressure in lactating rats after i.v. (A,B) or i.n. (C,D) administration of OXT (A,C) or CE (B,D). (A) 1: 2.75×10^{-7} mg, 2: 5.5×10^{-7} mg, 3: 1.1×10^{-6} mg; (B) 1: 4×10^{-5} mg; (C) 1: 2×10^{-4} mg, 2: 2×10^{-3} mg, 3: 2×10^{-2} mg; (D) 1: 1×10^{-4} mg, 2: 1×10^{-3} mg, 3: 1×10^{-2} mg.*

The half-life of CE determined in plasma of female rats in induced estrus after i.v. injection of radioactive CE $(5 \times 10^5 - 1 \times 10^6$ cpm) in six experiments was 8.1 min. A similarly enhanced value was found for deamino-dicarba-OXT (7.8 min) [4], whereas for OXT the value was ca. 4 min.

Penetration of CE into the blood stream was investigated using tritiated CE as a tracer. We followed the rate at which the radioactivity appeared in plasma after i.n. application and the time course of the galactogogic effect. The radioactive compounds appeared in plasma 2.5 min after application, and the plateau level was reached in 10 min. After this time, the radioactivity in plasma increased only very slowly (Fig. 2). The rate of absorption was relatively high and coincided with the appearance of biological effect. However, the galactogogic response disappeared from the plasma at a faster rate than the radioactivity.

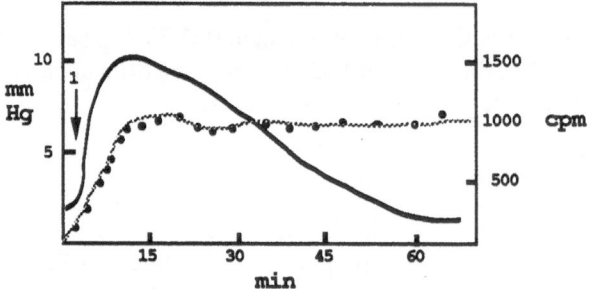

Fig. 2.   *Time-courses of the intramammary pressure (——) and the radioactivity (●---●) in plasma of a lactating rat after i.n. administration of tritiated CE. 1-CE (2.5 × 10⁻³ mg + 1 μCi of the tritiated compound) was given in a volume of 20 μl.*

Tritiated CE $(4 \times 10^{-5}$ mg) was administered subcutaneously to lactating female rats to establish its distribution among tissues and its possible passage into the young rats via feeding. The lactating rats were killed after 24 h and their young 24 or 48 h after peptide administration. Radioactivity was determined in blood, kidneys, liver, brain, reproductive organs and milk tissue. The highest radioactivity was found in kidneys of lactating rats, as well as of their young and in the liver (about half of the amount in kidneys).

References

1.   Frič, I., Kodíček, M., Procházka, Z., Jošt, K. and Bláha, K., Coll. Czech. Chem. Commun., 39 (1974) 1290.
2.   Lebl, M., Barth, T., Slaninová, J., Hrbas, P., Eichler, J. and Černý, B. In Jung, G. and Bayer, E. (Eds.) Peptides 1988, Walter de Gruyter, Berlin, 1989, p. 546.
3.   Slaninová, J. In Jošt, K., Lebl, M. and Brtník, F. (Eds.) Handbook of Neurohypophyseal Hormone Analogs, Vol. I, Part 1, CRC Press, Boca Raton, FL, 1987, p. 83.
4.   Vaněčková, J., Barth, T., Jošt, K., Rychlík, I., Fromageot, P. and Morgat, J.L., Coll. Czech. Chem. Commun., 41 (1976) 2124.

Structure-dependent behavior function of AVP analogs

Yu-cang Du, Chao Lin, Ling-fei Xu, Zi-xian Lu, Ben-xian Gu and Ning-ning Guo

Shanghai Institute of Biochemistry, Academia Sinica, 320 Yue-yang Road,
Shanghai 200031, China

AVP plays an important role in memory processes. In a previous paper, we reported that neonatal treatments with both [Mpr^1,D-Arg^8]AVP (DDAVP) and hypertonic saline facilitated acquisition and subsequent maintenance of brightness discrimination (BD) in rats [1]. Furthermore, we found that neonatal administration of [Mpr^1,D-Arg^8]des-(Gly^9-JAVP (DDDGAVP) or [PCA^4,Cyt^6]-JAVP$<$4–8$>$(ZNC(C)PR) both show significant facility for maintenance of BD in rats, but, unexpectedly, D-Arg-ZNC(C)PR did not. We synthesized more AVP analogs, [PCA^4,Cyt-Ome^6]JAVP$<$4–8$>$(ZNC(C-Ome)PR), its D-arginine isomer DA-ZNC(C-Ome)PR, and the covalent dimer of AVP$<$4–8$>$i.e. (ZNCPR)2. Rats administered with some of them showed significantly longer latency on the maintenance of shuttle-box-avoidance response 24 and 48 h post-trial; their ineffective doses are illustrated in Table 1. This indicated that ZNC(C-Ome)PR and ZNC(C)PR were 100 and 10 times respectively as effective as AVP on behavioral response, but the substitution with D-Arg dramatically decreased their behavioral activities to the level of AVP or DDDGAVP, and (ZNCPR)2 was totally ineffective in passive avoidance. Moreover, we found that ZNC(C-Ome)PR was the most effective stimulator of the accumulation of inositol phosphates in the hippocampus; the approximate concentrations resulting in half-maximal response were 0.05 μM for ZNC(C)PR and 1 μM for AVP. ^{35}S-labeled ZNC(C)PR synthesized in this laboratory showed a tight binding to the dentate gyrus and the hippocampal pyramidal cell layer in rat brain. Since the binding sites of

Table 1 *Effects of AVP analogs on retention in passive avoidance in adult rats*

	AVP Analogs	Effective doses* (ng)
AVP		300– 1 000
DDAVP	[Mpr^1,D-Arg^8]AVP	1 000– 3 000
DDDGAVP	[Mpr^1,D-Arg^8]AVP$<$1–8$>$	1 000– 3 000
ZNC(C)PR	[PCA^4,Cyt^6]AVP$<$4–8$>$	30– 3 000
ZNC(C-Ome)PR	[PCA^4,Cyt-Ome^6]AVP$<$4–8$>$	3–10 000
DA-ZNC(C)PR	[PCA^4,Cyt^6,D-Arg^8]AVP$<$4–8$>$	1 000–10 000
DA-ZNC(C-Ome)PR	[PCA^4,Cyt-Ome^6,D-Arg^8]AVP$<$4–8$>$	300– 3 000
(ZNCPR)$_2$	[PCA^4]AVP$<$4–8$>$ dimer	$>$10 000

* The effective dose with which each rat (130–150 g) was subcutaneously injected and the mean latency in 24-h retention test was significantly increased in comparison with that of controlled rats (p$<$0.05).

AVP or AVP<4–9>were reported to be CA1 and the molecular layer of the dentate gyrus or hilus only [2], the receptor of ZNC(C)PR in the membrane of rat hippocampus may be a new subtype.

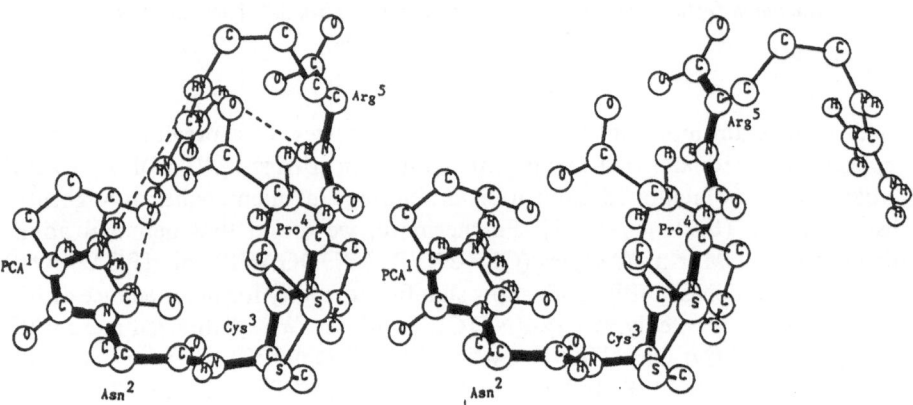

Fig. 1. *The NOE constraint molecular models of ZNC(C)PR containing L-(left) and D-arginine (right). Thick lines represent main peptide chain.*

In order to clarify the SAR of AVP analogs, conformation studies of ZNC(C)PR and its D-Arg isomer using NMR were performed on a Bruker 400 MHz spectrometer with an Aspect 3000 computer system at 25°C. All the resonance peaks on the one-dimensional spectra have been assigned by a combination of COSY and 2DJ spectra. 2D NOESY spectra of these two peptides in D_6-DMSO provided semi-quantitative information on proton-proton distances in the range 2–5 Å. The constraint distances of some protons could be estimated from the intensities of the cross-peaks in the NOESY spectra. Although some strong NOE connectivities between CαH-NH [dαN(i, i + 1)] indicated that proton-proton distances were shorter than 3.0 Å, some cross-peaks existed only in the former but not in the latter, such as Asn^2 αNH to Arg^5 guanidinyl ωNH and Asn^2 γNH$_2$ to Arg^5 εNH, which suggested a compact structure. It was also postulated that the carbonyl oxygen of beta carboxyl group of Asn^2 was linked with guanidine group of Arg^5 by a hydrogen bond in molecule ZNC(C)PR but not in the D-Arg isomer. The NOE constraint molecular models of these peptides are illustrated in Fig. 1.

Therefore, we concluded that Arg in short AVP analogs plays a critical role in the maintenance of a relatively steady structure necessary for its behavioral activity.

Acknowledgements

This study was supported by grants from the National Committee of Science

Foundation (9389007) and Academia Sinica (87-50-01). We thank Wang Peng-liang and Dr. Liu Ren-yi for assistance in deducing molecular models and bioassay.

References

1. Chen, X., Chen, Z., Leu, R. and Du, Y., Peptides, 9 (1988) 717.
2. Brinton, R.E., Gehlert, D.R., Wamsley, J.K., Wan, Y.P. and Yamamura, H.I., Life Sci., 38 (1986) 443.

Neokyotorphin from the brain of hibernating ground squirrels can participate in heart activation and arousal from hibernation

Boris V. Vaskovsky[a], Vadim T. Ivanov[a], Inessa I. Mikhaleva[a], Stella G. Kolaeva[b], Yuri M. Kokoz[b], Vladimir I. Svieryaev[a], Rustam H. Ziganshin[a], Galina S. Sukhova[c] and Dmitryi A. Ignatiev[b]

[a]*Shemyakin Institute of Bioorganic Chemistry, U.S.S.R. Academy of Sciences, Miklukho-Maklaya, 16/10, Moscow 117871, U.S.S.R.*
[b]*Institute of Biological Physics, U.S.S.R. Academy of Sciences, Pushino, U.S.S.R.*
[c]*Moscow State University, Moscow, U.S.S.R.*

Introduction

At present, a large number of peptides such as bombesin, opioid peptides, CCK and others are known as potential endogenous regulators of seasonal physiological functions of hibernating animals. The physiological spectrum of the activity of these peptides doesn't cover all the phenomenology of hibernation, therefore the search for new regulators in tissues of the hibernating animals is desirable. The totality of the endogenous regulators of hibernation must contain both the inhibitors of different physiological functions and the activators providing the restoration of normotropic state and arousal.

Results and Discussion

The acid extract from the brain of the hibernating ground squirrel *Citellus undulatus* was fractionated by ultrafiltration with Amicon PM-10 and UM-2 filters and gel-filtration on Sephadex G-25 column in 0.5 M acetic acid at 8°C. All fractions obtained were tested on hypothermic and antimetabolic activity; active 1000–2000 D fraction was further subjected to Nucleosil $7/C_{18}$ RPHPLC column. At this stage we employed a more sensitive test system based on substantial differences between electrophysiological parameters of functioning hearts of hibernating and nonhibernating animals. For instance, the action potential of hibernator heart has practically no plateau phase that is characteristic for nonhibernator heart; moreover, the former has a rather delayed phase of repolarization [1]. We found that fractions active in hypothermic and antimethabolic tests were also able to transform action potentials on frog atrial preparations to the form typical for hearts of hibernating animals. It is well known that the plateau in action potential is formed by Ca^{2+} and K^+-currents, and blockers of transport of these ions have antimetabolic and antiarrhythmic potency. We also tested the effect of all HPLC components on Ca^{2+}-currents in isolated

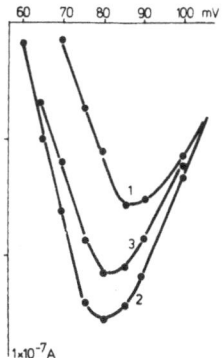

Fig. 1.   *Effect of neokyotorphin on Ca^{2+}-current in frog atrial fibers 1 = control, 2 = peptide 5 × 10^{-7} M, 3 = washout.*

atrial trabecula from frog *Rana ridibunda* at room temperature. Ca^{2+}-currents were recorded under voltage clamp. The sodium current was blocked by tetrodotoxin. Among the HPLC fractions, we were able to isolate components capable of both increasing and diminishing Ca^{2+}-current.

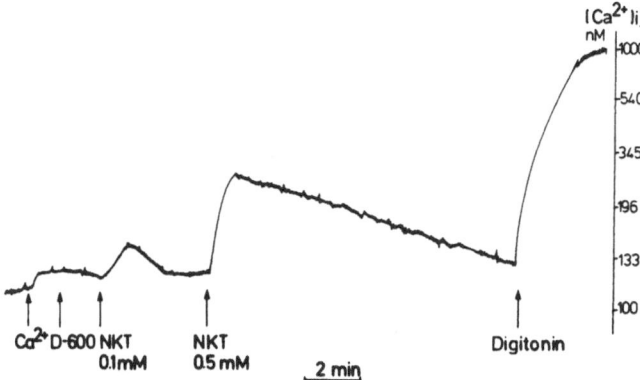

Fig. 2.   *Effect of neokyotorphin on intracellular Ca^{2+}-concentration in rat myocardial cells estimated by fluorescent method (D-600 – Ca^{2+}-blocker).*

The former compound was shown by automatic Edman degradation to have the amino acid sequence of neokyotorphin, an analgesic pentapeptide found earlier in bovine brain [2]. Its possible role as hibernation regulator has not yet been explored. Neokyotorphin was synthesized conventionally. Synthetic neokyotorphin, as well as the native component in 5 × 10^{-7} M concentration, activates the voltage-dependent calcium current in the frog atrial fibers (Fig. 1). At larger concentrations the peptide increases intracellular Ca^{2+} in rat heart ventricle cells (Fig. 2). In the experiments with hibernating ground squirrels,

303

the i.p. injection of 0.7 mg/kg neokyotorphin speeds up arousal and sharply increases the heart rate. Administration of the peptide to mice considerably accelerates transition of animals from the state of artificial hypothermy to the homeothermal state. Thus, neokyotorphin found in the brain of hibernating ground squirrels may be responsible for the arousal of the animals from hibernation, along with activation of metabolism and heart function in special states.

References

1. Duker, G.D., Sjoguist, P.-O., Swensson, G., Wolhlfart, B. and Johansson, B.W., in Heller, H.C., Musachia, X.G. and Wang, L.C.H. (Eds.) Living in the Cold, Elsevier, New York, 1986, p. 565.
2. Fukui, K., Shiomi, H., Takagi, H., Hayashi, K., Kiso, Y. and Kitagawa, K., Neuropharmacology, 22 (1983) 191.

Cyclic disulfide analogs of [Sar¹,Ile⁸]-angiotensin II

Elizabeth E. Sugg[a], Catherine A. Dolan[a], Arthur A. Patchett[a], Ray S.L. Chang[b], Kristie A. Faust[b] and Victor J. Lotti[b]

[a]*Department of Exploratory Chemistry and [b]Department of New Lead Pharmacology, Merck, Sharp and Dohme Research Laboratories, P.O. Box 2000, Rahway, NJ 07065, U.S.A.*

Introduction

[Sar¹, Ile⁸]-Angiotensin II (Ang II) is a potent antagonist [1] of the native pressor peptide, angiotensin II (Asp-Arg-Val-Tyr-Ile-His-Pro-Phe). In an attempt to define the bioactive conformation of Ang II antagonists, we have prepared seven novel cyclic analogs of [Sar¹, Ile⁸]-Ang II, in which positions 1 and 5, 3 and 5 or 5 and 8 have been replaced with L- or D-cysteine (Cys) or L-homocysteine (hCys).

Results and Discussion

Analogs were prepared by SPPS on Merrifield resin using standard Boc-methodology [2], cleaved by liquid HF, cyclized using $K_3Fe(CN)_6$, and purified by gel filtration (Sephadex G-15) and RPHPLC. All compounds were analytically pure and had appropriate ^1H NMR and FABMS. [Sar¹,Cys³,⁵,Ile⁸]-Ang II was examined by 1D and 2D ^1H NMR and found to exist as a mixture of conformers. Table 1 lists the receptor binding affinities and relevant pA_2 values for these new analogs.

Poor receptor affinity was observed with [Cys¹,Cys⁵,Ile⁸]-Ang II and [D-Cys¹,Cys⁵,Ile⁸]-Ang II. Both the L-Cys¹ and D-Cys¹ analogs were equipotent and the linear analog [Ser¹,⁵,Ile⁸]-Ang II had 10-fold better receptor affinity than the cyclic analogs. [Sar¹,Cys⁵,Cys⁸]-Ang II did not bind the Ang II receptor at concentrations up to 100 μM. However, [Sar¹,Cys⁵,D-Cys⁸]-Ang II had an IC_{50} of 18 μM. Expansion of the ring from 14 atoms to 15 atoms ([Sar¹,hCys⁵,D-Cys⁸]-Ang II gave a 20-fold improvement in receptor affinity. Further analogs in this series are under investigation.

[Sar¹,Cys³,⁵,Ile⁸]-Ang II was 50 times more potent than the linear analog [Sar¹,Ser³,⁵,Ile⁸]-Ang II. Expansion of the ring size from 11 atoms to 13 atoms produced an analog ([Sar¹,hCys³,⁵,Ile⁸]-Ang II) with nanomolar receptor affinity, although this compound is only 5-fold more potent than the corresponding linear analog, [Sar¹,Met³,⁵,Ile⁸]-Ang II. These two cyclic analogs suggest that the bioactive conformation has a β-structure in this region of the peptide.

Table 1 *Receptor binding affinities and pA$_2$ values for cyclic angiotensin II analogs in rabbit aorta*

Analog	IC$_{50}$ (nM[a])	pA$_2$[b]
[Sar^1,Val^5,Ile^8]Ang II	0.8	
[Cys^{1,5},Ile^8]Ang II	8 000	
[D-Cys^1,Cys^5,Ile^8]Ang II	9 000	
[Ser^{1,5},Ile^8]Ang II	1 100	
[Sar^1,Cys^{5,8}]Ang II	> 100 000	
[Sar^1,Cys^5,D-Cys^8]Ang II	18 000	
[Sar^1,hCys^5,D-Cys^8]Ang II	800	
[Sar^1,Cys^{3,5},Ile^8]Ang II	140	7.1
[Sar^1,Ser^{3,5},Ile^8]Ang II	7 900	5.3
[Sar^1,hCys^{3,5},Ile^8]Ang II	4	9.16
[Sar^1,Met^{3,5},Ile^8]Ang II	26	

[a] Determined by displacement of [^{125}I][Sar^1,Ile^8]-Ang II from solubilized rabbit aortic membranes.
[b] pA$_2$ = –log K$_B$ = log (IC$_{50}$ dose ratio) – log [antagonist]. IC$_{50}$ ratios for Ang II determined in the same tissues in the absence and presence of antagonists.

References

1. Turker, R.K., Hall, M.M., Yamamoto, M., Sweet, C.S. and Bumpus, F.M., Science, 177 (1972) 1203.
2. Stewart, J.M. and Young, J.M., Solid Phase Peptide Synthesis, Pierce Chemical Co., Rockford, IL, 1984.

The substitution of conformationally constrained amino acids into position 7 of angiotensin II (Ang II) agonist and antagonist analogs

J. Samanen[a], T. Cash[a], D. Narindray[a], T. Yellin[a] and D. Regoli[b]

[a]Smith Kline and French, Laboratories, P.O. Box 1539, King of Prussia, PA 19406-0939, U.S.A.
[b]Department of Pharmacology, U. Sherbrooke, Quebec, Canada J1H5N4

Introduction

The reduced agonist potency of analogs 2 [1], 3 [2], and 4 [3] suggested that [Pro7] in Ang II was critical for maintaining bioactive conformation. Position 7 conformational requirements in Ang II antagonists have not been investigated. In native [Pro7], Ang II torsion angle ϕ is constrained (ca. -60°), while ψ is unconstrained. Continuing the evaluation of conformational requirements for [Pro7] in *both* Ang II agonists and antagonists, [Pro7] in both [Sar1]Ang II and [Sar1,Ile8]Ang II was replaced with conformationally constrained residues Aib, (NMe)Ala, and Dtc (5,5-dimethylthiazolidine-4-carboxylic acid).

Results and Discussion

Peptides containing Dtc cannot adopt a C-7 conformation due to syn-βMe-C=O steric interaction [4,5]. For [Phe8]-agonists, both [Aib7] and [Dtc7] substitution dramatically reduced in vitro (rabbit aorta) activity, (see Table 1) [1–9]. For [Ile8]-antagonists, Aib, (NMe)Ala, Sar and Dtc substitution reduced both in vitro (rabbit aorta) and in vivo (rat blood pressure) activity (Table 1). These results disfavor α-helical conformations ($\phi = -60°$, $\psi = -60°$ preferred by Aib) or more extended conformations ($\phi = -60°$, $\psi = 150°$ preferred by Dtc) for [Pro7], suggesting that intramolecularly hydrogen-bonded C-7 conformations ($\phi = -60°$, $\psi = 90°$) are preferred for native Pro7 in both Ang II agonists and antagonists.

Table 1 Activities of Ang II analogs substituted in position seven

Peptide sequence: -A----Arg--Val--Tyr-B-----His-----C-----D (positions: 1 2 3 4 5 6 7 8)

No.	1 (A)	5 (B)	7 (C)	8 (D)	Agonist		Antagonist	
					In vitro[a] Ang II-like[c]	In vivo[b] Ang II-like[d]	In vitro[a] pA_2	In vivo[b] ID_{50}[e]
1	Asp	Ile	Ala	Phe	0.1 (RU)[f]	1.5[g],7.5[h]		
2	Asp	Val	(NMe)Ala	Phe	4.3 (GI)[i]	22[i]		
3	Asp	Ile	Cle[j]	Phe		1–1.5[j]		
4	Asp	Ile	Aib	Phe		1.0[h]		
5	Sar	Ile	Pro	Phe	180 ± 14	125 ± 15		
6	Sar	Ile	Sar	Phe	22 (RU)[k]			
7	Sar	Ile	Tpr[k]	Phe	238 (RU)[k]			
8	Sar	Ile	Dtc	Phe	4.0 ± 0.5			
9	Sar	Ile	Aib	Phe	18.0 ± 1.7			
10	Sar	Ile	Pro	Ile	0	10.2 ± 0.94	9.1	10.0 ± 0.71
11	Sar	Ile	(NMe)Ala	Ile	0	10.0 ± 1.2	8.4	20.0 ± 3.1
12	Sar	Ile	Sar	Ile	0		8.2 (RU)[k]; 8.1 (RU)[k]	
13	Sar	Ile	Aib	Ile	0	20.0 ± 1.9	8.0	60 ± 8.5
14	Sar	Ile	Tpr	Ile	0	15.0 ± 1.7	9.7	12.5 ± 1.9
15	Sar	Ile	(O)Tpr[l]	Ile	0	–	7.2 (RU)[k]; 6.7	–
16	Sar	Ile	Dtc	Ile	4.0	40 ± 6.3	7.5	250 ± 32

a In vitro agonist 'Ang II-like' activity and antagonist activity, pA_2, were measured in the rabbit aorta strip assay according to Rioux et al. [6] except where noted.
b In vivo residual 'Ang II-like' activity and antagonist activity, ID_{50}, were measured in vivo, in the rat blood pressure assay described by Regoli et al. [7] except where noted.
c Ang II-like activity in vitro = % activity rel. Ang II.
d Ang II-like activity in vivo = % activity rel. Ang II for [Phe8]-agonists, or the mm Hg of blood pressure increase produced by a one microgram bolus intravenous injection of compound, for [Ile8]-antagonists.
e ID_{50} in ng/rat/min (using 250 g rats).
g Ref. [8]; h Ref. [3]; i Ref. [1]; j Cle = cycloleucine, Ref. [2]; k (RU) = rat uterus, Tpr = thioproline, Ref. [9]; l (O)Tpr = thioproline sulfoxide.

References

1. Andreatta, R.H. and Sheraga, H.A., J. Med. Chem., 14 (1971) 489.
2. Park, W.K., Regoli, D. and Rioux, F., Br. J. Pharmacol., 48 (1973) 288.
3. Marshall, G.R., In Lande, S. (Ed.) Prog. Pept. Res., Gordon and Breach, New York, 1972, p. 15.
4. Samanen, J., Cash, T., Eggelston, D.S. and Saunders, M., In Marshall, G.R. (Ed.) Peptides: Chemistry and Biology (Proceedings of the 10th American Peptide Symposium), ESCOM, Leiden, 1987, p. 81.
5. Samanen, J., Zuber, G., Bean, J., Romoff, T., Kopple, K. and Saunders, M., Int. J. Pept. Prot. Res., in press.
6. Rioux, R., Park, W.K. and Regoli, D., Can. J. Physiol. Pharmacol., 51 (1973) 665.
7. Regoli, D. and Park, W.K., Can. J. Physiol. Pharmacol., 50 (1972) 99.
8. Park, W.K., Choi, C., Rioux, R. and Regoli, D., Can. J. Biochem., 52 (1974) 113.
9. Moore, G.J., In Rich, D.H. and Gross, E. (Eds.) Pierce Chemical Co., Rockford, IL, 1981, p. 245.

Role of protein kinase C and ion fluxes in the desensitization of angiotensin II receptors in smooth muscle cells

Antonio C.M. Paiva, Suma I. Shimuta, Celia A. Kanashiro, Maria E.M. Oshiro and Therezinha B. Paiva

Department of Biophysics, Escola Paulista de Medicina, Caixa Postal 20.388, 04034 Sao Paulo, SP, Brazil

Introduction

Angiotensin II (Ang II) induces a contraction in smooth muscles that is due to activation of phospholipase C. The response is composed of a fast transient contracture, due to inositol trisphosphate-induced Ca^{2+} release from intracellular stores, followed by a maintained tonus, whose mechanism is not well understood. A general model for agonists that induce phospholipase C-mediated responses proposes that the slower sustained phase of the response is due to a protein kinase C branch activated by diacylglycerol [1]. In vascular smooth muscles, Ang II induces a sustained increase in diacylglycerol production [2], but its role in the maintenance of the tonic response has not been demonstrated. In fact, it has been suggested that diacylglycerol activation of protein kinase C causes a negative feedback by inhibiting the phospholipase C-mediated responses triggered by the hormone–receptor interaction [3]. This inhibitory feedback mechanism might be responsible for the homologous desensitization that occurs during Ang II stimulation. This desensitization, also known as tachyphylaxis, is a property that may play an important role in the physiological regulation of vascular tone by Ang II.

In the present investigation, the effects of Ang II on Na^+ and Ca^{2+} fluxes in cultured intestinal smooth muscle cells from the guinea pig ileum were studied and correlated with the contraction and desensitization observed in whole muscles. The possible role of protein kinase C in desensitization was also studied through activation of that enzyme by treatment with phorbol-12-myristate-13-acetate (PMA).

Results and Discussion

The effects of Ang II were compared with those of acetylcholine (ACh), an agonist that acts at muscarinic receptors in the intestinal smooth muscle and does not induce desensitization. Ang II and ACh stimulated $^{24}Na^+$ influx upon addition to the cells (Table 1), and this stimulation persisted for at least 30 min (not shown). Both agonists also stimulated $^{45}Ca^{2+}$ uptake (Table 1), but

310

Ang II's effect was transcient, returning to control values within 30 min, whereas the effect of ACh persisted during that time. The decrease in Ang II-stimulated $^{45}Ca^{2+}$ influx in the presence of the hormone paralleled the relaxation (desensitization) observed in the whole muscle during prolonged treatment with Ang II.

Table 1 *Effect of PMA (100 nM) on Na^+ and Ca^{2+} influxes induced by Ang II (100 nM) and ACh (5 μM) in cultured guinea pig ileum smooth muscle cells*

Treatment	$^{24}Na^+$ influx	$^{45}Ca^{2+}$ influx
	(nmol/10^6 cells/15 s)[a]	
Control	21.3 ± 0.5	0.48 ± 0.04
PMA	16.6 ± 1.6	0.49 ± 0.03
ACh	36.3 ± 2.2	0.71 ± 0.07
ACh + PMA	37.7 ± 3.2	0.76 ± 0.07
Ang II	33.5 ± 2.1	0.75 ± 0.08
Ang II + PMA	37.8 ± 3.2	0.58 ± 0.06

[a]Means ± S.E.M.

Pretreatment of the muscle for 30 min with 100 nM PMA caused an abbreviation of the tonic response to Ang II that mimicked the desensitization phenomenon. Pretreatment with PMA also blocked Ang II's stimulating effect on $^{45}Ca^{2+}$, but not $^{24}Na^+$ influx (Table 1). The stimulating effects of ACh on $^{24}Na^+$ and $^{45}Ca^{2+}$ influxes were not affected by PMA (Table 1). Both Ang II (100 nM) and acetylcholine (5 μM) also induced marked transient increases in $^{45}Ca^{2+}$ efflux (ca. 425% and 550% of basal efflux, respectively). However, Ang II's effect reached a maximum 60–90 s after addition of the agonist and returned to control values after 90–120 s, whereas the effect of acetylcholine already attained a maximum at 15 s and returned to control values by 60–90 s. The effects of the two agonists on Ca^{2+} efflux were not affected by 30 min pretreatment with PMA. This suggests that protein kinase C-activation does not affect the initial phase of the response mediated by release of inositol trisphosphate by phospholipase C. Thus, our findings are in agreement with the hypothesis that protein kinase C has a modulating effect on the responses of the intestinal smooth muscle cells to Ang II, but this effect appears to be exerted at a step in the stimulus–response chain subsequent to phospholipase C-activation.

References

1. Rasmussen, H., Takuwa, Y. and Park, S., FASEB J., 1 (1987) 177.
2. Griendling, A.M., Lew, P.D. and Valottoon, M.J., Biol. Chem., 260 (1985) 7836.
3. Berk, B.C., Brock, T.A., Gimbrone, M.A. Jr. and Alexander, R.W., J. Biol. Chem., 262 (1987) 5065.

The role of the histidine nitrogens of angiotensin II in receptor binding

Richard S. Pottorf, William C. Ripka, Andrew T. Chiu and Pancras C. Wong

Medical Products Department, E.I. DuPont de Nemours & Co. Inc., Experimental Station, P.O. Box 80353, Wilmington, DE 19880-0353, U.S.A.

Introduction

The protonated imidazole ϵN of His[6] is suggested to be important for angiotensin II's (Ang II) biological activity [1–3]. More specifically, a conformational role has been proposed, such that the imidazole interacts intramolecularly with the Tyr[4] phenol and the α-carboxyl group [4]. To assess the individual role each imidazole nitrogen has as a hydrogen bond donor in receptor binding, His[6] was substituted with the nonbasic isosteres: Asn and Gln [5].

ε-Tautomer	Glutamine	Asparagine	δ-Tautomer

Results and Discussion

All peptides were synthesized by SPPS on PAM resins using standard Boc methodology [6], cleaved by liquid HF/anisole, and purified by RPHPLC. The compounds were judged homogenous by RPHPLC and were characterized by FABMS and AAA. The receptor binding affinities were determined by displacement of [³H]Ang II from rat adrenal cortical microsomes (Table 1).

Study of [Asn[6]] and [Gln[6]] analogs (**1–6**) indicates that while the imidazole ϵNH is *required* for both receptor binding and vasoconstrictive activity of Ang II agonists, the δNH is *not essential* for either. That a basic δN plays a minor role is seen from the 10-fold loss in binding of **3** vs. **1**, suggesting the loss of a hydrogen bond involving the δN. The potent binding and activity of nonbasic **3**, indicates that the imidazole of Ang II is not protonated during receptor binding.

Comparing the [Sar¹,Tyr(OCH₃)⁴] and [Sar¹,Phe⁴] Ang II series [4,7] (**7–9** and **10–12**), the 10-fold loss of binding (**7** vs. **4**) suggests that a phenol : imidazole hydrogen bond has been removed by substitution of the O-methyl in **7**. A substantial loss of binding results when His[6] of **7** is substituted by Asn (**8**) and Gln (**9**). One interpretation of this is that important interactions exist involving Tyr[4], His[6], and possibly some other group. However, the improved binding observed with the Phe[4] series (**10–12**) compared to the Tyr(OCH₃)⁴ series (**7–9**)

Table 1 *Receptor binding and contractile activity data for Ang II analogs*

No.	Ang II analog	Receptor binding (IC$_{50}$, nM)	Contractile activity, in vitro[a]
1	Ang II	4.2	100% at 3×10^{-8} M
2	[Asn^6]	2 200	None
3	[Gln^6]	67	90% at 10^{-5} M
4	[Sar^1]	2.4	–
5	[Sar^1,Asn^6]	110	None
6	[Sar^1,Gln^6]	25	22% at 10^{-6} M
7	[Sar^1,Tyr(OCH$_3$)^4][b]	82	–
8	[Sar^1,Tyr(OCH$_3$)^4,Asn^6]	3 000	–
9	[Sar^1,Tyr(OCH$_3$)^4,Gln^6]	2 000	–
10	[Sar^1,Phe^4][c]	30	–
11	[Sar^1,Phe^4,Asn^6]	740	–
12	[Sar^1,Phe^4,Gln^6]	200	–

[a] Measured by incubating analog with rabbit aortic strips, activity is expressed as % of the maximal response to Ang II. Maximum concentration tested: 10^{-5} M.
[b] Ref. 4.
[c] Ref. 7.

suggests that direct hydrogen bonding between the Tyr^4 and His^6 in Ang II is not essential. The weaker binding of **7–9** may be the result of steric interactions due to the methoxy that either prevents these analogs from attaining the proper conformation, or interacts with some functionality on the receptor. Further studies are aimed at distinguishing these possibilities as well as determining the effects of Asn^6 and Gln^6 on other Ang II peptide antagonists.

Acknowledgements

We gratefully acknowledge the technical assistance of Nicholas C. Caputo, Tam T. Nguyen and Gordon A. Slack for synthesis, receptor binding, and the in vitro functional assay. We also wish to thank Dr. Joe Lazar for the FABMS analysis and Prof. Roger W. Roeske (Indiana University) for the AAA.

References

1. Andreatta, R. and Hofmann, K., J. Am. Chem. Soc., 90 (1968) 7334.
2. Needleman, P., Marshall, G.R. and Rivier, J., J. Med. Chem., 16 (1973) 968.
3. Hsieh, K., Jörgensen, E.C. and Lee, T.C., J. Med. Chem., 22 (1979) 1199.
4. Scanlon, M.N., Matsoukas, J.M., Franklin, K.J. and Moore, G.J., J. Life Sci., 34 (1984) 317.
5. Lowe, D.M., Fersht, A.R., Wilkinson, A.J., Carter, P. and Winter, G., Biochemistry, 24 (1985) 5106.
6. Barany, G. and Merrifield, R.B., In Gross, E. and Meienhofer, J. (Eds.) The Peptides, Vol. 2, Academic Press, New York, NY, 1980, p. 1.
7. Goghari, M.H., Franklin, K.J. and Moore, G.J., J. Med. Chem., 29 (1985) 1121.

2D NMR studies of neuropeptide Y [NPY] in dimethyl sulfoxide-d_6

V. Renugopalakrishnan[a], S.G. Huang[b], M. Prabhakaran[c] and
A. Balasubramaniam[d]

[a]*Laboratory for the Study of Skeletal Disorders and Rehabilitation, Department of
Orthopaedic Surgery, Harvard Medical School and Children's Hospital,
300 Longwood Avenue, Boston, MA 02115, U.S.A.*
[b]*Harvard University, Cambridge, MA 02138, U.S.A.*
[c]*University of Illinois, Chicago, IL 60680, U.S.A.*
[d]*University of Cincinnati Medical Center, Cincinnati, OH 45267, U.S.A.*

Introduction

NPY, a 36-residue polypeptide rich in Tyr, was recently discovered in brain
[1,2] and peripheral nervous system [3]. Its primary structure was determined
to be: Tyr-Pro-Ser-Lys-Pro-Asp-Asn-Pro-Gly-Glu[10]-Asp-Ala-Pro-Ala-Glu-Asp-
Leu-Ala-Arg-Tyr[20]-Tyr-Ser-Ala-Leu-Arg-His-Tyr-Ile-Asn-Leu[30]-Ile-Thr-Arg-
Gln-Arg-Tyr-NH$_2$ [1]. NPY belongs to a class of peptide hormones that shares
sequence similarity with peptide YY (PYY) and pancreatic polypeptides. NPY
has been implicated in a number of central and peripheral effects in mammals
[4–7].
   Synthesis of NPY by SPPS has been reported [8]. We have recently investigated
the secondary structure of NPY and its 4-norleucine analog by CD and laser
Raman spectroscopy, augmented with Chou–Fasman predictions [9]. In this
preliminary communication, the results of 2D NMR studies of NPY are briefly
discussed.

Results and Discussion

*2D NMR studies of NPY in 4:1-H$_2$O:D$_2$O and dimethyl sulfoxide-*d$_6$
   1D and 2D NMR spectra of NPY in H$_2$O and dimethyl sulfoxide-d_6 were
obtained on a Bruker AM500 NMR spectrometer. 2D COSY and NOESY
experiments were performed using standard pulse sequences [10], as previously
described in 2D NMR studies of polypeptides and proteins from this laboratory
[11]. Magnetic field was locked on deuterium resonance of the solvent.
   NOESY spectrum of NPY in dimethyl sulfoxide-d_6 at a concentration of ~10
mg/ml is shown in Fig. 1. The assignments of proton resonances in the 500
MHz 1D ^{1}H NMR spectra of NPY in water and dimethyl sulfoxide-d_6 are now
in the process of final confirmation. Numerous cross peaks in the NOESY
spectrum of NPY in Fig. 1, with the assignments presently known, are suggestive
of a well-defined secondary structure, consistent with earlier CD and Raman

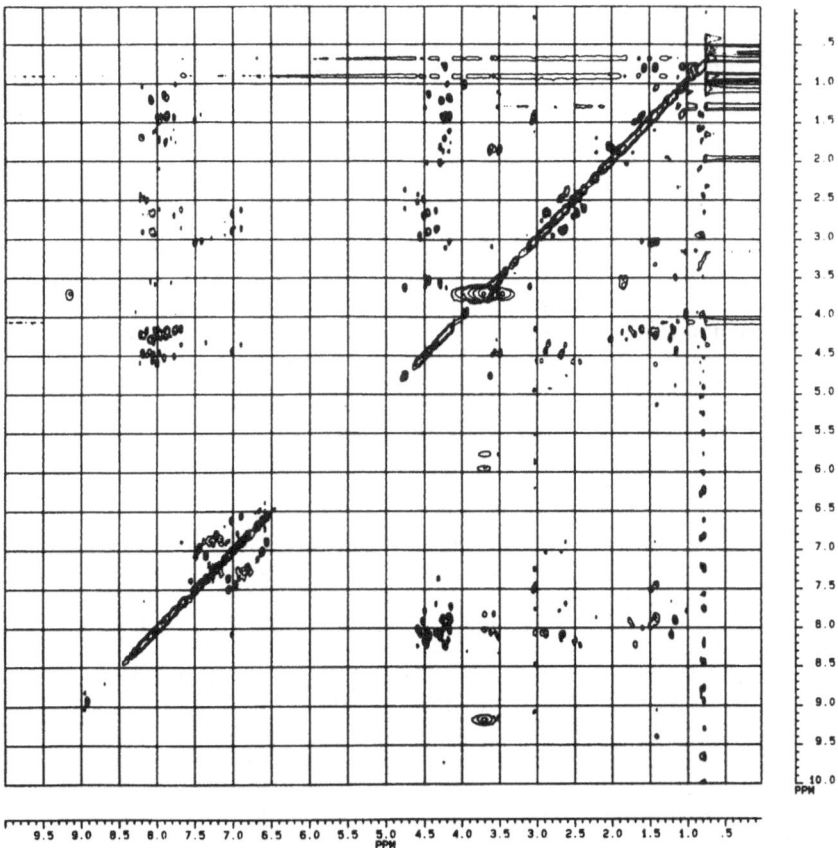

Fig. 1. NOESY spectrum of neuropeptide Y in dimethyl sulfoxide-d₆ at a concentration of ~10 mg/ml at a mixing time of 400 min. The NOESY spectrum shown is phase-sensitive.

studies of NPY [9]. The assignment of proton resonances was facilitated by 2D COSY studies of NPY fragments. The NOE contacts derived from NOESY studies were used as constraints in the molecular mechanics–dynamics studies that are currently in progess.

2D NOESY spectrum of NPY shown in Fig. 1, contains an inter-locking pattern generally observed in polypeptides and protein containing α-helical segments [12]. Especially noteworthy are short NH-NH distances (d_{NN}). In addition, amide NH protons also exhibited strong NOEs to C beta H proton and C beta proton of the neighboring residue. Furthermore the observation of the above NOEs was supplemented by the derivation of Ramachandran angle, ϕ, from the vicinal coupling constants. At the present time, the above observation relates to the C-terminal segment of NPY. 2D NMR evidence for the presence of β-turns at the N-terminal regions is not completely established.

The model derived from 2D NMR and molecular mechanics studies are consistent with previous spectroscopic studies [9].

Acknowledgements

The research reported in this study was supported partially by NIH Grant GM-38601.

References

1. Tatemoto, K., Carlquist, M. and Mutt, V., Nature (London), 296 (1982) 659.
2. Tatemoto, K., Proc. Natl. Acad. Sci. U.S.A., 79 (1982) 5485.
3. Emson, P.C. and DeQuidt, J.E., Trends Neurosci., 2 (1984) 31.
4. Balasubramaniam, A., Grupp, I., Matlib, M.A., Benza, R., Jackson, R.L., Fischer, J.F. and Grupp, M., Regulatory Peptides, 21 (1988) 289.
5. Clark, J.T., Sahu, A., Kalra, P.S., Balasubramaniam, A. and Kalra, S.P., Regul. Pept., 17 (1987) 31.
6. Gray, T.S. and Morley, J.E., Life Sci., 38 (1986) 389.
7. Allen, J.M. and Bloom, S.R., Neurochem. Int., 8 (1986) 1.
8. Balasubramaniam, A., Grupp, I., Srivastava, L., Tatemoto, K., Murphy, R.F., Joffe, S.N. and Fisher, J.E., Int. J. Pept. Prot. Res., 29 (1987) 78.
9. Balasubramaniam, A., Renugopalakrishnan, V., Rigel, D.F., Nussbaum, M.S., Rapaka, R.S., Dobbs, J.C., Carreira, L.A. and Fischer, J.E., Biochim. Biophys. Acta, 997 (1989) 176.
10. Wüthrich, K., NMR of Proteins and Nucleic Acids, John Wiley and Sons, New York, 1986.
11. Huang, S.G., Renugopalakrishnan, V. and Rapaka, R.S., Biochemistry, in press.
12. Englander, S.W. and Wand, A.J., Biochemistry, 26 (1987) 5953.

Neuropeptide Y[18-36] – An NPY fragment with hypotensive action in vivo: Structure-activity relationships

Jaroslav H. Boublik, Neal A. Scott[a], Robert D. Feinstein[b], Murray Goodman[b], Marvin R. Brown[a] and Jean E. Rivier[c]

[a]*University of California San Diego Medical Center, San Diego, CA 92103, U.S.A.*
[b]*Department of Chemistry, University of California San Diego, La Jolla, CA 92037, U.S.A.*
[c]*The Clayton Foundation Laboratories for Peptide Biology, The Salk Institute, La Jolla, CA 92037, U.S.A.*

Introduction

Neuropeptide Y (NPY) is a 36-amino acid, C-terminally amidated peptide that was first isolated from porcine brain by Tatemoto et al. [1] in 1982. NPY elicits a potent vasoconstriction upon peripheral administration and so is thought to act as a cardiovascular modulatory factor. We have been interested in defining SAR for the hypertensive action of NPY and have previously shown [2] that while modifications (such as deletion, *N*-acetylation and D-substitution) in the N-terminal region are tolerated C-terminal modifications result in inactive analogs. To define the minimum structure required for activity, we made substantial N-terminal truncations and showed that NPY[16-36] was the shortest NPY fragment with measurable hypertensive activity. Further deletion to give NPY[17-36] and NPY[18-36] resulted in compounds with hypotensive activity in vivo [3]. In the present study we have synthesized NPY[18-36] analogs with N- and C-terminal modifications to probe further the structural requirements for the hypotensive activity of this NPY fragment. We have also subjected NPY[18-36] and selected analogs to 2D NMR in order to define a solution conformation for the native fragment.

Results and Discussion

NPY, NPY[18-36] and N- and C-terminally modified NPY[18-36] analogs were synthesized by manual SPPS methods and purified by preparative HPLC (Table 1). The effect of intraarterial administration on mean arterial pressure and heart rate of these peptides was determined in a conscious rat bioassay. The results indicated that, while N-terminally modified analogs of NPY[18-36] had no significant hypotensive potency, C-terminal modifications gave analogs with hypotensive potencies similar to that of NPY[18-36]. These data are in contrast to findings with NPY analogs, in which C-terminal modifications resulted in a loss of hypertensive activity, whereas N-terminally modified analogs were essentially equipotent to NPY. N- and C-terminally modified NPY and NPY[18-36] analogs elicited parallel but more variable effects on heart rate (data not shown). These

317

Table 1   *HPLC retention times, purities, optical rotations and effects on mean arterial pressure (MAP) for NPY, NPY[18-36] and N- and C-terminally modified NPY[18-36] analogs*

No.	Compound	Iso RT @ % MeCN[a]	Purity (%)[b]	$[\alpha]_D^{25}$ (deg.)[c]	Effect on MAP[d]
1	NPY[1-36]	4.3 @ 33.6	>97	−58.2	+++
2	NPY[18-36]	3.4 @ 30.0	>98	−46.0	− − −
3	[Ac-Ala[18]] NPY[18-36]	4.3 @ 33.0	>98	−47.7	+
4	[NMe-Ala[18]] NPY[18-36]	3.4 @ 30.0	>97	−46.4	+
5	[des-NH$_2$-Ala[18]] NPY[18-36]	4.5 @ 32.4	>98	−44.2	+
6	[D-Ala[18]] NPY[18-36]	4.3 @ 30.0	>97	−46.9	+
7	[D-Arg[19]] NPY[18-36]	4.0 @ 30.0	>96	−43.4	+
8	[D-Arg[33]] NPY[18-36]	4.4 @ 28.8	>97	−39.4	− −
9	[D-Gln[34]] NPY[18-36]	3.8 @ 28.8	>97	−40.3	− − −
10	[D-Arg[35]] NPY[18-36]	3.8 @ 28.2	>98	−41.0	− − −
11	[D-Tyr[36]] NPY[18-36]	4.1 @ 28.2	>98	−48.9	− − −
12	NPY[18-36]-Tyr-OH	4.8 @ 28.8	>97	−45.9	− −

[a] Isocratic retention times at specified % MeCN-in min.
[b] Consensus of values from three analytical determinations.
[c] Determined in 1.0 M AcOH (c~0.5).
[d] Qualitative expression of the effect of maximal doses of compounds on MAP as determined in a conscious rat bioassay.

findings may suggest that the differing cardiovascular actions of NPY and NPY[18-36] occur via mechanisms involving distinct subclasses of NPY receptors with differing structural requirements for agonist ligands.

NMR assignment of NPY[18-36], the *N*-methyl and *N*-acetyl analogs of NPY[18-36] was performed, and detailed NOE data for NPY[18-36] in DMF-d_7 were obtained. Of interest was the observation that *N*-acetylation resulted in an obvious disruption of the N-terminal 5–6 residues as reflected by changes in the chemical shifts of all protons.

Acknowledgements

This work was supported in part by the N.I.H. (grant Nos. AM-26741, HL-41901, HL-43154 and HL-27716). J.B. is the recipient of a Neil Hamilton Fairley Postdoctoral Fellowship provided by the Australian National Health and Medical Research Council. N.S. is supported by the Robert Wood Johnson Foundation. We thank K. Carver, C. Miller and D. Pantoja, for excellent technical assistance.

References

1. Tatemoto, K., Carlquist, M. and Mutt, V., Nature, 296 (1982) 659.
2. Boublik, J.H., Scott, N.A., Brown, M.R. and Rivier J.E., J. Med. Chem., 32 (1989) 597.
3. Boublik, J.H., Scott, N.A., Taulane, J., Brown, M.R., Goodman, M.J. and Rivier, J.E., Int. J. Pept. Prot. Res., 33 (1989) 11.

Examination of an intramolecularly stabilized model for neuropeptide Y with centrally truncated and stabilized analogs

John L. Krstenansky, Thomas J. Owen, Stephen H. Buck, Karen A. Hagaman and Larry R. McLean

Merrell Dow Research Institute, 2110 E. Galbraith Road, Cincinnati, OH 45215, U.S.A.

Introduction

Spectroscopic studies of porcine neuropeptide Y (pNPY) [1] indicated that it had an intramolecularly stabilized structure like that of the homologous avian pancreatic polypeptide (APP), whose crystal structure was known [2]. Using the available APP structure, a model of pNPY was built (Fig. 1) [3]. The model consists of an N-terminal polyproline helix whose lipophilic face (proline residues) is in association with the lipophilic face of a C-terminal amphipathic α-helical region (residues 14–30).

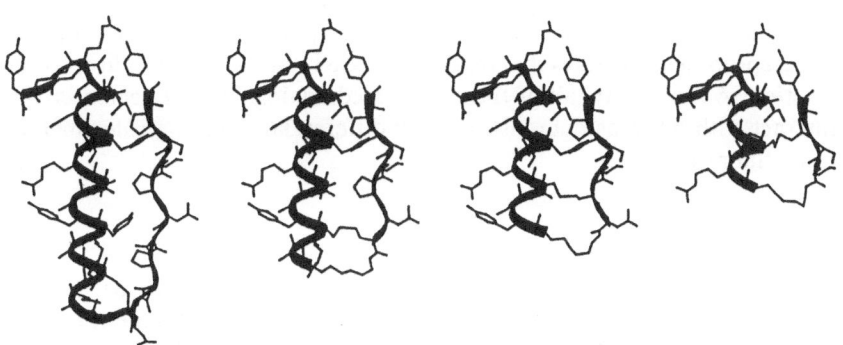

Fig. 1. Molecular models with ribbon diagram overlay of pNPY, C7-NPY, C5-NPY and C2-NPY (left to right). Tyr[1] is at the upper right of each molecule.

Results and Discussion

Previous work [4] had suggested that both the N-terminal Tyr and the C-terminal region were required for potent interaction with one (Y_1) of at least two types of NPY receptor (Y_1 and Y_2). The intramolecularly stabilized model of NPY places the Tyr[1] and the C-terminal region in close spacial proximity. If the purpose of residues 2–31 was to provide a stable template for presenting

319

residues 1 and 32–36 to the NPY receptor, then a peptide analog devoid of portions of the central residues should be able to bind potently to the NPY receptor, provided the N-terminal to C-terminal spatial relationship could be maintained. Such an analog would also lose its ability to interact with lipid, due to the loss of the amphipathic helical regions. NPY itself is able to disrupt liposomes to an extent even greater than that of glucagon, a hormone for which lipid interaction is thought to be important. Thus, if lipid interaction plays an important role in NPY receptor interaction, any interruption of that interaction should adversely affect the receptor interaction.

Analogs were designed which progressively removed portions of the N- and C-terminal helices. This was accomplished by removing residues from the central turn region between the helices and spanning the gap with an 8-aminooctanoic acid residue. A disulfide bridge spans the two helices to replace stabilization lost from the shortening of the interacting surfaces of the amphipathic helices. The analogs are [D-Cys^7,Aoc$^{8\text{-}17}$,Cys^{21}]-pNPY, [Cys^5,Aoc$^{7\text{-}21}$,D-Cys^{24}]-pNPY and [Cys^2,Aoc$^{5\text{-}24}$,D-Cys^{27}]-pNPY which are designated as C7-NPY, C5-NPY and C2-NPY, respectively. The molecular models of these analogs are shown in Fig. 1. C7-NPY (IC$_{50}$ = 3 nM) is equipotent to pNPY (IC$_{50}$ = 4 nM) in the mouse brain, a tissue where C-terminal fragments bind poorly, indicating that this represents the receptor type of interest; C5-NPY (IC$_{50}$ = 100 nM) and C2-NPY (IC$_{50}$ = 600 nM) are less potent. In porcine spleen, all of the analogs bind well, including the C-terminal fragments (IC$_{50}$ = <5 nM), indicating that the potency at the receptors in this tissue exhibited by C5-NPY and C2-NPY is probably due to the presence of C-terminal residues alone. The potency of C7-NPY on mouse brain receptors demonstrates that residues 7–17 are not required for direct receptor interaction. Additionally, all of these truncated analogs have a decreased ability to interact with lipid, as demonstrated by differential scanning calorimetry and tyrosine fluorescence.

Conclusion

Because C7-NPY and pNPY are equipotent on mouse brain tissue, we suggest that lipid interaction is not a crucial aspect of NPY receptor interaction, and that the amphipathic helical regions of NPY serve a structural purpose.

References

1. Krstenansky, J.L. and Buck, S.H., Neuropeptides, 10 (1987) 77.
2. Blundell, T.L., Pitts, J.E., Tickle, I.J., Wood, S.P. and Wu, C.W., Proc. Natl. Acad. Sci. U.S.A., 78 (1981) 4175.
3. Krstenansky, J.L., Owen, T.J., Buck, S.H., Hagaman, K.A. and McLean, L.R., Proc. Natl. Acad. Sci. U.S.A., 86 (1989) 4377.
4. Rioux, F., Bachelard, H., Martel, J.-C. and St-Pierre, S., Peptides, 7 (1986) 27.

μ to δ Selectivity of opioid peptides: A comparison of dermorphin and deltorphins

P.A. Temussi[a],*, D. Picone[a], T. Tancredi[b], R. Tomatis[c], S. Salvadori[c], M. Marastoni[c] and G. Balboni[c]

[a]*Dipartimento di Chimica, University of Naples, via Mezzocannone 4, I-80134 Napoli, Italy,*
[b]*Istituto Chimica M.I.B. del CNR, Arco Felice, Napoli, Italy*
[c]*Dipartimento di Scienze Farmaceutiche, University of Ferrara, Ferrara, Italy*

Introduction

The N-terminal tetrapeptide sequence of enkephalins, i.e. Tyr-Gly-Gly-Phe, is generally considered essential for high δ activity. On the other hand, the message domain (Tyr-D-Ala-Phe) of dermorphin [1], a natural μ opioid heptapeptide, has long been considered the main cause of the high μ selectivity of this peptide and of its analogs. The discovery, in the skin of South American tree frogs, of deltorphins [2–4] challenges this belief. The relative μ/δ activities (S. Salvadori, unpublished) range from 100 for dermorphin, Tyr-D-Ala-Phe-Gly-Tyr-Pro-Ser-NH$_2$ (**DR**), to ca. 1×10^{-3} for all deltorphins: deltorphin A, Tyr-D-Met-Phe-His-Leu-Met-Asp-NH$_2$ (**DA**), deltorphin B, Tyr-D-Ala-Phe-Glu-Val-Val-Gly-NH$_2$ (**DB**), and deltorphin C, Tyr-D-Ala-Phe-Asp-Val-Val-Gly-NH$_2$ (**DC**).

The reversal of μ/δ selectivity certainly cannot be explained on the basis of the message domain [5], but before drawing conclusions on the relevance of sequential changes, it is necessary to study the conformational preferences of deltorphins in comparison to those of dermorphin, since all these peptides are very flexible. Here we present a conformational analysis based on ^{1}H NMR studies in DMSO-d_6 and in a cryoprotective mixture [6].

Results and Discussion

All peptides were synthesized by classical solution methods. NMR spectra were run on Bruker AM-400 and WM-500 spectrometers, in 6 mM solutions of neat DMSO-d_6 at 300 K or of a 90:10 (v/v) DMSO-d_6:H$_2$O cryoprotective mixture at 277 K. Full proton assignments were achieved by means of 2D spectroscopy techniques. Most chemical shifts of the labile backbone protons in the two solvent systems employed have values consistent with standard literature values [7]. However, sequential NH-C$_\alpha$H, NH-NH and many long range intrachain NOEs indicate that the mixture of conformations present in solution is *not* made up solely of random extended chains. Besides, in both

---

*To whom correspondence should be addressed.

solvent systems, the C_β protons of the second residue have unusually high-field chemical shifts.

In particular, in the cryoprotective mixture, the β-CH$_2$ of D-Met^2 of **DA** are at 1.30 and 1.37 ppm; the methyl group of D-Ala^2 resonates at 0.69 ppm, both for **DR** and **DB**, and at 0.66 ppm for **DC**. It had already been noticed in dermorphin and its fragments [8] that the chemical shift of the methyl group of D-Ala^2 is unusually low with respect to the standard value for alanine [7], and that the high-field shift is paralleled by a high μ activity. Thus, although some parameters related to the backbone protons point to a disordered state, it is possible that most conformations present in solution have a similar arrangement of residues 2 and 3, representing a local relative energy minimum. It seems significant that H-Tyr-D-Ala-Phe-Gly-NH$_2$, in the type II′ β-turn conformation consistent with a receptor model [9] based on rigid opiates, has the methyl group of D-Ala^2 and the ring of Phe^3 at a short enough distance to induce a large diamagnetic shift in the resonance of the methyl group of D-Ala. Thus, also in the case of deltorphins, it seems reasonable to assume that the relative arrangement of residues 2 and 3 is similar to that present in the β-turn that fits the model receptor we proposed for μ and δ peptides [9]. In fact, it can be said that the sequence Tyr^1-D-Xxx^2-Phe^3, long believed to be characteristic of μ opioids only, is an ideal feature of both μ and δ opioid peptides. The cause of the reversal of selectivity in going from the sequence of dermorphin to those of deltorphins must be sought in the properties of residues 4–7.

We propose that the key factor causing μ to δ reversal is *charge in the message domain*: i.e. the picture of the δ receptor that emerges from the present work is similar to that of the μ receptor [9], with an important difference: the ability to tolerate positive and negative charges inside (or close to) the message domain.

References

1. Erspamer, V., Melchiorri, P., Erspamer, G.F., Montecucchi, P.C. and de Castiglione, R., Peptides, 6 (1985) 7.
2. Richter, K., Egger, R. and Kreil, G., Science, 238 (1987) 200.
3. Kreil, G., Barra, D., Simmaco, M., Erspamer, V., Erspamer-Falconieri, G., Melchiorri, P., Negri, L., Severini, C. and Corsi, R., Eur. J. Pharmacol., 162 (1989) 123.
4. Erspamer, V., Erspamer-Falconieri, G., Melchiorri, P., Negri, L., Corsi, R., Severini, C., Barra, D., Simmaco, M. and Kreil, G., Proc. Natl. Acad. Sci. U.S.A., (1989) in press.
5. Temussi, P.A. Picone, D., Tancredi, T., Tomatis, R., Salvadori, S., Marastoni, M. and Balboni, G., FEBS Lett., 247 (1989) 283.
6. Motta, A., Picone, D., Tancredi, T. and Temussi, P.A., J. Magn. Res., 75 (1987) 364.
7. Wüthrich, K., NMR In Biological Research: Peptides and Proteins, Elsevier, Amsterdam, 1976.
8. Pastore, A., Temussi, P.A., Tancredi, T., Salvadori, S. and Tomatis, R., Biopolymers 23 (1984) 2349.
9. Castiglione-Morelli, M.A., Lelj, F., Pastore, A., Salvadori, S., Tancredi, T., Tomatis, R., Trivellone, E. and Temussi, P.A., J. Med. Chem., 30 (1987) 2067.

DALDA: An extremely μ receptor-selective dermorphin analog capable of inducing peripheral antinociception

Peter W. Schiller[a],*, Thi M.-D. Nguyen[a], Nga N. Chung[a], Gervais Dionne[b] and René Martel[b]

[a] Clinical Research Institute of Montreal, 110 Pine Avenue West, Montreal, Quebec, Canada H2W 1R7
[b] IAF BioChem International, Laval, Quebec, Canada H7V 1B7

Introduction

The dermorphin-related tetrapeptide analog H-Tyr-D-Arg-Phe-Lys-NH$_2$ (DALDA) [1] shows μ receptor affinity comparable to that of H-Tyr-D-Ala-Gly-Phe(NMe)- Gly-ol (DAGO) and of H-Tyr-Pro-Phe(NMe)-Pro-NH$_2$ (PLO17) in the rat brain membrane binding assay. However, because of its extremely low δ receptor affinity ($K_i^\delta = 19.2$ M), DALDA displays considerably higher receptor selectivity ($K_i^\delta/K_i^\mu = 11\,400$) than DAGO ($K_i^\delta/K_i^\mu = 1050$) or PLO17 ($K_i^\delta/K_i^\mu = 1470$). Testing in the GPI assay revealed that DALDA is a potent μ agonist that does not significantly interact with κ receptors. Because of its high positive charge (3+), DALDA is a very polar molecule not expected to cross the blood–brain barrier (BBB) and, therefore, is of interest as a most likely peripherally acting, highly selective μ agonist.

Recently, it has been shown that in hyperalgesic or inflammatory conditions, systemically administered opioids without access to the brain can elicit an analgesic effect through interaction with peripheral opioid receptors. Thus, both N-methylmorphine [2] and the relatively polar enkephalin analog BW443 [3] were shown to produce an analgesic effect in the mouse writhing assay, a test model detecting peripheral antinociception, and the effect of these compounds could be reversed with N-methylnalorphine, a quaternized opiate antagonist unable to cross the BBB. In view of these findings, it was of interest to explore the potential of DALDA as a peripherally acting analgesic. DALDA was synthesized by the solid-phase method [1] and the analgesic assays were performed by following published procedures [3].

Results and Discussion

DALDA, in comparison with morphine, was tested in the mouse writhing assay, using phenyl-1,4-benzoquinone (PBQ) as irritant. Opioids were injected s.c. 5, 20 or 60 min prior to the administration of PBQ (2.5 mg/kg, i.p.). As indicated by the ED$_{50}$ values determined 5 or 20 min after s.c. administration,

---

*To whom correspondence should be addressed.

DALDA exerts about as potent an analgesic effect as morphine in this assay system (Table 1). However, comparison of the ED_{50} values obtained after a 60 min time interval reveals that the antinociceptive effect of the peptide analog is of somewhat shorter duration than that of morphine. The analgesic effect of DALDA could be antagonized by low doses of naloxone (5 mg/kg) or of the quaternized opiate antagonist N-methyllevallorphan (10 mg/kg) administered i.p. 20 min prior to the peptide analog. Since N-methyllevallorphan does not have access to the brain, its ability to reverse the analgesic effect of the subcutaneously administered dermorphin analog suggests that DALDA does indeed produce antinociception through interaction with peripheral opioid receptors. Administration (i.p.) of the same dose of N-methyllevallorphan had no effect on the analgesic activity of subcutaneously injected morphine, as it was to be expected for this centrally acting compound. DALDA also showed a potent analgesic effect in the writing assay after i.v. administration and was found to be even orally active at high doses (100–200 mg/kg). The analgesic activity of DALDA in the mouse hot-plate assay, a test system permitting the detection of centrally mediated analgesic effects only, was weak, as indicated by an ED_{50} value of 71.5 mg/kg determined 30 min after s.c. administration. Interestingly, a lower ED_{50} value (24.2 mg/kg) was obtained after a time interval of 1 h, suggesting that DALDA at high concentrations may slowly accumulate in the brain.

Table 1   *Mouse writing assay of DALDA and morphine*

Compound	Time (min)[a]	ED_{50} (mg/kg)
DALDA	5	0.48
	20	0.65
	60	2.80
Morphine	5	0.31
	20	0.43
	60	0.60

[a]Time interval between administration of opioid and PBQ.

Conclusion

Peripherally induced analgesia is of interest because some of the centrally mediated side effects (respiratory depression, dependence, etc.) are no longer implicated. The results presented here suggest that DALDA is capable of producing a peripherally mediated analgesic effect. For this reason, DALDA and other polar dermorphin analogs may have therapeutic potential.

Acknowledgements

This work was supported by operating grants from the MRCC (MT-5655), QHF and NIDA (DA-04443).

References

1. Schiller, P.W., Nguyen, T.M.-D., Chung, N.N. and Lemieux, C., J. Med. Chem., 32 (1989) 698.
2. Smith, T.W., Buchan, P., Parsons, D.N. and Wilkinson, S., Life Sci., 31 (1982) 1205.
3. Hardy, G.W., Lowe, L.A., Pang, Y.S., Simpkin, D.S.A., Wilkinson, S., Follenfant, R.L. and Smith, T.W., J. Med. Chem., 31 (1988) 960.

A highly selective in vitro μ-opioid agonist with atypical in vivo pharmacology

R.T. Shuman[a], M.D. Hynes[a], J.H. Woods[b] and P.D. Gesellchen[a]

[a]*The Lilly Research Laboratories, Eli Lilly and Company, Indianapolis, IN 46285, U.S.A.*
[b]*University of Michigan Medical School, Ann Arbor, MI 48109, U.S.A.*

Introduction

Evidence for multiple opioid receptors has caused a considerable amount of research to be focused on the development of ligands specific for each of the major opioid receptor sub-types (μ, δ and κ). Specific ligands can be used to determine differences between these receptors and to study the receptor roles in various biological processes.

Results and Discussion

The pharmacological evaluation (in vitro binding studies) of a new enkephalin tripeptide amide, MeTyr-D-Ala-Gly-(Et)N-CH(CH$_2$C$_6$H$_5$)-CH$_2$-N(CH$_3$)$_2$ (LY-164929) indicate that it is a potent and selective μ-receptor ligand (Table 1). The μ/δ selectivity ratio of 1500 makes LY164929 one of the most selective μ-receptor peptide ligands reported to date [1]. The data from other in vitro test systems (MVD and GPI, data not shown) indicate that LY164929 is a full opiate agonist with no antagonist properties. In vivo, LY164929 exhibits potent analgesic activity. Further pharmacology of LY164929 was studied in three assays that correlate with a drug's potential for producing physical dependence: (1) the mouse locomotor assay [2]; (2) the mouse withdrawal-jumping assay [3]; and (3) the monkey single-dose suppression assay [4] (SDS).

Table 1 *In vitro and in vivo opioid activities*

Compound	Receptor binding–IC$_{50}$ (nM)			Analgesic activity[a] ED$_{50}$ (mg/kg)	
	[³H]Naloxone	[³H]DADL	[³H]EKC	s.c.	p.o.
LY164929	0.6	900	>1000	6.6	50.4
Metkephamid	2.5	4.4	>1000	3.1	523
Morphine	5.7	74	167	0.93	7.4

[a]Mouse writing assay, ED$_{50}$ measured at 30 min.

In the locomotor assay, narcotic analgesics, such as morphine, will markedly stimulate running. However, metkephamid, an enkephalin analog that has been shown to have a low potential to produce physical dependence [5] exhibits low locomotor activity. The overall stimulation of locomotor activity produced by LY164929 was virtually identical to that of metkephamid. Based upon these results metkephamid and LY164929 should have a reduced liability for the production of physical dependence.

In the withdrawal-jumping assay, stereotypical jumping behavior can be induced by injection of the opiate antagonist, naloxone, in mice treated chronically with narcotic opiates, such as morphine. However, in mice treated chronically with metkephamid, the jumping behavior was not significantly different from that of control. Unexpectedly, in the LY164929 treated group, a very high degree of jumping was observed. Consequently, in this test system a prediction of high physical dependence liability can be made for LY164929.

Finally, in the monkey SDS assay, extremely large doses of metkephamid and LY164929 are required to bring about suppression of withdrawal symptoms in morphine dependent rhesus monkeys even when administered i.c.v. Metkephamid (s.c.) requires a 16-fold higher dose than morphine, and LY164929 (s.c.) requires approximately a 100-fold higher dose than morphine (Table 2). These data indicate that LY164929 has a weaker morphine-like effect in vivo than metkephamid.

Table 2   *Comparison of potency of opioids in suppressing morphine abstinence in rhesus monkeys [4]*

Drugs	Dose[a] (mg/kg, s.c.)	Drugs	Dose[a] (mg/kg, s.c.)
Etorphine	0.001	Methadone	3.0
Fentanyl	0.04	Meperidine	10.0
Azidomorphine	0.2	Codeine	20.0[b]
Heroin	1.0	Propoxyphene	15.0[b]
Levorphanol	1.0	Metkephamid	48.0
Morphine	3.0	LY164929	300[c]

[a] Dose required for complete suppression of abstinence.
[b] Incomplete suppression. A dose of this magnitude may induce convulsions.
[c] Extrapolation based on metkephamid s.c. and i.c.v. compared to LY164929 i.c.v. data.

Thus, LY164929, which was expected to be a typical morphine-like opioid, based on in vitro data, was found to exhibit contradictory in vivo pharmacology. The withdrawal-jumping assay predicts that LY164929 should possess a high degree of physical dependence liability. However, both the locomotor and the monkey SDS assays indicate that LY164929 should behave more like metkephamid than morphine. This atypical in vivo pharmacology may be attributable to interactions at peripheral opioid receptors [6] or specific binding to a low affinity [^{3}H]-DADL binding site [7].

References

1.   Recently a series of dermorphin analogs with high μ-receptor selectivity was reported.

Schiller, P.W., Nguyen, T.M.D., Chung, N.N. and Lemieux, C., J. Med. Chem., 32 (1989) 698.

2. Brase, D.A., Iwamoto, E.T., Loh, H.H. and Way, E.L., J. Pharmacol. Exp. Ther., 197 (1976) 317.

3. Way, E.L., Loh, H.H. and Shen, F., J. Pharmacol. Exp. Ther., 167 (1969) 1.

4. Woods, J.H., Proceedings of the Thirty-Ninth Annual Meeting, Committee on Problems of Drug Dependence, NAS-ARC, 1977, p. 420.

5. Frederickson, R.C.A., Smithwick, E.L., Shuman, R. and Bemis, K.G., Science, 211 (1981) 603.

6. Stein, C., Millan, M.J., Shippenberg, T.S., Peter, K. and Herz, A., J. Pharmacol. Exp. Ther., 248 (1989) 1269.

7. Rothman, R.B., Bykov, V., Ofri, D. and Rice, K., Neuropeptides, 11 (1988) 13.

Analgesic activity of new cyclic enkephalin peptides containing conformationally constrained tryptophan residues

D.E. Wright[a] and J.L. Vaught[b]

[a]R.W. Johnson Pharmaceutical Research Institute, San Diego, CA 92121, U.S.A.
[b]R.W. Johnson Pharmaceutical Research Institute, Spring House, PA 19477, U.S.A.

Introduction

New cyclic enkephalin peptides have been synthesized in which the disulfide bond used to form the cyclic peptide consists of either modified tryptophan residues (2-thioltryptophan) or a modified tryptophan residue and cysteine. Using the indole moiety as an integral part of the cyclic structure provides new constraints to cyclic peptides, which differ from those obtained using cysteine/ penicillamine or carboxylic acid-amine residues. An additional advantage in using modified tryptophan residues to form all or part of the disulfide bridge is the added constraint of an aromatic group (indole) in the cyclic structure. These constraints yield increased flexibility in designing new peptidomimetics. The biological activity and preliminary molecular modeling studies of [Cys^2,D-(2'-SH)Trp^5]enkephalin and [D-(2'-SH)Trp^2,D-(2'-SH)Trp^5] enkephalin are reported here.

Results and Discussion

The cyclic peptides have unique UV spectra. Cyclization via Cys-Trp gives an absorbance maximum around 310 nm; Trp-Trp cyclization shows an absorbance maximum at about 340 nm. This contrasts with linear enkephalin peptides containing unmodified tryptophan, which show the usual absorbance maximum around 280 nm.

The analgesic activity of these new cyclic enkephalin peptides has been determined but using the mouse tailflick assay (intra-cerebroventricular injection; 15 min after injection, the mouse is tested). Both of the peptides are inactive at 20 μg as is enkephalin at 150 μg.

From a preliminary examination of the conformation of the peptides with the molecular modeling program, Alchemy, the phenylalanine residue in position four of both peptides seems situated very close to the tyrosine residue, thereby sterically hindering the interaction of tyrosine with the receptor. In both peptides, tryptophan in position five appears to assume the aromatic position occupied by phenylalanine in the native peptide. The distance between the tyrosine and phenylalanine residues is about 3.5Å for both peptides. By comparison NMR studies of [D-Pen^2,D-Pen^5]enkephalin give a distance of 4.5 Å between the aromatic ring protons of Tyr^1 and Phe^4 [1]. In contrast, fluorescence energy-transfer measurements of [D-Cys^2,Trp^4,D-Cys^5]enkephalinamide in H_2O provide an estimated average intramolecular distance of 9.7 Å between the Tyr^1 and Trp^4 aromatic rings [2].

The two peptides reported here seem to possess different backbone configurations. Cyclization with Cys-Trp used to form the disulfide ring, results in a 16-member ring, whereas the Trp-Trp cyclic peptide has an 18-member ring. This contrasts with the 14-member rings found in previously reported cyclic disulfide bond enkephalin analogs. Although the two new cyclic enkephalin peptides reported here are inactive, we have synthesized cyclic enkephalin peptides that show good analgesic activity using this modified tryptophan procedure to form the disulfide bond. Comparing the conformations of the active cyclic peptides with the inactive cyclic peptides, it would appear that the structure of the peptide backbone, which differs greatly depending on where and how the cyclization is constructed, is not as important as the position of the phenolic side chain with respect to the aromatic (phenylalanine) residue. The appropriate distance between the aromatic groups in cyclic enkephalin analogs must be maintained to retain biological activity. The peptide backbone appears to serve as the scaffolding necessary for maintaining the suitable aromatic side-chain spacing.

References

1. Hruby, V.J., Lung-Fa, K., Pettitt, B.M. and Karplus, M., J. Am. Chem. Soc., 110 (1988) 3351.
2. Schiller, P.W., Biochem. Biophys. Res. Commun., 114 (1983) 268.

Synthesis and biological activities of linear and cyclic enkephalin analogs containing a Gly³ψ(E,CH=CH)Phe⁴ replacement

D. Tourwé[a], D. Meert[a], J. Couder[a], M. Ceusters[a], M. Elseviers[a], G. van Binst[a], G. Tóth[c], T.F. Burks[b], J.E. Schook[b] and H.I. Yamamura[b]

[a]*Organic Chemistry Laboratory, Vrije Universiteit Brussel, Pleinlaan 2, B-1050 Brussels, Belgium*
[b]*Department of Pharmacology and* [c]*Department of Chemistry, University of Arizona, Tucson, AZ 85721, U.S.A.*

Introduction

The Gly³-carbonyl in the enkephalins has been proposed to correspond to the C^6-OH in morphine [1] or to the C^{19}-OH in PEO [2], and its modification should, in that case, have a profound effect on the binding to the μ-opiate receptor. We have investigated the effect of the replacement of the peptide function by an E,CH=CH function at the 3–4 position in Tyr-D-Ala-Gly-Phe-D-Leu-OH (DADLE), Tyr-D-Cys-Gly-Phe-D-Cys-NH₂ (DCDCE-A) and Tyr-D-Pen-Gly-Phe-D-Pen-OH (DPDPE). In DADLE, the 2–3 position was also modified.

Results and Discussion

The synthesis of Boc-D-Alaψ(E,CH=CH)Gly-OH and Boc-Glyψ(E, CH=CH)D,L-Phe-OH was performed as described in [3] and [4]. Since after incorporation of the racemic dipeptide analog, two diastereomeric peptides are obtained, an enantioselective synthesis (Scheme 1) was used to determine the absolute stereochemistry of the pseudo-Phe residue. In all compounds, the replacement of the peptide function by an E,CH=CH function leads to a large drop in affinity for the μ-receptor, but even more for the δ-receptor (Table 1). Reduction of the double bond to the saturated analogs causes a further drop in potency. This is an indication for the importance of rigidity at these positions, and may explain the low potencies of other flexible peptide bond

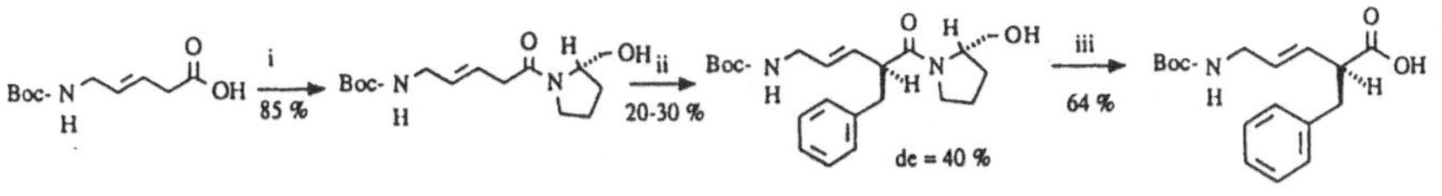

Scheme 1. Enantioselective synthesis.
i = S-prolinol, DCC/HOBt; ii = benzylbromide; iii = 6 N HCl, then Boc₂O.

Table 1 *Biological activities of enkephalin analogs* IC_{50} *(nM)*

	GPI	MVD	GPI/MVD	$[^3H]\mu^{*,s}$	$[^3H]DPDPE$
1 Tyr-D-AlaψGly-Phe-D-Leu-OH	51 522	10 592	4.86	>10000*	393
2 Tyr-D-Alaψ(sat)Gly-Phe-D-Leu-OH	30 000 (49% inhib)	30 000 (10% inhib)	–	>10000*	3 628
3 Tyr-D-Ala-GlyψPhe-D-Leu-OH I	403 364	~60 000	–	>10000*	1 023
4 Tyr-D-Ala-GlyψPhe-D-Leu-OH II	1 230	3 475	0.35	2 631	756
5 Tyr-D-Cys-GlyψD-Phe-D-Cys-NH₂	1 315	3 281	0.40	717* 711s	1 259
6 Tyr-D-Cys-GlyψL-Phe-D-Cys-NH₂	5.15	190	0.27	2.1* 12.1s	50
7 Tyr-D-Cys-Glyψ(sat)Phe-D-Cys-NH₂ I	–	–	–	3 835s	8 629
8 Tyr-D-Cys-Glyψ(sat)Phe-D-Cys-NH₂ II	–	–	–	325s	1 173
9 Tyr-D-Pen-GlyψPhe-D-Pen-OH I	100 000 (21% inhib)	10 000 (7% inhib)	–	>10000*	3 419
10 Tyr-D-Pen-GlyψPhe-D-Pen-OH II	~25 000	~25 000	–	>10000*	670

$\psi = \psi(E,CH=CH)$; $\psi(sat) = \psi(CH_2CH_2)$; $[^3H]\mu^{*,s} = [^3H]CTOP^*$ or $[^3H]$sufentanil[s]

modified analogs at these positions [5–7]. One analog: DCDCE-A **6** retains considerable binding potency and in vitro activity in the GPI test. The DPDPE analog **10** also binds to the δ-receptor, although its in vitro activity is rather low but still better than that of the flexible ψ(CH$_2$-S) or ψ(CH$_2$-NH) analogs [6]. Whereas the binding potency to the δ-receptor of the cyclic compound **10** is similar to that of the linear analog **4**, its potency in the bioassay is much lower. This is an unusual case where the 'efficacy' of the linear compound is higher than that of the cyclic one. These results confirm the importance of the Gly[3]-carbonyl, not only for interaction with the μ-receptor, but even more for the δ-receptor.

Acknowledgements

We thank Dr. J.E. Leysen (Janssen Research Foundation) for some of the binding experiments.

References

1. Maigret, B., Premilat, S., Fournié-Zaluski, M.C. and Roques, B.P., Biochem. Biophys. Res. Commun., 99 (1981) 267.
2. Hudson, D., Sharpe, R. and Szelke, M., Int. J. Pept. Prot. Res., 15 (1980) 122.
3. Miles, N.J., Sammes, P.J., Kennewell, P.D. and Westwood, R, J. Chem. Soc. Perkin Trans. I (1985) 2299.
4. Cox, M.T., Heaton, D.W. and Horbury, J., J. Chem. Soc. Chem. Commun., (1980) 799.
5. Spatola, A.F., In Weinstein B. (Ed.) Chemistry and Biochemistry of Amino Acids, Peptides and Proteins, Marcel Dekker, New York, 1983, p. 267.
6. Spatola, A.F. and Darlak, K., Tetrahedron, 44 (1988) 821.
7. Sherman, D.B., Porter, R.A. and Spatola, A.F., In Marshall, G.R. (Ed.) Peptides: Chemistry and Biology (Proceedings of the 10th American Peptide Symposium), ESCOM, Leiden, 1988, p. 84.

The effect of chirality and polarity within the peptide backbone

Arno F. Spatola[a], Krzysztof Darlak[a], William S. Wire[b] and Thomas F. Burks[b]
[a]Department of Chemistry, University of Louisville, Louisville, KY 40292, U.S.A.
[b]Department of Pharmacology, University of Arizona, Tucson, AZ 85824, U.S.A.

Introduction

DMSO is an excellent high dielectric solvent with an established propensity to form strong hydrogen bonds. Dialkyl sulfoxides are known to be configurationally stable, and thus able to produce a new stereogenic center when incorporated within a nonsymmetric environment. These properties have been investigated and exploited through the synthesis of a series of peptide analogs containing the $\psi[CH_2SO]$ amide bond surrogate.

Results and Discussion

We have previously reported the synthesis of $\psi[CH_2SO]$ analogs of linear GnRH [1,2], dermorphin [3], and enkephalin [4] peptides. We now report the synthesis (previously described [4]) and separation of $\psi[CH_2SO]$ analogs in the 4–5 position of the potent DiMaio-Schiller parent peptide, Tyr-cyclo[D-Lys-Gly-Phe-Leu-] [5]. As seen in Table 1, the biological potencies of the $\psi[CH_2SO]$ compounds reveal *increased potency*, a divergence in selectivity, and confirm the importance of chirality in drug design.

Both $\psi[CH_2SO]$ analogs (two diastereoisomers), as well as a $\psi[CH_2SO_2]$ replacement (the sulfone analog – which is achiral and thus furnishes only one isomer), can be prepared from the $\psi[CH_2S]$ pseudopeptides [4 and 6] by H_2O_2/HOAc oxidation. In Fig. 1, the RPHPLC chromatogram illustrates the baseline separation of all analogs (but now with a D-Leu[5] residue). The activities of the D-Leu series will be reported elsewhere.

The NMR and biological properties (Table 1) of the L-Leu analogs have been contrasted. No new hydrogen bonding patterns have been discerned in these analogs. Although it has not yet been possible to establish absolute configurations of the sulfoxides, their biological potencies are significantly different from that of the sulfides and from one another. The lack of opioid receptor selectivity (μ vs δ) of the $\psi[CH_2SO]$ analogs suggests that they also share the inherent flexibility of the thiomethylene unit, even within this cyclic framework. Nevertheless, these $\psi[CH_2SO]$ pseudopeptides are among the most potent surrogates that we have prepared, and further structural analysis, as well as comparison to the relatively unknown sulfone series, will be pursued.

Table 1 *Biological activities of backbone modified cyclic enkephalins*[a]

Compound	GPI		MVD		GPI:MVD ratio
	IC_{50} (nM)	Rel. potency	IC_{50} (nM)	Rel. potency	
H-Tyr-Gly-Gly-Phe-Leu-OH	246 ±37[b]	1	11.4 ± 1.1[b]	1	21.7
H-Tyr-cyclo[D-Lys-Gly-Phe-Leu-]	4.80± 1.77[b]	51.2	141 ±28[b]	0.081	0.034
H-Tyr-cyclo[D-Lys-Gly-Pheψ[CH$_2$S]-Leu-]	1.1 ± 0.5	227	2.1 ± 1.0	5.4	0.52
H-Tyr-cyclo[D-Lys-Gly-Pheψ[CH$_2$SO]-Leu-] (Isomer I)	0.75± 0.18	328	0.46± 0.1	24.8	1.63
H-Tyr-cyclo[D-Lys-Gly-Pheψ[CH$_2$SO]-Leu-] (Isomer II)	1.7 ± 0.06	145	3.8 ± ˙1.3	3.0	0.45

[a] GPI = guinea pig ileum; MVD = mouse vas deferens; $n = 3$; Leu-enkephalin = 1.
[b] Ref. [5].

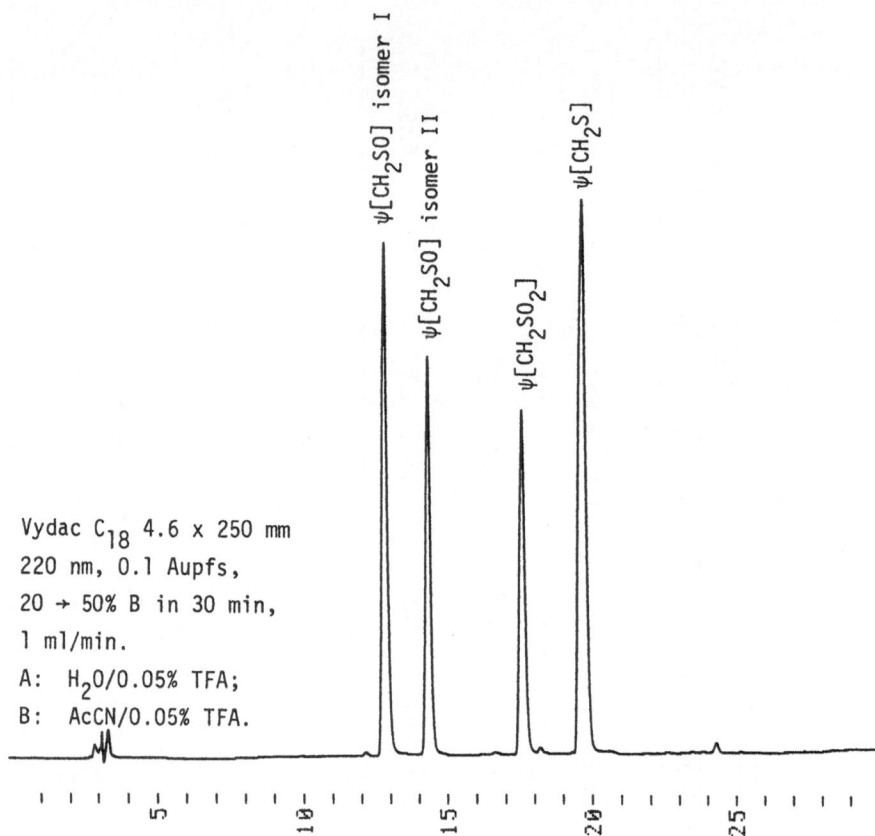

Fig. 1. *HPLC profile of H-Tyr-cyclo[D-Lys-Gly-Pheψ[xxxx]D-Leu] analogs.*

Acknowledgements

Supported by NIH GM-33376 and NIDA DA-04504.

References

1. Spatola, A.F., Agarwal, N.S., Bettag, A.L., Yankeelov Jr., J.A., Bowers, C.Y. and Vale, W.W., Biochem. Biophys. Res. Commun., 97 (1980) 1014.
2. Spatola, A.F., Bettag, A.L., Agarwal, N.S., Saneii, H., Vale, W.W., Bowers, C.Y., In Zatuchni, G.I., Shelton, J.D. and Sciarra, J.J. (Eds.) LH-RH Peptides as Female and Male Contraceptives, Harper and Row, New York, NY, 1981, p. 24.
3. Darlak, K., Grzonka, Z., Spatola, A.F., Benovitz, D.E., Burks, T.F. and Wire, W.S., In Jung, G. and Bayer, E. (Eds.) Peptides 1988, (Proceedings of the 20th European Peptide Symposium), De Gruyter, Berlin, 1989, p. 634.
4. Spatola, A.F. and Edwards, J.V., Biopolymers, 25 (1986) S229.
5. DiMaio, J. and Schiller, P.W., Proc. Natl. Acad. Sci. U.S.A., 77 (1980) 7162.
6. Edwards, J.V. and Spatola, A.F., J. Liquid Chromatogr., 9 (1986) 903.

Cyclic Dynorphin A analogs with high selectivities and potencies at κ opioid receptors

Andrew M. Kawasaki[a], Richard Knapp[b], William S. Wire[b], Thomas Kramer[b], Henry I. Yamamura[b], Thomas F. Burks[b] and Victor J. Hruby[a],*

[a]Department of Chemistry and [b]Department of Pharmacology, University of Arizona, Tucson, AZ 85721, U.S.A.

Introduction

We have designed and synthesized several cyclic disulfide-containing peptide analogs of Dynorphin A (Dyn A) that are conformationally constrained in the putative 'address' segment of the opioid ligand. These Dyn A analogs exhibit unexpected selectivities between the kappa (κ) opioid receptor(s) of the CNS vs. peripheral nervous system.

Results and Discussion

The analogs **1** and **2** displayed very high affinities for the κ receptor but low κ vs. mu(μ) selectivities in the GP brain binding studies (Table 1). Surprisingly, in contrast to the high κ and μ affinities shown by **1** and **2** in the binding assays, **1** and **2** exhibited very low potencies in the GPI (Table 2).

Table 1 *Opioid receptor binding affinities and selectivities of various cyclic disulfide Dynorphin analogs with guinea pig brain homogenate*

No.	Compound	IC$_{50}$ (nM)		
		[^{3}H]U-69593	[^{3}H]PL-17	[^{3}H]DPDPE
	Dyn A$_{1-17}$-OH	0.087	3.40	2.74
1	[Cys^{5,11}]Dyn A$_{1-11}$-NH$_2$	0.285	0.270	1.63
2	[Cys^{5,11},D-Ala^8]Dyn A$_{1-11}$-NH$_2$	0.391	2.30	18.6
3	[Cys^{8,13}]Dyn A$_{1-13}$-NH$_2$	0.076	0.986	3.97
4	[D-Cys^8,Cys^{13}]Dyn A$_{1-13}$-NH$_2$	1.76	10.3	104
5	[D-Cys^{8,13}]Dyn A$_{1-13}$-NH$_2$	0.110	0.362	14.3

U-69593 = (5α,7α,8β)-(-)-N-methyl-N-(7-(1-pyrrolidinyl)-1-oxaspiro(4,5)dec-8-yl)ben-zene-acetamide; PL-17 = Tyr-Pro-*N*-MePhe-D-Pro-NH$_2$;

DPDPE = Tyr-D-Pen-Gly-Phe-D-Pen-OH.

The analogs **3**, **4**, and **5** displayed very different pharmacological profiles in the peripheral bioassays when compared to **1** and **2**. Analogs **3** and **5** showed high potencies and displayed κ selectivity in the GPI bioassay. In the binding

---

*To whom correspondence should be addressed.

337

Table 2 *Bioassays with the smooth muscle tissue of the guinea pig ileum (GPI) and the mouse vas deferens (MVD)*

No.	Compound	Bioassay	IC_{50} (nM)	K_e (nM) for:		
				NLX	CTAP	ICI
	Dyn A_{1-17}-OH	GPI	2.5	66.7	no*	N.D.
		MVD	22.5	58.8	no*	no
1	[Cys^{5,11}]Dyn A_{1-11}-NH$_2$	GPI	1 080	66.7	no	N.D.
		MVD	421	80	2 000	86.2
2	[Cys^{5,11},D-Ala^8]Dyn A_{1-11}-NH$_2$	GPI	4 406	y	50	N.D.
		MVD	1 660	N.D.	500	21.3
3	[Cys^{8,13}]Dyn A_{1-13}-NH$_2$	GPI	1.3	38.9	no	N.D.
		MVD	20.1	95.2	175	156
4	[D-Cys^8,Cys^{13}]Dyn A_{1-13}-NH$_2$	GPI	2.27	32.3	256	N.D.
		MVD	24.5	35.6	370	667
5	[D-Cys^{8,13}]Dyn A_{1-13}-NH$_2$	GPI	1.75	233	no	N.D.
		MVD	19.5	18.2	1 430	1 000

NLX = Naloxone; CTAP = D-Phe-Cys-Tyr-D-Trp-Arg-Thr-Pen-Thr-NH$_2$; CTP = D-Phe-Cys-Tyr-D-Trp-Lys-Thr-Pen-Thr-NH$_2$; ICI = ICI-174864, *N,N*-diallyl-Tyr-Aib-Aib-Phe-Leu-OH; y = shifted DRC; no = did not shift DRC; * = CTP, not CTAP used as μ antagonist; N.D. = not determined.

assay, **3**, **4**, and **5** exhibited high affinities for the κ opioid receptor, but only low κ vs. μ selectivities.

Thus, incorporation of conformational constraint in the putative 'address' segment of Dyn A analogs has resulted in the κ opioid receptor ligands **1** and **2** possessing high κ opioid receptor affinities centrally (GP brain), but only weak activity at peripheral κ opioid receptors (GPI). On the other hand, **3** and **5** display high κ potencies *and* selectivities at the peripheral (GPI), but not at the central (GP brain) κ opioid receptor. However, in mice, **5** is a full agonist in vivo (ICV, $A_{50} = 10$ μg) in the hot-plate analgesia test. The lack of correlation between the pharmacological profiles observed in smooth muscle and in the brain binding assays suggests the existence of different subtypes of the κ receptor in the brain and peripheral nervous systems. Recent findings in the delta (δ) opioid receptor area have led to a similar proposal [1] (Yamamura, Hruby, Shimohigashi, et al., submitted).

Acknowledgements

This work was supported by grants from the U.S. Public Health Service DA04248 and GM12427.

Reference

1. Shimohigashi, Y., Costa, T., Pfeiffer, A., Herz, A., Kimura, H. and Stammer, C.H., FEBS Lett., 222 (1987) 71.

Conformational analysis of galactosyl enkephalin analogs showing high analgesic activity

J.L. Torres[a], H. Pepermans[b,*], G. Valencia[a], F. Reig[a], J.M. García-Antón[a] and G. Van Binst[b]

[a]*Laboratory of Peptides, C.I.D.-C.S.I.C., Jordi Girona 18-26, E-08034 Barcelona, Spain*
[b]*Organic Chemistry Laboratory, Vrije Universiteit Brussel, Pleinlaan 2, B-1050 Brussels, Belgium*

Introduction

In the present work we describe the conformational analysis by 2D NMR of a series of potent analgesic galactosyl enkephalins in DMSO-d_6 solutions. The parent compound was [D-Met^2,Pro^5] enkephalinamide [1] (**1**), and the glycosyl derivatives were [D-Met^2,Pro^5] enkephalin [N$^{1.5}$-β-galactopyranosyl] amide (**2**) and O$^{1.5}$-(β-D-galactopyranosyl) [D-Met^2,Hyp^5] enkephalinamide (**3**). **3** is 50 000-fold more potent than morphine and more than 1000-fold more potent than the parent compound **1** in rats, according to both the tail immersion and paw pressure tests of analgesia after intraventricular administration (10 animals) [2]. **2** is about 100-fold more potent than **1** in the same test of analgesia [3].

Results and Discussion

Table 1 *Significant NOEs observed by ROESY in DMSO-d_6 solutions*

		1	2	3 303 K	3 333 K
Intra-residual					
Gly NH-α_1H	D$_1$		+++	++	
Gly NH-α_2H	D$_2$		+++	-	
Inter-residual					
α_iH-NH$_{i+1}$					
TyrαH-D-Met NH	A	+++	+++	+++	-
D-MetαH-Gly NH	B	+++	+++	+++	+
Glyα_1H-Phe NH	C$_1$	+++	+++	-	-
Glyα_2H-Phe NH	C$_2$	+++	+++	+++	+

---
*Present address: Unilever Research Laboratories, O. van Noortlaan 120, 3133 AT Vlaardingen, The Netherlands.

The three compounds tested present two β-strands connected to each other by a fairly flexible glycyl residue at position 3. For **1** and **2**, experimental data are inconsistent with a single conformation of Gly^3 as indicated by both intra-residual D_1, D_2 ROESY cross-peaks and inter-residual C_1, C_2 ROESY cross-peaks (Table 1). In contradistinction, **3** seems to exist preferentially in a more rigid folded conformation as indicated by the absence of both D_1 and C_1 ROESY cross-peaks. From these data and the measured $^3J_{NH\alpha1\alpha2}$ coupling constants, we propose a conformation in which $\phi_3 \approx -30°$ and $\psi_3 \approx -120°$.

The difference in the analgesic activity of **1** and **2** did not relate to a conformational change in DMSO solution, whereas the increment for **3** with respect to **2** could relate to the adoption of the preferential folded conformation proposed in this work.

References

1. Bajusz, S., Rónai, A.Z., Székely, J.I, Gráf, L., Dunnai-Kovács, Z. and Berzétei, I., FEBS Lett., 76 (1977) 91.
2. Rodríguez, R.E., Rodríguez, F.D., Sacristán, M.P., Torres, J.L., Valencia, G. and García-Antón, T.I., Neurosci. Lett., 101 (1989) 89.
3. Rodríguez, R.E. et al., Psychopharmacology, in press.

Comparative theoretical conformational analysis of [D-Pen², D-Pen⁵]enkephalin

Brian C. Wilkes* and Peter W. Schiller

Clinical Research Institute of Montreal, 110 Pine Avenue West, Montreal, Quebec,
Canada H2W 1R7

Introduction

In an effort to identify low energy-conformations of the δ-selective enkephalin analog [D-Pen², D-Pen⁵]enkephalin (DPDPE), we carried out a theoretical conformational study using the software package SYBYL (Tripos) and methodology developed in our laboratory [1].

Results and Discussion

A systematic grid search of the bare ring structure of DPDPE devoid of the exocyclic Tyr[1] residue and the Phe[4] side chain generated 63 allowed conformations. After minimization, 23 ring conformations were retained for further analysis. The exocyclic atoms were added and a systematic search on the non-ring portions of the 23 completed structures was then carried out. In this manner, over 130 conformations were generated and finally minimized. Among these conformers, 12 had energies within 2 kcal/mol of that of the lowest energy conformation. Obviously, this conformationally restricted enkephalin analog still possessed some flexibility. The four lowest energy structures are presented in Fig. 1. In the lowest energy conformation, the Tyr[1] aromatic ring folded over the 14-membered ring in close proximity to the Gly[3] residue, whereas the Phe[4] side chain pointed away from the rest of the molecule. No strong hydrogen bonds were detected in this conformer.

Several groups performed theoretical conformational analyses of DPDPE using various programs including CHARMM [2], AMBER [3] and ECEPP [4,5]. Using published torsion angles, we constructed the various candidate conformers proposed in these studies and minimized them using the SYBYL force field, thus allowing us to compare the energies of these proposed conformations directly with our results. All the proposed candidate structures were considerably higher in energy than the lowest energy structure described above. The comparatively lowest energy structure was reported in one of the ECEPP studies [5], although it was still 4.4 kcal/mol higher in energy than the lowest energy conformer found in this study.

---

*To whom correspondence should be addressed.

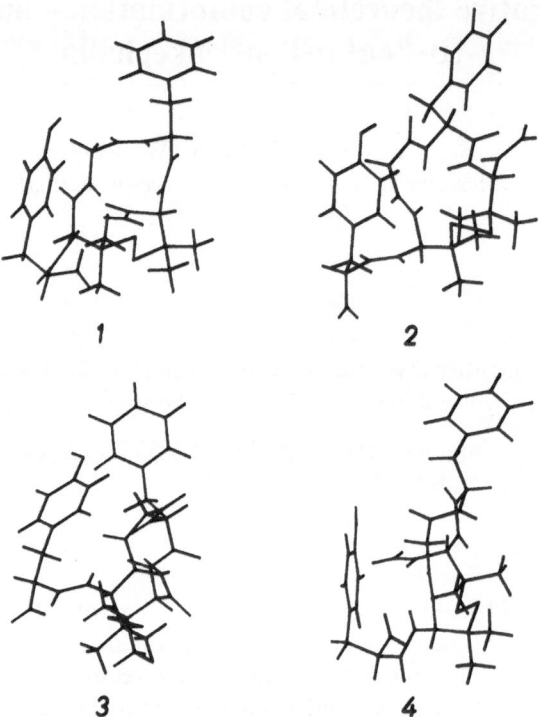

Fig. 1.   Low energy conformation of DPDPE. Energies are: 1, –22.35; 2, –21.90; 3, –21.41; 4, –21.33 kcal/mol.

Structural comparison revealed considerable differences between the various proposed candidate conformers and our lowest energy structure, both in terms of backbone conformation (hydrogen-bonding) and side-chain orientation. A particularly interesting structural feature is the disulfide dihedral angle, which in the present study was systematically altered (30° increments) and in the various low-energy conformers obtained was generally about 180°. In contrast to this result, two other analyses [2,3] had been restricted to conformers with a disulfide dihedral angle of $\pm 90°$.

Conclusion

We have identified conformers of DPDPE considerably lower in energy than structures proposed by others. The fact that the intramolecular distance between the Tyr-Phe aromatic rings in this conformer of DPDPE is considerably larger than that in the lowest energy structure of the μ-selective analog H-Tyr-D-Orn-Phe-Asp-NH$_2$ [1] may be relevant to the preference of these compounds for either δ or μ receptors.

Acknowledgements

This work was supported by the MRCC (MA-10131) and NIDA (DA-04443).

References

1. Wilkes, B.C. and Schiller, P.W., Biopolymers, 26 (1987) 1431.
2. Hruby, V.J., Kao, L.F., Pettit, B.M. and Karplus, M., J. Am. Chem. Soc., 110 (1988) 3351.
3. Froimowitz, M. and Hruby, V.J., Int. J. Pept. Prot. Res., 34 (1989) 88.
4. Keys, C., Payne, P., Amsterdam, P., Toll, L. and Loew, G., Mol. Pharmacol., 33 (1988) 528.
5. Nikiforovich, G.V. and Balodis, J., FEBS Lett., 227 (1988) 127.

Synthesis and bioactivities of [2,3-methanotyrosine[1]]-enkephalins

Charles H. Stammer, Claudio Mapelli and V.P. Srivastava

Department of Chemistry, School of Chemical Sciences, University of Georgia, Athens, GA 30602, U.S.A.

Introduction

Since the N-terminus of the family of enkephalins must be preserved if analgesic activity is to be maintained, the N-terminal tyrosine moiety has been compared to the tyramine grouping in morphine. In order to test the positional requirements of the tyramine moiety, we prepared the conformationally constrained 2,3-methanotyrosine, which has the dihedral angle, χ_1, fixed at 0° and 120° in the Z- and E-isomers, respectively. These two isomers of the derived tyrosine were prepared as racemates and incorporated into [D-Ala[2],Leu[5]]-enkephalin. The GPI and MVD assays are reported.

Results and Discussion

Table 1 *Biological properties of cyclopropyl enkephalins*

Analog	Assay (IC$_{50}$)			
	GPI (μM)		MVD (μM)	
	Agonist	Antagonist[a]	Agonist	Antagonist[b]
(+)∇ ^{E}Phe	I^c	1.0	30	0.01
(−)∇ ^{E}Phe	200	I	60	I
(−)∇ ^{Z}Phe	5	I	5	I
(+)∇ ^{Z}Phe	0.1	10	0.01	1.0
(±)∇ ^{E}Tyr	1.0	I	0.001	I
(±)∇ ^{Z}Tyr	I	I	0.01	I
(+)∇ ^{Z}Tyr	I	I	0.01	I
(−)∇ ^{Z}Tyr	I	I	0.01	I

[a] Tested to reverse effects of morphine (3×10^{-5} M).
[b] Test to reverse effect of morphine (1×10^{-6} M).
[c] Inactive (> 300 μM).

The biological results given in Table 1 show that the analog containing the racemic E-2,3-methanotyrosine showed an MVD/GPI ratio of 1000, while it was 100 times more active than the Z-isomer in the MVD assay. Based on literature data, the same E-compound is about 1/10 as active as [Leu[5]]-enkephalin. We have not been able to separate the two diastereoisomeric peptides formed from the racemic E-isomer, so that only the mixture was tested. The two peptides

344

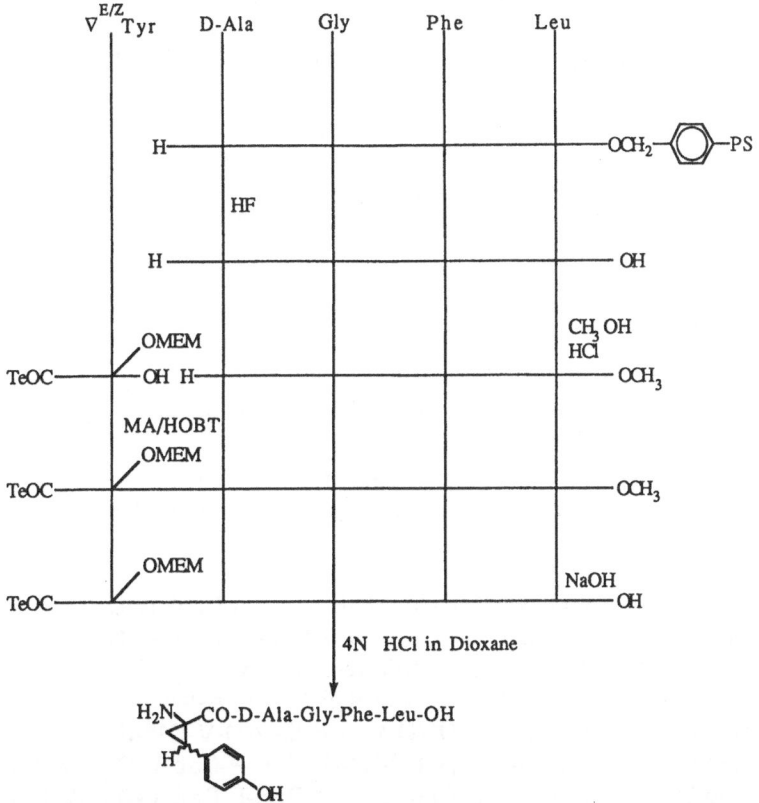

Fig. 1. Synthesis of [V^{E/Z}Tyr^1,D-Ala^2,Leu^5]enkephalin.

formed from the Z-isomer, however, were separated by HPLC with difficulty, and tested.

As in morphine, the dihedral angle corresponding to χ_1 in tyrosine is approximately 80°. This is about midway between the χ_1 angles, 0° and 120°, of the two 2,3-methanotyrosine isomers. These results confirm the stringent steric requirements of the N-terminal tyrosine residue for binding of the enkephalin molecule.

345

The design, synthesis and biological studies of synthetic peptide models of dynorphin A (1–17)*

Chi-Ching Yang** and **John W. Taylor**

Laboratory of Bioorganic Chemistry and Biochemistry, The Rockefeller University, 1230 York Avenue, New York, NY 10021, U.S.A.

Introduction

Dynorphin A (Fig. 1), an opioid peptide, is postulated to consist of a [Leu]enkephalin region (residues 1–5), a linker (arginine), a potential amphiphilic β-strand region (residues 7–15) and a disordered region [1] (residues 16–17). The functional significance of the potential amphiphilic β-strand region, in which Pro^{10} causes an interruption of an idealized β-strand peptide chain conformation, has been investigated.

Dynorphin:	Y-G-G-F-L-R-^{7}R-I-R-P-K-L-K-W-D^{15}-N-Q
Peptide 1:	Y-G-G-F-L-K-^{7}K-V-K-P-K-V-K-V-K^{15}-S-S
Peptide 2:	Y-G-G-F-L-K-^{7}K-V-K-V-K-V-K-V-K^{15}-S-S
Peptide 3:	Y-G-G-F-L-K-^{7}K-A-K-P-K-A-K-A-K^{15}-S-S
Peptide 4:	Y-G-G-F-L-O-^{7}O-V-O-P-O-V-O-V-O^{15}-S-S
Peptide 5:	Y-G-G-F-L-Dab-^{7}Dab-V-Dab-P-Dab-V-Dab-V-Dab^{15}-S-S
Peptide 6:	Y-G-G-F-L-Dap-^{7}Dap-V-Dap-P-Dap-V-Dap-V-Dap^{15}-S-S

Fig. 1. Dynorphin A and dynorphin model peptides.

Results and Discussion

Peptide **1** (Fig. 1), with Ser as the substitute for Asn^{16} and Gln^{17}, was designed to replace the potential amphiphilic β-strand region of the native sequence with a sequence of alternating Lys and Val residues, while conserving Pro^{10}. In **2** (Fig. 1), Pro^{10} was also replaced by Val, allowing us to investigate the role of the Pro residue. Additionally, the importance of hydrophobic residues in the 7–15 segment was studied by using **3** (Fig. 1), in which Ala substituted for each Val in **1**. Peptides **4**, **5** and **6** (Fig. 1), in which ornithine (O),L-2,4-diaminobutyric acid (Dab) and L-2,3-diaminopropionic acid (Dap) residues replaced each Lys in **1**, respectively, were synthesized as well and allowed us to study the functional importance of the length of the side chain in hydrophilic residues in the 7–15 segment.

As summarized in Table 1, **1** gives potent κ receptor affinity and high selectivity

---

*Dedicated to the memory of Dr. E.T. Kaiser.
**Present address: Protos Corporation, 4560 Horton Street, Emeryville, CA 94608-2916, U.S.A.

for κ receptors over μ and δ receptors. However, **3** containing Ala as the hydrophobic component in residues 7–15 exhibits even higher κ receptor affinity. The lower κ receptor affinity and selectivity for κ receptors over μ and δ receptors of **2** demonstrate that Pro^{10} is important to the bioactive conformation of dynorphin A. These results indicate that a nonhomologous segment from residues 7–15, such as that in **1**, can be used as a substitute for the same region in dynorphin A. Comparing the binding studies of **1, 4, 5** and **6, 4** with propylamine as the side chain for hydrophilic residues in positions 7–15 has the highest affinity for κ receptors. Peptide **5**, with ethyleneamine as the side chain, shows the highest μ receptor affinity. In contrast, **6** with the shortest side chain, methyleneamine, has much lower affinity for both κ and μ receptors. These findings support a crucial role for charge interactions in dynorphin-κ receptor binding [3].

Table 1 *Competitive binding of peptide models to opioid receptors in guinea-pig brain membranes*[a]

Peptide	IC_{50} + SEM (nM) (for radioligand displacement)		
	μ sites	δ sites	κ sites
Dynorphin A	3.16 + 0.62	13.1 + 2.90	0.63 + 0.19
1	4.82 + 1.11	39.1 + 3.20	0.53 + 0.04
2	4.97 + 1.27	12.5 + 1.80	1.86 + 0.63
3	5.52 + 0.85		0.29 + 0.02
4	2.65 + 0.42		0.32 + 0.04
5	1.07 + 0.16		0.62 + 0.14
6	19.11 + 1.56		15.31 + 3.66

All assays performed at 4°C in 50 mM Tris–HCl, pH 7.4.
[a]For experimental details see Ref. 2.

References

1. Tayor, J.W. and Kaiser, E.T., Pharmacol. Rev. 38 (1986) 291.
2. Tayor, J.W. and Kaiser, E.T., Int. J. Pept. Prot. Res., 34 (1989) 75.
3. Sargent, D.F., Bean, J.W., Kosterlitz, H.W. and Schwyzer, R., Biochemistry, 27 (1988) 4974.

Peptidomimetics in the study of opiate peptides

Dale F. Mierke[a], Odile E. Said-Nejad[a], Toshimasa Yamazaki[a], Eduard Felder[a],
Peter W. Schiller[b] and Murray Goodman[a]

[a]Chemistry Department, University of California, San Diego, CA 92093, U.S.A.
[b]Clinical Research Institute of Montreal, Montreal, Quebec, Canada H2W 1R7

Introduction

We are interested in examining the conformational requirements necessary
for biological activity of opiate peptides [1,2]. It has clearly been demonstrated
that the aromatic residues Tyr[1] and Phe[3] (dermorphin, morphiceptin) and Phe[4]
(enkephalin) are important for the opiate activity of peptides. To characterize
the role of these aromatic residues, we have synthesized a series of 14-membered
cyclic opiate analogs that contain phenylalanine at both positions three and
four (Fig. 1). In addition, we have initiated the synthesis of a series of cyclic
analogs of enkephalin incorporating a β-(1-naphthyl)alanine residue at the fourth
position in place of the phenylalanine.

Methods

The analogs were synthesized by solution and SPPS and purified by RPHPLC.
The conformational characterization of the molecules was carried out by [1]H
NMR and NOE-restrained molecular dynamics and flexible geometry energy
minimizations. The proton resonances were assigned using 2D COSY and
HOHAHA techniques. The NOEs were measured in the rotating frame (ROESY
experiments) at various mixing times, from 75 to 500 ms. Computer simulations
were carried out at the San Diego Supercomputer Center using the DISCOVER
program. The biological response of the analogs for two subclasses of opiate
receptors is shown in Table 1.

Results and Discussion

Multiple resonances were observed in the proton spectra of three of the eight
compounds shown in Fig. 1. The Tyr-c[D-A$_2$bu-Gly-D-Nal(1)-Leu] analog contains 22% of *cis* configurational isomers, and two additional isomers were
observed for Tyr-c[D-A$_2$bu-Phe-gPhe-R-mLeu] accounting for 21% and 13%.
The analog Tyr-c[D-Glu-Phe-gPhe-D-retro-Leu] is composed of only 28% of all
trans structure, with two *cis* amide-containing isomers accounting for 51% and
21%, respectively. We have been able to show with ROESY experiments that
the additional resonances arise from *cis/trans* isomerization about unsubstituted
amide linkages [3]. With saturation transfer techniques, the rate of iso-

348

Table 1 *Guinea pig ileum (GPI) and mouse vas deferens (MVD) assay of various opioid peptide analogs*

Compound	GPI IC$_{50}$ [nM]	MVD IC$_{50}$ [nM]	MVD/GPI IC$_{50}$-ratio
Tyr-c[D-A$_2$bu-Phe-Phe-Leu]	1.15 ± 0.17	2.86 ± 0.37	2.49
Tyr-c[D-A$_2$bu-Phe-Phe-D-Leu]	1.14 ± 0.12	13.9 ± 0.9	12.2
Tyr-c[D-A$_2$bu-Phe-gPhe-S-mLeu	0.518 ± 0.099	6.30 ± 1.11	12.2
Tyr-c[D-A$_2$bu-Phe-gPhe-R-mLeu	17.6 ± 2.8	117 ± 31	10.1
Tyr-c[D-Glu-Phe-gPhe-rLeu]	39.0 ± 9.3	3.64 ± 0.10	0.09
Tyr-c[D-Glu-Phe-gPhe-D-rLeu]	2.75 ± 0.63	49.1 ± 9.9	17.9
Tyr-c[D-A$_2$bu-Gly-βNal(1)-Leu]	5.80 ± 0.62	23.8 ± 1.4	4.10
Tyr-c[D-A$_2$bu-Gly-D-βNal(1)-Leu]	62.0 ± 6.7	266 ± 26	4.29
[Leu^5]enkephalin	246 ± 39	11.4 ± 1.1	0.0463

merization was measured and found to be similar to that measured for proline containing compounds.

The conformational analysis employing NOE constrained molecular dynamics has been completed for the analogs shown in Fig. 1. These results will be compared with the findings from our studies of cyclic enkephalin analogs and applied to our proposed model for opiate receptor selectivity.

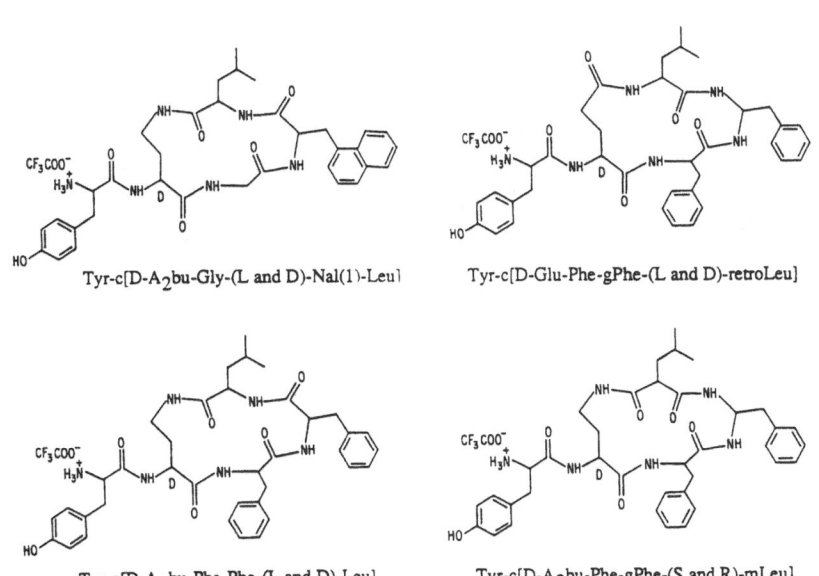

Tyr-c[D-A$_2$bu-Gly-(L and D)-Nal(1)-Leu]   Tyr-c[D-Glu-Phe-gPhe-(L and D)-retroLeu]

Tyr-c[D-A$_2$bu-Phe-Phe-(L and D)-Leu]   Tyr-c[D-A$_2$bu-Phe-gPhe-(S and R)-mLeu]

Fig. 1.   Structures of cyclic analogs related to enkephalin and dermorphin.

Acknowledgements

This project was supported in part by the NIH grant (DK-15410) and by the San Diego Supercomputer Center.

349

D.F. Mierke et al.

References

1. Goodman, M. and Mierke, D.F., J. Am. Chem. Soc., 111 (1989) 3489.
2. Mierke, D.F., Lucietto, P., Schiller, P.W and Goodman, M., Biopolymers, 26 (1987) 1600.
3. Mierke, D.F., Said-Nejad, O.E., Yamazaki, T., Felder, E. and Goodman, M., J. Am. Chem. Soc., 111 (1989) 6847.

Solvation state of Leu-enkephalin in phospholipid micellar environment as revealed by FTIR studies

N. Birlirakis[a], I. Gerothanassis[a], M. Sakarellos-Daitsiotis[a], C. Sakarellos[a], B. Vitoux[b] and M. Marraud[b]

[a]Department of Chemistry, University of Ioannina, Box 1186, 45110 Ioannina, Greece
[b]LCPM-ENSIC-INPL, CNRS-UA 494, B.P. 451, 54001 Nancy Cedex, France

Introduction

Transfer of an aqueous-soluble peptide hormone such as Leu-enkephalin, Tyr[1]-Gly[2]-Gly[3]-Phe[4]-Leu[5], to the micellar environment of phospholipids may convert the peptide into a conformation for eliciting biological activity. It is, therefore, reasonable that a variety of spectroscopic techniques have been applied in investigating the conformational state of peptide hormones in phospholipid environment [1–5]. However, the models of site-specific solvation state of hormones remain largely speculative. As a first step towards the solution of specific mode solvation-hydratation effects of peptide oxygens, we performed, for the first time, FTIR studies of [1-^{13}C-Gly[2],Leu[5]]- and [1-^{13}C-Gly[3],Leu[5]]-enkephalin in dodecylphosphorylcholin ($C_{12}PC$) micelles.

Results and Discussion

We have recently shown that the IR difference spectra, with reference to Leu-enkephalin, of the [1-C^{13}-Gly[2],Leu[5]]- and [1-^{13}C-Gly[3],Leu[5]]-enkephalins in CH_3CN/DMSO are similar, with a frequency shift of 36 cm^{-1} between the ^{12}C$=$O and ^{13}C$=$O stretching vibrations [6]. In aqueous solution, the Gly[2] and Gly[3] C$=$O amide I stretching vibrations of both [1-^{13}C-Gly[2], Leu[5]]- and [1-^{13}C-Gly[3],Leu[5]]-enkephalins are shifted to lower frequencies by 15 cm^{-1} and 25 cm^{-1}, respectively, thus indicating the formation of one hydrogen bond, and that the solvation species involved are largely monohydrated [6,7]. This is in agreement with extensive IR studies of simple model amides in which a characteristic high frequency shift of the ν(C$=$O) vibration is observed upon disruption of a hydrogen bond. It is, therefore, clear that the insertion or penetration of Gly[2] and Gly[3] peptide oxygens into the interior of the micelle and the concomitant effective dissolution would result in a significant shift to high frequency provided that no conformational changes occur that could induce intramolecular hydrogen bonding involving the peptide NH groups. However, it is found that the C$=$O vibration frequencies of both Gly[2] and Gly[3] remain unaffected upon successive addition of $C_{12}PC$ above the critical micelle concentration. Therefore it is clear that both Gly[2] and Gly[3] oxygens are exposed to the solvent. The solvation species involved are, as in aqueous solution, largely monohydrogen-bonded. The alternative interpretation that these two peptide oxygens are inserted within the interior of the micelle with the primary solvation (hydratation) state identical to that in aqueous solution should be excluded since, as shown from extensive FTIR studies on model amides, the dielectric constant of the medium should induce significant spectral modifications for a given primary solvation state that is not observed in our case. On the other hand, at least one of the other Tyr[1] or Phe[4] $C=O$ vibration frequencies is slightly affected by the addition of $C_{12}PC$, probably due to some interaction of this peptide carbonyl within the micelle.

Conclusion

The present results suggest that the solvation state of both Gly[2] and Gly[3] peptide oxygens remains essentially unaffected upon addition of $C_{12}PC$ micelles. It is also apparent that Leu-enkephalin does not fold into an intramolecular H-bonded $2 \leftarrow 5$ β-turn structure in the lipid environment. The above results show that FTIR of selectivily labeled ^{13}C analogs of small peptide hormones is a promising tool for studying specific solvation state of peptide oxygens in micellar environment.

Acknowledgements

This work was supported by C.E.C. (Grant ST2J-0184), the Greek General Secretary for Research and Technology, the Research Committee of the University of Ioannina, the Greek Foundation 'Leonidas Zervas' (Scholarship to N.B.), FEBS (summer fellowship to N.B.), and EMBO (short-term fellowship to IPG).

References

1. Behnam, B.A. and Deber, C.M., J. Biol. Chem., 259 (1984) 14935.
2. Deber, C.M. and Behnam, B.A., Proc. Natl. Acad. Sci. U.S.A., 81 (1984) 61.
3. Deber, C.M. and Behnam, B.A., Biopolymers, 24 (1985) 105.
4. De Marco, A., Zetta, L., Gariboldi, P. and Menegatti, E., In Hruby, V. and Rich, D. (Eds.) Peptides: Structure and Function, Pierce Chemical Company, Rockford, IL, 1985, p. 887
5. Zetta, L., De Marco, A., Zannoni, G. and Cestaro, B., Biopolymers, 25 (1986) 2315.
6. Sakarellos, C., Gerothanassis, I., Birlirakis, N., Karayannis, T., Sakarellos-Daitsiotis, M., Vitoux, B. and Marraud, M., In Bayer, E. and Jung, G. (Eds.) Peptides 1988, (Proceedings of the 20th European Peptide Symposium), De Gruyter, Berlin, 1989, p. 498.
7. Sakarellos, C., Gerothanassis, I.P., Birlirakis, N., Karayannis, T., Sakarellos-Daitsiotis, M. and Marraud, M., Biopolymers, 28 (1989) 15.

Session II
Enzymology

Chairs: Bruce W. Erickson
University of North Carolina
Chapel Hill, North Carolina, U.S.A.

and

Robert S. Hodges
University of Alberta
Edmonton, Alberta, Canada

Enzymatic synthesis of peptides in anhydrous organic solvents

Alexey L. Margolin* and Alexander M. Klibanov

Department of Chemistry, Massachusetts Institute of Technology, Cambridge, MA 02139, U.S.A.

Introduction

Enzymatic methods, with their unparalleled specificity, are an attractive alternative to the classical techniques of organic chemistry. The use of organic solvents as a reaction medium allowed for several transformations impossible in water, thus significantly broadening the synthetic repertoire of enzymes [1]. In this paper, we will consider our recent work on enzymatic peptide synthesis in anhydrous organic solvents.

The advantages of enzymatic peptide synthesis, such as mild reaction conditions, absence of racemization, minimal protection and activation requirements, are well recognized. Despite several successful examples, enzymatic – namely protease-catalyzed – peptide synthesis has yet to reach the versatility of well-known chemical methods. Three major problems should be solved before enzymatic techniques become general routine procedure: (1) unfavorable thermodynamics of peptide synthesis in water; (2) a narrow substrate specificity and enantioselectivity of enzymes; and (3) undesirable proteolysis of the growing polypeptide chain. The first drawback can be alleviated by using enzymes in biphasic aqueous organic mixtures, mixtures of water with water-miscible organic solvents, or nonaqueous media. In our research discussed herein, we addressed the remaining problems, namely secondary hydrolysis and strict substrate- and stereoselectivity of enzymes.

Results and Discussion

Since lipases can catalyze the reaction of aminolysis between carboxylic esters and aliphatic amines in anhydrous organic solvents [2], we decided to use these nonproteases for the formation of peptide bond. We hoped that lipases under such reaction conditions would catalyze the formation of the peptide bond. Indeed, we succeeded [2] in synthesizing a variety of dipeptides using lipases as catalysts in anhydrous toluene and tetrahydrofuran. The reaction procedure was very simple. In a representative experiment, 18.5 mmol of N-acetyl-L-Phe chloroethyl ester and 16.7 mmol of L-Leu-NH$_2$ were dissolved in 330 ml of

---

*Present address: Merrell Dow Research Institute, 9550 N. Zionsville Rd., Indianapolis, IN 46268, U.S.A.

355

anhydrous toluene. Then, porcine pancreatic lipase was added and the suspension was shaken at 45°C. The precipitated N-Ac-Phe-Leu-NH$_2$ was extracted with ethanol and then recrystallized to give 4.4 g (84% yield) of the pure dipeptide. This methodology was used for the synthesis of other dipeptides. The results obtained indicate that porcine pancreatic lipase catalyzes the reaction with other amino acids, such as Tyr, Met, and Ala. A number of both amide and ester derivatives of N-terminal amino acids can be used. Significantly, L and D isomers of Leu and Ala serve equally well as nucleophiles. Porcine pancreatic lipase was by no means a unique nonprotease to form a peptide bond. Lipases from *Chromobacterium viscosum, Pseudomonas* sp., *Mucor* sp. and *Candida cylindracea* catalyzed peptide synthesis, although their substrate specificity was different. It is important to note that lipases did not catalyze the hydrolysis of synthesized peptides. Thus, the use of organic solvents, instead of water, as a reaction medium solves the equilibrum problems, and lipases, in contrast to proteases, do not catalyze the secondary hydrolysis of peptides.

A number of biologically active peptides, including important antibiotics, synthetic vaccines, and hormones contain unnatural and D-amino acid residues. Enzymatic methods (as well as recombinant DNA technology) are not generally applicable to peptides involving D-amino acids because the L-specificity of proteolytic enzymes. The only exception has been the use of D-amino-acid derivatives as nonspecific nucleophiles competing with water for the acyl enzyme formed in the reaction between a protease and a protected L-amino-ester. This approach, however, is inherently limited to the enzymatic incorporation of a single D-amino-acid residue into a peptide's C-terminal position. We decided to attack this problem by using anhydrous organic solvents as a reaction medium and the microbial protease subtilisin as a catalyst. We have observed that, upon a transition from water to organic solvents, the enantioselectivity of several proteases dramatically relaxes [3]. We measured the ratio of specificity constants $(k_{cat}/K_m)_L/(k_{cat}/K_m)_D$, which reflects enantioselectivity of an enzyme, in water and in several organic solvents. The enzymatic reaction studied in water was the hydrolysis of 2-chloroethyl esters of N-acetyl-L- and D-amino acids. In organic solvents, we examined the enzymatic transesterification reaction between the same esters and propanol. For all nonaqueous solvents tested, this enantioselectivity factor was found to be 10- to 100-fold lower. Also, there is a linear correlation between the difference in free energy of activiation ($\Delta\Delta G^{\#}$) for subtilisin-catalyzed transesterification of L- and D-enantiomers of N-Ac-Ala-OEtCl in 13 different organic solvents, and the hydrophobicity of the solvent [3]. These results indicate that an enzyme's enantioselectivity decreases as the hydrophobicity of the solvent increases. To test the generality of the observed phenomenon, we investigated enantioselectivity of five other serine proteases in water and in butyl ether. Elastase, α-lytic protease, subtilisin BPN, α-chymotrypsin, and trypsin all exhibit striking enantioselectivity in water, but not in organic solvents: while the $(k_{cat}/K_m)_L/(k_{cat}/K_m)_D$ in water is on the order of 10^3–10^4, in butyl ether it does not exceed even one order of magnitude. These results demonstrate the predictable and rational control of enantioselectivity

Table 1 *Synthesis of peptides containing D-amino acids catalyzed by subtilisin in anhydrous tert-amyl alcohol*[a]

Substrates (mmol)		Product	Isolated yield of the product (%)
Amino acid ester	Nucleophile		
N-Ac-D-Phe-OEtCl (3.1)	L-Phe-NH$_2$ (3.1)	N-Ac-D-Phe-L-Phe-NH$_2$	67
N-F-D-Ala-OEtCl (5.1)	L-Phe-NH$_2$ (15.6)	N-F-D-Ala-L-Phe-NH$_2$	82
N-Ac-D-Asn-OEtCl (5.0)	L-Leu-NH$_2$ (7.5)	N-Ac-D-Asn-L-Leu-NH$_2$	71
N-Ac-D-Trp-OEtCl (8.4)	L-Phe-NH$_2$ (8.4)	N-Ac-D-Trp-L-Phe-NH$_2$	47
N-F-D-Ala-OEtCl (6.0)	D-Ala-NH$_2$ (6.0)	N-F-D-Ala-D-Ala-NH$_2$	65
N-F-D-Ala-OEtCl (16.0)	L-Phe-L-Leu-NH$_2$ (16.0)	N-F-D-Ala-L-Phe-L-Leu-NH$_2$	61
N-CBZ-L-Tyr-OEtCl (2.5)	D-Ala-NH-(CH$_2$)$_3$-Ph (7.8)	N-CBZ-L-Tyr-D-Ala-NH-(CH$_2$)$_3$-Ph	54

of enzymes by changing the reaction medium. The discovered phenomenon allowed us to solve the problem of enzymatic synthesis of peptides containing D-amino acid residues. Using subtilisin Carlsberg as a catalyst and *tert*-amyl alcohol as a reaction medium, we synthesized several peptides with D-amino acids (Ala, Phe, Trp, Asn) in the N-terminal position [4] (Table 1). It is worth noting that various amino acid derivatives, including both L- and D-isomers and dipeptides, were utilized as nucleophiles. This led us, for the first time, to the enzymatic synthesis of a dipeptide (N-F-D-Ala-D-Ala-NH$_2$) constructed of only D-amino acids. Thus, a radically altered stereospecificity of subtilisin in organic solvents affords facile enzymatic preparation of diverse peptides containing D-amino acids.

Another group of peptides representing a challenging and potentially useful target for enzymatic methods is the isopeptides. These compounds, with the bond formed with functional groups located in the amino acid side chains, may have a number of attractive pharmacological properties. For example, the peptide bond in N-Ac-L-Phe-(ϵ)-L-Lys-*tert*-Bu is much more stable toward proteolysis than its α-isomer [5]. We synthesized this isopeptide from N-Ac-L-Phe-OEtCl and L-Lys-O-*tert*-Bu in *tert*-amyl alcohol using subtilisin as a catalyst in an 85% isolated yield [5]. The regioselectivity of this process was remarkably high ($>99\%$). Similar data were obtained when a *Pseudomonas* sp. lipoprotein lipase was used as a catalyst of peptide synthesis. Therefore, subtilisin and lipase, when used in anhydrous organic solvents, preserve their regioselectivity and afford the synthesis of the unnatural ϵ peptide linkage.

One of the most important features of enzymatic peptide synthesis is an inherent chemoselectivity of enzymes. The property to choose exclusive α-carboxy and α-amino groups without protection and then deprotection of highly reactive side chains makes enzymes valuable practical catalysts. We studied a model reaction of N-Ac-L-Phe-OEtCl with 6-amino-1-hexanol in *tert*-amyl alcohol and have shown that subtilisin Carlsberg exhibits a 50-fold preference toward the NH$_2$ group over the OH group [6]. We applied this procedure for the synthesis of peptides to tri-functional amino acids (Tyr, Thr, Ser) [6]. As an illustration, we utilized both subtilisin Carlsberg and BPN to prepare protected dipeptide fragments (L-Tyr-L-Thr, L-Thr-L-Thr, and D-Ala-L-Ser) of the octapeptide T, which has been reported to have an anti-HIV activity.

Conclusion

The unique combination of broad stereoselectivity and strict chemo- and regioselectivity of enzymes in anhydrous organic solvents allowed the solution of several problems of enzymatic peptide synthesis, thus making this technique a valuable addition to chemical procedures.

Acknowledgements

This work was financially supported by NIH grant no. GM 39794.

References

1. Klibanov, A.M., Trends Biochem. Sci., 14 (1989) 141.
2. Margolin, A.L. and Klibanov, A.M., J. Am. Chem. Soc., 109 (1987) 3802.
3. Sakurai, T., Margolin, A.L., Russell, A.J. and Klibanov, A.M., J. Am. Chem. Soc., 110 (1988) 7236.
4. Margolin, A.L., Tai, D.F. and Klibanov, A.M., J. Am. Chem. Soc., 109 (1987) 7885.
5. Kitaguchi, H., Tai, D.-F. and Klibanov, A.M., Tetrahedron Lett., 29 (1988) 5487.
6. Chinsky, N., Margolin, A.L. and Klibanov, A.M., J. Am. Chem. Soc., 111 (1989) 386.

Site-directed labeling and identification of modified peptides in a catalytic antibody-combining site

Sheri Hunt, Kathy Bowdish, Nebojsa Janjic and Al Tramontano

*Department of Molecular Biology, Scripps Clinic and Research Foundation,
10666 N. Torrey Pines Rd., La Jolla, CA 92037, U.S.A.*

Introduction

The activity of monoclonal antibodies to specific haptens as enzymatic catalysts is a topic of much current interest [1]. One implication of this research is in the structural analysis of proteins with catalytic function. All antibodies have a common structural motif, and therefore the determination of the structure of an antibody–antigen complex can be approached by empirical modeling methods [2]. Knowledge of the interactions between antibody and ligands for the combining site can be used to construct an accurate model of the complex.

The amino-acid sequence of the esterolytic antibody [3] 50D8 has been deduced from its cDNA sequence. Its combining site complementarity-determining regions (CDRs) are suggested by homology to known antibody sequences (K. Bowdish and J. Hicks, unpublished results). This sequence can be used to define a preliminary model for the combining site and to predict the sequence of peptides containing residues essential for the binding and catalytic activity. We have begun to map the interactions between hapten and antibody by affinity-labeling techiques and to identify labeled residues in its tryptic peptides. This report describes our progress towards the characterizations of two labeling sites.

Results and Discussion

Two reagents were applied to active site-directed labeling. Tetranitromethane (TNM) is known to nitrate a functional tyrosine residue in the active site, with loss of catalytic activity [3]. A photoactivatable reagent, p-azidophenylglyoxal (APG) as its hydrate, is thought to mimic the phosphonyl structure of the hapten (Fig. 1). We report here that esterolytic activity of 50D8 is efficiently and irreversibly inactivated by photolysis in the presence of 10 mol equivalents of APG. Chemical modification of the antibody is conveniently assayed by loss of the esterolytic activity in a previously described spectrophotometric assay [3].

At concentrations of TNM in modest excess to protein (100 mol equiv.) under standard conditions [4], the antibody's catalytic activity is reduced by more than 60% in 20 min. Trypsin digest of the reduced (β-mercaptoethanol) and alkylated (iodoacetamide) protein produced a mixture of peptides detected by

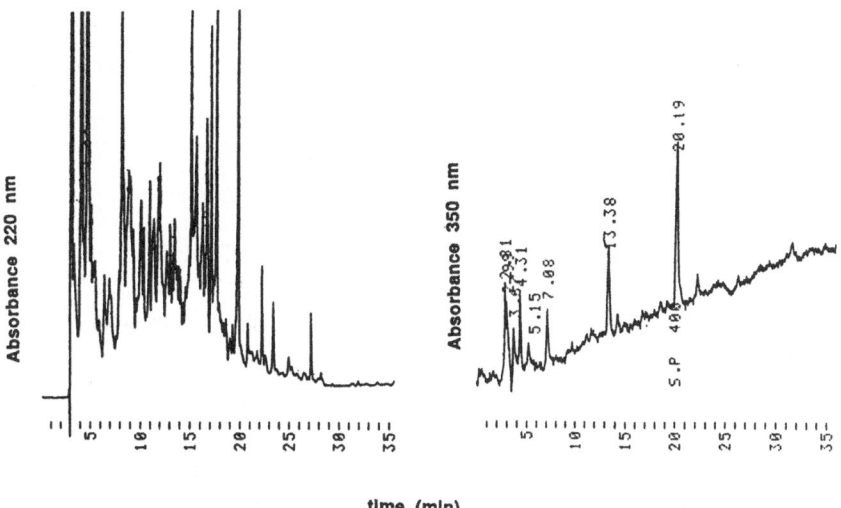

Fig. 1.   Azidophenylglyoxal (APG) hydrate analog for phosphonate hapten.

RPHPLC (C_{18} Vydac 201 TP54). A chromatogram of the mixture at 350 nm UV detection revealed a major peak at 20 min retention time (Fig. 2). The peak was collected and repurified by RPHPLC by a CH_3CN/H_2O 0.1% TFA 0–50% in 30-min gradient elution. The purified peptide, characterized by Edman sequencing, has in its first six positions the sequence S-G-G-I-I-Y_N. The residue in the sixth position was identified as nitrotyrosyl by correspondence with an authentic sample. The position of modification in the protein can, therefore, be assigned based on the presence of this unique sequence in the segment contained by trypsin cleavage sites in positions 52–53 and 64–65 on the H chain of 50D8. The sequence 52–64 reads R-S-G-G-I-I-Y-Y-P-D-S-M-K. This fragment falls within CDR2, and the hydroxyl group of Y^{58} can be proposed to contact the carbonyl oxygen of the substrate in the complex where it can participate in general base catalysis.

Fig. 2.   MAb 50D8 nitration by tetranitromethane. Tryptic peptides by RPHPLC mapping. Gradient elution with CH_3CN/H_2O, 2% NH_4OAc, pH 7, 3–30% CH_3CN in 30 min.

The modification by APG may be assumed to map a position of contact between the protein and the *para* substituent of a *p*-amidophenylacetate ester substrate. Thus, the antibody was photolyzed in the presence of a 10-fold molar excess of APG, reduced with sodium borohydride, alkylated and digested as before. The RPHPLC peptide map at 280 nm detection showed a large number of peaks. The 280-nm chromophore is therefore of little use in identification of labeled peptides. Since the glyoxal group in the affinity label can be reduced to a diol with incorporation of hydride, this serves as a useful method for specific radiolabeling of APG-modified peptides (Fig. 3). When the inactivated protein was reduced with tritiated borohydride (NaB_3H_4) and digested, the HPLC map contained a unique fraction, eluted at about 13–14 min, with significant tritium content. This mixture has been separated by HPLC and the radiolabeled material was collected. The purification is incomplete and a second gradient elution is being applied to obtain a homogeneous material. The sequence of the peptide obtained shall provide supplementary information for mapping of the combining site. The usefulness of the APG labeling shall depend on the sequencing through the modified residue, such that the position of modification may be inferred. It is expected that the identification of two sites contacting the substrate or hapten in a well-defined manner will provide sufficient constraints with which to orient the ligand in the combining site. Further refinement of the active site structure may then proceed from the analysis of adjacent sequences and side-chain interactions.

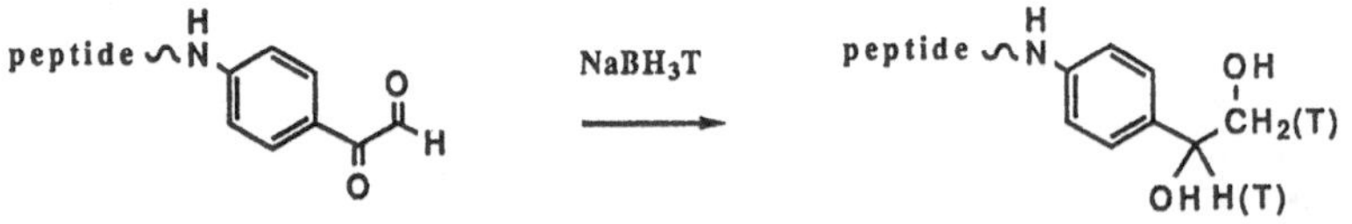

Fig. 3.   Radiolabeling of APG photo-inactivated antibody.

Acknowledgements

We are grateful to Diane Schloeder for production of antibody and to the National Institutes of Health for financial support. This is contribution 5979-MB from the Department of Molecular Biology, Scripps Clinic and Research Foundation, La Jolla, CA 92037.

References

1a. Lerner, R.A. and Tramontano, A., Trends Biochem. Sci., 12 (1987) 427.
 b. Tramontano, A., Janda, K., Napper, A.D., Benkovic, S.J and Lerner, R.A., Cold Spring Harbor Symp. Quant. Biol., 52 (1987) 91.
2a. Smith-Gill, S.J., Mainhart, C., Lavoie, T.B., Feldmann, R.J., Drohan, W. and Brooks, B.R., J. Mol. Biol., 194 (1987) 713.
 b. Rees, A.R. and de la Paz, P., Trends Biochem. Sci., 11 (1986) 144.
3.   Tramontano, A., Ammann, A.A. and Lerner, R.A., J. Am. Chem. Soc., 110 (1988) 2282.
4.   Sokolovsky, M., Riordan, J.F. and Vallee, B.L., Biochemistry, 5 (1966) 3582.

Synthetic substrates of the IgA1 proteinase from *Neisseria gonorrhoea*

Stephen G. Wood

Laboratory of Rational Drug Design, Evans Department of Clinical Medicine, University Hospital, Boston University Medical Center, Boston, MA 02118, U.S.A.

Introduction

The pathogenic bacteria, *Neisseria gonorrhoea*, secretes a high molecular weight enzyme (106 kDA) that cleaves and inactivates human secretory IgA1 [1]. The precursor of this enzyme, upon extracellular secretion, undergoes autoproteolysis at three sites to generate the mature enzyme [2]. The three sites all contain a proline residue in the P_1 position that is consistent with the cleavage sites of IgA1.

Results and Discussion

Three decapeptides based on the autoproteolytic cleavage sites, as well as two decapeptides containing the IgA cleavage sites were synthesized, along with their possible catabolites, by SPPS (Table 1). The synthetic peptides were digested with the neisserial enzyme and monitored by HPLC for cleavage. Two of the decapeptides have been proven to be substrates for the neisserial enzyme.

These are the first synthetic substrates of any IgA1 proteinase.

Table 1   *Synthetic peptides synthesized*

Autoproteolytic sites	
HRP-91	Ac-VKPAPSPAAN-NH$_2$
HRP-92	Ac-VVAPPSPQAN-NH$_2$
HRP-93	Ac-LPRPPAPVFS-NH$_2$
IgA1 cleavage sites	
HRP-94	Ac-TPPTPSPSCC-NH$_2$
HRP-95	Ac-PSTPPTPSPS-NH$_2$

The decapeptides were digested with the neisserial enzyme in Tris buffer (50 mM, pH 7.5) for 24 h and checked for cleavage. Upon HPLC two of the decapeptides, HRP-92 and HRP-93, produced two peaks corresponding to their respective catabolites. The peaks were collected and subjected to AAA and sequencing. The results confirmed the HPLC data that cleavage had occurred. No cleavage of the other decapeptides had occurred after 24 h incubation.

The two decapeptides were then synthesized using a high tritium label (1 Ci/mol) to effectively monitor the cleavage. The peptides were digested, as above, in five concentrations. The results are given in Table 2. The tetrapeptide inhibitor, HRP-48, was incubated along with a digest mixture containing HRP-93, and $<3\%$ cleavage had occurred after 6 h.

Boronate transition–state inhibitors also effectively inhibit the cleavage of the synthetic peptides, as does human colostrum.

Both HRP-92 and HRP-93 are cleaved by IgA(+) constructs from *Escherichia coli*. Neither synthetic substrate is cleaved by neisserial IgA(+) constructs.

Inhibitors of the cleavage of the synthetic substrates may also prevent maturation of the IgA1 proteinase.

Table 2 k_{cat}/K_m *of synthetic peptides*

	K_m (M)	k_{cat} (pmol/h/U)	k_{cat}/K_m ($\times 10^6$)
IgA1	6×10^{-6}	13.3	2.16
HRP-92	1.35×10^{-3}	280	0.21
HRP-93	3.43×10^{-3}	439	0.13

Acknowledgements

This research was supported by NIDR Grant DE-07257.

References

1. Plaut, A.G., Gilbert, J.V., Artenstein, M.S. and Capra, J.D., Science, 190 (1975) 1103.
2. Pohlner, J., Halter, R., Bevreuther, K. and Meyer, T.F., Nature, 325 (1987) 458.

Action of pepsin on peptide substrates containing N^{α}-methyl amino acids

Jan Pohl

Microchemical Facility, Winship Cancer Center, Emory University, Atlanta, GA 30322, U.S.A.

Introduction

In complexes of aspartic proteinases with inhibitors, peptides are bound at the active site in an extended beta-strand conformation with hydrogen bonds between the main chain of the inhibitor and the enzyme [1–3]. Hydrogen bonding in complexes of these enzymes with oligopeptide substrates may play an important role in their catalytic efficiency, k_{cat}/K_M, e.g., by promoting strain on the scissile peptide bond [4]. In fact, k_{cat}/K_M increases with increased substrate length (K_M remaining unchanged). We explored the effect of eliminating substrate amide protons (H-bond donor) on the hydrolytic efficiency of pepsin. In positions P_3 through P'_2 of the chromophoric octapeptide substrates [5], peptide bond -NH- groups were replaced by -N(CH$_3$)-, and the effect of these modifications was quantitated.

Results and Discussion

Peptides (Table 1) were synthesized by using SPPS on PAM-resin (0.1 mmol scale), and Boc-L-amino acids. Boc-N-methyl amino acids were incorporated into the peptides by the symmetric anhydride coupling protocol used for leucine and required double couplings (monitoring by quantitative ninhydrin test and by solid phase sequencing). The peptides were deprotected and cleaved from the resin in HF/anisole/DMS (85:10:5, 12°C, 90 min), and purified by RP-HPLC; their composition was confirmed by AAA.

Pepsin cleaved peptides between residues P_1 and P'_1 (Table 1). The kinetics are position-dependent, in the order of $P_1 \gg P'_1 > P_3, P'_2, P_2$ for effects on k_{cat} and $P_1 \gg P'_2 > P_3, P'_1 > P_2$ for effects on k_{cat}/K_M. The amide proton in P_1 is nearly essential for hydrolysis, however, with remarkably little effect on K_M. This correlates with the view [1–3] that hydrogen bonding of the P_1 amide to the carbonyl of Gly^{217} (pepsin numbering) is crucial for correct positioning of the scissile peptide bond. The amide proton in P'_1 is not an essential structural feature, since k_{cat} is reduced only 20-fold. Proline was placed in P'_1 for comparison, and **9** was also found to be a moderately good substrate. Replacements in secondary substrate positions, P_3 and P'_2, have negative effects on both k_{cat} and K_M (38-fold increase in P'_2). For the doubly substituted **12**, these changes are approximately

Table 1 *Kinetics of cleavage of peptides by pepsin*[a]

	P_5 P_4 P_3 P_2 P_1 P_1' P_2' P_3'	k_{cat} (s^{-1})	K_m (μM)	k_{cat}/K_m $(s^{-1} M^{-1})$
1	Lys-Pro-Ala-Lys-Phe-Nph-Arg-Leu	87	45	1 930 000
2	Lys-Pro-Ala-Lys-Phe-Nph-Arg-Leu	17	230	74 000
3	Lys-Pro-Ala-Ala-Phe-Nph-Arg-Leu	91	21	4 300 000
4	Lys-Pro-Ala-Ala-Phe-Nph-Arg-Leu	45	7	6 400 000
5	Lys-Pro-Ala-Pro-Phe-Nph-Arg-Leu	0.007	130	54
6	Lys-Pro-Ala-Lys-Phe-Nph-Arg-Leu[b]	0.004	20	200
7	Lys-Pro-Ala-Lys-Nph-Ala-Arg-Leu	0.8	280	2 900
8	Lys-Pro-Ala-Lys-Nph-Ala-Arg-Leu	0.04	130	300
9	Lys-Pro-Ala-Lys-Nph-Pro-Arg-Leu[b]	3.0	520	5 800
10	Lys-Pro-Ala-Glu-Phe-Nph-Ala-Leu	110	20	5 500 000
11	Lys-Pro-Ala-Glu-Phe-Nph-Ala-Leu	30	770	39 000
12	Lys-Pro-Ala-Lys-Phe-Nph-Ala-Leu	~2	>2 000	<1 000

[a] Nph = p-nitrophenylalanine; N-methyl amino acid residues are underlined; at 37°C, pH 4.5, I = 0.1 M; see Ref. 5 for methods.
[b] Compare with 1.

additive. As shown previously [6,7], occupation of these secondary binding subsites by small residues, e.g., Ala, despite rather large increase of k_{cat}, does not influence K_M at all [6,7]. Therefore, in subsites S_3 and S_2', the main chain amide hydrogen bonds (in pepsin to Ser[219] and Gly[34], respectively) are likely to contribute to the magnitude of K_M. The position least influenced is P_2 for which 4 has a slight increase in k_{cat}/K_M. Therefore, hydrogen bonding of P_2 amide to Thr[77] of the 'flap' is not catalytically important. In striking contrast with the effect of N-methyl-Ala, proline residue in P_2 (5) is not acceptable. This stresses the importance of substrate conformation in this subsite.

Acknowledgements

This work was supported in part by NIH grant AM-18865 to Dr. Ben M. Dunn, University of Florida, Gainesville, FL.

References

1. Kostka, V. (Ed.) Aspartic Proteinases and their Inhibitors, de Gruyter, Berlin, 1985.
2. Foundling, S.I., Cooper, J., Watson, F.E., Clesbay, A., Pearl, L.H., Sibanda, B.L., Hemmings, A., Wood, S.P., Blundell, T.L., Valler, M.J., Norey, C.G., Kay, J., Boger, J., Dunn, B.M., Leckie, B.J., Jones, D.M., Atrash, B., Hallet, A. and Szelke, M., Nature, 327 (1987) 349.
3. Suguna, K., Padlan, E.A., Smith, C.W., Carlson, W.D. and Davies, D.R., Proc. Natl. Acad. Sci. U.S.A., 84 (1987) 7009.
4. Pearl, L.H., FEBS Lett., 214 (1987) 8.
5. Pohl, J. and Dunn, B.M., Biochemistry, 27 (1988) 4827.
6. Fruton, J.S., Adv. Enzymol. Relat. Areas Mol. Biol., 33 (1976) 401.
7. Hofmann, T., Allen, B., Bendiner, M., Blum, M. and Cunningham, A., Biochemistry, 27 (1988) 1140.

Expression, refolding and characterization of recombinant human procathepsin D

Paula E. Scarborough[a], Gregory E. Conner[b] and Ben M. Dunn[a]

[a]*Department of Biochemistry and Molecular Biology, University of Florida, Gainesville, FL 32610, U.S.A.*
[b]*Department of Cell Biology and Anatomy, University of Miami School of Medicine, Miami, FL 33101, U.S.A.*

Introduction

Cathepsin D is a lysosomal enzyme that functions, primarily, in the degradation of intracellular and endocytosed proteins. A member of the aspartic protease family, cathepsin D shows a great deal of sequence homology with pepsin and renin. Cathepsin D is synthesized in the cell as a preproenzyme that must undergo several cleavages to generate the active enzyme. In the lysosome, cathepsin D exists in two enzymatically active forms: a single-chain form and, after further proteolytic cleavage, a predominant and stable two-chain form that lacks seven amino acids in the middle of the molecule [1]. To study the structural and enzymatic properties of the biosynthetic intermediates of this enzyme, a bacterial expression system has been developed to yield large quantitatives of procathepsin D. The proenzyme must be refolded and activated to recover proteolytic activity.

Results and Discussion

Human procathepsin D with an initiation codon preceding an authentic amino-terminus was expressed in *Escherichia coli* using the plasmid pTCPSD2, derived from the T7 phage promotor-driven cloning vector pET3a. The recombinant procathepsin D was identified in Coomassie-stained whole cell lysates on SDS–PAGE gels and confirmed on Western blots by anti-cathepsin D antibodies. The recombinant procathepsin D accumulated in insoluble intracytoplasmic inclusion bodies visible in electron micrographs of cells containing the pTCPSD2 construct. The insoluble inclusion bodies were isolated from lysed bacteria by centrifugation. It was determined that procathepsin D constituted greater than 95% of the total protein in the inclusion bodies.

The purpose of this work was to optimize the conditions under which the proenzyme material was solubilized, refolded and activated, in order to recover the maximum amount of active enzyme possible. All conditions were evaluated based on proteolytic activity assays of the activated enzyme using the nitro-phenylalanine (Nph)-containing synthetic substrate, Lys-Pro-Ile-Glu-Phe-Nph-Arg-Leu [2], which has an apparent K_m for native human cathepsin D of 329 μM and 298 μM for the refolded recombinant human cathepsin D.

The inclusion bodies were solubilized under strongly denaturing conditions (50 mM CAPS, 50 mM β-ME, 8 M urea, pH 10.7), then diluted to a concentration of 500 pmol/ml. As determined from a pH range of 7.20 – 10.20 and time courses up to 24 h, after dilution, it was most efficient to incubate the refolding solution for 2 h at a pH of 8.70–8.75. Ellman's sulfhydryl quantitation studies indicated that, within the first hour following dilution, all but a residual 5% of cysteine residues had formed disulfide bonds. The following millimolar ratios of reduced : oxidized glutathione were examined for their enhancement of correct folding, each at pH 8.75 and 3.50: 10:1, 5:1, 5:5, 2:1, 1:1, 1:0.1, 0.5:0.5, 0.1:1, 1:2, 1:5, 1:10. A 10:1 ratio at pH 3.50 with 3 mM cysteine present enhanced correct folding 4–5-fold. From a pH range of 3.35–4.09, the optimum pH for activation of the proenzyme was found to be between 3.50 and 3.90. However, proteolytic activity was detected no earlier than 24 h after the refolding solution had been adjusted to the acidic pH. Activity was seen to increase up to at least 96 h after the final pH adjustment.

The extensive activation times along with the observed disappearance of free sulfhydryl groups within the first hour of refolding indicated that the initial disulfide bonds formed might not be those of the active enzyme, which, according to the classic model for aspartic proteases, are believed to be Cys^{27}-Cys^{96}, Cys^{46}-Cys^{53}, Cys^{222}-Cys^{226} and Cys^{265}-Cys^{302}. The refolding process is believed to continue at the acidic pH. These statements are supported by changes in fluorescence emission spectra and HPLC analyses of tryptic digests of the procathepsin D at different times during the refolding and activation processes.

AAA and sequencing of collected HPLC peaks, combined with CD measurements of samples taken during the refolding and activation processes, should yield a clearer picture of changing disulfide bonds. Determining the optimal conditions for recovering correctly refolded, active cathepsin D is instrumental in realizing our ultimate goal of developing site-directed mutants to further our understanding of the cleavage mechanism.

References

1. Conner, G.E., Blobel, G. and Erickson, A.H., In Glaumann, H. and Ballard, F.J. (Eds.) Lysosomes – Their Role in Protein Breakdown, Academic Press Inc., London, 1987, p. 152.
2. Dunn, B.M., Jimenez, M., Parten, B.F., Valler, M.J., Rolph C.E. and Kay, J., Biochem. J., 237 (1986) 899.

In vitro and in vivo inhibition of human leukocyte elastase (HLE) by two series of electrophilic carbonyl containing peptides

M.M. Stein, R.A. Wildonger, D.A. Trainor, P.D. Edwards, Y.K. Yee, J.J. Lewis, M.A. Zottola, J.C. Williams and A.M. Strimpler

ICI Pharmaceuticals Group, A Business Unit of ICI Americas Inc., Wilmington, DE 19897, U.S.A.

Introduction

HLE is a serine protease believed to be involved in the pathogenesis of emphysema. In an attempt to find therapeutic agents for the treatment of emphysema, we have focused our efforts on discovering new, potent, selective inhibitors of HLE. We have previously reported on the activity of a number of peptides containing electrophilic carbonyl groups including aldehydes, trifluoromethylketones, difluoromethyleneketones and α-ketoamides. We report here on the synthesis and activity of peptidyl α,α-difluoro-β-ketoamides and peptidyl α-diketones, two new series of electrophilic carbonyl-containing peptides.

Results and Discussion

α-Diketones

A series of peptidyl α-diketones of general structure (1) (where R' = aryl or arylacylsulfonamido, R = aryl, alkyl) has been synthesized by two different routes both utilizing peptidyl acyloins as the penultimate intermediate. In the first route, the key step is the reaction of a peptidyl α-hydroxy-N-methyl-N-methoxy amide [1] with excess Grignard reagent to form the acyloin (Eqn. I). In the second route, a peptidyl trimethylsilylcyanohydrin was reacted with excess Grignard reagent [2] to form the same acyloin (Eqn. I). Oxidation of the acyloin to the α-diketone is best accomplished via a Swern oxidation.

369

These peptidyl α-diketones are potent, competitive, and reversible inhibitors of HLE in vitro. For example, for compound (**1a**) (R′ = Cbz, R = Et) the $K_{i(HLE)} = 2.4$ nM. They have comparable activity to a series of related peptides containing trifluoromethylketones ($K_{i(HLE)} = 1.9$ nM for (**2**) if R′ = Cbz).

These peptidyl α-diketones also inhibit HLE in vivo in our acute hemorrhagic model of lung injury. Compound (**1b**) (R′ = $C_6H_6SO_2NH$-terephthaloyl, R = *n*-Bu) has a significant retention time in the lung and protects the lung from HLE damage when the inhibitor is predosed intratracheally up to 6 h prior to the enzyme.

α,α-Difluoro-β-ketoamides

A series of α,α-difluoro-β-ketoamides of general structure (**3**) (where R′ = aralkoxy or arylacylsulfonamido, R = alkyl or aralkyl) has been synthesized and evaluated as inhibitors of HLE. In vitro these compounds are potent, selective, reversible and competitive inhibitors of HLE.

The potency of these compounds in vitro increases as the length of the peptidyl portion is increased from a dipeptide to a tripeptide. Structural variations to the R′ group are well tolerated and arylacylsulfonamido is preferred. Structural manipulation of R can effect potency by more than one order of magnitude with an aralkyl side-chain preferred.

The most potent HLE inhibitor of the series (**4**) ($K_{i(HLE)} = 0.04$ nM) was tested versus a wide variety of proteases and has exceptional selectivity for HLE.

This compound (**4**) also inhibits HLE in vivo in our acute hemorrhagic model of lung injury when predosed intratracheally up to 24 h prior to the enzyme.

Compound (**4**) is prepared using standard peptide coupling and protecting agents. The synthesis of fragment (**C**) involves a key Reformatskii reaction between Cbz-Valinal and ethyl bromodifluoroacetate. After displacement of the ester to form the amide and removal of the protecting group, one obtains the β-hydroxyde amide as a single crystalline diastereomer. The (**A,B**) fragment is obtained as a crystalline calcium salt that is neutralized and coupled to fragment (**C**). In the final step, the alcohol is oxidized via the Pfitzner-Moffat reaction to obtain the final crystalline ketone.

References

1.   Nahm, S. and Weinreb, S.M., Tetrahedron Lett., 22 (1981) 3815.
2.   Krepski, L.R., Heilmann, S.M. and Rasmussen, J.K., Tetrahedron Lett., 24 (1983) 4075.

Phosphinates as transition-state analog inhibitors of thermolysin: The importance of hydrophobic and hydrogen bonding effects

Bradley P. Morgan and Paul A. Bartlett

Department of Chemistry, University of California, Berkeley, CA 94720, U.S.A.

Introduction

Substitution of the phosphonamide NH with an ester oxygen in a series of identically bound [1] thermolysin inhibitors decreased the binding affinity by 4 kcal/mol (Table 1) [2]. This reduction was attributed to differences in H-bonding, solvation, and metal coordination between the inhibitors, an interpretation subsequently supported by a computational study [3]. The related series of phosphinate analogs, **1**, has been prepared to evaluate further the role of H-bonding and hydrophobic effects, as well as validate an independent computational study.

Table 1 *Tetrahedral phosphorus inhibitors of thermolysin*

1: Y = Ch$_2$ 2: Y = NH   3: Y = O

Inhib. X	K$_i$ (nM)		
	Y = CH$_2$	Y = NH	Y = O
NH$_2$	1 400	760	660 000
Gly	300	270	230 000
Phe	66	78	53 000
Ala	18.4	16.5	13 000
Leu	10.6	9.1	9 000

Results and Discussion

Across a similar series of inhibitors (Table 1), the phosphinates are bound an average of only 0.1 kcal/mol less tightly than the corresponding phosphonamidates. This trend is observed across two orders of magnitude in absolute binding affinity (Fig. 1), suggesting that the phosphinate and phosphonamidate inhibitors bind to the active site in a similar manner and, aside from those due to the NH group itself, with similar interactions to the protein.

The difference in binding energy of compounds **1**, **2**, and **3**, as shown in the thermodynamic cycle in Fig. 1, arise from steps 1 and 3. Step 1, desolvation of the inhibitor, is expected to be more difficult for the phosphonamidate than for the phosphinate, largely because of the hydrogen bonding capability of the former. Step 3, association of inhibitor and enzyme, is dependent on two major factors: the favorable interaction from the NH-enzyme hydrogen bond and the ligand strength of the phosphorus oxyanion with the enzyme-bound zinc atom. The difference in binding affinities can, therefore, be summarized by the following equation: $\Delta\Delta G = \Delta G_{solv} + \Delta G_{ligand} + \Delta G_{H\text{-}bond}$.

Assuming the differences in ligand affinities to be negligible [4], the difference in solvation energies of the two inhibitors is approximately equal to the difference of hydrogen-bonding. In other words, the magnitude of the hydrophobic effect of the phosphinate inhibitors is similar to the binding energy due to the hydrogen-bond of the phosphonamidate inhibitors.

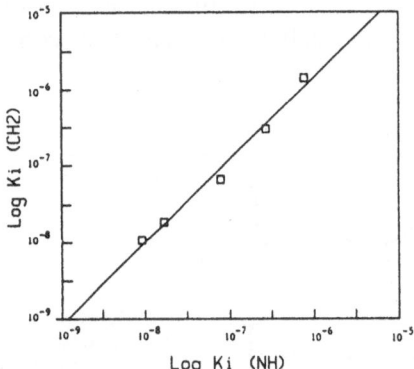

Fig. 1. *Comparison of inhibition constants for 1 and 2.*

$I_{(s)} + E_{(s)} \rightarrow EI_{(s)}$ (solution)

$1\downarrow \quad 2\downarrow \quad 4\uparrow$

$I_{(g)} + E_{(g)} \rightarrow EI_{(g)}$ (gas phase)

Step		Difference
1	Desolvation	ΔG_{solv}
2	Identical	0
3	Binding	$\Delta G_{H\text{-}bond} + \Delta G_{ligand}$
4	Similar	≈ 0

Fig. 2. *Thermodynamic Cycle.*

Acknowledgements

This research is supported by NIH grant CA-22747.

References

1. Matthews, B.W., Holden, H.M. and Tronrud, D.E., Science, 235 (1987) 571.
2. Bartlett, P.A. and Marlowe, C.K., Science, 235 (1987) 569.
3. Bash, P.A., Kollman, P.A., Singh, U.C., Brown, F.K. and Langridge, R., Science, 235 (1987) 574.
4. Wozniak, W. and Nowogrocki, G., Talanta, 26 (1979) 381.

The bisubstrate inhibitor concept: Application to protein kinases C and A

A. Ricouart[a], A. Tartar[b] and C. Sergheraert[a]

[a]*Institut Pasteur, SCBM, 1 rue Calmette, F-59019 Lille, France*
[b]*Faculté de Pharmacie, Laboratoires de Chimie Générale et Chimie Organique,*
3 rue Laguesse, F-59045 Lille, France

Introduction

Protein kinase C (PKC) is recognized as a major regulatory enzyme and has been involved in the control of a wide variety of physiological processes [1]. Different classes of compounds have been reported to inhibit PKC and some of them were shown to act as anti-tumor promotors [2–4]. Most inhibitors have been classified as interacting with the regulatory or the catalytic domain based on competitive or noncompetitive inhibition with respect to phospholipids, calcium, ATP and substrate.

Results and Discussion

With such an enzyme showing separated regulatory and catalytic domains and interacting with two substrates at the active site, a bisubstrate inhibitor can be designed that incorporates aspects of both substrates in the same molecule. Such constructs allow the taking advantage of the binding energy of each substrate and also adds an entropic contribution by having both binding moieties covalently linked in one molecule. We have developed such bisubstrates by covalently linking two competitive moieties:

(1) an isoquinoline sulfonamide (a common structure for several competitive inhibitors of the ATP binding site of different kinases: PKC, cAMP-dependent protein kinase (PKA) and cGMP-dependent protein kinase [5]); and
(2) a peptidic substrate based on the consensus recognition pattern of PKC or PKA.

As an estimate of the relative disposition of substrates in the catalytic site of PKC, we have used the X-ray structure of phosphoglycerate kinase crystallized with an ATP and the substrate molecule [6]. From these coordinates, we could calculate a 16.3 Å distance between the carbon atom bearing the phosphorylated hydroxyl and the adenine nitrogen bound to the ribose moiety (Fig. 1).

Several constructs were thus prepared by SPS. Their inhibiting potency for PKC and PKA were determined and compared to those of the separated moieties.

One of these bisubstrates, isoquinoline-SO$_2$-(βAla)$_2$-Ser-Arg^6, is a potent inhibitor of PKC ($K_i = 4$ μm). Surprisingly, isoquinoline-SO$_2$-(βAla)$_2$-Ser-Arg^6 is also a very potent inhibitor of PKA ($K_i = 170$ nM), strongly surpassing the

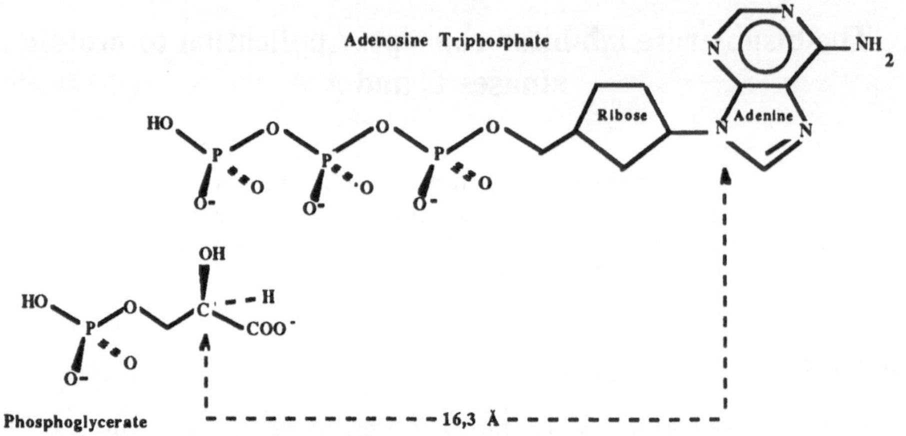

Fig. 1.   Active site of phosphoglycerate kinase: relative substrates disposition.

commercial product H7 ($K_i = 3$ μM). In contrast to the bisubstrate, each separated moiety: isoquinoline-SO_2-βAla and βAla-Ser-Arg[6] does not have significant inhibitory capacity, even when used together.

These results show that properly designed bisubstrate inhibitors can be valuable tools to attain inhibition of protein kinases in the nanomolar range.

References

1.   Nishizuka, Y., Nature, 308 (1984) 693.
2.   Huang, K.P. and Huang, F.L., Biochem. Biophys. Res. Commun., 139 (1986) 320.
3.   Lee, M.H. and Bell, R.H., J. Biol. Chem., 261 (1986) 14867.
4.   Matsui, T., Nakao, Y., Koisumi, T., Katakami, Y. and Fujita, T., Cancer Res., 46 (1986) 583.
5.   Hidaka, H., Inagaki, M., Kawamoto, S. and Sasaki, Y., Biochemistry, 23 (1984) 5036.
6.   Bryant, T.N., Watson, H.C. and Wendell, P.C., Nature, 257 (1974) 17.

Synthesis of cysteine proteinase inhibitors structurally based on the proteinase-binding center of human cystatin C

Z. Grzonka[a], R. Kasprzykowska[a], F. Kasprzykowska[a], A. Grubb[b],
M. Abrahamson[b], I. Olafsson[b] and J. Trojnar[c]

[a]Institute of Chemistry, University of Gdansk, 80-952 Gdansk, Poland
[b]Department of Clinical Chemistry, University of Lund, S-221 85 Lund, Sweden
[c]Ferring AB, P.O. Box 305, S-200 62 Malmö, Sweden

Introduction

Cysteine proteinases of microorganisms and mammalian cells participate in several important intra- en extra-cellular biological processes [1]. Several structurally related mammalian low molecular weight proteins, which inhibit the activities of cysteine proteinases, have recently been described. These natural cysteine proteinase inhibitors have been named cystatins and constitute the cystatin superfamily of proteins. Cystatin C is perhaps the physiologically most important extracellular human cystatin, and a part of the molecule involved in the interaction with the target enzymes has been localized [2]. This work describes the synthesis of peptide derivatives structurally related to the binding segment of cystatin C and their proteinase inhibitory properties.

Results and Discussion

Peptides were obtained by SPPS or in solution. Diazomethane derivatives were synthesized from the appropriate protected peptides, and diazomethane by the mixed anhydride method. Pseudopeptides containing $-CH=CH-$,

$$\overset{O}{\overset{\displaystyle\wedge}{}}$$

$-CHOH-CH_2-$, or $-CH-CH$-groups were obtained by coupling either the (ϵ)-5-aminopent-3-enoic or (RS)-3-hydroxy-5-aminovaleric acids to Z-Leu-Val-OH, and by oxidation of the olefinic precursor.

Inhibition or inactivation (for diazomethyl ketone inhibitors) rate constants were determined by the methods of Barrett et al. [3] and Nicklin and Barrett [4], respectively.

Cystatin C fragments comprising residues 4–21 (GKPPRLVGGPMDAS-VEEE), 7–16, 8–12, and 11–15, were synthesized and tested as substrates for papain. The three first were all good substrates and were all preferentially cleaved at the expected Gly^{11} - Gly^{12} bond, while the fourth was not cleaved at all. Therefore, peptide derivatives with the N-terminal peptide segment adjacent to the Gly-Gly bond of cystatin C were synthesized. The synthesized compounds were tested for inhibitory activity against papain, and four of them displayed

Table 1 *Calculated inactivation rate constants (K'_{+2}) for synthetic peptidyldiazomethylketones and inhibition constants (K_i) for reversible inhibitory derivatives*

Synthetic peptide	Papain	Streptococcal proteinase	Cathepsin B (bovine)
Boc-Val-Gly-CHN$_2$	$k'_{+2} =$ 5 000 M^{-1}s^{-1}	$k'_{+2} =$ 20 M^{-1}s^{-1}	$k'_{+2} =$ 230 M^{-1}s^{-1}
Z-Leu-Val-Gly-CHN$_2$	$k'_{+2} =$ 300 000 M^{-1}s^{-1}	$k'_{+2} =$ 102 M^{-1}s^{-1}	$k'_{+2} =$ 3 300 M^{-1}s^{-1}
Z-Leu-Val-NH-CH$_2$-CH=CH-CH$_2$-COOH	$K_i =$ 0.4 mM		
Z-Leu-Val-NH-CH$_2$-CH–CH-CH$_2$-COOH ⟍⟋ O	$K_i =$ 0.15 mM		

Thermolysin (metalloproteinase):   no inhibition.
Collagenase (metalloproteinase):   no inhibition.
Pepsin (aspartic acid proteinase):   no inhibition.
Trypsin (serine proteinase):   no inhibition.

inhibitory properties. Two of the diazomethyl ketones gave irreversible rapid inactivation of papain and were also found to inhibit irreversibly bovine cathepsin B and streptococcal proteinase. These inhibitors did not exert any inhibitory activity against the tested serine- and aspartic-, or metallo-proteinases (Table 1).

Preliminary studies demonstrate that some of the peptide derivatives are efficient inhibitors of several human cysteine proteinases, and Z-LVG-CHN$_2$ has recently been shown to block the replication of group A streptococci [1], that of the protozoan, *Entamoeba histolytica*, and that of the herpes simplex virus, probably by interfering with cysteine proteinases of fundamental importance of the replication processes. It has also been demonstrated in animal experiments that ZLVG-CHN$_2$ might be used as a life-saving antimicrobial agent in lethal infections caused by group A streptococci [1].

References

1. Björk, L., Åkesson, P., Bohus, M., Trojnar, J., Abrahamson, M., Olafsson, I., Grubb, A., Nature, 337 (1989) 385.
2. Abrahamson, M., Ritonja, A., Brown, M., Grubb, A., Machelidt, W., Barrett, A., J. Biol. Chem., 262 (1987) 9688.
3. Barrett, A.J., Kembhavi, A.A., Brown, M.A., Kirschke, H., Knight, C.G., Tamai, M., Hanada, K., Biochem. J., 201 (1982) 189.
4. Nicklin, M.J.H. and Barrett, A.J., Biochem. J., 223 (1984) 245.

Studies on the enzymatic degradation of peptides corresponding to the subunit 2 C-terminus of herpes virus ribonucleotide reductases

Pierrette Gaudreau[a], Hélène Paradis[b], Yves Langelier[b] and Paul Brazeau[a]

[a]Notre-Dame Hospital Research Center and [b]Montreal Cancer Institute, Notre-Dame Hospital, 1560 Sherbrooke St. East, Montreal, Quebec, Canada H2L 4M1

Introduction

HSV R2-(329–337) corresponds to the subunit 2 C-terminus of herpes virus ribonucleotide reductase (RR) and VZV R2-(298–306) to that of varicella-zoster virus RR. These two peptides as well as Ac-HSV R2-(329–337) and HSV R2-(332–337), but not [Ala^{336}]HSV R2-(329–337), specifically inhibit in vitro viral RR activity [1]. In this report, we present data on their proteolysis, using quantitative RPHPLC.

Results and Discussion

Under standard RR assay conditions (40 nmol of peptide and 250 μg of HSV-1 partially purified RR protein extract from BHK-21/C13-infected cells in 50 mM HEPES, pH 7.8, containing 4 mM NaF, 30 mM DTT and 50 μM CDP, 37°C) HSV R2-(329–337) was rapidly degraded, since only $17 \pm 1\%$ of the initial concentration remained after 30 min of incubation. The proteolytic resistances of VZV R2-(298–306), Ac-HSV-R2-(329–337), HSV R2-(332–337) and [Ala^{336}]-HSV R2-(329–337) relative to that of HSV R2-(329–337) (100%) were, respectively, 350%, 150%, 30% and 80% under conditions of linear degradation (100 nmol of peptide and 60 μg of protein extract). The beneficial effect of N^α-acetylation suggests that aminopeptidases play an important role in HSV R2-(329–337) processing. Isolation and characterization of this peptide's metabolites revealed that it was transformed into HSV R2-(330–337). This octapeptide, which possesses 13% of the nonapeptide inhibitory potency (IP) was further reduced to the heptapeptide 331–337 (IP, 16%). These results suggest that HSV R2-(329–337) assumes, in the RR assay conditions, a conformation that protects it against other peptidase activities. Hence, structure modifications such as amino-terminus shortening, internal monosubstitution (Ala^{336}) or disubstitutions (Thr^{332} with Ile^{334}) could either destabilize or stabilize such a conformation. Since degradation patterns of HSV R2-(329–337) fragments and analogs have not been elucidated, we cannot completely exclude the participation of other proteolytic activities in their degradation. The residual concentrations of HSV H2-(329–337), in this RR assay were, respectively, $49 \pm 1\%$, $44 \pm 2\%$ and $95 \pm 6\%$ when 2.0 mM amastatin, 2.0 mM bestatin (two aminopeptidase inhibitors) or 2.5 mM bacitracin (a broad spectrum inhibitor) were added to the incubation medium. Altogether these findings provide a rationale for the independent study of the affinity of RR inhibitory peptides and their metabolic resistance. They also suggest structure modifications that may improve HSV R2-(329–337) efficacy, Ac-Tyr-Ala-Gly-Thr-Val-Ile-Asn-Asp-Leu-OH being totally resistant to proteolysis by RR protein extracts.

References

1. Gaudreau, P., Michaud, J., Cohen, E.A., Langelier, Y. and Brazeau, P., J. Biol. Chem., 226 (1987) 12413.

Inhibitors of tissue kallikrein based on the structure of the porcine pancreatic kallikrein-aprotinin complex

Milind S. Deshpande[a], John Boylan[b] and James Hamilton[b]

[a]Laboratory of Rational Drug Design, Department of Medicine, University Hospital, Boston, MA 02118, U.S.A.
[b]Biophysics Institute, Boston University School of Medicine, Boston, MA 02118, U.S.A.

Introduction

The biologic role of tissue kallikrein is poorly understood. In vitro, this enzyme produces the hormone lysyl-bradykinin from kininogen [1] and other polypeptide hormones, such as insulin and atriopeptin, from their higher molecular weight precursors. Kinins are involved in blood pressure regulation, in perception of pain, and in the manifestation of rhinitis. Use of specific kallikrein inhibitors offers a means of determining which of the active kinins are produced by tissue kallikrein. Substrate analog inhibitors, designed around the Arg-Ser cleavage site of low molecular weight kininogen, have been shown to specifically inhibit tissue kallikrein [2]. The substrate analog inhibitors have K_i values in the μmolar range. In order to increase the affinity of the inhibitors, we have focused on synthesizing inhibitors based on the amino acid sequence of aprotinin, a potent inhibitor of tissue kallikrein. The crystal structure of porcine pancreatic kallikrein-aprotinin complex is known, thus facilitating the design of inhibitors.

Results and Discussion

Peptides were synthesized by SPPS, gel filtered on a Sephadex G-15 column, purified by RPHPLC, and characterized by TLC and AAA. Inhibition constants were determined in a chromogenic assay at 37°C by using 0.25 units of porcine pancreatic kallikrein (PPK) and D-Val-Leu-Arg-pNA (0.1 mM or 0.2 mM). Kinetic data was analyzed by Dixon plots. NMR spectra in D_2O were recorded on a Brucker WP-200 spectrometer. Two-dimensional experiments in D_2O included COSY, RELAY and NOESY. Delays of 25 ms were used between mixing pulses in RELAY experiments, and 50–300 ms mixing delays were used in NOESY experiments. For samples in H_2O, the spectra were recorded on a Brucker WM-400 spectrometer. Two-dimensional spectra included 2QF-COSY and NOESY. A mixing time of 100 ms was used in the NOESY experiments.

Table 1 lists the peptides synthesized and their K_i values for inhibiting PPK. The residue sequences are based on the amino acid sequence of aprotinin that is in contact with the active site of PPK. Note that α-aminobutyric acid was substitued for cysteine. A minimum sequence of five amino acids, spanning residues 12–16 of aprotinin, is necessary for inhibition of PPK. Addition of

Table 1 *Inhibition of porcine pancreatic kallikrein by peptides homologous with the binding segment of aprotinin*

Inhibitor	Sequence	K_i ($10^{-4}M$)
KKI-40	Ac-GP-NH$_2$	NI
KKI-39	Ac-GPX-NH$_2$	NI
KKI-38	Ac-GPXK-NH$_2$	NI
KKI-37	Ac-GPXKA-NH$_2$	2.3
KKI-36	Ac-GPXKAR-NH$_2$	2.3
KKI-35	Ac-GPXKARI-NH$_2$	5.0
KKI-34	Ac-GPXKARII-NH$_2$	1.2

Single letter abbreviations for amino acids are used: NI = no inhibition; X = 2 α-amino-butyric acid.

residues 17 and 18 results in a poorer inhibitor, while addition of residue 19 increases inhibition. Since the octapeptide, KKI-34, inhibits PPK, its solution structure was studied by one- and two-dimensional NMR methods for comparison with the known structure of the segment in aprotinin that contacts PPK. Since all chemical shifts could not be assigned from one-dimensional spectra, two-dimensional experiments (COSY, 2QF, and RELAY) were performed to aid in spectral assignments. For the purpose of assigning chemical shifts to the amide protons, one-dimensional and 2QF-COSY spectra were obtained in 90% H$_2$O at 400 MHz. In the 2QF-COSY spectrum of KKI-34, six clearly resolved cross-peaks were observed, corresponding to each of the seven expected NH-CαH couplings. Amide proton is absent in proline, and the amide protons of the two isoleucine co-resonate. Thus, only six cross-peaks were observed. The observed vicinal coupling constants (^{3}JNH-CαH) measured from one-dimensional spectra in H$_2$O were all between 5.5 and 8.0 Hz.

The range of observed coupling constants, and the absence of significant NOE interactions suggest that KKI-34 exists as a random coil, or it rapidly samples several restricted conformations leading to averaging of the observed coupling constants on the NMR time scale. Thus, inhibition by the octapeptide occurs because of its homology with residues 12–19 of aprotinin. Moreover, the absence of a stable solution conformation, similar to the binding segment of aprotinin, may explain the 150 000-fold increase in the K_i of KKI-34 compared to that of aprotinin.

References

1. Levinsky, N.G., Circ. Res., 44 (1979) 441.
2. Okunishi, H., Spragg, J. and Burton, J., Hypertension, 8 (1986) I114.

Synthesis of eglin c and related peptides and studies on the relationship between their structure and inhibitory effects on human leukocyte elastase and cathepsin G

Satoshi Tsuboi[a], Yuko Tsuda[a], Kazunori Nakabayashi[a], Yoshio Okada[a], Yoko Nagamatsu[b] and Junichiro Yamamoto[b]

[a]Faculty of Pharmaceutical Sciences and [b]Faculty of Nutrition, Kobe-Gakuin University, Nishi-ku, Kobe 673, Japan

Introduction

Eglin c, isolated from the leech *Hirudo medicinalis* [1], consists of 70 amino acid residues (Fig. 1) and effectively inhibits chymotrypsin and subtilisin, as well as leukocyte elastase and cathepsin G. It was revealed that eglin c present in the complex with subtilisin was shortened N-terminally by 7 amino acid residues [2]. This observation indicates that eglin c derivatives, shortened from the N-terminus by up to 7 amino acid residues, should be active as an inhibitor.

This paper deals with the synthesis of eglin c (8–70) and various peptide fragments related to eglin c, and the relationship between their structure and inhibitory activity against leukocyte elastase, cathepsin G and α-chymotrypsin.

Results and Discussion

Peptides (I–XIII) shown in Fig. 1 were prepared by the conventional solution method and their inhibitory effects on leukocyte elastase, cathepsin G and α-chymotrypsin were examined. The results are summarized in Table 1 with K_i values. Peptide (31–40) [V] inhibited cathepsin G ($K_i = 2.3 \times 10^{-4}$ M). Peptide (41–49) [VI] inhibited cathepsin G and α-chymotrypsin ($K_i = 4.0 \times 10^{-5}$ and 2.0×10^{-5} M, respectively), but not leukocyte elastase; while peptide (60–63) [IX]

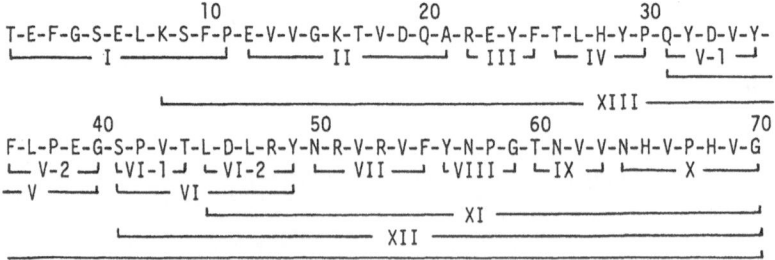

Fig. 1. Primary structure of eglin c and related peptides [I–XIII].

inhibited leukocyte elastase ($K_i = 1.6 \times 10^{-4}$ M) but not cathepsin G and α-chymotrypsin, indicating that the interacting site of eglin c with leukocyte elastase and cathepsin G and α-chymotrypsin might be different.

Table 1 K_i values of eglin c fragments and N^α-Ac-eglin c^a

	K_i (M)		
	Elastase	Cathepsin G	α-Chymotrypsin
H-(22–25)-OMe·2AcOH [III]	$-^b$	8.0×10^{-4}	7.0×10^{-5}
H-(31–35)-OMe·TFA [V-1]	2.1×10^{-3}	7.3×10^{-4}	–
H-(31–40)-OMe·TFA [V]	–	2.3×10^{-4}	–
H-(41–49)-OMe·2AcOH [VI]	–	4.0×10^{-5}	2.0×10^{-5}
H-(60–63)-OMe·HCl [IX]	1.6×10^{-4}	–	–
H-(45–70)-OH [XI]	5.6×10^{-5}	–	ND^c
H-(41–70)-OH [XII]	–	7.5×10^{-5}	ND
H-(8–70)-OH·9AcOH [XIII]	2.7×10^{-8}	1.2×10^{-7}	8.4×10^{-9}
N^α-Ac.eglin c	5.1×10^{-9}	1.5×10^{-8}	2.2×10^{-9}

[a] Substrate; Suc-Ala-Tyr-Leu-Val-pNA ($K_m = 0.13$ mM) for elastase, Suc-Ile-Pro-Phe-pNA ($K_m = 0.7$ and 0.05 mM) for cathepsin G and α-chymotrypsin.
[b] Not detectable.
[c] Not determined.

Previously, Bode et al. [2] reported that the nine residues of the binding loop (40–48) of eglin c are involved in direct contact with subtilisin, and the conformation of this loop is maintained by formation of electrostatic and hydrogen bonds with the C-terminal part of eglin c. XI inhibited leukocyte elastase with a K_i value of 5.6×10^{-5} M, but not cathepsin G; XII inhibited cathepsin G with a K_i value of 7.5×10^{-5} M, but not leukocyte elastase, indicating that the peptide part (41–49) in eglin c is very important for manifestation of inhibitory activity against cathepsin G; and the peptide part (41–44) in XII might block the access of the peptide part (60–63) to the active site of leukocyte elastase. It is also concluded that in XII, electrostatic and hydrogen bonds to maintain the rigid conformation of the peptide part (41–49) could not be formed.

Next, eglin c (8–70) [XIII] was prepared and its inhibitory effect on the enzymes was examined. XIII exhibited very potent inhibitory activity against leukocyte elastase, cathepsin G and α-chymotrypsin.

References

1. Seemüller, U., Meier, M., Ohlsson, K., Müller, H.P. and Fritz, H., Hoppe-Seyler's Z. Physiol. Chem., 358 (1977) 1105.
2. Bode, W., Papamokas, E., Musil, D., Seemüller, U. and Fritz, H., EMBO J., 5 (1986) 813.

Determinations of disulfide pairings and biological activities of synthetic cholecystokinin-releasing peptide (monitor peptide)

Yao-Zhong Lin, Daniel D. Isaac, Xiao-Hong Ke, Cathy Volin and James P. Tam

The Rockefeller University, 1230 York Avenue, New York, NY 10021, U.S.A.

Introduction

A 61-residue trypsin-sensitive cholecystokinin-releasing peptide (monitor peptide), purified from rat pancreatic juice on the basis of its stimulatory activity towards pancreatic enzyme secretion, was recently reported to inhibit bovine trypsin and possess epidermal growth factor (EGF) activities at a concentration of 10 nM [1,2]. This monitor peptide shares about 45% sequence homology with pancreatic secretory trypsin inhibitors (PSTIs; Kazal-type) but displays very limited sequence homology with the EGF and lacks the characteristic cysteinyl alignments of the EGF family. To confirm whether monitor peptide possesses putative EGF-like growth factor activity, we have chemically synthesized the

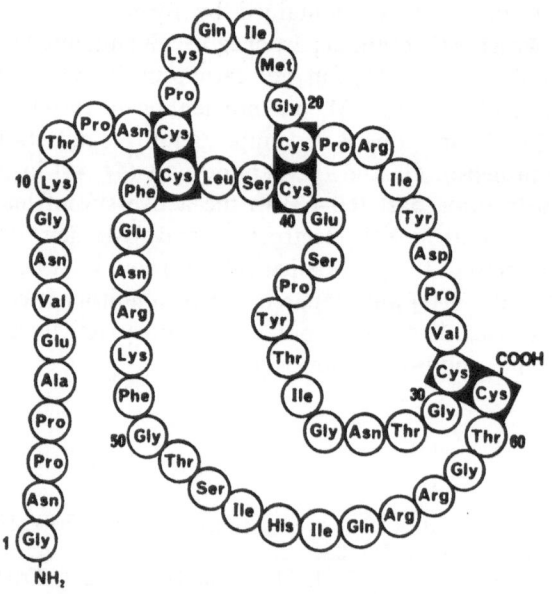

Fig. 1. *Structure of monitor peptide.*

61-residue peptide. Unlike the monitor peptide reported previously [1,2], the synthetic monitor peptide, structurally determined unambiguously by MS and thermolysin digestion, did not possess any putative EGF-like growth factor activity. However, we found that the monitor peptide is a trypsin inhibitor and residue Arg^{23} is crucial to the inhibitory activity.

Results and Discussion

Two peptides including monitor peptide (Fig. 1) and [Ala^{23},Ala^{47}]-monitor peptide were synthesized by a stepwise solid-phase method [3]. The reduced peptides after synthesis were oxidized in a combination of reduced and oxidized glutathiones [4,5]. The reaction mixture was easily purified to homogeneity by C$_{18}$ RPHPLC in about 10% overall yield.

The integrality of the purified peptides were characterized by AAA and Cf-252 fission fragment MS. Enzymatic digestion with thermolysin produced fragments corresponding to the disulfide pairings of Cys^{14}-Cys^{43} and Cys^{21}-Cys^{40}. These results suggest that the disulfide pairings of synthetic monitor peptide are similar to those of Kazal-type trypsin inhibitors [6] (Fig. 1). Trypsin inhibitory activity was determined by bovine trypsin at 2 µg/ml with the purified monitor peptide or its analog in the presence of substrate, benzoyl arginine *p*-nitroanilide. Synthetic monitor peptide showed full trypsin inhibition activity similar to that of the native product (Fig. 2). However, the analog, [Ala^{23},Ala^{47}]-monitor peptide, did not exhibit trypsin inhibition activity (Fig. 2). The reactive site P$_1$-P$'_1$ (residues 18–19) of pancreatic secretory trypsin inhibitor (Kazal-type) [7] was reported

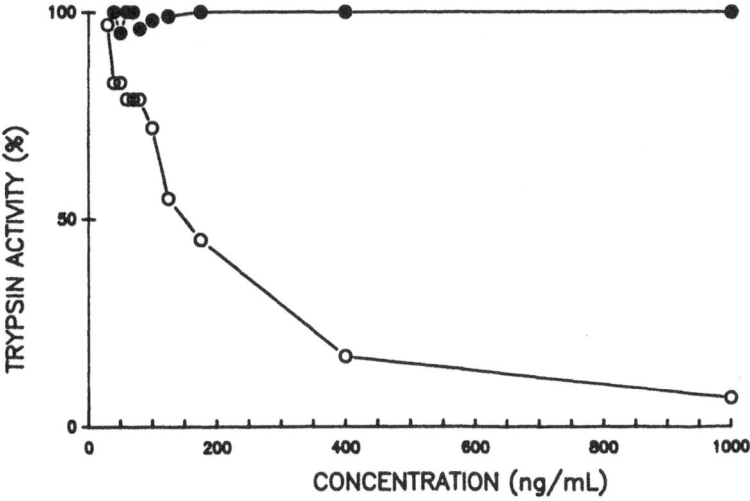

Fig. 2. *Inhibition of bovine trypsin by the monitor peptide (o) and [Ala^{23}, Ala^{47}]-monitor peptide (•).*

385

to be Lys-Ile for human and porcine PSTIs; Arg-Ile for bovine and ovine inhibitors. The Arg-Ile pair in the P_1-P'_1 site is conserved in the monitor peptide (residues 23–24). Thus, our results suggest that the Arg^{23} is required for the inhibitory activity. Inhibition of [^{125}I]EGF binding to the EGF-receptor was examined by subconfluent monolayer of formalin-fixed A431 cells after 1 h incubation at 22 °C with synthetic peptides. Incorporation of [^{3}H]thymidine in the mitogenic assay was measured in normal rat kidney fibroblasts, clone 49F or Swiss 3T3 cells. The synthetic monitor and its analog did not inhibit [^{125}I]EGF binding to A431 cells up to 0.1 μM concentration and showed no growth stimulation effect on both NRK 49F and Swiss 3T3 cells in the mitogenic assays. These results suggest that the monitor peptide does not possess any putative EGF-like growth factor activity.

Acknowledgements

This work was supported by USPHS Grant CA-36544, awarded by the National Cancer Institute, DHHA.

References

1. Iwai, K., Fukuoka, S.-L., Fushiki, T., Tsujikawa, M., Hirose, M., Tsunasawa, S. and Sakiyama, F., J. Biol. Chem., 262 (1987) 8956.
2. Fukuoka, S.-L., Fushiki, T., Kitagawa, Y., Sugimoto, E. and Iwai, K., Biochem. Biophys. Res. Commun., 145 (1987) 646.
3. Merrifield, R.B., J. Am. Chem. Soc., 85 (1963) 2149.
4. Lin, Y.-Z., Caporaso, G., Chang, P.-Y., Ke, X.-H. and Tam, J.P., Biochemistry, 27 (1988) 5640.
5. Tam, J.P., Sheikh, M.A., Salomon, D.S. and Ossowski, L., Proc. Natl. Acad. Sci. U.S.A., 83 (1986) 8082.
6. Guy, O., Shapanka, R. and Greene, L.J., J. Biol. Chem., 246 (1971) 7740.
7. Bartelt, D.C., Shapanka, R. and Greene, L.J., Arch. Biochem. Biophys., 179 (1977) 189.

Comparative molecular modeling analyses of endothiapepsin complexes as renin model templates

Elizabeth A. Lunney and Christine C. Humblet

Computer Assisted Drug Design Group, Parke-Davis Pharmaceutical Research Division, Warner-Lambert Company, 2800 Plymouth Road, Ann Arbor, MI 48105, U.S.A.

Introduction

Endothiapepsin, a fungal, aspartic protease enzyme homologous to human renin, presents a high level of residue conservation in the active site region [1]. In the absence of a 3D structure for human renin, endothiapepsin provides a tentative, but useful template for the elaboration of a human renin model.

Results and Discussion

The X-ray structures of various inhibitors cocrystallized with endothiapepsin [2], including PD 125967 and PD 126664 (Blundell, T.L. et al., unpublished results), were studied. The inhibitors contain a non-amide moiety that binds at the cleavage site. In addition, key hydrogen bonds anchor each backbone in the bound conformation [2].

The X-ray data provide information regarding steric allowances of the bound inhibitors. The crystal structure of PD 125967 bound to endothiapepsin shows the expanse of the P_3 and P_4 pockets, while both PD 125967 and PD 126664 illustrate how the cyclohexyl moiety increases the steric occupation of the P_1 pocket. The availability of multiple binding pockets is observed in the X-ray data, and illustrated in comparing the bound conformations of H-142C39 and L-363,564 [2].

A series of proposed renin inhibitors containing an isostere at the P_2–P_3 junction (Table 1) was synthesized at Parke-Davis [3]. Based upon data obtained from the X-ray crystal structures and the docking of compounds* from this series into the human renin model [1], we can explain the inhibitory potency effects of various substituents present in the compounds. One of the key interactions between the enzyme and the inhibitors is the hydrogen bond involving the P_3 carbonyl. The most potent isostere analogs have a hydrogen bond acceptor at the P_3 site. With **1**, the oxygen of the hydroxyethylene isostere can form a bifurcated hydrogen bond with the enzyme. This interaction, along with the presence of the ACHPA residue providing strong van der Waals interaction, supports the observation of increased potency. Compounds **5**, containing a double

---

*The modeling work was carried out using the Sybyl software package (Version 5.2, February 1989), Tripos Associates, Inc.

bond with Z geometry, and **7**, containing an epoxide derived from **5**, are both inactive with IC_{50} values of less than 10^{-4}. However, **6**, containing the double bond with E geometry, has activity in the μmolar range. Compounds with the E double bond can bind in the cleft in the extended conformation with the side chains at the P_2 and P_3 positions alternating on either side of the backbone. This is not possible with the Z double bond geometry.

Table 1 *Isostere analogs (3)*

Compound	Structure	IC_{50} (μM) renin
1	P_4 P_3 P_2 P_1–P_1' P_2' P_3' Boc-Phe[CHOHCH$_2$]Gly-Achpa-Leu-NHCH$_2$—⬡—CH$_2$NH$_2$	0.022
5	Boc-Phe[Z – CH=CH]Gly--Sta--Leu-NHCH$_2$Ph	> 100
6	Boc-Phe[E – CH=CH]Gly--Sta--Leu-NHCH$_2$Ph	7.0
7	Boc-Phe[CH—CH]-Gly--Sta--Leu--NHCH$_2$Ph	> 100

Although the majority of the compounds in this series have no side chain at the P_2 site, μmolar activity is often observed. In analyzing the crystal structure of endothiapepsin in the absence of an inhibitor and filled with water molecules, virtually no bound water molecule is found in the general area where the P_2 side chain of an inhibitor would be positioned. Therefore, P_2 side chains do not seem critical, and compounds with P_2 glycine or P_2 glycine derivatives can be potent renin inhibitors.

Conclusion

The human renin model derived from the crystal structure of endothiapepsin, combined with the X-ray data of various endothiapepsin-bound inhibitors [5], provide experimental data that are useful in describing and analyzing hydrogen bonding patterns, steric requirements, van der Waals interactions and multiple modes of binding.

Acknowledgements

We acknowledge Drs. Tom Blundell and Jon Cooper for their contribution to the present work.

References

1. Blundell, T.L., Sibanda, B.L., Hemmings, A., Foundling, S.F., Tickle, I.J., Pearl, L.H. and Wood, S.P., In Burgen, A.S.V., Roberts, G.C.K. and Tute, M.S. (Eds.)

Molecular Graphics and Drug Design, Elsevier Science Publishers, Amsterdam, 1986, p. 324.
2.   Blundell, T.L., Cooper, J. and Foundling, S.I., Biochemistry, 26 (1987) 5585.
3.   Kaltenbronn, J.S., Hudspeth, J.P., Lunney, E.A., Nicolaides, E.D., Repine, J.T., Roark, W.H., Stier, M.A., Tinney, F.J., Woo, P.K.W. and Essenburg, A.D., J. Med. Chem., in press.

Angiotensin-converting enzyme-like gastrin-degrading activity in fundic mucosal cells with release of the C-terminal dipeptide

Philibert Dubreuil[a], Marc Rodriguez[a], Pierre Fulcrand[a], Hélène Fulcrand[a], Jeanine Laur[a], Jean-Pierre Bali[b] and Jean Martinez[a]

[a]Centre CNRS-INSERM de Pharmacologie-Endocrinologie, rue de la Cardonille, F-34094 Montpellier Cedex, France
[b]Faculté de Pharmacie, 15 avenue C. Flahaut, F-34060 Montpellier, France

Introduction

We have reported that the tetragastrin analog, Boc-Trp-Leu-Asp-Phe-NH$_2$, was degraded by a vesicular membrane fraction from rat gastric mucosa, yielding the two dipeptides, Boc-Trp-Leu and Asp-Phe-NH$_2$ [1]. We have recently shown that various gastrin analogs, including Boc-Trp-Leu-Asp-Phe-NH$_2$, were hydrolyzed in vitro by angiotensin-converting enzyme (ACE) from rabbit lung (Sigma), with release of the C-terminal dipeptide Asp-Phe-NH$_2$ [2]. We report here on the degradation of gastrin analogs by a gastric mucosal cell preparation from rabbit, enriched in parietal cells ($\approx 50\%$) and containing gastrin receptors.

Results and Discussion

Rabbit (or rat, dog) fundic mucosal cells (5 millions/ml) were prepared according to the collagenase/EDTA procedure [3,4]. They were incubated at 37°C, at different times, with the tetragastrin analogs Boc-Trp-Leu-Asp-Phe-NH$_2$, Boc-Gly-Trp-Leu-Asp-Phe-NH$_2$, and Boc-βAla-Trp-Leu-Asp-Phe-NH$_2$. As shown in Fig. 1A, degradation is time-dependent, the gastrin analogs being significantly degraded within the first 7 min, degradation being almost complete after 1 h incubation. The hydrolysis appeared to be also pH- and temperature-dependent, maximal enzyme activity occurring at pH 7.4 and 37°C. Degradation of these analogs produced the dipeptide Asp-Phe-NH$_2$ as the main metabolite, which was identified by HPLC. CCK-8 was also hydrolyzed by the same cell preparation, with release of the dipeptide Asp-Phe-NH$_2$. Hydrolysis of the gastrin analogs by fundic mucosal cells from different species (rabbit, rat or dog) was inhibited by metalloprotease inhibitors, such as EDTA and 1–10-phenantroline, and more efficiently and specifically by the ACE inhibitor captopril (IC$_{50} \approx 50$ nM) (Fig. 1B). Other specific enzyme inhibitors, such as phosphoramidon, thiorphan, bestatin, amastatin, p-chloromercuribenzoate (PCMB), pepstatin A and phenylmethylsulfonyl fluoride (PMSF) were not very efficient. Interestingly, gastrin analogs synthesized in our laboratory [5–9], and acting as antagonists,

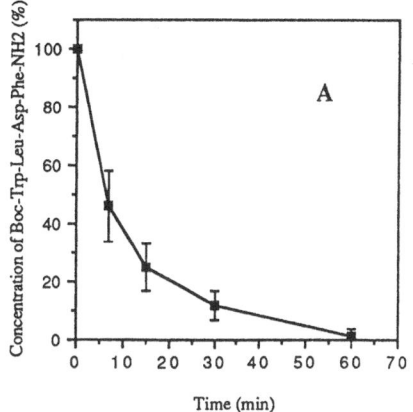
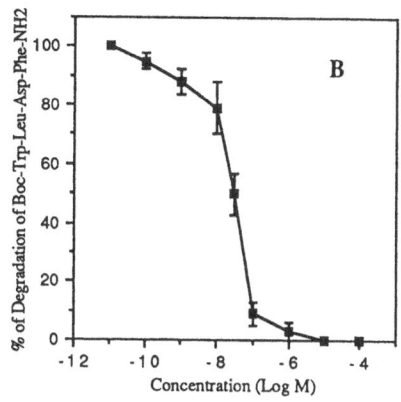

Fig. 1. (A) Degradation of Boc-Trp-Leu-Asp-Phe-NH$_2$ (0.1 mM) by a gastric mucosal cell preparation (5 millions/ml) from rabbit stomach. Incubations were performed at 37°C, pH 7.4; Concentration of Boc-Trp-Leu-Asp-Phe-NH$_2$ was evaluated by HPLC (detection at 279 nm). Each value (± SEM) represents results of at least 5 separate experiments. (B) Effects of captopril on degradation of Boc-Trp-Leu-Asp-Phe-NH$_2$ after 1 h incubation in the same conditions. Each value (± SEM) represents results of at least 5 separate experiments.

such as Boc-Trp-Leuψ(CH$_2$NH)Asp-Phe-NH$_2$, Boc-Trp-Leu-Asp-2-phenylethyl-amide, Boc-Trp-Leu-Asp-2-phenylethylester, Boc-Trp-Leu-Asp-NH$_2$, Boc-Trp-Leuψ(NHCO)Asp-Phe-NH$_2$, Boc-Trp-βhomo-Leu-Asp-Phe-NH$_2$, and Boc-Trp-Gly-Asp-Phe-NH$_2$, remained completely unaffected when incubated with fundic mucosal cells under the same conditions. On the other hand, comparison of hydrolysis of various gastrin analogs by a rabbit fundic mucosal cell preparation with their ability to bind to gastrin receptor, showed a close correlation between the extent of cleavage and the apparent binding affinity. However, correlation between the gastrin receptor and the enzymatic cleavage, as well as the participation of the enzymatic system in the mechanism of action of gastrin remains to be determined. Nevertheless, the existence of an ACE-like degrading enzyme, able to hydrolyze tetragastrin analogs by release of the amidated C-terminal dipeptide in the fundic mucosal cells of various species, has clearly been demonstrated.

References

1. Dubreuil P., Lignon, M.F., Magous, R., Rodriguez, M., Bali, J.P. and Martinez, J., Drug Design and Delivery, 2 (1987) 49.
2. Dubreuil, P., Fulcrand, P., Rodriguez, M., Fulcrand, H., Laur, J. and Martinez, J., Biochem. J., (1989) in press.
3. Soll, A.H., J. Clin. Invest., 61 (1978) 370.
4. Magous, R. and Bali, J.P., Eur. J. Pharmacol., 82 (1982) 47.

5. Martinez, J., Bali, J.P., Magous, R., Laur, J., Lignon, M.F., Briet, C., Nisato, D. and Castro, B., J. Med. Chem., 28 (1985) 273.
6. Martinez, J., Bali, J.P., Rodriguez, M., Castro, B., Magous, R., Laur, J. and Lignon, M., J. Med. Chem., 28 (1985) 1874.
7. Martinez, J., Rodriguez, Bali, J.P., and Laur, J., J. Med. Chem., 29 (1986) 2201.
8. Rodriguez, M., Dubreuil, P., Bali, J. P. and Martinez, J., J. Med. Chem., 30 (1987) 758.
9. Rodriguez, M., Fulcrand, P., Laur, J., Aumelas, A., Bali, J. P. and Martinez, J., J. Med. Chem., 32 (1989) 522.

Design of rigid heterocyclic phenylalanine replacements for incorporation into renin inhibitors

Dale J. Kempf, Stephen L. Condon, Jacob J. Plattner, Herman H. Stein, Jerome Cohen, Ed de Lara, William R. Baker and Hollis D. Kleinert

Pharmaceutical Products Division, Abbott Laboratories, Abbott Park, IL 60064, U.S.A.

Introduction

The incorporation of rigid analogs of amino acids into peptides with potential therapeutic importance can provide information of both theoretical and practical significance. In the course of our studies directed toward the discovery of novel inhibitors of human renin, we have investigated a number of rigid phenylalanine analogs in order to probe the active conformation of the P_3 side chain. Recently, we reported a potent, proteolytically stable renin inhibitor in which penylalanine was replaced by (Z)-dehydrophenylalanine [1]. Here we describe our further efforts in this area.

Results and Discussion

The structures and inhibitory potencies of the rigidified inhibitors against purified human renin (pH 6), expressed as IC_{50} values, are shown in Table 1. Compounds 1–7, containing dehydro-Phe at the P_3 position with a variety of N-blocking groups, are nearly comparable in activity to corresponding Phe-containing inhibitors [2], indicating that the position occupied by the side chain of dehydro-Phe resembles that preferred for the natural substrate.

In an effort to achieve added rigidity, we eliminated the N-blocking group of 1 and attached the nitrogen of the dehydro-Phe to the ortho-position of the phenyl group. The resulting bicyclic inhibitor (8) retains nanomolar potency with significantly reduced molecular weight [3]. SAR of other bicyclic inhibitors (9–13) indicated that the indole-2-carbonyl group of 8 is optimal for binding to the enzyme.

We next wondered whether further rigidification of 8 might be feasible by cyclizing from the 3-position of the indole ring to the nitrogen of the P_2 amino acid, since it had previously been established that the P_2 NH was not involved in hydrogen bonding to the enzyme [1]. The inhibitory potencies of the resulting tricyclic compounds 14–21 are given in Table 1. Comparison of 14 to 20 and 15 to 21 shows that the pyrrolidine ring of 14 apparently directs the phenylring to a more favorable position for lipophilic binding than does the piperidine ring of 20. Interestingly, while other nitrogen heterocycles can replace imidazole in the P_2 site in this series (16 and 17), substitution with branched lipophilic

Table 1 *Rigid inhibitors of human renin*

RCONH ... His-R' Ph		RCO-His-R'	...COR'
1–7		**8–13**	**14–21**

R' = (1-cyclohexyl-3(R),4(S)-dihydroxy-6-methylhept-2(S)-yl)amino

No.	R	n	IC_{50} (nM)
1	CH_3	–	2.4
2	$Z\text{-}NHC(CH_3)_2CH_2$	–	3.3
3	$H_2NC(CH_3)_2CH_2$	–	2.9
4	$Z\text{-}NH(CH_2)_5$	–	1.3
5	$(CH_3)_2N$	–	1.2
6	$(C_2H_5)_2N$	–	1.6
7	morpholin-4-yl	–	0.6
8	indol-2-yl	–	4
9	indol-3-yl	–	1 500
10	1-methylindol-2-yl	–	160
11	benzofuran-2-yl	–	10
12	bezothiazol-2-yl	–	32
13	3-methylinden-2-yl	–	200
14	$imidazol\text{-}4\text{-}yl\text{-}CH_2$	1	11
15	$CH_3(CH_2)_2CH_2$	1	20
16	$pyrazol\text{-}3\text{-}yl\text{-}CH_2$	1	10
17	$thiazol\text{-}4\text{-}yl\text{-}CH_2$	1	8
18	$thien\text{-}2\text{-}yl\text{-}CH_2$	1	68
19	$(CH_3)_2CHCH_2$	1	>100
20	$imidazol\text{-}4\text{-}yl\text{-}CH_2$	2	67
21	$CH_3(CH_2)_2CH_2$	2	66

side chains (cf. **18** and **19**) results in substantial loss of activity, in contrast to the corresponding peptide-based inhibitors [2].

Conclusion

Heterocyclic mimics of phenylalanine have been incorporated into inhibitors of renin without serious loss of potency. These rigid analogs provide insight into the preferred conformation of the P_2 side chain.

References

1. Plattner, J.J., Marcotte, P.A., Kleinert, H.D., Stein, H.H., Greer, J., Bolis, G., Fung, A.K.L., Bopp, B.A., Luly, J.R., Sham, H.L., Kempf, D.J., Rosenberg, S.H., Dellaria, J.F., De, B., Merits, I. and Perun, T.J., J. Med. Chem., 31 (1988) 2277.

2.   Luly, J.R., BaMung, N., Soderquist, J., Fung, A.K.L., Stein, H.H., Kleinert, H.D., Marcotte, P.A., Egan, D.A., Bopp, B.A., Merits, I., Bolis, G., Greer, J., Perun, T.J. and Plattner, J.J., J. Med. Chem., 31 (1988) 2264.
3.   For a similar approach, see Buhlmayer, P., Rasetti, V., Fuhrer, W., Stanton, J.L. and Goschke, R., U.S. Patent 4,727,060, Feb. 23, 1988.

Orally active renin inhibitors containing a novel aminoglycol dipeptide (Leu-Val) mimetic

Gunnar J. Hanson*, John S. Baran, Michael Clare, Kenneth Williams, Maribeth Babler, Stephen E. Bittner, Mark. A. Russell, S.E. Papaioannou, Po-Chang Yang and Gerald M. Walsh

G.D. Searle & Co., 4901 Searle Parkway, Skokie, IL 60077, U.S.A.

Introduction

The development of highly potent renin inhibitors has come about through the replacement of angiotensinogen (Ang) residues with ingenious amino acid and dipeptide surrogates. For example, we have reported dipeptide glycol renin inhibitors in which the Leu-Val (P_1-P_1' position) of Ang has been replaced with the non-hydrolyzable (2S,3R,4S)-2-amino-1-cyclohexyl-3,4-dihydroxy-6-methyl-heptane (1) dipeptide mimic [1]. Oral activity in marmosets, as measured by plasma renin activity (PRA) reduction assayed at pH 6, was achieved by replacing the P_4-P_3 position with an N-morpholinocarbonyl-phenyllactic acid surrogate [1].

Results and Discussion

The renin inhibitor SC-46944 [1] displays favorable biochemical characteristics [IC_{50} human renin = 5 nM; rhesus renin = 28 nM; cynomolgus renin = 36 nM; dog renin = 310 nM; endothiapepsin = 100 nM] and exhibits PRA lowering activity in salt-depleted marmosets [$95 \pm 3\%$ reduction @ 20 mg/kg after 1 h, i.g., $n = 5$, assayed at pH 7.4]. There is great interest in ascertaining the mode of binding of SC-46944 to human renin as an important piece of information towards de novo rational design. Because the 3D structure of human renin was not available, we chose to use as a model of human renin a related aspartic acid protease, *endothiapepsin*, whose high resolution X-ray crystal structure had been published [2]. In addition, a published [3] X-ray structure of rhizopuspepsin complexed to a 'reduced bond' inhibitor was used as an alignment template to position SC-46944. The P_3-P_2 region of SC-46944 was mapped onto the native endothiapepsin structure, aligning the strong hydrogen bonds from the P_3 and P_2 inhibitor carbonyls to residues Thr[219] and Asp[77] of endothiapepsin in analogy to the cited rhizopuspepsin-inhibitor template. The corresponding side chains were similarly fitted and well accommodated by the endothiapepsin S_3 and S_2 subsites. The fitting of the critical P_1-P_1' region was aided by the use of a simplified model of the glycol moiety, (2S,3R,4S)-2-*t*-butyloxycarbonylamino-1-cyclohexyl-

---

*To whom correspondence should be addressed.

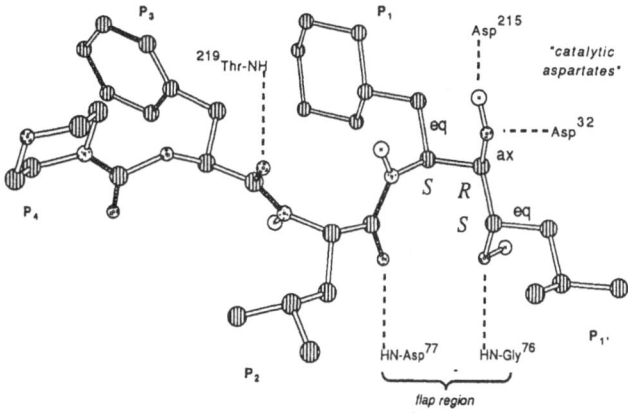

Fig. 1.   *Model of SC-46944 complexed to endothiapepsin.*

3,4-dihydroxypentane. This structure was subjected to an exhaustive conformational analysis using the MacroModel V2.5 MM2 force field. Examination of the low energy conformers revealed a common conformational feature: *an intramolecular H-bonded 8-membered ring array, reminiscent of the C7eq (gamma) turn found in proteins.* Positioning this simplified model with the (3R)-hydroxyl at the location of the key Asp-bound water in the native endothiapepsin, and with the P_1 side chain fitted into the S_1 subsite, the (4S)-hydroxyl was found to fall within hydrogen bonding distance of the NH of 'flap' residue Gly[76]. Combining the above alignment criteria, the complete inhibitor SC-46944 was positioned and the entire inhibitor-enzyme complex was subjected to further energy refinement. The final conformation of SC-46944 in the endothiapepsin complex is shown in Fig. 1. The inhibitor P_4, P_3, P_2, P_1 residues adopt an extended conformation. Particularly interesting is the spatial arrangement of the P_1-P_1' atoms with the large side chains, cyclohexylmethyl (P_1) and isobutyl (P_1') adopting an antiperiplanar diequatorial conformation, the (3R) and (4S) hydroxyls being positioned anti- and diaxial. The P_1-P_1' mimic (**1**), adopts a conformation ($\phi = -93°$, $\psi = 59°$) resembling that of an inverse gamma turn ($\phi = -80°$, $\psi = 65°$). At the outset of our discovery of glycol-based inhibitors, it was unclear exactly what role the (4S)-hydroxyl played. Our modeling results indicate that the glycol (4S)-hydroxyl of SC-46944 forms a hydrogen bond to the 'flap' NH of Gly[76] in endothiapepsin; in rhizopuspepsin [3], the P_1' carbonyl of a substrate analog exhibits this same interaction. The role of the (4S)-hydroxyl in glycol-based inhibitors may very well be to mimic the function of the P_1' carbonyl of angiotensinogen.

References

1. Hanson, G.J., Baran, J.S., Lowrie, H.S., Russell, M.A., Sarussi, S.J., Yang, P.-C., Babler, M., Bittner, S.E., Papaioannou, S.E. and Walsh, G.M., Biochem. Biophys. Res. Commun., 160 (1989) 1.
2. Pearl, L.H., Sewell, B.T. Jenkins, J.A., Cooper, J.B. and Blundell, T.L., Brookhaven Data Base, 1986.
3. Suguna, K., Padlan, E.A., Smith, C., Carlson, W. and Davies, D.R., Proc. Natl. Acad. Sci. U.S.A., 84 (1987) 7009.

Dog renin inhibitors: Structure-activity and in vivo studies

Kwan Y. Hui[a],* and Helmy M. Siragy[b]

[a]*Massachusetts General Hospital, Boston, MA 02114, U.S.A.*
[b]*University of Virgina School of Medicine, Charlottesville, VA 22908, U.S.A.*

Introduction

Dog models are frequently used to study the role of the renin-angiotensin system in the maintenance of blood pressure and in a variety of experimental forms of diseases. The availability of potent and long-acting dog renin inhibitors will facilitate such studies. We have synthesized a series of substrate analogs in an effort to identify SAR.

Results and Discussion

The IC_{50} values of the peptides against dog plasma renin were determined at pH 7.4 [1]. The structures of the renin inhibitors and the results of the in vitro studies are summarized in Table 1.

A potent dog plasma renin inhibitor, compound I, was infused intravenously (i.v.) into sodium-depleted, conscious, normotensive dogs following a published method [2]. The in vivo activities are presented in Fig. 1.

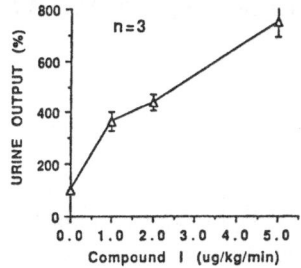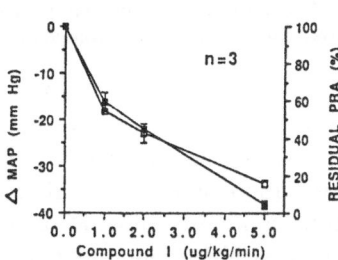

Fig. 1. Dose–response curves for changes in mean arterial pressure (□), residual plasma renin activity (■) and urine output (Δ) after 45 min i.v. infusion of compound I into conscious, Na-depleted, normotensive dogs (n = 3). Prior to each i.v. infusion, the dog was injected i.v. with compound I at a bolus dose 4 times the per min infusion rate [2]. Each value is the mean ± SEM.

*Present address: Lilly Research Laboratories, Lilly Corporate Center, Indianapolis, IN 46285, U.S.A.

Table 1   *In vitro activities of renin inhibitors of dog plasma renin*

Compound	Sequence P_6 P_5 P_4 P_3 P_2 P_1 P_1' P_2' P_3' P_4'	IC_{50} (nM)
A	Ac-His-Pro-Phe-Val-Statine-Leu-Phe-NH$_2$	5.2
B	Ac-Phe-Pro-Phe-Val-Statine-Leu-Phe-NH$_2$	24
C	Ac-Tyr-Pro-Phe-Val-Statine-Leu-Phe-NH$_2$	17
D	Ac-YOM-Pro-Phe-Val-Statine-Leu-Phe-NH$_2$	14
E	Ac-Trp-Pro-Phe-Val-Statine-Leu-Phe-NH$_2$	113
F	Ac-CHA-Pro-Phe-Val-Statine-Leu-Phe-NH$_2$	87
G	Ac-NA-Pro-Phe-Val-Statine-Leu-Phe-NH$_2$	58
H	Ac-His-Pro-Phe-Val-Statine-Leu-Phe-PAF-NH$_2$	3.4
I	Ac-PAF-Pro-Phe-Val-Statine-Leu-Phe-PAF-NH$_2$	1.7
J	Ac-His-Pro-Phe-Val-ACHPA-Leu-Phe-PAF-NH$_2$	3.5
K	Ac-PAF-Pro-Phe-Val-ACHPA-Leu-Phe-PAF-NH$_2$	33
L	Pro-His-Pro-Phe-Val-ACHPA-Leu-Phe-PAF-NH$_2$	4.7
M	Pro-PAF-Pro-Phe-Val-ACHPA-Leu-Phe-PAF-NH$_2$	24
N	Ac-His-Pro-Phe-His-Statine-Leu-Phe-NH$_2$	53
O	Ac-His-Pro-CHA-His-Statine-Leu-Phe-NH$_2$	70
P	Ac-His-Pro-NA-His-Statine-Leu-Phe-NH$_2$	74
Q	Ac-PAF-Pro-PAF-Val-Statine-Leu-Phe-PAF-NH$_2$	31
R	Ac-His-Pro-Phe-His-ACHPA-Leu-Phe-NH$_2$	32
S	Ac-His-Pro-Phe-Val-ACHPA-Leu-Phe-NH$_2$	0.96
T	Ac-His-Pro-Phe-His-ACHPA-Leu-Phe-NH$_2$	32
U	His-Pro-Phe-Val-Statine-Leu-Phe-PAF-NH$_2$	290
V	Pro-His-Pro-Phe-His-Statine-Leu-Phe-NH$_2$	7.8
W	Pro-His-Pro-Phe-His-Statine-Ile-Phe-NH$_2$	50
X	Pro-His-Pro-Phe-His-Statine-Ile-His-NH$_2$	160
Y	Ac-His-Pro-Phe-Val-ACHPA-Leu-Phe-Phe-NH$_2$	0.78
Z	Ac-His-Pro-Phe-Val-ACHPA-Leu-Phe-His-NH$_2$	3.2

NA = 3 (1'naphthyl)alanine; PAF = p-aminophenylalanine; YOM = O-methyl-tyrosine; Statine = (3S,4S)-4-amino-3-hydroxy-6-methylheptanoic acid; ACHPA = (3S,4S)-4-amino-3-hydroxy-5-cylcohexylpentanoic acid.

The following observations can be made:

(1) At the P_5 position, a side chain with a 6-member aromatic ring or smaller is important for optimal activity (A, B, C and D vs. E, F and G).

(2) Substitution of PAF for His at P_5 results in a mild increase in potency of the statine-containing substrate analog (H vs. I), but causes a marked decrease in potency of ACHPA-containing analogs (J vs. K; L vs. M).

(3) At P_3, Phe is preferred over CHA, NA and PAF (N vs. O and P; I vs. Q).

(4) Substitution of Val for His at the P_2 position enhances potency by an order of magnitude (A vs. N; R vs. S).

(5) The difference between statine and ACHPA in their contribution to potency is moderate (N vs. T; A vs. S; H vs. J).

(6) N-terminal protection of the P_5 position is important for optimal activity (U vs. H). The use of acetyl or Pro as the protecting residue at the P_6 position results in equivalent potency (J vs. L; K vs. M).

(7) Substitution of Leu for Ile at the P_2' position has a marked increase in

potency (V vs. W and X). Extension of the C-terminus to the P_4' position has a mild influence on potency (A vs. H; S vs. J, Y and Z).

(8) Compound I, a potent dog plasma renin inhibitor in vitro, exhibits a dose-related hypotensive effect associated with progressive reduction of plasma renin activity and increase in urine output when infused into conscious, sodium-depleted dogs. These results suggest that compound I is a potent and selective agent for the study of the renin-angiotensin system in dogs.

Acknowledgements

Supported in part by grants from the AHA (Massachusetts Affiliate) and NASA.

References

1. Hui, K.Y., Carlson, W.D., Bernatowicz, M.S. and Haber, E., J. Med. Chem., 30 (1987) 1287.
2. Boger, J., Payne, L.S., Perlow, D.S., Lohn, N.S., Poe, M., BLaine, E.H., Ulm, E.H., Schorn, T.W., LaMont, B.I., Lin, T.Y., Kawai, M., Rich, D.H. and Veber, D.F., J. Med. Chem., 28 (1985) 1779.

Renin inhibitors containing a C-terminal heterocycle

Saul H. Rosenberg, Joseph F. Dellaria, Dale J. Kempf, Hing L. Sham,
Keith W. Woods, Ed de Lara, Robert G. Maki, Kenneth P. Spina,
Herman H. Stein, Jerome Cohen, Jacob J. Plattner, William R. Baker,
Hollis D. Kleinert and Thomas J. Perun

Cardiovascular Research Division, Abbott Laboratories, Abbott Park, IL 60064, U.S.A.

Introduction

Dipeptide glycols [1] and azido glycols [2] are representatives of a new class of renin inhibitors that contain a novel erythro-1,2-diol, functionality designed to mimic the tetrahedral transition state of the enzyme-mediated amide bond hydrolysis. SAR have delineated the optimum stereochemical [1] and regiochemical [2] relationships for the two hydroxyls; however, the exact nature of their interactions with the active site of the enzyme is not known. While molecular modeling studies [1] indicated that the second hydroxyl [3] acts as a donor in a hydrogen bond, SAR within the azido-glycol series suggested that this hydroxyl was, in fact, acting as an acceptor [2].

Results and Discussion

While examining a series of compounds related to the azido-glycols, we realized that these compounds could be used to probe the interactions of the second hydroxyl. Specifically, cyclizing this group and the amine derived from the azide through a carbonyl linkage resulted in oxazolidinone **9**, in which this oxygen can only participate as an acceptor. Compound **9** is slightly less potent than the parent azido-glycol **8**, perhaps because it lacks a lipophilic C-terminal residue.

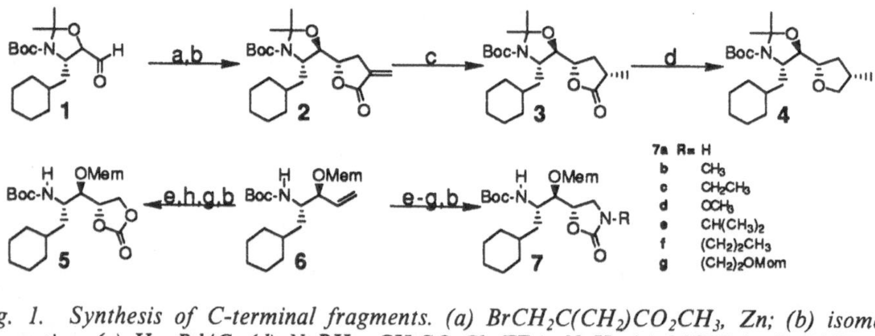

Fig. 1. Synthesis of C-terminal fragments. (a) $BrCH_2C(CH_2)CO_2CH_3$, Zn; (b) isomer separation; (c) H_2, Pd/C; (d) $NaBH_4$; CH_3SO_2Cl, TEA; NaH; (e) MCPBA; (f) RNH_2 or NaN_3; $HS(CH_2)_3SH$, TEA; (g) Cl_2CO, TEA; (h) OsO_4.

Table 1 *In vitro potencies*

Boc-Phe-His-N — (structure with H, OH, and O–X, Y groups on cyclohexylmethyl oxazolidinone scaffold)

No.	X	Y	IC$_{50}$ (nM) HR[b]	IC$_{50}$ (nM) PR[c]	% Inhibition[a] CD[d]	% Inhibition[a] PP[e]
8[f,g]	H	N$_3$	0.4	9.0	0	0
9	CO	NH	4.7	25		
10	CO	NCH$_3$	1.4	8.8		
11	CO	NCH$_2$CH$_3$	0.63	2.8	21	0
12	CO	NOCH$_3$	0.90	9.2	0	0
14	CO	NCH(CH$_3$)$_2$	0.84	8.2		
15	CO	N(CH$_2$)$_2$CH$_3$	1.0	12		
16	CO	N(CH$_2$)$_2$OH	0.70	7.4		
17	CO	O	9.8	130		
18	CO	CCH$_2$	0.71	20	34	7
19	CO	(S) CHCH$_3$	0.32	2.3		
20	CH$_2$	(S) CHCH$_3$	1.2	3.0	9	2
21[g,h]	H	CH(CH$_3$)$_2$	1.5	–	0	0

[a] At 10^{-5} M.　　[d] Bovine cathepsin D, pH 3.　　[g] Acyclic.
[b] Purified human renin, pH 6.0.　　[e] Porcine pepsin, pH 2.　　[h] Ref. 1.
[c] Human plasma renin, pH 7.4.　　[f] Ref. 2.

A series of oxazolidinones that contains this lipophilic group was prepared from protected allylic alcohol 6 and these compounds were incorporated into inhibitors 10–16. Compound 11 is more potent than both unsubstituted oxazolidinone 9 and azide 8 against human plasma renin, demonstrating that an ethyl group is optimal. That even a C-terminal proton contributes to binding is evident upon comparison of the potencies of 9 and carbonate 17.

Although 11 possessed considerable potency, it was not clear which oxazolidinone oxygen was contributing to binding. Cyclic ether 4 was prepared from lactone 3 that was derived from aldehyde 1. Cyclic ether containing inhibitor 20 is equipotent with lactone 19, strongly suggesting that the ring carbonyl plays no role in binding to the active site. Compound 20 is also equipotent with diol 21, indicating that the ether oxygen does contribute to binding and must do so acting as a hydrogen bond acceptor.

References

1. Luly, J.R., BaMaung, N., Soderquist, J., Fung, A.K.L., Stein, H., Kleinert, H.D., Marcotte, P.A., Egan, D.A., Bopp, B., Merits, I., Bolis, G., Greer, J., Perun, T.J. and Plattner J.J., J. Med. Chem., 31 (1988) 2264.
2. Rosenberg, S.H., Woods, K.W., Kleinert, H.D., Stein, H., Nellans, H.N., Hoffman, D.J., Spanton, S.G., Pyter, R.A., Cohen, J., Egan, D.A., Plattner, J.J. and Perun, T.J., J. Med. Chem., 32 (1989) 1371.
3. The first hydroxyl refers to that hydroxyl lying closest to the N-terminus and occupying the same position as the single hydroxyl in statine containing inhibitors.

Isostere containing renin inhibitors

James S. Kaltenbronn, James P. Hudspeth, Elizabeth A. Lunney,
Ernest D. Nicolaides, Joseph T. Repine, W. Howard Roark, Michael A. Stier,
Francis J. Tinney, Peter K.W. Woo and Arnold D. Essenburg

*Parke-Davis Pharmaceutical Research Division, Warner-Lambert Company,
Ann Arbor, MI 48105, U.S.A.*

Introduction

As one aspect of our renin inhibitor strategy, we chose to prepare modified compounds based on the potent renin inhibitor (I) reported by Bock et al. [1]. In this way we sought to improve stability by preventing enzymatic hydrolysis at the altered amide bond.

$$\text{BOC} - \text{PHE} - \text{HIS} - \text{STA} - \text{LEU} - \text{NHCH}_2\text{Ph} \qquad \text{(I)}$$

$$\underbrace{}_{P_4} \underbrace{}_{P_3} \underbrace{}_{P_2} \underbrace{}_{P_1\text{-}P_1'} \underbrace{}_{P_2'} \underbrace{}_{P_3'}$$

We describe the results of replacing the amide linkage connecting the P_3 and P_2 sites, the only linkage in the above structure connecting two natural amino acids, with 14 different isosteres. While several reports [2–4] have described isosteric replacements at the P_1-P_1' site, only isolated reports have appeared describing isosteric replacement at the P_3-P_2 site and the P_4-P_3 site [5,6].

Results and Discussion

Table 1 gives the structure and IC_{50} value of the most potent derivative prepared from each of the 14 isosteric classes cited in this paper. While none of the isostere-containing compounds matched the high potency of peptide model **1**, the most potent of these, **11**, did approach it, being 4-fold less potent. When **2**, which has a Gly in P_2, was compared to isostere-containing compounds having a Gly equivalent in P_2, several compounds showed enhanced potency.

Compounds **9** en **11** were the most potent. However, not all compounds having a hydroxyl-bearing isostere at this position were highly active, as evidenced by the diminished potency of **8** and **10**. Compound **3**, containing ϵ-double bond isostere, and epoxide **5** derived from it, had IC_{50} values in the μM range, while **4**, containing the Z-double bond isostere, and epoxide **6** derived from it, were essentially inactive.

Table 1 *In vitro renin inhibitory activity*

Compd.	Structure	IC$_{50}$ (μM)[a]
1	Boc-Phe-His-Sta-Leu-NHCH$_2$Ph	0.0057[b]
2	Boc-Phe-Gly-Sta-Leu-NHCH$_2$Ph	1.1
3	Boc-Pheψ[E – CH = CH]Gly-ACHPA-Leu-NHCH$_2$Ph[c]	1.3
4	Boc-Pheψ[Z – CH = CH]Gly-Sta-Leu-NHCH$_2$Ph	> 100
5	Boc-Pheψ[CH – CH]Gly-ACHPA-Leu-NHCH$_2$Ph $\overset{O}{\overset{\diagdown}{}}$	0.53
6	Boc-Pheψ[CH – CH]Gly-Sta-Leu-NHCH$_2$Ph $\overset{O}{\overset{\diagdown}{}}$ (derived from Z-4)	> 100
7	Boc-Pheψ[CH$_2$CH$_2$]Gly-ACHPA-Leu-NHCH$_2$–C$_6$H$_4$–CH$_2$NH$_2$	0.63
8	Boc-Pheψ[CHOHCHOH]Gly-Sta-Leu-NHCH$_2$Ph	1.4
9	Boc-Pheψ[CHOHCH =CHCO]ACHPA-Leu-NHCH$_2$Ph	0.09
10	Boc-Pheψ[CHOHCHOHCHOHCO]Sta-Leu-NHCH$_2$Ph	11.0
11	Boc-Pheψ[CHOHCH$_2$]Gly-ACHPA-Leu-NHCH$_2$–C$_6$H$_4$–CH$_2$NH$_2$	0.02
12	Boc-Pheψ[COCH$_2$]Gly-Sta-Leu-NHCH$_2$Ph	21.0
13	Boc-Pheψ[CH$_2$NH]His-ACHPA-Leu-NHCH$_2$Ph	0.2
14	Boc-Pheψ[CH$_2$NOH]His-Sta-Leu-NHCH$_2$Ph	6.5
15	IVA-Pheψ[CH$_2$S]Phe-Sta-Ala-Sta-NHCH$_2$Ph	7.4
16	IVA-Pheψ[CH$_2$SO]Phe-Sta-Ala-Sta-NHCH$_2$Ph	2.0
17	IVA-Pheψ[CH$_2$SO$_2$]Phe-Sta-Ala-Sta-NHCH$_2$Ph	1.5

[a] IC$_{50}$ values were determined using a standard RIA for angiotensin I. Human plasma with a plasma renin activity of 0.68 –4.04 μg/AI/mL/h was used in this assay.
[b] Ref. 4 gives the IC$_{50}$ as 0.026 μM when tested against human plasma renin.
[c] ACHPA is 4(*S*)-amino-3(*S*)-hydroxy-5-cyclohexylpentanoic acid.

References

1. Bock, M.G., DiPardo, R.M., Evans, B.E., Rittle, K.E., Boger, J., Poe, M., LaMont, B.I., Lynch, R.J., Ulm, E.H., Vlasuk, G.P., Greenlee, W.J. and Veber, D.F., J. Med. Chem., 30 (1987) 1853.
2. Smith, C.W., Saneii, H.H., Sawyer, T.K., Pals, D.T., Scahill, T.A., Kamdar, B.V. and Lawson, J.A., J. Med. Chem., 31 (1988) 1377.
3. Bülmayer, P., Caselli, A., Fuhrer, W., Göschke, R., Rasetti, V., Rüedger, H., Stanton, J.L., Criscione, L. and Wood, J.M., J. Med. Chem., 31 (1988) 1839.
4. Luly, J.R., Bolis, G., BaMaung, N., Sonderquist, J., Dellaria, J.F., Stein, H., Cohen, J., Perun, T.J., Greer, J. and Plattner, J.J., J. Med. Chem., 31 (1988) 532.
5. Evans, B.E., Rittle, K.E., Ulm, E.H. Veber, D.F., Springer, J.P. and Poe, M., In Deber, C.M., Hruby, V.J. and Kopple, K.D. (Eds.) Peptides: Structure and Function (Proceedings of the Ninth American Peptide Symposium), Pierce Chemical Co., Rockford, IL, 1985, p. 743.
6. TenBrink, R.E., Pals, D.T., Harris, D.W. and Johnson, G.A., J. Med. Chem., 31 (1988) 671.

Renin inhibiting azapeptides

Joachim Gante*, Harald Kahlenberg, Peter Raddatz and Reinhard Weitzel
*E. Merck Darmstadt, Pharmaceutical Research Department, Frankfurter Strasse 250,
D-6100 Darmstadt, F.R.G.*

Introduction

'Azapeptides' were described for the first time in 1962 [1]. In these peptide analogs the α-CH of one or more amino acid residues of a peptide chain is replaced with nitrogen (Fig. 1).

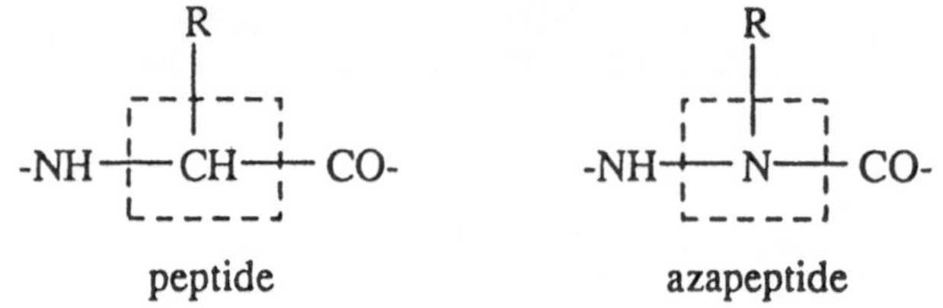

Fig. 1.

This backbone modification is attractive, since from a purely structural point of view there is only a minimal change in the peptidic structure: the basic 'skeleton' is preserved and the side chains remain identical. In this respect, the synthesis of biologically active azapeptides has become more and more important during the last few years. Many of them exert stronger activity and – due to reduced biodegradability – extended duration of action compared with the original compounds [2].

In recent years renin inhibitors have been investigated as effective tools in combating high blood pressure [3]. While a renin-inhibiting azatetrapeptide, Azaleu-Leu-Val-Phe-OMe, with higher biological activity than the original compound was synthesized some years ago by others [4], we designed corresponding compounds of the 'transition-state' type. This class of compounds, at present under intense investigation in many laboratories, contains molecules incorporating either statine or one of its analogs [5].

Results and Discussion

Exchanging one amino acid each at various positions within the peptide chain for an azaamino acid, we synthesized analogs of the highly active renin inhibitor, Boc-Phe-Gly-ACHP-Ile-3-pyridylmethylamide, recently synthesized in our com-

---

*To whom correspondence should be addressed.

406

pany [6] with Azaphe, Azagly and Azaile (Compounds A–C) residues. Thus, we should be able to compare the biological activity of the analogs with that of the original compound with respect to the position of the altered residue.

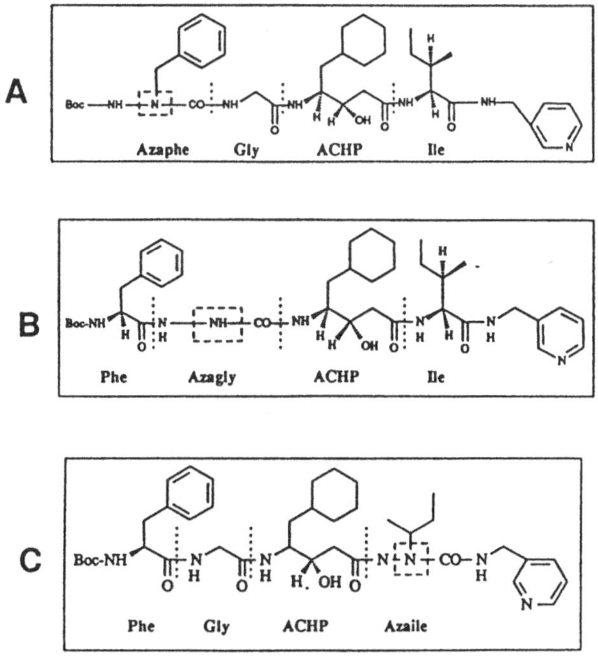

Fig. 2.

The synthesis of these compounds was carried out by a combination of semicarbazide forming reactions and peptide couplings (see reviews, Ref. 2).

Biological activity

Inhibition of renin (pepsin, cathepsin D):

A: $IC_{50} = 1.5 \times 10^{-6}$ M* $(> 10^{-4}$ M, $> 10^{-4}$ M$)$

B: $IC_{50} = 2.45 \times 10^{-7}$ M $(> 10^{-4}$ M, $> 10^{-4}$ M$)$

C: $IC_{50} = 8.5 \times 10^{-7}$ M $(> 10^{-4}$ M, $> 10^{-4}$ M$)$

Boc-Phe-Gly-ACHP-Ile-
-3-pyridylmethylamide: $IC_{50} = 2.8 \times 10^{-9}$ M $(4 \times 10^{-5}$ M, 6.5×10^{-6} M$)$

As can be seen, the analogs are significantly less potent than the original compound. They have, however, the same high inhibitory specificity for renin when compared with that for pepsin and cathepsin D. The relative potency of the compounds A, B and C is about 1:6:2, i.e. that the Azagly residue in the 'middle' of the molecule causes the strongest rening inhibition, this molecule being about six times more potent than that with N-terminal Azaphe.

Enzymatic degradation experiments with chymotrypsin

Interesting results were provided by enzymatic stability experiments. Because chymotrypsin is known to cleave -Phe-CO-NH- peptide bonds preferentially, the Azaphe (A) and Azagly (B) derivatives with an unnatural residue in front of and behind that amide bond, respectively, were chosen as candidates for corresponding degradation experiments.

Whereas the original compound, Boc-Phe-Gly-ACHP-Ile-3-pyridylmethyl-amide, exhibited significant degradation by chymotrypsin, the Azagly analog (B) was degraded even more rapidly. On the other hand, no significant enzymatic breakdown could be detected with the Azaphe analog (A). The experiments show that Azagly in combination with 'normal' Phe surprisingly enhances the breakdown whereas 'altered' Phe in this case protects the adjacent amide bond against degradation. These observations are highly interesting in connection with the development of orally active azapeptides.

Conclusions

It could be shown that the Aza compounds A, B and C are specific renin inhibitors, but with significantly lower activity than the original compound. On the other hand, chymotryptic degradation of the Azaphe derivative (A) – in contrast to the original compound – is almost completely inhibited.

Acknowledgements

We thank Dr. A. Jonczyk (E. Merck) for running the enzymatic stability experiments.

References

1. Gante, J., Thesis, Freie Universität, Berlin, 1962.
2. Reviews: (a) Gante, J., Synthesis, (1989) 405;
   (b) Gante, J., Fortschr. Chem. Forsch., 6 (1966) 358;
   (c) Thamm, P., In Houben-Weyl (Ed.) Methoden der organischen Chemie, Vol. XV/1, 4th edn., Georg Thieme Verlag, Stuttgart, 1974, p. 894 and Vol. XV/2, p. 388.
3. Greenlee, W.J., Pharm. Res., 4 (1987) 364.
4. Parry, M.J., Russell, A.B. and Szelke, M., Chemistry and Biology of Peptides (Proceedings of the 3rd American Peptide Symposium), 1972, p. 541.
5. Boger, J., Payne, L.S., Perlow, D.S., Lohr, N.S., Poe, M., Blaine, E.H., Ulm, E.H., Schorn, T.W., LaMont, B.I., Lin, T.Y., Kawai, M., Rich, D.H. and Veber, D.F., J. Med. Chem., 28 (1985) 1779.
6. Merck Patent GmbH (Raddatz, P., Sombroek, J., Gante, J., Schmitges, C.J., Minck, K.O., Jonczyk, A. and Hölzemann, G., Inv.), Eur. Pat. Appl. 249 096 (Dec. 16, 1987) [Chem. Abstr., 108 (1988) 112 960t].

The synthesis and use of 3-amino-4-phenyl-2-piperidones and 4-amino-2-benzazepine-3-ones as conformationally restricted phenylalanine isosteres in renin inhibitors

S.E. de Laszlo[a], B.L. Bush[a], J.J. Doyle[b], W.J. Greenlee[b], D.G. Hangauer[b], T.A. Halgren[b], R.J. Lynch[a], T.W. Schorn[a] and P.K.S. Siegl[b]

[a]*Department of Exploratory Chemistry, Merck Sharp & Dohme Research Laboratories, Rahway, NJ 07065, U.S.A.*
[b]*Department of Pharmacology, Merck Sharp & Dohme Research Laboratories, West Point, PA 19486, U.S.A.*

Introduction

The goal of developing an orally effective renin inhibitor has been hindered by the low bioavailability and sensitivity to peptidases of the inhibitors developed to date. As part of our program in this area, we designed, through the use of the Merck renin computer graphic model (CGM), two classes of novel conformationally restricted renin inhibitors (compounds **1–5** and **6–9**), in order to investigate the conformation of inhibitors bound in the active site of the enzyme, and to improve the potency of these inhibitors as well as their stability to proteinases.

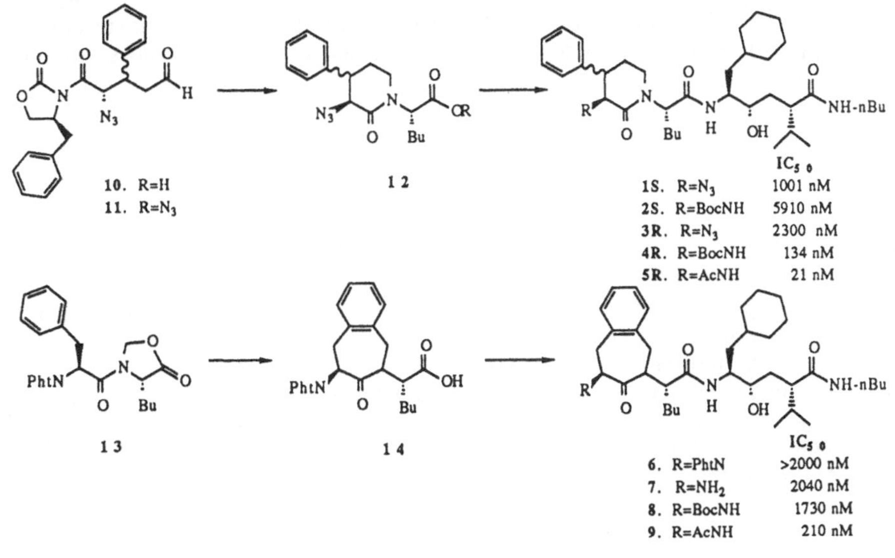

Fig. 1. Synthetic approach and potencies of conformationally restricted renin inhibitors.

409

Results and Discussion

The synthesis of the piperidones was achieved by a novel route via the aldehyde **10**. The aldehyde **10** was prepared, via Evans methodology, from the fully separated diastereomeric oxazolidinones **11**, protected as the dioxolane acetals derived from 3-phenyl glutaric acid via a 7-step synthesis. The absolute configurations of the azide and 3-phenyl substituents were determined by X-ray crystallography of an intermediate. The aldehyde **11** was reductively alkylated with norleucine *t*-butyl ester. On warming, the intermediate amines cyclized to the piperidones **12**. Deprotection of the ester and condensation with (2S,4S,5S)-5-amino-6-cyclohexyl-4-hydroxy-2-isopropyl-hexanoic acid gave **1S** and **3R** [1]. Conventional functional group conversions gave the derivatives illustrated.

The benzazepinones were prepared from the oxazolidinone **13** derived from *N*-phthalyl-phenylalaninyl-norleucyl carboxylic acid. Treatment of **13** with triflic acid gave the benzazepinone **14** with only 10% racemization [2]. The acid **14** was coupled and derivatized, as described above, and gave the inhibitors **6–9**.

The peptides were assayed as inhibitors of human renin [3]. The most potent inhibitor **5R** was 20 times less potent than the control compound with phenylalanine at P_3. This result was the product of two changes to the control: alkylation of the P_2–P_3 amide bond and introduction of a conformational restriction into two rotatable bonds. The low potency evident in the series **6–9** is the result of the introduction of a restriction into a further bond. Evident from this work was the increased potency of the smaller amino derivatives. This result had been expected from our analysis of the CGM. Although both series of structures overlay model inhibitors in the CGM well, we were surprised to find that the 2S, 3R diastereomers **3–5** were more potent than the diastereomeric compounds **1–2**. The CGM had indicated, at the outset of the work, that the 2S, 3S diastereomer should be more potent due to its superior overlay with the model inhibitor. Adjustments to the renin model to accomodate these results are in progess.

References

1. Chakravarty, P.K., de Laszlo, S.E., Sarnella, C.S., Springer, J.P. and Schuda, P.F., Tetrahedron Lett., 30 (1989) 415.
2. Flyn, G.A., European Patent Application 0 249 223, 1987.
3. The human plasma IC_{50} values were determined at pH 7.4, as in Boger, J., Payne, L.S., Perlow, D.S., Lohr, N.S., Poe, M., Blaine, E.H., Ulm, E.H., Schorn, T.W., Lamont, B.I., Lin, T.Y., Kawai, M., Rich, D.H. and Veber, D.F., J. Med. Chem., 28 (1985) 1779.

Approaches to human renin inhibitors with improved bioavailability

William J. Greenlee[a,*], Jan tenBroeke[a], Stephen E. de Laszlo[a],
Prasun K. Chakravarty[a], Valerie J. Camara[a], Kenneth J. Fitch[a],
Carol S. Sarnella[a], Arthur A. Patchett[a], Peter D. Williams[b], Debra S. Perlow[b],
Daniel F. Veber[b], Robert J. Lynch[b], John J. Doyle[b], Terry W. Schorn[b],
John F. Strouse[b] and Peter K.S. Siegl[b]

[a]*Exploratory Chemistry Department, Merck Sharp and Dohme Research Laboratories,
Rahway, NJ 07065, U.S.A.*
[b]*Department of Medicinal Chemistry and Department of Pharmacology, Merck Sharp and
Dohme Research Laboratories, West Point, PA 19486, U.S.A.*

Introduction

As a part of a program directed toward orally-active renin inhibitors, we
studied a series of inhibitors based on the statine analog 'ACHPA' (4S-amino-
5-cyclohexyl-3S-hydroxypentanoic acid) that incorporate amino-terminal N-
alkylated phenylalanine analogs (as the P_4 and P_3 elements). It was hoped that
they would show improved aqueous solubility and stability toward both chymotrypsin and aminopeptidases.

Results and Discussion

Removal of the amino-terminal Boc group from the inhibitor Boc-Phe-His-
ACHPA-Ile-NHCH$_2$(pyridin-2-yl) (IC$_{50}$ = 0.8 nM against human plasma renin)
[1] led to a less-potent inhibitor (IC$_{50}$ = 33 nM). While N-methylation of
phenylalanine gave no improvement, large N-substituents gave more-potent
inhibitors. Particularly interesting was inhibitor 1, which incorporates a 3S-
quinuclidinyl substituent, and has in vitro potency nearly equal to that of the
Boc-Phe analog. Inhibitor 1 had aqueous solubility (>20 mg/ml in 0.1 M citric
acid) greatly improved relative to that of neutral inhibitors, and was stable to
the action of homogenates of liver, kidney, intestine and pancreas [2].

Intravenous administration of inhibitor 1 to sodium-depleted rhesus monkeys
produced a drop in plasma renin activity (PRA) that persisted (>50% inhibition)
for 2 h, and lowered blood pressure (BP) for over 1 h. Oral administration
of 1 to monkeys (10 mg/kg) resulted in nearly complete inhibition of PRA
(>97%) for 6 h with an accompanying drop in blood pressure (max. drop = −17
mm Hg at 3 h). Inhibitors 2 and 3, that incorporate an hydroxyethylene isostere
or a lactam analog of ACHPA [3], eliminate the carboxy-terminal isoleucine

---

*To whom correspondence should be addressed.

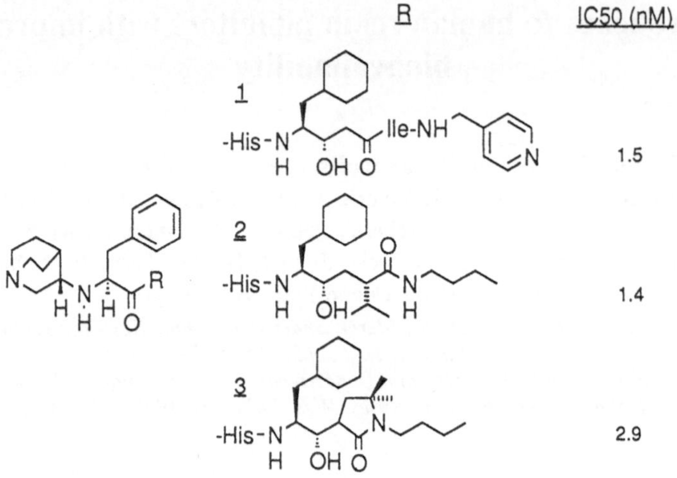

Fig. 1. Inhibitors of human renin.

while maintaining high in vitro potency, but neither showed oral activity matching that of **1**. It is hoped that other modifications at the carboxy-terminus in this series will lead to further improvements in bioavailability.

References

1. The human plasma renin IC_{50} values were determined at pH 7.4 following the procedure described in Boger, J., Payne, L.S., Perlow, D.S., Lohr, N.S., Poe, M., Blaine, E.H., Ulm, E.H., Schorn, T.W., Lamont, B.I., Lin, T.Y., Kawai, M., Rich, D.H. and Veber, D.F., J. Med. Chem., 28 (1985) 1779.
2. These experiments were carried out by Dr. P. Krieter, MSDRL, Rahway, NJ.
3. Williams, P.D., Payne, L.S., Perlow, D.S., Lundell, G.F., Gould, N.P., Siegl, P.K.S., Schorn, T.W., Lynch, R.J., Doyle, J.J., Strouse, J.F., Sweet, C.S., Arbegast, P.T., Stabilito, I.I., Vlasuk, G.P., Freidinger, R.M. and Veber, D.F., In Rivier, J.E. and Marshall, G.R. (Eds.) Peptides: Chemistry, Structure and Biology (Proceedings of the 11th American Peptide Symposium), ESCOM, Leiden, 1990, p. 43.

Session III
Analytical methodologies

Chairs: Robert S. Hodges
University of Alberta
Edmonton, Alberta, Canada

and

Bruce W. Erickson
University of North Carolina
Chapel Hill, North Carolina, U.S.A.

Chromatotopography: Peptide dynamics at solid-liquid interfaces

Milton T.W. Hearn

Department of Biochemistry and Centre for Bioprocess Technology Monash University,
Clayton, Victoria, Australia 3168

Introduction

Chromatographic methods have been extensively applied in recent years to study the conformational behaviour of proteins in solution, particularly in the presence of dissociating reagents. Typically, size exclusion chromatography, most recently in the high performance mode (e.g. SEC-HPLC), has been employed to evaluate the effect of mobile phases containing increasing concentrations of urea, guanidine hydrochloride or other dissociating chaotropic reagents. Changes in the hydrodynamic volume, Stokes radius of gyration, intrinsic viscosity and solvation state of the protein can be directly monitored from variation in the permeation coefficient, average retention and/or band broadening of the peak zone. Similar methods, again with dissociants present in the eluent, have been applied to characterize the bulk changes in the conformation of proteins with ion-exchange and RPHPLC techniques (for a selected compendium of recent applications see Refs. 1–6). However, such methods, as routinely employed, provide little or no information on the molecular mechanism of the unfolding/refolding trajectory or the location of structural regions, motifs or domains of the protein (or peptide) which associate with their complementary ligand. In particular, these methods, as generally employed, do no indicate structural regions within continuous or noncontinuous sequence frameworks, which are more or less accessible to the surrounding solvent.

Recent studies [7,8] have, however, demonstrated that modern HPLC techniques, when operated in appropriate adsorption modes, can provide information and directives that address different aspects of these fundamental question. Although the mosaic of experimental results is far from complete, the available information indicates that the chemical ligand, irrespective of whether it is a *n*-butyl or a *n*-octadecyl group as used in RPHPLC, a quaternary ammonium, sulphate or carboxylate group as employed in ion-exchange HPLC, or another type of chemical functionality as used in biomimetic and ligand exchange HPLC, can function as a molecular probe to unravel the conformational vagaries and interactive contact regions of peptides and proteins. This publication reports further progress towards the routine use of such rapid high-resolution chromatographic methods to probe, and to analyze, induced changes in the structural hierarchy of peptides and proteins – an application of modern HPLC for which the evocative term chromatotopography has been coined. In particular, this study documents further examples of the use of RPHPLC to characterize transient amphipathic conformations of synthetic peptides related to the (6–13)-fragment of human growth hormone and an approach to characterize conformational changes associated with regions of high electrostatic potential of proteins.

Results and Discussion

Methodology, instrumentation and calculation of parameters for RPHPLC and ion-exchange HPLC were described earlier [8–10].

Over the past several years our understanding of the molecular basis of protein function and antigen recognition in the normal and pathological state has been dependent on the application of a variety of experimental approaches involving (i) monoclonal antibody (MAb) mapping methods; (ii) development of computer-aided predictive algorithms; (iii) analysis of immunological or biological properties of cleavage fragments of a relevant protein; (iv) arbitrary synthesis of large numbers of overlapping peptides; (v) analysis of the effect of amino acid substitutions and deletions in variant or homologous proteins, and (vi) spectroscopic and X-ray crystallographic analysis. All of these experimental methods have as their common goal the elucidation of the structure of the recognitopic regions of a protein or polypeptide and to relate these data to information encoded within the primary sequence. For different reasons, however, all of these methods currently have a number of significant limitations, either in terms of the restricted applicability of the derived data to unrelated proteins, or their cost and time required for execution of the investigations. Clearly, improved and expanded data bases on the conformational and accessibility status of sequential, as well as spatially assembled or discontinuous, recognitopes are required if experimental methods of wide applicability are to be used to correlate the segmental/domain behavior of a specific protein or polypeptide (in terms of its primary, secondary or higher order structures) with its biological function as manifested in immune response or receptor binding phenomena.

In order to allow molecular definition of a particular topographic region, it is necessary to determine the composition and orientation of amino acids which comprise the binding determinant. Early availability of predictive data on the preferred conformational limits of a designated peptide in a particular environment, and directives on the possible topographic contact regions of a protein capable of interacting with relevant chemical or biological ligands would considerably assist this molecular definition. The underlying thermodynamic and kinetic processes observed in peptide and protein interactions with chemical as well as biological ligands involve common physicochemical considerations. The concept that chromatographic sorbents can be used to access and assess surface features such as regional hydrophobic moments or electrostatic ionotopes, thus has the capacity to transform current qualitative understanding of interfacial behavior of peptides and proteins into more quantitative and mechanistic relationships. Interestingly, this concept is not new, but rather has evolved over the past forty years or so, and forms the basis of all algorithms for structure-retention prediction in partition chromatography, as well as in paper, thin-layer and reversed-phase chromatography. Iteration of the derived retention data can be used to derive amino acid functional group coefficients [11–13], which then find use in the analysis of the presence of accessible hydrophobic stretches along a polypeptide amino acid sequence by residue averaging methods. Derivation of these coefficients and their assembly into structural motifs of defined retention characteristics has its origin in the relationship between the thermodynamic parameter, known in chromatography as the capacity factor k', and the compositions of the protein (or peptide), the surrounding solvent and the interfacial microenvironment.

Knowledge of the relationship between the logarithmic capacity factor, e.g. log k', of a peptide or protein and the mole fraction, φ, or logarithmic concentration (log [c]) of the displacing species, e.g. solvent, salt, etc. employed with a particular adsorption HPLC mode, can be used to evaluate in thermo-dynamic as well as empirical terms, preferred conformational pathways involved in biosolute folding. Related LFER and structure–retention relationships can be used to suggest possible molecular regions within the total structure involved in solute–ligand interactions. Similarly, relationships between peak broadening and chromatographic parameters can be used to provide insight into the kinetic processes associated with solute–ligand interaction. The forms of these retention dependencies most commonly used [7,10] to analyze data obtained from reversed phase (hydrophobic interaction) and ion-exchange (coulombic) chromatographic experiments are:

for reversed phase

$$\log k' = \log K_0 - S\varphi \tag{1}$$

or

$$\log k' = C + \gamma \frac{N\Delta A_h + 4.836 N^{\frac{1}{3}} (\kappa^e - 1) V_m^{\frac{2}{3}}}{RT} \tag{2}$$

and alternatively for ion exchange

$$\log k' = \log K_0 - Z \log C \tag{3}$$

or

$$\log k' = C - \frac{2.303}{RT} \left[\frac{Z_c Z_i \epsilon^2 r_p}{2D \; A_c} - \frac{\kappa}{1 + \kappa a} \right] \tag{4}$$

where S and Z are the slopes of the log k' vs. φ or log k' vs. log [ionic concentration]; log K_0 is related to the capacity factor of the peptide and protein in neat water,

417

ΔA_h is the relative surface area of the molecule in hydrophobic contact with the nonpolar surface, and A_c is the ionotopic surface area encompassing the vectorial charges at the surface of the protein which are in electrostatic contact with the coulombic surface.

In the case of ionic interactions as the electrostatic potential per unit area increases, i.e. as the number of charge groups per unit area within the contact region increases, retention in ion-exchange HPLC will increase for a designated buffer and salt system of dielectric constant D, charge Z_i, temperature T and effective protein ion radius, r_p. The remaining terms in Eqn. (4) relate to the distance of closest approach, **a**, of the contact region to the charged ligand and the composition of the solution, the so called κ-term which is an approximate measure of the thickness of the ionic atmosphere or the distance over which the electrostatic field of an ion extends with appreciable strength. Continuous linear dependencies of log k vs. log [1/c] or log k' vs. 1/D are thus predicted by Eqn. (4) provided the electrostatic potential per unit area, i.e. the $Z_c r_p/A_c$ term remains constant for different values of [c] with a defined sorbent. Furthermore, numerical evaluation of Z_c, r_p or A_c can be carried out from appropriate experimental measurements. In the case of hydrophobic van der Waals/Lifshitz or London force interactions as the hydrophobic contact area ΔA_h, or the energy required to create a cavity with the molecular dimensions of the solute (the κ^e term) increase, retention will also increase. Thus, reversed phase and coulombic modes of HPLC independently permit assessment of hydrophobic or electrostatic patches at the surface of a protein or peptide accessible to an appropriate ligand. Under conditions dictated by the thermo-dynamic near equilibrium assumptions, these contact regions will correspond to sites of the highest hydrophobic moments or highest electrostatic charge potential.

When changes in the contact area, electrostatic potential or charge density arise from specific interfacial processes mediated by ion condensation of a particular anion or cation to the protein (or sorbent) surface or via unfolding phenomena then non-linear plots are predicted. Figure 1 illustrates one such case observed for hen egg white lysozyme (HEWL). Knowledge of the value of the $Z_c Z_i \epsilon^2 r_p/2DA_c$ term and the slope and intercept of the log k' versus log 1/[c] can be used to suggest potential regions of the protein surface which may be involved in the interfacial contact. For example, in the case of HEWL it has been proposed [14] that the topographical region encompassing the amino acid residues Asp[48], Asp[52], Asp[101] and Glu[35] represent the dominant electrostatic binding domain (ionotope) recognized by the quaternary ammonium ligand found in Q-type sorbents. Involvement of this topographical region, which is located around the catalytic cleft, would also be consistent with the known influence of salts, e.g. NaI, on the hinge-bending of the two domains of HEWL. Similar bimodal dependencies of log k vs. log [1/c] as shown in Fig. 1 have also been described [8] for the cation exchange behavior of subtilisin muteins where a salt bridge transition influencing the ionisation state of His[64] has been proposed. Interestingly, His[64] is part of the active site triad and, based on earlier

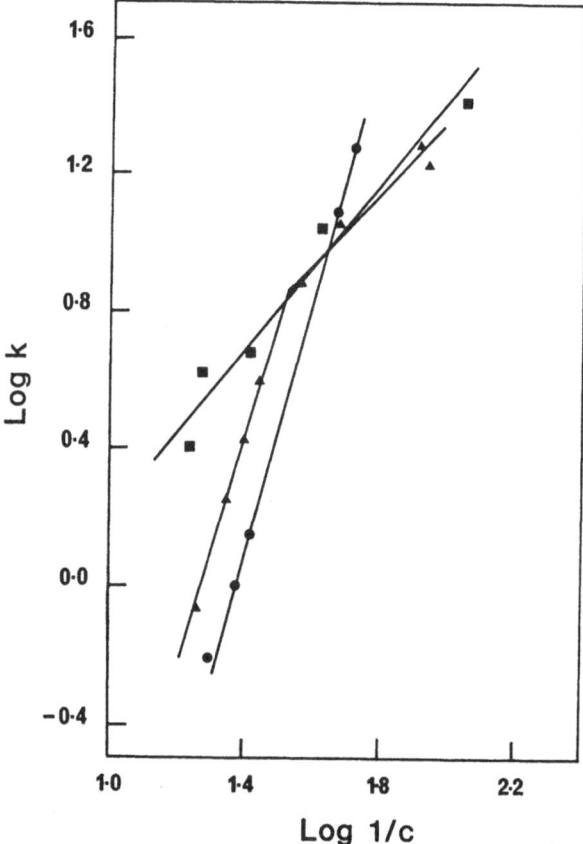

Fig. 1. Retention plots for hen eggwhite lysozyme eluted from a Mono-Q sorbent under different displacer salt conditions and column residency times. The displacer salts used were: (●) KF; (■) KCl; and (▲) KBr.

studies [15] showing the importance of the active site Ser in cation exchange retention, the His[64] has been proposed to form part of the chromatographic contact region. What these studies illustrate is the potential for HPIEC to rapidly provide, with the current generation of instrumentation, very useful insight into the charge distribution of clusters of amino acid residues accessible at the surface of folded or partially folded peptides or proteins in a particular chemical micro-environment.

Previously, similar concepts have been utilized in RPHPLC of peptides to derive retention coefficients for individual amino acid residues. These retention coefficients have been subsequently employed in various predictive algorithms to anticipate the retention behaviour of different peptides, or used in structural classifications such as hydropathy plots or the prediction of antigenic sites. What is now evident is that RPHPLC is a potent tool to aid the characterization of amphipathic and non-amphipathic features of peptides. Because of the dependency of log k' on the hydrophobic contact area, ΔA_h, (Eqn. 2), the effect of sequential changes in the secondary and tertiary structure of a specified peptide can be readily monitored by RPHPLC from plots of log k' vs. φ (or γ) and from van 't Hoff plots of log k' vs. 1/T. Such approaches have been successfully applied to studies with peptides related to growth hormone, myoglobin, tropomyosin, paracelsins, β-endorphins as well as various peptide polymers with repeating units or amino acid residue replacement/migration along a defined sequence.

Illustrative of these studies are the results shown in Table 1 for a series of synthetic peptides related to the (6–13) human growth hormone sequence.

Table 1 *Sequence (S) and log k_0 values for peptides related to hGH(6–13)*

Peptide No.	Sequence	Link	C_{18}/0.1% TFA		C_4/0.1% TFA		C_{18}/Bicarb	
			S	log K_0	S	log K_0	S	log K_0
1	LSRLFDNA	β	13.1	2.3	12.8	1.8	10.1	2.5
2	LSRLFDNA	imide	7.2	3.6	7.8	3.5	6.3	3.0
3	LSRLFENAG	β	12.0	2.1	13.1	1.9	8.1	2.1
4	LSRLFENAG	α	11.4	3.0	11.1	2.5	–	–
5	LSRLfDNA	imide	8.2	4.0	7.3	3.2	7.4	3.7
6	LFDNAG	α	11.2	1.8	12.7	1.4	13.0	2.5
7	LSRLFDNG	β	12.8	1.7	10.6	1.0	8.3	1.8
8	LSRLFENG	β	12.8	1.9	13.0	1.5	9.5	1.8

f=D-phenylalanine.

Associated studies had demonstrated [16] that the parent peptide sequence Leu-Ser-Arg-Leu-Phe-Asp-Asn-Ala is characterized by the configuration of the side chain at position Asp[6], with three alternative structures, derived from the β-Asp rearrangement, occurring following synthesis. Only the imide form is active as a hypoglycemic and insulin-potentiating peptide in in vitro and in vivo assays. The smaller S value but larger log K_0 value suggests a smaller hydrophobic contact area but one of higher binding affinity to the hydrophobic surface. These requirements would be achieved if the imide structure was more conformationally stabilized (compared to the α- or β-forms) as a nascent amphipathic structure involving the amino-terminal residues. Measurement of the standard entropy change, ΔS°_{assoc}, associated with the transfer of the peptide analogs to the sorbent surface indicated that the open chain α- and β-peptides do in fact exist in more flexible conformations while the helical structure generated with the imide form is more constrained. For example, the ΔS°_{assoc} value of the imide peptide **2** was 3.8 ± 0.6 J mol^{-1} whilst the corresponding ΔS°_{assoc} values for the α- and β-forms were -39.7 ± 1.2 J mol^{-1} and -32.8 ± 5.5 J mol^{-1}, respectively. The enthalphic change, ΔH°_{assoc}, for these peptides was -3.3 ± 0.6, -8.7 ± 0.3 and -6.1 ± 1.0 kJ mol^{-1}, respectively, which indicates for the α-, β- and imido-forms related to peptide **2**, that heat is released upon adsorption of the peptide to the hydrocarbonaceous surface. Additional spectroscopic measurements, e.g.

COSY and NOESY NMR studies, have provided [16] further supportive evidence for amphipathic helical stabilization to be enhanced through hydrogen bonding between the guanidino protons of Arg^3 and the imide structure.

This behavior for a folded peptide structure to exhibit smaller hydrophobic contact areas (i.e. smaller S-values) compared to the fully unfolded structure represents a trend, often noted anecdotally in the scientific literature. Besides leading to divergencies in the prediction of peptide retention in RPHPLC when algorithms are used which assume linear additivity of retention coefficients of amino acids as defined by the primary sequence, inappropriate definition of the contact area, as determined from Eqn. 2 also forms the basis for the limited correlations of the predictive algorithms in the definition of antigenic determinant sites based on the use of the same or similar coefficient values. A new approach under development in this laboratory to remedy this situation allows multi-sequence frames to be read as topographical coefficient motifs, which defines more accurately the hydrophobicity/hydrophilicity and accessibility of surface regions of a peptide or protein. This approach has lead to a data base of over 3000 peptide sets being accumulated and analyzed in terms of structure–retention dependencies. Based on these data, refined coefficients which describe the transfer function of individual amino acids within a reading frame of 3,5,7,9..... amino acid residues in different conformational motifs have been developed (the Matwindow program). This approach has been employed in these laboratories to characterize potential continuous and non-continuous regions of biological interest within the structures of the glycoprotein hormone subunits, inhibin subunits and the somatotrophin family. The definition of the region (6–13)hGH as a relevant, surface-accessible region was predicted using this Matwindow approach. Details of these investigations will be described elsewhere.

Acknowledgements

Investigations described in this publication have been supported by the National Health and Medical Research Council of Australia, the Australian Research Council, Monash University Special Research Grants Committee, the Potter Foundation, the Helen Schutt Trust and the Buckland Foundation. The author wishes to acknowledge the dedicated participation of Dr. M.I. Aguilar, and Messrs. A.N. Hodder, M. Wilce and A. Purcell in this research effort.

References

1. Wetlaufer, S.B. and Koenigbauer, M.R., J. Chromatogr., 259 (1986) 55.
2. Corbett, R.J. and Roche, R.S., Biochemistry, 23 (1984) 1988.
3. Benedek, K., Doug, S. and Karger, K.B., J. Chromatogr., 317 (1984) 227.
4. Hearn, M.T.W., Hodder, A.N. and Aguilar, M.I., J. Chromatogr., 327 (1985) 47.
5. Kunitani, M.G., Cunico, R.L. and Staats, S.J., J. Chromatogr., 443 (1988) 205.
6. Saito, Y. and Wada, A., Biopolymers, 9 (1983) 2123.

7. Hearn, M.T.W. and Aguilar, M.I., In Neuberger, A. and Van Deenen, L.L.M. (Eds.) Modern Physical Methods in Biochemistry, Part B, Elsevier, Amsterdam, 1988, p. 107.
8. Chicz, R.M. and Regnier, F.E., J. Chromatogr., 443 (1988) 193.
9. Hearn, M.T.W. and Aguilar, M.I., J. Chromatogr., 359 (1986) 31.
10. Hodder, A.N., Aguilar, M.I. and Hearn, M.T.W., J. Chromatogr., 476 (1989) 391.
11. C.T. Mant, N.E. Zhou and Hodges, R.S., J. Chromatogr., 476 (1989) 363.
12. Parker, J.M., Gou, D. and Hodges, R.S., Biochemistry, 25 (1986) 5425.
13. Hearn, M.T.W., Wilce, M. and Aguilar, M.I., submitted.
14. Hodder, A.N., Machin, K., Aguilar, M.I. and Hearn, M.T.W., in press.
15. Polgar, L. and Bender, M.L., Biochemistry, 8 (1969) 136.
16. O'Donoghue, M.F., Hearn, M.T.W., Ng, F., Rae, I.D. and Cross, B.A., In Jung, G. and Bayer, E. (Eds.) Peptides 1988 (Proceedings of the 20th European Peptide Symposium), De Gruyter, Berlin, 1989, in press.

Induced conformational effects during RPHPLC: Prediction of immunodominant T_H-cell antigenic sites

K. Büttner[a], J.M. Ostresh[b] and R.A. Houghten[c]

[a]*Scripps Clinic and Research Foundation, 10666 N. Torrey Pines Road, La Jolla, CA 92037, U.S.A.*
[b]*Multiple Peptide Systems, 10955 John Jay Hopkins Drive, San Diego, CA 92121, U.S.A.*

Introduction

RPHPLC is currently the method of choice for the rapid estimation of the homogeneity of synthetic peptides. While mostly based on hydrophobic interactions, the exact mode of binding of peptides with the stationary lipid phase is not fully understood. This becomes evident due to the inability of retention coefficients (which are based solely on the amino acid composition of the relevant peptides [1] to predict accurately peptide retention times during RPHPLC, as we have shown in earlier work [2]. We believe this is due to the induction of specific secondary structures in peptides by the lipid of the solid support, or, alternatively, by the increasing organic component of the mobile phase, analogous to α-helicity induced by organic solvents observed by circular dichroism. The increased retention times of peptides having α-helical amphipathic structures is used here to predict immunodominant T_H-cell sites in proteins.

Results and Discussion

In an ongoing study to investigate induced conformational effects determined by RPHPLC, we have prepared a wide range of synthetic peptides analogs. In earlier studies [3] we found that 'walking' a lysine through H-(Ala)[14]-NH₂ yielded a series of peptides having a periodicity of their retention times that is compatible with a α-helical conformation of the peptide as it interacts with the C-18 of the stationary phase. In a different series, a proline was 'walked' through the alanines of H-Lys-(Ala)[18]-Lys-NH₂ and the resulting 18 analogs were examined by RPHPLC at pH 2.1 (Fig. 1).

The variation in the retention times corresponds well with the disruption and/or destabilization of an α-helical array by proline. Thus, we believe proline causes a bend in the α-helix, with the carbonyl of the proline and the adjacent disrupted amide bonds now hydrogen bonding with H_2O [4]. This additional bound water decreases the effective hydrophobicities, and therefore, the retention times of the peptides [also see Ref. 9].

In another series of peptides, derived by 'walking' a lysine through H-(Pro)[18]-NH₂, we found a periodicity strongly indicative of a type II helix for polyproline (3_{10}-helix, trans amide bonds – 3 residues per turn). Surprisingly, this series

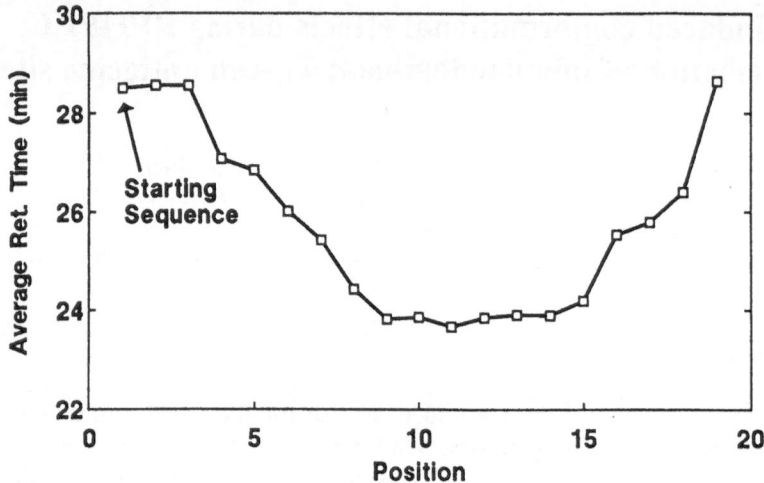

Fig. 1.   Proline 'walked' through polyalanine.

eluted approximately 15 min earlier than would be predicted with the use of retention coefficients. Modeling of this peptide as a 3_{10}-helix indicates that it is twice the length of its α-helical polyalanine counterpart, and the carbonyl groups are oriented approximately perpendicular to the axis of the helix. Both the carbonyl orientation and overall peptide length would facilitate hydrogen bonding with water, thereby imparting greatly reduced hydrophobicities, and also decreased retention times for this series, relative to that predicted by retention coefficients.

Peptides having increased retention times relative to those predicted, point to a structural stabilization of the peptide in the two-phase partitioning RP-HPLC system. We have found this to be associated with an overall amphipathic secondary structure. The work of De Lisi and Berzofsky [5] correlates amphipathicity of peptides with immunodominant T_H-cell antigenicity. If our RPHPLC results are correct and of a general nature, we should be able to predict these epitopes. As a model protein, sperm whale myoglobin (SWM) was examined, for which the antigenic T_H-cell epitopes are known. A nested series of 29 overlapping 15-residue peptides spanning the sequence of (SWM) were synthesized using the method of simultaneous multiple peptide synthesis [6], followed by RPHPLC at pH 7.2. Using this approach we were able to confirm the three immunodominant antigenic T_H-cell epitopes of (SWM) by their increased retention times on RPHPLC, relative to those predicted through the use of retention coefficients (Fig. 2). As can be deduced from the (SWM) crystal structure [7], these epitopes are all associated with helical segments of the protein, and we believe they adopt this secondary structure during RPHPLC. The same approach was applied to the circumsporozoite protein of *Plasmodium falciparum*, and all of the known immunodominant T_H-cell sites [8] were confirmed.

424

We conclude that RPHPLC offers an excellent means to study induced conformational effects of peptides at aqueous/lipid interfaces.

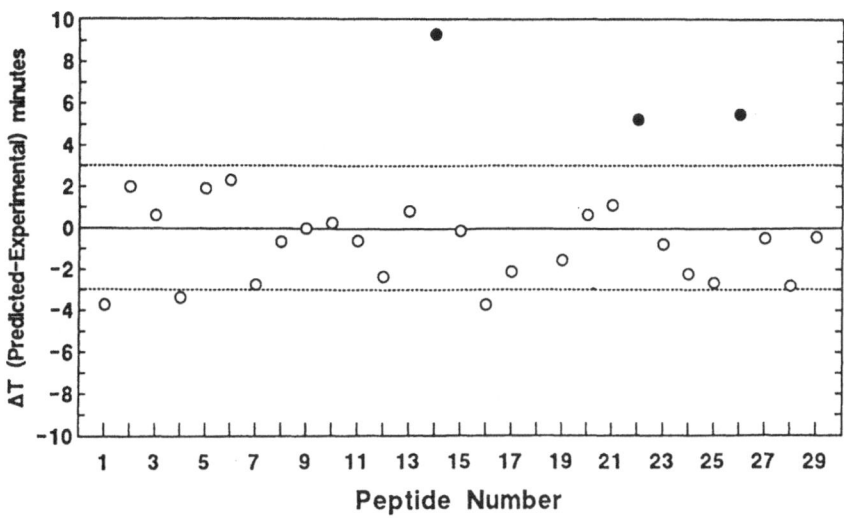

Fig. 2. Sperm whale myoglobin.

References

1. Meek, J.L., Proc. Natl. Acad. Sci. U.S.A., 77 (1981) 1632.
2. Houghten, R.A. and DeGraw, S.T., J. Chromatogr., 386 (1987) 223.
3. Houghten, R.A. and Ostresh, J.M., Biochromatography, 2 (1987) 80.
4. Houghten, R.A. and Ostresh, J.M., In Shiba and Skakibara (Eds.) Peptide Chemistry 1987, Protein Research Foundation, Osaka, Japan, 1988, p. 101.
5. De Lisi, C. and Berzofsky, J.A., Proc. Natl. Acad. Sci. U.S.A., 82 (1985) 7048.
6. Houghten, R.A., Proc. Natl. Acad. Sci. U.S.A., 82 (1985) 5131.
7. Kendrew, J.C., Dickerson, R.E., Strandberg, B.E., Hart, R.G., Davies, D.R., Phillips, D.C. and Shore, V.C., Nature, 185 (1960) 442.
8. Good, M.F., Pombo, D., Quakyi, I.A., Riley, I.M., Houghten, R.A., Menon, A., Alling, D.W., Berzofsky, J.A. and Miller, L.H., Proc. Natl. Acad. Sci. U.S.A., 85 (1988) 11995.
9. Karl, I.L. and Flippen-Anderson, J.L., Biopolymers, 28 (1989) 773.

Determination of the sequence of peptides and proteins by tandem mass spectrometry

Klaus Biemann, James E. Biller, James A. Hill, Richard S. Johnson,
Stephen A. Martin, Ioannis A. Papayannopoulos and James E. Vath

Department of Chemistry, Massachusetts Institute of Technology, Cambridge, MA 02139, U.S.A.

Introduction

Over the past few years, we have developed high-performance tandem mass spectrometry (MS/MS) [1] into a technique for the determination of the structure of peptides up to 20–25 amino acids in length [2]. This is just the size of peptides generated by cleavage of proteins with various proteolytic enzymes, and thus represents the basis of a strategy for the determination of the primary structure of proteins. The most important characteristic of MS/MS for peptide sequencing is its speed (it takes only 1–2 min of instrument time for each peptide to collect all the data required), and the fact that this can be accomplished with a mixture of peptides, i.e., the digest does not have to be separated into the individual components.

In high-performance, MS/MS [3], the sample mixture of peptides is dissolved in a matrix, such as glycerol or thioglycerol, and ionized in the ion source of the first double focusing MS (MS-1) by irradiation with a beam of xenon atoms [4] or cesium ions, which causes the formation of protonated peptide molecules, $(M_1 + H)^+$, $(M_2 + H)^+$, ... These are then separated by mass in MS-1, resulting in the determination of the molecular weights (accurate to $\pm < 0.3$ Da) of the individual peptides present in the mixture. Alternatively, it permits passing any selected ion beam (the ^{12}C-only specie of each $M + H$ ion cluster) into a collision cell filled with 10^{-3} torr helium, where collision-induced dissociation (CID) into fragments takes place. Fragmentation occurs along the peptide bonds, producing ions differing in mass by the mass increment of an amino acid [2]. These fragment ions (F_1, F_2, ...) are then mass-analyzed in the second MS (MS-2). The CID spectrum of a peptide can thus be interpreted by following the spacing (in mass) between peaks from low mass to high mass, which reveals the sequence from the N-terminus to the C-terminus, if the fragmentation leads to retention of the charge at the N-terminus, and vice versa for fragments retaining the charge at the C-terminus.

A detailed study of the CID spectra of hundreds of peptides of known or originally unknown sequence led to a good understanding of these fragmentation processes and their dependence upon the presence and location of certain amino acids, especially the basic arginine and lysine. Unique fragmentation processes

426

that involve the side chain of the amino acid located at the site of peptide bond cleavage allow the differentiation of the isomeric pair leucine and isoleucine [5]. To aid in the interpretation of the CID spectra of peptides produced under the conditions of our experiments (monoisotopic fragment ions produced at kilovolt collision energies), we have written a computer program (SEQPEP) that incorporates all this information and takes an average of only 2–3 min to generate a ranked list of sequences that are compatible with the data [6].

Since the identity and location of an amino acid is deduced from the mass difference of consecutive peaks due to cleavage at successive, analogous peptide bounds, this MS approach also permits the identification and location of modified amino acids that may be produced by posttranslational modification, planned or accidental chemical conversions, or other processes. Acylation at the amino-terminus is just one such modification, which not only does not interfere with the methodology but also reveals the size (and usually the identity) of the acyl group.

Results

We have developed this strategy in the course of the determination of the primary structure of a number of thioredoxins [7], and present here the work on the structure of a glutaredoxin isolated from a mammalian source, rabbit bone marrow. In contrast to thioredoxins, eukaryotic glutaredoxins tend to have a blocked N-terminus which frustrated earlier attempts to determine the sequence by the conventional Edman degradation. Thus, MS/MS was the obvious choice for the determination of the structure of this protein.

A tryptic digest of the glutaredoxin was partially separated by HPLC into a few fractions containing 1–5 peptides each. The $(M + H)^+$ ions produced in MS-1 ranged from m/z 542.0 to 2242.8, and of all but the last one, complete CID spectra were obtained. One of them clearly indicated the sequence Ac-AQEFVNSK and must thus represent the N-terminus. Two other digests, utilizing chymotrypsin and *S. aureus* V8 protease, generated two entirely different sets of peptides that were also analyzed by MS and MS/MS. The resulting individual peptide sequences exhibited considerable overlap and redundancy, making it possible to arrive at the complete 106-amino acid sequence of the rabbit bone marrow glutaredoxin with a high degree of confidence [8]. The final structure exhibited a large degree of homology with the related thioltransferase from pig liver [9], but differed from the published structure of calf thymus glutaredoxin [10] in two significant aspects. The latter was believed to have an N-terminal pyroglutamic acid and lacked a 4-amino acid segment (TVPR), present in both the rabbit glutaredoxin and the thioltransferase. A reinvestigation [11] by MS/MS of the chymotryptic digest of the calf thymus protein clearly revealed that it also had an N-terminal Ac-AQ... sequence, and the four amino acids TVPR were indeed present at positions analogous to the other two examples.

Discussion

These two examples clearly demonstrate that MS/MS has reached the point that it can be used for the determination of the hitherto unknown primary structure of a protein, as well as for quick and reliable checks of previously proposed structures. In the latter case – as well as in other instances, such as structures deduced from DNA sequences, mutants, synthetically modified sequences, etc. – it often suffices to determine only the molecular weights of peptides generated by specific enzymatic or chemical cleavage. Wherever an uncertainty remains, it can be solved by acquiring the CID spectrum of the particular peptide(s) to determine the sequence in question.

More recent improvements were directed toward an increase in sensitivity of the instrumentation, to make it possible to make these measurements on the picomole, rather than nanomole level, and to eliminate losses during the manipulation of the material between the digestion step and the introduction of the peptides into the ion source of the MS. For these reasons, we have designed and constructed a focal plane detector [12] that allows the simultaneous recording of up to 30% of the CID spectrum [13]. The complete spectrum can thus be recorded in a number of consecutive steps, requiring only a few seconds in total. This integrating mode increases sensitivity about 100-fold. However, in order to translate this sensitivity increase into the capability of sequencing smaller amounts of proteins, transfer losses have to be eliminated. Injection of the digest into a packed capillary HPLC column that is interfaced directly or indirectly to the first of the two MS, permits the efficient handling of digests of 50–100 pmol of protein [14].

With these improvements, peptide and protein sequencing by MS/MS should become a useful alternative to the automated Edman degradation. The relatively high capital investment required for the former is more than compensated by its much higher efficiency and speed, which should be particularly important for laboratories and organizations where a large amount of sequencing work has to be accomplished in a timely fashion. In addition, MS/MS is a generally applicable method suitable for many other areas of biochemistry and biotechnology beyond the specific protein structure problems outlined above.

Acknowledgements

We are grateful to S. Hopper for the glutaredoxin from rabbit bone marrow and for financial support from the National Institutes of Health (RR00317 and GM05472) and the National Institute of Environmental Health Sciences (ES04675).

References

1. Sato, K., Asada, T., Ishihara, M., Kunihiro, F., Kammei, Y., Kubota, E., Costello, C.E., Martin, S.A., Scoble, H.A. and Biemann, K., Anal. Chem., 59 (1987) 1652.

2. Biemann, K., Biomed. Environ. Mass Spectrom., 16 (1988) 99.
3. Biemann, K. and Scoble, H.A., Science, 237 (1987) 992.
4. Barbar, M., Bordoli, R.S., Sedgwick, R.D. and Tyler, A.N., J. Chem. Soc. Chem. Commun., (1981) 325.
5. Johnson, R.S., Martin, S.A., Biemann, K., Stults, J.T. and Watson, J.T., Anal. Chem., 59 (1987) 2621; Johnson, R.S., Martin, S.A. and Biemann, K., Int. J. Mass Spectrom. Ion Processes, 86 (1988) 137.
6 Johnson, R.S. and Biemann, K., Biomed. Environ. Mass Spectrom., 18 (1989) 94S.
7 Johnson, R.S. and Biemann, K., Biochemistry, 26 (1987) 1209; Johnson, R.S., Mathews, W.R., Biemann, K. and Hopper, S., J. Biol. Chem., 263 (1988) 9589; Mathews, W.R., Johnson, R.S., Cornwell, K.L., Johnson, T.C., Buchanan, B.B. and Bieman, K., J. Biol. Chem., 262 (1987) 7537; Johnson, T.C., Yee, B.C., Carlson, D.E., Buchanan, B.B., Johnson, R.S., Mathews, W.R. and Biemann, K., J. Bacteriol., 170 (1988) 2406.
8. Hopper, S., Johnson, R.S., Vath, J.E. and Biemann, K., J. Biol. Chem., 264 (1989) 20438.
9 Yang, Y., Gan, Z.-R. and Wells, W.W., Gene, 83 (1989) 339.
10. Klintrot, I.-M., Hoog, J.-O., Jornvall, H., Holmgren, A. and Luthman, M., Eur. J. Biochem., 144 (1984) 417.
11. Papayannopoulos, I.A., Gan, Z.-R., Wells, W.W. and Biemann, K., Biochem. Biophys. Res. Commun., 159 (1989) 1448.
12. Hill, J.A., Martin, S.A., Biller, J.E. and Biemann, K., Biomed. Environ. Mass Spectrom., 17 (1988) 147.
13. Hill, J.A., Biller, J.E., Martin, S.A., Biemann, K. and Ishihara, M., presented at the 37th ASMS Conf. on Mass Spectrometry and Allied Topics, Miami Beach, FL, May 1989.
14. Cappiello, A., Palma, P., Papayannopoulos, I.A., Biemann, K., Crescentini, G. and Bruner, F., presented at the 37th ASMS Conf. on Mass Spectrometry and Allied Topics, Miami Beach, FL, May 1989.

An examination of the potential of capillary zone electrophoresis for the analysis of polypeptide samples

John Frenz, John Battersby and William S. Hancock*

Department of Medicinal and Analytical Chemistry, Genentech Inc.,
460 Point San Bruno Blvd., South San Francisco, CA 94080, U.S.A.

Introduction

Capillary zone electrophoresis (CZE) is a promising approach to the analysis of biopolymers [1,2] that combines the instrumental convenience of HPLC with the selectivity afforded by electrophoretic processes. Theoretical plate counts in excess of one million have been reported for separations of oligonucleotide mixtures in gel-filled capillaries [1]. Zone electrophoretic separations of proteins and peptides have not yielded efficiencies as high as these, although resolution of charge variants of human growth hormone and insulin have been reported [3,4]. In other cases, larger proteins and glycoproteins have yielded poor peak shapes or have not been eluted from the capillary (unpublished results). The adverse behavior of these species may be linked to sample microheterogeneity, e.g., due to the carbohydrate moieties in glycoproteins [2], internal wall interactions with the capillary [5], or to Joule heating of the electrophoresis buffer [6]. Therefore, it may be more convenient to analyze the peptides resulting from enzymatic digestion, rather than the intact molecule. Such analyses are well known in HPLC, where the tryptic map has become a standard tool of analytical protein chemistry [7].

The electropherograms of the peptides resulting from digestion of human growth hormone (hGH) with trypsin have been reported previously [3,4]. Trypsin has the disadvantage that it can yield peptides with little differences in charge, with many of the resulting peptides carrying a net charge of +2 at low pH, and therefore little basis for separation in CZE. Yet acidic conditions may be desirable in CZE because, under these conditions, interactions between the sample and the silanols at the capillary wall are minimized, electroosmotic flow is negligible and all the peptides electrophorese in the same direction, toward the cathode. Therefore, in this report we will explore capillary electrophoretic maps of peptides generated by digestion of hGH with chymotrypsin and *S. aureus* V8, as well as with trypsin.

Results and Discussion

The CZE maps resulting from digestion of hGH with the three proteolytic

---

*To whom correspondence should be addressed.

430

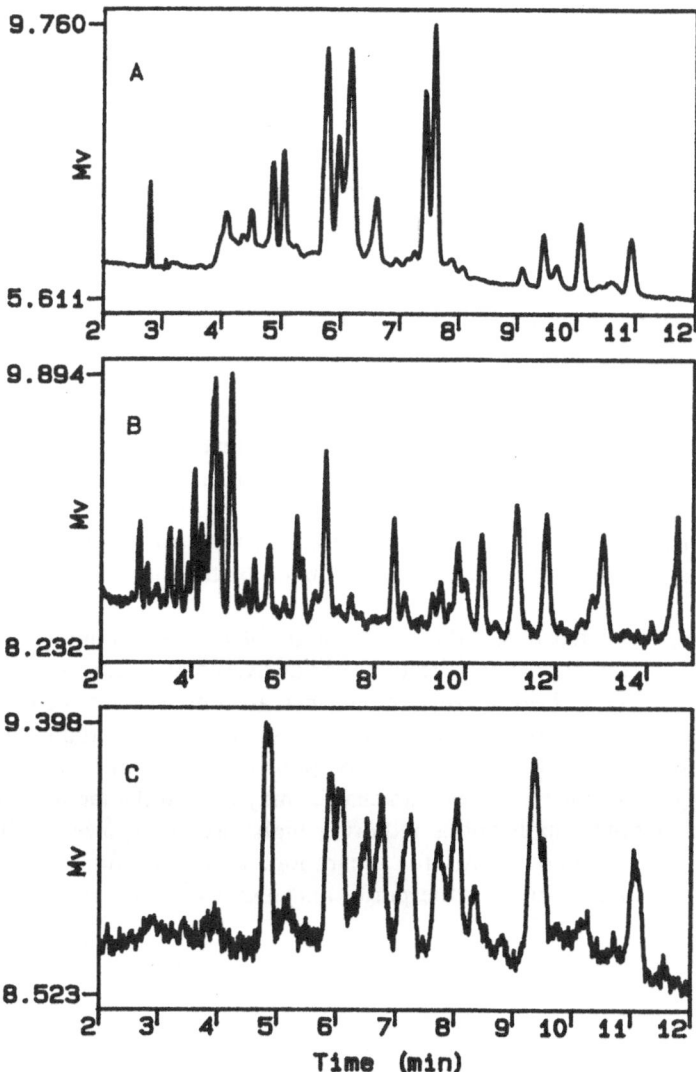

Fig. 1. Electropherograms of hGH digested with (A) trypsin, (B) chymotrypsin and (C) S. aureus V8. CZE was carried out on a Bio-Rad HPE 100, in 20 cm × 25 μm capillaries, in a 100 mM phosphate buffer, pH 2.5, at 8 kV with detection at 200 nm. Tryptic and chymotryptic digestions were carried out in 100 mM Tris–acetate buffers, pH 8.3, with enzyme:hGH ratios of 1% for 4 h at 37°C. S. aureus V8 digestion was performed in 200 mM sodium phosphate, pH 7.8, for 18 h at 37°C with an enzyme:hGH ratio of 30%.

enzymes are shown in Fig. 1. The identities of individual peaks in the tryptic map, Fig. 1A, were determined by injection of fractions from the standard reversed phase HPLC map of this mixture, as reported earlier [4]. In the CZE tryptic map, there is a cluster of peaks eluting around 6 min in Fig. 1A that are poorly resolved.

The chymotryptic map of hGH is shown in Fig. 1B. Chymotrypsin cleaves the protein after hydrophobic residues, so does not exhibit the charge-bias of trypsin. The chymotryptic map shows a large number of peaks, most of them well-resolved. Some clustering of the peaks around 4 min is evident in Fig. 1B. This result indicates that chymotrypsin provides the altered selectivity that is expected to yield a higher resolution map in electrophoresis. The specificity of chymotrypsin, however, is poor relative to trypsin, making it difficult to predict where in the protein cleavage occurs. For example, assuming that chymotrypsin cleaves after phenylalanine, tyrosine and tryptophan, there should be 22 peptides in the tryptic map. Figure 1B , and the reversed phase map of this mixture (data not shown) both indicate that many more peptides are produced by proteolysis under the conditions employed here.

The V8 digest map is shown in Fig. 1C, and reveals the relatively smaller number of peaks, well spaced in the electropherogram, that are produced by cleavage after acidic residues. The low concentrations of peptides shown in Fig. 1C are indicative of the problems associated with the low yields obtained with this protease, however, the enzyme shows much better specificity than chymotrypsin.

The results reported here indicate that some of the shortcomings of trypsin in enzymatic digest mapping applications can be overcome by the use of different proteolytic enzymes. Two alternative enzymes are not without their own disadvantages, however. Since trypsin is the protease of choice for studies of protein structure, and yields peptides that may be poorly resolved in CZE, separation based on hydrophobicity, rather than charge, may remain the method of choice for tryptic mapping applications requiring high resolving power. CZE, since it provides an alternative basis for separation, is a powerful supplementary technique for confirmation of the purity and identity of the peptides isolated by HPLC.

References

1. Karger, B.L., Cohen, A.S. and Guttman, A., J. Chromatogr., in press.
2. Grossman, P.D., Colburn, J.C., Lauer, H.H., Nielsen, R.G., Riggin, R.M., Sittampalam, G.S. and Rickard, E.C., Anal. Chem., 61 (1989) 1186.
3. Nielsen, R.G., Sittampalam, G.S. and Rickard, E.C., Anal. Biochem., 177 (1989) 20.
4. Frenz, J., Wu, S.-L. and Hancock, W.S., J. Chromatogr., 481 (1989) 379.
5. Hjertén, S., J. Chromatogr., 347 (1985) 191.
6. Grushka, E., McCormick, R.M. and Kirkland, J.J., Anal. Chem., 61 (1989) 241.
7. Hancock, W.S., Bishop, C.A., Partridge, R.L. and Hearn, M.T.W., Anal. Biochem., 89 (1978) 203.

Chemical recognition of terminally amidated peptides

Gordon Marc Loudon[a], Hee-Sung Choi[a] and Philip C. Andrews[b]

[a]*Department of Medicinal Chemistry and Pharmacognosy, School of Pharmacy and Pharmacal Sciences, and* [b]*Department of Biochemistry, Purdue University, West Lafayette, IN 47907, U.S.A.*

Introduction

Peptides amidated at their carboxy termini invariably have neuropeptide or hormonal activity [1]. Although such peptides can, in principle, be identified by a bioassay-directed search, a strategy involving chemical recognition of the C-terminal amide feature would be more general, and could well lead to the isolation of previously unknown peptide hormones [2,3]. We have developed nonenzymic chemistry that should be usable to determine whether a sample peptide is terminally amidated. The reactions take advantage of the reagent *1,1*-bis(trifluoroacetoxy)iodobenzene (PIFA, **1**), which has been used for other unique chemistry of peptides [4–7] (Scheme 1).

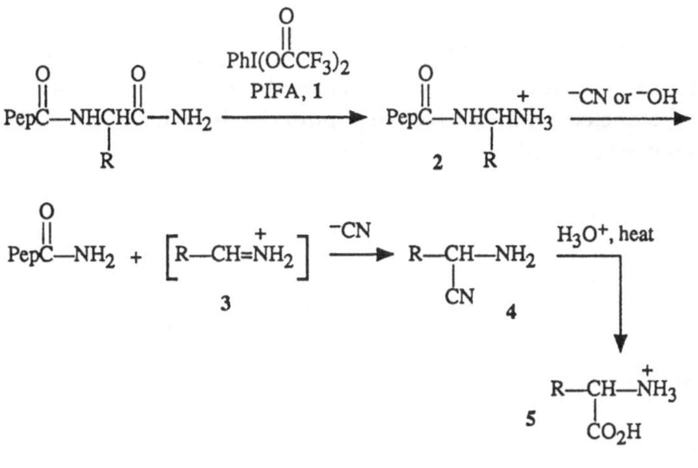

Scheme 1.

Results and Discussion

The formation of a cyanide-trapable fragment **3** is what differentiates C-terminal amides from both nonamidated C-terminal residues and the side chains of Gln and Asn. Although the latter residues do undergo the PIFA-induced rearrangement [7], they do not fragment further under basic conditions [8].

After testing this scheme on a number of *N*-acetyl amino acid amides with

433

promising results, it was applied to peptide amides, with the results shown in Table 1. In each control, an identical sample was subjected to the same procedure, but without cyanide. Hydrolysis of the reaction mixture gave the appropriate amino acid **5** corresponding to the C-terminal residue; in the control, this residue was largely absent. The controls show that rearrangement takes place in nearly quantitative yield.

Table 1   *Reactions of peptides*[a]

Peptide (nmol used)	PIFA (amount)	NaCN (amount)	Recovery of C-term. residue (%)	
			+CN⁻	no CN⁻ (control)
α-MSH[b] (89.5)	50 equiv.	98 equiv.	58.0	3.4
N-R-K-R-H-amide (141.1)	30 equiv.	49 equiv.	49.1	2.5
N-I-F-N-Q-amide (563.3)	21 equiv.	41 equiv.	62.5	6.9

[a] Peptides were treated with large excess of PIFA in DMF/H_2O; excess PIFA was extracted, and the reaction mixture was treated with excess cyanide or no cyanide (control) at 140°C, 18 h in DMSO.
[b] N-Ac-S-Y-S-M-E-H-P-R-W-G-K-P-V-amide.

Having validated the proposed chemistry, we envision that, in future studies, the use of radiolabeled cyanide will afford enhanced sensitivity, and will permit us to determine whether a peptide of unknown structure is terminally amidated as follows: we hypothesize that the trapping of imine **3** (or aldehyde hydrolysis product) will be the only cyanide-fixing reaction observed, and thus, that radioactivity will be irreversibly fixed in the sample only when it is a terminally amidated peptide. Moreover, hydrolysis of the reaction mixture should yield a single radiolabeled amino acid corresponding to the C-terminal residue. Further work is required to validate this hypothesis.

Acknowledgements

We gratefully acknowledge support of this work by NIH grant GM-37500.

References

1. Eipper, B.A. and Mains, R.E., Ann. Rev. Physiol., 50 (1988) 333.
2. Treston, A.M., Cuttitta, F., Yergey, A.L., Mulshine, J.L., and Kaspryzk, P.G., In Rosen, S.T., Mulshine, J.L., Cuttitta, F. and Abrams, P.G. (Eds.) Biology of Lung Cancer – Diagnosis and Treatment, Marcel Dekker, New York, 1988, p. 92.
3. Tatemoto, K. and Mutt, V., Proc. Natl. Acad. Sci. U.S.A., 75 (1978) 4115.
4. Parham, M.E. and Loudon, G.M., Biochem. Biophys. Res. Comm., 80 (1978) 437.
5. Pallai, P.V., Richman, S., Struthers, R.S. and Goodman, M., Int. J. Pept. Prot. Res., 21 (1983) 84.
6. Kemp, D.S. and Stites, W.S., Tetrahedron Lett., 29 (1988) 5057.
7. Blodgett, J.K. and Loudon, G.M., J. Am. Chem. Soc., 111 (1989) 6813.
8. Loudon, G.M., Almond, M.R. and Jacob, J.N., J. Am. Chem. Soc., 103 (1981) 4508.

Affinity methods for purifying large synthetic peptides

H. Ball[a], C. Grecian[a], S.B.H. Kent[b] and P. Mascagni[a]

[a]School of Pharmacy, 29–39 Brunswick Square, London WC1N 1AX, U.K.
[b]California Institute of Technology, Pasadena, CA 91125, U.S.A.

Introduction

The principal impurities arising from the stepwise nature of peptide synthesis are truncated peptides and deletion peptides. For peptides larger than 40 residues, the purification of the target peptide from these impurities, using conventional chromatographic and electrophoretic methods, becomes a difficult task.

Results and Discussion

We have devised a purification scheme based on the acid-stable Fmoc-4-carboxylic acid **1a** which improves on previous work involving either attachment of a Sulph-Fmoc group to the peptide [1], or a covalent biotin-peptide derivative [2]. **1a** can be derivatized with one of the groups shown in Fig. 1, and introduced via the urethane moiety onto the last amino acid of the sequence being synthesized. After HF cleavage, the purification of the peptide, thus derivatized, should be carried out using either reverse phase, ion-exchange or affinity chromatography. If this purification scheme is combined with capping procedures after each coupling step of the synthesis, then the purified peptide should also be free of deletion peptides.

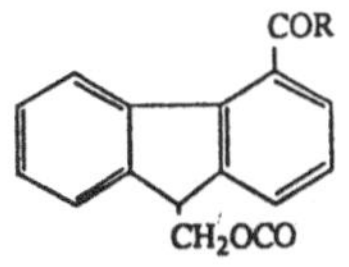

1b   R1 = $NH_2CH(CH_2)_7CH_3COOtBu$

1c   R2 = $NH_2CH(CH_2)_7CH_3CONHCH(CH_2)_7CH_3COOCH_3$

1d   R3 = NH_2-(2ClCBZ)Lys-(2ClCBZ)Lys-(CBZ)Lys-OCH_3

Fig. 1. Fmoc derivatives used in this study.

To probe our hypothesis, we synthesized FMDV VP1-143-160 using the Boc-amino acid chemistry, and introducing the last residue (Gly) as the Fmoc derivative **1b**. After HF cleavage, the crude peptide was eluted through a C-18 reverse phase column. Figure 2 shows the HPLC chromatogram of the FMDV derivative; the peptide containing the Fmoc derivative eluted 8 min later than the underivatized peptide, thus confirming our hypothesis.

435

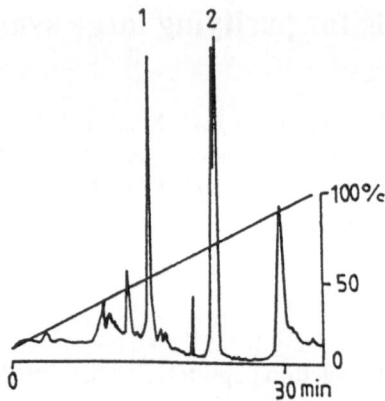

*Fig. 2. C-18 RPHPLC of the FMDV(143–160) derivative with **1b**. 1 = underivatized peptide; 2 = derivatized peptide.*

Next we synthesized the 39–86 HIV-1 TAT fragment using the Boc strategy. However, this time, capping with 20% acetic anhydride in DCM containing 5% DIEA was performed after each coupling. The resin-peptide was stored in DMF at –70°C, and subsequently treated with TFA to remove the Boc group. Only 20% of the peptide chains were deprotected, as indicated by a quantitative ninhydrin test conducted after two TFA treatments.

Using a manual shaker, the Gly derivative of **1c** was then added as the HOBt active ester. The incorporation was monitored by taking small aliquots of peptide-resin and treating them with 20% piperidine in DMF. Readings of fluorene UV absorbance at 270 nm were then taken. The reaction was stopped when two successive such readings were found to be identical. After HF cleavage, the crude peptide was eluted on a C-4 column with a AcCN linear gradient. The derivatized peptide eluted at 68% AcCN, which contrasted with 33% for the uncoupled peptide. RPHPLC of 15 mg of crude material allowed the separation of these two peptides. Their molecular weight was assessed by SDS-PAGE and, in both cases, found to correspond to the expected masses.

Conclusions

These experiments have shown that the proposed scheme can be successfully applied to the purification of peptides of about 50 residues. The application of this procedure to 100-residue peptides using **1d** is currently under investigation.

References

1. Merrifield, R.B. and Bach, A.E., J. Org. Chem., 43 (1978) 4808.
2. Lobl, T.J., Deibel, M.R. and Yen, A.W., Anal. Biochem., 170 (1988) 502.

RPHPLC resolution of enantiomeric N-methyl amino acids by GITC (2,3,4,6-tetra-O-acetyl-β-D-glucopyranosyl isothiocyanate) derivatization

Rainer Albert and François Cardinaux

Preclinical Research, Sandoz Ltd., CH-4002 Basle, Switzerland

Introduction

N-Methyl amino acids (NMe-AA) are known to occur in natural products (e.g. cyclosporin) and are often used in synthetic peptide analogs. For the quality control of starting materials, as well as for the analysis of racemization during peptide synthesis, a simple and rapid method for determining enantiomeric purity of NMe-AA is required. Existing methods [1] for the separation of NMe-AA are unsatisfactory. The stable reagent GITC [2] was found to react rapidly with a DL-NMe-AA forming diastereomeric thiourea derivatives [3], that are easily separable by RPHPLC.

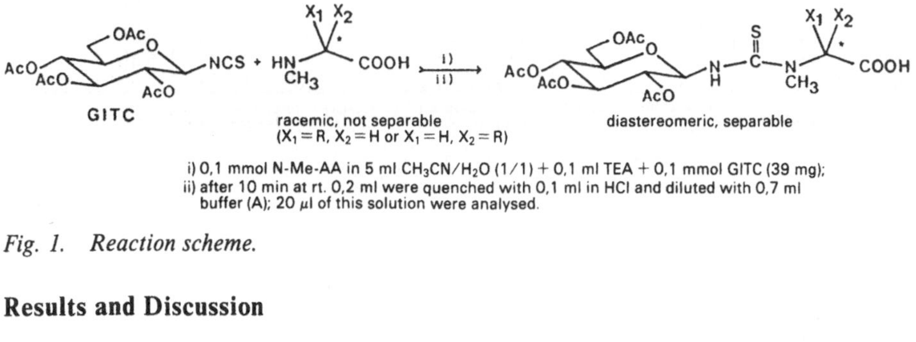

i) 0,1 mmol N-Me-AA in 5 ml CH₃CN/H₂O (1/1) + 0,1 ml TEA + 0,1 mmol GITC (39 mg);
ii) after 10 min at rt. 0,2 ml were quenched with 0,1 ml in HCl and diluted with 0,7 ml buffer (A); 20 µl of this solution were analysed.

Fig. 1. Reaction scheme.

Results and Discussion

Using this method we were able to determine the enantiomeric purity of a variety of NMe-AA in starting materials (Fig. 2), as well as in natural and synthetic peptides after acid hydrolysis (Fig. 3).

This simple and rapid method is recommended for the racemization analysis of NMe-AA by RPHPLC on a standard support and common buffer systems. Detection at 250 nm gives the best signal-to-noise ratio.

Acknowledgements

We wish to thank Miss U. Zweifel for developing the separation conditions and for many analyses.

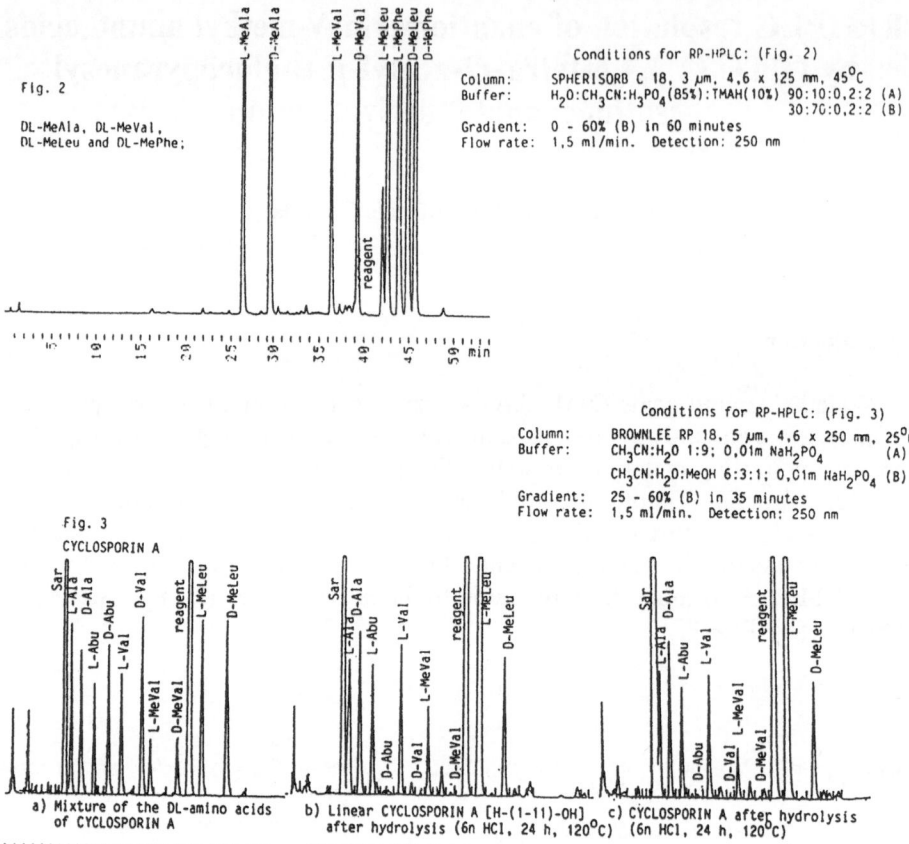

Fig. 2

DL-MeAla, DL-MeVal,
DL-MeLeu and DL-MePhe;

Conditions for RP-HPLC: (Fig. 2)

Column: SPHERISORB C 18, 3 μm, 4,6 x 125 mm, 45°C
Buffer: H₂O:CH₃CN:H₃PO₄(85%):TMAH(10%) 90:10:0,2:2 (A)
                                              30:70:0,2:2 (B)

Gradient: 0 - 60% (B) in 60 minutes
Flow rate: 1,5 ml/min. Detection: 250 nm

Conditions for RP-HPLC: (Fig. 3)

Column: BROWNLEE RP 18, 5 μm, 4,6 x 250 mm, 25°C
Buffer: CH₃CN:H₂O 1:9; 0,01m NaH₂PO₄   (A)
             CH₃CN:H₂O:MeOH 6:3:1; 0,01m NaH₂PO₄ (B)

Gradient: 25 - 60% (B) in 35 minutes
Flow rate: 1,5 ml/min. Detection: 250 nm

Fig. 3
CYCLOSPORIN A

a) Mixture of the DL-amino acids of CYCLOSPORIN A

b) Linear CYCLOSPORIN A [H-(1-11)-OH] after hydrolysis (6n HCl, 24 h, 120°C)

c) CYCLOSPORIN A after hydrolysis (6n HCl, 24 h, 120°C)

Figs. 2 and 3.

References

1. Brueckner, H., Chromatographia, 24 (1987) 725.
    Porter, J., Dykert, J. and Rivier, J., Int. J. Pept. Prot. Res., 30 (1987) 13.
    Lindner, W.F. and Hirschboeck, I., J. Liq. Chromatogr., 9 (1986) 551.
    Koenig, W.A., Steinbach, E. and Ernst, K., J. Chromatogr., 301 (1984) 129.
    Koenig, W.A., Benecke, I., Lucht, N., Schmidt, E., Schulze, J. and Sievers, S., J. Chromatogr., 279 (1983) 555.
2. GITC is commercially available from Aldrich Chemical Company, Milwaukee, WI.
3. Nimura, N., Toyama, A. and Kinoshita, T., J. Chromatogr., 316 (1984) 547.

Comparison of capillary electrophoresis of pentigetide and related peptides with HPLC, TLC and paper electrophoresis

Thomas J. Lobl[a], Jo-Lynne S. Boone[a], Awaz H. Ahmed[a], Thomas J. Stolzer[a] and
Joel Colburn[b]

[a]Quality Control/Analytical Chemistry Department, Immunetech Pharmaceuticals,
11045 Roselle Street, San Diego, CA 92121, U.S.A.
[b]Applied Biosystems Inc., 3745 North First Street, San Jose, CA 95131, U.S.A.

Introduction

Homologous polar peptides often represent a difficult RPHPLC chromatographic problem in terms of both retention and resolution. Pentigetide, a polar peptide (Asp-Ser-Asp-Pro-Arg; DSDPR) derived from immunoglobulin E (IgE) Fc region, is an especially good example of this problem [1]. All the likely contaminants in a synthetic mixture would be expected to have similar retention properties to pentigetide, and therefore, would be difficult to resolve.

Pentigetide has been shown to inhibit hypersensitivity in the Prausnitz-Küstner reaction [2] and is being tested for the treatment of allergic rhinitis, allergic conjunctivitis, asthma, and other inflammatory conditions [3,4]. The need to measure synthetic impurities and decomposition products prompted us to evaluate capillary electrophoresis (CE) as an alternate technique for RPHPLC. CE is a sensitive, high resolution technique that separates molecules based on charge, molecular size and shape [5–7]. It is now possible to evaluate the utility of CE in analytical chemistry, where reproducibility and variability constraints are rigorous. We describe now the CE studies of very similar hydrophilic peptides, and compare them to RPHPLC, TLC and high voltage paper electrophoresis (HVE).

Materials and Methods

Capillary electrophoresis: Capillary electrophoresis was done on a Model 270A from Applied Biosystems Inc., using the following standard conditions: 0.1 M NaOH wash, 20 mM sodium citrate running buffer, pH 2.5, 20 000 V, 30°C, and monitored at 200 nm. Depending on the experiment, the voltage, temperature, buffer, and pH were varied. Each run contained an internal standard (RKRSRKE). The fundamental property, electrophoretic mobility (μ_{ep}), was calculated for each peptide from the equation:

$$\mu_{ep} = [(L_D)(L_t)]/[(\text{voltage})(\text{time})] - \mu_{eo}$$

where L_D and L_t are capillary length to detector and total length, respectively, and μ_{eo}(electroendoosmotic flow) was calculated from the internal marker in each run, based on the equation that $\mu_{eo} = \mu_{apparent} - \mu_{ep}$, where the $\mu_{(apparent\ marker)} = (L_D)(L_t)/(voltage)(time)$. μ_{ep} for the marker is 3.95×10^{-4} cm²/V·s.

HPLC: HPLC k′ data was generated as reported earlier [1]. *TLC*: TLC was done on silica gel (E. Merck), developed in *n*-butanol/acetic acid/water/ethyl acetate (1:1:1:1) and visualized using ninhydrin. *HVE*: HVE was done on Whatman 3 MM chromatography paper, and developed with 0.2 M Na₂HPO₄, pH 6.5 for 1 h at 2 kV, 20–25°C and ~35 mA. Spots were visualized using ninhydrin and are relative to L-aspartic acid. All peptides were synthesized at Immunetech, as described earlier [1].

Results and Discussion

Our initial objective was to show that CE could give equivalent, or better resolution and retention than RPHPLC. The chromatogram (Fig. 1a) shows that DSDPR, D(DS)DPR, β-DSDPR, DPR, PR and SDPR are resolved. The very polar DSD comes out with the solvent front. CE is able to separate the same compounds (Fig. 1b). Unlike HPLC, where it was difficult to achieve retention with polar compounds such as DSD, CE elutes it clearly and slowly. The μ_{ep} for DSD is 1.23×10^{-4} cm²/V·s (migration time (t_m) = 26 min) under standard conditions. Even the polar peptide, DSD, had long elution times due to its low charge with these varieties of pH and buffer conditions in CE (data not shown) [8]. The data shown in Table 1 and in Fig. 1 shows that HPLC and CE are comparable in resolution. CE was able to separate structural isomers with identical charge and molecular weight to pentigetide. The ability to manipulate mobility with increasing voltage or temperature during CE of pentigetide gives more flexibility to the methods development than HPLC [8].

Table 1 *Comparison of CE with HPLC, TLC and HVE*

Compound	CE μ_{ep} ($\times 10^{-4}$ cm²/V·s)	HPLC k′	TLC R_f	HVE mobility
β-DSDPR	1.52	4.69	0.15	0.40
DSDPR	1.91	4.54	0.18	0.43
D(D-Ser)DPR	1.95	6.80	0.18	0.39

A second objective was the demonstration that CE was superior to TLC and HVE in resolution capability. The electropherogram (Fig. 1b) and the data in Table 1, demonstrate that the narrow peaks and high resolution for CE show clear superiority over the low resolution information obtained from TLC and HVE, and the run times are much shorter (frequently 15–30 min versus more than 1 h).

A third objective of this study was to investigate the potential of using CE for quantitative analysis of small percentage impurities, frequently below one

a

b

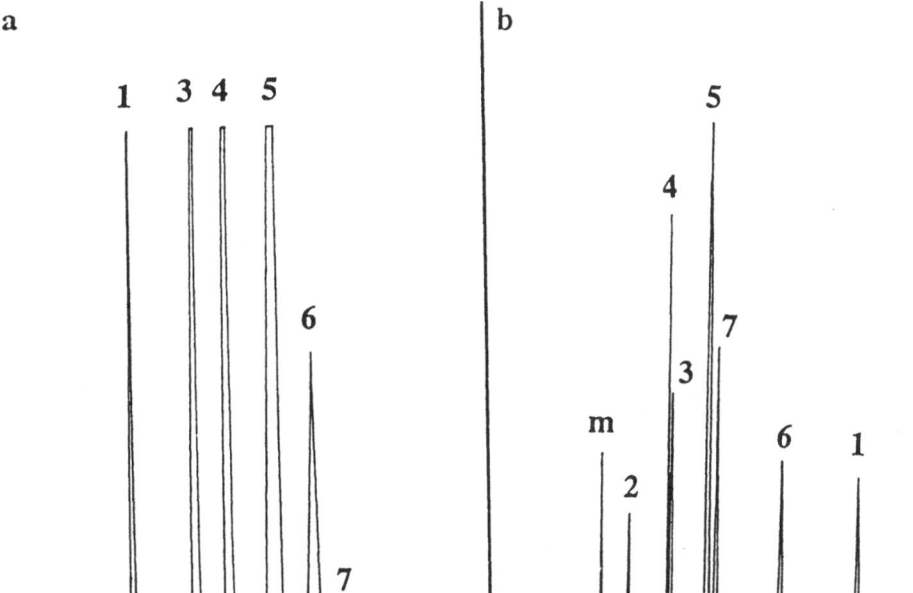

Fig. 1. (a) RPHPLC chromatogram and (b) electropherogram of pentigetide and analogs (m = marker, 1 = DSD, 2 = PR, 3 = DPR, 4 = SDPR, 5 = DSDPR, 6 = β-DSDPR, 7 = D(D-S)DPR.

percent. The linearity of the response to different concentrations of pentigetide, within and between run variation and day to day variations, show that CE has the necessary high degree of reproducibility [8]. Further, it is possible to obtain quantitative results of small percentage impurities. The runs of bulk manufactured pentigetide (not shown) showed no impurity peaks not seen previously by HPLC. Further studies are needed to optimize the quantitative analysis methodology.

Conclusion

Capillary electrophoresis was investigated as an alternative high resolution, quantitative technique to RPHPLC, TLC and paper electrophoresis with a series of polar peptidic analogs of pentigetide. The results show that the technique is superior to TLC and paper electrophoresis, and equivalent, if not better than

HPLC. CE is able to separate different structural isomers of pentigetide that have the same charge and molecular weight. Finally, no impurities were observed by CE that were not seen previously by HPLC.

References

1.   Nagarajan, G.R., Boone, J.S., Stolzer, T.J. and Richieri, S.P., In Hugli, T.E. (Ed.) Techniques in Protein Chemistry, Academic Press, San Diego, 1989, p. 357.
2.   Hamburger, R. N., Science, 189 (1975) 389.
3.   Cohen, G.A., O'Conner, R.D. and Hamburger, R.N., Ann. Allergy, 52 (1984) 83.
4.   Hahn, G.S., Nature, 324 (1986) 283.
5.   Gutzman, M.A., Hernandez, L. and Hoebel, B.G., BioPharm, 2 (1989) 22.
6.   Grossman, P.D., Colburn, J.C., Lauer, H.H., Nielsen, R.G., Riggin, R.M., Sittam-palam, G.S. and Rickard, E.C., Anal. Chem., 61 (1989) 1186.
7.   Gordon, M.J., Huang, X., Pentoney Jr., S.L. and Zare, R.N., Science, 242 (1988) 244.
8.   Lobl, T.J., Boone, J.-L., Ahmed, A., Stolzer, T. and Colburn, J., Manuscript in preparation (1989).

Identification of a thioether by-product in the synthesis of a cyclic disulfide peptide by tandem mass spectrometry

Mark F. Bean[a], Steven A. Carr[a], Emanuel Escher[b],*, Witold Neugebauer[b] and James Samanen[a]

[a]*Smith Kline & French Laboratories, King of Prussia, PA 19406, U.S.A.*
[b]*Université de Sherbrooke, Sherbrooke, Quebec, Canada J1H 5N4*

Introduction

It is now well established that MS/MS on a four-sector instrument allows much enhanced spectral information from peptides [1]. MS/MS with high-energy, collision-induced fragmentation of peptides is able, even in complex mixtures to sequence through N-terminally-blocked peptides, handle unusual amino acid modifications, resolve complex amino acid insertion/deletion problems, sequence through cystine bridges, and distinguish positional isomers of amino acids. Clarification of these structural problems is difficult by any other method. Here we illustrate the utility of the technique in uncovering an unusual peptide modification in which the internal disulfide bond of a somatostatin analog has been converted to a thioether link with retained activity.

Results and Discussion

Synthesis of small conformationally-constricted photoaffinity labels related to sandostatin was undertaken preparatory to isolation of somatostatin receptors. This new approach involving the di-alcohol analog of threonine as a starting material and built up by the Boc-TFA procedure is described elsewhere in these Proceedings [2]. Sandostatin and its labeling analogs were obtained as anticipated, but in every case, an unidentified, active by-product was isolated as a significant impurity.

Initial FABMS analysis showed these minor components to have masses of 32 Da lower than expected; this could be explained either by amino acid substitutions or by a hard-to-explain loss of sulfur. High-resolution exact mass measurements indicated that the mass difference indeed corresponded to a sulfur loss. The putative disulfide link could not be reduced, and a thioether bridge was suspected, although the nature of the bridge was unclear.

The tandem mass spectrum of the intact sandostatin yielded compositional data from the R-group losses and immonium ion fragments, as well as the identity of the terminal amino acids, but no internal sequence information (Fig. 1). The lack of sequence ions below m/z 840 from the intense parent ion was suggestive

---

*To whom correspondence should be addressed.

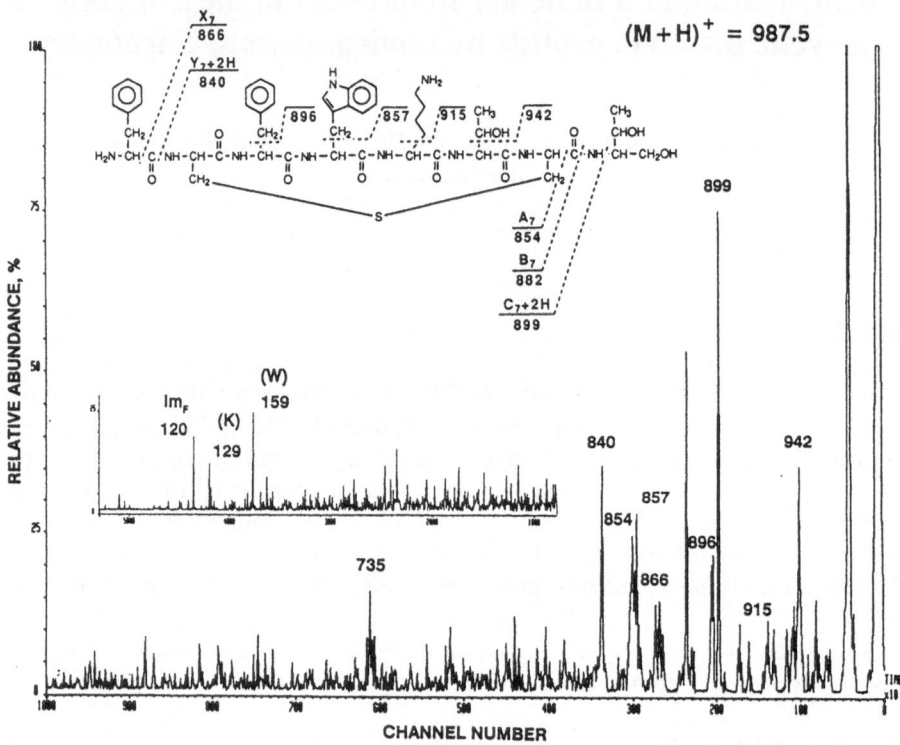

Fig. 1. *MS/MS of the intact somatostatin analog. VG ZAB-SE 4F mass spectrometer; C.E. = 10 keV; 75% attenuation of parent by He; matrix: 3-nitrobenzylalcohol/DTT/DTE (9 : 3 : 1) with 0.1% TFA; Cs ion gun operated at 35 kV.*

of a cyclic peptide that might be sequenced if it could first be opened. Although the compound was completely resistant to tryptic cleavage, proteolysis was achieved with chymotrypsin (24 h, 35% yield). The tandem mass spectrum of the ring-opened sandostatin is rich with sequence information (Fig. 2), and the sulfur loss is shown to have occurred in the cystine bridge to form the unusual thioether. We treat the smaller of the linked peptide sequences as a modification of the cysteine of the larger peptides, and use the established ABCD/VWXYZ nomenclature to indicate the collisionally induced fragment ions [1]. Reversing the peptides, we employ 'A'B'C'D/'V'W'X'Y'Z.

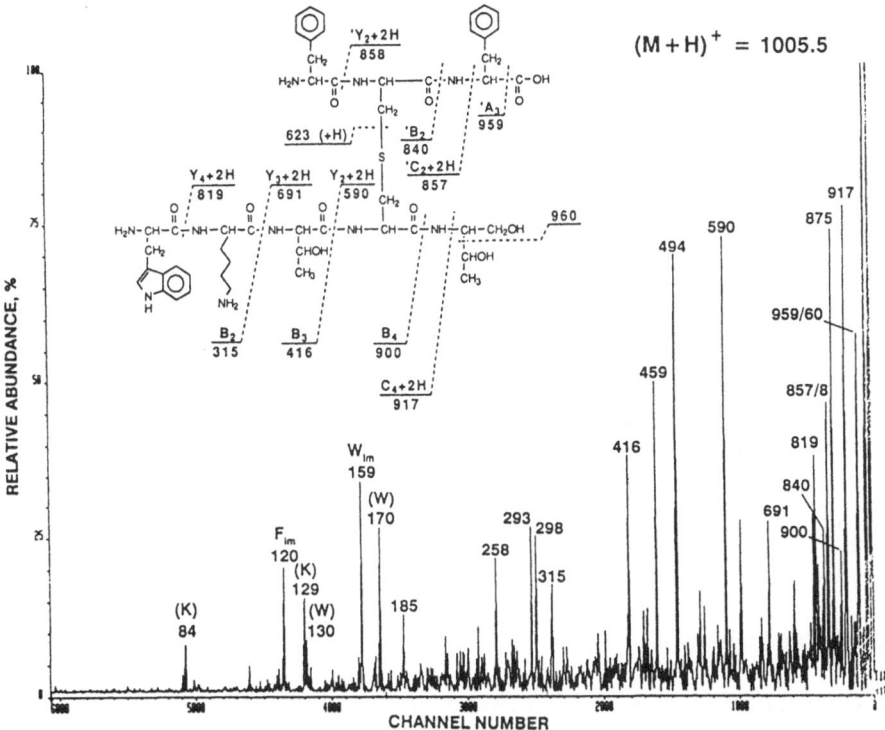

Fig. 2. MS/MS of the somatostatin analog after chymotryptic proteolysis (refer to Fig. 1 for experimental details).

References

1. Biemann, K., Biomed. Environ. Mass Spectrom., 16 (1988) 99.
2. Neugebauer, W., Lefebvre, M.R., Laprise, R. and Escher, E., In Rivier, J.E. and Marshall, G.R. (Eds.) Peptides: Chemistry, Structure and Biology (Proceedings of the 11th American Peptide Symposium), ESCOM, Leiden, 1990, p. 1020.

Methods for the automated amino acid sequencing of peptide-resin conjugates

Doris A. Sparrow and James T. Sparrow

Department of Medicine, Baylor College of Medicine and The Methodist Hospital, 6535 Fannin St., Houston, TX 77030, U.S.A.

Introduction

We have shown that peptide-resin conjugates are useful for the production of antibodies [1,2]. In order to confirm the integrity of the amino acid sequence of the peptide on the resin after HF or TFA deprotection, methods for the sequencing of the peptide still attached to the hydrophilic resin were needed. The programs supplied with the Applied Biosystems 477A Sequencer were used initially. Since these programs led to poor results with the deprotected resins, improvements in the methodology were needed.

Results and Discussion

The peptidyl-resins were synthesized on an Applied Biosystems 430A Synthesizer using a beaded glycyl-polyacrylamide cross-linked with *bis*-acryloyl-1,3-diaminopropane. The peptide was deprotected at –20°C for 3 h with anhydrous HF containing 10% anisole and 1% ethanedithiol. The resin was washed with ether, methanol, water, 1% acetic acid, water, 0.1 M Tris, water, 1% acetic acid and water before drying in a vacuum desiccator. The peptidyl-resin (1 mg) was suspended in TFA (0.5 ml) and a 5 μl sample immediately applied to the glass filter disc of the sequencer. With resins containing 0.7 meq/g of starting amino acid, this procedure resulted in the consistent transfer of approximately 1000 pmol of peptidyl-resin to the disc. The PTH-amino acids were separated on the Model 120 PTH Analyzer using a buffer of trimethylamine (1.5 ml)/ sodium acetate (37 ml, pH 3.8; 8.4 ml, pH 4.6) in THF-water (1 liter) and acetonitrile in the gradient in Table 1. The retention times of the PTH-derivatives of protected amino acids [3] are given in Table 2.

When various ABI programs were used to sequence the deprotected peptides attached to polyamide resins, we found that large amounts of PTH-amino acid were carried over to the next cycle (Fig. 1A). Since the programs for the sequencer contain wash steps using hydrophobic solvents that do not swell the hydrophilic resin, we reasoned that elimination of the apolar solvents from the program should improve recovery and reduce carryover. Subsequently, we found that using butyl chloride for all washes and extractions results in improved cleavage yields and greatly reduced carry-over (Fig. 1B). When these programs are used

446

Table 1 *Gradient for separation of protected PTH-amino acids*

Time	% Acetonitrile	Flow rate (μl/min)	Time	% Acetonitrile	Flow rate (μl/min)
0.0	10[a]	210	31	90	210
18	44[a]	210	31.5	90	300[c]
25	44	210	34.8	90	300
28[b]	90	210	35	90	5

[a] Depending on column and temperature it may be necessary to increase or decrease these percentages to separate normal PTH-amino acids in the same gradient with protected PTH-derivatives.

[b] The slope of this increase in acetonitrile concentration may need to be changed in order to separate some of the later protected PTH-derivatives.

[c] The increase in flow is needed to elute the protected PTH-tyrosines quickly.

Table 2 *Retention times of various protected PTH-amino acids*

PTH-amino acid	Retention time (min)	PTH-amino acid	Retention time (min)
AcM-Cysteine	10.8	Chx-Aspartic	26.1
TOS-Arginine	19.2	Benzyl-Threonine	27.5
Formyl-Tryptophan[a]	20.9	Cl-Z-Lysine	27.6
MTS-Arginine	23.2	Chx-Glutamic	28.3
Benzyl-Aspartic	23.4	Me-Benzyl-Cysteine	29.2
BOM-Histidine	24.1	Br-Z-Tyrosine	31.5
Benzyl-Glutamic	25.2	Cl$_2$-Benzyl-Tyrosine	31.6
MeO-Benzyl-Cysteine	25.3		

[a]PTH-formyl-tryptophan co-elutes with PTH-phenylalanine.

to sequence protected peptides on polystyrene or polyamide, we also find about a two-fold increase in yield and very low carryover.

The ABI programs use a cleavage temperature of 55°C that can lead to significant side-chain protecting-group cleavage [4,5]. With the butyl chloride washes and extractions, we found that the temperature could be lowered to 48°C greatly reducing side-chain deprotection. However, with both procedures, we found that the recovery of PTH-serine and -threonine was poor.

We have sequenced numerous protected and deprotected peptides on polyamide resin using the methodology described above with good results. The peptides showed little if any pre-view and were cleanly deprotected at –20°C with HF. The only side reaction of significance was β-aspartimide formation when β-benzyl aspartic acid was used for the synthesis. In conclusion, we find that in the sequencing of peptides on polymer supports it is desirable to keep the polymer in a swollen state to facilitate cleavage and minimize carry-over.

Acknowledgements

This research was funded by Grant No. 4107 from the Texas Advanced Technology Program and by NIH Grants HL-30064 and HL-27341.

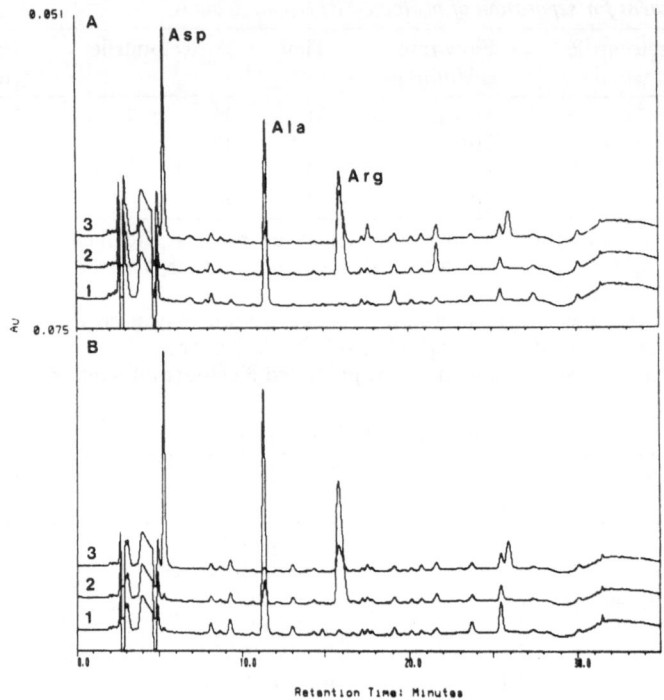

Fig. 1. The analysis of the PTH-amino acids from the first 3 cycles of the peptide,
ARDCEVHKSWFYRCTLMG, synthesized on polyamide and deprotected with anhydrous
HF for 3 h at –20°C. (A) Cycles using ABI programs. (B) Cycles using butyl chloride
for all washes and extractions. Note the decreased carryover and increased yield of PTH-
derivatives in cycle 2 and 3 compared to (A).

References

1. Chanh, T.C., Dreesman, G.R., Kanda, P., Linette, G.P., Sparrow, J. T., Ho, D.D.
    and Kennedy, R.C., EMBO J., 5 (1986) 3065.
2. Kennedy, R.C., Dreesman, G.R., Chanh, T.C., Boswell, R.N., Allan, J.S., Lee, T.-
    H., Essex, M., Sparrow, J.T., Ho, D.D. and Kanda, P., J. Biol. Chem., 262 1987) 5769.
3. Steiman, D.M., Ridge, R.J. and Matsueda, G.R., Anal. Biochem., 145 (1985) 91.
4. Kochersperger, M., Blacher, R., Kelly, P., Pierce, L. and Hawke, D.H., Am. Biotech.
    Lab., March (1989).
5. ABI User Bulletin No. 13, (1985) 1.

Structural modification of recombinant consensus α-interferon produced in *E. coli*

Hsieng S. Lu, Michael L. Klein, David W. Whiteley and Timothy D. Bartley

Amgen Inc., 1900 Oak Terrace Lane, Thousand Oaks, CA 91320, U.S.A.

Introduction

Mammalian proteins can be produced in high levels in bacterial expression systems using recombinant DNA technologies. Alterations in molecular structure of bacteria-synthesized recombinant proteins frequently occur. The most frequent alterations may be due to incomplete processing of the N-terminal amino acids [1], post-translational modicification, or by chemical and physical modifications during the purification processes. In this report, the analysis of such molecular modification occurring in recombinant consensus α-interferon (rIFN-Con1) is described.

Results and Discussion

Recombinant rIFN-Con1, a molecule whose sequence corresponds to the consensus sequence of 14 human α-interferon subtypes, was expressed in *E. coli* [2]. The purified protein contained two intramolecular disulfide bonds and exhibited full antiviral activity [3,4]. When rIFN-Con1 was subjected to analytical isoelectric focusing (IEF) in the presence of urea, three distinct subforms with apparent pIs of 6.1, 6.0 and 5.7 could be observed (Fig. 1, lane 1). Each of these different pI forms of rIFN-Con1 exhibited an apparent molecular weight of 20 kDa, indistinguishable from that of the original rIFN-Con1 starting material. Preparative IEF using Immobiline gel (1 mm thickness with a pH range of 5.5–6.5) was performed to isolate each of the individual species. Approximately 12 mg of the protein was applied, and a total of 3.8 mg of protein was recovered: 0.83 mg of the pI 5.7 band, 2.29 mg of the pI 6.0 band, and 0.71 mg of the pI 6.1 band. The purity of each isolated pI subform was further confirmed by analytical IEF as shown in Fig. 1 (lanes 2–4). N-terminal sequencing indicated that the pI 6.1 subform was methionyl rIFN-Con1 (M-C-D-L-P-E-T-H-S-L-G---), whereas the pI 6.0 subform corresponded to des-methionyl rIFN-Con1. Sequence analysis of the pI 5.7 subform showed that it contained no N-terminal sequence signal, indicating that the N-terminus of this subform was blocked. Bioassay results indicated that all three subforms had equivalent antiviral activity; CD analysis revealed indistinguishable bands in the near-UV region (250–320 nm), suggesting that there were no major differences among subforms in terms of their tertiary structure.

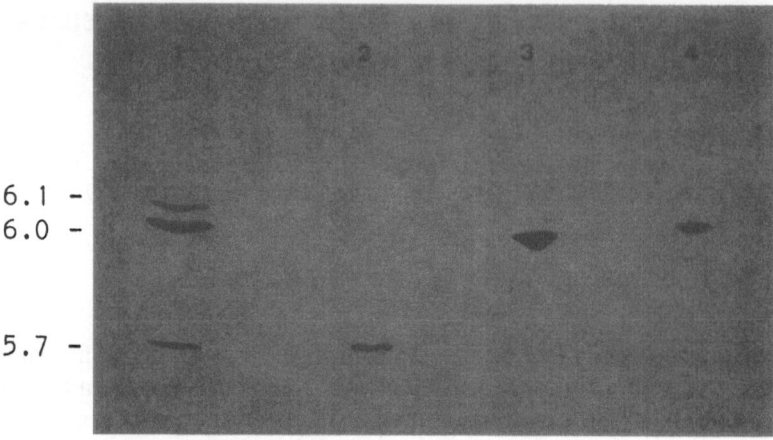

Fig. 1. *IEF of recombinant IFN-Con1 before and after isolation by preparative IEF. Lane 1: rIFN-Con1; Lanes 2–4: pI 5.7, 6.0 and 6.1 subforms, respectively.*

Recombinant IFN-Con1 was further subjected to a preparative C_4 RPHPLC (2.25×25 cm) by 0.1% TFA-acetonitrile gradient elution. The pI 5.7 subform was separated from pI 6.0 and 6.1 subforms. Carboxymethylated [Cys(CM)] derivatives of these subforms were prepared and subjected to trypsin digestion and RPHPLC peptide mapping using C_4 columns. Two maps were obtained: map A for the pI 6.0 and 6.1 subforms and B for the pI 5.7 subform. Both maps were identical except that two peptides (a and b, retention time of 33.5 and 33.7 min) in map A were absent from map B. Peptide c (37.5 min) was found in map B, and also in small quantity in map A. When subjected to sequence analysis, peptide a yielded the sequence corresponding to the N-terminal tryptic peptide of methionyl rIFN-Con1 [M-C-A-L-P-G-T-H-S-L-G-D-R], and peptide b yielded the sequence of the des-Met N-terminal tryptic peptide. Peptide c isolated from map B was sequenced and confirmed to be N-terminally blocked.

Peptides a, b and c were then subjected to AAA using the precolumn PITC derivatization method [4]. Peptides a and b yielded compositions consistent with data obtained from sequencing results (see above). The composition of peptide c was identical to that of peptide b, indicating that this blocked peptide was also a des-Met N-terminal tryptic fragment. When the blocked N-terminal peptide was subjected to FABMS, a quasi-molecular ion [M + H], was observed at m/z 1441. A strong signal was also observed at m/z 1382. The difference between these two values corresponded to removal of the carboxymethyl group at the cysteinyl residue during analysis. Since the expected mass of the des-Met N-terminal tryptic peptide of Cys(CM)-rIFN-Con1 is 1398, the difference between this value and the [M + H] value of peptide c is 43, a mass equivalent to an acetyl group. It was thus concluded that the pI 5.7 subform present in rIFN-Con1 preparation was acetylated, at its N-terminal cysteinyl residue. Acetylation occurs in many proteins in vivo by processing during and after translation [5,6].

Acetylation at the N-terminal cysteinyl residue of rIFN-Con1 seems to be unique, since similar modification of other types of recombinant α-interferons has not been reported in bacteria expression systems.

Acknowledgements

We are indebted to Drs. P.H. Lai, L. Goldstein (posthumously) and P.G. Righetti for their significant contribution to this work.

References

1. Sherman, F., Stewart, J.W. and Tsunasawa, S., BioEssays, 3 (1985) 27.
2. Alton, K., Stabinsky, Y., Richards, R., Ferguson, B., Goldstein, L., Altrock, B., Miller, L. and Stebbing, N. In De Maeyer, E. and Schellekens, H. (Eds.) The Biology of the Interferon System, Elsevier, Amsterdam, 1983, p. 119.
3. Klein, M.L., Bartley, T.D., Lai, P.-H. and Lu, H.S., J. Chromatogr., 454 (1988) 205.
4. Lu, H.S. and Lai, P.-H., J. Chromatogr., 368 (1986) 215.
5. Wold, F., Annu. Rev. Biochem., 50 (1981) 783.
6. Aofin, S.M. and Bradshaw, R., Biochemistry, 27 (1988) 7979.

Session IV
Recent bioactive peptides and biology

Chairs: Irwin M. Chaiken
Smith Kline & French Laboratories
Swedeland, Pennsylvania, U.S.A.

Maurice Manning
Medical College of Ohio
Toledo, Ohio, U.S.A.

Roger Acher
Université de Paris VI
Paris, France

and

Andrew Baird
Whittier Institute
La Jolla, California, U.S.A.

Synthesis and biological activities of peptides from marine organisms

Takayuki Shioiri, Yasumasa Hamada and Kohfuku Kohda

Faculty of Pharmaceutical Sciences, Nagoya City University, Tanabe-dori, Mizuho-ku, Nagoya 467, Japan

Introduction

Lipophilic peptides from marine organisms constitute a growing class of naturally occurring cytotoxic and/or antineoplastic substances. Since these peptides are structurally unique, have quite interesting biological activities, and can be obtained in only small quantities in nature, we have had a keen interest in the exploitation of versatile synthetic methods applicable to their large scale production.

Using diethyl phosphorocyanidate [DEPC, $(C_2H_5O)_2P(O)CN$] and diphenyl phosphorazidate [DPPA, $(C_6H_5O)_2P(O)N_3$] mainly for the peptide-chain elongation and for the cyclization, respectively, we have succeeded in the total synthesis of the cytotoxic cyclic peptides (1–9) of marine origin, shown in Chart 1 [1]. Their biological activities have also been investigated [2]. We describe here the total synthesis of real dolastatin 3 (2), and cytotoxic studies on ulithiacyclamide (7) and ulicyclamide (3).

Results and Discussion

(1) Dolastatin 3 revisited

Dolastatin 3 was isolated from an Indian Ocean sea hare, *Dolabella auricularia*, by Pettit and co-workers [3]. We and others have already revealed by synthetic studies that both the proposed structure (1) and its reversed isomer were unaccepted [1a]. Recently, the revised structure (2) was proposed for dolastatin 3 by Pettit et al. and confirmed by synthesis [4].

We also accomplished the synthesis of (2) by the $3+2$ fragment condensation approach, as shown in Chart 2. Instead of the L-(gln)Thz derivative, the corresponding cyano derivative (10), prepared from Boc-L-Gln-OH, was utilized. DEPC was used for the peptide-chain elongation, and DPPA was used for the cyclization. The cyclized product (11) was treated with alkaline hydrogen peroxide to give dolastatin 3 (2). Although the 400 MHz NMR spectrum of our synthetic specimen was completely identical with that of natural dolastatin 3, the cytotoxic activity of our specimen has been quite low ($IC_{50} > 80 \mu g/ml$) against both L1210 and P388 mouse leukemia cells, as compared to the high toxicity ($ED_{50} = 0.16$ vs 0.17 $\mu g/ml$) against the PS leukemia of native dolastatin 3 [4]. Although

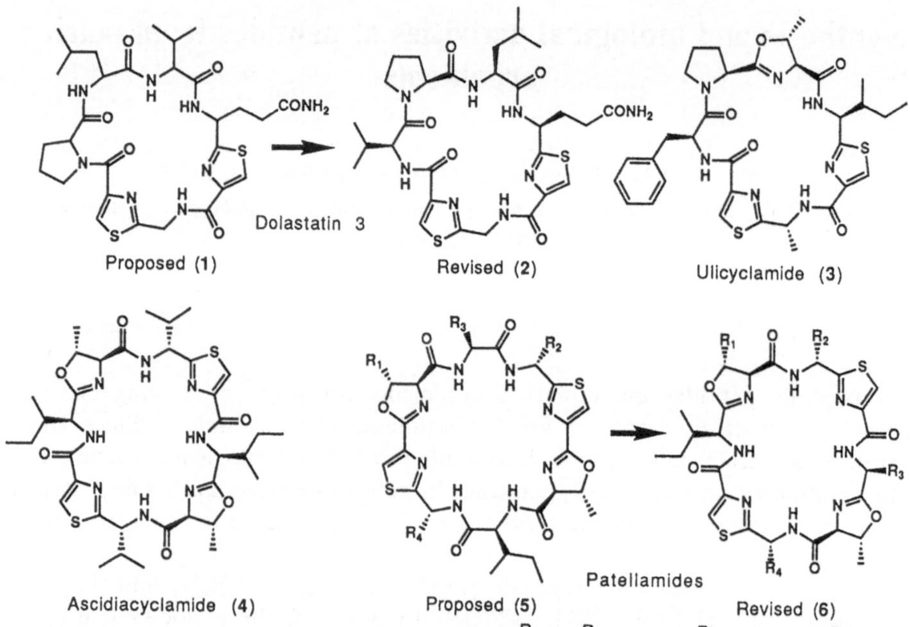

Proposed (1)

Dolastatin 3

Revised (2)

Ulicyclamide (3)

Ascidiacyclamide (4)

Proposed (5)

Patellamides

Revised (6)

	R_1	R_2	R_3	R_4
a: Patellamide A	H	$(CH_3)_2CH$	$(S)\text{-}CH_3CH_2CH(CH_3)$	$(CH_3)_2CH$
b: Patellamide B	CH_3	$PhCH_2$	$(CH_3)_2CHCH_2$	CH_3
c: Patellamide C	CH_3	$PhCH_2$	$(CH_3)_2CH$	CH_3

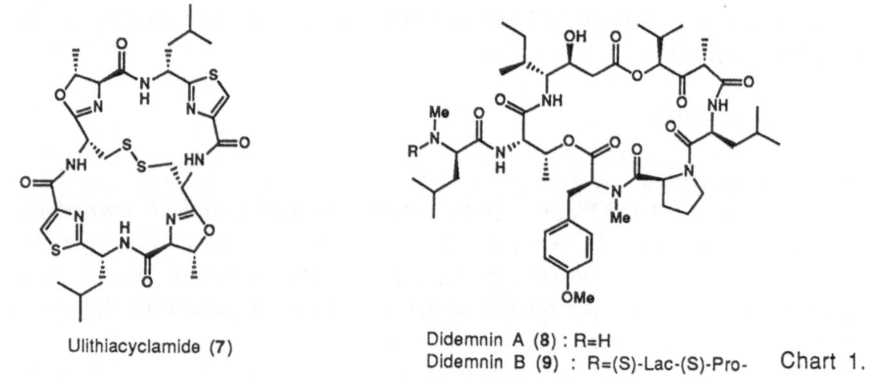

Ulithiacyclamide (7)

Didemnin A (8) : R=H
Didemnin B (9) : R=(S)-Lac-(S)-Pro-

Chart 1.

clarification of this discrepancy still remains to be done in the future, apparent lack of cytotoxicity of our synthetic dolastatin 3, having no oxazoline function, coincides with our observation [2] that the oxazoline function is important for cytotoxicity in the cyclic peptides (1–7) and their synthetic intermediates.

(2) Cytotoxic mechanism of ulithiacyclamide

Among the cyclic peptides (1–7) and their synthetic intermediates tested, ulithiacyclamide (7) was found to have the most potent cytotoxic activity against mouse leukemia L1210 cells in vitro, $IC_{50} = 0.04$ μg/ml, whose value is comparable

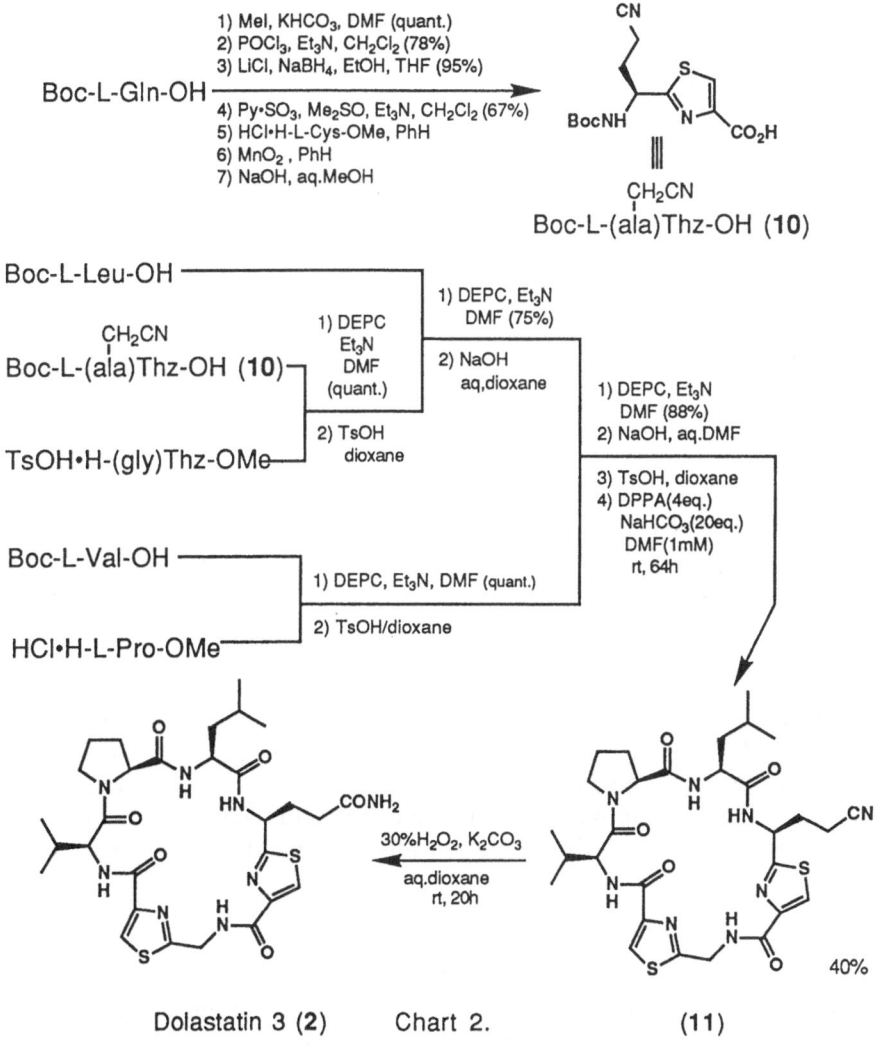

Chart 2.

Dolastatin 3 (2)   Chart 2.   (11)

or superior to those of some clinically useful anticancer drugs [2]. We now investigated the mechanistic aspects of the cytotoxicity of ulithiacyclamide and its effect on the cytotoxicity of several anticancer drugs using L1210 cells in vitro.

Exposure of ulithiacyclamide (7) on mouse leukemia L1210 cells resulted in the steep decrease in cell growth, suggesting that ulithiacyclamide might directly interact with certain cell constituent(s) to produce lethal damage. Furthermore, inhibition of cell growth by ulithiacyclamide was revealed to be self-destructive. In binding experiments, it was found that ulithiacyclamide might not specifically interact with cell membrane constituents.

Examination of the effect of ulithiacyclamide on macromolecular biosynthesis has revealed that ulithiacyclamide inhibits protein biosynthesis most effectively, $IC_{50} = 0.21$ μg/ml. Inhibition of RNA biosynthesis follows protein biosynthesis, but DNA biosynthesis is only slightly inhibited even at the maximum concentration, 1 μg/ml, tested.

Such a selective inhibitory effect prompted us to examine the potentiation effect of ulithiacyclamide on the cytotoxicity of several clinically useful anticancer drugs when used in combination. Among the drugs tested, only bleomycin and its analog pepleomycin showed synergistic cytotoxicity with ulithiacyclamide, while others were additively or antagonistically toxic.

(3) Cytotoxicity of ulicyclamide

Ulicyclamide (3) is also cytotoxic against mouse L1210 leukemia cells, $IC_{50} = 13$ μg/ml. It is interesting to note that the synthetic cyclic precursor, cyclo[L-aThr-L-(ile)Thz-D-(ala)Thz-L-Phe-L-Pro], also exhibits cytotoxicity, $IC_{50} = 35$ μg/ml, though it has no oxazoline function. However, the linear synthetic intermediate, H-L-aThr-L-(ile)Thz-D-(ala)Thz-L-Phe-L-Pro-OH, has no cytotoxicity. Unlike ulithiacyclamide (7), ulicyclamide (3) and its cyclic precursor do not inhibit protein biosynthesis, but produce a remarkable inhibitory effect on both DNA and RNA biosynthesis.

Conclusion

Lipophilic cyclic peptides (1–9) are structurally unique and biologically interesting. Among them, didemnin B (9) is quite promising as a new antitumor agent, and already is now in Phase II clinical trials. Since ulithiacyclamide (7) also shows a strong cytotoxicity comparable to that of didemnin B, a new useful antitumor drug may be derived from this after suitable structural modifications.

References

1a. Hamada, Y. and Shioiri, T., J. Synth. Org. Chem. Japan, 45 (1987) 957.
 b. Hamada, Y., Kondo, Y., Shibata, M. and Shioiri, T., J. Am. Chem. Soc., 111 (1989) 669.
2. Shioiri, T., Hamada, Y., Kato, S., Shibata, M., Kondo, Y., Nakagawa, H. and Kohda, K., Biochem. Pharmacol., 36 (1987) 4181.
3. Pettit, G.R., Kamano, Y., Brown, P., Gust, D., Inoue, M. and Herald, C.L., J. Am. Chem. Soc., 104 (1982) 905.
4. Pettit, G.R., Kamano, Y., Holzapfel, C.W., van Zyl, W.J. Tuinman, A.A., Herald, C.L., Baczynskyj, L. and Schmidt, J.M., J. Am. Chem. Soc., 109 (1987) 7581.

Comparison and organization of the vasotocin- and isotocin-encoding genes from a teleost fish (*Catostomus commersoni*)

Steven D. Morley[a], Jörg Heierhorst[a], Jaime Figueroa[a], Christiane Schönrock[a], Karl Lederis[b] and Dietmar Richter[a],*

[a]*Institut für Zellbiochemie und klinische Neurobiologie, UKE, Universität Hamburg, Martinistr. 52, D-2000 Hamburg 20, F.R.G.*
[b]*Department of Pharmacology and Therapeutics, Health Science Centre, The University of Calgary, 3330 Hospital Drive N.W., Calgary, Alberta, Canada T2N 4N1*

Introduction

Vasopressin and oxytocin are the predominant mammalian homologs of a larger, structurally conserved, family of neuropeptide hormones, whose members are distributed throughout the animal kingdom [1]. These peptides all consist of 9 amino acids with several completely conserved residues, including the two cysteine residues in positions 1 and 6 that form a disulfide bridge, Asn^5, Pro^7 and the C-terminal glycine amide. Amino acid variations occur at positions 2, 3 and 4 and most significantly, at position 8 where the change from Leu or a related lipophilic amino acid to Arg or Lys accounts for a vasopressin-type (control of water retention), as opposed to an oxytocin-type (smooth muscle contraction) function. In all vertebrates other than mammals, vasotocin has been chemically identified in place of vasopressin, while in non-mammalian tetrapods and bony fish, oxytocin is replaced respectively by mesotocin and isotocin [1].

Vasopressin and oxytocin are encoded by distinct, but structurally related, genes and synthesised as part of larger precursor molecules [2]. These contain, in addition to the hormone moiety, a cysteine-rich protein termed neurophysin, a carrier protein that is involved in transporting the hormone from the hypothalamus (site of synthesis) to the neurohypophysis (site of storage and release). The vasopressin precursor includes an additional entity, a glycopeptide or copeptin [1] that may be the conjectured hypothalamic prolactin-releasing factor [3].

We report here the presence of vasotocin and isotocin precursor-encoding mRNAs in the hypothalamic region of the teleost fish, *Catostomus commersoni* (white sucker), showing striking differences in their gene and precursor organization, as compared to those for their mammalian counterparts vasopressin and oxytocin.

---

*To whom correspondence should be addressed.

459

Results and Discussion

When a λgt11 cDNA library, constructed from mRNA of sucker brain hypothalamic region, was screened with fully degenerate pools of oligonucleotides deduced from the first 7 amino acids of vasotocin or isotocin [4], cDNAs clones encoding several members of the vasotocin-isotocin hormone precursor gene family were obtained. The DNA sequence analysis predicts precursors of similar structures, each consisting of a putative signal peptide connected to the hormone moiety, followed by a cysteine-rich protein with features of the neurophysin family. Each hormone is linked to the rest of the precursor by the residues Gly-Lys-Arg (potential signals for C-terminal hormone amidation and precursor processing, respectively). The predicted vasotocin and isotocin precursors all contain neurophysin sequence, that are longer at their C-termini by a tract of some 30 amino acids compared to their mammalian counterparts. Each of these sequences contains a leucine-rich core segment resembling that found in the copeptin, a glycopeptide moiety present in mammalian vasopressin precursors. They differ, however, distinct in respect to processing signals and consensus sequence for glycosylation, raising the possibility that the mammalian copeptin arose by the introduction of a functional cleavage signal into a larger ancestral neurophysin.

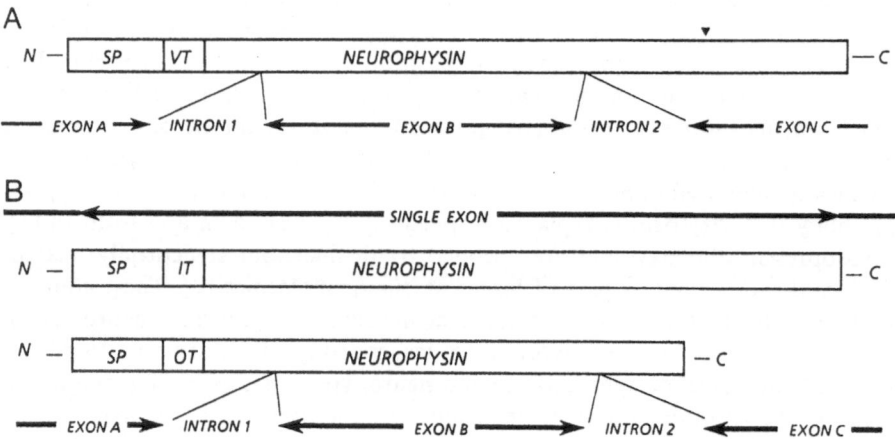

Fig. 1. *Structural organisation of the* Catostomus *vasotocin (A) and isotocin (B) genes compared to the mammalian oxytocin gene (2). SP, signal peptide; IT isotocin; OT, oxytocin; VT, vasotocin; the arrow head points to the position corresponding to the basic amino acid separating the neurophysin from the copeptin moiety of the mammalian vasopressin precursor.*

Each hormone is encoded by two non-allelic genes, all of which are expressed, as indicated by Northern blot analysis. Genomic DNA amplified by the polymerase chain reaction has been used to define exon–intron boundaries. Both of the vasotocin genes contain introns in positions corresponding to those found in the gene of their mammalian counterpart, vasopressin. Thus, the hormone and neurophysin moieties are encoded on distinct exons, although the latter is encoded jointly by the second and third exons (Fig. 1).

In contrast, however, the isotocin genes lack introns in their protein–coding regions, a feature that is unprecedented for genes of the vasopressin–oxytocin family of neuropeptide hormones. This absence of introns is puzzling since isotocin and vasotocin are conjectured to be members of distinct lineages that arose originally from the duplication of a single gene [1]. If this is, indeed, the case, one must suppose either that the vasotocin genes have gained introns, or that introns have been lost from the genes for isotocin by gene conversion during subsequent evolution. An alternative possibility is that the isotocin genes have arisen by a mechanism involving reverse transcription and reinsertion of a vasotocin mRNA into the *Catostomus* genome, eliminating introns in the process, followed by duplication and genetic drift. To clarify this point, it is essential to obtain further sequence and structural information for genes of the vasopressin–oxytocin family from phylogenic groups that arose prior to the separation, approximately 400 million years ago [5], of the evolutionary lines leading to bony fish and mammals.

References

1. Acher, R. and Chauvet, J., Biochemie, 70 (1988) 1197.
2. Richter, D., In C.W. Smith (Ed.) The Peptides, Academic Press, New York, 1987, Vol. 8, p. 41.
3. Nagy, G., Mulchahey, J.J., Smyth, D.G. and Neill, J.D., Biochem. Biophys. Res. Commun., 151 (1988) 524.
4. Heierhorst, J., Morley, S.D., Figueroa, J., Krentler, C., Lederis, K. and Richter, D., Proc. Natl. Acad. Sci. U.S.A., 86 (1989) 5242.
5. Acher, R., In Maruani, J. (Ed.) Molecules in Physics, Chemistry and Biology, Kluwer Academic Publishers, 1989, Vol. IV, p. 13.
6. Uyeno, T. and Smith, G.R., Science, 125 (1972) 644.

NRP-11, a novel neurotensin-resembling peptide from bovine brain

V.T. Ivanov[a], V.I. Tsetlin[a], V.V. Ul'yashin[a], A.A. Karelin[a], E.V. Karelin[a],
K.V. Sudakov[b], V.V. Sherstnev[b], O.N. Dolgov[b], V.E. Klusha[c], S.E. Severin Jr.[d]
and D.G. Mikeladze[e]

[a]Shemyakin Institute of Bioorganic Chemistry, U.S.S.R. Academy of Sciences,
Moscow, U.S.S.R.
[b]Anokhin Institute of Normal Physiology, U.S.S.R. Academy of Medical Sciences,
Moscow, U.S.S.R.
[c]Institute of Organic Synthesis, Latvian S.S.R. Academy of Sciences, Riga, U.S.S.R.
[d]Lomonosov Institute of Fine Chemical Technology, Moscow, U.S.S.R.
[e]Beritashvili Institute of Physiology, Georgian S.S.R. Academy of Sciences,
Tbilisi, U.S.S.R.

Introduction

Understanding the diversity of biological effects manifested by endogenous
peptides requires a search for and SAR of new members of various peptide
families. Here a novel neurotensin-like peptide is described.

Results and Discussion

Acid extraction of bovine brain homogenate followed by several ion-exchange
and adsorbtion chromatography steps with final RPHPLC (Fig. 1) afforded the
title peptide, NRP-11, whose structure was established by gas-phase sequencing
and confirmed by FAB MS. The sequence obtained was novel, with neurotensin
being the closest analog among the known peptides.

Biological activity was assayed using NRP-11 and its 1–7 fragment (P7) prepared
by SPPS. Intraperitoneal injection of both peptides elicited no behavioral
responses in rats. When administered i.c.v. in doses over 20 μg, NRP-11 and
P7 induced many behavioral reactions such as convulsions, rotations, grooming,
and also affected food consumption. In general, the observed effects are similar
to those of neurotensin [1]. The activity of NRP-11 and P7 in the passive avoidance
test depends on the individual features of the experimental animals.

NRP-11 and P7, contrary to NT, are practically devoid of vasopressor and
contractile activity, but are much more potent in releasing histamine from rat
mast cells. Noteworthy, considerable differences in peripheral activity were
observed among the NT congeners [2].

When assayed with several protein kinases, NRP-11, but not P7, proved to
be a highly specific substrate for Ca^{2+}-dependent protein kinase C, Km 3μM,
V_{max} 30 nm/min/mg, being phosphorylated at Ser^7. The preliminary results
implicate NRP-11 in the activation of the phosphoinositide pool.

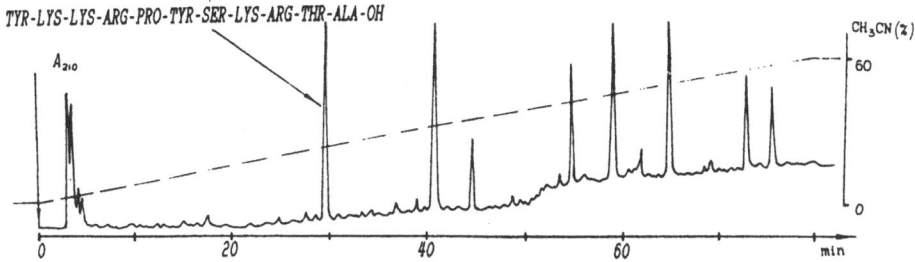

Fig. 1. Isolation of NRP-11 with Ultrasphere-ODS column 4.6×250 mm, 0.1% TFA.

The NRP-11 and P7 effects on the brain muscarinic acetylcholine receptor, as well as glutamate, β-adrenergic, and D2-dopamine receptors were examined. Both peptides were shown selectively to modulate ligand binding to the D2-dopamine receptors, NRP-11 affecting similarly to NT [3] the [^{3}H]-spiroperidol interaction with the rat brain striatal membranes, and contrary to NT suppressing [^{3}H]-haloperidol binding to the *n. accumbens* membranes. P7 acted in a different manner, inhibiting the antagonists binding to striatum and potentiating [^{3}H]-haloperidol binding to the *n. accumbens* membranes.

The data for NRP-11 and P7 are generally consonant with the essential role of NT fragment 8–13 in the manifestation of its peripheral activities [4]. Obviously, other SAR should hold for the peptide modulation of the dopaminergic system, since either NRP-11 or P7 having substitutions or deletions in this fragment, are equipotent with NT in the respective assays. Sharing a number of common features with the NT family, NRP-11 might, in addition, have some distinct functions associated with its phosphorylation.

References

1. Martin, G.E. and Naruse, T., Regul. Pept., 81 (1982) 97.
2. Carraway, R.E., Reynecke, M., In Falkmer, S., Hakanson, R. and Sundler, F., (Eds.) Evolution and Tumor Pathology of the Neuroendocrine System, Elsevier Science Publishers, Amsterdam, 1984, p. 245.
3. Yoshida, T., Kito, S., Matsuboyashi, H. and Mioshi, R., Neurochem. Res., 13 (1988) 255.
4. Kitabgi, P., Carraway, R., Van Reitschoten, J., Granier, C., Morgat, J.L., Menez, A., Leeman, S.E. and Freycget, P., Proc. Natl. Acad. Sci. U.S.A., 74 (1977) 1848.

Purification and sequence analysis of rat melanin concentrating hormone

W.H. Fischer, J. Vaughan, J.-L. Nahon, F. Presse, C. Hoeger, J.E. Rivier and W. Vale

The Clayton Foundation Laboratories for Peptide Biology, The Salk Institute, 10010 N. Torrey Pines Road, La Jolla, CA 92037, U.S.A.

Introduction

Salmon melanin concentrating hormone (sMCH) is a cyclic heptadeca-peptide first isolated from chum salmon pituitaries and characterized by Kawauchi et al. [1]. It causes aggregation of melanin granules within melanocytes of teleosts [1,2]. In some other fish species, reptiles and amphibians, it has the opposite effect, i.e., it causes dispersion of melanin granules within melanocytes [3,4]. sMCH has been shown to be an inhibitor of CRF-induced ACTH and α-MSH secretion by teleost pituitary glands in vitro [5] and is also reported to have a similar, albeit, weaker activity on the rat pituitary [6]. An activity similar, but not identical to sMCH was detected in several higher vertebrates including mammals [7,8]. The role of this MCH-like activity has been difficult to study, mainly because its structure was unknown.

Results and Discussion

Using a rabbit anti-sMCH antibody, we were able to detect sMCH-like immunoreactivity in extracts from 60 000 rat hypothalami. These extracts were previously used in this laboratory to isolate, among others, corticotropin releasing hormone [9] and growth hormone releasing factor [10]. Throughout our purification, a radioimmunoassay utilizing a rabbit anti-sMCH antibody was used to monitor activity. The purification was achieved by gel filtration on Sephadex G-50, immunoaffinity chromatography using the rabbit anti-sMCH antibody bound covalently to Sepharose, FPLC on Superose 12B and RPHPLC. In the last HPLC purification step, several immunoreactive species were observed. These were subjected separately to Edman degradation in a gas phase sequencer. The same N-terminal amino acid sequence was obtained for all fractions extending to up to 19 residues. The sequence of rat MCH (rMCH) is shown in Fig. 1 in comparison to that of sMCH [11]. A high degree of homology is observed, especially in the central and C-terminal portion of the molecules. The sequence between the two cysteines exhibits only one conservative substitution of valine for leucine. This region of the molecule is believed to be responsible for melanin concentrating activity in the fish skin assay [12,13]. The N-terminal region

Rat MCH	Asp	Phe	Asp	Met	Leu	Arg	Cys	Met	Leu	Gly	Arg	Val	Tyr	Arg	Pro	Cys	Trp	Gln	Val
Salmon MCH			Asp	Thr	Met	Arg	Cys	Met	Val	Gly	Arg	Val	Tyr	Arg	Pro	Cys	Trp	Glu	Val

Fig. 1. Sequence of rMCH in comparison to that of sMCH. Conserved residues are boxed.

shows an extension by two amino acids and two other substitutions. The corresponding part of the salmon hormone was found to be essential for MSH-like (MCH antagonistic) activity [12,13].

The peptide corresponding to the experimentally determined sequence was subsequently synthesized by SPPS [14]. After HF cleavage, the material was purified by preparative RPHPLC [15] and the intramolecular disulphide bond was formed by oxidation with ferricyanide.

Synthetic rMCH and fresh extracts from rat hypothalami exhibited a parallel displacement curve in the RIA (Fig. 2). The HPLC elution position of the hypothalamic extracts coincides with synthetic rMCH and is considerably

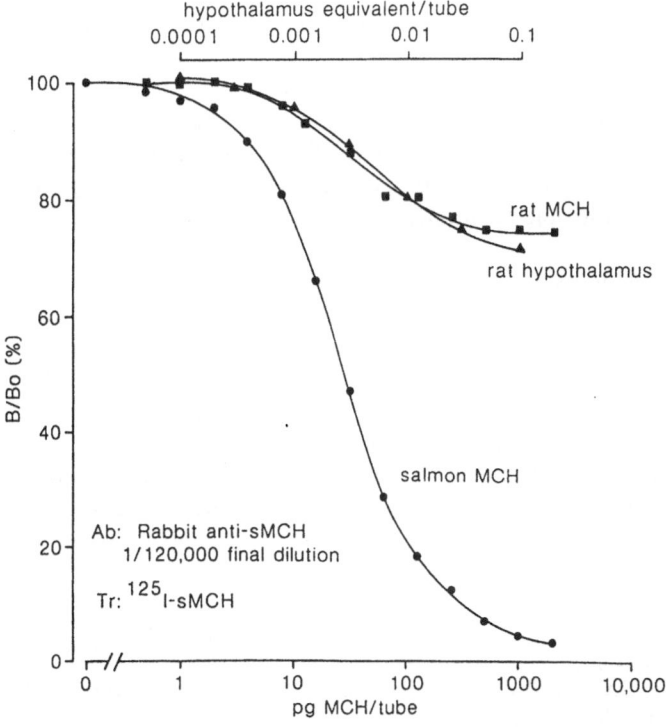

Fig. 2. RIA of synthetic rMCH. A rabbit anti-sMCH antibody was used. The displacement curves of fresh rat hypothalamic extracts and synthetic sMCH are shown for comparison.

different from that of sMCH [11]. The immunoreactive material from our initial hypothalamic extracts exhibited a different behavior in both the RIA and its HPLC elution position. This was probably due to deamidation of glutamine in position 18 and oxidation of sulfur containing amino acids.

Based on the N-terminal portion of the experimentally determined amino acid sequence, an oligonucleotide was designed. Using this oligonucleotide as a probe, we were able to identify a DNA in a rat hypothalamic cDNA library coding for the peptide precursor. The DNA was isolated, cloned and sequenced. From the DNA sequence, a 165-amino acid precursor is predicted [16]. The organization of the precusor is shown in Fig. 3. The MCH peptide sequence is found at the C-terminus of the precursor. It is followed by a stop codon. Two putative neuropeptides (NEI and NGE following Tatemoto and Mutt's convention [17]) are flanked by either single or paired basic residues.

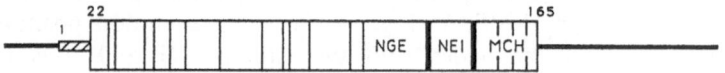

Fig. 3. Organization of the rMCH precursor. Thin lines represent basic residues (arginine and lysine), bold lines paired basic residues. The hatched box indicates the putative signal peptide. Bold horizontal lines correspond to 5' and 3' untranslated DNA regions.

We have explored the actions of salmon and rat MCH on the basal and CRF-stimulated secretion of ACTH by cultured rat anterior pituitary cells. No effect is observed over a 4-h period. The location of MCH in the rat brain in the dorsolateral hypothalamus suggests a role as a neuroregulator rather than as a hypophysiotropic hormone.

Acknowledgements

We thank C. Douglas, J. Miller, D. Karr, R. Kaiser, J. Porter and C. Miller for excellent technical assistance. Research was supported by NIH program grants DK-26741 and HD-13527, NATO, Philippe Foundation, la Ligue Française contre le Cancer, l'Association pour la Recherche contre le cancer, and the Adler Foundation. Research conducted in part by the Clayton Foundation for Research, California Division; Wylie Vale is a Clayton Foundation Investigator.

References

1. Kawauchi, H., Kawazoe, I., Tsubokawa, M., Kishida, M. and Baker, B.I., Nature, 305 (1983) 321.
2. Wilkes, B.C., Hruby, V.J., Sherbrooke, W.C., de L. Castrucci, A.M. and Hadley, M.E., Biochem. Biophys. Res. Commun., 122 (1984) 613.
3. Wilkes, B.C., Hruby, V.J., de L. Castrucci, A.M., Sherbrooke, W.C. and Hadley, M.E., Science, 224 (1984) 1111.
4. Ide, H., Kawazoe, I. and Kawauchi, H., Gen. Comp. Endocrinol., 58 (1985) 486.
5. Baker, B.I., Bird, D.J. and Buckingham, J.C., Gen. Comp. Endocrinol., 63 (1986) 62.

6. Baker, B.I., Bird, D.J. and Buckingham, J.C., J. Endocrinol., 106 (1985) R5.
7. Zamir, N., Skofitsch, G., Bannon, M.J. and Jacobowitz, D.M., Proc. Natl. Acad. Sci. U.S.A., 83 (1986) 1528.
8. Sekiya, K., Ghatei, M.A., Lacoumenta, S., Burnet, P.W.J., Zamir, N., Burrin, J.M., Polak, J.M. and Bloom, S.R., Neuroscience, 25 (1988) 925.
9. Rivier, J., Spiess, J. and Vale, W., Proc. Natl. Acad. Sci. U.S.A., 80 (1983) 4851.
10. Spiess, J., Rivier, J. and Vale, W., Nature, 303 (1983) 532.
11. Vaughan, J.M., Fischer, W.H., Hoeger, C., Rivier, J. and Vale, W., Endocrinology, 125 (1989) 1660.
12. Hadley, M.E., Zechel, C., Wilkes, B.C., de L. Castrucci, A.M., Visconti, M.A., Pozo-Alonso, M. and Hruby, V.J., Life Sci., 40 (1987) 1139.
13. Kawazoe, I., Kawauchi, H., Hirano, T. and Naito, N., Int. J. Pept. Prot. Res., 29 (1987) 714.
14. Rivier, J., J. Amer. Chem. Soc., 96 (1974) 2986.
15. Hoeger, C., Galyean, R., Boublik, J., McClintock, R. and Rivier, J., Biochromatography, 2 (1987) 134.
16. Nahon, J.-L., Presse, F., Bittencourt, J.C., Sawchenko, P.E. and Vale, W., Endocrinology, 125 (1989) 2056.
17. Tatemoto, K. and Mutt, V., Proc. Natl. Acad. Sci. U.S.A., 78 (1981) 6603.

Isolation and structure determination of neurosecretory peptides from the *corpora cardiaca* of the insect *Locusta migratoria*

Hélène Hietter[a],*, Alain Van Dorsselaer[a], Francine Goltzene[b], Daniel Zachary[b], Jules Hoffmann[b] and Bang Luu[a]

[a]*Laboratoire de chimie organique des substances naturelles associé au CNRS, 5 rue B. Pascal, F-67084 Strasbourg, France*
[b]*Laboratoire de biologie générale associé au CNRS, 12 rue de l'Université, F-67000 Strasbourg, France*

Introduction

Insect development is controlled by peptide hormones synthesized in the neurosecretory cells of the brain and released by retrocerebral neurohaemal organs. In *Locusta*, the storage and release site of these neurohormones are the neurohaemal lobes of the *corpora cardiaca*. The insect *corpora cardiaca* are of dual anatomical composition. They contain most of the axon endings from the neurosecretory cells of the brain, serving as storage and release sites for the brain neurohormones, and they have intrinsic secretory cells which synthesize and release peptide hormones. In the characterization of the developmental neurohormones of *Locusta*, we have been hampered by the lack of reliable microscale bioassays. We have, therefore, started a systematic structural study of the major peptides contained with the *corpora cardiaca*. The complete characterization of these peptides will enable chemical synthesis, necessary for probing their physiological function. In addition, physicochemical studies will establish the secondary and tertiary structures of the synthetic peptides, necessary for understanding their mode of action and for designing analogs and inhibitors. The amino acid sequence of the isolated neuropeptides can also serve to devise oligonucleotide probes to screen cDNA or genomic libraries.

In the present paper, we report the isolation and full chemical characterization of two 6-kDa dimeric peptides contained within the secretory granules of the glandular lobes, and of a novel 5-kDa peptide from the neurohaemal lobe of the *corpora cardiaca*.

Results and Discussion

The peptides were isolated from aqueous extracts of *corpora cardiaca*, purified by RPHPLC and characterized by automated microsequencing and LSIMS.

---

*To whom correspondence should be addressed.

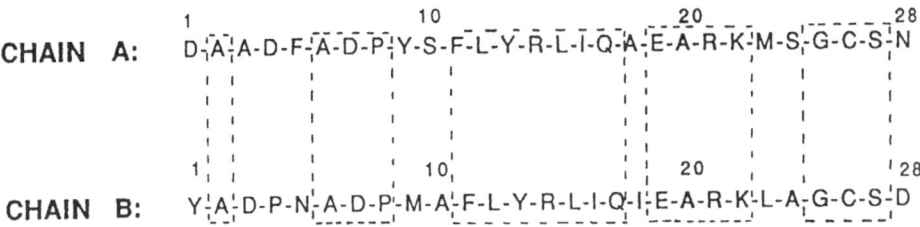

Fig. 1. *A and B chains of the A-A and A-B 6 kDa peptides: 6K-I and 6K-II.*

Identification of the 6-kDa peptides

Glandular lobes were extracted and subjected to C_{18}-RPHPLC. Two major compounds were isolated and shown to be structurally related dimers: one being a homodimer (A-A chains linked via a disulfide bridge at Cys^{26}, the second being a heterodimer (A-B chains linked via a disulfide bridge at Cys^{26}). A 60% similarity exists between the two chains (Fig. 1). Wide scan and narrow scan spectra were performed and allowed the mass determination: the protonated molecular ion of the homodimer (6K-I) was observed at m/z = 6279.8 (expected value 6280.0) and that of the heterodimer (6K-II) at 6282.2 (expected value 6282.1). Both peptides have been chemically synthesized and the synthetic compounds appeared to be identical to the native 6K-I and 6K-II. Polyclonal antibodies raised against each of these peptides demonstrated that they were contained within the secretory granules of the intrinsic cells of the *corpora cardiaca* glandular lobes. The physiological significance of 6K-I and 6K-II is unknown. We are currently probing the biological role of 6K-I and 6K-II [1] with the synthetic peptides.

Identification of the 5-kDa peptide

Neurohaemal lobes of *corpora cardiaca* were extraced and subjected to C_{18}-RPHPLC. The predominant absorption peak was further purified by C_1-RP-HPLC. The major compound was then characterized as a 50-residue peptide (5K-peptide, 4978 Da). Its sequence (Fig. 2) is remarkable by its large number of alanine residues which make up about one-fourth of the 50 amino acids; in particular, the sequence contains a penta-alanine motif. This raises the possibility that the 5K-peptide could be a spacer peptide. This hypothesis was indeed confirmed by Lagueux et al. [2] who showed that the coding sequence

```
1                    10                    20
A-S-Q-D-V-S-D-S-E-S-E-D-N-Y-W-S-G-Q-S-A-D-E-A-A-E-


          30                    40                    50
A-A-A-A-A-L-P-P-Y-P-I-L-A-R-P-S-A-G-G-L-L-T-G-A-V
```

Fig. 2. *Structure of the 5 kDa peptide.*

for the 5K-peptide in the mRNA is flanked by sequences exhibiting significant homology to the A and B chains of the superfamily of insulins. In the conventional terminology used for insulins, the 5K-peptide would correspond to the C-peptide. Interestingly, a computer search of protein sequence data banks did not reveal any significant homology of 5K to other identified proteins [3].

References

1. Hietter, H., Luu, B., Goltzene, F., Zachary, D., Hoffmann, J. and Van Dorsselaer, A., Eur. J. Biochem., 182 (1989) 77.
2. Lagueux, M., Lwoff, L., Meister, M., Goltzene, F. and Hoffmann, J., Eur. J. Biochem., (1989) in press.
3. Hietter, H., Van Dorsselaer, A., Green, B., Denoroy, L., Hoffmann, J. and Luu, B., Eur. J. Biochem., (1989) in press.

Interaction of selected peptide fragments of tubulin with MAP-2: A structure-function relationship

John R. Cann*, Eunice J. York and John M. Stewart

Department of Biochemistry, Biophysics and Genetics, University of Colorado Health Sciences Center, B-121, 4200 E. 9th Avenue, Denver, CO 80262, U.S.A.

Introduction

The cytoskeleton (the internal scaffolding) of mammalian and other eukaryotic cells is a dynamic structure giving the cell its distinctive shape and enabling it to migrate and change shape, transport vesicles, and to divide into daughter cells [1]. The cytoskeleton of animal cells is composed of three principal types of protein filaments: microfilaments that are 70 Å in diameter and made of actin; intermediate filaments, 70–110 Å in diameter, formed from a family of proteins that have common α-helical coiled-coil core but are otherwise diverse; and the ubiquitous microtubules (MT). This latter element of the cytoskeleton is the focus of this investigation.

The MT [1–3] are hollow cylindrical structures, 300 Å in external diameter and 140 Å internal diameter. They are composed of an helical array of dimeric subunit molecules called tubulin. The MT play an essential role in several intracellular functions, including modulation of surface receptors and secretory activities, cellular division, and fast axoplasmic transport.

The subunit protein tubulin has a molecular weight of 100 000 and consists of two nonidentical subunits designated α- and β-tubulin. It reassembles into MT in vitro at 37°C via a GTP-dependent polymerization process that requires either an assembly promoter, such as glycerol, or one of the micromolecule-associated proteins (MAPs), in particular, either MAP-2 or tau. A large number of biologically active substances interact with tubulin. These include the mitotic poison colchicine [4], the antineoplastic drug vinblastine [5–7], the neurotransmitter substance P [8], and the neuroleptic chlorpromazine [9]. Each of these substances interferes with the assembly of MT and disassembles preformed MT.

Chlorpromazine (CPZ) exerts its action on tubulin and MT by binding to a single site on the tubulin molecule with concomitant decrease in the α-helical content of the protein. The many biological activities of CPZ include blockade of axonal transport, reduction in the number of MTs in spinal ganglion cells and neuroblastoma cells, mitotic arrest, and disorganization of the organized microtubular structure produced in cells by cAMP. We suggest that the psychotropic action of CPZ may involve interaction with tubulin as one of a set of orchestrated mechanisms. In another vein, work in this area illustrates the

---

*To whom correspondence should be addressed.

J.R. Cann, E.J. York and J.M. Stewart

Table 1 A. Peptide fragments of the C-terminal moiety of α- and β-tubulin[a]

α(430–441)	Nα-Acetyl-Lys430-Asp-Tyr-Glu-Glu-Val-Gly-Val-Asp-Ser-Val-Glu441-NH2
β(422–434)	Nα-Acetyl-Tyr422-Gln-Gln-Tyr-Gln-Asp-Ala-Thr-Ala-Asp-Glu-Gln-Gly434-NH2
α(401–410)	Nα-Acetyl-Lys401-Arg-Ala-Phe-Val-His-Trp-Tyr-Val-Gly410-NH2
β(391–400)	Nα-Acetyl-Arg391-Lys-Ala-Phe-Leu-His-Trp-Tyr-Thr-Gly400-NH2

B. Predicted conformation and binding constants to MAP-2

4-kDa peptide	Chou–Fasman in situ conformation[b]	k_a (M^{-1})[c]	Number of sites on MAP-2[c]
α(430–441)	one α-helical turn at N-terminus	2×10^6 1.1×10^4	2
β(422–434)	one α-helical turn at C-terminus	$4.0(\pm 1.5) \times 10^5$	2
α(401–410)	predominately β-structure	no binding	
β(391–400)	predominately β-structure	no binding	

[a] The peptides were synthesized by solid-phase method, purified and characterized as described [15].
[b] The MSEQ computer program was used for the secondary structure analysis.
[c] Data from Refs. 12–14 and personal communication from R.B. Maccioni (1989).

472

unanticipated and rewarding direction that basic research can take. Thus, Seebeck and Gehr [10], noting the CPZ-tubulin interaction, were led to discover the potent lethal activity of CPZ and certain other phenothiazines on *Trypanosoma brucei*, with complete disruption of pellicular MT. Disruption is apparently mediated through binding of CPZ by a protein component of the link between the MT array and the cell membrane [11]. Different MAb against this protein can, in turn, distinguish between different trypanosome species, which may prove of diagnostic and epidemiologic value [11].

The study reported below probes the interaction of tubulin with MAP-2, which binds to an essential site on the regulatory C-terminal 4-kDa moiety of the heterodimer [12,13]. Limited proteolysis of tubulin with subtilisin produces the cleaved tubulin heterodimer and the two C-terminal 4-kDa peptide fragments of α- and β-tubulin, thereby rendering assembly of tubulin into MT independent of MAPs. In order to define more precisely the structure of the essential binding site for MAPs, four synthetic peptide fragments of tubulin were examined in this context. Their amino acid sequences are given in Table 1A. It seemed likely that the two anionic peptides α(430–441) and β(422–434), from the low homology region within the 4-kDa domain, might interact with MAP-2, while the cationic peptides α(401–410) and β(391–400), of the high homology region adjacent to the 4-kDa domain, might be inactive. The binding characteristics of these peptides to MAP-2 have already been reported [12–14, R.B. Maccioni, personal communication, 1989]. Their solution conformations are the subject of this communication. A SAR has emerged from these combined studies.

Results and Discussion

The binding data summarized in Table 1B confirmed that the peptide fragments from the 4-kDa domain of tubulin bind to MAP-2, while those from the adjacent region do not. Preferential binding of β(422–434) compared to α(430–441) suggests a possible composition of the binding site on the tubulin molecule, the affinity of the β-fragment for MAP-2 accounting for the affinity of the 4-kDa peptide. Thus, the standard free energy of binding of the β-fragment to the first site on MAP-2 is $\Delta G° = -RT\ln 2k_a = -8.1$ kcal mol^{-1}, which is to be compared to the standard free energy of binding of the 4-kDa peptide, $\Delta G° = -RT\ln k_a = -8.6$ kcal mol^{-1}. The Chou–Fasman predictions of the in situ conformation of the four fragments, Table 1B, suggests a SAR, prompting CD and NMR studies of their solution conformations.

Comparison of the CD of each peptide in two solvents is presented in Fig. 1, from which it is apparent that each has a disordered structure in water as solvent. However, in 95% ethanol/5% water, the MAP-2-binding peptides α(430–441) and β(422–434) show characteristic α-helical spectra that analyze [16] 60% and 82% helix, respectively. Peptide α(430–441) was also examined in 80% methanol/20% water, in which solvent it showed (Fig. 2 in Ref. 17) a partial α-helical conformation (30%), as confirmed by 2D ROESY ¹H NMR [18], the NMR conformation being described as an α-helical turn at the N-terminus,

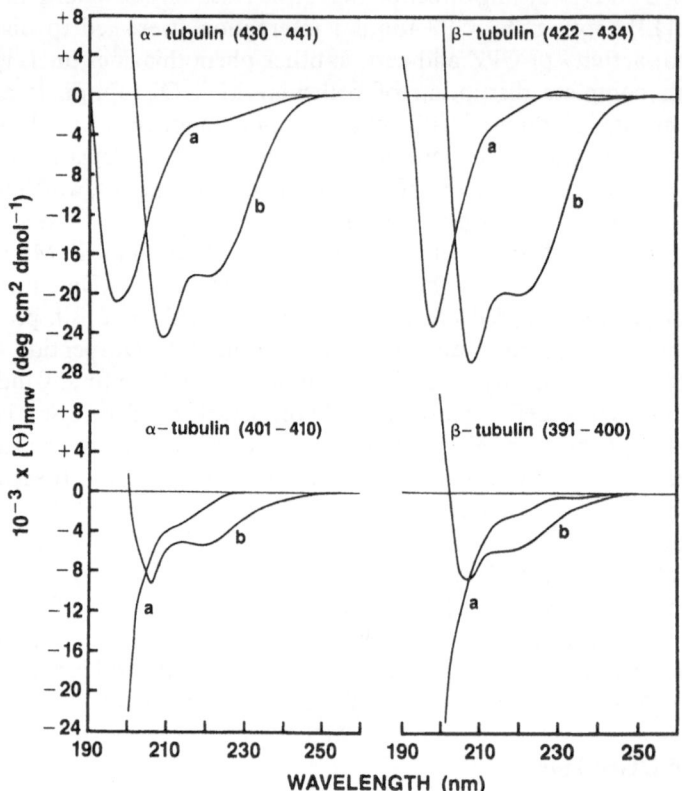

Fig. 1. *Far-UV CD spectra of four selected peptide fragments of tubulin; (a) water at pH 5 as solvent; (b) 95% ethanol/5% water. Upper panels, 10°C; lower panels, 27°C.*

followed by a kinked extended section. These results are in striking agreement with the in situ conformation predicted by the Chou–Fasman algorithm (Table 1B).

The CD and NMR of the nonbinding peptides α(401–410) and β(391–400) is contrastive. Although superficial reading of their CD in ethanol (or methanol)/ H_2O (Fig. 1) might suggest a degree of hydrogen-bonded structure in alcohol/ H_2O as solvent, 1D and 2D ^1H NMR of α(401–410) speak against hydrogen bonding; none of the amide proton resonances were shifted downfield, and there were no significant changes in coupling constants in 90% CD_3OH/10% H_2O [G. Kotovych, personal communication, 1988]. Measurements on model compounds indicate that the change in CD of the peptide on going from H_2O to CH_3OH/H_2O is attributable to nonspecific perturbation of the CD of the four aromatic side groups and the peptide bonds. These results point up the hazards of superficial assignment of CD to a particular conformation.

Conclusion

The interaction of the peptide fragments with MAP-2 conforms to a SAR; namely, whereas the two anionic fragments from the 4-kDa domain of tubulin with their potential for adopting an α-helical conformation bind to MAP-2, the two cationic fragments from the adjacent region, which are nonhydrogen-bonded in solution, do not bind. The specificity of the interactions has been established immunologically [19].

References

1. Alberts, B., Bray, D., Lewis, J., Raff, M., Roberts, K. and Watson, J.D., Molecular Biology of the Cell, 2nd ed., Garland Publishing Inc., New York, 1989, Chap. 11 and 13.
2. Dustin, P., Microtubules, 2nd ed., Springer-Verlag, New York, 1984.
3. Soifer, D. (Ed.) Dynamic Aspects of Microtubule Biology, Ann. N.Y. Acad. Sci., 466 (1986).
4. Hastie, S.B., Williams, Jr., R.C., Puett, D. and Macdonald, T.L., J. Biol. Chem., 264 (1989) 6682.
5. Lee, J.C., Harrison, D. and Timaseff, S.N., J. Biol. Chem., 250 (1975) 9276.
6. Na, G.C. and Timasheff, S.N., Biochemistry, 19 (1980) 1355.
7. Na, G.C. and Timasheff, S.N., Biochemistry, 25 (1986) 6214.
8. Maccioni, R.B., Cann, J.R. and Stewart, J.M., Eur. J. Biochem., 154 (1986) 427.
9. Appu Rao, A.G. and Cann, J.R., Mol. Pharmacol., 19 (1981) 295.
10. Seebeck, T. and Gehr, P., Mol. Biochem. Parasitol., 9 (1983) 197.
11. Stieger, J. and Seebeck, T., Mol. Biochem. Parasitol., 21 (1986) 37.
12. Maccioni, R.B., Rivas, C.I. and Vera, J.C., EMBO J., 7 (1988) 1957.
13. Rivas, C.I., Vera, J.C. and Maccioni, R.B., Proc. Natl. Acad. Sci. U.S.A., 85 (1988) 6092.
14. Cann, J.R., York, E.J., Stewart, J.M., Vera, J.C. and Maccioni, R.B., Anal. Biochem., 175 (1988) 462.
15. Stewart, J.M. and Young, J.D., Solid Phase Peptide Synthesis, 2nd edn., Pierce Chemical Company, Rockford, IL, 1984.
16. Chang, C.T., Wu, C.-S.C. and Yang, J.T., Anal. Biochem., 91 (1978) 13.
17. Sugiura, M., Maccioni, R.B., Cann, J.R., York, E.J., Stewart, J.M. and Kotovych, G., J. Biomol. Struct. Dyn., 4 (1987) 1105.
18. Otter, A. and Kotovych, G., Can. J. Chem., 66 (1988) 1814.
19. Vera, J.C., Rivas, C.J. and Maccioni, R.B., Proc. Natl. Acad. Sci. U.S.A., 85 (1988) 6763.

Interleukin 1: Biological effects, structures, and gene regulation

Burton D. Clark and Charles A. Dinarello

*Tufts University School of Medicine and the New England Medical Center,
Boston, MA 02111, U.S.A.*

Introduction

Interleukin 1 (IL-1) has been shown to mediate a wide variety of immunologic and inflammatory responses. It is a polypeptide hormone, predominantly synthesized by circulating, activated monocytes or resident macrophages that have entered tissues and differentiated, depending on their site of localization. Some examples are Kupffer cells, alveolar macrophages, and microglia. The name interleukin, meaning 'between leukocytes', is misleading since many non-leukocytic cells also synthesize this hormone, including synovial fibroblasts, keratinocytes, and mesangial, endothelial, and smooth muscle cells.

IL-1 was first described as leukocytic, granulocytic, or endogenous pyrogen (reviewed in Ref. 1). It was purified and biochemically characterized in the 1970s, and in the last several years, two genes and their respective cDNAs have been cloned. These genes encode the two known forms designated IL-1α and IL-1β. The expression of these cDNAs has supplied abundant quantities of pure recombinant IL-1 protein for further study of its biological properties and for confirming previous reports that used natural sources.

Results and Discussion

(1) Biological effects

One of the first effects found to be associated with IL-1 was its ability to act as an endogenous pyrogen and produce monophasic fever in rabbits [2,3]. A multitude of other dramatic biological effects include the ability to induce the acute phase response [4,5], augment T-cell responses to antigens and mitogens [6], stimulate B-cell proliferation [7,8], decrease plasma iron and zinc levels, and activate endothelial cells and fibroblasts to synthesize and release PGE_2 [9–11].

IL-1, as well as TNFα, IFNα and IL-6, mediate the febrile response as a component of the acute phase reaction [12]. Currently there are two hypotheses to explain this activity. In the first hypothesis, it is thought that plasma IL-1, produced in response to injury, infection, or inflammation, enters the brain through the *organum vasculosum laminae terminalis* (OVLT), which is the site of IL-1-responsive neurons involved in changing the temperature set point.

Increased PGE_2 is found in this area during fever, but it is not juxtaposed to thermosensitive neurons; however, it may stimulate interneurons to mediate an effect on thermoregulation. The exact site of action of IL-1 has not been determined, but this hypothesis is consistent with the destruction of the OVLT in guinea pigs, which leads to abolishment of the febrile and acute phase response [13,14]. In a recent paper, a second hypothesis arises. Histochemical staining of the human brain using anti-IL-1β shows that the most significant staining is in the preoptic (PO) and paraventricular nuclei (PVN), both some distance from the OVLT, the so-called main gateway for IL-1 entrance into the brain [15]. It is unlikely that these neurons are affected by peripheral IL-1. It could be that the IL-1 in the PO and PVN serves as its own intermediate messenger and that these neurons may use IL-1 as a neuromodulator. This cascade may then stimulate autocrine and endocrine pathways, resulting in the cerebral component necessary to activate the acute phase response. We have also observed similar histochemical staining of rat brain taken during endotoxemia [16]. Using anti-rat IL-β, immunoreactive IL-1β is seen in cell bodies of the anterior and posterior hypothalamus, and the anterior and posterior lobes of the pituitary. In all histochemical studies, binding could be inhibited with exogenous IL-1. Other regions of the brain, such as the median eminence, may also serve as a gateway for IL-1, possibly working in concert with the OVLT.

The ability of IL-1 to up-regulate the cellular metabolism of hepatocytes in the acute phase response is very pronounced. IL-1 induces a dramatic increase of many acute phase proteins, including C-reactive protein, serum amyloid A, antiproteases, complement components C3 and factor B, α_1-major acute phase protein, α, β, and γ fibrinogen, and α_1-acid glycoprotein [17]. The regulation of these proteins is at the level of transcription [4,18]. Other proteins expressed during this response in the liver, such as actin, are unchanged, whereas albumin and transferrin synthesis are decreased. Many of the acute phase protein genes have been cloned [19] and studies are underway to investigate the signal mechanisms by which IL-1 enhances the transcription of these specific genes.

The liver also responds to IL-1 by synthesizing metalloproteins which bind serum zinc and iron. The removal of iron and zinc seems to be a basic host defense mechanism, since bacteria and tumor cells require large amounts of iron for cell metabolism. It is believed that the acute phase proteins induced during infection and fever are important for removing bacterial toxins and oxygen radicals. Furthermore, elevated temperatures during fever may inhibit bacterial growth.

Other important biological activities of IL-1 are the augmentation of T-cell and B-cell proliferation. In an early report, IL-1 was shown to be a comitogen in stimulating thymocyte proliferation [20]. IL-1 will not stimulate these cells alone, but, in combination with lectins, such as Con A, it induces IL-2 and IL-2 receptors [21]. De novo IL-2 can then work in combination with IL-1 to stimulate more IL-2 receptors and thymocyte proliferation [22]. The effect of IL-1 on B-cells is similar to that of T-cells; it acts as a cofactor with IL-4 to activate B-cells and increase antibody formation [23,24].

(2) Structure and gene expression of IL-1

In the last several years, great advances have been made in understanding the structure and expression of IL-1. There are two forms, IL-1α and IL-1β, that have isoelectric points of 5 and 7, respectively. The genes and their respective cDNAs have been isolated and sequenced for both human forms and cDNA counterparts have been sequenced for murine, bovine, rat and rabbit [25–27]. Both forms are translated as prointerleukin-1 polypeptides of 31 kDa before being processed to mature 17.5-kDa peptides. The exact pathway of processing the propeptide is unclear. In macrophages, the intracellular mass of IL-1 is 31 kDa, and this is thought to be enzymatically processed when extracellular. The IL-1α form is predominantly cell associated, whereas IL-1β is 'secreted'. Neither form has traditional signal sequences of hydrophobic amino acids to facilitate secretion.

There is strong amino acid homology (75%) for the same form of IL-1 when comparing between species, and there is approximately 30% amino acid homology between IL-1α and IL-1β. Despite the lack of amino acid homology between the two IL-1 forms, they share nearly identical biological activities. The two forms do have five small regions of conserved sequences designated A–E, and it has been suggested that regions C, D and E may represent 'active sites' [28]. Regions C + D are encoded by the entire VIth exon conserved in the two human genes [29]. Furthermore, both IL-1α and β recognize the same receptor and have similar binding affinities [30,31]. Thus, even though the two forms are structurally distinct, they do possess conserved regions that may be responsible for their similar biological actions.

Although a considerable amount of information is known about the biological activities of the two IL-1s, less is known about the expression of the genes. Endotoxin-induced expression of IL-1β in the monocytic cell line, THP-1, is transient and similar to several proto-oncogenes and competence factors [32–34]. There is rapid transcriptional activation after addition of endotoxin, followed by rapid repression (expression is non-transient when stimulated with phorbol myristic acetate). This expression is controlled at the level of transcription as shown by nuclear run-off experiments [32]. Similar transient expression is seen in human monocytes, endothelial, and smooth muscle cells with the peak levels of mRNA at 2–3 h [33,35, 36]. There is differential expression of IL-1α and β when cultured monocytes are compared to peripheral blood monocytes and alveolar macrophages [37]. In general, there is more IL-1β mRNA relative to IL-1α in response to endotoxin, suggesting that there are separate mechanisms regulating each gene. The ratio of β:α ranges from 3.5 for aveolar macrophages to 21 for cultured monocytes.

Both human IL-1 genes have been sequenced [29,38]. They each contain seven exons and have very similar exon boundaries and inclusive amino acids [29]. To better understand the regulation of the IL-1β gene, *cis*- and *trans*-acting elements of the human prointerleukin-1β gene have been studied using two approaches [39]. The first approach makes use of chimeric plasmids containing proIL-1β DNA sequences fused to the promoter-less bacterial CAT gene to investigate important *cis*-acting elements. The second approach involves the binding of proteins to important sequences of this gene, as visualized by gel mobility shift assays. From these studies, the proIL-1β upstream sequences are able to direct CAT expression, which is cell type specific, being active in monocytic cell lines THP-1 and U937, but inactive in non-monocytic HeLa or Colo16 cells. Only the first 131 base pairs of upstream proIL-1β DNA sequence are essential for orchestrating CAT expression, and this can be enhanced by having the first 513 or 1097 bp of upstream sequence relative to the transcriptional initiation site.

Electrophoretic band shift assays have identified regions of the proIL-1β DNA sequences that bind nuclear factors (*trans*-acting) that may mediate expression of the gene. A specific factor from endotoxin-stimulated THP-1 and human monocytes binds to the *Hind*III-*Taq*I fragment whose sequence is located in the vicinity of the TATA promoter box. One possibility is that this factor is important for endotoxin-induced expression of IL-1β. This factor is not found in nuclear protein extracts of U937 cells, however, these monocytic-like cells do not express IL-1β in response to endotoxin.

Acknowledgements

The authors wish to thank Drs. Philip E. Auron, Matthew J. Fenton, Andrew C. Webb and Steven P. Sirko.

References

1. Dinarello, C.A. and Wolff, S.M., Am. J. Med., 72(1982)799.
2. Dinarello, C.A., Renfer, L. and Wolff, S.M., Proc. Natl. Acad. Sci. U.S.A., 74(1977)4624.
3. Murphy, P.A., Chesney, P.J. and Wood, W.B.J., J. Lab. Clin. Med., 83(1974)310.
4. Ramadori, G., Sipe, J.D., Dinarello, C.A., Mizel, S.B. and Colten, H.R., J. Exp. Med., 162(1985)930.
5. Ghezzi, P., Saccardo, B., Villa, P., Rossi, V., Bianchi, M. and Dinarello, C.A., Infect. Immun., 54(1986)837.
6. Smith, K.A., Lachman, L.B., Oppenheim, J.J. and Favata, M.F., J. Exp. Med., 151(1980)1551.
7. Lipsky, P.E., Thompson, P.A., Rosenwasser, L.J. and Dinarello, C.A., J. Immunol., 130(1983)2708.
8. Matsushima, K., Procopio, A., Abe, H., Scala, G., Ortaldo, J.R. and Oppenheim, J.J., J. Immunol., 135(1985)1132.
9. Schmidt, J.A., Oliver, C.N., Lepe, Z.J.L., Green, I. and Gery, I., J. Clin. Invest., 73(1984)1462.
10. Rossi, V., Breviario, F., Ghezzi, P., Dejana, E. and Mantovani, A., Science, 229(1985)174.
11. Dejana, E., Breviario, F., Erroi, A., Bussolino, F., Mussoni, L., Gramse, M., Pintucci, G., Casali, B., Dinarello, C.A. and Van, D.J., Blood, 69(1987)695.
12. Dinarello, C.A., Cannon, J.G. and Wolff, S.M., Rev. Infect. Dis., 10(1988)168.

13. Blatteis, C.M., Bealer, S.L., Hunter, W.S., Llanos, Q.J., Ahokas, R.A. and Mashburn, T.A.J., Brain Res. Bull., 11 (1983) 519.
14. Blatteis, C.M. and Banet, M., Pflüg. Arch. Eur. J. Physiol., 406 (1986) 480.
15. Breder, C.D., Dinarello, C.A. and Saper, C.B., Science, 240 (1988) 321.
16. Lechan, R.M., Toni, R., Clark, B.D., Cannon, J.G., Shaw, A.R., Dinarello, C.A. and Reichlin, S., 71st Annual Meeting of The Endocrine Society, 1989.
17. Koj, A., In Reutter, W.R. and Popper, H. (Eds.) Modulation of Liver Cell Expression (Falk Symposium, No. 43), MTP Press, Lancaster, 1986, p. 331.
18. Liao, W.S.L., Ma, K.-T., Woodworth, C.D. and Isom, H.C., Mol. Cell. Biol., 9 (1989) 2779.
19. Liao, W.S.L. and Stark, G.R., Inflamm. Res., 10 (1985) 220.
20. Gery, I. and Waksman, B.H., J. Exp. Med., 136 (1972) 143.
21. Simic, M.M. and Stosic, G.S., Folia Biol. (Praha), 31 (1985) 410.
22. Mannel, D.N., Mizel, S.B., Diamantstein, T. and Falk, W., J. Immunol., 134 (1985) 3108.
23. Falkoff, R.J., Butler, J.L., Dinarello, C.A. and Fauci, A.S., J. Immunol., 133 (1984) 692.
24. Falkoff, R.J., Muraguchi, A., Hong, J.X., Butler, J.L., Dinarello, C.A. and Fauci, A.S., J. Immunol., 131 (1983) 801.
25. Auron, P.E., Webb, A.C., Rosenwasser, L.J., Mucci, S.F., Rich, A., Wolff, S.M. and Dinarello, C.A., Proc. Natl. Acad. Sci. U.S.A., 81 (1984) 7907.
26. Lomedico, P.T., Gubler, U., Hellmann, C.P., Dukovich, M., Giri, J.G., Pan, Y.C., Collier, K., Semionow, R., Chua, A.O. and Mizel, S.B., Nature, 312 (1984) 458.
27. Cannon, J.G., Clark, B.D., Wingfield, P., Schmeissner, U., Losberger, C., Dinarello, C.A. and Shaw, A.R., J. Immunol., 142 (1989) 2299.
28. Auron, P.E., Rosenwasser, L.J., Matsushima, K., Copland, T., Dinarello, C.A., Oppenheim, J.J. and Webb, A.C., J. Mol. Cell Immunol., 2 (1985) 169.
29. Clark, B.D., Collins, K.L., Gandy, M.S., Webb, A.C. and Auron, P.E., Nucleic Acids Res., 14 (1986) 7897.
30. Kilian, P.L., Kaffka, K.L., Stern, A.S., Woehle, D., Benjamin, W.R., Dechiara, T.M., Gubler, U., Farrar, J.J., Mizel, S.B. and Lomedico, P.T., J. Immunol., 136 (1986) 4509.
31. Sims, J.E., March, C.J., Cosman, D., Widmer, M.B., MacDonald, H.R., McMahan, C.J., Grubin, C.E., Wignall, J.M., Jackson, J.L., Call, S.M., Friend, D., Alpert, A.R., Gillis, S., Urdal, D.L. and Dower, S.K., Science, 241 (1988) 585.
32. Fenton, M.J., Clark, B.D., Collins, K.L., Webb, A.C., Rich, A. and Auron, P.E., J. Immunol., 138 (1987) 3972.
33. Fenton, M.J., Vermeulen, M.W., Clark, B.D., Webb, A.C. and Auron, P.E., J. Immunol., 140 (1988) 2267.
34. Greenberg, M.E. and Ziff, E.B., Nature, 311 (1984) 433.
35. Libby, P., Ordovas, J.M., Auger, K.R., Robbins, A.H., Birinyi, L.K. and Dinarello, C.A., Am. J. Pathol., 124 (1986) 179.
36. Libby, P., Ordovas, J.M., Birinyi, L.K., Auger, K.R. and Dinarello, C.A., J. Clin. Invest., 78 (1986) 1432.
37. Smith, F.M., Kueppers, F.R., Yound, P.R. and Lee, J.C., In Powanda, M.C., Oppenheim, J.J., Kluger, M.J. and Dinarello, C.A. (Eds.) Monokines and Other Non-lymphocytic Cytokines, Alan R. Liss, New York, 1988, p. 79.
38. Furutani, Y., Notake, M., Fukui, T., Ohue, M., Nomura, H., Yamada, M. and Nakamura, S., Nucleic Acids Res., 14 (1986) 3167.
39. Clark, B.D., Fenton, M.J., Rey, H.L., Webb, A.C. and Auron, P.E., In Powanda, M.C., Oppenheim, J.J., Kluger, M.J. and Dinarello, C.A. (Eds.) Monokines and Other Non-lymphocytic Cytokines, Alan R. Liss, New York, 1988, p. 47.

The TGF-β family of multifunctional peptides

Anita B. Roberts* and Michael B. Sporn

Laboratory of Chemoprevention, National Cancer Institute, Bethesda, MD 20892, U.S.A.

Introduction

The largest known family of growth factors comprises peptides structurally related to transforming growth factor-β (TGF-β), a name that was first coined in 1983 to describe a disulfide-linked homodimeric peptide of 25 000 MW that, based on its ability to induce anchorage-independent growth of cells, was thought to play a role in the expression of the transformed phenotype (for review see [1]). However, in the ensuing 5 years it has become apparent that this peptide is probably one of the most highly conserved and multifunctional of all the known growth factors. Platelets and bone represent the most abundant sources of the peptide, suggesting that it plays a central role in healing and repair of both soft and hard tissues. It is both secreted by and acts on a wide variety of cell types, often functioning as a 'master control' to regulate the activity of other growth factors.

Results and Discussion

The family of peptides related to TGF-β presently comprises 5 highly conserved functional homologs, called TGF-βs 1 to 5, which are between 62% and 84% homologous to each other, and 8 other functionally distinct peptides, which are only 30–40% homologous to the TGF-βs [1]. These more distantly related peptides include the mammalian inhibins, activins, Mullerian inhibitory substance, and the bone morphogenetic proteins, as well as the putative products of the *Drosophila* DPP-C gene complex and the *Xenopus* Vg1 gene. Although each member of the TGF-β family is first synthesized as a large precursor molecule and then processed to the bioactive C-terminal peptide, the familial relationships of these more distantly related peptides to the TGF-βs are restricted to positional conservation of the cysteine residues within the C-terminal mature form of TGF-β.

In contrast, within the set of 5 TGF-βs, homologies extend throughout the entire precursors, which range in size from 382 to 414 amino acids; all are processed at a tetrabasic site to release the C-terminal 112 amino acids (except TGF-β-4, which has 2 additional amino acids). This C-terminal peptide, in dimeric form, constitutes the biologically active form of the peptide. Each of the different types of TGF-β is highly conserved between species, and the mature forms of the peptides are greater than 99% conserved, either between different mammalian

---

*To whom correspondence should be addressed.

481

species or between mammalian and avian species [1]. Thus, for example, the amino acid sequences of human, bovine, porcine, simian, and chicken TGF-β-1 are identical, and that of murine TGF-β-1 differs by one amino acid. Although TGF-βs-4 and -5, cloned from embryonic chicken [2] and frog cDNA libraries [3], respectively, have not yet been identified in mammals, it is likely that these peptides represent new forms of TGF-β that also will be highly conserved between species.

Practically speaking, this high degree of conservation means that a particular type of TGF-β can be used with equivalent effectiveness in almost all species, regardless of its source. On the other hand, one must ask why so many different TGF-βs exist and whether they will be found to have distinct biological activities. Recent data demonstrate that all TGF-βs, with the exception of TGF-β-4 that has not yet been expressed, act interchangeably in most in vitro assay systems, such as inhibition of cell growth or differentiation, stimulation of matrix protein synthesis, and monocyte chemotaxis ([4], unpublished). Thus one can often speak in a generic sense of the activities of the various TGF-βs. However, a limited number of specific activities are also beginning to be defined; thus TGF-β-1 is far more potent in inhibiting the growth of endothelial cells than is TGF-β-2 [5], and TGF-β-2, but not TGF-β-1 or TGF-β-5, is active in the induction of amphibian mesoderm from ectodermal explants ([6]; F. Rosa and I. Dawid, personal communication). This suggests that in vivo assay of the specific TGF-β types might reveal additional selective activities.

TGF-βs are unique among the cytokines in that they are released from platelets and secreted from cells in a high molecular weight latent, biologically inactive form that is unable to bind to cellular receptors [7,8]. This latent form is now known to be a non-covalent complex between three peptides: mature TGF-β, the remainder of its precursor, called the latency-associated peptide, and a third protein of about 150 kDa. Physiological mechanisms of activation of latent TGF-β probably include the action of proteases or of local acidic microenvironments, such as might be found in the vicinity of a healing wound or of activated osteoclasts in remodeling bone. Since many different cell types secrete TGF-β and have receptors for TGF-β, it is clear that activation of the latent form of the peptide is an important control point in regulation of the biological activity of the peptide.

The TGF-βs as a class are some of the most potent of all known growth factors; ED_{50}s for in vitro activities typically range from 10 to 100 pg/ml (0.4–4 pM), but for certain activities, such as chemotaxis of monocytes, they are in the femtomolar concentration range [1,9] Recent immunohisto-chemical analysis of expression of the TGF-βs in vivo [10,11] as well as in situ hybridization studies [12] demonstrate that TGF-β is an endogenous mediator of cell growth and differentiation, and that it is expressed in a developmentally regulated fashion by specific cell types at specific times, not only during early development, but also in adult tissues, and in certain pathological states.

Evidence suggests that expression of the different TGF-βs is independently regulated. For example, in the sources from which TGF-βs are typically obtained, platelets or bone, human platelets exclusively contain TGF-β-1, but porcine

platelets and bovine bone contain both TGF-β-1 and -2 in approximately a
4 : 1 ratio [1,4]. Moreover, certain cell lines have been identified that secrete
predominantly one or the other form of the peptide [13]. Relative expression
of the various TGF-β mRNAs is also dependent on the species. Thus, levels
of expression of TGF-βs-1, -2 and -3, and -5 mRNAs are most prominent in
mouse, chicken, and frog embryos, respectively ([3,10]; S. Jakowlew, personal
communication). Specific probes and antibodies are now available to quantitate
the different types of TGF-β [13]; the promoters for the different forms of TGF-
β are being isolated [14]. Used together, these type-specific reagents will allow
investigation of the pattern of differential expression of the TGF-β genes in
vivo.

   The action of TGF-β is dependent on many factors including the cell type,
the state of differentiation of the target cell, and the set of other growth factors
with which it is acting [15]. As such, TGF-β can be considered to be the
prototypical multifunctional growth factor. Numerous examples could be cited
where it inhibits a process under one set of conditions and stimulates the same
process under a different set of conditions [1]. The diversity of effects of TGF-
β, together with the almost universal ability of cells to respond to TGF-β, place
it in a unique position with regard to regulation of normal and pathologic
physiology [9,16]. Thus it plays a central role in control of many physiological
processes, including control of connective tissue function, suppression of immune
responses, regulation of growth and differentiation of both epithelial and
mesenchymal tissues, control of formation and remodeling of mineralized tissues,
as well as control of hematopoiesis and steroidogenesis. Aberrant expression
of TGF-β is found in many connective tissue diseases, as in proliferative
vitreoretinopathy [17], and in carcinogenesis [18], where the peptide is thought
to stimulate formation of tumor stroma by mechanisms analogous to those that
stimulate formation of granulation tissue in a healing wound [9].

References

1. Roberts, A.B. and Sporn, M.B., In Sporn, M.B. and Roberts, A.B. (Eds.) Peptide
   Growth Factors and Their Receptors, Handbook of Experimental Pharmacology,
   Springer-Verlag, Heidelberg, 95-1 (1990) 419.
2. Jakowlew, S.B., Dillard, P.J., Sporn, M.B. and Roberts, A.B., Mol. Endocrinol.,
   2 (1988) 1186.
3. Kondaiah, P., Sands, M.J., Smith, J.M., Fields, A., Roberts, A.B., Sporn, M.B.
   and Melton, D.A., J. Biol. Chem., (1989) in press.
4. Cheifetz, S., Weatherbee, J.A., Tsang, M.L.S., Anderson, J.K., Mole, J.E., Lucas,
   R. and Massague, J., Cell, 48 (1987) 409.
5. Jennings, J.C., Mohan, S., Linkhart, T.A., Widstrom, R. and Baylink, D.J., J. Cell.
   Physiol., 137 (1988) 167.
6. Rosa, F., Roberts, A.B., Danielpour, D., Dart, L.L., Sporn, M.B. and Dawid, I.B.,
   Science, 239 (1988) 783.
7. Wakefield, L.M., Smith, D.M., Flanders, K.C. and Sporn, M.B., J. Biol. Chem.,
   263 (1988) 7646.

8. Miyazono, K., Hellman, U., Wernstedt, C., and Heldin, C.-H., J. Biol. Chem., 263 (1988) 6407.
9. Roberts, A.B., Flanders, K.C., Kondaiah, P., Thompson, N.L., Van Obberghen-Schilling, E., Wakefield, L., Rossi, P., Crombrugghe, B. de, Heine, U.I. and Sporn, M.B., Rec. Prog. Horm. Res., 44 (1988) 157.
10. Heine, U.I., Flanders, K., Roberts, A.B., Munoz, E.F. and Sporn, M.B., J. Cell Biol., 105 (1987) 2861.
11. Thompson, N.L., Flanders, K.C., Smith, M., Ellingsworth, L.R., Roberts, A.B. and Sporn, M.B., J. Cell Biol., 108 (1989) 661.
12. Lehnert, S.A. and Akhurst, R.J., Development, 104 (1988) 263.
13. Danielpour, D., Dart, L.L., Flanders, K.C., Roberts, A.B. and Sporn, M.B., J. Cell. Physiol., 138 (1989) 79.
14. Kim, S.-J., Jeang, K.-T., Glick, A., Sporn, M.B. and Roberts, A.B., J. Biol. Chem., 264 (1989) 7041.
15. Sporn, M.B. and Roberts, A.B., Nature, 332 (1988) 217.
16. Wakefield, L.M., Smith, D.M., Masui, T., Harris, C.C. and Sporn, M.B., J. Cell Biol., 105 (1987) 965.
17. Connor, T.B., Roberts, A.B., Sporn, M.B., Danielpour, D., Dart, L.L., Michels, R.G., deBustros, S., Enger, C. and Glaser, B.M., J. Clin. Invest., 83 (1989) 1661.
18. Roberts, A.B., Thompson, N.L., Heine, U.I., Flanders, K. C. and Sporn, M.B., Br. J. Cancer, 57 (1988) 594.

Synthetic peptides reveal specificity of molecular recognition for acetylation by N^α-acetyltransferase

John A. Smith[a,b,d],* and Fang-Jen S. Lee[a,c]

[a]Department of Molecular Biology and [b]Department of Pathology, Massachusetts General Hospital, Boston, MA 02114, U.S.A.
[c]Department of Genetics and [d]Department of Pathology, Harvard Medical School, Boston, MA 02114, U.S.A.

Introduction

N^α-Acetylation is a common co-translational modification seen in eukaryotic proteins, but its physiological significance remains unknown. However, it is presumed that N^α-acetylation plays a role in protein secretion and degradation. A large number of proteins in various organisms (animal and plant) are known to be N^α-acetylated (e.g. $>80\%$ of soluble cytoplasmic proteins in mouse L-cells and Ehrlich ascites cells [1]). Recently, the first N^α-acetyltransferase (AcT) was purified to homogeneity from *Saccharomyces cerevisiae* and partially characterized [2], the encoding yeast gene (*AAA1*) was cloned and sequenced [3], and a null mutation (*aaa1-1*) was created by gene replacement [4]. In comparison to wild-type yeast cells, *aaa1-1* cells cannot enter stationary phase, are sporulation-defective, are sensitive to heat-shock, and display reduced mating functions for a-type cells.

Results and Discussion

AcT catalyzes the transfer of an acetyl group to various synthetic peptides, including human ACTH(1–24) and its [Phe²] analog, yeast alcohol dehydrogenase I(1–24) and II(1–24), yeast phosphoglycerate kinase (1–24), human superoxide dismutase (SOD) (1–24), and human enolase (1–24) (Table 1). These peptides contain Ser or Ala as their NH_2-terminal residues, which together with Met are the most commonly acetylated NH_2-terminal residues [5]. The enzyme does not acetylate other synthetic peptides, including ACTH(11–24), ACTH(7–38), ACTH(18–39), human β-endorphin, yeast SOD(1–24), and yeast enolase(1–24).

In general, proteins derived from mitochondrial DNA are not acetylated. However, certain proteins (e.g. bovine cytochrome c oxidase, subunit VI) derived from nuclear DNA and imported into the mitochondria are acetylated. At present, it is unclear whether acetylation of such important proteins occurs within the mitochondria, and if so, why endogenously synthesized proteins remain refractory to acetylation. Interestingly, two 24-residue peptides mimicking two nuclear

---

*To whom correspondence should be addressed.

Table 1 *Relative activity of yeast acetyltransferase for the* Nα-*acetylation of synthetic peptides*

Substrate	Activity (%)[a]
ACTH(1–24)	100 ± 5
Ser-Tyr-Ser-Met-Glu-His-Phe-Arg-Trp-Gly-Lys-Pro-Val-Gly-Lys-Lys-Arg-Arg-Pro-Val-Lys-Val-Tyr-Pro	
[Phe²]ACTH(1–24)	90 ± 9
Ser-Phe-Ser-Met-Glu-His-Phe-Arg-Trp-Gly-Lys-Pro-Val-Gly-Lys-Lys-Arg-Arg-Pro-Val-Lys-Val-Tyr-Pro	
ACTH(11–24)	0
Lys-Pro-Val-Gly-Lys-Lys-Arg-Arg-Pro-Val-Lys-Val-Tyr-Pro	
ACTH(7–38)	0
Phe-Arg-Trp-Gly-Lys-Pro-Val-Gly-Lys-Lys-Arg-Arg-Pro-Val-Lys-Val-Tyr-Pro-Asn-Gly-Ala-Glu-Ser-Ala-Glu-Ala-Phe-Pro-Leu-Glu	
ACTH(18–39)	0
Arg-Pro-Val-Lys-Val-Tyr-Pro-Asn-Gly-Ala-Glu-Asp-Glu-Ser-Ala-Glu-Ala-Phe-Pro-Leu-Glu	
β-Endorphin(1–24) (Human)	2 ± 2
Tyr-Gly-Gly-Phe-Met-Thr-Ser-Glu-Lys-Ser-Gln-Thr-Pro-Leu-Val-Thr-Leu-Phe-Lys-Asn-Ala-Ile-Ile-Lys-Asn-Ala-Tyr-Lys-Lys-Gly-Glu	
Alcohol dehydrogenase I(1–24) (Yeast)	101 ± 5
Ser-Ile-Pro-Glu-Thr-Gln-Lys-Gly-Val-Ile-Phe-Tyr-Glu-Ser-His-Gly-Lys-Leu-Glu-Tyr-Lys-Asp-Ile-Pro	
Alcohol dehydrogenase II(1–24) (Yeast)	102 ± 4
Ser-Ile-Pro-Glu-Thr-Gln-Lys-Ala-Ile-Ile-Phe-Tyr-Glu-Ser-Asn-Gly-Lys-Leu-Glu-His-Lys-Asp-Ile-Pro	
Phosphoglycerate kinase (1–24) (Yeast)	101 ± 6
Ser-Leu-Ser-Ser-Lys-Leu-Ser-Val-Gln-Asp-Leu-Asp-Leu-Lys-Asp-Lys-Arg-Val-Phe-Ile-Arg-Val-Asp-Phe	
Cytochrome c oxidase(1–24) (Yeast, mitochondrial, polypeptide VI)	60 ± 5
Ser-Asp-Ala-His-Asp-Glu-Glu-Thr-Phe-Glu-Glu-Phe-Thr-Ala-Arg-Tyr-Glu-Lys-Glu-Phe-Asp-Glu-Ala-Tyr	
ATPase inhibitor(1–24) (Yeast, mitochondrial)	76 ± 6
Ser-Glu-Gly-Ser-Thr-Gly-Thr-Pro-Arg-Gly-Ser-Gly-Ser-Glu-Asp-Ser-Phe-Val-Lys-Arg-Glu-Arg-Ala-Thr	
Superoxide dismutase(1–24) (Human)	86 ± 6
Ala-Thr-Lys-Ala-Val-Cys-Val-Leu-Lys-Gly-Asp-Gly-Pro-Val-Gln-Gly-Ser-Ile-Asn-Phe-Glu-Gln-Lys-Glu	
Superoxide dismutase(1–24) (Yeast)	0
Val-Gln-Ala-Val-Ala-Val-Leu-Lys-Gly-Asp-Ala-Gly-Val-Ser-Gly-Val-Val-Lys-Phe-Glu-Gln-Ala-Ser-Glu	
[Ala⁻¹-Thr¹]Superoxide dismutase(1–24) (Yeast)	14 ± 4
Ala-Thr-Gln-Ala-Val-Ala-Val-Leu-Lys-Gly-Asp-Ala-Gly-Val-Ser-Gly-Val-Val-Lys-Phe-Glu-Gln-Ala-Ser	
[Ala⁻¹-Thr¹-Lys²]Superoxide dismutase(1–24) (Yeast)	54 ± 5
Ala-Thr-Lys-Ala-Val-Ala-Val-Leu-Lys-Gly-Asp-Ala-Gly-Val-Ser-Gly-Val-Val-Lys-Phe-Glu-Gln-Ala-Ser	
[Ala⁻¹-Thr¹-Lys²-Ala³-Val⁴-Cys⁵]Superoxide dismutase(1–24) (Yeast)	80 ± 5
Ala-Thr-Lys-Ala-Val-Cys-Val-Leu-Lys-Gly-Asp-Ala-Gly-Val-Ser-Gly-Val-Val-Lys-Phe-Glu-Gln-Ala-Ser	

Table 1   *(continued)*

Substrate	Activity (%)[a]
Enolase(1–24) (Human) Ser-Ile-Leu-Lys-Ile-His-Ala-Arg-Glu-Ile-Phe-Asp-Ser-Arg-Gly-Asn-Pro- Thr-Val-Glu-Val-Asp-Leu-Phe	54 ± 5
Enolase(1–24) (Yeast) Ala-Val-Ser-Lys-Val-Tyr-Ala-Arg-Ser-Val-Tyr-Asp-Ser-Arg-Gly-Asn- Pro-Thr-Val-Glu-Val-Glu-Leu-Thr	4 ± 2
[Tyr^2]Enolase(1–24) (Yeast) Ala-Tyr-Ser-Lys-Val-Tyr-Ala-Arg-Ser-Val-Tyr-Asp-Ser-Arg-Gly-Asn- Pro-Thr-Val-Glu-Val-Glu-Leu-Thr	2 ± 1
[Ala^3]Enolase (1–24) (Yeast) Ala-Val-Ala-Lys-Val-Tyr-Ala-Arg-Ser-Val-Tyr-Asp-Ser-Arg-Gly-Asn- Pro-Thr-Val-Glu-Val-Glu-Leu-Thr	0
[Ser^1-Ile^2]Enolase(1–24) (Yeast) Ser-Ile-Ser-Lys-Val-Tyr-Ala-Arg-Ser-Val-Tyr-Asp-Ser-Arg-Gly-Asn- Pro-Thr-Val-Glu-Val-Glu-Leu-Thr	80 ± 5
[Ser^1-Ile^2-Leu^3]Enolase(1–24) (Yeast) Ser-Ile-Leu-Lys-Val-Tyr-Ala-Arg-Ser-Val-Tyr-Asp-Ser-Arg-Gly-Asn- Pro-Thr-Val-Glu-Val-Glu-Leu-Thr	4 ± 5
[Ser^1-Ile^2-Leu^3-Ile^5-His^6-Glu^9-Ile^{10}](1–24) (Yeast) Ser-Ile-Leu-Lys-Ile-His-Ala-Arg-Glu-Ile-Tyr-Asp-Ser-Arg-Gly-Asn-Pro- Thr-Val-Glu-Val-Glu-Leu-Thr	1 ± 0.5

[a]Data reported as mean activity $\pm$ SD ($n = 3$–5).

genome-derived yeast mitochondrial proteins (i.e. cytochrome c oxidase, subunit VI and ATPase inhibitor) are acetylated in vitro (Table 1), although in vivo these proteins are not acetylated.

Since human, but not yeast, SOD(1–24) is acetylated by AcT, NH_2-terminal residues from human were progressively substituted into the yeast sequence. After substituting 6 residues from the human sequence into yeast, the human-yeast hybrid peptide is acetylated as efficiently as the human peptide (Table 1). In addition, yeast [Tyr^2] and [Ala^3] enolase (1–24), whose three NH_2-terminal residues are identical to those of acetylated parvalbumins from carp and coelacanth, respectively, are not acetylated. All higher eukaryote enolases have blocked NH_2-termini, and the NH_2-terminus of the human enzyme is Ac-Ser. In contrast, the endogenous yeast enzyme (approx. 60% identical to human enolase in its NH_2-terminal 24 residues) is not N^α-acetylated. Synthetic peptides were prepared that progressively substituted the NH_2-terminal sequence from the human enzyme into that of the yeast enzyme. Although the substrate specificity of N^α-acetyltransferase is highly dependent on the NH_2-terminal residues, residues distal to the NH_2-terminal ten residues of enolase clearly affect N^α-acetylation (Table 1). These results contrast with the findings of Tsunasawa et al. [6], who suggested that only the first three NH_2-terminal residues play a critical role in determining the specificity of N^α-acetyltransferase and suggest that conformation and/or sequences located more C-terminally also modulate N^{α}-acetylation.

Acknowledgements

Supported by a grant from Hoechst Aktiengesellschaft (F.R.G.).

References

1. Brown, J.L. and Roberts, W.K., J. Biol. Chem., 254 (1979) 1447.
2. Lee, F.-J., Lin, L.-W. and Smith, J.A., J. Biol. Chem., 263 (1988) 14948.
3. Lee, F.-J., Lin, L.-W. and Smith, J.A., J. Biol. Chem., 264 (1989) 12339.
4. Lee, F.-J., Lin, L.-W. and Smith, J.A., J. Bacteriol., 171 (1989) 5795.
5. Persson, B., Flinta, C., von Heijne, G. and Jornvall, H., Eur. J. Biochem., 152 (1985) 523.
6. Tsunasawa, S., Stewart, J.W. and Sherman, F., J. Biol. Chem., 260 (1985) 5382.

Conformation-driven, stepwise processing of three-domain neurohypophysial hormone/neurophysin/copeptin precursors

R. Acher*, J. Chauvet, M.T. Chauvet, B. Levy, G. Michel and Y. Rouille

Laboratory of Biological Chemistry, University of Paris VI, Paris, France

Introduction

Like most secreted polypeptides, neurohypophysial hormones are biosynthesized as protein precursors that are processed into two or three fragments prior to secretion (for reviews see Refs. 1 and 2). Multistep processing occurs along the pathway from the entrance of the primordial precursor into the rough endoplasmic reticulum to the exit of mature products, packaged into secretory granules, through exocytosis. At least four factors should be taken into consideration in this dynamic mechanism: (1) the conformations of the primordial and intermediate precursors that serve as substrates; (2) the primary specificities of the processing enzymes; (3) the relative mobilities of the processing enzymes, luminal or membrane-bound, in the compartments; and (4) the successive compartments where enzyme-substrate interactions can occur: endoplasmic reticulum, Golgi apparatus, secretory granules, and plasma membrane. The three-domain organization of the vasopressin precursor (vasopressin/neurophysin/copeptin) is revealed by the two-step proteolytic processing.

Results and Discussion

The first system: The dibasic endopeptidase

Many secreted peptides have, as neurohypophysial hormones, an amidated C-terminal end, and this physiologically essential feature results from the successive actions of three enzymes on a Gly-Lys-Arg sequence: (1) an endopeptidase splitting at the level of a pair of basic amino acids [3–5]; (2) a carboxypeptidase B-like enzyme that removes the two basic residues [6,7], and (3) a peptidyl-glycine α-amidating monooxygenase that oxygenates a C-terminal glycine, which in turn, by dismutation, generates the amide group on the penultimate residue [8,9].

It is assumed that a first cleavage occurs between hormone and neurophysin as a consequence of both an accessible processing site, Lys-Arg, in the precursor, and the 'dibasic' specificity of an endopeptidase (Fig. 1). It is important to note that another potential dibasic processing site, Arg-Arg, present usually near

---

*To whom correspondence should be addressed.

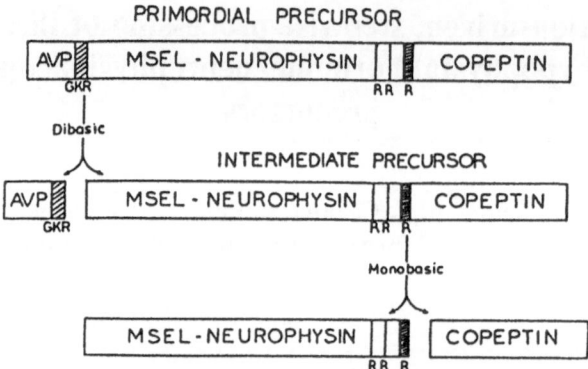

Fig. 1.   The first cleavage of the primordial precursor occurring at the dibasic site (KR) releases vasopressin (AVP) and the intermediate precursor. The second cleavage at the monobasic site (R) leads to free MSEL-neurophysin and copeptin.

the neurophysin-copeptin junction, is never split, showing that this site is inaccessible, probably buried in the primordial precursor. No intermediate, comprising only hormone and neurophysin domains, has ever been isolated, whereas, in contrast, an intermediate neurophysin-copeptin has been occasionally identified [10]. So, the first cleavage occurs in an early compartment.

The first processing systems leading to active carboxyamidated neurohormone could be regulated at the level of each of its three enzymes. In amphibians, along with mature vasotocin, the usual antidiuretic hormone of nonmammalian tetrapods, processing intermediates, termed *Hydrins* because of their action on water permeability of skin and bladder, have been identified. *Hydrin-1* is vasotocinyl-Gly-Lys-Arg, found in the aquatic *Xenopus laevis*, and *hydrin-2* is vasotocinyl-Gly, detected in semi-aquatic or terrestrial frogs and toads [11].

The second system: The monobasic endopeptidase

After vasopressin is trimmed off, a conformational change in the neurophysin-copeptin fragment develops, making the second cleavage site accessible. This latter probably occurs late in granules. An intermediate encompassing MSEL-neurophysin and copeptin (132 residues), representing about 20% of the primordial precursor, has been characterized in guinea pig posterior pituitary gland [10] (Fig. 2).

To ascertain whether the slower cleavage in guinea pig was due to a variation in the amino acid sequence or to a peculiar conformational hindrance, the intermediate has been passed through a column of trypsin immobilized onto Sepharose in order to mimic the passage of the intermediate on a membrane-bound endopeptidase. Only two cleavages, identified through microsequencing of the products, were found in the inter-domain region between neurophysin and copeptin (Fig. 2). These results show, on the one hand, that neurophysin and copeptin domains are not attacked by trypsin when attached to one another

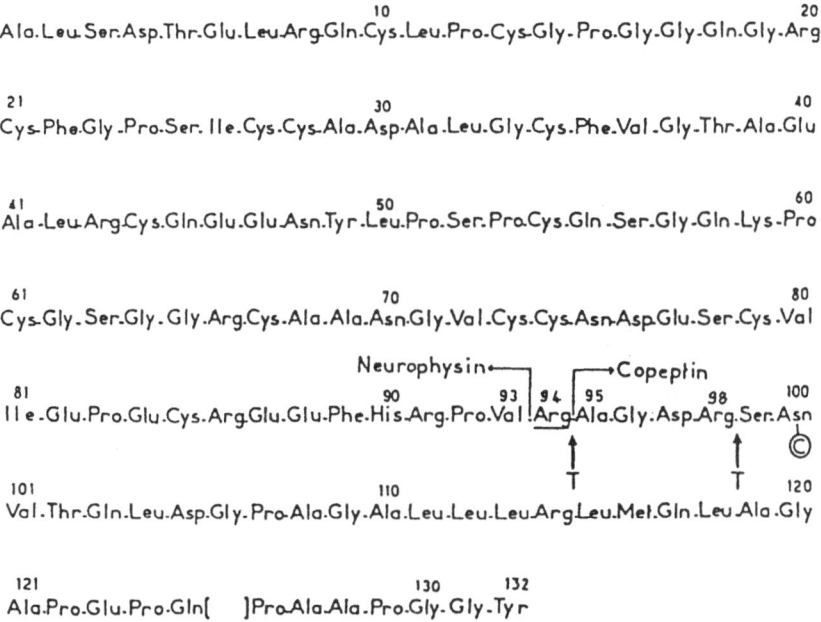

10   20
Ala.Leu.Ser.Asp.Thr.Glu.Leu.Arg.Gln.Cys.Leu.Pro.Cys.Gly.Pro.Gly.Gly.Gln.Gly.Arg

21   30   40
Cys.Phe.Gly.Pro.Ser. Ile.Cys.Cys.Ala.Asp.Ala.Leu.Gly.Cys.Phe.Val.Gly.Thr.Ala.Glu

41   50   60
Ala.Leu.Arg.Cys.Gln.Glu.Glu.Asn.Tyr.Leu.Pro.Ser.Pro.Cys.Gln.Ser.Gly.Gln.Lys.Pro

61   70   80
Cys.Gly.Ser.Gly. Gly.Arg.Cys.Ala.Ala.Asn.Gly.Val.Cys.Cys.Asn.Asp.Glu.Ser.Cys.Val

Neurophysin←───┐   ┌──→Copeptin
81   90   93 │94│95   98   100
Ile.Glu.Pro.Glu.Cys.Arg.Glu.Glu.Phe.His.Arg.Pro.Val│Arg│Ala.Gly.Asp.Arg.Ser.Asn
                                        ↑   ↑
                                        T   T

101   110   120
Val.Thr.Gln.Leu.Asp.Gly.Pro.Ala.Gly.Ala.Leu.Leu.Leu.Arg.Leu.Met.Gln.Leu.Ala.Gly

121   130   132
Ala.Pro.Glu.Pro.Gln[   ]Pro.Ala.Ala.Pro.Gly.Gly.Tyr

Fig. 2. Artificial processing of the guinea pig neurophysin-copeptin intermediate precursor by trypsin-Sepharose. Arg⁹⁴ is the physiological processing site. The two tryptic cleavages are shown by arrows (T).

in the intermediate precursor, whereas they are split when subjected separately, on the other hand results also show that the linking arginine between the two moieties (the second processing site) becomes accessible in the intermediate precursor [12].

In nonmammalian terrestrial vertebrates, vasopressin is replaced by vasotocin and oxytocin by mesotocin, but the organization of the precursors is similar [1,2]. However, in the vasotocin precursor processing, the second cleavage ('monobasic') between neurophysin and copeptin no longer occurs, so that a 'big' neurophysin resembling the guinea pig intermediate precursor is produced [13]. Again, when this 'big' MSEL-neurophysin is passed through a column of trypsin-Sepharose, cleavages occur only in the putative inter-domain region; apparently there is no conformational hindrance for monobasic endopeptidases.

References

1. Acher, R., In Kobayashi, H., Bern, H.A. and Urano, A. (Eds.) Neurosecretion and the Biology of Neuropeptides (Proceedings of the 9th International Symposium on Neurosecretion), Japan Sci. Soc. Press, Tokyo/Springer-Verlag, Berlin, 1985, p. 11.

491

2. Acher, R., Chauvet, J., Chauvet, M.T., Levy, B., Michel, G. and Rouillé, Y., In Yoshida, S. and Share, L. (Eds.) Recent Progress in Posterior Pituitary Hormones 1988, Elsevier Science Publishers (Biomedical Division), 1988, p. 419.
3. Bond, J.S. and Butler, P.E., Annu. Rev. Biochem., 56 (1987) 333.
4. Parish, D.C., Tuteja, R., Alstein, M., Gainer, H. and Loh, Y.P., J. Biol. Chem., 261 (1986) 14392.
5. Julius, D., Brake, A., Blair, L., Kunisawa, R. and Thorner, J., Cell, 37 (1984) 1075.
6. Hook, V.Y.H. and Loh, Y.P., Proc. Natl. Acad. Sci. U.S.A., 81 (1984) 2276.
7. Fricker, L.D., In 8th International Congress of Endocrinology, Kyoto, Japan, 1988, p. 21 (Abstract S-43.)
8. Bradbury, A.F., Finnie, M.D.A. and Smith, D.G., Nature, 298 (1982) 686.
9. Eipper, B.A., Mains, R.E. and Glembotski, C.C., Proc. Natl. Acad. Sci. U.S.A., 80 (1983) 5144.
10. Chauvet, M.T., Chauvet, J., Acher, R., Sunde, D. and Thorn, A.N., Mol. Cell. Endocrinol., 44 (1986) 243.
11. Rouillé, Y., Michel, M., Chauvet, M.T., Chauvet, J. and Acher, R., Proc. Natl. Acad. Sci. U.S.A., 86 (1989) 5272.
12. Chauvet, J., Chauvet, M.T. and Acher, R., FEBS Lett., 217 (1987) 180.
13. Michel, G., Chauvet, J., Chauvet, M.T. and Acher, R., Biochem. Biophys. Res. Commun., 149 (1987) 538.

Chemical-enzymatic synthesis of emerimicin IV and III, membrane-active pentadecapeptide antibiotics

Miroslaw T. Leplawy[a],*, Karol Kociolek[a], Adam S. Redlinski[a], Urszula Slomczynska[a], Janusz Zabrocki[a], James B. Dunbar Jr.[b] and Garland R. Marshall[b]

[a]Institute of Organic Chemistry, Politechnika, P-90-924 Lodz, Poland
[b]Department of Pharmacology, Washington University School of Medicine, St. Louis, MO 63110, U.S.A.

Introduction

Emerimicin IV (1, Fig. 1), the principal component, and emerimicin III (2), the minor component of emerimicins, belong to the family of peptaibol antibiotics that form voltage-gated ionic membrane channels. Synthesis of peptaibols is a challenge to peptide chemists due to difficulties arising from low reactivity of α,α-disubstituted amino acids, presence of acid-labile EtA(MeA)-Hyp(Pro) bonds, and increased racemization risk of protein amino acids involved in the

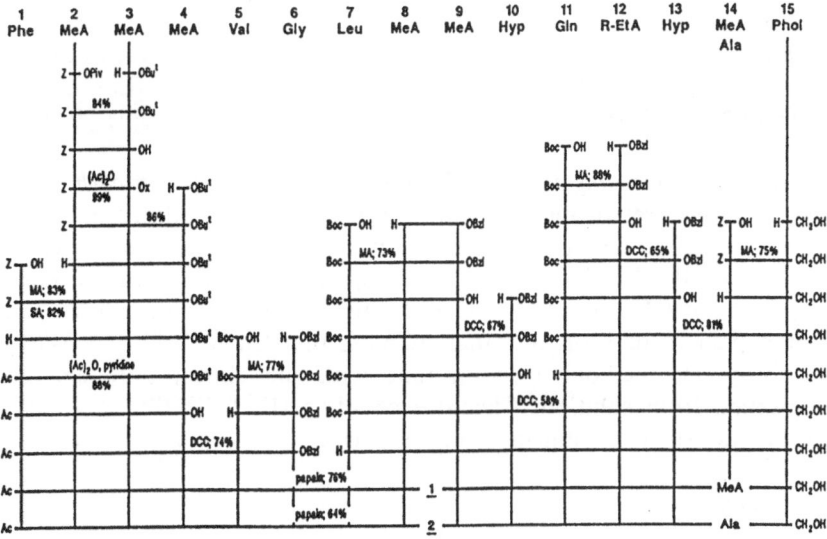

Fig. 1. Chemical-enzymatic synthesis of emerimicin IV (1) and III (2). Abbreviations: MeA = α-methylalanine (or Aib), EtA = α-ethylalanine (or Iva), OPiv = pivaloyl mixed anhydride, MA = mixed anhydride with isobutyl chloroformate, SA = symmetrical anhydride, Ox = oxazolone, DCC = DCC and HOBt as additive.

---

*To whom correspondence should be addressed.

synthesis. Several synthetic emerimicin segments have been described since 1981 [1,2], and we report on the total synthesis in which an enzymatic approach to final 6 + 9 coupling plays a crucial role (Fig. 1).

Results and Discussion

Initial, inefficient 9 + 6 coupling strategy

Ac-Phe-(MeA)₃-Val-Gly-Leu-(MeA)₂-OBzl, segment 1–9. The 9 + 6 strategy was chosen because it provides access to the N-terminal nonapeptide with the high content of MeA (55%) needed for structural studies. This segment was successfully synthesized by 4 + 5 coupling (53%, DCC/HOBt). Boc/OBzl protection was used throughout the synthesis. Crystalline nonapeptide 1–9 proved to be an interesting example of a natural continuous α-helix.

Z-Hyp-Gln-R-EtA-Hyp-Ala-Phol, segment 10–15. Two synthetic routes were explored. At first, hydroxyl-protected Boc-Hyp(Bzl)-OH and 1 + n coupling strategy were used [2] to suppress racemization risk. However, this approach had to be discontinued because stepwise lengthening of Boc-R-EtA-Hyp(Bzl)-Ala-Phol could not be accomplished beyond the tetrapeptide step. This failure may be attributed to spontaneous cleavage of N-deprotected tetrapeptide, an analogy to the known case of H-MeA-Pro-Trp-OH. In a successful approach, the key segment was Boc-Gln-R-EtA-Hyp-OBzl. C-elongation (+ 2) followed by N-elongation (+ 1, OH-unprotected Z-Hyp) by means of DCC/HOBt furnished hexapeptide 10–15 in moderate yield.

9 + 6 coupling. Final coupling performed by means of DCC/HOBt over many hours furnished a complex mixture, which after extensive purification afforded pure (HPLC, FABMS, NMR) emerimicin III in an unacceptable yield of 5%.

Serviceable 6 + 9 coupling strategy: Significance of enzymatic approach

The main features of the successful synthetic scheme outlined in Fig. 1 are: the oxazolone procedure for preparation of -MeA-MeA-MeA- sequence; the key segment Boc-Gln-R-EtA-Hyp-OBzl enabling assembly of C-terminal nonapeptide; the use of OH-unprotected Hyp throughout the whole synthesis; and the use of papain to perform the final coupling that led to 76% and 64% yield, respectively, in one-hour reaction time with the almost pure emerimicins precipitating from solution. Chemical coupling (DCC/HOBt) of these same segments furnished emerimicin III in 12% yield.

Acknowledgements

Supported by the National Institutes of Health, Grants GM-24483 and GM-33918, as well as the Polish Academy of Sciences, Grants CPBP-01.13.2 and CPBR-3-13-4.

References

1. Raj, P.A., Das, M.K. and Balaram, P., Biopolymers, 27 (1988) 683 (and references cited therein).
2. Marshall, G.R., Clark, J.D., Dunbar, Jr., J.B., Smith, G.D., Zabrocki, J., Redlinski, A. and Leplawy, M.T., Int. J. Pept. Prot. Res., 32 (1988) 544 (and references cited therein).

Functional dissection of C3a using peptide agonists

M. Gier[a], D. Ambrosius[a], M. Casaretto[a], B. Gilge[a], D. Saunders[b], M. Federwisch[c] and H. Höcker[a]

[a]*Deutsches Wollforschungsinstitut, Veltmanplatz 8, D-5100 Aachen, F.R.G.*
[b]*Grünenthal GmbH, Forschungszentrum, Zieglerstr. 6, D-5100 Aachen, F.R.G.*
[c]*Institut für Biochemie, RWTH Aachen, Pauwelsstr., D-5100 Aachen, F.R.G.*

Introduction

The anaphylatoxin C3a (77 residues) is generated during activation of the serum complement cascade. By interacting with pharmacologically defined receptors, C3a induces the release of histamine and other vasoactive amines, supporting acute inflammatory reactions.

Earlier studies with synthetic peptides showed that the C-terminal pentapeptide (73–77) is fully, if weakly, active, whereas the sequence 57–77 has nearly full C3a potency in gp-ileum assay [1]. These results were developed further using the 'amphipathic helix' principle propagated by Hügli (1) and Kaiser and Kézdy [2]. There is indeed reason to believe that helical elements play a significant role in the function of native C3a. Our goal, however, is to synthesize simple analogs in which complex natural elements have been replaced by simpler elements possessing similar properties.

Results and Discussion

Based on SAR of more than 70-peptides, synthesized by SPPS [3] and tested by an ATP-release assay from gp-platelets [4], we have developed a model describing the functional elements required for high potent C3a analogs. From the N-terminus, these are: MEMBRANE ANCHOR – SPACER – MESSAGE SEQUENCE.

In general, we found two elements to enhance the biological activity of short C3a analogs. Both effects, N-terminal elongation by spacers and N-terminal acylation by hydrophobic groups (e.g., Fmoc-, Fmoc-Ahx-, Nap-Ahx), supplement each other. Cationic spacers, based on appropriate arginine arrangements [3], produced higher potentiation than neutral spacers (e.g., [α,ω-aminocarboxylic acids or unspecific amino acids) [4].

The degree of potentiation is maximal in short sequences (up to 4000-fold). Augmenting the potency by a membrane anchor, and using the very sensitive ATP-release assay, allows the design of very short, active peptides to characterize the essential active binding site of C3a. It has been possible to reduce the previously assumed binding site from 5 (LGLAR) to 3 amino acids (LAR). The tripeptide LAR is only able to desensitize, but not to activate gp-platelets. N-terminal acylation with Fmoc-α-aminohexanoic acid generates a full agonist (0.4% C3a activity), with a potency 40 times higher than that of the pentapeptide (0.01% C3a activity). Unlike others, we find no relevance of amphipathic helix, other than its ability to reduce the association event with the receptor from 3 to 2 dimensions.

Conclusions

In our studies, the C3a analogs were made amphipathic by adding hydrophilic cationic spacers and hydrophobic residues to the N-terminus. However, we expect a 'linear' amphipathy for our peptides, and not a 'circular' amphipathy, as with the helical peptides of Hügli [1]. The high potency of our peptides results from electrostatic, as well as hydrophobic, interactions with the membrane. This approach is supported by receptor binding studies with platelets, as well as fluorescence lifetime experiments using phosphatidyl choline liposomes.

Acknowledgements

This research was supported by BMFT grant No. 01VM86048. M. Gier holds a scholarship from the Graduiertenförderung des Landes Nordrhein-Westfalen.

References

1. Hügli, T.E., In Müller-Eberhard, H.J. and Miescher, P.A. (Eds.) Complement, Springer-Verlag, Berlin, 1984, p. 73.
2. Kaiser, E.T. and Kézdy, F.J., Science, 223 (1984) 249.
3. Ambrosius, D., Casaretto, M., Gerardy-Schahn, R., Saunders, D., Brandenburg, D. and Zahn, H., Biol. Chem. Hoppe-Seyler, 370 (1989) 217.
4. Gerardy-Schahn, R., Ambrosius, D., Casaretto, C., Grötzinger, J., Saunders, D., Wollmer, A., Brandenburg, D. and Bitter-Suermann, D., Biochem. J., 255 (1988) 209.

The design and synthesis of nonpeptide mimetics of erabutoxin B

Michael Kahn and Susanne Wilke

*University of Illinois at Chicago, Department of Chemistry, 829 W. Taylor Street,
P.O. Box 4348 (m/c 111), Chicago, IL 60680, U.S.A.*

Introduction

Peptides and proteins play critical roles in the regulation of virtually all biological processes. However, the understanding of the relationship between the conformation and the activity of bioactive peptides and proteins remains one of the critical goals of contemporary biochemistry. In this regard, we are engaged in the design and synthesis of conformationally restricted nonpeptide mimetics of bioactive peptides and proteins. We have designed an 11-membered ring bis-lactam (1) as a conformationally stable nonpeptide mimic of the ubiquitously distributed β-turn [1]. Turns, being surface-localized and containing predominantly polar potentially reactive side chains, are implicated as recognition sites that trigger complex immunologic, metabolic and endocrinologic mechanisms [2].

(1)   (2)

One target of our studies is erabutoxin B, a low molecular weight neurotoxin (MW = 7.2 kDa, 62 AA residues), that acts through the blockade of the post-synaptic nicotine acetylcholine receptor (kDa ~ 10^{-11} M) in a nondepolarizing curare-like manner [3]. It is isolated from the broad-banded sea snake. Venoms from both *Hydrophiidae* (sea snakes) and *Elapida* families (cobras, kraits, mambas) show a common mode of action and strong structural homology. The highly conserved region, Asp-Phe-Arg-Gly (residues 31–34), is contained within a β-turn region at the end of the longest of three loops of β-pleated sheet, and has been proposed to contain the residues critical for binding to the acetylcholine receptor protein [4]. It is this region we have chosen to mimic. Based on molecular modeling, we have designed a type I β-turn mimetic that exquisitely mimics the backbone of the 'neurotoxic' β-turn loop and appropriately orients the neurotoxic side chains (2).

Results and Discussion

For the first generation mimetics, introduction of the charged amino acid side chains Arg and Asp is desired. A retrosynthetic strategy for the construction of these mimetics is outlined in Scheme 1.

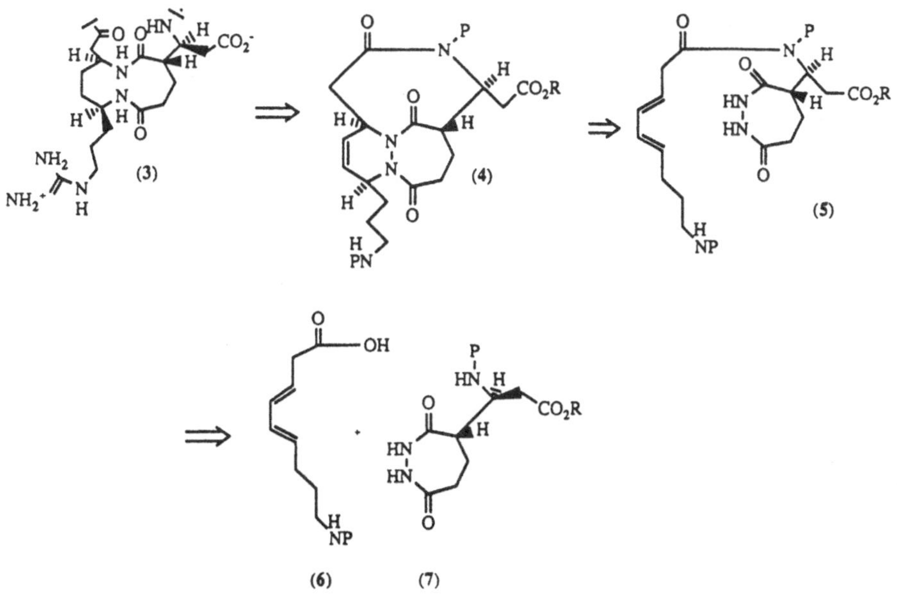

Scheme 1.

The key transformation in this scenario involves an intramolecular cycloaddition reaction of the diazodicarbonyl system (5) via in situ oxidation of the diacylhydrazide, that subsequently undergoes cycloaddition through the less strained exo transition state [5]. The cycloaddition reaction was modeled successfully, and proceeded in moderate yield. The Arg guanidino moiety is incorporated by means of the *N*-protected E,E diene (6), the Asp moiety through the optically active diacylhydrazide (7). Essential for the success of this mimetic strategy was the ease by which the resulting tricyclic lactam (4) was opened by methanol to generate an N- as well as a C-terminus (3). Consequently, this synthetic turn can be incorporated into a Merrifield solid phase peptide synthesis to generate conformationally restricted mimetic chimeras [6]. The enantioselective synthesis of (3) is nearing completion; biological studies will be published in due course. In summary, an efficacious strategy for the generation of nonpeptide mimetics of the proposed neurotoxide loop is outlined, providing a unique system to explore the relationship between peptide structure and function.

Acknowledgements

M.K. wishes to acknowledge the generous financial support of the Camille and Henry Dreyfus Foundation, the Searle Scholars Program/The Chicago Community Trust, the NSF (Presidential Young Investigators Award), Monsanto, Procter and Gamble, Schering, Searle and Syntex for matching funds, the American Cancer Society (Junior Faculty Fellowship), and the NIH (GM-38260). Additionally we wish to thank the NSF for funds for purchase of a 400 MHz NMR spectrometer.

References

1. Kahn, M., Wilke, S., Chen, B. and Fujita, K., J. Am. Chem. Soc., 110 (1988) 1638; Kahn, M., Lee, Y., Wilke, S., Chen, B., Fujita, K. and Johnson, M., J. Biomol. Recogn., 1 (1988) 75.
2. Rose, G.D., Gierasch, L.M. and Smith, J.A., Adv. Protein Chem., 37 (1985) 1.
3. Lee, C.Y., Ann. Rev. Pharmacol., 12 (1972) 265.
4. Ryden, L., Gabel, D. and Eaker, D., Int. J. Pept. Prot. Res., 5 (1973) 261.
5. Medina, J.C., Cadilla, R. and Kyler, K.S., Tetrahedron Lett., 28 (1987) 1059.
6. Kahn, M. and Bertenshaw, S., Tetrahedron Lett., 30 (1989) 2317.

Active domains of staphylococcal enterotoxin

Amrit K. Judd, Robert C. Humphres* and Mark A. Winters

*Life Sciences Divison, SRI International, 333 Ravenswood Avenue,
Menlo Park, CA 94025, U.S.A.*

Introduction

Staphylococcal enterotoxins are proteins produced in food or in culture media
by various strains of staphylococci, resulting in acute food poisoning outbreaks
in humans and a limited number of other mammalian hosts. Staphylococcal
enterotoxin B (SEB) is a very potent mitogen for mouse and human lymphocytes
[1].

We report here the studies on synthetic peptides derived from the SEB sequence
[2]. The selection of peptide sequences was based on computer graphic studies
of secondary structure prediction [3], hydropathicity [4], and published results
of chemical modifications of different amino acid residues [5]. We synthesized
seven peptides (Table 1). Free peptides were evaluated for their ability to stimulate
mitogenesis of murine spleen cells and antisera to free peptides, and peptide
carrier conjugates were evaluated for binding to SEB and for the inhibition
of SEB-stimulated mitogenesis.

Table 1   *Amino acid sequences of the peptides synthesized*

No.	Sequence
1.	$\overset{1}{E}$ S Q P D P K P D E L H K S S $\overset{16}{K}$
2.	$\overset{90}{Y}$ Q C Y F S K K T N N I D S H E N T K R K T $\overset{112}{C}$
3.	$\overset{116}{C}$ G V T E HGNNQ L D K Y Y R $\overset{131}{S}$
4.	$\overset{148}{V}$ Q T N K K K V T A E Q L D $\overset{162}{Y}$
5.	C G $\overset{199}{D}$ MM P A P G N K F D Q S K $\overset{213}{Y}$
6.	$\overset{216}{M}$ Y N N D K M V D S K D V $\overset{229}{Y}$
7.	$\overset{225}{D}$ S K D V K I E V Y L T T K K $\overset{239}{K}$

Note: Underlined regions predict the probable location of reverse or β-turns.

---

*Deceased.

501

Results and Discussion

Antiserum against one peptide, SEB (224–239), showed significant binding activity against native SEB after two immunizations (Day + 35, Fig. 1). The binding activity of this serum was significantly greater than that of normal serum. The amount of binding of SEB (224–239)-BSA to SEB (Fig. 2) was smaller than that of free peptide (Fig. 1). This is probably because of the lower number of (224–239) antigenic determinants available for binding in equal concentrations of peptide and peptide–BSA conjugate. Another peptide, SEB (116–131), also showed binding activity against SEB, but to a lesser extent.

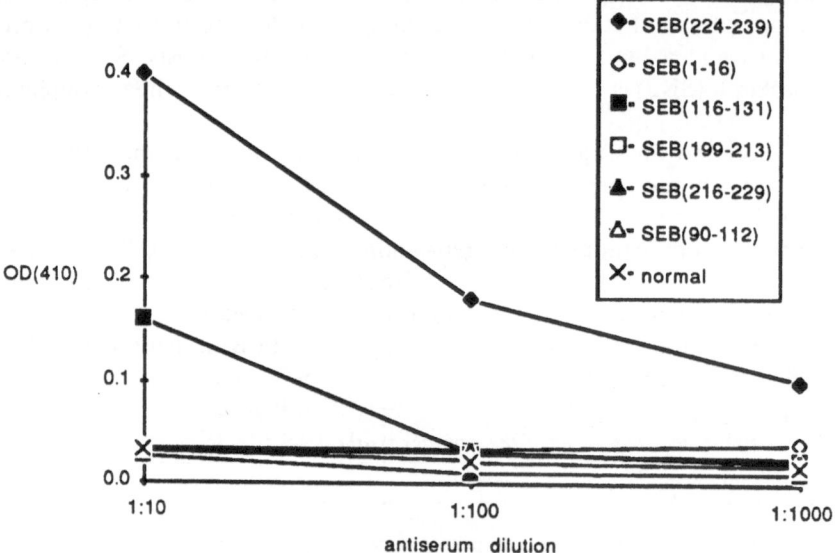

Fig. 1. *Binding of antiserum from mice immunized with 100 μg SEB peptides to SEB.*

Results of mitogenic activity of SEB peptides are shown in Fig. 3. It was observed that SEB (90–112) had appreciably more mitogenic activity compared with that of control (media + spleen cells), which was negligible. SEB (116–131) showed some mitogenic activity.

Antisera of those SEB peptides that had binding activity against native SEB, i.e., SEB (224–239), SEB (116–131) and SEB (116–131)-KLH, were tested for the inhibition of SEB-induced mouse spleen cell mitogenesis (Fig. 4). Anti-SEB (224–239) serum showed some inhibitory activity.

Conclusion

The results of binding and mitogenic activities of SEB peptides raise the question of whether these activities derive from different sites on the SEB molecule. Our

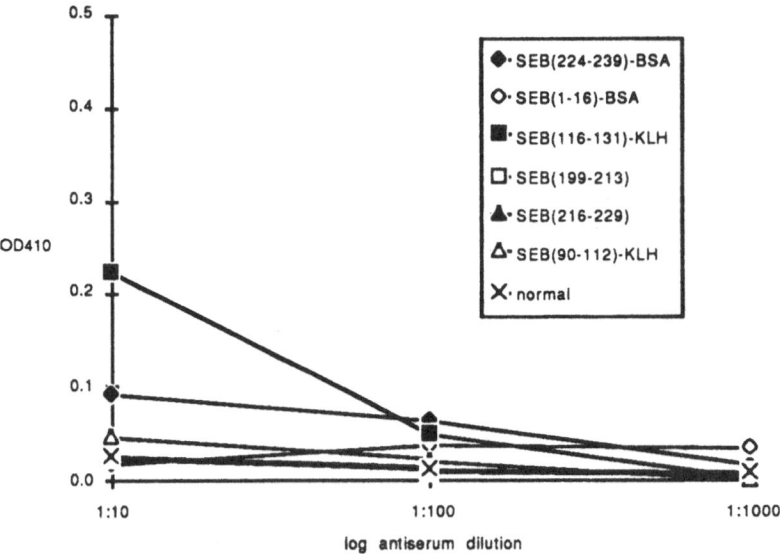

Fig. 2. Binding of antiserum from mice immunized with 100 µg of peptide–carrier conjugates to SEB.

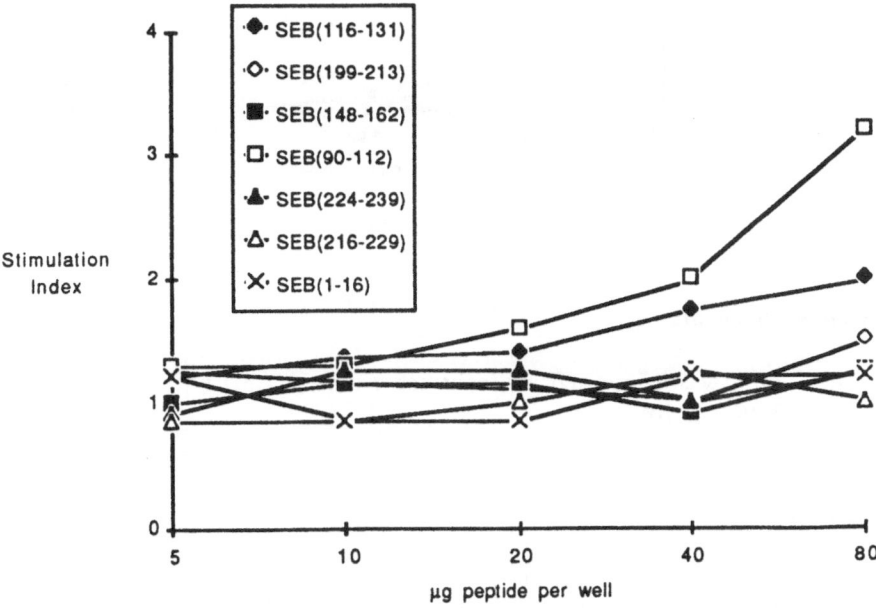

Fig. 3. Induction of mouse spleen cell mitogenesis by SEB peptides.

studies are not complete and a few more peptides need to be synthesized and evaluated. However, from the results obtained so far, it is plausible that these two properties of SEB might involve different sites. Our results show that SEB (224–239) had no mitogenic activity, but antisera to these peptides bind to SEB. On the other hand, SEB (90–112) had mitogenic activity, but antisera to it did not bind to SEB.

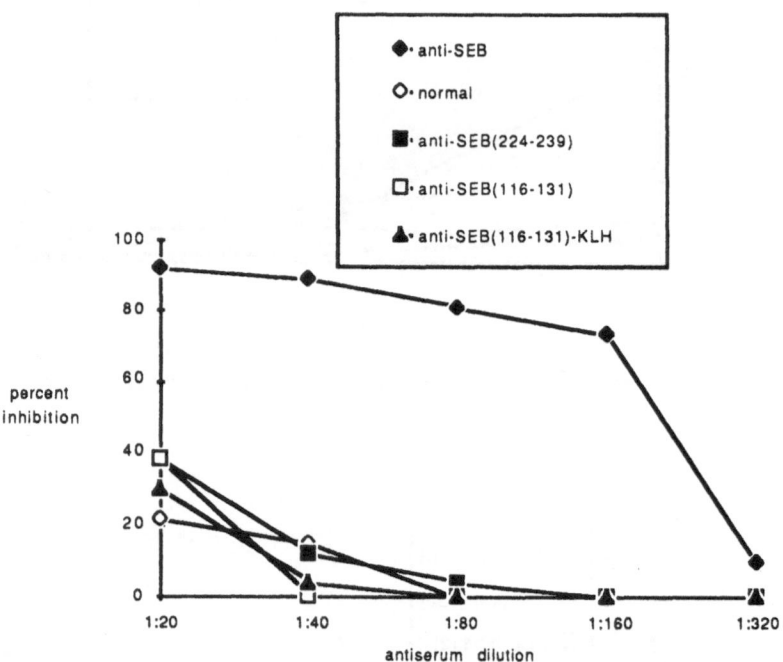

Fig. 4. *Inhibition of SEB-induced spleen cell mitogenesis by serum from mice immunized with SEB peptides.*

Acknowledgements

This work was supported by contract from USAMRDC (Contract No. DAMD17-86-C-6110).

References

1. Spero, L., Leatherman, D.L. and Adler. W.H., Infect. Immun., 12 (1975) 1018.
2. Huang, I.-Y. and Bergdoll, M.S., J. Biol. Chem., 245 (1970) 3518.
3. Chou, P.Y. and Fasman, G.D., Annu. Rev. Biochem., 47 (1978) 251.
4. Kyte, J. and Doolittle, R.F., J. Mol. Biol., 157 (1982) 105.
5. Chu, F.S., Crary, E. and Bergdoll, M.S., Biochem., 8 (1969) 2890.

Development of antagonists of des-Glu^1-Conotoxin GI

Ronald G. Almquist, Srinivasa R. Kadambi, Dennis M. Yasuda,
Frederick L. Weitl, Willma Polgar, Lawrence R. Toll and Edward T. Uyeno
Life Sciences Division, SRI International, Menlo Park, CA 94025, U.S.A.

Introduction

Marine snails of the genus *Conus* produce a series of peptide neurotoxins that include conotoxin GI [1]. Conotoxin GI is known to antagonize the binding of acetylcholine (ACh) to nicotinic receptor sites and thereby causes paralysis of skeletal muscle [2,3]. This process can be lethal to animals. In our work we wanted to develop analogs of conotoxin GI that would antagonize the lethal effects of conotoxin. We hoped to accomplish this goal by synthesizing analogs of conotoxin GI that would bind to the ACh receptor site and displace conotoxin without displacing ACh.

Results and Discussion

A series of 21 analogs of des-Glu^1-Conotoxin GI were synthesized by solid phase synthesis on MBHAR. Since conotoxin GI and des-Glu^1-Conotoxin GI have comparable activities [1], we chose to make analogs of the shorter des-Glu^1-Conotoxin GI to simplify the syntheses. All analogs were tested initially for their abilities to inhibit contractions in a mouse-diaphragm-with-phrenic-nerve assay. Replacement of Asn with Ala, Gly with D-Ala, His with either Phe or Leu, or cyclization (3–13) of des-Glu^1-[Asp^3,Dpr^{13}] Conotoxin GI gave analogs with comparable IC$_{50}$s to that of des-Glu^1-Conotoxin GI. Replacement of Tyr with Leu lowers paralytic activity by 20-fold. As can be seen in Table 1, total loss of paralytic activity occurs when Gly replaces Pro, D-Tyr replaces Tyr, or D-Phe replaces Gly. In most cases, loss or change in length of one of the disulfide rings eliminates paralytic activity, except with analog **15** which is weakly active, and analog **17**, which is quite paralytic. Replacement of the Cys^2-Cys^7 disulfide bond with an amide bond (analog **20**) greatly lowers paralytic activity. Analog **15** (at doses of 10 μM) was the only analog that antagonized (by 24%) the paralytic effect of conotoxin GI in this assay.

Table 1 gives the results of testing all analogs with low or no paralytic activity in the mouse-diaphragm-with-phrenic-nerve assay for their abilities to antagonize the lethal effects of conotoxin GI in mice. Analogs **15** and **16** with equal activity are the best antagonists in this assay. Extending the C-terminal chain of **16** or reducing it in **15** reduces antagonistic activity. It appears that the small disulfide loop of des-Glu^1-Conotoxin GI is more important than the large loop for

Table 1 Activities of selected conotoxin GI analogs

Analog	Amino acid sequence (changed residues from Ctxn GI are underlined)	In vitro IC$_{50}$ (μM)[a]	In vivo % Antag.[b] (dose, μmol/kg)
Ctxn GI	Cys-Cys-Asn-Pro-Ala-Cys-Gly-Arg-His-Tyr-Ser-Cys-NH$_2$	0.27	Toxic(0.031)
1	Cys-Cys-Asn-Gly-Ala-Cys-Gly-Arg-His-Tyr-Ser-Cys-NH$_2$	>100	33(6.0)
5	Cys-Cys-Asn-Pro-Ala-Cys-Gly-Arg-His-DTyr-Ser-Cys-NH$_2$	>100	11(6.0)
8	Cys-Cys-Asn-Pro-Ala-Cys-DPhe-Arg-His-Tyr-Ser-Cys-NH$_2$	>100	0(5.7)
9	Ala-Cys-Asn-Pro-Ala-Ala-Gly-Arg-His-Tyr-Ser-Cys-NH$_2$	>100	50(6.0)
10	Cys-Asn-Pro-Ala-Ala-Gly-Arg-His-Tyr-Ser-Cys-NH$_2$	>100	25(5.1)
11	Cys-Ala-Asn-Pro-Ala-Cys-NH$_2$	>100	0(6.0)
12	Cys-Ala-Asn-Pro-Ala-Cys-Gly-NH$_2$	>100	17(6.0)
13	Cys-Ala-Asn-Pro-Ala-Cys-Gly-Arg-NH$_2$	>100	11(7.4)
14	Cys-Ala-Asn-Pro-Ala-Cys-Gly-Arg-His-NH$_2$	>100	86(10.8) 33(6.5)
15	Cys-Ala-Asn-Pro-Ala-Cys-Gly-Arg-His-Tyr-NH$_2$	70	100(7.3) 88(5.1) 22(3.7)
16	Cys-Ala-Asn-Pro-Ala-Cys-Gly-Arg-His-Tyr-Ser-NH$_2$	≥100	88(5.1)
17	Cys-Ala-Asn-Pro-Ala-Cys-DAla-Arg-His-Tyr-Ser-NH$_2$	3.9	Toxic(5.1) 0(0.51)
18	Cys-Ala-Asn-Pro-Ala-Cys-Gly-Arg-His-Tyr-Ser-Ala-NH$_2$	>100	22(6.4)
20	Asp-Cys-Asn-Pro-Ala-Dpr-Gly-Arg-His-Tyr-Ser-Cys-NH$_2$	37	60(6.0)

[a] Concentration of analog that inhibits the contraction of the mouse-diaphragm-with-phrenic-nerve by 50% at the 10-min time point following analog addition to the tissue bath.
[b] A toxic dose (0.031 μmol/kg) of conotoxin GI is given with and without the analog.
% Antagonism = [1−(mice dead with analog/mice dead without analog)] × 100.

binding conotoxin to the portion of the ACh receptor that does not interfere with binding ACh. The antagonist activity of **20** is probably caused by the effect of the amide loop on the conformation of some of the binding residues that are important in conotoxin GI for displacing ACh from its receptor. Substitution of D-Ala for Gly in **16** to yield **17** probably changes the conformation of this peptide in such a way that it now displaces ACh from its receptor site and causes toxicity.

Acknowledgements

This research was supported by the U.S. Army Medical Research Acquisitions Activity Contract No. DAMD17-86-C-6107.

References

1. Gray, W.R., Olivera, B.M. and Cruz, L.J., Ann. Rev. Biochem., 57 (1988) 665.
2. Cruz, L.J., Gray W.R., and Olivera, B.M., Arch. Biochem. Biophys., 190 (1978) 539.
3. Gray, W.R., Luque, A., Olivera, B.M., Barrett, J. and Cruz, L.J., J. Biol. Chem., 256 (1981) 4734.

Dimeric interaction of Ca-binding photoprotein aequorin

Kozo Nagano[a,*] and Frederick I. Tsuji[b,c]

[a]Faculty of Pharmaceutical Sciences, University of Tokyo, 7-3-1 Hongo Bunkyo-ku,
Tokyo 113, Japan
[b]Department of Enzymes and Metabolism, Osaka Bioscience Institute, 6-2-4 Furuedai,
Suita, Osaka 565, Japan
[c]Marine Biology Research Division, Scripps Institution of Oceanography, University of
California, San Diego, La Jolla, CA 92093, U.S.A.

Introduction

Aequorin is a small monomeric protein ($M_r = 21\,400$) from the jellyfish *Aequorea victoria*, that emits blue light in the presence of Ca^{2+} by the following intramolecular reaction [1]:

$$\text{aequorin} \xrightarrow{3\ Ca^{2+}} \text{apoaequorin} + \text{coelenteramide} + CO_2 + h\nu$$

Aequorin consists of coelenterazine (an imidazopyrazine compound) and molecular oxygen bound noncovalently to apoaquorin (apoprotein). Aequorin has three Ca^{2+}-binding sites and three cysteine residues. When three Ca^{2+} binds to aequorin, a conformational change takes place and the protein is converted to an oxygenase, catalyzing the oxidation of coelenterazine by the bound oxygen with light emission. Aequorin may be regenerated from apoaequorin by incubation with coelenterazine, EDTA, 2-mercaptoethanol (2-ME), and dissolved oxygen. The process involves the removal of Ca^{2+} by EDTA and the dissociation of coelenteramide from apoaequorin, followed by binding of fresh coelenterazine and molecular oxygen. The cDNA for apoaequorin has been cloned [2].

Results and Discussion

Dimeric interactions appear to play an important role in aequorin bioluminescence. First, the color of light from aequorin is blue, whereas, in the jellyfish it is green. The reason for the green light emission is that there is an energy transfer from aequorin to a hydrophobically bound green fluorescent protein that serves as the emitter in the reaction. Second, dimeric interactions involving intermolecular disulfide bonds appear, in general, to interfere with the regeneration of aequorin. Little is known about the function of 2-ME during regeneration, but is is presumed to reduce disulfide bonds. From studying molecular models and analyzing the results of site-directed mutagenesis, we find

---

*To whom correspondence should be addressed.

508

that the role of 2-ME is to disrupt both intra- and inter-molecular disulfide bonds and promote disulfide exchange. No regeneration of wild-type aequorin occurs in the absence of 2-ME, and replacement of one or two cysteine residues with serine only serves to decrease activity, whereas substitution of all three cysteine residues by serine leads to full regeneration of activity, albeit at a slower rate [3]. In the case of aequorin with one cysteine and two serine residues, the effect of an intermolecular disulfide bond cannot be neglected, since an intra-disulfide bond is absent. Thus, all three cysteine residues play an essential role in aequorin regeneration. The replacement of the highly conserved glycine in Ca^{2+}-binding sites 1, 2, and 3 by arginine leads to 0, 49, and 97% in relative activity respectively [4]. These results also aid in the interpretation of dimeric interactions of aequorin.

References

1. Johnson, F.H. and Shimomura, O., In DeLuca, M.A. (Ed.) Bioluminescence and Chemiluminescence (Methods in Enzymology, Vol. 57), Academic Press, New York, 1978, p. 271.
2. Inouye, S., Noguchi, M., Sakaki, Y., Takagi, Y., Miyata, T., Iwanaga, S., Miyata, T. and Tsuji, F.I., Proc. Natl. Acad. Sci. U.S.A., 82 (1985) 3154.
3. Kurose, K., Inouye, S., Sakaki, Y., and Tsuji, F.I., Proc. Natl. Acad. Sci. U.S.A., 86 (1989) 80.
4. Tsuji, F.I., Inouye, S., Goto, T. and Sakaki, Y., Proc. Natl. Acad. Sci. U.S.A., 83 (1986) 8107.

Purification and sequence of a new hydrophobic polypeptide from cardiac muscle

Evelyne Terzi, Philippe Boyot, Alain Van Dorsselaer, Bang Luu and
Elisabeth Trifilieff

*Laboratoire de Chimie Organique des Substances Naturelles associé au CNRS,
5, rue Blaise Pascal, F-67084 Strasbourg, France*

Introduction

We have developed a method for the purification of hydrophobic polypeptides
that is characterized by the intensive use of organic solvents and exclusion of
detergent. This method was successful for the purification of cardiac muscle
phospholamban (the supposed regulatory protein of the sarcoplasmic reticulum
pump) [1], the H^+-ATPase proteolipid, subunits VIIIa and b of the cytochrome-*c*

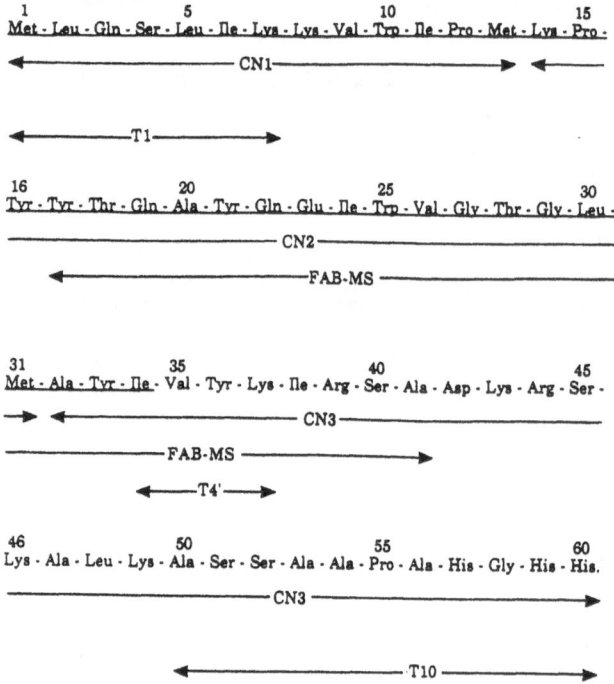

Fig. 1. Amino acid sequence of the new hydrophobic polypeptide. —— *= automated Edman
degradation, CN = cyanogen bromide peptides; T = tryptic peptides.*

510

oxidase and the A6L protein of ATP synthase [2]. We now describe the purification and the primary sequence of a new organic solvent-soluble mitochondrial polypeptide of 60 residues.

Results and Discussion

The hydrophobic polypeptide was purified from an acidic chloroform-methanol extract of bovine cardiac muscle by a combination of gel permeation and ion-exchange chromatographies in organic solvents as described in [2].

The molecular weight of the HPLC-purified hydrophobic polypeptide was mesured by FABMS and gives a protonated molecular ion of m/z 6834.1, indicating that the polypeptide contains about 60 residues.

According to AAA, automated Edman degradation on the intact polypeptide and on cyanogen bromide and tryptic peptides, we could determine the primary sequence as described in Fig. 1. Sequence 17–41 was confirmed by fragmentation peaks observed in the FABMS spectrum. The C-terminal sequence was deduced by analysis of the tryptic T10 peptide and especially by the measurement of its molecular mass by FABMS ($[M + H^+]$: 1042.6).

To localize this new polypeptide, antibodies against it were raised in rabbits. Western blots were run with a 1/1000 dilution of the serum. The pure polypeptide was indeed detected. The antibodies also reacted specifically with a band of the same molecular weight contained in total brain and liver mitochondrial proteins.

Conclusion

Our method of extraction and purification characterized by the intensive use of organic solvent, allowed the purification of a new mitochondrial hydrophobic peptide of 60 residues.

References

1. Boyot, P., Luu, B., Jones, L.R. and Trifilieff, E., Arch. Biochem. Biophys., 269 (1989) 639.
2. Boyot, P., Trifilieff, E., Van Dorsselaer, A. and Luu, B., Anal. Biochem., 173 (1988) 75.

Isolation of a tumor-derived 186-residue peptide amide related to human chromogranin A and its in vitro conversion to human pancreastatin-48

Susumu Funakoshi[a], Hirokazu Tamamura[a], Nobutaka Fujii[a], Haruaki Yajima[a], Akihiro Funakoshi[b] and Mitsuhiro Ohta[c]

[a]*Faculty of Pharmaceutical Sciences, Kyoto University, Sakyo-Ku, Kyoto 606, Japan*
[b]*National Kyushu Cancer Center, Fukuoka 815, Japan*
[c]*Clinical Research Center, Utano National Hospital, Narutaki, Kyoto 616, Japan*

Introduction

Pancreastatin (PS), a 49-residue peptide amide, was first isolated from porcine pancreas by Tatemoto et al. in 1986 [1]. This peptide has been shown to inhibit glucose-induced insulin release from the isolated perfused pancreas.

Using C-terminal peptide-amide specific porcine pancreastatin antibody [2] as a guide, we isolated a C-terminal α-amidated peptide related to human pancreastatin and chromogranin A from liver metastasis of a patient with insulinoma. The established NH_2-terminal amino acid sequence (116–165) of the isolated protein hPS-186 was identical to the corresponding region of its cDNA-derived sequence of human chromogranin A (hCgA). The results are consistent with the proposal that the isolated protein consists of 186 amino acid residues corresponding to the human chromogranin A-116–301 (hCgA-116–301).

Results and Discussion

Using our newly developed human pancreastatin RIA, we identified and isolated three peptides containing the C-terminal glycine amide of hCgA from the tryptic digestion of the 186 residues polypeptide. These peptides overlapped with the 29-residue peptide isolated by Sekiya et al. [3,4], and the 92-residue peptide isolated by Schmidt et al. [5], that were shown to have the pancreastatin activity. Figure 1 shows the positions of each of these peptides within the structure of human chromogranin A deduced from its cDNA sequence.

The isolation and characterization of hPS-29 (273–301), hPS-48 (254–301), and hPS-92 (210–301) generated either in vivo or in vitro, suggests the potential proteolytic processing sites of human pancreastatin to yield smaller fragments with retention of biological activities. The biosynthesis of various sizes of human pancreastatin from chromogranin A (hCgA-1–439) could have initially initiated with the synthesis of hPS-301 (hCgA-1–301) (Fig. 1), although the protein of this size has as yet to be isolated or detected from physiological sources. hPS-301 undergoes proteolytic processing to yield fragments of various sizes that retain biological activities. The significance of the presence of polypeptides of different sizes with similar biological activities is not clear.

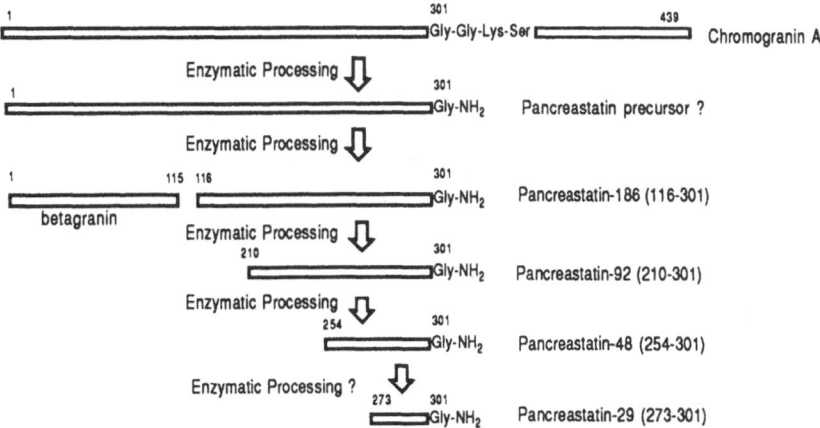

Fig. 1. Schematic representation of the structures of human chromogranin A-related peptides generated either in vivo or in vitro.

While hPS-186 was identified as the major form of pancreastatin in a liver metastasis from a patient with insulinoma, we are not certain that hPS-186 is the major active form of human pancreastatin in circulation. Analysis of human sera, derived from a patient with insulinoma, indicated the presence of an immunoreactive polypeptide(s) eluting at or near hPS-186, and a peptide(s) that co-eluted with a synthetic, 48-residue peptide amide hPS-48. The hPS-48 peptide was the main proteolytic product (42%) with biological activity when hPS-186 was treated with trypsin. Finally, the corresponding major species of pancreastatin in porcine is a 49-residue peptide amide. It would seem, therefore, that hPS-48 would be an important candidate for consideration of naturally occurring pancreastatin in human sera.

Conclusion

Biological studies on synthetic peptides, hPS-29, -48 and -52, indicated that the molar potency of hPSs is almost equivalent to that of porcine PS-49. The potent effect of the C-terminal fragment of pancreastatin on pancreatic secretion suggested that this portion of the molecule is important for the biological activity and that the C-terminal active fragments of pancreastatin may be generated in vivo at the site where it exhibits its physiological functions.

References

1. Tatemoto, K., Efendic, S., Mutt, V., Makk, G., Feistner, G.J. and Barchas, J.D., Nature, 324 (1986) 476.
2. Shimizu, F., Ikei, N., Iwanaga, T. and Fujita, T., Biomed. Res., 8 (1987) 457.
3. Sekiya, K., Ghatei, M.A., Minamino, N., Bretherton-Watt, D., Matsuo, H. and Bloom, S.R., FEBS Lett., 228 (1988) 153.
4. Sekiya, K., Ghatei, M.A., Minamino, N., Bretherton-Watt, D., Matsuo, H. and Bloom, S.R., FEBS Lett., 231 (1988) 451.
5. Schmidt, W.E., Siegel, E.G., Kratzin, H. and Creutzfeldt, W., Proc. Natl. Acad. Sci. U.S.A., 85 (1988) 8231.

Isolation and properties of insulin-like materials from crystalline porcine insulin

Shang-quan Zhu, Ying-gao Xu, Xin-tang Zhang and Da-fu Cui

Shanghai Institute of Biochemistry, Academia Sinica, People's Republic of China

Introduction

When monopeak-insulin, obtained from crystalline porcine insulin by chromatography on Sephadex G-50 column, was purified on DEAE-Sepharose CL-6B column, peaks more basic than insulin were observed. Among them, only peak 3 (called AP3) showed a single band on polyacrylamide gel electrophoresis at pH 8.3. It was found that AP3 possessed insulin-like structure and activity. For example, AP3 contains two peptide chains with glycine and phenylalanine as its N-terminal residues and possesses receptor binding capacity and in vivo biological activity of insulin. However, AP3 could be separated into three peaks on HPLC C-8. We report here on the preparation and properties of AP3.

Methods

Crystalline porcine insulin was purchased from the Shanghai Biochemical Factory. The monopeak porcine insulin was prepared by gel-filtration on Sephadex G-50 column with 1 N acetic acid elution. The procedures of ion-exchange chromatography and HPLC are described in Figs. 1 and 2. The insulin–receptor binding assay on human placenta membrane was carried out at 4°C for 20 h and the in vivo activity was determined by mouse convulsion method.

Results

Figure 1 shows the separation of monopeak porcine insulin on DEAE-Sepharose CL-6B column. Four peaks ahead of the insulin peak (peak 5) can be identified. Among them, only peak 3 (called AP3) was homogeneous on PAGE. The N-terminal amino acids of AP3 determined by DABITC method [1] were glycine and phenylalanine. AP3 was hydrolyzed by trypsin more easily than insulin under the same conditions, and tryptic hydrolysate of AP3 was still more basic than desoctapeptide (B23-30) insulin. However, 4 peaks were observed on HPLC C-8 after sulfitolysis of AP3, in which 3 peaks have the same N-terminal amino acid, glycine, and another peak has phenylalanine as its N-terminal amino acid. AP3 was separated by HPLC C-8 into 3 components, AP3-I, AP3-II and AP3-III (see Fig. 2). AP3-I and AP3-II were collected and rechromatographed on HPLC C-8 indicating that they are homogeneous. The

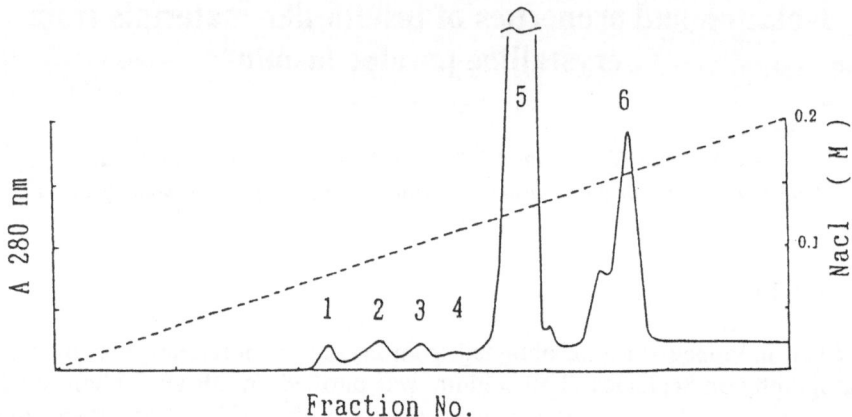

Fraction No.

Fig. 1.   Isolation of the insulin-like materials on DEAE-Sepharose CL-6B column (2.2×24 cm) chromatography. Eluent: 0.05 M Tris buffer, pH 8.0, containing 40% isopropanol with salt gradient. The peak 1 and peak 6 were named AP1 and AP6, respectively.

in vivo activity and insulin-receptor binding capacity of AP3-I and AP3-II are shown in Table 1.

Table 1   Biological properties of AP3-I and AP3-II

Sample	Activity in vivo[a]	Receptor binding capacity[b]
Insulin	100	100
AP3-I	70	85.8
AP3-II	50	73.6

[a] Determined by mouse convulsion method.
[b] See Methods.

Discussion

Insulin-like materials prepared by us are unlikely to be proinsulin or its converting intermediate materials for the following reasons.

(1) Proinsulin had been removed from the starting material by Sephadex G-50 column.

(2) The N-terminal AAA of AP3 showed the presence of only glycine and phenylalanine, but no basic amino acids as its N-terminus.

(3) PAGE of tryptic hydrolysate of AP3 showed that the difference between insulin and AP3 is not in the C-terminus of the β-chain, but perhaps in the α-chain.

The mutant insulins with only one amino acid change were found in some diabetic persons [2–4]. It would be interesting to know if a mutant insulin might also be found in some animals, such as the pig. Therefore, it is necessary to further elucidate the structure and biological significance of insulin-like materials.

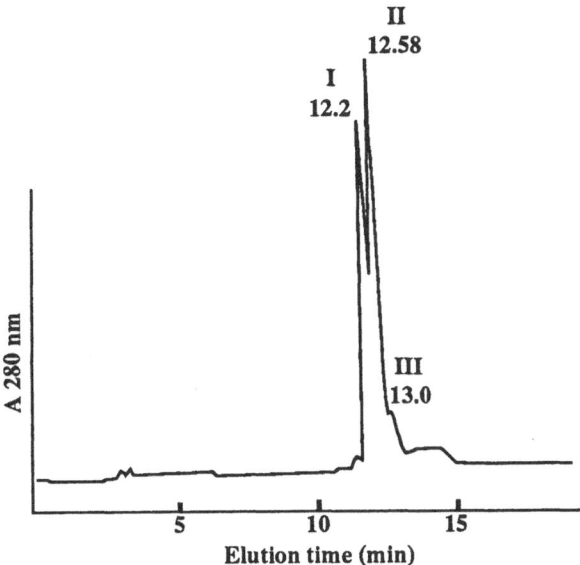

Fig. 2. HPLC separation of AP3 on C-8. Solution A: 30% acetonitrile/70% 0.03 M ammonium bicarbonate/0.2% TFA. Solution B: 80% acetonitrile, 0.1% TFA. Column was developed, at a flow rate of 1 ml/min, with a linear 20-min gradient from 0 to 100% B. The column temperature was kept at 30°C and the fractions were monitored by absorbance at 280 nm. Each peak was collected manually. The peaks from left to right in part A were named AP3-I, AP3-II and AP3-III, respectively.

References

1. Chang, J.Y., FEBS Lett., 91 (1978) 63.
2. Tager, H.B., Given, B., Baldwin, D., Mako, M., Markese, J., Rubenstein, A., Olefsky, J., Kobayashi, M., Kolterman, O. and Poucher, R., Nature, 281 (1979) 122.
3. Shoelson, S., Haneda, M., Blix, P., Nanjo, A., Sanke, T., Inouye, K., Steiner, D., Rubenstein, A. and Tager, A., Nature, 302 (1983) 504.
4. Vinik, A. and Bell, G., Horm. Metab. Res., 20 (1988) 1.

Isolation and characterization of pituitary adenylate cyclase activating polypeptides (PACAP) from ovine hypothalamic extract

Atsuro Miyata[a], Paul E. Gottschall[a], Raymond R. Dahl[a], Masahiko Fujino[c], David H. Coy[b] and Akira Arimura[a,b]

[a]U.S.-Japan Biomedical Research Laboratories, Tulane University Hebert Center, Belle Chasse, LA 70037, U.S.A.
[b]Department of Medicine, Tulane University School of Medicine, New Orleans, LA 70112, U.S.A.
[c]Tsukuba Research Laboratories, Takeda Chemical Industries, Ltd., Tsukuba, Ibaraki 300-42, Japan

Introduction

All the hypothalamic-releasing hormones identified so far, more or less stimulate the pituitary adenylate cyclase and increase accumulation of cyclic AMP. Increase of intracellular, as well as extracellular cyclic AMP in cultured pituitary cells is most strikingly demonstrated by GHRH, and to lesser extent by CRF [1]. GnRH and TRH also increase cyclic AMP accumulation, though it does not seem to be linked with secretion of these hormones [1]. It is, therefore, likely that any hypophysiotropic-stimulating hormone, whether its target pituitary cells are known or not, would increase intra- and extra-cellular cyclic AMP levels in pituitary cell cultures. In this line of thought, we considered it possible to discover novel hypophysiotropic-releasing peptides by monitoring the pituitary adenylate cyclase-stimulating activity (ACSA) in hypothalamic extracts. The use of ACSA in pituitary cell cultures for screening hypophysiotropic-stimulating substances may have an additional advantage that an inhibiting factor specific for a hormone, such as somatostatin for GH and TSH, does not significantly interfere with the ACSA.

Results and Discussion

In the present study, we used the rat pituitary cell cultures for screening ACSA. By determining the content of cyclic AMP in culture media with RIA, accumulation of cyclic AMP in the culture media for a 3-h incubation period was compared with the control value and used as a response parameter for ACSA of the test substance. Ovine hypothalami fractions (2400 g) were extracted as indicated in Fig. 1. On the step of C-18, C and D showed considerable ACSA. Since ACSA derived from fraction D had GHRH/CRF immunoreactivity on further purification steps, it appeared to represent GHRH/CRF-related materials. Our efforts were then concentrated on isolating the most potent ACSA

518

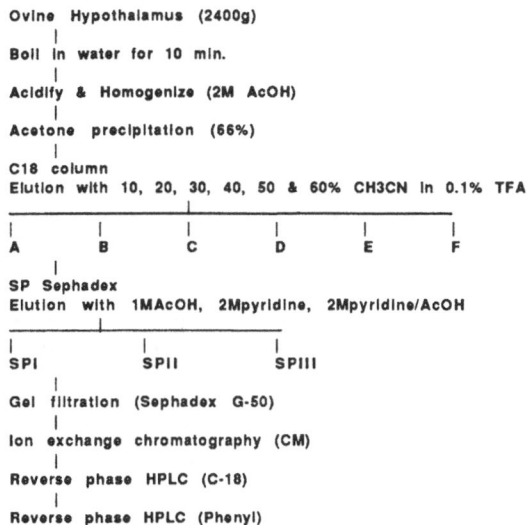

Fig. 1.   *Isolation procedure of PACAP.*

substances in fraction C. After SP-Sephadex, fraction C SPIII, which had potent ACSA, was subjected to gel-filtration chromatography on a Sephadex G-50 column. Since the fractions corresponding to molecular weights 3000–4000 showed the greatest ACSA, these fractions were pooled and subjected to cation exchange chromatography on a CM-52 cellulose column (Fig. 2), followed by two steps of RPHPLC. A peptide with ACSA was isolated in a pure form.

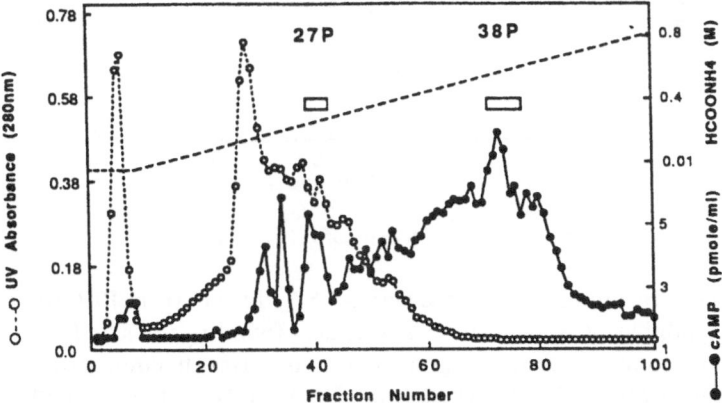

Fig. 2.   *Cation exchange chromatography of the ACSA fractions obtained from gel filtration of ovine hypothalamic extract. Column: CM-52 Cellulose (Whatman, 1×20 cm). Fraction size 4 ml/fraction. Solvent system: A linear gradient of ammonium formate (pH 6.5) from 10 mM to 0.8 M.*

519

The primary structure, which was determined by automated Edman degradation, was revealed to be: His-Ser-Asp-Gly-Ile-Phe-Thr-Asp-Ser-Tyr-Ser-Arg-Tyr-Arg-Lys-Gln-Met-Ala-Val-Lys-Lys-Tyr-Leu-Ala-Ala-Val-Leu-Gly-Lys-Arg-Tyr-Lys-Gln-Arg-Val-Lys-Asn-Lys-NH$_2$. The peptide consists of 38 amino acids with amidated C-terminus. Its N-terminal sequence, 1–28, shows significant homology with VIP (68%) [2]; however, the C-terminal sequence after the 29th residue is different from any other known peptide, as revealed by a computer-assisted search for homology. We previously reported the isolation and partial structure of an ovine VIP-like GH-releasing factor that stimulates GH release from rat pituitary fragments in vitro [3]. The two peptides appear to be very closely related. The peptide with 38 residues was synthesized by SPPS and the synthetic peptide elicited ACSA to the same extent as the native peptide (Fig. 3). This peptide was thus named PACAP38 (pituitary adenylate cylase activating polypeptide with 38 residues). The ACSA of PACAP38 was comparable to that of CRF and approximately 1000 times greater than that of VIP. PACAP38 stimulated release of GH, PRL, and ACTH from superfused rat pituitary cells in a dose-dependent manner, in a dose range between 10^{-10} M to 10^{-8} M. The responses declined as the dose increased and showed 'bell-shaped' dose–response curves. On the other hand, LH response was linear in a dose range of 10^{-9}–10^{-6} M. The release of FSH and TSH was not significantly altered under the same condition.

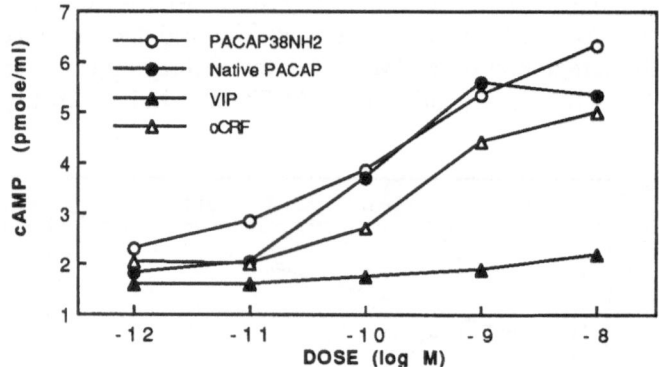

Fig. 3.   *Adenylate cyclase-stimulating activity of PACAP in rat pituitary cell culture.*

Furthermore, another substance with ACSA, distinct from PACAP38 or other known hypophysiotropic hormones, was also found in the side fractions of the CM-cation exchange chromatography of ovine hypothalamic extract obtained during purification of PACAP38 (Fig. 2). After two steps of RPHPLC, this peptide was isolated in a pure form. The amino-acid sequence of this peptide revealed it to be a (1–27)-NH$_2$ of PACAP38. PACAP38 contains a signal sequence for processing and amidation [-Gly^{28}-Lys^{29}-Arg^{30}-], which also suggested the presence of PACAP27NH$_2$. This peptide was also synthesized and found to exhibit a similar ACSA as PACAP38. Several oligonucleotides corresponding to various portions of PACAP38 were synthesized and are being used for cloning cDNA for the precursor protein of PACAP38. A broad range of physiological experiments are also now underway to unveil the physiological significance of these novel peptides.

Acknowledgements

Research was supported in part by NIH grant DK-09094. We thank Mrs. L. Jiang, P. Portilla and M.C. Noriega for their technical assistance.

References

1. Negro-Villar, A. and Conn, P.M. (Eds.) Peptide Hormones – Effect and Mechanism of Action, Vol. III, CRC Press Inc., Boca Raton, FL, 1988.
2. Mutt, V. and Said, S.I., Eur. J. Biochem., 42 (1974) 581.
3. Arimura, A., Matsumoto, K., Culler, M.D., Kanda, M., Itoh, Z., Murphy, W., Shively J.E. and Palkovits M., Peptides, 5 (Suppl. 1) (1984) 41.

Tyrphostins: A new class of tyrosine kinase inhibitors that selectively blocks EGF-dependent cell proliferation

Chaim Gilon[a], Aviv Gazit[a], Pnina Yaish[b] and Alexander Levitzki[b]

[a]Department of Organic and [b]Department of Biological Chemistry, The Hebrew University of Jerusalem, Jerusalem 91904, Israel

Introduction

We have recently reported on a new class of low molecular weight protein tyrosine kinase inhibitors, termed 'tyrphostins' (Fig. 1), which are based on a benzylidene malono nitrile nucleus [1]. We now describe (a) the molecular requirements for inhibition, derived from SAR studies of a large series of molecules with increased affinity towards the substrate site of the EGF receptor kinase (EGFRK) domain, a comparison between the (b) inhibitory effect of some tyrphostins on the phosphorylation catalyzed by EGFRK and insulin receptor kinase (InsRK), and (c) the inhibition of EGF-dependent proliferation of A431/clone 15 cells.

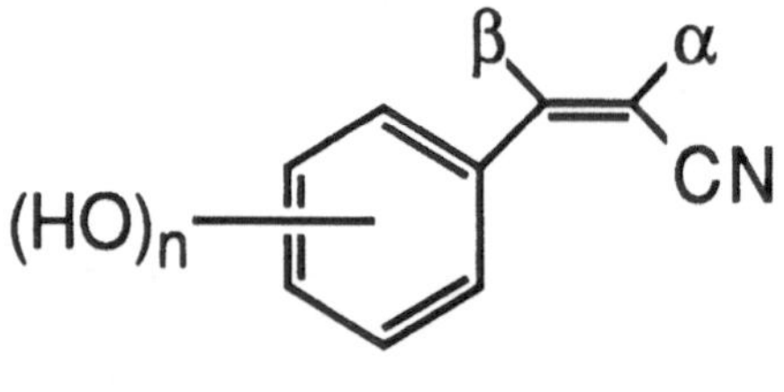

Fig. 1.   Structure of tyrphostins.

Results and Discussion

Most tyrphostins were prepared by a straightforward Knoevenagel condensation of benzaldehydes or acetophenones with malononitriles, cyanoacetamides or malononitrile dimer. The substructure responsible for the potency of the tyrphostins appears to be polyhydroxy cis cinnamonitrile. Thus p-hydroxy trans cinnamonitrile is inactive, and saturation of the double bond abolished activity. A strong increase in potencies is seen when hydroxyl groups are added on the aromatic ring, with 10–15-fold increase in potency for one OH added, and 3–5-fold for the other OH. Addition of electron withdrawing or donating groups on the aromatic ring makes relatively little difference. When trans substitution is added to the cis cinnamonitrile nucleus (the α-position), improvement is obtained. Replacement of H by OH at the β-position increases potency, whereas

522

replacement by Me lowers it. The best competitive inhibitors of the phosphorylation of the substrate poly(Glu$_6$-Ala$_3$-Tyr) catalyzed by purified EGF receptor were 3,4-di-OH, α-carboxamido *cis* cinnamonitrile (K$_i$ = 2.3 μM), and 3,4-di-OH, α-thiocarboxamido *cis* cinnamonitrile (K$_i$ = 0.85 μM). These inhibitors poorly inhibited the phosphorylation of poly(Glu$_4$-Tyr) catalyzed by insulin receptor [2] (K$_i$ = 410 μM for the carboxamide and K$_i$ = 640 μM for the thiocarboxamide). These tyrphostins also inhibited the EGF-dependent proliferation of A431/clone 15 epidermoid carcinoma cells, with little or no effect on EGF-independent cell growth.

References

1. Yaish, P., Gazit, A., Gilon, C. and Levitzki, A., Science, 242 (1988) 933.
2. Braun, S., Raymond, W.E. and Racker, E.J., J. Biol. Chem., 259 (1984) 2051.

Mitotoxins: Tools for the determination of the function of growth factors

Douglas A. Lappi[a], Darlene Martineau[a], Emelie Amburn[a], Robert Fox[b] and Andrew Baird[a]

[a]Department of Molecular and Cellular Growth Biology, The Whittier Institute for Diabetes and Endocrinology, 9894 Genesee Avenue, La Jolla, CA 92037, U.S.A.
[b]Department of Immunology, Research Institute of Scripps Clinic, La Jolla, CA 92037, U.S.A.

Introduction

Basic fibroblast growth factor (FGF) is mitogenic for cells derived from the primary mesenchyme and the neuroectoderm. Basic FGF has a strong affinity for its receptor (20–50 pM) and is one of the most powerful mitogens characterized. Antagonists for the activities of basic FGF would be particularly useful as tools to determine the physiological role that it plays as a neurotrophic factor [1], as an angiogenic factor [2], as a mitogen [3] and as a differentiation factor [4] in vivo. Because it has also been recently implicated in the development of Kaposi's sarcoma in AIDS [5,6] and the progression of melanocytes to melanomas [7], there is considerable potential for an anti-FGF.

In an effort to characterize the importance of basic FGF in these roles, we have synthesized a mitotoxin [8] by conjugating the ribosome-inactivating protein saporin SAP (SAP or SO-6) [9,10]) to basic FGF. The presence of the specific FGF receptor on cells permits the entry of SAP into the cell. SAP, in turn, catalyzes the cleavage of nucleotide base A_{4324} of ribosomal RNA and causes target cell death. Therefore, the binding and internalization of the mitotoxin then results in cell death.

Results and Discussion

Microscopic examination of BHK cells 48 h after plating at 5000 cells per ml in 24-well plates reveals nearly confluent cells. In wells that have been treated with 10 nM mitotoxin, there are very few attached, morphologically normal cells. The majority of those present have become unattached, indicating cell death.

The treatment of bovine corneal endothelial cells with FGF-SAP (Fig. 1) elicits a powerful cytotoxic effect at picomolar concentrations (ED_{50} of 400 pM). At low concentrations of mitotoxin, there appears to be a slight proliferative effect, presumably because basic FGF has a strong mitogenic effect on these cells. At no concentration does SAP alone have a mitogenic or cytotoxic effect. It is interesting to note that a mixture of basic FGF and SAP has an effect similar

524

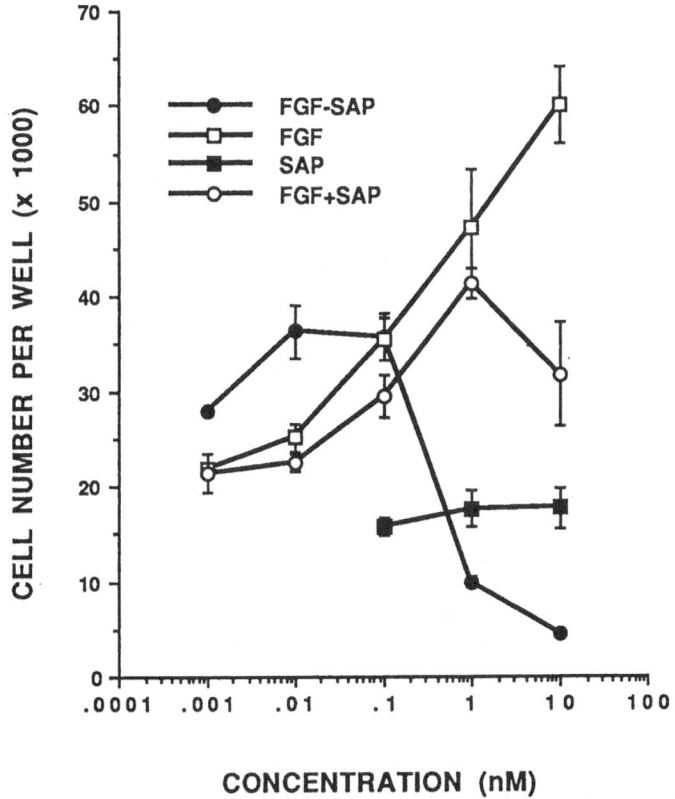

Fig. 1. The effect of FGF-SAP on bovine corneal endothelial cells. The procedure was as described in Lappi et al. [8].

to basic FGF alone except at the highest dose when slight inhibition of cell number is seen. These data indicate that the mitotoxin can be a potent inhibitor of endothelial cell proliferation and an antagonist of basic FGF for an important target cell population of the growth factor, endothelial cells.

The efficacy of the mitotoxin, as measured by the ED_{50}, is dependent on the number of basic FGF receptors (Table 1). Five cell types were treated with the mitotoxin and for these same 5 cell types, the number of FGF receptors per cell was determined. The higher the number of FGF receptors on the target cell, the greater the potency of the mitotoxin. These data indicate that, if tissues express the receptor, the mitotoxin could be an effective tool to target FGF-dependent processes in vivo.

Because basic FGF has several in vitro and in vivo activities [11], we propose that the mitotoxin can be used as a probe to establish its physiological role in regulating cell growth and function. Clearly, it should also be evaluated as a potential therapeutic anti-FGF.

Table 1   *Receptor number per cell and ED₅₀ of FGF-SAP for various cell types*

Cell type[a]	ED$_{50}$ (pM)[b]	Receptor number per cell[c]
ACE	1 200	2 200
PFHR9	3 400	2 700
DC3/3	350	14 000
3T3	62	23 000
BHK	25	42 600

[a] ACE (adult capillary endothelial cells) were obtained from primary culture according to the method of Gospodarowicz et al. [12]. PFHR9 cells were a kind gift of Dr. E. Engvall of the La Jolla Cancer Research Foundation. DC3/3 cells were from a primary culture from tissue obtained from surgery for Dupuytren's contracture (R. Fox et al., manuscript in preparation). 3T3 and BHK (baby hamster kidney) cells were obtained from the American Type Culture Collection.
[b] ED$_{50}$s were calculated as described in Lappi et al. [8].
[c] Receptor number was determined as described by Moscatelli [13].

Acknowledgements

Research was funded by grants from the National Institutes of Health (DK-18811) and The Whittier Institute Angiogenesis Research Program. We also acknowledge Drs. Pamela Maher, Robert Z. Florkiewicz, and Marino Buscaglia for helpful discussion, and Denise Higgins for preparation of the manuscript.

References

1. Walicke, P., Cowan, W.M., Ueno, N., Baird, A. and Guillemin, R., Proc. Natl. Acad. Sci. U.S.A., 83 (1986) 3012.
2. Montesano, R., Vasalli, J.D., Baird, A., Guillemin, R. and Orci, L., Proc. Natl. Acad. Sci. U.S.A., 83 (1986) 7297.
3. Böhlen, P., Baird, A., Esch, F., Ling, N. and Gospodarowicz, D., Proc. Natl. Acad. Sci. U.S.A., 81 (1984) 5364.
4. Kimelman, D. and Kirschner, M., Cell, 51 (1987) 869.
5. Salahuddin, S.Z., Nakamura, S., Biberfeld, P., Kaplan, M.H., Markham, P.D., Larsson, L. and Gallo, R.C., Science, 242 (1988) 430.
6. Nakamura, S., Salahuddin, S.Z., Biberfeld, P., Ensoli, B., Markham, P.D., Wong-Staal, F. and Gallo, R.C., Science, 242 (1988) 426.
7. Halaban, R., Kwon, B., Ghosh, S., Delli Bovi, P. and Baird, A., Mol. Cell. Biol., 8 (1988) 2933.
8. Lappi, D.A., Martineau, D. and Baird, A., Biochem. Biophys. Res. Commun., 160 (1989) 917.
9. Stirpe, F., Gasperi-Campani, A., Barbieri, L., Falasca, A., Abbondanza, A. and Stevens, W.A., Biochem. J., 216 (1983) 617.
10. Lappi, D.A., Esch, F.S., Barbieri, L., Stirpe, F. and Soria, M., Biochem. Biophys. Res. Commun., 129 (1985) 934.
11. Baird, A., Esch, F., Mormède, P. et al., Rec. Prog. Horm. Res., 42 (1986) 143.
12. Gospodarowicz, D., Massoglia, S., Cheng, J. and Fujii, D.K., J. Cell. Physiol., 127 (1986) 121.
13. Moscatelli, D. J., Cell. Physiol., 131 (1987) 123.

Importance of Ca^{2+} in agonist-receptor interaction: Structural studies on ligand binding site peptide of β_2-adrenergic receptor

V.S. Ananthanarayanan* and C. Horne

Department of Biochemistry, McMaster University, Hamilton, Ontario, Canada L8N 3Z5

Introduction

Two fundamental questions concerning the action of hormones, drugs and other stimulants are: (a) What is the bioactive conformation of the stimulant (agonist)? and (b) What is the nature of the interaction between these agents and their respective receptors in the cell membrane? Based on our recent finding that many peptide hormones [1] and drugs [2] exhibit definitive binding of Ca^{2+} in non-polar solvents, we proposed [1] that the Ca^{2+}-bound form of these and other stimulants may represent the biologically relevant conformation that interacts with the receptor. Based on our further observation [1] that many agonists are also capable of transversing the lipid bilayer with the bound Ca^{2+}, we envisaged the interaction of the agonist with the receptor to be mediated by Ca^{2+}. We expected that the receptor would, itself, bind Ca^{2+} as part of this interaction. Here, we provide evidence for Ca^{2+} binding by two synthetic peptides representing the transmembrane ligand-binding domain of β_2-adrenergic receptor (β-AR).

Results and Discussion

Using covalent incorporation of a ligand, Dohlman et al. [3] have identified the sequence 83–96 of β-AR to be the ligand binding site: G L A V V P F G A S H I L M. We synthesized, by solid-phase techniques, the above 14-mer (termed KC-14). Since Asp^{79} in β-AR has been implicated in agonist binding [3,4], we also synthesized the sequence 79–96 of β-AR (termed KC-18) with D L V M added to the N-terminal of KC-14. The peptides were purified by preparative HPLC on C-18 columns.

We obtained CD spectra of KC-14 and KC-18 in TFE in the absence and presence of Ca^{2+}. Figure 1 shows the KC-18 data. Both the peptides assume α-helical conformation in the non-polar solvent. Ca^{2+} addition significantly destabilizes the helix. No major change in the helix content of KC-14 or KC-18 was seen on the addition of phenylephrine, an adrenegic agonist, or propranolol, a β-adrenergic blocker (data not shown). Interestingly, the structure

---

*To whom correspondence should be addressed.

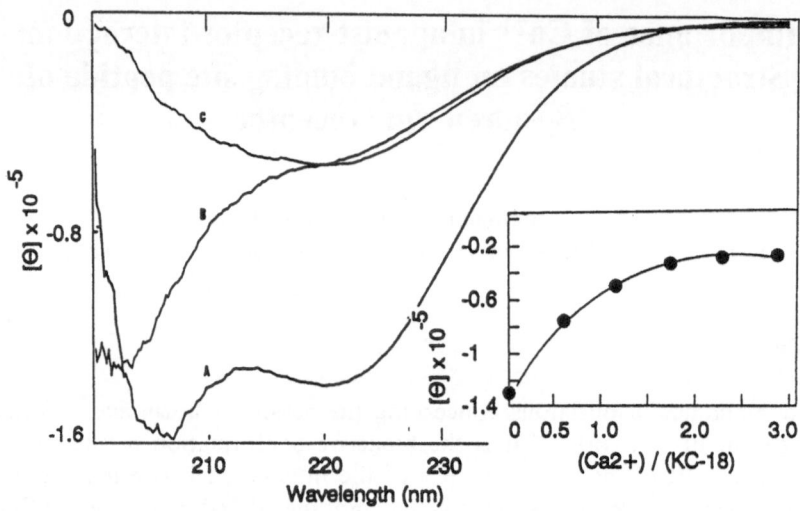

Fig. 1. CD spectra of KC-18 in TFE: (A) peptide alone; (B) in 2 molar excess of Ca(ClO$_4$)$_2$; (C) 2 molar excess of Ca^{2+} and 2.0 molar excess of phenylephrine. Inset: Ca^{2+} binding isotherm (22°C) of KC-18.

of the peptide is different in the presence of saturating concentrations of Ca^{2+} and drug (Fig. 1). The Ca^{2+} binding isotherm of KC-18 (Fig. 1 inset) showed a 1:1 Ca^{2+}/peptide complex; detailed analysis of the data will be reported elsewhere.

Figure 2 shows plots of the fluorescence emission intensity of phenylephrine.

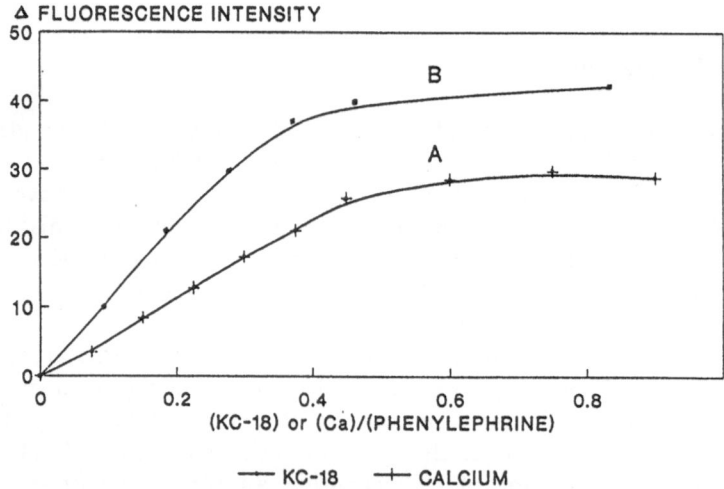

Fig. 2. Changes in fluorescence emission intensity of phenylephrine (excited at 270 nm) on addition of (A) Ca^{2+} and (B) peptide KC-18.

528

Substantial changes in the drug structure are seen in the presence of Ca^{2+} and KC-18. Two moles of drug complex with 1 mole of Ca^{2+} or peptide.

The peptide sequences studied here have been identified as crucial for ligand binding by β-AR; deletion of Asp^{79} results in a 10-fold decrease in agonist binding affinity [4]. The important implications of this study are: (a) β-AR binds Ca^{2+}, possibly in the same domain where the ligand binds; (b) the ligand (phenylephrine) also binds Ca^{2+} stoichiometrically; (c) the formation of the putative ternary complex: receptor–Ca^{2+}–ligand, may trigger the subsequent events in signal transduction. (Mg^{2+} may compete with or replace Ca^{2+} in certain cases).

On the basis of our earlier and present data, we propose that the interaction of Ca^{2+} with both the stimulant and the receptor, possibly as a 'cofactor' could be a general requirement and may represent a key event in signal transduction.

Acknowledgements

This work was supported by the Medical Research Council of Canada.

References

1. Ananthanarayanan, V.S., J. Cell Biol., 107 (1988) 500a (Abstr.).
2. Ananthanarayanan, V.S. and Taylor, L.B. In Proceedings of the Symposium on Cellular Variations in Ca^{2+} Signalling, Toronto, August, 1988.
3. Dohlman, H.G., Caron, M.C., Strader, C.D., Amlaiky, N. and Lefkowitz, R.J., Biochemistry, 27 (1988) 1813.
4. Strader, C., Segal, I.S. and Dixon, R.A.F., FASEB J., 3 (1989) 1825.

Antisense peptides of melanocyte-stimulating hormone (MSH): Surprising results

Fahad A. Al-Obeidi[a], Victor J. Hruby[a], Shubh D. Sharma[a], Mac E. Hadley[b] and Ana M. De L. Castrucci[b]

[a]Department of Chemistry and [b]Department of Anatomy, University of Arizona, Tucson, AZ 85721, U.S.A.

Introduction

Melanotropic peptides (α-MSH and β-MSH) are generated from a precursor protein (pro-opiomelanocortin) whose mRNA nucleotide sequence was previously reported [1]. These hormones stimulate melanin pigment synthesis and distribution in vertebrates. Both α-MSH and β-MSH have the same pentapeptide message sequence (EHFRW), but the N- and C-terminal extensions differ in primary structure.

It has recently been suggested that antisense peptides (peptides encoded by a mRNA complementary to the mRNA for a specific peptide hormone) might act as 'receptors' for the hormones (e.g., ACTH) [2]. We have designed and synthesized several antisense peptides for β-MSH and α-MSH based on their mRNAs [1], and have obtained some unexpected biological results.

Results and Discussion

Eight antisense peptides related to β-MSH, and two related to α-MSH (Table 1) were synthesized by SP methods. They were purified by HPLC and characterized by AAA, FABMS, TLC, and analytical HPLC. The primary sequences and design of various peptides antisense to α- and to β-MSH are described in Table 1. The possible interaction of these antisense peptides with α-MSH and MCH was studied in the in vitro frog and fish skin (melanocyte) bioassays, respectively.

Peptides 1B, 2A and 2B exhibited agonistic activity (melanin, dispersion, skin darkening) in the frog skin bioassay.

Peptides 3B and 4B were full but weak agonists (melanin aggregation, skin lightening) in the fish skin bioassay; whereas, 1B and 4A were not fully effective antagonists (Table 1). Recently, we have suggested that α-MSH and MCH (melanin-concentrating hormone) may be evolutionarily related [3].

Affinity HPLC on a [Nle^4]α-MSH affinity column was used to investigate the binding of various antisense peptides to it. Peptides 2A and HSM1 were the most active in binding. Testing the recognition ability of these antisense peptides against β-MSH is in progress in our laboratory.

Table 1 *Melanotropin antisense peptides: structures and biological activities*

Antisense peptide	Melanotropic activity[a]			
	α-MSH (bioassay) Frog skin bioassay		MCH (bioassay) Fish skin bioassay	
	Agonist	Antagonist	Agonist	Antagonist
β-MSH related[b]				
1A Ac-VLGRVAPAEVLHPVGPLV-NH$_2$	–	+	–	–
1B Ac-VLGRVAPAEVLHPVGPRV-NH$_2$	+	–	–	+
2A Ac-VLPGVPHLVEAPAVRGLV-NH$_2$	+	–	–	–
2B Ac-VRPGVPHLVEAPAVRGLV-NH$_2$	+	–	–	–
3A Ac-LLPGMAHLVKATSLGGFL-NH$_2$	–	–	–	–
3B Ac-LSPGMAHLVKATSLGGFL-NH$_2$	–	–	+	–
4A Ac-LFGGLSTAKVLHAMGPLL-NH$_2$	–	–	–	+
4B Ac-LFGGLSTAKVLHAMGPSL-NH$_2$	–	–	+	–
α-MSH related[b]				
HSM1 Ac-HRLAPAEVFHGVR-NH$_2$	Strongly binds to [Nle^4]α-MSH affinity column			
HSM2 Ac-RVGHFVEAPALRH-NH$_2$	Weakly binds to [Nle^4]α-MSH affinity column			

a Skins were first incubated in the presence of an antagonist (10^{-4} – 10^{-7} M), then followed by the addition of MSH (10^{-10} M) or MCH (10^{-10} M). The agonistic or antagonistic activity of the antisense peptides in each bioassay (frog, α-MSH, or fish, MCH) is indicated as positive (+) or negative (–).

b Shaded area represents sequence antisense to the message sequences in α-MSH and β-MSH (EHFRW). Peptides were generated from both 5' to 3' reading frame of anti-mRNA for β-MSH (human and rat) and α-MSH (1A, 1B and HSM1) and 3' to 5' reading frame (3A and 3B). All others represent the reverse amino acid sequences corresponding to these peptides.

531

Acknowledgements

This work was supported in part by grants from the U.S. Public Health Service and NSF.

References

1. Nakanishi, S., Inoue, A., Kita, T., Nakamura, M., Chang, A.C.Y., Cohen, S.N. and Numa, S., Nature 278 (1979) 423.
2. Bost, K.L., Smith, E.M. and Blalock, J.E., Proc. Natl. Acad. Sci. U.S.A., 82 (1985) 1372.
3. Castrucci, A.M.L., Hadley, M.E., Lebl, M., Zechel, C. and Hruby, V.J., Regul. Pept., 24 (1989) 27.

Session V
Peptide/protein folding

Chairs: Arthur M. Felix
Hoffmann-La Roche Inc.
Nutley, New Jersey, U.S.A.

Lila M. Gierasch
University of Texas Southwestern Medical Center
Dallas, Texas, U.S.A.

William A. Gibbons
University of London
London, U.K.

and

Evaristo Peggion
University of Padua
Padua, Italy

Small peptide models for side-chain–backbone interactions

Michel Marraud[a], Samuel Prémilat[b] and André Aubry[c]
[a]UA-CNRS-494, ENSIC-INPL, BP 451, F-54001 Nancy Cedex, France
[b]UA-CNRS-494, University of Nancy I, BP 239, F-54506 Vandoeuvre Cedex, France
[c]UA-CNRS-809, University of Nancy I, BP 239, F-54506 Vandoeuvre Cedex, France

Introduction

In crystallized proteins, a quarter of the hydrogen-bonded side chains interact with peptide atoms, and half of these main-chain–side-chain interactions are short-range interactions that could stabilize local conformations [1]. Simple model peptides are particularly suited to studying these local conformations, and we have carried out a conformational analysis, using ^{1}H NMR and FTIR spectroscopy in solution, and X-ray diffraction on crystals, of model dipeptides reproducing the β- and Asx-turns in order to elucidate the role of polar side chains in these two structures.

Results and Discussion

It is known that short polar residues (Ser, Thr, Asp, Asn, His) are preferentially incorporated in β-turns, a structure that concerns a sequence of four residues presenting a short contact between a peptide carbonyl and the peptide nitrogen three residues further down the sequence [2]. The stability of this structure was studied on the tBuCO-Pro-X-NHMe dipeptides in relation with the nature of the X residue with the aim of looking for possible main-chain–side-chain attractive interactions.

From the NH and $C=O$ stretching frequencies in IR spectroscopy, and the solvent-induced variation of the NH proton NMR signals, it appears that the tBuCO-Pro-X-NHMe dipeptides are effectively β-folded in solution [3]. However, aliphatic X residues give smaller β-turn ratios than polar residues or the weakly polar Cys and Met and aromatic Phe and Tyr residues. The only exception is the protonated His$^+$ residue (Table 1).

IR and NMR data show the polar X side chain to be the proton-accepting site from the X-NH bond, which imposes the βI-folding mode on the peptide backbone. This interaction, contributing to the stability of the βI-turn, favors the gauche$^+$ rotamer ($\chi^1 = 60°$) of the X C^α-C^β bond, the percentage of which varies like that of the β-turn ratio (Table 1).

In the solid state, most polar residues induce the βI-turn conformation with retention of the main-chain–side-chain interaction found in solution (Table 1). However, several βII-turn conformations are also observed, mainly for apolar residues, showing that this latter folding mode is also accessible to homochiral sequences.

535

Table 1 *Conformational data for the tBuCO-Pro-X-NHR dipeptides (R = Me, iPr) in CHCl$_3$ solution and in crystals*

X	Solution		Crystal		
	βI-turn (%)[a]	g⁺ rotamer (%)[b]	β-turn type	Cα-Cβ rotamer	mc.–sc. interaction[c]
Leu	56	30	II	g⁻	
Phe	62	54	II	g⁻	
Tyr	65	60	II	g⁻	
Met	67	52			
Cys(Me)	74	75	II	g⁻	
Cys(StBu)	74	67	II	g⁻	
Cys	73	88	I	g⁻	
Ser	90	100	I	g⁺	+
Thr	80	100	I	g⁺	+
Asn	94	88	II	g⁻	
Asp	87	78	II	t	
Asp(OMe)	83	82	I	g⁺	+
Asp⁻.NMe₄⁺	100	100			
His	85	84	I	g⁺	+
His(Nτ-Me)	90	85	I	g⁺	+
His(Nπ-Me)	65	47	I	g⁺	
His⁺.PF₆⁻	35	88	d	g⁻	d

[a] See Ref. 4.
[b] See Ref. 5.
[c] Main-chain–side-chain interaction between the X nitrogen and the X side chain (see text).
[d] The imidazolium is the proton-donating site to the pivaloyl carbonyl (see Ref. 6).

A clear conformational transition due to the side chain is given by the His derivative. In the neutral form, it assumes the βI-folding mode with a His-NH to His-Nπ interaction, but protonation of the imidazole ring results in a new conformation stabilized by a His⁺-NπH to tBuCO interaction [6].

In proteins, the gauche⁺ rotamer appears to be favored for the short polar residues in βI-turns (39% occurrence against 11% for the other residues). This suggests that the main-chain–side-chain interaction found in these dipeptides could be responsible for the preferential incorporation of the short polar residues in βI-folded sequences, and could contribute to the stability of this conformation in proteins.

The Asx-turn (Asx = Asn or Asp) is much less documented than the β-turn. It concerns a shorter sequence of three residues in which the Asx side carbonyl is hydrogen-bonded to the peptide nitrogen two residues further down the sequence [1]. This conformation is found in β-turn and α-helix inducing sequences [1,7], and could be involved in post-translational modifications, such as Asn-deamidation [8] and N-glycosylation [9].

In crystallized proteins, nearly 18% of the Asx residues are involved in Asx-turns. A Monte Carlo analysis of the Gly-Asn-Ala-Gly sequence reveals the existence of four Asx-turn conformers (Table 2), among which the three most stable types, I, II and III, exist in proteins with 70, 10 and 20% occurrences, respectively. They differ by the trans ($\chi^1 = 180°$) or gauche+ ($\chi^1 = 60°$) conformation of the Asx C^α-C^β bond, and/or by a 180° rotation of the Asx-following amide plane.

Table 2 *Conformational angles of the Asx-turn conformers predicted by a Monte Carlo analysis of the Gly-Asn-Alα-Gly sequence*

Asx-turn	Asn				Ala		Enthalpy[a] (kcal/mol)
	ϕ	ψ	χ^1	χ^2	ϕ	ψ	
Type I	−70	121	176	−152	−69	−29	0
Type II	−70	−14	66	150	59	30	1.5
Type III	−70	166	65	−98	−105	12	1.4
Type IV	−70	−37	−175	101	70	17	2.8

[a]The most stable Asn-turn conformer was taken as a reference.

IR and ^{1}H NMR experiments have been carried out on the Boc-Asx-X-NHMe dipeptides with Asx = Asn, Asp, Asp(OBzl), Asp⁻.NMe₄⁺ (Asp⁻ standing for the anionic form of Asp), and X = Pro or Phe. The Pro-derivatives are the shortest molecules that clearly assume the Asx-turn in solution. The distribution of the Asx C^α-C^β rotamers shows that the type I Asx-turn is more frequent than type III, with, respectively, 60 and 35% occurrences in CHCl₃ for Asx = Asn. Furthermore, the Asx-turn ratio decreases in the order Asn > Asp⁻ > Asp.

The Asx-turn also contributes to the molecular conformation of the Boc-Asn-X-Ser-NHMe tripeptides (X = Ala, Pro) deriving from the marker sequence for *N*-glycosylation of proteins [9]. In solution, X = Pro favors the type I Asx-turn, and X = Ala favors both II and III types. Four crystal structures of these tripeptides, one Ala- and three Pro-derivatives, have been solved so far [10].

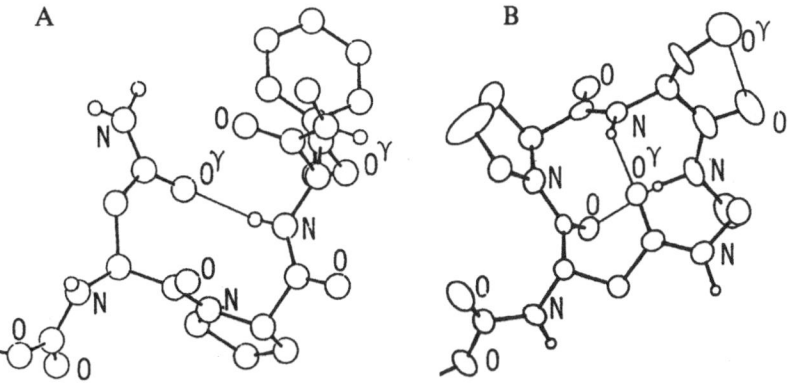

A   B

Fig. 1. *Crystal molecular structures of the Boc-Asn-Pro-Ser (Bzl)-NHMe (A) and Boc-Asn(Me)-Pro-Ser-NHMe (B) tripeptides (tBu omitted).*

All present the type I Asx-turn with a Ser-N to Asn-O$^\gamma$ distance ranging from 2.88 to 3.10 Å, but both Ala-ϕ and ψ angles differ by 50° from their Pro-counterparts. The Asx-turn is the unique intramolecular interaction in one case, and it is followed by a βI-folded Pro-Ser sequence in two cases (Fig. 1).

The study of simple model peptides is a convenient way of collecting information on local conformations in larger peptides and in proteins. The β- and Asx-turns are intrinsically stable and do not require additional long-range stabilizing interactions. They illustrate the importance of the side chains for orienting the conformation of the peptide chain. However, both are flexible structures with possible transition from one conformer to another, due to long-range interactions.

References

1. Baker, E.N. and Hubbard, R.E., Progr. Biophys. Mol. Biol., 44 (1984) 97.
2. Rose, G.D., Gierasch, L.M. and Smith, J.A., Adv. Protein Chem., 37 (1985) 1.
3. Aubry, A., Cung, M.T. and Marraud, M., J. Am. Chem. Soc., 107 (1985) 7640.
4. Boussard, G. and Marraud, M., J. Am. Chem. Soc., 107 (1985) 1825.
5. Cung, M.T. and Marraud, M., Biopolymers, 21 (1982) 953.
6. Aubry, A., Vlassi, M. and Marraud, M., Int. J. Pept. Prot. Res., 28 (1986) 637.
7. Richardson, J.S. and Richardson, D.C., Science, 240 (1988) 1648.
8. Kossiakoff, A.A., Science, 240 (1988) 191.
9. Abbadi, A., Boussard, G., Marraud, M., Pichon-Pesme, V. and Aubry, A., In Aubry, A., Marraud, M. and Vitoux, B. (Eds.) Second Forum on Peptides, INSERM Colloquium, John Libbey Eurotext, Paris, 1989, p. 375.
10. Pichon-Pesme, V., Aubry, A., Abbadi, A., Boussard, G. and Marraud, M., Int. J. Pept. Prot. Res., 32 (1988) 175.

Spectroscopic studies of a chemically synthesized zinc finger peptide: Domain structure and folding

Grace Párraga[a], Suzanne Horvath[b], Jon Herriott[a], Leroy Hood[b]
and Rachel E. Klevit[a],*

[a]*Department of Biochemistry, SJ-70 University of Washington, Seattle, WA 98195, U.S.A.*
[b]*Division of Biology, California Institute of Technology, Pasadena, CA 91125, U.S.A.*

Introduction

About four years ago, 2 groups noted the presence of repeated amino acid sequences in the *Xenopus* transcription factor TFIIIA [1,2]. At that time, the sequences were designated as zinc fingers, based on limited proteolysis, zinc content and sequence analysis of the TFIIIA–5SRNA complex [1,2]. Both groups proposed that metal binding via conserved cysteine and histidine residues stabilized the domains and that the repeated zinc finger structures were responsible for the DNA binding properties of the protein.

Subsequent to the identification of nine zinc finger sequences in TFIIIA, the motif has been found in a plethora of deduced amino acid sequences of putative and demonstrated eukaryotic DNA binding proteins [3–6]. The zinc finger model has also been directly supported by genetic and deletion analyses of zinc finger proteins [7,8], as well as physical analysis and footprinting of the TFIIIA–5SRNA complex [9–11], and spectroscopic studies of zinc finger peptides [12–15].

Because of the implied importance of this protein motif in sequence specific DNA binding, several groups have been actively pursuing the structural details of zinc finger proteins and peptides. Zinc finger models have been built, based on conserved substructures from crystallized metalloproteins [16], and using interactive model building followed by molecular dynamics refinement [17]. In order to obtain structural information, our group has concentrated on applying 2D ^{1}H NMR techniques to synthetic peptides consisting of zinc finger sequences from the yeast transcription factor ADR1 [13–15].

We have synthesized the wild type and mutant single and double zinc finger peptides (outlined in Fig. 1) by the stepwise solid-phase method [18,19]. To determine the structure of a zinc finger domain, it was first necessary to ascertain whether zinc finger peptides would fold in vitro into a single conformational form upon the addition of metal ions. Once the folding and unfolding of single finger peptides could be demonstrated and monitored by spectroscopy [12,13], specific questions regarding zinc finger folding, structure, and stability could be addressed. Models for two single finger peptides – ADR1a [13] and ADR1b [14] – have been determined from 2D NMR studies. The wild type zinc finger

---

*To whom correspondence should be addressed.

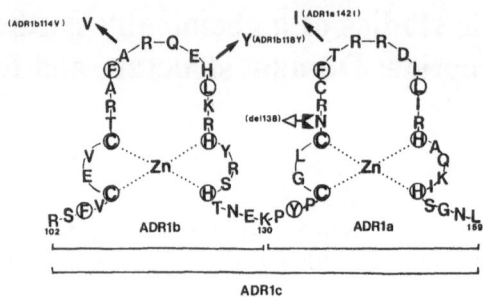

Fig. 1. Amino acid sequence of zinc fingers in ADR1, and the sequences of the finger peptides synthesized. The sequence of ADR1b is underlined; missense mutants and a deletion mutant peptide are denoted by arrows.

structures that we have determined consist of an N-terminal loop involving the two conserved Cys residues, a C-terminal amphiphilic α-helix, and a tightly packed hydrophobic core. In this report, we address the thermal unfolding properties of ADR1b monitored by 1D ^{1}H NMR spectroscopy.

Results and Discussion

In a previous study [15], we reported the effects of cysteine thiol alkylation and histidine imidazole protonation on the metal binding and folding characteristics of ADR1a and ADR1b. We also described the inability of a mutant zinc finger peptide (del138, Fig. 1) to bind zinc tetrahedrally and to fold properly. We were interested in the temperature stability of wild type single finger peptides because it has been shown that zinc finger domains from proteins [1] and peptides [12] are protease-resistant structures, implying that the domains are stable and compact in their zinc-reconstituted form.

The aromatic and upfield aliphatic regions of the 1D NMR spectra of ADR1b from 17°C tot 87°C are shown in Figs. 2A and 2B, respectively. When reconstituted in the presence of zinc, ADR1b folds into a single conformation, characterized by downfield-shifted CαH resonances, histidine C2 and C4 proton resonances (3 His residues, therefore, 3 separate C2H and C4H resonances, Fig. 2A) and upfield-shifted methyl proton resonances (Fig. 2B). The spectrum of ADR1b reconstituted in the absence of zinc is shown in Fig. 2C as an example of the 'unfolded' ADR1b spectrum.

The thermal unfolding of ADR1b, as monitored by NMR, is completely reversible from temperatures up to 87°C (the highest temperature tested). The spectra obtained at increasing temperatures exhibit several interesting properties. One feature worth noting is the differential rates of deuteration (protons exchange for deuterons and disappear from the ^{1}H NMR spectrum) of the imidazole His C2 protons (denoted by arrows). In folded ADR1b, the C2 protons of His^{122} and His^{118} are deuterated at a slower rate (qualitatively) than His^{126} which is

540

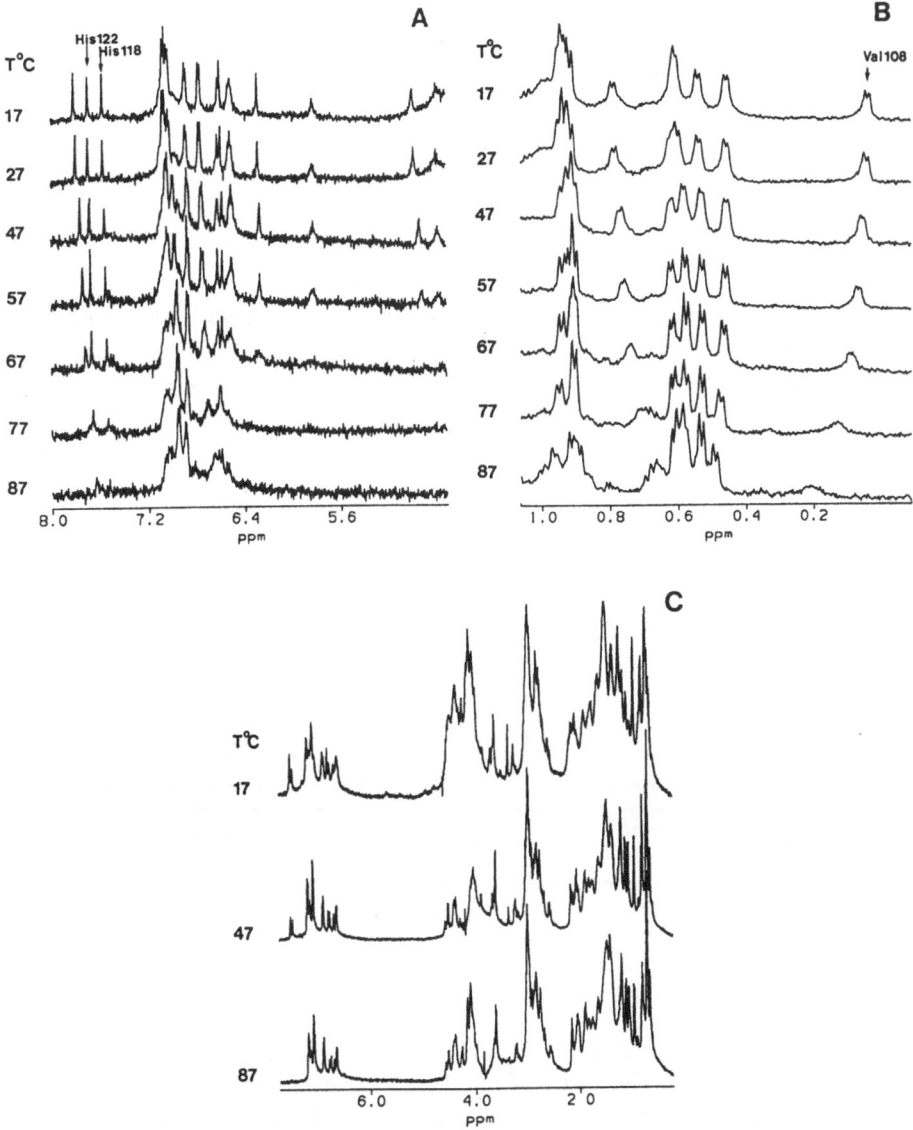

Fig. 2. *Temperature effects on 1D ¹H NMR spectra of ADR1b. D₂O spectra are shown of 0.5 mM ADR1b in 50 mM deuterated Tris reconstituted in (A) the presence of 0.5 mM ZnCl₂ (aromatic proton resonances), (B) the presence of 0.5 mM ZnCl₂ (aliphatic proton resonances), and (C) the absence of ZnCl₂. Methods are described in Ref. 13.*

near the extended C-terminal tail of the peptide [14]. Furthermore, the resonance of C2H of His[126] also exhibits a linear dependence of its chemical shift position on temperature that is likely to be a consequence of its rapid exchange with solvent. All three His C2 protons are deuterated more slowly in folded ADR1b (Fig. 2A) than in the unfolded peptide (Fig. 2C).

The behavior observed for proton resonances in the temperature-dependent spectra of Zn^{2+}–ADR1b is complex. Some resonances exhibit slow exchange behavior over the temperature range tested. The intensities of such resonances gradually decrease with increasing temperature, while their chemical shift positions remain unchanged. This type of behavior usually indicates the existence of two species in solution, each with lifetimes that are long on the NMR timescale. There are a variety of resonances that exhibit fast-exchange behavior; their chemical shift positions change with increasing temperature. The thermal transition curves for the His[118] C2H and His[122] C2H are shown in Fig. 3A, and for Val[108] γ-methyl protons in Fig. 3B. The baselines for the thermal denaturation curves are quite flat until about 60°C, when the resonance positions of protons change in the direction of the positions of unfolded ADR1b. These curves are all qualitatively similar and indicate that the peptide is not completely unfolded at 87°C (Fig. 3). Thus, it is clear that this 30-residue peptide folds into a remarkably heat-stable structure when reconstituted with zinc.

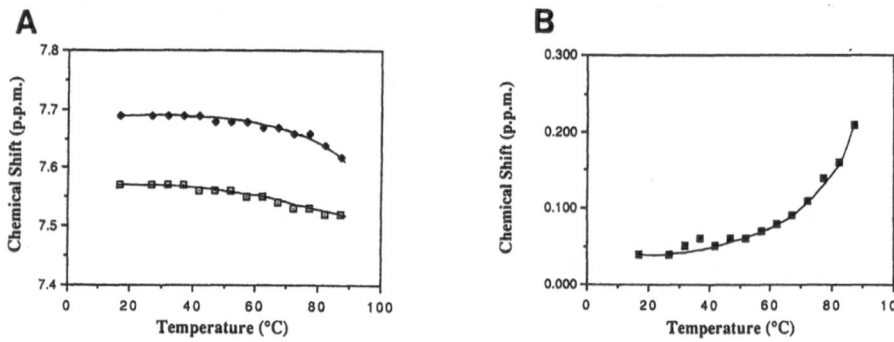

Fig. 3. Temperature dependence of chemical shift changes for selected protons of ADR1b. Proton resonances are indicated using nomenclature for the corresponding residue in intact ADR1, and assignments are detailed in Ref. 14. (A) His[118] C2H, □ = approximate random coil position = 7.46 p.p.m. and His[122] C2H, ♦ = 7.51 p.p.m. (B) Val[108] methyl protons, ■ approximate random coil position = 0.66 p.p.m.

These results contrast with those observed for $P\alpha P\beta$ – a 30-residue disulphide-bonded peptide pair designed to mimic the native tertiary structure of a folding intermediate of bovine pancreatic trypsin inhibitor [20]. For the $P\alpha P\beta$ dipeptide, folding is thermodynamically linked to disulphide bridge formation. Thermal denaturation curves are quite broad, and unfolding is complete at about 60°C. The secondary and tertiary structure of ADR1b is thermodynamically and kinetically coupled to tetrahedral zinc binding, and unfolding is not completely achieved at 87°C. Indeed, the thermal unfolding of a zinc finger peptide is also quite different from its pH-dependent unfolding [15]. The pH-induced unfolding behavior is fairly simple, with all resonances exhibiting slow exchange, indicating that the folded and unfolded species interconvert slowly at low pH. The pH-dependent unfolding of a zinc finger domain can be described as a two-state process between the zinc-bound, unprotonated, folded form and the zinc-free, protonated, unfolded form. In contrast, the thermal unfolding of a zinc finger peptide does not appear to be a simple two-state process. The complication may arise from an intermediate form that has (partially or completely) unfolded secondary structure, but still has a zinc ion bound. Further analysis should allow us to distinguish between the complex pH-dependent and temperature-dependent unfolding mechanisms of zinc finger domains.

The remarkable heat stability of the zinc finger structure has important implications for protein and peptide modeling and design. It may be possible to engineer small heat and protease-stable intracellular DNA-binding modules using a metal binding site at the core of these domains.

Acknowledgements

This work was supported by NIH 2 PO1 GM32681.

References

1. Miller, J., McLachlan, A.D. and Klug, A., EMBO J., 4 (1985) 1609.
2. Brown, R.S., Sander, C. and Argos, P., FEBS Lett., 186 (1985) 271.
3. Hartshorne, T.A., Blumberg, H. and Young, E.T., Nature, 328 (1986) 443.
4. Rosenberg, U.B., Shröder, C., Preiss, A., Kienlin, A., Côté, S., Riede, J. and Jäckle, J., Nature, 319 (1986) 335.
5. Ruiz i Altaba, A., Perry-O'Keefe, H. and Melton, D.A., EMBO J., 6 (1987) 3065.
6. Page, D.C., Mosher, R., Simpson, E.M., Fisher, E.M.C., Mardon, G., Pollack, J., McGillivray, B., De la Chapelle, A. and Brown, L.G., Cell, 51 (1987) 1091.
7. Blumberg, H., Eisen, A., Sledziewski, A., Bader, D. and Young E.T., Nature, 328 (1987) 443.
8. Thukral, S.K., Tavianini, M.A., Blumberg, J. and Young, E.T., Mol. Cell. Biol., 9 (1989) 2360.
9. Diakun, G.P., Fairall, L. and Klug, A., Nature, 324 (1986) 698.
10. Fairall, L., Rhodes, D. and Klug, A., J. Mol. Biol., 192 (1986) 577.
11. Vrana, K.E., Churchill, M.E.A., Tullius, T.D., Brown, D.D., Mol. Cell. Biol., 8 (1988) 1684.
12. Frankel, A.D., Berg, J. and Pabo, C.O., Proc. Natl. Acad. Sci. U.S.A., 84 (1987) 4841.
13. Párraga, G., Horvath, S.J., Eisen, A., Taylor, W.E., Hood, L., Young, E.T. and Klevit, R.E., Science, 241 (1988) 1489.
14. Klevit, R.E., Horvath, S.J. and Herriott, J.R., (submitted).
15. Párraga, G. Horvath, S.J., Hood, L., Young, E.T. and Klevit, R.E., (submitted).
16. Berg, J., Proc. Natl. Acad. Sci. U.S.A., 85 (1988) 99.
17. Gibson, T.J., Postma, J.P.M., Brown, R.S. and Argos, P., Protein Eng., 2 (1988) 209.
18. Roise, D., Horvath, S.J., Tomich, J.M., Richard, J.J. and Schatz, G., EMBO J., 5 (1986) 1327.
19. Bruist, M.F., Horvath, S.J., Hood, L.E., Steitz, T.A. and Simon, M.I., Science, 235 (1987) 777.
20. Oas, T.G. and Kim, P.S., Nature, 336 (1988) 42.

Hydration of helix backbones of hydrophobic peptides

Isabella L. Karle[a], Judith L. Flippen-Anderson[a], Kuchibhotla Uma[b] and Padmanabhan Balaram[b]

[a]*Laboratory for the Structure of Matter, Naval Research Laboratory, Washington, DC 20375-5000, U.S.A.*
[b]*Molecular Biophysics Unit, Indian Institute of Science, Bangalore, India*

Introduction

Crystal structure analyses have shown that 7–16 residue peptides, composed of only apolar residues and at least one Aib residue, form α-helices or mixed $3_{10}/\alpha$-helices. All of these helical peptides form intermolecular head-to-tail hydrogen bonds directly between NH and CO moieties, or through a water or alcohol molecule intermediary. The columnar surface of the peptide is covered with hydrocarbon side chains, and the peptide columns pack with van der Waals' attractions. In a number of these peptide structures, water has been found in unexpected places, that is, other than the head-to-tail region. Apolar helices have acquired mini areas of polar surface or even amphiphilicity by the action of water. Water molecules have been found to interact with the backbone atoms of an apolar helical molecule in 3 different modes, as illustrated in Fig. 1.

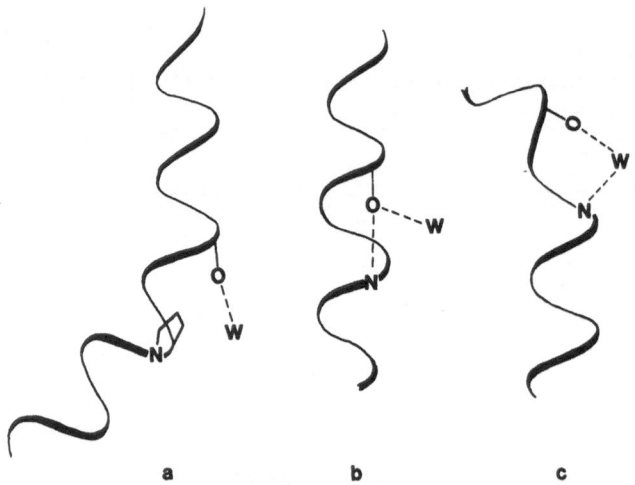

Fig. 1. Three modes of hydrating α-helix backbone atoms (W = water).

Results and Discussion

When a Pro residue is incorporated into an α-helix (Fig. 1a), there is a loss of a hydrogen atom, thereby leaving a carbonyl group without a hydrogen bond. The extra bulk of the pyrrolidine group forces a bend in the helix that exposes the carbonyl group to the outside environment, and allows it to attract a water molecule. The schematic diagram represents the backbone of a 16-residue peptide (actually with 3 Pro residues [1]). A ring of water molecules surrounds the middle of the apolar molecule in the crystal.

The second mode of hydration (Fig. 1b), has been found in the structures of Boc-(Aib-Val-Ala-Leu)$_n$-Aib-OMe, where n = 2 and 3 [2,3]. The α-helix is straight. In one of the normal NH···O=C bonds that is formed in the helix, the carbonyl participates in another hydrogen bond with a water molecule. The surrounding apolar side chains in that part of the peptide are small, or directed away, so that the carbonyl group is exposed to the outside environment and allows the approach of a water molecule.

Figure 1c represents the most intriguing action of water in which a water molecule severs a normal hydrogen bond in an α-helix and inserts itself to form

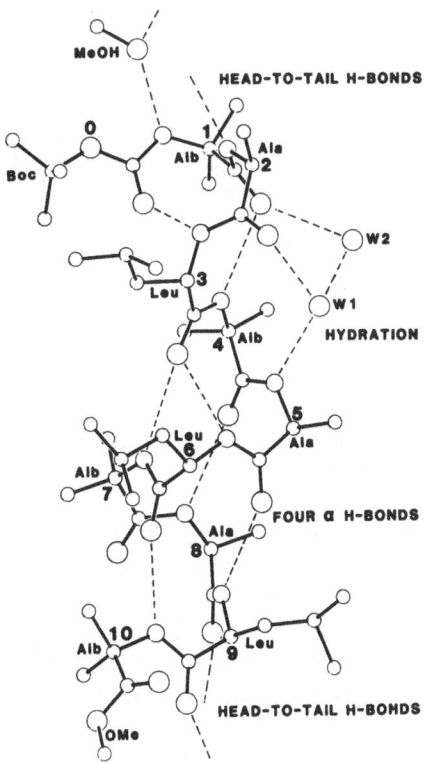

Fig. 2. Water insertion into backbone of Boc-(Aib-Ala-Leu)$_3$-Aib-OMe in the crystal [4].

two new hydrogen bonds to the NH and CO groups. The decapeptide Boc-(Aib-Ala-Leu)₃-Aib-OMe, when crystallized from ethylene glycol, forms a nearly ideal α-helix (to be published). When the solvent is changed to methanol/water, water insertion occurs in the manner shown in Figs. 1c and 2 [4]. The entire distortion of the helix is confined to rotation of the N-C^α and C^α-C' bonds at Leu³ (Fig. 2), where the values of the torsion angles ϕ and ψ become $-102°$ and $+15°$, respectively. In this case, the hydration of the apolar peptide results in an amphiphilic mimic. In the crystal packing, on one side of the peptide, the leucyl residues of one peptide interdigitate with the leucyl residues of a neighboring peptide to form only hydrophobic contacts. On the other side of the peptide, the water molecules and the exposed carbonyl groups form a hydrophilic surface and make hydrophilic contacts with a neighboring peptide molecule.

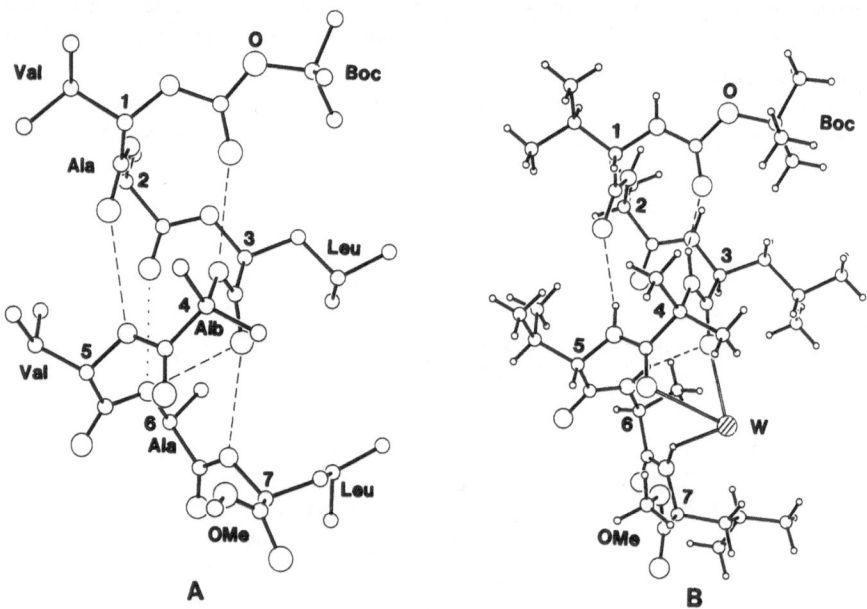

Fig. 3. Two conformers of Boc-Val-Ala-Leu-Aib-Val-Ala-Leu-OMe occurring in the same crystal. Water is inserted into conformer B.

A similar insertion of a water molecule into the backbone of an apolar peptide, but with a hydrophobic environment for the water molecule, is shown in Fig. 3 (in publication). Two different conformers of Boc-Val-Ala-Leu-Aib-Val-Ala-Leu-OMe cocrystallize, side-by-side in a unit cell of the crystal. Molecule A occurs as an α-helix, whereas in molecule B, the N(7)H···O(3) hydrogen bond has been broken by the insertion of water (W) that forms new hydrogen bonds W···O(3) and N(7)···W, 2.77 Å and 2.97 Å, respectively. Again, the entire conformational change for the accommodation of the water molecule occurs solely by rotations at C^α for Val⁵, where $\phi = -91°$ and $\psi = +2°$. These conformers may represent two stages in helix unfolding or folding.

Acknowledgements

This research was supported in part by National Institutes of Health Grant GM-30902, and in part by a grant from the Department of Science and Technology, India.

References

1. Karle, I.L., Flippen-Anderson, J.L., Sukumar, M. and Balaram, P., Proc. Natl. Acad. Sci. U.S.A., 84 (1987) 5087.
2. Karle, I.L., Flippen-Anderson, J.L., Uma, K. and Balaram, P., Int. J. Pept. Prot. Res., 32 (1988) 536.
3. Karle, I.L., Flippen-Anderson, J.L., Uma, K. and Balaram, P., Biochemistry, in press.
4. Karle, I.L., Flippen-Anderson, J.L., Uma, K. and Balaram, P., Proc. Natl. Acad. Sci. U.S.A., 85 (1988) 299.

NMR analysis of structural stability and specific residue conformations in wild-type and mutant *lamB* signal sequences

M.D. Bruch, C.J. McKnight and L.M. Gierasch

Departments of Pharmacology and Biochemistry, University of Texas Southwestern Medical Center at Dallas, 5323 Harry Hines Blvd., Dallas, TX 75235-9041, U.S.A.

Introduction

Signal sequences are relatively short (15–30 amino acids), highly hydrophobic, N-terminal sequences that constitute the most general requirement for protein export and are essential for selective targeting of nascent protein chains [1]. The lack of homology among signal peptides despite performance of similar cellular functions raises the question of the relationship between their primary sequences and their mode of action. In previous work from this and other laboratories, a strong preference for α-helical conformation was associated with functional signal sequences [2]. We now report detailed conformational analysis of synthetic peptides corresponding to the wild-type and several mutant signal sequences of the *lamB* gene product (*lamB*) from *E. coli*. The wild-type signal peptide, MMITLRKLPLAVAVAAGVMSAQAMA, contains a characteristic core of hydrophobic residues (Leu[8]-Ala[16]), which is a hallmark of signal sequences. When four residues (10–13) are deleted from this hydrophobic core, the resultant deletion mutant is completely nonfunctional; no protein is exported. However, partial activity is restored in two peptides that have point substitutions in addition to the deletion of four residues. The first of these pseudo-revertants (R1, G17 → C) enables export of 50% of the protein, and the other revertant (R2, P9 → L) has 90% of the activity of wild-type.

Results and Discussion

CD spectra of the wild-type, deletion mutant and revertant R1 signal peptides in 50/50 TFE/H$_2$O (by volume) at 5, 25, 50°C (Fig. 1) show significant helical content at 25°C. The deletion mutant, which is completely inactive, has a much lower helical content than the wild-type, while the revertant R1, which has 50% of wild-type activity, is intermediate in helical content. In all three peptides, decreased helical content is observed at 50°C, while cooling to 5°C results in an increase in helical content. The isodichroic point observed in each case is indicative of a two-state random coil $\leftrightarrow$ α-helix interconversion.

CD results yield a description of the behavior of the peptides as a whole, but give no insight into the conformational states of specific residues within

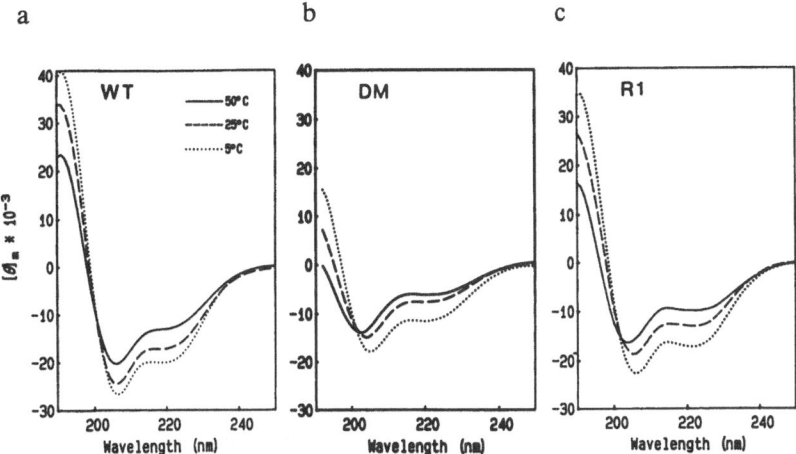

Fig. 1. *CD spectra of* lamB *signal peptides in 19 mol% TFE as a function of temperature; peptide concentration 10μM. (a) wild-type, (b) deletion, (c) revertant R1.*

the peptide. 2D NOESY yields a map of protons that are spatially close irrespective of bonding networks, and the type and pattern of the observed short distances (<4.5A) reveals elements of secondary structure at the level of specific residues. In a regular α-helix: d_{NN} (i, i + 1) ~ 2.8 Å, $d_{αN}$ (i, i + 1) ~ 3.5 Å, $d_{αN}$ (i, i + 3) ~ 3.4 Å, $d_{αN}$ (i, i + 4) ~ 4.2 Å and d_{NN} (i, i + 2) ~ 4.2 Å [3].

The NOESY spectrum of the wild-type signal peptide (Fig. 2a) yields all sequential NH (i)/NH (i + 1) interactions but one between residues 10 and 24, implying a significant population of α-helical conformation in this region. In addition, the interactions observed between the Pro[9] δ-methylene protons and the amide protons of both Leu[8] and Leu[10] suggest that the helix extends from residue 8 to the C-terminal end of the peptide. This interpretation is confirmed by the large number of α(i)/NH (i + 3,4) interactions observed in this region. By contrast, there are no interactions indicative of an α-helix observed in the N-terminal part of the peptide. The observed interactions at 25°C are summarized in Fig. 2b.

There also is a network of weak NH(i)/NH(i + 2) interactions from residues 11–18, but no interactions of this type are observed outside the hydrophobic core. These observations suggest that the helix is more stable in the hydrophobic core, which is confirmed by results obtained at 50°C. Both NH(i)/NH(i + 1) and α(i)/NH(i + 3,4) interactions occur in the region from residues 8–21 at the higher temperature with a significantly greater density of these interactions in the hydrophobic core. At 5°C, some NH(i)/NH(i + 1) interactions are now observed in the N-terminal part of the molecule, but the absence of other interactions expected for a helix in this region shows that the helix is much less stable in the N-terminal region, even at low temperature.

In the deletion mutant, NH(i)/NH(i + 1), interactions are observed from Val[14] (the first residue after Pro) to Ser[20]. The absence of α(i)/NH(i + 3,4) interactions

549

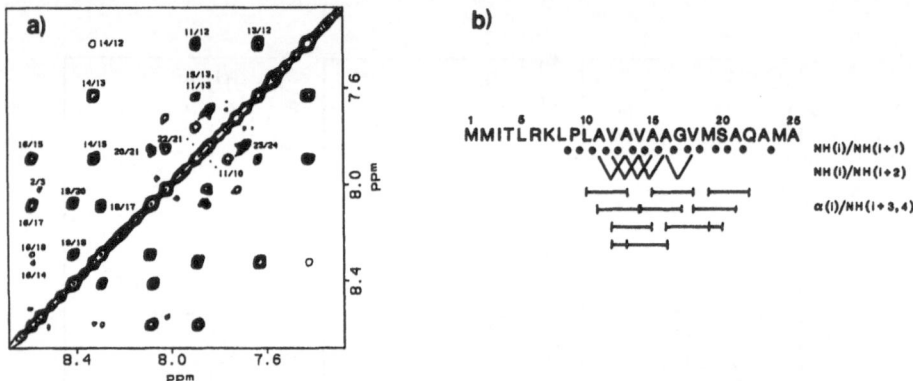

Fig. 2. (a) NH region of the NOESY spectrum of lamB *wild-type signal peptide in 19 mol% TFE at 25°C; peptide concentration 2 mM; mixing time 300 ms. (b) Summary of interactions observed in the NOESY spectra of* lamB *wild-type signal peptide at 25°C.*

suggests that the helix formed in the hydrophobic core of this peptide is only marginally stable, which is supported by the complete loss of structure seen by nuclear magnetic resonance (NMR) at 50°C. It should be noted that the helix from Val[14] to Ser[20] comprises 7 residues, which is approximately the minimal number for helix initiation.

By contrast, the revertant R1 is more similar to the wild-type in its behavior. Most NH(i)/NH(i + 1) interactions are observed from Val[14] to Ala[23], and some α(i)/NH(i + 3,4) interactions also occur in this region. In addition, the Pro[9] δ-methylene protons and the amide protons of both Leu[8] and Val[14] interact. These results suggest that the revertant R1 has a reasonably stable helix from residues 8–23. Unlike data for the deletion mutant, NMR results at 50°C show that R1 has retained some structure at the higher temperature, but less than in the wild-type at 50°C.

Conclusions

Our results show that even in this flexible, linear peptide, helix content can be determined in specific portions of the molecule by NMR. In both the wild-type and the mutant peptides, the most stable part of the helix is in the hydrophobic core. Furthermore, the functional signal peptides have more stable helices than the nonfunctional peptides. Since the existence of a hydrophobic core is a hallmark of functional signal peptides, these results suggest that helix formation in the hydrophobic core may be required for proper function. Finally, our results suggest that proline has a stronger disrupting effect on helix propagation than glycine in these peptides. Glycine appears to be readily accomodated in the helix, whereas proline may destabilize its N-terminal neighbors.

Acknowledgements

Supported in part by grants from the NIH (GM-34962) and the NSF (DCB-8896144). LMG gratefully acknowledges the support of the Robert A. Welch Foundation.

References

1. Gierasch, L.M., Biochemistry, 28 (1989) 923.
2. McKnight, C.J., Briggs, M.S. and Gierasch, L.M., J. Biol. Chem., in press.
3. Wüthrich, K., Billeter, M. and Braun, W., J. Mol. Biol., 180 (1984) 715.

Solution conformation of endothelin

Yuji Kobayashi

Institute for Protein Research, Osaka University, Suita, Osaka 565, Japan

Introduction

An endothelial cell-derived 21-peptide, endothelin, (ET) shows potent and long-lasting vasoconstricting activity [1,2]. This activity, related to a cardiovascular control system, has triggered the curiosity of biochemists and endocrinologists. Unusual in endothelin are the two pairs of disulfide bridges in its rather short chain length. Bioactive peptides, of mammalian origin, do not usually have such a high content of intramolecular disulfide bridges. But some patterns in amino acid sequences similar to that of ET have been found in toxic peptides produced by animals of lower orders.

The amino acid sequences of such homologous peptides are shown in Fig. 1. They reveal that four Cys residues are located in identical positions with two kinds of disulfide bridge formation. Among them, the tertiary structures of apamin [3–5] and variant-3 scorpion toxin [6] have been determined by NMR and X-ray analyses. It is interesting to compare the conformations of ET with that of these peptides having different activities to elucidate the SAR.

The structure determination of porcine (human) ET in solution was done by the combined use of NMR measurements and our distance geometry algorithm [7].

Results and Discussion

ET was synthesized and purified as previously reported [8] to provide a sufficient amount for NMR measurements. The solvent was a 10% aqueous acetic acid solution and the concentration of solute was 5 mM. Sedimentation equilibrium experiments were carried out and showed that ET exists in a monomer state under the conditions of the NMR measurements.

Two-dimensional NMR spectra of NOESY, DQF-COSY and HOHAHA were recorded at 500 MHz. Whole peak assignments of the spectra were done by the sequence specific assignment procedure introduced by Wüthrich [9].

The regular secondary structure elements can be elucidated by analyzing the sequential and medium range NOEs. By investigating the NOEs, as found in the NOESY spectra of ET and summarized in Fig. 2, α-helical conformation, indicated by $d_{\alpha N}(i, i+3)$ and $d_{\alpha\beta}(i, i+3)$, was revealed in the region from Lys^9 to Cys^{15}.

The distance geometry algorithm used to calculate the tertiary structure of ET consists of constructing an initial conformation with random dihedral angles

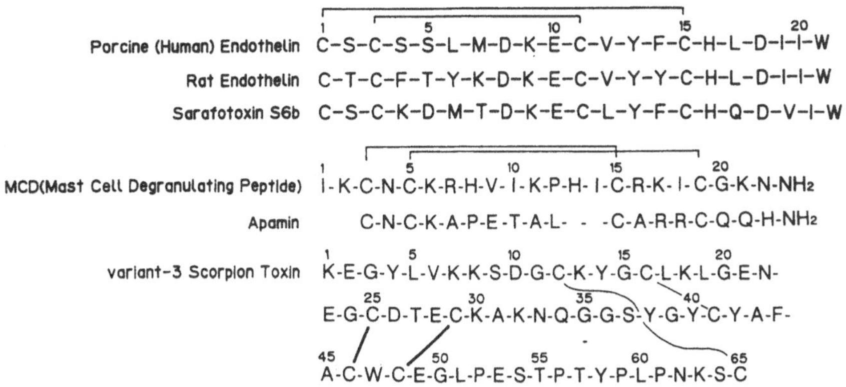

Porcine (Human) Endothelin C-S-C-S-S-L-M-D-K-E-C-V-Y-F-C-H-L-D-I-I-W

Rat Endothelin C-T-C-F-T-Y-K-D-K-E-C-V-Y-Y-C-H-L-D-I-I-W

Sarafotoxin S6b C-S-C-K-D-M-T-D-K-E-C-L-Y-F-C-H-Q-D-V-I-W

MCD(Mast Cell Degranulating Peptide) I-K-C-N-C-K-R-H-V-I-K-P-H-I-C-R-K-I-C-G-K-N-NH₂

Apamin C-N-C-K-A-P-E-T-A-L- - -C-A-R-R-C-Q-G-H-NH₂

variant-3 Scorpion Toxin K-E-G-Y-L-V-K-K-S-D-G-C-K-Y-G-C-L-K-L-G-E-N-

E-G-C-D-T-E-C-K-A-K-N-Q-G-G-S-Y-G-Y-C-Y-A-F-

A-C-W-C-E-G-L-P-E-S-T-P-T-Y-P-L-P-N-K-S-C

Fig. 1. Amino acid sequences of porcine ET, rat porcine ET, sarafotoxin[12], MCD (mast cell degranulating peptide)[13], apamin and variant-3 scorpion toxin.

and minimizing the sum of the squared differences between mutual atomic distances in the calculated conformation, and the corresponding distances obtained by the interpretation of NOE intensities. The details of this procedure and examples of its application have been given elsewhere [7,10].

Porcine (Human) Endothelin: .

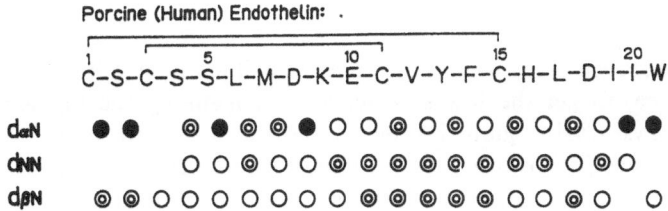

Fig. 2. Amino acid sequence of porcine ET with a summary of sequential assignments. Sequential connectivities $d_{\alpha N}$, d_{NN} and $d_{\beta N}$ manifested by strong, medium and weak NOESY cross peaks are indicated by filled, doubled and open circles, respectively.

Starting from a hundred initial conformations, chosen individually and randomly, an assembly of a hundred conformers was obtained by repetition of the calculations. Five conformers best satisfying the distance constraints experimentally deduced from NOEs were selected for analysis of convergence. The root mean square distance (rmsd) value of the backbone atoms among these five conformers was 4.1 Å which indicates rather poor convergence. But the analysis of rmsd values of individual parts of the peptide demonstrated that convergences are good in three regions. These are the Cys^1-Ser^4, Ser^5-Asp^8 and Lys^9-Cys^{15} regions in which averaged rmsd values are 1.1, 0.9 and 1.0Å, respectively. The backbone structures of these three regions are illustrated, respectively, in Fig. 3.

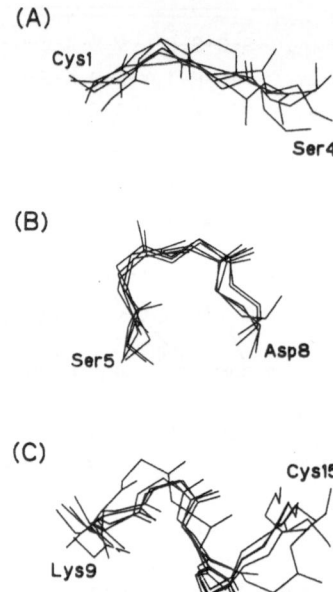

(A)

Cys1

Ser4

(B)

Ser5   Asp8

(C)

Cys15

Lys9

Fig. 3. *Backbone folding of five calculated ET structures superimposed in the region (A) Cys¹-Ser⁴, (B) Ser⁵-Asp⁸ and (C) Lys⁹-Cys¹⁵ by least-squares fit of backbone atoms, respectively.*

The convergence of the whole molecule is also illustrated in Fig. 4, by arranging the backbone, to get the lowest value of rmsd among the segment of Cys¹-Cys¹⁵. The Lys⁹-Cys¹⁵ region showed a typical helical conformation with rather nice convergence. This is consistent with the result of the sequential NOE analysis. Furthermore, the helical structure in this region was confirmed by the deuteron

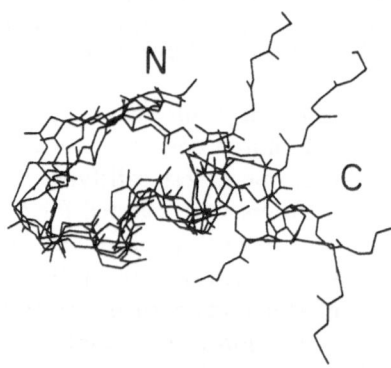

N

C

Fig. 4. *Backbone folding of five calculated endothelin structures superimposed in the region between Cys¹ and Cys¹⁵ by least-squares fit of backbone atoms.*

exchange rate measurements of NH protons in D_2O solution. The Ser[5]-Asp[8] region also appears well converged and shows Type I β-turn structure. The Cys[1]-Ser[4] region appears not very well converged, but still indicates some defined secondary structure.

Conformational resemblances were observed among the structures of apamin, scorpion-toxin and ET. Those are found especially in helical regions where Ala[9]-Gln[17], Glu[23]-Lys[32] and Lys[9]-Cys[15] appear, respectively. It is quite interesting that, whichever amino acid X is (Fig. 5), these helical cores are located in the characteristic sequence Cys-X-X-X-Cys, which is linked through two Cys residues to the other two Cys in the sequence Cys-X-Cys. Zell et al. [6] have recognized such a pattern between apamin and scorpion toxin in spite of the different arrangements of the Cys residues as shown in Fig. 5A and 5B, respectively. Here we recognize the same pattern in ET with another type of arrangement of the Cys residues as shown in Fig. 5C. We could expect a helical conformation of the same pattern in another possible arrangement (shown in Fig. 5D) which has not been found yet. We conclude that such a disulfide bridging pattern of four Cys residues should be a motif to promote a helical formation regardless of the differences among the topological arrangements shown in Fig. 5.

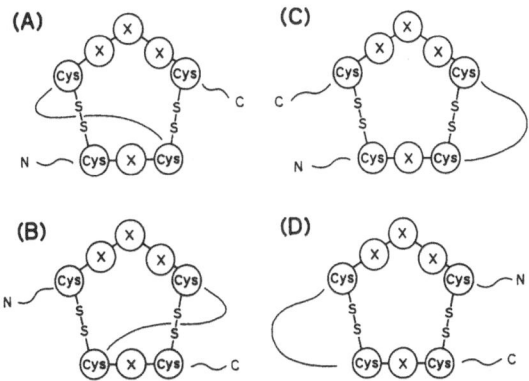

Fig. 5. Schematic drawing of four kinds of disulfide bridging patterns.

We have synthesized an analog of ET with a different pattern of disulfide formation with Cys[1]-Cys[11] and Cys[3]-Cys[15] as shown in Fig. 5A, and compared its activity with that of ET [11]. The activity of the analog was less than 1% of that of native ET. The conformational analysis showed that this analog takes a helical conformation in almost the same region as that of native ET.

It is interesting that even though there are large differences among the biological activities of these peptides with such a helical core, all of them are considered to be related to ion channels.

References

, 1. Yanagisawa, M., Kurihara, H., Kimura, S., Tomobe, Y., Kobayashi, M., Mitsui, Y., Yazaki, Y., Goto, K. and Masaki, T., Nature, 332 (1988) 411.
2. Yanagisawa, M., Inoue, A., Ishikawa, T., Kasuya, Y., Kimura, S., Kumagaye, S., Nakajima, K., Watanabe, T.X., Sakakibara, S., Goto, K. and Masaki, T., Proc. Natl. Acad. Sci. U.S.A., 85 (1988) 6964.
3. Bystrov, V.F., Okhanov, V.V., Miroshnikov, A.I. and Ovchinnikov, Y.A., FEBS Lett., 119 (1980) 113.
4. Wemmer, D. and Kallenbach, N.R., Biochemistry, 22 (1983) 1901.
5. Pease, J.H.B. and Wemmer, D., Biochemistry, 27 (1988) 8491.
6. Zell, A., Ealick, S.E. and Bugg, C.E., In Bradshaw, R.A. and Tang, J. (Eds.) Molecular Architecture of Proteins and Enzymes, Academic Press, New York, 1985, p. 65.
7. Ohkubo, T., Kobayashi, Y., Shimonishi, Y., Kyogoku, Y., Braun, W. and Go, N., Biopolymers, 25 (1986) S 123.
8. Kumagaye, S., Kuroda, H., Nakajima, K., Watanabe, T.X., Kimura, T., Masaki, T. and Sakakibara, S., Int. J. Pept. Prot. Res., 32 (1988) 519.
9. Wüthrich, K., NMR of Proteins and Nucleic Acids, John Wiley & Sons, New York, 1986.
10. Kobayashi, Y., Ohkubo, T., Kyogoku, Y., Koyama, S., Kobayashi, M. and Go, N., J. Biochem., 104 (1988) 322.
11. Nakajima, K., Kumagaye, S., Nishio, H., Kuroda, H., Watanabe, T.X., Kobayashi, Y., Tamaoki, H., Kimura, T. and Sakakibara, S., J. Cardiovasc. Pharmacol., 13 (Suppl. 5) (1988) S8.
12. Takasaki, C., Tamiya, N., Bdolah, A., Wollberg, Z. and Kochva, E., Toxicon, 26 (1988) 543.
13. Gauldie, J., Hanson, J.M., Rumjanek, F.D., Shipolini, R.A. and Vernon, C.A., Eur. J. Biochem., 83 (1978) 405.

Independent folding of domains and subdomains of thermolysin

Angelo Fontana

*Department of Organic Chemistry, Biopolymer Research Centre of CNR,
University of Padua, Via Marzolo 1, I-35131 Padua, Italy*

Introduction

In recent years, emphasis has been given to the fact that globular proteins are assemblies of compactly folded substructures, usually called domains. These domains are almost invariably seen with proteins made up of more than about 100 amino acids, and often their existence can be recognized simply by visual inspection of 3D models of protein molecules [1,2], or by the aid of computer alogorithms utilizing the Cα-coordinates given by X-ray analyses [3–11]. These algorithms, based on principles such as interface area minimization, plane cutting, clustering, distance mapping, specific volume minimization and compactness, allow a description of globular proteins in terms of a 'hierarchic' architecture ranging from elements of secondary structure (helices, strands), subdomains (supersecondary structures or folding units), domains, and whole protein molecule [3,7].

Wetlaufer [1] was first to emphasize the structural role of protein domains and, in addition, to propose that these protein substructures could represent intermediates in the folding process of globular proteins. It is conceivable that specific segments of an unfolded polypeptide chain first refold to individual domains, that then associate and interact to give the final tertiary structure of the globular protein, much the same as do subunits in oligomeric proteins. The major implication of this model of protein folding by a mechanism of modular assembly is that isolated protein fragments corresponding to domains in the intact protein are expected to fold into a native-like structure, independently from the rest of the polypeptide chain, thus resembling, in their properties, a small globular protein. Indeed, independent folding of relatively large protein fragments excised from the parent protein by limited proteolysis or chemical cleavage has been demonstrated in a number of cases [12].

In recent years, we have addressed the question of properties of protein domains and subdomains, examining the folding and stability characteristics of fragments of the thermolysin molecule, a well-characterized metallo-protease isolated from *B. thermoproteolyticus* [13–15]. This protein is a single polypeptide chain of 316 amino acid residues without thiol or disulfide groups, and containing a functional zinc ion and four calcium ions. The 3D structure of thermolysin was determined at 1.6 Å resolution [15], allowing establishment that this protein is constituted by two structural domains of equal size, with the active site located at the interface between them [1,14]. Moreover, computer algorithms have been used by several investigators to describe the hierarchic structural organization of the thermolysin molecule, identifying the size and location of its domains and subdomains [3,5–8, 10]. On this basis, it appears that the thermolysin molecule is a most suitable model to test the general hypothesis of an independent folding of protein fragments corresponding to domains and, perhaps, subdomains.

Results and Discussion

Cyanogen bromide cleavage of thermolysin at the level of the two methionine residues in position 120 and 205 of the polypeptide chain produces fragments 1–120, 121–205 and 206–316 [13,16]. By using a low amount of reagent, and shorter reaction times it was also possible to prepare the 'overlapping' fragments 1–205 and 121–316 [16,17]. All these fragments have been isolated to homogeneity, and their conformational properties investigated, but the most detailed studies were carried out on fragment 121–316 [18], entirely comprising the 'all-α' COOH-terminal domain 158–316, as well as fragment 206–316 [16,19–21]. Both fragments are able to refold in aqueous solution at neutral pH into a structure of native-like characteristics, as judged from CD measurements, and immunochemical properties using rabbit anti-thermolysin antibodies [22]. The figures of α-helix content of fragment 121–316 and 206–316, calculated from far-ultraviolet CD spectra, were in essential agreement with those expected for a native-like structure of the fragments, and calculated from the known 3D structure of thermolysin (see Table 1). Furthermore, the fact that both fragments recognize anti-thermolysin antibodies, as observed by precipitin analysis and competitive binding experiments [22], are indicative of a close structural relationship between fragments and intact native thermolysin. In fact, antibodies elicited towards a native globular protein are specific for antigenic determinants that are located in the more exposed regions of the protein molecule, and thus able to probe similarities of important details (loops, corners) of the 3D structure in the isolated fragments and the native parent protein [23].

The independent folding of fragment 206–316 correlates well with predictions of location of subdomains in the COOH-terminal structural domain (sequence 158–316) of the thermolysin molecule [3,5–8, 10], since, for example, a subdomain comprising residues 212–316 was computed by interface area scans [6]. Rashin [5,24] computed the location of *stable* subdomains with native-like conformation in the COOH-terminal portion of the thermolysin molecule on the basis of surface area measurements, and predicted that subfragments of the cyanogen bromide fragment 206–316 should have a good chance to show independent folding. The calculated stabilities of these subfragments, starting at the COOH-terminus of the fragment molecule are shown in Fig. 1 (all fragments corresponding to stability minima include the COOH-terminus). Thus, these calculations predict that fragments (approximately) 214–316, 235–316, and 255–316 should be able, in

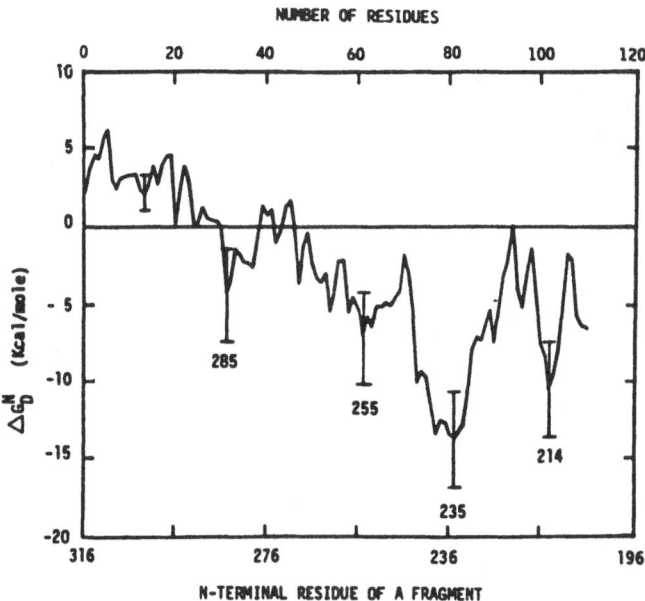

Fig. 1. Calculated stabilities of subfragments of the cyanogen bromide fragment 206–316 including the COOH-terminus of the thermolysin molecule. Numbers indicate NH_2-terminal residues of subfragments corresponding to stability minima. Vertical bars indicate standard errors (taken from Ref. 24).

aqueous solution, to acquire a stable and native-like structure. Shorter COOH-terminal fragments of 20–30 residues are, instead, predicted not to fold.

In order to study the effect of polypeptide chain length on fragment folding and stability, fragment 206–316 was selectively cleaved with hydroxylamine at the level of the peptide bond Asn[227]-Gly[228], and from the reaction mixture, homogeneous fragment 228–316 was isolated [25]. In agreement with predictions (Fig. 1), removal of a 22-residue segment from the 111-residue chain of fragment 206–316 does not impair the stable and native-like fragment fold [25]. However, since the sizes of both fragment 206–316 and 228–316 were dictated by the location of specific amino acid residues sensitive to specific fragmentation by chemical reagents (cyanogen bromide, hydroxylamine), the use of limited proteolysis was anticipated as a most suitable approach in order to simplify fragment 206–316. The idea was that proteolytic enzymes could easily remove chain segments that are flexible and exposed and thus, easily attacked by the protease, leaving in solution the more rigid, compact and folded part of fragment 206–316, and thus, more resistant to further proteolysis (see Ref. 26 for a discussion on the molecular mechanism of limited proteolysis of globular proteins).

Experiments of limited proteolysis of fragment 206–316 were carried out using four different proteases (trypsin, chymotrypsin, thermolysin and subtilisin) [27]. In each case it was found that the fragment is easily digested, but a fragment of approximate molecular weight of 6–7000 Da was always observed when aliquots of the individual reaction mixtures were analyzed by SDS-PAGE. This was taken as a clear indication that a 'core' folded fragment, and thus resistant to further proteolysis, was produced by proteolysis from the 111-residue chain of fragment 206–316. The majority of fragments produced by proteolysis of fragment 206–316 were isolated and characterized, allowing establishment of the sites of proteolytic cleavage along the polypeptide chain 206–316. Figure 2 shows location of the sites of cleavage in the 3D structure of this fragment (assumed in a native-like conformation as in intact thermolysin). Of note, only the NH$_2$-terminal portion of the fragment is attacked by the four different proteases, leaving in solution a 'core' fragment encompassing the three COOH-terminal helices. A peculiar aspect shown in Fig. 2 is that subtilisin, thermolysin and chymotrypsin, despite their broad substrate specificity, cleave at the same segment 250–255 corresponding to an exposed loop.

From the subtilisin digest of fragment 206–316, the 'core' fragment was isolated by chromatography and shown to correspond to sequence 255–316 [27]. The conformational and stability properties of this fragment were investigated using CD and immunochemical measurements, allowing establishment of its folded and native-like structure (see figures for its helical content in Table 1) [27], signifying that indeed it is possible to isolate stable supersecondary structures (subdomains or folding units) from globular proteins of native-like characteristics. Moreover, as expected from typical properties of globular proteins, fragment 255–316 shows cooperative and reversible thermal unfolding, with a T_m 65°C (conc. 0.1 mg/ml). This thermal stability is quite remarkable, if one considers the small size of the fragment, being a simple polypeptide chain of 62 amino acid residues, lacking disulfide bridges, cofactors, strongly bound ions, all characteristics well known to contribute significantly to the folding and stability of protein structures.

From the calculated stabilities of subfragments of the cyanogen bromide fragment 206–316 shown in Fig. 1, it can be anticipated that a COOH-terminal fragment of ~20 residues, and comprising the COOH-terminal helix (see Fig. 2), should not be able to maintain a stable native-like fold in aqueous solution. To test this hypothesis, a 21-residue peptide corresponding to sequence 296–316 of thermolysin, and thus encompassing entirely the COOH-terminal helical segment 301–312 of the native protein, was synthesized by solid-phase methods and purified to homogeneity by RPHPLC. The homogeneous peptide 296–316 was cleaved with trypsin at Lys[307] and with *Staphylococcus aureus* V8 protease at Glu[302], producing the additional fragments 296–307, 308–316, 296–302 and 303–316 (see Fig. 3). Far-ultraviolet CD measurements were carried out on all these peptides, establishing that they are unable to acquire a helical conformation in aqueous solution to a measurable extent and thereby, an independent folding to a stable native-like structure. CD spectra were those of typical random-coil polypeptides, with figures for $[\theta]_{222\,nm}$ ranging from –500 to –2000 deg cm^2 dmol^{-1} [31]. On the other hand, significant helicity can be induced by dissolving the peptide 296–316 in aqueous TFE or ethanol solution. Considering that the curves

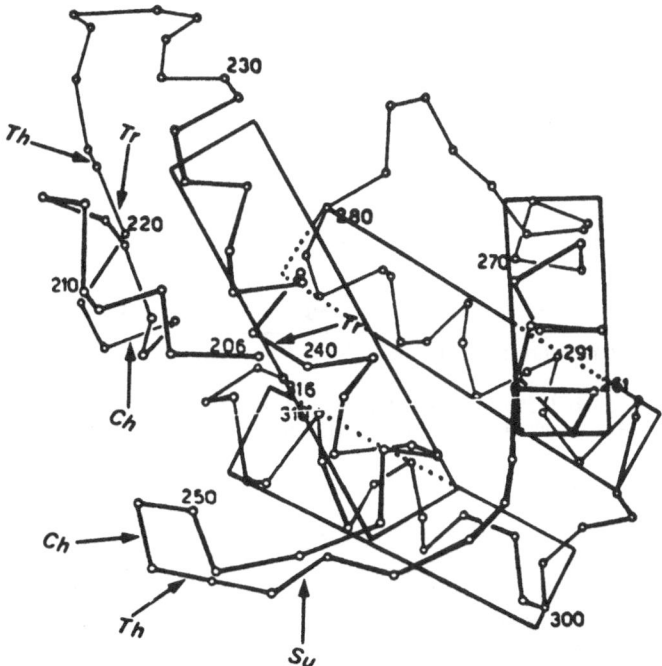

Fig. 2. Polypeptide backbone conformation of the cyanogen bromide fragment 206–316 in the intact thermolysin (adapted from Ref. 14). The four helices are indicated. Arrows indicate sites of proteolytic cleavage by trypsin (Tr), chymotrypsin (Ch), thermolysin (Th) and subtilisin (Su). These sites have been deduced by isolation and characterization of the individual peptide fragments obtained by limited proteolysis [27].

of CD spectra obtained by increasing TFE or ethanol concentration show an isodichroic point at 203 nm, and that the spectrum observed in the presence of 97% alcohol is typical of α-helical polypeptides with minima near 208 and 222 nm [32], it can be concluded that alcohol induces a two-state random to helix conformational transition of peptide 296–316. From the figures listed in Table 1, it is seen that this peptide attains in 97% TFE the same amount of helix of the corresponding chain segment in native thermolysin. The structure-promoting effect of alcohol is also seen with subfragments of peptide 296–316 (see Fig. 3), but the most significant conformational changes (induction of helicity) are observed with peptide SP2(303–316) and T1(296–307) only, i.e., with sub-fragments encompassing at least, in part, the helical segment 301–312 of the intact protein, whereas with subfragments SP1(296–302) and T2(308–316), ethanol has marginal effect. The far-ultraviolet CD spectra of subfragments of peptide 296–316 dissolved in aqueous ethanol were evaluated in terms of content of secondary structure, and the figures for the per cent helical structure are given in Table 1. However, these estimates could be problematic in the present case, since the reference CD spectra for helix and unordered form in aqueous alcohol

Table 1 *Helical content of COOH-terminal fragments of thermolysin as determined from far-ultraviolet CD spectra[a]*

Fragment	Number of residues	Solvent[b]	Per cent helix		Reference
			X-ray[c]	CD[d]	
121–316	196	A	45	41	[18]
206–316	111	B	49	45	[16]
228–316	89	B	62	53	[25]
255–316	62	B	69	63; 77[e]	[27]
296–316	21	C	57	59[f]	[31]
303–316	14	D	71	47[f]; 63[g]	[31]
296–307	12	D	58	32[f]; 42[g]	[31]

[a] CD measurements were carried out at 20–22°C at peptide concentrations of 0.05–0.1 mg/ml.

[b] A, 10 mM Tris–HCl/5 mM CaCl$_2$ buffer, pH 7.5; B, 10 mM Tris–HCl/0.1 M NaCl buffer, pH 7.5; C, 97% TFE in 5 mM Tris–HCl/0.1 M NaCl buffer, pH 7.5; D, 97% ethanol in Tris/0.1 M NaCl buffer, pH 7.5.

[c] Calculated on the basis of the crystallographically determined structure of thermolysin [14].

[d] Calculated using the method of Chen et al. [28].

[e] Calculated using the method of Siegel et al. [29].

[f] Calculated using the formula of Wu et al. [30]: $f_H \times 100 = ([\theta]_{222\,nm} + 2000)/([\theta]_H + 2000)$, where f_H represents the fraction of helix and $[\theta]_H$ the reference $[\theta]_{222\,nm}$ for the helix, taken as –28 400 deg cm^2 dmol^{-1} for a helix having an average length of 10 residues.

[g] Calculated taking as reference $[\theta]_{222\,nm} = -21\,800$ deg cm^2 dmol^{-1} for a helix having an average length of 6 residues [30].

is not firmly established, and since the CD signal in the far-ultraviolet region is strongly influenced by the length of helical segments [28]. In the present case, the helix content was calculated from CD data, taking as reference $[\theta]_{222\,nm}$ for the helix, the figure given for a helical segment of average length of both 10 and 6 residues [30]. The figures for per cent helical content of fragments SP2 and T1, thus estimated (see Table 1), are lower than those predicted from X-ray data, perhaps due to end-effects [33]. At any rate, the ranking of CD-

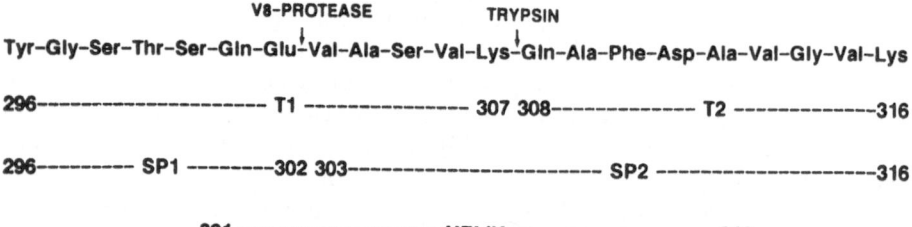

Fig. 3. Amino acid sequence of peptide 296–316 of thermolysin prepared by solid-phase methods. Cleavages at Lys[307] with trypsin and at Glu[302] with V8-protease produce peptides T1 and T2 and peptides SP1 and SP2, respectively. The helical segment 301–312 in native thermolysin [14, 15] is also indicated.

calculated figures is % helix of SP2 > % helix T1, as predicted from X-ray data.

Inspection of the 3D model of thermolysin reveals that the COOH-terminal helix 301–312 runs almost parallel to helix 281–296, and that the interface between the two helices is characterized by a cluster of adjacent hydrophobic residues. It appears, therefore, that in the absence of this hydrophobic interaction, peptide 296–316 is unable to attain a helical structure, whereas in a more hydrophobic environment (such as aqueous TFE and ethanol), likely amphiphilic helix can be formed and stabilized. In addition, the C-terminal helix 301–312 in the intact protein is linked to the nearby helical segments by two salt bridges, namely Glu^{302}...Lys^{262} and Asp^{311}...Arg^{285} [14]. Likely, absence of suitable hydrophobic, as well as charge–charge interactions, could destabilize the helical secondary structure in the isolated fragment 296–316 in aqueous solution.

The fact that the isolated COOH-terminal helix of thermolysin is not stable in aqueous solution is in agreement with predictions of helix stability based on the Zimm–Bragg equation [34], and host–guest data [35] that indicate that helix formation should be negligible for all short peptides. However, it cannot be excluded that peptide 296–316, or some of its subfragments, could attain, in aqueous solution, a conformational ensemble of 'nascent helices' of extremely low stability. It could well be that by using NMR, a more powerful analytical technique than CD, it will be possible to identify such transient secondary structures. In this ambit, mention should be given to recent studies showing that the 'nascent helices' of the myohemerythrin C-helix peptide [36] or of protein fragments of ribonuclease [37,38] that were identified by NMR could not be detected by CD measurements (see Wright et al. [39] for references and a recent discussion on folding of small protein fragments in aqueous solution).

Conclusion

In summary, the results of the experimental studies of folding of thermolysin fragments, using classical methods of protein chemistry (e.g., fragmentation), indicate that there is a minimum size of a fragment of the COOH-terminal large structural domain of thermolysin (second half of the molecule) capable of folding by itself into a stable and native-like structure, and that this folded 'core' is a polypeptide chain of ~60 residues comprising the three COOH-terminal helices (see Fig. 2). The stable and native-like folding of this thermolysin fragment was predicted by computing the buried surface area of amino acid residues [24]. At the same time, from these calculations it was inferred that the synthetic peptide 296–316 should be unable to fold, as experimentally substantiated. In this context, it is also of interest to recall the prediction advanced by Dill [40], based on statistical dynamics calculations, that there is an optimum chain length for globularity and stability of protein molecules, and that polypeptide chains of less than 70 residues should not fold. This chain length of 50–70 residues is also that required for attainment of a functional property, such as binding of cofactor, a metal ion, etc. (functional modules) [41]. Thus, overall, the experimental results on folding and stability properties of COOH-terminal fragments of thermolysin of different chain length are in line with these predictions. As a final comment, it should be noted that several experiments described here were prompted by the results and models provided by theoretical studies. This kind of interaction between experimentalists and theoreticians is likely vital for evaluating more fully the rules governing the hierarchic 3D organization of globular proteins.

Acknowledgements

This work was supported in part by the Commission of the European Communities, Biotechnology Action Program (contract BAP-0249-I). The author wishes to thank his coworkers who contributed to the various papers reviewed here and, in particular, D. Dalzoppo, V. De Filippis, C. Vita and M. Zambonin.

References

1. Wetlaufer, D.B., Proc. Natl. Acad. Sci. U.S.A., 70 (1973) 697.
2. Rossmann, M.G. and Liljas, A., J. Mol. Biol., 85 (1974) 177.
3. Crippen, G.M., J. Mol. Biol., 126 (1978) 315.
4. Rose, G.D., J. Mol. Biol., 134 (1979) 447.
5. Rashin, A.A., Nature, 291 (1981) 85.
6. Wodak, S.J. and Janin, J., Biochemistry, 20 (1981) 6544.
7. Janin, J. and Wodak, S.J., Prog. Biophys. Molec. Biol., 42 (1983) 21.
8. Sander, C., In Balaban, M. (Ed.) Structural Aspects of Recognition and Assembly in Biological Macromolecules, Balaban Int. Science Services, Rehovot, Israel, 1981, p. 183.
9. Zehfus, M.H. and Rose, G.D., Biochemistry, 25 (1986) 5759.
10. Zehfus, M.H., Proteins, 2 (1987) 90.
11. Kikuchi, T., Nemethy, G. and Scheraga, H.A., J. Protein Chem., 7 (1989) 427.
12. Wetlaufer, D.B., Adv. Protein Chem., 34 (1981) 61.
13. Titani, K., Hermodson, M.A., Ericsson, L.H., Walsh, K.A. and Neurath, Biochemistry, 11 (1972) 2472.
14. Colman, P.M., Jansonius, J.M. and Matthews, B.W., J. Mol. Biol., 70 (1972) 701.
15. Holmes, M.A. and Matthews, B.W., J. Mol. Biol., 160 (1982) 623.
16. Vita, C., Fontana, A., Seeman, J.R. and Chaiken, I.M., Biochemistry, 18 (1973) 3023.
17. Vita, C., Dalzoppo, D., Patti, S. and Fontana, A., Int. J. Pept. Prot. Res., 24 (1984) 104.
18. Vita, C., Dalzoppo, D. and Fontana, A., Int. J. Pept. Prot. Res., 21 (1983) 49.
19. Vita, C. and Fontana, A., Biochemistry, 21 (1982) 5196.
20. Fontana, A., Vita, C. and Chaiken, I.M., Biopolymers, 22 (1983) 69.
21. Dalzoppo, D., Vita, C. and Fontana, A., Biopolymers, 24 (1985) 767.
22. Vita, C., Fontana, A. and Chaiken, I.M., Eur. J. Biochem., 151 (1985) 191.
23. Crumpton, M.J., In Sela, M. (Ed.) The Antigens, Vol. 2, Academic Press, New York, 1974, p. 1.
24. Rashin, A.A., Biochemistry, 23 (1984) 5512.
25. Vita, C., Dalzoppo, D. and Fontana, A., Biochemistry, 23 (1984) 5512.
26. Fontana, A., In Kotyk, A., Skoda, J., Paces, V. and Kostka (Eds.) Highlights of Moderm Biochemistry, Vol. 2, VSP, Utrecht, The Netherlands, 1989, p. 1711.
27. Dalzoppo, D., Vita, C. and Fontana, A., J. Mol. Biol., 182 (1985) 331.
28. Chen, Y.-H., Yang, J.T. and Chau, K.H., Biochemistry, 13 (1974) 3350.

29. Siegel, J.B., Steinmetz, W.F. and Long, G.L., Anal. Biochem., 104 (1980) 160.
30. Wu, C.-S.C.,Ikeda, K. and Yang, J.T., Biochemistry,20 (1981) 566.
31. Vita, C., Dalzoppo, D., De Filippis, V., Longhi, R., Manera, E., Pucci, P. and Fontana, A., Int. J. Pept. Prot. Res., (1989) submitted.
32. Greenfield, N.J. and Fasman, G.D., Biochemistry, 8 (1969) 4108.
33. Hol, W.G.J., Prog. Biophys. Molec. Biol., 45 (1985) 149.
34. Zimm, B.H. and Bragg, J.K., J. Chem. Phys., 31 (1959) 526.
35. Sueki, M., Lee, S., Powers, S.P., Denton, J.B., Konishi, Y. and Scheraga, H.A., Macromolecules, 17 (1984) 148.
36. Dyson, H.J., Rance, M., Houghten, R.A., Lerner, R.A. and Wright, P.E., J. Mol. Biol., 201 (1988) 201.
37. Jimenez, M.A., Rico, M., Herranz, J., Santoro, J. and Nieto, J.L., Eur. J. Biochem., 175 (1988) 101.
38. Jimenez, M.A., Nieto, J.L., Herranz, J., Rico, M. and Santoro,J., FEBS Lett., 221 (1987) 320.
39. Wright, P.E., Dyson, H.J. and Lerner, R.A., Biochemistry, 27 (1988) 7167.
40. Dill, K.A., Biochemistry, 24 (1985) 1501.
41. Traut, Th.W., Mol. Cell. Biochem., 70 (1986) 3.

565

Synthetic α-helical model proteins: Contribution of hydrophobic residues to protein stability

P.D. Semchuk, C.M. Kay and R.S. Hodges

Department of Biochemistry and The Medical Research Council of Canada Group in Protein Structure and Function, University of Alberta, Edmonton, Alberta, Canada T6G 2H7

Introduction

Proteins, as we all know, are complex molecules, and it is difficult to determine the features in their sequences that are responsible for protein folding and protein stability. Our approach to this problem was to carry out the de novo design and synthesis of very simplistic model proteins [1–4] meeting the following requirements. The protein must have a defined secondary, tertiary and quaternary structure, with only one type of secondary structure present. The secondary structure should be stable in aqueous solution at neutral pH and easily monitored by such techniques as circular dichroism. The protein must have sufficient stability to tolerate a wide range of amino acid substitutions, and still be of appropriate size to allow the easy chemical synthesis of a large number of analogs by SPPS. Having selected the α-helix, a minimum of two interacting α-helices would be required to introduce tertiary and quaternary structure into the molecule. The simplest model protein would then be a two-stranded α-helical coiled-coil. The advantages of studying coiled-coils are many. There is no other type of structure present, other than the α-helix, to complicate interpretation of the results, that is, no turns, β-sheet, or undefined structure. They are highly charged, soluble and stable molecules in aqueous conditions of neutral pH. The hydrophobic side-chain interactions that stabilize α-helices in coiled-coils, stabilize α-helices in globular proteins as well.

Results and Discussion

The amino acid sequence of the 70-residue disulfide-linked two-stranded α-helical coiled-coil is shown in Fig. 1. The major factor contributing to the folding and stabilization of proteins is the burying of hydrophobic side chains. Coiled-coils are no exception, with hydrophobic residues occurring between the two α-helices as a result of a repeating pattern, X-N-X-X-N-X-X-X-N-X-X-N-X-X-X-N ..., where N is a non-polar residue. This hydrophobic repeat, first identified by Hodges et al. [5], if often referred to as a 3–4 repeat, and results in a coiled-coil structure as first described by Crick [6] (the α-helices are in-register; one α-helix is rotated 180° to the other; the two α-helices are tilted at an angle near 20° to one another, and a small deformation along their entire length results

SEQUENCE

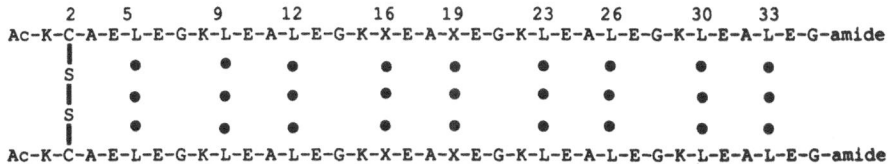

Fig. 1. *The amino acid sequence of the synthetic 70-residue disulfide-linked model protein. The L_o-protein refers to the sequence where position X is occupied by leucine residues. Similarly, I_o, V_o, F_o, Y_o and A_o denote the substitution of leucine residues at positions 16 and 19 in each chain (position X) by isoleucine, valine, phenylalanine, tyrosine and alanine, respectively. The subscript o refers to the oxidized protein (disulfide-linked). The hydrophobic interactions in each heptad responsible for the formation and stabilization of the two-stranded α-helical coiled-coil are indicated by the dots.*

in the α-helices wrapping around each other like a two-stranded rope). The criteria used for the selection of the amino acid sequence was described previously [2]. Considering the results of previous studies [3], a peptide length of 35-residues with only leucine residues occupying all the hydrophobic positions was chosen to meet our objectives. A disulfide bridge was introduced at the N-terminal end of the polypeptide to maintain the two α-helices in a parallel alignment. The disulfide bridge was also considered an important feature to stabilize further the protein model and allow the model to tolerate the effects of destabilizing substitutions. The amino acid substitutions were made at the hydrophobic positions denoted by X (positions 16 and 19) in Fig. 1. The hydrophobic residues substituted at position X were Leu, Ile, Val, Phe, Tyr and Ala. The effect these substitutions have on protein stability were monitored by circular dichroism during guanidine-hydrochloride denaturation studies at pH 2 and pH 7 (Fig. 2 and Table 1). These model proteins have demonstrated a wide range of stability depending upon pH, ranging from instability in the absence of denaturant, to complete stability in the presence of 6 M Gn·HCl. The L_o, I_o, V_o and F_o proteins are essentially completely α-helical at pH 2 and 7 (Table 1) while the substitution of Tyr or Ala in the central heptad results in a decrease in α-helical content (65% for Y_o and 29% for A_o at pH 7). There is an increased stability of all the proteins at pH 2 compared to pH 7. In fact, the L_o and I_o proteins are essentially 100% α-helical at pH 2 in the presence of 6 M Gn·HCl (Fig. 2). This increased stability at pH 2 is probably the result of protonation of the glutamic acid residues and removing the charge repulsion from negatively charged residues at pH 7 in the i and i + 4 positions in the sequence (e.g., Glu[6,10]; Glu[13,17] etc.). This effect must outweigh the potential ionic interactions at pH 7 between the Glu and Lys residues (Lys[1] in chain 1 to Glu[6'] in chain 2; Glu[8,13'] etc.). The coiled-coil stability varies, depending upon the hydrophobic residue in the 3–4 repeat positions, with leucine providing the most stable coiled-coil, and

Table 1 Circular dichroism results of the synthetic model proteins

Protein	$[\theta]_{220}$				Transition midpoint[d] (M)		$\Delta G_D^{H_2O}$* (kcal mol^{-1})	
	Benign[a] pH 2	% Helix[b]	Benign[c] pH 7	% Helix[b]	pH 2	pH 7	pH 2	pH 7
L_o	−31,750	100	−30,250	95	ND	5.3	ND	7.3
I_o	−33,050	104	−29,750	94	ND	4.1	ND	5.6
V_o	−33,900	107	−28,900	91	5.9	2.9	8.0	4.0
F_o	−32,500	102	−30,550	96	4.6	2.4	6.2	3.2
Y_o	−28,500	90	−20,650	65	3.4	1.1	4.7	1.5
A_o	−23,350	74	− 9,050	29	3.1	ND	4.3	ND

[a] A 0.1% aqueous TFA solution, pH 2.4. Temperature was 25°C for ellipticity measurements.
[b] The % α-helix is calculated based upon the native 70-residue disulfide-bridged model protein (L_o) having 100% α-helix in benign conditions [3] at pH 2.4.
[c] A 0.1 M KCl, 0.05 M PO$_4$ buffer, pH 7.0. Temperature was 25°C for ellipticity measurements.
[d] The transition midpoint is the value of guanidine hydrochloride (M) required to give a 50% decrease in ellipticity.
* $\Delta G_D^{H_2O}$* is the free energy of unfolding in the absence of guanidine hydrochloride and is estimated by extrapolating the free energy of unfolding at each individual concentration of guanidine hydrochloride (ΔG_D) to zero concentration assuming that they are linearly related [7].
ND = Not determined for the L_o and I_o proteins since they were not denatured with 6 M guanidine hydrochloride (Fig. 2). Not determined for the A_o protein due to the low α-helical content at pH 7 in the absence of guanidine hydrochloride.

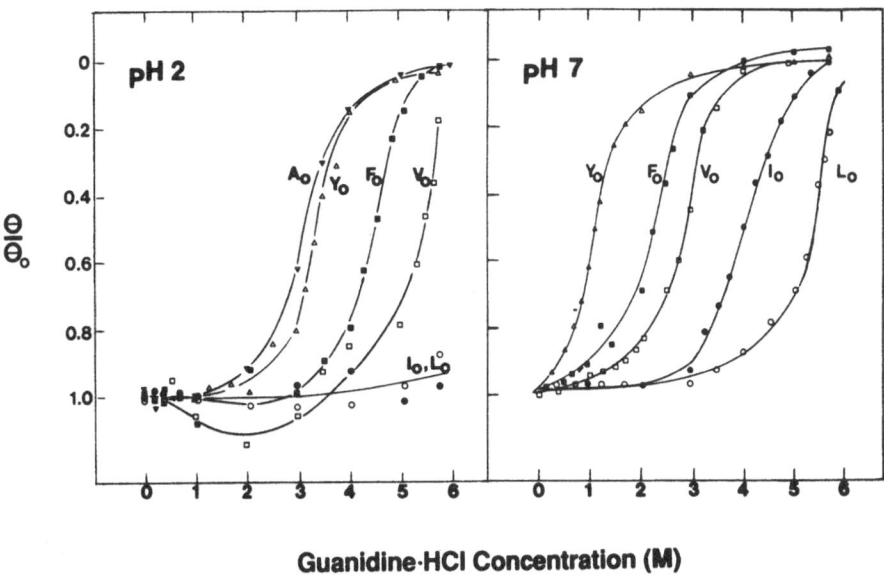

Fig. 2.   *Guanidine hydrochloride denaturation profiles at pH 2 and pH 7 of the 70-residue disulfide-linked model proteins as described in Fig. 1. θ/θ_o represents the ratio of ellipticity at 220 nm at the indicated molarity of guanidine hydrochloride to the ellipticity without denaturant.*

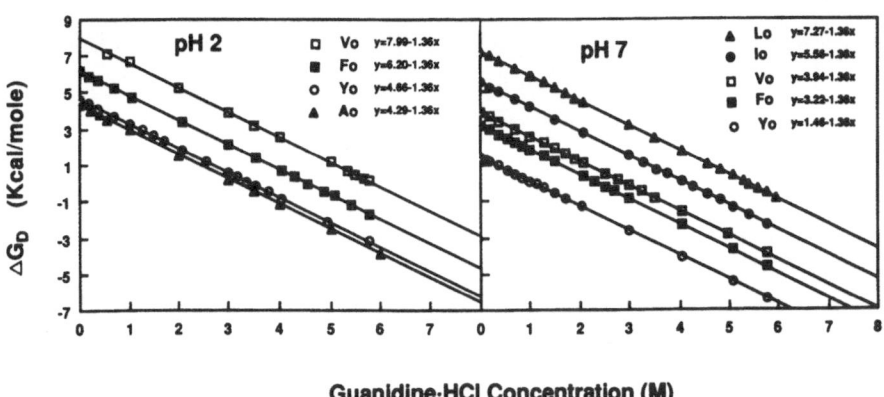

Fig. 3.   *Linear dependence of ΔG_D on the concentration of guanidine hydrochloride. The free energy of unfolding in the absence of guanidine hydrochloride ($\Delta G_D^{H_2O}$) is estimated by extrapolating the free energy of unfolding at each individual concentration of denaturant (ΔG_D) to zero concentration, assuming that they are linearly related ($\Delta G_D = \Delta G_D^{H_2O} - m$-[guanidine hydrochloride]), where m is the slope of the line. ΔG_D can be determined from the experimental data by using the equation, $e^{-\Delta G_D/RT} = ([\theta]_N - [\theta])/([\theta] - [\theta]_D)$, where $[\theta]$ is the observed ellipticity at any particular guanidine hydrochloride concentration, and $[\theta]_N$ and $[\theta]_D$ are the ellipticities of the native and denatured states, respectively [7-9].*

569

alanine providing the least stable molecule. This order of stability (Leu > Ile > Val > Phe > Tyr > Ala) is, as expected, independent of pH. From the Gn·HCl denaturation curves, we can calculate the free energy of unfolding at each concentration of denaturant, and by extrapolation to zero concentration, we can determine the free energy of unfolding in the absence of denaturant (Fig. 3, Table 1).

In summary, we have demonstrated the validity of our model protein to quantitate the effects of various hydrophobic residues on stabilizing α-helices in aqueous solution, and these results should be generally applicable to the de novo design of proteins in general.

References

1. Talbot, J.A. and Hodges, R.S., Acc. Chem. Res., 15 (1982) 224.
2. Hodges, R.S., Saund, A.K., Chong, P.C.S., St.-Pierre, S.A. and Reid, R.E., J. Biol. Chem., 256 (1981) 1214.
3. Lau, S.Y.M., Taneja, A.K. and Hodges, R.S., J. Biol. Chem., 259 (1984) 13253.
4. Hodges, R.S., Semchuk, P.D., Taneja, A.K., Kay, C.M., Parker, J.M.R. and Mant, C.T., Peptide Res., 1 (1988) 19.
5. Hodges, R.S., Sodek, J., Smillie, L.B. and Jurasek, L., Cold Spring Harbor Symp. Quant. Biol., 37 (1972) 299.
6. Crick, F.H.C., Acta Crystallogr., 6 (1953) 689.
7. Pace, C.N., Methods Enzymol., 131 (1986) 266.
8. Kellis, Jr., J.T., Nyberg, K., Sali, D. and Fersht, A.R., Nature, 333 (1988) 784.
9. McCubbin, W.D., Oikawa, K., Sykes, B.D. and Kay, C.M., Biochemistry, 21 (1982) 5948.

Molecular characterization of bGH self-association

S. Russ Lehrman, Jody L. Tuls and Marilyn Lund

The Upjohn Company, Kalamazoo, MI 49001, U.S.A.

Introduction

Bovine growth hormone (bGH), a 4-α-helix bundle containing 191 amino acids, undergoes self-association when partially denatured at high protein concentrations (>10 μM). This self-associated form of bGH has been identified using a number of hydrodynamic and spectral characteristics [1,2]. A central region of this protein, spanning residues 96–133, appears to be involved in this process, since the corresponding peptide inhibits bGH self-association and itself self-associates to form an α-helix [3,4]. Recent efforts in our laboratory have been directed to the study of bGH self-association using bGH(96–133) analogs as a model (Fig. 1). This model system would permit examination of the effect of peptide structure on self-association. Validation of the model requires that characteristics of the peptide analogs parallel protein self-association. Previous efforts have shown that [Leu[112]]-bGH self-associates to a greater extent [5] and human growth hormone (hGH) self-associates to a lesser extent than bGH (Lehrman, Tuls and Havel, manuscript in preparation). The experiments

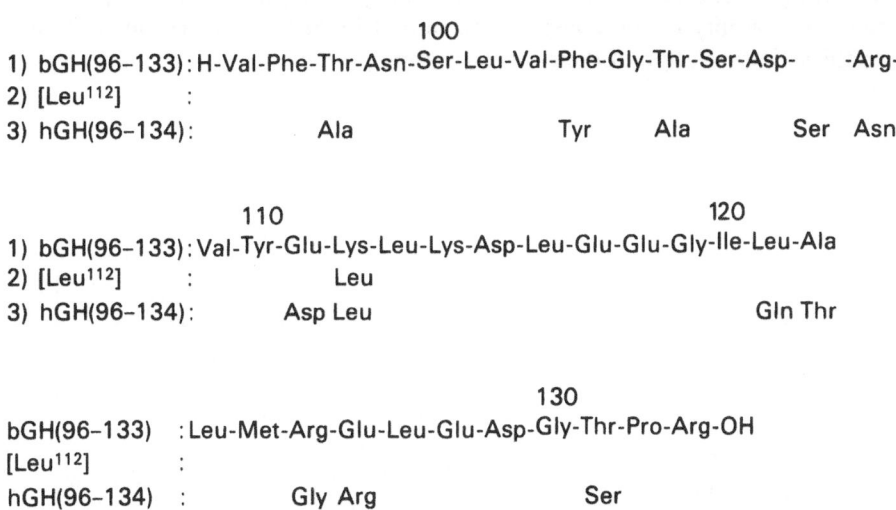

Fig. 1. Primary sequences of bGH. Numbering corresponds to the bGH sequence. Note that hGH(96–134) contains an insertion of one amino acid residue between positions 107 and 108 of the bGH primary sequence.

571

presented here compare the ability of bGH(96–133), [Leu112]-bGH(96–133) and hGH(96–134) to inhibit bGH self-association with the self-association of the corresponding protein analogs. In addition, we compare the ability of these peptides to form an α-helix as a function of pH and concentration. Our results confirm that these peptides provide a model system for bGH self-association.

Results and Discussion

bGH(96–133) (peptide 1), [Leu112]-bGH(96–133) (peptide 2), and hGH(96–134) (peptide 3) were synthesized using standard solid-phase procedures. These peptides were determined to be at least 95% pure by RPHPLC, and their structures were verified by liquid SIMS mass spectroscopy (m/z 4343, 4328 and 4362 daltons, respectively) and AAA.

Self-associated bGH and related proteins precipitate to varying extents on the addition of this material into dilute aqueous buffers [5]. When bGH and [Leu112]-bGH are treated in this manner, 72 and 85 ± 4% precipitation is observed, respectively [5]. More recent studies have shown that no significant precipitation of hGH occurs when this protein is subjected to the same conditions. We have compared the relative self-association of these proteins with the ability of their corresponding peptides to inhibit bGH self-association. Previous studies had shown that preincubation of bGH with a 5-molar excess of bGH(96–133), cleaved from the intact protein using mild trypsin digestion, inhibits the formation of precipitate [3]. Using this system, peptides 1, 2, and 3 were compared for their ability to inhibit the precipitation of self-associated bGH [3]. These peptides reduce the precipitation of self-associated bGH by 68, 83, and 5 ± 4%, respectively. The relative ability of these peptides to inhibit bGH self-association correlates with the self-association of their corresponding proteins.

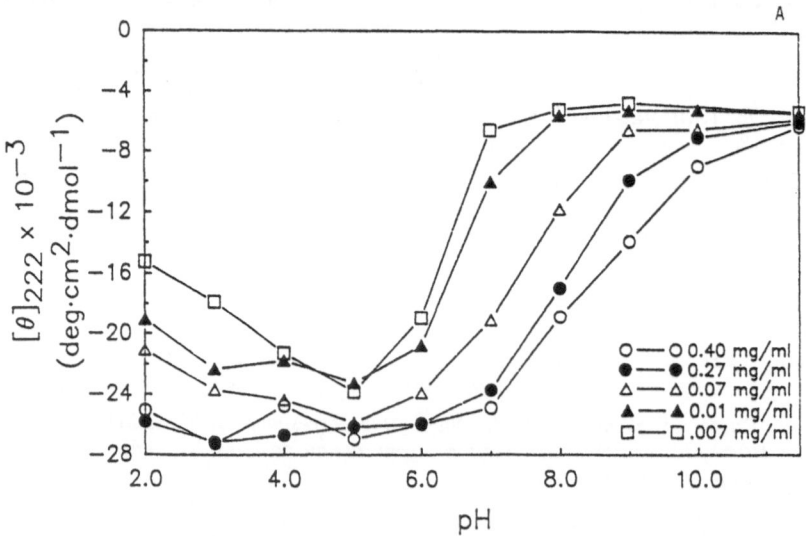

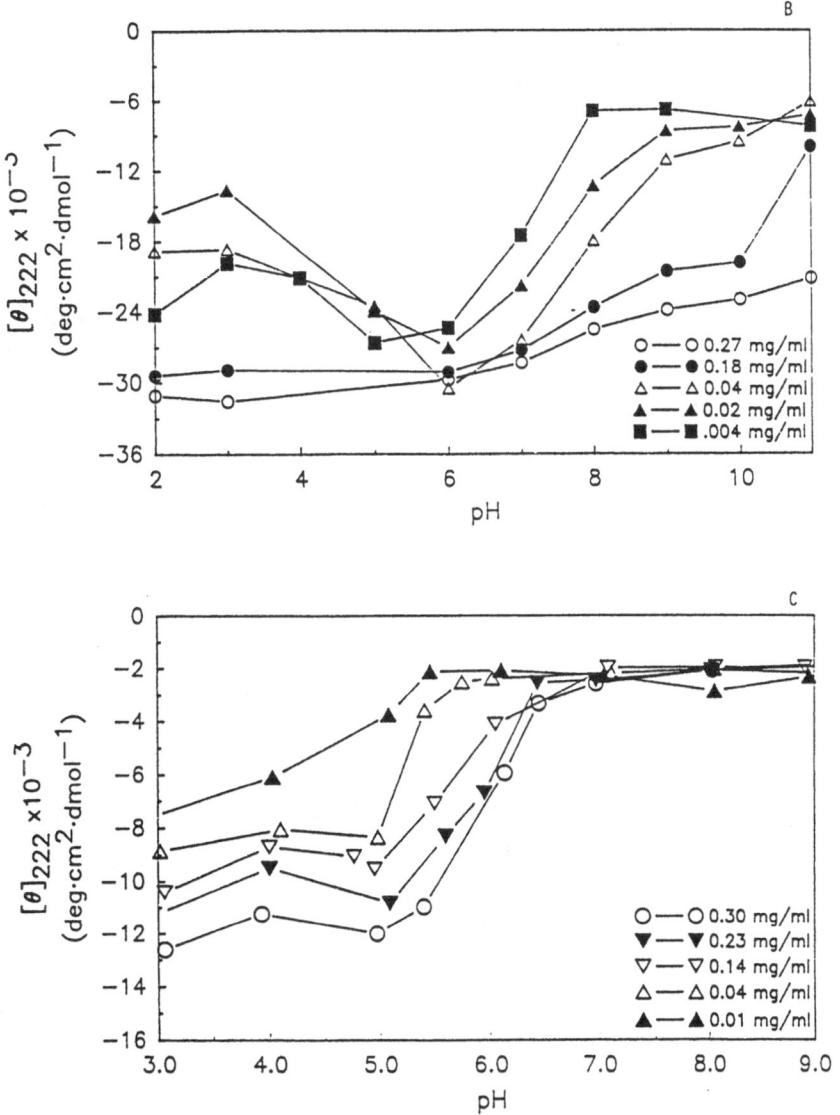

Fig. 2. *pH and concentration dependence of α-helicity in peptides 1 (panel A), 2 (panel B), and 3 (panel C). The peptides were dissolved in 1 mM sodium citrate, sodium borate and sodium phosphate. Buffer pH was adjusted by the addition of concentrated sodium hydroxide or hydrochloric acid. All spectra were obtained at room temperature.*

α-Helix formation in bGH(96–133) has been previously shown to occur as the result of peptide self-association [4]. The α-helicity of peptides 1, 2, and 3 was, therefore, determined using circular dichroism at 222 nm, as a function of pH and peptide concentration (Fig. 2). The mean residue ellipticity, $[\theta]_{222}$, was calculated for each peptide using an average amino acid molecular weight of 114 daltons and peptide concentrations that were determined using the bicinchoninic acid (BCA) method (Pierce Chemical Co., Rockford, IL) [6]. Although the α-helicity of all three peptides increases significantly with increasing peptide concentration and decreasing pH, the maximum α-helicity of peptides decreases in the following order: $2>1>3$. Under conditions where bGH self-association was studied, (pH 8.5), the relative α-helicity of these peptides is maintained. Therefore, the conformations of these peptides also parallel the self-association of their corresponding proteins.

These experiments indicate that the observed characteristics of bGH(96–133) analogs provide a useful model of bGH self-association. Experiments are in progress to further validate this model and determine molecular features which are responsible for bGH self-association.

Acknowledgements

The authors thank Wayne Duholke for FABMS, Lynn Snyder for AAA, and Patricia Barr for her secretarial assistance.

References

1. Havel, H.A., Kauffman, E.W., Plaisted, S.M. and Brems, D.N., Biochemistry, 25 (1986) 6533.
2. Brems, D.N., Plaisted, S.M., Kauffman, E.W. and Havel, H.A., Biochemistry, 25 (1986) 6539.
3. Brems, D.N., Biochemistry, 27 (1988) 4541.
4. Brems, D.N., Plaisted, S.M., Kauffman, E.W., Lund, M.E. and Lehrman, S.R., Biochemistry, 26 (1987) 7774.
5. Brems, D.N., Plaisted, S.M., Havel, H.A. and Tomich, C.-S.S., Proc. Natl. Acad. Sci. U.S.A., 85 (1988) 3367.
6. Smith, P.K., Krohn, R.I., Hermanson, G.T., Mallia, A.K., Gartner, F.H., Provenzano, M.D., Fujimoto, E.K., Goeke, N.M., Olson, B.J. and Klenk, D.C., Anal. Biochem., 150 (1985) 76.

Conformational analysis of bioactive analogs of growth hormone-releasing factor (GRF)

Vincent S. Madison, David C. Fry, David N. Greeley, Voldemar Toome,
Bogda B. Wegrzynski, Edgar P. Heimer and Arthur M. Felix

Roche Research Center, Hoffmann-La Roche Inc., Nutley, NJ 07110, U.S.A.

Introduction

There are a number of potential therapeutic indications for GRF in human and animal health. Generally, a linear GRF analog will have a large ensemble of conformations because of flexible segments. Non-polar solvents and covalent restrictions can be used to induce order in specific segments that may be recognized by the receptor. Conformational analysis via CD, NMR and computations has been performed on a series of GRF analogs to determine the effects of $i - (i + 4)$ lactams on peptide folding, and to provide a basis for further analog design.

Results and Discussion

Analogs of GRF(1–29)-NH$_2$ with replacements for Asn^8 and Gly^{15}, and with lactams formed between the side chains of Asp^8 to Lys^{12} and/or Lys^{21} to Asp^{25} were investigated in aqueous solution at pH 3 and in 75% methanol/water at pH 6. The CD spectra of these analogs are independent of pH within the range of pH 3–7. The percentage α-helix deduced from the CD spectra as well as the in vitro bioactivity are reported in Table 1.

Table 1 *Relative potency and percentage α-helix for GRF analogs*

Analog	Potency	% α-Helix from CD	
	(in vitro)	In water	In 75% MeOH
GRF(1–44)-NH$_2$	1.0	–	–
GRF(1–29)-NH$_2$	0.8	20	90
[Sar^{15}]-GRF(1–29)-NH$_2$	0.04	15	50
[Ala^{15}]-GRF(1–29)-NH$_2$	4.0	25	90
Cyclo^{8,12}[Asp^8,Ala^{15}]-GRF(1–29)-NH$_2$	0.8	35	90
Cyclo^{21,25}[Ala^{15}]-GRF(1–29)-NH$_2$	1.3	35	90
Dicyclo$^{8,12;21,25}$[Asp^8, Ala^{15}]-GRF(1–29)-NH$_2$	0.7	45	90

Helical segments were apparent for the GRF analogs from the pattern of consecutive $i - (i + 3)$NOEs in 2D spectra. For each analog, about 200 pairwise NOEs were observed. The NOEs were converted into distance constraints using assumptions and a classification scheme that we have outlined [1]. These distances were used to calculate a penalty function in a structural optimization scheme using molecular dynamics and energy minimization in the CHARMM program package [2], as has been described [1,3] using $0.01 - 3.0$ for the weighting factor (WNOE) and 2 or 4 for the exponent (ENOE) in the energy term: $E = WNOE$ $(kT/2) [(R - R_0)/\delta R]^{ENOE}$. For each peptide, the optimization yielded 30 (49 for the dicyclic analog in 75% methanol/water) low-energy structures that are consistent with the experimental data. In a typical fit to the data, 75% of the distances in the models fit the experimental ones within the estimated error, and the remaining 25% show an additional RMS deviation of 0.1 Å.

The i–(i+4) lactams are fully compatible with the α-helical conformation. In aqueous and methanol/water solutions, the optimization reveals well-formed helical segments that include the region of the lactam, and extend toward the carboxyl terminus. In 75% methanol/water, all of the peptides (except the Sar[15] analog) feature a single helical segment including most of the molecule, except for fraying of about two residues at each end. The high degree of order in this segment is seen from the low RMS deviations from the average structure for consecutive segments of four Cα's (Fig. 1A). Nevertheless, the linear peptides show a tendency to kink near residues 16 and 25 (see the spikes in Fig. 1A). The fraction of kinked structures is 6/30 for GRF(1–29)-NH$_2$, and 9/30 for the Ala[15] analog. For comparison, this fraction is reduced to 2/49 for the dicyclic analog, and the two monocyclic analogs had no sharply kinked structures, but only small bends. For the cyclic analogs, the segment RMSD is large only at the ends.

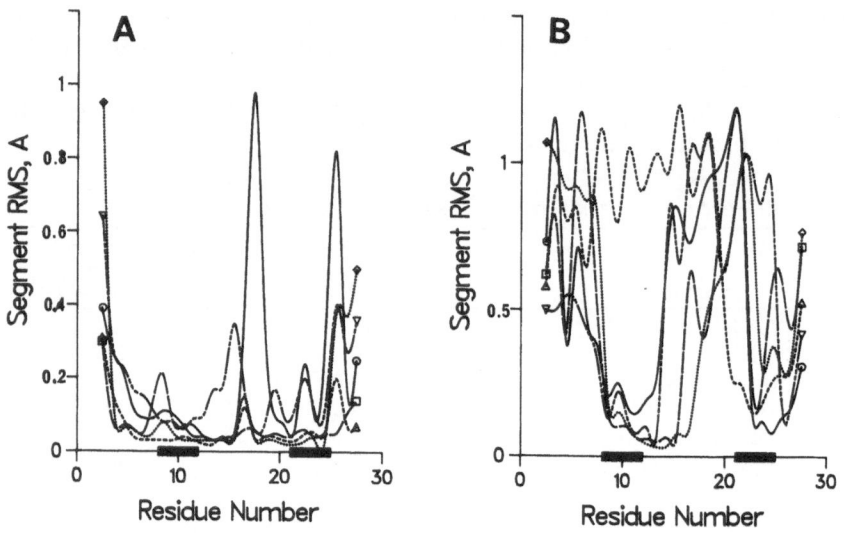

Fig. 1.   RMS deviation from the average over the ensemble of 30 optimized structures for segments of four consecutive Cα's. The bars on the abscissa mark the position of lactams in the cyclic analogs. The symbols and line types for the analogs are: inverted triangles, (long dash, three short dashes) = GRF(1–29)-NH$_2$; circles (solid line) = Ala[15]; squares (long dashes) = cyclo[8,12]; triangles (short dashes) = cyclo[21,25]; and diamonds (dots) = dicyclo[8,12;21,25] (A) In 75% methanol/water and (B) in aqueous solution.

In aqueous solution, the linear peptides have two short helical segments centered near residues 11 and 25 (see Fig. 1B). In the region of residues 8–14, the degree of order is very similar for GRF(1–29)-NH$_2$ and its Ala[15] analog, but, surprisingly, the native analog is less well-ordered near the carboxyl terminus. The Asp[8]-Lys[12] lactam extends the helical segment to include residues 7–17 in the monocyclic analog, and 7–19 in the dicyclic analog. The Lys[21]-Asp[25] lactam stabilizes a helical segment in the 18–28 region for the monocyclic analog, but destabilizes the 8–14 segment, which was helical for the linear peptides. In the dicyclic analog, the C-terminal helical segment is shorter and less well-ordered than in the Ala[15] or the monocyclic analogs.

The i – (i + 4) lactams stabilize α-helical segments in aqueous solution, as seen by the larger helix content determined via CD, and the longer helical segments obtained from the optimization constrained by NOE-derived distances. Further, there is negative cooperativity between segments in the peptide chain: extending one helical segment destabilizes the other.

For the biologically active peptides investigated herein, the lactam-stabilized helical segments that are retained in aqueous solution and in 75% methanol/ water should also be retained at the receptor. Thus, we propose that the bioactive conformer is compatible with helical segments in the vicinity of residues 8–12 and 21–25.

References

1. Fry, D.C., Madison, V.S., Bolin, D.R., Greeley, D.N., Toome, V. and Wegrzynski, B., Biochemistry, 28 (1989) 2399.
2. Brooks, B.R., Bruccoleri, R.E., Olafson, B.D., States, D.J., Swaminathan, S. and Karplus, M., J. Comp. Chem., 4 (1983) 187.
3. Madison, V., Berkovitch-Yellin, Z., Fry, D., Greeley, D. and Toome, V., In Tam, J.P. and Kaiser, E.T. (Eds.) Synthetic Peptides – Approaches to Biological Problems, Alan R. Liss Inc., New York, 1989, p. 109.

Conformational studies on bombolitin III in aqueous solution and in the presence of surfactant micelles

Eleni Bairaktari, Carol A. Bewely*, Stefano Mammi and Evaristo Peggion
Biopolymer Research Center, Department of Organic Chemistry, University of Padua, Via Marzolo 1, I-35131 Padua, Italy

Introduction

Bombolitins are five structurally related heptadecapeptides that were recently isolated from bumblebee venom and characterized in terms of biological properties by Argiolas and Pisano [1]. Similar to mellitin and mastoparan isolated from bee venom, bombolitins lyse erythrocytes and liposomes and enhance the activity of phospholipase A_2 from different sources. All of these peptides, in spite of the remarkable difference in the amino acid sequence, have a strong tendency to fold into an amphiphilic α-helical structure. The hypothesis has been made that the amphiphilic nature of these peptides might account for their affinity for membranes and for their biological activity [1].

In this paper, we present preliminary results of conformational studies by CD and ^{1}H NMR on Bombolitin III (IKIMDILAKLGKVLAHV) in aqueous solution and in the presence of sodium dodecylsulfate (SDS) micelles.

Results and Discussion

Bombolitin III was synthesized by the solid phase Merrifield procedure and purified by HPLC. In dilute aqueous solution (1×10^{-4} M or lower), the CD spectrum indicated that the peptide is essentially, but not completely in a random structure (Fig. 1). This is revealed by the presence of a negative shoulder at 222 nm and by the low intensity of the negative band below 200 nm. The CD pattern is scarcely dependent upon pH changes up to 8. At strongly basic pH, there is an increase of ordered structure (data not shown), possibly due to deprotonation of the lysine side chains.

When SDS is added to aqueous solution at a concentration above the critical micellar concentration, the typical CD pattern of the right-handed α-helix is formed (Fig. 1). The helix content at a 1:1 ratio of peptide to micelles is of the order of 60%. In the micelle-bound state, the peptide conformation is completely pH-independent up to pH 9.

The characterization of the peptide conformation in the presence of detergent micelles was carried out by 2D homonuclear ^{1}H NMR experiments at 40°C using perdeuterated SDS and appropriate solvent resonance suppression

---

*On leave of absence from Bachem Fine Chemicals, Torrance, CA, U.S.A.

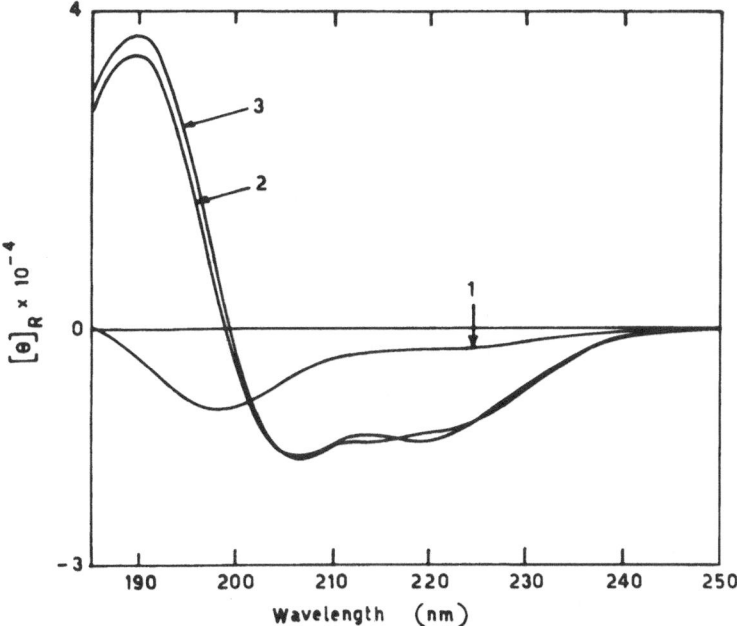

Fig. 1. CD spectra of Bombolitin III in aqueous solution at pH 4.9 in the absence [1] and in the presence of SDS micelles. Spectra 2 and 3 refer to solutions with a peptide to micelles molar ratio of 1 : 1 and 1 : 10, respectively.

techniques. Using standard COSY, NOESY and sequential assignment procedures, the spin systems of all individual amino acid residues were identified,
and all proton resonances assigned. In the NOESY spectra, connectivities were
observed between NH_i and NH_{i+1} protons spanning the entire amino acid
sequence. In addition to sequential connectivities, NOESY cross peaks were found
between residues in position i and $i+3$, notably between $C^\alpha H_i$ and $C^\beta H_{i+3}$ protons
(Fig. 2) and also between $C^\alpha H_i$ and NH_{i+3} protons. These connectivities are
diagnostic for the presence of an α-helical structure [2] and indicate that the
helical segment in Bombolitin III comprises the central part of the peptide chain
from Met in position 4 to Val in position 13 (Fig. 3). Its length is consistent
with the helix content estimated from the amplitude of the negative CD band
at 222 nm. Because of the amphiphilic character of the helix, it is conceivable
that its axis is oriented parallel to the micelle surface, with the hydrophobic
surface in contact with the hydrocarbon side chain inside the micelle, and the
hydrophilic portion at the water–micelle interface.

The NMR results also show that there is a single set of resonance lines for
each spin system. Taking into account that in our experimental conditions the
peptide is completely in the micelle-bound state (Fig. 1), these data indicate
that a single conformation is assumed by Bombolitin III upon interaction with
SDS micelles [3].

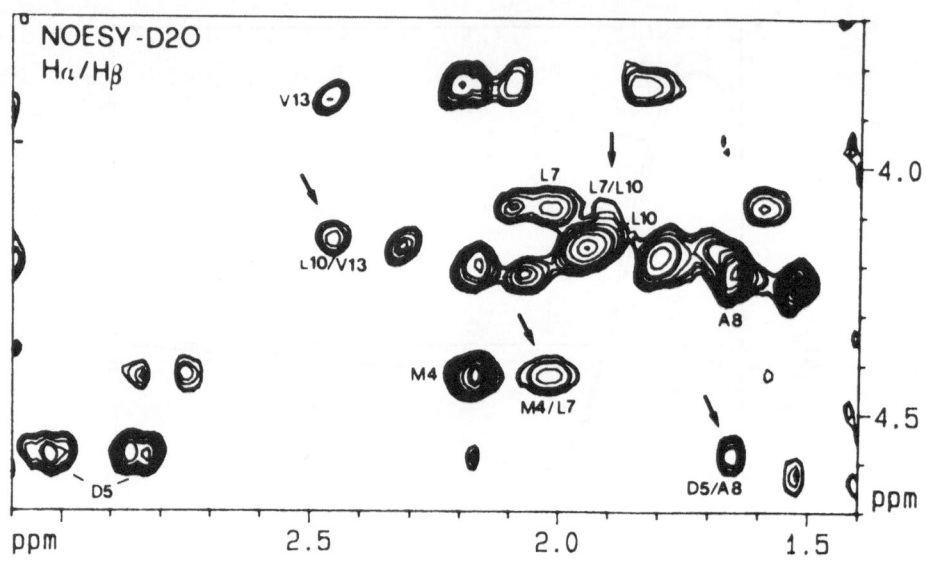

Fig. 2. *Portion of a NOESY spectrum in D_2O, in the presence of SDS micelles. Arrows indicate connectivities between $C^\alpha H_i$ and $C^\beta H_{i+3}$ protons.*

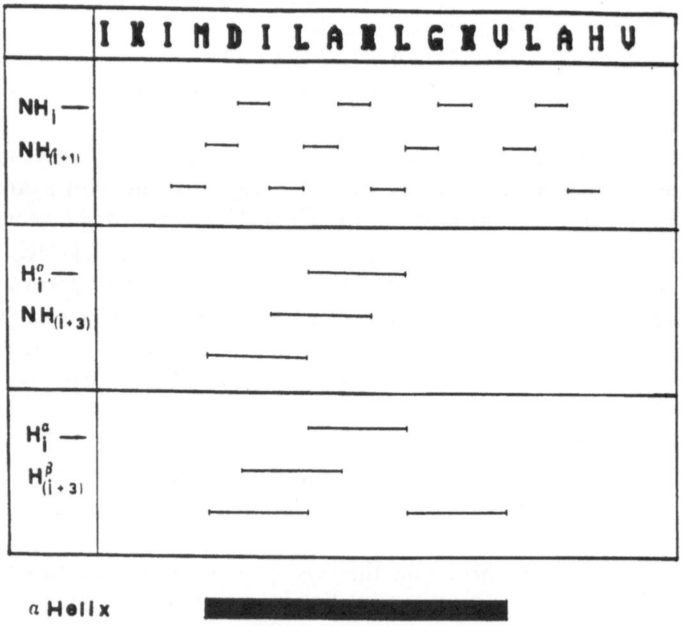

Fig. 3. *Summary of connectivities found in NOESY experiments in aqueous solution and in the presence of SDS micelles. The length of the helical segment is also indicated.*

580

Similar to other amphiphilic peptides [4], Bombolitin III tends to form intermolecular aggregates at sufficiently high concentration in aqueous solution and in the absence of detergent. At pH 4.5, the CD spectra show an increase of helical content upon increasing the peptide concentration above 1 mM. Thus, aggregation favors the formation of the helical structure in individual monomeric units, very likely stabilized by intermolecular hydrophobic interactions. Under these conditions, different from the micelle-bound state, the CD spectrum is strongly pH-dependent. Aggregation and consequent helix formation is substantially reduced at pH $\leqslant 3.0$. Since the only event taking place at this pH is the protonation of the carboxylate function of the Asp residue in position 5, it is possible that in the aggregates this group is involved in a salt bridge stabilizing the helical form.

References

1. Argiolas, A. and Pisano, J.J., J. Biol. Chem., 260 (1985) 1437.
2. Wüthrich, K., NMR of Proteins and Nucleic Acids, J. Wiley and Sons, New York, 1986, p. 162.
3. Lee, K.H., Fitton, J.E. and Wüthrich, K., Biochim. Biophys. Acta, 911 (1987) 144.
4. Kaiser, E.T. and Kezdy, F.J., Proc. Natl. Acad. Sci. U.S.A., 80 (1983) 1137.

3-Substituted prolines: Preparation and assignment of absolute configuration to optically pure *trans*-3-phenyl- and *trans*-3-*n*-propylprolines

Mark W. Holladay, John Y.L. Chung, William A. Arnold, Catherine S. May, James T. Wasicak and Alex M. Nadzan

Neuroscience Research Division, Pharmaceutical Discovery, Abbott Laboratories, Abbott Park, IL 60064, U.S.A.

Introduction

An important approach toward understanding the bioactive conformation of a peptide involves the study of conformationally constrained analogs. Proline is a readily available conformationally constrained amino acid that is often incorporated into a peptide in the place of other residues as a way of introducing conformational restrictions [1–3]. However, loss of activity with this modification could be due to loss of important interactions by the side chain of the original amino acid. Therefore, we have investigated the use of proline analogs that retain the side chain moieties of standard amino acids. One series in this effort consists of 3-substituted proline analogs (Fig. 1). Here we describe our results at providing convenient preparations of optically pure *trans*-3-phenyl- (**1a**) and *trans*-3-*n*-propylprolines (**1b**) (analogs of Phe and Nle, respectively) of known absolute configuration.

a. R = Ph, P = Ac
b. R = n-propyl, P = Boc

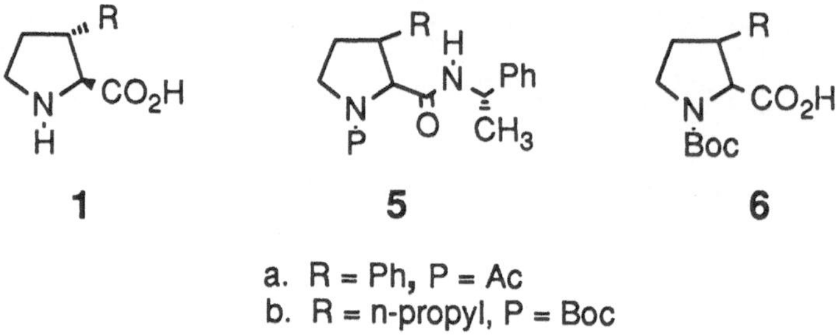

Fig. 1. Structure of trans-3-substituted prolines and derivatives.

Results and Discussion

The preparation of racemic *N*-acetyl-*trans*-3-phenylproline is illustrated in Fig. 2. The sequence is based on a literature route [4], but several modifications lead to improved yields and simplified procedure. In particular, the acid-catalyzed silane reduction [5] of hydroxylactam **2** affords the pyrrolidine **3** cleanly in a single step; performing the reduction at this stage allows subsequent steps to proceed more cleanly and in higher yields. Base-catalyzed epimerization of ester **4** improves the *trans : cis* ratio, and the ability to separate the isomers by selective saponification is retained. The 3-*n*-propyl series can be prepared analogously; however, the separation of *cis* and *trans* isomers is best achieved with the selective saponification carried out on *N*-Boc protected esters [6].

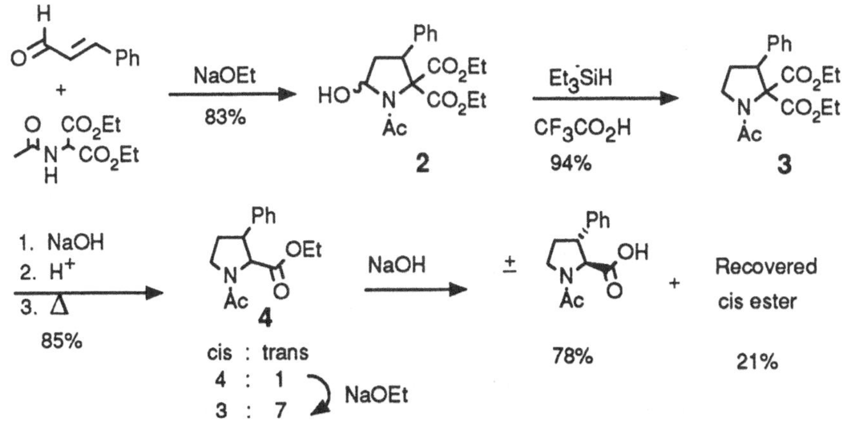

Fig. 2. Synthetic route to N-acetyl-3-phenyl-D,L-proline.

The racemic *N*-protected trans acids were resolved by coupling with S(−)-α-methylbenzylamine (EDAC, HOBt) to give diastereomers of **5**, which were separated by silica gel chromatography. Total hydrolysis under acid conditions, followed by *N*-protection with Boc, provided derivatives **6** for peptide synthesis. Table 1 shows physical data by which the isomers of **5** and **6** can be identified. The absolute stereochemistry in the 3-phenyl series was assigned by decarboxylation (2-cyclohexen-1-one, cyclohexanol, Δ) [7] of the free amino acids to 3-phenylpyrrolidine, for which absolute configuration has been correlated with optical rotation [8]. The stereochemistry of 3-*n*-propylproline was assigned by stereospecific synthesis of the L-isomer from 4-hydroxy-L-proline (M.W. Holladay and C.S. May, unpublished results). The key step in this sequence was the reaction of the pyrrolidine enamine derived from Z-4-oxo-L-Pro-OMe with allyl bromide

Table 1 *Physical data for derivatives of 3-substituted prolines*

Stereoisomer	R	5	6
		TLC R_f (solvent)[a]	$[\alpha]_D^{24}$ (c1, CHCl$_3$)
2S,3R (L-*trans*)	Ph	0.24 (2.5% HOAc/EtOAc)	+ 33.7°
2R,3S (D-*trans*)	Ph	0.40 (2.5% HOAc/EtOAc)	− 35.9°
2S,3S (L-*trans*)	*n*-Pr	0.39 (1 : 2 EtOAc/hexane)	− 38.3°
2R,3S (D-*trans*)	*n*-Pr	0.45 (1 : 2 EtOAc/hexane)	+ 43.2°

[a]E. Merck Silica Gel-60 precoated glass plates.

M.W. Holladay et al.

to give, after hydrolysis, the 3-allyl-ketone in fair yield without racemization at the α-center. Catalytic hydrogenation of the olefin, followed by reduction of the 4-keto function to methylene (Barton-type deoxygenation of the intermediate secondary alcohol) gave a mixture of *cis*- and *trans*-*N*-Cbz-3-*n*-propyl-L-proline methyl esters that, after conversion to the Boc derivatives, were separated by selective saponification as described above.

In summary, convenient procedures for the preparation of Boc-*trans*-3-phenyl- and Boc-*trans*-3-*n*-propylproline in optically pure form with defined stereochemistry were developed. Full experimental details are forthcoming [6].

References

1. Momany, F.A. and Chuman, H., Methods Enzymol., 124 (1986) 3.
2. Marshall, G.R., In Creighton, A.M. and Turner, S. (Eds.) Chemical Recognition in Biological Systems, The Chemical Society, London, 1982, p. 279.
3. Arison, B.H., Hirschmann, R. and Veber, D.F., Bio-org. Chem., 7 (1978) 447.
4. Sarges, R. and Tretter, J.R., J. Org. Chem., 39 (1974) 1710.
5. Auerbach, J., Zamore, M. and Weinreb, S.M., J. Org. Chem., 41 (1976) 725.
6. Chung, J.Y.L., Wasicak, J.T., Arnold, W.A., May, C.S., Nadzan, A.M. and Holladay, M.W., J. Org. Chem., in press.
7. Hashimoto, M., Eda, Y., Osanai, Y., Iwai, T. and Aoki, S., Chem. Lett., 893 (1986).
8. Tseng, C.C., Terashima, S. and Yamada, S.-I., Chem. Pharm. Bull., 25 (1977) 166.

Conformations of proline-containing segments in membrane environments

Charles M. Deber, Mira Glibowicka and G. Andrew Woolley

Research Institute, Hospital for Sick Children, Toronto, Ontario, Canada M5G IX8
Department of Biochemistry, University of Toronto, Toronto, Ontario, Canada M5S IA8

Introduction

Although noted as hydrophilic residues with helix-breaking potential, proline residues are widely observed in the putatively α-helical transmembrane (TM) segments of many channel-forming integral membrane proteins. Because of the recognized property of X-Pro peptide bonds (where X = any amino acid) to occur in *cis* as well as *trans* isomeric states, intramembranous Pro residues have been proposed as conformational participants in channel regulatory events [1]. However, the tertiary amide character of the X-Pro bond [2] also confers increased propensity for involvement of its carbonyl group in liganding interactions with positively-charged species, and/or in specific H-bonded structures (e.g., β- and γ-turns) [3]. We have been investigating this last property of the X-Pro bond in further detail through synthesis and conformational analysis of Pro-containing peptides with membrane affinity. Inspection of sequence triads X-Pro-Y, occurring in TM regions of transport proteins [1], led to the choice of Leu-Pro-Phe as a consensus sequence typifying the environment of intramembranous Pro residues. To model the behavior of such membrane-occurring Pro-containing segments, we are synthesizing a series of non-polar peptides, of which t-Boc-(Ala)$_3$-Leu-Pro-Phe-OH is an initial compound.

Results and Discussion

Effects on ^{13}C NMR spectra (75 MHz), upon incorporation of t-Boc-AAAL*P*F-OH (*: ^{13}C=O isotopic enrichment) into unilamellar phosphatidylcholine (PC) vesicle bilayers, are presented in Fig. 1. The spectrum of the free peptide dissolved to saturation (3 mM) in D$_2$O displays single resonances for both Pro and Leu C=O carbonyls, likely corresponding to an all-*trans* conformation at the Leu-Pro bond [4]. Stepwise additions of PC vesicles produced a well-defined trend: the Leu C=O resonance remains essentially invariant, while the Pro C=O resonance moves selectively upfield, where it coincides with the Leu resonance at a PC/peptide mole ratio of 2.6. These initial results suggest that (in lieu of any apparent lipid-induced population of *cis* peptide bonds) the membrane appears instead to induce a specific conformation of t-Boc-AAAL*P*F-OH, as reflected by the altered local magnetic environment of the Pro residue.

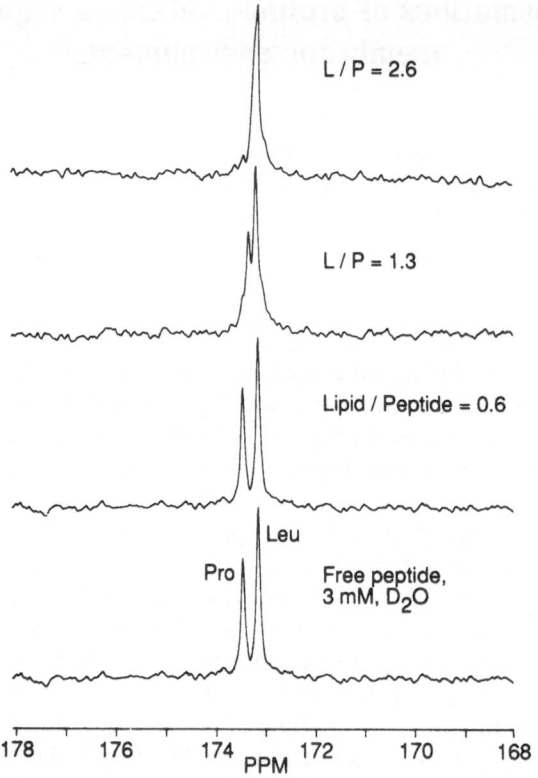

Fig. 1. *^{13}C NMR spectra (75 MHz) of Boc-(Ala)$_3$-Leu*-Pro*-Phe-OH (3 mM, D$_2$O, 27°C) in the region of Pro and Leu carbonyl carbons. The peptide was enriched with ^{13}C-atoms during synthesis at both Pro (40%) and Leu (60%) *C=O sites. The bottom spectrum is the free hexapeptide. The upper three spectra result from successive addition of unilamellar phosphatidylcholine (PC) vesicles at the indicated lipid/peptide mole ratios. Lipid concentrations were 1.89, 3.90, and 7.80 mM. Spectra were recorded on a Bruker 300 MHz spectrometer, using 32 K data points, with 500 accumulations per spectrum, and a 4.0-s pulse delay. Chemical shifts are referenced to external dioxane at 66.50 ppm. The peptide was prepared by solution-phase techniques, using the fragment condensation approach; see Ref. 5 for further details of synthesis and spectroscopy.*

In conjunction with observation of corresponding ^{13}C NMR spectral effects displayed by this peptide in monolayers of mixed (50/50, w/w) lyso-phosphatidylcholine/lyso-phosphatidylglycerol micelles, and from analysis of its ^{1}HNMR (300 MHz) spectra in perdeuterated sodium dodecylsulfate (SDS) micelles [5], the overall spectral results can be interpreted in terms of bilayer-stabilization of an intramolecularly H-bonded structure, possibly an inverse γ-turn conformation involving the Leu C=O and the Phe N-H in a hydrogen bond. The upfield movement of Pro C=O likely arises from changes in local magnetic environment due to backbone dihedral angle rotations (minimally Pro ψ and

Phe ϕ angles) that produce eclipsing interactions of Pro C=O with Pro C_β [6], as are requisite for conversion from the initially (aqueous-solvated) inter-molecularly H-bonded conformation to the inverse γ-turn. The π-electron density of the Phe aromatic ring may also participate in creating an electron-rich pocket [5].

Preliminary molecular modeling experiments indicate that insertion of an inverse γ-turn into a longer [e.g., 20 residue $(Ala)_m$-Leu-Pro-Phe-$(Ala)_n$ idealized TM] α-helical segment does impart a 'kink' to the chain, but that such a structure may be incorporated into an α-helix without a large deviation in overall chain direction. Alteration in domain polarity can thus lead to an alternative category of local conformation of Pro-containing segments, a property that may be of functional importance in membrane proteins for which channel 'opening-closing' involves an aqueous/membrane re-location of a critical segment.

Acknowledgements

This work was supported, in part, by a grant to C.M.D. from the Medical Research Council of Canada (MT-5810). G.A.W. held an Ontario Government Scholarship.

References

1. Brandl, C.J. and Deber, C.M., Proc. Natl. Acad. Sci. U.S.A., 83 (1986) 917.
2. Veis, A. and Nawrot, C.F., J. Am. Chem. Soc., 92 (1970) 3910.
3. Smith, J.A. and Pease, L.G., C.R.C. Crit. Rev. Biochem., 8 (1980) 315.
4. Fraser, P.E. and Deber, C.M., Biochemistry, 24 (1985) 4593.
5. Deber, C.M., Glibowicka, M. and Woolley, G.A., Biopolymers, 28 (1989) in press.
6. Deber, C.M., Madison, V. and Blout, E.R., Acc. Chem. Res., 9 (1976) 106.

2D NMR spectroscopy study of the spatial structure of kinins: Bradykinin, kallidin and their cyclic analogs

J.B. Saulitis, E.E. Liepins, I.P. Sekacis, F.K. Mutulis and G.I. Chipens

Institute of Organic Synthesis, Latvian S.S.R. Academy of Sciences, Riga, U.S.S.R.

Introduction

A new methodology for quantitative evaluation of interproton distances in peptides using the cross-peak intensities in 2D NOE spectra was developed. The method is based on the measurement of the intensities of the cross-peaks in the pure-phase 2D NOE spectra; and the ratios of the cross-peak intensities $I_{N\alpha}/I_{\alpha N} = (d_{\alpha N}/d_{N\alpha})^6$ and $I_{NN}/I_{\alpha N} = (d_{\alpha N}/d_{NN})^6$ enable the determination of the corresponding interproton distances $d_{N\alpha} = d(N_iH, C_i^{\alpha}H)$, $d_{\alpha N} = d(C_i^{\alpha}H, N_{i+1}H)$ and $d_{NN} = d(N_iH, N_{i+1}H)$ for several amino acid residues. The spatial structures of bradykinin (BK), kallidin (KL) and their cyclic analogs [(9–1$^\epsilon$)Lys^1,Gly^6]BK (CBK), cyclo(10–1$^\epsilon$)KL (CKL) in H_2O and DMSO were established by this approach, and using routine ^1H NMR conformationally dependent parameters: $^3J(NHC^{\alpha}H)$ spin-spin coupling constants and temperature coefficients of amide proton resonances $\Delta\delta_{NH}/\Delta T$. Further, these structures are used as a crude model in the refinement by the restraint minimization of the conformational energy.

Results and Discussion

In order to elucidate the role of the Coulombic interaction between charged groups in the BK and KL molecules, the study under conditions with charged (BK-I and KL-I) and neutral (BK-II and KL-II) C-terminal carboxyl groups (peptide lyophilized from the H_2O at pH 7.8 and 1.5, respectively), is carried out in DMSO. The different NOEs, $^3J(NHC^{\alpha}H)$ and $\Delta\delta_{NH}/\Delta T$ values were observed for BK-I and BK-II. The neutralization of the C-terminal carboxyl group caused essential change of the $C^{\alpha}H$-Arg^1 proton chemical shifts ($\Delta\delta_{BK-I,BK-II} = 0.20$ ppm), while the chemical shifts of other protons, excepting NH protons, changed negligibly, approximately ± 0.05 ppm. For KL, the passage from the KL-I to KL-II state also causes the changes of the proton chemical shifts, NOEs, coupling constants and $\Delta\delta_{NH}/\Delta T$ values. These data suggest that the BK-I conformation is stabilized in DMSO by the Coulombic interaction of the N-terminal α-amino and C-terminal carboxyl groups. The BK-II molecules do not exhibit such interaction. For KL-I, the interaction of the C-terminal carboxyl group with the guanidino group of Arg^2 and Arg^{10} residues occurs in DMSO at concentrations of 70% and 30%, respectively. As follows from the study, the kinin BK-I and KL-I molecules form two β-turns: Pro-Pro-Gly-

Phe (type II) and Ser-Pro-Phe-Arg (type I), and those structures are stabilized by intramolecular Coulombic interaction. Note that the second β-turn appears only in 70% of all KL-I molecules. The BK-II and KL-II molecules do not adopt any dominant conformation in DMSO and the structures of BK and KL in H_2O are also disordered. The majority (86%) of CBK molecules with trans-configuration of all peptide bonds also possesses these β-turns, whose orientation is determined by H-bond Phe[5]-CO...N$^\epsilon$H-Lys[1]. The ^{1}H NMR data show the three slowly exchangeable conformers of CKL molecules: CKL-I with trans-configuration of all peptide bonds constitute 25%, CKL-II with *cis*-Pro[3]-Pro[4] – 35%, whereas CKL-III with the *cis*-Arg[2]-Pro[3] bond – 40%. In H_2O, the populations of these conformers are 45, 25 and 30%, respectively. The assignment of CKL conformers has been done from the 2D NOE spectra taken at various temperatures, 303, 323 and 353 K. From the intensities of the exchange cross-peaks in 2D NOE spectra, the Gibbs free energy $\Delta G^{\neq}_T$ for the cis-trans isomerization of the peptide bond Pro[3]-Pro[4] is $\Delta G^{\neq}_{303\ K} = 73.2$ kJ/mol·K, and for the peptide bond Arg[2]-Pro[3] isomerization $\Delta G^{\neq}_{353\ K} = 81.6$ kJ/mol·K. The dominant conformation of CKL-I has β-turn type III in sequence Ser-Pro-Phe-Arg and second H-bond Pro[3]-CO...HN-Ser[7]. The most stable conformer of CKL–CKL-III has a similar type II β-turn. The CKL-II has the β-turn encompassing residues Ser-Pro-Phe and the additional H-bond Arg[2]-CO...HN-Phe[6]. In H_2O the CKL-I and CKL-III keep β-turns, while CKL-II has an unordered structure. In conclusion, linear kinins – bradykinin and kallidin can be stabilized by intramolecular interaction between charged groups, therefore having the ability to form quasicyclic structures in DMSO.

Prothrombin fragment 1: Conformational studies and the role of residues 35–46

Lisa M. Balbes, Lee G. Pedersen and Richard G. Hiskey*

Department of Chemistry, University of North Carolina, CB 3290, Chapel Hill, NC 27599-3290, U.S.A.

Introduction

The three dimensional crystal structure of residues 36–156 of bovine pro-thrombin fragment 1 (residues 1–156, BF1) has recently been solved by Tulinsky et al. [1]. The γ-carboxyglutamic acid (Gla) domain was disordered in the absence of metal, but folded into a specific tertiary structure upon crystallization in the presence of Ca[II] ions (A. Tulinsky, personal communication). Tulinsky et al. noted three turns of α-helix located in the 38–47 region of the protein. This hydrophobic helical region is conserved in the vitamin K-dependent proteins, and may stabilize the final folded conformation of the Ca[II]-Gla domain complex. Pollock et al. [2] have reported the isolation of the bovine Gla domain peptide, residues 1–45. The [^{125}I]1–45 peptide was shown [3] to bind to phosphatidylserine/phosphatidylcholine vesicles (25 : 75, PS/PC) in the presence of 2.0 mM Ca[II] ions, $K_d = 11.8$ μM (BF1, $K_d = 0.6$ μM). The peptide dimerized readily; increasing the NaCl concentration favored dimer formation and decreased PS/PC binding. Ca[II] concentrations below 2.0 mM had no effect on the monomer-dimer equilibrium; higher Ca[II] concentrations caused the peptide to precipitate. The dimerization of the 1–45 peptide might result from interactions of the hydrophobic portions of two 1–45 molecules. The peptide corresponding to residues 35–46 was synthesized using a standard PAM resin protocol (Applied Biosystems 430A). The synthetic 35–46 peptide (purified by HPLC), was also acetylated at the amino terminus (Leu^{35}) and Lys^{44} using [^{14}C]-Ac$_2$O, nor reductively methylated at these residues using [^{14}C]-HCHO and NaCNBH$_4$. The effect of these peptides on the PL binding of 1–45 and BF1 was evaluated. The conformations of 35–46, 1–45 and BF1 were also investigated by circular dichroism (CD) spectroscopy [4] and by molecular modeling. Modeling was carried out using Macromodel 1.5, AMBER 3.0 and SYBYL+ 5.05. Three distinct homodimers were modeled: one associated via hydrophobic interactions, and two through electrostatic interactions.

Results and Discussion

Addition of the 35–46 peptide (10-fold molar excess) to a mixture of 1–45

---

*To whom correspondence should be addressed.

590

peptide, PS/PC vesicles, and Ca[II] inhibited the PS/PC binding of the radioiodinated 1–45 peptide (Table 1). The acetylated peptide had no effect on the PS/PC binding of [^{125}I]1–45, while the methylated peptide led to a slight decrease in binding. Neither the acetylated, nor the methylated peptide bound to PS/PC, in the presence or absence of Ca[II] ions. The hydrophobic peptide (86-fold molar excess) does not appear significantly to affect the PS/PC binding of BF1, although solubility problems prevented the use of higher concentrations of 35–46. The results of the modeling study indicated that, regardless of the program used, the dimer conformations of the 35–46 peptide modeled were more stable than two isolated monomers, with the head-to-tail charge dimer as the most stable. CD spectroscopy showed that a 340 μM solution of the 35–46 peptide contains 42 ± 5% helix at 25°. The CD spectra of BF1 and 1–45 and the effect of metal ions and NaCl concentration on the spectra were examined. From the studies described above, we conclude that: (1) Binding of the 35–46 peptide to the radioiodinated 1–45 peptide, presumably via the hydrophobic C-terminal region, contributes to the loss in Ca[II]-dependent PS/PC binding of the 1–45 peptide. (2) The 35–46/1–45 interaction also requires some contribution from the amines at positions 35 and/or 44. (3) The conformational change induced in BF1 by Ca[II] ions and monitored by CD differs from that induced by Mg[II] or Sr[II].

Table 1   *Effect of 35–46 and derivatives on PL binding of [^{125}I]1–45*

Inhibitor	Molar Excess	ΔPL bound[a]	ΔDimer[a]	ΔMonomer[a]
None[b] [^{125}I]1–45	0.0	36.7%	19.7%	43.6%
35–46	1.1	+ 0.5%	+ 0.2%	− 0.7%
35–46	10.7	− 31.0%	+21.9%	+ 9.0%
35–46	20.3	− 33.5%	+27.2%	+ 6.2%
[^{14}C]AC-35–46	4.4	+ 1 %	+ 4 %	− 5 %
[^{14}C]Me-35–46	17	− 8 %	− 5 %	+14 %

[a] Δ refers to the observed difference as compared to the 1–45 control.
[b] No inhibitor. Numbers reported are actual percent of counts in each peak.

Acknowledgements

We would like to gratefully acknowledge the contributions of Dr. Lila Gierasch and Dr. W. Curtis Johnson to this study.

References

1.   Tulinsky, A., Park, C.H. and Skrzypczak-Jankun, E., J. Mol. Biol., 202 (1988) 885.
2.   Pollock, J.S., Shepard, A.J., Weber, D.J., Olson, D.L., Klapper, D.G., Pedersen, L.G. and Hiskey, R.G., J. Biol. Chem., 263 (1988) 14216.
3.   Weber, D.J., Pollock, J.S., Pedersen, L.G. and Hiskey, R.G., Biochem. Biophys. Res. Comm., 155 (1988) 230.
4.   Johnson Jr., W.C., Ann. Rev. Biophys. Biophys. Chem., 17 (1988) 145.

Conformation induction in amphiphilic peptide hormones bound to model interfaces

John W. Taylor

The Rockefeller University, Laboratory of Bioorganic Chemistry and Biochemistry, 1230 York Avenue, New York, NY 10021, U.S.A.

Introduction

Amphiphilic structures are likely to be a common feature of the active conformations of small, flexible peptides [1]. These active conformations are identified more readily through studies of these peptides at model interfaces than through more conventional analyses of their solution conformations. The compression isotherms of the monolayers formed by several potentially amphiphilic peptide hormones at the air–water interface have previously been analyzed to determine their surface-bound conformations [2]. However, interpretation of the results is less straightforward than for idealized amphiphilic model peptides [see, for example, Ref. 3], because it is usually unclear which segments of the native peptides occupy surface space and which segments are in the aqueous subphase. To obtain more direct spectroscopic information, a simple method has been developed that involves the adsorption of such peptides from aqueous buffer onto siliconized quartz slides and subsequent measurement of their CD spectra.

Results and Discussion

Four peptides, all with disordered structures in aqueous solution, were selected as examples: β-endorphin (β-EP, 31 residues), which has a potential amphiphilic α-helical structure in residues 13–31; salmon calcitonin (32 residues), which has a potential amphiphilic α-helical structure in residues 8–22; gonadotropin-releasing hormone (GnRH), all 10 residues of which can form an amphiphilic β-strand, and dynorphin A(1–17), which has a potential amphiphilic β-strand segment in residues 7–15 [1]. Each peptide was adsorbed onto four siliconized quartz slides by soaking the slides for 5 min in an aqueous solution of the peptide (20–30 μM) in 20 mM NaH_2PO_4/NaOH, pH 7.5, containing 160 mM KCl, and then rinsing them briefly in distilled, deionized water to remove excess peptide solution. CD spectra were measured for each set of slides in eight slide orientations about the lightpath. These spectra were then averaged to eliminate linear dichroism artefacts [4] and a blank spectrum, obtained in the same way for the same set of slides before peptide adsorption was subtracted. The results, shown in Fig. 1, indicate that the potential amphiphilic α-helical structures

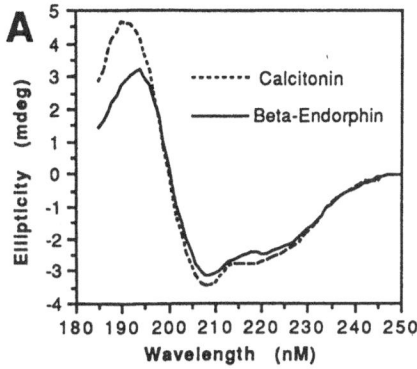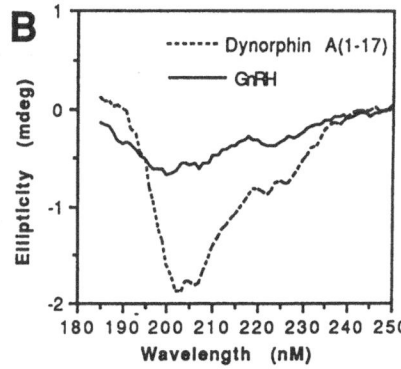

Fig. 1.   CD spectra of peptides adsorbed onto siliconized quartz slides.

in β-EP and calcitonin, were indeed, induced upon binding to the siliconized slides, in agreement with the behavior of these peptides at the air–water interface [2]. Furthermore, a comparison of these spectra with those obtained for α-helical, β-sheet and disordered poly-L-lysine films transferred from the air–water interface [5] suggests, interestingly, that these helices might be propagated through most of the β-EP and calcitonin structures. In contrast, GnRH was too poorly adsorbed onto the siliconized slides for a useful spectrum to be obtained. This hormone also failed to form monolayers at the air–water interface, suggesting that it has a preferred, nonamphiphilic, folded conformation in aqueous solution. Dynorphin A(1–17) adsorbed onto the siliconized slides moderately well, in a conformation that was mostly disordered, but with a small helical component indicated by the shoulder in the CD spectrum at 222 nM [3,5]. This is also consistent with the behavior of dynorphin monolayers at the air–water interface, which have a low collapse pressure (5 dyn/cm) and a limiting area (29-Å²/residue) that is larger than that observed for model amphiphilic α-helical and β-sheet peptides [3]. Studies of dynorphin analogs indicate that the presence of a proline residue in position 10 in the dynorphin sequence inhibits the amphiphilic β-strand formed by residues 7–15 from aggregating in the form of β-sheets at these planar model interfaces (unpublished data). There are no examples to date of peptide hormones that form amphiphilic β-sheets at these model interfaces.

Acknowledgements

The assistance of Teddy Liu and Adam Profit in performing these experiments is gratefully acknowledged. This research was supported by P.H.S. grant GM-38811.

J.W. Taylor

References

1. Taylor, J.W. and Kaiser, E.T., Pharmacol. Rev., 38 (1986) 291.
2. Taylor, J.W., In Dhawan, B.N. and Rapaka, R.S. (Eds.) Recent Progress in Chemistry and Biology of Centrally Acting Peptides, Central Drug Research Institute, Lucknow, India, 1988, p. 25.
3. DeGrado, W.F. and Lear, J.D., J. Am. Chem. Soc., 107 (1985) 7684.
4. Cornell, D.G., J. Colloid Interface Sci., 70 (1978) 167.
5. Stevens, L., Townend, R., Timasheff, S.N., Fasman, G.D. and Potter, J., Biochemistry, 7 (1968) 3717.

Structural studies of endothelin by CD and NMR

S.C. Brown, M.E. Donlan and P.W. Jeffs

Department of Structural and Biophysical Chemistry, GLAXO Inc., 4117 Emperor Blvd., Morrisville, NC 27560, U.S.A.

Introduction

Endothelin is a recently discovered peptide [1] that is the most potent vasoconstricting substance known, though its physiological role is unclear. 'Human endothelin' (ET-1) is one of a class of 21-residue peptides containing two disulfide bonds. Two other endothelins, ET-2 and ET-3 ('Rat'), have been cloned from a human cDNA library and have similar biological effects, while also having nearly identical primary sequence homology and disulfide bond arrangements. Recent results from our laboratory on the structure of ET-1 in various solvent systems, and bound to lipid vesicles, indicate that a helical structure exists within the sequence K9-H16 in DMSO. ET-1 exhibits a higher percentage of helicity in either TFE, or bound to sonicated dimyristoyl phosphatidyl glycerol (DMPG) vesicles, but not when bound to dimyristoyl phosphatidylcholine (DMPC) vesicles.

Methods

Standard 2D NMR experiments (DQF-COSY, DQF-RELAY (30 ms), HO-HAHA (80 ms), and NOESY (200 ms, 400 ms)) were done on the GN-500 NMR spectrometer at the Duke University Biomedical NMR Center. We were not able to achieve a stable solution of ET-1 (Peptides International, Louisville, KY) in aqueous/co-solvent systems (acetic acid, acetonitrile, methanol) for concentrations above 1 mM. Though one-dimensional spectra were obtainable, these samples ET-1 invariably precipitated after several hours, so that 2D NMR spectra could not be done. DMSO was selected as a solvent system for NMR structural studies due to its high dielectric constant and the stability of ET-1 solutions. All NMR experiments were done at 30°C in DMSO-d_6, at a concentration of 3 mM ET-1. CD spectra were done at (10–100 μM) peptide in various solvents at 20°C on an Aviv 6D spectrophotometer. Lipid vesicles were prepared by careful sonication (50% duty cycle) at 0°C under nitrogen. ET-1 solutions were added to stock solutions of sonicated lipid, vortexed gently, and observed for several hours to ensure the attainment of equilibrium

Results and Discussion

Proton resonances of endothelin were assigned in DMSO-d_6 by the standard sequential technique using 2D NMR data from DQF-COSY, DQF-RELAY ($\tau_m = 30$ ms), HOHAHA ($\tau_m = 80$ ms), and NOESY($\tau_m = 200$, 400 ms) experiments. Interproton distances were calculated from the NOESY data using the proportional ratio method and the fixed internal H_β-H_β (1.81 Å) and tyrosine H_δ-H_ϵ (2.51 Å) distances. In this regard, we discovered that the C-terminal tryptophan (W21) aromatic ring was rotating rapidly, exhibiting inter-proton NOEs much reduced in intensity, compared to those of tyrosine (Y13) and phenylalanine (F14). Twenty-one non-sequential inter-residue NOEs were confidently assigned, all of which were $i, i+2$ or $i, i+3$ in nature. Complete sequential amide-amide and amide-alpha NOEs were observed, but showing relative intensities incompatible with a single type of stable secondary structural element. Structures were generated by the distance geometry program DSPACE (Hare Research) using the conformational constraints derived from the NMR data (van der Waals radii–2.5 Å, 2.0 Å–3.2 Å, 3.0 Å–4.5 Å). The ten structures generated indicated a helical structure between Lys^9 and His^{16} driven by the $i, i+2$ and $i, i+3$ constraints, but a 3_{10} or α-helix could not be distinguished for this region due to incomplete data. These helical types can be distinguished only by the relative intensities of these characteristic NOEs [2], most of which could not be assigned in this case, due to overlap with strong intraresidue NOEs. Amide resonance temperature dependencies and D_2O exchange rates were also measured, but did not yield information distinct enough to assign hydrogen-bonded or solvent-inaccessible amide protons. The carboxyl region [H16-W21] of ET-1 is more dynamic and shows no non-sequential NOEs, indicating no stable structure in DMSO. A loose type I turn is found for residues S5-K9.

A distinct negative CD band at 222 nm was only observed in 100% TFE, a notorious helix inducer, and for ET-1 bound to DMPG vesicles (1 ET: 55 lipid). Relative helical content was measured by relative specific molar ellipticities at 222 nm vs. the omnipresent band at 208 nm. CD experiments in aqueous solutions indicated a slight pH dependence of the CD signal for ET-1, minimum at pH = 6.6, and increasing at either higher or lower pH values. Measurement in a series of alcohols indicated increasing helical content proportional to solvent dielectric, common for helices stabilized by electrostatic interactions. Helical content was higher in all alcohols studied vs. aqueous solutions. Neither proendothelin nor ET-1 demonstrated a concentration-dependent CD signal (5–500 μM). Finally, ET-1 showed markedly enhanced helicity in the presence of dimyristoylphosphatidylglycerol (DMPG) vesicles, but not in the presence of DMPC vesicles. The maximal attainable lipid/peptide ratios were found to be 75:1, so peptide–peptide interactions on the lipid bilayer are expected to be minimal. We are currently measuring the binding properties of ET-1 to various lipid vesicles, so it cannot be determined, as yet, whether the lack of CD spectral change in the presence of DMPC vesicles is due merely to lack of binding.

References

1. Yanagisawa, M., Kurihara, H., Kimura, S., Tomobe, Y., Kobayashi, M., Mitsui, M., Yasaki, Y., Goto, K. and Masabi, T., Nature, 332 (1988) 411.
2. Wagner, G., Neuhaus, D., Wörgötter, E., Vašak, M., Kägi, J.H.R. and Wüthrich, K., J. Mol. Biol., 187 (1986) 131.

Synthesis of shortened analogs of ANP: Evidence for a reverse turn in the C-terminal 'tail'

S.F. Brady, T.M. Ciccarone, T.M. Williams, D.F. Veber and R.F. Nutt

Merck Sharp and Dohme Research Laboratories, West Point, PA 19486, U.S.A.

Introduction

Various investigators have established a key role for the arginine residues in atrial natriuretric peptide, ANP(1–28) Ile^{12} = rat sequence [1]. Comparison of bioactivities of compounds in which the arginine (Arg^{27}) in the shortened sequence **Ia** has been deleted or replaced [2] indicates that the guanido group contributes about 20-fold to potency and that it may be supplied by either the N- or C-terminal 'tail' section. In addition, the reduced biological activity seen in cyclic analogs having both 'tails' been deleted (e.g., structure **IIa**) has been ascribed to lack of the critical Arg guanido function. Replacement of certain residues in **IIa** by arginine, for example Gly^{22} (**IIb**), has led to recovery of significant potency [3].

(H)-Cys^{13}-FGGRIDRIGAQSGLG-Cys^{23}-NSF-Arg^{27}-Y-(OH)   **Ia**

Cyclo-(Pro-FGGRIERIGAQSGL-X^{22}-Phe^{23})   **IIa**: X = Gly

     IIb: X = Arg

In the present work, we sought to define the steric positioning of the Arg cationic moiety realtive to the hydrophobic binding elements of the ring by preparing analogs of **Ia**. In these analogs, a proline residue was introduced independently in place of each of the 'tail' residues (Asn^{24},Ser^{25},Phe^{26}) to provide the constraint of a reverse turn.

Results and Discussion

Compounds were prepared by a general route [1] in which resin-bound precursor was carried through HF cleavage. Oxidative cyclization using iodine in 80% HOAc was followed by final purification using preparative RPHPLC, essentially according to procedures previously described [1]. Each analog was characterized by FABMS and AAA.

Biological results, summarized in Table 1, are consistent with the likelihood of a reverse turn at Phe^{26}. The optimal Pro^{26} analog **Ie** is 4 times more potent than reference analog **Ib**. Analogs **Ic** and **Id**, although less active than **Ib**, retain sufficient potency to suggest that more than one backbone conformation is capable of giving active structures.

Thus, substitutions in ANP that promote a turn conformation in the C-terminus result in highly active analogs. A molecular model based on these findings,

Table 1   *Relative potencies of pro-substituted analogs*

(H)-Cys-F G G R I D R I G A Q S G L G-Cys-Asn-Ser-Phe-Arg-(X)
7   23 24 25 26 27

No.	X	Residue substitution	Potency (%)[a]
Ia	Tyr-OH	–	100[b]
Ib	NH_2	–	55
Ic	NH_2	Pro[24]	72
Id	NH_2	Pro[25]	13
Ie	NH_2	Pro[26]	211

[a] Vasorelaxant activity on rabbit renal arterial rings determined as described in R.J. Winquist et al., Eur. J. Pharm., 102 (1984) 169.
[b] Ref. 3.

in which the C-terminus is folded back to place the Arg[27] side chain near the critical hydrophobic groups of the cyclic core, is depicted in Fig. 1. This model was constructed by setting the dihedral angles of the Phe-Arg segment equivalent to those of Pro-X with proline in the $i+1$ position of a Type I β-turn. This structure places the guanidine close enough to the α-carbon of Gly[22] to support the possibility of superposition by a guanidine supplied by the Arg[22] in structure **IIb**.

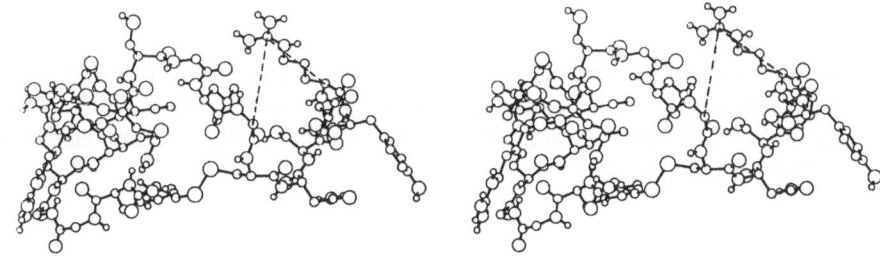

Fig. 1.   *Molecular model of ANP(7–28) (IIa) with Phe[26] and Arg[27] as the $i+1$ and $i+2$ residues, respectively, of a Type I β-turn; energy-minimized using CHARMm to remove any bad contacts. Distances between the guanido function and each of the α-carbons of the Gly[22] and Arg[27] are 6.28 Å and 6.34 Å, respectively.*

References

1. Nutt, R.F. and Veber, D.F., Endocrinol. Metab. Clin. of N. Am., 16(1)(1987) 19.
2. Nutt, R.F., Brady, S.F., Lyle, T.A., Ciccarone, T.M., Colton, C.D., Paleveda, W.J., Veber, D.F. and Winquist, R.J., In Peeters, H. (Ed.) Protides of the Biological Fluids (Proceedings of the 34th Colloquium), Vol. 34, Pergamon Press, Oxford, 1986, p. 55.
3. Nutt, R.F., Ciccarone, T.M., Brady, S.F., Williams, T.M., Colton, C.D., Winquist, R.J. and Veber, D.F., In Jung, G. and Bayer, E. (Eds.) Peptides 1988 (Proceedings of the 20th European Peptide Symposium), De Gruyter, Berlin, 1989, p. 572.

¹H NMR assignments and conformational studies of melanin-concentrating hormone in water using NOE constraints and molecular modeling algorithms

Terry O. Matsunaga, Catherine A. Gehrig and Victor J. Hruby

Department of Chemistry, University of Arizona, Tucson, AZ 85721, U.S.A.

Introduction

Salmon melanin-concentrating hormone (sMCH) is a 17-amino acid peptide hormone that stimulates melanosome aggregation in teleost fish. The molecule is a disulfide-linked peptide of the following sequence:

D T M R C M V G R V Y R P C W E V. Recent evidence has suggested that peptides of this size possess more structural integrity than previously thought [1,2]. As a part of attempts to determine the conformational structure requirements for sMCH at its receptor, we report the conformational analysis of sMCH in water using 2D NMR and molecular modeling methods.

Results and Discussion

Absolute assignments and NOE data were acquired at 500.13 MHz. Assignments were made according to the sequential assignment method (Fig. 1) [3]. A total of 60 NOEs were established in 90% H_2O/10% ^2H_2O. Distances were based upon the formula [4]:

$$r_{ij} = r_{kl} \left(\frac{\sigma_{kl}}{\sigma_{ij}} \right)^{1/6}$$

and quantitated with the use of FTNMR (Hare Research Inc.). Conformational analysis was determined by using a combination of distance geometry (DISGEO) [5] and restrained molecular mechanics, using the CHARMM program [6]. Of significance was the observance of the medium range dipolar interactions $d_{\alpha N}$ between the $i, i+2$ residues of Val^{10} and Gly^8. This interaction, coupled to the strong short-range d_{NN} interaction between the $i, i+1$ residues of Arg^9 and Val^{10}, as well as the moderate d_{NN} between Gly^8 and Arg^9 are indicative of a Type I β-turn [6] encompassing residues 7–10 (Fig. 2). In addition, a transannular NOE is observed between the 2′, 6′ protons of Tyr^{11} to the CH_α of Met^6 and the NH of Val^7.

All molecular modeling algorithms lead to a β-turn in the 7–10 region [7]; however, independently, the schemes do not as yet appear to converge upon a common structure, suggesting some flexibility in other parts of the structure.

600

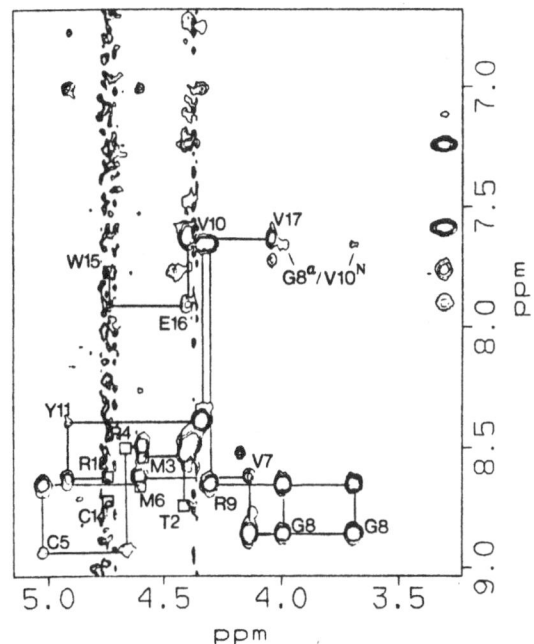

Fig. 1. *500.13 MHz ¹H phase-sensitive NOESY spectrum of the fingerprint region of MCH, illustrating the sequential assignments, as well as NOE secondary assignments.*

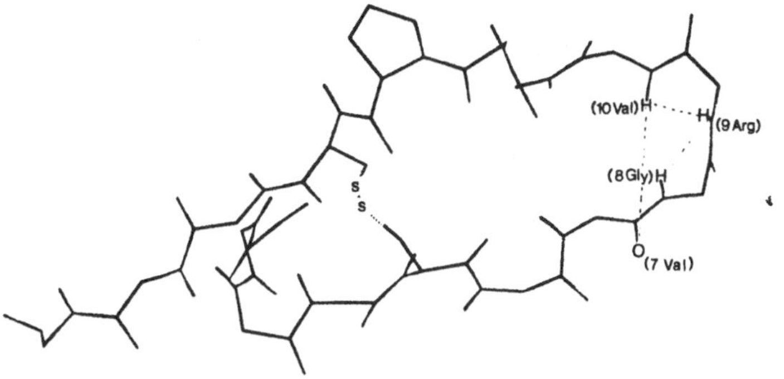

Fig. 2. *Backbone conformation of MCH after applying NOE constraints into the distance geometry (DISGEO) modeling program.*

Acknowledgements

This work was supported by grants from the U.S. Public Health Service (DK-17420), the National Science Foundation, and by a NIDA Fellowship to T.O.M. (DA-05371). A travel grant from ABI is gratefully acknowledged.

601

References

1.   Tsou, C.-L., Biochemistry, 27 (1988) 1809.
2.   Wright, P.E., Dyson, J.H. and Lerner, R.A., Biochemistry, 27 (1988) 7168.
3.   Billeter, M., Braun, W. and Wüthrich, K., J. Mol. Biol., 155 (1982) 321.
4.   Olejniczak, E.T., Gampe, Jr., R.T., Rockway, W.W. and Fesik, S.W., Biochemistry, 27 (1988) 7124.
5.   Havel, T. and Wüthrich, K., Bull. Math. Biol., 46 (1984) 673.
6.   Brooks, B.R., Bruccoleri, R.E., Olafson, B.D., States, D.J., Swaminathan, S. and Karplus, M., J. Comput. Chem., 4 (1983) 187.

A prediction of conformational domains defined by NOEs: Human transforming growth factor alpha

J.C. Hempel[a], F.K. Brown[b], S.C. Brown[c], L. Mueller[b], K.D. Kopple[b] and
P.W. Jeffs[c]

[a]Biosym Technologies, San Diego, CA 92121, U.S.A.
[b]Smith Kline & French Laboratories, Swedeland, PA 19479, U.S.A.
[c]Glaxo Research, Chapel Hill, NC 27514, U.S.A.

Introduction

TGF-α, a 50-residue protein, regulates cell growth and replication. The ^{1}H NMR spectrum has been previously reported by our group [1]. A question of interest in modeling studies incorporating distance constraints is how to assign confidence levels to conformational features observed in modeling studies. In this study, conformational domains, sets of residues linked by clustered NOEs, are defined using an ab initio analysis of the NOE assignments [2,3]. The more highly correlated the NOEs, the more constrained will be the conformation of the domain in modeling studies. The orientation of domains relative to one another will reflect the number and placement of the NOEs linking the domains.

Results and Discussion

NOEs are summarized by residue in Fig. 1. Domains defined by clustered NOEs are mapped by row in Fig. 2 (filled squares). Each residue of each domain is linked by NOEs to two (or more) other residues of the domain [2,3]. The protein is divided into two major domains and a loop region of seven residues. Three residues at the N-terminus are linked by sequential NOEs as are three residues at the C-terminus. The major domains are linked by relatively few NOEs.

NOE-derived distance constraints were used to define 3D structures.* These are similar to structures reported in two independent studies [4,5]. Conformational domains predicted by the ab initio domain analysis superimpose well while the orientation of domains, with respect to one another, varies. This is consistent with the ab initio analysis of the NOEs. The ab initio analysis can be done before modeling studies to identify where in the molecule additional experimental information could significantly increase confidence levels in modeling studies. The analysis also facilitates error checking of NOE assignments and distance bounds.

---

*Using distance geometry (DSPACE) and energy refinement (AMBER).

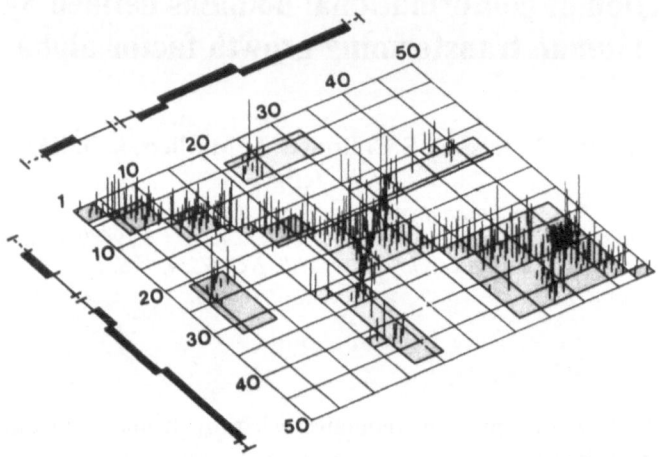

Fig. 1.   NOEs observed for TGF-α are grouped by residue. Peak heights are proportional to the number of NOEs-linking residues indexed by row and column position. Shaded squares group NOEs that define conformational domains (Fig. 2).

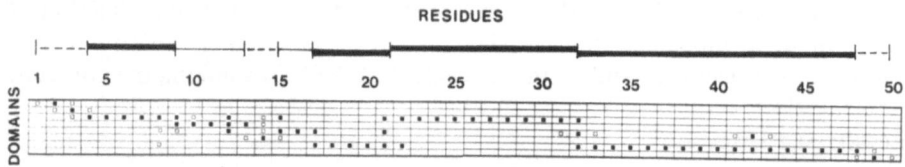

Fig. 2.   Conformational domains defined by the NOEs are mapped by row. Filled squares identify domains. Open squares indicate pendant residues (linked by NOEs to one residue of the domain).

References

1.   Brown, S.C., Mueller, L. and Jeffs, P.W., Biochemistry, 28 (1989) 593.
2.   Hempel, J.C., J. Am. Chem. Soc., 111 (1989) 491.
3.   Hempel J.C. and Brown, F.K., J. Am. Chem. Soc., in press.
4.   Kohda, D., Shimda, I., Miyake, T., Fuws, T. and Inagaki, F., Biochemistry, 28 (1989) 953.
5.   Campbell, I.D., Cooke, R.M., Baron, M., Harvey, T.S. and Tappin, M.J., Progr. Growth Factor Res., 1 (1989) 13.

Expression of a reactive molecular perspective within the triple helical sequence of collagen

P.V. Scaria, Keith R. Sorensen and Rajendra S. Bhatnagar*

Box 0650, University of California, San Francisco, CA 94143-0650, U.S.A.

Introduction

The structural protein collagen exists as fibers under physiological conditions. Its interactions, including self-association, contribute to tissue biomechanical properties, and it is a regulatory protein with profound effects on the behavior of cells. The interactions of collagen with cells are facilitated by denaturation and collagen-derived peptides bind to cells [1]. These observations suggest that the cell binding domains of collagen may be released from the triple helical conformation, for recognition and allosteric binding. The unique conformation of collagen is related to a large content of imino residues that impart a polyproline-like character to each chain, and to the presence of glycine in every third residue facilitating the supercoiling of three chains into the long, rod-like triple helix [2]. In this structure, all side chains are present on the surface and the backbone moieties are available for interaction with the solvent. The major stabilizing interactions in this conformation are based on stereochemical properties of the imino peptide bonds [2], and on non-bonded interactions between imino side chains in adjacent polypeptides [3]. Substitution of α-amino residues for imino residues lowers the stability of the triple helix [4]. The imino-deficient domains of collagen can be expected to be conformationally labile, and therefore they are potential participants in interactions of collagen.

Results and Discussion

Although collagen in aqueous solutions is denatured at 41°C and fibers undergo thermal transition above 60°C, it can be expected that the molecules on the surface of the fiber may be destabilized because of their location at the interface between a hydrophobic core and surrounding water. Anisotropic environments, in which most interactions of proteins with cells occur, can be expected to induce distinct conformations for recognition and binding. We have examined the conformation of a 15-residue peptide, GTPGPQGIAGQRGVV (P-15), based on the sequence around the collagenase-susceptible peptide bond located in an imino-deficient region of α1(I) chain. The structure was examined in aqueous solutions and in the presence of methanol, ethylene glycol and trifluoroethanol, and in the presence of detergents by circular dichroism and ^{1}H and ^{13}C NMR

---

*To whom correspondence should be addressed.

spectroscopy. These conditions have been used as models for anisotropic environments [5]. Only the CD data are presented here. P-15 did not show ordered conformations in aqueous solutions. In the presence of MeOH, Etglycol, TFE, and in sodium dodecyl sulfate (SDS), a clear shift in the spectrum towards a β-strand-like conformation was seen (Fig. 1). A Chou-Fasman analysis of this sequence showed a high preference for the β-strand conformation, with 11 out of 15 residues favoring this conformation. Similar analyses on imino-deficient domains showed that large regions of collagen may share this preference. We examined the conformation of rat tail tendon type I collagen in SDS. As seen in Fig. 2, the conformation of collagen in SDS is quite different from its conformation in the native or heat-denatured states. The CD spectrum of collagen in the presence of SDS shows a high content of β-strand structure suggesting the induction of this conformation in certain domains. SDS has been shown to induce this conformation in the random coil forms of polylysine [6].

Other studies in our laboratory showed that P-15 may be recognized by fibroblast collagen receptors since it inhibits the binding of cells to collagen [7], induces collagen-degrading metalloproteinases and inhibits collagen synthesis in fibroblasts in culture (in preparation). P-15 also binds stoichiometrically to fibronectin and to the 40 kDa collagen-binding domain of fibronectin, inducing major conformational changes in these proteins (submitted for publication).

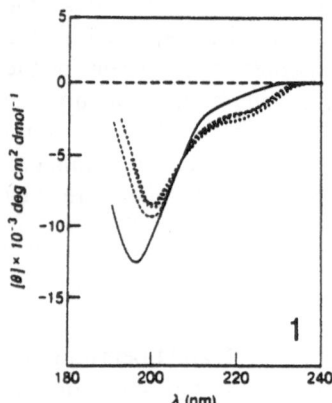 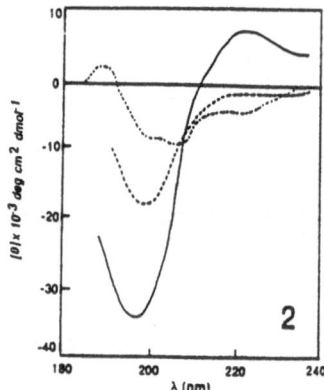

Fig. 1. CD spectra of P-15 in aqueous solution (——), in 75% MeOH in H_2O (-----), in 50% TFE in H_2O (-·-··-), and in 8.27×10^{-3} M SDS (·······).

Fig. 2. CD spectra of pepsin-solubilized rat tail tendon type I collagen, in H_2O at 20°C, in H_2O at 45°C, and in 8.27×10^{-3} M SDS at 20°C.

Conclusion

Our studies suggest that a reactive peptide that mimics some of the interactions of collagen has a greater preference for β-strand conformation than for collagen conformation. An examination of the Ramachandran map shows that the energy barrier between collagen–polyproline conformations and the β-strand conformation is quite low. It is possible that under appropriate conditions, a tautomerism may be induced between collagen–polyproline conformation and β-strand conformation. Since the latter conformation segregates polar and non-polar residues, it is an amphiphilic structure favoring inter-molecular interactions [8]. Thus, local induction of this conformation may be a mechanism for the expression of reactive molecular perspectives within the triple helix of collagen.

Acknowledgements

These studies were supported by NIH Grant AR-37267.

References

1.   Yamada, K.M., Annu. Rev. Biochem., 52 (1983) 761.
2.   Ramachandran, G.N. and Ramakrishnan, C., In Ramachandran, G.N. and Reddi, A.H., (Eds.) Biochemistry of Collagen, Plenum, New York, 1976, p. 45.
3.   Bhatnagar, R.S., Pattabiraman, N., Sorensen, K.R., Langridge, R., MacElroy, R.D. and Renugopalakrishnan, V., J. Biomol. Struct. Dynamics, 6 (1988) 223.
4.   Bhatnagar, R.S. and Rapaka, R.S., In Agris, P.F., (Ed.) Biomolecular Structure and Function, Academic Press, New York, 1978, p. 429.
5.   Mammi, S., Mammi, N.J., Toffani, M.T., Peggion, E., Moroder, L. and Wunsch, E., Biopolymers, 26 (1987) S1.
6.   Mattice, W.L. and Harrison III, W.H., Biopolymers, 15 (1976) 559.
7.   Bhatnagar, R.S. and Qian, J.J., (1989) in press.
8.   Kaiser, E.T. and Kedzy, F.J., Annu. Rev. Biophys. Biophys. Chem., 16 (1987) 561.

Determination of a precise interatomic distance in a helical peptide by REDOR NMR

Garland R. Marshall[a], Denise D. Beusen[a,*], Karol Kociolek[b], Adam S. Redlinski[b],
Miroslaw T. Leplawy[b], Yong Pan[c] and Jacob Schaefer[c]

[a]Department of Pharmacology, Washington University School of Medicine,
St. Louis, MO 63110, U.S.A.
[b]Institute of Organic Chemistry, Politechnika, PL 90-924 Lodz, Poland
[c]Department of Chemistry, Washington University, St. Louis, MO 63130, U.S.A.

Introduction

Recent advances in NMR based on nuclear Overhauser effects have allowed the determination of the solution structure of small proteins [1]. These methods for distance determination suffer from the need to approximate correlation times in the context of a motional model and can have large error ranges. A new solid-state spectroscopic technique, rotational-echo double-resonance (REDOR) NMR [2], measures the heteronuclear dipolar coupling between pairs of labeled nuclei and allows interatomic distances to be directly and accurately determined without invoking simplifying assumptions. We have applied REDOR to measure an interatomic distance in fragment 1–9 of the peptide antibiotic, emerimicin. The crystal structure of this peptide [3] [Ac-Phe-MeA-MeA-MeA-Val-Gly-Leu-MeA-MeA-OBzl, MeA = α-methylalanine] shows it to be α-helical. Emerimicin (2–9), only one residue shorter, has a 3_{10}-conformation in the crystal [4], emphasizing the ease of transition between these two helix types. The measurement by REDOR of an interatomic distance previously established validates the NMR technique and demonstrates its utility in studying α-helix-3_{10}-helix transitions in peptides containing multiple α,α-dialkyl amino acids.

Results and Discussion

Molecular modeling studies suggested that placement of ^{13}C in the carbonyl carbon of residue i and ^{15}N at the amino group of residue i + 4 would yield an interatomic distance that was maximally different [1.74Å] between the α and 3_{10} conformations while maintaining the absolute interatomic distances within the experimentally observeable range. The i to i + 4 carbon-nitrogen distance for oligo-MeA in an α-helix was estimated at 4.13Å vs. 5.87Å in a 3_{10}-helix. The availability of [^{15}N]-glycine and the ease of synthesis of MeA[1-^{13}C] led to the incorporation of ^{13}C at position 2 and ^{15}N at position 6 of the nonapeptide [5]. This pair of atoms is involved in intramolecular hydrogen bonding in the

---

*To whom correspondence should be addressed.

crystal, and the actual measured distance between them in the crystal structure was 4.128 Å.

In a solid with ^{13}C-^{15}N dipolar coupling, the ^{13}C rotational spin echoes, which form each rotor period following a ^{1}H-^{13}C cross-polarization transfer, can be prevented from reaching full intensity by inserting two ^{15}N π pulses per period. The difference between spectra aquired with and without the ^{15}N pulses is related to the C-N dipolar coupling by $\Delta S/S = K(N_c D/n_r)^2$ where $\Delta S/S$ is the experimental ratio of the carbonyl-carbon REDOR difference signal to the full echo, N_c is the number of rotor cycles, D is the dipolar coupling constant to be determined, n_r is the spinning speed, and K is a dimensionless constant equal to 1.066. In this study, $N_c = 8$, $n_r = 3205$ Hz, and the calculated value of D was 44.1 Hz. Using this coupling constant, an experimental standard (the C-N distance in alanine [6] for which the coupling is known), and the r^3 distance dependence for dipolar coupling, a distance of 4.07 Å was calculated between the emerimicin labels. In a second experiment with $N_c = 4$ and $n_r = 2222$ Hz, the distance was determined to be 4.05 Å.

Errors in the REDOR measurement due to differences in motional averaging for alanine and emerimicin, lack of consideration of the J-coupling contribution to the directly-bonded C-N coupling in alanine, and the presence of intermolecular contributions to ^{13}C-^{15}N dipolar coupling in emerimicin are estimated to be less than 0.1 Å.

Our validation of REDOR and its inherent precision suggest that this technique can be used to map intermolecular distances such as those seen in drug–receptor, inhibitor–enzyme, or antigen–antibody complexes. Membrane-bound receptors and enzymes that resist crystallization may be approached using this technique.

Acknowledgements

Support for this work was received from NIH (GM-24483 and GM-33918, GRM, and GM-40634, J.S.), NSF (DIR-8720089, J.S.), and the Polish Academy of Sciences (Grant CPBP 01.13.2.5, A.S.R., K.K., M.T.L.). A travel stipend for Denise Beusen provided by Applied Biosystems Inc. is gratefully acknowledged.

References

1. Wüthrich, K., Science, 243 (1989) 45.
2. Gullion, T. and Schaefer, J., J. Magn. Res., 81 (1989) 196.
3. Marshall, G.R., Hodgkin, E.E., Langs, D.A., Smith, G.D., Zabrocki, J. and Leplawy, M.T., Proc. Natl. Acad. Sci. U.S.A., in press.
4. Toniolo, C., Bonora, G.M., Bavoso, A., Benedetti, E., DiBlasio, B., Pavone, V. and Pedone, C., J. Biomol. Struct. Dynamics, 3 (1985) 585.
5. Marshall, G.R., Beusen, D.D., Kociolek, K., Redlinski, A.S., Leplawy, M.T., Pan, Y. and Schaefer, J., J. Am. Chem. Soc., in press.
6. Stejskal, E.O., Schaefer, J. and McKay, R.A., J. Magn. Res., 57 (1984) 471.

Systematic search in the analysis of conformational constraints derived from NMR: Cyclosporin and other small peptides

Hiroshi Iijima[a], Denise D. Beusen[b],* and Garland R. Marshall[b]

[a]*Kirin Brewery Company, Maebashi, Gunma 371, Japan*
[b]*Department of Pharmacology, Washington University School of Medicine, St. Louis, MO 63110, U.S.A.*

Introduction

Structural determination of peptides and small proteins made possible by advances in NMR has facilitated attempts to correlate biological activity with solution conformation and dynamics. Computional methods incorporating NOE distance constraints have been shown to be robust tools for the elucidation of the solution conformation of small proteins at low resolution, as determined by comparison of NMR-derived structures with their X-ray homologs [1,2]. Ideally, any effort to couple biological effects and solution conformation would consider all possible conformations present in solution. Commonly used structure generation methods, such as distance geometry and molecular dynamics (MD), have not been shown to meet this criterion. The application of systematic search to simulated and actual NMR datasets suggests that this approach does have the potential to determine the entire ensemble of conformations consistent with the data and may reveal conformations not found by other methods.

Results and Discussion

Solution conformations of cyclosporin A (CSA), an 11-residue immunosuppressive cyclic peptide, were determined from an NMR dataset in DMSO [3]. The crystal structure of iodocyclosporin [4] was used to generate a model for the backbone of CSA, and the SEARCH module of the SYBYL software package** was used to generate conformers by scanning ϕ and ψ angles in 10° increments. Conformers were screened for unfavorable van der Waals contacts using radii previously calibrated for proteins and peptides [5]. To speed the calculation, the CSA model was analyzed initially as several small subproblems. Two or three residue fragments were evaluated to determine ϕ,ψ angle ranges consistent with 55 short-range NOEs and vicinal coupling constants. Ultimately, the molecule was analyzed in two major fragments sharing common amide bonds. The loop fragment consisted of Ala[7], D-Ala[8], MeLeu[9], MeLeu[10], and MeVal[11],

---

*To whom correspondence should be addressed.
**Tripos Associates Inc., 1699 S. Hanley Road, St. Louis, MO 63144, U.S.A.

and the sheet fragment included MeBmt[1], Abu[2], Sar[3], MeLeu[4], MeVal[5], and MeLeu[6]. Torsional ranges for ϕ and ψ determined in small subsearches, five long-range NOEs, and four hydrogen bonds were used to constrain the search of these major fragments. Distances linking the atoms at the ends of the loop fragment common to the sheet fragment were monitored for every valid conformation. The resulting distance map described all possible relative orientations between terminal atoms for the loop fragment and was used as a constraint in the subsequent analysis of the sheet fragment. In a final calculation, the distance map generated from the sheet fragment was used to constrain a search of the loop fragment. Use of a distance map from one fragment to constrain the search of the other fragment ensured that not only the NMR-derived constraints were met, but also that the ring closure requirement was satisfied. The sheet and loop fragments were then combined to generate complete CSA structures by matching conformer pairs that possessed identical distance map points.

A total of two and half million conformations were found to be available to the 11 CSA backbone ϕ,ψ pairs. Two distinct conformational families were found to be available to Sar[3]. One family was a βII'-turn ($\phi3 = 60°$, $\psi3 = -120°$) and constituted ~94% of the valid conformations. This is equivalent to the βII' structures reported for CSA [3] and its iodo-derivative [4] in the crystalline state. The remaining 6% corresponded to a βI-turn ($\phi3 = -50°$, $\psi3 = -60°$). Both families were found to be energetically acceptable. Residues other than Sar[3] were conformationally less variable, having similar ϕ,ψ values regardless of the Sar[3] conformation.

The results of this study contrast with those of two recent structural studies based on the same NMR dataset. A 30-ps molecular dynamics simulation of CSA [6] at 300 K found no βI-structure, and the authors concluded that the backbone solution conformation of CSA was similar to that of the crystal. Both the MD simulation and the search calculations unexpectedly suggest that the sheet region of CSA is more flexible than the loop. A second study using distance geometry [7] alone and in combination with MD also failed to find the alternative conformation.

In anticipation of applying systematic search to larger systems, we have developed a hybrid approach combining grid searching with an analytical solution, which significantly speeds the calculation. This new algorithm has been applied to a simulated NMR dataset derived from the crystal structure of cyclo(Gly-Pro-D-Phe-Gly-Val) [8] and to a real data set on cyclo(Tyr-Ile-Gly-Ser-D-Arg), an analog of the cell adhesion pentapeptide YIGSR of laminin [9].

Acknowledgements

Support for this work was received from NIH (GM-24483). A travel stipend for Denise Beusen provided by Applied Biosystems Inc. is gratefully acknowledged.

References

1. Wüthrich, K., Science, 243 (1989) 45.
2. Clore, G.M. and Gronenborn, A.M., Protein Eng., 1 (1987) 275.
3. Loosli, H.R., Kessler, H., Oschkinat, H., Weber, H.-P., Petcher, T.J. and Widmer, A., Helv. Chim. Acta, 68 (1985) 682.
4. Petcher, T.J., Weber, H.-P. and Ruegger A., Helv. Chim. Acta, 59 (1976) 1480.
5. Iijima, H., Dunbar, Jr., J.B. and Marshall, G.R., Proteins: Struct. Funct. Genet., 2 (1987) 330.
6. Lautz, J., Kessler, H., Kaptein, R. and van Gunsteren, W.F., J. Comput.-Aided Mol. Design, 1 (1987) 219.
7. Lautz, J., Kessler, H., Blaney, J.M., Scheek, R.M. and van Gunsteren, W.F., Int. J. Pept. Prot. Res., 33 (1989) 281.
8. Stroup, A.N., Rheingold, A.L., Rockwell, A.L. and Gierasch, L.M., J. Am. Chem. Soc., 109 (1987) 7146.
9. Graf, J., Ogle, R.C., Robey, R.A., Sasaki, M., Martin, G.R., Yamada, Y. and Kleinman, H.K., Biochemistry, 26 (1987) 6896.

Sequence-specific resonance assignment and solution conformational analysis of nisin by ^{1}H 2D NMR spectroscopy

Weng C. Chan[a],*, Barrie W. Bycroft[a], Lu-Yun Lian[b] and Gordon C.K. Roberts[b]

[a]*Department of Pharmaceutical Sciences, University of Nottingham, University Park, Nottingham NG7 2RD, U.K.*
[b]*Biological NMR Centre and Department of Biochemistry, University of Leicester, University Road, Leicester LE1 9HN, U.K.*

Introduction

The peptide antibiotic nisin is produced by strains of *Lactococcus lactis*, and possesses potent antimicrobial activity against a spectrum of gram-positive organisms. Nisin contains several atypical residues, namely dehydroalanine, dehydrobutyrine, *meso*-lanthionine and (2*S*,3*S*,6*R*)-3-methyllanthionine; the latter two residues introduce thioether bridges at various locations in the molecule, resulting in a series of cyclic units [1,2]. We report here to complete resonance assignment of the ^{1}H NMR spectrum of nisin in aqueous solution, and some preliminary conformation models based on constraints derived from a series of NOESY experiments.

Results and Discussion

Nisin was purified to homogeneity by semi-preparative RPHPLC [3]. Complete sequence-specific resonance assignment of nisin was achieved by application of several 2D NMR techniques [4], and the results are summarized in Table 1. The H$_2$O resonance was suppressed by using either the SCUBA-pulse [5] or the jump-and-return [6] sequence.

Resonances were first assigned to individual types of amino acids by observation of relayed scalar connectivities from the peptide backbone amide protons to the side-chain aliphatic protons (by analysis of HOHAHA spectra). The sequential assignment that followed was based on a systematic search for short- and medium-range NOESY cross-peaks, and confirmed by the results obtained from the relayed-NOESY experiment, which is composed of a NOESY pulse sequence ($\tau_m = 350$ ms), followed by a MLEV-17 pulse sequence [7] ($\tau_m = 40$ ms). The occurrences of the five thioether ring systems were confirmed by the observation of NOE connectivities across the sulphur atom (e.g. D-Abu^{13}CβH–Ala$_s^{19}$CβH NOE and D-Abu^{13}CβH–Ala$_s^{19}$CαH NOE-*J* connectivities).

---

*To whom correspondence should be addressed.

Table 1 ^{1}H NMR (500 MHz) chemical shifts of nisin (5 mM) in aqueous solution (100 mM) sodium phosphate buffer; pH 2.25; $H_2O,85:D_2O,15$) at 303 K

	NH	CαH	CβH	CγH	CδH	CεH
			δ (ppm)			
Ile[1]	8.22	4.19	2.16	1.37,1.62 1.16	0.95	
Δ^{z}Abu[2]	9.93		6.65	1.87		
D-Ala$_s$[3a]	8.24	4.68	3.18,3.32			
Ile[4]	7.89	4.39	2.12	1.22,1.45 0.99	0.88	
ΔAla[5]	9.85		5.49,5.61			
Leu[6]	8.94	4.47	1.78	1.69	0.94,1.00	
Ala$_s$[7]	8.27	4.60	2.99,3.11			
D-Abu[8b]	8.88	5.16	3.61	1.40		
Pro[9]		4.48	1.77,2.50	2.22,1.90	3.50	
Gly[10]	8.70	3.66,4.45				
Ala$_s$[11]	7.96	4.07	3.07,3.70			
Lys[12]	8.64 7.60 (NεH$_3$$^+$)	4.39	1.86	1.44,1.54	1.75	3.05
D-Abu[13]	8.29	4.60	3.66	1.37		
Gly[14]	8.35	4.12,4.16				
Ala[15]	8.57	4.30	1.51			
Leu[16]	8.47	4.37	1.78	1.70	0.95,0.98	
Met[17]	7.86	4.68	2.15,2.32	2.53,2.70	2.18	
Gly[18]	8.12	3.90,4.18				
Ala$_s$[19]	7.72	4.55	3.01,3.04			
Asn[20]	8.55	4.75	2.86	6.94,7.60 (βCONH$_2$)		
Met[21]	8.27	4.57	2.04,2.17	2.56,2.64	2.17	
Lys[22]	8.43 7.60 (NεH$_3$$^+$)	4.37	1.90	1.48,1.55	1.77	3.06
D-Abu[23]	8.80	5.05	3.60	1.40		
Ala[24]	8.24	4.72	1.50			
D-Abu[25]	9.09	4.85	3.57	1.48		
Ala$_s$[26]	7.87	3.93	2.72,3.48			
His[27]	8.74	4.98	3.13,3.42	8.65 (H2)	7.35 (H5)	
Ala$_s$[28]	7.90	4.47	2.75,3.73			
Ser[29]	8.47	4.54	3.90			
Ile[30]	8.13	4.24	1.87	1.20,1.38 0.90	0.90	
His[31]	8.65	4.85	3.21,3.31	8.63 (H2)	7.32 (H5)	
Val[32]	8.32	4.24	2.17	1.01		
DAla[33]	9.71		5.74			
Lys[34]	8.50 7.60 (NεH$_3$$^+$)	4.51	1.91,2.01	1.54	1.77	3.07

[a]Alanine and [b]α-aminobutyric acid moieties of the lanthionine and 3-methyllanthionine residues.

Preliminary computational analysis, with input of distance constraints derived from over 150 observed NOE data revealed that the peptide backbone adopts a rather compact conformation. In addition, the low NMR temperature coefficient ($-\Delta\delta/\Delta T$, x10^{-3} ppm/K) values of the α-amide NHs of Ile[4], Ala$_s$[7], Ala$_s$[11], Gly[14],

Met[17], Ala$_s$[19] and Ala$_s$[26] (1.2–3.1) suggest that these N*H*s are either solvent shielded or involved in intramolecular H-bonding; in contrast, the solvent exposed amide N*H*s generally displayed $-\Delta\delta/\Delta T$ in the range of 8.0–16.5.

Further work is currently in progress in structural refinement, as well as correlating the solution structure with the structure–activity data of several nisin-derived peptides.

Acknowledgements

This work is supported by a grant from Science and Engineering Research Council, U.K. We thank Aplin & Barrett, Beaminster, U.K. for generous supply of nisin.

References

1. Hurst, A., Adv. Appl. Microbiol., 27 (1981) 85.
2. Gross, E. and Morrell, J.L., J. Am. Chem. Soc., 93 (1971) 4634.
3. Chan, W.C., Bycroft, B.W., Lian, L.-Y and Roberts, G.C.K., FEBS Lett., 252 (1989) 29.
4. Chan, W.C., Lian, L.-Y., Bycroft, B.W. and Roberts, G.C.K., J. Chem. Soc., Perkin Trans. I., (1989) in press.
5. Brown, S.C., Weber, P.L. and Mueller, L., J. Magn. Reson., 77 (1988) 166.
6. Plateau, P. and Gueron, M., J. Am. Chem. Soc., 104 (1982) 7310.
7. Bax, A. and Davis, D.G., J. Magn. Reson., 65 (1985) 355.

Conformational studies of nisin and its fragments by NMR and computer simulations

Darryl E. Palmer[a], Dale F. Mierke[a], Christian Pattaroni[a], Kenichi Nunami[a], Seonggu Ro[a], Ziwei Huang[a], Tateaki Wakamiya[b], Koichi Fukase[b], Manabu Kitazawa[b], Hiroshi Fujita[b], Tetsuo Shiba[b] and Murray Goodman[a]

[a]*Department of Chemistry, B-043 University of California, San Diego, CA 92093, U.S.A.*
[b]*Department of Chemistry, Osaka University, Japan*

Introduction

We report recent progress in elucidating the structure of the peptide antibiotic, nisin. Nisin contains α,β-unsaturated amino acids and five lanthionine (monosulfide) bridged rings [1]. Similar features appear in other naturally occurring antibiotics such as subtillin, epidermin and gallidermin, possibly playing a role in antibiotic activity [2]. We have characterized segments containing each of these constrained rings, (with acetyl and *N*-methyl amide end groups) and the tripeptide region linking rings C and D, (with adamantoyl and adamantamide end groups) through NMR and MD simulations [3]. In a molecular *aufbau* approach, we are assembling the resulting conformations to aid in structural determination of the whole molecule.

Methods

Our NMR studies included HOHAHA, NOESY, ROESY, as well as t_1, homonuclear J-coupling, and temperature coefficient measurements [4–6]. Additionally, we have examined the metal-binding properties of whole nisin. To date we have worked primarily in DMSO-d_6, but we have also used chloroform and water. We use a GE Nicolet 500 MHz spectrometer.

Computer simulations began with high temperature (750–1000 K) dynamics. At ps intervals, conformations were extracted and energy-minimized, with or without NOE constraints. We used NOE constraints as a short-cut in our computational search, and not to fit the experimental data. All constrained structures were subsequently 'relaxed' through unconstrained minimizations. Starting from the lowest energy conformations encountered (within 3 kcal/mol of the lowest energy conformation), we proceeded with MD at 300 K. Again at ps intervals, we carried out unconstrained minimizations to generate families of the lowest energy conformations. Properties time-averaged over the course of these low temperature dynamics agreed better with NMR results than did minimum energy conformations. All calculations were carried out on the San Diego Supercomputer Center Cray X-MP computer using DISCOVER Ver. 2.21 software.

616

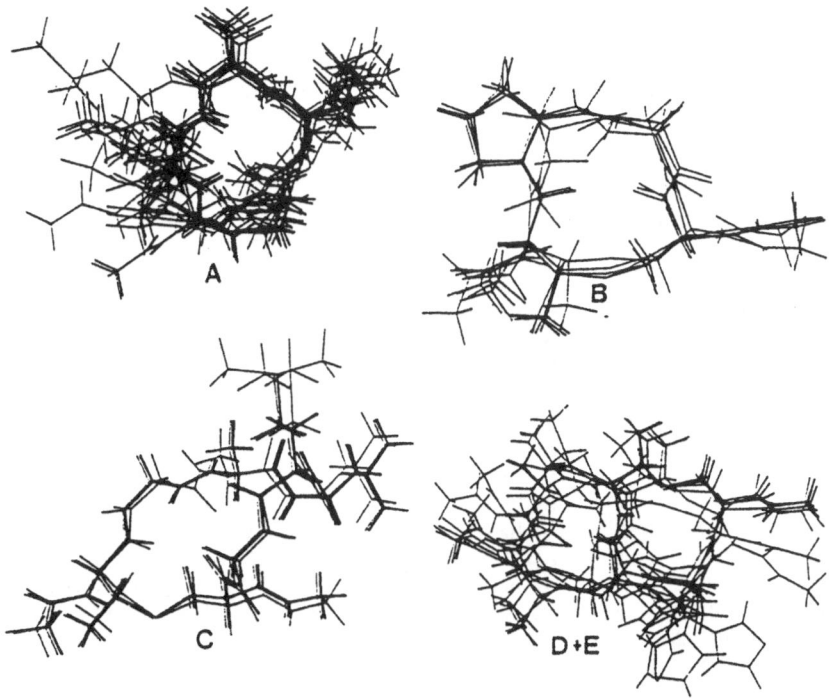

Fig. 1. The lowest energy conformations determined for the rings of nisin.

The novel α,β-unsaturated amino acid residues required new force constants and geometric parameters that we have developed through calculations, IR spectroscopy, and X-ray diffraction studies. Using these parameters, we have calculated a minimum energy conformation for acetyl, alanyl *N*-methyl amide, which agrees to within 2% with bond lengths and angles from X-ray diffraction. The calculated carbonyl stretching frequencies agree to within 0.5% with IR spectra collected in chloroform.

Results and Discussion

In such a manner we have characterized each ring of nisin (Fig. 1). The simulations of rings B, C, and D + E, and the C-D linker region, were in overall agreement with the NMR observations. Ring B shows one stable and one weak transannular hydrogen bond. Ring C, surprisingly rigid for a heptapeptide, shows two stable hydrogen bonds. Ring D + E shows no stable hydrogen bonding, and the histidine side chain swings from one lanthionine to the other, as indicated by the NOE to each. However, simulations of ring A indicate that the psi torsion of dehydroalanine is *cis*, and a strong NOE from DhA HB (*trans*) to Leu HN indicates a *trans* conformation. Thus, we are continuing to refine the force constants for dehydroalanine.

The entire nisin molecule as well as nisin residues 1–32 have been assigned. Our assignments deviate somewhat from results recently reported by another group, perhaps because of TFA, which they added to increase peak dispersion [7]. Table 1 lists interresidue NOEs (ROESY, 300 ms mixing time) observed for the entire molecule and demonstrates how well these were reflected by NOEs found in the isolated fragments.

Table 1   *Observed NOE of nisin in DMSO-d_6*

Ile^{1}HA-DhB^{2}HN		Ala^{11}HA-Lys^{12}HN		Ala^{26}HA-His^{27}HN	P
Ile^{1}HG-DhB^{2}HG		Lys^{12}HA-D-Abu^{13}HN		His^{27}HA-Ala^{28}HN	A
Ile^{1}HD-DhB^{2}HG		D-Abu^{13}HB-Ala^{19}HB2	P	Ala^{28}HA-Ser^{29}HN	
DhB^{2}HB-D-Ala^{3}HN		Ala^{15}HA-Leu^{16}HN	P	Ser^{29}HA-Ile^{30}HN	
Ile^{4}HN-DhA^{5}HN	P	Leu^{16}HA-Met^{17}HN	P	Ile^{30}HA-His^{31}HN	
Ile^{4}HA-DhA^{5}HN	P	Met^{17}HA-Gly^{18}HN	P	Ile^{30}HG-His^{31}HN	
Ile^{4}HB-DhA^{5}HN	P	Ala^{19}HA-Asn^{20}HN	S	Ile^{30}HG-His^{31}HD	
DhA^{5}HB(*cis*)-Leu^{6}HN	N	Asn^{20}HA-Met^{21}HN	P	His^{31}HA-Val^{32}HN	
DhA^{5}HB(*trans*)-Leu^{6}HN	P	Met^{21}HA-Lys^{22}HN	P	Val^{32}HN-DhA^{33}HN	
Leu^{6}HA-Ala^{7}HN	P	Lys^{22}HA-D-Abu^{23}HN	S	Val^{32}HA-DhA^{33}HN	
Ala^{7}HA-D-Abu^{8}HN	S	D-Abu^{23}HA-Ala^{24}HN	P	Val^{32}HB-DhA^{33}HN	
D-Abu^{8}HA-Pro^{9}HD1	N	Ala^{24}HA-D-Abu^{25}HN	P	DhA^{33}HN-Lys^{34}HN	
D-Abu^{8}HA-Pro^{9}HD2	A	D-Abu^{25}HN-Ala^{26}HN	P	Dha^{33}HB(*cis*)-Lys^{34}HN	N
D-Abu^{8}HA-Ala^{11}HB2	N	D-Abu^{25}HA-Ala^{26}HN	P	DhA^{33}HB(*trans*)-Lys^{34}HN	
D-Abu^{8}HB-Ala^{11}HB2	A	D-Abu^{25}HA-Ala^{28}HB2	N		
Gly^{10}HN-Lys^{12}HD		D-Abu^{25}HB-Ala^{28}HB2	A		

P = present in fragment spectra.
A = absent from fragment spectra.
S = suggested in fragment spectra by NOE to end group.
N = indirect NOE, not observed in fragment spectra.

Acknowledgements

This work was supported in part by NIH grant GM-19694 and a grant from the SDSC.

References

1. Gross, E., In Walter, R. and Meienhofer, J. (Eds.) Proceedings of the 4th American Peptide Symposium, Ann Arbor Science Publishers, Ann Arbor, MI, 1975, p. 31.
2. Jung, G., In Rivier, J.E. and Marshall, G.R. (Eds.) Peptides: Chemistry, Structure and Biology (Proceedings of the 11th American Peptide Symposium), ESCOM, Leiden, 1990, p. 865.
3. Palmer, D.E., Mierke, D.F., Pattaroni, C., Goodman, M., Wakamiya, T., Fukase, K., Kitazawa, M., Fujita, H. and Shiba, T., Biopolymers, 28 (1989) 397.
4. States, D.J., Haberkorn, R.A. and Ruben, D.J., J. Magn. Reson., 48 (1982) 286.
5. Bax, A. and Davis, D.G., J. Magn. Reson., 65 (1985) 355.
6. Bax, A. and Davis, D.G., J. Magn. Reson., 63 (1985) 207.
7. Chan, W.C., Lian, L.-Y., Bycroft, B.W. and Roberts, G.C.K., J. Chem. Soc. Perkin Trans. 1, (1989) in press.

Conformation of nikkomycin X in aqueous solution

Eduardo Krainer[a], L.D.S. Yadav[a], Fred Naider[a] and Jeffrey M. Becker[b]

[a]*Department of Chemistry, The College of Staten Island, CUNY, Staten Island,
NY 10301, U.S.A.*
[b]*Department of Microbiology, University of Tennessee, Knoxville, TN 37996, U.S.A.*

Introduction

Nikkomycin X (Fig. 1), a dipeptidyl nucleoside antibiotic produced by
Streptomyces tendae, is a strong inhibitor of chitin synthetase from pathogenic
fungi [1]. We undertook the study of its conformation in aqueous solution as
a first step toward understanding the molecular interactions of this antibiotic
with large membrane-associated proteins, such as peptide permeases and chitin
synthetase, from pathogenic yeasts.

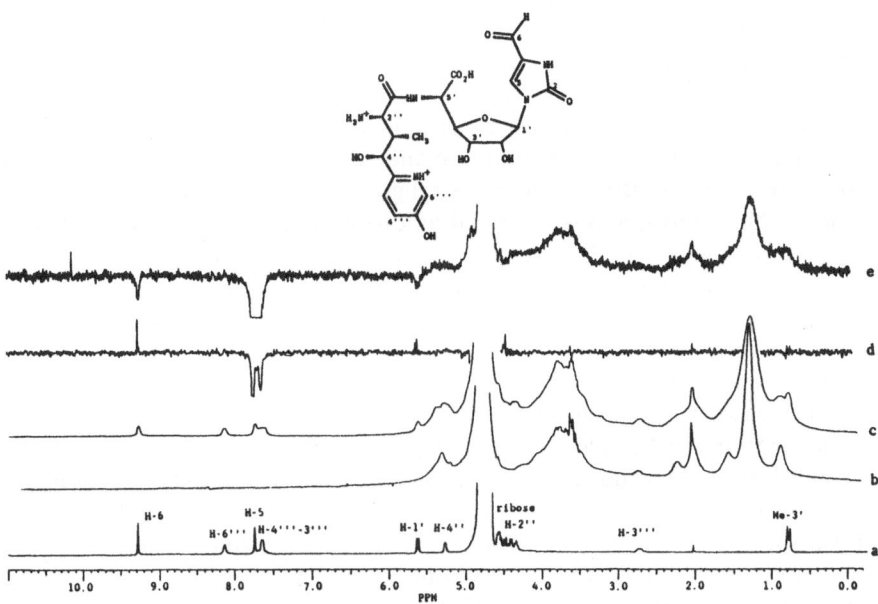

Fig. 1. *^{1}H NMR spectra in D_2O of (a) Nikkomycin X, and (b) membrane fractions from*
Candida albicans *containing chitin synthetase; (c) Nikkomycin X plus the membrane fractions;*
(d) NOE difference spectrum, same conditions as in (a), H-5 irradiated (3 s); (e) NOE
difference spectrum, same conditions as in (c), H-5 irradiated (500 ms).

Results and Discussion

200-MHz ^{1}H NMR spectra were run directly in H_2O, using the 1331 solvent-suppression pulse sequence proposed by Hore [2]. We measured ^{1}H NMR chemical shifts and coupling constants in the pH range from 1 to 6, and NOEs upon irradiation of the resolved proton signals.

The ^{1}H NMR titration data were fit to a one-proton titration curve,

$$\delta = \frac{\delta_{AH} + \delta_A 10^{(pH-pKa)}}{1 + 10^{(pH-pKa)}}$$

using a non-linear, least-squares curve-fitting iterative computer algorithm. The amide proton curve showed an upfield titration shift $\delta_A - \delta_{AH} = -0.61$ ppm, and an inflection point at $pKa = 2.95 \pm 0.05$ corresponding to the ionization of the carboxyl group. The pKa for the pyridine moiety, deduced from the H-3'''-4''' and H-6''' curves, was 4.34 ± 0.02. The curves for H-5 and H-1' showed small titration shifts (+0.08 and –0.03 ppm, respectively) and inflections at the pKa of the carboxyl group. This implies an intramolecular through-space interaction between those protons and the carboxyl group, and is consistent with a conformation where the carboxyl group is situated above the plane of the ribose ring and close to the imidazolinone heterocycle. The imidazolinone ring adopts an *anti* conformation which brings H-5 close to the carboxyl group. It is possible that an indirect interaction occurs between the carboxyl group and H-5 in a *syn* conformation. However, we view this interaction as unlikely since both H-6 and NH-3 exhibited significantly smaller changes (0.02 and ≤ 0.04 ppm, respectively) upon titration of the carboxyl group.

The coupling constant $J_{1'2'}$ increased gradually during titration, from a value of 4.3 Hz at pH 1.0 to 5.9 Hz at pH 5.85. This variation corresponds to a shift of the conformational equilibrium from $X_N \simeq 0.6$ to $X_N \simeq 0.4$ as the pH increased from 1 to 6 (X_N is the mole fraction of N-conformer [3]).

We measured NOEs in D_2O solution using long irradiation times (3.0 s). All the enhancements were small and positive. Through-space interactions were only observed between hydrogen atoms within a given amino acid. This supports a conformationally flexible structure in water. NOEs were observed between H-5 and H-2' and H-5 and H-3', supporting the existence of the *anti* conformation in the nucleoside moiety. NOEs measured between H-5 and H-1' demonstrate [4] that the *syn* conformation is also significantly populated (Fig. 1d).

We also carried out NOE measurements in the presence of *Candida albicans* membranes containing activated chitin synthetase (protein: 16 mg/ml; activity: 0.95 nmol/mg.min; pD* 6.45). As seen in Fig. 1c, the protons of the inhibitor are clearly resolved in the presence of the membranes. Most significantly (Fig. 1e), in contrast to the NOEs found in D_2O, all NOEs in the presence of membranes were negative. This strongly suggests that we are observing the inhibitor bound to the enzyme. However, it is also possible that nikkomycin binds non-specifically to lipids or other membrane proteins, or that changes in the solution viscosity cause the change in sign of the NOEs. Strains of *C. albicans* unable to synthesize chitin are not viable. We therefore ran control experiments by measuring NOEs for nikkomycin X in the presence of DPPC vesicles (16 mg/ml, diameter 1000 Å) and DPPC vesicles (16 mg/ml) plus BSA (16 mg/ml). In both experiments, the sign of the Overhauser effect was identical to that measured for free nikkomycin in solution. This result supports our conclusion that the negative NOEs provide information on the conformation of nikkomycin bound to chitin synthetase.

Acknowledgements

The authors acknowledge support of grants from the NIH (AI-14387) and the American Cancer Society (BC 626).

References

1. Hagenmaier, H., Keckeisen, A., Zahner, H. and Konig, W.A., Liebigs Ann. Chem., (1979) 1494.
2. Hore, P.J., J. Magn. Res., 55 (1983) 283.
3. Altona, C. and Sundaralingam, M., J. Am. Chem. Soc., 95 (1973) 2333.
4. Schirmer, R.E., Davis, J.P., Noggle, J.H. and Hart, P.A., J. Am. Chem. Soc., 94 (1972) 2561.

Site-directed chemical modifications as an aid for the three-dimentional structure studies of the toxic site of a cardiotoxin using proton NMR and distance geometry calculations

Christian Roumestand, Eric Gatineau, Bernard Gilquin, André Ménez and Flavio Toma

Service de Biochimie, Département de Biologie, CEN-Saclay, F-91191 Gif-sur-Yvette, France

Introduction

Toxin γ, a cardiotoxin from *Naja nigricollis*, is a protein containing 60 residues with four disulfide bridges in a unique polypeptide chain [1]. Single-site chemically modified derivatives of toxin γ were prepared for residues at different positions (Leu[1]; Lys[2,12,16,18,23,35]; Trp[11]; Tyr[22,51]) in the attempt to delineate the toxic site of the protein. It followed from this work that (i) Lysine[12], and to a lesser extent Lys[35], probably play a direct role in the toxic activity, and (ii) other residues, i.e., Trp[11], Tyr[22] and Lys[23], might be rather involved in the conformational properties of the toxic site of the protein [2,3]. This paper deals with the comparison of the 3D structures of toxin γ and the three derivatives, i.e., (NPS-Trp[11])-toxin γ, [(3-nitro)-Tyr[22]]-toxin γ and [(3-nitro)-Tyr[51]]-toxin γ.

Results and Discussion

The proton NMR spectra at 500 or 600 MHz (Bruker WM500 and AM600 spectrometers) of toxin γ and of the derivatives were fully assigned in sequence-specific terms. The NOEs, backbone-coupling constants and temperature coefficients of the amide protons thus obtained, were used for generating 3D structures by means of the DISGEO algorithm. Ten structures of toxin γ compatible with the NMR experimental data (381 distance constraints for proton pairs at <0.45 nm) were computed (Fig. 1). Root-mean-square differences for Cα positions of the calculated structures vary from 0.275 to 0.418 nm. All of them are close to the crystal structure of the 74% sequence-homologous cardiotoxin of *Naja mossambica mossambica V[II]4* [4] (Fig. 1) and, respectively, to the solution structure of the 95% sequence-homologous cardiotoxin of *Naja mossambica mossambica IIb* [5].

Three adjacent loops are present, stabilized by antiparallel β-sheets formed in both the solid and solution states by strands involving respectively, residues 2–4/11–13 (loop I), residues 18–24/34–40 (loop II) and residues 20–26/49–55 (loop II and III). Chemical modifications of Trp[11] in loop I, Tyr[22] in loop II

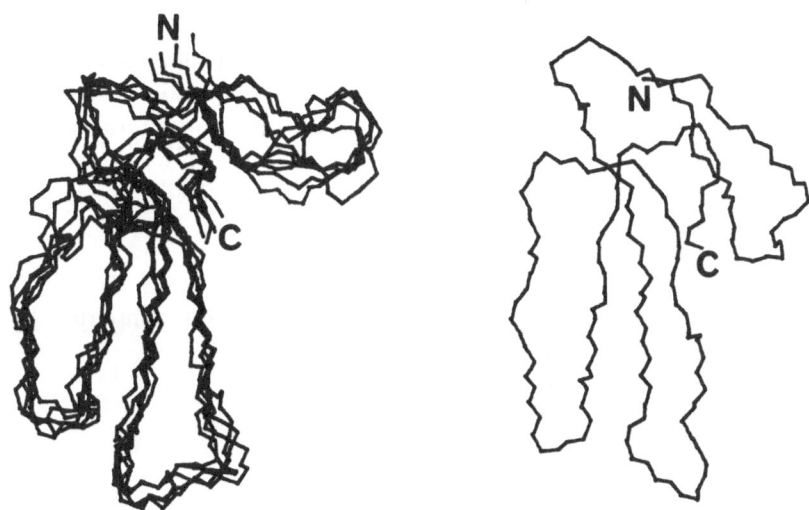

Fig. 1. Five structures of toxin γ computed with DISGEO (left) and projected view of the crystal structure of cardiotoxin V[II]4 [4] (right).

and Tyr[51] in loop III have no influence on the folding. On the contrary, local or medium-range perturbations of the 3D structure appear in the derivatives through variations in the NMR parameters (NH and H_α chemical shifts, non-sequential inter-residue NOEs). These were analyzed in detail in the Trp[11] derivative. In this, large concomitant variations of the NH chemical shifts of residues at position 3–13 (Fig. 2) and, respectively, of the Hα of Leu[1] (+0.23 ppm), Asn[4] and Lys[12] (–0.15 ppm) were found relative to toxin γ. Moreover, interactions of the indole side chain of Trp[11] with the backbone (positions 7, 10, 12) and other side chains (positions 2, 8, 9) disappear in the derivative.

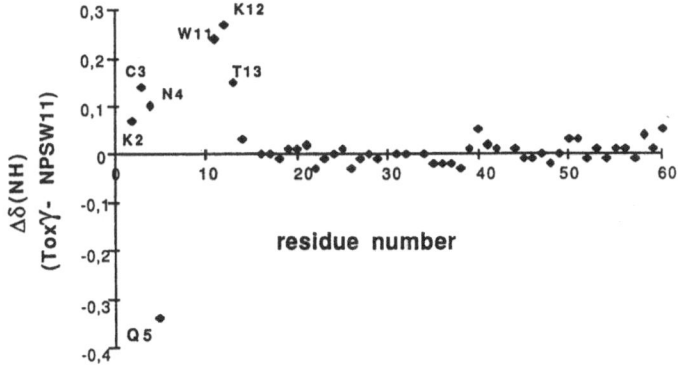

Fig. 2. Effect of Trp[11] modification on NH chemical shifts (ppm) of residues 1–60 (toxin γ vs. derivative).

Altogether, these results indicate a strong perturbation of the 3D organization of the antiparallel β-sheet present in loop I.

In conclusion, this structural change appears as the major factor related to the decrease of the toxic activity in the Trp[11] derivative of toxin γ. Interestingly, the antiparallel β-sheet structure in loop I is conserved in natural mutants of toxin γ [4,5] that all belong to the same cardiotoxin family [1].

Acknowledgements

We thank Drs. I. D. Kuntz and J. Thomason (UCSF, San Francisco) for providing us with a copy of VEMBED. This work was supported by the Commissariat à l'Energie Atomique.

References

1. Grognet, J.-M., Ménez, A., Drake, A., Hayashi, K., Morrison, I.E.G. and Hider, R.C., Eur. J. Biochem., 172 (1988) 383.
2. Gatineau, E., Toma, F., Montenay-Garestier, Th., Takechi, M., Fromageot, P. and Ménez, A., Biochemistry, 26 (1987) 8046.
3. Gatineau, E., Takechi, M., Harvey, A.L., Montenay-Garestier, Th. and Ménez, A., Biochemistry, submitted.
4. Rees, B., Samama, J.-P., Thierry, J.-C., Gillibert, M., Fisher, J., Schweitz, H., Ladzdunski, M. and Moras D., Proc. Natl. Acad. Sci. U.S.A., 84 (1987) 3132.
5. Steinmetz, W.E., Bougis, P.E., Rochat, H., Redwine, O.D., Braun, W. and Wüthrich, K., Eur. J. Biochem., 172 (1988) 101.

NMR and preliminary modeling studies on the variant-3 neurotoxin from *Centruroides sculpturatus* Ewing

N. Rama Krishna[a], Carolynn H. Moore[b], Sudha Narasimhan[a] and Dean D. Watt[c]

[a]*Comprehensive Cancer Center and Department of Biochemistry, University of Alabama at Birmingham, Birmingham, AL 35294, U.S.A.*
[b]*Health Data Systems, University of Alabama at Birmingham, Birmingham, AL 35294, U.S.A.*
[c]*Department of Biochemistry, Creighton University, Omaha, NB 68178, U.S.A.*

Introduction

The variant-3 neurotoxin from the venom of the scorpion *Centruroides sculpturatus* Ewing (range southwestern U.S.A.) is a small basic protein composed of 65 residues [1]. It binds to the sodium channels in a voltage-dependent manner and prolongs the inactivation of the sodium current. The crystallographic structure for this neurotoxin has been characterized in great detail [2]. The present study reports some preliminary results of our model building efforts for this protein based on NMR and distance geometry calculations.

Results and Discussion

The sequence specific assignments for this protein based on 2D NMR spectroscopic methodology have been reported by us recently [3,4]. A number of proton spatial contacts among the different amio acids in the sequence have been established by 2D NOESY experiments on spectrometers operating at 400, 500 and 600 MHz (Bruker WH-400, WM-500 and AM-600). From an analysis of the sequential NOESY connectivities and some NOESY contacts between non-neighboring residues, certain general conclusions were deduced about the secondary structure of this protein in aqueous solutions. An α-helical conformation for residues 23–31 was suggested by the observation of $N(i)$–$N(i+1)$ and $\alpha(i)$–$\beta(i+3)$ contacts among these residues. Similarly, an antiparallel β-sheet formed by residues 1–4 and 47–50 was indicated by the presence of cross-strand α-α NOESY contacts between residues 2 and 49 and between 4 and 47. Evidence for the participation of a third strand formed by residues 36–41 in the antiparallel β-sheet was also indicated by the presence of α-α NOESY contacts between residues 41 and 46 and between 36 and 50, with a bulge involving residues 37 and 38. The *cis* conformation for Pro[59] was identified from the α-α contacts between residues 59 and 58. A number of interesting contacts among the aromatic side chains were also detected.

Distance geometry calculations [5,6] were initiated to define further the overall conformation of the protein based on the NMR data. The calculations were performed on a CRAY computer using the UCSF distance geometry algorithm. In defining the constraints, a data set consisting of 237 NOESY contacts obtained primarily from 400 and 500 MHz spectra was employed. A uniform distance range of 2–4.2 Å was used for these contacts. Some hydrogen bond distance constraints in the α-helix and β-sheet regions as well as some non-NOE constraints (distances $\geqslant 4.5$ Å) to keep some atoms apart were also introduced into the calculations. The results of the preliminary distance geometry calculations are shown in Fig. 1. The structures in this figure are arranged to minimize the rms deviation with respect to the average structure. The rms deviations varied from 1.09 Å to 1.86 Å. The α-helix and the antiparallel β-sheet structures are well-represented in the distance geometry structures. A comparison of the backbone conformations deduced from crystallography and our preliminary distance geometry calculations shows that they are qualitatively very similar. The side chains for some of the residues, however, undergo a minor rearrangement. Refinement of these distance geometry structures by inclusion of additional NOESY constraints obtained from 600 MHz data, and by restrained MD and energy minimization calculations, are in progress.

Fig. 1. *Variant-3 neurotoxin structures obtained from distance geometry calculations.*

Acknowledgements

The authors thank Drs. I.D. Kuntz and J. Thomason (UCSF, San Francisco) for the distance geometry program, and for advice on its use. The work was supported by grants DMB-8705496 and BBS-8611303 from the National Science Foundation, and by grants CA-13148 and RR-03373 from the National Institutes of Health.

References

1. Watt, D.D. and Simard, J.M., J. Toxicol., Toxin Rev., 3 (1984) 181.
2. Fontecilla-Camps, J.C., Almassy, R.J., Ealick, S.E., Suddath, F.L., Watt, D.D., Feldman, R.J. and Bugg, C.E., Trends Biochem. Sci., 6 (1981) 291.
3. Nettesheim, D.G., Klevit, R.E., Drobny, G., Watt, D.D. and Krishna, N.R., Biochemistry, 28 (1989) 1548.
4. Krishna, N.R., Nettesheim, D.G., Klevit, R.E., Drobny, G., Watt, D.D. and Bugg, C.E., Biochemistry, 28 (1989) 1556.
5. Havel, T.F., Kuntz, I.D. and Crippen, G.M., Bull. Math. Biol., 45 (1983) 665.
6. Braun, W., Quart. Rev. Biophys., 19 (1987) 115.

Proton-detected ^{15}N NMR studies of little gastrin

Stefano Mammi[a], Henriette Molinari[b] and Evaristo Peggion[a]

[a]*Biopolymer Research Center, Department of Organic Chemistry, University of Padova, Via Marzolo 1, I-35131 Padova, Italy*
[b]*Dipartimento di Chimica Organica ed Industriale, Universita' di Milano, Via Golgi 19, I-20133 Milano, Italy*

Introduction

The nitrogen atom in the peptide linkage is a potential source of information on secondary structure. The application of ^{15}N NMR to biological systems was made possible by the relatively recent development of reverse detection methods for observation of low sensitivity nuclei via the coupled protons. The crucial problem in these experiments is the suppression of large unwanted signals from the protons attached to ^{14}N. In the case of experiments in water, or other non-deuterated solvents, an additional and even more critical source of undesired resonances is the solvent itself. Although some methods have been suggested, water suppression in the reverse mode still lacks a general solution.

In the reverse experiment, selection of the desired signals can be achieved in two ways. The suppression can occur after digitization, by means of phase cycling, thus imposing severe stability and dynamic range requirements on the spectrometer [1]. These methods are unsuitable for experiments in water even with the insertion of efficient water suppression schemes such as DANTE [2] or selective excitation (Redfield 2-1-4, Jump-Return) [3,4].

A second kind of experiment achieves selection of the desired signals at each scan, using either BIRD pulse sandwiches [5] or spin-lock pulses [6,7]. These types of sequences ar potentially more favorable for experiments in water.

Results and Discussion

We have compared the two different classes of sequences for water suppression in reverse experiments. Better results were obtained with sequences from the second type. Among these, the use of the BIRD scheme in conjunction with presaturation of the water signal is unfavorable because of the different relaxation times of solvent and solute protons. We obtained best results employing the sequence NEMESIS [7]. The pulse sequence is a modified version of the reverse INEPT [8] in which the water suppression is accomplished by means for randomizing spin-lock pulses after the initial refocused INEPT sequence. We implemented this sequence on our Bruker AM-400 and modified it with the insertion of a 180° refocusing pulse in the middle of the evolution in order to eliminate heteronuclear coupling in F1.

We applied these methods to [Nle[15]]-little gastrin (pEGPWLEEEE-EAYGWXDF-amide) at natural abundance ^{15}N. This peptide was shown to have a random structure in water [9,10], but to adopt an ordered conformation in SDS micelles [11,12]. We performed experiments using the modified NEMESIS sequence at concentrations as low as 10.8 mM in 90% H_2O, both in water and in SDS micelles in order to detect possible effects of structural variations on the ^{15}N chemical shift.

The proton spectrum of little gastrin in water was fully assigned by conventional 2D techniques. Our results agree with and complete the work of Torda et al. [10]. The complete assignment of the proton spectrum in SDS was previously accomplished [12]. From these results, we were able to assign all the ^{15}N resonances in water and many of the ^{15}N resonances in SDS. In both cases, the Gly resonances are shifted upfield relative to the other residues. From water to SDS micelles, the biggest differences in chemical shift are found for Nle and Leu.

References

1. Bax, A., Griffey, R.H. and Hawkins, B.L., J. Magn. Reson., 55 (1983) 301.
2. Esposito, G., Gibbons, W. and Bazzo, R., J. Magn. Reson., 80 (1988) 523.
3. Griffey, R.H., Poulter, C.D., Bax, A., Hawkins, B.L., Yamaizumi, Z. and Nishimura, S., Proc. Natl. Acad. Sci. U.S.A., 80 (1983) 5895.
4. Sklenar, A. and Bax, A., J. Magn. Reson., 71 (1987) 379.
5. Bax, A. and Subramanian, S., J. Magn. Reson., 67 (1986) 565.
6. Otting, G. and Wüthrich, K., J. Magn. Reson., 76 (1988) 569.
7. Lee, K.S. and Morris, G.A., J. Magn. Reson., 78 (1988) 156.
8. Bodenhausen, G. and Ruben, D.J., Chem. Phys. Lett., 69 (1980) 195.
9. Peggion, E., Jaeger, E., Knof, S., Moroder, L. and Wuensch, E., Biopolymers, 20 (1981) 633.
10. Torda, A.E., Baldwin, G.S. and Norton, R.S., Biochemistry, 24 (1985) 1720.
11. Wu, C.C., Hachimori, A. and Yang, J.T., Biochemistry, 21 (1982) 4556.
12. Mammi, S. and Peggion, E., Biochemistry, submitted.

Conformationally restricted formyl-methionyl tripeptide chemoattractants

Claudio Toniolo[a,*], Marco Crisma[a], Giovanni Valle[a], Gian Maria Bonora[a],
Stefano Polinelli[a], Elmer L. Becker[b], Richard J. Freer[c] and
Padmanabhan Balaram[d]

[a]*Biopolymer Research Center, CNR, Department of Organic Chemistry, University of Padova, Via Marzolo 1, I-35131 Padova, Italy*
[b]*Department of Pathology, University of Connecticut Health Center, Farmington, CT 06032, U.S.A.*
[c]*Department of Pharmacology, Medical College of Virginia, Richmond, VA 23298, U.S.A.*
[d]*Molecular Biophysics Unit, Indian Institute of Science, Bangalore 560012, India*

Introduction

The tripeptide CHO-L-Met-L-Leu-L-Phe-OH is known to induce chemotaxis and selective release of lysosomal enzymes in neutrophils. The extended β-sheet conformation has been proposed as the receptor-bound conformation from spectroscopic analyses in solution. More recently, the X-ray diffraction structure of the bioactive derivative CHO-L-Met-L-Leu-L-Phe-OMe indicated the preference of the tripeptide for an 'open' folded conformation in the crystal state, helical at the central Leu residue but extended at the terminal Met and Phe residues. In this communication, we describe the results of a systematic, detailed biological study (using the release of the neutrophil granule enzyme β-glucosaminidase) and conformation analysis (using X-ray diffraction and ^{1}H NMR) of the tripeptide CHO-L-Met-Xxx-L-Phe-OMe (Xxx = Aib, Ac$_3$c, Ac$_5$c, Ac$_6$c, and Ac$_7$c) and their Boc-protected synthetic precursors as well. The α-amino acids dialkylated at the α-carbon listed above are known to be conformationally restricted and to favor strongly intramolecularly H-bonded forms of the β-bend type.

Results and Discussion

The N^{α}-formylated Ac$_5$c, Ac$_6$c, and Ac$_7$c analogs are approximately 2-, 5-, and 7-fold more active than the prototypical CHO-L-Met-L-Leu-L-Phe-OH (ED$_{50}$ = 1.3–3.5×10^{-10}). Conversely, the Aib and Ac$_3$c analogs are 4- and 150-fold less active, respectively. Thus, in general, the conformational restrictions in the CHO-L-Met-Xxx-L-Phe-OMe peptides are compatible with quite high activity. In addition, the activity increases as the hydrophobic pocket in the receptor for position 2 is more completely occupied. In the N^{α}-Boc-protected tripeptides, a marked fall in activity is observed.

---

*To whom correspondence should be addressed.

Our crystal-state X-ray diffraction analysis indicates that the peptide backbone of Boc-L-Met-Xxx-L-Phe-OMe (Xxx = Aib and Ac$_5$c) adopts a β-bend conformation at the -L-Met-Xxx- sequence stabilized by an intramolecular H-bond between the Phe NH and the Boc C=O groups. The ϕ,ψ values for the L-Met and Xxx residues are in reasonable agreement with those expected for a type-II β-bend conformation. The L-Phe residue is semi-extended. The only relevant conformational difference between the two structures is found in the L-Met side-chain disposition, (t, t, g^-) in the Aib tripeptide while (t, t, g^+) in the Ac$_5$c analog.

The involvement of the C-terminal (Phe) NH group of the tripeptides in intramolecular H-bonding in CDCl$_3$ solution (concn. 2×10^{-3} M) was determined on the basis of the modest variation in chemical shift experienced upon addition of DMSO to the CDCl$_3$ solution. Conversely, the behavior of the NH resonances of the Met and Xxx residues strongly favors the conclusion that these NH groups are solvent-accessible. In the different NOE spectra, obtained by irradiation of either the Xxx NH or the Met C$^\alpha$H resonance, significant intensity enhancements are observed on the Met C$^\alpha$H and Xxx NH resonances, respectively. These findings support the view that in CDCl$_3$ solution these peptides are folded in an intramolecularly H-bonded type-II β-bend conformation, as observed in the crystal state.

Our results establish that the conformationally restricted, β-bend forming formyl-methionyl tripeptides incorporating, at position 2, an α-carbon dialkylated residue, are able to induce granule enzyme secretion from rabbit peritoneal neutrophils. Therefore, this type of folded conformation may allow highly favorable interaction with the neutrophil formylpeptide receptor. The significant enhancements in activity observed by an increase of the bulkiness of the side chain of central residue indicates that peptide–receptor interactions involving this specific site are important modulators of biological effects.

Evidence for the tendency of the highly active CHO-L-Met-L-Leu-L-Phe-OH and its methyl ester to adopt the type-II β-bend conformation has not been found so far, either in the crystal state or in solution. In this connection, the possibility that the interaction of the formylpeptides with their neutrophil receptor involves either multiple sites or an induced-fit mechanism is one that cannot be ignored in view of the present findings and of the published results.

Evaluation of side-chain contributions to α-helix stability

Ping C. Lyu, Luis A. Marky and Neville R. Kallenbach

Department of Chemistry, New York University, New York, NY, U.S.A.

Introduction

Studies with natural and synthetic models show that peptides with as few as 13 amino acids have detectable helicity at low temperature [1–3]. Co-operative helix formation in short peptides provides an important possible model for very early intermediates in folding natural proteins.

Factors that have, so far, been identified as important in stabilizing helical structure in isolated peptides are: (1) The identity of the side chains of groups within helices [4–7]. (2) The arrangement of charged side chains [8–10]. (3) Participation of certain side chains in helix nucleation and termination [11]. (4) Temperature: helical structure in short peptides is strongly T-dependent [6, 7,10].

Results and Discussion

To evaluate quantitatively the free energy contributions of different side chains in stabilizing a helix, we have designed and synthesized a series of peptides (Fig. 1) of 21 amino acids that possess partial helical structure in water at low temperature. The parent structure consists of two adjacent blocks of glutamic acid and lysine residues [12], providing a polar and soluble matrix into which we insert different side chains. The approach is similar to the host–guest studies of Scheraga's group [13] but involves short peptides and only naturally occurring side chains. By estimating the change in helix content for different side chains, we can assess their relative helical forming propensities.

The CD spectra of the peptides in Fig. 1 at 4°C are shown in Fig. 2, and the results of a two-state analysis of the helix content in these peptides are

EAK: Su-YSEEEEKKKKAAAEEEEKKKK-NH$_2$

ELK: Su-YSEEEEKKKKLLLEEEEKKKK-NH$_2$

EGK: Su-YSEEEEKKKKGGGEEEEKKKK-NH$_2$

Fig. 1. Sequences of the three substituted peptides. Su, succinyl; Y, tyrosine; S, serine; E, glutamic acid; A, alanine; K, lysine.

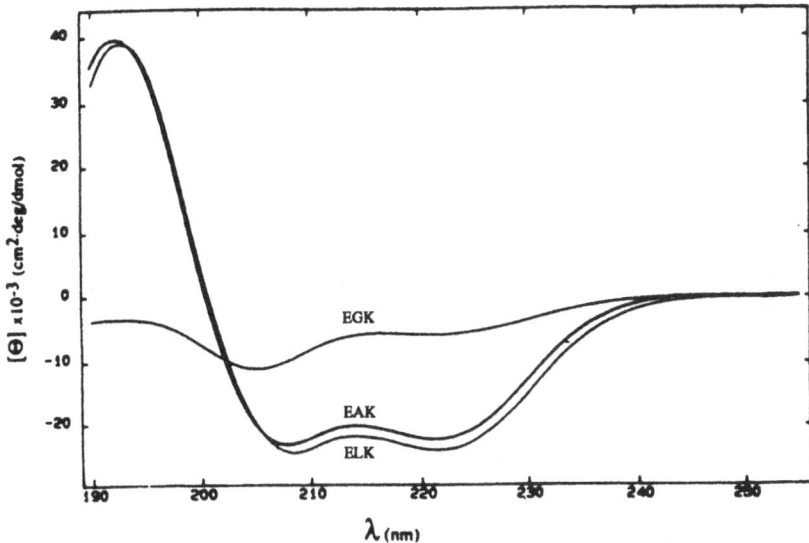

Fig. 2. Circular dichroism spectra of the peptides EAK, ELK and EGK in 10 mM KF (4°C, pH7).

given in Table 1. So far, we can conclude that the stability constant for alanine, S_{Ala}, is greater than that for leucine, S_{Leu}, and both are well above S_{Gly}. Our estimates of *relative* helix-forming potential thus differ from Scheraga's values. In addition, the absolute values are much larger for Ala and Leu in the context of these peptides, and the apparent enthalpies are also greater, since the helix structure is T-sensitive. This study shows that introducing sets of test residues into a soluble, partially helical matrix formed by repeats of E_4K_4 permits assessment of relative helix-forming potential. We are extending the series to other amino acids to develop a more complete set.

Table 1 *Helix content and the free energy of substitution of the three peptides*

Peptide	$-[\theta]_{222}$[a]	f[b]	K_h[c]	ΔG[d]	$\Delta\Delta G$[e]
EAK	28 300	0.79 ± 0.06	3.98 ± 1.28	−0.73 ± .18	−1.41 ± .23
ELK	23 370	0.81 ± 0.06	4.85 ± 1.85	−0.83 ± .23	−1.50 ± .27
EGK	5 600	0.23 ± 0.02	0.30 ± 0.03	0.68 ± .05	0

[a] Mean residue ellipticity (cm²deg/dmol) of the peptides at 222 nm (4°C, pH 7).
[b] $f = ([\theta] - [\theta]^0 / [\theta]^{100} - [\theta]^0)$, the fraction of helix. The maximum value calculated from $[\theta]^0 = 200$ cm²deg/dmol, $[\theta]^{100} = -27\,000$ cm²deg/dmol [14], and the minimum value from $[\theta]^0 = 2\,500$ cm²deg/dmol, $[\theta]^{100} = -32\,000$ cm²deg/dmol [15].
[c] $K_h = f/1-f$.
[d] $\Delta G = -R\,T \ln K_h$ (kcal/mol).
[e] $\Delta\Delta G = \Delta G_{EAK,ELK \text{ or } EGK} - \Delta G_{EGK}$.

Acknowledgements

This work is supported by grants GM-40746 from the NIH and NSF DIR 8722895.

References

1. Brown, J.E. and Klee, W.A., Biochemistry, 10(1971)470.
2. Bierzynski, A., Kim, P.S. and Baldwin, R.L., Proc. Natl. Acad. Sci. U.S.A., 79(1982)2470.
3. Merutka, G. and Stellwagen, E., Biochemistry, 28(1989)352.
4. Creighton, T.E., Proteins, W.H. Freeman, New York, NY, 1984.
5. Kim, P.S. and Baldwin, R.L., Nature, 307(1984)329.
6. Shoemaker, K.R., Kim, P.S., Brems, D.A., Marqusee, S., York, E.J., Chaiken, I.M., Stewart, J.M. and Baldwin, R.L., Proc. Natl. Acad. Sci. U.S.A., 82(1985)2349.
7. Shoemaker, K.R., Kim, P.S., York, E.J., Stewart, J.M. and Baldwin, R.L., Nature, 326(1987)563.
8. Chou, P.Y. and Fasman, G.D., Biochemistry, 13(1974)211.
9. Hol, W.G.J., Van Duijnen, P.T. and Berendsen, H.J.C., Nature, 273(1978)443.
10. Marqusee, S. and Baldwin, R.L., Proc. Natl. Acad. Sci. U.S.A., 84(1987)8898.
11. Presta, L.G. and Rose, G.D., Science, 240(1988)1632.
12. Lyu, P.C., Marky, L.A. and Kallenbach, N.R., J. Am. Chem. Soc., 111(1989)2733.
13. Sueki, M., Lee, S., Powers, S.P., Denton, J.B., Konishi, Y. and Scheraga, H.A., Biopolymers, 17(1984)148.
14. Shoemaker, K.R., Fairman, R., Schultz, D.A., Robertson, A., York, E.J., Stewart, J.M. and Baldwin, R.L., Biopolymers, in press.
15. Marqusee, S., Robbins, V.H. and Baldwin, R.L., Proc. Natl. Acad. Sci. U.S.A., in press.

Effects of a disulfide bridge on the helix in C-peptide analogs

Eunice J. York[a], John M. Stewart[a], Robert Fairman[b] and Robert L. Baldwin[b]

[a]Department of Biochemistry, University of Colorado School of Medicine, Denver, CO 80262, U.S.A.
[b]Department of Biochemistry, Stanford University Medical Center, Stanford, CA 94305, U.S.A.

Introduction

Modifications of the C-peptide of RNase A (residues 1–13) have led to analogs of increased helicity as measured by CD at 222 nm [1]. The reference peptides RN-21 (Ac-A-E-T-A-A-A-K-F-L-R-A-H-A-NH$_2$) and RN-80 (RN-21 : Phe^8 → Tyr) are about 50% helical in H$_2$O (pH 5.3, 3°C, 0.1 M NaCl). Introduction of an intramolecular disulfide bond across one turn of the helix might impart increased stability to these structures; computer modeling indicated that substitution of homocysteine (Hcy) in positions 7 and 11 of RN-21 with formation of a disulfide bond would not distort the helix and would not interfere with either the Glu^{2-}...Arg^{10+} ion pair or the Phe^8...His^{12+} interaction.

RN-83 (Ac-A-E-T-A-A-A-Hcy-F-L-R-Hcy-H-A-NH$_2$) was synthesized, and the bridge was formed both by oxidation of the free SH peptide and by use of the nitrophenylsulfenyl (Npys) group for disulfide-bond formation [2].

Results and Discussion

Synthesis was performed by SPPS methods using Boc chemistry; peptides were purified and characterized by standard methods [3]. Boc-Hcy(MeB) (Chemical Dynamics) was a gift from Peter S. Kim (MIT), and Npys-Cl was prepared by Wieslaw Klis in our laboratory. In the N-terminal region 1–6, 0.4 M KSCN was used in the coupling mixture in order to disrupt secondary structure of the growing peptide resin and increase the coupling rate [4]. Cleavage of Hcy(MeB) peptides employed a low-high HF method using anisole as the scavenger [5]. The free SH peptide, synthesized with the HF-labile MeB protection on Hcy, was reduced with DTT prior to air oxidation or was oxidized with K$_3$Fe(CN)$_6$. Alternately, use of the HF-stable Npys group for protection of Hcy made possible the purification of the peptide before formation of the disulfide bond. Npys was introduced by displacement of MeB on Hcy by Npys-Cl in the completed peptide resin. Formation of the disulfide bond proceeded by reaction of triphenylphosphine with Npys, giving either one free SH that could displace the second Npys with concomitant formation of the S—S bond, or two free SH groups

635

with S—S bond formation by air oxidation. Subsequently, RN-92 (RN-83: Phe → Tyr) was synthesized in order to use the tyrosine absorbance as a measure of peptide concentration for CD measurements. CD was performed as previously described [1] on an AVIV 60DS spectropolarimeter.

The methods used to form the disulfide bond in RN-83 all gave an identical major product, albeit in low yield, that was ascertained to be the desired product by AAA and FABMS, RN-92 was synthesized by the air oxidation method. The CD of RN-92 shows a higher helix content ($\theta_{222} = -15\,800$, pH 5.3, 3°C, 0.1 M NaCl) than the reference peptide RN-80 ($-12\,700$). Partial reduction of RN-92 causes a sharp decrease to $-12\,400$. It is not yet clear if RN-92 is purely an α-helical structure in the oxidized form. The 222-nm minimum is shifted slightly to lower wavelength vs. the reduced peptide. Although the temperature dependence of θ_{222} shows a high T_m of 32°C, indicating a highly stable structure (the T_m for RN-80 is 0–5°C), it extrapolates to $-19\,000$, lower than the expected $-27\,000$. TFE titration gives a maximum θ_{222} of $-19\,000$ at 12.5 mol% TFE, consistent with the temperature dependence data, whereas the reduced form (which is probably not fully reduced as shown by the presence of S—S bonds in the 260–280 nm region of the CD spectrum) is as high as $-22\,000$. Also, the difference spectrum of the oxidized and reduced forms shows a miminum between 211 and 215 nm, which is not typical of an α-helix. The disulfide bond may have produced a kink or a twist in the backbone, and further experiments will be needed to characterize fully the secondary structure of these peptides.

Acknowledgements

This research was supported by NIH Grant GM-31475 to R.L.B. We thank Thanh Le for AAA and Leland V. Miller of the NIH Clinical Mass Spectrometry Resource, University of Colorado, supported by Grant RR-01152, for the FABMS. E.J.Y. thanks Applied Biosystems Inc. for a travel grant.

References

1. Shoemaker, K.R., Kim, P.S., York, E.J., Stewart, J.M. and Baldwin, R.L, Nature, 326 (1987) 563.
2. Bernatowicz, M.S., Matsueda, R. and Matsueda, G.R., Int. J. Pept. Prot. Res., 28 (1986) 107.
3. Stewart, J.M. and Young, J.D., Solid Phase Peptide Synthesis, 2nd edn. Pierce Chemical Comp., Rockford, IL, 1984.
4. Klis, W. and Stewart, J.M., In Rivier, J.E. and Marshall, G.R. (Eds.) Peptides: Chemistry, Structure and Biology (Proceedings of the 11th American Peptide Symposium), ESCOM, Leiden, 1990, p. 904.
5. Ploux, O., Chassaing, G. and Marquet, A., Int. J. Pept. Prot. Res., 29 (1987) 162.

Identifying secondary structure in mastoparan and in gramicidin S by IR and IR-ATR spectroscopies

John W. Bean, Gary E. Zuber and Kenneth D. Kopple

Smith Kline & French Laboratories, Department of Physical and Structural Chemistry, L-940, P.O. Box 1539, King of Prussia, PA 19406-0939, U.S.A.

Introduction

The ability of IR and IR-attenuated total internal reflection (ATR) spectroscopies to identify secondary structure in peptides was tested using two conformationally well characterized peptides, mastoparan and gramicidin S. Mastoparan, which has little ordered structure in aqueous solution, adopts an α-helical conformation in methanol and upon binding to lipid membranes, as determined by CD, by the vicinal NH-αH coupling constants in the NMR spectra, and by TRNOE analysis [1,2]. Deuterium NMR spectroscopy of mastoparan bound to dilaurylphosphatidylcholine vesicles indicates that the helical axis is oriented parallel to the plane of the membrane, and that the charged residues of mastoparan are exposed to the aqueous solvent [2,3]. Gramicidin S has been shown to have a β-pleated sheet conformation in a variety of solvent systems, and has been used in the past as a probe to test structure determination methodologies [4]. Our IR and IR-ATR results agree with these prior results concerning the conformations of mastoparan and gramicidin S.

Results and Discussion

The amide-I band of mastoparan in methanol and on palmitoyloleoyl-phosphatidylcholine (POPC) multilayers at a lipid:peptide molar ratio of 50:1 has absorption maxima at 1659 cm^{-1} and 1656 cm^{-1}, respectively. The amide-I band of gramicidin S on POPC multilayers at a lipid:peptide molar ratio of 50:1 has an absorption maximum at 1638 cm^{-1}. These amide-I band maxima concur with experimentally determined values for α-helical and β-sheet types of secondary structure in 21 globular proteins [5].

The deuterium-shifted amide-I band of mastoparan in D_2O was a broad band centered at 1645 cm^{-1} with shoulders at 1650 cm^{-1} and 1638 cm^{-1}. The transition at 1650 cm^{-1} may be assigned to α-helical secondary structure based on the band maximum of mastoparan in CD_3OD, but it constitutes only a fraction of the overall amide-I band area. This result supports prior evidence indicating that mastoparan has little ordered secondary structure in aqueous solution [1].

Polarized IR-ATR spectra of mastoparan and gramicidin S on the POPC multilayers cast on germanium crystals indicated that the transition dipoles giving rise to their amide-I bands were aligned parallel to the plane of the POPC multilayer. For mastoparan, $R_{ATR} = 0.81$, this indicates that the α-helical axis is oriented parallel to the plane of the lipid multilayer, and for gramicidin S, $R_{ATR} = 0.86$, this indicates that the plane of the β-sheet is parallel to the plane of the lipid multilayer. Parallel orientation of transition dipoles, with respect to the lipid multilayer, is not always the case, however. Dynorphin A 1–13 adopts an α-helical conformation oriented perpendicular to the plane of the membrane, $R_{ATR} = 1.44$ [6]. This is an experiment and a result which we were able to replicate.

We were unable to resolve the transitions and their dipole orientations comprising the turns in gramicidin S. This may reflect a limit on the ability to discern discrete structural motifs in regular secondary structures by the IR-ATR technique, as applied by us to peptides on lipid multilayers at ratios less than 1:25.

References

1. Higashijima, T., Wakamatsu, K., Takemitsu, M., Fujino, M., Nakajima, T. and Miyazawa, T., FEBS Lett., 152 (1983) 227.
2. Wakamatsu, K., Higashijima, T., Fujino, M., Nakajima, T. and Miyazawa, T., FEBS Lett., 162 (1983) 123.
3. Saito, K., Higashijima, T., Fujino, M., Nakajima, T. and Miyazawa, T., In Shiba, T. and Sakakibara, S. (Eds.) Peptide Chemistry 1987, Protein Research Foundation, Osaka, 1988, p. 689.
4. Ovchinnikov, Y.A. and Ivanov, V.T., In Neurath, H. and Hill, R.T. (Eds.) The Proteins, 3rd ed., Academic Press, New York, 1982, p. 547.
5. Byler, M.D. and Susi, H., Biopolymers, 25 (1986) 469.
6. Erne, D., Sargent, D.F. and Schwyzer, R., Biochemistry, 24 (1985) 4261.

Conformational energy calculations on aminoisobutyric acid containing peptides and thiopeptides

V.N. Balaji, Salvatore Profeta Jr.*, A. Mobasser and M. Garst

Allergan Inc., 2525 Dupont Drive, Irvine, CA 92715, U.S.A.

Introduction

α-Aminoisobutyric acid (Aib) is an unusual, sterically hindered amino acid with predicted $(\phi,\psi)=\pm(50°,40°)$ [1–4], confirmed from crystal-structure data analysis [5–8]. The conformational restriction of peptide, N-methyl peptide, thioamide, N-methyl thioamide containing Aib residues is likely to yield stereochemically interesting structures. While oligopeptide containing Aib residues favor the formation of 3_{10} or α-helical structures [3,4,8], the presence of thioamide [9] in the helical structure will disrupt the formation of 3_{10} or α-helical structures. We have investigated energetics (methods as described in [9]) of dipeptide analogs (Fig. 1) and polymers of Aib containing residues.

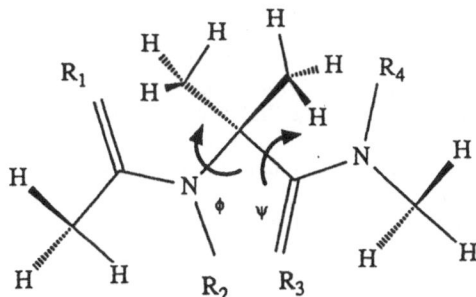

Fig. 1. Model Aib compounds $(R_2=R_4=H)$ (I) dipeptide $(R_1=R_3=O)$, (II) dithiopeptide $(R_1=R_3=S)$, (III) peptide-thiopeptide $(R_1=O,R_3=S)$ and (IV) thiopeptide-peptide $(R_1=S,R_3=O)$.

Results and Discussion

Ac-(Aib)$_2$-NHMe (I) (Fig. 1) has minima at $(\phi,\psi)=\pm(50°,40°)$ and local minima, 0.5, 0.9 and 1.4 kcal/mol above the global at $\pm(-60°,160°)$, $\pm(-170°,60°)$, and $(\pm180°,\pm180°)$, respectively. The barrier from the left-handed (50°,40°) to right-handed (−50°,−40°) conformation is about 2.0 kcal/mol. Ac-(Aib-S)$_2$-NHMe

---

*Present address: Glaxo Inc., Research Triangle Park, NC, U.S.A.

(II) has global minima at $\pm(-60°,140°)$ with local minima, 0.2 and 2.4 kcal/mol above the global minima, at $\pm(50°,40°)$ and $\pm(-170°,60°)$ respectively. The conformational energy contours for both mixed peptides, Ac-(Aib-Aib-S)- NHMe (III) and Ac-(Aib-S-Aib)- NHMe (IV), show more similarities to I than II, consistent with the observed NMR data [10] on model compounds (details to be published elsewhere). For Aib-di(N-methylpeptide), the global minima occur at $\pm(50°,40°)$. The barrier to transition from these two conformers is more than 5 kcal/mol compared to 2 kcal/mol for Aib-dipeptide. Poly(Aib)peptide adopts both 3_{10} and α-helical structure, the former being slightly more favored than the latter, consistent with previous reports. Poly(Aib)- thiopeptide adopts 3- and 4-fold polyglycine type II helical structures [11] with $C=S$ and NH groups perpendicular to the helix axis. In such a structure, $C=S$ groups point away from the helix axis, whereas NH groups point toward the helix axis.

References

1. Marshall, G.R. and Bosshard, H.E., Circ. Res., Suppl. 30 and 31 (1972) 143.
2. Burgess, A.W. and Leach, S.J., Biopolymers, 12 (1973) 2599.
3. Prasad, B.V.V and Sasisekharan, V., Macromolecules, 12 (1979) 1107.
4. Paterson, Y., Rumsey, S.M., Bendetti, E., Nemethy G. and Scheraga, H.A., J. Am. Chem. Soc., 103 (1981) 2947.
5. Smith, G.D., Pletnev, V.Z., Duax, W.L., Balasubramanian, T.M., Bosshard, H.E., Czerwinski, E.W., Kendrick, N.E., Mathews F.S. and Marshall, G.R., J. Am. Chem. Soc. 103 (1981) 1493
6. Toniolo, C., Bonora, G.M., Bavoso, A., Benedetti, E., Blasio, B.D., Pavone, V. and Pedone, C., Biopolymers, 22 (1983) 205.
7. Karle, I.L., Sukumar, M. and Balaram, P., Proc. Natl. Acad. Sci. U.S.A. 83 (1986) 9284.
8. Karle, I.L., Anderson, J.F., Sukumar, M. and Balaram, P., Proc. Natl. Acad. Sci. U.S.A. 84 (1987) 5087.
9. Balaji, V.N., Profeta Jr., S. and Dietrich, S.W., Biochem. Biophys. Res. Commun., 145 (1987) 834; 146 (1987) 1531.
10. Jensen O.E. and Senning, A., Tetrahedron, 42 (1986) 6555.
11. Balaji, V.N., Int. J. Quantum Chem., 20 (1981) 347.

Turn tendency of α-aminoisobutyric acid-containing sequences studied by HPLC and chiroptical methods

Takashi Yamada[a], Masayuki Nakao[a], Toshifumi Miyazawa[a], Shigeru Kuwata[a], Masao Kawai[b,*], Yasuo Butsugan[b] and Makiko Sugiura[c]

[a]Department of Chemistry, Faculty of Science, Konan University Okamoto, Higashinada-ku, Kobe 658, Japan
[b]Department of Applied Chemistry, Nagoya Institute of Technology Gokiso-cho, Showa-ku, Nagoya 466, Japan
[c]Kobe Women's College of Pharmacy, Okamoto, Higashinada-ku, Kobe 658, Japan

Introduction

The introduction of α, α-dialkylated α-amino acids, the prototype of which is α-aminoisobutyric acid (Aib), into peptide chains has proven to restrict the available range of backbone conformation [1]. The conformational difference between peptide diastereomers has a definite influence on separation by HPLC.

We recently reported that diastereomers of protected tetrapeptides **1**,

Z-(L-D)-Val-X-Y-L-Phe-OMe   **1**

containing two achiral amino acids (X and Y), exhibit marked separation in reversed-phase HPLC, when X is such an α,α-dialkylated glycine as Aib, Deg, Dpg, Ac$_5$c and Ac$_6$c [2], and Y is not bulky, e.g., Gly [3,4]. Excellent correlation between the separation and ¹H NMR chemical shift difference of Val-NH between diastereomers of **1** was found; β-turn conformation with two parallel intramolecular H-bonds was proposed for the L-L isomers from NMR study, and HPLC separation of diastereomers could be explained by the difference in the turn tendency [4].

Results and Discussion

In this work we examine the turn tendency of diastereomeric tetrapeptides **2**

Dnp-Val-X-Y-Leu-pNA   **2**

containing the X-Y sequences, Aib-Gly, Aib-Aib and Gly-Aib. Val-Aib-Gly-Leu is the central sequence of alamethicin.

CD spectra of **2** show the intense Cotton effects due to the interaction between the terminal chromophores of the L-L isomers (Fig. 1) and, thus, indicates their

---

*To whom correspondence should be addressed.

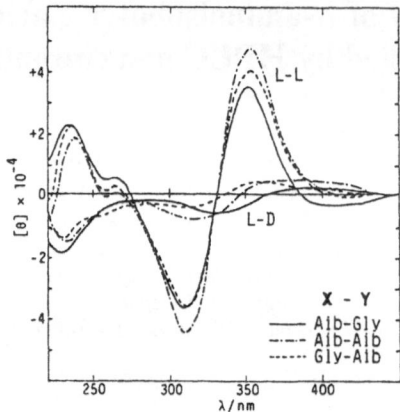

Fig. 1. *CD Spectra of Dnp-L-Val-X-Y-(L-D)-Leu-pNA in MeOH.*

strong turn tendency [5,6]. However, none of the L-D isomers unexpectedly show distinct Cotton effects. Marked differences between diastereomers are also observed in HPLC, as shown in Fig. 2. Every diastereomeric pair of **2** is much more significantly separated than that of **1** containing the corresponding X-Y sequence, particularly in the case of X-Y = Aib-Aib. Noticeably, the retention order of diastereomers of **2** is reversed in comparison with that of **1**.

The conformational investigation has been extended to ^{1}H NMR, mainly to the temperature dependence of NH resonances of **2** in DMSO-d_6. Both Leu-NH and pNA-NH of the L-L isomers show much lower values of dδ/dT (ppm × 10^{-3}/K) than those of X-NH and Y-NH, which are exposed to solvent (–3.2~–5.2), as follows: Aib-Gly, –0.5, –2.0; Aib-Aib, –0.5, –0.2; Gly-Aib, –1.0, +1.0 for Leu-NH and pNA-NH, respectively. Val-NH is strongly H-bonded to the o-nitro group of Dnp (–0.8~–1.8). These results indicate that the L-L isomers adopt 3_{10} helices with two intramolecular H-bonds in which Leu-NH and pNA-NH participate, and that the stability of the conformation may be in the order of Aib-Aib > Gly-Aib > Aib-Gly, corresponding to the order of magnitude of Cotton effects.

On the other hand, three L-D isomers show various temperature dependences: Aib-Gly, –1.9, –3.0; Aib-Aib, –1.2, –1.0; Gly-Aib, –2.9, –0.5 for Leu-NH and pNA-NH, respectively. Thus, these L-D isomers also adopt turn conformations, although the conformations differ from each other. The L-D/D-L isomer of **2** seems to prefer incorporation of an Aib residue into the left-hand corner (N-terminal side) of type II and II′ β-bend, contrary to the findings by Toniolo et al. [7]. Two chromophores of the L-D isomer may be separated from one another in the turn conformation, in contrast to those of the L-L isomer, which may be considerably closer to one another. This means that the L-D isomer has a larger hydrophobic surface than the L-L one, and thus, the latter elutes much faster than the former in RPHPLC.

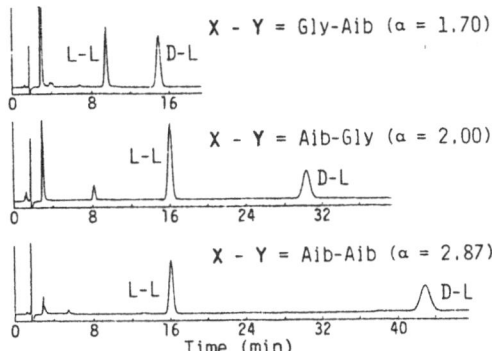

Fig. 2. Separation of diastereomers of Dnp-(L-D)-Val-X-Y-L-Leu-pNA (Cosmosil 5C₁₈ (4.6 I.D. × 150 mm), 70% MeOH aq.).

Similar study on Dnp-Gly-Aib-L-Pro-Gly-pNA and Dnp-Gly-Aib-L-Pro-OCH₂CO-pNA have shown the larger turn tendency of the former compound being consistent with the conformational difference found in solid state [8].

References

1. Toniolo, C., Br. Polymer J., 18(1986)221.
2. We had proposed the abbreviation, Ac$_n$c, for 1-aminocycloalkanecarboxylic acid where the subscript 'n' means the number of carbon atoms in cycloalkane ring. See, Yamada, T., Nakao, M. and Kuwata, S., In Kiso, Y. (Ed.) Peptide Chemistry 1985, Protein Research Foundation, Osaka, 1986, p. 333.
3. Yamada, T., Nakao, M., Tsuda, K., Nonomura, S., Miyazawa, T., Kuwata, S. and Sugiura, M., In Shiba, T. and Sakakibara, S. (Eds.) Peptide Chemistry 1987, Protein Research Foundation, Osaka, 1988, p. 97.
4. Yamada, T., Nakao, M., Yanagi, T., Miyazawa, T., Kuwata, S. and Sugiura, M., In Jung, G. and Bayer, E. (Eds.) Peptides 1988 (Proceedings of the 20th European Peptide Symposium), De Gruyter, Berlin, 1989, p. 301.
5. Sato, K., Kawai, M. and Nagai, U., Biopolymers, 20(1981)1921.
6. Kawai, M., Maekawa, M., Saito, Y., Butsugan, Y., Taga, T., Itoh, M. and Machida, K., In Ueki, M. (Ed.) Peptide Chemistry 1988, Protein Research Foundation, Osaka, 1989, p. 241.
7. De Pieri, G., Signor, A., Bonora, G.M. and Toniolo, C., Int. J. Biol. Macromol., 6(1984)35.
8. Kawai, M., Butsugan, Y., Fukuyama, K. and Taga, T., Biopolymers, 26(1987)83.

Influence of asparagine on turn formation in cyclic pentapeptides

Sarah J. Stradley, Josep Rizo, Martha D. Bruch, Zhi-Ping Liu and
Lila M. Gierasch

*Department of Pharmacology and Department of Biochemistry, University of Texas
Southwestern Medical Center, 5323 Harry Hines Boulevard, Dallas, TX 75235-9041, U.S.A.*

Introduction

To explore the influence of asparagine on reverse turn conformations, we have designed and synthesized model cyclic pentapeptides that incorporate Asn residues in positions likely to reside within β- and γ-turns. In this paper, we focus on the β-turn region and compare conformational preferences of cyclo(Gly-Pro-Asn-D-Ala-Pro) and cyclo(Gly-Pro-Ala-D-Phe-Pro), which incorporate Asn or Ala, respectively, in position $i+2$ of a β-turn. We find in the model peptide system co-existing conformers that contain type I or type II β-turns in dynamic equilibrium; Asn in position $i+2$ of the β-turn leads to a higher proportion of a type II β-turn than when Ala occupies this position.

Results and Discussion

From NMR data such as temperature dependences of NHs, ^{13}C chemical shifts and coupling constants, the overall backbone structures for the two cyclic pentapeptides consist essentially of fused β- and γ-turns as previously found for other related peptides [1]. NOE analysis revealed an interaction between the Pro H^α and the residue following it in the β-turn, which usually indicates of a type II β-turn, since this interproton distance is 3.5 Å in a standard type I β-turn and 2.1 Å in a type II β-turn (Fig. 1) [2]. To characterize these turns more fully, we carried out quantitative NOE analysis using the build-up rates of 1D NOEs (Fig. 2). From the build-up rates, one can obtain cross-relaxation rates (σ), which then can be used to determine accurate interproton distances with equation (1) [3]:

$$\frac{r_{ij}}{r_{kl}} = \left(\frac{\sigma_{kl}}{\sigma_{ij}}\right)^{1/6} \tag{1}$$

Here, σ_{kl} was between the methylene protons of the Gly residue, whose interproton distance can be assumed to be 1.75 Å. The calculated interproton distance, r_{ij}, is from the proton (j) that is irradiated to the proton (i), whose transient NOE build-up is observed. The distance determined from the Pro^2 H^α to the $i+2$ NH is 2.9 Å for the Pro-Ala peptide and 2.45 Å for Pro-Asn.

644

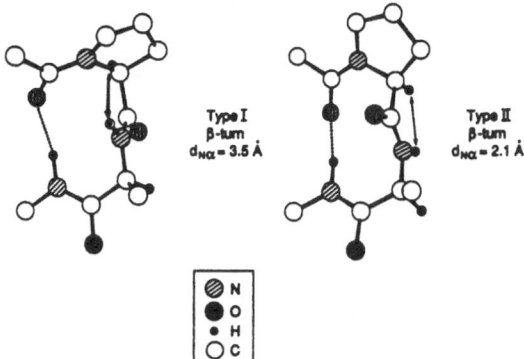

Fig. 1.   Structures of type I and type II β-turns, showing the Hα (i + 1) to NH (i + 2) distances.

The distances consistent with observed NOE build-up rates could arise either from a single conformation with ϕ,ψ angles intermediate between type I and type II β-turns or from interconverting species. Torsional forcing of the Pro^2 ψ angle in 5° intervals for a total rotation of 360° allowed us to calculate an energy profile corresponding to a type I/II transition (data not shown). These results showed that an intermediate structure was of relatively high energy (ca. 6 kcal/mol above the low-energy regions) and, hence, was unlikely to exist as a single conformer. Therefore, we conclude that these peptides are in dynamic equilibrium between conformers containing type I and type II β-turns.

We can thus calculate percentages of the populations of the two turn types for each peptide that would be consistent with the measured distances between

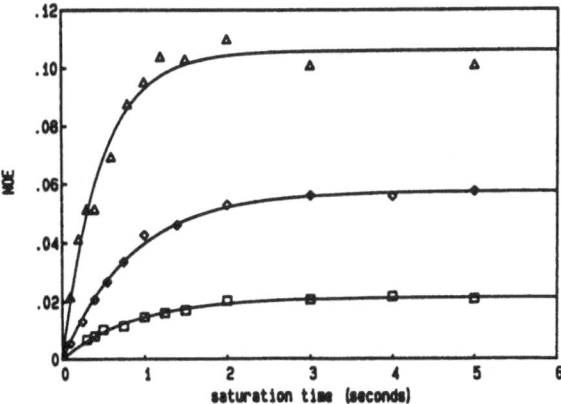

Fig. 2.   NOE as a function of irradiation time for: cyclo(Gly-Pro-Ala-D-Phe-Pro) (open squares); cyclo(D-Ala-Pro-Asn-Gly-Pro) (open diamonds); cyclo(Gly-Pro-D-Ala-Gly-Asn) (open triangles). NH resonance irradiated, Hα observed in all cases. For detailed analysis see Ref. 3.

645

the Pro H$^\alpha$ and the NH of the residue following. The Pro-Ala peptide would be in the type I conformation 80% of the time and in type II 20%, compared to the Pro-Asn peptide 65% type I, 35% type II. The introduction of asparagine in position $i+2$ of the β-turn has shifted the equilibrium significantly towards type II. Interestingly, this shift parallels the behavior of Asn residues in proteins [4,5].

Acknowledgements

We thank Adam N. Stroup for help with NMR data analysis. This work was supported in part by a grant from the NIH (GM-27616). J.R. thanks the Ministerio de Educacion y Ciencia, Spain. L.M.G. appreciates the support of the Robert A. Welch Foundation.

References

1. Stradley, S.J., Rizo, J., Bruch, M.D., Stroup, A.N. and Gierasch, L.M., Biopolymers, in press.
2. Narasinga Rao, B.N., Kumar, A., Balaram, H., Ravi, A. and Balaram, P., J. Am. Chem. Soc., 105 (1983) 7423.
3. Noggle, J.H. and Schirmer, R.E., The Nuclear Overhauser Effect, Chemical Applications, Academic Press, New York, 1971.
4. Wilmot, C.M. and Thornton, J.M., J. Mol. Biol., 203 (1988) 221.
5. Richardson, J.S. and Richardson, D.C., In Fasman, G.D. (Ed.) Prediction of Protein Structure and the Principles of Protein Conformation, Plenum Press, New York, 1989.

Low temperature rotating frame relaxation studies of cyclic peptides

Cynthia A. D'Ambrosio and Kenneth D. Kopple

Smith Kline and French Laboratories, L-940, P.O. Box 1539, King of Prussia, PA 19406-0939, U.S.A.

Introduction

The cyclic peptides c(D-Phe-D-Pro-Ala-Pro) (**I**), c(Phe-D-Trp-Lys-Thr-Phe-Pro) (**II**), gramicidin S (**III**) and α-amanitin (**IV**) all give narrow line NMR spectra in methanol down to –60°C. In principle, these 4 peptides could be conformationally stable or, alternatively, they could be exchanging among widely different conformations at rates exceeding 10^3 s^{-1}.

Rotating frame relaxation studies ($T_{1\rho}$) can be used to extend the frequency range of recognizable low frequency conformation exchange two orders of magnitude, to $< 10^5$ s^{-1}. A conformation exchange contribution to the relaxation rate $R_{1\rho}$ of a nucleus is dependent on the rate of the exchange, and on the square of the chemical shift difference in the different conformations [1]. To see if internal motions of the 4 peptides have been brought into this 10^5 s^{-1} range at lower temperature, we carried out 400 MHz NMR relaxation studies of peptide α-protons on methanol solutions at –60°C.

Results and Discussion

For these carbon-bound protons, the only important contributions to the rotating frame relaxation rate are those from interproton dipole–dipole (d-d) coupling and chemical shift modulation by any chemical (conformation) exchange:

$$R_{1\rho} \text{ (observed)} = R_{1\rho}(\text{d-d}) + R_{1\rho}(\text{exchange})$$

The dipole–dipole contribution is estimated from measurements of the laboratory frame spin-lattice relaxation rate (R_1), containing only a dipole–dipole contribution and knowledge of a rotational correlation time τ_c. In these studies we have obtained τ_c from the ratio:

$$R_1(\text{selective, initial rate}) / R_1(\text{nonselective})$$

where the selective rate is the initial rate of relaxation of a selectively inverted proton resonance (30 ms pulse) and the nonselective rate is that from the usual

647

hard pulse inversion recovery experiment. These two rates have different dependences on the rotational correlation time [2]. The advantages of this approach over ^{13}C relaxation for obtaining τ_c are use of lower peptide concentrations and the fact that correlation times obtained refer specifically to the interproton vectors directly involved in the dipolar relaxation. For the α-protons of **III** in DMSO at 20°C, we obtain values of τ_c near 0.8 ns in this work; using $^{13}C\alpha$ data, we earlier obtained values near 0.6 ns. The shorter τ_c values from the ^{13}C work could lead to overestimation of exchange contributions to $R_{1\rho}$.

Representative results are given in Table 1.

Table 1   *Relaxation data for α-protons of cyclic peptides, methanol, 213 K*

	I, D-Phe	II, Trp	III, Orn	IV, Cys	IV, Gly
R_1 (s^{-1})	1.41	0.67	1.68	1.54	1.06
R_1, sel. init. (s^{-1})	1.64	2.3	3.3	2.7	4.4
τ_c (s)	0.57	1.6	1.0	1.7	1.8
$R_{1\rho}$ ($^{-1}$)	5.9±.5	8.0±.4	14.6±.4	14.5±1.	43±5.
$R_{1\rho}$, d-d, est. (s^{-1})	4.5±.4	8.9±.9	11.3±.8	10.7±1.	17.7±1.
$R_{1\rho}$, exch. contr. (s^{-1})	~1	~0	~3	~4	large

Rotating frame field dependence, above the error level, was not observed (γH_1 in the range 2.9–8 kHz), indicating the motions involved are at the upper end of the range affecting rotating frame relaxation in these experiments. Exchange contributions to $R_{1\rho}$ were not observed for all of the protons examined. However, spectra of **II, III,** and **IV,** measured at –90°C, 30°C lower, show differential line broadening, confirming that conformation exchange is occurring at –60°C, even though narrow line spectra are obtained. These measurements do not suggest specific details of the conformational motions.

References

1. Kopple, K.D., Wang, Y.-S., Cheng, A.G. and Bhandary, K.K., J. Am. Chem. Soc., 110 (1988) 4168.
2. Mirau, P.A., Behling, R.W. and Kearns, D.R., Biochemistry, 24 (1985) 6200.

Comparison between the conformation of renin inhibitory peptides determined by cryogenic CD and 2D exchange NMR and molecular modeling based upon the crystal structure of proteases

Graeme J. Anderson, Colin H. James and William A. Gibbons

The London School of Pharmacy, Brunswick Square, London WC1N 1AX, U.K.

Introduction

Although linear peptides had been thought previously to possess a wide spectrum of conformations in solution, recent experiments using cryogenic CD, solvent perturbation CD and 2D NMR have shown this not to be the case, e.g., polylysine [1] and rabies virus epitopes [2]. Here we report similar studies of a renin inhibitory peptide, RIP [3]:

H-Pro-His-Pro-Phe-His-Phe-Phe-Val-Tyr-Lys-OH

Renin inhibitory peptides were thought to have many solution conformations, and their conformation derived from crystallography and molecular graphics was predicted to be an extended chain conformation. Our results show RIP to have two principal solution conformations, neither of which is an extended conformation.

Results and Discussion

CD spectra recorded as a function of solvent polarity, pH, and temperature revealed the presence of a two-state conformational equilibrium. The low temperature state consisted of *cis* amide bonds at Pro^3. The room-temperature spectra indicated a mixture of *cis* and *trans* isomers, the ratio varying with the polarity of solvent.

2D ^{1}H NMR studies confirmed and refined the conclusions from CD results, and were used to construct an approximate structure for RIP in both conformational states.

Both solution conformations differed from the enzyme-bound conformation predicted by Blundell et al. [4]. Our energy-minimized structures have been used to design conformationally constrained inhibitors of renin.

649

G.J. Anderson, C.H. James and W.A. Gibbons

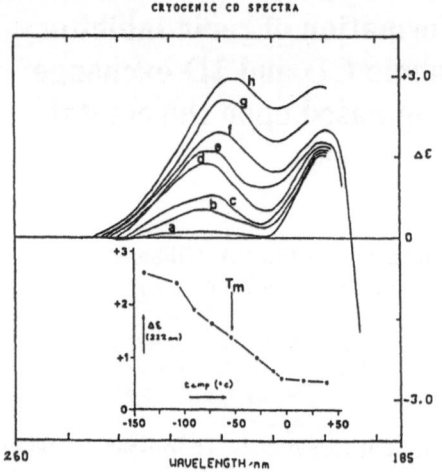

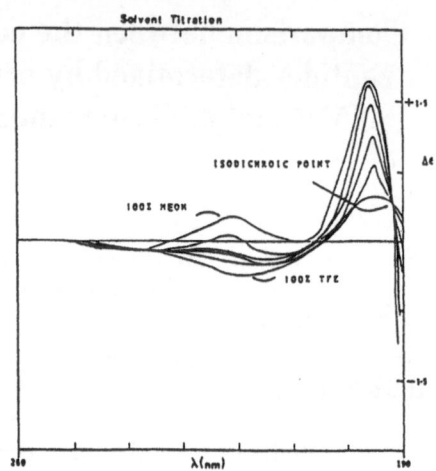

Fig. 1. Circular dichroism spectra of RIP in methanol/glycerol solution (9:1), recorded as a function of temperature.

$a = +20°C;$ $b = -13°C;$ $c = -30°C;$ $d = -54°C;$ $e = -73°C;$ $f = -90°C;$ $g = -108°C;$ $h = -142°C.$

The plot of $\Delta\epsilon$ versus temperature showed a 'leveling off' of the curve at $-142°C$, consistent with the 'freezing out' of the lowest temperature state, in this instance a structure involving cis peptide bonds. At room temperature the curve was also level, indicating a high temperature conformational state.

Fig. 2. CD solvent titration between 100% MeOH and 100% TFE. The presence of an isodichroic point was indicative of a two-state system for RIP and represented a further step in resolving the conformational components of RIP in solution.

References

1. Drake, A.F., Siligardi, G. and Gibbons, W.A., Biophys. Chem., 31 (1988) 143.
2. Siligardi, G., Drake, A.F., Mascagni, P., Neri, P., Lozzi, L., Niccolai, N. and Gibbons, W.A., Biochem. Biophys. Res. Commun., 143 (1987) 1005.
3. Burton, J., Cody Jr., R.J., Herd, J.A. and Haber, E., Proc. Natl. Acad. Sci. U.S.A., 77 (1980) 5476.
4. Blundell, T.L., Sibanda, B.L. and Pearl, L., Nature, 304 (1983) 273.

Conformational analysis of peptides using time-resolved fluorescence resonance energy transfer

J.F. Hedstrom[a,*], F.G. Prendergast[b] and L.J. Miller[a]

[a]Gastroenterology Unit and [b]Department of Biochemistry, Mayo Clinic, Rochester, MN 55905, U.S.A.

Introduction

Conformation determines the pharmacophoric selectivity between a ligand and a putative binding site. Consequently, many techniques have been employed attempting to establish receptor preferences for specific ligand conformations. Steady state methods can only provide static structural information about peptides in solution. Recently developed time-resolved fluorescence resonance energy transfer (TR-FRET) techniques [1] have paved the way for more insightful dynamical information to be garnered from fluorescence data. TR-FRET not only yields information about the average separation between two chromophores, but also indicates whether a distribution of peptide conformations is more plausible. Using TR-FRET, we investigated the conformation of a dansylated analog of cholecystokinin octapeptide (CCK-OP): Dansyl-Asp-Tyr(SO$_3$H)-Met-Gly-Trp-Met-Asp-Phe-NH$_2$.

SAR were performed to ensure that the derivatization of the α-amino group of CCK with dansyl (DNS) would not alter the biological activity of the peptide.

Results and Discussion

10-μM samples of both the unmodified and dansyl-derivatized CCK-OP were prepared in (i) 25 mM N-(2-acetamido)iminodiacetic acid (ADA) buffer with 120 mM KCl at pH 6.8, and (ii) 6 M guanidine·HCl (known to denature polypeptides).

Fluorescence lifetimes were measured using time-correlated single photon counting [2]. Intensity decays for tryptophan in most polypeptides is multi-exponential. Lakowicz et al. [3] have shown that the donor fluorescence for a donor-acceptor (tryptophan-DNS) pair separated by a distance r may be expressed as

$$I_{da}(r,t) = \sum_i \alpha_{di} \exp\left[-t/\tau_{di} - t/\tau_{di} (R_o/r)^6 \right] \tag{1}$$

and the observed decay given by

---

*To whom correspondence should be addressed.

J.F. Hedstrom, F.G. Prendergast and L.J. Miller

$$I_{da}(t) = \int_0^\infty f(r) \, I_{da}(r,t) \, dr \tag{2}$$

with α_{di} defined as the pre-exponential factor and τ_{di} the associated decay time for component i of the donor fluorescence (with no acceptor). The Förster radius R_o is 15 Å for the tryptophan-DNS pair in water [4], and f(r) is the distance distribution function.

Simulated and experimentally acquired TR-FRET decay files were fit to discrete and Gaussian-distributed distance functions. The results presented in Table 1 emphasize the utility of TR-FRET techniques in measuring distances in polypeptides. The technique is not only extremely sensitive (permitting use of micromolar peptide concentrations), but also yields information about the dynamics of the system. For CCK-OP, the implication is that, at physiological pH, the peptide has a broad range of conformations.

Table 1 *Recovered parameters for TR-FRET decays assuming distributed or discrete distance model functions*

Sample	Length (Å)	Width (Å)	χ^2_r
Gaussian simulation	12.04[a]	5.72[a]	–
distribution	12.24	5.48	1.04
discrete	14.18	–	25.17
Discrete simulation	14.19[a]	–	–
distribution	14.19	0.02	0.96
discrete	14.19	–	0.96
CCK in ADA buffer			
distribution	10.39	5.06	17.22
discrete	12.88	–	92.77
CCK in guanidine/HCl			
distribution	12.48	5.60	12.17
discrete	14.37	–	68.32

[a]Values used in simulation.

Acknowledgements

We thank Steve Powers and Delia Pinon for preparing the dansylated CCK. J.H. was supported by NIH Training Grant DK-07198.

References

1. Amir, A. and Haas, E., Biochemistry, 26 (1987) 2162.
2. O'Connor, D.V. and Phillips, D., Time-correlated Single Photon Counting. 1st edn., Academic Press, Orlando, FL, 1984, p. 36.
3. Lakowicz, J.R., Gryczynski, I., Cheung, H.C., Wang, C.K., Johnson, M.L. and Joshi, N., Biochemistry, 27 (1988) 9149.
4. Steinberg, I.Z., Ann. Rev. Biochem., 40 (1971) 83.

Synthesis of phenyl-substituted BTD (bicyclic-turned dipeptide) and its incorporation into bioactive peptides

Ukon Nagai and Rika Kato

Mitsubishi Kasei Institute of Life Sciences, Minamiooya 11, Tokyo 194, Japan

Introduction

BTD has been developed as a tool to elucidate the conformation of bioactive peptides when it binds to the receptor sites [1]. It is characterized by the fixed backbone conformation simulating the central part of type II′ β-turn. It can be incorporated into a peptide chain and the BTD-containing peptide always takes a folded conformation at the point of BTD substitution, since BTD cannot take on the other conformation due to the rigid bicyclic structure. The BTD-containing analogs of gramicidin S and GnRH were synthesized, and they retained high biological activity demonstrating that their bioactive conformation is, indeed, turned at the position of BTD incorporation [2,3]. During the course of investigation, introduction of side-chain functionality on the BTD skeleton was found necessary in order to extend its applicability to peptides containing a functional group at the turning point which is essential for biological activity. 8-Phenyl BTD has been selected as the first target for the substituted BTDs.

Results and Discussion

The BTD skeleton can be constructed from Glu and Cys derivatives. Therefore, β-phenylcysteine (Fcys) is required as the building block for the synthesis of 8-phenyl BTD. Although the synthesis of chiral Fcys was first reported by O. Ploux et al., we devised an original route to obtain all the four diastereomers of chiral Fcys [4]. First, the addition of a mercaptan to acylamidocinnamic ester was followed by removal of N- and S-protecting groups. The Fcys ester obtained was resolved to threo- and erythro-isomers by recrystallization. Treatment of each of the stereoisomers with acetone afforded the corresponding 2,2-dimethylthiazolidine derivative quantitatively, which was then resolved into chiral forms by passing through a chiral HPLC column. The γ-carbonyl compound (aldehyde or ketone) of the N-protected Glu derivative was used as the second component. Actually, two experiments were carried out using (1) Pht = Glu-(CHO)-OMe and H-Fcys-OMe·HCl and (2), Boc-Glu(CO-Me)-OH and H-Fcys-OH·HCl as sets of building blocks on the basis of preliminary experiments with model compounds. Although the first experiment gave poor yield of the desired product, the second experiment gave good yield of the product, Boc-6-Me-8-Ph-BTD-OH (see the structure in Fig. 1). It showed no melting point up to

300°C, but colored gradually yellow to dark brown above 220°C, and had specific rotation value of –28.1° (c 1.0, MeOH).

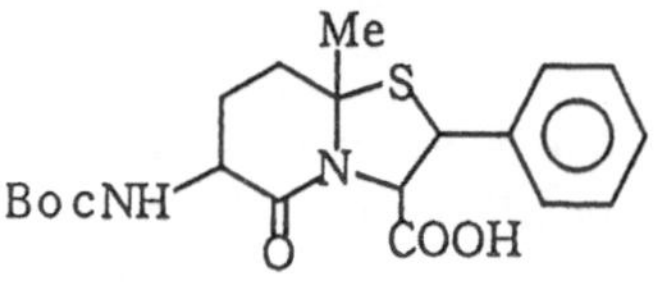

Fig. 1.

Its ¹H NMR spectrum proved the structure unambiguously except for the stereochemistry of angular methyl group. This compound has a convenient form of Boc-protected dipeptide that can be used directly for incorporation into a peptide chain by the usual procedure.

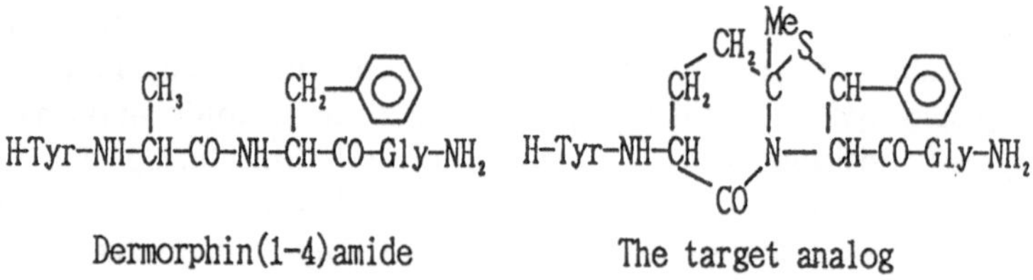

Dermorphin(1–4)amide   The target analog

Fig. 2.

The 8-phenyl BTD can be used as a substitute for the β-turn having Phe at the third position. So, dermorphin has been selected as the first target peptide to incorporate the 8-phenyl BTD unit. The experiment to synthesize the 6-Me-8-Ph-BTD-containing analog of dermorphine(1–4)-amide is in progress (Fig. 2). Dermorphin is an opioid peptide isolated from frog skin by an Italian group, and it is known that even the short-chain analog, Tyr-D-Ala-Phe-Gly-amide retains high potency [5]. If our analog would exhibit potent activity, valuable information on the bioactive conformation would be obtained.

References

1. Nagai, U. and Sato, K., Tetrahedron Lett., 26 (1985) 647.
2. Sato, K. and Nagai, U., J. Chem. Soc., Perkin Trans. I (1986) 1231.
3. Nagai, U., Nakamura, R., Sato, K. and Ying, S.-Y-., In Miyazawa, T. (Ed.) Peptide Chemistry 1986, Protein Research Foundation, Osaka, 1987, p. 295.
4a. Nagai, U. and Pavone, P., Heterocycles, 28 (1989) 589.
  b. Ploux, O., Caruso, M., Chassaing, G. and Marquet, J., J. Org. Chem., 53 (1988) 3154.
5. Broccardo, M., Erspamer, V.G., Falconieri-Erspamer, G., Impota, G., Linari, G., Melchiorri, P. and Montecucchi, P.C., Brit. J. Pharmacol., 73 (1981) 625.

Long-range effects of single site mutations in bacteriorhodopsin

Martin Engelhard[a], Benno Hess[a], Günther Metz[b], Fritz Siebert[b], Jörg Soppa[c] and Diether Oesterhelt[c]

[a]*Max Planck Institut für Ernährungsphysiologie, Rheinlanddamm 201, D-4600 Dortmund, F.R.G.*
[b]*Institut für Biophysik und Strahlenbiologie, Albertstr. 23, D-7800 Freiburg, F.R.G.*
[c]*Max Planck Institut für Biochemie, Am Klopferspitz 18a, D-8033 Martinsried, F.R.G.*

Introduction

The modification of biologically active peptides and proteins has been extensively utilized in structure-function studies. However, the altered function could not always be unequivocally associated with the modified amino acid because the peptide accommodated the modification by long-range conformational changes. The introduction of site-directed mutagenesis into this field together with X-ray structural analysis provided the possibility to study systematically how a modification affects stability and function (for an overview see Ref. 1). However, this approach fails if crystal structures are not available or if the proteins are not susceptible to solution NMR techniques. In the following, it is shown that solid-state NMR techniques can provide data of mutational effects on the structure of a membrane protein. As an example, bacteriorhodopsin (bR), the light-driven proton pump form *Halobacterium halobium* was chosen (for a recent review see Ref. 2).

Results and Discussion

Bacteriorhodopsin was biosynthetically modified with [4-^{13}C]Asp [3]. The isotope label was also found in C-11 of the Trp ring system [3]. The ^{13}C solid-state magic angle sample spinning (MASS) NMR of this sample revealed four classes of Asp residues distinguished by their chemical environment. Of the nine Asp contained in bR, five residues are on the cytoplasmic surface of the membrane protein. Two of them experience an ionic environment that is changed by removing the C-terminal tail. Also, two protonated internal Asp could be discerned, and two internal deprotonated Asp could be tentatively correlated with a signal at 173 ppm. Five different C-11 Trp resonances were found between 110 ppm and 115 ppm.

Three mutant strains of *Halobacterium spec.* GRB, with the site of mutation in the bacteriorhodopsin gene (PM 326: Asp^{96}-Asn; PM 374: Asp^{96}-Gly; PM 384: Asp^{85}-Glu) [4,5], were grown in a synthetic medium containing [4-^{13}C]Asp. The mutant bacteriorhodopsins, labeled with [4-^{13}C]Asp (37–45%) and due to

655

the metabolic reaction with [11-^{13}C]Trp (50–100%), were isolated as purple membranes and solid-state ^{13}C NMR spectra of the samples were taken.

The two Asp[96] mutants lacked the signal at 171.3 ppm that was previously assigned to a protonated internal Asp. From this observation one can conclude that Asp[96] is protonated in the ground state. The NMR spectrum of PM 384 (Asp[85]-Glu) differs dramatically from the wild type. However, one can assign the resonance at 173 ppm in the wild-type spectrum to Asp[85].

The distribution of the ^{13}C isotope provides an internal antenna that monitors the effects of single-site mutations on Trp, internal Asp residues and Asp residues from the cytoplasmic surface. External surface Asp residues of both classes of mutant bR are better available for hydrogen bonds as compared to the wild type. A similar effect was also observed in a spectrum of the wild type that lacked the C-terminal tail [3]. Apparently, the C-terminus folds over the cytoplasmic surface, thereby protecting certain parts from the bulk phase [6]. This interaction is partially disrupted in the mutant proteins, indicating structural changes as far reaching as from the interior of the protein to the cytoplasmic surface.

Whereas the Asp[96]-Asn and Asp[96]-Gly mutant proteins display only little structural changes, the functional properties of these proteins are drastically impaired [7]. One can conclude from these observations, that the perturbations are primarily locally confined. It is, therefore, reasonable to assume that Asp[96] is functionally involved in the reaction cycle of bR and an integral member of the proton transfer chain.

On the contrary, the Asp[85]-Glu mutant alters not only the chromophore–protein interaction, as indicated by the shift of the absorption maximum from 570 nm to 610 nm and the chemical environment of the surface Asp residues, but also two of the Trp resonances and the chemical shift of Asp[115]. In this case, the altered function [7] might not necessarily be due only to the mutated site.

In this communication, it was shown that conservative mutations like the replacement of Asp by Glu in bacteriorhodopsin can have dramatic effects on its structure. On the other hand, a substitution of Asp by Gly, which leaves a gap at the location of the side chain of Asp, can be compensated locally by the protein.

References

1. Alber, T., Annu. Rev. Biochem., 58 (1989) 765.
2. Lanyi, J.K., In Ernster, L. (Ed.) Bioenergetics, North-Holland, Amsterdam, 1984, p. 315.
3. Engelhard, M., Hess, B., Emeis, D., Metz, G., Kreutz, W. and Siebert, F., Biochemistry, 28 (1989) 3967.
4. Soppa, J. and Oesterhelt, D., J. Biol. Chem., 264 (1989) 13043.
5. Soppa, J., Otomo, J., Straub, J., Tittor, J., Meessen, S. and Oesterhelt, D., J. Biol. Chem., 264 (1989) 13049.

6. Engelhard, M., Pevec, B. and Hess, B., Biochemistry, 28 (1989) 5432.
7. Butt, H.J., Fendler, K., Bamberg, E., Tittor, J. and Oesterhelt, D., EMBO J., 8 (1989) 1657.

Membrane-spanning bent helix modeling receptor protein channel function

Günther Boheim[a], Inge Helbig[a], Sabine Meder[a], Horst Vogel[b], Lennart Nilsson[c], Rudolf Rigler[c], Brigitte Franz[d] and Günther Jung[d],*

[a]Department of Cell Physiology, Ruhr-Universität Bochum, Bochum, F.R.G.
[b]Ecole Polytechnique Fédérale de Lausanne, Lausanne, Switzerland
[c]Karolinska Institutet, Stockholm, Sweden
[d]Institut für Organische Chemie, Universität Tübingen, Tübingen, F.R.G.

Introduction

Putatively all hydrophobic and amphiphilic polypeptides induce ion permeation effects in planar bilayers as far as they interact with the lipids and adopt transmembrane structures [1]. Experiments with a series of simple helical rods P_n [Boc-(Ala-Aib-Ala-Aib-Ala)$_n$-OMe, n = 1–4] reveal a strongly voltage-dependent conductivity at high positive and negative voltages, but sharply distributed open channel conductances could not be resolved [2]. This differs from ion channels of alamethicin, which are switched on at positive and negative voltages and display a sequence of long-lived channel states of non-integral conductance values.

Results and Discussion

In order to impose channel-stabilizing structural elements to the icosapeptide P_4, we added aromatic residues to the termini (N-terminal dansyl-group and C-terminal Trp-OMe), and a Pro residue was inserted ⅓ rod length away from the C-terminus to yield Dns-(Ala-Aib-Ala-Aib-Ala)$_3$-Pro-Ala-Aib-Ala-Aib-Ala-Trp-OMe (I). Besides a strongly voltage-dependent conductivity at high voltages, this α-helical polypeptide forms a channel type of weak voltage-dependence at moderate voltages [3]. Open channel events appear in bursts (Fig. 1, upper traces), similar to the gating characteristics of the acetylcholine receptor channel WTBγ (bottom traces in Fig. 1). Surprisingly, the model channel exhibits only one open conductance state of 63 pS, independent of voltage. Mean lifetime analyses of the open and the closed state within burst fluctuations reveal that its ratio, the equilibrium distribution function, depends weakly on membrane voltage. The formal gating charge is 0.19 elementary charge.

The synthetic α-helix I forms a bursting channel, the kinetic properties of which resemble those of the nicotinic acetylcholine receptor. This important result reveals that bursting behavior already is an intrinsic feature of simple

---

*To whom correspondence should be addressed.

Dans-P21-Trp

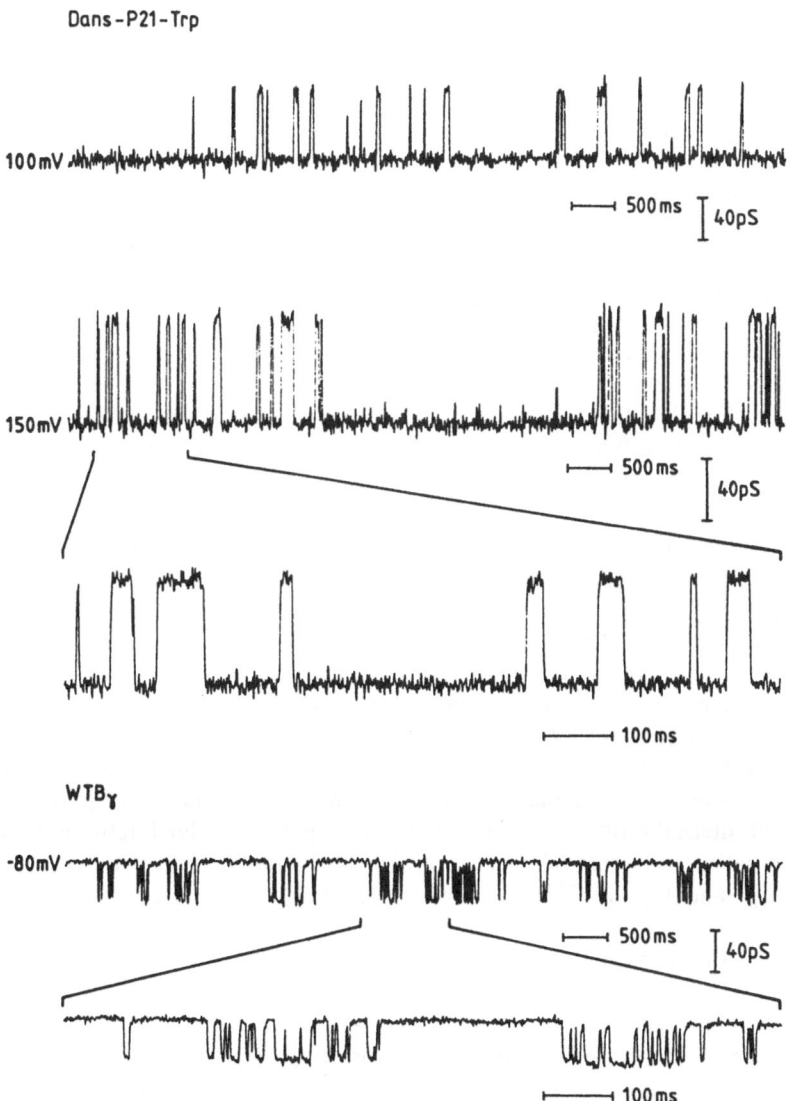

Fig. 1. Single channel current fluctuation traces of bursting ion channels. Upper traces: *bent helix I (Dns-P21-Trp) 50 μg/ml in* cis *compartment; BLM: POPC/DOPE/Chol., 85:9:6; 1.0 M NaCl, 10 mM Tris/HCl, pH 7.0, 22°C.* Lower traces: *Patch-clamping on skinned* Xenopus *oocytes after injection of m-RNA of fetal nicotinic acetylcholine receptor of bovine skeletal muscle (WTBγ from S. Numa, Kyoto) and expression of translated receptor protein; 10 μM acetylcholine in pipette, 21°C, conductance 40 pS.*

molecules that form stable aggregates (probably 4 or 5 monomers for I) in lipid bilayers. The weak voltage dependence of channel lifetime of I is found with ligand-gated channels, too, indicating similar structural prerequisites. An interesting question is that of the origin of the two closed states, one short-lived within a burst, and another long-lived between the bursts of activity. Lateral rearrangements of the helices within the membrane plane and a helix movement at the Pro residue might be feasible. Pro insertion into C-terminal half of the polypeptide chain leads to an asymmetric molecular shape. Thus, compact ion channels with maximum hydrophobic intermolecular interactions are formed in the parallel helix orientation. Experiments with Pro-free analogs of alamethicin [1] and melittin indicate that a break and bend in helix axis improves aggregate stability.

Straight α-helical voltage-dependent channel formers such as Boc-(Ala-Aib-Ala-Aib-Ala)$_4$-OMe [2] are lacking the stabilizing elements leading to resolved single channel events in lipid bilayer experiments, whereas Dns-(Ala-Aib-Ala-Aib-Ala)$_3$-Pro-Ala-Aib-Ala-Aib-Ala-Trp-OMe (I) exhibits the well resolved traces shown in Fig. 1. Fluorescence quenching and time-resolved fluorescence anisotropy of Trp in different positions along 21-peptide helices gave very valuable information about position in the membrance, and side chain and backbone motions. In order to investigate the influence of the Pro residue on the internal flexibility, the fluorescence energy transfer (FET) from Trp to Dns was measured in I and Dns-(Ala-Aib-Ala-Aib-Ala)$_4$-Trp-OMe (II). For the bent helix I the FET measurements in methanol at RT indicated an average distance Dns-Trp of 25 Å and a halfwidth of P(r) of 12–13 Å, whereas the corresponding values for the straight helix II are: Dns-Trp 30 Å and P(r) 9–10 Å. Measurements in fluid lipid membranes show a similar trend, however, with a 20–30% reduced halfwidth of P(r) for both I and II.

According to molecular dynamics simulations for 600 ps, the broader distance distribution between the helix ends in I compared to that in II is due to an increased internal mobility of the bent helix. Regular angular fluctuations found for two helical segments (characterized by the angle between two vectors from C_α^6 to C_α^{13} and C_α^{13} to C_α^{20}) for the straight helix II are around ± 6 degrees. For I, a similar type of motion (± 6 degrees) is superimposed by jumps of 15–20 degrees that appear to occur preferentially in a plane defined by the average positions of the two helix segments. Future experiments correlating the fluorescence with the electrical signals will show if the motions of the bent helices described here are responsible for channel opening and closing.

References

1. Boheim, G., Gelfert, S., Jung, G. and Menestrina, G., In Yagi, K. and Pullman, B. (Eds.) Ion Transport through Membranes, Academic Press, Tokyo, 1987, p. 131.
2. Menestrina, G., Voges, K.-P., Jung, G. and Boheim, G., J. Membrane Biol., 93 (1986) 111.
3. Boheim, G., Helbig, I., Meder, S., Franz, B. and Jung, G., In Maelicke, A. (Ed.) Molecular Biology of Neuroreceptors and Ion Channels, Springer-Verlag, Berlin, 1989, p. 401.

Conformational control in cyclic hexapeptides

Nathan Collins and Nigel G.J. Richards*

Department of Chemistry, The University, Southampton S09 5NH, U.K.

Introduction

The use of dipeptides and their analogs to control conformation in cyclic peptides remains the subject of much investigation. We report here the synthesis and preliminary conformational analysis of two cyclic hexapeptides, cyclo(-Pro[1]-D-Phe[2]-Gly[3]-Val[4]-Tyr[5]-Gln[6]-) **1** and cyclo(-Pro[1]-Phe[2]-Gly[3]-Val[4]-Tyr[5]-Gln[6]-) **2**. These two sequences were chosen in order to investigate the effect upon peptide conformation and flexibility of changing the dipeptide sequence Pro-D-Phe to Pro-L-Phe. The former sequence, when incorporated into cyclic hexapeptides, has been shown to restrict their conformational space through its preference for adopting a type II β-turn [1]. The other residues were chosen because they comprise the turn region of the 'helix-turn-helix' motif in *cro* repressor protein [2]. The synthesis of **1** and **2** was achieved by solid phase methods using the Fmoc N-protection strategy [3], followed by cyclization of the linear precursors using DPPA.

Results and Discussion

We examined the conformational properties of **1** and **2** using both amide temperature-dependence measurements [4] and ROESY experiments [5] using d6-DMSO as solvent. The mixing time for the ROESY pulse sequence was usually 300 ms. Peptide **1** adopted a single preferred conformation, although peaks from a second minor conformer could be detected. However, the ratio of these was at least 30:1. The Pro-D-Phe segment was determined to adopt a type II β-turn, while a type I β-turn was present about the Val-Tyr region. In addition, the glycine residue was distorted such that the NH resonance was observed as a doublet, the assignment being based upon a TOCSY experiment. An intriguing feature of the modeled conformation consistent with experiment was the alignment of 3 N–H bonds upon one face of the molecule in the type I turn region. From statistical studies upon proteins [6] in the Brookhaven database [7] we expected that Pro-L-Phe would not favor the formation of a type II β-turn, and therefore that **2** would prefer an alternate conformation to that observed for **1**. Although peptide **2** was found to exist in predominantly two conformations (in a ratio of 3:1) in DMSO solution, the observed temperature dependence of the amide chemical shifts in the major conformation was identical to that determined for **1**.

---

*To whom correspondence should be addressed.

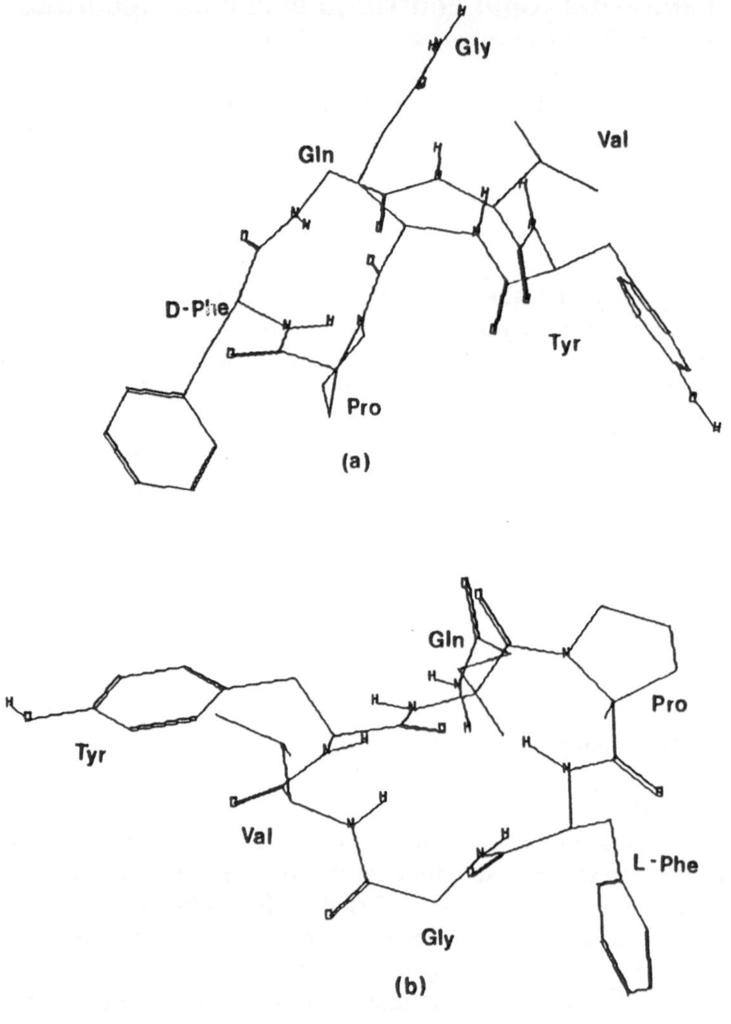

Fig. 1. Modeled solution phase conformations of 1 and 2 based upon coupling constant, temperature dependence of NH chemical shift and rotating frame NOEs. (a) Major conformation for 1 showing the turn structures and the alignment of the NH bonds about the type I β-turn. A similar structure has been determined for the preferred conformation of 2. (b) Minor conformation of 2 showing the type VI β-turn about the Gln-Pro residues.

Moreover, the glycine NH was again a doublet, and a similar set of rotating frame NOEs was determined for this conformer of **2** when compared to those of peptide **1**. The gross backbone conformation of the peptide must therefore be similar for the major conformations of both **1** and **2**. Similar experiments have also allowed us to establish that the minor conformation of **2** contains a *cis*-proline linkage, and therefore possesses a type VI β-turn about the Gln-

Pro segment. Chemical exchange peaks in the ROESY spectrum of **2** have established that the major and minor conformers are interconverting in solution. Indeed, the magnitude of these peaks is comparable to those observed for linear peptide sequences under study in our laboratory (N.G.J. Richards et al., manuscript in preparation). Our current models of the major conformation of **1** and the minor conformer of **2** are shown in Fig. 1. Refinement of these structures is underway using constrained Monte Carlo searching methods [8] and MD calculations.

Conclusion

From these results, the ability of the Pro-D-Phe sequence to control the shape of cyclic hexapeptides does not seem to lie wholly in a structural preference for a type II β-turn, since **2** also exists in a similar conformation in its most populated form. However, the substitution of D-Phe for L-Phe in the Pro-Phe region does increase the ratio of the two most highly populated conformers by an amount corresponding to ca. 5.7 kJ/mol in free energy, at 298 K. This substitution also appears to decrease the conformational flexibility of the cyclic hexapeptide. Whether the Pro-D-Phe unit exerts its influence by stabilizing the type I β-turn about Val-Tyr, or by destabilizing the type VI turn about Gln-Pro, relative to the Pro-L-Phe dipeptide is the subject of current molecular modeling studies.

Acknowledgements

This work was supported by the Science and Engineering Research Council (U.K.). We thank Dr. D. Turner for aid in the interpretation of the 2D NMR spectra of **1** and **2**.

References

1. Spatola, A.F., Anwer, M.K., Rockwell, A.L. and Gierasch, L.M., J. Am. Chem. Soc., 108 (1986) 825.
2. Anderson, W.F., Ohlendorf, D.H., Takeda, Y. and Matthews, B.W., Nature (London), 290 (1981) 754.
3. Atherton, E., Caviezel, M., Fox, H., Harkiss, D., Over, H. and Sheppard, R.C., J. Chem. Soc., Perkin Trans. 1 (1983) 65.
4. Kopple, K.D., Ohnishi, O. and Go, N., J. Am. Chem. Soc., 91 (1969) 4264.
5. Bothner-By, A.A., Stephens, L.R., Lee, J., Warren, C.D. and Jeanloz, R.W., J. Am. Chem. Soc., 106 (1984) 811.
6. Creighton, T.E., Proteins, W.H. Freeman, New York, 1983.
7. Bernstein, F.C., Koetzle, T.F., Williams, G.J.B., Meyer, E.F., Brice, M.D., Rodgers, J.R., Kennard, O., Schimanouchi, T. and Tasumi, M., J. Mol. Biol., 112 (1977) 535.
8. Still, W.C., Guida, W. and Chang, G., J. Am. Chem. Soc., 112 (1989) 4379.

Physical and conformational properties of a synthetic leader peptide from bovine serum albumin*

Ann Eisenberg Shinnar[a,**], Jennifer Howitt Anolik[a], Daniel A. Johnson[a] and Thomas. J. Lobl[b,***]

[a]Department of Chemistry, Swarthmore College, Swarthmore, PA 19081, U.S.A.
[b]Biopolymer Chemistry, The Upjohn Co., Kalamazoo, MI 49001, U.S.A.

Introduction

In an effort to understand how the conformational and physical properties of leader (signal) peptides contribute to their function in protein translocation across membranes, we have chemically synthesized the 24-residue amino terminal sequence of prepro bovine serum albumin (BSA): MKW(CHO)VTFISLLLLFS-SAYSRGVFRR(NH$_2$). The primary structure of preproBSA LP exemplifies features commonly observed in both eukaryotic and prokaryotic leader peptides, including: (1) an N_t-region bearing a charged residue (lysine), (2) a hydrophobic core region of 11 residues, and (3) a C_t-region with Ala and Ser occupying the –3, –1 positions proximal to the cleavage site [1]. PreproBSA LP is also predicted to have secondary structural features typical of leader peptides [2]. Application of Chou–Fasman rules [3] suggests the strong potential of forming either α-helix or β-sheet secondary structures in the N_t and hydrophobic core regions, and the signficant probability of a β-turn near the cleavage site region. These structural features are thought to be essential for the functional interaction of leader peptides with membranes. In the CD studies described below, we report the ability of preproBSA LP to undergo major changes in backbone structure that are dependent upon conditions of solvent choice and peptide concentration and thus illustrate the conformational flexibility of this synthetic leader peptide.

Results and Discussion

Size exclusion chromatography studies provide evidence that the peptide solubilized in 1% acetic acid undergoes a monomer-oligomer equilibrium. The minimum aggregation number is 50 and the critical micelle concentration (cmc) is estimated to be between 10^{-5} and 10^{-4} M. CD spectra show that preproBSA LP adopts β-sheet structure above the cmc and aperiodic conformation below the cmc (Fig. 1A). Addition of TFE to this system shifts the backbone

---

*This work is dedicated in memory of Professor Emil Thomas Kaiser and Eleanor Howitt.
**Present address: Department of Biochemistry and Biophysics, University of Pennsylvania, Philadelphia, PA 19104 U.S.A.
***Present address: Immunotech Pharmaceuticals, San Diego, CA 92121, U.S.A.

conformation to α-helix, whether the peptide is monomeric or oligomeric. In simple buffers such as phosphate or acetate, preproBSA LP exhibits CD spectra characteristic of a mixture of α-helix and β-sheet structures. In solvent mixtures of acetonitrile, methanol and phosphate buffer, used in C4 RPHPLC purification, the peptide shows a CD spectrum typical of α-helix (Fig. 1B).

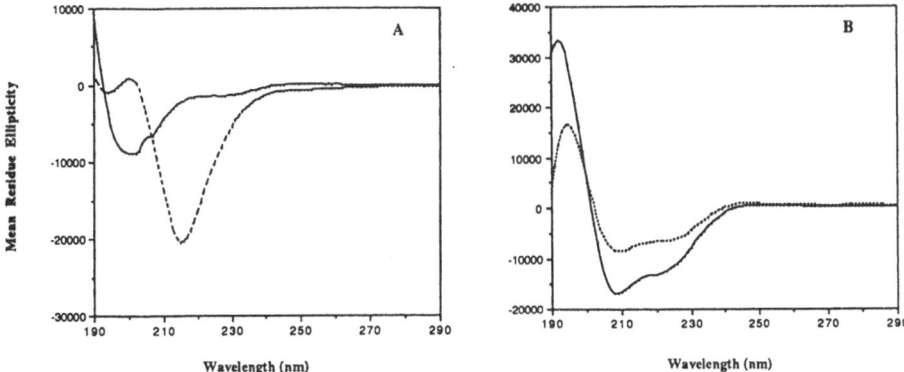

Fig. 1. (A) CD spectra of preproBSA in 1% acetic acid. Above the cmc, the peptide adopts > 80% β-sheet structure (-----); below the cmc, the peptide shows > 60% aperiodic structure (——) [4]. (B) α-Helix is promoted by TFE and organic solvents. Addition of 40% TFE to monomeric preproBSA in 1% acetic acid changes the backbone structure from aperiodic to > 45% α-helix content (——). In aqueous mixtures of acetonitrile and methanol, the peptide has > 25% α-helix (·······).

PreproBSA LP gives rise to a monomer-oligomer equilibrium, an aggregation property that typifies surfactants, i.e., molecules exhibiting interfacial activity. This micellization in 1% acetic acid is accompanied by a major change in conformation from predominantly aperiodic to β-sheet structure. The ability of this synthetic leader peptide to assume a more ordered structure upon encountering another peptide or protein surface could be important for receptor-mediated contact with biological membranes. Transition to an α-helical structure occurs easily in organic solvents, such as HPLC mixtures of acetonitrile and methanol, and also in hydrogen bond promoting solvents such as TFE. This suggests that an α-helical conformation would be favored for leader peptides inserting into the interior of a biological membrane.

Acknowledgements

Grants from Swarthmore College and The Upjohn Company are gratefully acknowledged.

References

1. Gierasch, L.M., Biochemistry, 28 (1989) 923.
2. Perlman, D. and Halvorson, H.O., J. Mol. Biol., 167 (1983) 391.
3. Chou, P.Y. and Fasman, G.D., Annu. Rev. Biochem., 47 (1978) 251.
4. Greenfield, N. and Fasman, G.D., Biochemistry, 8 (1969) 4108.

β-Ala residues for molecular design of cyclic peptides containing H-bonding turns

B. Di Blasio, V. Pavone*, X. Yang, A. Lombardi, E. Benedetti and C. Pedone

Department of Chemistry, University of Naples, Via Mezzocannone 4, 80134 Napoli, Italy

Introduction

The continuous and growing interest for cyclic peptides derives from the reduced conformational flexibility that can be achieved when a topological constraint is introduced in a linear peptide. For this reason, cyclic peptides have been used to develop highly potent biologically active molecules. Several cyclic peptides have been synthesized using S-S bond, amide or ester linkage. δ-Aminovaleric acid and ε-aminocaproic acid have been used to freeze naturally occurring amino acids in a typical β-turned conformation. With the same aim, we have undertaken the synthesis of a series of peptides containing the βAla–βAla linkage dipeptide as putative cyclization arm to force the remaining residues in a β-turned conformation.

Results and Discussion

As a part of our efforts on the conformational properties of cyclic peptides, we report the synthesis and X-ray single crystal analysis of two cyclic peptides: cyclo(Pro-Phe-βAla-βAla) and cyclo(Pro-Pro-Phe-βAla-βAla) (Fig. 1). The choice of these α-amino acids residues was based on their occurrence in bioactive molecules.

The peptides were synthesized in solution using classical methods. Cyclization of the free tetra- and pentapeptides was achieved in DCM using DCCI as activating agent in dilute solution (≈ 4.2 mM) in a reasonable yield (12%).

The cyclo(Pro-Phe-βAla-βAla) crystallizes in the monoclinic P2$_1$ space group from hot water, with two tetrapeptide and seven water molecules as an independent moiety of the unit cell. This compound contains, in the solid state, an intramolecular hydrogen bond between the CO group of the βAla[4] and the NH group of the βAla[3] residues stabilizing a type I β-turn conformation in which Pro[1] and Phe[2] occupy the relative position 2 and 3 of the turn, respectively. A rather complex network of 18 hydrogen bonds involving all the remaining CO and NH groups and water molecules is present in the crystal.

The cyclo(Pro-Pro-Phe-βAla-βAla) crystallizes in the monoclinic space group P2$_1$ from hot water with five solvent molecules. The Pro[1]-Pro[2] peptide bond is *cis* and the molecular conformation is stabilized by an intramolecular hydrogen

---

*To whom correspondence should be addressed.

667

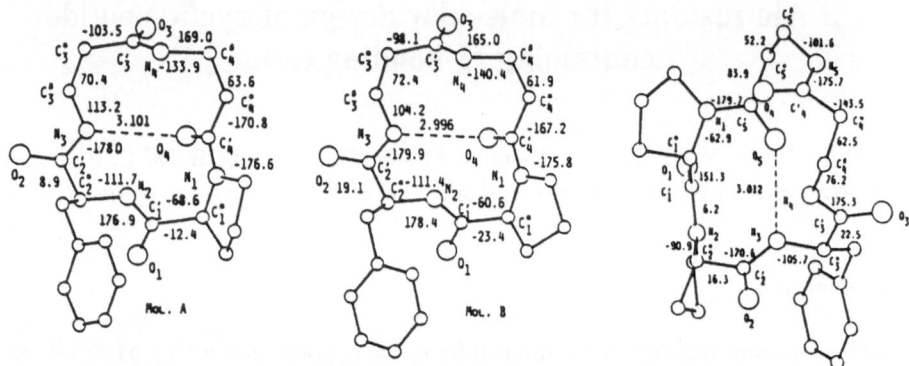

Fig. 1. Single crystal X-ray structure of cyclo(Pro-Phe-βAla-βAla) and cyclo(Pro-Pro-Phe-βAla-βAla).

bond between the CO group of β-Ala^5 and the NH of the Phe^3 residue. The Pro^1-Pro^2 segment occupies the relative positions 2 and 3 of a type VIa β-turn, while the β-Ala-β-Ala-Phe segment is incorporated in a C$_{13}$-like ring structure. The crystal packing is characterized by a network of 12 intermolecular hydrogen bonds involving all the remaining CO, NH and the water molecules.

Determination of peptide α-helicity in aqueous trifluoroethanol and its use in studies of protein secondary structure

S. Russ Lehrman, Jody L. Tuls and Marilyn Lund

The Upjohn Company, Kalamazoo, MI 49001, U.S.A.

Introduction

Linear peptides that lack well-defined secondary structure in aqueous solution often become α-helical in less polar solvents such as TFE [1]. The relevance of TFE-induced α-helicity to protein secondary structure is unclear. For example, it is unknown if TFE-induced α-helicity correlates with predicted α-helicity. This type of correlation would suggest that the α-helicity of peptides, dissolved in aqueous TFE, can be used as an indicator of their α-helical propensity, and would enhance the utility of peptide chemistry in studies of protein structure.

Bovine growth hormone (bGH) is a four-antiparallel-helix-bundle protein [2], which has been shown to fold in a manner consistent with the framework and molten-globule hypotheses [3,4,5]. Chou-Fasman analysis predicts the presence of several helical regions in this protein [6]. We can, therefore, determine the conformation of bGH fragments in aqueous TFE, and compare these with the known and predicted secondary structure of this protein. We have synthesized a series of eleven bGH fragments which span the complete primary structure of this protein (Fig. 1), and have determined the α-helical content of these peptides in aqueous TFE. In this report, we compare these results with the predicted α-helicity of each peptide.

Results and Discussion

The bGH fragments were synthesized by SPPS and characterized by FABMS and AAA. The percent α-helical content of each peptide was calculated from the mean residue ellipticity at 222 nm ($[\theta]_{222}$) of each peptide dissolved in 10 mole percent TFE. For these calculations, we used a value of 31 500 deg·cm^2·dmol^{-1}, as the $[\theta]_{222}$ for 100% α-helix [7]. Peptide concentrations were determined using BCA analysis (Pierce Chemical Co., Rockford, IL) [8]. The average variability of the $[\theta]_{222}$ determinations is ± 5%. Predictions of the α-helical content of each peptide were determined using the Chou-Fasman method [6].

We find that most of these peptides become increasingly α-helical as a function of TFE concentration up to 10 mole percent (data not shown). At 10 mole percent TFE, the amount of α-helicity developed by these peptides varies from 0 to 74%. These values have been compared with the predicted α-helicity for

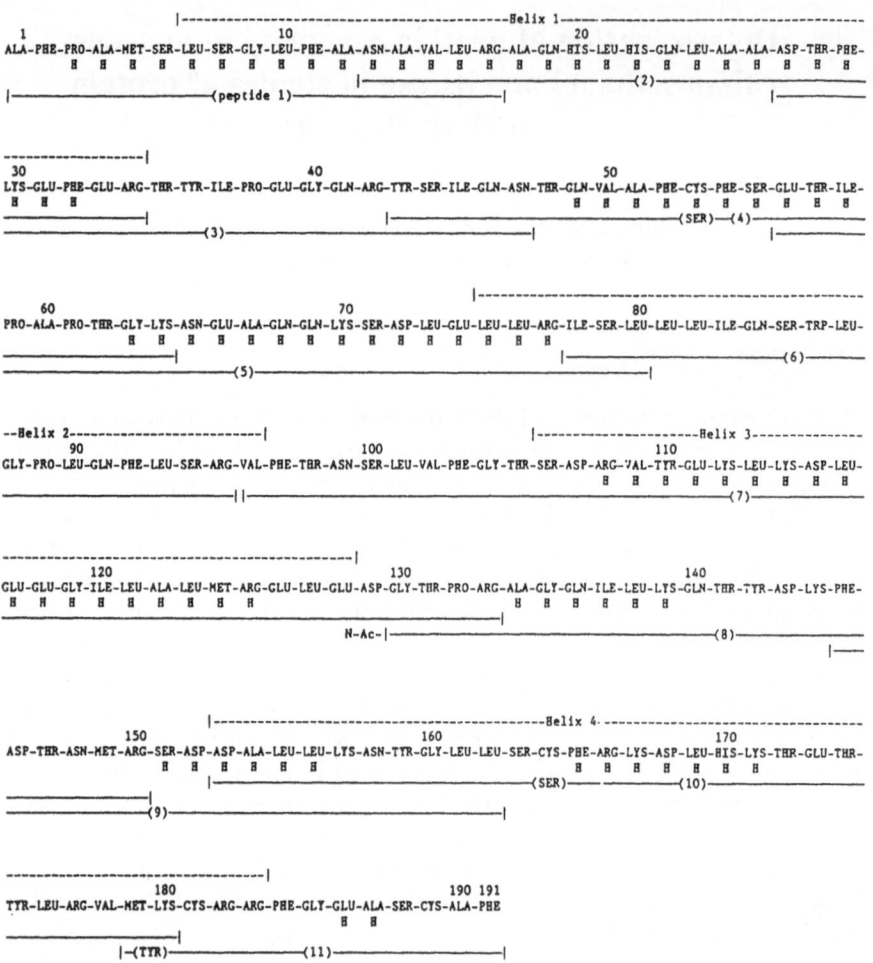

Fig. 1.   *The primary sequence of bovine growth hormone. α-Helical regions are indicated by dotted lines above the sequence. Regions of predicted α-helicity using the Chou-Fasman algorithm are noted by the letter H immediately below the primary sequence. Peptide fragments which were synthesized for use in this study are indicated by solid lines below the sequence.*

each primary sequence (Table 1). For eight of the eleven peptides studied, the predicted and TFE-induced α-helicity agree to within 20%, and are statistically correlated ($r = 0.9$). Therefore, for peptides which span 75% of the bGH primary sequence, it appears that the TFE-induced α-helicity reflects a propensity for α-helix formation. Peptides 1, 6, and 10 do not behave in this manner, and inclusion of these peptides in this analysis significantly lowers the observed correlation ($r = 0.6$). Peptides 1 and 6 are significantly more hydrophobic than the other peptides included in this study, suggesting that any relationship between predicted and TFE-induced α-helicity may only be valid for hydrophilic peptides.

670

Table 1   *Comparisons of predicted and TFE-induced peptide α-helicity*

	Peptide	TFE-induced α-Helicity (%)	Predicted α-Helicity (%)*
1	bGH(1–17)	46	88
2	bGH(12–34)	74	91
3	bGH(27–47)	16	29
4	[Ser^{53}]-bGH(43–64)	27	45
5	bGH(56–80)	51	60
6	bGH(78–95)	37	0
7	bGH(96–133)	60	47
8	[N-Acetyl]-bGH(130–150)	32	29
9	bGH(145–162)	27	33
10	[Ser^{164}]-bGH(153–80)	62	25
11	[Tyr^{179}]-bGH(179–191)	0	0

*See Ref. 6.

Conclusions

The results reported here show that, for a series of eight bGH fragments, TFE-induction of α-helicity parallels predicted α-helicity. However, three of the bGH fragments, two of which contain significant hydrophobic character, do not follow this pattern. These results suggest that investigations of peptide conformation in aqueous TFE are useful in studies of protein structure, but with limitations to be defined.

Acknowledgements

The authors thank Wayne Duholke for FABMS analyses, Lynn Snyder for amino acid analyses, Drs. Francois Kezdy and Henry Havel for helpful discussions, and Pat Barr for secretarial assistance.

References

1. DeGrado, W.F., Adv. Protein Chem., 39 (1988) 51.
2. Abdel-Meguid, S.S., Shieh, H.-S., Smith, W.W., Dayringer, H.E., Violand, B.N. and Bentle, L.A., Proc. Natl. Acad. Sci. U.S.A., 84 (1987) 6434.
3. Kim, P.S. and Baldwin, R.L., Annu. Rev. Biochem., 51 (1982) 459.
4. Brems, D.N., Plaisted, S.M., Dougherty, J.J. and Holzman, T.F., J. Biol. Chem., 262 (1987) 2590.
5. Brems, D.N. and Havel, H.A., Proteins, 5 (1989) 93.
6. Chou, P.Y. and Fasman, G., Annu. Rev. Biochem., 47 (1978) 251.
7. Chen, Y.H., Yang, J.T. and Martinez, H.M., Biochemistry, 11 (1972) 4120.
8. Smith, P.K., Krohn, R.I., Hermanson, G.T., Mallia, A.K., Gartner, F.H., Provenzano, M.D., Fujimoto, E.K., Goeke, N.M., Olson, B.J. and Klenk, D.C., Anal. Biochem., 150 (1985) 76.

Positively charged amino acid residues, because of their amphipathic nature, can increase lipid affinities of amphipathic helixes

Y.V. Venkatachalapathi, Kiran B. Gupta, Hans DeLoof, Jere P. Segrest and G.M. Anantharamaiah

Departments of Medicine and Biochemistry and the Atherosclerosis Research Unit, UAB Medical Center, Birmingham, AL 35294, U.S.A.

Introduction

The amphipathic helical (AH) domains of apolipoproteins (apo), compared to other biologically active AH domains such as polypeptide hormones and toxins, are unique. They possess a clustering of positively-charged (+) residues at the polar–nonpolar interface and negatively-charged (−) residues in the center of the polar face. Synthetic peptide analogs of the AH with the charged residue positions reversed, i.e., + residues clustered in the center of the polar face and − residues at the polar-nonpolar interface, have significantly decreased lipid affinity compared to apo-mimicking AH peptides [1]. From these studies, we hypothesized that the + residues, because of their amphipathic nature, can increase the lipid affinity of the AH.

As part of a study to understand the molecular properties of apo A-I, the major protein component of high density lipoproteins, we synthesized analogs for the consensus lipid-associating domain with the sequence PVLDEFREKL-NEXEEALKQKLK (A-I_{con}) [2]. The 13th position (denoted by X) indicates the marked polymorphism that occurs at this position in the eight 22mers of apo A-I and is located on the nonpolar face 40 degrees from the polar-nonpolar interface of the AH. We now describe studies of three A-I_{con} analogs, ([Lys^{13}]A-I_{con}, [Glu^{13}]A-I_{con} and [Haa^{13}]A-I_{con}) with X at the 13th position = Glu, Lys and L-homoaminoalanine (Haa, an unnatural + amino acid having the same acyl chain length as Glu) to test our hypothesis. [Haa^{13}]A-I_{con} serves as a control peptide to study the effect of acyl-chain length on the lipid affinity.

Results and Discussion

The peptides were compared for their ability to interact with multilamellar vesicles of dimyristoyl phosphatidylcholine (DMPC), and the extent of interaction was compared by their ability to clarify turbid DMPC suspension and to form stable discoidal complexes. We have shown earlier that the size of the discoidal complexes varies inversely with the lipid affinity of the peptide. As measured by the EM and non-denaturing gradient gel electrophoresis of peptide : DMPC

672

complexes (1 : 1 weight ratio), the [Lys^{13}]A-I$_{con}$ forms the smallest discoidal complexes of the three (Stoke's diameters: [Lys^{13}]A-I$_{con}$: 110 ± 30 Å [Glu^{13}]A-I$_{con}$: 200 ± 50 Å and [Haa^{13}]A-I$_{con}$: 310 ± 30 Å). The results of the dye leakage experiments also suggest that [Lys^{13}]A-I$_{con}$ interacts with the carboxyl fluorescein-entrapped egg PC to release the dye at least twice as effectively as the other two. That is, [Lys^{13}]A-I$_{con}$ at 200 µg equals Triton X-100. The other two, at the same concentration, released about 50% of the entrapped dye.

Polar face

Nonpolar face

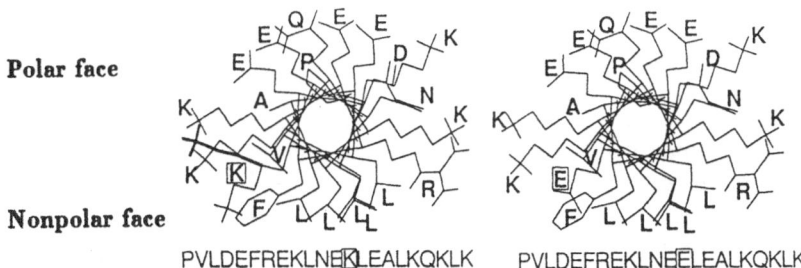

PVLDEFREKLNEKLEALKQKLK   PVLDEFREKLNEELEALKQKLK

Fig. 1. Idealized helix representation of two peptides. The mutated Lys residue [Lys^{13}]A-I$_{con}$ is shown in the original and in snorkel orientation (bold).

Based on these results, we propose the (Snorkel) model shown in Fig. 1 for the association of the apo class AH with lipids. The bulk of the van der Waals surface areas of the + residues are hydrophobic. We propose that the amphipathic basic residues, when associated with phospholipid, extend toward the polar face of the helix to insert their charged moieties into the aqueous milieu for solvation. We suggest that essentially all of the uncharged van der Waals surface of the AH of the apos can be buried within the interior of a phospholipid monolayer. Compared to the other classes of the AH, the arrangement of charged residues found in the apos provides for a deeper helix insertion into a monolayer and thus possess a greater lipid affinity.

References

1. Anantharamaiah, G.M., Jones, J.L., Brouillette, C.G., Schmidt, C.F., Chung, B.H., Hughes, T.A., Bhown, A.S. and Segrest, J.P., J. Biol. Chem., 260 (1985) 10248.
2. Anantharamaiah, G.M., Brouillette, C.J., Venkatachalapathi, Y.V. and Segrest, J.P., Arteriosclerosis, (1989) in press.

Diacylaminoepindolidiones as models for β-sheet structure

**D.S. Kemp*, Christopher C. Muendel, Daniel E. Blanchard and
Benjamin R. Bowen**

*Department of Chemistry, Massachusetts Institute of Technology, Room 18-584,
Cambridge, MA 02139, U.S.A.*

Introduction

Earlier, we introduced peptide conjugates of 1,8-diaminoepindolidiones as
models for antiparallel β-sheet formation [1,2]. The epindolidione nucleus mimics
the H-bonding sites of a polypeptide backbone in its extended conformation
as noted in Fig. 1. Acylation with the dipeptide Pro-D-Ala that has a strong
tendency towards β-turn formation, followed by a chain-reversing urea spacer
linked to a dipeptide dimethylamide (Fig. 2), permits exploration of biases of
amino acids for and against antiparallel sheet formation. Figure 3 shows a 1,8-
diacylaminoepindolidione bearing an acetyltetrapeptide residue terminated by
the turn-forming Pro-D-Ala. This system permits exploration of the nucleation
of parallel sheets.

Results and Discussion

The structures of the epindolidione derivatives were studied by NMR in DMSO-
d_6. For some of the derivatives containing the urea linker, NH J-values, NH
$\Delta\delta$'s, and Gly methylene δ's imply that ordered conformations exist. NH
$\Delta\delta/\Delta T$'s indicate that the amides positioned for H-bonding are sequestered from
solvent. NOE data show proximity of the entire peptide to the epindolidione
core.

In addition to NH $\Delta\delta$'s, shifts are observed in H3 on the polyheterocycle.
We have attributed this to deshielding from the adjacent anilidic carbonyl as
it is forced in proximity to the proton when a second H-bond is formed. To
test this point, three series of derivatives were made varying the fourth residue
between series and the third residue within a series. Each group reveals a linear
relationship between H3 δ and NH δ of residue 4, the NH involved in the
second hydrogen bond, and the same sequence of increasing $\Delta\delta$ in all three
series.

The simple tetrapeptide derivatives (Fig. 3) incorporating permutations of the
same four hydrophobic residues exhibit similar evidence for parallel sheet
formation in DMSO. The most compelling indications are shift anisotropies,
here of H3 and the αH of residue 4. The magnitude of the last anisotropy

---

*To whom correspondence should be addressed.

674

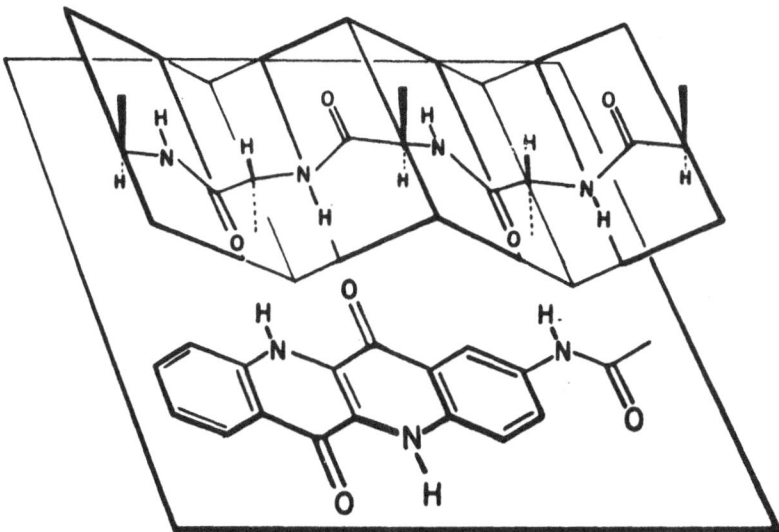

Fig. 1. *Alignment of epindolidione nucleus with β-sheet peptide.*

in particular, ranging from zero in the Ala-Ala derivative to 0.55 ppm for the Val-Val case, closely reflects the bulk of the residues comprising the sheet. Further evidence includes the high NH J-values of residue 4 (up to 9.5 Hz) and the increasingly stabilized β-turn (low $\Delta\delta/\Delta T$ and large aryl NH $\Delta\delta$). Finally, difference and 2D NOE experiments of the Val-Val compound reveal a number of spatially proximate hydrogens, notably H4 and the acetyl, and Val4 αH and the pyridone NH, that confirm an extended structure adjacent to the epindolidione.

We have thus demonstrated the ability of a rigid polyheterocycle to direct β-sheet formation in both the parallel and antiparallel sense, and we have identified parameters that report consistently on the robustness of the structure.

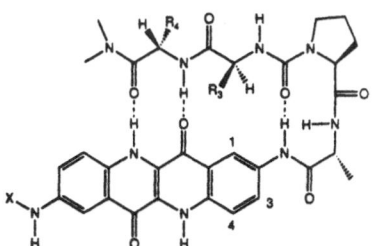

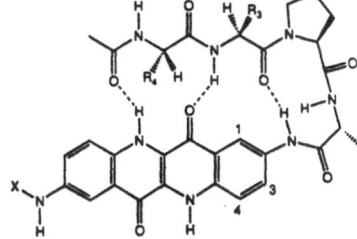

Fig. 2. *Dipeptide conjugate linked through urea spacer to Pro-D-Ala-epindolidione system.*

Fig. 3. *Peptide directly coupled to Pro-D-Ala-epindolidione system.*

675

Future work will focus on NMR and CD studies in water, including a quantitative assessment of the sheet-forming propensities of the different amino acids.

References

1.   Kemp, D.S. and Bowen, B.R., Tetrahedron Lett., 29 (1986) 5077.
2.   Kemp, D.S. and Bowen, B.R., AAAS Protein Folding Conference, in press.

Solution phase conformation of templated proteins: Progress in the modeling of α-helix initiation and propagation

D.S. Kemp and James G. Boyd

Department of Chemistry, Massachusetts Institute of Technology, Room 18-584, Cambridge, MA 02139, U.S.A.

Introduction

As reported elsewhere in these proceedings [1], owing to conformational flexibility at the acetyl amide and 8,9 C-C bonds, peptide conjugates of template **1** (T) exist in water as a mixture of two conformers. Here, we present evidence that concerted with conversion of the template to its trans, eclipsed form, a helix-coil transition occurs in the linked peptide that is readily measurable by H NMR. Peptide conjugates of **1** (Fig. 1) thus constitute a laboratory for studying helix-coil transitions in water for small peptides.

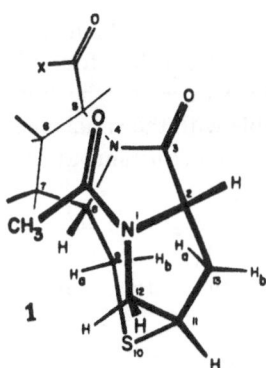

1

Fig. 1. *trans, Eclipsed conformation of the helix-nucleating template 1.*

Results and Discussion

Scrutiny of the 500 MHz H NMR spectrum of TA_6OH (Fig. 2) reveals two peptide conformers present in a helix/coil ratio (h/c) of approximately 2:1. The thiomethylene protons 9a,b (Fig. 1) and the two acetyls also integrate to 2:1, demonstrating the cooperative transition described above. The striking feature of the NH spectrum of TA_6OH is the pronounced difference in NH chemical shift (δ) between the two forms of the peptide. The helical conformer

677

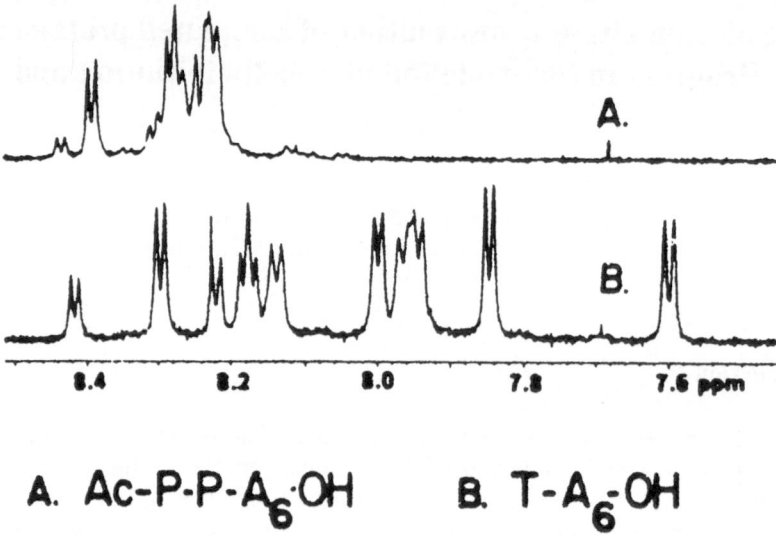

A.

B.

8.4   8.2   8.0   7.8   7.6 ppm

A. Ac-P-P-A$_6$·OH   B. T-A$_6$·OH

Fig. 2.   NH resonances of hexa Ala conjugates.

has δ's consistent with those of helical residues in native peptides [2]. The lower field NH resonances, however, are the same as for AcPro$_2$Ala$_6$OH (Fig. 2). The change in δ with respect to temperature ($\Delta\delta/\Delta T$) is a test for shielding from solvent. We observe $\Delta\delta/\Delta T$ values for the helical amides of 3.5–5.4×10^{-3} ppm/K. These values are consistent with values found for internally H-bonded protons [3]. Conversely, the $\Delta\delta/\Delta T$ values for the low field NHs range from 7.7 to 8.4×10^{-3} ppm/K, comparable with the value measured for random coil peptides. In addition, the $J_{\alpha CH-NH}$ coupling constants are 5–6 Hz, consistent with the range predicted for helical residues [2].

To test the effect of chain length on h/c ratio, we prepared two series of templated Ala oligomers. The first series: TA$_{4,6,8}$ONa, shows ratios of 1.1, 1.5 and 1.7, respectively. The second series, containing a C-terminal lysine amide: TA$_{4-8}$KNH$_2$ ϵ-TFA salt, has ratios of 2.0, 2.3, 2.5, 2.6 and 2.7. From these values, several points are evident. First, the template responds to small changes in peptide structure. Second, the effect is approximately that predicted from the Scheraga [4] Zimm–Bragg's value for alanine. Third, negative charge at the C-terminal is destabilizing (compare TA$_4$KNH$_2$ with TA$_6$ONa).

We have also prepared a templated alanine peptide series with glycine inserted at each position. The h/c values are shown in Fig. 3 and are in agreement with what we predict using the accepted s value of 0.60. Finally, a striking difference in the chemical environments of the Gly methylene protons is seen for TGA$_5$OH. The helical form in D$_2$O has two chemically non-equivalent protons at δ 3.97 and 3.90 (J = 16 Hz). The random coil Gly methylenes, however, are chemically equivalent and show up as a singlet at δ 3.94.

678

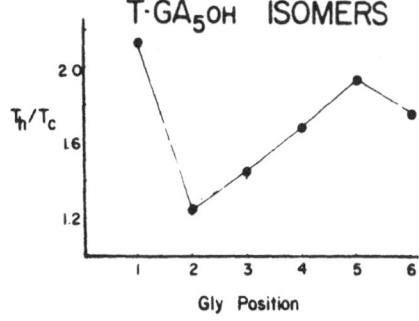

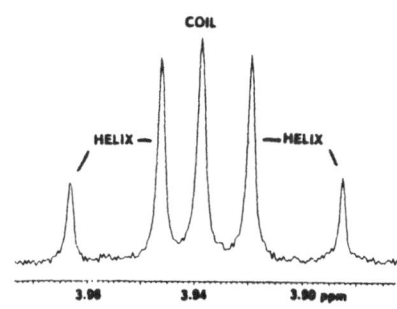

Fig. 3. h/c Ratios for positional isomers of TGA₅OH from integration of ¹H NMR spectra.

Fig. 4. ¹H NMR resonances for the Gly methylene group of TGA₅OH.

Conclusion

Three observations were made: (1) our helix initiating template induced large amounts of helical structure in attached peptides, (2) the bias that amino acid residues have towards forming a helix is detectable by H NMR, and (3) this system can be used for the assignment of helix-forming and breaking parameters.

References

1. Kemp, D.S., Boyd, J.G., Curran, T.P. and Fotouhi, N., In Rivier, J.E. and Marshall, G.R. (Eds.) Peptides: Chemistry, Structure and Biology (Proceedings of the 11th American Peptide Symposium), ESCOM, Leiden, 1990, p. 861.
2. Pardi, A., Wagner, G. and Wüthrich, K., Eur. J. Biochem., 137 (1983) 445.
3. Dyson, H.J. et. al., J. Mol. Biochem., 201 (1988) 161.
4. Scheraga, H.A., Pure Appl. Chem., 50 (1978) 315.

Synthetic approaches to the azole peptide mimetics

Thomas D. Gordon, Philip E. Hansen, Barry A. Morgan and Jasbir Singh
*Department of Medicinal Chemistry, Sterling Research Group, Rensselaer,
NY 12144, U.S.A.*

Introduction

The rational design of molecules that mimic the conformation of physiologically important peptides is a major goal of medicinal chemists. The strategy of conformational restriction we have adopted involves a cyclization between the carbonyl of the residue n and the N, Cα and Cβ of the residue n + 1. We refer to these thiazoles, imidazoles and oxazoles, etc., as 'dipeptide azoles'.

Our initial synthesis [1] of 5-substituted thiazole dipeptide mimics using a modification of the Hantzsch [2] synthesis led to nearly complete racemization. We envisioned that a suitably protected β-oxo-dipeptide could serve as a key intermediate for the synthesis of a variety of azole dipeptides of known absolute chirality.

Results and Discussion

A search of the literature revealed that α-amino-β-oxo-esters, such as (**2**), were prepared rather laboriously via hydrolysis of oxazoles [3]. Consequently, we have developed a general, one pot high yield synthesis of α-amino-β-oxo-esters (**2**), by addition of the potassium salt of Schiff's base (**1**) at –78°C to a pre-cooled solution of the acyl chloride, followed by in-situ acid hydrolysis and crystallization.

Coupling of (**2**) with Boc- or Z-D-Trp-OH by the mixed anhydride method, under conditions where the free base of α-amino-β-oxo-ester was gradually generated in low concentrations in situ, provided the β-oxo-dipeptide (**3a**) in good yield. Treatment of (**3a**) with Lawesson's reagent gave the crystalline thiazole ester (**4a**) [4]. After several attempts, the 'formal' dehydration of (**3a**) to the desired oxazole (**4c**) was accomplished using the conditions shown in Scheme 1. In addition, imidazole (**4b**) was obtained from the β-oxo-dipeptide intermediate (**3a**). The synthesis of azole dipeptide mimics (**4a–c**), as shown in Scheme 1, proceeded with retention of chiral integrity. To the best of our knowledge, this procedure constitutes the first synthesis of the 5-substituted dipeptide azoles. Dipeptide mimics (**4a–c**) were introduced into [Pro[6]-D-Trp[7,9]]SP(6–11) analogs. Their pharmacology is reported elsewhere in these proceedings [5].

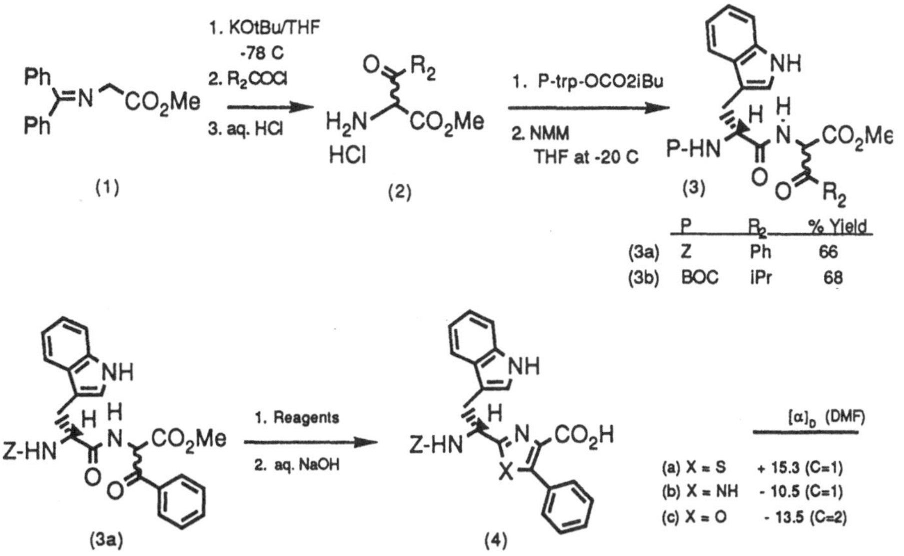

Scheme 1. *Reagents: (a) Lawessons' reagent/THF, reflux, (66%); (b) NH₄OAc/HOAc,* reflux, (62%); (c) Ph₃P/DBU/CCl₄ in Py/CH₃CN, (43%).

$$\textit{Scheme 1. Reagents: (a) Lawessons' reagent/THF, reflux, (66\%); (b) } NH_4OAc/HOAc,$$

reflux, (62%); (c) $Ph_3P/DBU/CCl_4$ in Py/CH_3CN, (43%).

Conclusion

We have developed a simple and versatile method for the synthesis of optically pure 'azole' dipeptide mimics via β-oxo-dipeptides as key intermediates. These dipeptide mimics can be successfully elaborated both at the amino and the carboxyl termini without racemization.

References

1. Gordon. T.D., Hansen, P.E., McKay, F.C., Morgan, B.A., Singh, J., Baizman, E.R., Kiefer, D. and Ward, S.J., In Jung, G. and Bayer, E. (Eds.) Peptides 1988 (Proceedings of the 20th European Peptide Symposium), De Gruyter, Berlin, 1989, p. 292.
2. Dean, B.M., Mijouic, M.P.V. and Walker, J., J. Chem. Soc., (1961) 3394.
3. Suzuki, M., Iwasaki, T., Mutsumoto, K. and Okumura, K., Syn. Comm., 2 (1972) 237.
4. Schmidt, U., Gleich, P., Greisser, H. and Utz, R., Synthesis, (1986) 992 (and references cited therein).
5. Gordon, T.D., Hansen, P.E., Morgan, B.A., Singh, J., Baizman, E.R. and Ward, S.J., In Rivier, J.E. and Marshall, G.R. (Eds.) Peptides: Chemistry, Structure and Biology (Proceedings of the 11th American Peptide Symposium), ESCOM, Leiden, 1990, p. 877.

Protein engineering of betabellins 9, 10 and 11

Robert D. McClain[a], Scott B. Daniels[a], Robert W. Williams[b], Arthur Pardi[c],
Michael Hecht[d], Jane S. Richardson[d], David C. Richardson[d] and
Bruce W. Erickson[a]

[a]Department of Chemistry, University of North Carolina, Chapel Hill, NC 27599, U.S.A.
[b]Department of Biochemistry, Uniformed Services University of the Health Sciences,
Bethesda, MD 20814, U.S.A.
[c]Department of Chemistry and Biochemistry, University of Colorado, Boulder,
CO 80309, U.S.A.
[d]Department of Biochemistry, Duke University, Durham, NC 27710, U.S.A.

Introduction

We are pursuing the solid-phase synthesis and structural characterization of
betabellin, a *beta*-barrel *bell*-shaped prot*ein*. Versions 3–7 of betabellin had two
identical 32-residue peptide chains cross-linked by both a symmetric diamino
acid and a disulfide bond [1–5]. Each chain was designed to fold into an
antiparallel β-pleated sheet consisting of four β-strands joined by 3 β-turns,
as shown below for betabellin 9. Two amphiphilic sheets should adopt a
conformation in water involving noncovalent interactions between their hydrophobic faces. Exposure of their hydrophilic faces should render the folded
structure quite water-soluble. But these early versions of betabellin were very
hydrophobic and may not have folded properly due to the constraints of two
covalent cross-links. Betabellin 9, which lacked the diamino-acid cross-link of
betabellin 7, was much more hydrophilic than betabellin 7. Biophysical studies
showed that betabellin 9 may adopt a β-structure. Engineering of betabellins
10 and 11 used insights from molecular mechanics calculations to optimize the
geometry of the β-turns.

Results and Discussion

The hydrophilicity of betabellin 9 was measured as the upper phase/lower
phase solubility ratio in butanol/acetic acid/water (4:1:5, v/v) as quantitated
by analytical RPHPLC. The partition ratio of betabellin 9 was 0.5, or 26 times
lower than that of betabellin 7. The increased hydrophilicity of the disulfide-
bridged betabellin 9 suggests that it folds into a more compact structure than
did the doubly cross-linked versions. The CD spectrum of betabellin 9 resembled
those of β-sheet proteins but was dissimilar to those of α-helical models;
denaturation removed these features. Laser Raman analysis of betabellin 9
indicated that the peak phase, intensity, and frequency shift of the amide-I region
is most like that of the β-barrel protein, concanvalin A, and is very different

from that of the α-helical peptide, melittin. 2D NMR analysis revealed that the NOESY spectrum of betabellin 9 contains several Cα-Cα cross peaks absent from the COSY spectrum. These cross peaks may represent NOE interactions between adjacent strands of an antiparallel β-sheet.

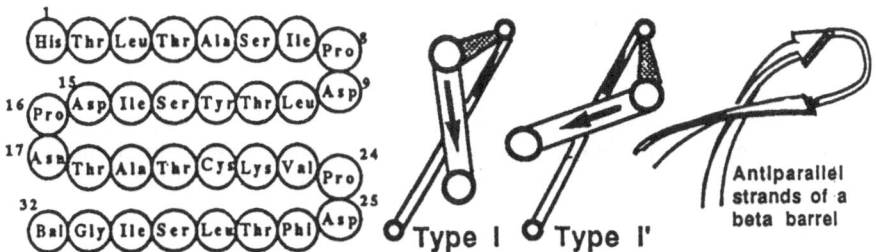

Our molecular modeling calculations indicated that the use of D-amino acids at the i + 1 and i + 2 positions of the six β-turns of betabellin might enhance folding into a β-barrel by favoring the Type-I' β-turn conformation. As shown above, the twist of the Type-I' turn is closer than the twist of the Type-I turn to the inherent right-handed twist of antiparallel strands of a β-barrel [5]. MACROMODEL calculations of model hexapeptides containing an Ala-Pro-Asx-Ala turn indicated that the D-Pro-D-Asx chirality is more likely to preserve the dihedral geometry of the Type-I' turn conformation than L-Pro-L-Asx. On energy minimization, the D-Pro-D-Asn model adopted a Type-I' turn conformation and was about 9 kcal/mol more stable than the L-Pro-Asn model. In addition, SYBYL caluculations by energy minimization with a Kollman force field indicated that the D-Pro-D-Asn turn is much better than the L-Pro-Asn turn at maintaining the average cross-turn distances for the Type-I' β-turns found in six proteins (actinidin, elastase, staphylococcal nuclease, γ-chymotrypsin, α-lytic protease, streptococcal protease A).

Thus, betabellin 10 was synthesized with D-amino acids at all six i + 1 and i + 2 turn positions of each sheet D-Pro⁸,D-Asp⁹,D-Pro¹⁶,D-Asn¹⁷,D-Pro²⁴,D-Asp²⁵). Denaturation and reduction of the *S*-sulfonated 32-residue peptide under conditions used for betabellin 9 (0.1% TFA, 25% 2-mercaptoethanol, 100°C, 10 min) resulted in substantial cleavage of the Asp¹⁵-D-Pro¹⁶ peptide bond. The peptide fragments were identified by FABMS and Edman sequencing with assistance from Dean Marbury and Russ Henry. The presence of D-Pro¹⁶ in betabellin 10 might induce a local conformation that orients the carboxylate side chain of Asp¹⁵ in a way that more efficiently catalyzes cleavage of the Asp¹⁵-D-Pro¹⁶ peptide bond.

Betabellin 11 was designed to avoid this chain cleavage by replacing the second turn sequence Asp¹⁵-D-Pro¹⁶-D-Asn¹⁷-Thr¹⁸ of betabellin 10 with the isomeric sequence Asn¹⁵-D-Pro¹⁶-D-Asp¹⁷-Thr¹⁸, which lacks a labile Asp-Pro peptide bond but preserved the net charge. The structural integrity of betabellin 11 is being studied.

683

Acknowledgements

This work was supported in part by grants from NIH, ONR, and ARO.

References

1. Reddy, P.A. and Erickson, B.W., In Deber, C.M., Hruby, V.J. and Kopple, K.D. (Eds.) Peptides: Structure and Function, (Proceedings of the 9th American Peptide Symposium), Pierce Chemical Co., Rockford, IL, 1985, p. 453.
2. Erickson, B.W., Daniels, S.B., Reddy, P.A., Unson, C.G., Richardson, J.S. and Richardson, D.C., In Zoller, M. and Fletterick, R. (Eds.) Computer Graphics and Molecular Modeling, Cold Spring Harbor Laboratory, Cold Spring Harbor, NY, 1986, p. 53.
3. Daniels, S.B., Reddy, P.A., Albrecht, E., Richardson, J.S., Richardson, D.C. and Erickson, B.W., In Marshall, G.R. (Ed.) Peptides: Chemistry and Biology, (Proceedings of the 10th American Peptide Symposium), ESCOM, Leiden, 1988, p. 383.
4. Erickson, B.W., Daniels, S.B., Reddy, P.A., Higgins, M.L., Richardson, J.S. and Richardson, D.C., ICSU Short Rep., 8 (1988) 2.
5. Richardson, J.S. and Richardson, D.C., In Oxender, D.L. and Fox, C.F. (Eds.) Protein Engineering, Alan R. Liss, New York, NY, 1986, p. 149.

684

A polypeptide based hemeprotein model

Tomikazu Sasaki* and Emil T. Kaiser

*Laboratory of Bioorganic Chemistry and Biochemistry, Rockefeller University,
1230 York Avenue, New York, NY 10021, U.S.A.*

Introduction

De novo design of functional proteins requires basic understanding of the stability and dynamics of the folded state of polypeptide chains. Although predicting a final folded conformation of a peptide from its primary sequences is still very difficult, and largely depends on empirical rules, such as Chou-Fasman parameters, several artificial proteins have been successfully synthesized [1] based on designed tertiary structures. We have recently reported [2] the synthesis and hydroxylase activity of an artificial hemeprotein (helichrome, Fig. 1) by utilizing amphiphilic α-helices as a building unit. We now wish to report the structural stability of helichrome in the presence of guanidine hydrochloride (Gn·HCl) as a denaturant.

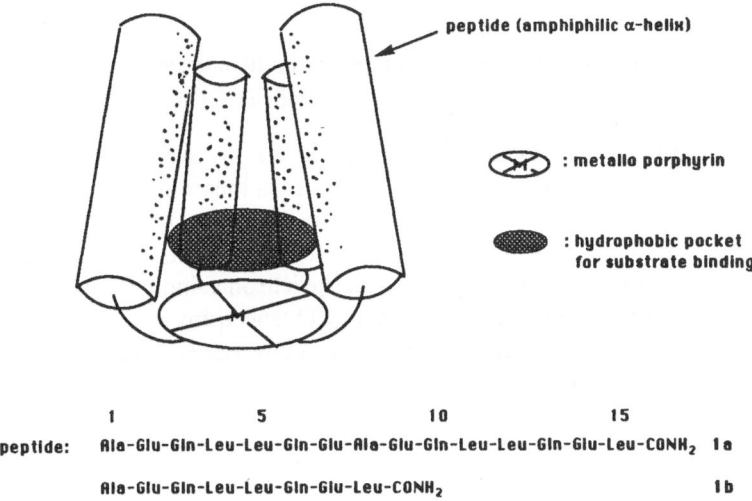

peptide (amphiphilic α-helix)

: metallo porphyrin

: hydrophobic pocket for substrate binding

	1	5	10	15

peptide: Ala-Glu-Gln-Leu-Leu-Gln-Glu-Ala-Glu-Gln-Leu-Leu-Gln-Glu-Leu-CONH₂ 1a

Ala-Glu-Gln-Leu-Leu-Gln-Glu-Leu-CONH₂ 1b

Fig. 1. Design concept of helichrome as an artificial hemeprotein.

---
*Present address: Department of Chemistry, BG-10, University of Washington, Seattle, WA 98195, U.S.A.

685

T. Sasaki and E.T. Kaiser

Results and Discussion

Helichrome (**1a**) and mini-helichrome (**1b**) were synthesized via a segment synthesis-condensation approach as reported previously [2]. Figure 2 shows the CD spectrum of helichrome in an aqueous buffer solution. The helical content was calculated to be 70% and 32% for (**1a**) and (**1b**), respectively. On the other hand, single peptide showed the typical CD pattern for a disordered confirmation under the same experimental condition. Helichrome was found to be monomeric based on the gel filtration chromatography and sedimentation equilibrium experiments, indicating an intramolecularly folded state of helichrome.

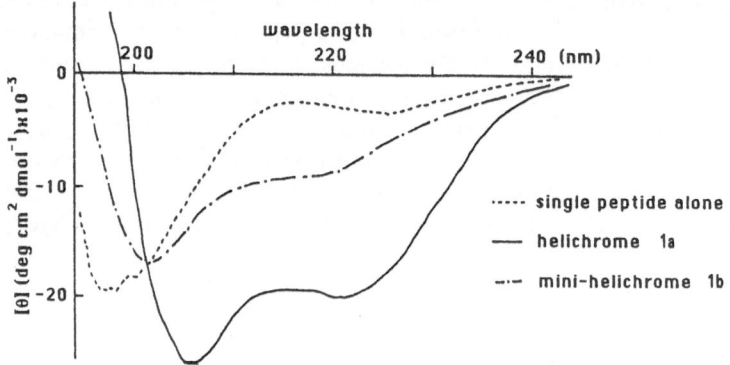

Fig. 2. *CD spectra of single peptide, 1a and 1b in 20 mM phosphate, 0.16 M KCl pH 7.5 at 27°C.*

An unfolding experiment was conducted to investigate the stability and dynamics of the folded structure of helichrome. In the presence of 7.2 M Gn·HCl, helichrome exhibited a very similar spectrum to that of the single peptide in a buffer solution, indicating a disordered conformation of the denatured helichrome. The unfolding transition of helichrome induced by Gn·HCl was followed by observing changes in the CD spectrum as a function of the concentration of denaturant. The result was analyzed according to the published procedure [3]. The folded structure of (**1a**) appeared to have a comparable stability ($C_{0.5} = 5.2$ M, m = 0.84) to that of native globular proteins.

References

1a. Regan, L. and De Grado, W.F., Science, 241 (1988) 976.
  b. De Grado, W.F., Wasserman, Z.R. and Lear, J.D., Science, 243 (1989) 622.
  c. Mutter, M., Altmann, E., Altmann, K.-H., Hersperger, R., Kozieg, P., Nebel, K., Tuchscherer, G., Vuilleumier, S., Gremlich, H.-U and Muller, K., Helv. Chim. Acta, 71 (1988) 835.

d. Richardson, J.S., Richardson, D.C. and Erickson, B.W., Biophys. J., 45 (1984) 25A.
e. Oxender, D.L. and Fox, C.F., (Eds.) Protein Engineering, Alan. R. Liss Inc., New York, 1987.
2a. Sasaki, T. and Kaiser, E.T., J. Am. Chem. Soc., 111 (1989) 380.
b. Sasaki, T. and Kaiser, E.T., Biopolymers, (1989) in press.
3a. Ahmad, F. and Bigelow, C.C., Biopolymers, 25 (1986) 1623.
b. Ahmad, F. and Bigelow, C.C., J. Biol. Chem., 257 (1982) 12935.

Minimally designed ion channels

Arlene L. Rockwell, James D. Lear and William F. DeGrado

E.I. du Pont de Nemours and Co., Central Research and Development Dept.,
P.O. Box 80328, Wilmington, DE 19880-0328, U.S.A.

Introduction

In order to investigate a simple model of large ion channel proteins, we have synthesized a series of 21-residue amphiphilic peptides with a repeating pattern of Ser and Leu [1]. H_2N-(LSLLLSL)$_3$-CONH$_2$ formed a channel only large enough to conduct protons, consistent with an aggregate of four α-helical monomers. Changing one Leu in each heptad to a Ser, (LSSLLSL)$_3$, increased the size of the central pore to allow passage of larger monovalent cations (Li$^+$, Na$^+$, K$^+$, Cs$^+$, guanidinium$^+$), and modeling suggested this peptide formed a hexameric aggregate. To establish further the sequence requirements for ion channel activity, we prepared a series of 21-residue peptides containing the two heptads LSSLLSL and LSLLLSL in each of the eight possible combinations.

Results and Discussion

The heptads were synthesized on a *p*-nitrobenzophenone oxime resin, as described previously [2]. The protected hexamer Fmoc-Leu-Ser(Bzl)-Ser(Bzl)-Leu-Leu-Ser(Bzl)-O-Polymer was cleaved from the resin by nucleophilic displacement with the acetate salt of Leu-O-t-But. After precipitation from DCM:MeOH, the ester group was removed by TFA/anisole 9:1. Chromatographically pure protected peptides were obtained in multigram quantities with typical yields being 40–65%. The heptamers were then coupled to 4-MeBHA resin in a 1.2 M excess with 1.6 equiv. HOBT and 1.6 equiv. DIC in NMP overnight (Scheme 1). The extent of reaction was assessed by titration with picric acid [3], and a second coupling done if necessary. The resin was then acetylated with acetic anhydride (0.6 M in DCM) to cap unreacted sites, and deprotected with 50% piperidine/DMF. This coupling/deprotection procedure was repeated twice more with the appropriate heptamers, and the peptide cleaved from the resin. Crude yields were typically 45–55%, based on the loading of the first heptad onto the resin.

The 21-mers were purified by RPHPLC, using a semi-preparative PRP-1 (Hamilton) column eluted isocratically with from 63% to 78% (depending on the peptide) aqueous acetonitrile containing 0.1% TFA at a flow rate of 4 ml/min. Isocratic elution and careful avoidance of contamination were necessary to obtain peptides pure enough to give channel records suitable for analysis. FABMS confirmed the identity of the parent ions.

Synthesis of (LSSLLSL)3

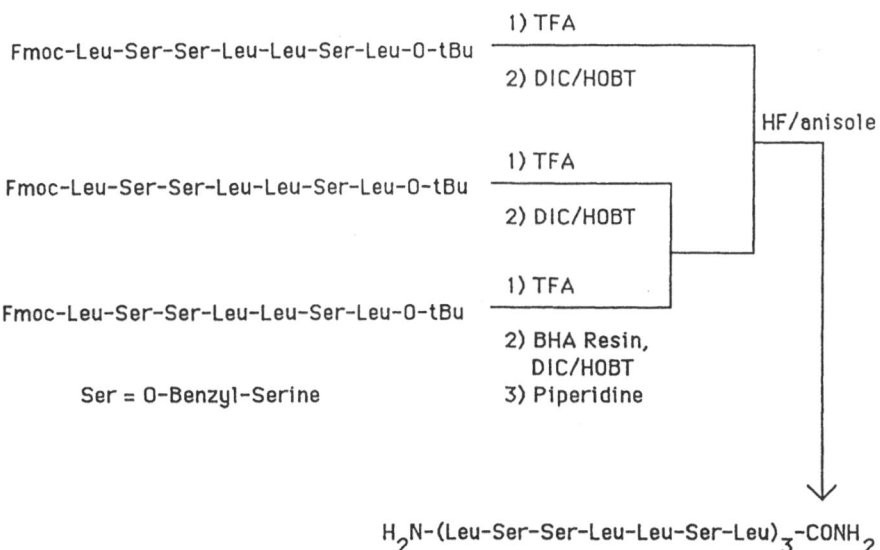

Scheme 1.

The purified peptides were analyzed for ion channel formation as described previously [1], and the selectivity for conduction of protons versus other cations was determined. The peptides indicated as 'proton channels' (Table 1) were conductive to protons but not larger cations whereas the peptides indicated as 'ion channels' were capable of passing guanidinium cations. The heptad '332' formed approximately equal numbers of channels of both types. These data indicate that both the position and the type of the heptamer repeat in the peptides are important for determining selectivity. Complete biophysical and computational analyses of the peptides are in progress.

Table 1   *'Mixed heptamers' results*

	Proton channels	Ion channels
333 H_2N(LSSLLSL)-(LSSLLSL)-(LSSLLSL)-$CONH_2$		×
233 H_2N(LSLLLSL)-(LSSLLSL)-(LSSLLSL)-$CONH_2$		×
332 H_2N(LSSLLSL)-(LSSLLSL)-(LSLLLSL)-$CONH_2$	×	×
323 H_2N(LSSLLSL)-(LSLLLSL)-(LSSLLSL)-$CONH_2$	×	
222 H_2N(LSLLLSL)-(LSLLLSL)-(LSLLLSL)-$CONH_2$	×	
322 H_2N(LSSLLSL)-(LSLLLSL)-(LSLLLSL)-$CONH_2$	×	
223 H_2N(LSLLLSL)-(LSLLLSL)-LSSLLSL)-$CONH_2$	×	
232 H_2N(LSLLLSL)-(LSSLLSL)-(LSLLLSL)-$CONH_2$	×	

689

References

1. Lear, J.D., Wasserman, Z.R. and DeGrado, W.F., Science, 240 (1988) 1177.
2. DeGrado, W.F. and Kaiser, E.T., J. Org. Chem., 47 (1982) 3258.
3. Gisin, B.F. and Merrifield, R.B., J. Am. Chem. Soc., 94 (1972) 3102.

Peptides which stabilize membrane bilayers inhibit insulin-promoted 2-deoxyglucose uptake in rat adipocytes

Richard M. Epand, Alan R. Stafford, Laura Abbott and Louise Gorton

Department of Biochemistry, McMaster University, 1200 Main Street West, Hamilton, Ontario, Canada, L8N 3Z5

Introduction

A number of Z-dipeptide-amides have been shown to inhibit insulin-activated glucose uptake in adipocytes [1]. Several of these peptides inhibit membrane fusion, and there is a relationship between the membrane bilayer stabilizing potency of these peptides and their inhibitory effect on insulin-activated glucose uptake [2]. It is thus possible that peptides can modulate the effects of insulin on adipocytes through alterations in the bulk biophysical properties of the membrane. We have investigated whether this is a general effect of membrane-stabilizing peptides, and have begun to evaluate the mechanism of action of these agents.

Results and Discussion

In addition to the Z-dipeptide-amides which have been previously shown to inhibit insulin-promoted glucose uptake [1], Z-D-Phe-Phe-Gly and Z-Phe-Gly are also effective inhibitors of this process in isolated rat adipocytes (Table 1). This indicates that a number of small peptides with different amino-acid composition, sequence, and charge, can affect insulin-promoted activity. However, not all peptides have this effect. The inhibition caused by these peptides is related to their bilayer stabilizing potency (Table 1); this activity is determined by measuring the effect of the peptide in shifting the bilayer-to-hexagonal-phase transition temperature of dielaidoylphosphatidyl-ethanolamine to higher temperatures [2,3]. The peptides which are the best bilayer stabilizers are also those which show the most inhibition of insulin-stimulated glucose uptake (Table 1). Cyclosporin A has been shown to stabilize the bilayer at low concentrations, but to promote hexagonal-phase formation at high concentration [5]. We find a similar biphasic effect on its inhibition of insulin-stimulated 2-deoxyglucose uptake in adipocytes. At 100 nM, cyclosporin is very effective at inhibiting insulin-promoted glucose uptake, but this inhibition decreases at both higher and lower concentrations of cyclosporin. This effect of cyclosporin may be responsible for the impaired glucose-tolerance observed in cyclosporin-treated renal graft recipients [6]. Alkanes [7] and threodihydrosphingosine [8] have the opposite effect to the bilayer-stabilizing peptides. These hydrophobic substances lower

691

Table 1 *Bilayer stabilization and glucose uptake*

Compound	Slope[a]	% Basal[b]	% Insulin[c]
Z-D-Phe-Phe-Gly	84 ± 5	112 ± 15	12 ± 15
Z-Gly-Phe-NH$_2$	52 ± 7	78 ± 10	42 ± 15
Z-Phe-Gly	34 ± 3	107 ± 12	49 ± 9
Z-Gly-Phe	15 ± 3	98 ± 10	70 ± 16
Phe-Met-Arg-Phe-NH$_2$	12 ± 3	141 ± 21	70 ± 19
Phe-Gly	1 ± 2	108 ± 14	119 ± 24

[a] Slope of plot of L_α – H_{II} phase transition temperature vs. mol fraction of peptide additive. Units are degrees/mol fraction.

[b,c] Ratio of 2-deoxy-glucose uptake in rat adipocytes with and without the addition of 10 μM peptide in the absence (b) or in the presence (c) of 1 μM porcine insulin. Glucose uptake assay similar to a previously described procedure [4].

the bilayer-to-hexagonal-phase transition temperature of dielaidoylphosphatidyl-ethanolamine, i.e. they have negative slopes. Both hexane and threodihydro-sphingosine are somewhat cytotoxic at higher concentrations, but at 10 μM they stimulate the basal activity by 20 and 30%, respectively. These results suggest that there is a relationship between the hexagonal-phase-forming tendency of the membrane and the regulation of glucose uptake in adipocytes.

There are many possible sites of action of these modulators of the biophysical properties of membranes. One known effect of certain Z-dipeptide amides is on the activity of the glucose transporter. The adipocyte glucose transporter, reconstituted into liposomes, is inhibited non-competitively by Z-Gly-Phe-NH$_2$ with a K_i of 1.5-2 mM [9]. This is about 100 to 1000-fold higher than the concentration we find to be required for the inhibition of glucose transport in adipocytes. Thus, there are likely to be other sites of action of these peptide inhibitors.

Another effect of insulin in adipocytes is the stimulation of protein synthesis. It has recently been shown that amino acid transport is not insulin-dependent [10]. Using the protocol of Marshall [10] to measure the incorporation of [^{3}H]-Leu into protein, we find that 10 μM Z-D-Phe-Phe-Gly inhibits insulin-promoted protein synthesis to an extent comparable to the effect of this peptide on glucose transport. Thus, two different targets of insulin-signaling, i.e. glucose transport and protein synthesis, are affected to similar extents by a bilayer-stabilizing inhibitor. This suggests that the insulin signaling mechanism is affected. In agreement with this, the insulin-stimulated activity, but not the basal activity, is inhibited by the bilayer-stabilizing peptides (Table 1).

We have directly demonstrated that insulin receptor binding and subsequent receptor internalization in adipocytes is inhibited by Z-D-Phe-Phe-Gly. At a concentration of 10 μM Z-D-Phe-Phe-Gly, the binding of [^{125}I]-insulin to its receptor is inhibited approximately 50%, while initial receptor internalization into the adipocyte is inhibited almost completely. Thus, insulin receptor function is altered by Z-D-Phe-Phe-Gly.

Conclusions

Insulin signaling is modulated by the bulk biophysical properties of the membrane. Bilayer stabilizers inhibit insulin function, while hexagonal-phase promoters may, to a small extent, mimic the effects of insulin. The molecular basis of these effects is currently under further investigation.

Acknowledgements

This work was supported by NSERC and by the Ontario Heart and Stroke Foundation.

References

1. Aiello, L.P., Wessling-Resnick, M. and Pilch, P.F., Biochemistry, 25 (1986) 3944.
2. Epand, R.M., Lobl, T.J. and Renis, H.E., Biosci. Rep., 7 (1987) 745.
3. Epand, R.M., Biosci. Rep., 6 (1986) 647.
4. Hyslop, P.A., Kuhn, C.E. and Sauerheber, R.D., Biochem. Pharmacol., 36 (1987) 2305.
5. Epand, R.M., Epand, R.F. and McKenzie, R.C., J. Biol. Chem., 262 (1987) 1526.
6. Ost, L., Tydén, G. and Fehrman, I., Transplantation, 46 (1988) 370.
7. Epand, R.M., Biochemistry, 24 (1985) 7092.
8. Epand, R.M., Chem.-Biol. Interact., 63 (1987) 239.
9. Wheeler, T.J., Biochem. Biophys. Acta, 979 (1989) 331.
10. Marshall, S., J. Biol. Chem., 264 (1989) 2029.

Biophysical and structural properties of a pulmonary surfactant peptide in phospholipid dispersions

Alan J. Waring[a], Lorne Taylor[b], Gary Simatos[b], K.M.W. Keough[b] and H.W. Taeusch[a]

[a]Department of Pediatrics, King/Drew-UCLA, Los Angeles, CA 90059, U.S.A.
[b]Department of Biochemistry, Memorial University of Newfoundland, St. John's, Nfld., Canada A1B 3X9

Introduction

SP-C is a hydrophobic protein isolated from mammalian pulmonary surfactant. One function of SP-C is to increase the rate of delivery of amphipathic surfactant lipids to air–water interfaces in the lung. SP-C is a 35 amino acid residue protein with several unique features apparent from its amino acid sequence (Table 1). The N-terminal domain has several hydrophobic residues with prolines flanking vicinal cysteine residues that have a low probability of forming intramolecular disulfide linkages [1]. These residues are followed by several polar amino acids (histidine, lysine and arginine). The more polar N-terminal segment is adjacent to a long hydrophobic sequence toward the carboxyl end of the protein.

Table 1 *Sequence of native SP-C [2] and the synthetic peptide of the N-terminal domain of SP-C used for biophysical studies*

Native Human SP-C
LIPCCPVHLKRLLIVVVVVVLIVVVIVGALLMGLH
[Trp¹]SP-C(1–10)-OH
NH₂-Trp-Ile-Pro-Cys-Cys-Pro-Val-His-Leu-Lys-COOH

Results and Discussion

In order to better characterize the structure and function of the N-terminal domain, we synthesized this sequence (Table 1) by manually using the SPPS method of Merrifield employing a Boc strategy (UCLA Peptide Synthesis Facility) [3]. A tryptophan residue was substituted for leucine[1] to provide a fluorescent marker of this peptide domain.

The interaction of the synthetic peptide [Trp¹]SP-C(1–10)-OH with lipid dispersions of various compositions was monitored by fluorescence spectroscopy, electron paramagnetic resonance (EPR) with nitroxide spin-labeled lipids, circular dichroism (CD), and by measurement of surface tension by Langmuir–Wilhelmy surface balance and King/Clements surface adsorption device.

The fluorescence emission of the N-terminal tryptophan of the synthetic peptide was observed at 355 nm in 100 mM saline, 352 nm in DMPC, 355 nm in DPPC, and 337 nm in dioleoyl phosphatidylglycerol (DOPG) at 37°C. Mixtures of lipids consisting of DPPC:DOPC:palmitic acid (PA) (66:22:2, wt/wt) had a tryptophan emission peak at 348 nm. The blue shift of tryptophan fluorescence in this mixture indicates that the N-terminus is located in the more hydrophobic domains of this amphipathic lipid mixture.

The addition of the synthetic peptide also influenced the molecular ordering of several lipid dispersions. For example, using the nitroxide spin-labeled phospholipid 5-doxyl stearyl PC as a probe of the lipid domains adjacent to the phospholipid polar head groups, there was an increase in the molecular order parameter of 0.012 upon addition of 3% wt/wt synthetic peptide to the dispersion containing DMPC, DOPG, and lipid mixtures consisting of DPPC:DOPG:PA (66:22:9, wt/wt). In contrast, with DPPC alone at 37°C, there was no difference in molecular order in vesicles with and without peptide. The absence of change in the DPPC lipid molecular order in the presence of peptide may be explained from the fluorescence emission of the peptide N-terminal residue. Since the fluorescence emission for the peptide is the same in DPPC and normal saline solution, the tryptophan residue of the peptide may remain in a relatively polar environment and does not localize in hydrophobic regions near the lipid nitroxide probe.

The specific nature of the interaction of the [Trp^1]SP-C(1–10)-OH with lipid dispersions is indicated by the circular dichroism spectra of lipid–peptide mixtures. CD spectra of the peptide in 2-chloroethanol, DMPC, DOPG, and the lipid mixture DPPC:DOPG:PA (66:22:9, wt/wt), under non-reducing conditions have a negative ellipticity, with a maximum from 213 to 220 nm. Addition of DTT to the lipid–peptide dispersions and to peptide solutions (peptide in 2-chloroethanol) resulted in a reduction in the magnitude of the ellipticity in the 213–220 nm range, suggesting that the peptide conformations that are associated with lipid under oxidizing conditions are largely disulfide-linked oligomers.

Surface tension measurements with lipid–peptide mixtures on the Langmuir-Wilhelmy balance indicated that dispersions of DPPC:DOPG:PA (66:22:9, wt/wt) with 3% [Trp^1]SP-C(1–10)-OH fell to a minimum surface tension below 10 mN/m when the surface area was compressed by 50%. DPPC:DOPG:PA mixtures without peptide required much higher surface compression to achieve these surface tensions. The movement of the lipid–peptide mixture from the aqueous hypophase to the air–liquid interface was monitored by the King-Clements adsorption device [4]. Rates of adsorption for the DPPC:DOPG:PA-synthetic peptide mixture were greater than rates for lipids alone, but did not duplicate rates of adsorption found using mixtures containing native protein.

Acknowledgements

We thank NIH (HL-40666), the Medical Research Council of Canada, and the California American Lung Association for financial support.

References

1. Ovchinnikov, Y.A., Lipkin, V.M., Shuvaeva, T.M., Bogachuk, A.P. and Shemyakin, V.V., FEBS Lett., 179 (1985) 107.
2. Possmayer, F., Am. Rev. Respir. Dis., 138 (1988) 990.
3. Stewart, J.M. and Young, J.D., Solid Phase Peptide Synthesis, 2nd edn., Pierce Chemical Co., Rockford, IL, 1984.
4. King, R. and Clements, J., Am. J. Physiol., 223 (1972) 727.

Conformational study of synthetic MIR-decapeptides bound to the anti-acetylcholine receptor antibodies

Constantinos Sakarellos[a], Ioannis Hadjidakis[a], Eleni Bairaktari[a],
Vassilios Tsikaris[a], Socrates J. Tzartos[b], Irene Papadouli[b], Spyros Potamianos[b],
Michel Marraud[c] and Manh Thong Cung[c]

[a]Department of Chemistry, University of Ioannina, Box 1186, 451 10 Ioannina, Greece
[b]Hellenic Pasteur Institute, 127 Vas. Sofias Ave., 115 21 Athens, Greece
[c]LCPM-ENSIC, CNRS-UA 494, BP 451, F-54001 Nancy-Cedex, France

Introduction

We have recently reported [1] the RIA binding properties of anti-MIR antibodies (MAbs) to some synthetic decapeptides corresponding to the main immunogenic region (MIR: α67–76 acetylcholine receptor (AChR) fragment of *Torpedo* fish electric organs (residues W[67] NPADYGGIK[76])) [2]. Some of them enhanced MAb binding, suggesting that the construction of a very antigenic MIR is feasible. We now report on the NMR conformational study (NOESY) of the decapeptide analogs bound to the anti-MIR MAbs. These results are briefly analyzed in terms of mobility restrictions in the decapeptide conformation in order to yield some insight into the structure of the α67–76 binding site of AChR.

Results and Discussion

The MIR decapeptide of *Torpedo* and two analogs with particular behavior, [A[69]] (important change in the free decapeptide conformation) and [A[76]] (enhancement of binding capacity to the anti-MIR MAbs), are examined in the presence of the anti-MIR MAbs. The concentrations used in all experiments were 5 mM for the decapeptides and 0.1 mM for the MAbs in D_2O. The NOESY experiments were run in phase-sensitive mode with a mixing time of 200 ms. The bound state of a ligand to MAb ($\omega\tau_c \gg 1$) exhibits opposite signs for the cross-relaxation rates and give rise to negative NOEs. None of the three decapeptides studied in this work showed NOEs in aqueous solution in the absence of MAb. However, in the presence of MAb, a progressive increase in the NOE intensities, more or less following the MAb binding potencies, was observed. Indeed, the [A[76]] analog (300% of binding with reference to *Torpedo* decapeptide) showed the strongest NOE cross-peaks and the [A[69]] analog (30% of binding) the weakest ones. In the case of the latter, only few cross-peaks are observed. The cross-peaks between geminal protons, which exist in the case of *Torpedo* and [A[76]] analog, are absent in the NOESY map of [A[69]]. The most striking point is the presence of strong interresidue cross-peaks in [A[76]] analog between the aromatic protons of W[67] and Y[72] side chains and protons of the P[69], A[70] and D[71] residues, respectively. In addition, the comparison between COSY spectra of decapeptides that were obtained in the presence and absence of MAb showed a drastic decrease, due to dynamical filtering [3], in intensity of the signals for the five N-terminal residues; in particular, those of P[69] are absent, whereas the signals of the C-terminal part are only moderately decreased. According to [3], the most perturbed signals denote the rigid parts of molecule.

In conclusion, the above experiments argue for the existence of a folded structure of the decapeptide in the presence of MAb, and suggest that the structure of the N-terminal sequence is requisite for full molecular recognition.

Acknowledgements

This work was supported by grants from the Association Française contre les Myophathies to M.T.C., M.M., S.J.T. and C.S. and the Muscular Dystrophy Association of America to S.J.T., and EMBO to E.B. and V.T.

References

1. Tzartos, S.J., Papadouli, I., Potamianos, S., Hadjidakis, I., Bairaktari, E., Tsikaris, V., Sakarellos, C., Cung, M.T. and Marraud, M., In Maelicke, A. (Ed.) Molecular Biology of Neuroreceptors and Ion Channels, NATO ASI series, Springer Verlag, Vol. H 32, 1989, p. 361.
2. Tzartos, S.J. and Lindstrom, J.L., Proc. Natl. Acad. Sci. U.S.A., 77 (1980) 755.
3. Weiss, M.A., Eliason, J.L. and States, D.J., Proc. Natl. Acad. Sci. U.S.A., 81 (1984) 6019.

A conformationally constrained 'RGD' analog specific for the vitronectin receptor: A model for receptor binding

Teruna Siahaan[a], Laura R. Lark[b], Michael Pierschbacher[a], Erkki Ruoslahti[a] and Lila M. Gierasch[b]

[a]*La Jolla Cancer Research Foundation, 10901 N. Torrey Pines Rd., La Jolla, CA 92037, U.S.A.*
[b]*Department of Biochemistry and Department of Pharmacology, University of Texas Southwestern Medical Center at Dallas, Dallas, TX 75235, U.S.A.*

Introduction

Integrins are a family of cell surface proteins that include the fibronectin (FN) or vitronectin (VN) receptor and that mediate the adhesion of cells to extracellular matrices or to other cells [1]. Integrins play an important role in cell anchorage, migration, proliferation and differentiation and influence a wide variety of biological processes ranging from tumor metastasis to blood clotting. In many cases, the critical binding site in the matrix protein (i.e., FN and VN) centers around a common tripeptide, arginine-glycine-aspartate (RGD), and short peptides containing this sequence inhibit binding of the normal ligand [2,3]. The integrin receptors, however, are specific for ligand. This binding specificity may arise from adjacent sequences and/or the conformation of 'RGD' in each protein. In an effort to study the conformational requirements for RGD binding and for specificity, we have studied by ^{1}H NMR in DMSO and by MD simulations in vacuo a conformationally constrained 'RGD' peptide analog [c(2–9)GPenGRGDSPCA, Pen = penicillamine] that demonstrates a 10 000-fold higher affinity for the VN receptor than for the FN receptor [3].

Results and Discussion

NMR data suggest that this peptide exists in one preferred all-trans conformational family. An interesting finding is a series of strong NH to NH NOE connectivities between consecutive residues from Gly[3] to Ser[7] (see Fig. 1). The amide protons of Arg[4], Asp[6], and Ala[10], as well as the ϵNH of Arg[4] are sequestered from solvent, since their resonances do not shift with temperature or broaden upon titration with a paramagnetic probe, TEMPO (tetramethyl-1-piperidinyloxy). One conformation that is consistent with these observations contains a β-turn around Gly[5] and Asp[6]; in this model, the NHs of Arg[4], Asp[6], Ser[7] and the ϵNH of Arg[4] are hydrogen bonded to C=O of Ser[7], Asp[6] side chain, Arg[4] and the Asp[6] side chain, respectively. Coupling constants support a type I β-turn.

To supplement the NMR data and to refine our model, MD simulations and energy minimizations were carried out, both constrained by a harmonic potential

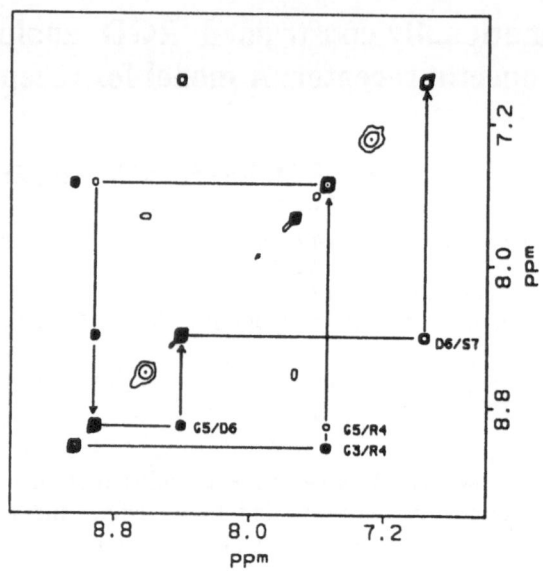

Fig. 1. The NH to NH region of the 500-MHz ^{1}H NOESY spectrum of c(2–9)GPenGRGDSPCA in DMSO-d$_6$; the mixing time is 300 ms, concentration about 5 mg/ml, temperature was 25°C. The downfield residue is labeled first in each case.

to fit observed NOEs, and non-constrained. The lowest energy conformation for the constrained dynamics (20.3 kcal/mol) has consecutive type I β-turns around Gly^5/Asp^6 and Asp^6/Ser^7, and an inverse γ-turn at Pro^8 (see Fig. 2).

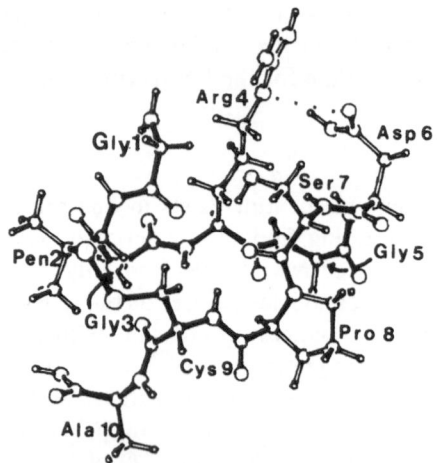

Fig. 2. The lowest energy structure from the constrained molecular dynamics simulations. Note consecutive β-turns around Gly^5/Asp^6 and Asp^6/Ser^7.

In this structure, the ϵNH of Arg[4] is salt-bridged to the Asp[6] side chain. The root mean square deviation of interproton distances in this structure from the NOE distances is 1.01 Å. A similar low energy structure was also found in the unconstrained dynamics. These structures resemble the structure that was recently published for a linear analog, GRGDSP [4].

Our working model for the binding site conformation in VN is a structure with a type I β-turn around Gly[5] and Asp[6]. Alternative conformations may be available since preliminary NMR data in water show cis/trans isomerization, suggesting that this peptide has some degree of conformational flexibility despite cyclization. Subsequent work will focus on the requirements for selective binding to the FN receptor.

Acknowledgements

We are grateful to David Kitson and Arnold Hagler for their help with molecular dynamics. In addition, we thank Applied Biosystems Inc. for a travel award to L.R.L. and the San Diego Computing Center for access to the system. This work was supported in part by grants from the NIH (GM-27616 to L.M.G., CA-28896 to E.R. and HL-38417 to M.D.P.). L.M.G. gratefully acknowledges the Robert A. Welch Foundation.

References

1. Ruoslahti, E. and Pierschbacher, M.D., Science, 238 (1987) 491.
2. Ruoslahti, E. and Pierschbacher, M.D., Cell, 44 (1986) 517.
3. Pierschbacher, M.D. and Ruoslahti, E., J. Biol. Chem., 262 (1987) 17294.
4. Reed, J., Hull, W.E., von der Lieth, C.W., Kubler, D., Suhai, S. and Kinzel, V., Eur. J. Biochem., 178 (1988) 141.

Conformational analysis of a peptide-DNA interaction by Fourier transform infrared spectroscopy (FTIR)

Lee Walters[a],* and S.B. Dev[b],**

[a]*Research Institute of Scripps Clinic, MB-15, 10666 N. Torrey Pines Road, La Jolla, CA 92037, U.S.A.*
[b]*Department of Applied Biological Sciences, Massachusetts Institute of Technology, Cambridge, MA, U.S.A.*

Introduction

The applicability of FTIR to peptide– and protein–DNA interactions was explored and demonstrated in a model system consisting of the complex that formed in aqueous solution between heterogeneous cellular DNA and a 29-residue synthetic peptide. The peptide was designed, de novo: (a) to bind to B-DNA through putative ionic contacts, hydrogen bonds and hydrophobic forces; and (b) to adopt a regular, ordered secondary structure when bound to DNA.

Results and Discussion

The sequence of model Peptide-1 is: EERAL RESVY RQRAP ERQRE ISARV AREE. Peptide-1 bound to DNA with a nonspecific dissociation constant of 90×10^{-6} M, as determined by electrophoretic separation of complexes from free peptide and DNA [1–4]. FTIR spectra were acquired on peptide, DNA (calf thymus), and peptide–DNA complexes under the following conditions: Digilab FTS-15 spectrometer; HgCdTe detector; 8 cm^{-1} resolution; 2000 scans per sample; Spectratech microcircle cell with ZnSe crystal; samples were in aqueous 5 mM sodium phosphate at pH 7.0.

Second derivative methods [5,6] resolved the conformationally sensitive Amide I band in the spectra of native peptide into components located at 1627 (β-strand), 1658 (α-helix) and 1681 (turn) cm^{-1} with a distinct shoulder at 1647 cm^{-1} (disordered structure) [7–11]. Upon interacting with DNA, the band at 1681 cm^{-1} (turn) was no longer seen; a new band appeared at 1675 cm^{-1}; the 1627 cm^{-1} band (β-strand) was considerably reduced in intensity; the position of the α-helical (1658 cm^{-1}) component remained unchanged; the shoulder at 1647 cm^{-1} (disorder) disappeared. The new vibration at 1675 cm^{-1} was characteristic of β-strand structures. The asymmetric stretch (ν_{AS}) of the DNA phosphates shifted from 1223 cm^{-1} (unbound) to 1229 cm^{-1} (bound) and the

---

*To whom correspondence should be addressed.
**Current address: Biotechnologies and Experimental Research Inc., 3742 Jewell Street, San Diego, CA 92109, U.S.A.

realtive intensities of ν_{AS} and the phosphate symmetric stretch (ν_S) were altered upon peptide binding.

The data was consistent with the following conclusions: (a) DNA-binding changed the secondary structure of the peptide; (b) disordered region(s) were only observed in free peptide, i.e., DNA-binding stabilized and increased order in the peptide secondary structure; (c) turn(s) changed into β-strand and/or α-helical conformation(s) when peptide bound to DNA; (d) a β-strand conformation that was characterized by a 1627 cm^{-1} vibration was present in free and bound peptide; (e) there was a particular β-strand vibration/conformation that was only present in the bound peptide; (f) α-helical region(s) existed in both free and bound peptide; and (g) DNA phosphates participated in peptide binding.

In summary, FTIR has revealed singificant molecular details about a peptide–DNA interaction. These results provide a basis for extending this approach to protein–DNA interactions.

Acknowledgements

Some of the work reported here was done at the National Center for Biomedical Infrared Spectroscopy, Battelle Memorial Institute, Columbus, OH (S.B.D.) and in the laboratory of E.T. Kaiser at The Rockefeller University, New York, NY (L.W.). This work was supported in part by National Institutes of Health Grant RR-013676 to the National Center for Biomedical Infrared Spectroscopy. We also thank C.K. Rha for hospitality in her laboratory at MIT (S.B.D.).

References

1. Garner, M.M. and Revzin, A., Nucleic Acids Res., 9 (1981) 3047.
2. Fried, M.G. and Crothers, D.M., Nucleic Acids Res., 9 (1981) 6505.
3. Fried, M.G. and Crothers, D.M., J. Mol. Biol., 172 (1984) 241.
4. Spassky, A., Rimsky, S., Garrean, H. and Buc, H., Nucleic Acids Res., 12 (1984) 5321.
5. Susi, H. and Byler, D.M., Biochem. Biophys. Res. Commun., 115 (1983) 391.
6. Susi, H. and Byler, D.M., In Hirs, C.H.W. and Timasheff, S.N. (Eds.) Methods in Enzymology, Academic Press, New York, 1986, p. 290.
7. Byler, D.M. and Susi, H., Biopolymers, 25 (1986) 469.
8. Koenig, J.L., and Tabb, D.L., In Durig, J.R. (Ed.) Analytical Applications of FTIR to Molecular and Biological Systems, D. Reidel, Boston, 1980, p. 241.
9. Krimm, S. and Bandekar, J., Adv. Protein Chem., 38 (1986) 181.
10. Kawai, M. and Fasman, G., J. Am. Chem. Soc., 100 (1978) 3630.
11. Bandekar, J. and Krimm, S., Proc. Natl. Acad. Sci. U.S.A., 76 (1979) 774.

Studies on protein folding using the bis-cysteinyl hinge peptide 225–232/225′–232′ of human IgG1 as a model

L. Moroder, G. Hübener*, S. Göhring-Romani, W. Göhring and E. Wünsch

Max-Planck-Institute of Biochemistry, Department of Peptide Chemistry,
D-8033 Martinsried, F.R.G.

Introduction

Studies on thiol-disulfide interchanges have clearly shown that no simple alkane or arenedithiols prefer the dimeric bis(disulfide)-structure, and that large-ring cyclic disulfides are not intrinsically highly stable structures [1].

Results and Discussion

Selective disulfide bridging of the bis-cysteinyl-octapeptide **1** to the antiparallel dimer **3** was found to be accompanied by substantial disproportionation to the parallel structure **2**, thus leading to the discovery of the unexpectedly predominant parallel dimerization via the thermodynamically controlled thiol-disulfide interchange occurring in the slow air oxidation process (Scheme 1) [2,3].

H-Thr(tBu)-Cys-Pro-Pro-Cys-Pro-Ala-Pro-OH **1**

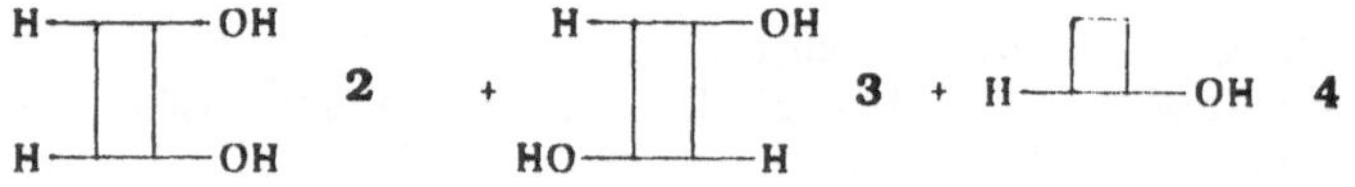

Scheme 1.

The rate of air oxidation, as shown in Fig. 1, is influenced by the addition of oxidized thioredoxin, but not the final product distribution (appr. molar ratio of $2:3:4 = 90:8:2$). Conversely, the kinetically controlled oxidation of **1** with azodicarboxylic acid dimorpholide at an oxidant/SH ratio of 0.5 : 1 leads, within minutes, to quantitative consumption of **1** with generation of the parallel and antiparallel dimer in nearly statistical distribution (2.0 : 1.7), and of the cyclic monomer to a greater extent (concentration factor); by decreasing the oxidant to SH ratio, the intervening thiol–disulfide interchange again allows for the thermodynamic equilibrium between the 3 species. Similarly, a thiol–disulfide

---

*To whom correspondence should be addressed.

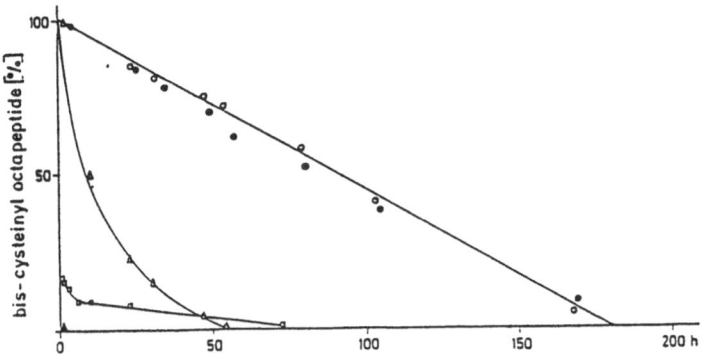

*Fig. 1.   Oxidation of the bis-cysteinyl-peptide **1** (3 × 10⁻³ M in 0.1 M ammonium acetate, pH 6.8) at room temperature by (i) air oxygen in absence (o—o) and in presence of oxidized thioredoxin at a ratio of **1**/thioredoxin of 100:1 (●—●) and 10:1 (△—△); (ii) azodicarboxylic acid dimorpholide at ratios of SH/oxidant of 1:0.5 (▲—▲) and 1:0.25 (□—□). The oxidation was followed by HPLC (column: μ-Bondapak C-18; eluent: CH₃CN/0.1 M sodium phosphate, pH 3.5, linear gradient: 13–40% CH₃CN in 50 min; flow rate 2.0 ml/min; UV: 210 nm); the product distribution was determined by approximate procedures from the peak areas.*

interchange of the antiparallel dimer **3** (Fig. 2) with the bis-cysteinyl-peptide **1** (molar ratio = 5:1), in the presence or absence of oxidized thioredoxin, leads to thermodynamic equilibrium (again **2**:**3**:**4** = 90:8:2). Since concomitant air oxidation was not excluded for technical reasons, the equilibration end-point was assessed with an additional amount of **1**, but without any effect on the product distribution. The identical results were obtained with DTT, whereby the concurring air oxidation leads to fast consumption of free thiols, thus requiring repeated addition of reductant to reach the thermodynamic equilibrium.

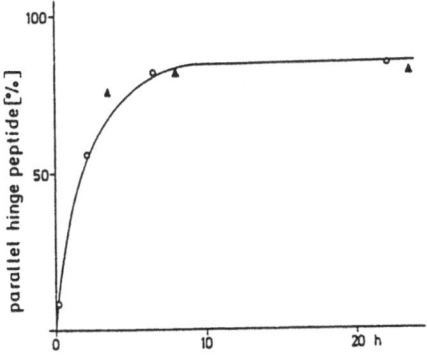

*Fig. 2.   Thiol-disulfide interchange of the antiparallel dimer **3** (2.5 × 10⁻³ M; 0.1 M ammonium acetate, pH 7) with the bis-cysteinyl-peptide **1** (molar ratio of 5:1) in the absence (o—o) and in the presence of oxidized thioredoxin; ratio of **1**/thioredoxin = 5:0.5 (▲—▲).*

Conclusion

The unique thermodynamically preferred parallel dimerization of the IgG1 hinge fragment can only be explained on the basis of sequence-dependent structural information. This local information may also explain why the initial step in covalent assembly of immunoglobulins is the disulfide bridging between a nascent heavy chain and a completed chain [4]. The protein-disulfide isomerase may, in this context, play a negligible role in terms of correct pairing, if its involvement is not claimed in terms of speeding up the correct folding.

References

1. Houk, J. and Whitesides, G.M., J. Am. Chem. Soc., 109 (1987) 6825.
2. Moroder, L., Bovermann, G. and Wünsch, E., In Shiba T. and Sakakibara, S. (Eds.) Peptide Chemistry 1987, Protein Research Foundation, Osaka, 1988, p. 759.
3. Wünsch, E., Moroder, L., Göhring-Romani, S., Musiol, H.-J., Göhring, W. and Bovermann, G., Int. J. Pept. Prot. Res., 32 (1988) 368.
4. Bergmann, L.W. and Kuehl, W.M., J. Biol. Chem., 254 (1979) 5690.

Session VI
Immunology

Chair: John A. Smith
Harvard Medical School
Boston, Massachusetts, U.S.A.

Predetermined supersecondary motifs in the de novo design of protein antigenic determinants

Pravin T.P. Kaumaya*, Anne Van Buskirk, Erwin Goldberg and Susan K. Pierce

Northwestern University, Evanston, IL 60208, U.S.A.

Introduction

The immune response to lactate dehydrogenase C_4 (LDH-C_4) is of medical interest because immunization with LDH-C_4 has been shown to reduce fertility in mice, rabbits and baboons [1]. One strategy for developing this protein antigen as a contraceptive vaccine requires mapping the antigenic structure with sequential peptides [2]. However, as has been observed for other protein antigens, immunization with linear LDH-C_4 synthetic peptides, containing antigenic determinants, coupled to carrier proteins, elicits antibodies that bind poorly, with greatly reduced affinity, to the native protein [3]. Recent determination of the structure of Fab bound to their antigens show antigen-antibody binding is lock-and-key-like, and highly conformation-dependent, involving amino acid side-chain interactions over 700 $Å^2$ on discontinuous chains of the antigen [4]. It is, perhaps, not surprising that antibodies, elicited by immunization with sequential epitopes having little or no structure, do not bind strongly to native protein. Thus, an effective peptide immunogen must more closely mimic the structure of the antigenic determinant in the native protein. In addition, synthetic peptides by themselves are poorly immunogenic and require coupling to carrier proteins that are undesirable for use in humans, as there is little control over the antigenic determinants. This obstacle could be circumvented by incorporation of synthetic antigenic T-cell epitopes along with a B-cell determinant, as indicated by recent studies [5]. Our ongoing studies are directed towards understanding the conformation dependency of protein epitopes in the induction of antibody responses through peptide engineering, in an attempt to mimic the molecular features of a protein antigenic site [6,7]. The basic idea of our experimental approach uses the framework model of protein folding and a set of rules about protein structure and stability to design a peptide able to fold into a native-like structure. Here we show that various conformational peptides of LDH-C_4, even in the absence of carrier protein, do indeed yield far superior immune responses than classical strategies in use at present.

---

*To whom correspondence should be addressed at The Ohio State University, College of Medicine, Department of Obstetrics and Gynecology, 5th Floor, Means Hall, 1654 Upham Drive, Columbus, OH 43210-1228, U.S.A.

Results and Discussion

(1) Peptide engineering

A surface-accessible α-helical segment (sequence 310-327, α_N) of mouse lactate dehydrogenase C_4 (LDH-C_4) was chosen based on properties known to enhance peptide and protein stability (e.g., electrostatic ion pairs, salt bridges, helix-dipole and amphiphilicity). The amphiphilicity of the sequence was idealized without affecting the spatial orientation of the antibody-accessible residues, as measured with a spherical probe of radius 10 Å. Mutations in the sequence were maximized to promote hydrophobic interaction and folding of the peptide into an $\alpha\alpha$ supersecondary structure (Fig. 1). The 'heptad repeat', a 4-residue β-turn and a synthetic strategy to allow the display of an anti-parallel α-helix to permit a crossover of 20° between the helices to yield a structure comparable to 4-helix bundles is displayed (Fig. 1). A parallel α-helical coiled-coil structure was designed by adding a cysteine residue at the N-terminal of α_1, and the peptide was dimerized in solution. An $\alpha\beta$ supersecondary peptide was synthesized by connecting α_1 via a 3-residue turn to a (Leu-Ser)$_5$ sequence (Fig. 1).

(2) Synthesis and characterization

All synthetic peptides were assembled semi-manually by stepwise Fmoc-*t*-butyl strategy on MBHAR support. The purity of crude synthetic peptides was in excess of 80%. Peptides were further purified by semi-preparative RPHPLC. Chemical characterization was by AAA, Edman sequencing and FABMS. The conformational properties of the model peptides were investigated by various biophysical techniques. CD spectra, typical of α-helices were obtained for α_1 and α_3, whereas α_N was typically random coil [6]. The concentration dependence of the mean residue ellipticity at 222 nm indicates that α_1 is a tetramer (−20310 deg cm^2/dmol) and α_3 is a dimer (−22485 deg cm^2/dmol). The free energy of tetramerization (α_1) is −20.9 kcal, and dimerization (α_3) is −7.8 kcal. FTIR spectroscopic studies indicate a high degree of secondary structure for α_3 (60%), α_1 (45%), and α_N (14%). The SAXS data strongly suggest that the helices are arranged in a 4-helix bundle, since the radius of gyration of 17.2 Å and the vector distribution function are indicative of a prolate ellipsoid of axial dimensions and molecular weight appropriate for the 4-helix bundle.

(3) Immune response

Both mice and rabbits were immunized with the peptides initially in complete Freund's adjuvant, and subsequently in phosphate-buffered saline at four-week intervals. Sera obtained at various times were tested for their ability to bind respective peptides, heterologous peptides, and native protein by ELISA. Results obtained in mice demonstrate that the conformational peptide α_3 is a far more potent immunogen than the linear unstabilized sequence α_1 and α_N [7]. Antisera from rabbits immunized with the conformational peptides α_3 and $\alpha\beta$ conformational peptides (Fig. 2) showed high titered antibody response, recognizing both native LDH and respective peptides. The high IgG titers obtained in response

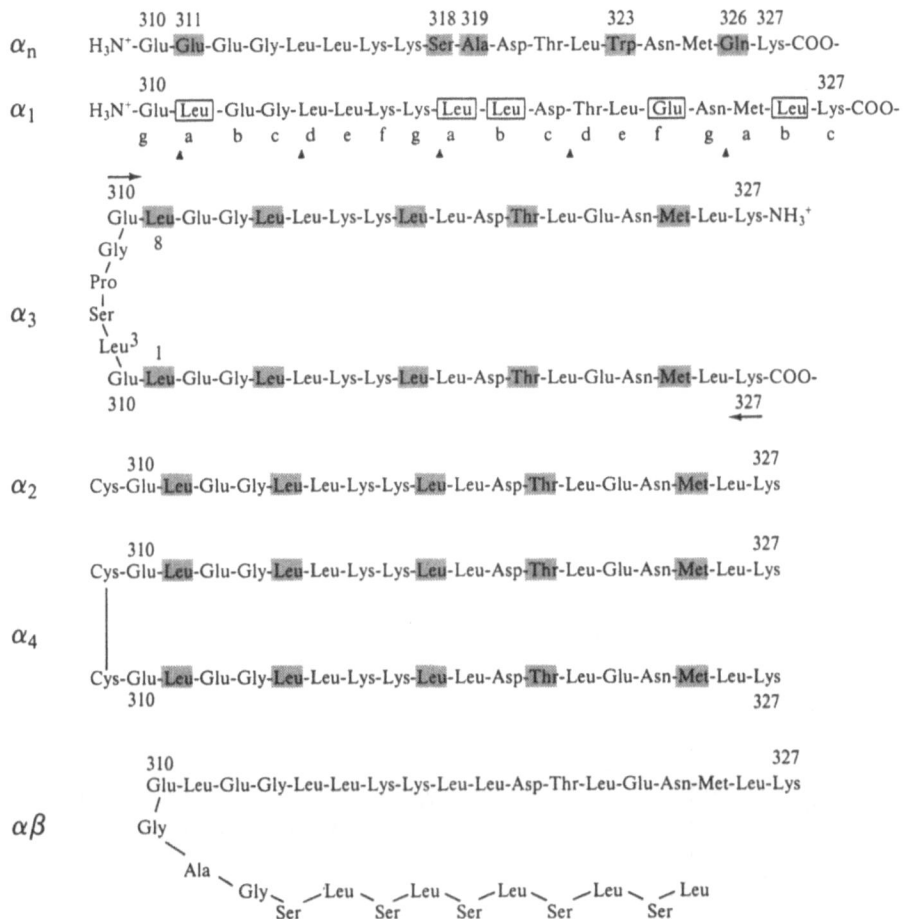

Fig. 1. Amino acid sequences of LDH-C₄ peptides: natural sequence, α_N, idealized sequence, α_1, and conformational peptides $\alpha_3(\alpha\alpha$-fold), α_4 (disulphide bonded-dimer) and $\alpha\beta$ ($\alpha\beta$-fold). Residue substitutions are indicated in open and shaded boxes, heptad repeat indicated in positions a and d.

to α_3 and $\alpha\beta$, that increase with successive immunizations, suggest that the conformational peptides are capable of stimulating memory cells. The disulphide dimer (α_4) is no better in eliciting an immune response than the unstabilized natural and idealized sequences. The antipeptide antibodies revealed specificity for their anticipated structure, as well as sequence in the native protein, and point to the importance of tertiary conformation in antigen recognition, as they did not react to unrelated α-helical proteins. The immune response to the conformational peptides shows that buried hydrophobic residues are most likely not involved in antigen recognition per se, but merely serve to contribute to

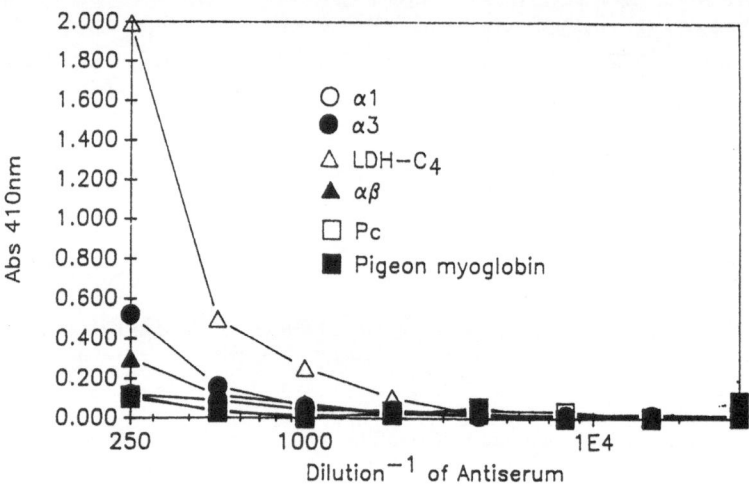

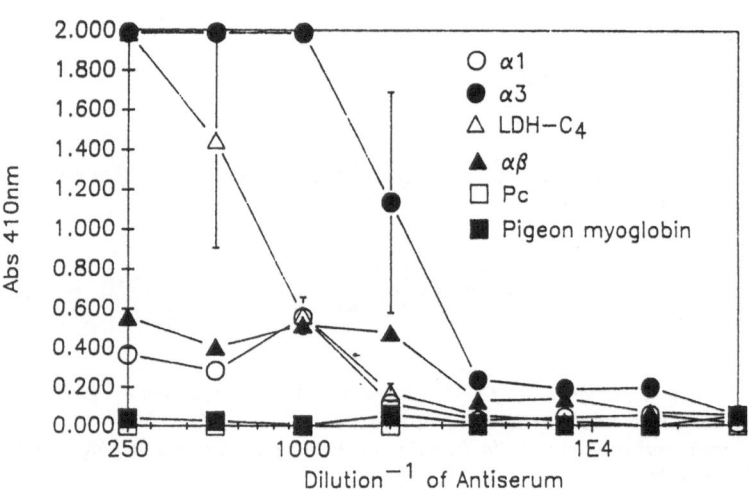

Fig. 2. Secondary immune response of rabbits immunized with conformational peptides α_3, $\alpha\beta$ assayed to homologous peptides, heterologous peptides, native protein (LDH-C$_4$) and unrelated α-helical proteins by ELISA.

the overall stability of the folded structure. This conclusion bears from the fact that the mutations introduced within the natural sequence did not result in abolition of these peptides to induce antibodies that bind to the native protein.

References

1. Goldberg, E., Wheat, T.E. and Gonzales-Prevatt, V., In Weggmann, T.G. and Gill, T.J. (Eds.) Immunology of Reproduction, Oxford University Press, New York, NY, 1983, p. 491.
2. Lerner, R.A., Nature, (London), 299 (1982) 592.
3. Hogrefe, H.H., Kaumaya, P.T.P. and Goldberg, E., J. Biol. Chem., 264 (1989) 105.
4. Amit, A.G., Mariuzza, R.A., Philips, S.E.V. and Poljak, R.J., Science, 233 (1986) 747.
5. Milich, A.R., McLachlan, A., Thornton, G.B. and Hughes, J., Nature, 329 (1987) 547.
6. Kaumaya, P.T.P., Berndt, K.D., Heidorn, D.B., Trewhella, J., Kezdy, F.J. and Goldberg, E., Biochemistry, (1990) in press.
7. Kaumaya, P.T.P., O'Hern, P., and Goldberg, E., manuscript submitted.

The augmentation of antibody production using peptide-resin conjugates

James T. Sparrow[a], Doris A. Sparrow[a], Zhang LiXin[a], Merry Kovar[a], Wanjun Li[b] and Ralph B. Arlinghaus[b]

[a]Department of Medicine, Baylor College of Medicine and The Methodist Hospital, 6535 Fannin St., Houston, TX, U.S.A.
[b]Department of Pathology, U.T-M. D. Anderson Hospital and Tumor Institute, Houston, TX 77030, U.S.A.

Introduction

We have previously shown [1] that good titers of antibodies to native proteins can be produced by injecting into animals a deprotected synthetic peptide attached to the hydrophilic polymer on which it was synthesized. The polymer is swollen by polar organic solvents or aqueous buffers making it ideal for use in biological systems. Peptide–resin conjugates have been used to produce neutralizing antibodies to HBsA and HIV [2]. The corresponding peptide–resin conjugate can also be used to detect circulating antibodies to these viruses [2]. In investigating the general utility of this methodology and to devise methods to increase the antigenic response, we chose to attach an amphipathic T-cell epitope [3–5] from Staph. A. nuclease to the peptide–resin conjugate.

Results and Discussion

The peptides were synthesized automatically using Boc or Fmoc chemistry on a beaded polydimethylacrylamide resin cross-linked with bis-acryloyl-1,3-diaminopropane containing 0.7 meq/g of amino group from acryloyl-diaminopropane or -diaminohexane [6]; Boc-glycine was first attached to the resin with DCC/DMAP and deprotected with 50% TFA. After completion of the synthesis, the peptidyl–resin was deprotected at –20°C for 3 h with anhydrous HF containing 10% anisole and 1% ethanedithiol. The peptidyl–resin conjugate was washed with ether, methanol, water, 1% acetic acid, water, 0.1 M Tris, water, 1% acetic acid, and water to neutrality; the resin was dried in a vacuum desiccator over P_2O_5. The amino acid composition and peptide sequence [7] were determined before the peptidyl–resin was homogenized in buffer and mixed with Freund's adjuvant for injection into mice or rabbits; each animal was given approximately 500 μg of peptidyl–resin conjugate per injection. Antibody titers against the native protein were determined by an ELISA; the antibodies produced were IgG.

The amphipathic T-cell epitope was residues 64–70 of Staph. A. nuclease and has the sequence KMVENAK. The position of the T-cell epitope with respect

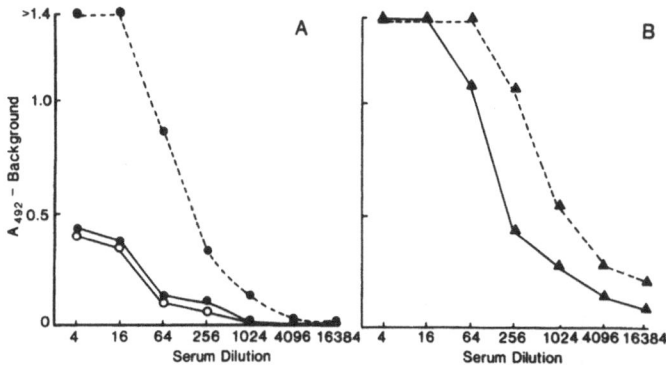

Fig. 1. *The antibody titers obtained from injection of peptide–resin conjugates into rabbits. (A): Residues 386–403 of v-abl oncogene protein attached to a polyamide resin.* ●——● *peptide–resin;* o——o *peptide–T-cell epitope–resin; and* ●-----● *T-cell epitope–peptide–resin. (B): Residues 22–34 of rat fatty acid binding protein.* ▲——▲ *peptide–resin;* ▲-----▲ *T-cell epitope–peptide–resin.*

to the resin was found to be important to the antigenic response in rabbits (Fig. 1A). If the T-cell epitope was first synthesized on the resin followed by the peptide, the anti-peptide titer, determined by an ELISA using the purified v-*abl* oncogene peptide residues 389–403 as antigen, was similar to the titer obtained from the v-*abl* peptide-resin alone, approximately 1:256. However, the antibodies obtained did not effectively precipitate the native proteins encoded by the *gag-abl* genes of Abelson murine leukemia virus. When the T-cell epitope was synthesized on the amino terminus of the peptide, a dramatic increase in antibody titer was obtained (1:2000) and the antibody precipitated the native proteins (data not shown). When the same T-cell epitope was attached to the amino terminus of residues 22–34 of rat fatty acid binding protein, the antibody titer increased more than 5-fold (Fig. 1B). Therefore, in subsequent experiments, the T-cell epitope was synthesized on the amino terminus of the peptide.

In Fig. 2 are a number of examples of this methodology as applied to the serum apolipoproteins. Except for apoB, it has been difficult to obtain antibodies against these smaller amphipathic lipid-binding proteins. As illustrated here, antibody titers are increased significantly by inclusion of the T-cell epitope from Staph. A. nuclease. With low density lipoprotein (LDL) as the antigen in the ELISA, the antibody titer against the receptor binding region of apoB contained in residues 3357–3369 increased 5-fold over that found with the peptidyl–resin (Fig. 2A). ApoC-III (33–49) alone on the resin was not antigenic in mice (Fig. 2B). However, attachment of the T-cell epitope leads to significant levels of antibody against apoC-III alone (Fig. 2B), or in human serum very low density lipoproteins (VLDL) that contain apoC-II and apoC-III (data not shown). With residues 68–79 of apoC-II, a 5-fold increase in antibody titer was found with the T-cell-peptidyl–resin (Fig. 2C). With the lipoprotein lipase (LPL) activating region

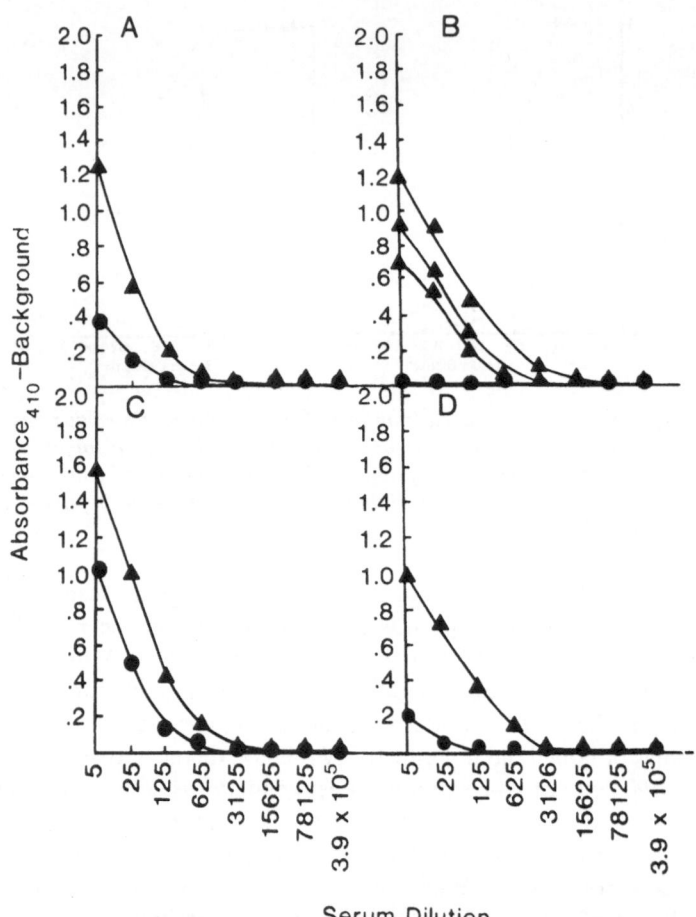

Fig. 2.   ELISA of anti-sera obtained after injection of apolipoprotein synthetic peptide–resin conjugates into rabbits and mice. (A): ApoB residues 3357–3369 with (▲) and without (●) a T-cell epitope. (B): ApoC-III residues 33–49 with (▲) and without (●) a T-cell epitope injected into three sets of mice. (C): ApoC-II residues 68–79-resin conjugate injected into rabbits. (D): Apo C-II residues 56–69-resin conjugate injected into rabbits.

between residues 56 and 69 of apoC-II, a dramatic increase in titer against apoC-II (Fig. 2D) or VLDL was observed by including the T-cell epitope. In addition, the attachment of the Staph. A. nuclease T-cell epitope described above has permitted us to obtain significant antibody titers to numerous proteins in rabbits, sheep, and mice. The spleens from the mice have been used to generate specific monoclonal antibody-producing hybridomas by fusion with myeloma cells.

Conclusion

These data provide new insights into the role of primary protein sequences in the generation of a useful humoral immune response to a specific antigen. Our results raise the possibility that generic T-cell-specific epitopes exist in defined synthetic peptides, and that these generic T-cell sites can be used to provide T-cell help to a variety of antigenic sites in proteins. We believe that these sites are contained in amphipathic regions of restricted length and can be used for the preparation of antibodies of predetermined specificity by eliminating the carrier protein. This approach could lead to methods for generating defined reagents useful for diagnostic test and vaccines.

Acknowledgements

A major part of this work was made possible by Grant No. 4107 from the Texas Advanced Technology Program to J.T.S., and by NIH Grants HL-30064 and HL-27341.

References

1. Chanh, T.C., Dreesman, G.R., Kanda, P., Linette, G.P., Sparrow, J.T., Ho, D.D. and Kennedy, R.C., EMBO J., 5 (1986) 3065.
2. Kennedy, R.C., Dreesman, G.R., Chanh, T.C., Boswell, R.N., Allan, J.S., Lee, T.-H., Essex, M., Sparrow, J.T., Ho, D.D. and Kanda, P., J. Biol. Chem., 262 (1987) 5769.
3. DeLisi, C. and Berzofsky, J.A., Proc. Natl. Acad. Sci. U.S.A., 82 (1985) 7048.
4. Finnegan, A., Smith, M.A., Smith, J.A., Berzofsky, J., Sachs, D.H. and Hodes, R.J., J. Exp. Med., 164 (1986) 897.
5. Margalit, H., Spouge, J.L., Cornette, J.L., Cease, K.B., DeLisi, C. and Berzofsky, J.A., J. Immunol., 138 (1987) 2213.
6. Sparrow, J.T., Kennedy, R.C. and Kanda, P., J. Org. Chem., (1989) submitted.
7. Sparrow, D.A. and Sparrow, J.T., In Rivier, J.E. and Marshall, G.R. (Eds.) Peptides: Chemistry, Structure and Biology (Proceedings of the 11th American Peptide Symposium), ESCOM, Leiden, 1990, p. 446.

Modular branched peptides: Synthesis of potential branched peptide vaccines

Elisabeth Albrecht, Marisa Engel, Laura G. Melton, Brian M. Peek and Bruce W. Erickson

Department of Chemistry, University of North Carolina, Chapel Hill, NC 27599-3290, U.S.A.

Introduction

A restricted immune response has been induced to a branched lysine octamer [1] bearing multiple copies of a B-cell epitopic peptide ((NANP)$_4$ from the circumsporozoite protein of the malarial parasite (*Plasmodium falciparum*). But effective branched peptide vaccines should include both B-cell and T-cell epitopic peptides in order to trigger both humoral and cell-mediated immunity.

Results and Discussion

We have used a modular approach to synthesize branched peptide structures designed to overcome genetic restriction of the immune response. For example, structure **5** is an asymmetric dialkyl sulfide in which one alkyl group is a branched module having two copies of a T-cell epitopic peptide and the other alkyl group is an unbranched module containing a B-cell epitopic peptide.

We have synthesized two derivatives of 3-(Aminomethyl)-4-aminobutanoic acid (Aab) for assembling branched peptide modules: the symmetrically protected derivative **9**, Boc-Aab(Boc), and the asymmetrically protected derivative **13**, Boc-Aab(Z). The latter was designed for making branched peptide modules containing different peptides. Both protected forms of Aab were derived from benzyl 3,3-dicyanopropionate (**6**), which was produced by alkylation of the malononitrile anion with benzyl bromoacetate. Following crystallization to remove the undesired dialkylated product **7**, ester **6** was simultaneously hydrogenated and hydrogenolyzed to yield Aab (**8**). Silica-catalyzed cyclization of **8** furnished 4-(aminomethyl)-2-pyrrolidinone (**10**), which was protected as the Z derivative **11** and was converted into Boc-Aab(Z) (**13**) by reaction with di(*tert*-butyl) dicarbonate followed by regioselective hydrolysis [2]. A related synthesis of 3,3-bis(aminomethyl)-4-aminobutanoic acid (Tab) is in progress. A branched peptide module such as **4** could use the symmetric triamino acid Tab in place of Aab.

Synthesis of the branched peptide module **4** involved parallel assembly of two identical peptide chains on Aab-Cys(4-MeBzl)-NH-resin by solid-phase methods. It was S-alkylated in solution in 53% yield with the N-bromoacetylated peptide module **3** to furnish the potential peptide vaccine **5**, which contains two copies of a T-cell epitopic peptide (Tpep = F-E-R-F-E-I-F-P-K-E, from influenza hemagglutinin [3]) and one copy of a B-cell epitopic peptide (Bpep = (N-A-N-P)₃, from *P. falciparum* [4]). A variation of modular branched peptide **5** could have a branched B-cell epitopic module and a unbranched T-cell epitopic module.

We are also pursuing alternate approaches to producing modular branched peptides by reversing the branching roles of the sulfide and amide bonds. For example, succinimido 3,5-bis(bromomethyl)benzoate (Br₂Dmb-OSu, mp 171–172°C), obtained by NBS bromination of succinimido 3,5-dimethylbenzoate followed by extensive recrystallization, could be used to synthesize a Br₂Dmb-peptide, which could be used to S-alkylate two copies of a thiol-containing peptide.

E. Albrecht et al.

Acknowledgements

Marisa Engel was supported as a La Caixa Fellow. This work was supported in part by NIH grant GM 42031.

References

1. Tam, J.P., Proc. Natl. Acad. Sci. U.S.A., 85 (1988) 5409.
2. Flynn, D.L., Zelle, R.E. and Grieco, P.A., J. Org. Chem., 48 (1983) 2424.
3. Hackett, C.J., Dietzschold, B., Gerhard, W., Ghrist, B., Knorr, R., Gillessen, D. and Melchers, F., J. Exp. Med., 158 (1983) 294.
4. Zavala, F., Tam, J.P., Hollingdale, M.R., Cochrane, A.H., Quakyi, I., Nussenzweig, R.S. and Nussenzweig, V., Science, 228 (1985) 1436.

Characterization of hardware for peptide synthesis using Geysen's method, and epitope scanning of a malarial protein

J. Mark Carter and Jeffrey A. Lyon

Walter Reed Army Institute of Research, Division CD&I, Department of Immunology, Washington, DC 20307-5100, U.S.A.

Introduction

A method for synthesis of peptides on blocks of polypropylene pins was described by Geysen [1], who used them to perform epitope mapping of sera and MAb via ELISA. We obtained hardware and protocols marketed by Cambridge Research Biochemicals (CRB) for this technique, and have performed chemical characterization of the pins and their attached peptides.

Results and Discussion

To determine the quality of the peptides obtained by this synthesis method, we performed AAA on 10 of the control pins supplied by CRB. These pins bore the sequence Ac-Pro-Leu-Ala-Gln-Bal. The analyses (Beckman 6300) showed wide variation in the amount of peptide between pins, but fairly good stoichiometry of amino acids on the same pin. This suggests variable loading during derivatization with good efficiency in synthesis.

When we synthesized peptides on the pins using the Fmoc protocol recommended by CRB [2], we experienced difficulty in washing away the piperidine from the pins after deprotection, as judged from the alkaline pH of an aqueous solution of the wash, so we added a wash step with methanol, containing 1% acetic acid. This resulted in complete neutralization (if not removal) of the piperidine, thereby preventing premature deprotection of the next amino acid added in the sequence, and potential tandem coupling.

AAA were performed on a few random pins after completion of peptide synthesis. Results confirmed the success of the synthesis. However, successful ELISA required prior 'disruption' of the peptides with a high-power (500 W) ultrasonic cleaning bath; a low-power cleaner was not effective. Sonication is apparently necessary to disrupt the unreactive conformation assumed by the peptides after synthesis. We also observed that AAA showed higher peptide content for the pins after sonication.

ELISA was used to identify epitopes in two malarial proteins (*Plasmodium falciparum* Camp strain): gp195 [3] and p126 (J.L. Weber, personal communication, 1989). A number of *Aotus* monkey sera and mouse monoclonal Ab

721

were screened, showing prominent recognition. Even weakly reacting epitopes were reliably detected. Some of the strongest reacting epitopes were predicted by searching for hydrophilic stretches in the sequence. However, most contained a mixture of hydrophilic and hydrophobic amino acids in series with an aromatic amino acid – usually Tyr.

AAA of pins after binding serum showed the appearance of 2 nmol per pin of lysine, as well as smaller amounts of other amino acids. This implies binding of about 0.2 nmol (30 μg as IgG) antibody (Ab) per pin. These additional amino acids were not detectable after ultrasonic disruption, suggesting that this treatment quantitatively removes Ab from the pins.

Next, we synthesized 8 identical pins containing the epitope TGESQTGN, the peptide from p126 most reactive in ELISA. The pins were allowed to bind to serum (diluted 1/200), and then they were washed four times with PBS containing 0.1% Tween 20 (PBST). Ab was then eluted from the pins with 0.2 M glycine·HCl, pH 2.8, (200 μl/well) for 10 minutes, and the eluate was neutralized with Tris base (solid). This solution was then used without further treatment to probe 4-mm wide strips cut from nitrocellulose blots of 8% PAGE of cultured *P. falciparum* merozoites. Strips were washed three times with PBST and then amplified by incubation for 2 h with affinity purified rabbit IgG versus *Aotus* IgG (2 ml, 1 μg/ml in PBS). After four washes in PBST, the strips were incubated with ^{125}I protein A (2 ml, 1 ng/ml in PBS, 41 mCi/mg, Amersham) for 1 h. Finally, strips were washed five times with PBST and autoradiographed for 72 h. Results indicated that the desorbed Ab recognized the protein on the blot.

From the overlapping octamer peptides constituting gp195, we selected 11 pins bearing permutations of the repeated tripeptide SGT (SGTSGTSG, GTSGTSGT, TSGTSGTS), which is the strongest reacting epitope in this protein. After adsorbing and eluting these pins, the Ab was diluted and used to probe blots as before. Results clearly showed that the blots can detect Ab eluted from a single pin. However, when other epitopes were tested, only those reacting very strongly in ELISA yielded detectable amounts of Ab.

Conclusion

Peptide synthesis on polypropylene pins was found to be an effective approach for defining the continuous epitopes recognized by monoclonal and polyclonal Ab. In some cases, these peptides may be used for purification of small amounts of Ab for immunoblot experiments. Successful application of this technique probably requires thorough removal of piperidine during peptide synthesis as well as adequate 'disruption' of the pins prior to ELISA. The method, however, is probably not appropriate for quantitative evaluation of Ab behavior, because the pins are derivatized with variable amounts of peptide.

Acknowledgements

We gratefully acknowledge the technical assistance of Eddie Wright for preparation of nitrocellulose blots, and Larry Loomis, for AAA, as well as Carolyn Deal for administrative help.

References

1. Geysen, H.M., Meloen, R.H. and Barteling, S.J., Proc. Natl. Acad. Sci. U.S.A., 81 (1984) 3998.
2. Epitope Mapping and the Determination of Antibody Specificities, Cambridge Research Biochemicals, Valley Stream, NY, pp. 37–46.
3. Weber, J.L., Leininger, W.M. and Lyon, J.A., Nucl. Acids Res., 14 (1986) 3311.

Vaccine engineering: Chemically defined hepatitis vaccine models

Yi-An Lu and James P. Tam

The Rockefeller University, 1230 York Avenue, New York, NY 10021, U.S.A.

Introduction

An important objective in the development of synthetic vaccines is the design of chemically defined and multivalented vaccines with the relevant B and T cell epitopes, but unaided by carriers. The multiple antigen peptide (MAP) system is designed with these purposes in mind [1]. To test the importance of B and T epitopes covalently linked in the MAP system, we designed two synthetic peptide-vaccine models of hepatitis B virus (HBV). The models contain two epitopes of the HBV surface protein: the a determinant of the S region consisting of tandemly repeating residues 140–146 (TKPTDGN, designated as TN-14), and the pre-S(2) region consisting of residues 12–26 (LQDPRVRGLYFPAGG, designated as LG-15).

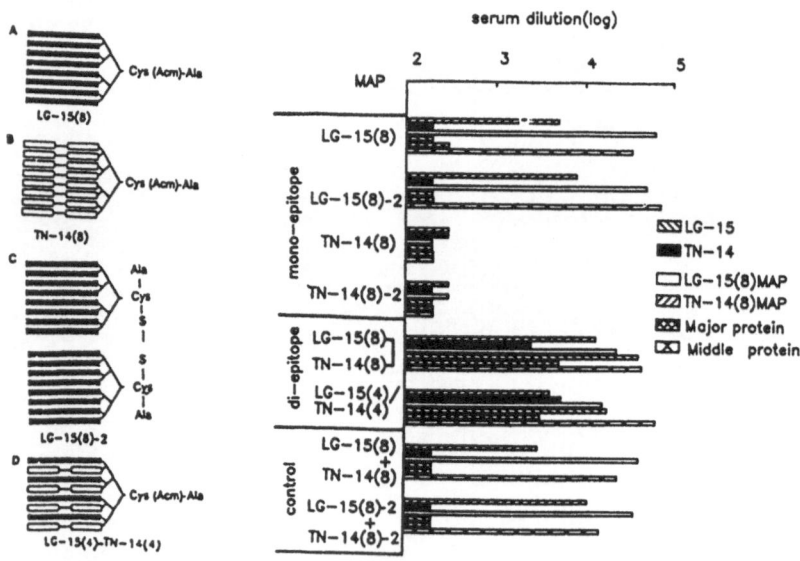

Fig. 1. MAP models.   Fig. 2.   Immunological responses of different antisera.

Results and Discussion

Two models (Fig. 1) referred as homologous (A, B, C) and heterologous (D) branching epitopes were designed to produce MAPs containing two different peptide epitopes. In the homologous branching, each epitope, TN-14 or LG-15, was synthesized as an individual, mono-epitope, octameric MAP (A and B). Oxidative linkage by iodine formed a disulfide bond, producing a heteromeric di-epitope MAP containing TN-14 and LG-15 (C). In the heterologous branching, both epitopes were synthesized in one octameric MAP system utilizing the α-amino groups for one epitope and the ϵ-amino groups for another (D). The synthesis of MAP was accomplished manually by a stepwise SPPS [1,2] on Boc-Ala-OCH$_2$-Pam resin. In synthesizing the core matrix of a MAP containing two different peptide sequences, Fmoc-Lys(Boc) was used in the third level to give Fmoc-Lys(Boc) end groups. The synthesis of the first peptide epitope used the Boc chemistry. The second epitope used the Fmoc chemistry. Finally, cleavage was mediated by the low-high HF method [3] to remove the MAP from resin support. Antibodies from New Zealand white rabbits were immunized with each of the 8 MAP models. The secondary antibody response (Fig. 2) showed that the mono-epitope MAP of TN-14 was a poor immunogen and produced little antibody response, either against the peptide antigen or the cognate protein (major protein). The di-epitope MAPs (model C and D) elicited high antibody response to the cognate protein of the S region (major protein) and pre-S(2) region (middle protein), as well as to the peptides LG-15 and TN-14. The controls containing non-covalent mixture of mono-epitopes did not elicit the desired response for TN-14. Thus, the di-epitope MAPs produce enhanced immunological responses in animals and provide defined models for the design and engineering of synthetic vaccines.

Acknowledgements

This work was supported in part by grants from AID and Grant CA-36544.

References

1. Tam, J.P., Proc. Natl. Acad. Sci. U.S.A., 85 (1988) 5409).
2. Posnett, D.N., McDrath, H. and Tam, J.P., J. Biol. Chem., 263 (1988) 1719.
3. Tam, J.P., Heath, W.F. and Merrifield, R.B., J. Am. Chem. Soc., 105 (1983) 6442.

New approaches to malaria vaccines: Synthesis and immunological evaluation of a multiple (NANP)$_3$ peptide comprising the immunodominant epitope of *P. falciparum* CS protein

Edgar P. Heimer[a], Arthur M. Felix[a], Timothy McGarty[a], Theodore Lambros[a], Mushtaq Ahmad[a], Howard Etlinger[b] and Fidel Zavala[c]

[a]*Peptide Research Department, Hoffmann-La Roche Inc., Nutley, NJ 07110, U.S.A.*
[b]*Central Research Department, F. Hoffmann-La Roche & Co., Ltd., Basle, Switzerland*
[c]*Department of Medical and Molecular Parasitology, New York University School of Medicine, New York, NY 10016, U.S.A.*

Introduction

Plasmodium falciparum, the parasite that is the major cause of malaria infection, continues to be responsible for approximately one million deaths each year. Failures to control the spread of the disease by traditional methods, including insecticides and drugs, have caused a wide interest in the development of a synthetic vaccine. The immunodominant epitope of the *P. falciparum* circumsporozoite protein (CSP) has a repetitive sequence consisting of Asn-Ala-Asn-Pro (i.e. NANP) [1,2]. Antibodies directed against this sequence have been shown to neutralize CSP in vitro [3]. Recently, a synthetic tridecapeptide conjugate consisting of Ac-Cys(Asn-Ala-Asn-Pro)$_3$-OH linked to tetanus toxoid (TT) was shown to be potentially useful as a first generation malaria vaccine [4]. The peptide-conjugate evoked high anti-(NANP)$_3$ antibody response in mice, but the titers in human volunteers were relatively low, possibly due to a dampening of the immune response by TT.

Recently, Tam has developed a novel approach in which an immunogenic peptide was multiply-linked to a branching trifunctional amino acid and assembled by SPPS [5]. The resultant multiple antigenic peptides, that are chemically unambiguous, have been shown to elicit high titers of antipeptide antibodies without the need for conjugation of peptide to carriers. To explore the relationship between the molar ratio and configuration of (NANP)$_3$ in the conjugate, and its ability to elicit antipeptide antibody, a novel derivative, **4**, was prepared consisting of multiple (NANP)$_3$ units covalently linked to Lys and containing a spacer/diagnostic residue aminocaproic acid. This material was also (a) conjugated to TT and (b) dimerized for immunization studies in mice.

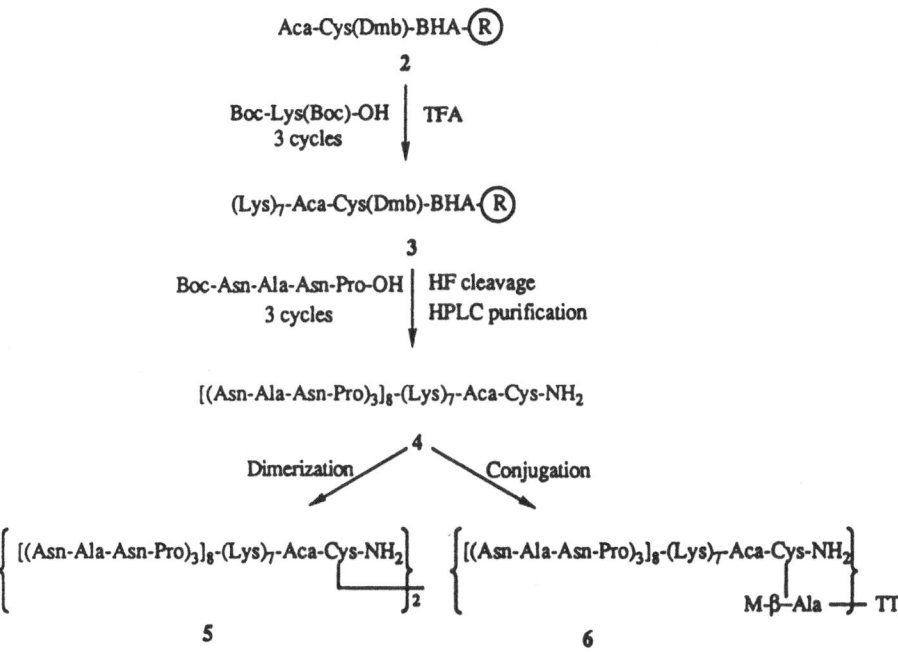

Fig. 1. Synthesis, conjugation and dimerization of multiple-(NANP)₃ peptide.

Results and Discussion

The synthetic procedure is outlined in Fig. 1. Boc-Lys(Boc)-OH was used in 3 cycles of SPPS and the final deprotection with TFA yielded the heptalysyl peptide-resin, **3**, containing 8 free amino groups. After 3 more cycles of SPPS using the fragment Boc(NANP)-OH, it was cleaved and purified by HPLC. The synthetic multiple (NANP)₃ peptide, **4**, was homogeneous by HPLC and characterized by amino acid composition, sequence analysis, Ellman test and FAB-MS. Compound **4** was dimerized (pH 8) to form **5** and conjugated to TT (**6**).

In two initial comparative studies (Table 1), it was observed that both the multiple (NANP)₃ peptide, **4**, as well as the dimer, **5**, elicited at least comparable antibody response as that observed for the (NANP)₃ peptide conjugate, **1**, which was used for the clinical study. The multiple (NANP)₃ peptide conjugated to TT, **6**, gave no significant enhancement in antibody response. Therefore, these novel mutiple (NANP)₃ peptide antigens may lead to malaria vaccine candidates that would not require conjugation to a carrier protein.

E.P. Heimer et al.

Table 1 *Immunization results for (NANP)$_3$ peptide conjugate, 1, multiple (NANP)$_3$ peptide, 4, dimer, 5, and conjugate, 6, using different adjuvants and strains of mice*

Antigen	Antibody response					
	15 days	30 days	30 days	41 days	63 days	88 days
1	2 000	128 000	400[b]	275 000	191 000[b]	229 000[b]
			8 000[c]	476 000[c]	330 000[c]	228 000[c]
			2 000[d]	226 000[d]	397 000[d]	687 000[d]
4	8 000	64 000	92 000[e]	229 000[e]	159 000[e]	191 000[e]
5	8 000	256 000	275 000[f]	229 000[f]	110 000[f]	132 000[f]
6	8 000	128 000	53 000[c]	999 000[c]	476 000[d]	397 000[d]
			15 000[d]	572 000[d]	476 000[d]	330 000[d]
			3 000	275 000[b]	275 000[b]	191 000[b]

[a] Groups of 5 mice (C57-BL/10) were immunized. Each mouse received 20 μg of antigen i.p. (emulsified in CFA). After 15 days mice were boosted (10 μg of antigen in incomplete Freund's adjuvant). IRMA (recombinant protein).
[b] BALB/C mice without adjuvant.
[c] C57BL/6 mice using Al(OH)$_3$ adjuvant.
[d] BALB/C mice using Al(OH)$_3$ adjuvant.
[e] C57BL/6 mice using IFA.
[f] C57BL/6 mice without adjuvant.

References

1. Zavala, F., Cochrane, A.H., Nardin, E.H., Nussenzweig, R.S. and Nussenzweig, V., J. Exp. Med., 157 (1983) 1947.
2. Enca, V., Ellis, J., Zavala, F., Arnot, D.E., Asanavitch, A., Masula, A., Quakyi, I. and Nussenzweig, R.S., Science, 225 (1984) 628.
3. Zavala, F., Tam, J.P., Cochrane, A.H., Quakyi, I., Nussenzweig, R.S. and Nussenzweig, V., Science, 228 (1985) 1436.
4. Herrington, D.A., Clyde, D.F., Losonsky, G., Cortesia, M., Murphy, J.R., Davis, J., Baqar, S., Felix, A.M., Heimer, E.P., Gillessen, D., Nardin, E., Nussenzweig, R., Nussenzweig, V., Holligsdale, M.R. and Levine, M.M., Nature 328 (1987) 257.
5. Tam, J.P., Proc. Natl. Acad. Sci., U.S.A., 85 (1988) 5409.

Immunogenicity of synthetic peptides specific for the major immunogenic determinant of hepatitis B surface antigen (HBsAg)

Kwang-Soon Shin, Joong-Ho Choi and Kyong-Ho Kim

Mogam Biotechnology Research Institute, 341 Pojung-ri Koosung-myon Yongin-kun Kyonggi-do, 449-910, Korea

To investigate the antigenicity and immunogenicity of synthetic peptides specific for the major immunogenic determinant of HBsAg, a nonapeptide, H_2N-Cys-Thr-Lys-Pro-Thr-Asp-Gly-Asn-Aba-COOH, which corresponds to HBsAg amino acid residues 139–147, was synthesized by SPPS with a slight modification. This peptide was identical in amino acid composition to the theoretical value. The degree of purification and molecular weight were ascertained by HPLC and MS. Using m-maleimidobenzoyl-N-hydroxysuccinimide ester as a conjugating agent, the synthetic peptide was conjugated to rabbit albumin and γ-globulin, tetanus and diphtheria toxoids, and keyhole limpet hemocyanin. The conjugation yields were 8.3, 9.5, 15.8, 13.5 and 11.2%, respectively. The antigenicity and immunogenicity of this specific peptide were then compared with those of purified plasma-derived natural HBsAg. Antigenicity of the synthetic peptide and the plasma-derived natural HBsAg was determined by competition radioimmunoassay using ^{125}I-labeled natural HBsAg. The 50% inhibition was 90.00 µg/ml and 0.12 µg/ml for the synthetic peptide and the natural HBsAg, respectively, or about 750-fold less antigenicity for the latter (Fig. 1).

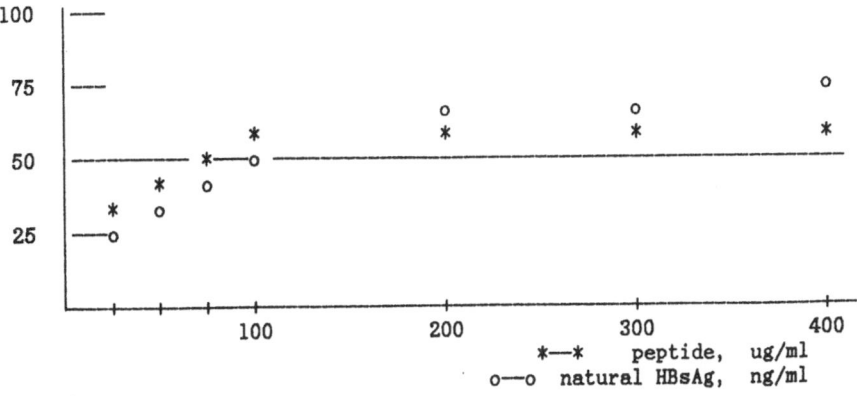

Fig. 1. Competition radioimmunoassay for synthetic peptide and natural HBsAg.

Table 1 *Immune responses in rabbits against synthetic peptide*

Carrier protein	Rabbit albumin	Rabbit γ-globulin	Tetanus toxoid	Diphtheria toxoid	KLH	Free peptide
with FCA	3/3	3/3	2/3	3/3	3/3	1/3
without FCA	3/3	3/3	1/3	2/3	3/3	1/3

Responding/immunized.

Immunogenicity was determined by administering the synthetic peptide-carrier conjugates into rabbits with and without Freund's complete adjuvant (FCA). The 50% immunizing dose in mice of the various peptide-carrier conjugates was 5.47, 6.00, 65.16, 31.25 and 13.03 μg/dose, respectively, for rabbit albumin, γ-globulin, tetanus and diphtheria toxoids and keyhole limpet hemocyanin. Natural HBsAg required 0.65 μg/dose. So we postulated that homologous proteins were preferable to heterologous ones as the carriers.

Despite the carrier protein used and the presence or absence of adjuvant, positive immune responses to the synthetic peptide were observed (Table 1 and Fig. 1). Antibody titers were higher, however, in the groups given FCA.

Mapping of human T-cell epitopes, occurring in the meningococcal class 1 outer membrane protein

Emmanuel J.H.J. Wiertz[a], Jacqueline A.M. van Gaans[a,*], Peter Hoogerhout[a],
Robert C. Seid Jr.[b], John E. Heckels[c], Geziena M.Th. Schreuder[d] and
Jan T. Poolman[a]

[a]National Institute of Public Health and Environmental Protection, P.O. Box 1,
3720 BA Bilthoven, The Netherlands
[b]Praxis Biologics, Rochester, NY, U.S.A.
[c]University of Southampton, Southampton, U.K.
[d]Immunohaematology, Academic Hospital Leiden, Leiden, The Netherlands

Introduction

Immunity to disease caused by *Neisseria meningitidis* is associated with the presence of bactericidal and opsonic antibodies to the capsular polysaccharides, lipopolysacccharides and outer membrane proteins. The capsular polysaccharides of group A and C meningococci are proven efficacious vaccines, although the immunogenicity in infants is poor and the immunity is of short duration. The combination with T-helper epitopes from outer membrane proteins will certainly improve the immunogenic properties of these T-cell independent antigens. For this reason we started an investigation to identify the location of T-cell epitopes in the class 1 outer membrane protein.

Results and Discussion

Six MHC-typed volunteers were immunized with an experimental meningo-coccal vaccine. They received a boost after 4–6 weeks. One dose of vaccine contained 25 μg of B outer membrane proteins (15 : P1.16), 25 μg of C polysaccharide, 25 μg of thiomersal and 0.25 mg ALPO$_4$. Six months after the first immunization, peripheral blood mononuclear cells were isolated and cultured with the outer membrane protein for six days. The proliferative response of T-cells was measured by a [^{3}H]thymidine incorporation assay. Cells from vaccinees 1–3 proliferated strongly, whereas cells from vaccinees 4 and 5 showed an intermediate response. Vaccinee 6 was found to be a low responder.

The recent elucidation of the sequence of the *N. meningitidis* class 1 outer membrane protein [1] enables a detailed mapping of its T-cell epitopes. Possible epitopes were selected using the prediction methods of Margalit et al. [2], and Rothbard and Taylor [3].

Peptides corresponding to the sequences 16–34, 47–59, 57–71, 103–121, 143–

---

*To whom correspondence should be addressed.

184, 153–158, 190–202, 215–228, 241–261, 304–322, and 352–366 were synthesized by continuous flow SPPS, applying Fmoc-polyamide chemistry [4].

The antigenicity of the peptides was measured by a similar assay as described for the whole protein. Six out of eleven peptides induced proliferation of T-cells from one or more volunteers (Table 1). The location of a number of T-cell epitopes on the meningococcal class 1 outer membrane protein, in relation with different MHC-types of human vaccinees, is therefore established. We intend to continue the investigation by testing more synthetic peptides and using human class 1 specific T-cell clones.

Table 1 *Stimulation index (SI)[a] of synthetic peptides on the proliferation of peripheral blood monocular cells from MHC-typed vaccinees*

Vaccinee	MHC specifity	Peptides					
		16–34	47–59	143–184	153–158	241–261	304–322
1	DR: 2 W6 W13 W15 W52 DQ: W1 W6	++	++	++	++	++	±
2	DR: 5 W11 W10 W52 DQ: W1 W3 W5 W7	ND[b]	+	++	+	ND	ND
3	DR: 3 W52 DQ: W2	–	+	±	±	+	±
4	DR: 3 7 W52 W53 DQ: W2	–	–	+	–	–	++
5	DR: 5 W11 7 W52 W53 DQ: W2 W3 W7	–	+	±	–	–	–
6	DR: 4 W10 W53 DQ: W1 W3 W5 W8	±	–	++	+	–	–

[a] Determined by [³H]thymidine incorporation (SI = cpm antigen/cpm background). ++: $SI > 6$; +: $3 < SI < 6$; ±: $2 < SI < 3$; –: $SI < 2$.
[b] Not (yet) determined.

References

1. Barlow, A.K., Heckels, J.E. and Clarke, I.N., Mol. Microbiol., 3 (1989) 131.
2. Margalit, H., Sponge, J.L., Cornette, J.L., Cease, K.B., Delisi, C. and Berzofsky, J.A., J. Immunol., 138 (1987) 2213.
3. Rothbard, J.B. and Taylor, W.R., EMBO J., 7 (1988) 93.
4. Dryland, A. and Sheppard, R.C., J. Chem. Soc. Perkin Trans. I., (1986) 125.

Synthetic antigens: Synthesis of several active immunogens representing epitopes of two picorna viruses and of fibrin

J.M. Peeters[a], P. Kretschmann[a], W.J.G. Schielen[a,b], M. Voskuilen[b],
W. Nieuwenhuizen[b], W.M.M. Schaaper[c], W.C. Puijk[c], H. Lankhof[c], A. Thomas[c],
R.H. Meloen[c], E.C. Beuvery[d], T.G. Hazendonk[d] and G.I. Tesser[a]

[a]*Laboratory of Organic Chemistry, Catholic University Nijmegen, Nijmegen,
The Netherlands*
[b]*Gaubius Institute, Leiden, The Netherlands*
[c]*Central Veterinary Institute, Lelystad, The Netherlands*
[d]*National Institute Public Health & Environmental Protection (RIVM), Bilthoven,
The Netherlands*

Introduction

Polio and foot-and-mouth disease are caused by picornaviridae carrying immunodominant antigenic sites on their capsid (VP1) proteins. The structural framework of the capsid proteins is similar for all picornaviridae, consisting of an eight-stranded antiparallel β-barrel and two flanking helices [1]. In fibrinogen, proteolytic removal of two small peptides by thrombin causes polymerization of the resulting fibrin monomers. The proteolysis exposes a new site on the Aα-chain (148–160), which causes degradation of a blood clot (by plasmin), since the plasminogen-to-plasmin conversion by tissue type plasminogen activator is enhanced [2] by this site.

Results and Discussion

Antigenic determinants
Polio virus: The spatial form of the antigenic site was mimicked by the synthesis of compound **A**. The peptide, which contains three spatially adjacent fragments of VP1, also contains two disulfide bridges to limit the number of possible conformations.

Compound A: Ac-CKY·TSK·DYC·NPC·STT·NKD·KCF·AYP·YPE·KWD-OH

Ac-C-(252-256)-D-(92-C-94-95-C-97-103-C-105-107)-(165-171)-OH

Foot-and-mouth disease virus: VP1-(119–211)-heneicosapeptide (i.e., the C-terminus) was prepared as a linear peptide or as a derivative (compound **B-1** and **B-2**) that could be cyclized (Ata = *S*-thioacetyl, Mal = ε-maleinimido-caproyl, Sic = ε-succinimidocaproyl):

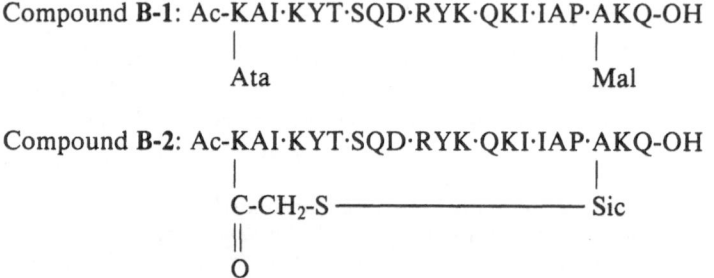

Compound **B-1**: Ac-KAI·KYT·SQD·RYK·QKI·IAP·AKQ-OH
　　　　　　　　|　　　　　　　　　　　|
　　　　　　　Ata　　　　　　　　　　Mal

Compound **B-2**: Ac-KAI·KYT·SQD·RYK·QKI·IAP·AKQ-OH
　　　　　　　　|　　　　　　　　　　　|
　　　　　　　C-CH₂-S ——————————— Sic
　　　　　　　‖
　　　　　　　O

Fibrin: The sequence Aα(148–160), assumed to be exposed solely in fibrin, is presumably also a unique determinant in fibrin. It contains a sequence of alternating polar–apolar aminoacyl residues. Compound C was prepared as a determinant:

Compound C: Ata-Nle-K·RLE·VDI·DIK·IRS-OH

Syntheses

Compound **A** was prepared by SPPS on a SASRIN resin and **B** and **C** were prepared on a *p*-alkoxybenzyl alcohol ('Wang') resin, using Fmoc for temporary protection and Boc for side chain protection. Arg was protected with Pmc or Pms; condensations were by the DCC/HOBt method.

Antigens

The antigenic determinants were condensed with protein carriers as thioethers using a linear linker in the maleimido method. This procedure was preferable, since defined conjugation took place and formation of antibodies directed against the linker was greatly diminished [3].

Polio virus: Conjugates of synthetic peptide derivatives comprising sequences of site 1 of poliovirus type 1, (i.e., compound **A**) elicited type-specific neutralizing antibodies when forced into the indicated loop structure. The peptides themselves showed no priming effect for a subsequent injection with trivalent vaccine.

Foot-and-mouth disease virus: Neutralizing antibodies were formed in rabbits upon immunization with the conjugate of compound **B-1**. Guinea pigs were fully protected from foot-and-mouth disease type A10 upon a single immunization with the adduct (to tetanus toxoid) of the linear sequence **B-1**.

Fibrin: A monoclonal antibody (anti-Fb-1/2) was obtained on immunization with an antigen consisting of BSA, repeatedly condensed with compound C, [Nle¹⁴⁷]-fibrinogen-Aα-(147–160). Anti-Fb-1/2 constitutes a viable diagnostic for fibrin concentration in human plasma, since fibrin can be detected in the presence of a 1000-fold molar excess of fibrinogen [4].

SPPS on Sasrin-resins [5] lends itself well to the preparation of protected peptides that are to be processed in the protected form. Here, protection was wanted to facilitate the iodine oxidation of *S*-protected (Trt and Acm) cysteine residues to disulfides [6].

The unexpected complete protection of guinea pigs (5 out of 5), which are (in contrast to rabbits) susceptible to foot-and-mouth disease, supports the feasibility of using synthetic vaccines for immunization. Additionally, the delicate balance between blood coagulation and fibrinolysis can be monitored with the monoclonal antibody described here that detects 2.5 μg/ml of fibrin in the presence of 1–5 mg of fibrinogen.

References

1. Hogle, J.M., Chow, M. and Filman, D.J., Science, 229 (1985) 1358.
2. Nieuwenhuizen, W., Vermond, A., Voskuilen, M., Traas, W.D. and Verheijen, J.H., Biochim. Biophys. Acta, 748 (1983) 86.
3. Peeters, J.M., Hazendonk, T.G., Beuvery, E.C. and Tesser, G.I., J. Immunol. Methods, 120 (1989) 133.
4. Schielen, W.J.G., Voskuilen, M., Tesser, G.I. and Nieuwenhuizen, W., Proc. Natl. Acad. Sci. U.S.A, in press.
5. Mergler, M., Nyfeler, R., Tanner, R., Gosteli, J. and Grogg, P., Tetrahedron Lett., 29 (1988) 4009.
6. Kamber, B., Hartmann, A., Eisler, K., Riniker, B., Sieber, P. and Rittel, W., Helv. Chim. Acta, 63 (1980) 899.

The use of bovine serum albumin in the production of antibodies to peptides: Application to the tachykinin family

Robert Benoit[a], Yin-Zeng Liu[b] and Solange Lavielle[c]

[a]*Montreal General Hospital, 1650 Avenue Cedar, Montreal, Canada*
[b]*Shanghai Institute of Organic Chemistry, 345 Lingling Lu, Shanghai 200032, China*
[c]*Laboratoire de Chimie Organique Biologique, CNRS UA 493, Université Paris VI,*
4 place Jussieu, F-75005 Paris, France

Introduction

Methylated bovine serum albumin (BSA-OCH$_3$) has been used for the preparation of antibodies against basic polypeptides [1]. Since no covalent coupling is involved before immunization, it has been proposed that the peptide is adsorbed on the ionized surface of BSA-OCH$_3$. Although no systematic study has yet been done, co-injections of BSA-OCH$_3$ with the positively charged peptides GnRH and TRH which are peptides lacking a nucleophilic amine function, i.e. a free N-terminus or a lysine side chain, did not elicit antibodies formation (R.B., unpublished results). The specificity patterns of previously described antibodies to somatostatin (1–14) [1] suggested that a covalent transamination reaction might occur in vivo between the nucleophilic amino groups of the Lys residues and the methyl ester of BSA-OCH$_3$. Therefore, we speculated that such a covalent linkage might also exist between the methyl ester of a peptide and BSA. Indeed, co-injection of Substance P methyl ester (SP-OCH$_3$) and BSA to rabbits led to high-affinity antibodies to Substance P whose reading frames were interestingly shifted toward the center and the N-terminal region of SP.

Results and Discussion

Co-injection of SP with BSA-OCH$_3$ yielded, as expected, high-titer, high-affinity anti-SP whose antigenic determinant resides in the C-terminal sequence of SP, i.e. SP (5–11). In agreement with previous negative results obtained with GnRH and TRH, it seemed that a nucleophilic amino group in the sequence of the peptide was a prerequisite for inducing antibodies formation with BSA-OCH$_3$ suggesting that a covalent amide linkage could be formed in vivo between the peptide and BSA-OCH$_3$, i.e. BSA-OCH$_3$ + peptide → BSA-peptide + CH$_3$OH. However, no transamination reaction could be evidenced in vitro when tritiated SP was incubated at 37°C with BSA-OCH$_3$ for 1–4 weeks with or without incomplete Freund's adjuvant (data not shown). This negative result does not exclude that such a reaction might occur in vivo.

As speculated, SP-OCH$_3$ co-injected with BSA induced the formation of high-affinity antibodies which recognize both SP and SP(1–9) on an equimolar basis,

Table 1 *Antigenic determinants of anti-SP obtained after co-injection, without covalent coupling, of SP with BSA-OCH$_3$, SP with BSA and SP-OCH$_3$ with BSA*

Peptides	Methods (percentage of cross-reactivity, %)			
	SPA/BSA-OCH$_3$	SP/BSA		SP-OCH$_3$/BSA
	(a)	(b)	(c)	(d)
SP	100	100	100	100
SP(5–11)	46.5	100	<0.01	n.d.
SP(6–11)	2.1	n.d.	n.d.	n.d.
SP(7–11)	n.d.	n.d.	<0.01	<0.01
SP(1–9)	<0.01	0.01	100	100
SP-COOH	0.1	n.d.	n.d.	n.d.
[Met^7]SP	10	n.d.	0.03	0.03
[Pro^9]SP	<0.01	n.d.	n.d.	n.d.
NKA	0.02	0.1	<0.01	<0.01
NKB	0.3	0.5	0.02	n.d.

The final dilutions used in the RIA performed at 4°C were: (a) 1:800 000; (b) 1:150 000; (c) 1:22 000 and (d) 1:10 000. The iodinated tracers were [^{125}I][Tyr^0]SP for (a), (b) and (c) experiments and [^{125}I][Tyr^{12}]SP for (d) experiments. EAch value is in percentage and represents (ED$_{50}$ analog/ED$_{50}$ SP) × 100. Calculations are done on a weight basis. n.d. = not determined.

whereas SP(5–11) does not cross-react at all (Table 1). As a control experiment, SP co-injected with BSA yielded high-titer anti-SP whose antigenic determinant resides in the C-terminal sequence of SP(5–11), and SP(1–9) does not cross-react with these antibodies. The condition, rendering a given region of a peptide more immunogenic than another, remains elusive even if we already know that a consistent pattern of response exists when the same conditions of immunization are used. For example, utilization of BSA-OCH$_3$ and antrin leads to antibodies directed essentially toward the COOH terminus of the decapeptide [2]. The same method applied to somatostatin-14 leads to antibodies directed mostly towards the central region of the tetradecapeptide [1]. Whatever the exact mechanisms operating in the selection of antigenic determinants are, this preliminary study has already shown that: (1) a small basic amphiphilic peptide such as SP could generate anti-SP antibodies when co-injected without coupling with BSA, and (2) SP methyl ester co-injected without coupling with BSA yielded high-affinity anti-SP whose antigenic determinant has been shifted toward the center and the N-terminal region of SP. Both results have to be further analyzed with other peptides. It would be worthwhile to apply this strategy for raising antibodies to members of the tachykinin family for which N-terminal directed antibodies are required.

References

1. Benoit, R., Ling, N., Brazeau, P., Lavielle, S. and Guillemin, R., In Boulton, A.A., Baker, G.B. and Pittman, Q.J. (Eds.) Neuromethods, Vol. 6, The Humana Press Inc., Clifton, NY, 1987, p. 43.
2. Ravazzola, M., Benoit, R., Ling, N. and Orci, L., J. Clin. Invest., 83(1989) 362.

Immunostereochemistry of peptides related to the flap of human renin

C.F. Liu[a], J.A. Fehrentz[a], A. Heitz[a], B. Castro[a], F. Heitz[b], C. Carelli[c], F.X. Galen[c], J.B. Michel[c], P. Corvol[c]

[a]CCIPE, rue de la Cardonille, F-34094 Montpellier Cedex 2, France
[b]Laboratoire de Physicochimie des Systèmes Polyphasés, CNRS, BP 5051, route de Mende, F-34033 Montpellier, France
[c]INSERM U36, 17 rue du Fer à Moulin, F-75005 Paris, France

Introduction

We have shown in earlier studies that it is possible to inhibit the renin-angiotensin system by immunization against renin [1] or peptides [2] related to renin and that specifically designed peptides could be recognized by anti-renin antibodies. These peptides [3] (flaps 10 SS and 14 SS) were cyclized through disulfide bridge and were able to bind respectively to five and six of our seven anti-renin monoclonal antibodies, showing first that the flap of human renin presents a conformation similar to that of these peptides (a β-hairpin), and second, that the flap of human renin seems to be an immunopotent region of the enzyme.

Results and Discussion

For this study, we have synthesized numerous peptides closed by amide bond or disulfide bridge and correlated their conformation to their ability to bind

C Y S T G C	Flap 6 SS
C R Y S T G T C	Flap 8 SS
C L R Y S T G T V C	Flap 10 SS
C T L R Y S T G T V S C	Flap 12 SS
C L T L R Y S T G T V S G C	Flap 14 SS
L T L R Y S T G T V S G	Flap 12 cyclo
T L R Y S T G T V S Y S T G	Flap 14 cyclo
C E L T L R Y S T G T V S G C	Flap 15 SSE
C L T L R Y S T G A T V S G C	Flap 15 SSA

Fig. 1. Sequences of the synthesized peptides.

738

antirenin monoclonal antibodies	flap 6 SS	flap 8 SS	flap 10 SS	flap 12 SS	flap 14 SS	flap 12 cyclo	flap 15 SSB	flap 15 SSA
4 G 1	.	.	0.6	.	0.3	.	.	.
F 15	0.3	0.3	0.3	0.3	0.3	0.3	0.6	0.3
4 B 11	.	.	.	.	.	.	.	0.3
3 E 8	.	.	0.3	.	0.3	.	.	.
1A12	.	.	.	.	0.6	.	.	.
2 D 12	.	.	0.6	.	0.6	0.6	0.6	.
4 B 1	0.6	.	0.6	.	0.6	0.6	.	0.6

Fig. 2. Recognition of the peptides by seven anti-renin monoclonal antibodies. Results are expressed in µg of monoclonal antibody giving an optical density three times as high as the negative control serum.

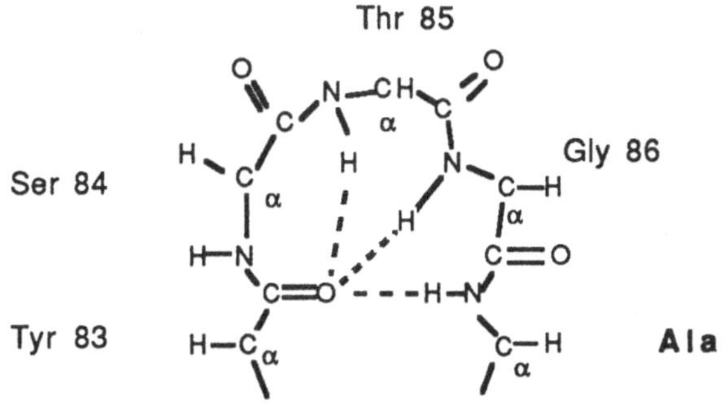

Fig. 3. Proposed model of the reversal of flap 15 SSA.

anti-renin monoclonal antibodies (Figs. 1 and 2). One non-natural peptide (flap 15 SSA) with an extra residue (with regard to the deletion that occurs when going from the pepsins to the renins) was also synthesized. It presents the good anti-parallel hydrogen bonds pattern as the flaps 10 SS and 14 SS, but the reversal was composed of five residues with three NHs engaged in hydrogen bonds with the same carbonyl (Fig. 3). The affinity of this peptide for several monoclonal antibodies was measured, compared to that of flaps 10 SS and 14 SS, and was found better.

It is of great interest in our search of immuno-control of renin activity to have a non-natural peptide related to the renin sequence that presents an important antigenic property. This peptide may be a candidate for the basis of a synthetic vaccine against renin and is actually studied for its immunogenic properties.

References

1. Michel, J.-B., Guettier, C., Philippe, M., Galen, F.-X., Corvol, P. and Menard, J., Proc. Natl. Acad. Sci. U.S.A., 84 (1987) 4346.
2. Bouhnik, J., Galen, F.-X., Menard, J. and Corvol, P., J. Biol. Chem., 262 (1987) 2913.
3. Fehrentz, J.-A., Heitz, A., Seyer, R., Fulcrand, P., Devilliers, R. and Castro, B., Biochemistry, 27 (1988) 4071.

A pair of peptides encoded by the same RNA but read in the 5'-3' or 3'-5' direction are antigenically similar as determined using monoclonal antibodies

David W. Pascual and Kenneth L. Bost

Department of Physiology and Biophysics, University of Alabama at Birmingham,
UAB Station, Birmingham, AL 35294, U.S.A.

Introduction

RNA sequences translated in the 5'-3' or 3'-5' direction often vary in their respective amino acid sequence, but will usually have similar hydropathicity [1]. This results from the middle nucleotide of the codon, which determines the hydropathic nature of the amino acid. To ascertain the antigenic relatedness of a given pair of hydropathically similar peptides, two peptides were synthesized based on the RNA sequence CAU CAA UCC AAA GAA CUG CUG AGG CUU GGG UCG. Reading this RNA sequence in the 5'-3' or 3'-5' directions resulted in two peptides, HQSKELLRLGS (termed 5'-3' peptide) and AGFGVVKKPEY (termed 3'-5' peptide), with different primary structures, but similar hydropathicity. Mice were immunized with either peptide conjugated to keyhole limpet hemocyanin, and monoclonal antibodies were produced to determine if these peptides represented cross-reactive antigens.

Results and Discussion

Using an enzyme-linked immunosorbent assay, we found that the IgM monoclonal 2F8 anti-5'-3' peptide antibody bound to microtiter wells coated with the 5'-3' peptide in a dose-dependent and saturable fashion (Fig. 1). This binding could be blocked with an 8-fold excess of BSA conjugated 5'-3' peptide, demonstrating the specificity of this interaction. Similar results were obtained with IgM monoclonal 8D2 anti-3'-5' peptide antibody's binding to 3'-5' peptide with regard to dose-dependence, saturability and specificity (Fig. 2). The IgM monoclonal 2F8 anti-5'-3' peptide antibody also recognized the 3'-5' peptide, and this binding could be blocked with either peptide conjugate (Fig. 1). Furthermore, the IgM monoclonal 8D2 anti-3'-5' peptide antibody also recognized the 5'-3' peptide, and this interaction could be blocked with either peptide conjugate (Fig. 2). Examination of the slopes for the curves depicting 2F8 IgM antibody binding to the 5'-3' and 3'-5' peptides at 50% saturation demonstrated that the extent of cross-reactivity was 96%. By the same criteria, the extent of cross-reactivity for the 8D2 IgM antibody with the two peptides was 84%. Clearly from the evidence provided, the two monoclonal antibodies, 2F8 and

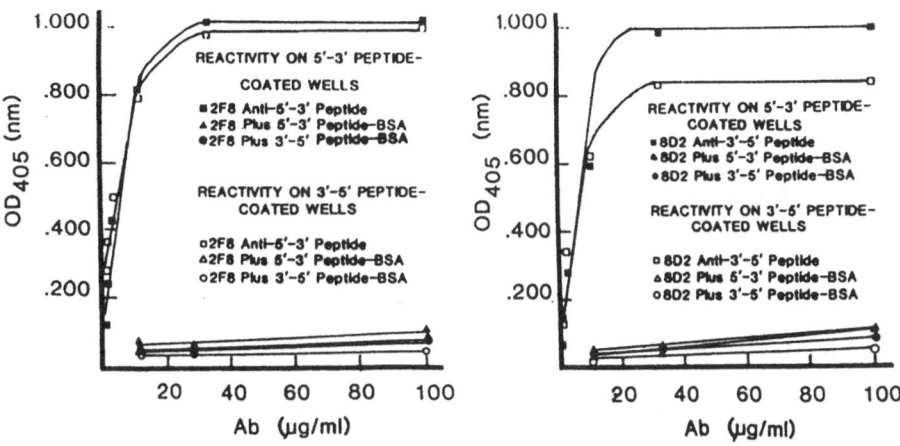

Fig. 1.   *2F8 Cross-reactivity.*   Fig. 2.   *8D2 Cross-reactivity.*

8D2, demonstrated similar abilities to recognize either peptide. Thus, these results support the notion that peptides read from the same RNA sequence either in the 5'–3' or 3'–5' directions are antigenically-related despite having different amino acid sequences.

References

1.   Blalock, J.E. and Bost, K.L., Biochem. J., 234 (1986) 679.

A blocking micro-ELISA for detection of antimalarial antibodies in human blood based on the synthetic peptide (NANP)$_{20}$

Carlo A. Nuzzolo[a],*, Adriano Bernardi[a], Antonio S. Verdini[b] and Antonello Pessi[b]

[a]Bioscience Department, Eniricerche S.p.A. and [b]Peptide Synthesis Department, Sclavo S.p.A., Via E. Ramarini 32, I-00015 Monterotondo, Rome, Italy

Introduction

Malaria from *Plasmodium falciparum* is widely diffused in many tropical and subtropical regions of the world, causing great morbidity and mortality. Immunological methods for detection of antibodies to different life stages of the parasite have recently been developed, as they are very useful tools for epidemiological studies. Anti-sporozoite antibodies have been determined by IRMA (immunoradiometric assay) or ELISA (enzyme-linked immunosorbent assay) using the synthetic NANP = Asn-Ala-Asn-Pro repeated sequence as supported antigen [1–4]. Since ELISA based on the use of two microtiter plates, with and without the supported antigen, may give false results owing to disuniformities of both the plate and the adsorbed materials, we have overcome this difficulty by developing an 'inhibition monoplate ELISA', based on a single (NANP)$_{20}$-coated plate: the specific value of a sample is given as the difference between the absorbance values of the uninhibited and the fully inhibited sample; (NANP)$_{20}$ is used as specific inhibitor.

Results and Discussion

All the parameters of this 'monoplate ELISA' have been optimized. The best results were obtained using: Dynatech M 129A plate, (NANP)$_{20}$ coated without postcoating; 1 μg (NANP)$_{20}$/ml TBS, pH 7.8, as coating solution; TBS, pH 7.8, 0.5% casein, 0.05% Triton X-100 and 0.005% Thimerosal, as diluent/eluent buffer; TBS, pH 7.8, 0.05% Triton X-100 and 0.005% Thimerosal, as washing buffer; 1 μg (NANP)$_{20}$ per well (9 μg/ml), as inhibitor. The uninhibited and (NANP)$_{20}$-inhibited samples are placed in adjacent wells.

Different kinds of samples were assayed: serum, plasma, whole blood, eluted bloodspot, and mosquito bloodmeal as well. Horseradish peroxidase (HRP) and alkaline phosphatase (AP) conjugates of goat antihuman IgG were used. The best results were obtained with the HRP conjugate and TMB or ABTS as substrates: in this case the results (as positive/negative detection) could also

---

*To whom correspondence should be addressed.

be seen 'by eye', as difference/equality of color between the uninhibited and the inhibited sample wells.

Highly correlating results were obtained for a number of African sera and bloodspots assayed by this monoplate ELISA and by the commercial ELISA kit from Sclavo S.p.A., Siena, Italy [4]. A 'rapid' monoplate ELISA has also been developed, using an affinity-purified HRP conjugate and TMB as substrate, in which the incubation times for sample and conjugate are shortened to 15 min, instead of the 60 min of the normal procedure: thus the overall test duration is reduced to about 1 h.

A preliminary study on human bloodspots from an endemic area of Comores (G. Sabatinelli et al., in preparation) has yielded results in agreement with the known epidemiological picture, lending support to the usefulness of this test as a sensitive and highly specific epidemiological tool.

Acknowledgements

We wish to thank Drs. G. Sabatinelli and G. Maiori, Istituto Superiore di Sanità, Rome, Italy, for providing the African sera and bloodspots, and Dr. F. Esposito, Università di Camerino, Italy, for the gift of the anti-NANP MAb.

References

1. Young, J.F., Hockmeyer, W.T., Gross, M., Ballou, W.R., Wirtz, R.A., Trosper, J.H., Beaudoin, R.L., Hollingdale, M.R., Miller, L.H., Diggs, C.L. and Rosenberg, M., Science, 228 (1985) 958.
2. Zavala, F., Tam, J.P. and Masuda, A., J. Immunol. Methods, 93 (1986) 55.
3. Del Giudice, G., Verdini, A.S., Pinori, M., Pessi, A., Verhave, G.P., Tougne, C., Ivanoff, W., Lambert, P.-H. and Engers, H.D., J. Clin Microbiol., 25 (1987) 91.
4. Esposito, F., Fabrizi, P., Provvedi, A., Tarli, P., Habluetzel, A. and Lombardi, S., Acta Tropica, (1989) in press.

Molecular characterization of two types of murine interleukin-1 receptors

Ueli Gubler[a,*], Patricia L. Kilian[b], Alvin S. Stern[c] and Richard Chizzonite[a]

[a]Department of Molecular Genetics, [b]Department of Immunopoharmacology and
[c]Department of Protein Biochemistry, Roche Research Center, Hoffmann-La Roche Inc.,
Nutley, NJ 07110, U.S.A.

Introduction

Labeled recombinant interleukin-1 alpha and beta bind to a number of different murine cell types in a specific and saturable manner, indicating the presence of IL-1 receptors. Depending on the cell type, the binding can be classified as either high affinity (kDa = 20 – 100 pM, T-cells, fibroblasts, keratinocytes and hepatocytes) or intermediate affinity (kDa = 200 – 500 pM, B-cells and macrophages). The number of sites per cell is uniformly low (200–6000) [1]. Affinity cross-linking experiments corroborate these apparent differences. Labeled IL-1 cross-links to a receptor protein of about 100 kDa on cells with high-affinity receptors, and to an 85 kDa protein on cells with intermediate-affinity receptors. A monoclonal antibody, specific for the high-affinity IL-1 receptor, inhibits IL-1 binding to T-cells and fibroblasts, but not to B-cells and macrophages, thus, also differentiating between the two types of receptors. The IL-1-induced proliferation of D10 helper T-cells or thymocytes is also blocked by this antibody, further indicating its specificity by inhibiting the biological response of IL-1 at its receptor. The antibody studies thus indicate the possible existence of two types of IL-1 receptors.

Results and Discussion

The availability of a cloned cDNA for the murine T-cell/fibroblast (high affinity) type receptor mRNA [2] allowed us to determine directly the presence of this transcript in different cell types. An S1 nuclease protection assay showed that the T-cell/fibroblast (high affinity) type receptor mRNA is, indeed, not present in the intermediate affinity receptor-bearing cells (B-cells and macrophages), corroborating the data obtained with the antibody.

Since the levels of receptor transcripts in total RNA are low, we decided to use a much more sensitive detection method based on the polymerase chain reaction. Total cellular RNA was reverse transcribed into cDNA. The cDNA was subsequently amplified with primers specific for the high-affinity receptor mRNA. Preliminary evidence suggests that coexpression of high- and interme-

---

*To whom correspondence should be addressed.

744

diate-affinity receptor mRNAs occurs in the cells tested. Both high- and intermediate-affinity receptor-bearing cells contain detectable, steady state levels of the high-affinity receptor mRNA, only at very different levels. The maximum levels detected in the intermediate-affinity receptor-bearing cells constitute only a few percent of the levels found in the high-affinity type T-cells (EL-4 cells). It is not known whether these low amounts of receptor mRNA actually give rise to functional protein, since the levels of protein expressed would be too low for detection by ligand binding or antibody inhibition studies. The availability of a cDNA probe for the low-affinity receptor will help to further clarify the questions concerning putative receptor coexpression and its possible biological significance.

References

1. Chizzonite, R., Truitt, T., Kilian, P.L., Stern, A.S., Nunes, P., Parker, K.P., Kaffka, K.L., Chua, A.O., Lugg, D.K. and Gubler, U., Proc. Natl. Acad. Sci. U.S.A., 86 (1989) 8029.
2. Sims, J.E., March, C.J., Cosman, D., Widmer, M.B., MacDonald, H.R., McMahan, C.J., Grubin, C.E., Wignall, J.M., Jackson, J.L., Call, S.M., Friend, D., Alpert, A.R., Gillis, S., Urdal, D.L. and Dower, S.K., Science, 241 (1988) 585.

Recombinant interleukin-2 analogs: Modulation of bioactivity by conformational perturbation

B.E. Landgraf[a], W.G. Berndt[a], D.P. Williams[b], J.R. Murphy[b], F.E. Cohen[c], K.A. Smith[a] and T.L. Ciardelli[d]

[a]*Dartmouth Medical School, Hanover, NH 03756, U.S.A.*
[b]*The University Hospital, Boston, MA 02118, U.S.A.*
[c]*University of California, San Francisco, CA 94143, U.S.A.*
[d]*Veterans Administration Hospital, White River Jct., VT 05001, U.S.A.*

Introduction

Understanding the immunomodulatory role of Interleukin-2 (IL-2) and its primary mechanism of action on the growth of T lymphocytes is a topic of wide interest [1]. One aspect of this research receiving particular attention is the SAR of IL-2 [2]. If receptor binding residues can be identified, selective engineering at these sites may produce molecules possessing agonist/antagonist activity. Despite considerable efforts in this area, knowledge of the exact binding residues remains largely obscure. The development of IL-2 agonists/antagonists, however, is not necessarily restricted by information regarding receptor contact sites. It may be possible to circumvent the search for binding residues through a design approach affecting conformation.

Results and Discussion

Using a semi-synthetic model system and the multiple simultaneous residue replacement technique pioneered by Kaiser et al. [3], we demonstrated that IL-2 contained an amphiphilic C-terminal helix prior to the availability of X-ray diffraction data [4,5]. Furthermore, we determined that the integrity of the C-terminal helix was critical for biologic activity. [6]. Although useful, the semi-synthetic two-chain derivatives displayed decreased biologic potency when compared to the authentic single-chain protein. Therefore, we have employed cassette mutagenesis to prepare fully recombinant single-chain proteins to explore the applicability of multiple residue replacement for protein design. The effect of these substitutions on biologic activity, receptor binding, conformation, and stability are examined.

X-ray diffraction data reveals a core tertiary structure stabilized by a left-handed, four-fold α-helical bundle. The C-terminal segment of IL-2, positions 117–133 (Helix F), participates in stabilizing this structure. In this study, helix F (FLNRWITFCQSIISTLT) was mutated to incorporate modifications that would be helix-stabilizing in Mutant I (FL*QKWIQFAQQLLQALK*), or helix-destabilizing in Mutant II (FL*PRWITFAQPIISTLT*). These mutants were expressed, purified, and refolded from *E. coli* inclusion-body extracts prior to characterization. Biologic activity for both mutants was compared to the activity exhibited by wild-type IL-2, using human peripheral blood T lymphocytes. Mutant I retained an activity indistinguishable from that of authentic IL-2, despite the incorporation of 10 simultaneous substitutions, while Mutant II displayed a > 100-fold decrease (Fig. 1). Assessment of both mutants' receptor binding characteristics was evaluated by a competitive radioligand receptor binding assay for the high affinity IL-2 receptor. As with the biologic activity, Mutant I displayed an affinity similar to that of the natural ligand, and Mutant II demonstrated a > 100-fold decrease in binding capability. Analysis of the structural integrity of each mutant was based on far-UV CD spectra obtained at 25°, 50° and 80°C. These spectra indicate that Mutant I, possessing helical secondary structure virtually identical to that of the authentic protein at 25°C, is more stable toward thermal denaturation. Mutant II, although exhibiting a noticeable decrease in helical secondary structure, remains stable at all temperatures tested.

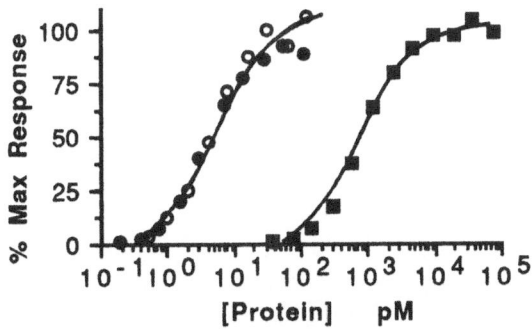

Fig. 1.   The ability of IL-2 (o), Mutant I (●), and Mutant II (■) to stimulate the growth of human T lymphocytes.

Conclusion

From this investigation, we observed that the properties of the semi-synthetic variants were predictive of those properties possessed by the fully recombinant mutants. We found that ten simultaneous helix-stabilizing replacements can be made within a 17-residue segment to generate a biologically identical molecule. Yet, only two proline substitutions are enough to reduce bioactivity and receptor binding by > 100-fold. CD spectra suggest that secondary structural stability is enhanced for Mutant I. Furthermore, CD confirms that Mutant II, while possessing less helix, remains thermally stable. We conclude that, through rational manipulation of secondary and tertiary conformation, it may be possible to modulate bioactivity and receptor binding of protein hormones without knowledge of the exact receptor contact sites.

B.E. Landgraf et al.

Acknowledgements

This work is sponsored in part by the National Cancer Center and NIH grant No. A-123398.

References

1. Smith, K.A., Science, 240 (1988) 1169.
2. Ciardelli, T.L., Holley, H., Smith, K.A., Cohen, F.E., Butler, L. and Gadski, R., In Smith, K. (Ed.) Interleukin-2, Academic Press, New York, 1988, p. 67.
3. Kaiser, E.T. and Kezdy, F.J., Science, 223 (1984) 249.
4. Ciardelli, T.L., Cohen, F.E., Gadski, R., Butler, L., Landgraf, B. and Smith, K.A., In Marshall, G.R. (Ed.) Peptides: Chemistry and Biology, (Proceedings of the 10th American Peptide Symposium), ESCOM, Leiden, 1988, p. 364.
5. Ciardelli, T.L., Landgraf, B., Gadski, R., Strnad, J., Cohen, F.E. and Smith, K.A., J. Mol. Rec., 1 (1988) 42.
6. Landgraf, B.E., Cohen, F.E., Smith, K.A., Gadski, R. and Ciardelli, T.L., J. Biol. Chem., 264 (1989) 816.

Structure-function studies of interleukin-3

Ian Clark-Lewis

The Biomedical Research Centre, 2222 Health Sciences Mall, UBC, Vancouver, B.C., Canada V6T 1W5

Introduction

Interleukin-3 (IL-3) is a T-lymphocyte-derived glycoprotein growth factor that stimulates a broad spectrum of hemopoietic cells [1]. Its biological activities suggest that it is a pivotal regulator of hemopoiesis, and clinical uses have been proposed. The broad aim of this on-going study is to understand the structural basis for the activities of IL-3 to permit engineering of IL-3 analogs with novel properties.

Previous studies established that the chemically synthesized and folded 140 residue polypeptide chain of mouse IL-3 had similar activities to that of the native or recombinant forms indicating that the carbohydrate moieties were not essential for function [2]. Synthesis of cysteine-modified analogs demonstrated that a single disulfide bridge between cysteines 17 and 80 was essential for full activity, but cysteines 79 and 140 could be replaced by alanine without loss of function [3].

Results and Discussion

The specific questions that were addressed in these studies were: (1) What is the minimal structure required for full activity? (2) How can this active fragment be purified to homogeneity for structural studies? (3) Which residues/regions of IL-3 are essential for function?

All the analogs were synthesized as described previously [2] but with Cys^{79} replaced by alanine so that they folded unambiguously [3]. The analogs were purified by RPHPLC and examined by AAA, HPLC, SDS gel electrophoresis and isoelectric focusing.

To determine the minimal structure required for activity, a series of analogs were constructed that were shortened at the C-terminus. IL-3 (1–127) and (1–118) had potencies that were equivalent to that of IL-3 (1–140) (50% of maximal response about 1 ng/ml); however, (1–112) was 100-fold less potent and (1–108) lacked activity at concentrations up to 50 μg/ml. When the corresponding analogs were synthesized without the N-terminal 16 residues, i.e., IL-3 (17–127), (17–118), (17–112) and (17–108), the pattern was similar; however, the N-terminally shortened analogs had about 5-fold lower potency. This is consistent with previous data showing that IL-3 (17–140) was about 5 times less potent than (1–140) [2]. The results demonstrate that residues 109–118 are essential for activity and that the 102 residue fragment (17–118), represents the minimal structure with high biological activity. This fragment was the focus for further studies.

For structural studies, homogeneous material is required. Experience with IL-3 (17–118) and analogs was that a narrow cut on RPHPLC gave material between 80% and 90% pure as judged by isoelectric focusing and capillary electrophoresis. An additional orthogonal technique was required to reproducibly obtain large amounts of pure material. Isoelectric focusing using the 'Rotofor' (BioRad) system was found to have the required capacity, resolution and recovery. Urea (8 M) was required to maintain solubility, and refocusing selected fractions improved resolution further. Two RPHPLC steps resulted in apparently homogeneous material that will be assessed for structural characterization. Isoelectric focusing and RPHPLC represent a powerful combination for the purification of synthetic proteins (>80 residues).

Although knowledge of the 3-dimensional structure is essential for rational design of analogs, this information alone will not tell us which residues/regions are essential for binding to the cell surface receptor and biological activity. To begin to address this question a series of 12 analogs of IL-3 (17–118) were synthesized in which *all* the residues of one particular type were substituted in a conservative manner (e.g., all the Asp for Glu). Based on these results, further analogs can be designed to locate individual residues that are sensitive to modification by a process of elimination. This strategy is based on the hypothesis that residues that are highly sensitive to minimal modification (i.e., result in inactive analogs) are important for function. This approach requires no prior assumptions regarding the 3D structure.

All the IL-3 (17–118) analogs synthesized had lower activity than the wild type and none demonstrated inhibitory effects. Analogs with every Ser converted to Thr, Thr to Ser, Asn to Gln, Gln to Asn, Tyr to Phe and Lys to Arg had readily detectable activity ranging between 10 ng/ml and 1 μg/ml giving 50% maximal response. Particularly remarkable was the lack of activity of an analog with the single His at position 111 replaced by Lys. An analog with this residue replaced by D-His was also inactive. However, when His[111] was replaced by Phe, the analog had activity equivalent to that of the wild type. This result suggests that His is probably essential for structure but is not involved directly in receptor binding. However, distinguishing between these two possibilities will require further structural analysis. Analogs in which all Arg were converted to Lys, Asp to Glu or Glu to Asp were inactive and, therefore, of particular interest. Design of further analogs aims to locate the individual residues that are sensitive to modification.

References

1. Schrader, J.W. Lymphokines 15: Interleukin-3; the Pan-Specific Hemopoietin, Academic Press, New York, NY, 15 (1988) 1.
2. Clark-Lewis, I., Aebersold, R.A., Ziltener, H.J., Schrader, J.W., Hood, L.E. and Kent, S.B.H., Science, 231 (1986) 134.
3. Clark-Lewis, I., Hood, L.E. and Kent, S.B.H., Proc. Natl. Acad. Sci. U.S.A., 85 (1988) 7897.

Mapping of the neutralization epitopes of pertussis toxin S1 subunit using synthetic peptides

Pele Chong*, Margaret Sydor, Gloria Zobrist, Heather Boux and Michel Klein

Connaught Centre for Biotechnology Research, 1755 Steeles Ave. West, Willowdale, Ont., Canada M2R 3T4

Introduction

Pertussis toxin (PT) is a major virulence factor of *Bordetella pertussis*. Antibodies raised against PT have been shown to protect mice against both intracerebral and respiratory challenges with virulent *B. pertussis* [1]. Thus, PT is the prime candidate for an acellular pertussis vaccine. Pertussis toxin is an oligomeric A-B type toxin in which the S1 subunit is an ADP-ribosyltransferase, whereas the B-oligomer mediates the binding of the holotoxin to target cell receptors [2,3]. The polycistronic gene coding for PT has recently been sequenced, and the primary amino-acid sequences of the five different subunits have been deduced from the corresponding DNA sequences [4,5]. Two regions of S1 molecule have structural similarity with the A subunit of both cholera and *E. coli* heat-labile toxins (Table 1). These conserved regions may play an important role in the NAD-binding and/or enzymatic activities of S1. We have established, by photocrosslinking experiments, that residue Glu^{129} plays a critical role in NAD-binding [6]. A recent study by Burns et al. [7] has indicated that the C-terminal residues of S1 are necessary for its interaction with the B-oligomer. In this paper, we describe the immunological properties of four sets of synthetic peptide analogs corresponding to these potentially critical functional regions of S1 subunit.

Results and Discussion

Four sets of peptides were synthesized using SPPS and purified by RPHPLC. These peptides include: the N-terminal N18-S1 (residues 1–18), and S1-P4 (residues 41–65) peptides corresponding to the first and second conserved regions of the bacterial ADP-ribosyltransferases; the NAD-S1 (residues 121–138) which contains the contact residue (Glu^{129}) of the NDA active site; the C-terminal C35-S1 peptide (residues 201–235) is postulated to be the binding site for the B-oligomer (Table 1). Peptides were conjugated to KLH or BSA at a 10:1 molar ratio of peptide over carrier, as described by Liu et al. [8]. Anti-peptide antisera raised in rabbits (Table 2) were tested for their immunological properties (a) in the Chinese hamster ovary (CHO) cell antitoxin neutralization assay [9]; (b) by PT- and peptide-

---

*To whom correspondence should be addressed.

Table 1 *Protein sequences of potentially critical functional regions of pertussis toxin S1 subunit*

Peptides	Sequences[a]
Region 1	
	1 18
PT-S1	DDPPATVYRYDSRPPEDV
CT-A	N DDKL YRADSRPPDEI
E. coli HLT-A	YRADSRPP
Region 2	
	41 65
PT-S1	CQVGSSNSAFVSTSSSRRYTEVYL
CT-A	VSTSI SLR
E. coli HLT-A	VSTSLSLR
Region 3	
PT-S1	121 #[b] 138
NAD-binding site	GALATYQSEYLAHRRIPP
Region 4	
PT-S1	201 235
B-oligomer	CMARQAESSEAMAAWSERAGEAMVLVYYESIAYSF
Binding site	

[a] Amino acid sequences of functionally critical regions of pertussis toxin S1 subunit. Two regions (residues 1–18 and 41–65) are structurally similar to both A subunits of cholera and E. coli heat-labile toxins.

[b] The symbol # points to residue Glu[129] which has been shown to be involved in NAD binding.

specific ELISAs [10]; and (c) immunoblot analysis [11]. The results of these experiments are summarized in Table 2.

Rabbit antisera were shown to be monospecific for their respective immunizing peptides except for the rabbit antiserum raised against NAD-S1-KLH, which cross-reacted with N18-S1-BSA in peptide-ELISA (data not shown). Both N18-S1-specific and NAD-S1-specific antibodies recognized PT in PT-ELISA and the S1 subunit by immunoblot analysis (Table 2). These antibodies, as well as antibodies raised against C35-S1, were capable of neutralizing PT toxicity in the CHO cell clustering assay (Table 2). Although C35-S1-specific antibodies recognized the S1 subunit in the PT-ELISA, they failed to react with S1 on immunoblots. S1-P4-specific antibodies failed to recognize S1, as judged by both PT-ELISA and immunoblot analysis. These data indicate that this region is either buried inside S1 or its native structural conformation is not recognized by the anti-peptide antibodies. In contrast, NAD-S1, N18-S1, and C35-S1 sequences are exposed. Indeed, antibodies specific for these regions were shown to interact with native S1. These three regions also contain PT-neutralizing epitopes. In addition, the C35-S1 peptide expresses a potent T-cell epitope, since the unconjugated peptide is able to elicit a strong antibody response.

Table 2   *Biochemical and immunological properties of rabbit antisera raised against S1 peptides or peptide-KLH conjugates*

Immunogens[a]	Antibody titer		Recognition of S1 by immunoblot analysis
	PT-neutralization CHO cell assay[b]	PT-specific ELISA[c]	
N18-S1(1-18-Cys)	1/64	1/10000	Yes
N4-18-S1(4-18-Cys)	0	1/8000	Yes
N7-17-S1(7-18-Cys)	0	1/8000	Yes
N10-18-S1(10-18-Cys)	0	1/2000	Yes
S1-P4(41-65)	0	0	No
NAD-S1(Cys-121-138-NH$_2$)	1/64	1/32000	Yes
NAD-S1-AN1(128-138-Cys)	0	0	Yes
NAD-S1-AN2(124-138-Cys)	0	0	Yes
NAD-S1-AN3(121-138-Cys)	0	0	Yes
C12-S1(223-235)	0	1/200	Yes
C23-S1(212-235)	0	1/800	Yes
C35-S1(201-235)	1/32	1/10000	No

[a] Two rabbits were immunized im with individual peptide-KLH conjugates, except for C35-S1 and its analogs which were injected as free peptides in complete Freund's adjuvant. After two weeks, each rabbit received two booster injections with the same immunogens in incomplete Freund's adjuvant. Between 100 and 500 μg of immunogens were used for each injection. Rabbit sera were collected two weeks after the final booster injection.
[b] The PT toxicity neutralization assay was performed according to Gillenius et al. [9].
[c] The specificity of rabbit antisera was determined as previously described [10].

To further identify the residues critical for the expression of PT-neutralizing epitopes, peptide analogs were synthesized, HPLC-purified, and coupled to KLH. All N-terminal deleted N18-S1 peptide analogs were able to induce antibodies against S1, as judged by both PT-ELISA and immunoblot analysis (Table 2). However, these antibodies did not neutralize PT toxicity in the CHO-cell clustering-inhibition assay. These data suggest that the first three N-terminal residues of S1 are essential for the induction of neutralizing antibodies. Similarly, the peptide analogs of NAD-S1 coupled to KLH via a C-terminal cysteine residue were capable of inducing antibodies reacting with S1 on immunoblots. In contrast to antibodies raised against the NAD-S1 sequence linked to the carrier protein via an N-terminal cysteine residue, these antibodies failed to neutralize PT in the CHO-cell inhibition assay, and did not recognize PT in the PT-ELISA (Table 2). This observation leads to the hypothesis that the N-terminal end of the NAD-S1 sequence is buried in the holotoxin, whereas its C-terminus is exposed. Two N-terminal-truncated analogs of C35-S1 peptide were capable of inducing S1-specific antibodies, but in contrast to the anti-C35-S1 antibody, they had no PT-neutralizing activity (Table 2). This indicates that the N-terminal sequence of C35-S1 (residues 201–212) contains neutralization epitope(s).

Conclusion

Three exposed B-cell neutralization epitopes (residues 1–18, 121–138, 201–235) on the enzymatic S1 subunit of pertussis toxin have been mapped. The identification of these epitopes represents a first step towards the rational design of a synthetic whooping-cough vaccine.

References

1. Manclark, C.R. and Cowell, J.L., In Germanier, R. (Ed.) Bacterial Vaccine, Academic Press, New York, 1984, p. 69.
2. Tamura, M., Nogimori, K., Yajima, M., Ito, K., Katada, T., Ui, M. and Ishii, S., Biochemistry, 21 (1982) 5512.
3. Burns, D.L., Kenimer, J.G. and Manclark, C.R., Infect. Immun., 55 (1987) 24.
4. Locht, C. and Keith, J.M., Science, 232 (1986) 1258.
5. Nicosia, A., Perugini, M., Franzini, C., Casagli, C., Borri, M., Antoni, G., Almoni, M., Neri, P., Ratti, G. and Rappuoli, R., Proc. Natl. Acad. Sci. U.S.A., 83 (1986) 4631.
6. Cockle, S., FEBS Lett., 249 (1989) 329.
7. Burns, D.L., Hausman, S.Z., Linder, W., Robey, F.A. and Manclark, C.R., J. Biol. Chem., 262 (1987) 17677.
8. Liu, F.-T., Zinnecker, M., Hamaok, T. and Kazt, D.H., Biochemistry, 18 (1979) 690.
9. Gillenius, P., Jaattmaa, E., Askelof, P., Granstrom, M. and Tiru, M., J. Biol. Stand., 13 (1985) 61.
10. Chong, P. and Klein, M., Biochem. Cell. Biol., 67 (1989) 387.
11. Towbin, H., Staehelin, T. and Gordon, J., Proc. Natl. Acad. Sci. U.S.A., 76 (1979) 4350.

Heterotypic protection induced by synthetic peptides corresponding to three serotypes of foot-and-mouth disease virus (FMDV)

R.D. DiMarchi[a,*], G. Mulcahy[b], C.M.C.F. DoAmaral[b], G.S. Brooke[a], C. Gale[a] and T. Doel[b]

[a]Lilly Research Laboratories, Eli Lilly & Comp., Indianapolis, IN 46285, U.S.A.
[b]AFRC Institute, Pirbright, Woking, Surrey, GU24 ONF, U.K.

Introduction

Foot-and-mouth disease (FMD) is a highly contagious disease of cloven-hooved animals and, as such, is of great concern to the agricultural economies of the world. The measures employed to control the disease include extensive and frequent prophylactic campaigns based on killed virus vaccines. A number of laboratories have attempted to develop alternative vaccines based either on the viral coat protein, VP_1 [1], or synthetic peptides equivalent to specific regions of the same protein [2,3]. The 141–160 region of VP_1 has been clearly identified as a major site for induction of virus neutralizing antibody. We have used it in the specific 40-residue peptide Cys-Cys-(200–213)-Pro-Pro-Ser-(141–158)-Pro-Cys-Gly to protect cattle against intradermolingual challenge with the virulent O_1BFS 1860 strain of FMDV [3].

A common expectation with synthetic peptide vaccines is the generation of highly specific antibodies, such that vaccinates will not be protected against variant viruses possessing minor mutations in their protein sequences. Indeed, the heterogeneity of FMDV is such that it is necessary to produce conventional killed virus vaccines against each of the seven serotypes of the virus. It is recognized that there is little protective enhancement between virus strains included in conventional trivalent vaccines [4]. The protective capacity of three peptides, based on the VP_1 sequences of three FMDV serotypes, with particular reference to heterotypic protection, was evaluated.

Results and Discussion

We have extended our previous FMDV studies by demonstrating that peptides bearing two specific regions of the VP_1 protein in the A_{24} Cruzeiro, C_3 Indaial, and O_1 Kaufbeuren serotypes can confer protection against viral challenge (Table 1). The A- and O-peptides were found capable of providing complete

---

*To whom correspondence should be addressed.

protection at all dose levels against their respective virus challenge. The C-peptides provided a high level of protection, but less than complete.

Table 1 *Protection of guinea pigs to homotypic FMDV challenge (28 dpv)*

Peptide of immunization	Peptide dose (mg)		
	1	0.2	0.04
A40	5/5[a]	5/5	5/5
C40	2/4	5/5	3/5
O40	4/4	5/5	5/5

[a]Number of guinea pigs protected/challenged.

It was demonstrated that peptide vaccines are capable of inducing a broad level of reactivity (Table 2). Cross-reactive antiviral antibody, as measured by ELISA, was observed with anti-O40-peptide sera and a range of O-viruses. The best experimental evidence for heterotypic protection was with guinea pigs, although one cow given two doses of O40-peptide was also protected against challenge with A_{24}-virus. In general, high levels of heterotypic virus neutralizing antibody (VNA) were not observed, which contrasts to the level of heterotypic antipeptide antibody.

Table 2 *Protection of guinea pigs to heterotypic FMDV challenge (28 dpv)*

Peptide of immunization	Dose (mg)	Challenge virus		
		A-serotype	C-serotype	O-serotype
A40	0.2	5/5[a]	0/5	4/5
C40	0.2	1/5	3/5	3/5
O40	0.2	1/5	0/5	5/5
A40	5.0	5/5	1/5	3/5
C40	5.0	1/5	5/5	3/5
O40	5.0	4/5	1/5	5/5

[a]Number of guinea pigs protected/number challenged.

The lack of correlation of protection with VNA was reported in a previous paper from this laboratory [5] and suggests a significant qualitative difference between the antibodies induced by peptide and viral vaccines. Cattle immunized with virulent or inactivated FMDV mount a humeral immune response in which the isotype profile differs significantly from that which occurs in cattle immunized with synthetic peptide. The fact that an IgG_1 response is favored in virus-immunized, and IgG_2 in peptide-immunized animals may partially explain the lower protective capacity of antipeptide antibodies against FMDV challenge. It would appear that our peptide vaccines do not exactly mimic the immune response of cattle or guinea pigs to virus, but, nonetheless, achieve the objective of protection. Clearly, peptides can be used to induce a unique immune response, in our case, heterotypic protection, that cannot be achieved with the whole virus. This argues well for FMD and other potential peptide vaccines.

References

1. Kleid, D.G., Yansura, D., Small, B., Dowbenko, D., Moore, D.M., Grubman, M.J., McKercher, P.D., Morgan, D.O., Robertson, B.H. and Bachrach, H.L., Science, 214 (1981) 1125.
2. Bittle, J.L., Houghten, R.A., Alexander, H., Shinnick, T.M., Sutcliffe, J.G., Lerner, R.A., Rowlands, D.J. and Brown, F., Nature, 298 (1982) 30.
3. DiMarchi, R., Brooke, G., Gale, C., Cracknell, V., Doel, T. and Mowat, N., Science, 232 (1986) 639.
4. Black, L., Nicholls, M.J., Rweyemamu, M.M., Ferrari, R. and Zunino, M.A., Res. Vet. Sci., 40 (1986) 303.
5. Doel, T.R., Gale, C., Brooke, G. and DiMarchi, R., J. Gen. Virol., 69 (1988) 2403.

Epitope mapping of human transforming growth factor α

P.D. Hoeprich Jr.[a],*, S. Martin[a], B. Langton[a], R. Harkins[a], C. Volin[b] and J.P. Tam[b]

Triton Biosciences, Alameda, CA 94501, U.S.A.
bThe Rockefeller University, New York, NY 10021, U.S.A.

Introduction

TGFα is a mitogenic peptide that stimulates growth in a variety of cell types [1–3]. First isolated from medium conditioned by malignant cells in culture, it is a single-chain polypeptide consisting of 50 amino acids, six of which are cysteine residues [4,5]. The latter form three disulfide bonds that are required for full biological activity and divide the molecule into three subdomains, each defined as a disulfide 'loop' region (Fig. 1). A similar disulfide bond pattern is found in homologous mitogenic polypeptides including EGF, vaccinia growth factor and Shope fibroma growth factor [6,7]. Although TGFα shares 33–40% sequence homology with EGF, antibodies against one do not crossreact with the other. Mitogenic effects of TGFα are apparently mediated through the EGF receptor. Addition of TGFα to certain nontransformed cells in culture results in aberrant growth typical of a transformed phenotype. Furthermore, it has been reported that the overexpression of cell surface receptors capable of binding TGFα increases in certain tumor cell lines [8].

Because of the apparent importance of TGFα in the role of cellular growth control, coupled with potential therapeutic value (e.g., wound healing), an understanding of its interaction with the EGF receptor will provide useful information. An approach to this problem is to use monoclonal antibodies to identify structural determinants that compete for receptor binding. We have raised a panel of 11 monoclonal and four polyclonal antibodies to human TGFα, some of which neutralize/prevent receptor binding. We now report strategies and results of mapping studies undertaken to identify the epitopes recognized by each of these antibodies. The mapping strategies involve two different approaches, the method of Geysen et al. [9], using synthetic peptides immobilized on polyethylene pins containing sets of overlapping peptides based on the sequence of hTGFα, and an improved conventional synthetic approach that yields free, soluble peptides with a *p*-hydroxybenzhydrylamine moiety at the COOH terminus to facilitate surface association of peptides for solid-phase immunoassays [10]. Two immunodominant regions are identified unambiguously; they contain the amino terminus residues 1–9 and residues 22–31 of the second domain or 'B' loop (Fig. 1).

---

*Present address: Affymax Research Institute, Palo Alto, CA 94304, U.S.A.

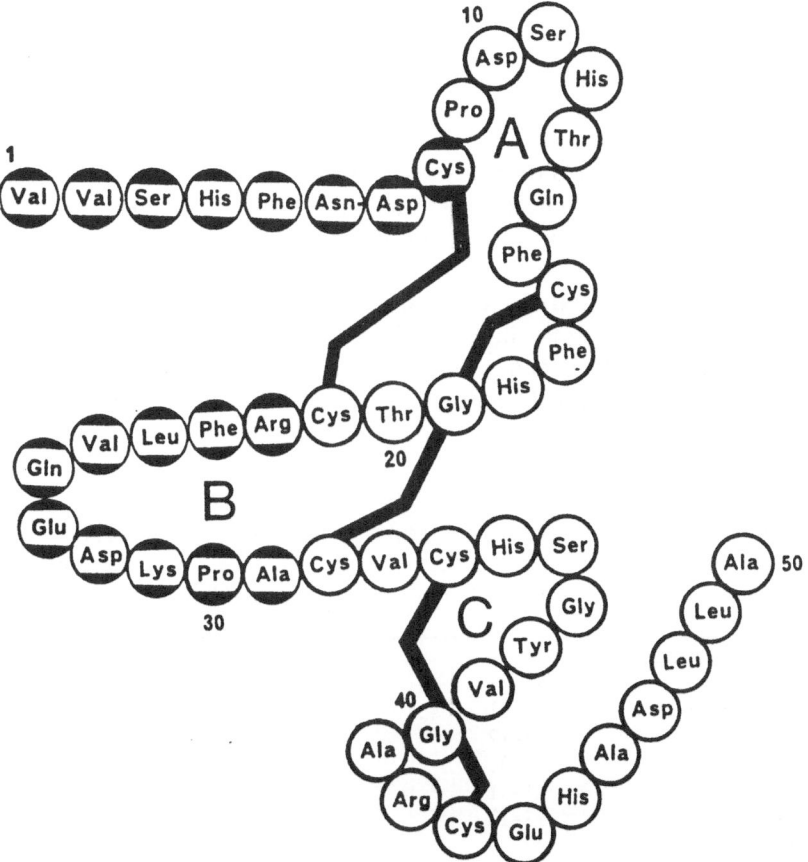

Fig. 1. Schematic representation of human TGFα. Three disulfide bonds conform the peptide into a biologically active structure containing three 'loop' regions, designated A, B and C, respectively. The shaded residues indicate the two immunodominant regions.

Results and Discussion

The epitope mapping results of the panel of TGFα-specific antibodies are shown in Table 1. Of the 11 monoclonal antibodies studied, four reacted specifically with the amino terminal region, four recognized a portion of the 'B' loop covering a putative β-turn region and three did not show any reactivity when studied by the Geysen method. In every case, the antibodies were exposed to families of peptides, i.e., 4-mers, 6-mers and 8-mers, that were synthesized based on the primary amino acid sequence of hTGFα. Members of each family of peptides overlapped one another by a single residue; as such, the minimal size of an epitope could be determined. For the monoclonal Tab 2, a minimal binding epitope containing residues 4–7, His-Phe-Asn-Asp, was detected. As

759

expected, the response of the four polyclonal antibodies was heterogeneous. They reacted with both the regions identified in the case of the monoclonal antibodies, as well as showing some reactivity with the C-terminal region containing residues 42–50 (Table 1). The results obtained using the soluble synthetic peptides essentially paralleled the observations made with the Geysen method. Three polyclonal antisera showed some disagreement in the two approaches, but this discrepancy may reflect differences in conformational states between covalent and non-covalent immobilization. Monoclonal antibodies Tab 1, 8 and 10 are unique in that they prevent interaction of hTGFα with the EGF receptor and they did not react with any synthetic peptide representing a continuous epitope. These results suggest that the two immunodominant regions identified above are not involved in receptor binding, since none of the antibodies that reacted to these regions affect ligand-receptor interaction. It is tempting to speculate that part of the 'A'- and 'C'-loops, as well as the C-terminal region, may be involved in receptor binding.

Table 1 *Primary antigenic sites of human TGFα-specific antibodies*

Antibody	Major antigenic epitope (residues)	
	Peptide-pin ELISA[a]	Soluble peptide ELISA[b]
Monoclonal antibodies[c]		
A1.5	21–31	23–30
A9.1	21–31	23–30
A13.5	2–12	2–12
B4.5	21–31	23–30
B6.1	21–31	23–30
Tab 2	1–12	1– 8
Tab 3	2– 9	1– 8
Tab 5	2–12	1– 8
Tab 1, 8, 10	NR[d]	ND[d]
Polyclonal antisera		
39–74	1– 9, 21–31	1– 8, 23–30
39–11[e]	1– 9, 21–32, 42–59	43–50
R8	2–12, 21–31	17–24
R9	2–12, 21–31	17–24, 37–44

[a] Determined by the method of Geysen et al. [9].
[b] Assay done using soluble octamer peptide epitopes covering the entire hTGFα sequence and overlapping by six residues.
[c] A1.5, A9.1, A13.5, B4.5, B6.1 and Tab 1–3, 5 are derived from synthetic human TGFα.
[d] NR, no reaction, ND, not determined.
[e] 39–11 derived from synthetic rat TGFα.

Conclusion

Our results show that the major immunodominant epitopes of human TGFα are located at the amino terminus (residues 1–9) and the β-turn of the second disulfide loop domain (residues 21–31). This conclusion is derived from results obtained by two different approaches to systematic mapping of all possible $^{\bullet}$ continuous or linear epitopes occurring in the hTGFα sequence.

Acknowledgements

This work was in part supported by Triton Biosciences, Inc and the U.S. Public Health service Grant CA-36544.

References

1. DeLarco, J.E. and Todaro, G.J., Proc. Natl. Acad. Sci. U.S.A., 75 (1978) 4001.
2. Ozanne, B., Fulton, J. and Kaplan, P.L., J. Cell Physiol., (1982) 163.
3. Moses, H.L. and Robinson, R.A., Fed Proc., (1982) 3008.
4. Marquadt, H., Hunkapiller, M., Hood, L.E., Twardzik, D.R., DeLarco, J. and Todaro, G.J., Science, 220 (1983) 1079.
5. Derynck, R., Roberts, A.B., Winkler, M.E., Chen, E.Y. and Goeddel, D.V., Cell, 41 (1984) 287.
6. Savage, C.R., Inagami, J. and Cohen, S., J. Biol. Chem., 235 (1972) 7612.
7. Stroobant, P., Rice, A.P., Gullick, W.J., Cheng, D.J., Kerr, I.M. and Waterfield, M.D., Cell, 42 (1985) 383.
8. Slamon, D.J., Clark, G.M., Wong, S.G., Levin, W.J., Ullrich, A. and McGuire, W.L., Science, 235 (1987) 177.
9. Geysen, H.M., Meloen, R.H. and Barteling, S.J., Proc. Natl. Acad. Sci. U.S.A., 81 (1984) 3998.
10. Vergera, U. and Tam, J.P., Peptide Res., 2 (1989) 134.

761

Conformation analysis of antigenic variant epitopes of the foot-and-mouth disease virus (FMDV)

Giuliano Siligardi[a], Alex F. Drake[b], Fred Brown[c], David Rowlands[c],
Nero Niccolai[d], William A. Gibbons[a], Colin H. James[a] and Paolo Mascagni[a]

[a]The School of Pharmacy, 29/39 Brunswick Square, London WC1N 1AX, U.K.
[b]Birkbeck College, 20 Gordon Street, London WC1H OAJ, U.K.
[c]Department of Virology, Wellcome Biotech, Beckenham, Kent, U.K.
[d]University of Siena, Department of Chemistry, I-53100 Siena, Italy

Introduction

The FMDV viral capsid consists of 60 copies of each of four proteins: VP-1, -2, -3, -4. Four FMDVs (A, B, C and USA) of serotype A-subtype 12 and the synthetic peptides corresponding to the major antigenic site [1] (VPI 141-160) are distinguishable in cross-neutralization and cross-precipitation tests [2]. According to these tests the four viruses and peptides were grouped into two distinct pairs: A with C and B with USA.

141		148		153		160	
Gly-Ser-Gly-Val-Arg-Gly-Asp-Ser-Gly-Ser-Leu-Ala-Leu-Arg-Val-Ala-Arg-Gln-Leu-Pro							A
———————————	Ser	————	Ser	————————			C
———————————	Leu	————	Pro	————————			B
———————————	Phe	————	Pro	————————			USA

In light of these serological tests, peptides A and USA were selected as representatives of the two classes for solution conformation studies. These were particularly relevant since the recently published [3] three-dimensional structure of the virus of serotype O revealed that, although the major antigenic site was exposed on the surface of the virus, its structure could not be specified because of poor definition of the electron density map in this region.

Results and Discussion

CD spectra in H_2O or the cryogenic substitute ethanediol–H_2O 2:1 (Fig. 1) showed that the structure of peptide USA was similar to that of poly-L-lysine [4]. At 20°C, no more than 20% of peptide USA (Fig. 1) was in the left-handed helical form (in poly-L-lysine it was 50%) [4]. Computer simulations [5] of the CD spectra of peptide A in ethanediol–H_2O (2:1) and at 20°C indicated a 50:20:30 (left-helix, α-helix, β-turn I/III-3_{10}) [6] ratio. Low temperatures showed an increase in the α-helical content with a 44:28:28 ratio present at −100°C. At room temperature and in TFE, a solvent known for supporting H-bonded

762

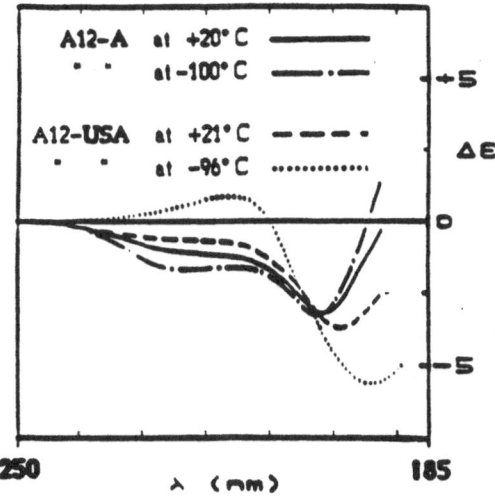

Fig. 1.   *CD spectra of peptides A and USA in ethanediol–H_2O 2:1.*

structures [7], the CD spectra of peptide A and USA were qualitatively similar with α-helix and 3_{10}-helix contributions. At lower temperatures the differences between the two peptides were emphasized. The CD intensity of peptide A at 220 m doubled on cooling from +40° to –33°C as the fraction of α-helix component increased, whereas peptide USA was less temperature sensitive. Computer simulations [6] indicated that, at +20°C, the ratio of left-helix, α-helix, β-turn I/III was 37:33:30 and that there was a substantial γ-turn [8] of the α-helix

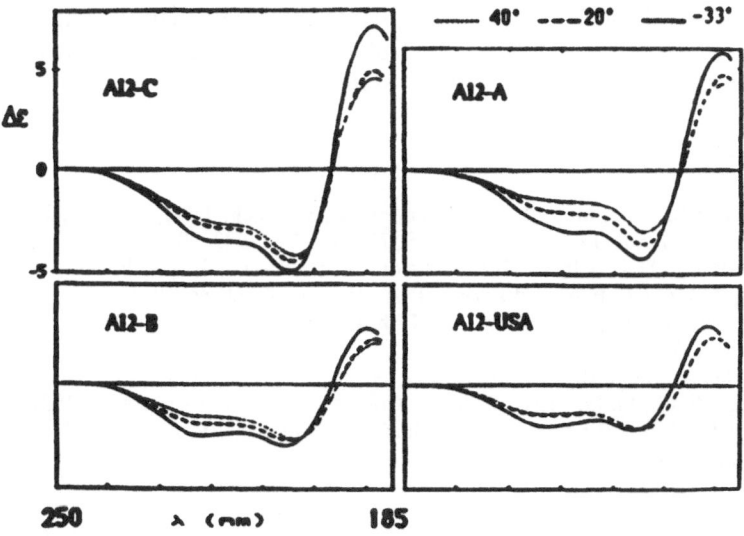

Fig. 2.   *CD spectra of peptides A, B, C and USA in TFE.*

763

A12–A

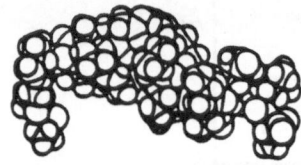

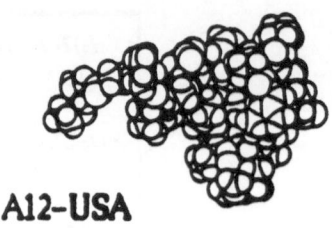

A12–USA

Fig. 3.   MD simulations for peptides A (left) and USA (right).

component. When peptides B and C were studied in TFE (Fig. 2), the CD spectra showed that the four peptides clearly fell into the same two groups deduced from the serological tests. For peptide A, NMR temperature studies indicated that the N-H of residues 150, 151, 152, 153 and 157, 158, 159 were likely to be H-bonded. J(NH-α) coupling constants and NH-NH NOEs between adjacent residues were consistent with three highly structured regions (140–145 left-handed helix; 146–153, α-helix; and 154–159, 3_{10}-helix. However, peptide USA, which had one less α-helical turn (Pro replaces Leu[153]), showed an additional H-bond for Arg[154] and Ala[152] NH Pro[153δ] and Arg[154α]-Pro[153δ] NOEs that were consistent with a 'γ-turn'. The NMR parameters were used as constraints for computer molecular modeling and energy minimization calculations. The structures of the two peptides (Fig. 3) were then refined by molecular dynamics simulations over 10 ps.

Conclusion

The hypothesis stemming from immunological studies that the four viruses owe their serological differences to conformational changes attributable to mutations of positions 148 and 153 of the VP_1 protein were correlated and confirmed by extensive CD and NMR studies of all four peptides.

References

1. Bittle, J.L., Houghten, R.A., Alexander, H., Shinnick, T.M., Sutcliffe, J.G., Lerner, R.A., Rowlands, D.J. and Brown, F., Nature, 298 (1982) 30.
2. Rowlands, D.J., Clarke, B.E., Carroll, A.R., Brown, F., Nicholson, B.H., Bittle, J.L., Houghten, R.A. and Lerner, R.A., Nature, 306 (1984) 694.
3. Acharya, R., Fry, E., Stuart, D., Fox, G., Rowlands, D. and Brown, F., Nature, 337 (1987) 709.
4. Drake, A., Siligardi, G. and Gibbons, W.A., Biophys. Chem., 31 (1988) 143.
5. Siligardi, G., Ph.D. thesis, Birkbeck College, University of London, (1989).
6. Woody, R.W., Peptides, Polypeptides and Proteins, Blout, Bovey, Goodman and Lotan, New York, 1974, p. 338.
7. Woody, R.W., In Udenfriend, S. Meienhofer, J. and Hruby, V.J. (Eds.) The Peptides – Analysis, Synthesis and Biology, Vol. 1, Academic Press, New York, 1985, p. 5.
8. Rose, G.D., Gierasch, L.M. and Smith, J.A., Adv. Protein Chem., 37 (1985) 1.

Fine-specificity of antisera raised against a multiple antigenic peptide from foot-and-mouth disease virus

W.M.M. Schaaper[a], Y.-A. Lu[b], J.P. Tam[b] and R.H. Meloen[a]

[a]*Central Veterinary Institute, Edelhertweg 15, 8219 PH Lelystad, The Netherlands*
[b]*The Rockefeller University, 1230 York Avenue, New York, NY 10021, U.S.A.*

Introduction

For an optimal production of antibodies against low molecular weight synthetic peptides, large carrier proteins like KLH or combinations with lipopeptides, for instance, have been used. An attractive alternative is a peptide bound to a branching lysine core [1]. Three successive lysine couplings lead to an octameric structure of the peptide. This multiple antigenic peptide (MAP) can be synthesized directly in SPPS, and the well-defined high molecular weight product can be used as immunogen without further conjugation procedures.

Results and Discussion

In this study, antibodies were raised in rabbits against a MAP containing the sequence VP1(141-160) of foot-and-mouth disease virus (FMDV) type O, as well as against VP1(141-160)-Cys coupled by MBS or GDA with KLH. The MAP induced a very high antipeptide response (see serum dilutions in PEPSCAN, Table 1); the peptides conjugated with KLH induced a lower response. However, neutralization titers of all antisera were similar.

We tested the specificity of the antisera in PEPSCANs. In the PEPSCAN, we used overlapping peptides from the region 134–165 of VP1 of FMDV type O,

Table 1 *Titers of rabbit antisera, 8 weeks after immunization*

No.	Antiserum	PEPSCAN[a]	PE[a,b]	MNT[a,c]	Core sequences[d]
a	MAP[e]	4.3	4.4	1.5	VPN
b		5.0	5.0	2.1	VPN
c	peptide[f]/MBS/KLH	3.5	4.1	1.8	NLRGDL, DLQVLA, VART
d		3.0	4.3	2.1	VPN, RGDLQVL, TLP
e	peptide[f]/GDA/KLH	4.0	4.8	2.7	PNLR, DLQVLA
f		3.5	4.4	2.4	DLQVL, VARTLP

[a]titers are expressed as – log(serum dilution).
[b]Peptide ELISA.
[c]Micro Neutralization Test.
[d]Shortest reactive peptides from PEPSCANs.
[e]Ac-VP1(141-160)$_8$-Lys$_4$-Lys$_2$-Lys-βAla.
[f]Ac-VP1(141-160)-Cys-amide.

ranging from tri- to nonapeptides (Fig. 1 shows scans of octapeptides). The results reveal that a large population of antibodies raised against the MAP is specific to the 'free' N-terminal amino acids, and that both rabbits seem to give a similar response (Fig. 1a,b). In contrast, antibodies raised against the peptides conjugated with KLH react with different sequences, and each individual rabbit gives a different response (Fig. 1c,d or 1e,f). Thus, the MAP produced a better controlled antipeptide response.

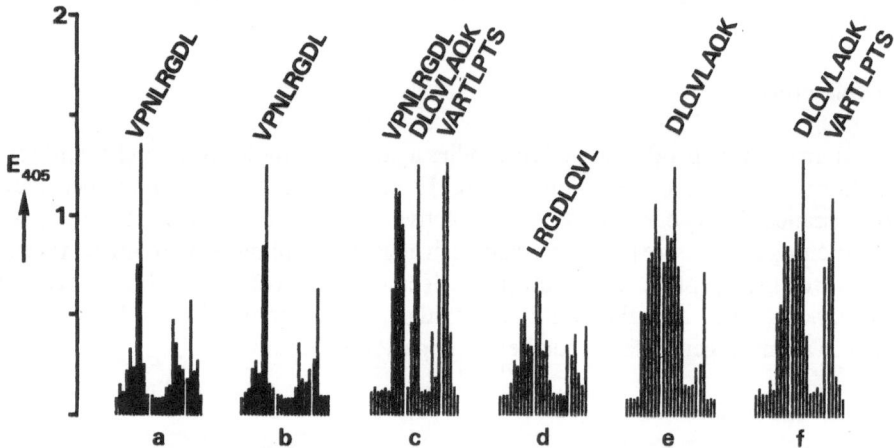

Fig. 1. *PEPSCANs of antipeptide sera. Vertical lines represent extinctions at 405 nm in the antibody binding ELISA. Each group of lines represents a set of overlapping octapeptides synthesized on polyethylene rods and used in the ELISA. The sequence scanned VP1(134-165): CRYNRNAVPNLRGDLQVLAQKVARTLPTSFNY (FMDV, type 0) a and b: anti-MAP; c and d: anti-VP1(141-160)Cys/MBS/KLH; e and f: anti-VP1(141-160)Cys/GDA/KLH.*

In earlier experiments [2], we found that antiviral antibodies reacted with the sequences GDLQVL and DLQVLA. Similar reactivities were found in the anti-peptide/KLH sera (Fig. 1c,d,e,f) but not in the anti-MAP sera. Therefore, we assume that the antibodies raised against the MAP block the RGD cell-attachment site of FMDV by reaction with the sequence VPN, which is very close to the RGD sequence. A careful selection of the sequence to be used in the MAP may result in antisera with even higher virus-neutralizing activity.

References

1.   Tam, J.P., Proc. Natl. Acad. Sci. U.S.A., 85 (1988) 5409.
2.   Meloen, R.H. and Barteling, S.J., Virology, 149 (1986) 55.

Synthesis of *Agaricus bisporus* metallothionein (MT) and related peptides and examination of their heavy metal-binding properties

Yasuhiro Nishiyama[a], Yutaka Matsuno[a], Yoshio Okada[a], Kyong-Son Min[b], Satomi Onosaka[b] and Keiichi Tanaka[b]

[a]Faculty of Pharmaceutical Sciences and [b]Faculty of Nutrition, Kobe-Gakuin University, Nishi-ku, Kobe 673, Japan

Introduction

Metallothioneins (MTs) are low-molecular-weight, cysteine-rich and heavy metal-binding proteins. They participate in heavy metals (Zn, Cu) metabolism and detoxification of heavy metals (Cd, Hg), although their precise biological role is not well understood. To help clarify MTs' functions, we synthesized various kinds of MTs and related peptides, and examined their heavy metal-binding properties. This paper deals with the synthesis of *Agaricus bisporus* MT and related peptides by conventional solution method, and examination of their heavy metal-binding properties.

Results and Discussion

A. bisporus MT consists of 25 amino acid residues (Fig. 1). The amino acid sequence is similar to the N-terminal region of mammalian MTs [1]. For the construction of the pentacosapeptide, 7 kinds of peptide fragments were prepared using newly developed β-2-adamantyl aspartate [2] in order to suppress the imide formation. Final deprotection was achieved by HF method. Metal-binding activities of various peptides obtained were assessed by measuring the increase in absorbance of mercaptide at 250 nm or at 265 nm as a function of the concentration of Cd^{2+} ($CdCl_2$) or Cu^{2+} ($CuCl_2$) and Cu^+ [$Cu(CH_3CN)_4ClO_4$], respectively.

H-Gly-Asp-Cys-Gly-Cys-Ser-Gly-Ala-Ser-Ser-Cys-Thr-Cys-Ala-Ser-Gly-Gln-Cys-Thr-Cys-Ser-Gly-Cys-Gly-Lys-OH

Fig. 1. Structure of Agaricus bisporus *metallothionein.*

Metal-binding properties of various peptides are illustrated in Fig. 2a,b,c. Their Cu^+ or Cu^{2+}-binding properties were similar to each other (Fig. 2a and b), whereas Cd^{2+}-binding properties of peptides were fairly dependent on their structures (Fig. 2c). These results are compatible with those in the case of *Neurospora crassa* MT [3] and would provide an answer to the role of the additional C-terminal residues in further evolved MTs.

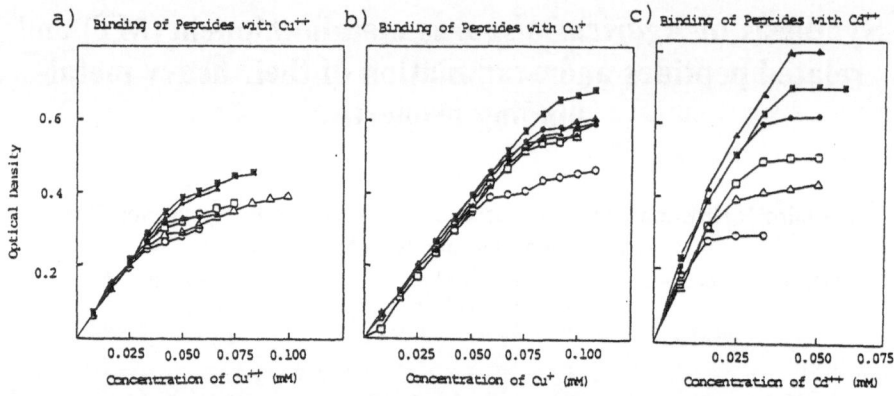

Fig. 2. Binding properties of peptides with heavy metals (a) with Cu^{2+}, (b) with Cu^+, (c) with Cd^{2+}. Peptide, 0.15 mM as SH in 3 ml of Tris-HCl (10 mM, pH 7.0). (■) A. bisporus MT (1–25), (▲) (7–25), (□) (14–25), (o) (4–25), (●) (11–25), (△) (16–25).

References

1. Munger, K. and Lerch, K., Biochemistry, 24 (1985) 6751.
2. Okada, Y. and Iguchi, S., J. Chem. Soc. Perkin Trans., 1 (1988) 2129.
3. Matsuno, Y., Okada, Y., Nishiyama, Y., Tanaka, K., Onosaka, S. and Min, K.-S., In Ueki, M. (Ed.) Peptide Chemistry (1988), Protein Research Foundation, Osaka, 1989, p. 295.

768

Solid phase synthesis and immunological properties of cyclic peptides of the hemagglutinin of influenza virus

S. Plaué[a],*, S. Muller[a], J.P. Briand[a], S. Valette[b], M. Aymard[b] and
M.H.V. Van Regenmortel[a]

[a]I.B.M.C., 15, rue René Descartes, F-67000 Strasbourg, France
[b]Université Claude Bernard, Lyon, France

Introduction

In recent years, the potential use of synthetic peptides as a new generation of vaccines, has received considerable attention. With viruses such as hepatitis B, foot-and-mouth disease, poliovirus and influenza, it has been shown that synthetic peptides are capable of inducing partial protective immunity [1]. Influenza virus appeared to be an excellent model for studying the potential role of synthetic peptides in vaccination since the 3D structure of the hemagglutinin is known [2] as well as the location of the major antigenic sites [3]. In order to mimic a loop structure corresponding to site A, eight cyclic peptides derived from the region 139–147 of Strain V_3A and X31 were synthesized.

Results and Discussion

Chain elongation was performed using SPPS and cyclization was achieved with the peptide still attached to the resin prior to final cleavage. Ring closure was performed in two different ways providing peptides of different loop size. In the first case, the ring closure was made between the N-terminal amino group and the β-carboxyl group of an aspartic acid placed after residue 147; these peptides were called small loop peptides (SLP) (see Table 1). In the second case, ring closure was achieved by a succinimidyl linker between the N-terminus and the ϵ-amino function of an additional lysine placed at position 148; these peptides were called large loop peptides (LLP) (see Table 1). We studied the rate of cyclodimerization as a function of resin substitution (0.1 – 0.6 mmol/g), and we found that the proportion of cyclodimer increased from 8% to 19% in the crude product, while the rate of polymer remained stable. These results were obtained when DIC/HOBt was used for the cyclization reaction. However, when the BOP reagent was used for the cyclization (in the same conditions), up to 25% of polymeric material was found. In the case of N-α-Fmoc protection, we have observed that the blocking of the side chain of aspartic acid with benzyl ester led to 30% of an imide side product. When cyclohexyl ester was used for side-chain protection, no imide was detected.

---

*Present address: NEOSYSTEM S.A., 7, rue de Boulogne, F-67100 Strasbourg, France.

Table 1 *Formulae of cyclic analogs of site A of hemagglutinin of influenza virus corresponding to strains V_3A and X31.*

			139	147
(I)	V_3A	SLP	Cys-Lys-Arg-Gly-Pro-Gly-Ser-Asp-Phe-Asp-Tyr-NH$_2$	
(II)	V_3A (Ser139)	SLP	Ser-Lys-Arg-Gly-Pro-Gly-Ser-Asp-Phe-Asp-Tyr-NH$_2$	
(III)	V_3A	LLP	Cys-Lys-Arg-Gly-Pro-Gly-Ser-Asp-Phe-Lys-Tyr-NH$_2$ — CO-CH$_2$-CH$_2$-CO	
(IV)	V_3A (Ser139)	LLP	Ser-Lys-Arg-Gly-Pro-Gly-Ser-Asp-Phe-Lys-Tyr-NH$_2$ — CO-CH$_2$-CH$_2$-CO	
(V)	X31	SLP	Cys-Lys-Arg-Gly-Pro-Gly-Ser-Gly-Phe-Asp-Tyr-NH$_2$	
(VI)	X31(Ser139)	SLP	Ser-Lys-Arg-Gly-Pro-Gly-Ser-Gly-Phe-Asp-Tyr-NH$_2$	
(VII)	X31	LLP	Cys-Lys-Arg-Gly-Pro-Gly-Ser-Gly-Phe-Lys-Tyr-NH$_2$ — CO-CH$_2$-CH$_2$-CO	
(VIII)	X31(Ser139)	LLP	Ser-Lys-Arg-Gly-Pro-Gly-Ser-Gly-Phe-Lys-Tyr-NH$_2$ — CO-CH$_2$-CH$_2$-CO	

The ability of these peptides to react in ELISA with anti-virus antibodies was measured. Anti-peptide antibodies were raised in rabbits and tested for their ability to react with the viruses. An antiserum to whole virus (strain NT 60/68) recognized the cyclic peptides when they were conjugated to ovalbumin through the N-terminal tyrosine via bisdiazobenzidine, but not when they were conjugated or coupled through the cysteine[139] or lysine[140].

In ELISA inhibition experiments, we observed that the unconjugated SLP, but not the LLP, was able to inhibit the binding of anti-virus antibodies up to 60%. The capacity of the peptides to induce a protective immune response was tested in 10 mice (strain OF1) that received 4 injections at 10 days interval with 50 μg of conjugated or unconjugated cyclic peptides. After 6 weeks, the mice were challenged with an infective dose sufficient to kill 50% of the mice. The unconjugated SLP had little protective effect (20%), whereas the LLP and SLP, when coupled through the tyrosine, protected 70% and 80% of the mice respectively.

References

1. Arnon, R., Synthetic Vaccines, Vol. I and II, CRC Press Inc., Boca Raton, FL, 1987.
2. Wilson, I.A., Skehel, J.J. and Wiley, D.C., Nature, 289 (1981) 366.
3. Wiley, D.C., Wilson, I.A. and Skehel, J.J., Nature, 289 (1981) 373.

Biologically active retro-inverso analogs of thymopentin

Alessandro Sisto[a], Sabina Mariotti[a], Antonio Groggia[a], Giordana Marcozzi[a], Luigi Villa[b], Luciano Nencioni[b], Sergio Silvestri[b], Antonio S. Verdini[a] and Antonello Pessi[a]

[a]Peptide Synthesis Laboratory, Sclavo Spa & Bioscience Department, Eniricerche Spa, I-00015 Monterotondo-Rome, Italy
[b]Sclavo Research Center, Via Fiorentina 1, I-53100 Siena, Italy

Introduction

The pentapeptide thymopentin [TP5: Arg-Lys-Asp-Val-Tyr (**1**)], corresponding to the active fragment 32–36 of the thymic hormone thymopoietin, retains all the biological activities of the parent polypeptide [1]. TP5 has shown immunomodulatory effects on several immune system components: it promotes differentiation in immature prothymocites and exerts a regulatory influence on more mature effector T-cells in immune disbalance states.

TP5 has already found clinical use in the treatment of immune-depressed conditions, autoimmune diseases and as adjuvant in vaccinations; the clinical potential of TP5 is, however, reduced by the critical design of the therapeutic regimen, due to the difficult determination of effective dosing [2]. This result is probably related to the fast breakdown by plasma and membrane peptidases: the half-life of TP5 in human plasma is about 1 min [3].

We applied the approach of retro-inverso modification of the peptide bond to increase the stability of TP5 toward enzymatic degradation and investigated the effect of a prolonged half-life on immunostimulatory activity.

Results and Discussion

The synthesis of the peptides was performed according to previously described procedures [4] or by means of a new methodology based on the use of Meldrum's acids, as activated derivatives of the corresponding malonic acids, for the condensation with the gemdiamino compound [5].

The dose eliciting the maximum immunostimulatory activity of the analogs in the plaque-forming cells assay and the relative potency expressed as percent of the control response of mice receiving the antigen alone is shown in Table 1.

The half-life ($t_{1/2}$) and the degradative pattern of the analogs in human plasma were monitored chromatographically.

The increase in immunostimulating effect parallels the augmented half-life in plasma. The results agreed with the study of Tischio et al. [3] who located the main sites of enzymatic breakdown at the bonds adjacent to Arg and Lys, resulting in inactive compounds, while the hydrolysis of the Val-Tyr bond

Table 1 *Biological activity and enzymatic stability*

	Peptide	Immunostimulating activity		$t_{1/2}$ (min)
		(dose in ng)	% control	
	Control	–	100	–
1	Arg-Lys-Asp-Val-Tyr	100	137	1.5
2	gArg-(R,S)mLys-Asp-Val-Tyr	100	237	22
3	Arg-Lys-Asp-gVal-(R,S)mTyr	100	190	2.5
4	Arg-Lys-gAsp-(R,S)mVal-Tyr	100	132	10
5	gArg-(R,S)mLys-Asp-gVal-(R,S)mTyr	1	231	>360
6	gArg-(R,S)mLys-gAsp-(R,S)mVal-Tyr	100	196	>360
7	gArg-D-Lys-D-Asp-D-Val-(R,S)mTyr	100	220	>360
8	gArg-(R,S)mLys-Asp	100	137	>360

proceeded more slowly. As expected, the retro-inverso modification at the N-terminal amide bond (2) slowed the enzymatic cleavage yielding an analog with prolonged half-life in human plasma and an increased activity. Moreover, 2 showed an alteration of its degradative pathway: the more susceptible site becoming the Asp^3-Val^4 bond. This breakdown gave rise to 8, as the main, still active, metabolite. Analog 6, bearing a second retro-inverso modification at the Asp^3-Val^4 bond to block the enzymatic activity at this site, resulted in the most potent compound, over 200-fold more potent than the parent compound.

Further studies are planned to test the immunomodulatory effects of these analogs in several pathological states.

References

1. Goldstein, G., In Goldstein, G., Bach, J.F. and Wigzell, H. (Eds.) Immune regulation by characterized polypeptides, A.R. Liss, New York, 1987, p. 51.
2. Bolla, K., Duchateau, J., Delespesse, G. and Servais, G., Int. J. Clin. Pharm. Res. IV, 6 (1984) 431.
3. Tischio, T.P., Patrick, J.E., Weintraub, H.S., Chasin, M. and Goldstein, G., Int. J. Pept. Prot. Res., 14 (1983) 479.
4. Berman, J.M. and Goodman, M., Int. J. Pept. Prot. Res., 23 (1984) 610.
5. Pinori, M., Centini, F. and Verdini, A.S., It. Patent. no. 23098 A/88.

The immune response of haptenic peptides is dictated by the specific attachment position and carrier protein

Clemencia Pinilla*, Jon Appel and Richard A. Houghten

Scripps Clinic and Research Foundation, 10666 N. Torrey Pines Road, La Jolla, CA 92037, U.S.A.

Introduction

A sequence derived from the hemagglutinin protein of the influenza virus, YPYDVPDYASLRS (HA1 : 98–110) was used to study peptide-specific immune responses. Two different proteins, KLH and tetanus toxoid (TT), were used as carrier molecules for this peptide attached to the carrier through an additional cysteine at either the N-terminal or C-terminal position. This resulted in four different conjugates. These conjugates were used to generate polyclonal antisera in rabbits in order to study the effect of the attachment point and/or carrier protein on the immune response of haptenic peptides.

To study the effect of peptide conjugation efficiency on the immune response, three concentrations of the above peptide were conjugated through its C-terminal cysteine to KLH and antisera raised separately in rabbits against each conjugate. These antisera were tested for their titers against the peptide, and the antigenic determinants of the peptide were located using a series of omission analogs of the original peptide. For the preparation of the conjugates, the heterobi-functional, crosslinking reagent, SPDP, was used [1]. This reagent allows the monitoring of both the protein modification reaction and the peptide conjugation reaction.

Results and Discussion

The antisera raised against the four different peptide-carrier immunogens were assayed by ELISA using omission analogs of the peptide to locate the antigenic determinants (Table 1). Each of the antisera was found to recognize the native protein from which this peptide was derived. The position of the cysteine in the peptide, which is used to conjugate the peptide to the carrier protein, plays an important role in the overall specific peptide immune response. This can be seen by the location of the antigenic determinants recognized by the antisera. No significant difference, however, was found in the immune response as a result of the two different proteins used as carriers. The antisera for the two N-terminal cysteine conjugates recognized the same antigenic determinants. The two antisera

---

*Present address: Torrey Pines Institute for Molecular Studies, 10955 John Jay Hopkins Drive, San Diego, CA 92121, U.S.A.

Table 1 *Identification of antigenic determinants using antisera raised against different peptide-carrier conjugates*

Carrier protein	Location of peptide attachment to carrier	Peptide[a]
KLH	C-terminal	YPYDVPDYASLRS
	N-terminal	YPYDVPDYASLRS
TT	C-terminal	YPYDVPDYASLRS
	N-terminal	YPYDVPDYASLRS

[a]Residues in the antigenic determinant are underlined.

for the C-terminal cysteine conjugates, while not identical, shared four common residues. The antisera raised against the C-terminal cysteine peptide conjugated to TT had a 10-fold lower titer to the peptide than the other three conjugate antisera.

A tritiated form of this C-terminal cysteine peptide facilitated the quantitation of the amount of peptide conjugated to KLH. Three concentrations (40, 100 and 200 μg peptide/mg KLH) were prepared. The resulting antisera displayed significantly different titers to the peptide (Fig. 1). The difference in the observed titers was a direct result of the amount of peptide injected (all animals were injected with identical amounts of KLH). The low titers found when 40 μg of haptenic peptide was injected was not the result of poor immunization or low responsiveness by those animals, since the titers to KLH were the same to the 100 and 200 μg peptide antisera. Also, these antisera did not recognize a clear antigenic determinant when assayed with omission analogs by ELISA, whereas the other two antisera recognized a clear, four residue antigenic determinant (data not shown).

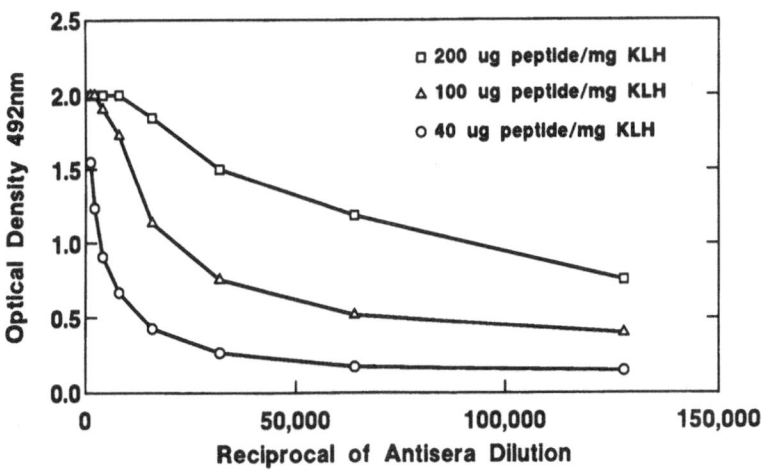

Fig. 1. *Antisera titer curve for Ac-YPYDVPDYASLRS-NH₂.*

775

C. Pinilla, J. Appel and R.A. Houghten

We conclude that the immune response to haptenic peptides is dictated by their attachment position regardless of the carrier protein used and that antisera having low titers to haptenic peptides, are often the result of an insufficient amount of the peptide actually coupled to the carrier protein.

References

1.   Carlsson, J., Drevin, H.L. and Axen, R., Biochem. J., 173 (1978) 723.

Peptidic and pseudopeptidic inhibitors of erythrocyte invasion by *Plasmodium falciparum* merozoites

Roger Mayer[a], Isabelle Picard[a], Philippe Lawton[b], Philippe Grellier[b], Joseph Schrével[b] and Michel Monsigny[a]

[a]Centre de Biophysique Moléculaire, CNRS, F-45071 Orléans, France
[b]Laboratoire de Biologie Cellulaire, Université de Poitiers, F-86022 Poitiers, France

Introduction

Malaria constitutes a major cause of morbidity in most tropical areas in the world. In view of its dramatic worldwide resurgence due to the spreading of drug-resistant *P. falciparum* parasite strains and to the absence of antimalarial vaccines, we developed a new therapeutic approach based upon the mechanism of the molecular interactions between erythrocytes and merozoites. By use of peptidic substrates with a fluorogenic 3-amino-9-ethylcarbazole group (AEC), a specific parasite neutral endopeptidase has been detected, isolated and characterized [1]. This enzyme, which specifically recognizes the Val-Leu-Gly-Lys (VLGK) sequence, is a cysteine protease assumed to play a key role in the invasion process of erythrocytes [2]. In order to block this crucial step during malarial infection, various specific peptidic and pseudopeptidic inhibitors were synthesized. In this report, we present preliminary data on their inhibitory activity on the purified protease and on the reinvasion of erythrocytes by merozoites in in vitro experiments.

Results and Discussion

Peptidic inhibitors were synthesized by the classical DCC method. Their N-terminal end was acylated by a hydrosolubilizing gluconoyl group (GlcA) [3] that also prevents them from aminopeptidase degradation. Reversible inhibitors were prepared by replacing the AEC group of the GlcA-Val-Leu-Gly-Lys-AEC substrate by ethyl- or diethylamine, (R)-2-amino-1-butanol (AB) or 2-amino propanediol (AP), and *N*-methyl glycine (Sar) or 2-amino-L-isobutyric acid. In order to increase the enzyme resistance of the peptidyl-ethylamide substrate, its amide bond, at the proteolytic cleavage site, was modified by a pseudopeptidic $\psi[CH_2NH]$ bond. A peptidyl-nitrile was also synthesized.

As the *P. falciparum* neutral endopeptidase has been characterized as a thiol protease [2], we also synthesized a peptidyl-diazomethyl ketone and a peptidyl-fluoromethyl ketone, derivatives known as specific inactivators of this class of enzyme.

All the substrates and inhibitors were purified by preparative RPHPLC.

777

Experiments were performed with cultured *P. falciparum* FRC3 Gambian strain synchronized at the schizont stage. The purification of the specific neutral endoprotease was achieved as already described [1].

The activity of the competitive inhibitors and suicide substrates was determined in the presence of purified protease. With compounds GlcA-VLGK-R, where R is ψ[CH$_2$NH]Et, CHN$_2$, Sar-OH, CH$_2$F or NHEt, the K_i and the IC$_{50}$ values increased from 0.05 to 0.5 mM and from 0.08 to 0.5 mM, respectively.

In the in vitro tests of erythrocytes reinvasion by *P. falciparum* merozoites, the % of inhibition was expressed as % inhibition $= 100 - (R/R_o \times 100)$, where R is the number of ring stage that appeared in the presence of inhibitor, and R_o the number in the absence of inhibitor. At a 1 mM concentration, GlcA-VLGK-R compounds, where R is ψ[CH$_2$NH]Et, AB, AP, Sar-OH or NHEt, the percentage of inhibition was in the range from 100% to 72%; 0.1 mM of GlcA-VLGK-ψ[CH$_2$NH]Et still gave 60% inhibition.

On the basis of the in vitro data, and in order to increase the apparent affinity of these inhibitors for the enzyme by an avidity effect [4], the synthesis of hydrosoluble macromolecular poly-inhibitors expected to have high in vivo lifetime and efficiency is in progress.

References

1. Grellier, P., Picard, I., Bernard, F., Mayer, R., Heidrich, H., Monsigny, M. and Schrével, J., Parasitol. Res., 75 (1989) 455.
2. Schrével, J., Grellier, P., Mayer, R. and Monsigny, M., Biol. Cell, 64 (1988) 233.
3. Monsigny, M. and Mayer, R., (1987) U.S. Patent No. 4, 703, 107.
4. Derrien, D., Midoux, P., Petit, C., Nègre, E., Mayer, R., Monsigny, M. and Roche, A.-C., Glycoconjugate J., 6 (1989) 241.

Murine major histocompatability complex polymorphism and T cell specificities: Mapping I-A^{d,k} and I-E^{d,k} restricted T-cell epitopes of staphylococcal nuclease

Zhuoru Liu[a,c] and John A. Smith[a,b,d],*

[a]*Department of Molecular Biology and* [b]*Department of Pathology, Massachusetts General Hospital, Boston, MA 02114, U.S.A.*
[c]*Department of Genetics and* [d]*Department of Pathology, Harvard Medical School, Boston, MA 02114, U.S.A.*

Introduction

Major histocompatability complex (MHC) genes are exceedingly polymorphic. Such genes code for receptors for 'processed' peptides that, after they are bound, are displayed to specific T-cell receptors. Balb/c (I-A^d/I-E^d) and A/J (I-A^k/I-E^k) mice were immunized with purified recombinant staphylococcal nuclease (Nase), and murine T-cell hybridomas were prepared and screened for T-cell proliferation when stimulated by Nase or a Nase peptide, according to established procedures [1].

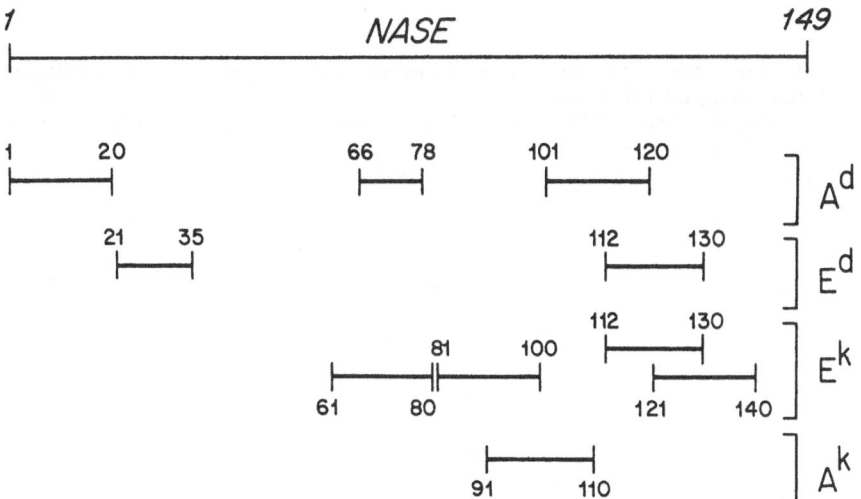

Fig. 1. Location of I-A^{d,k}- and I-E^{d,k}-restricted T-cell antigenic sites in staphylococcal nuclease.

---

*To whom correspondence should be addressed.

Results and Discussion

Among 167 hybrids (2 fusions) derived from Balb/c mice, 86% were I-A^d-restricted, and a nested set of synthetic Nase peptides identified three T-cell epitopes: Nase(61–80)(84%), Nase(1–20) (15%), and Nase(101–120) (1%) (Fig. 1). Additional Nase peptides were used to identify two discrete I-E^d-restricted epitopes: Nase(21–35) (83%) and Nase(112–130) (17%).

Among 80 hybrids (2 fusions) derived from A/J mice, 96% were I-E^k-restricted (77 hybridomas), and a nested set of synthetic peptides identified four T-cell epitopes: Nase(81–100) (70%), Nase(112–130) (21%), Nase(121–140) (6%), and Nase(61–80) (3%). Four I-A^k-restricted clones recognized Nase(91–100).

Finnegan et al. [2] have also localized an I-E^k epitope to residues 81–100 of Nase, as well as localized an I-A^b epitope to residues 91–110.

These data regarding the precise location of these disparate T-cell antigenic sites in Nase provide a starting point from which to dissect certain of the biochemical mechanisms that underlie protein antigen processing [3] and presentation [4].

Acknowledgements

Supported by a grant from Hoechst Aktiengesellschaft (F.R.G.).

References

1. Lai, M.-Z., Ross, D.T., Guillet, J.-G., Briner, T.J., Gefter, M.L. and Smith, J.A., J. Immunol., 139 (1987) 3973.
2. Finnegan, A., Smith, M.A., Smith, J.A., Berzofsky, J.A., Sachs, D.H. and Hodes, R.J., J. Exp. Med., 164 (1986) 897.
3. Unanue, E.R. and Allen, P.M., Science, 236 (1987) 551.
4. Guillet, J.-G., Lai, M.-Z., Briner, T.J., Buus, S., Sette, A., Grey, H.M., Smith, J.A. and Gefter, M.L., Science, 235 (1987) 865.

Mapping of epitopes and agretopes on an HLA A$_2$ synthetic hexadecapeptide

Fabrice Cornille[a], Philippe Laugaa[a], Frédéric Martinon[b], Marie-Claude Fournié-Zaluski[a], Ada Prochnicka-Chalufour[c], Elisabeth Gomard[b], Jean-Paul Levy[b] and Bernard P. Roques[a]

[a]Chimie Organique, U266 INSERM, UA498 CNRS, UER des Sciences Pharmaceutiques et Biologiques, 4 avenue de l'Observatoire, F-75006 Paris, France
[b]Laboratoire d'Immunologie et d'Oncologie des Maladies Rétrovirales, U152 INSERM, UA628 CNRS, Hôpital Cochin, 27 rue du Fbg. Saint Jacques, F-75014 Paris, France
[c]Unité d'Informatique Scientifique, Institut Pasteur, 25 rue du Docteur Roux, F-75015 Paris, France

Introduction

The class-I molecules of the major histocompatibility complex (H-2 in mouse, HLA in humans) are membrane proteins composed of a polymorphic heavy chain associated with β2 microglobulin. Extensive studies have shown that class-I molecules present peptides derived from processed antigens to the receptor of cytolytic T cells [1,2]. Particularly in the H-2^d haplotype, the synthetic HLA-A2 170–185 peptide (RYLENGKETLQRTDAP) can be recognized on K^d-bearing target cells by K^d-restricted cytolytic T-cell clones specific for the 170–185 region [3]. This peptide bears determinants which interact either with the TCR ('epitopes') or with the MHC class I molecules ('agretopes').

Using synthetic substituted analogs of this peptide, we have investigated the agretopic and the epitopic residues. The structural requirements of our CTL clones have been elucidated from a combination of biological studies, molecular modeling, and comparative studies by ^{1}H NMR spectroscopy (400 MHz) of the native peptide and its substituted D-Ala[175] analog.

Results and Discussion

Several substituted analogs of HLA A2 170–185 were tested on two CTL clones lytic for P 815 cells presenting this peptide. By cytolysis and competition tests, we have distinguished between residues important for presentation by H-2 K^d from those important for CTL recognition, and those whose substitution had no effect. The results show that Glu[173] and Glu[177] are epitopic determinants for the clone 332/2A, and only Glu[177] for the clone 223/14. For both clones, Thr[178] and Gln[180] belong to the agretopic region, and the last four residues Thr[182] to Pro[185] are not implicated in the recognition phenomenom. Then we substituted Gly[175] by D-Ala[175] in this peptide in order to verify the importance of relative spatial disposition of Glu[173] and Glu[177] side chains. In this substituted

781

analog, only the epitope is modified because this peptide is inactive in cytolysis test, but competition test reveals its good affinity for H-2 K^d.

Then NMR experiments and molecular modeling were used to investigate the agretopic and epitopic components at the molecular level. ^1H NMR experiments were performed in H_2O and D_2O using two-dimensional homonuclear spectroscopy (HOHAHA, relayed COSY). Results obtained from titration of the HLA A2 170–185 protons were in good agreement with those obtained by molecular modeling of the peptide conformation derived from the alpha carbons coordinates of the published structure of the HLA A2 molecule [4]. This peptide has a helical structure from Arg^{170} to Arg^{181}, followed by a random coil part from Thr^{182} to Pro^{185}. In the helical part, which is the only important region for CTL and H2 K^d recognition, the two epitopic determinants Glu^{173} and Glu^{177} are on the same side of the peptide. On the opposite side there are three hydrophobic side chains, Tyr^{171}, Leu^{172} and Leu^{179}, surrounded by Thr^{178} and Gln^{180}.

^1H NMR experiments were then performed on the D-Ala^{175}-substituted analog. We found that the helical and linear part is conserved, but the most important difference occurs at Glu^{177} level. In the native peptide, the side chain of Glu^{177} was pointing up, far from other side chains. In contrast, in the D-Ala substituted analog there is a spatial proximity between the Asn^{174} and Glu^{177} side chains. So the relative movement of Glu^{177} and the consecutive steric hindrance of Asn^{174} modify the epitopic determinant, aborting the CTL recognition without modification of the agretope.

References

1. Babbitt, B.P., Allen, P.M., Matsueda, G., Haber, E. and Unanue, E.R., Nature, 317 (1985) 359.
2. Bouillot, M., Choppin, J., Cornille, F., Martinon, F., Papo, T., Gomard, E., Fournié-Zaluski, M.C. and Levy, J.P., Nature, 339 (1989) 473.
3. Martinon, F., Cornille, F., Gomard, E., Fournié-Zaluski, M.C., Abastado, J.P., Roques, B.P. and Levy, J.P., J. Immunol. 142 (1989) 3489.
4. Bjorkman, P.J., Saper, M.A., Samraoui, B., Bennett, W.S., Strominger, J.L. and Wiley, D.C., Nature, 329 (1987) 506 and 512.

Session VII
Protein/DNA interactions

Chair: Robert Schwyzer
Swiss Federal Institute of Technology
Zürich, Switzerland

Targeting DNA sites with simple metal complexes

Jacqueline K. Barton*

Department of Chemistry, Columbia University, New York, NY 10027, U.S.A.

Introduction

In my laboratory we have been interested in exploring those factors which govern the specific recognition of DNA sites. How can one target chemistry to a particular site along the DNA helix? Proteins which bind to DNA and regulate gene expression are able to discriminate among DNA sites with high specificity. Restriction enzymes, whose discovery marked the beginnings of recombinant DNA technology, bind to their 4–8 base pair recognition sequences of DNA with 10^4 times greater affinity than to other sequences. In each case, it is likely that a discrete peptide domain is responsible for the site-specific recognition. Yet the principles which govern this recognition are still far from well understood. In order to begin to design peptide mimics or other small molecular structures which bind DNA sites with high specificity, so as to develop rationally based DNA-binding pharmaceuticals, we need to determine those factors which contribute to the recognition of a particular site along the DNA helix.

What are the factors which are likely to be important in the discrimination of a particular site? Originally it was proposed that some sort of code might exist, involving the matching of individual amino acid residues, or perhaps pairs of residues, to the different nucleic acid bases [1]. Through a series of hydrogen bonding interactions, one might then imagine a one-dimensional readout of the particular DNA sequence by the DNA-binding peptide. It has, however, become quite clear, from the few crystallographic determinations of proteins bound to specific DNA sequences, that such a code does not exist [2,3]. Indeed double helical DNA is not one-dimensional. Instead DNA is quite polymorphic in its 3D structure, [4,5] and the variations in structure along the strand, both those that are subtle and perhaps not so subtle, provide a basis also to discriminate among different DNA sites. In addition to hydrogen bonding, one might then consider that another basis for site recognition could be found in both the stabilizing van der Waals interactions and destabilizing steric clashes which result from associations of peptide and nucleic acid structures one with another. Another component of recognition could therefore be considered to be that based upon the recognition of the local shape or conformation associated with a nucleic acid sequence.

*Present address: Division of Chemistry and Chemical Engineering, California Institute of Technology, Pasadena, CA 91125, U.S.A.

A major focus of our laboratory has been to explore the recognition of DNA sites based upon their local variations in shape [5]. Can molecules be designed which target DNA sites based upon their local conformation and, if so, with what level of selectivity? Can molecules so designed be applied also to probe the variations in structure that arise along the strand so as to develop some rules which govern the 3D structures associated with different nucleic acid sequences? From this perspective, we have designed a series of simple transition metal complexes of well-defined shapes and symmetries which target different local nucleic acid conformations. Complexes have been constructed which target the various conformations of double helical DNA, and even subtle variations within the most common B-form family. Remarkably, high levels of recognition, based solely upon considerations of shape, may be achieved.

Results and Discussion

Features of the metal complexes

There are several features which the transition metal complexes we will describe share and which have been proven to be extremely useful to this investigation. First, and importantly, the complexes all are coordinatively saturated octahedral complexes which are inert to substitution. The metal center does not coordinate directly to the polynucleotide. Instead it is the matching of the shape of the metal complex to that of the DNA which is important. The complexes, therefore, bind DNA through an ensemble of weak, non-covalent interactions. From that standpoint, another important feature shared by the metal complexes under study which is important is that the complexes are quite rigid in structure. Then, if we know some aspect of the binding mode and orientation of the complex on the DNA helix, we know in detail the orientation of all the atoms on the complex with respect to the double helical structure. In addition, another feature that appears important to establish specific interactions with DNA is that all the complexes are chiral. Thus the diastereomeric interactions of the complexes with the DNA helix provide a basis for site discrimination and means also to explore symmetry-dependent structures along the helix [6].

Perhaps most critical is that the complexes all contain a transition metal at their center. The metal serves several functions. First, as described above, it is the metal that defines the structure, symmetry, and shape of the molecule. But the metal also provides the means to probe the binding interaction. Ruthenium(II) complexes have been extremely valuable as spectroscopic probes for binding, owing to their intense metal to ligand charge transfer transitions and to the fact that these transitions are perturbed on DNA binding [7–9]. More recently, nickel(II) and chromium(III) analogs have been employed as paramagnetic broadening agents in ^{1}H NMR experiments [10]. Lastly we have taken advantage of the photoredox properties of cobalt and rhodium complexes to develop complexes which upon photoactivation cleave the DNA strand at their binding site [11,12]. This conversion of a DNA-binding molecule to a DNA-cleaving molecule has been extremely useful in analyzing the recognition characteristics of the various metal complexes. The cleavage reaction, in breaking the phosphodiester backbone at the binding site, sensitively marks the site of binding. This technique permits our detection of where the complex has bound along the strand to single nucleotide resolution using biochemical methods.

Shape-selective recognition

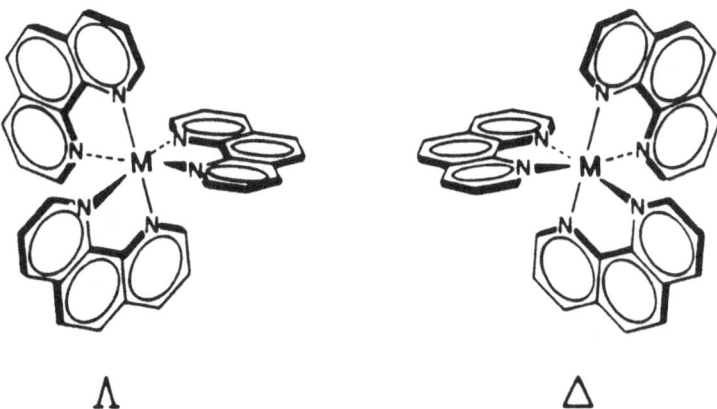

Λ Δ

Fig. 1. *The tris(phenanthroline)–metal complexes. Shown on the left is the Λ-isomer, a left-handed propeller-like structure, and on the right, the Δ-enantiomer, a right-handed propeller-like structure.*

Complexes were first designed which recognize and distinguish the A- and Z-conformations of DNA. Other complexes that have been prepared recognize substantial deviations in DNA tertiary structure and still others detect small variations in twist and tilt within an ostensibly B-form helix.

The compelexes we have explored for the most part represent derivatives of the parent tris(phenanthroline)-metal complex [6,13], shown in Fig. 1. As determined by photophysical [7,8] and NMR [10] studies, the tris(phenanthroline)-metal complexes bind to DNA predominantly through two non-covalent binding modes: (i) intercalation and (ii) surface or groove binding. The groove binding interaction, a common mode of association of small molecules with DNA, is likely stabilized by hydrophobic and electrostatic interactions and occurs in the minor groove of the helix. For this binding mode, the Λ-isomer, with a symmetry complementary to the helix, is preferred. The other binding mode, again a common association of small molecules and DNA, is an intercalative interaction, where likely one ephenanthroline ligand inserts partially and stacks in between the base pairs, anchoring the complex relative to the helix. For this binding mode it is the Δ-isomer that is favored and the intercalation appears to occur from the major groove of the helix.

Based upon the intercalative interaction, the probe for Z-DNA was derived. The chiral discrimination favoring the Δ-isomer for intercalative binding arises because of the matching of the symmetry of the metal complex to that of the helix. With the Λ-isomer bound intercalatively into a right-handed helix, there are steric clashes between the ancillary non-intercalated phenanthroline ligands, disposed in a left-handed fashion, and the sugar-phosphate backbone, having the right-handed helicity. Hence to probe Z-DNA, a lefthanded helix, the Λ-isomer would be an appropriate choice. The complex $Ru(DIP)_3^{2+}$ (DIP = 4,7-diphenylphenanthroline), with bulkier ancillary ligands to completely preclude binding of the Λ-isomer to a right-handed helix, is shown in Fig. 2. This complex, which binds avidly to Z-DNA and shows no detectable binding to the B-form, was our first spectroscopic probe for left-handed helical structures [14].

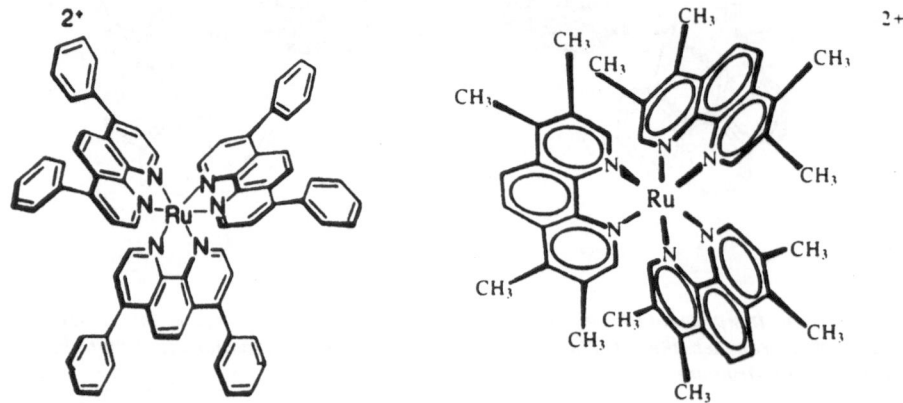

Fig. 2. *Probes for Z- and A-DNAs. Shown on the left is $Ru(DIP)_3^{2+}$. The Λ-isomer provides a spectroscopic probe for left-handed helical structures. $Ru(TMP)_3^{2+}$, a photoreactive probe for A-form helical structures, is shown on the right.*

The probe for A-form helices was derived, based upon a matching of shapes rather than of symmetries. The A-form helix is characterized by a shallow minor groove surface and an inaccessible major groove pulled deeply into the interior of the right-handed helix. The complex, $Ru(TMP)_3^{2+}$ (TMP = 3,4,7,8-tetramethylphenanthroline), also shown in Fig. 2, was designed so as to target preferentially the A-form helix [15,16]. The methyl groups, situated about the periphery of the phenanthroline ligands, limit intercalative stacking, owing to their bulk, and should promote surface binding, owing to their increased hydrophobicity. But the surface of the $Ru(TMP)_3^{2+}$ is simply too big to fit well against the grooved structure of a B-form helix. The surface is instead well-matched to the shallow A-form groove. Consistent with this model, preferential binding of $Ru(TMP)_3^{2+}$ to A-form polymers is indeed observed.

A striking example of the specificity that may be associated with shape-selective recognition is evident in photoactivated cleavage reactions with $Rh(DIP)_3^{3+}$ [17,18]. This complex, like its ruthenium(II) congener, recognizes altered conformations such as Z-DNA. Furthermore, upon photoactivation, the rhodium

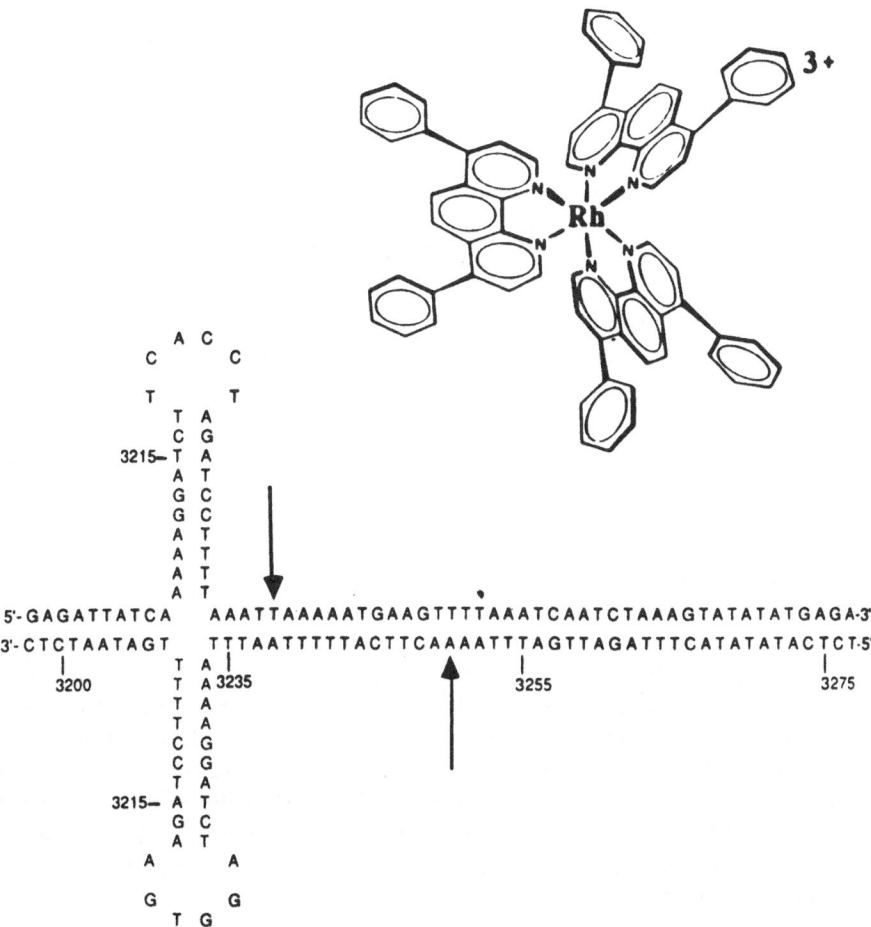

Fig. 3. Rh(DIP)₃³⁺, shown above, targets cruciform structures through double-stranded cleavage upon photoactivation. A cruciform from pBR322 is illustrated below and the arrows mark the sites of cleavage by Rh(DIP)₃³⁺ adjacent to the cruciform.

and cobalt [19,20] analogs promote strand cleavage at these conformationally distinct sites. Interestingly, among the sites targeted have been cruciform structures. In particular $Rh(DIP)_3^{3+}$ targets cruciform structures through double-stranded cleavage [17].

Cruciforms, one of which is illustrated schematically in Fig. 3, correspond to unique tertiary structures of DNA which may arise at palindromic sequences under supercoiled stress; at such palindromes, each strand of DNA may extrude out so as to form an intrastrand base-paired segment, relieving some torsional strain associated with the supercoiling. Such structures may be important elements in gene recombination. Although we depict cruciforms easily as a 2D base-paired array, our understanding of the complex 3D structure at such cruciform sites is poor. Certainly, the cruciform represents a grossly altered shape extruded along the double-helical polymer and one which is targeted by Rh(DIP)$_3^{3+}$ based upon its shape with high specificity.

Plasmids containing extruded cruciforms show site-specific double-stranded fragmentation after photolysis in the presence of Rh(DIP)$_3^{3+}$ at sites directly adjacent to the extruded cruciform (Fig. 3) [17]. Cruciforms of different sequences and lengths from different supercoiled DNAs have been targeted; it appears only that it is the conformation, not the sequence directly, which is important. Moreover, if a supercoiled DNA containing the cruciform is first linearized, relieving the torsional strain and hence the cruciform, and then incubated and irradiated in the presence of Rh(DIP)$_3^{3+}$, no specific cleavage is evident. What the complex appears therefore to recognize is the *shape* associated with a cruciform, its folded tertiary structure. Possibly, cleavage occurs adjacent to the cruciform owing to the formation of a hydrophobic pocket for the rhodium complex generated by two close, almost coaxial helices. Clearly the specificity of the targeting is one governed by *shape selection* and one therefore indirectly encoded by the sequence, since it is the sequence that determines the local conformation.

If cruciform targeting is an example of the recognition of grossly altered structures, then an example of a more subtle discrimination of sites based also upon shape selection may be found in experiments conducted with rhodium complexes of phenanthrenequinone diimines (ϕ) [21]. The ϕ-ligand was designed so as to maximize its planar surface for intercalation between the base pairs, and metal complexes containing the ϕ-ligand appear to bind DNA quite avidly by intercalation [9]. One can furthermore compare site-selectivities of rhodium complexes containing the ϕ-ligand but differing in their ancillary ligands. Shown in Fig. 4 are Rh(phen)$_2\phi^{3+}$ and Rh(ϕ)$_2$bpy^{3+}. These complexes both bind to DNA with affinities $\geq 10^7$ M^{-1} and, upon photoactivation, efficient DNA strand cleavage is observed [21].

What are the differences in site-selectivities associated with these complexes? For Rh(phen)$_2\phi^{3+}$, it is the ϕ-ligand that intercalates and two phenanthrolines, which lack hydrogen bonding substituents, maintain ancillary positions. For Rh(ϕ)$_2$bpy^{3+}, with one ϕ intercalated, the remaining ϕ, occupying an ancillary site, is well poised for hydrogen bonding between the imine hydrogen and the guanine O6 oxygen atom (based upon molecular modeling). On the basis of hydrogen bonding considerations, then, one might argue that, of the two complexes, Rh(ϕ)$_2$bpy^{3+} ought to display the greater (though not high) sequence selectivity. In fact what is oberved is the contrary. Rh(ϕ)$_2$bpy^{3+} cleaves DNA in essentially a sequence-neutral fashion. Indeed the complex serves as an efficient and widely applicable reagent for photofootprinting experiments [22]. Instead it is Rh(phen)$_2\phi^{3+}$ that shows some level of sequence-selectivity, and one which necessarily must be dependent upon a recognition of shape rather than hydrogen bonding. Indeed we find that the sequences cleaved preferentially by Rh(phen)$_2\phi^{3+}$ are those which may be characterized as being more open in the major groove [21]. This shape-selection is understandable since if the ϕ-ligand is to intercalate

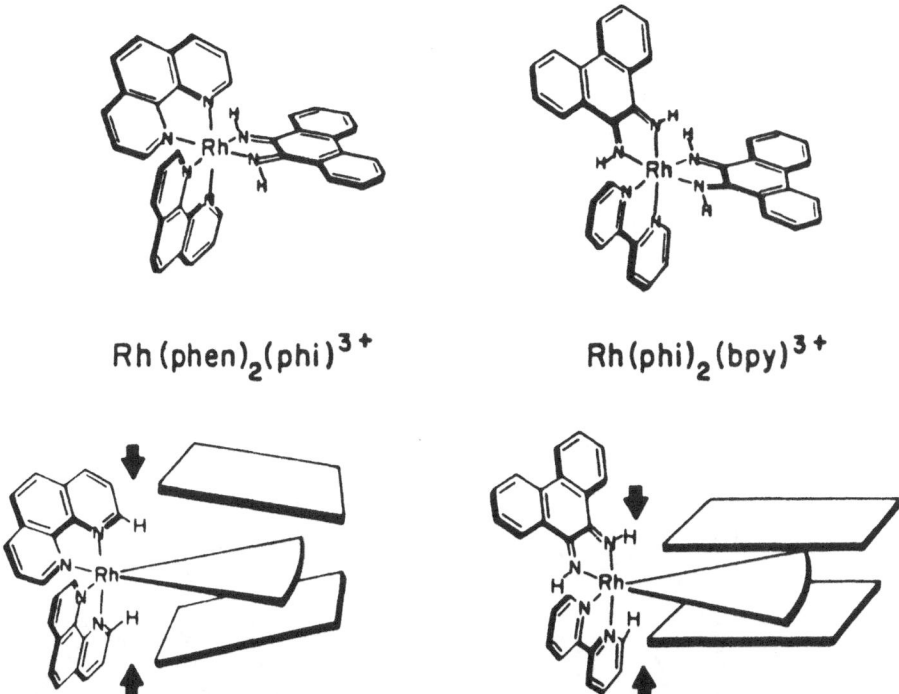

$$Rh(phen)_2(phi)^{3+} \qquad\qquad Rh(phi)_2(bpy)^{3+}$$

Fig. 4. Shape-selective cleavage by phenanthrenequinone (ϕ) complexes of rhodium(III). Shown on the left is Rh(phen)$_2\phi^{3+}$, which targets open sites in the major groove of B-form helices on the right is shown Rh(ϕ)$_2$bpy^{3+}, which cleaves DNA in essentially a sequence-neutral fashion and provides a versatile reagent for photofootprinting. Both complexes bind DNA avidly by intercalation and in the presence of light induce strand cleavage. The basis for the differences in recognition characteristics is illustrated schematically below. For Rh(phen)$_2\phi^{3+}$, steric clashes between the base pairs and the ancillary phenanthroline ligands preclude intercalation of the ϕ-ligand unless the base-pair site is open, while with Rh(ϕ)$_2$bpy^{3+}, the steric bulk of the complex is back from the axial coordination axis (shown by the arrows), permitting a facile sampling of all intercalation sites.

intimately from the major groove direction, steric clashes may ensue between ancillary phenanthroline hydrogen atoms and bases above and below the intercalation site if the site is not one which is more 'open' (owing to base-pair unwinding, tilting, or propeller twisting). For Rh(ϕ)$_2$bpy^{3+}, in contrast, the steric bulk of the molecule is predominantly pulled back relative to the axial coordination axis, and thus a similar basis for site-discrimination is not available; Rh(ϕ)$_2$bpy^{3+} can sample all sites equally well. Subsequent experiments revealing chiral discrimination associated with the site-selection by Rh(phen)$_2\phi^{3+}$ has allowed us, using considerations of shape and symmetry selection, to distinguish DNA sites which are open based upon propeller twisting from those opened owing to base tilting [23]. Such schemes for recognition based upon these subtle as well as the more gross distinctions between local DNA conformations are surely ones which Nature may exploit in targeting DNA-binding proteins to a specific site.

Metal–peptide complexes

We might also consider whether shape-selection might govern at least in part the interaction of peptide residues and complexes with DNA. Perhaps such considerations regarding shape are important only in the absence of appropriate hydrogen bonding groups on the DNA binding molecule, such as with phenanthroline ligands. A series of cobalt complexes containing coordinated dipeptides was therefore prepared and its recognition of DNA, explored [24].

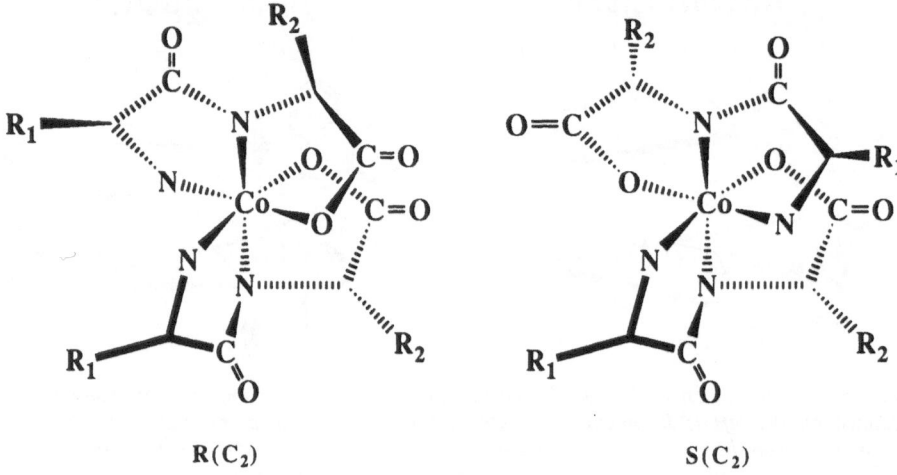

$R(C_2)$ $S(C_2)$

Fig. 5. The R-(left) and S-(right) isomers of bis(dipeptidato) cobalt(III). Each dipeptide forms a tridentate ligand about the octahedral metal center. R_1 and R_2 indicate peptide side chains.

Shown in Fig. 5 are two isomers of bis(dipeptide) complexes of cobalt(III). In these complexes, the metal is again coordinatively saturated and inert to substitution. Each dipeptide provides a tridentate ligand for the metal with coordination through the amino terminus, the deprotonated peptide nitrogen atom and carboxylate oxygen. In such complexes, the metal serves then to constrain the peptide; the orientation of each of the side chains is fixed by the coordination. We restricted our consideration to complexes of the form $Co(Arg-X)_2^+$ and $Co(X-Arg)_2^+$, (where X = Phe, Ile, Leu, Ser, Gly and Tyr) so as to preserve an overall positive charge on the complex and therefore some electrostatic interaction with the helix. The question of interest was whether the recognition characteristics on the helix would be determined primarily by X, the varied peptide residue.

Upon photoactivation, these bis(dipeptide) complexes of cobalt all induce DNA strand cleavage, albeit inefficiently. Most interesting, however, is the recognition characteristics associated with these cleavage reactions. Complexes of the form $Co(Arg-X)_2^+$ all appear essentially sequence-neutral in their cleavage of DNA. The binding to the helix is low, irrespective of the side chain. The shape of the complex, highly spherical with polar substituents dispersed equally about the metal center, may be poorly suited to maximize contacts with the helix. Complexes of the form $Co(X-Arg)_2^+$, in contrast, all showed some sequence preferences and the sites were equivalent among the series; the complexes cleaved primarily at AT-rich sequences with a 3'-cleavage asymmetry that is characteristic of association from the minor groove. Importantly, again the recognition characteristics appeared essentially the same for the different complexes in the series $Co(X-Arg)_2^+$ irrespective of the side chain. The recognition therefore appeared to depend less upon the nature of the X-residue and instead more upon the shape of the resultant coordinated complex (the overall molecular shape is of course shared by members of the family, independent of the nature of X). For the family of $Co(X-Arg)_2^+$ complexes, the guanidinium groups are disposed axially in the molecule, perhaps neatly poised for association with the sugar-phosphate backbones, and the X-residues are disposed in an isohelical fashion out from the center of the molecule, likely generating a surface which maximizes associations in the groove.

In sum, then, shape selection appears to govern also the recognition characteristics of these simple oriented peptide complexes. Certainly for these simple families of metal–peptide complexes, it appears that the first order interactions may result more from the nature of the molecular shape and how well that shape matches that of the DNA structure, than from the nature of the individual peptide residues themselves.

Conclusions

Simple transition metal complexes have been prepared which contain rigid, well-defined structures and symmetries and these have been targeted to DNA based upon *shape selection*. The complexes bind to DNA through non-covalent interactions which appear to depend upon the shape associated with the complex, and how well that shape is matched to that of the local conformation of DNA.

The transition metal complexes distinguish variations in DNA secondary and tertiary structure, and they may become useful in probing the sequence-dependent variations in structure along the strand. Not only grossly altered local DNA conformations but also the subtleties of base tilting and propeller twisting may be detected and selectively targeted based upon shape and symmetry considerations. Indeed a high level of site discrimination may be accomplished through *shape selection*. The same considerations apply in the targeting of metal–peptide complexes to sites along the double helix. Bis(dipeptide)-metal complexes have been prepared which target DNA, and here as well the pattern of recognition appears to depend more upon the molecular shape associated with the peptide complex than upon the individual peptide residues.

It is tempting to consider, given the polymorphic nature of DNA, that such factors play a role also in determining the recognition of DNA sites by proteins. Surely in the design of constrained peptides and smaller molecular analogs for DNA binding proteins, the contributions from shape-selective targeting might be considered.

Acknowledgements

I am grateful to the National Institutes of Health and the National Foundation for Cancer Research for their financial support. In addition I thank all my coworkers and collaborators, mentioned in the individual references, for their valuable contributions.

References

1. Rich, A., Seeman, N.C. and Rosenberg, J.M., Proc. Natl. Acad. Sci. U.S.A., 73 (1976) 804.
2. Matthews, B.W., Nature, 335 (1988) 294.
3. Otwinowski, R.W., Schevitz, R.W., Zhang, R.-G., Lawson, C.L., Joachimiak, A., Marmorstein, R.Q., Luisi, B.F. and Sogler, P.B., Nature, 335 (1988) 321.
4. Saenger, W., Principles of Nucleic Acid Structure, Springer-Verlag, New York, 1984 (and references therein).
5. Barton, J.K., Chem. Eng. News, 66 (39) (1988) 30.
6. Barton, J.K., Science, 233 (1986) 727.
7. Kumar, C.V., Barton, J.K. and Turro, N.J., J. Am. Chem. Soc., 107 (1985) 5518.
8. Barton, J.K., Goldberg, J.M., Kumar, C.V. and Turro, N.J., J. Am. Chem. Soc., 108 (1986) 2081.
9. Pyle, A.M., Rehmann, J.P., Meshoyrer, R., Kumar, C.V., Turro, N.J. and Barton, J.K., J. Am. Chem. Soc., 111 (1989) 3051.
10. Rehmann, J.P. and Barton, J.K., Biochemistry, in press.
11. Fleisher, M.B., Waterman, K.C., Turro, N.J. and Barton, J.K., Inorg. Chem., 25 (1986) 3549.
12. Fleisher, M.B., Mei, H.Y. and Barton, J.K., Nucleic Acids and Mol. Biol., 2 (1988) 65.
13. Barton, J.K., Danishefsky, A.T. and Goldberg, J.M., J. Am. Chem. Soc., 106 (1984) 2172.
14. Barton, J.K., Basile, L.A., Danishefsky, A. and Alexandrescu, A., Proc. Natl. Acad. Sci., U.S.A., 81 (1984) 1961.
15. Mei, H.Y. and Barton, J.K., J. Am. Chem. Soc., 108 (1986) 7414.
16. Mei, H.Y. and Barton, J.K., Proc. Natl. Acad. Sci. U.S.A., 85 (1988) 1339.
17. Kirshenbaum, M.R., Tribolet, R. and Barton, J.K. Nucleic Acids Res., 16 (1988) 7948.
18. Kirshenbaum, M.R., Ph.D. Dissertation, Columbia University, 1989.
19. Barton, J.K. and Raphael, A.L., Proc. Natl. Acad. Sci, U.S.A., 82 (1985) 6460.
20. Muller, B.C., Raphael, A.L. and Barton, J.K., Proc. Natl. Acad. Sci, U.S.A., 84 (1987) 1764.
21. Pyle, A.M., Long, E.C. and Barton, J.K., J. Am. Chem. Soc., 111 (1989) 4520.
22. Uchida, K., Pyle, A.M., Morii, T. and Barton, J.K. Nucleic Acids Res., (1989) in press.
23. Pyle, A.M., Morri, T. and Barton, J.K., submitted for publication.
24. Levy, E. and Barton, J.K., submitted for publication.

Biomimetic design of ion-carriers

Shneior Lifson[a], Yitzhak Tor[b], Jacqueline Libman[b] and Avi Shanzer[b]

[a]Department of Chemical Physics and [b]Department of Organic Chemistry,
Weizmann Institute of Science, Rehovot 76100, Israel

Introduction

No living cell exists without possessing mechanisms for uptake or release of metal ions. Nature invented many ways to perform this function, including ion-transport across the cell membrane by molecular vehicles named ion-carriers. Indeed, a large number of ion-carriers with specific and highly optimized properties are found in nature. They all function by similar principles, although they vary widely in details. They all bind a particular metal ion in a well-suited internal cavity by virtue of polar ligating groups, and generate a hydrophobic envelope. In the case of Fe^{3+}-carriers, siderophores, the ion uptake is receptor-driven and the envelope matches the receptor surface. The design and synthesis of synthetic ion-carriers should incorporate the above-listed desired properties into otherwise simple and easy to synthesize molecules.

Molecular Design

The general concept of design of the families of ion-carriers discussed here is that of binding 3 identical short chains to a common anchor at their one end, and to 3 identical bifunctional ion-binding groups at their other end. The ion-binding groups are chosen according to their known ion-specificity. The branches, linked to the common anchor by peptide or ester bonds, are designed to endow the molecule with the desired hydrophobicity, shape and chirality, and to bring the 3 ion-binding groups together to form an ion-binding octahedral cavity suited to the particular ion. A similar design with 2 branches is chosen for ions that prefer a tetracoordination. After choosing the general outlines of a family of molecules, the empirical force field method is used to calculate the equilibrium conformations of its individual members and the conformational distortions and strain energy imposed on the molecule by binding the target ion [1].

Results and Discussion

The two families of molecules described here mimic the two natural siderophores (Fe^{3+}-carriers), enterobactin and ferrichrome, respectively.

1. Enterobactin analogs (Fig. 1). Enterobactin **1** is the most powerful natural siderophore [2–4]. It is composed of 3 L-serines, linked to each other by ester

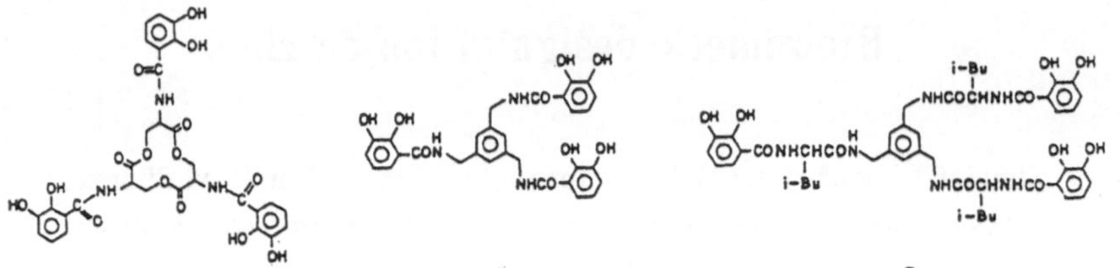

Fig. 1.   Enterobactin analogs 1, 2 and 3.

bonds, and to 3 catecholes by amide bonds. The ferric ion strips the catechols of their hydroxylic hydrogens, and the catecholates surround the ion in a right-handed (Δ-*cis*) propeller-like conformation [5]. Only the ferric complex of native enterobactin, and not its synthesized enantiomer, is biologically active, being taken up by *Escherichia coli* [6].

A combined experimental-theoretical analysis of enterobactin [7] showed that the Δ-*cis* propeller conformation is also the preferred conformation of the free enterobactin. It is stabilized by H-bonds between the oxygens of the ring and their closest amide hydrogens. Upon complexation, these weak H-bonds are broken, and much stronger H-bonds are formed between the amide hydrogen and the closest catecholate oxygens.

The potency of enterobactin as a siderophore has stimulated the synthesis of a large number of analogs, with 1,3,5-mesitylene-triscatecholamide **2**, being the most efficient synthetic binder [8], although it still lags behind enterobactin by several orders of magnitude of its binding constant. Our calculations [7] indicated that the difference is only partly due to the strain imposed by complexation, higher in **2** than in **1** by about 2 kcal/mol. The rest may be due to entropy differences, since the uncomplexed mesitylene analog, being achiral, has a larger conformational freedom than enterobactin. We modified **2** by interjecting one amino-acid residue between the mesitylene ring and the ends of the side chains. A combined experimental-theoretical analysis of members of the family of these tripeptides indicated that, since the side chains carry two amides each, they tend to form a circularly organized structure stabilized by inter-chain H-bonds between different amides [9]. When the end groups are catechols and the amino acids are L-leucines, a siderophore **3** is obtained that binds ferric ions somewhat better than **2**, with a preferred Δ-*cis* configuration [10], similar to enterobactin. Whether **3** or other members of the family possess biological activity comparable to enterobactin has still to be investigated.

2. Ferrichrome analogs (Fig. 2). Among the natural siderophores, the hydroxmate-based binders are the most abundant [2–4,11]. The microbial ferrichrome Fe^{3+}-carrier **4**, resembles the enterobactin siderophore by being a macrocyclic molecule with three L-amino acids and three ligating side chains [2–4]. It differs, however, from the latter by lacking C_3 symmetry, by using hydroxamates instead of catecholates as binding sites, and by forming Fe^{3+}-complexes of opposite absolute configuration, Λ-*cis*.

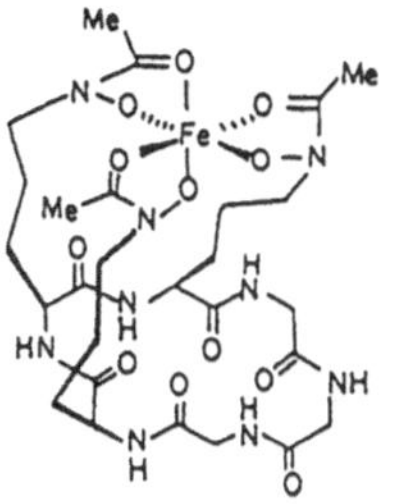

Fig. 2.   Ferrichrome analogs **4** and **5**.

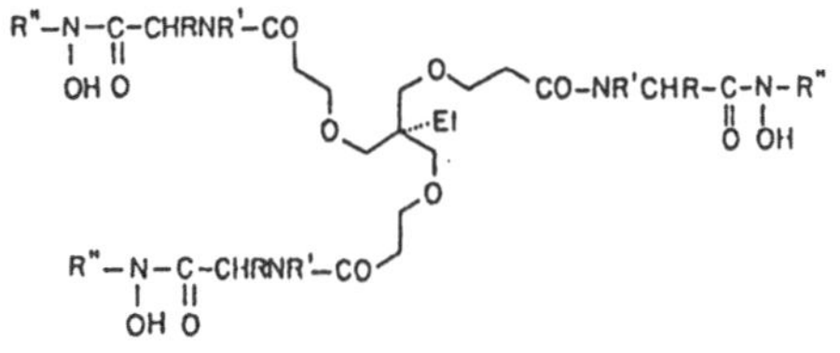

Our ferrichrome analogs have the general formula **5** [12]. The three chains are here again found by IR and NMR studies to be circularly organized through inter-chain hydrogen bonds. The ferric complexes of ligands **5** are found by CD to prefer, like ferrichrome, the Λ-*cis* configuration.

The L-amino-acid residues incorporated in **5** were leucyl, propyl, sarcosyl and alanyl. Their biological activity as growth promoters was tested by Dr. T. Emery, Utah State University, on *Arthrobacter flavescens*. This bacterium possesses ferrichrome-specific receptors, but lacks the capability to produce ferrichrome, and its growth depends on externally supplied ferrichrome. Some of these synthetic analogs proved to match the activity of ferrichrome as growth promoters [13].

References

1. Lifson, S., Felder, C.E. and Shanzer, A., J. Am. Chem. Soc., 105 (1983) 3866.
2. Nielands, J.B., Structure and Bonding, 58 (1984) 1.
3. Raymond, K.N., Mueller, G. and Matzanke, B.F., Top. Current Chem., 123 (1984) 49.
4. Hider, R.C., Structure and Bonding, 58 (1984) 25.
5. McArdle, J.V., Sofen, S.R., Cooper, S.R. and Raymond, K.N., Inorg. Chem., 17 (1978) 3075.
6a. Neilands, J.B., Erickson, T.J. and Rastetter, W.H., J. Biol. Chem., 256 (1981) 3831.
  b. Neilands, J.B., Ann. Rev. Microbiol., 36 (1982) 285.
7. Shanzer, A., Libman, J., Lifson, S. and Felder, C.E., J. Am. Chem. Soc., 108 (1986) 7609.
8. Harris, W.R. and Raymond, K.N., J. Am. Chem. Soc., 101 (1979) 6534.
9. Tor, Y., Libman, J., Shanzer, A., Felder, C.E. and Lifson, S., J. Chem. Soc. Ser. Chem. Commun., (1987) 749.
10. Tor, Y., Libman, J., Shanzer, A. and Lifson, S., J. Am. Chem. Soc., 109 (1987) 6517.
11. Emery, T., Met. Ions Biol. Syst., 7 (1978) 77.
12. Tor, Y., Libman, J. and Shanzer, A., J. Am. Chem. Soc., 109 (1987) 6518.
13. Shanzer, A., Libman, J., Lazar, R., Tor, Y. and Emery, T., Biochem. Biophys. Res. Commun., 157 (1988) 389.

The use of linear gramicidins as model peptides for the investigation of peptide-lipid interactions: A monolayer study

N. Van Mau, A. Benayad and F. Heitz*

Laboratoire de Physicochimie des Systèmes Polyphasés, CNRS, BP 5051, Route de Mende F., F-34033 Montpellier, France

Introduction

When incorporated into a lipid bilayer, gramicidin A (HCO-Val^1-Gly^2-Ala^3-D-Leu^4-Ala^5-D-Val^6-Val^7-D-Val^8-Trp^9-D-Leu^{10}-Trp^{11}-D-Leu^{12}-Trp^{13}-D-Leu^{14}-Trp^{15}-NH-C$_2$H$_4$OH) forms an ionic channel through a head-to-head dimerization process. Therefore, gramicidin-containing monolayers can be considered as representing half a bilayer, thus providing a good model for the investigation of membrane peptide–lipid interactions. Information is obtained through two types of measurement that are, on the one hand, the surface pressure and, on the other hand, the surface potential.

Results and Discussion

Several gramicidins, containing either tyrosine (Tyr), phenylalanine (Phe), or naphtylalanine (Nal), instead of the Trp residues in all four positions, or containing both Nal and Trp residues in various numbers or positions ('hybrides'), have been studied at the air–water interface. All the gramicidins show the same trend, characterized by a phase transition around 220–250 Å^2 and is detected in the surface pressure (inflexion of plateau depending on the nature of the aromatic side chain) and in the surface potential measurements (Fig. 1). Variations of the surface potential with the molecular density can be divided into three different domains. The first one (a), below the phase transition, is characterized by an increase of the potential V with the molecular density n, i.e., when the molecular area is reduced. When no charged groups are involved, as is the case for linear gramicidin, V is given by the relationship: $V = K \mu n \cos\theta$ with $K = 12 \pi 10^3$, μ in Debye, n is the number of molecules per Å^2, V in mV and $\cos\theta$ the vertical component of the dipole moment. Comparison of the experimental and theoretical (calculated using Urry's [1] model) values leads to $\theta = 84$, 67, 26 and 80° for GA, GM, GN and the 'hybrides' respectively, suggesting that the Trp-containing gramicidins have their helical axis parallel to the air–water interface, while GN and GM, which are more hydrophobic, have their axis perpendicular to this

---

*To whom correspondence should be addressed.

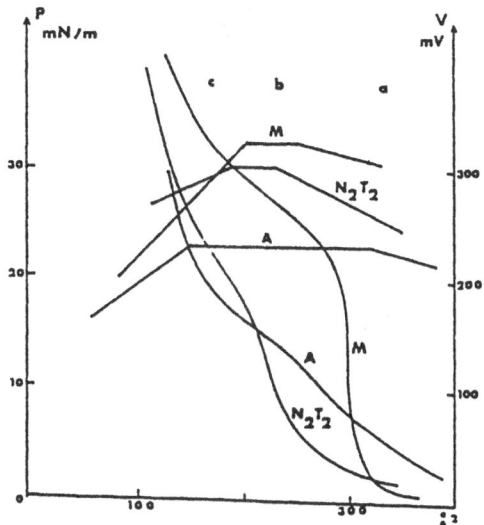

Fig. 1. Variations of the surface pressure and surface potential for GA, GM and GN[9,15] (GN₂T₂).

direction. The second domain (b), corresponding to the inflexion or to the plateau of the pressure-area curves, is attributed to a phase transition of the monolayers. The last domain (c), obtained at high molecular density, is characterized by a decrease of V when n is increased. Such a trend, corresponding to a general behavior for polypeptides, is attributed to a compensation of the dipoles by multilayer formation on compression. The same behavior is observed for mixed (GA- or GM-DOPC) monolayers which would also indicate, from pressure-area curves, that both components are not miscible [2].

References

1. Urry, D.W., Proc. Natl. Acad. Sci. U.S.A., 68 (1971) 672.
2. Van Mau, N., Trudelle, Y., Daumas, P. and Heitz, F., Biophys. J., 54 (1988) 563.

Glyco-β-turns anchor the antennae systems in glycoproteins

Miklós Hollósi[a], András Perczel[a], Gyula Szókan[a], Péter Sándor[b] and Gerald D. Fasman[c],*

[a]*Institute of Organic Chemistry, L. Eötvös University, H-1088 Budapest, Hungary*
[b]*Central Research Institute of Chemistry, Budapest, Hungary*
[c]*Graduate Department of Biochemistry, Brandeis University, Waltham, MA 02254-9110, U.S.A.*

Introduction

In *O*-glycoproteins, the first sugar moeity is linked to serine, threonine, hydroxylysine or hydroxyproline residues. These amino acids are well known to be frequently involved in reverse turn structures of proteins. In an attempt to study the effect of glycosylation on the conformation of the peptide backbone, a series (**1**) of peracetylated D-gluco-, D-galacto-, and D-mannopyranoside derivatives of Boc-X-L,D-Ser-NHCH$_3$ dipeptides (**2**, X = L-Pro,L-Val or Gly) has been synthesized (Fig. 1).

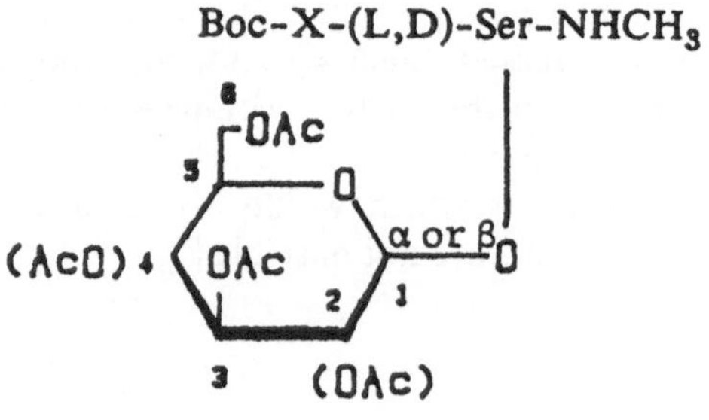

Fig. 1. 2e,3e,4e,6e-tetra-O-acetyl-D-glucopyranosyl (TAc-Glcp), 2e,3e,4a,6e-tetra-O-acetyl-D-galactopyranosyl (TAc-Galp), 2a,3e,4e,6e-tetra-O-acetyl-D-mannopyranosyl (TAc-Manp).

Results and Discussion

Boc-dipeptide *N'*-methylamides (**2**) were prepared using classical methods of peptide synthesis. 2,3,4,6-tetra-O-acetyl-1-α-bromo-D-glucopyranose, -1-α-bromo-D-galactopyranose, and -1-α-chloro-D-mannopyranose were used to convert the protected dipeptides into their peracetylated pyranosides [1]. According to HPLC and 400 MHz ¹H NMR studies, the method resulted in practically pure

---

*To whom correspondence should be addressed.

β-anomers of D-gluco- and D-galactopyranosides and 4:1 mixtures of the α- and β-anomers of D-mannopyranosides.

Circular dichroism (CD) spectra of the parent peptides and their glycosylated derivatives were measured in CH_3CN, TFE and water. In acetonitrile, the glycosylated Pro-L-Ser and Val-L-Ser models were found to show typical C spectra according to the classification of Woody [2], while glycosylated Pro-D-Ser and Val-D-Ser derivatives gave B or C' spectra [2]. In some cases, class C, B or C' spectra were also obtained in aqueous media. This suggests that peracetylated glycosides of Pro-L-Ser and Val-L-Ser dipeptides have a tendency to form type I(III) β-turns, while the Pro-D-Ser and Val-D-Ser models will likely adopt type II or distorted type II β-turn conformations. Based on CD data, the configuration at C-2 of the carbohydrate residue [gluco, (galacto) or manno] has no effect on the turn-forming ability of the peptides.

Infrared (IR) spectra of the glycosylated models were measured in CCL_4 and CH_2Cl_2 at different concentrations. The intensity ratios of the free and intra-H-bonded NH bands showed that models **1** are present as mixtures of single- and double H-bonded conformers in dilute nonpolar solution, and the gluco- and galactoconjugates (equatorial acetoxy group at C-2) have a higher tendency to form double H-bonded conformers than the mannoconjugates (axial acetoxy group at C-2). The peracetylated models (**1**) have a great variety of weak proton acceptor groups (glycosidic and ring oxygens, as well as acetoxy groups) which, based on molecular mechanical calculations and experiments with Dreiding models, have the potential to form C_5, C_7, C_{10} or C_{12} glyco-turns in addition to the peptide intramolecular H-bonds. Our studies indicate that the C=O of an equatorial acetoxy (or acetamido) group at C-2 is a favored acceptor, and the formation of an H-bonded glyco-turn does not affect the backbone conformation and intramolecular H-bond system of the parent peptide. Further studies are needed to explore the possibility of the formation of the other types of glyco-turns which are fixed by an intramolecular H-bond between a peptide C=O and the NH of the 2-acetamido group.

Acknowledgements

This work was supported in part by NSF Grant No. DMB-8713193.

References

1. Paulsen, H. and Paal, M., Carbohyd. Res., 135 (1984) 71.
2. Woody, R.W., In Hruby, V.J. (Ed.) The Peptides, Vol. 7, Academic Press, New York, NY, 1985, p. 38.

Derivatization of amino acids with 2,3,4,6-tetra-O-acetyl-β-D-glucopyranosyl isothiocyanate and resolution of their diastereomers by RPHPLC

Doreen R. Pangilinan[a], Michael L. Moore[a], John Hughes[b] and David Stevenson[b]

[a]Department of Peptide Chemistry and [b]Department of Synthetic Chemistry, Smith Kline & French Laboratories, Swedeland, PA 19479, U.S.A.

Introduction

To synthesize large quantities of high purity peptides useful as drug substances for evaluation in human clinical trials, it is essential to determine the chiral purity of amino acid derivatives to be used in the synthesis. This is especially true of the D-amino acid derivatives or of any other amino acids that are not derived from natural sources. This knowledge is especially critical when determining if racemization may have occurred during synthesis. This problem was recently presented to us during the synthesis of clinical supplies of the vasopressin antagonist, SK&F 105494 (1). This synthesis required supplies of Boc-D-Arg(Tos), Boc-D-Tyr(Et) and Boc-ω-benzyl-6,6-pentamethylene-2-amino-L-suberic acid [Boc-L-Pas(Bzl)] of known chiral purity with an accuracy greater than obtainable from specific rotation data. Chiral capillary gas chromatography was not applicable to Boc-D-Arg(Tos) because of difficulty in preparing volatile derivatives.

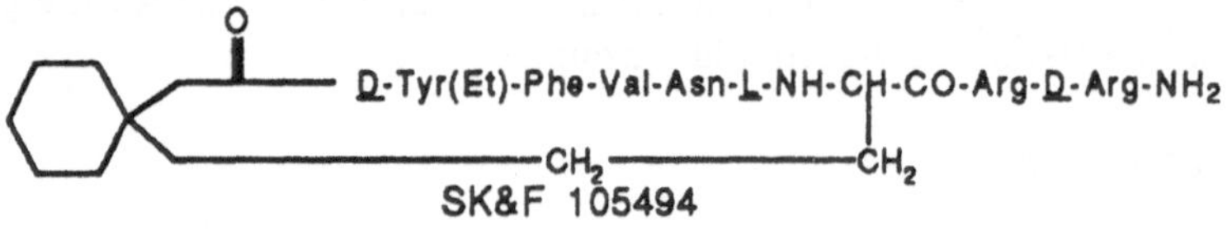

Scheme 1.

Methods

Boc-amino acid (5 mg) was deprotected with 1.0 ml 1:1 (v/v) TFA/CH$_2$Cl$_2$ for 30 min. The reaction mixture was concentrated by a rotary evaporator and the residue was washed several times with CH$_2$Cl$_2$ to remove excess TFA. The remaining oil was redissolved in aqueous CH$_3$CN (1:1) containing 0.4% (w/v) triethylamine to give a final volume of 10 ml. To a 100 μl aliquot of this solution was added 100 μl of 0.2% (w/v) GITC in CH_3CN. The mixture was allowed to stand at room temperature for 30 min. A 5 μl aliquot of this reaction mixture was injected into the HPLC.

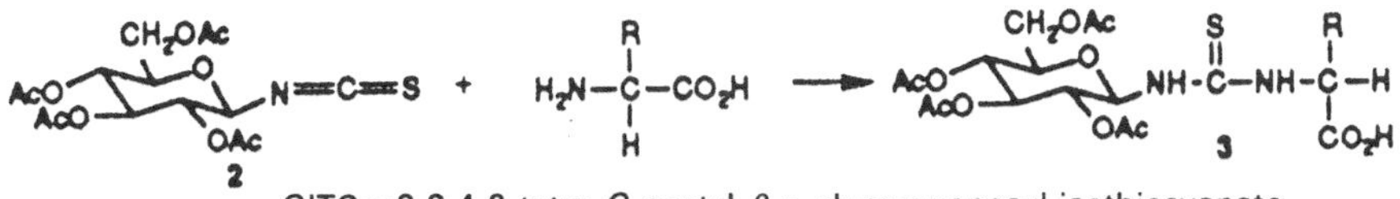

GITC = 2,3,4,6-tetra-*O*-acetyl-β-D-glucopyranosyl isothiocyanate

Scheme 2.

Results and Discussion

Enantiomeric peptides often have different pharmacological properties. In peptide synthesis, racemization is occasionally unavoidable. SK&F 105494 has 7 residues including D-Tyr(Et), L-Pas and a C-terminal D-Arg. D-Enantiomers of amino acids are made synthetically and may contain trace amounts of the L-enantiomers. Without prior knowledge of the chiral purity of Boc-D-Arg(Tos), it is very difficult to determine how much racemization may have occurred during SPPS. Optical rotation is insufficiently sensitive for detecting small amounts (below 10%) of the wrong enantiomer and it requires Boc-D-Tyr(Et) and Boc-L-Pas(Bzl) of high chemical purity. Boc-L-Pas(Bzl) is an unnatural amino acid synthesized via a Kolbe reaction [3] and standard reference material was not available. It has been reported in the literature [1,2] that diastereomeric thiourea derivatives (3) of simple amino acids can be formed using GITC (2) and separated by RPHPLC. We have implemented this method to determine the chiral purity of Boc-D-Arg(Tos), Boc-D-Tyr(Et) and Boc-ω-benzyl-6,6-pentamethylene-2-amino suberic acid [Boc-L-Pas(Bzl)]. Resolution of the resulting diastereomers on RPHPLC using a C_{18} Ultrasphere Beckman column does not require isolation of the pure derivatives. Resolving diastereomers of these amino acids provided information of how much of the wrong enantiomer was present. This knowledge helped to set up analytical specifications on large batches of incoming amino acids to be used for synthesis of peptides for clinical trials. Minimizing the presence of the wrong enantiomer simplifies the purification of the target peptide since fewer impurities are present in the crude peptide. All amino acid derivatives analyzed by this method gave baseline separation allowing quantitation of enantiomeric impurities in the 0.1% range.

References

1. Nimura, N., Ogura, H. and Kinoshita, T., J. Chromatogr., 202 (1980) 375.
2. Kinoshita, T., Kasahara, Y. and Nimura, N., J. Chromatogr., 210 (1981) 77.
3. Callahan, J.F., Newlander, K.A., Bryan, H.G., Huffman, W.F., Moore, M.L. and Yim, N.C.F., J. Org. Chem., 53 (1988) 1527.

Unprotected *O*-glycosylated Fmoc-threonine derivatives in the synthesis of glycopeptides

Fernando Filira, Laura Biondi, Franco Cavaggion, Barbara Scolaro and Raniero Rocchi

Centro di Studio sui Biopolimeri del C.N.R., Dipartimento di Chimica Organica, Universita' di Padova, Via Marzolo 1, I-35131 Padova, Italy

Introduction

The synthesis of glycopeptides normally involves, as the last step, the deblocking of protective groups both on the peptide and carbohydrate portions without affecting the chemically sensitive glycosidic linkage. During the synthesis, sugar hydroxyl functions are usually protected as ethers or esters. Benzyl ether-protecting groups can be cleaved by catalytic hydrogenolysis and acyl-protecting groups are removed under basic conditions. Side reactions such as incomplete removal, racemization, and/or β-elimination may occur, and unsuccessful attempts to obtain fully deblocked synthetic glycopeptides have been reported. In order to simplify the final deblocking steps and to avoid possible side reactions, the use of *O*-glycosylated Fmoc-threonine derivatives, unprotected at the sugar hydroxyl functions, in the synthesis of glycopeptides was investigated.

Results and Discussion

Four tuftsin (Thr-Lys-Pro-Arg) derivatives containing a D-glycopyranosyl (Glc) or a D-galactopyranosyl (Gal) unit covalently linked to the hydroxy side-chain function of the threonine residue through an either α, or β, *O*-glycosylated Fmoc-threonine derivative were synthesized.

To explore the possibility of utilizing the sugar-unprotected, *O*-glycosylated threonine derivatives in the solid phase glycopeptide synthesis, β-galactosylated tuftsin was also prepared by the continuous flow variant of the Fmoc-polyamide method.

Z-Thr[α-Glc(OBzl)$_4$]-OBzl and Z-Thr[α-Gal-(OBzl)$_4$]-OBzl were prepared from the tetra-*O*-benzylated sugar, and Z-Thr-OBzl by the trichloroacetimidate method [1,2] in the presence of TMSOTf, and converted into Fmoc-Thr(α-Glc)-OH and Fmoc-Thr(α-Gal)-OH by catalytic hydrogenation followed by acylation with Fmoc-OSu.

β-Glycosylation was carried out by reacting the proper per-*O*-acetylated sugar with Z-Thr-OBzl and boron trifluoride ethyl etherate by the neighboring group assisted procedure [3]. Catalytic hydrogenation of the β-glycosylated threonine derivatives, followed by acylation with Fmoc-OSu and deacetylation with methanolic hydrazine yielded Fmoc-Thr(β-Glc)-OH and Fmoc-Thr(β-Gal)-OH, respectively. The *O*-glycosylated threonine derivatives were reacted with H-Lys(Z)-Pro-Arg(NO$_2$)-OBzl [4] by the DCC-HOBt procedure, and the resulting glycosylated tuftsin derivatives were fully deblocked by catalytic hydrogenation, purified by HPLC or ion exchange chromatography, and characterized by optical rotation, AAA, and ^{1}H NMR.

In the SPS of β-galactosylated tuftsin, the symmetrical anhydride obtained from Fmoc-Arg(Mtr)-OH and Fmoc-Lys(Boc)-Pro-OH was used for introducing the first 3 amino acid residues. The last residue was incorporated into the peptide chain by using Fmoc-Thr(β-Gal)-OPfp as the acylating agent. After completion of the synthesis, the Fmoc protecting group was removed, the final peptide was cleaved from the 4-hydroxymethyl-phenoxyacetyl-norleucyl derivatized Kieselguhr-supported polydimethylacrylamide resin with 95% TFA, and purified by ion-exchange chromatography, yielding a β-galactosylated tetrapeptide analytically indistiguishable from that obtained by the solution synthesis.

The procedure used for the solution synthesis was satisfactorily applied to the preparation of the α- and β-glycosylated tuftins (Iα and Iβ). In the case of both the α- and β-galactosylated derivatives, two main products were formed by catalytic hydrogenation. One of them (IIα and IIβ) showed the expected amino acid composition after acid hydrolysis, while, apparently, the threonine residue is missing in the other one (IIα' and IIβ'). Surprisingly, in all cases ^{1}H NMR data were consistent with the synthesized compound, and threonine protons were unambigously assigned, even in those galactosylated derivatives (IIα' and IIβ') that according to the AAA, are lacking this residue. With the exception of a slight upfield shift in the Thr Hα resonances (IIα 4.03δ, IIα' 3.91δ; IIβ 3.89δ, IIβ' 3.80δ), the increase in the proton–proton vicinal coupling constants $J_{\alpha-\beta}$ (IIα 4.58 Hz, IIα' 5.49 Hz; IIβ 3.89 Hz, IIβ' 3.80 Hz), and the appearance of a new signal integrating three protons (IIα' 2.67, IIβ' 2.64), the spectra of IIα and IIa' or IIβ and IIβ' are practically superimposable. According to FABMS, peptide IIα' provided a molecular mass 14 daltons higher than that of IIα.

Further experiments are under way to explain these data that could indicate that a partial modification (methylation?) occurred at the threonine amino group during the final deblocking of the galactosylated tuftsins.

References

1. Schmidt, R.P., Angew. Chem. Int. Ed. Engl., 25 (1986) 212.
2. Wegmann, B. and Schmidt, R.P., J. Carbohydr. Chem., 6 (1987) 357.
3. Paulsen, H., Chem. Soc. Rev., 13 (1984) 15.
4. Rocchi, R., Biondi, L., Filira, F., Gobbo, M., Dagan, S. and Fridkin, M., Int. J. Pept. Protein Res., 29 (1987) 2501 (and refs. cited therein).

Carbohydrate derivatives of insulin: Preparation of Phe(B1) monoglycated and Gly(A1),Phe(B1) diglycated human insulin by reaction with glucose in non-aqueous solvents

John B. Halstrøm[a], Klavs H. Jørgensen[a], Gustav Bojesen[b], Mogens Christensen[a] and Anders R. Sørensen[a]

[a]*Novo Research Institute, Novo Allé, DK-2880 Bagsvaerd, Denmark*
[b]*Odense University, Department of Chemistry, Campusvej 55, DK-5230 Odense M, Denmark*

Introduction

Monomeric, rapidly absorbed insulin mutants have been obtained recently [1] using a strategy of introducing charge repulsion into the monomer–monomer interface by protein engineering.

The present approach aims at achieving a similar rapid absorption by covalent attachment of 1-deoxy-D-fructose residues to the N-termini of the A and B chains of human insulin using a new method of glycation.

Results and Discussion

In order to suppress substitution at the epsilon amino group of Lys(B29), we looked for conditions which would direct glycation preferentially to the more weakly basic α-amino groups. Extending a study of non-aqueous glycation of peptides [2] to insulin, we found that in methanol/acetic acid without added water, insulin is converted by excess D-glucose into 3 main products, of which 2 (mono- and diglycated insulin) successively predominate and may be isolated in substantial (30–60%) yields. The time course is followed by RPHPLC, allowing the point of maximum formation of each to be determined. From two preparative glycations at 40°C for 8 and 16 h, the major components of each reaction mixture were purified by semi-prep RPHPLC, using an H_3PO_4–acetonitrile system [3]. After desalting, they were characterized by FABMS, quantitative AAA before and after borohydride reduction, and bioassays (see Table 1). Gly(A1) monoglycated insulin does not appear to be a major product, presumably because of its rapid conversion to the A1,B1 diglycated species. Triglycated insulin forms very slowly (<10% after 10 h), probably due to protonation of the epsilon amino group. Thus, the amino groups react in the order B1 > A1 >> B29, in contrast to conventional glycation in water at pH 7–8 [4], where Lys(B29) is also extensively glycated.

The rates of s.c. absorption in pigs of ^{125}I-labeled Gly(A1), Phe(B1) diglycated

Table 1 *Analysis and bioassay of human insulin glycation products*

	HPLC R.T. relative to insulin	FABMS $(MH)^+$ observed (calculated)	AAA[a] Before (upper) and after (lower) $NaBH_4$			Potency[a] (% of insulin)	
			Gly	Phe	Lys	Free fat cell	Blood glucose
B1 monoglycated insulin	0.80	5970.4 (5970.7)	3.95 3.94	2.81 1.95	1.05 0.99	86	99
A1,B1 diglycated insulin	0.67	6132.0 (6132.9)	3.86 3.00	2.76 1.96	1.04 0.93	23	66
A1,B1,B29 triglycated insulin	0.60	6294.6 (6295.0)	3.73 2.94	2.75 1.95	0.59[b] 0.15	22	53

[a]For method see Ref. 1.
[b]Low due to characteristic dehydration of N-ε-glycated lysine [2].

807

human insulin and human Actrapid insulin were compared (see Fig. 1). The result: mean $T_{50\%} \pm SEM$: 1.47 h$\pm$0.14 h vs. 2.33 h$\pm$0.24 h shows that it is possible by appropriate hydrophilic substitution of human insulin to increase significantly the rate of s.c. absorption relative to a conventional rapid acting insulin. The potential utility of such derivatives in the treatment of diabetes is being investigated.

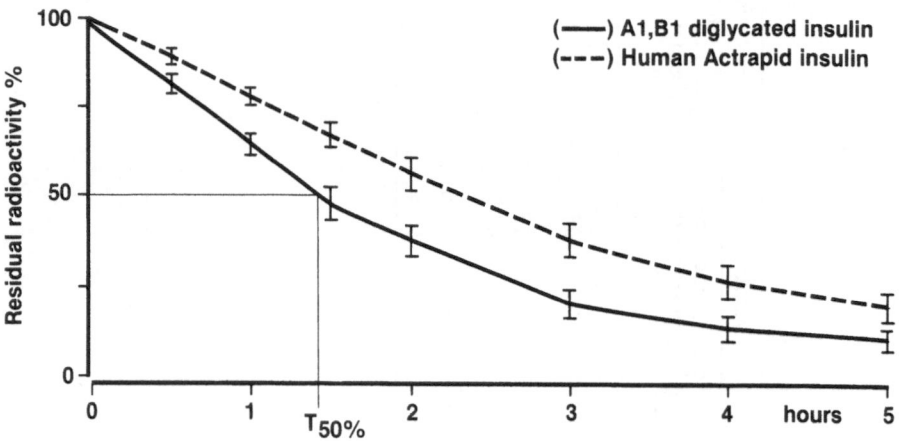

Fig. 1.   Absorption study in pigs (for methodology see Ref. 1).

References

1. Brange, J., Ribel, U., Hansen, J.F., Dodson, G., Hansen, M.T., Havelund, S., Melberg, S.G., Norris, F., Norris, K., Snel, L., Sørensen, A.R. and Voigt, H.O., Nature, 333 (1988) 679.
2. Halstrøm, J. and Kruse, V., In Deber, C.M., Hruby, V.J. and Kopple, K.D. (Eds.) Peptides: Structure and Function, Pierce Chemical Co., Rockford IL, 1985, p. 479.
3. Snel, L., Damgaard, U. and Mollerup, I., Chromatographia, 24 (1987) 329.
4. Lapolla, A., Tessari, P., Poli, T., Valerio, A., Duner, E., Iori, E., Fedele, D. and Crepaldi, G., Diabetes, 37 (1988) 787.

Conformational approach to the *N*-glycosylation of the Asn-Xaa-Ser/Thr marker sequence

Guy Boussard[a], Abdelilah Abbadi[a], Michel Marraud[a], Virginie Pichon-Pesme[b] and André Aubry[b]

[a]CNRS-URA 494, ENSIC-INPL, BP 451, F-54001 Nancy Cedex, France
[b]CNRS-URA 809, Université de Nancy I, BP 239, F-54506 Vandoeuvre, France

Introduction

N-Glycosylation, a co-translational event, requires the signal sequence Asn-Xaa-Ser/Thr on the nascent polypeptide chain. This enzymatically controlled process [1], that takes place at the luminal face of ER, consists in the block transfer of a preformed oligosaccharidic chain on to the Asn side chain.

Since Ac-Asn-Ala-Thr-NH$_2$, the simplest tripeptide that has ever been glycosylated in vitro, fulfills most of the necessary conditions, and since Xaa can be any of the coded amino acids, possibly excepting Pro, we believe that conformational features are of great importance in giving preference to one of the two proposed hypothetical enzymatic mechanisms of *N*-glycosylation [2, 3].

Results and Discussion

A series of Boc-Asx-Xaa-Yaa-R tripeptide derivatives was carefully designed and synthesized with Asx = Asn, Asn(Me), Asn (GlcNAc), Xaa = Pro, Ala, Yaa = Ser, Ser (Bzl), Thr and R = OMe, NHMe in order to obtain a maximum of informative and clear-cut conclusions concerning their conformational features, either in organic solution by means of NMR and FTIR spectroscopies or, whenever possible, in the crystalline state by X-ray diffraction, and finally to afford arguments for what the active conformation in the enzymatic process would be.

All the compounds were studied in organic solvents (DCM, DMSO, chloroform, acetonitrile), that were supposed to imitate, as closely as possible, the highly lipophilic microenvironment in which *N*-glycosylation takes place.

The chemical nature of the required tripeptide sequence suggests that, in addition to the usual main chain–main chain interaction, the polar Ser and Asn side chains may give rise to main chain–side chain and/or side chain–side chain interactions resulting in a complex hydrogen bonding network. Thus, both Ala- and Pro- unglycosylated families exhibit a variable percentage of a well-structured conformer characterized by a β-turn with Xaa and Ser as corner residues and imbricated in an Asx-turn stabilized by the intramolecular side chain–main chain (Asn) C$^\gamma$O$^\gamma$...$\underline{H}$N(Ser) hydrogen bond. The existence of this particular disposition is confirmed by the crystal structure of two derivatives of the Pro-family [4].

Nevertheless, in solution such a conformer is less frequent for Ala-sequences because of the larger number of accessible Ala-ϕ values, as reflected in the greater flexibility of the Ala family. For example, the spectroscopic properties and the geometrical dimensions of the Asx-turn show that an Asn-side chain t-disposition ($\chi_1 = 180°$) is more frequently adopted by Pro-derivatives while a g+-disposition prevails in the case of the Ala-family. Moreover, if we refer to all $^3J_{N\alpha}$ and $^3J_{\alpha\beta}$ measured coupling constants, an Ala-ϕ angle of $\approx +80°$ is compatible with a certain type of Asx-turn which excludes the presence of an imbricated β-turn of any type.

The side chain–side chain (Asn) CγOγ...$\underline{H}$Oγ(Ser) hydrogen bond, closing a 13-membered ring, has been evidenced in the crystal structure of Boc-Asn(Me)-Ala-Ser-OMe [5]; an interaction put forward by Marshall with reference to the enzymatic process, and also in a theoretical conformational analysis of Ac-Asn-Ala-Thr-NH$_2$.

None of our results suggest that a (Ser)HOγ...$\underline{H}$NHδ(Asn) type of hydrogen bond exists to a significant extent in any of the compounds studied.

The chemically prepared N-glycosylated compounds do not seem to be affected in the general conformation of their main polypeptidic backbone, although the Asx-turn vanishes. In addition, some NMR results (NOESY) seem to indicate the presence of new interactions between the Ser side chain hydroxyl function and the O^{6}H and O^{3}H alcoholic functions of the GlcNAc monosaccharide. Thus, the glucosidic part would bend over the main oligopeptidic backbone leading to shielding and masking of the polypeptide chain, as is the case in proteins.

Taking into account these last results, one may argue that the role of the (Asn) CγOγ...$\underline{H}$N(Ser) hydrogen bond, which stabilizes the Asx-turn, would seem to be catalytic rather than conformational. In addition to the (Asn) CγOγ...$\underline{H}$Oγ(Ser) intramolecular interaction, this bond facilitates the heterolysis of one of the two (Asn) NδH$_2$ protons, as suggested by Marshall's enzymatic mechanism. However, neither of the two putative mechanisms [3] refers to the occurrence of the Asx-turn stabilized by the (Asn) CγOγ...$\underline{H}$N(Ser) intramolecular hydrogen bond.

References

1. Kaplan, H.A., Welply, J.K. and Lennarz, W.J., Biochim. Biophys. Acta, 906 (1987) 167 (and references cited therein).
2. Marshall, R.D., Biochem. Soc. Symp., 40 (1974) 17.
3. Bause, E., Biochem. Soc. Trans., 12 (1984) 514.
4. Aubry, A., Boussard, G., Abbadi, A. and Marraud, M., New J. Chem., 11 (1987) 739.
5. Pichon-Pesme, V., Aubry, A., Abbadi, A., Boussard, G. and Marraud, M., Int. J. Pept. Prot. Res., 32 (1988) 175.

'Sticky fingers' for peptides

L. Moroder and H.-J. Musiol

Max-Planck-Institute of Biochemistry, Department of Peptide Chemistry,
D-8033 Martinsried, F.R.G.

Introduction

Attachment of lipid-derived hydrophobic moieties to peptides and proteins is expected to affect strongly their association to outer lipid bilayers of cell membranes and related translocation processes, as well as their insertion into liposomes or proteosomes.

Results and Discussion

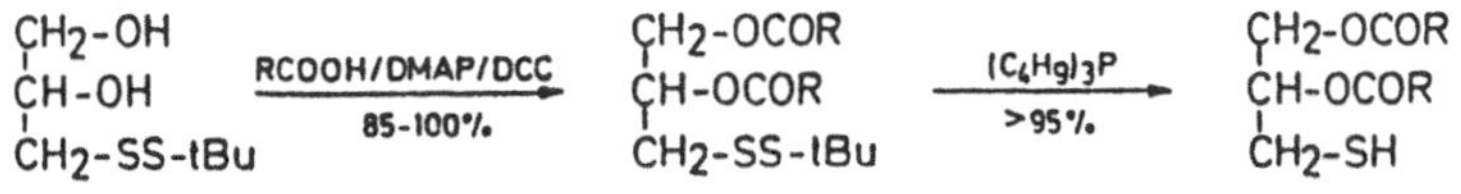

$$RCO = palmitoyl, myristoyl, retinoyl$$

Fig. 1. Synthetic scheme for compounds 2 and 3.

The observation that the maleimide function is sufficiently stable under certain conditions of peptide synthesis to allow for its insertion at preselected chain positions [1,2], compelled us to develop thiol-functionalized lipid derivatives for the selective linkage of 'sticky fingers' to peptides via the maleimide-thiol principle. For this purpose, 3-mercaptoglycerol was selected as starting compound and protected at the thiol function as unsymmetrical disulfide using 1-*tert*-butylthio-hydrazine-1,2-dicarboxylic acid dimorpholide as *tert*-butylthio donor [3]. Compound **1** (Fig. 1) was then acylated with palmitic, myristic and retinoic acid via the DCC/DMAP procedure to generate in high yields the homogeneous, symmetrical 1,2-di-O-acyl-3-*tert*-butyldithio-3-deoxy-*rac*-glycerols **2**. For the synthesis of the unsymmetrical diglycerides (Fig. 2), compound **1** was tritylated at the primary hydroxyl function and then selectively acylated with palmitic acid in position 2. By detritylation of **5**, via silicic acid/boric acid [4] or $ZnBr_2$ [5], acyl migration was suppressed to minimal extents ($\leqslant 4\%$ according to ^{1}H NMR). Subsequent acylation of **6** with myristic and retinoic acid produced the mixed-acid diglyceride derivatives **7**. Reduction of **2** and **7** with $(C_4H_9)_3P$ in aqueous trifluoroethanol/ether proceeded quantitatively without noticeable isomerization

811

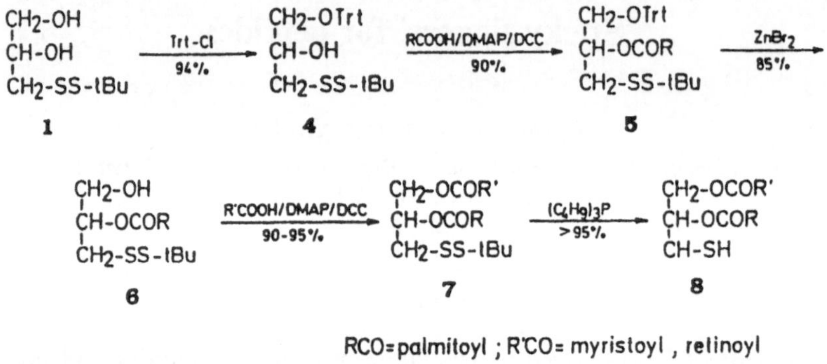

CH2-OH   Trt-Cl   CH2-OTrt   RCOOH/DMAP/DCC   CH2-OTrt   ZnBr2
CH-OH   ——————→   CH-OH   ————————————→   CH-OCOR   ——————→
CH2-SS-tBu   94%   CH2-SS-tBu   90%   CH2-SS-tBu   85%

    1   4   5

CH2-OH   R'COOH/DMAP/DCC   CH2-OCOR'   (C4H9)3P   CH2-OCOR'
CH-OCOR   ——————————————→   CH-OCOR   ——————→   CH-OCOR
CH2-SS-tBu   90-95%   CH2-SS-tBu   >95%   CH-SH

    6   7   8

RCO=palmitoyl ; R'CO= myristoyl , retinoyl

Fig. 2. Synthetic scheme for compounds 4 to 8.

(O → S acyl migration), yielding the symmetrical and unsymmetrical 1,2-di-O-acyl-3-mercapto-rac-glycerols 3 and 8, respectively, as suitable reagents for their addition to maleoyl-peptide derivatives (Fig. 3). Their reaction with 3-maleimido-propionic acid as model compound produced the lipophylic addition products in good yields. ^{1}H NMR spectra revealed the presence of 2 diastereoisomers resulting from the additional chiral center formed in the succinimide moiety.

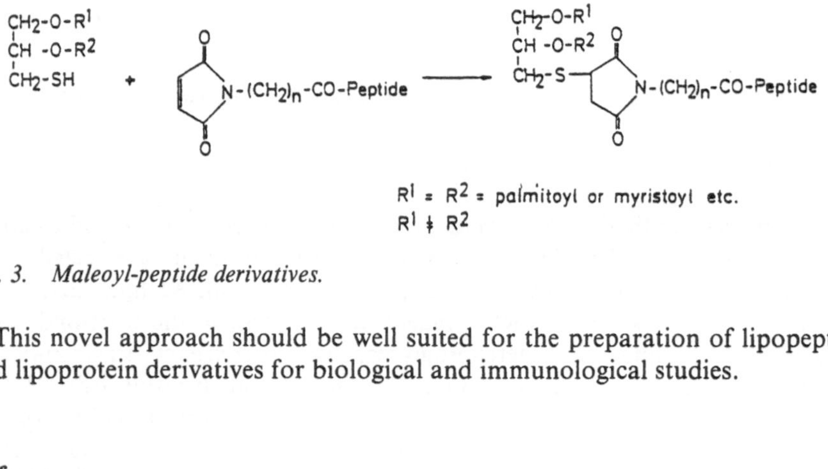

R^1 = R^2 = palmitoyl or myristoyl etc.
R^1 ≠ R^2

Fig. 3. Maleoyl-peptide derivatives.

This novel approach should be well suited for the preparation of lipopeptide and lipoprotein derivatives for biological and immunological studies.

References

1. Moroder, L., Nyfeler, R., Gemeiner, M., Kalbacher, H. and Wünsch, E., Biopolymers, 22 (1983) 481.
2. Wünsch, E., Moroder, L., Nyfeler, R., Kalbacher, H. and Gemeiner, M., Biol. Chem. Hoppe-Seyler, 366 (1985) 53.
3. Wünsch, E., Moroder, L. and Romani, S., Hoppe-Seyler's Z. Physiol. Chem., 363 (1982) 1461.
4. Buchnea, D., Lipids, 9 (1974) 55.
5. Kohli, V., Blöcker, H. and Köster, H., Tetrahedron Lett., 21 (1980) 2683.

Session VIII
HIV and related areas

Chairs: John M. Stewart
University of Colorado Health Sciences Center
Denver, Colorado, U.S.A.

and

Stephen B.H. Kent
Bond University
Gold Coast, Queensland, Australia

Molecular targets of HIV that are candidates for rational drug design

John J. McGowan*, Nava Sarver, H.S. Allaudeen and Margaret M. Johnston

Developmental Therapeutics Branch, AIDS Program, National Institute of Allergy and Infectious Diseases, National Institute of Health, Bethesda, MD 20892, U.S.A.

Introduction

The last three years have witnessed a revolution in the discovery, development and approval process for drugs to treat people infected with HIV. In particular, many laboratories within the private, academic and government sectors are utilizing the knowledge and reagents generated through basic research in the rational design of drugs and the development of selective biochemical assays to facilitate the search for anti-HIV drugs by random drug screening [1].

Examination of the life cycle of HIV reveals a number of steps that appear sufficiently distinct to serve as molecular targets for antiviral therapy (Fig. 1). These targets include the cellular CD4 receptor, virus-encoded enzymatic activities (reverse transcriptase, RNase H, integrase, proteinase), regulatory proteins critical for activation of viral transcription (tat, rev, nef), processing of viral mRNAs, processing of viral proteins (e.g., glycosylation, myristoylation), assembly at the plasma membrane, and maturation and budding of virus particles. Already, drugs have been identified through screening and rational drug design that inhibit these steps in replication (Fig. 1).

In this article, the focus will remain on the rational approaches to the discovery and design of anti-HIV drugs that will be blocking viral attachment or processing of viral proteins.

Results and Discussion

CD4-based therapies

Basic research on the HIV viral glycoprotein (gp 120) and the identification of the receptor for the virus on T-cells led to the design and biologic production of recombinant soluble forms of the receptor (sCD4) as a drug. CD4 is a membrane glycoprotein that protrudes from the surface of glial, muscle and fibroblastic cells. The extracellular region of the glycoprotein has four domains that share homolgy to immunoglobulin variable domains. CD4 has been used to identify a sub-population of T-cells (CD4+) that interacts with class II MHC molecules [2,3]. The evidence that CD4 is the receptor for HIV includes the following: high binding affinity of gp 120 for CD4 ($K_D \sim 10^{-9}$ M; [4]), antibodies to CD4

---

*To whom correspondence should be addressed.

Attack Points to Stop Life Cycle of HIV

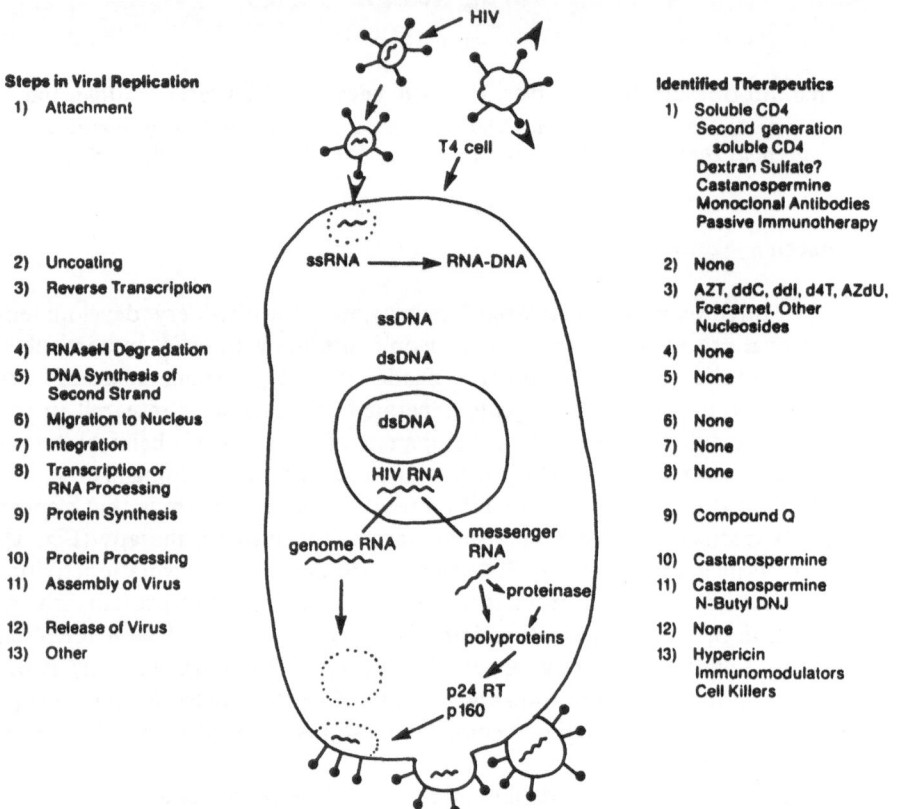

Steps in Viral Replication
1) Attachment
2) Uncoating
3) Reverse Transcription
4) RNAseH Degradation
5) DNA Synthesis of Second Strand
6) Migration to Nucleus
7) Integration
8) Transcription or RNA Processing
9) Protein Synthesis
10) Protein Processing
11) Assembly of Virus
12) Release of Virus
13) Other

Identified Therapeutics
1) Soluble CD4
   Second generation soluble CD4
   Dextran Sulfate?
   Castanospermine
   Monoclonal Antibodies
   Passive Immunotherapy
2) None
3) AZT, ddC, ddI, d4T, AZdU, Foscarnet, Other Nucleosides
4) None
5) None
6) None
7) None
8) None
9) Compound Q
10) Castanospermine
11) Castanospermine N-Butyl DNJ
12) None
13) Hypericin Immunomodulators Cell Killers

Fig. 1. Life cycle of HIV.

block HIV infection of T-cells [5], cells that are not infectable with HIV become susceptible if CD4 is expressed, by transfection, on the cell surface, and soluble forms of CD4 bind gp 120, block HIV binding, syncytia formation and HIV infectivity [2,3,6–9].

At present, Biogen, Genentech and Smith Kline and French have developed a first-generation sCD4 therapeutic. Each company has sponsored a phase I evaluation of their product. No untoward toxicities have been observed in the studies, nor have auto-antibodies to CD4 been reported from patients who received sCD4 for 8 weeks.

The first generation products of sCD4 were approximately 370 nucleotides in length. A number of second generation sCD4 being examined as possible clinical candidates include:

- sCD4 linked to toxins that prolong serum half life and selectively kill HIV-infected cells;
- sCD4 fused with antibodies (immunoadhesins) that prolong serum half-life may have other immune effector functions (e.g., C1q and Fc γ-binding activities);
- shorter versions (~ 118 nucleotides in length);
- cells genetically engineered to produce sCD4 in vivo;
- chemically modified synthetic peptides (11–12 amino acids in length) that are stable and more potent than those already published.

Of particular interest to this conference, benzyl-derivatives synthetic peptides of sCD4 (sCD4 region 81–92; [10]) that inhibit HIV are not coincident with the gp 120 binding site identified by others for CD4 (region 40–56; [10–12]). The data suggest that the binding, attachment and entry of the virus into a CD4-positive cell may be more complicated than originaly thought.

Each of the second-generation sCD4 identified raises some concern. Changes in the size of the product (shorter or longer) may cause antigenic recognition of newly exposed epitopes. Antibodies are likely to be generated with a short period of time to those sCD4 molecules with toxins attached. The long-term effect of the toxins is unknown. sCD4-toxin linked may kill a significant number of HIV-infected cells resulting in a substantial increase in the viral load in the patient. More data are needed to quantitate the amount of HIV released from cells killed by toxins to address this point. The large size, site of protein fusion and potential folding of immunoadhesins may lead to increased antibody formation. Finally, the bioavailability of sCD4 beyond the vascular system may be extremely limited. Despite these potential pitfalls, the sCD4 may prove to be an important, non-toxic therapy that will be useful in the successful management of HIV in those people infected.

HIV protease inhibitors

Another viral target that has received considerable attention recently is the HIV-encoded proteinase [13]. The proteinase mediates the post-translational cleavage of HIV polyproteins that are critical for virus assembly. In particular, protease processing of viral gag (p55) and gag-pol (p160) precursors has been characterized (Fig. 2; [14]). It has been suggested that the proteinase may mediate the cleavage of mature nucleocapsid protein in the early steps of HIV infection; the implications of this event in HIV replication are not known [15].

The HIV proteinase is an excellent target for rational drug design. It is essential for viral replication [14,16,17]; responsible for the cleavage of the HIV polyproteins at specific sites (cellular proteinases cannot cleave at the unique sites present in retroviral proteins [14,18]; and amenable to rational drug design without knowing the structure of the protein. Further, the structure and base of information established on known retroviral and aspartyl proteinases are being used to solve the structure of the HIV proteinase [19–23]. The HIV proteinase has been cloned, chemically synthesized, sequenced and characterized

[13,16,17,23–25]. Several groups have designed and are using rapid assays for evaluating agents for their ability to inhibit this enzyme.

The HIV proteinase is structurally related to other aspartyl proteinases studied to date. The active site of other retroviruses (e.g., Rous Sarcoma Virus), HIV and other aspartyl proteinases are superimposable within 0.5 Å [26,27]. Navia and coworkers of Merck Sharp and Dohme Research Laboratories have recently published the crystal structure of the enzyme to a 3 Å resolution [28].

Though similar in structure, significant differences exist between the published crystal structure of HIV and that of other aspartyl proteinases [27]. HIV proteinase appears to possess:

- a strict 2-fold axis of symmetry where other aspartyl proteinases have an approximate 2-fold axis;
- unique topologies at both the N- and C-terminus;
- a 4-stranded β-sheet where other aspartyl proteinases have a 6-stranded β-sheet;
- no strict counterpart of the highly conserved 'flap' structure of other aspartyl proteinases;
- an active site that is the interface of two homologous peptide chains (99 aa in length), rather than of two domains along a single peptide chain (~ 200 aa in length).

These differences require confirmation of the crystal structure to a higher resolution. [Note in proof: See Ref. 29 for more detailed structure and corrections.]

Already, computer-generated docking models are being used to search drug repositories and data bases for compounds that 'fit' into the enzyme-active site, and classical approaches of using synthetic peptides coupled with chemical groups have led to the discovery of potential proteinase inhibitors that may reach clinical trials next year [20,21,6]. Small peptide inhibitors are being synthesized that contain a reduced amide bond at the cleavage site, or amino acid substitutions on either side of the cleavage site, or other modifications that confer resistance to proteinase-mediated cleavage while maintaining a strong affinity for the active site.

The design and development of an inhibitor of HIV proteinase require the identification of an active prototype inhibitor in a biochemical screening assay and confirmation of antiviral activity in cell culture followed by animal and human studies. The inhibitor must have a sufficient 1/2 in serum to establish an effective intracellular concentration to block the viral proteinase. While oral administration of the antiviral is the preferred route, many proteinase inhibitors are peptide analogs and may not survive digestion by gastric enzymes or be poorly absorbed.

The HIV proteinase is active at the interface of the viral/cellular membrane and remains active in the cleavage of the gag-pol precursor during the process of virus budding and assembly. Evidence suggests that the viral proteinase associated with the membrane is critical in budding and assembly of the virus particle by transforming the local high concentration of polyproteins to form virus particles [22,25,30]. Proteinase cleavage continues after the incomplete virus particle buds from the host cell membrane until a complete infectious virion is formed. Elucidation of the structure and inhibitor binding characteristics of the gag-pol precursor may yield valuable information on viral structure, viral assembly and cellular processes that take place within the cellular/viral membrane.

References

1. McGowan, J. and Hoth, D., JAIDS, (1989) in press. .
2. Clarke, S., Vogel, J.P., Deschenes, R.J. and Stock, J., Proc. Natl. Acad. Sci. U.S.A., 85 (1988) 4643.
3. Maddon, P.J., Dalgleish, A.G., McDougal, J.S., Clapham, P.R., Weiss, R.A. and Axel, R., Cell, 47 (1986) 333.
4. Smith, D.H., Byrn, R.A., Marsters, S.A., Gregory, T., Groopman, J.E. and Capon, D.J., Science, 238 (1987) 1704.
5. Dalgleish, A.G., Beverley, P.C.L., Clapham, P.R., Crawford, D.H., Greaves, M.F. and Weiss, R.A., Nature, 312 (1984) 763.
6. Capon, D.J., Chamow, S.M., Mordenti, J., Marsters, S.A., Gregory, T., Mitsuya, H., Byrn, R.A., Lucas, C., Wurm, F.M., Groopman, J.E., Broder, S. and Smith, D.H., Nature, 337 (1989) 525.
7. Till, M.A., Ghetie, V., Gregory, T., Patzer, E.J., Porter, J.P., Uhr, J.W., Capon, D.J. and Vitetta, E.S., Science, 242 (1988) 1166.
8. Chao, B.H., Costopoulos, D.S., Curiel, T., Bertonis, J.M., Chisholm, P., Williams, C., Schooley, R.T., Rosa, J.J., Fisher, R.A. and Maraganore, J.M., J. Biol. Chem., 264 (1989) 5812.
9 Ibegbu, C.C., Kennedy, M.S., Maddon, P.J., Deen, K.C., Hicks, D., Sweet, R.W. and McDougal, J.S., 142 (1989) 2250.
10. Lifson, J.D., Hwang, K.M., Nara, P.L., Fraser, B., Padgett, M., Dunlop, N.M. and Eiden, L.E., Science, 241 (1988) 712.
11. Lasky, L.A., Nakamura, G., Smith, D.H., Fennie, C., Shimasaki, C., Patzer, E., Berman, P., Gregory, T. and Capon, D.J., Cell, 50 (1987) 975.
12. Palker, P.J., Matthews, T.J., Clark, M.E., Cianciolo, G.J., Randall, R.R., Langlois, A.J., White, G.C., Safai, B., Snyderman, R., Bolognesi, D.P. and Haynes, B.F., Proc. Natl. Acad. Sci. U.S.A., 84 (1987) 2479.
13. Kräusslich, H.-G., Shoog, M.T., Tallai, P.V., Carter, C.A. and Wimmer, E. In Kräusslich, H.-G., Oroszlan, S. and Wimmer, E. (Eds.) Current Communications in Molecular Biology: Viral Proteinases as Targets for Chemotherapy, vol. 28, Cold Spring Harbor Laboratory Press, Cold Spring Harbor, NY, 1989, p. 147.
14. Kräusslich, H.-G. and Wimmer, E., Ann. Rev. Biochem., 57 (1988) 701.
15. Roberts, M.M. and Oroszlan, S., Biochem. Biophys. Res. Comm., 160 (1989) 486.
16. Kohl, N.E., Emini, E.A., Schleif, W.A., Davis, L.J., Heimbach, J.C., Dixon, R.A.F., Scolnick, E.M. and Sigal, I.S., Proc. Natl. Acad. Sci. U.S.A., 85 (1988) 4686.
17. Le Grice, S.F.J., Mills, J. and Mous, J., EMBO J., 7 (1988) 2547.
18. Seelmeier, S., Schmidt, H., Turk, V. and von der Helm, K., Proc. Natl. Acad. Sci. U.S.A., 85 (1988) 6612.
19. Sham, H.L., Bolis, G., Stein, H.H., Fesik, S.W., Marcotte, P.A., Plattner, J.J., Rempel, C.A. and Greer, J., J. Med. Chem., 31 (1988) 284.
20. Billich, S., Knoop, M.-T., Hansen, J., Strop, P., Sedlacek, J., Mertz, R. and Moelling, K., J. Biol. Chem., 263 (1988) 17905.

819

21. Kotler, M., Katz, R.A., Danho, W., Leis, J. and Skalka, A.M., Proc. Natl. Acad. Sci. U.S.A., 85 (1988) 4185.
22. Kräusslich, H.-G., Ingraham, R.H., Skoog, M.T., Wimmer, E., Pallai, P.V. and Carter, C.A., Proc. Natl. Acad. Sci. U.S.A., 86 (1989) 807.
23. Loeb, D.D., Hutchison III, C.A., Edgell, M.H., Farmerie, W.G. and Swanstrom, R., J. Virol., 63 (1989) 111.
24. Debouck, C., Gorniak, J.G., Strickler, J.E., Meek, T.D., Metcalf, B.W. and Rosenberg, M., Proc. Natl. Acad. Sci. U.S.A., 84 (1987) 8903.
25. Farmerie, W.G., Loeb, D.D., Casavant, N.C., Hutchison III, C.A., Edgell, M.H. and Swanstrom, R., Science, 236 (1987) 305.
26. Pearl, L.H. and Taylor, W.R., Nature, 329 (1987) 351.
27. Erickson, J. et al., (1989) submitted.
28. Navia, M.A., Fitzgerald, P.M.D., McKeever, B.M., Leu, C.-T., Heimbach, J.C., Herber, W.K., Sigal, I.S., Darke, P.L. and Springer, J.P., Nature, 337 (1989) 615.
29. Wlodawer, A., Miller, M., Jaskólski, M., Sathyanarayana, B.K., Baldwin, E., Weber, I.T., Selk, L.M., Clawson, L., Schneider, J. and Kent, S.B.H., Science, 245 (1989) 616.
30. Witte, O.N. and Baltimore, D., J. Virol., 26 (1978) 750.

Diagnostic and antiviral applications of synthetic HIV-1 peptides

Bruce E. Kemp[a], Kim M. Wilson[a], Dennis B. Rylatt[b], Colin M. House[a], Dale A. McPhee[c], Peter G. Bundesen[b], Carmel J. Hillyard[b] and Richard R. Doherty[b]

[a]St. Vincent's Institute of Medical Research, Fitzroy, Vic., Australia 3065
[b]Agen Biomedical Ltd., Acacia Ridge, Qld., Australia 4110
[c]NH&MRC Special Unit for AIDS Virology, Macfarlane Burnet Centre for Medical Research, Fairfield Hospital, Fairfield, Australia 3078

Introduction

Diagnostic test

The measurement of circulating antibodies to HIV-1 has played a vital role in detecting carriers of HIV-1 as well as monitoring contamination of blood products. Several rapid agglutination assays for antibodies to HIV-1 have been developed including those based on latex beads [1], gelatin beads [2] and turkey red cells [3]. We have developed a whole blood rapid agglutination test for antibodies to HIV-1 that uses the patient's own red cells as the indicator [4]. This is analogous to the procedure pioneered by Coombs et al. [5], except that autologous red cells are used rather than exogenous ones. A non-agglutinating monoclonal antibody against human red blood cells is chemically cross-linked to the immunodominant synthetic peptide epitope of gp41. When this reagent is added to a drop of blood, the patient's red blood cells are coated with the synthetic viral antigen antibody conjugate. Agglutination of the patient's own red blood cells occurs if the blood contains antibodies to the immunodominant viral antigen.

The monoclonal antibody to human red cells was selected using the criterion that it was non-agglutinating when added to whole blood but caused agglutination in the presence of anti-mouse antibody. The monoclonal antibody recognized glycophorin as an abundant antigen on red blood cells.

Antiviral peptides

We have identified conserved regions of gp120 and gp41 that may be important in forming contacts between these proteins. Synthetic peptides corresponding to these regions have antiviral activity as measured by reverse transcriptase activity.

Results and Discussion

Blood samples from a series of HIV-infected patients, normal healthy blood donors and hospital patients with unrelated conditions were tested. The auto-

821

logous red cell agglutination test detected 42/43 known HIV-1-positive individuals, giving a sensitivity of 98% compared to 100% (43/43) using the Abbott
commercial test. With hospital patients there were 3/66 false positive tests, giving
a specificity of 95%. The number of false positive results with blood donors
was only 1/873 giving a specificity of 99.9% comparable to the regular EIA
test. These results indicated that the red cell agglutination test principle could
be used to develop a test with the sensitivity and specificity approaching the
commerical EI test [4]. In order to reduce the level of false positive tests among
hospital patients as well as reduce the amount of blocking monoclonal antibody
required, we investigated the use of Fab fragments of the anti-red blood cell
antibody. The antibody was treated with pepsin to generate Fab_2 fragments
and the inter heavy chain disulphides were reduced with mercaptoethylamine
and protected with Ellman's reagent, as described previously by Brennan et
al. [6]. The resultant TNB-protected Fab fragments were purified by chromatography on S-200 (Pharmacia) and reacted with reduced synthetic peptide. The
residual sulphydryl groups were blocked with iodoacetamide and the conjugate
purified by cation-exchange chromatography on mono-S (Pharmacia). The Fab-
peptide conjugate could be used without the addition of blocking antibody.
Blocking the residual sulphydryl groups with iodoacetamide greatly enhanced
the stability of the reagent.

The antigen used in any test is a vital determinant of the sensitivity and
specificity of the test. Recombinant proteins are replacing viral lysate material
because of the need for batch-to-batch consistency and inclusion of adequate
amounts of cricital epitopes. The autologous red cell agglutination test described
here used synthetic peptide antigens for ease in cross-linking to the antibody.
It was initially thought that recombinant antigens would be needed for a
autologous red cell agglutination test kit suitable for clinical use. However, recent
findings from a number of laboratories indicate that synthetic peptide antigens,
particularly those including the immunodominant epitope on the envelope
glycoprotein gp41, may be adequate for a 'front line test' [7–11]. They indicated
that sensitivities of 98–99% and a false positive rate of $<1\%$ (specificity $>99\%$)
are attainable.

Antiviral peptides

The functional regions of the HIV envelope glycoproteins have been studied
in a number of laboratories [12,13]. The binding site for CD4 is located in
the carboxyl-terminal region of gp120, whereas the amino-terminal 300 residues
of gp120 are important for the association between gp120 and gp41. We found
that the peptide gp120(105–129) interfered with virus replication as indicated
by reverse transcriptase activity. Since this is located in the region involved
in the interaction between gp120 and gp41, we inspected the sequence of gp41
for a possible complementary structure. This revealed a region in gp41(571–
591) with four matching charged residues spaced in the same juxtaposition as
corresponding residues in the region gp120(99–119). The alignment of these two
sequences is illustrated below. The matched charged residues are underlined.

```
HIV-1 gp41    W G I K⁵⁷⁴ Q L Q A R⁵⁷⁹ I L A V E R⁵⁸⁵ Y L K D⁵⁸⁹ Q Q
(571-591)

HIV-1 gp120   D M V E¹⁰² Q M H E D¹⁰⁷ I I S L W D¹¹³ Q S L K¹¹⁷ P C
(99-119)

HIV-2 gp41    W G T K⁵⁷³ N L Q A R⁵⁷⁸ V T A I E K⁵⁸⁴ Y L Q D⁵⁸⁸ Q A
(570-590)

HIV-2 gp120   T V T E⁸¹ Q A I E D⁸⁶ V W H L F E⁹² T S I K⁹⁶ P C
(78-98)
```

Both regions, gp120(99–117) and gp41(571–591), are highly conserved across all known isolates of HIV-1 and HIV-2 and have a propensity to form amphipathic α-helices with three complementary charged residues and complementary hydrophobic residues. The synthetic peptide corresponding to the region gp41(571–591) was prepared and found to have antiviral activity analogous to that seen for the complementary gp120 peptide [14]. One interpretation of these results is that the antiviral action of these peptides is due to their disturbing the interactions of the envelope proteins, perhaps analogous to ribonucleotide reductase from HSV where a 9-residue peptide from the carboxyl terminal of the β-subunit inhibited the enzyme [15].

Acknowledgements

This work was supported by the NH&MRC, a Commonwealth AIDS Research Grant and an Industrial Research and Development Grant to Agen Biomedical Ltd.

References

1. Riggen, C.H., Beltza, G.A., Hung, C.H., Thorn, R.M. and Marciani, D.J., J. Clin. Microbiol., 25 (1987) 1772.
2. Yoshida, T., Matsui, T., Kobayashi, S., Harada, S., Kurimiera, T., Hiriama, Y. and Yamamoto, N., Jpn. J. Cancer Res., 77 (1986) 1211.
3. Spielberg, F., Kabeya, C.M., Ryder, R.W., Kifuani, N.K., Harris, J., Bender, T.R., Heyward, W.L. and Quinn, T.C., Lancet, i (1989) 580.
4. Kemp, B.E., Rylatt, D.B., Bundesen, P.G., Doherty, R.R., McPhee, D.A., Stapleton, D., Cottis, L.E., Wilson, K.M., John, M.A., Khan, J.M., Dinh, D.P., Miles, S. and Hillyard, C.J., Science, 241 (1988) 1352.
5. Coombs, R.R.A., Scott, M.L. and Grange, M.P., J. Immunol. Methods, 101 (1987) 1.
6. Brennan, M., Davidson, P.F. and Paulus, H., Science, 229 (1985) 81.
7. Wang, J.J.G., Steel, S., Wisniewolski, R. and Wang, C.Y., Proc. Natl. Acad. Sci. U.S.A., 83 (1986) 6159.

823

8. Gnann, J.W., Schwimmbeck, P.L., Nelson, J.A., Traux, A.B. and Oldstone, M.B., J. Infec. Dis., 156 (1987) 261.

9. Smith, R.S., Naso, R.B., Rosen, J., Whalley, A., Hom, Y.-L., Hoey, K., Kennedy, C.J., McCutchan, A., Spector, S.A. and Richman, D.D., J. Clin. Microbiol., 25 (1987) 1498.

10. McPhee, D.A., Kemp, B.E., Cumming, S., Stapleton, D., Gust, I.D. and Doherty, R.R., FEBS Lett., 233 (1988) 393.

11. Narvanen, A., Korkolainen, M., Suni, J., Korpela, J., Kontio, S., Partanen, P., Vaheri, A. and Huhtala, M.-L., J. Med. Virol., 26 (1988) 111.

12. Kowalski, M., Potz, J., Basiripour, L., Goh, T., Terwilliger, E., Dayton, A., Rosen, C., Haseltine, W. and Sodroski, J., Science, 237 (1987) 1351.

13. Lasky, L.A., Nakamura, G., Smith, D.H., Fennie, C., Shimasaki, C., Patzer, E., Berman, P., Gregory, T. and Capon, D.J., Cell, 50 (1987) 975.

14. McPhee, D.A., Cummung, S.A., Pavuk, N.C. Doherty, R.R, Stapleton, D.I. and Kemp, B.E., Vaccines, 89 (1989) 185.

15. McClements, W., Yamanaka, G., Garsky, V., Perr, H., Bacchetti, S., Colonno, R. and Stein, R.B., Virology, 162 (1988) 270.

Chemical synthesis of a 99-residue HIV-1 protease with high enzymatic activity

R.F. Nutt, T.M. Ciccarone, S.F. Brady, P.L. Darke and D.F. Veber

Merck, Sharp & Dohme Research Laboratories, West Point, PA 19486, U.S.A.

Introduction

The human immunodeficiency virus (HIV-1), a retrovirus, has been identified as the causative agent of acquired immune deficiency syndrome (AIDS). A critical step in the life cycle of retroviruses is the proteolytic cleavage of initially translated polyproteins to mature and functional proteins. This processing step is carried out by a virally encoded protease belonging to the aspartyl protease family. For HIV-1, a region that encodes a 99-peptide with sequence homology to other retroviral proteases has been identified in the viral genome [1]. A single point mutation in the putative active site of this protease eliminates proteolytic activity of the protease as well as production of noninfectious virus [2]. These findings support the rationale that inhibitors of this enzyme are potentially useful therapeutic agents for the treatment of AIDS. To set up screening efforts for inhibitors, the need for adequate supplies of the protease became a critical issue for us. Therefore, the chemical synthesis of the 99-residue protease was carried out [3]. Additional studies have focused on the preparation and isolation of this synthetic enzyme with increased purity and full biological activity.

```
                Bzl      For     Tos              Bzl      ClZ      15
      Pro-Gln-Ile-Thr-Leu-Trp-Gln-Arg-Pro-Leu-Val-Thr-Ile-Lys-Ile-

                      ClZ Chx              Chx Bzl          Chx Chx 30
      Gly-Gly-Gln-Leu-Lys-Glu-Ala-Leu-Leu-Asp-Thr-Gly-Ala-Asp-Asp-

      Bzl         Chx Chx      Bzl              Tos For ClZ      ClZ 45
      Thr-Val-Leu-Glu-Glu-Met-Ser-Leu-Pro-Gly-Arg-Trp-Lys-Pro-Lys-

                                      ClZ      Tos      BrZ Chx 60
      Met-Ile-Gly-Gly-Ile-Gly-Gly-Phe-Ile-Lys-Val-Arg-Gln-Tyr-Asp-

                      Chx      Pmb     Bom ClZ              Bzl     75
      Gln-Ile-Leu-Ile-Glu-Ile-Cys-Gly-His-Lys-Ala-Ile-Gly-Thr-Val-

                      Bzl                              Tos         90
      Leu-Val-Gly-Pro-Thr-Pro-Val-Asn-Ile-Ile-Gly-Arg-Asn-Leu-Leu-

      Bzl              Pmb Bzl              99
      Thr-Gln-Ile-Gly-Cys-Thr-Leu-Asn-Phe-PAM RESIN
```

Fig. 1. Protection scheme for protease assembly.

Results and Discussion

The 99-peptide sequence of the postulated HIV-1 protease was synthesized by the SPPS [3]. Successful assembly of peptide-resin as shown in Fig. 1 was ascertained by chemical characterization of products at intermediate and final stages of the synthesis before and after HF cleavage.

Cleavage of peptide from the resin was accomplished by reaction with HF. The cleaved reaction product was purified by size permeation chromatography using Sephadex G-50F and G-75F in 50% HOAc. Fractions that contained a major peak when analyzed by HPLC (Fig. 2) were combined and used for preparation of the folded enzyme. Proteolytic activity could be obtained upon folding under reducing conditions, in the presence of ethylene glycol, glycerol and bovine serum albumin, and at the slightly acidic pH of 5.5 [3].

In an attempt to purify the synthetic enzyme further, the active enzyme was

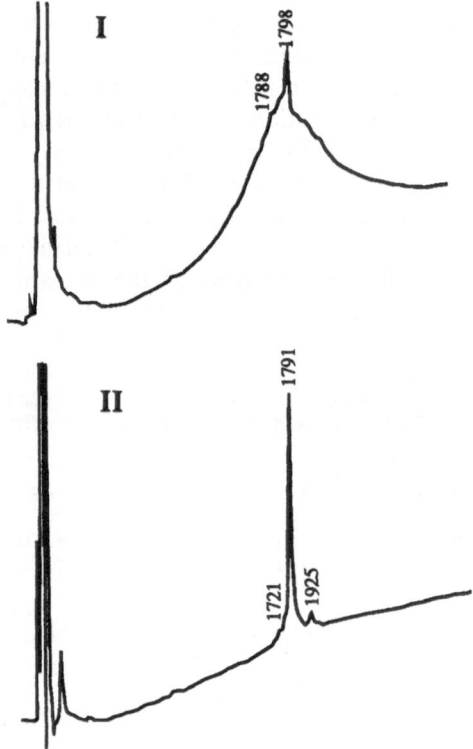

Fig. 2. HPLC of HIV-1 protease. I: After HF cleavage and size exclusion chromatography; II: after folding and ion-exchange chromatography. Conditions: Vydac C_4, 10 μm, 300 A; Abs = 210 nm; Buffers: A = 0.1% TFA, B = 0.1% TFA-CH$_3$CN; Gradient: 70% A to 50% A over 30 min.

isolated under non-denaturing conditions. Upon size permeation chromatography of the folded material, it was discovered that enzymatic activity was associated with a product of dimeric molecular weight. More efficient isolation of active material could be accomplished by ion-exchange chromatography. Use of a silica based carboxyethyl column (Baker CBX) and a solvent gradient of 0–0.5 M NaCl at pH 5.5 under reducing conditions afforded material with specific enzymatic activity of 6800 units (nmol of substrate VSQNYPIV cleaved/min/ mg of enzyme). This activity represented a 10-fold enhancement of potency over that previously reported [3] and was comparable to the activity observed for microbially expressed protease (P. Darke, personal communication). HPLC analysis (Fig. 2) of the isolated product indicated essentially a single peak, and preview sequence analysis confirmed the improvement in purity. At cycle 21 of the analysis, only 0.5% preview of Ala was observed which was indistinguishable from background. This result compares favorably with 4.6% preview observed at the same position for the product analyzed directly afte size permeation chromatography.

Conclusions

In summary, the chemical synthesis of a 99-peptide HIV-1 protease has been accomplished. A synthetic product of high purity and of full biological activity was isolated. Synthetic material was used to characterize the protease as to substrate and inhibitor specificity. A diversity of substrate cleavage sites was shown to be intrinsic to the HIV-1 protease. The high speed of synthesis and the availability of active material allowed the expedient establishment of screens for inhibitors as potential therapeutic agents for the treatment of AIDS.

References

1. Ratner, L., Haseltine, W., Patarca, R., Livak, K.J., Starcich, G., Josephs, S.F., Doran, E.R., Rafalski, J.A., Whitehorn, E.A., Baumeister, K., Ivanoff, L., Petteway, Jr., S.R., Pearson, M.L., Lautenberger, J.A., Papas, T.S., Ghrayeb, J., Chang, N.T., Gallo, R.C. and Wong-Staal, F., Nature, 313 (1985) 277.
2. Kohl, N.E., Emini, E.A., Schleif, W.A., Davis, L.J., Heimbach, J.C., Dixon, R.A.F., Scolnick, E.M. and Sigal, I.S., Proc. Natl. Acad. Sci. U.S.A., 85 (1988) 4686.
3. Nutt, R.F., Brady, S.F., Darke, P.L., Ciccarone, T.M., Colton, C.D., Nutt, E.M., Rodkey, J.A., Bennett, C.D., Waxman, L.H., Sigal, I.R., Anderson, P.S. and Veber, D.F., Proc. Natl. Acad. Sci. U.S.A., 85 (1988) 7129.

Studies on the synthesis and biological activity of HIV tat protein

Sara Biancalana[a], Derek Hudson[a],* and Alan D. Frankel[b]

[a]*Milligen/Biosearch, 81 Digital Drive, Novato, CA 94949, U.S.A.*
[b]*Whitehead Institute for Biomedical Research, Cambridge, MA 02142, U.S.A.*

Introduction

The HIV tat protein transactivates genes that are expressed from the viral long terminal repeat, and tat is essential for viral replication. The sequence of the 86-residue protein is rich in arginines, cysteines and glutamines and presents a complex test for synthetic methodology.

Results and Discussion

Peptides from tat were synthesized to determine the requirements for biological activity. Noteworthy features of the syntheses include automated Fmoc-methodology (Milligen/Biosearch 9600 synthesizer), with the BOP + HOBt coupling method [1], protection of the 8 Gln residues with trimethoxybenzyl (Tmob) [2], protection of Arg side chains with 2,2,5,7,8-pentamethylchroman-6-sulfonyl (Pmc) [3] and the use of PAL polystyrene resin, to prepare the peptides as their C-terminal amides. Peptides were cleaved and deprotected using a novel cocktail, dubbed Reagent R, formulated from TFA/ethanedithiol/thioanisole/anisole in the ratios of 90 : 3 : 5 : 2.

To test the use of trityl groups for protection of the 7 cysteines of tat, (contained within the sequence 22–37) and for histidine, an 18-residue peptide, 21–38, was prepared as its C-terminal amide using PAL resin. This peptide had previously been prepared by Boc-synthesis and found to be capable of forming metal-linked dimers [4], similar to those formed by the natural protein [5]. The presence of adjacent His(Trt) and Cys(Trt) residues was expected to provide a stringent test because of possible steric hindrance. The high purity of the 18-mer product showed that this selection of protecting groups could be successfully used with the BOP + HOBt coupling method.

To determine which regions of the 86-residue tat protein are important for transactivation, several peptides were synthesized (Table 1). Two syntheses were carried out, each starting at residue 58, and aliquots of resin were removed at appropriate points. The final products from both syntheses were identical. Also shown are 5 peptides, prepared by sampling a synthesis started at Phe[38], and a single peptide, 37–62 prepared from Ser[62]. After cleavage and deprotection

---

*To whom correspondence should be addressed.

with Reagent R for 4 h, each peptide was isolated by precipitation with cold anhydrous ether.They were then reduced with dithiothreitol, purified by preparative RPHPLC, and carefully characterized. Each peptide was tested for its ability to transactivate the HIV-1 promoter by scrape-loading peptides into HeLa cells (HL3T1) that contained an integrated HIV LTR controlling expression of the chloramphenicol acetyltransferase (CAT) gene [6].

Table 1   *Biological activity of tat peptides*

Origin	Sequence	Activity[a]	Origin	Sequence	Activity[a]
Bacterial	1–86	700	Synthetic	37–62	N.D.
Bacterial	1–72	350	Synthetic	1–38	N.D.
Synthetic	1–58	34	Synthetic	4–38	N.D.
Synthetic	5–58	10	Synthetic	10–38	N.D.
Synthetic	10–58	N.D.[b]	Synthetic	15–38	N.D.
Synthetic	15–58	N.D.	Synthetic	21–38	N.D.; inhib.[c]
Synthetic	21–58	N.D.	Chymotryptic fragment	1–47	N.D.
Synthetic	38–58	N.D.	Chymotryptic fragment	48–86	N.D.

[a] Activity is expressed as the -fold increase in HIV promoter activity in the presence of peptide, as measured by chloramphenicol acetyltransferase activity.
[b] Not detectable.
[c] 21–38 showed nonspecific inhibition of promoter activity at high peptide concentrations (see text).

Of the synthetic peptides, only tat 1–58 showed high levels of transactivation. Deletion of as few as 4 residues from the amino-terminus significantly decreased the activity, and removal of the first 9 residues abolished activity. After we had prepared tat 38–58 and found it to be inactive, a similar peptide, tat 37–62, was reported to retain most of the activity of the full protein [7]. We resynthesized this peptide, and now report it to be completely inactive in our assay system, even at a concentration 100-fold higher than that used with 1–58.

Since tat is essential for replication of HIV, an effective inhibitor of tat might be used to treat AIDS. Consequently, all peptides shown in Table 1 were assayed for possible inhibitory activity. At high peptide concentrations, 21–38 did inhibit transactivation. Using the recommended procedures and a manual simultaneous synthesis technique [1], a series of 7 peptide analogs of 21–38 were made in which each cysteine was systematically replaced with serine. Further variants replacing Cys^{22} with Cys(Acm), Cys(Me) and Cys(tBu) were made, as well as variants replacing His^{33} and Tyr^{26} with Phe. None of these analogs possessed higher inhibitory activity than 21–38. Indeed, further study showed that the inhibition observed with 21–38 was nonspecific in that it also blocked expression from the SV40 early promoter at similar concentrations.

S. Biancalana, D. Hudson and A.D. Frankel

Conclusions

The synthetic methodology reported provides a reliable recipe for the synthesis of complex peptides and small proteins. The structure activity data assembled on the tat sequence show that the N-terminal, the cysteine-rich region, and the basic region are all necessary for transactivation [8]. No partial sequence possesses specific inhibitory activity.

References

1. Hudson, D., J. Org. Chem., 53 (1988) 617.
2. Milligen-Biosearch Technical Bulletin, 9000-01 (1988).
3. Ramage, R. and Green, J., Tetrahedron Lett., 28 (1987) 2287.
4. Frankel, A.D., Chen, L., Cotter, R.J. and Pabo, C.O., Proc. Natl. Acad. Sci. U.S.A., 85 (1988) 6297.
5. Frankel, A.D., Bredt, D.S. and Pabo, C.O., Science, 240 (1988) 70.
6. Frankel, A.D. and Pabo, C.O., Cell, 55 (1988) 1189.
7. Green, M. and Loewenstein, P., Cell, 55 (1988) 1179.
8. Frankel, A.D., Biancalana, S. and Hudson, D., Proc. Natl. Acad. Sci. U.S.A., 86 (1989) 7397.

Development of fluorescent substrates for viral proteinases

Jeffrey R. Weidner, Raymon F. Roberts and Ben M. Dunn

Department of Biochemistry & Molecular Biology, University of Florida, Gainesville, FL 32610, U.S.A.

Introduction

Proteolytic processing of virally encoded polyproteins is an essential step in the maturation of many RNA viruses including picornaviruses and retroviruses. Picornaviruses (e.g., polio) synthesize their entire genome as a single polyprotein that is cleaved by two viral cysteine proteinases, 2A and 3C, whereas the retrovirus human immunodeficiency virus (HIV) uses a viral aspartic proteinase (HIV-PR) to process its gag and gag-pol polyproteins. These proteinases appear to be highly specific and present an attractive target for antiviral therapies. The development of a convenient, continuous spectrophotometric assay should be useful for characterizing these enzymes.

Fluorescence energy transfer is a useful principle for the design of substrates for continuous assay of proteinase activity [1]. Cleavage of the substrate results in an increase in the distance between the fluorescence donor and acceptor pair and a relief of fluorescence quenching, which can then be used to study the kinetics of the reaction.

Results and Discussion

Peptides based upon all eight known poliovirus 3C cleavage sites were examined for their ability to act as substrates for 3C in an HPLC assay. One of the best substrates, POLIO 1.2: (Ac-Arg-Cys-Nle-Glu-Ala-Leu-Phe-Gln-Gly-Pro-Leu-Gln-Tyr-Lys-Asp), was selected for modification to develop a fluorescence assay. This peptide was readily cleaved with a K_m of approximately 1 mM. The cysteine and lysine moieties also presented excellent sites for attachment of reporter groups.

α-N-acetyl POLIO 1.2 was derivatized with two equivalents of maleimidyl coumarin (Molecular Probes, Eugene) at pH 7.5 in 20% DMF for 1 hour and purified by RPHPLC. The resulting peptide, POLIO 1.2-C, had a coumarin moiety attached to the cysteine, as verified by AAA and fluorescence and UV-VIS spectroscopy.

DABITC (4-dimethylaminophenylazophenyl-4'-isothiocyanate) (Molecular Probes, Eugene) was added to the ϵ-NH$_2$ of the lysine at pH 9.5 in 50% acetonitrile to yield POLIO 1.2-CD. The DABTC-lysine quenched the coumarin's fluorescence and added an appropriate absorbance peak at 480 nm. POLIO 1.2-CD was soluble in aqueous solution but only to 150 μM, much less than the 1 mM K_m for the parent substrate peptide.

831

A large, time-dependent increase in the fluorescence of a POLIO 1.2-CD solution was observed following the addition of polio 3C proteinase. This approximate ten-fold increase in fluorescence occurred concomitantly with cleavage of the substrate peptide, as evident by the appearance of two new HPLC peaks and a decrease in the size of the substrate peak. These peaks represented the expected cleavage product peptides as confirmed by AAA and fluorescence and UV-VIS spectra.

The rate of increase in fluorescence was directly proportional to the amount of 3C proteinase added and was also linear with substrate concentration below 3 μM; however, the rate of fluorescence change fell off above this concentration. Cleavage still occurred at the higher substrate concentrations as evident from HPLC; therefore, the decrease in the rates of fluorescence change could be attributed to a decrease in fluorescence yield, perhaps due to inter-filter effects. This non-linear increase in fluorescence yield at concentrations above 3 μM was also noted in solutions of POLIO 1.2-CD alone and in cleaved reaction mixtures. This problem might be avoided by the choice of different donor–acceptor pairs with less spectral overlap or by decreasing the path length in the cuvette.

In addition, POLIO 1.2-CD has been used to evaluate several potential inhibitors of the 3C proteinase, including chicken egg white cystatin, a cysteine proteinase inhibitor. An apparent K_i of 42 μM was determined by assuming that $[S] \ll K_m$ and $K_i \approx IC_{50}$. This is close to the value of 18 μM determined by HPLC assay. Thus, despite the limited concentration range, this assay should be useful for screening potential inhibitors of the proteinase.

Principles used in the design of this substrate are currently being applied to the development of similar assay systems for HIV-PR. In addition, other reporter pairs are being evaluated for their suitability in this approach. Efforts are also underway to increase the solubility of these derivatives and to increase the useful concentration range of the assay.

References

1.   Yaron, A., Carmel, A. and Katchalski-Katzir, E., Anal. Biochem., 95 (1979) 228.

Tetrazole analogs of HIV-1 protease substrate peptide: The role of *cis*-proline in enzyme recognition

Assunta Garofalo, Céline Tarnus, Jean-Marc Remy, Raymond Leppik, François Piriou, Bruce Harris and John Tom Pelton
Merrell Dow Research Institute, 16 rue d'Ankara, F-67084 Strasbourg, France

Introduction

The HIV-1 retrovirus, the causative agent of acquired immune deficiency syndrome (AIDS), codes for a virus-specific protease known to be essential for retrovirus maturation and replication. An unusual feature of this aspartyl protease is its ability to cleave on the N-terminal side of proline. To further understand the role of proline in the cleavage site of this enzyme, we have recently expressed the HIV-1 protease in *E. coli* and examined its ability to cleave synthetic peptides. The heptapeptide Ser-Gln-Asn-Tyr-Pro-Ile-Val-NH$_2$, corresponding to one of the native sequences in the gag precursor protein processed by the retroviral protease, was found to be a substrate for the enzyme. NMR studies of this heptapeptide indicated that it exists in two major conformations: (1) 70% as an extended peptide with all *trans* peptide bonds; and (2) 30% with a *cis*-proline peptide bond. Molecular modeling of the heptapeptide with *cis*-proline yields a molecule with a pronounced 'kink' that may be important for protease recognition. To examine the role of the *cis*-proline bond in enzyme recognition and binding, we have synthesized the tetrazole analog, Ser-Gln-Asn-Pheψ[CN$_4$]Ala-Ile-Val-NH$_2$. This peptide bond surrogate has been proposed as a means of mimicking the *cis* configuration [1].

Results and Discussion

The tetrazole dipeptide Pheψ[CN$_4$]Ala-OH was synthesized as previously described [2] except that alanine was protected as the *t*-butyl ester so as to avoid the base-promoted racemization of the C-terminal residue; benzyloxycarbonyl was used for phenylalanine. The protecting groups were removed with freshly prepared TFMSA-TFA and the amino terminus reprotected with the Boc group. The tetrazole analog was coupled to the peptide resin with HOBt/DCC and the synthesis completed by standard methods. Gel filtration and preparative HPLC afforded the pure peptide as determined by TLC and analytical C$_{18}$ RPHPLC.

AAA after acid hydrolysis gave the expected molar ratios ($\pm7.0\%$) of the constituent amino acids. The FABMS molecular ion ([MH$^+$] = 802.4 expt; 802.1 calc.) and fragmentation patterns were in agreement with the amino acid sequence and composition.

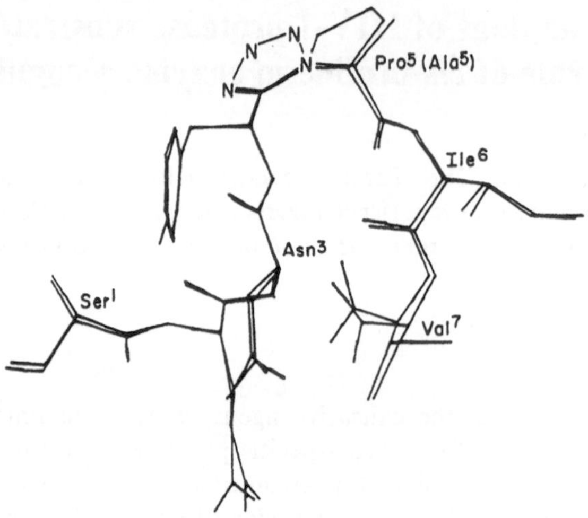

Fig. 1. Overlay of the tetrazole peptide and the substrate peptide Ser-Gln-Asn-Tyr-Pro-Ile-Val-NH₂ which contains a cis-peptide bond.

NMR studies of the tetrazole analog in D_2O indicate a random coil structure. Chemical shifts and coupling constants of the α and β protons were found to be similar to those observed in the cis-heptapeptide, except for the Phe and Ala αCH protons that were observed at lower field (5.40 and 5.25 ppm, respectively).

Both the substrate heptapeptide and the tetrazole analog were modeled using the SYBYL [3] program. A suitable starting structure for each molecule was generated by sequentially adding amino acid building blocks in a random coil conformation, followed by energy minimization. Overlay of the two peptides through a 'fit' program shows a good correlation between the conformation of the tetrazole analog and that of the cis proline substrate (Fig. 1). Studies with the expressed HIV-1 protease indicate that the tetrazole analog is not a substrate for, nor an inhibitor of, this enzyme. This result may indicate that the cis-amide bond is not important for enzyme recognition. Alternatively, the increased steric bulk of the tetrazole analog at the cleavage site may prevent effective binding of the peptide to the protease.

References

1. Marshall, G., Humblet, C., Van Opdenbosch, N. and Zabrocki, J., In Rich, D.H. and Gross, E. (Eds.) Proceedings of the 7th American Peptide Symposium, Pierce Chemical Company, Rockford, IL, 1981, p. 669.
2. Yu, K.-L. and Johnson, R.L., J. Org. Chem., 52 (1987) 2051.
3. SYBYL Molecular Modeling Software, Version 5.1, Tripos Associates Inc., St. Louis, Missouri.

Inhibitors of HIV protease based on modified peptide substrates

Mihaly V. Toth[a], Francis Chiu[a],*, George Glover[b],**, Stephen B.H. Kent[c], Lee Ratner[d], Nancy Vander Heyden[d], Jeremy Green[e], Daniel H. Rich[e] and Garland R. Marshall[a],***

[a]*Department of Pharmacology, Washington University School of Medicine, St. Louis, MO 63110, U.S.A.*
[b]*Monsanto Company, Chesterfield, MO 63198, U.S.A.*
[c]*Department of Biology, Bond University, Gold Coast Mail Centre, Queensland, Australia 4217*
[d]*Department of Medicine, Washington University School of Medicine, St. Louis, MO 63110, U.S.A.*
[e]*School of Pharmacy, University of Wisconsin, Madison, WI 53706, U.S.A.*

Introduction

Inhibition of the HIV protease as a means of therapeutic intervention in the treatment of AIDS is a logical strategy. HIV protease cleaves the virally encoded polyprotein at several sites to liberate the *gag* proteins (p17, p24, p15), as well as the protease (PR) itself, and reverse transcriptase (p66). Schneider and Kent [1] have prepared peptide substrates of 20 or more residues corresponding to all of these cleavage sites and demonstrated that their synthetic enzyme cleaved each at the appropriate site. Darke et al. [2,3] showed that both expressed and synthetic enzyme cleaved larger synthetic peptides at the proposed processing sites. The minimal length peptide substrate that was shown to be cleaved (at 90% of the rate of the octapeptide) was a heptapeptide, SQNYPIV. Either SQNYPI, QNYPIV, or acetyl-QNYPIV failed to be cleaved. Billich et al. [4] have also reported that heptapeptide substrates were cleaved by HIV-1 protease. In studies on the protease from the retrovirus, avian sarcoma-leukosis virus (ASLV), Kotler et al. [5,6] examined decapeptide substrates and suggest that the minimal size for cleavage with ASLV protease is longer than six residues. In contrast, we show that hexapeptides still retain sufficient binding affinity to function as substrates and serve as a basis for inhibitor design.

In an analogous approach to that of Szelke et al. [7], we have prepared shortened peptide substrates containing dipeptide analogs of Met-Met, Leu-Pro, Tyr-Pro and Phe-Pro, known cleavage sites of retroviral proteases, in which the amide bond has been reduced to give the $\psi[CH_2NH]$ linkage. A report of decapeptide

---

*Present address: Department of Chemistry, La Trobe University, Bundoora, Victoria, Australia 3083.
**Present address: Smith Kline & French Laboratories, King of Prussia, PA 19406-0939, U.S.A.
***To whom correspondence should be addressed.

substrate analogs, VSFNF-ψ[CH$_2$N]-PQITL and CTLNF-γ[CH$_2$N]-PISPI, and of HIV substrates (N- and C-terminal sites of the protease) containing the reduced amide bond that showed inhibitory activity (50% inhibition around or below 125 μM concentration) has appeared [4] since this work was completed.

Experimental

Solid phase synthesis of HIV protease substrates and inhibitors

Acetyl hexapeptide amides, with and without reduced internal amide bonds, were prepared by conventional SPPS using the p-MBHAR. The dipeptide analogs having reduced peptide bonds were prepared, either in solution and incorporated as a unit in the solid phase protocol, or in situ by reductive alkylation with 4 equivalents of Boc-amino acid aldehyde prepared by the procedure of Fehrentz and Castro [8] and coupled according to Sasaki et al. [9].

HIV protease assay

The HIV protease assay was conducted using either synthetic HIV protease [1] in which the two Cys residues (Cys[67], Cys[95]) had been replaced by the isosteric α-aminobutyric acid to eliminate the complications of free sulfhydryl groups (Clawson and Kent, unpublished), or cloned material expressed in $E.$ $coli$ supplied by Dr. George Glover of the Monsanto Company. In all cases examined, the cleavage patterns and inhibition results were identical. Synthetic protease (1 mg/ml) was dissolved in a buffer (20 mM PIPES, 0.5% NP-40 detergent, 1 mM DTT, pH 6.5). Substrates were also dissolved in the same buffer at 1 mg/ml concentration. Five microliters of substrate and 5 μl of HIV protease were added to an Eppendorf tube that was centrifuged and then incubated at 25°C for the desired time. The reaction was stopped by the addition of 50 μl of 10% TFA. The sample was diluted with 200 μl of water and applied to a C$_4$ HPLC column developed with 0.1% TFA for 5 min, followed by a gradient of 0–50% acetonitrile in 50 min. For inhibitor studies, 5 μl of the protease solution was preincubated for 10 min with the 5 μl of inhibitor (dissolved in DMSO and diluted to 1 mg/ml with buffer). Then 5 μl of test substrate, Ac-Thr-Ile-Met-Met-Gln-Arg-NH$_2$, was added in order to determine inhibition of cleavage. Reactions were stopped and cleavage rates were monitored by HPLC, according to procedures described above.

Results and Discussion

In contrast to the reports of Darke et al. [2,3] on HIV protease, and the studies of Kotler et al. [5,6] on ASLV protease, hexapeptides function as good substrates, provided that their charged amino and carboxyl termini are blocked, although both are not essential. The p24/p15 hexapeptide, Ac-Thr-Ile-Met-Met-Gln-Arg-NH$_2$, was characterized with a $K_m = 1.4$ mM and a $V_{max} = 725$ nmol/min/mg. This compares favorably with the data of Darke et al. [2,3] for most of the longer peptide substrates.

The significant point is that the reduced amide transition-state inhibitor strategy has yielded inhibitors in all four substrate sequences to which it has been applied (Table 1).

Table 1 *Reduced amide bond inhibitors of HIV protease*

Cleavage site	Peptide	K_I[a]
p17/p24	Ac-Gln-Asn-Tyr-ψ[CH$_2$N]-Pro-Ile-Val-NH$_2$	17.2 μM
(HIV-2)	Ac-Gly-Asn-Tyr-ψ[CH$_2$N]-Pro-Val-Gln-NH$_2$	++
p24/p15	Ac-Thr-Ile-Met-ψ[CH$_2$NH]-Met-Gln-Arg-NH$_2$	+++
	Ac-Thr-Ile-Nle-ψ[CH$_2$NH]-Nle-Gln-Arg-NH$_2$	789 nM
	Ac-Thr-Ile-Leu-ψ[CH$_2$NH]-Leu-Gln-Arg-NH$_2$	734 nM
	Ac-Thr-Ile-Nle-ψ[CH$_2$NH]-Leu-Gln-Arg-NH$_2$	+++
	Ac-Thr-Ile-Nle-ψ[CH$_2$NH]-Phe-Gln-Arg-NH$_2$	+++
	Ac-Thr-Ile-Nle-ψ[CH$_2$NH]-Nle-Gln-D-Arg-NH$_2$	++
	Ac-Thr-Ile-Nle-ψ[CH$_2$NH]-Ile-Gln-Arg-NH$_2$	+++
	Ac-D-Thr-Ile-Nle-ψ[CH$_2$NH]-Nle-Gln-Arg-NH$_2$	++
	Ac-Thr-Ile-Nle-ψ[CH$_2$NH]-Nle-Gln-NH$_2$	++
	Ac-Ile-Nle-ψ[CH$_2$NH]-Nle-Gln-Arg-NH$_2$	++
p15/PR	Ac-Phe-Asn-Phe-ψ[CH$_2$N]-Pro-Gln-Ile-NH$_2$	++
	Ac-Ser-Phe-Asn-Phe-ψ[CH$_2$N]-Pro-Gln-Ile-NH$_2$	++
(HIV-2)	Ac-Leu-Ala-Ala-ψ[CH$_2$N]-Pro-Gln-Phe-NH$_2$	+
PR/p66	Ac-Leu-Asn-Phe-ψ[CH$_2$N]-Pro-Ile-Ser-OH	++
	H$_2$N-Thr-Leu-Asn-Phe-ψ[CH$_2$N]-Pro-Ile-Ser-OH	++
(HIV-2)	Ac-Leu-Asn-Leu-ψ[CH$_2$N]-Pro-Val-Ala-NH$_2$	++
	H$_2$N-Ser-Leu-Asn-Leu-ψ[CH$_2$N]-Pro-Val-Ala-OH	13.1 μM

[a] Under these assay conditions, pepstatin A had a $K_I = 5.46$ μM. The compounds whose affinities have not been fully characterized are classified into four categories based on comparative inhibition studies: +++ = submicromolar; ++ = 1–100 micromolar; + = greater than 100 micromolar; – = inactive.

Acknowledgements

This work has been supported in part by the National Institutes of Health (Grants GM-24483 and AI-27302) as well as Monsanto (Grant 44353K). Analytical data were obtained at the Washington University High Resolution NMR Service Facility (NIH Grant RR-02004) and the Washington University Mass Spectroscopy Resource (RR-00945).

References

1. Schneider, J. and Kent, S.B.H., Cell, 54 (1988) 363.
2. Darke, P.L., Nutt, R.F., Brady, S.F., Garsky, V.M., Ciccarone, T.M., Leu, C.T., Lumma, P.K., Freidinger, R.M., Veber, D.F. and Sigal, I.S., Biochem. Biophys. Res. Comm., 156 (1988) 297.

3. Darke, P.L., Leu, C.T., Davis, L.J., Heimbach, J.C., Diehl, R.E., Hill, W.S., Dixon, R.A.F. and Sigal, I.S., J. Biol. Chem., 264(1989)2307.
4. Billich, S., Knoop, M.T., Hansen, J., Sedlacek, J., Mertz, R. and Moelling, K., J. Biol. Chem., 263(1988)17905.
5. Kotler, M., Katz, R.A., Danho, W., Leis, J. and Skalka, A.M., Proc. Natl. Acad. Sci. U.S.A., 85(1988)4185.
6. Kotler, M., Danho, W., Katz, R.A., Leis, J. and Skalka, A.M., J. Biol. Chem., 264(1989)3428.
7. Szelke, M., Leckie, B., Hallett, A., Jones, D.M., Sueiras, J., Atrash, B. and Lever, A.F., Nature, 299(1982)555.
8. Fehrentz, J.A. and Castro, B., Synthesis, (1983)676.
9. Sasaki, Y., Murphy, W.A., Heiman, M.L., Lance, V.A. and Coy, D.H., J. Med. Chem., 30(1987)1162.

Three-dimensional structure of an hexapeptide-derived inhibitor of human T-lymphotropic virus HTLV1

G. Precigoux, S. Geoffre, M. Hospital and P. Picard

Laboratoire de Cristallographie, Université de Bordeaux I, 351 cours de la Libération,
F-33450 Talence, France

Introduction

Retroviruses such as human T-cell leukemia virus (HTLV1) code for a virus-specific protease that is essential for the proteolytic cleavage yielding the mature viral protein. A 115-residue polypeptide, encoded at the 5′ end of *pol* gene, has been proposed as the putative HTLV1 protease. Analysis of the retroviral protease sequence led to the suggestion that this enzyme was a member of the aspartyl protease family, on the basis of the conserved characteristic Asp-Thr-Gly active side sequence and the observed in vitro inhibition by the protease-specific inhibitor, pepstatin A [1]. The recently described structures of the Rous sarcoma virus (RSV) protease [2] and of the human immunodeficiency virus (HIV) protease [3] show, for such enzymes, a dimeric structure with a high structural homology to other aspartyl proteases [4]. On examining the cleavage sites of the viral polyprotein *gag* and *pol* precursors, a preference for the Leu-Pro sequence can be noted. Nevertheless, the proline residue is not essential.

In order to investigate the requirements for the protease inhibition, synthetic peptides with either reduced bond at the cleavage site or a substitution by statine were synthesized. The crystal structure of one of them, Pro-Gln-Val-Sta-Ala-Sta, is reported.

Results and Discussion

The hexapeptide was synthesized according to the Merrifield solid phase procedure [5], and using a Boc-statine prepared as described by Rich et al. [6]. Crystallization was carried out in Corning depression glass plates by the vapor diffusion method [7]. All diffraction data were collected from one crystal obtained from a solution containing 30 mM hexapeptide, 30 mM sodium cacodylate (pH 6.0), 20% (g/ml) 1,6-hexamethylenediol by equilibration with a reservoir of 100% 2-methyl-2,4-pentanediol. The crystal structure was solved by direct-phase determination using the random-start multi-solution procedure in the SHELXS-86 computer program [8]. The molecule crystallizes as a pentahydrate in the $P2_1$ monoclinic space group, with $a = 10.874(3)$, $b = 9.501(2)$, $c = 21.062(6)$ Å, $\beta = 103.68°$ and $Z = 2$. The molecular structure shown in Fig. 1 can be succinctly described as extended. The crystal-state conformation is characterized by a N-terminal β-pleated sheet conformation followed, at the hydroxyl group level, by another extended structure.

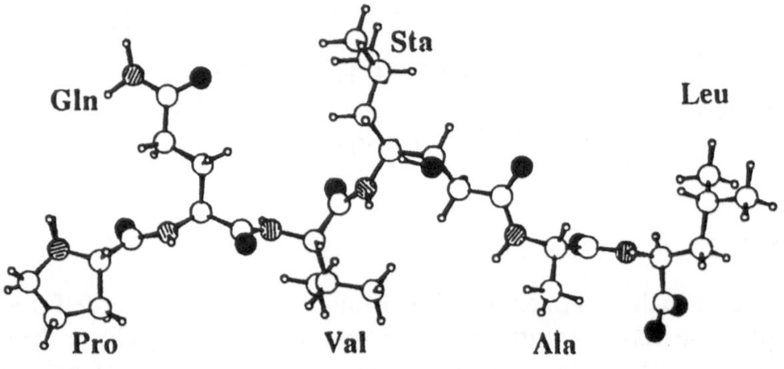

Fig. 1. Molecular structure for Pro-Gln-Val-Sta-Ala-Sta.

The comparison with other acidic-protease inhibitors described previously (inhibitor–endothiapepsin complexes) [9] shows a good agreement in the general shapes of the molecules. The orientations of the main chains and side chains of the first 4 residues are very similar. The hydrogen bonding arrangement, as far as these residues are concerned, corresponds to the classical antiparallel pleated-sheet scheme for the isolated molecule as well as for the molecular complexes. Therefore, the conformation of the first four residues observed for Pro-Gln-Val-Sta-Ala-Leu should provide a good model for the design of the N-terminal end of HTLV1-protease inhibitors. On the other hand, the Ala and Leu side chains are extended in opposite directions compared to that of the corresponding inhibitors' side chains in the site of endothiapepsin.

References

1. Katoh, I., Yasunaga, T., Ikawa, Y. and Yoshinaka, Y., Nature, 329 (1987) 654.
2. Miller, M., Jaskolski, M., Mohana Rao, J.K., Leis, J. and Wlodawer, A., Nature, 337 (1989) 576.
3. Navia, M.A., Fitzgerald, P.M.D., McKeever, B.M., Leu, C.-T., Heimbach, J.C., Herber, W.K., Sigal, I.S., Darke, P.L. and Springer, J.P., Nature, 337 (1989) 615.
4. Pearl, L.H. and Taylor, W.R., Nature, 329 (1987) 351.
5. Merrifield, R.B., J. Am. Chem. Soc., 77 (1963) 1067.
6. Rich, D.H., Sun, E.T. and Boparai, A.S., J. Org. Chem., 43 (1978) 3624.
7. McPherson, A., Preparation and Analysis of Proteins Crystals, Wiley, San Francisco, 1982.
8. Sheldrick, G.M. SHELXS-86, Program for crystal structure determination, Univ. of Göttingen, F.R.G., 1986.
9. Foundling, S.I., Cooper, J., Watson, F.E., Cleasby, A., Pearl, L.H., Sibanda, B.L., Hemmings, A., Wood, S.P., Blundell, T.L., Valler, M.J., Norey, C.G., Kay, J., Boger, J., Dunn, B.M., Leckie, B.J., Jones, D.M., Atrash, B., Hallett, A. and Szelke, M., Nature, 327 (1987) 349.

Early development of anti-HIV IgG

Britta Wahren[a,*], Per Anders Broliden[a,b], Jerzy Trojnar[c], Ellen Sølver[a], Jan Albert[a,b], Eric Sandström[d], Madeleine von Sydow[e] and Hans Wigzell[f]

[a]*Department of Virology, National Bacteriological Laboratory, S-105 21 Stockholm, Sweden*
[b]*Department of Virology, Karolinska Institute, S-104 01 Stockholm, Sweden*
[c]*Ferring AB, P.O. Box 30561, S-200 62 Malmö, Sweden*
[d]*Department of Venereology, South Hospital, S-100 64 Stockholm, Sweden*
[e]*Central Microbiological Laboratory, Stockholm County Council, P.O. Box 70470, S-107 26 Stockholm, Sweden*
[f]*Department of Immunology, Karolinska Institute, S-104 01 Stockholm, Sweden*

Introduction

A critical event during HIV infection is the binding of virion ligand protein to the cellular receptor. An important step in the viral infection involves binding of envelope glycoprotein gp120 to the CD4 complex [1]. The presence of this receptor on T-helper cells and many monocytes/macrophages make these cells the primary HIV-1 targets. Virion cell adherence or syncytium formation between infected cells can be inhibited by animal sera to an envelope region [2]. The variability of these regions explains why neutralization-escape is a common finding with individual isolates.

By modification of the natural amino acid sequence of a part of the CD4 binding region, we discovered a peptide to which the majority of both HIV-1 and HIV-2 infected persons have a strong seroreactivity. Previous experiences with cyclic peptides in virology have sometimes shown enhanced serological reactivities to peptides where the natural sequence permits cyclization. In the present case, we introduced an artificial sequence in order to enhance antibody reactivity and to acquire a stable peptide.

Results and Discussion

Several peptides synthesized according to the linear sequence of HIV-1 (strain HTLV-III$_B$) [3] were assayed for seroreactivity in ELISA. The strongest and most frequent reactivity was seen with the peptide JB-8p (Table 1). Other linear or cyclic peptides covering the C-terminal of the CD-4 binding region displayed weaker reactivity with sera. The N-terminus of the CD4 binding region also revealed some reactive sera. This reactivity was not obviously enhanced when cyclization was performed.

---

*To whom correspondence should be addressed.

Table 1 *Peptide JB-8p characteristics*

aa	non-variable (15 strains) aa 427–448	Chou, Fasman	amphipatic index	antigenic index	homology HIV-1/HIV-2	HIV-2
427	NH₂-W					W
	Q				x	M
	E		29		x	K
	V	H	30			V
	G	H	34	+		G
	K	H	27	+	x	R
	A	H	26		x	N
	C	H	27		x	V
	Y	H	14			Y
	A	H	10		x	L
	P	b	08			P
	P	b	14			P
	I	b	15		x	R
	S	b	13		x	E
	G	b	07			G
	Q	b	13	+	x	E
	I	b	13	+	x	L
	R		14	+		S
	C		14			C
	S	T	16		x	N
	S	T	17			S
448	N -COOH	T	16		x	T

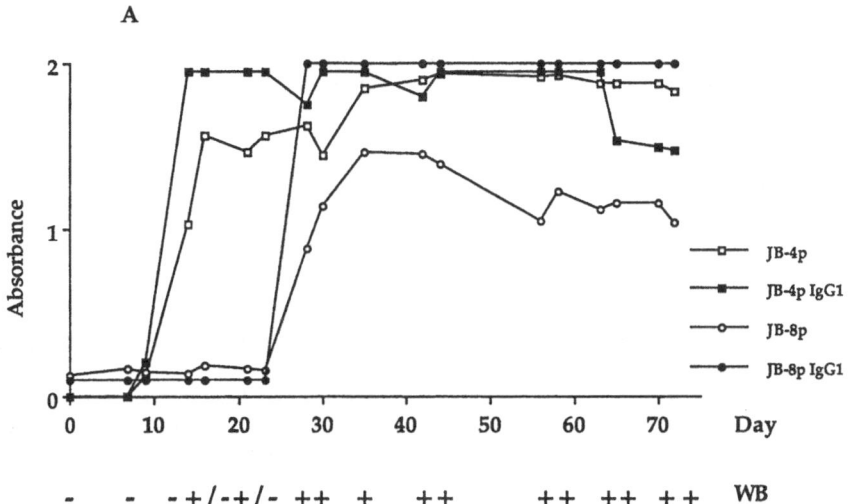

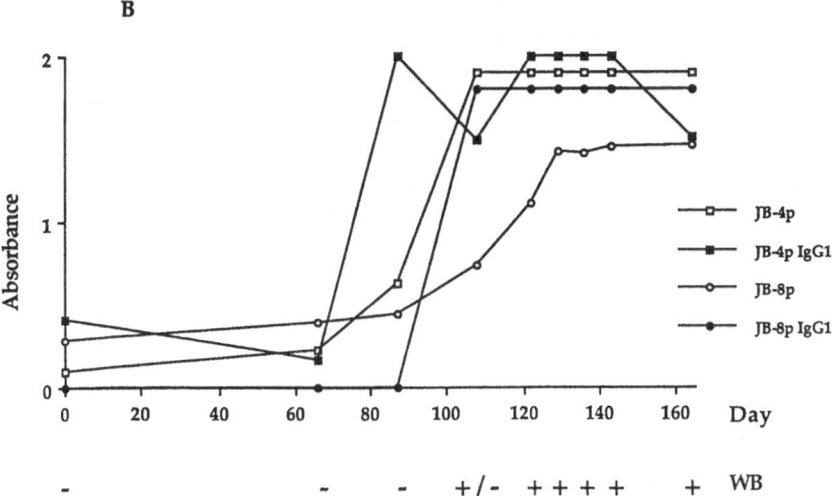

Fig. 1. Development of anti-JB-8p and anti-JB-4p antibodies by testing sequential sera from two primary HIV-infections (A and B). WB = reactivity in Western blot.

Predicted structure [4–6] of JB-8p is shown in Table 1. The peptide has hydrophobic and hydrophilic stretches, as well as two cysteines or a methionine and prolines. These amino acids should participate in a strongly structured compound. By exchanging M to C (position 434) it was possible to form an artificial loop between positions 434 and 445. When the resulting JB-8p peptide

843

was used with HIV-1 seropositive sera, 93% of the patients showed a clear seroreactivity. The figure for HIV-2-infected persons [7] was 73%. Anti-JB-8p antibodies developed in primary HIV-1 infection (Fig. 1). The time course was closely related to that of antibodies to a peptide representing gp 41 [8]. Obviously the conformational change imposed on the peptide JB-8p revealed and exposed stretches recognized as immunogens in the patient, but not well discovered by linear peptides.

References

1. Lasky, L.A., Nakamura, G., Smith, D.H., Fennie, C., Shimasaki, C., Patzer, E., Berman, P., Gregory, T. and Capon, D.J., Cell, 50 (1987) 975.
2. Rusche, J.R., Javaherian, K., McDanal, C., Petro, J., Lynn, D.L., Grimaila, R., Langlois, A., Gallo, R.C., Arthur, L.O., Pischinger, P.J., Bolognesi, D.P., Putney, S.D. and Matthews, T.J., Proc. Natl. Acad. Sci. U.S.A., 85 (1988) 3198.
3. Database of human retroviruses and AIDS, 3rd ed., Los Alamos, New Mexico, 1989.
4. Margalit, H., Spouge, J.L., Cornette, J.L., Cease, K.B., Delisi, C. and Berzofsky, J.A., J. Immunol., 138 (1987) 2213.
5. Wolf, H., Modrow, S., Motz, M. et al., Cabios, 4 (1988) 187.
6. Chou, P.Y. and Fasman, G.D., Biochem., 13 (1974) 222.
7. Broliden, P.A., Outtara, A.S., Sølver, E., Trojnar, J., Norrby, E. and Wahren, B., Abstract P-308, 11th American Peptide Symposium, La Jolla, U.S.A., 1989.
8. Rudén, U., Trojnar, J., Sølver, E. and Wahren, B., Abstract P-299, 11th American Peptide Symposium, La Jolla, U.S.A., 1989.

Synthesis of a peptide comprising the complete preS2 region of the hepatitis B virus (HBV) that confers protection in chimpanzees against HBV challenge

Victor M. Garsky[a,*], Patricia Conard[a], Jorg Eichberg[b], Emilio A. Emini[a], Roger M. Freidinger[a], Vivian Larson[a], Kenneth J. Stauffer[a] and Daniel F. Veber[a]

[a]Merck Sharp & Dohme Research Laboratories, Departments of Medicinal Chemistry and Virus and Cell Biology, West Point, PA 19486, U.S.A.
[b]Southwest Foundation for Biomedical Research, Virology and Immunology Department, San Antonio, TX 78284, U.S.A.

Introduction

Recent studies have emphasized the immunological role of the preS2 region of the hepatitis B virus (HBV) surface antigen in mediating responses to the S region [1]. In addition, studies have shown that truncated versions of preS2 conjugated to appropriate carriers can produce protective antibodies in chimpanzees [2]. Our studies extend these observations by demonstrating that alum adsorbed unconjugated synthetic peptide containing the full 55-amino acid sequence of preS2 (Fig. 1) can be used to immunize and protect chimpanzees from an infectious HBV challenge [3]. This report describes the synthesis, purification, and characterization of preS2.

NH_2-Met-Gln-Trp-Asn-Ser-Thr-Ala-Phe-His-Gln-Ala-Leu-Gln-Asp-Pro-Arg-Val-Arg-Gly-Leu-Tyr-Leu-Pro-Ala-Gly-Gly-Ser-Ser-Ser-Gly-Thr-Val-Asn-Pro-Ala-Pro-Asn-Ile-Ala-Ser-His-Ile-Ser-Ser-Ile-Ser-Ala-Arg-Thr-Gly-Asp-Pro-Val-Thr-Asn-OH

Fig.1. Sequence of preS2.

Results and Discussion

The chemical synthesis of the 55-amino acid-containing preS2 was the primary objective of our studies. Assembly of the peptide was achieved automatically. Selection of side-chain protection was made in an effort to minimize the formation of by-products [e.g., Asp(cHex), Glu(cHex), His(Bom)] during synthesis and HF deprotection. A double-coupling protocol was used for peptide resin assembly. Coupling efficiency was monitored by ninhydrin analysis and found to be greater than 98.5% in all cycles. The final 55-peptide resin was cleaved from the resin using the high HF procedure in the presence of *p*-cresol and *p*-thiocresol (1:1)

---

*To whom correspondence should be addressed.

as scavengers. The crude product was purified by preparative HPLC on a C_4 silica reverse-phase column using a 0.1% aqueous trifluoroacetic acid-acetonitrile gradient. After a single pass, the product appeared to be about 90% pure, as determined by analytical HPLC. A second purification using a 0.1% ammonium bicarbonate-acetonitrile gradient on C_4 gave a product that was >98% pure in a 5% overall yield.

The product was characterized for confirmation of structure and homogeneity. Amino acid composition following acid hydrolysis showed ratios within ±5% of expected values. Sequence analysis of preS2 carried out for 45 cycles produced no detectable preview. Proton NMR in D_2O established that the methionine had not oxidized to methionine oxide. The NMR also indicated that alkylation of side-chain functionalities had not occurred. Evidence of correct molecular weight (5734) was confirmed by FABMS that gave an M+H of 5735. Analytical RPHPLC in several buffer systems indicated that the product was >98% pure.

Following adsorption into alum, the preS2 peptide was inoculated into four chimpanzees, as described earlier [3]. Each animal was subsequently challenged with infectious doses of HBV. All of the peptide-inoculated chimpanzees were found to be fully protected.

Acknowledgements

The authors wish to acknowledge L. Wassel for AAA, J. Rodkey for sequence analyses, H. Ramjit for MS, J. Murphy for NMR studies, and V. Finley for typing the manuscript.

References

1. Milich, D.R., Immunol. Rev., 99 (1987) 71.
2. Itoh, Y., Takai, E., Ohnuma, H., Kitajima, H., Tsuda, F., Machida, A., Mishiro, S., Nakamura, T., Miyakawayk, Y. and Mayumi, M., Proc. Natl. Acad. Sci. U.S.A., 83 (1986) 9174.
3. Emini, E.A., Larson, V., Eichberg, J., Conrad, P., Garsky, V.M., Lee, D.R., Ellis, R.W., Miller, W.J., Anderson, C.A. and Gerety, R.J., J. Med. Virol., 28 (1989) 7.

Prevalence of anti-*nef* positive sera in HIV-infected patients: Mapping of the major epitopes of *nef*-protein using synthetic peptides

J.-M. Sabatier[a], J. van Rietschoten[a], G. Fontan[a], B. Clerget-Raslain[b],
E. Fenouillet[b], C. Granier[a], J.-C. Gluckman[b], L. Montagnier[c], H. Rochat[a]
and E. Bahraoui[a]

[a]UDC, CNRS UA1179 – INSERM U172, Laboratoire de Biochimie, Faculté de Médicine
Nord, Bd Pierre Dramard, F-13326 Marseille Cédex 15, France
[b]Laboratoire de Biologie et Génétique des Déficits Immunitaires, CERVI,
UER Pitié-Salpêtrière, F-75013 Paris, France
[c]Unité d'Oncologie Virale, CNRS UA1157, Institut Pasteur, 25 rue du Docteur Roux,
F-75724 Paris Cédex 15, France

Introduction

In addition to encoding the classic structural proteins of retroviruses (*gag, pol, env*), the human immunodeficiency virus (HIV) [1] genome encodes regulatory proteins (*tat, rev, nef*,...), one of which, the *nef* (negative regulatory factor) [2,3], is secreted and plays a keyrole in the pathophysiology of AIDS. This protein of 205 amino acid residues for HIV-1 is involved in the down-regulation of HIV replication, thus establishing and maintaining virus latency [4–6].

As a negative factor of HIV replication, *nef* appears to be produced before any detectable expression of *gag/env* proteins. *Nef* may thus be the target of antibodies that appear early in the course of HIV infection [7,8].

Results and Discussion

To investigate the immune response to *nef*, the prevalence of anti-*nef*-positive sera was assessed in HIV-infected patients, and immunodominant epitopes were mapped with synthetic peptides [9,10]. Human sera were tested for the presence of anti-*nef* antibodies by RIA with recombinant radiolabeled [^{125}I]*nef* expressed in *E. coli*. Of the 300 HIV-positive sera tested by RIA, $70 \pm 5.3\%$ were found to be anti-*nef*-positive at a dilution of 1:200. Anti-*nef* antibodies bound to *nef* with a high affinity ($K_{0.5} = 2.2 \times 10^{-9}$ M). The specificity of anti-*nef* antibodies was further analyzed on 31 of the anti-*nef*-positive sera by ELISA with large synthetic peptides ranging from 31 to 66 amino acid residues and spanning the total sequence of *nef* from HIV-1 (LAV bru isolate).

Stepwise assembling of the peptide chains was carried out automatically on 4-(oxymethyl)-PAM resin using Boc-/benzyl-based protecting groups. Two elongation protocols were used depending on the peptide chain length: a standard cycle, mainly characterized by a single coupling step (Boc-amino acid symmetrical anhydride in DMF) and an optimized one for large peptides, based on a double coupling strategy (Boc-amino acid symmetrical anhydride first in DMF, then in DCM). Final cleavage of the peptide from the solid support was with anhydrous HF, and crude peptides were purified by C_{18} medium pressure LC. The purified peptides were characterized by analytical HPLC, AAA and CD both in polar (H_2O) and non-polar (TFE) environments.

Of the 31 anti-*nef*-positive sera tested by ELISA, 26 responded against the C-terminal peptides (i.e., sequences 141–205 and 171–205), and 16 reacted with the N-terminal peptide (i.e., sequence 1–66). Three sera elicited a positive reaction with sequence 65–109, whereas 4 and 10 sera reacted with sequences 118–167 and 93–122, respectively. All 31 anti-*nef*-positive sera reacted strongly with a mixture of the N- and C-terminal peptides.

These results showed some heterogeneity of the recognized epitopes in the sera tested and the presence of at least four major antigenic sites within *nef* (i.e., sequences 1–31, 32–64, 93–122 and 171–205), the immunodominant epitopes being located close to the N- and C-termini of the molecule.

Besides indicating the possible usefulness for vaccination [11], this study provides a means for early diagnosis of HIV infection based on the detection in human sera of anti-*nef* antibodies with synthetic peptides.

References

1. Barré-Sinoussi, F., Chermann, J.-C., Rey, F., Nugeyre, M.T., Chamaret, S., Gruest, J., Dauguet, C., Axler-Blin, C., Vezinet-Brun, F., Rouzioux, C., Rozenbaum, W. and Montagnier, L., Science, 220 (1983) 868.
2. Allan, J.S., Coligan, J.E., Lee, T.H., Mc Lane, M.F., Kanki, P.J., Groopman, J.E. and Essex, M., Science, 230 (1985) 810.
3. Arya, S.K. and Gallo, R.C., Proc. Natl. Acad. Sci. U.S.A., 83 (1986) 2209.
4. Terwilliger, E., Sodroski, J.G., Rosen, C.A. and Haseltine, W.A., J. Virol., 60 (1986) 754.
5. Luciw, P.A., Cheng-Mayer, C. and Levy, J.A., Proc. Natl. Acad. Sci. U.S.A., 84 (1987) 1434.
6. Ahmad, N. and Venkatesan, S., Science, 241 (1988) 1481.
7. Ranki, A., Krohn, M., Allain, J.P., Franchini, G., Valle, S.L., Antonen, J., Leuther, M. and Krohn, K., Lancet, ii (1987) 589.
8. Ameisen, J.-C., Guy, B., Chamaret, S., Loche, M., Mouton, Y., Neyrinck, J.-L., Khalife, J., Leprevost, C., Beaucaire, G., Boutillon, C., Gras-Masse, H., Maniez, M., Kieny, M.-P., Laustriat, D., Berthier, A., Mach, B., Montagnier, L., Lecocq, J.-P. and Capron, A., AIDS Res. and Human Retrov. (1989) in press.
9. Sabatier, J.-M., Clerget-Raslain, B., Fontan, G., Fenouillet, E., Rochat, H., Granier, C., Gluckman, J.-C., Van Rietschoten, J., Montagnier, L. and Bahraoui, E., AIDS, 3 (1989) 215.
10. Sabatier, J.-M., Fontan, G., Loret, E., Mabrouk, K., Rochat, H., Gluckman, J.-C. J., Montagnier, L., Granier, C., Bahraoui, E. and Van Rietschoten, J., Int. J. Pept. Prot. Res. (1989) in press.
11. Bahraoui, E., Yagello, M., Billaud, J.-N., Berthault, S., Sabatier, J.-M., Guy, B., Muchmore, E., Girard, M. and Gluckman, J.-C. (manuscript submitted).

Delineation of the binding regions of CD4 for HIV and HLA class II antigens

Pradip K. Bhatnagar[a,*], Michael Busch[c,d], William F. Huffman[a], Joanne Silvestri[a], Alemseged Truneh[b], Paul Ulrich[d] and Girish N. Vyas[d]

[a]*Department of Peptide Chemistry and [b]Department of Immunology, Smith Kline & French Laboratories, King of Prussia, PA 19479, U.S.A.*
[c]*Irwin Memorial Blood Centers, San Francisco, CA 94118, U.S.A.*
[d]*Department of Laboratory Medicine, University of California, San Francisco, CA 94143-0134, U.S.A.*

Introduction

A 55 kDa T-cell surface glycoprotein, CD4, has been suggested as the cellular receptor for human immunodeficiency viruses (HIVs). This molecule exhibits high affinity for the surface glycoprotein (gp120) of these viruses [1]. CD4 also seems to interact with non-polymorphic regions of MHC class II antigen, thus mediating the interactions of T-helper cells and antigen presenting cells [2]. Through extensive genetic analyses and epitope mapping studies, the site of interaction of gp120 has been located in the V_1 domain of CD4 which is known to have homology with the $V\kappa$ domain of immunoglobulins (Ig) [3]. Recent studies have implicated several different residues from this domain to be responsible for the binding to gp120. Additionally, there is evidence that the sequences analogous to the complementarily-determining region 2 (CDR2) of Ig $V\kappa$ domains are most likely responsible for this interaction. We have attempted to delineate the regions of CD4 responsible for the binding to gp120 and HLA class II molecules using PEPSCAN technology, as developed by M. Geysen et al. [4] and commercialized by Cambridge Research Biochemical.

Results and Discussion

Overlapping hexameric sequences, covering the entire V_1 (1–92), J_1 (93–103), V_2 (104–164) and J_2 (165–181) domains of CD4 (1–189) were synthesized on polystyrene pins. Three different sets of pins were made. The first set of pins was incubated with cultured HIV (2×10^6 reverse transcriptase units), washed, and incubated with the pool of human anti-160, -120, -40 and -24 IgGs in human serum. The binding of IgGs was monitored using a Vector Elisa kit suitable for the detection of human IgG. The second set of pins was incubated with Raji B-lymphoblastoid cells (HLA class II[+]) which were previously labeled with ^{125}I anti-HLA class I antibodies. The binding of these pins to the cells was

---

*To whom correspondence should be addressed.

monitored using a gamma counter. The third set of pins was used to map OKT4a antibody using Geysen et al.'s method. Results of these experiments are shown in Fig. 1. Pins containing peptides from the regions 26–72, 105–134 and 175–189 reacted with HIV; pins from the regions 52–90 and 141–183 reacted with HLA class II-positive cells; and pins from the regions 57–65, 68–76 and 80–92 reacted with OKT4a. We have synthesized several peptides of varying length from these regions and found that none of these peptides showed activity in competition assays.

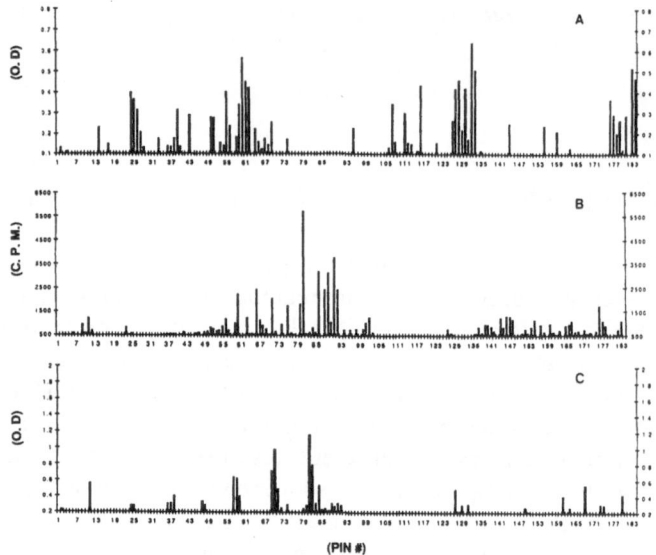

Fig. 1. *Binding of CD4 peptides to HIV (A), Raji B-lymphoblastoid cells (B) and OKT4a antibody (C).*

Consistent with several recent studies reporting binding sites for gp120, HLA class II and epitopes of OKT4A, our results indicate that the first immunoglobulin-like domain of CD4 (V_1) contains several binding regions for these molecules. However, other domains of CD4 may also participate in the interactions. Landau and Warton [5] and Mizukami et al. [6] have also implicated the residues from V_2 domains. Our attempts to find soluble peptides that will block the binding of CD4 to gp120 and HLA class II failed, although Jameson et al. [7] have reported that a peptide encompassing residue 23–56 blocked syncytium formation. Similarly, Mazerolles et al. [8] have reported that the peptide from the region 54–65 exhibited specific dose-dependent inhibitory effects on antigen-induced HLA class II-restricted T-cell proliferation and in vitro antibody synthesis, thus implicating this region to be responsible for interaction with HLA class II molecules.

Acknowledgements

Michael Busch, Paul Ulrich and Girish N. Vyas were supported by the NIH Program Project Grant PO1-HL-36589.

References

1. Sattentau, Q.J., Clapham, P.R., Weiss, R.A., Beverley, P.C.L., Montagnier, L., Alhalabi, M.F., Gluckmann, J.C. and Klatzmann, D., AIDS, 2 (1988) 101.
2. Doyle, C. and Strominger, J.L., Nature, 330 (1987) 256.
3. Arthos, J., Deen, K., Chaikin, M.A., Fornwald, J.A., Sathe, G., Sattentau, Q.J., Clapham, P.R., Weiss, R.A., McDougal, J.S., Pietropaolo, C., Maddon, P.J., Truneh, A., Axel, R. and Sweet, R.W., Cell, 57 (1989) 469 (and references cited therein).
4. Geysen, H.M., Rodda, S.J., Mason, T.J., Tribbick, G. and Schoofs, P.G., J. Immunol. Meth., 102 (1987) 259.
5. Landau, N.R., Warton, M. and Littman, D.R., Nature, 334 (1988) 159.
6. Mizukami, T., Fuerst, T.R., Berger, E.A. and Moss, B., Proc. Natl. Acad. Sci. U.S.A., 85 (1988) 9273.
7. Jameson, B.A., Rao, P.E., Kong, L.I., Hahn, B.H., Shaw, G.M., Hood, L.E. and Kent, S.B.H., Science, 240 (1988) 1335.
8. Mazerolles, F., Durandy, A., Piatier-Tonneau, D., Charron, D., Montagnier, L., Auffray, C. and Fischer, A., Cell, 55 (1988) 497.

Follow-up of HIV-infected persons with the help of synthetic peptides derived from *gag* and *env* sequences of HIV-1

Rüdiger Pipkorn[a], Elke Bernath[a], Jonas Blomberg[b], Per-Johan Klasse[b],
Bengt Ljungberg[c], Bertil Christensson[c] and Bo Ursing[c]

[a]*Center for Molecular Biology, University of Heidelberg, Im Neuenheimer Feld 282,*
D-6900 Heidelberg 1, F.R.G.
[b]*Section of Virology, Department of Medical Microbiology, Sölvegatan 23,*
S-223 62 Lund, Sweden
[c]*Section of Virology, Department of Infectious Diseases, Sölvegatan 23,*
S-223 62 Lund, Sweden

Introduction

In order to evaluate the development of the immune response to epitopes simulated by synthetic peptides during the course of HIV-1 infection, we synthesized the following peptides from HIV-1 *gag* (1–20, 320–338, 335–351) and from *env* (503–520, 581–599, 583–589, 586–606, 602–622, 848–863). The peptides were synthesized by SPPS, purified by preparative HPLC and characterized by analytical HPLC and AAA.

Results and Discussion

In a follow-up study of 34 (16 asymptomatic and 18 symptomatic) HIV-1 seropositive persons with the above synthetic peptides, over a time period of 0.5 – 3 years, we found that the antibody reactivity with a peptide deduced from the immunodominant region HIV-1 *env* (600–625) did not decrease with increasing degree of HIV-related illness, whereas reactivities with several other peptides deduced from *env* and *gag* sequences did. The temporal serological patterns from four patients is presented. The first patient seroconverted simultaneously with an acute encephalitis, probably due to HIV-1. The patient recovered without sequelae. Antibodies to sequences *env* (606–620) and *env* (848–863) rose most rapidly to high values, later followed by *env* (581–599, 583–599 and 586–606) [1]. The reactivity with peptide *env* (503–520) rose first, and then declined to zero. We have noted that antibodies to this sequence vary more than antibodies to other HIV-1 peptide sequences. Among the *gag* sequences, antibodies to peptide *gag* (1–20) rose rapidly and then stabilized at a lower level. The second person was a healthy HIV-1 seropositive person clinically judged as being in a stable condition. Conventional serological parameters (HIV-1 EIA and EIB) did not change during the period. However, in peptide serology we see that the situation was not stable. Although there were relatively minor changes

852

in *env* peptide serology, a more significant change occurred in *gag* serology, where the patient seroconverted to the sequence of *gag* (335–351), without any change in antibodies to the other *gag* sequences.

The condition of the third patient, who was referred to the ARC (AIDS related complex) category already in 1985, slowly deteriorated during the period. All peptide serological parameters, except antibodies to *env* (606–620), were low. The *gag* peptide serology was completely negative, and it is not shown. This serological situation is rather typical of the later stages of HIV-related disease. However, the last person (No. 4) also in the presence of advanced disease (AIDS with pneumocystis carinii pneumonia) exhibited an increasing synthesis of IgG reactive with only one of the HIV *env* peptides: i.e. *env* (586–606). Antibodies to *gag* peptides were not detected.

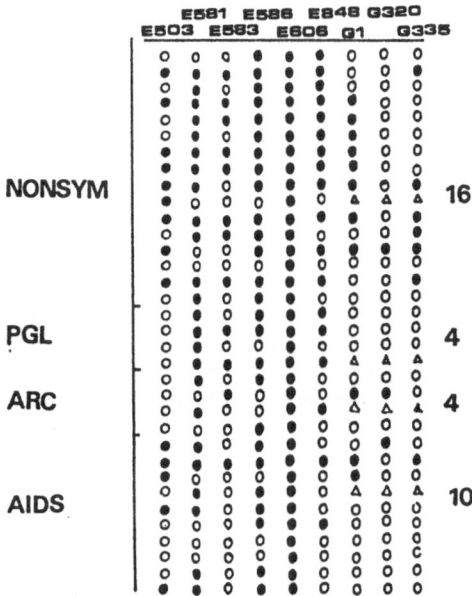

Fig. 1. Schematic representation of the results obtained with the 34 patients in the follow-up study. Results prevalent at the end of the study are shown. Δ, not done; o, unreactive.

We interpret these serological results in the following way. Absorption of *gag* and *env* specific antibodies by an increasing production of HIV and removal from the circulation during progression to AIDS is probably a major mechanism. However, the high and stable reactivity with peptides from HIV *env* (600–625), even in advanced disease, indicates that these B-cell clones produce much more antibody than other clones commited to HIV antigens. Even in patients with advanced disease, we have observed increases in antibody activity to peptides in or in the vicinity of this region. Healthy patients, who are 'stable' in conventional HIV serological tests, can have drastic changes in IgG reactivity with single peptides.

The mechanism behind antibody variation during the progression of HIV-related diseases, thus, is complex. Synthetic peptides derived from HIV *env* and *gag* proteins can be used to gain further insights into these events.

References

1. Pipkorn, R., Bernath, E., Blomberg, J. and Klasse, P.-J., In Jung, G. (Ed.) Peptides 1988, (Proceedings of the 20th European Peptide Symposium), Tübingen, F.R.G., De Gruyter, Berlin, 1989.

Design, structure-activity and specificity of highly potent P₁–P₁′-modified pseudopeptidyl inhibitors of HIV-1 aspartyl protease

Tomi K. Sawyer*, Alfredo G. Tomasselli, Roger A. Poorman, John O. Hui, Jessica Hinzmann, Douglas J. Staples, Linda L. Maggiora, Clark W. Smith and R. Heinrikson

Biopolymer Chemistry Unit, The Upjohn Company, Kalamazoo, MI, 49001, U.S.A.

Introduction

Human immunodeficiency virus (HIV), a member of the retrovirus family, is recognized as the causative agent in acquired immunodeficiency syndrome (AIDS). Molecular organization of the HIV genome comprises *gag, pol* and *env* as genes necessary for viral replication. During viral replication, these genes are expressed as polyproteins that undergo enzymatic cleavage to generate the functional proteins of the mature virus. Genetic and biochemical studies have demonstrated that a virally encoded aspartyl protease is responsible for the release of itself, reverse transcriptase, integrase and other proteins from the *gag–pol* fusion proteins [1]. Recently, the HIV-1 aspartyl protease has been proven to be essential for viral maturation; site-directed mutagenesis of this enzyme leads to the loss of viral infectivity [2]. Crystallographic studies on HIV-1 aspartyl protease have also recently contributed to our knowledge of the structural and mechanistic properties of the enzyme [3,4]. In this report we document recombinant HIV-1 aspartyl protease inhibition by pepstatin and several GAG[128-135]-based octapeptide derivatives, of which one provided a highly potent and selective inhibitor based on comparative studies performed on human renin.

Results and Discussion

Recombinant HIV-1 aspartyl protease was investigated by a series of synthetic substrate analogs of the generic formula H-Val-Ser-Gln-Asn-Xaa-Yaa-Ile-Val-OH (GOP, GAG[128-135] octapeptide). Specifically, several GOP derivatives having Xaa-Yaa substitutions, including Tyr-Pro (GAG-derived substrate peptide, GSP) or one of several nonhydrolyzable pseudodipeptides (i.e., Sta, Pheψ[CH₂N]Pro or Leuψ[CH(OH)CH₂]Val), were prepared using previously described methods [5] of pseudodipeptide and SPPS, purification and physicochemical analysis. The synthetic substrate H-Val-Ser-Gln-Asn-Tyr-Pro-Ile-Val-OH (GSP; $V_{max} = 1.5$ mol/min/mg, $K_m = 1.7$ mM at pH 5.5) was used to determine the HIV-1 aspartyl

---

*To whom correspondence should be addressed.

Table 1 *HIV-1 aspartyl protease inhibition SAR*

Entry	Compound	K_I(nM)
Pepstatin	Iva-Val-Val-Sta-Ala-Sta-OH	362
U-85549E	H-Val-Ser-Gln-Asn-Sta-Ile-Val-OH	3690
U-84645E	H-Val-Ser-Gln-Asn-Pheψ[CH$_2$N]Pro-Ile-Val-OH	3520
U-85550E	H-Val-Ser-Gln-Asn-Pheψ[CH$_2$N]Pro-NH$_2$	>10000
U-85072E	Ac-Pheψ[CH$_2$N]Pro-Ile-Val-OH	>10000
U-85548E	H-Val-Ser-Gln-Asn-Leuψ[CH(OH)CH$_2$]Val-Ile-Val-OH	< 10

Table 2 *Human renin inhibition SAR*

Entry	Compound	K_I(nM)
Pepstatin	Iva-Val-Val-Sta-Ala-Sta-OH	4450
U-85549E	H-Val-Ser-Gln-Asn-Sta-Ile-Val-OH	>20000
U-84645E	H-Val-Ser-Gln-Asn-Pheψ[CH$_2$N]Pro-Ile-Val-OH	>20000
U-85548E	H-Val-Ser-Gln-Asn-Leuψ[CH(OH)CH$_2$]Val-Ile-Val-OH	17000

protease inhibition kinetic activities. The angiotensinogen-derived substrate peptide, H-Pro-His-Pro-Phe-His-Leu-Val-Ile-His-D-Lys-OH (RSP; $k_{cat}=94$ s^{-1}, $K_m=25$ μM at pH 5.5), was used as a reference substrate for the measurement of renin inhibition kinetic activities. As summarized in Tables 1 and 2, pepstatin competitively inhibited the HIV-1 aspartyl protease ($K_I=362$ nM), as well as human renin ($K_I=4450$ nM). The GOP inhibitors varied markedly in their selectivity vs. the HIV-1 aspartyl protease and renin. The P_5-P_2 and $P_2'-P_3'$ sequences of the P_1-P_1' Pheψ[CH$_2$N]Pro substituted GOP derivative (U-84645E) were requisite for inhibition of the HIV-1 protease. Noteworthy was the discovery of a highly potent and selective inhibitor, H-Val-Ser-Gln-Asn-Leuψ[CH(OH)CH$_2$]Val-Ile-Val-OH (U-85548E; $K_I<10$ nM, HIV-1 protease; $K_I=17000$ nM, human renin). A structurally-related P_1-P_1' Leuψ[CH(OH)CH$_2$]Val-containing compound, Boc-His-Pro-Phe-His-Leuψ[CH(OH)CH$_2$]Val-Ile-His-OH(H-261), has also been recently reported [6] to be a potent inhibitor ($K_I=15$ nM) against HIV-1 aspartyl protease. Interestingly, H-261 is a well-known [7] angiotensinogen-based inhibitor of renin $K_I=0.1$ nM), as well as certain other aspartyl proteases. Thus, the selectivity of H-261 to inhibit HIV-1 aspartyl protease versus renin is approximately 100000-fold inferior to that of U-85548E. In summary, the inhibitor U-85548E is the first reported HIV substrate-based inhibitor having a Leuψ[CH(OH)CH$_2$]Val moiety at the P_1-P_1' site.

References

1. Darke, P.L., Nutt, R.F., Brady, S.F., Garsky, V.M., Ciccarone, T.M., Leu, C.-T., Lumma, P.K., Freidinger, R.M., Veber, D.F. and Sigal, I.S., Biochem. Biophys. Res. Commun., 156(1988) 297.
2. Kohl, N.E., Emini, E.A., Schleif, W.A., Davis, L.J., Heimbach, J.C., Dixon, R.A.F., Scolnick, E.M. and Sigal, I.S., Proc. Natl. Acad. Sci. U.S.A., 85(1988) 4686.

3. Navia, M.A., Fitzgerald, P.M.D., McKeever, B.M., Leu, C.-T., Heimbach, J.C., Herber, W.K., Sigal, I.S., Darke, P.L. and Springer, J.P., Nature, 337 (1989) 615.
4. Wlodawer, A., Miller, M., Jaskolski, M., Sathyanarayana, B.K., Baldwin, E., Weber, I.T., Selk, L.M., Clawson, L., Schneider, J. and Kent, S.B.H., Science, 245 (1989) 616.
5. Sawyer, T.K., Pals, D.T., Mao, B., Staples, D.J., DeVaux, A.E., Maggiora, L.L., Affholter, J.A., Kati, W., Duchamp, D., Hester, J.H., Smith, C.W., Saneii, H.H., Kinner, J.H., Handschumacher, M. and Carlson, W., J. Med. Chem., 31 (1988) 18.
6   Richards, A.D., Roberts, R., Dunn, B.M., Graves, M.C. and Kay, J., FEBS Lett., 247 (1989) 113.
7. Blundell, T.L., Cooper, J., Foundling, S.I., Jones, D.M., Atrash, B. and Szelke, M., Biochemistry, 26 (1987) 5585.

Session IX
Synthetic methodologies and
peptide bond mimetics

Chairs: R.B. Merrifield
The Rockefeller University
New York City, New York, U.S.A.

Victor J. Hruby
University of Arizona
Tucson, Arizona, U.S.A.

Peter W. Schiller
Clinical Research Institute of Montreal
Montreal, Quebec, Canada

and

Michael Szelke
Ferring Research Institute
Chilworth, Southampton, U.K.

Approaches to the protein folding problem through study of peptide-template conjugates

D.S. Kemp, James G. Boyd, Timothy P. Curran and Nader Fotouhi

Department of Chemistry, Room 18-584, Massachusetts Institute of Technology, Cambridge, MA 02139, U.S.A.

Introduction

Despite recent progress with the protein folding problem, the field lacks a workable description of the rules and processes that underlie the transformation of a freely orientable random coil into the compact, well-defined structure of a native protein. We have approached this problem unconventionally with the ultimate aim of synthesizing and studying small chimeric proteins in which sequences of natural amino acid residues are bridged at key sites with non-peptide functions that bear structure-inducing and reporting functions. With this long range in mind, several immediate goals can be set.

First, it must be possible to achieve reliable, convergent chemical synthesis of proteins in the 60–130 amino acid size range, using semisynthesis as a realistic option, and with a higher state of product homogeneity than appears to be possible using existing methodologies.

Second, it must be possible to prepare conjugates between peptides and chimeric functions that stabilize the common regional elements of protein structure, and structural and thermodynamic information must be available from their study. Although we have projects under development to model tertiary structure in this manner, the obvious starting point is the modeling of the three global types of protein secondary structure – parallel β-sheets, antiparallel β-sheets, and α-helices.

Results and Discussion

(1) Progress toward reliable protein synthesis by thiol capture

Recently we have reported application of thiol capture ligation to the synthesis of the C-terminal 29-peptide sequence of BPTI [1], and completion of the BPTI synthesis, as well as that of the 63-peptide ROP are in progress. The potential versatility of thiol capture is revealed by the clean coupling of a side-chain blocked octapeptide with a 21-peptide CRAKRNNFKSAEDCMRTCGGA, for which only the Cys thiol residues were blocked. The method thus meets a primary requirement for semisynthesis, compatibility with peptide fragments that lack side-chain protection, and we are currently exploring applications of the methodology to semisynthesis. Generalizing from a half-dozen couplings of medium-sized peptides, we find that the thiol capture methodology has worked well; the major complication observed thus far has been incomplete and side-reaction prone couplings of Arg and His residues during the solid phase synthesis of precursor fragments.

(2) Template–peptide conjugates with helix-forming properties

The premise that a preexisting sequence of secondary structure is likely to nucleate and stabilize the structure in a linked peptide of proper amino acid sequence underlies and defines our interest in template–peptide conjugates. We define a nucleating template as a rigid molecule that embodies one or more of the key structural features that stabilize adjacent regions of a particular type of secondary structure. Oriented C=O and N–H functions capable of polar or hydrogen bonding interactions have dominated our choices of templates thus far. Earlier [2], and elsewhere in this volume we have reported results demonstrating the nucleation of both parallel and antiparallel β-sheets by 2,8-diacylamino-epindolidione templates. Here we outline aspects of our studies of helix nucleation.

We have recently synthesized structure **1** with a tricyclic constraint [3] and structure **2** with a macrocyclic constraint (Fig. 1). Each orients a peptide backbone in the geometry characteristic of an α-helix, and **1** can be regarded as a conformationally locked form of Ac-Pro-Pro. Peptide conjugates of **1** have been

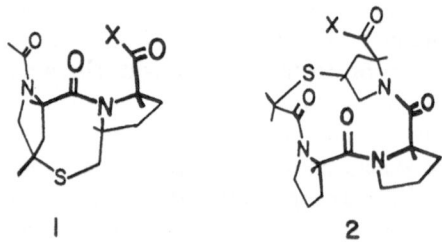

1   **2**

*Fig. 1.   Structures **1** and **2**.*

prepared, and their conformations have been studied by ¹H NMR in a variety of organic solvents as well as in water.

The template portions of conjugates of **1** are found to exist (Fig. 2) in two conformations that undergo an interconversion that is slow on the NMR time scale, thereby allowing determination of their concentration ratio. Figure 3 illustrates the conformational change that occurs in the series Temp-Ala$_n$-OtBu in CDCl$_3$. COSY and NOE analyses establish that Temp-OMe adopts the s-*cis*, staggered conformation of Fig. 2b, and members of the series Temp-Ala$_n$-OtBu(n = 3,4,5) adopt the s-*trans*, eclipsed conformation of Fig. 2a. For Temp-Ala$_n$-OtBu(n = 1,2), mixtures of these two states are observed.

The s-*trans*, eclipsed conformation has the correct orientation for helical nucleation. Examination of the peptide JαCH-NH values, and the solvent and

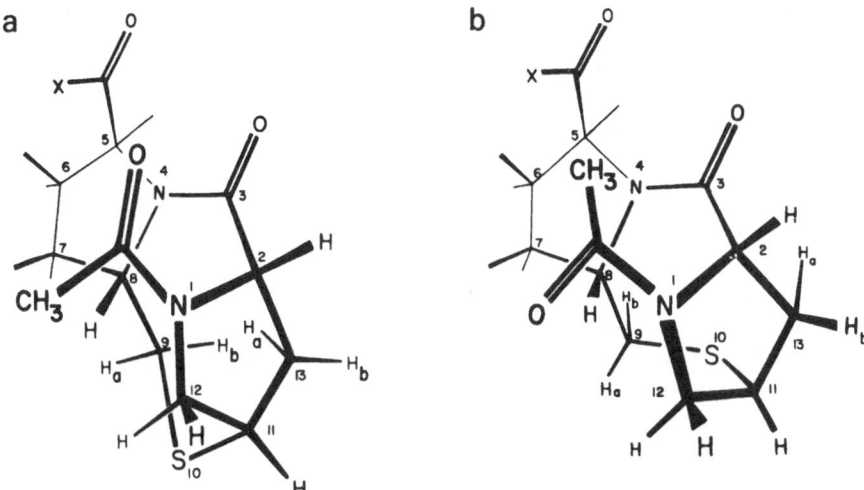

Fig. 2. *Two major conformations of **1** that differ in the orientations about the acetyl amide and the C-8, C-9 C-C bonds. The template nucleating conformation (a) has an s-trans amide and 8–9 eclipsed orientation; the nonnucleating conformation (b) has an s-cis amide and an 8–9 staggered orientation.*

temperature dependence of NH δ's establish that helical conformations of a hybrid 3_{10}-α type are adopted in the n = 3, 4, 5 cases, as shown in Fig. 4. Selective

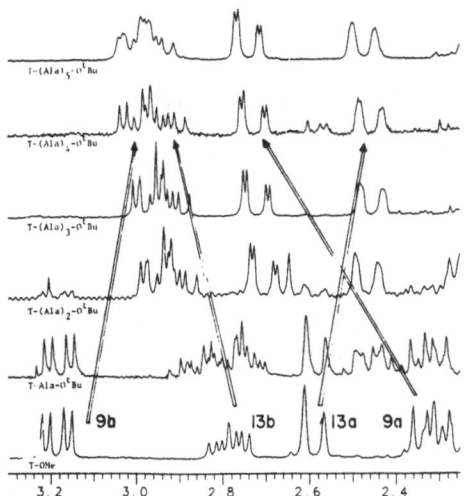

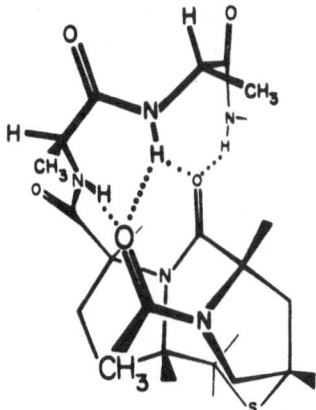

Fig. 3. *^1H NMR spectra in CDCl$_3$ of Temp-Ala$_n$-OtBu (n = 0–5). The resonances of H-9 and H-13 are shown.*

Fig. 4. *Proposed hybrid 3_{10}-α template-helix conjugate, showing bifurcated hydrogen bonds.*

863

irradiation of the three NH resonances of Temp-Ala$_3$-OtBu show the following strong NOE interactions: NH 1: Ala 2 NH, H-2, H-5, H-6, H-8, Ac-CH$_3$; NH 2: Ala 1 NH, Ala 3 NH, H-2, H-5, Ac-CH$_3$; NH 3: Ala 2 NH, H-5. These interactions establish the existence of a tight, helical coupling between the amino acid residues and the template.

Similar results are observed in other solvents, with the notable complication for the polar solvents water, DMF, and DMSO, that the template–peptide conjugate always exists as a mixture of s-*cis*, staggered and s-*trans*, eclipsed states with the former bearing a peptide with spectroscopic features expected for a random coil, and the latter bearing a helical peptide. From these studies, the order of helix stabilization has been found to be: CDCl$_3$ > CD$_3$CN > DMF-d_7 > H$_2$O > DMSO-d_6. Lengthening the peptide chain or strengthening its helical bias results in a shift in the measured ratio of s-*trans*/s-*cis* template populations. The template 1, when linked to peptides, can therefore be regarded as a two-state helix/coil switch that selectively nucleates helix in the s-*trans* state and also reports the presence of helical peptide at the template–helix junction. Studies in water are reported elsewhere in this volume and establish that peptide–template conjugates provide a very promising approach for resolving fundamental questions concerning the initiation and stabilization of secondary structure. When the studies now in progress are complete, it should be possible to quantitate the contributions of the relevant energetic factors to the stabilization of short helices in solution and, ultimately, by a kind of biophysical *aufbau* principle, to the stabilization of helices and sheets in proteins.

References

1. Fotouhi, N., Galakatos, N.G. and Kemp, D.S., J. Org. Chem., 54 (1989) 2803.
2. Kemp, D.S. and Bowen, B.R., Tetrahedron Lett., 29 (1988) 5077.
3. Kemp, D.S. and Curran, T.P., Tetrahedron Lett., 29 (1988) 4931.

Peptides with sulfide bridges and dehydroamino acids: Their prepropeptides and possibilities for bioengineering

Günther Jung

Institut für Organische Chemie, Universität Tübingen, D-7400 Tübingen, F.R.G.

Introduction

We proposed the collective name *lantibiotics* [1] for a group of complex polycyclic peptide antibiotics which contain the thioether amino acids *meso*-lanthionine (Lan) and (2S,3S,6R)-3-methyllanthionine (MeLan). Besides Lan and MeLan, many of the lantibiotics contain dehydroalanine (Dha) and dehydro-butyrine (Dhb), and some have β-hydroxyaspartic acid, lysinoalanine or *S*-aminovinyl-L-cysteine as unusual constituents. Lantibiotic-producing strains have the interesting capacity to synthesize prepropeptides on ribosomes which, via a number of post-translational modification steps, are transformed to rather complex peptide structures containing several sulfide bridges.

A well-known lantibiotic is the food preservative, nisin [2]. The pentacyclic 34-peptide structure has been elucidated by E. Gross and J. Morell [3] and the synthesis performed by the group of T. Shiba [4]. The earlier chemical work on nisin and other lantibiotics was done by E. Gross and coworkers, e.g., on subtilin [5], cinnamycin and duramycin [6]. Recently, the sequences of epidermin [7] and gallidermin [8] have been determined; both compounds may have therapeutic potential for the treatment of acne due to their high and specific antimicrobial activity against *Propionibacterium acne*. Other lantibiotics may act as enzyme inhibitors, e.g., ancovenin [9], or be effective against *Herpes simplex*, e.g., lanthiopeptin [10]. Ro 09-0198 [11] has been reported to be effective as immunopotentiator. Recently, lanthiopeptin, Ro 09-0198 [11], and cinnamycin [6] were found to be identical [10]. Furthermore, duramycin was reported to be [Lys2]-cinnamycin [10] and gallidermin [8] can be called [Leu6]-epidermin. This suggests that although many more lantibiotics might be found in the near future, most of them will belong to the ring structure groups already known. Multicomponent mixtures, as known for other large polypeptide antibiotics, e.g., alamethicin, are not produced in the ribosomal lantibiotic biosynthesis. The structure elucidation of the largest lantibiotic found so far, Pep5 [12,13], revealed a novel structure of a particularly basic 34-residue-peptide with membrane modifying properties. The occurrence of several sulfide rings and dehydroamino acid residues prevents routine sequencing unless chemical modifications are made [3,5–14]. In addition, for Pep5 we found the N-terminal 2-oxobutyryl [12,13], and for epidermin and gallidermin, an *S*-aminovinyl group as the C-terminal amide [7,8,14].

Results and Discussion

Our interest in the possible exploitation of lantibiotic-producing microbial strains and enzymes involved in the post-translational modification concerns two major aspects. Firstly, the construction of new analogs of lantibiotics with enhanced antimicrobial activities, and secondly, the novel biosynthetic possibilities which may enable the design and fermentative production of peptides with unusual constituents. For the realisation of these goals, the isolation of the enzymes responsible for the conversion of the ribosomally synthesized prelantibiotics is an essential step, followed by studies of the structure, mechanisms, and assembly of the enzymes. The versatility of the enzymes catalyzing the transformation of precursor polypeptides to lantibiotics is remarkable, and we may expect great potential for achieving useful biotransformations not only in vivo but also in vitro. Based on its chemical structure, the biosynthetic precursor of epidermin has been proposed [14], and using synthetic wobbled DNA probes, the corresponding structural gene for the lantibiotic epidermin could be isolated from a plasmid, and sequenced [1]. Within a very short period, the structural genes of subtilin [15], nisin [16,17], gallidermin [18], and Pep5 [19] were sequenced from plasmids or chromosomal DNA. The sequencing of the structural gene of preproepidermin [1] from *Staphylococcus epidermidis* was the first experimental confirmation of the hypothesis that lantibiotics originate from ribosomally synthesized precursors, such as those postulated for epidermin [14]. These precursors consist of an N-terminal leader sequence followed by a prolantibiotic part containing Ser and Thr residues, which are dehydrated to Dha and Dhb. The sulfide rings are formed via stereoselective addition of cysteine thiol groups to the double bonds of Dha and Dhb. (An intermediate phosphorylation of Ser, Thr must also be taken into consideration.) Thereby, the original L-configuration of the C_α-atom of Ser or Thr is changed to the opposite configuration. Thus, from meso-Lan and (2S,3S,6R)-MeLan in lantibiotics, one can derive corresponding Ser, Thr and Cys positions in the prelantibiotics for constructing synthetic oligonucleotides (Fig. 1).

In contrast to 'typical' signal peptides [20], the leader regions of lantibiotics contain many charged amino acids but no cysteine, and α-helix propensity is predicted, except for a short amphiphilic β-sheet in the middle, which may interact with a signal recognition particle (SRP). The prolantibiotic part is rich in turns and more hydrophobic. The cleavage sites for signal peptidases of prelantibiotics are highly conserved: residue (−1) is positively charged or polar, (−2) is always Pro, (−3) negatively charged or polar, (−4) hydrophobic, and (−5) charged or polar. Prelantibiotics do not exhibit hydrophobic stretches longer than 3 residues within the signal region. We assume a specific cooperation of the leader region with the prolantibiotic part during modification and export.

A reasonable working hypothesis is that, after having been anchored via SRP to a multienzyme complex, the prelantibiotics are modified at their C-termini. Thus, the prolantibiotic part with its high turn propensity moves along the phosphorylating, dehydrating, ring forming, etc., active enzymic sites and across

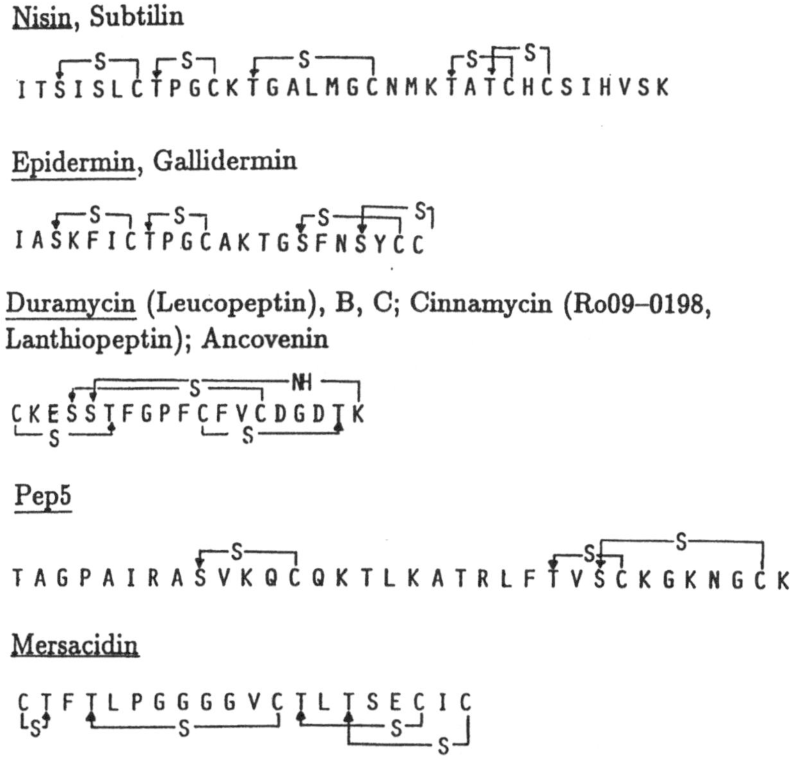

Nisin, Subtilin

Epidermin, Gallidermin

Duramycin (Leucopeptin), B, C; Cinnamycin (Ro09–0198, Lanthiopeptin); Ancovenin

Pep5

Mersacidin

Fig. 1. *Sulfide ring formation via dehydration of Ser and Thr followed by addition of Cys-SH groups*

of the membrane. The modification imposes a highly hydrophobic and compact C-terminus with membranophilic properties and a high dipole moment. The leader region may act as a protective principle for the various cysteine thiol groups before ring formation takes place.

Detailed 2D ^{1}H NMR experiments (COSY, NOESY, HOHAHA) at 500 MHz in DMSO and H_2O on epidermin and gallidermin, as well as on their tryptic fragments and molecular dynamic simulations with NOE constraints, show a rather compact and stable solution structure, regardless of whether free MDS or the approaches of van Gunsteren/Behrendsen or Karplus/Brünner are used. This conformational rigidity is in agreement with CD spectra, which are almost independent of temperature and solvent.

The tetracyclic gallidermin (and epidermin) folds into two wide helical turns. Three positive charges are accumulated on one side, and lipophilic residues on the other side of the helix. This amphiphilic structure is in close agreement with the proposed modification mechanisms and the membrane channel-forming activities (see below). The enzymic cleavage site (Lys[13]-Dhb[14]) [7,8] is exposed on the polar face of the helix and accessible by trypsin. The pronounced dipole of 53 Debye units for epidermin is comparable with that of voltage-dependent

867

channel formers such as alamethicin (about 70 D). This result explains the voltage-gated membrane pore formation of the cationic lantibiotics epidermin and gallidermin, which is very similar to that described for Pep5 [21].

Practical targets for possible applications of the biosynthetic machinery of lantibiotics are analogs of hormones, inhibitors, immunomodulators, vaccines, and antibiotics of higher chemical and enzymic stability with modified activities. Sulfide rings may be formed having any ring size from 4 to 8 residues from C-terminal L-cysteine to N-terminal Ser, Thr (resp. Dha, Dhb) residues. Using the biosynthetic possibilities of cinnamycin producers, sulfide bridges with reverse orientation (C-terminal D-configuration of Lan) and lysinoalanine bridges may also be incorporated. In addition, the incorporation of Dha, Dhb, hydroxyaspartic acid, as well as N-terminal and C-terminal blocking groups, should be achievable. Molecular dynamic simulations with NOE constraints have already been performed for the neutral, globular, cinnamycin-type lantibiotic (Ro 09-0198, lanthiopeptin) and the amphiphilic, positively charged, helical type gallidermin and its [Ile6]-analog epidermin. The nisin, subtilin, and Pep5 solution structures will soon be available. All of these structures may then serve as a basis for bioengineering novel metabolites. Cooperation between microbiologists and peptide chemists has opened a challenging new field with hopefully few limitations with respect to both structural goals and genetic possibilities.

Acknowledgements

The structural and conformational analytical work has been carried out by my highly-engaged coworkers H. Allgaier, R. Kellner, and S. Freund. Molecular dynamics simulations were made in cooperation with O. Gutbrod and G. Folkers. The excellent collaboration with the microbiology groups is acknowledged: H. Zähner, K.D. Entian (Frankfurt), F. Götz, H.G. Sahl (Bonn), and their outstanding research groups. We are indebted to R. Werner (Thomae) for his support and to the Deutsche Forschungsgemeinschaft (SFB 323, project C2-Jung). Finally, I thank my friend W.A. Gibbons and his group for having introduced S. Freund to 2D NMR.

References

1. Schnell, N., Entian, K.-D., Schneider, U., Götz, F., Zähner, H., Kellner, R. and Jung, G., Nature, 333 (1988) 276.
2. Hurst, A., Adv. Appl. Microbiol., 27 (1981) 85.
3. Gross, E. and Morell, J.L., J. Am. Chem. Soc., 93 (1971) 4634.
4. Fukase, K., Kitazawa, M., Sano, H., Shimbo, K., Fujita, H., Horimoto, S., Wakamiya, T. and Shiba, T., In Shiba, T. and Sakakibara, S. (Eds.) Peptide Chemistry 1987, Protein Research Foundation, Minohshi, Osaka, Japan, 1988, p. 337.
5. Gross, E., Kiltz, H. and Nebelin, E., Hoppe-Seyler's Z. Physiol. Chem., 354 (1973) 810.
6. Gross, E., Adv. Exp. Med. Biol., B86 (1977) 131.

7. Allgaier, H., Jung, G., Werner, R.G., Schneider, U. and Zähner, H., Eur. J. Biochem., 160 (1986) 9.
8. Kellner, R., Jung, G., Hörner, T., Zähner, H., Schnell, N., Entian, K.-D. and Götz, F., Eur. J. Biochem., 1977 (1988) 53.
9. Wakamiya, T., Ueki, Y., Shiba, T., Kido, Y. and Motoki, Y., Tetrahedron Lett., 26 (1985) 665.
10. Fukase, K., Wakamiya, T., Naruse, N., Konishi, M. and Shiba, T., In Ueki, M. (Ed.) Peptide Chemistry 1988, Protein Research Foundation, Minohshi, Osaka, Japan, 1989, p. 221.
11. Kessler, H., Steuernagel, S., Will, M., Jung, G., Kellner, R., Gillessen, D. and Kamiyama, T., Helv. Chim. Acta, 71 (1988) 1924.
12. Kellner, R., Jung, G., Josten, M., Kaletta, C., Entian, K.-D. and Sahl, H.-G., Angew. Chem., 101 (1989) 618; Angew. Chem. Int. Ed. Engl., 28 (1989) 616.
13. Kellner, R., Jung, G., Kaletta, C., Entian, K.-D., Reis, M. and Sahl, H.-G., In Jung, G. and Bayer, E. (Eds.) Peptides 1988, de Gruyter, Berlin, 1989, p. 369.
14. Allgaier, H., Jung, G., Werner, R.-G., Schneider, U: and Zähner, H., Angew. Chem. Int. Ed. Engl., 24 (1985) 1051.
15. Banerjee, S. and Hansen, J.N., J. Biol. Chem., 263 (1988) 9508.
16. Buchman, G.W., Banerjee, S. and Hansen, J.N., J. Biol. Chem., 263 (1988) 16260.
17. Kaletta, C. and Entian, K.-D., J. Bacteriol., 171 (1989) 1597.
18. Schnell, N., Entian, K.-D., Götz, F., Hörner, T., Kellner, R. and Jung, G., FEMS Microbiol. Lett., 58 (1989) 263.
19. Kaletta, C., Entian, K.-D., Kellner, R., Jung, G., Reis, M. and Sahl, H.-G., Arch. Microbiol., 152 (1) (1989) 16.
20. Gierasch, L.H., Biochemistry, 28 (1989) 923.
21. Kordel, M., Benz, R. and Sahl, H.-G., J. Bacteriol., 170 (1988) 84.

The substitution of an amide-amide backbone hydrogen bond in an α-helical peptide with a covalent hydrogen bond mimic

Thomas Arrhenius and Arnold C. Satterthwait

*Department of Molecular Biology, Research Institute of Scripps Clinic,
La Jolla, CA 92037, U.S.A.*

Introduction

We have proposed that the replacement of putative amide-amide hydrogen bonds with covalent mimics may serve as a general method for the conformational restriction of peptides to bioactive forms [1]. On average, every other amino acid in globular proteins engages in this bonding and, furthermore, different hydrogen bonding patterns define different structures. The replacement of a weak hydrogen bond at selected positions along the peptide chain could yield a variety of conformers determined by the position of substitution and the amino acid sequence.

Results and Discussion

To test this proposal, a hydrazone-ethylene bridge (N-N=CH-CH$_2$-CH$_2$) was substituted for an (i, i+4) backbone hydrogen bond (N-H···O=CR-NH) in one turn of an α-helix to give cyclic peptide A as outlined in Fig. 1. Conformational analysis of A in deuterated trifluoroethanol [2] utilizing NMR was consistent with a structure perhaps best described as a 'relaxed' helical turn that locates the three carbonyl oxygen atoms on the same face but lacks pitch. To confirm this conformation, a pentapeptide was added to the carboxyl terminus of A to give B (Fig. 1) with the expectation that, if correct, it would serve as a nucleation site in deuterated trifluoroethanol and fold the appended pentapeptide from a relatively disordered structure into an α-helix.

The conformations of the pentapeptide before and after appending it to the potential nucleation site A were compared by examining three properties of the amide NH protons using NMR experiments which are summarized in Table 1. All NH signal assignments were made on the basis of 2D COSY and ROESY spectra [3], except for those of the three final residues of the acetyl-peptide, which were made by analogy [4].

Spin-spin coupling constants $^3J_{HN\alpha}$ which measure ϕ angles, all decrease in B as predicted for the formation of an α helix [5]. Significantly, the $^3J_{NH\alpha}$ values for B show a regular increase from A1 to E5 indicating little or no distortion of the presumed helix in its interaction with the nucleation site [4]. Nuclear

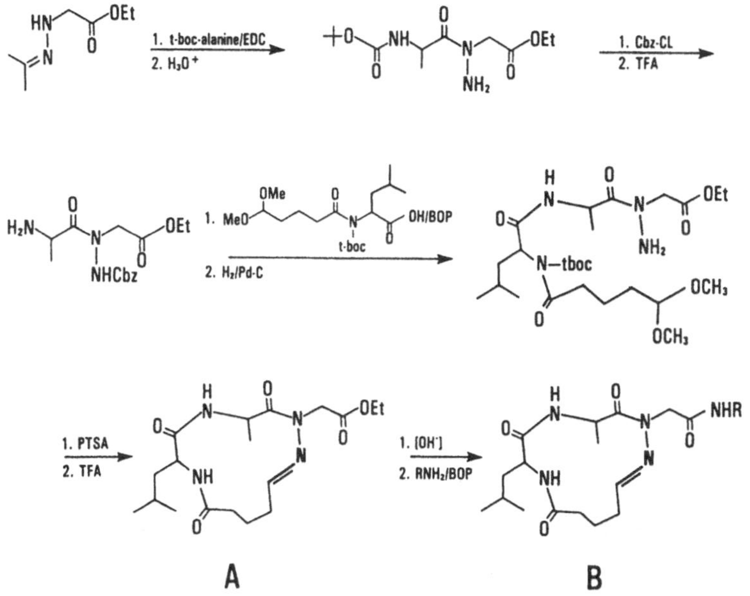

Fig. 1. Synthetic scheme. Pd-C = 10% palladium on carbon; BOP = Castro's reagent; PTSA = p-toluene sulfonic acid; RNH₂ = -L-alanine-(L-glutamic acid-γ-ethyl ester)₄-ethyl ester; other abbreviations are standard.

Table 1   NMR Experiments on amide protons

	Pentapeptide sequence				
	A1	E2	E3	E4	E5
1. $^3J_{NH\alpha}$, cps(δ)[a]					
Acetyl-pentapeptide	5.7(7.22)	6.1(7.84)	7.0(7.78)	7.8(7.67)	7.8(7.57)
Peptide B	3.8(7.42)	4.6(8.24)	5.9(7.72)	7.3(7.73)	7.5(7.62)
2. NOEs, dnn(i, i+1)					
Acetyl-pentapeptide		——			
Peptide B	——	——		[b]	——
3. H/D Exchange Rates[c]					
Acetyl-pentapeptide	>10	>10	>10	>10	10
Peptide B	1	1	2	2	5

[a] Chemical shifts relative to CF_3CDHOH at 3.88 ppm.
[b] Overlapping NH signals prevent observation.
[c] Relative rates determined at ambient temperature on mixtures of the two peptides, wihich showed no apparent aggregation. Rates are approximate due to overlapping signals.

Overhauser enhancements (NOEs) between NHs of adjacent residues identified in 2D ROESY spectra indicate their close spatial proximity ($\leqslant 3.5$ Å) as expected of an α-helix [5]. Each of the NH protons in the appended peptide shows the predicted reduction in hydrogen-deuterium exchange rates for the hydrogen bond network characteristic of an α-helix.

The NMR experiments provide good evidence that cyclic peptide A can serve as a nucleation site. Since α-helix formation requires several interactions between the appended peptide and the cyclic peptide, the cyclic peptide is probably close to being helical as well, establishing the hydrazone link as a functional covalent hydrogen bond mimic.

References

1. Arrhenius, T., Lerner, R.A. and Satterthwait, A.C. In Oxender, D. (Ed.) Protein Structure and Design. (UCLA Symposium on Molecular and Cellular Biology, New Series 69), 1987, p. 453.
2. Satterthwait, A.C., Arrhenius, T., Hagopian, R.A., Zavala, F., Nussenzweig, V. and Lerner, R.A., Trans. R. Soc. Lond., 323 (1989) 565.
3. Dyson, H.J., Rance, M., Houghten, R.A., Lerner, R.A. and Wright, P.E., J. Mol. Biol., 201 (1988) 161.
4. Ribeiro, A.A., Saltman, R. and Goodman, M., Biopolymers, 24 (1985) 2469.
5. Wüthrich, K. NMR of Proteins and Nucleic Acids., John Wiley and Sons, New York, NY, 1986, p. 162.

Helical preferences of peptides containing multiple α,α-dialkyl amino acids

Garland R. Marshall[a],*, Denise D. Beusen[a], John D. Clark[a], Edward E. Hodgkin[a], Janusz Zabrocki[b] and Miroslaw T. Leplawy[b]

[a]Department of Pharmacology, Washington University School of Medicine, St. Louis, MO 63110, U.S.A.
[b]Institute of Organic Chemistry, Politechnika, 90-924 Lodz, Poland

Introduction

The presence of α,α-dialkyl amino acids in microbial natural products, such as the peptaibol antibiotics, requires novel biosynthetic pathways to produce and incorporate these unusual amino acids, which argues strongly for a special role related to function. One aspect is the conformational restrictions imposed by these amino acids as first pointed out by Marshall and Bosshard [1]. Despite the variety of conformations theoretically available to α,α-dialkyl amino acids, which are also observed experimentally [2], the impact of multiple substitutions of this type of amino acid on the overall conformation of a peptide is dramatic. The crystal structure of alamethicin [3], which contains eight α-methylalanine (MeA) residues out of 20, is predominantly α-helical, with NMR data [4] supporting a similar solution conformation in methanol. A review [5] of crystal structures of 28 tri-, tetra- and pentapeptides containing MeA indicates that all but one show characteristics of the 3_{10}-helix. Crystal structures of longer MeA-containing peptides show significant α-helical content, while other MeA-containing peptides of seven or more residues are predominantly 3_{10}-helices. Much conjecture regarding the factors determining either α- or 3_{10}-helical conformations has appeared in the literature [6–9]. Recently, Karle et al. [10] have reported two crystalline forms of a decapeptide containing three MeA residues that switch from a predominantly α-helical hydrogen bonding pattern to a mixed $3_{10}/α$-helical pattern, depending on the solvent of crystallization. What is needed is a theoretical basis for understanding the relative stability of α- and 3_{10}-helical structures.

Results and Discussion

We can argue that the free energy difference between the α-helix and 3_{10}-helix conformations for a given peptide can be approximated by the following equation:

---

*To whom correspondence should be addressed.

$$\Delta G(3_{10} \text{ to } \alpha) = \Delta G(\text{end effects}) + n(\Delta G(\text{MeA})) + m(\Delta G(\text{AA}))$$

where n = number of MeA residues, m = number of normal amino acid residues, and end effects = difference at both N- and C-terminals between two conformations.

All ΔG terms consist of $\Delta H - T\Delta S$. While it is feasible to estimate the ΔH components through differences in energy minimizations, the $T\Delta S$ components are more problematic.

Estimation of ΔH of the end effect and ΔH per residue was achieved by calculating the energy of each type of helix for $(\text{MeA})_n$ and $(\text{Ala})_n$ over a range of values of n from 1 to 15. The difference in the slopes gives the relative helix stability per residue. The intercept with the energy axis represents the termination end effect, I_t. The calculations reveal that the α-helix of $(\text{MeA})_n$ gains approximately 1.94 kcal/mol/residue in stability over the 3_{10}-helix in the gas phase, as compared with the estimate of between 0.3 and 3.6 kcal/mol/residue by Prasad and Sasisekharan [11]. The equivalent number for the alanine peptides is 2.36 kcal/mol/residue, with previous [11] estimates between 3.3 and 3.6 kcal/mol/residue. These calculations are consistent with the small number of observations of 3_{10}-helix in proteins [12] as well as the increased probability of 3_{10}-helix with increased content of MeA residues. The intersections, I_t, should reflect the fact that the 3_{10}-helix has one more hydrogen bond due to 4-to-1 hydrogen bonding, rather than the 5-to-1 hydrogen bonding of the α-helix. The difference, $I_t(\alpha)-I_t(3_{10})$, for a distant-dependent dielectric of 4 r is 8.60 for MeA. Substituting these values into the equation gives:

$$\Delta G(3_{10} \text{ to } \alpha) = [8.60 - T\Delta S(\text{end effects})] + n[-1.94 - (T\Delta S(\text{MeA})] + m[-2.36 - (T\Delta S(\text{AA})]$$

If one assumes that the $T\Delta S$ differences are small (preliminary molecular dynamics calculations suggest 0.5 kcal/mol/residue at 300 K in favor of the 3_{10}-helix), then a plausible explanation of the experimental data emerges. In solution, a length of peptide is reached where the inherent increased stability of the α-helix dominates the energy contribution of the single additional hydrogen bond formed in a 3_{10}-helix, as compared with an α-helix of the same length. While literature estimates show considerable variation, 7 kcal/mol was the average hydrogen bond strength in a simulation of liquid N-methylacetamide by Jörgensen and Swenson [13]. Consequently, in short peptides, a 3_{10}-helix conformation is favored in solvents of low dielectric constant due to one additional hydrogen bond that can be satisfied. As the peptide length increases, the inherent stability of the α-helix compensates for the lost hydrogen bond, and the α-helix becomes dominant in solution. While the exact transition length will depend on the solvent and peptide sequence, we can estimate that most nonpolar solvents would require a length longer than seven residues for the α-helix to be favored over the 3_{10}-helix.

Evidence from solution NMR studies may support the conjectures regarding helix stabilities and transitions as a function of solvent polarity. Pentapeptides containing repetitive MeA-Ala or MeA-Val sequences showed [14] three intra-molecular hydrogen bonds in both $CDCl_3$ and $(CD_3)_2SO$ consistent with 3_{10}-helical structure. Heptapeptides of similar sequence [14] had five hydrogen bonds in $CDCl_3$ (implying a 3_{10}-helix), but only four hydrogen bonds in the more polar $(CD_3)_2SO$, consistent with α-helical conformation in accord with the considerations above. Balaram et al. [15] have reported that two decapeptides, Boc-MeA-Val-MeA-MeA-Val-Val-Val-MeA-Val-MeA-OMe and Boc-MeA-Leu-MeA-MeA-Leu-Leu-Leu-MeA-Leu-MeA-OMe, show the presence of eight in-tramolecular hydrogen bonds in $CDCl_3$, but only seven hydrogen bonds in $(CD_3)_2SO$, consistent with a transformation from 3_{10}-helix to α-helix upon increasing the polarity of the solvent and decreasing the hydrogen bond strength. While these studies are suggestive, the absence of the assignment of the amide protons in any of the above studies to specific residues leaves the helix-transformation interpretation of the data speculative, particularly due to the known propensity of these peptides to aggregate in a head-to-tail fashion.

While the factors that dictate helical preference in solution undoubtedly play a role in the crystal, intermolecular interactions become important and may actually dominate. The predicted enhanced stability of the α-helix in solution, once a critical length is reached, stands in contrast to the observation of 3_{10}-helical structure in crystals for numerous oligomers of MeA, as well as poly-MeA [16] itself. One factor governing the crystal structures in poly-MeA is an improved packing for the 3_{10}-helix as compared with the α-helix. This arises from the interdigitation of the methyl side chains, which is accommodated by the 3_{10}-helix and an increased stabilization by van der Waals interaction, estimated at approximately 0.5 kcal/mol/residue per helix–helix dimer interaction. Another major aspect is electrostatic, as the 3_{10}-helix has a smaller radius (1.9 vs. 2.3 Å) than the α-helix, and a similar dipole moment, resulting in a stronger electrostatic interaction during antiparallel helical stacking. The calculations of Hol and De Maeyer [17] on the α-helical crystals of Boc-Ala-MeA-Ala-MeA-Ala-Glu(OBzl)-Ala-MeA-Ala-MeA-Ala-OMe show the electrostatic interaction energy [17] between a helix and its neighbors was approximately –23 kcal/mol. With a distance between α-helices of 9.4 Å, the poly-MeA 3_{10}-helix has [16] a shorter packing distance than expected between helices of 7.5 Å due to interleaving of MeA side chains. The approximate 20% reduction in distance would result in an increase of approximately 73% in electrostatic energy, as dipole–dipole interactions have an r^{-3} dependency. If the undecapeptide could pack as efficiently as a 3_{10}-helix, it would gain a maximum of –16.7 kcal/mol, or –1.52 kcal/mol/residue, in electrostatic stabilization, which is not significantly different from the estimate of the enthalpic stabilization of the α-helix. If one considers both the estimated entropic stabilization, as well as increased van der Waals stabilization of the 3_{10}-helix in the crystal of poly-MeA, then the 3_{10}-form appears more energetically favored. The α-helical peptides containing multiple α,α-dialkyl amino acids seen in crystals, therefore, arise when the sequence of residues leads to unfavorable packing of the 3_{10}-helix. The correlation seen between MeA content and 3_{10}-helicity reflects the decreased bulk of the side chain. The replacement of the pBrBz group by acetyl-Phe at the N-terminus and replacement of the methyl ester by the more bulky benzyl ester in emerimicin 1–9 causes crystallization in the α-helix, rather than the 3_{10}-helix for the emerimicin 2–9 analog [18]. It is of interest that parallel packing of α-helical peptides occurs and has also been reported with 3_{10}-helices [8]. While it is certainly true that α-helices often pack in true parallel arrays (for a discussion of helical packing, see Ref. 19), the same observation for 3_{10}-helices is infrequent. Packing within a particular layer may be parallel, but alternate layers pack in an antiparallel fashion. In reality, the helices are packed antiparallel in a hexagonal array with four antiparallel and two parallel nearest neighbors.

Acknowledgements

This research was supported in part by the National Institutes of Health, grants GM-24482 and GM-33918, as well as the Polish Academy of Sciences, grant CPBP 01.13.2.5.

References

1. Marshall, G.R. and Bosshard, H.E., Circ. Res., 30/31 (1972) (Suppl. II) 143.
2. Marshall, G.R., Clark, J.D., Dunbar, Jr., J.B., Smith, G.D., Zabrocki, J., Redlinski, A.S. and Leplawy, M.T., Int. J. Pept. Prot. Res., 32 (1988) 544.
3. Fox, Jr., R.O. and Richards, F.M., Nature, 300 (1982) 325.
4. Esposito, G., Carver, J.A., Boyd, J. and Campbell, I.D., Biochemistry, 26 (1987) 1043.
5. Toniolo, C., Bonora, G.M., Bavoso, A., Benedetti, E., Di Blasio, B., Pavone, V. and Pedone, C., Biopolymers, 22 (1983) 205.
6. Bosch, R., Jung, G., Schmitt, H. and Winter, W., Biopolymers, 24 (1985) 979.
7. Karle, I.L., Sukumar, M. and Balaram, P., Proc. Natl. Acad. Sci. U.S.A., 83 (1986) 9284.
8. Toniolo, C., Bonora, G.M., Bavoso, A., Benedetti, E., Di Blasio, B., Pavone, V. and Pedone, C., J. Biomol. Struct. Dynamics, 3 (1985) 585.
9. Bavoso, A., Benedetti, E., Di Blasio, B., Pavone, V., Pedone, C., Toniolo, C., Bonora, G.M., Formaggio, F. and Crisma, M., J. Biomol. Struct. Dynamics, 5 (1988) 803.
10. Karle, I.L., Flippen-Anderson, J.L., Sakumar, M. and Balaram, P., Int. J. Pept. Prot. Res., 31 (1988) 567.
11. Prasad, B.V.V. and Sasisekharan, V., Macromolecules, 12 (1979) 1107.
12. Barlow, D.J. and Thornton, J.M., J. Mol. Biol., 201 (1988) 601.
13. Jörgensen, W.L. and Swenson, C.J., J. Am. Chem. Soc., 107 (1985) 569.
14. Vijayakumar, E.K.S. and Balaram, P., Tetrahedron, 39 (1983) 2725.
15. Balaram, H., Sukumar, M. and Balaram, P., Biopolymers, 25 (1986) 2209.
16. Malcolm, B.R., Biopolymers, 16 (1977) 2591; 22 (1983) 319.
17. Hol, W.G.J. and De Maeyer, M.C.H., Biopolymers, 23 (1984) 809.
18. Marshall, G.R., Hodgkin, E.E., Langs, D.A., Smith, G.D., Zabrocki, J. and Leplawy, M.T., Proc. Natl. Acad. Sci. U.S.A., (1990) in press.
19. Karle, I.L., Biopolymers, 28 (1989) 1.

Peptide azoles: A new class of biologically active dipeptide mimetics

Tom D. Gordon, Philip E. Hansen, Barry A. Morgan, Jasbir Singh, Eugene Baizman[a] and Susan Ward[a]

Department of Medical Chemistry and [a]Department of Pharmacology, Sterling Research Group, Rensselaer, NY 12144, U.S.A.

Introduction

The rational design of therapeutic agents that attenuate the pharmacologic effects of peptides is currently a major focus of medicinal chemistry. In addition to possessing appropriate pharmacologic activity in vitro, these 'peptide-mimetics' must be chemically and enzymatically stable and should possess acceptable bioavailability and pharmacokinetics. A wide variety of peptide mimetic strategies have been described in recent years, ranging in complexity from single atom substitution in the amide group, to ambitious de novo design of multiple residue mimetics. In general, the success of the design strategy has been limited, and advances have usually originated from empirical approaches, such as the screening of natural products [1]. In an attempt to utilize the structural variety exhibited by peptide-related secondary metabolites as a source for idea generation, we examined these structures [2] for repeating motifs. Some of these features, such as chiral inversion or N-methylation have been used to good effect for some time in the design of peptide analogs; however, others, such as the thiazole moieties embedded in molecules such as bleomycin [3] and dolostatin 3 [4], had not been incorporated into the design of peptide analogs. This feature, in which the carbonyl group of residue n and the N, Cα and Cβ of residue $n+1$ are constrained in a thiazole ring, met our criteria for consideration as potential peptide mimics: conformationally constrained moieties that could be synthesized by flexible routes from readily available chiral materials. Thus, we envisioned that the dipeptide (**1**) could be formally replaced by the thiazole (**2**). In addition, unlike the naturally occurring thiazole containing peptides, we would not be limited to the substitution R_2=H. We chose to apply this strategy to the hexapeptide neurokinin antagonist [5] [Pro^6,D-Trp^{7,9}]SP(6–11).

Results and Discussion

Chemistry

Our initial targets included the 'D-Trp-Phe' thiazole (**4b**, X=S). The Hantzsch thiazole synthesis [6], involving, in this case, condensation of an N-protected amino acid thioamide with a β-halo-α-keto-ester yielded a product with low optical activity that was further elaborated [7] to yield the desired hexapeptide mimic as a mixture of two diastereoisomers (**8** and **9**) that could be separated chromatographically. The corresponding Phe-D-Trp thiazole (**4a**, X=S) could not be synthesized by a Hantzsch approach, as indole-3-carboxaldehyde could not be converted into the required β-halo-α-keto-ester. However, based on the work of Oikawa and Yonemitsu [8], we envisioned that Xxx-Trp thioamides should cyclize to thiazoles on treatment with DDQ. Consequently, we were pleased to observe that treatment of the thioamide (**3a**, X=S) with DDQ in tetrahydrofuran provided the desired thiazole (**4a**, X=S) in excellent yield. Furthermore, the product had a significant optical rotation ($[\alpha_D] = -33.2°$, C=1, DMF) and gave no indication of diastereoisomer content on elaboration to the hexapeptide mimic (**12**), suggesting that this thiazole synthesis was free of racemization. Unfortunately, this route was applicable only to Xxx-Trp thiazoles, and we required a more general synthesis. We presumed that cyclization of thioamide (**3a**) with DDQ proceeded by a mechanism that did not give rise to equilibria involving sp² hybridization at the α-carbon bearing the R_1 substituent, and envisioned that β-oxo-dipeptide thioamides such as (**6**, X=S) should cyclize by an analogous mechanism *with retention of chiral integrity*.

(**3a,4a**) Y = R_1 = CH₂Ph, R_2 = 3-indolyl, R_3 = Me
(**3b,4b**) Y = CH₂Ph, R_1 = CH₂-3-indolyl, R_2 = Ph, R_3 = Me

    The synthesis of suitable thioamides from β-oxo-dipeptides (**6**, X=O) via hydrolysis and acylation of the Schiff's base (**5**) is described in detail elsewhere in these proceedings [9]. We were pleased to find that these compounds, that could not be isolated, cyclized spontaneously to yield the desired thiazoles with excellent retention of chiral integrity, confirming that intermediates such as (**6**) could serve as flexible synthons for the target thiazoles. Under other conditions

[9] the β-oxo-dipeptides (**6**, X=O) could be converted into the corresponding imidazoles (**4**, X=NH) or oxazoles (**4**, X=O).

(**5**)   (**6**)

Pharmacology

Guinea-pig myenteric plexus longitudinal muscle strips were prepared and pA_2 values for antagonism of substance P (SP) calculated as previously described [5]. The data for azole substitution at positions 7–8(trp-Phe) and 8–9(Phe-trp) of the hexapeptide (**7**) are shown in Table 1.

Discussion

For convenience, we have extended the ψ[...] nomenclature [10] to represent the azole modification: thus, ψ[thzl] implies that the fragment (**2**, X=S) has been substituted in place of the corresponding dipeptide. Incorporation of the thiazole moiety into the hexapeptide (**7**) at residues 7–8 to give analog (**8**) resulted in an increased potency of one pA_2 unit: interestingly, the diastereoisomer (**9**) was also more potent than (**7**). Insertion of the imidazole and oxazole moieties yielded analogs (**10**) and (**11**). The relative potency for the series at this position was [thzl](**8**)>[imzl](**10**)>[oxzl](**11**)=(**7**). Insertion of the thiazole moiety at the 8–9 dipeptide yielded (**12**), with an accompanied reduction in pA_2 relative to (**7**).

Table 1

			Structure								pA_2vs SP
(7)	Pro	–	trp	–	Phe	–	trp	–	Leu	–	Phe-NH$_2$ 6.4
(8)	Pro	–	trp	ψ[thzl]	Phe	–	trp	–	Leu	–	Phe-NH$_2$ 7.4
(9)	Pro	–	Trp	ψ[thzl]	Phe	–	trp	–	Leu	–	Phe-NH$_2$ 6.9
(10)	Pro	–	trp	ψ[imzl]	Phe	–	trp	–	Leu	–	Phe-NH$_2$ 6.7
(11)	Pro	–	trp	ψ[oxzl]	Phe	–	trp	–	Leu	–	Phe-NH$_2$ 6.3
(12)	Pro	–	trp	–	Phe	ψ[thzl]	trp	–	Leu	–	Phe-NH$_2$ 5.9

Conclusions

We have shown that an appropriately substituted azole moiety (**2**) can mimic a dipeptide, and, on substitution into a series of SP antagonist mimetics (**8–12**), can display up to 10 times the potency of the unmodified hexapeptide (**7**). We have developed a general racemization-free synthesis of azoles (**4**, X=S,O,NH), and shown that they can be elaborated into peptide-like structures such as (**8–12**). In addition, this methodology can be adapted to yield a facile route to the thiazole-containing natural products such as dolastatin **3**.

References

1. Morgan, B.A. and Gainor, J.A., In Allen, R.C. (Ed.), Annual Reports in Medicinal Chemistry, Academic Press, Orlando, 1989, in press.
2. For a review see Lewis, J.R., Natural Product Reports, 5 (1988) 351 and earlier reports.
3. Takita, T., Muraoka, Y., Yoshioka, T., Fujii, A., Maeda, K. and Umezawa, H., J. Antibiot., 25 (1972) 755.
4. Pettit, G.R., Kamano, Y., Brown, P., Gust, D., Inoue, M. and Herald, C.L., J. Am. Chem. Soc., 104 (1982) 905.
5. Baizman, E.R., Bentley, H., Gordon, T.D., Hansen, P.E., Keifer, D., McKay, F.C., Perrone, M.H., Morgan, B.A., Rosi, D., Singh, J., Stevenson, J. and Ward, S.J., In Ragnarsson, U. (Ed.) Peptides 1984, Almqvist and Wiksell International, Stockholm, 1984, p. 359.
6. Hantzsch, A., Annalen, 249 (1888) 1.
7. Gordon, T., Hansen, P., McKay, F., Morgan, B., Singh, J., Baizman, E., Keifer, D. and Ward, S., In Jung, G. (Ed.) Proceedings of the 20th European Peptide Symposium, in press.
8. Oikawa, Y. and Yonemitsu, O., J.O.C., 42 (1977) 1213.
9. Gordon, T.D., Hansen, P.E., Morgan, B.A. and Singh, J., In Rivier, J.E. and Marshall, G.R. (Eds.) Peptides: Chemistry, Structure and Biology (Proceedings of the 11th American Peptide Symposium), ESCOM, Leiden, 1990, p. 680.
10. IUPAC-IUB Nomenclature recommendations, Eur. J. Biochem., 138 (1984) 9.

Design and synthesis of a peptidomimetic employing β-D-glucose for scaffolding

Kyriacos C. Nicolaou[a], Joseph M. Salvino[a], Karen Raynor[c], Sherrie Pietranico[a], Terry Reisine[c], Roger M. Freidinger[b] and Ralph Hirschmann[a],*

[a]Department of Chemistry, University of Pennsylvania, Philadelphia, PA 19104-6323, U.S.A.
[b]Merck Sharp & Dohme Research Laboratories, West Point, PA 19486, U.S.A.
[c]Department of Pharmacology, University of Pennsylvania, Philadelphia, PA 19104, U.S.A.

Introduction

The search for peptidomimetics has been the subject of intensive investigations for several years. Successful mimetics range from agonists known since antiquity, to antagonists discovered through screening, such as morphine and asperlicin [1], respectively. Approaches to the *design* of novel mimetics have also varied greatly, including pseudopeptides [2], retro-inverso peptides [3], mimics of β-turns [4,5], and others. The principal objectives have been the discovery of molecules with improved transport properties and longer half-lives.

S-8307 [6,7], an angiotensin antagonist apparently discovered by screening, is an example of a peptidomimetic that has no amide backbone. We have initiated a program that seeks to discover such compounds *by design*.

Results and Discussion

Investigations in serveral laboratories [8–10] have shown that not all of the 14 amino acids of somatostatin (SRIF) are required for biological activity. For example, c-(Pro-Phe-D-Trp-Lys-Thr-Phe) (1) is about $1.7 \times$ as potent as SRIF [9] in vitro, and the 'super-active' MK-678 (2) is about $50 \times$ as potent. These studies also focused attention on the importance of the β-turn for activity [9]. SARs indicated no specific requirement for a *Thr* side chain in the $i+3$ position of the β-turn. With 1 as a guide, we designed (Fig. 1) and synthesized 2-indol-3-ylethyl 6-O-(5-aminopentyl)-2,3,4-tri-O-benzyl-β-D-glucopyranoside (3) in 10 steps from glucose. Modeling suggested that the spatial relationships of the C-1 and C-6 substituents of 3 resembled those of the Lys and Trp side chains in 1, and that the 2-O-benzyl group mimics Phe[7] (SRIF numbering throughout). The 4-O-Bzl was included to mimic that Phe of 1, located between Thr and Pro. The 3-O-Bzl was included only to simplify the synthesis. 3 completely displaced [^{125}I]CGP 23996 (des-Ala[1],Gly[2]-desamino-Cys[3]-[Tyr[11]]-dicarba[3,14]-

---

*To whom correspondence should be addressed.

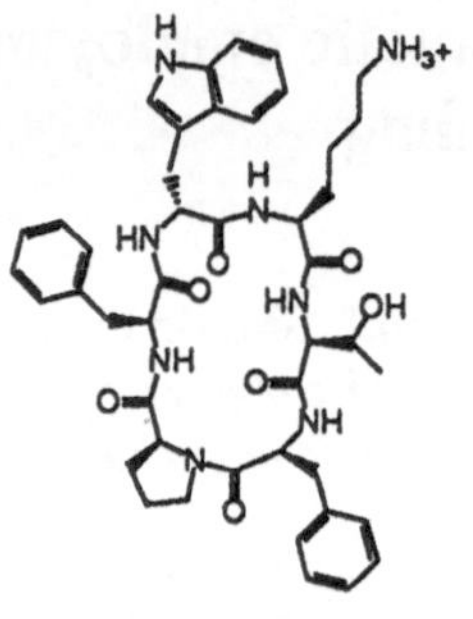

1

Cyclic Hexapeptide L-363,301

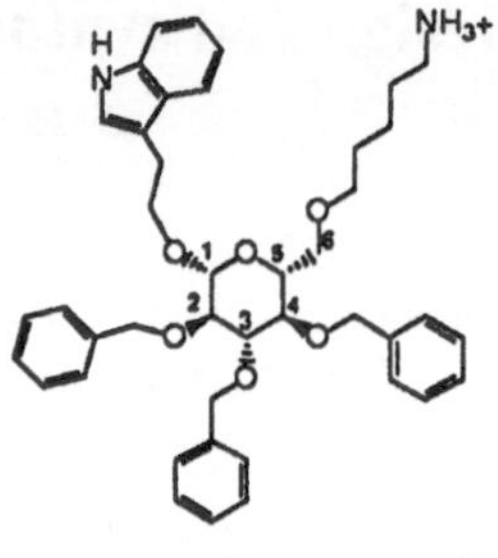

3

2-Indol-3-ylethyl 6-O-(5-aminopentyl)-
2,3,4--tri-O-benzyl-β-D-glucopyranoside

Fig. 1. Peptide template 1, and mimetic 3.

somatostatin) from SRIF receptors on membranes of pituitary, cerebral cortex and AtT-20 cells. Subsequent studies were carried out only with the AtT-20 cells, where **3** had an IC_{50} of about 9.5×10^{-6} M. MK-678 had an IC_{50} of about 60 nM and SRIF ca. 9.3 nM.

The following results encourage us to believe that the SRIF receptors do indeed recognize **3** as a mimic of **1**: (1) Elimination of the lysine-like substituent gave a compound that did not bind, and the 4-des-Bzl analog had an IC_{50} of about 1×10^{-4} M. By contrast, the analog **4** (Fig. 2), lacking the 'superfluous' 3-Bzl, had an IC_{50} of about 1.3 μM (Fig. 3). (2) The chemical shift of the 'γ' CH_2 in the aminopentyl group of **3** ($\delta = 1.28$ ppm), when compared with that of 5-aminopentanol ($\delta = 1.40$ ppm) or that of the des indole analog of **3**, suggests that the indole side chain of **3** can assume a position shielding the aminopentyl side chain of **3**.

It is of considerable interest that the H_6 protons of **3** are dramatically shifted upfield (δ ca. 2.6–3.0 ppm) by the 4-Bzl substituent. Molecular modeling based on this information and subsequent energy minimizations suggested that the 4-Bzl group, though still contributing some hydrophobic binding, does not occupy a position comparable to that of the side chain of Phe next to Thr in **1**. Thus, **3** may more closely resemble c-(Pro-Phe-D-Trp-Lys-Thr-*Ala*), which is 30× less potent than **1** in vitro [9]. This observation has important implications for future design.

An advantage of the AtT-20 cells is that they contain both SRIF and β-adrenergic receptors, facilitating study of the specificity of our compounds. **3** also bound to the latter receptor, but at a 5-fold higher concentration. We are encouraged that the more potent **4** had an IC_{50} for the β-receptor nearly one order of magnitude higher than that for the SRIF receptor, suggesting that further enhancing of potency will also further enhance specificity. This would not be without precedent.

Our results suggest that only three of the four β-turn side chains of SRIF

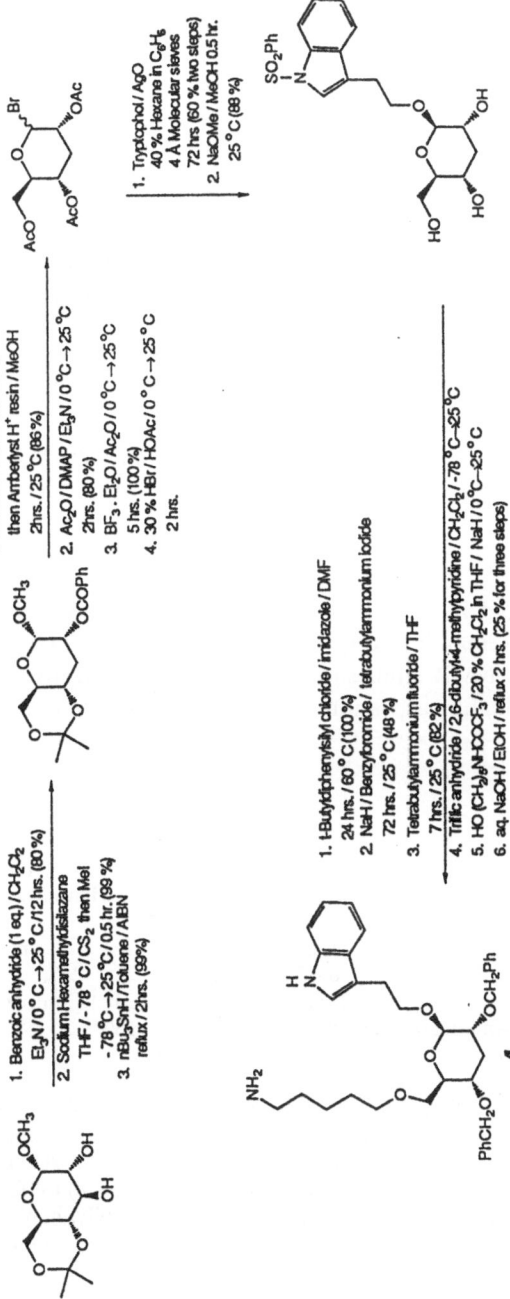

Fig. 2. Synthesis of 2-indol-3-ylethyl 6-O-(5-aminopentyl)-3-deoxy-2,4-di-O-benzyl-β-D-glucopyranoside, 4.

are needed for binding, and that there is no requirement for H-bonding between the SRIF receptors and the peptide backbone.

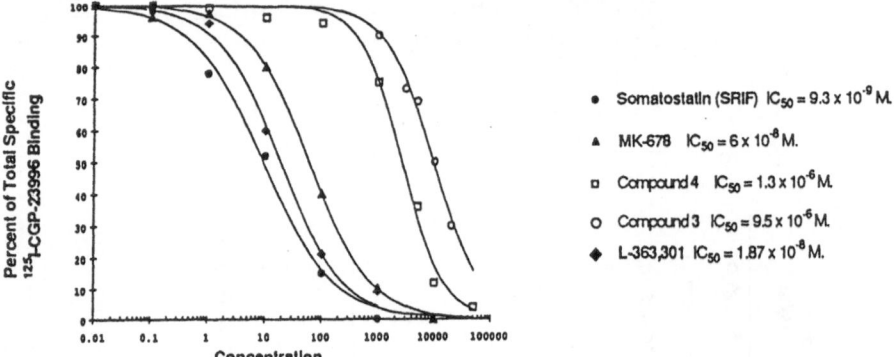

Fig. 3. Displacement of [^{125}I]CGP-23996 by compounds **3**, **4**, MK-678, L-363,301 and SRIF.

Acknowledgements

This work was supported by the National Institutes of Health, by Bachem Inc., and by the Sterling Research Group. We are indebted to Dr George Furst for his guidance with the NMR studies. R.H. is grateful to Dr Byron Arison (Merck, Rahway) for a valuable discussion on the NMR spectrum of **3**.

References

1. Chang, R.S.L., Lotti, V.J., Monaghan, R.L., Birnbaum, J., Stapley, E.O., Goetz, M.A., Albers-Schonberg, G., Patchett, A.A., Liesch, J.M., Hensens, O.D. and Springer, J.P., Science, 230 (1985) 177.
2. Spatola, A.F. In Weinstein, B. (Ed.) Chemistry and Biochemistry of Amino Acids, Peptides and Proteins, Marcel Decker, New York, 1983, Vol. 7, p. 267.
3. Freidinger, R. and Veber, D., In Vida, J. and Gordon, M. (Eds.) Conformationally Directed Drug Design, American Chemical Society, 1984, p. 169.
4. Kahn, M., Wilke, S., Chen, B. and Fujita, K., J. Am. Chem. Soc., 110 (1988) 1638.
5. Kemp, D.S. and Stites, W.E., Tetrahedron Lett., 29 (1988) 5057.
6. Furnkawa, Y., Kishimoto, S. and Nishikawa, K., U.S. Patent, 4,355,040 (1982).
7. Wong, P.C., Chiu, A.T., Price, W.A., Thoolen, M.J., Carini, D.J., Johnson, A.L., Taber, R.I. and Timmermans, P.B., J. Pharmacol. Exp. Ther., 247 (1988) 1.
8. Vale, W., Rivier, J., Ling, N. and Brown, M., Metabolism, 27 (1978) 1391.
9. Veber, D.F., Freidinger, R.M., Perlow, D.S., Paleveda, W.J., Holly, F.W., Strachan, R.G., Nutt, R.F., Arison, B.H., Homnick, C., Randall, W.C., Glitzer, M.S., Saperstein, R. and Hirschmann, R., Nature, 292 (1981) 55.
10. Bauer, W., Briner, U., Doepfiner, W., Haller, R., Huguenin, R., Marbach, P., Petcher, T.J. and Pless, J., Life Sci., 31 (1982) 1133.

BOP and congeners: Present status and new developments

Jacques Coste, Marie-Noëlle Dufour, Dung Le-Nguyen and Bertrand Castro

Centre CNRS-INSERM de Pharmacologie-Endocrinologie, Rue de la Cardonille,
F-34094 Montpellier, France

Introduction

BOP **1** (Fig. 1) was designed as a peptide coupling reagent 15 years ago [1]. It has been repeatedly used in our group since that time, in solution [2] or in solid phase [3] synthesis. In our group, strategies using Boc/TFA deprotection were almost exclusively used. Recently, it has been more and more widely used in other laboratories [4], often with Fmoc/morpholine deprotection system. BOP's high effectiveness is now widely recognized.

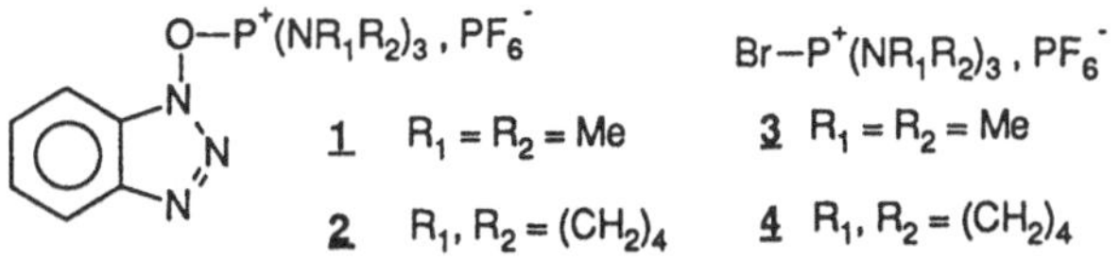

O—$P^+(NR_1R_2)_3$, PF_6^- Br—$P^+(NR_1R_2)_3$, PF_6^-

1 $R_1 = R_2 = Me$ **3** $R_1 = R_2 = Me$

2 $R_1, R_2 = (CH_2)_4$ **4** $R_1, R_2 = (CH_2)_4$

Fig. 1. Structure of the different coupling reagents.

Results and Discussion

BOP in peptide coupling

BOP reacts exclusively on carboxylate salts ; mixtures of BOP and a carboxylic acid remain indefinitely stable. In usual Boc synthesis, to the mixture of the new Boc-amino acid, the C-terminal protected peptide TFA salt and BOP is added a tertiary base, preferably DIEA in a minimal quantity of solvent.

In every case, the reaction mixture has to be kept rather basic in order to insure a fast coupling. Three equivalents of base are necessary to neutralize the carboxylic acid, the amine salt, and the acidic hydroxybenzotriazole. In such conditions, the coupling rate is so high that racemization is negligible in urethane-protected amino acid couplings, and fairly low, though not negligible, in segment coupling [5]. The excess of acid and BOP is typically 1.1 molar equivalent in solution synthesis; in solid phase synthesis, excesses from 1.5 to 3 equivalents are commonly used.

It is very characteristic that this procedure allows the skipping of a separate neutralization step, either in solution or in solid phase synthesis using the Boc/TFA strategy. This is not only time-saving but also useful to minimize diketopiperazine formation during the third step of a synthesis.

Trifluoroacetate anion is not activated by BOP; hence trifluoroacetylation side reaction is never observed.

Another facility offered by BOP is the possibility to couple DCHA salts, due to the high rate ratio of amino acid coupling towards dicyclohexylamine. This feature is especially useful in SPPS where the small amount of dicyclo-hexylamide formed is easily washed out. The main application of that procedure lies in the introduction of histidine as the cheap and easily accessible Boc-His(Boc)-OH. Up to three histidine residues could be introduced in the synthesis of human renin substrate tetradecapeptide.

Finally, the water- and DCM-soluble byproducts, HMPA and hydroxyben-zotriazole, allow very easy washing of reaction product, contrasting with DCC condensation where elimination of insoluble urea can be a problem.

New insights on mechanism

The breakthrough made by BOP in certain difficult coupling circumstances, where it appears to be really superior to DCC/HOBt, indicated that a completely different mechanism was followed, contrasting with the formerly proposed mechanisms involving the intermediacy of a benzotriazolyl active ester in both cases.

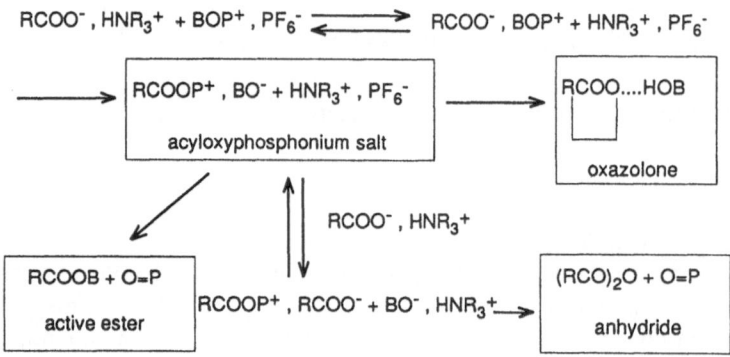

Fig. 2. Postulated reaction mechanism for BOP.

Various intermediates can be postulated in the case of BOP, as summarized in Fig. 2, including the acyloxyphosphonium intermediate, the symmetrical anhydride, the HOBt active ester and the oxazolone. In order to discriminate between these intermediates, we have realized competition experiments: one equivalent of Z-Val-OH, activated by various means, was opposed to a mixture of 1 equivalent of H-Val-OMe and 1 equivalent of H-Phe-OMe. The selectivity ratio, Z-Val-Val-OMe/Z-Val-Phe-OMe was measured in each case and reported in Fig. 3, together with the result obtained for DCC/HOBt.

It is clear from these data that BOP coupling differs from the other activations; the higher ratio of the more hindered product is indicative of the importance of an entropic barrier. The direct intermediacy of the acyloxyphosphonium salt, presumably through the cyclic complex suggested in Fig. 4, is likely.

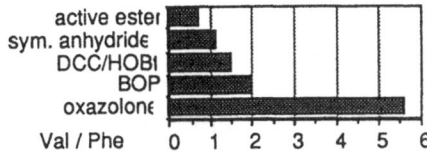

Fig. 3. Val to Phe ratios in competition experiments

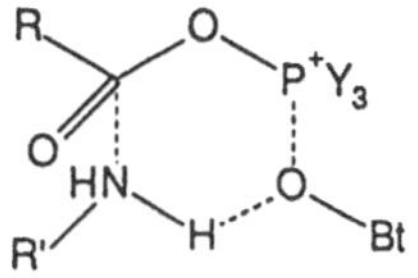

Fig. 4. Structure of a prepared intermediate.

New analogs of BOP

PyBOP Manufacturing BOP, as well as its use, involves hexamethylphosphoric triamide (HMPA), the respiratory toxicity (carcinogenicity) of which has been the subject of numerous reports [6]; HMPA has been recently included in the 'Seveso list' of dangerous chemicals [7]; its industrial availability is now limited and its rejection in the environment when using BOP may cause problems. Hence it was essential to find a replacement product for BOP. Two main criticisms could be made about HMPA, its relative volatility and the presence of numerous methyl groups that may confer to it alkylating properties in metabolization conditions.

The replacement of the phosphorous atom by a carbon could be attractive; however, the corresponding benzotriazolyl-tetramethyl-uronium salt (HBTU) and its analogs [8] are related to tetramethylurea, which *is* devoid of neither volatile and *has* methyl groups. Furthermore, some cytotoxic properties were reported for tetramethylurea [9].

The very straightforward replacement of methyl groups by ethyl, that we first tried, afforded an EtBOP analog with very poor coupling properties: the coupling of Boc-Phe-OH with H-Gly-OEt requires 5 h instead of a few minutes. The same behavior was observed in BOP analogs where dimethylamino groups were replaced by morpholino or piperidino groups.

However, the use of pyrrolidino groups on phosphorus afforded a new reagent nicknamed 'PyBOP®' **2**, allowing coupling rates as good as, if not better than those observed with BOP. Coupling tests in solution show that even better results are obtained in racemization tests. For a SPPS test, we used two model peptides, the rat renin substrate tetradecapeptide (RRS) and the acyl carrier protein (65-74) (ACP(65-74)) decapeptide. The excellent results obtained will be reported in another paper.

BroP It was striking to observe that BOP is not well suited for N-methyl amino acid coupling, whereas other reagents such as BOP-Cl and Dpp-Cl are more effective for this purpose. We have recently found that the corresponding halophosphonium salts, particularly the bromoderivatives such as BroP® **3** or

PyBroP® **4**, are especially well suited for such coupling reactions. For example, the coupling of Z-(N-Me)Val-OH on H-(NMe)Val-OMe yielded 80% of the dipeptide at room temperature within 1 h; the racemization was undetectable.

References

1. Castro, B., Dormoy, J.-R., Evin, G. and Selve, C., Tetrahedron Lett., (1975) 1219.
2a. Cumin, F., Evin, G., Fehrentz, J.A., Seyer, R., Castro, B., Ménard, J. and Corvol, P., J. Biol. Chem., 260 (1985) 9154.
  b. Le Nguyen, D., Seyer, R., Heitz A. and Castro, B., J. Chem. Soc. Perkin Trans. I (1985) 1025.
3a. Audousset-Puech, M.P., Dufour, M., Kervran, A., Jarrousse, C., Castro, B., Bataille, D. and Martinez, J., FEBS Lett., 200 (1986) 181.
  b. Bouhnik, J., Galen, F.-X., Ménard, J., Corvol, P., Seyer, R., Fehrentz, J.-A., Le Nguyen, D., Fulcrand P. and Castro, B., J. Biol. Chem., 262 (1987) 2913.
  c. Le Nguyen, D., Heitz A. and Castro, B., J. Chem. Soc. Perkin Trans. I (1987) 1915.
  d. Liu, C.F., Fehrentz, J.A., Heitz, A., Le Nguyen, D., Castro, B., Heitz, F., Carelli, C., Galen F.X. and Corvol, P., Tetrahedron, 44 (1988) 675.
  e. Fehrentz, J.A., Heitz, A., Seyer, R., Fulcrand, P., Devilliers, R., Castro, B., Heitz, F. and Carelli, C., Biochemistry, 27 (1988) 4071.
4a. Hudson, D., J. Org. Chem., 53 (1988) 617.
  b. Fournier, A., Wang C.-T. and Felix, A.M., Int. J. Pept. Prot. Res., 31 (1988) 86.
  c. Felix, A.M., Wang, C.T., Heimer, E.P. and Fournier, A., Int. J. Pept. Prot. Res., 31 (1988) 231.
  d. Fournier, A., Danho, W. and Felix, A.M., Int. J. Pept. Prot. Res., 33 (1988) 133.
5   Le-Nguyen, D., Dormoy, J.R., Castro, B. and Prévôt, D., Tetrahedron, 37 (1981) 4229.
6a. Schmidt, R.Z., Gesamte Hyg. Ihre Grenzgeb., 25 (1979) 662.
  b. Lloyd, J.W., J.Am. Ind. Hyg. Assoc., 36 (1975) 917.
  c. Steere, N.V., Chem. Educ., 53 (1976) A12.
  d. Shott, L.D., Borkovec, A.B., Knapp Jr., W.A., Toxicol. Appl. Pharmacol., 18 (1971) 499.
7. Journal Officiel de la République Française, Paris, November 24, 1988.
8a. Dourtoglou, V., Ziegler, J.C. and Gross, B., Tetrahedron Lett., (1978) 1269.
  b. Dourtoglou, V., Gross, B., Lamproglou, V. and Ziodrou, C., Synthesis, (1984) 572.
  c. Knorr, R., Tzeciak, A., Bannwarth W. and Gillessen, D., Tetrahedron Lett., 30 (1989) 1927.
9a. Rowell, R.M., Appl. Biochem. Biotechnol., 9 (1984) 447.
  b. Oustrin, M.L., Moisand, C., Cros M.L. and Bonnefoux, J., Ann. Pharmacol. Fr., 30 (1972) 685.
  c. Moisand, A., Moisand C. and Pitet, G., Ann. Pharmacol. Fr., 28 (1970) 575.

Unexpected dimerization in the reactions of activated Boc-amino acids with oxygen nucleophiles in the absence of tertiary amine

N. Leo Benoiton and Francis M.F. Chen

Department of Biochemistry, University of Ottawa, Ottawa, Ont., Canada K1H 8M5

Introduction

As part of continuing studies on the chemistry of activated forms of amino acids, we have examined the reaction of the mixed anhydride (MxAn) of Boc-valine which had been purified by edulcoration (removal of soluble salts by washing with aqueous solutions) [1] with methanol, and the EDAC-mediated reaction of Boc-valine with *p*-nitrophenol (HONp), in the absence of tertiary amine. Extensive dimerization was encountered in both cases. Recall that unisolated MxAn's, in the presence of the strong base DMAP and alcohols or phenols, generate esters [2,3], that the MxAn method was used to prepare *N*-succinimidyl esters years ago [4], and that activated esters are usually obtained by the use of DCC in pyridine [5] or inert solvents.

Results and Discussions

Boc-Val-O-COOMe [1] was left in methanol for 72 h, the solvent was removed, and the neutral fraction was examined by proton NMR spectroscopy after edulcoration. Two products, Boc-Val-OMe and Boc-Val-Val-OMe, had been formed in the ratio of 1:9. Since MxAn's react slowly with alcohols and decompose by cylization to the 2-alkoxy-5(4*H*)-oxazolone (AlkOx) [1,6], we had expected that the AlkOx would react with the alcohol to give Boc-Val-OMe. In order to confirm our suspicions that the AlkOx had decomposed, we examined its reaction with 1.5 equiv. of methanol in CDCl$_3$. One-half of the AlkOx remained after 72 h; the reaction was complete after 6 days. Products were estimated to be present as follows: Boc-Val-OMe, Boc-Val-Val-OMe, tBu-O-Me, Val-*N*-carboxyanhydride and polymer, in the ratio 7:2:2:1. Our interpretation of the results of the reaction of Boc-Val-O-COOMe with methanol is that the AlkOx initially formed underwent methanolysis to give Val-NCA and tBu-O-Me (Scheme 1, A). Val-NCA then reacted with methanol to give H-Val-OMe which captured intact MxAn or AlkOx to generate Boc-dipeptide ester. Methanolysis at the alkoxy group of the AlkOx is consistent with the finding that hydrolysis of 2-tBuO-4-iPr-5(4*H*)-oxazolone does not give Boc-Val-OH, but *tert*-butanol and Val-NCA instead [6]. This contrasts with 2-PhCH$_2$O-4-Bzl-5(4*H*)-oxazolone which undergoes hydrolysis at the carbonyl function to give Z-Phe-OH which is

889

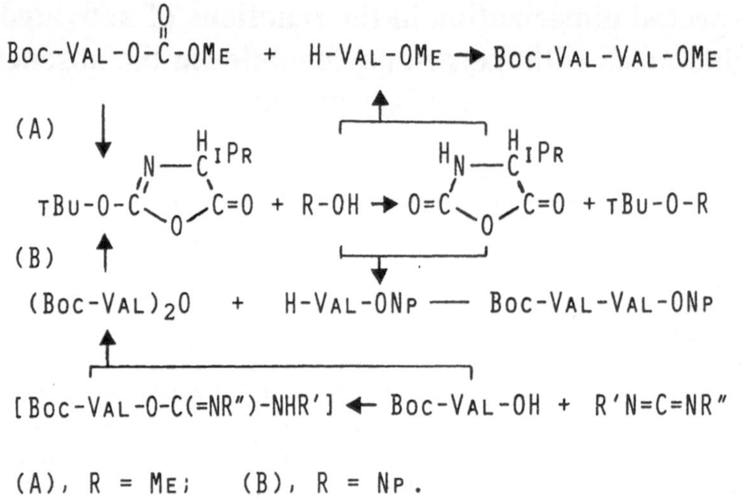

$$
\text{Boc-Val-O-}\overset{\overset{\text{O}}{\|}}{\text{C}}\text{-OMe} \quad + \quad \text{H-Val-OMe} \rightarrow \text{Boc-Val-Val-OMe}
$$

(A) ↓   ↑

$$
\text{TBu-O-C} \overset{\overset{\text{N—C IPr}}{\text{N'}}}{\underset{\text{O}}{\diagup}} \text{C=O} + \text{R-OH} \rightarrow \text{O=C} \overset{\overset{\text{HN—C IPr}}{}}{\underset{\text{O}}{\diagup}} \text{C=O} + \text{TBu-O-R}
$$

(B) ↑   ↓

$$
(\text{Boc-Val})_2\text{O} \quad + \quad \text{H-Val-ONp} \;\text{---}\; \text{Boc-Val-Val-ONp}
$$

↑

$$
[\text{Boc-Val-O-C(=NR'')-NHR'}] \leftarrow \text{Boc-Val-OH} + \text{R'N=C=NR''}
$$

(A), R = Me;   (B), R = Np.

Scheme 1.

captured by AlkOx generating symmetrical anhydride [6]. In accordance with this, Z-Val-O-COOMe in methanol gave the expected Z-Val-OMe, and not Z-Val-Val-OMe.

Boc-Val-OH, HONp, NMM and EDAC were left in DCM for 24 h, and the neutral fraction was purified by edulcoration. The product was Boc-Val-ONp (70%), with the expected ^{1}H NMR spectrum. On the other hand, the same reaction in the absence of NMM produced Boc-Val-ONp and Boc-Val-Val-ONp in a 1:3 ratio. A 2:1 ratio was obtained using DCC. 2-tBuO-4-iPr-5(4H)-oxazolone and 0.5 equiv. of HONp were left in DCM for 24 h, and the neutral fraction was examined after edulcoration. It contained Boc-Val-ONp, Boc-Val-Val-ONp, Val-NCA, tBu-O-Np and another component. The ether was isolated by reacting the mixture with dimethylaminopropylamine, followed by edulcoration. The observation was made that a solution of AlkOx and a deficiency of HONp is yellow until consumption of AlkOx is complete, at which time the color disappears. We attribute the yellow color to the nitrophenoxide anion released by protonation of the oxazolone, which would facilitate its reaction with the nucleophile. No color was generated when HONp was added to a MxAn. Our interpretation of the results of the EDAC-mediated reaction of Boc-Val-OH with HONp is similar to that for the MxAn reaction above, namely the intermediate AlkOx is decomposed by acidolysis by the HONp, successively generating Val-NCA, H-Val-ONp and the dimer (Scheme 1,B). The difference is that the electrophile in the aminolysis reaction is the symmetrical anhydride and not the MxAn. The generality of the reactions remains to be established.

The results demonstrate the possibility of side reactions associated with the coupling of *N*-alkoxycarbonylamino acids bearing an acid-sensitive *N*-α-protecting group when the activated form is not immediately captured by an incoming nucleophile. The side reaction issues from the formation and decomposition of the AlkOx by acidolysis. Evidence for the decomposition of Boc-amino acids during activation has been reported [7]. It is apparent that the side reaction is suppressed by tertiary amine. On this basis, it would seem that BOP, requiring a tertiary amine [8], is a most appropriate reagent for coupling Boc-amino acids.

Acknowledgement

This research was financially supported by the Medical Research Council of Canada.

References

1. Chen, F.M.F. and Benoiton, N.L., Can. J. Chem., 65 (1987) 619.
2. Kim, S., Kim, Y.C. and Lee, J.I., J. Org. Chem., 50 (1985) 560.
3. Jouin, P., Castro, B., Zeggaf, C., Pantaloni, A., Senet, J.P., Lecolier, S. and Sennyey, G., Tetrahedron Lett., 28 (1987) 1661.
4. Anderson, G.W., Callahan, F.M. and Zimmerman, J.E., J. Am. Chem. Soc., 89 (1967) 178.
5. Bodanszky, M., Funk, K.W. and Fink, M.L., J. Org. Chem., 38 (1973) 3565.
6. Benoiton, N.L. and Chen, F.M.F., Can. J. Chem., 59 (1981) 384.
7. Bodanszky, M., Klausner, Y.S. and Bodanszky, A., J. Org. Chem., 40 (1975) 1507.
8. Castro, B., Dormoy, J.R., Evin, G. and Selve, C., Tetrahedron Lett., (1975) 1219.

Solid-phase synthesis of ubiquitin

R. Ramage, J. Green and O.Y. Ogunjobi

Department of Chemistry, University of Edinburgh, West Mains Road, Edinburgh EH9 3JJ, Scotland, U.K.

Introduction

In the area of polypeptides and small proteins, the chemical approach to the synthesis of biologically significant systems has much to offer, particularly in analog design employing unnatural α-amino acids, thus complementing site-specific mutagenesis. For polypeptides made of up to 40 α-amino acids there is an excellent chance of success using existing methods of synthesis [1,2]. However, for small proteins of 75–150 α-amino acids the results to date indicate a real need for more efficient chemical methods of amide formation, protecting group strategy and purification. Recent examples of chemical synthesis of large polypeptides are interleukin-3 [3], porcine cardiodilatin 88 [4] and the 99 residue protein corresponding to the putative HIV-1 protease [5]. All of these syntheses employed a Merrifield strategy based upon strong acid final deprotection to reveal the final peptide. In our work at Edinburgh we have investigated the fundamental chemistry of N^α-protecting groups related to the Fmoc group of Carpino [6], which is the basis of an alternative strategy [7] that does not necessitate the use of strong acid during the synthesis.

Results and Discussion

Although synthesis of proteins by gene-based methodology has led to the production of many proteins and their analogs via site-specific mutagenesis, we felt that there was a role for chemical synthesis of small proteins and large polypeptides in studying the thermodynamic factors involved in controlling tertiary structure. The initial target selected for investigation was ubiquitin [8], which is a polypeptide made up of 76 amino acids having a stable tertiary structure and therefore may be classified as a small protein. As the name implies, ubiquitin is thought to be present in all eukaryotic cells and exhibits a most remarkable evolutionary sequence conservation. The crystal structure at 1.8 Å resolution [9] shows it to be a compact, globular protein with a hydrophobic core and a high degree of secondary structure, which accounts for the high resistance of ubiquitin to denaturation by heat, acid or alkali. This stable structure is achieved without the aid of intramolecular S–S links.

Ubiquitin is an interesting target for chemical synthesis from two complementary standpoints. Firstly, it is a highly defined target against which to measure the efficiency of new reagents in SPS of polypeptides so that this can be extended

Ubiquitin (human, bovine)

```
          10                  20                  30
M Q I F V K T L T G K T I T L E V E P S D T I E N V K A K I
          40                  50                  60
Q D K E G I P P D Q Q R L I F A G K Q L E D G R T L S D Y N
          70
I Q K E S T L H L V L R L R G G
```

to larger systems of ca. 15,000 Da molecular weight. Secondly, the chemical synthesis of ubiquitin can be adapted to incorporate unnatural α-amino acids and also to afford a range of large fragments of ubiquitin. In this way, 3-dimensional structural comparisons can be made between a segment of ubiquitin studied in isolation and when incorporated in the entire system. It is hoped that this will lead to a measure of the intramolecular interactions associated with the primary structure that integrate to the controlled folding, leading to the tertiary structure.

The approach to the SPS of ubiquitin utilized the N^{α}-Fmoc/TFA-labile side-chain protection strategy of Sheppard [7]. However, polystyrene was used as the resin to which was appended the TFA-labile linker devised by Wang [10]. In the course of this work we designed a TFA-labile N^{G}-protecting group for arginine that can be used for the synthesis of peptide sequences containing several arginines [11]. During the repetitive cycles of the synthesis, we designed a method for observing the deprotection of N^{α}-Fmoc groups that was continuously monitored by UV at 313 nm, giving a composite assessment of the coupling and deprotection stages. The amide-forming steps were accomplished by first coupling with the Fmoc amino acid, symmetric anhydride, then followed by a second coupling using the N-hydroxybenzotriazole (HOBt) activated ester of the Fmoc amino acid, both in two-fold excess. Exceptions to this protocol were the incorporation of Fmoc-Gly (single symmetric anhydride); Fmoc-Arg(Pmc)-OH, Fmoc-Asn-OH and Fmoc-Gln-OH employed only the DIC/HOBt method of activation in a double coupling. After each coupling phase, any unreacted NH_2 function was capped using acetic anhydride. After final cleavage of the side-chain protecting groups, using 5% H_2O in TFA containing thioanisole and ethylmethylsulphide as carbocation scavengers, the crude product was successively purified by G50 Sephadex gel filtration (under denaturing conditions), CM-cellulose followed by DEAE and FPLC on a mono-P column. Throughout the purification stages, the solution conditions were varied to encourage renaturation of the synthetic product. The eluants from the columns were scrutinized by Phast isoelectric focusing, by comparison with authentic bovine ubiquitin. Finally the synthetic product was shown to have identical behavior with natural material on HPLC and was finally purified on a C_8 reversed-phase column. The synthetic ubiquitin was sequenced by Dr. Linda Gilmore and Linda Kerr (WELMET Unit, Edinburgh University), and was also subjected to oxidation of Met-1, followed by tryptic digestion. The fragments so obtained were isolated, whereupon a portion was dansylated and hydrolyzed and the remainder subjected to hydrolysis and amino acid identification. It was found that the synthetic ubiquitin gave the same tryptic pattern as bovine ubiquitin; however, with the additional fragment (64–72). This interesting divergence would suggest that the synthetic product did not have the identical tertiary structure as the erythrocyte-derived material, with respect to the C-terminal region, which is also the case with the ubiquitin component of the lymphocyte-homing receptor [12].

Acknowledgements

We thank SERC, Applied Biosystems Inc., and Merck Sharp & Dohme for financial support; B. Whigham for technical support and K. Shaw for his major contribution to development of the monitor system for the ABI 430A instrument.

References

1. Merrifield, R.B., Angew. Chemie Int. Ed., 24 (1985) 799.
2. Kent, S.B.H., Annu. Rev. Biochem., 57 (1988) 957.
3. Clark-Lewis, I., Aebersold, R., Ziltener, H., Schrader, J.W., Hood, L.E. and Kent, S.B.H., Science, 231 (1986) 134.
4. Nokihara, K. and Semba, T., J. Am. Chem. Soc., 110 (1988) 7847.
5a. Schneider, J. and Kent, S.B.H., Cell, 54 (1988) 363.
  b. Nutt, R.F., Brady, S.F., Darke, P.L., Ciccarone, T.M., Colton, C.D., Nutt, E.M., Rodkey, J.A., Blunett, C.D., Waxman, L.H., Sigal, I.S., Anderson, P.S. and Weber, D.F., Proc. Natl. Acad. Sci. U.S.A., 85 (1988) 7130.
6. Carpino, L.A. and Han, G.Y., J. Org. Chem., 37 (1972) 3404.
7. Atherton, E., Gait, M.J., Sheppard, R.C. and Williams, B.J., Bioorg. Chem., 8 (1979) 351.
8a. Goldstein, G., Scheid, M., Hammerling, U., Boyse, E.A., Schlesinger, D.H. and Niall, H.D., Proc. Natl. Acad. Sci. U.S.A., 72 (1975) 11.
  b. Schlesinger, D.H., Goldstein, G. and Niall, H.D., Biochemistry, 14 (1975) 2214; Schlesinger, D.H. and Goldstein, G., Nature, 155 (1975) 423.
  c. Lund, P.K., Moats-Staats, B.M., Simmons, J.G., Hoyt, E., D'Ercoles, A.J., Martin, F. and Van Wyk, J.J., J. Biol. Chem., 260 (1985) 7609.
9. Vijay-Kumar, S., Bugg, C.E. and Cook, W.J., J. Mol. Biol., 914 (1987) 531.
10. Wang, S.S., J. Am. Chem. Soc., 95 (1973) 1328.
11a. Ramage, R. and Green, J., Tetrahedron Lett., 28 (1987) 2287.
  b. Green, J., Ogunjobi, O.Y., Ramage, R., Stewart, A.S.J., McCurdy, S. and Noble, R., Tetrahedron Lett., 29 (1988) 4341.
12. St. John, T.P., Gallatin, W.M., Siegelman, M., Smith, H.T., Fried, V.A. and Weissman, I.L., Science, 231 (1986) 845.

Orthogonal solid-phase synthesis of human gastrin-I under mild conditions

Nancy Kneib-Cordonier, Fernando Albericio and George Barany*

Department of Chemistry, University of Minnesota, Minneapolis, MN 55455, U.S.A.

Introduction

The advantages of segment condensation procedures for the synthesis of large peptides in solution are well known [1], but to date there have been difficulties in establishing analogous polymer-supported methods [2]. This report showcases several anchoring linkages recently developed in our laboratory, which were applied together with N^α-Fmoc and side-chain *tert*-butyl protection for the preparation of the 17-peptide amide human gastrin-I (Scheme 1). The chosen target [3] provides a stringent test for mild methodology because of the high density of acid-sensitive residues, including tryptophans (positions 4 and 14) and five consecutive glutamates (residues 6–10).

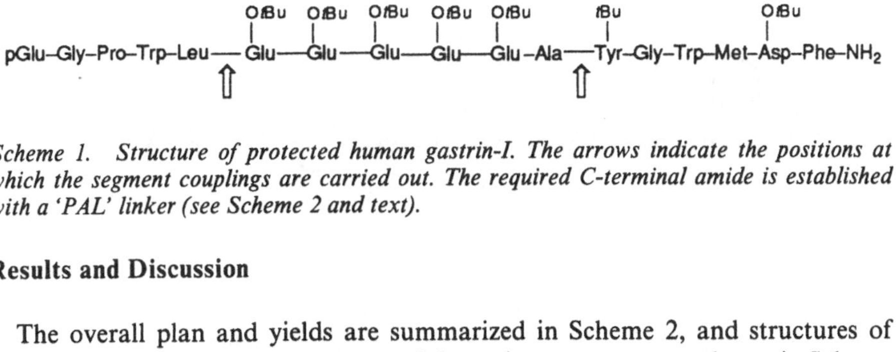

Scheme 1.   Structure of protected human gastrin-I. The arrows indicate the positions at which the segment couplings are carried out. The required C-terminal amide is established with a 'PAL' linker (see Scheme 2 and text).

Results and Discussion

The overall plan and yields are summarized in Scheme 2, and structures of the handles used to initiate syntheses of the various segments are drawn in Scheme 3. The N-terminal pentapeptide pGlu-Gly-Pro-Trp-Leu-OH was prepared using a *p*-alkoxybenzyl (PAB) ester linkage created by our published, preformed-handle strategy (Scheme 3, top structure) [4]. Cleavage was with the new Reagent M: TFA-DCM-βME-anisole (70:30:2:1), which was optimized to preserve the labile tryptophan residue.

A new, preformed-handle procedure expedited SPPS of the protected 'middle' hexapeptide, Fmoc-(Glu(O-tBu))$_5$-Ala-OH, anchored as an o-nitrobenzyl (ONb) ester. The key derivative (Scheme 3, middle structure) facilitated quantitative introduction of the C-terminal alanine onto amino-group-containing supports,

---

*To whom correspondence should be addressed.

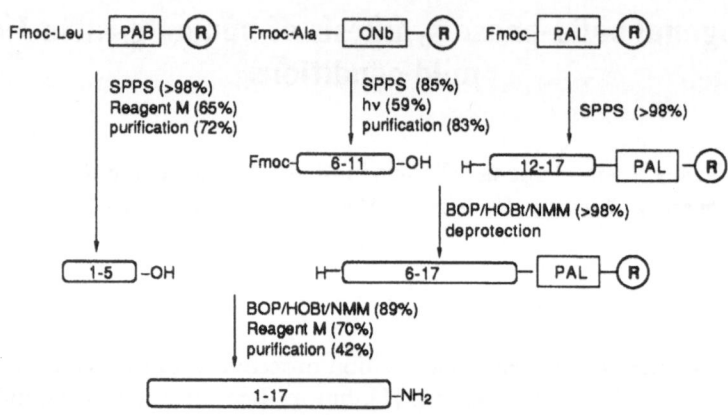

Scheme 2. Overview of segment synthesis of human gastrin-I.

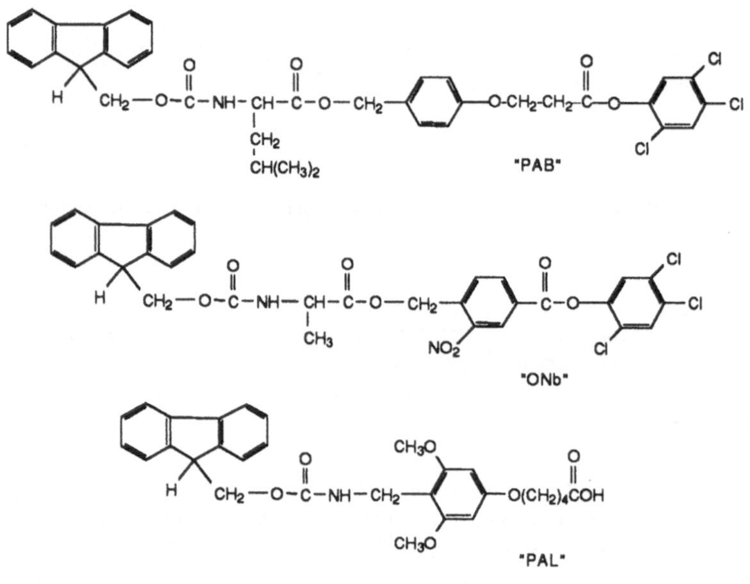

Scheme 3. Three preformed handles used for the preparation of protected peptide segments in orthogonal solid-phase synthesis of human gastrin-I.

while obviating separate protection/deblocking steps involving the handle carboxylate. Chain assembly with this Fmoc/ONb combination entailed relatively little loss of chains from the support ($>85\%$ retained throughout the synthesis, as judged by AAA and comparison to an 'internal reference' amino acid). These results are more promising than those found by Atherton et al.[5], who reported 30% diketopiperazine formation in this system. Final photolytic cleavage (350 nm) in toluene-TFE (4 : 1) gave the protected peptide in good yield. DCM must be avoided as a co-solvent, since small amounts of HCl are released under the photolysis conditions and serve partially to remove *tert*-butyl esters.

Both protected intermediates were purified by simple gel filtration, and comprised a single component by analytical HPLC. Proton NMR and FABMS data were entirely consistent with the expected structures. Lastly, the C-terminal hexapeptide, Tyr(*t*Bu)-Gly-Trp-Met-Asp(*O-t*Bu)-Phe, was assembled on our new tris(alkoxy)benzylamide support (starting handle shown in Scheme 3, bottom structure) [6]. For the polymer-supported segment condensation, the middle and N-terminal pieces were respectively added in >98% and 89% yields (judged by AAA and solid-phase sequencing), by overnight BOP/HOBt/NMM couplings in DMF. Racemization was 4% and 11%, respectively, at Ala and Leu. Cleavage with Reagent M and chromatography gave pure gastrin-I in an overall isolated yield of approximately 30%. These results compare favorably with those from stepwise assembly.

Acknowledgements

This contribution is taken in part from the Ph.D. thesis of N.K.C., University of Minnesota, June, 1989. We thank Ms. Darci L. Knowlton for expert technical assistance, and appreciate gifts of materials and collaborative insights from Dr. Derek Hudson of MilliGen/Biosearch. We are grateful to the NIH (GM 28934 and 42722) and the Graduate School of the University of Minnesota for financial support. F.A. is a faculty member in the Department of Organic Chemistry, University of Barcelona, Spain. His travel and living expenses while carrying out this research in Minnesota are covered by a NATO Collaborative Research Grant 0841/88, and he further thanks Applied Biosystems Inc. for a travel award to the Eleventh American Peptide Symposium.

References

1. Finn, F.M. and Hofmann, K., In Neurath, H. and Hill, R.L. (Eds.) The Proteins, 3rd Ed., Vol. 2, Academic Press, New York, NY, 1976, p. 105.
2. Barany, G., Kneib-Cordonier, N. and Mullen, D.G., Int. J. Pept. Prot. Res., 30 (1987) 705.
3. Gregory, R.A. and Tracy, H.J., Gut, 5 (1964) 103.
4. Albericio, F. and Barany, G., Int. J. Pept. Prot. Res., 26 (1985) 92.
5. Atherton, E., Brown, E., Priestley, G., Sheppard, R.C. and Williams, B.J., In Rich, D.H. and Gross, E. (Eds.) Peptides – Synthesis, Structure and Function: Proc. Seventh American Peptide Symposium, Pierce Chemical Company, Rockford, IL, 1981, p. 163.
6. Albericio, F. and Barany, G., Int. J. Pept. Prot. Res., 30 (1987) 206.

Mapping of a monoclonal antibody to FeLV gp70

Kuldip Singh, Patrick Sheridan, Lois Aldwin and Danute E. Nitecki

Cetus Corporation, 1400 53rd Street, Emeryville, CA 94608, U.S.A.

Introduction

Recently, Geysen et al. [1] have introduced a method for rapid synthesis and binding assay of large numbers of peptides. The approach involves SPPS on the surface of polyethylene rods arranged in a microtiter plate format. The *N*-acetylated, deprotected peptides, still on the rods, are assayed for binding of an antibody using ELISA methods. We have used this method and the commercially available kit from Cambridge Research Biochemicals (CRB, Cambridge, U.K.) to confirm a determinant in the FeLV coat protein gp70. A synthetic 14-mer, having sequence 213–226, MGPNLVLPDQKPPS, had been found to inhibit neutralization by a virus-neutralizing mouse monoclonal antibody cl.25 [2] at micromolar concentration.

Results and Discussion

We have synthesized 4 sets of 18 overlapping hexapeptides, shifted by one amino acid, starting with amino acid 208 and finishing with amino acid 230 from the gp70 sequence. These deprotected peptides, as well as 10 control hexapeptides, were compared in ELISA format for binding of monoclonal antibody cl.25. The synthesis, using minor variations of CRB protocols, utilized *N*-Fmoc-pentafluorophenyl ester chemistry. Many control hexapeptides were hydrolyzed and analyzed for amino acid composition. Representative analyses are shown in Table 1.

Table 1   *Examples of amino acid analyses of hexapeptide pins (nmol)*

Peptide	1	2	3	4	5	6	β-Ala
A D G K L N	2.07	1.42	4.58	1.36	4.05	1.42	9.04
G P N L V L	1.69	2.36	0.78	4.65	4.37	4.65	8.95
T V V A D L	0.8	1.89	1.89	3.41	2.3	8.95	20.64
P Q R S V S	2.93	1.06	0.99	3.85	7.18	3.85	24.73
M N P Q R V	1.17	0.66	9.58	1.19	1.18	16.6	28.98
K L M N P R	1.61	4.21	2.56	1.0	9.23	1.65	23.65
D G K L M P	1.27	3.63	1.66	2.98	1.88	10.61	18.79
G K L M N Q	2.8	1.01	2.61	1.81	1.03	2.61	21.70

Internal ratios were rather unsatisfactory, ranging several-fold within a given peptide; nor were we able to achieve the levels of peptides close to that of

898

the anchoring β-alanine. Nevertheless, ELISA assay with cl.25 of all 4 sets of hexapeptides indicated binding of the same region, albeit maximally binding pins were not identical in each set and the binding epitope shifted slightly. The 'consensus' epitope was found to reside in the sequence 212–220. We believe these inconsistencies to be a reflection of poor homogeneity of the peptide sequences on an individual pin. The antibody cl.25 binding to the hexapeptide pins could be inhibited by a soluble 9-mer peptide 211–219, with sequence Ac-QAMGPNLVL-OH but not by shorter peptides. We found relatively low ratios of binding to background in ELISA, as compared to those found in [1]; purification of monoclonal antibody from ascites fluid through a Sepharose–Protein A column improved the signal to noise ratio considerably, yet our best assays showed only a 6:1 ratio.

Mimotope mapping of monoclonal cl.25 did not show binding above background to any dipeptides from the linear epitope in the region indicated. We synthesized 3 sets of pins containing 400 dipeptides and some control pins for hydrolysis; a fourth set was obtained from CRB. All four sets indicated binding of highly hydrophobic and aromatic dipeptides. Table 2 shows the 5 highest binding dipeptides from all 4 sets. The striking feature is the presence of aromatic residues in all but one of the dipeptides.

Table 2 *Dipeptides showing maximum binding in each set*

SET					
1	IF	IY	IR	GY	WN
2	YY	LY	YL	YL	YC
3	YY	YR	LY	YC	WY
4	WI	YL	YY	YF	FY

The region 208–230 that binds this monoclonal cl.25 does not contain a single aromatic residue. It is entirely possible that these mimotope mapping attempts are, in fact, binding to some region outside of the antigen binding site. This is not entirely unexpected, because antibody is a busy polyfunctional molecule involved in a variety of binding events in a living animal (complement binding, Fc binding, placental transport, etc.).

References

1. Geysen, H.M., Rodda, S.J., Mason, T.J., Tribbick, G. and Schoofs, P.G., J. Immunol. Methods, 102 (1987) 259.
2. Nunberg, J.M., Rodgers, G., Gilbert, J.H. and Snead, R.M., Proc. Natl. Acad. Sci. U.S.A., 81 (1984) 3675.

BROP: A new coupling reagent for *N*-methyl amino acids

B. Castro, J. Coste, M.-N. Dufour and A. Pantaloni

CCIPE, Rue de la Cardonille, F-34094 Montpellier Cedex 2, France

Introduction

The usual peptide-coupling reagents often fail when coupling *N*-methylated amino acids [1]. For example, BOP **1** [2a] has been used by Wenger [3], but was occasionally not very effective [1,4]. It is noteworthy that BOP-Cl [1,5] and Dpp-Cl [6], which are effective reagents, produce this kind of coupling without HOBt, the usual additive for peptide coupling [7]. This led us to study BroP **2** [2b] (bromo tris (dimethylamino) phosphonium hexafluorophosphate), a compound which, unlike BOP, does not contain the oxybenzotriazole residue (see Scheme 1). This reagent proved to be excellent for coupling *N*-methylated amino acids.

Results and Discussion

The results summarized in Table 1 show that the reagent BroP is very effective, since it rapidly (1 h) produced high yields of non-racemized peptides. The absence of appreciable racemization was shown using ^{1}H NMR (360 MHz) to compare isomeric compounds obtained by the same pathway (e.g. see entries **4** and **5**, **11** and **12**, **13** and **15**). In contrast, the results with BOP are variable.

The yields were high with BroP even when the steric constraints were very strong (entries **7**, **9**, **11**, **12**, **13** and **15**). With BOP, however, when the steric factor was too great (entries **6** and **10**), the main compound formed was the 'active ester' [7] of oxybenzotriazole of the *N*-protected amino acid. This intermediate reacted very slowly: in entry **10**, for instance, only 40% dipeptide had been produced after 16 h of reaction, which was, moreover, 70% racemized. The formation of this derivative explains the poor efficacity of BOP.

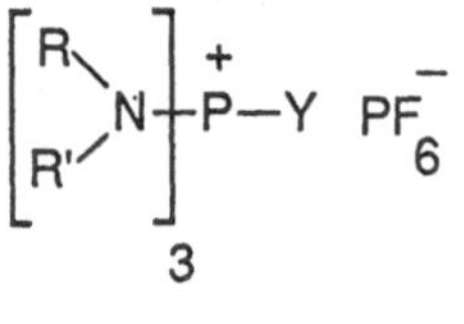

$$\left[\begin{array}{c} R \\ \diagdown \\ N-P-Y \\ \diagup \\ R' \end{array} \right]^{+} PF_6^{-}$$

3

1 BOP :   R = R' = Me   ; Y = OBt
2 BroP: R = R' = Me   ; Y = Br
3 PyBroP: R,R' = (CH$_2$)$_4$; Y = Br

Scheme 1.

Table 1 *Coupling reactions^a with reagents 1, 2 and 3*

Entry	Peptide	Reagent	Yield %
1	Z-Val-(Me)ValOMe	BOP	67
2		BroP	71
3	Z-(Me)Val-ValOMe	BOP	87
4		BroP	90
5	Z-(Me)DVal-ValOMe	BroP	85
6	Boc-Pro-(Me)ValOMe	BOP	25
7		BroP	82
8		PyBroP	91
9	Boc-Pro-(Me)LeuOMe	BroP	75
10	Z-(Me)Val-(Me)ValOMe	BOP	5
11		BroP	70
12	Z-(Me)DVal-(Me)ValOMe	BroP	80
13	Z-(Me)Leu-(Me)ValOMe	BroP	89
14		PyBroP	84
15	Z-(Me)DLeu-(Me)ValOMe	BroP	89

^{a} One-pot reactions in DCM for 1h at room temperature, with one equivalent of reagent and of N-protected amino acid, 1,1 equivalent of C-protected amino acid and three equivalents of DIEA.

Conclusion

Thus, BroP is an excellent reagent for coupling *N*-methylated amino acids. It gives results comparable to those of BOP-Cl [5] and Dpp-Cl [6], but in much shorter times. Another advantage of the reagent is its stability, which facilitates storage and use. The homologous compound 3 (PyBroP) gives identical results (entries 8 and 14).

References

1. Tung, R.T. and Rich, D.H., J. Am. Chem. Soc., 107 (1986) 4342.
2a. Castro, B., Dormoy, J.-R., Evin G. and Selve, C., Tetrahedron Lett., (1975) 1219;
 b. Castro, B., and Dormoy, J.-R., Tetrahedron Lett., (1973) 3243.
3. Wenger, R.M., Ang. Chem. Int. Ed., 24 (1985) 77.
4. Jouin, P., Poncet, J., Dufour, M.-N., Pantaloni, A. and Castro, B., J. Org. Chem., 54 (1989) 617.
5   Van Der Auwera, C. and Anteunis, M.J.O., Int. J. Pept. Prot. Res., 29 (1987) 574.
6. Galpin, I.J., Mohammed A.K. and Patel, A., Tetrahedron, 44 (1988) 574.
7. König, W. and Geiger, R., Chem. Ber., (1970) 788.

Use of BOP in SPPS: Application to incorporation of N^G-unprotected arginine

Marie-Line Ceccato[a], Jacques Chenu[a], Jacques Demaille[b], Bernard Calas[b] and Anne Laurent[b]

[a]*Sanofi-Recherche, 195 Route d'Espagne, F-31036 Toulouse Cedex, France*
[b]*CNRS, UPR 8402 et INSERM U.249, BP 5051, F-34033 Montpellier Cedex, France*

Introduction

At this time, BOP is used in SPPS at basic pH, but we observed that couplings may be conducted in slightly acidic conditions. With a protocol (Table 1) in which the basic washes used to deprotonate the α-NH$_2$ group after BOC cleavage were avoided, it became possible to build peptides easily with N^G-unprotected arginine.

Table 1 *Protocol for a synthetic cycle using BOP at pH 5–7*

Deprotection:	50% TFA in DCM 1×2 min	
	50% TFA in DCM 1×30 min	
Washes:	DCM	3×1 min
	DMF	3×1 min
Coupling:	2 equiv. amino acid, 2 equiv. *N*-methylmorpholine in DMF, add to the resin, shake for 2 min.	
	2 equiv. BOP in DMF, add to the resin, shake for 2 min.	
	Adjust pH to 6 with *N*-methylmorpholine, shake for 30 min.	
Washes:	DMF	2×1 min
	DCM	4×1 min

Results and Discussion

Using the protocol reported in Table 1, we have synthesized the tridecapeptides **1** and **2**: RFARKGXLRQKNV (X=Ser **1**, X=Ala **2**). **1** corresponds to the amino acid sequence between residues 19 and 31 of protein kinase C (PKC); **2** is a specific inhibitor of PKC. The resin used was 'Expansin' (Societé Expansia, BP 6, Aramon, 30390, France). C-terminal Val was linked to the support through a glycolamidic anchorage [1]. Side chains were protected as follows: Lys, 2 ClZ; Ser, Bzl and Arg, H$^+$. t-Boc was used to protect the α-NH$_2$. After completion of the synthesis, the partially protected peptides were cleaved from the resin by using 5 equivalents of NaOH in iPrOH-H$_2$O (70–30). Side chains were deprotected with 1 M TFMSA in TFA.

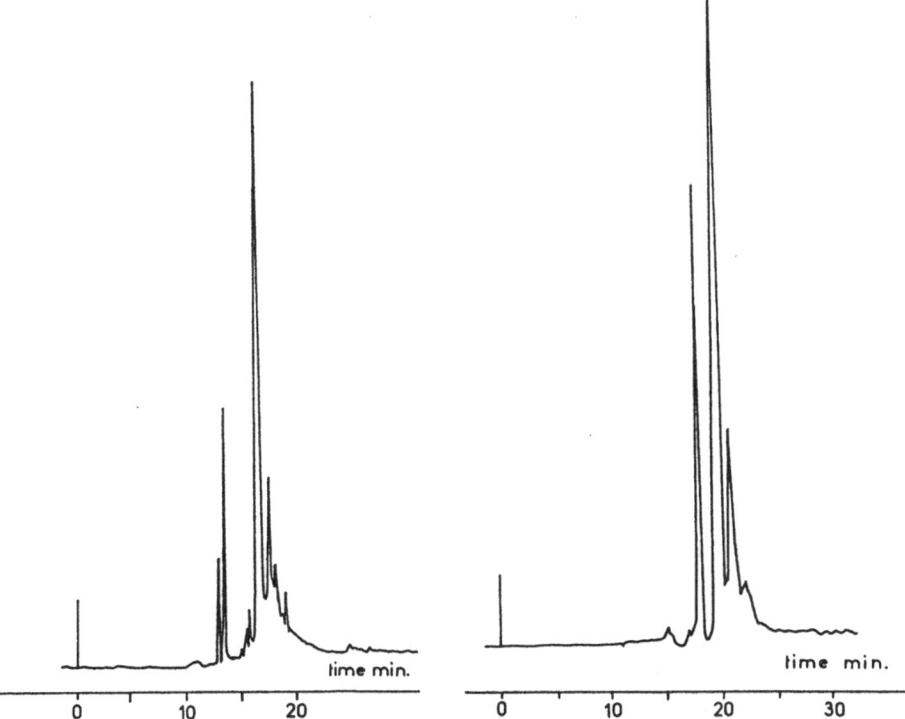

Fig. 1. HPLC of **1**. Aquapore RP 300 (0.46×20 cm). Elution was at 1 ml/min in a linear gradient from 0% buffer B to 60% buffer B in 30 mm. (Buffer B: 0.1% TFA in CH₃CN; buffer A: 0.1% TFA in water.)

Fig. 2. HPLC of **2**. Aquapore RP 300 (0.46×20 cm). Elution was at 1 ml/min in a linear gradient from 0% buffer B to 60% buffer B in 30 min. (Buffer B: 0.1% TFA in CH₃CN; buffer A: 0.1% TFA in water.)

HPLC of crude peptides indicated good homogeneity (Figs. 1 and 2). AAA of peptides **1** and **2** are in good agreement with the proposed structure; no ornithine was detected indicating the absence of acylation on the guanidino groups of arginines. Peptides were purified by ion-exchange chromatography followed by semi-preparative RPHPLC (Nucleosil 300, 5 μM, C₈, 25×2 cm).

With t-Boc strategy, the guanidino group of arginine is routinely protected by Tos or Mts. These groups are difficult to cleave and require drastic acidic conditons (HF or TFMSA). Use of our protocol with Boc-Arg-(HCl)-OH greatly simplifies the synthesis of peptides including in their sequences several arginyl residues.

References

1. Calas, B., Méry, J., Parello, J. and Cavé, A., Tetrahedron, 41 (1985) 5331.

Chaotropic salts improve SPPS coupling reactions

Wieslaw A. Klis and John M. Stewart

*Department of Biochemistry, University of Colorado Medical School, Denver,
CO 80262, U.S.A.*

Introduction

Many peptide chemists have found it difficult to obtain complete coupling in SPPS of certain sequences in the length range of 8–16 residues from the resin. In one case, secondary structure of the growing peptide on the resin was found to be a source of trouble. Such structures can make the N-terminal inaccessible for peptide synthesis. Deber et al. [1] confirmed the existence of β-sheet/β-turn structure in a 'difficult coupling' of a 16-residue peptidyl-resin fragment of human growth hormone releasing hormone. Complicated, stable, three-dimensional protein structures often cause difficulties for biological and biochemical research. Traditionally, those difficulties have been overcome by the use of salts and other chaotropic reagents. Possibly chaotropic salts decrease secondary structure formation in growing peptide chains and also perturb potential interactions between the peptide and the resin, thus exposing the N-terminal to the coupling reaction. We thought that perturbing the secondary structure might make synthesis easier. This idea had been tested in SPPS by using 1.5 M urea in DMF when adding glutamine as an active ester [2]. But urea is not a favorable additive in DCC types of peptide synthesis.

Results and Discussion

An analog of so-called C-peptide, the N-terminal fragment of ribonuclease A, was chosen as a model peptide. This peptide, Ac-AETAAAKFLRAHA-NH$_2$ has a strong tendency to adopt α-helical structure in cold water and can be expected to show a greater tendency for structure formation in non-polar solvents used in peptide synthesis. When synthesizing this peptide, difficult couplings were encountered in the coupling of alanines 6–4 and threonine 3. We sought to determine how chaotropic salts (KSCN, NaClO$_4$, LiBr and others) would affect the coupling of those residues. We used presynthesized heptapeptidyl resin (Boc-K(ClZ)FLR(Tos)AH(Bom)A-MBHA) as our precursor to study the coupling reaction. To achieve a satisfactory salt concentration DMF : DCM = 1 : 1 was selected as the working solvent system. The progress of coupling was monitored by the quantitative ninhydrin test [3]; completion of peptide coupling was verified by the qualitative Kaiser test. Figure 1 shows the progress of coupling Boc-Thr(Bzl) to AAAK(ClZ)FLR(Tos)AH(Bom)A-MBHA under several conditions, all with 10-fold excess DCC and Boc-AA. Under standard conditions,

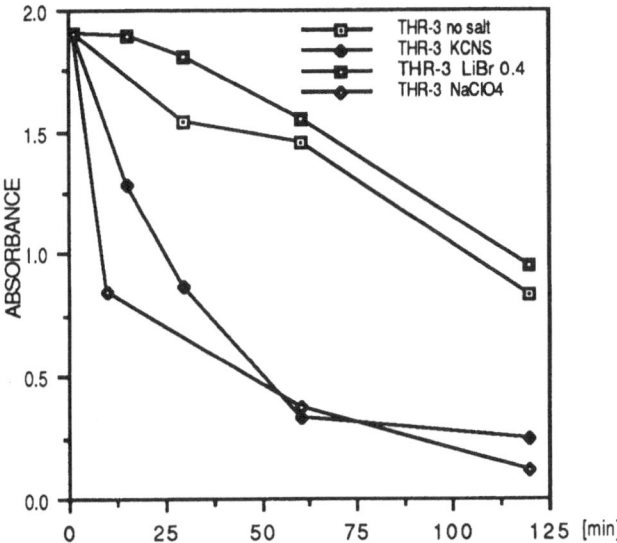

Fig. 1. Progress of DCC-mediated coupling of Boc-Thr(Bzl) (10-fold excess) to AAAK-(ClZ)FLR(Tos)AH(Bom)A-MBHA in DMF:DCM (1:1) as shown by absorbance in the quantitative ninhydrin test.

the reaction rate was very slow, and coupling was not complete. When the reaction was run in 0.4 M KSCN or 0.4 M NaClO$_4$ the rate accelerated and coupling was complete in 1 h. LiBr solution seemed to slow down the reaction. For difficult couplings of other residues tested, the picture was even more clear. With salt coupling, the reaction was complete in 30 min, whereas double coupling was required under standard conditions. In alanine coupling reactions, LiBr was as effective as KSCN. Salt coupling was extended to typical 2.5-fold Boc-AA excess and worked very well. To obtain complete coupling of Thr^3 under these conditions, DCC/HOBt or DCC symmetric anhydride preactivation was used; BOP reactions were also accelerated.

The series of experiments presented clearly shows that chaotropic salts are useful in SPPS to disrupt secondary structures and increase the rate of difficult coupling reactions. The effectiveness of salt synthesis seems to increase with the length of the peptide involved. The salts interact with the peptide chain and disturb or disrupt intramolecular interactions; prewashing the peptide-resin with salt solution is effective. The most effective salt concentration appears to be 0.4 M. The appropriate salt should contain a large anion, such as perchlorate or thiocyanate and a cation that does not form complex compounds easily, such as sodium or potassium. In some cases, KSCN has a tendency to catalyze *O*-acylisourea rearrangement, but perchlorate does not have such activity. Salts with small anions essentially do not enhance synthesis. Similarly, cations that form coordination compounds easily (Zn^{2+}, Ba^{2+}, Co^{3+}) inhibit the coupling

W.A. Klis and J.M. Stewart

reaction for Boc-Amino acids, which possess strong complexing activity due to side chain functional groups.

Acknowledgements

Aided by Biosearch Inc.

References

1. Deber, C.M., Lutek, M.K., Haimer, E.P. and Felix, A.M., Peptide Res., 2 (1989) 184.
2. Westall, F.C. and Robinson, A.B., J. Org. Chem., 35 (1970) 2842.
3. Merrifield, R.B., Vizioli, L.D. and Boman, H.G., Biochemistry, 21 (1982) 5020.

Peptide synthesis using bis(2-oxo-3-oxazolidinyl)-phosphinate esters of N-hydroxy compounds: Remarkable effects of solvent and varying amounts of tertiary amine on racemization suppression

Kusuo Horiki

Shionogi Research Laboratories, Shionogi & Co., Ltd., Fukushima-ku, Osaka 553, Japan

Introduction

We reported previously that bis(2-oxo-3-oxazolidinyl)phosphinate (**1**) of ethyl 2-hydroximino-2-cyanoacetate is very effective in the synthesis of simple peptides [1]. Diphenyl phosphate (**2**) of 3-hydroxy-3,4-dihydroquinazoline-4-one has been found to be equally useful (Fig. 1).

Fig. 1.   Structures of reagents used in this study.

Results and Discussion

This paper describes the application of the highly sensitive Young test (the synthesis of Bz-Leu-Gly-OEt, **3**) to **1** and **2**. Selected examples obtained by direct coupling under various conditions are shown in Table 1.

Polar solvents, or excess amounts of triethylamine (TEA) in nonpolar solvents, may reduce racemization by increasing the effective concentration of free Gly-OEt via large dissociation of hydrogen-bonded ion pairs of Gly-OEt•HCl and TEA. The increasing amount of free Gly-OEt would preclude the accumulation of a reactive species, whatever it may be, by rapid aminolysis, giving the less racemized **3**.

Next, the synthesis of **3** was attempted by the preactivation method using **1** in acetonitrile at RT. Surprisingly, **1** did not afford **3**, and many uncharacterized byproducts formed, which sharply contrasted with the results of the direct coupling. This result indicates that in an aprotic dipolar solvent, the less solvated carboxylate anion generated from Bz-L-Leu is very reactive, giving rise to degradation of **1**. At the initial activation stage of Bz-L-Leu in acetonitrile in the pre-activation method, degradation of **1** would occur exclusively, probably due to a slower collapse of the trigonal bipyramidal intermediate, [A] or [B]

907

Table 1 *Selected examples of the Young test using 1 and 2 by the direct coupling procedure*[a]

Reaction solvent	Amount of TEA (equiv.)	Yield (%)		L-Isomer (%)[b]	
		1	2	1	2
CH₂Cl₂	2	36.3	79.1	66.5	65.9
	3	18.1	71.9	90.6	82.4
Ethyl acetate	2	21.3	80.0	47.9	55.6
	3	26.9		83.5	
Acetonitrile	2	27.8	76.3	92.1	87.4
	3	18.8	70.0	97.4	82.9
DMF	2	40.6	68.4	97.9	92.9
	2	43.8[c]		98.2[c]	

[a] All reactions were carried out at RT for 16 h. Initial concentration: 1 or 2 = Bz-L-Leu = Gly-OEt•HCl = 0.1 M in DMF and 0.04 M in other solvents.
[b] Excluding L-isomer present as the racemate.
[c] Initial concentration: 1 = Bz-L-Leu = Gly-OEt•HCl = 0.04 M.

(formed after pseudorotation or other), either to the mixed anhydride, **4**, or to the active ester, **5** (Fig. 2). Alternatively, in the direct coupling method, the formation of either an intermediate or a transition state, [**C**], involving Gly-OEt, may be responsible for the formation of **3**. If this mechanism is correct, the low level of racemization with **1** may also be reasonably explained.

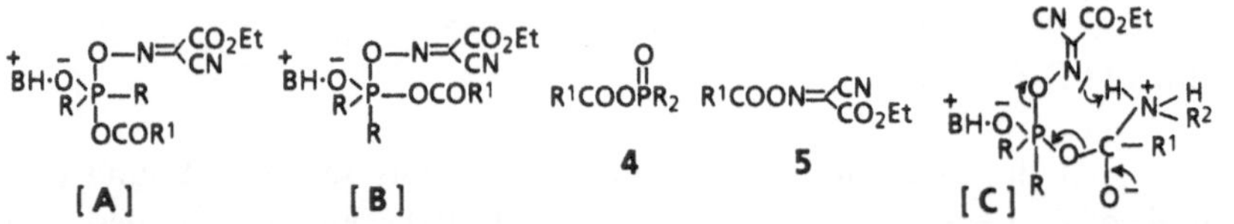

Fig. 2. Schematic illustration of either intermediates ([A], [B], 4, 5 and [C] or a transition state [C]. B = TEA; R = 2-oxo-3-oxazolidinyl; R¹ = Bz-Leu moiety; R² = Gly-OEt moiety.

In the case of **2**, the formation of either an intermediate or a transition state, like [**C**], seems to be sterically unfavorable. However, it is still uncertain which of the two alternative pathways (i.e., the mixed anhydride pathway or the active ester pathway) is predominant.

References

1. Horiki, K., In Shiba, T. and Sakakibara, S. (Eds.) Peptide Chemistry 1987, Protein Research Foundation, Osaka, 1988, p. 239.

908

Increased coupling efficiency in solid-phase peptide synthesis using elevated temperature

David H. Lloyd[a], Gordon M. Petrie[a], Richard L. Noble[a] and James P. Tam[b]

[a]Applied Biosystems Inc., Foster City, CA 94404, U.S.A.
[b]The Rockefeller University, New York, NY 10021, U.S.A.

Introduction

Despite many advances in SPPS, sequence-dependent coupling inefficiencies continue to be a problem. Approaches to defeat these have included multiple couplings and the addition of co-solvents. During the synthesis of big gastrin and TGF-α, it was found that difficult couplings could be driven to completion by re-coupling at 50°C for 1–2 h [1,2]. It was felt that this effect was primarily due to thermal disruption of inter-chain aggregates that normally inhibited complete coupling. This communication describes preliminary data on the use of elevated temperatures in automated peptide synthesis.

Results and Discussion

Syntheses were carried out using an ABI 430A synthesizer. Preformed tBoc-amino acid-HOBt esters (4 equiv.) were coupled in NMP. A reaction vessel heater and controller (activated and de-activated by a relay on the 430A) were used to maintain an elevated temperature during the coupling. The reaction mixture attained the desired temperature within a few minutes of heating. At the completion of the coupling phase, the peptide-resin was cooled by evaporation of DCM.

The peptide QFFYCNTTQLFNN (Q13N) provided a good example of a sequence-dependent coupling problem beginning at the ninth cycle of synthesis. Thirty-minute couplings (roughly half the duration of standard cycles) were used at room temperature, 40, 50 and 60°C. Yields, determined by quantitative ninhydrin assay (Fig. 1), indicate a temperature-dependent increase in reaction completion. When the peptide-resins were cleaved (HF/anisole/thiocresol, 60 min, –3°C) and analyzed by RPHPLC, it appeared that the 60°C synthesis provided the cleanest product. However, FABMS analysis of this sample indicated the presence of significant amounts of dehydrated material not present in the standard ambient temperature synthesis. Clearly, a compromise exists between improved couplings and dehydration that occurs at elevated temperatures.

Rat ANF(1–28) contains a commonly described [3] failure sequence and appeared to be another good model to test the effect of heated reaction cycles. Syntheses were carried out using the same cycles as used for Q13N at 60°C, as well as using standard (unheated) cycles for comparison. Coupling data (Fig. 2) showed significant improvements for the heated reactions in the failure region (cycles 14–17). In addition, it is interesting to note that heating did not markedly improve the incomplete couplings in the amino-terminal region. HPLC analysis of crude cleaved materials indicate fewer side products for the heated synthesis, and FABMS analysis of the purified peptide revealed no discernable dehydration products for this (heated) synthesis.

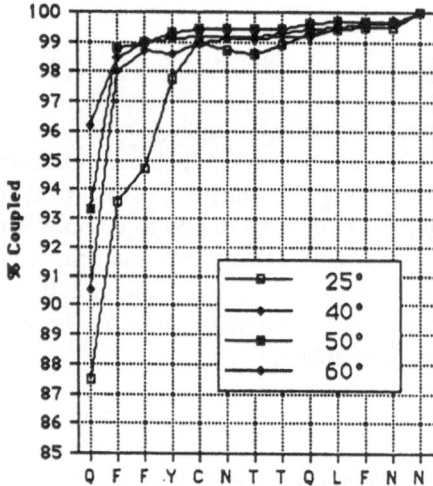

Fig. 1. *Temperature effect on coupling efficiency for Q13N.*

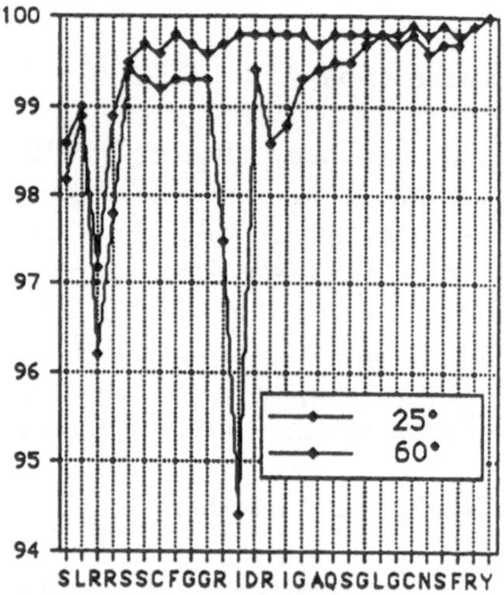

Fig. 2. *Temperature effect on coupling efficiency for ANF(1–28).*

Conclusions

Coupling efficiencies of some difficult sequences increase in a temperature-dependent manner, allowing shorter coupling cycles. In the case of one peptide (Q13N), significant amounts of dehydration also occurred with heated couplings. Characterization of side-reactions (racemization, aspartimide and pGlu formation, and other side-chain modifications) and their temperature-dependence will be necessary before heated couplings can be routinely used in SPPS.

References

1. Tam, J.P., In Deber, C.M., Hruby, V.J. and Kopple, K.D. (Eds.) Peptides: Structure and Function (Proceedings of the 9th American Peptide Symposium), Pierce Chemical Co., Rockford, IL, 1985, p. 423.
2. Tam, J.P., Int. J. Pept. Prot. Res., 29 (1987) 421.
3. Otteson, K.M., McCurdy, S.N., Harrison, J.L., Farnam, W.E., and Pierce, L.G., 13th Annual Lorne Conference on Protein Structure and Function, 1988.

BOP reagent for the coupling of pGlu and Boc-His(Tos) in solid-phase peptide synthesis

Mylène Forest and Alain Fournier

Institut National de la Recherche Scientifique-Santé, Université du Québec,
245, boul. Hymus, Pointe-Claire, Québec, Canada H9R 1G6

Introduction

The BOP reagent proposed by Castro et al. [1] is ideally suited for SPPS and the rate of coupling using BOP compares favorably to other methods of activation [2,3]. In the course of a previous study [3], it has been demonstrated that BOP couplings are practically racemization-free. However, in this study, the model peptides synthesized with BOP reagent were not containing histidine, a residue known to racemize easily. Racemization is favored in highly polar solvents such as DMF, and is influenced by the amount and chemical character of the organic base added to the reaction mixture. Therefore, we investigated the effects of the coupling conditions with BOP (i.e. DMF or NMP as solvent and DIEA as organic base) for the incorporation of histidine in the model peptide TRH. The derivative Boc-histidine(tosyl) was chosen in order to verify the feasibility of using the tosyl side-chain protecting group with BOP reagent. Indeed, according to the activation mechanism proposed by Castro et al. [4], BOP would generate hydroxybenzotriazole in the reaction mixture. It has been reported that histidine detosylation occurs when this reagent is used, for instance, as an additive for DCC-mediated couplings [5a]. We also investigated the compatibility of pyroglutamic acid (pGlu), the first residue of our model peptide TRH, with BOP coupling conditions. pGlu is usually introduced in peptide chain as a benzyloxycarbonyl (Z) derivative (ZpGlu), primarily because of its superior solubility characteristic in a non-polar solvent, such as DCM [5b]. The solubility problem of pGlu is not encountered in DMF or NMP.

Results and Discussion

A comparative racemization study for the synthesis of TRH and [D-His^2]TRH using either BOP or symmetrical anhydride-mediated couplings was carried out. With BOP reagent (3 equiv.) and DIEA (3 equiv.), crude products contained approximately 0.5% of diastereoisomer as estimated from HPLC. On the other hand, the crude material obtained when using the symmetrical anhydride procedure for the peptide bond formation showed the presence of up to 1% of racemate suggesting that at least 0.5% of C$^\alpha$-proton exchange occurred during the coupling of Boc-His(Tos), assuming that up to 0.5% of epimer might have been already present in the starting material.

We did not verify if detosylation occurred during Boc-histidine(tosyl) (3 equiv.) couplings with BOP (3 equiv.) in DMF, in presence of an equimolar amount (3 equiv.) of DIEA. However, the low racemization levels observed with BOP suggest that detosylation probably did not occur, or only to a very low extent, even if these conditions might allow the formation in the mixture of a little amount of the conjugate acid of HOBT. The conjugate acid is the chemical form of HOBt that is believed to be responsible for the tosyl removal from the imidazole moiety. Moreover, the extraction of the resin with TFA after HF cleavage gave a strongly absorbing side-product (retention time: 10.7 min) that has been identified after isolation by HPLC. NMR characterization (AA'BB': 7.713, 7.686, 7.243, 7.217 ppm; -CH$_3$: 2.362 ppm) and MS (m/z: 172 [M$^+$.]; 91 [tropylium cation]) showed this to be *p*-toluenesulfonic acid, thus confirming the presence of tosyl protection on the histidine side chain when the completed peptide was cleaved from the resin. This material was still present in the polymeric support, even after numerous washings with ether. When using 30% acetic acid for the extraction, most tosyl by-product is absent from the crude material suggesting that care must be taken when interpreting the peptide content and purity of post HF products when extractions are made with TFA.

We checked the effect of an increased amount of organic base by carrying out the coupling reaction of Boc-His(Tos) (3 equiv.) with BOP (3 equiv.) in DMF in presence of 9 equiv. of DIEA. The HPLC of the resulting crude TRH showed approximately 3% of [D-His^2]-TRH, clearly indicating that a racemization problem is encountered when Boc-His(Tos) coupling is carried out with BOP in presence of excess DIEA. This situation would be particular to histidine since previous investigations of the coupling conditions for other residues in different peptides did not reveal any major racemization problem when adding excess DIEA as organic base to the reaction mixture.

Since BOP couplings are favored in DMF, a polar aprotic solvent, rather than in DCM [3], we checked the possibility of using NMP in replacement of DMF for BOP-mediated couplings of Boc-amino acids. NMP containing 20% DMSO was already described by McCurdy and Otteson [6] as a useful solvent mixture in SPPS, with the use of the DCC/HOBT coupling method. The major advantage of this solvent system, or of NMP itself, is its very effective ability to swell the peptide-resin during coupling. Thus, a synthesis of TRH was carried out entirely with BOP (3 equiv.) in NMP, but this time in presence of 5 equiv. of DIEA. Indeed, as proposed by Le-Nguyen et al. [7,8], a protocol involving the concomittant steps of 'neutralization' and coupling was used for this synthesis that explains the use of a larger amount (5 equiv.) of DIEA in the coupling mixture. As observed on RPHPLC the crude material obtained after HF cleavage and 30% AcOH extraction was of excellent quality. The extent of racemization for the histidine residue was also estimated to be less or equal to 0.5%. These results confirmed the compatibility of BOP with NMP as solvent, and the practicality, as well as the usefulness, of the simultaneous TFA salt displacement and Boc-amino acid coupling. Moreover, although no precise kinetic studies were carried out, all couplings were completed in time periods equivalent to those observed with DMF.

The first residue of TRH is pGlu. This amino acid is usually introduced into the peptide chain as ZpGlu, primarily because of its superior solubility in a non-polar solvent, such as DCM [5b]. This solubility problem is not encountered in DMF or in NMP, and thus, we verified the feasibility of BOP-mediated couplings of pGlu under its free form, in DMF, as well as in NMP. We observed that the peptide bond formation with this derivative was satisfactorily achieved in both solvents in a single step of coupling. Therefore, its use in conjunction with BOP reagent in solvents such as DMF or NMP represents a real advantage over the conventional coupling procedures carried out in DCM, and requiring the much more expensive ZpGlu.

References

1. Castro, B., Dormoy, J.R., Evin, G. and Selve, C., Tetrahedron Lett., 1219 (1975) 1222.
2. Fournier, A., Wang, C.-T. and Felix, A.M., In Marshall, G.R. (Ed.) Peptides: Chemistry and Biology, ESCOM, Leiden, 1988, p. 212.
3. Fournier, A., Wang, C.-T. and Felix, A.M., Int. J. Pept. Prot. Res., 31 (1988) 86.
4. Castro, B., Dormoy, J.-R., Evin, G. and Selve, C., J. Chem. Res., (S) (1977) 182.
5a. Barany, G. and Merrifield, R.B., In Gross, E. and Meienhofer, J. (Eds.) The Peptides: Analysis, Synthesis, Biology, Vol. 2, Academic Press, New York, NY, 1980, p. 108.
  b. Ibid, p. 202.
6. McCurdy, S.N. and Otteson, K.M., Documentation from Applied Biosystems, 1987.
7. Le-Nguyen, D., Seyer, R., Heitz, A. and Castro, B., J. Chem. Soc. Perkin Trans 1 (1985) 1025.
8. Le-Nguyen, D., Heitz, A. and Castro, B., J. Chem. Soc. Perkin Trans 1 (1987) 1915.

New logically developed active esters for solid-phase peptide synthesis

Derek Hudson

MilliGen/Biosearch, 81 Digital Drive, Novato, CA 94949-5728, U.S.A.

Introduction

Although of adequate reactivity, currently available active esters (e.g., PFP, Dhbt) fall short of ideality.

Results and Discussion

Numerous active esters of Fmoc-valine were prepared and used in a modification of the competition and comparison assays previously used to assess direct coupling methods [1]. The results showed the following order of reactivity: Dnp > Dhbt > Pfp = TDO > Mpp. All other derivatives were much less active. Mpp is the abbreviation given to the enol ester of 3-methyl-1-phenyl-2-pyrazolin-5-one [2] (Fig. 1, R^2 = a = b = c = b' = a' = H; R^1 = CH_3). Its activity was unexpectedly high. Since Mpp itself is made simply by heating phenylhydrazine with ethyl acetoacetate [3], variants modified at R^1, R^2, and a, b and c are immediately accessible, and several have been prepared (see Table 1).

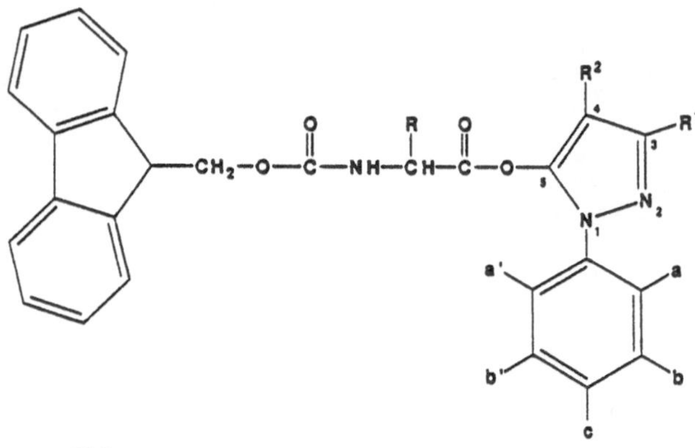

Fig. 1. Structure of Mpp.

Npc (**10**), derived from 2-carbethoxycyclohexanone, is of dramatically reduced activity, compared to Mpp, presumably a result of distortion of the pyrazoline ring. The monochloro derivative, PClp (**5**), is markedly more active than the corresponding dichloro derivative, DClpp (**8**). With Npp (**3**), a very high level

914

of activity is obtained, further increased in Pnp (**1a**) by the extra conjugation provided when R^1 is a phenyl ring (compare also Mpp with Dpp). Although highly reactive towards aminolysis, Npp esters are unreactive to hydroxyls. Anchiomeric assistance really may contribute to the aminolytic reactivity of these compounds. With Pnp and Npp esters, coupling can be facilely monitored. Ionized Npp, formed in the presence of unreacted amino groups, absorbs at 488 nm, a wavelength free from interference. During coupling, a blood red color appears on the support, and this rapidly fades, leaving a residual golden appearance. Ten Fmoc-amino acid Npp esters have been prepared in high yield and crystallinity. Several relatively simple peptides have been made in good yield, and visual observation of completion, normally occurring within 5 min, was found to be reliable.

Table 1 *The relative efficiency of several phenyl-pyrazolinone enol esters compared to other selected active esters*

Rank	Ester	R^1	R^2	1-subst.	M.p.	% Val inc.
1a	Pnp	Ph	H	$c=NO_2$	168–169	95[a]
1b	Dnp	–	–	–	112–113	95
3	Npp	Me	H	$c=NO_2$	195–197	92
4	Dhbt	–	–	–	106–109	82
5	PClp	Ph	H	$c=Cl$	112–113	77
6	Dpp	Ph	H	all H	147–149	59
7	PFP	–	–	–	117–118	56
8	DClpp	Ph	H	$a=c=Cl$	146–149	50
9	Mpp	Me	H	all H	148–149	32
10	Npc	$-(CH_2)\,4-$		$c=NO_2$	183–185	7.4

[a]Identical values obtained with tBoc-Val-OPnp (m.p. 120–122).

Conclusions

Novel enol esters of substituted 1-phenyl-pyrazolinones have been developed and found to possess remarkable properties. They possess a unique blend of reactivity and stability, are highly crystalline, do not require catalysts, and are 'self-indicating'. They are prepared from easily accessible intermediates that are non-toxic, non-carcinogenic and environmentally safe. Indeed, they have therapeutic applications [4]. The new active ester derivatives described promise to be widely applicable.

References

1. Hudson, D., J. Org. Chem., 53 (1988) 617.
2. Losse, G., Hoffmann, K. and Hetzer, G., Liebig's Ann. Chem., 684 (1965) 236.
3. Knorr, L., Liebig's Ann. Chem., 238 (1887) 137.
4. Nishi, H., Watanabe, T., Yuki, S., Morinaka, Y., Iseki, K. and Sakurai, H., European Patent Application (1987), 0 208 874.

Thio derivatives of amino acids as a carboxyl component in peptide synthesis catalyzed by papain

Yuri V. Mitin[a], Nina P. Zapevalova[a], Hans D. Jakubke[b] and Anne K. Patt[b]
[a]*Institute of Protein Research, U.S.S.R. Academy of Sciences, 142292 Pushchino, Moscow Region, U.S.S.R.*
[b]*Karl-Marx University, Leipzig, G.D.R.*

Introduction

Papain, a cysteine protease, can be successfully used for the synthesis of various peptides starting from alkyl esters of the *N*-protected amino acids and the corresponding amino component. Thanks to the wide papain specificity, this approach permits the synthesis of peptides consisting of any amino acids except proline [1]. A drawback of this method is the low solubility of the mentioned alkyl esters in water.

Results and Discussion

In attempting to find suitable esters for this kind of peptide synthesis, we studied thio derivatives: Z(Boc)NHCHRCOSR' (R' is H or SCH_2COOH). These substances can be easily synthesized starting from appropriate *N*-hydroxysuccinimide esters and Li_2S (thio amino acids) or $HSCH_2COOH$ (carboxymethyl thio ester). The carboxymethyl thio esters (R' is SCH_2COOH) are highly soluble

Table 1 *Peptides synthesized from thio derivatives of* N-*protected amino acids in the presence of papain at pH 8–8.5*

Synthesized peptide	Time (min)	Yield (%)
From thio amino acids		
Z-Ala–Val-NH$_2$[a]	60	99
Z-Ala–Gly-Phe-Leu-OH	30	98
Z-Glu(OBu^t)–Val-NH$_2$	60	88
Z-Tyr(Bzl)–Gly–Ala-NH$_2$	20	77
From carboxymethyl thio esters		
Z-Lys(Boc)-Val-NH$_2$	30	74
Z-Glu(OBu^t)–Ala-Glu(OBu^t)-Asn-OH	20	83
Z-Val–Glu(OBu^t)–Glu(OBu^t)-Ala-Glu(OBu^t)-Asn-OH	60	85
Z-Asp(OBu^t)–Val-NH$_2$	30	36
Z-Phe–Leu-NH$_2$[b]	5	83

[a] Longer hyphens indicate the formed peptide bond.
[b] Chymotrypsin was used at pH 10.

in water at pH>6, and stable at storage. In contrast, thio amino acids (R′ is H) are very easily oxidized to corresponding SS derivatives. But in the form of potassium salts, they are very stable. The synthesized thio derivatives of amino acids readily interact with papain in a water medium and form an intermediate active substance (acyl enzyme) that can react with the amino component, producing peptide. Peptides can be synthesized with a sufficiently high yield if a two-fold excess of thio derivatives is used. A maximal peptide yield (80–95% after 15–60 min) was obtained at pH 8–8.5. No organic solvents are required for this reaction. Thio amino acids with bulky side groups, for example, Z-IleSH, Z-Asp(OBu^t)SH, give peptides with a low yield. Thio amino acids readily interact only with papain, while in the case of carboxymethyl thio esters, not only papain, but also other enzymes with esterase activity are suitable for peptide synthesis.

References

1. Mitin, Yu.V., Zapevalova, N.P. and Gorbunova, E.Yu., Int. J. Pept. Prot. Res., 23 (1984) 528.

Optical resolution of amino acids using lipases

Toshifumi Miyazawa[a], Hitoshi Iwanaga[a], Shinichi Ueji[b], Takashi Yamada[a] and Shigeru Kuwata[a]

[a]Department of Chemistry, Faculty of Science, Konan University, Higashinada-ku, Kobe 658, Japan
[b]Department of Chemistry, College of General Education, Kobe University, Nada-ku, Kobe 657, Japan

Introduction

Enantiomerically pure (or enriched), unusual amino acids are useful building blocks for the synthesis of analogs of biologically active peptides. The enzymes commonly used for the optical resolution of amino acids, e.g., acylase I, may not always be applicable in the case of unusual amino acids. This report concerns the results of our extended research on the utilization of lipases for the resolution of amino acids via the asymmetric hydrolysis and transesterification of their N-protected esters.

Results and Discussion

Enzymatic hydrolyses were carried out in phosphate buffer (pH 7) at 25°C. The Z group was found to be suitable for the N-protection of an amino acid [$H_2NCH(R)CO_2H$] compared with other protecting groups. The 2-chloroethyl, 2,2,2-trifluoroethyl, or 2,2,2-trichloroethyl esters of the Z-derivatives of a number of amino acids were substrates for several commercially available lipases. A part of the results of experiments on the first derivatives of several unusual amino acids has already been reported [1], describing the high enantioselectivities shown by *Aspergilus niger* lipase. This was confirmed further by subsequent investigations on several aromatic amino acids: R, % conversion, % enantiomeric excess (e.e.) of the resulting acid; Ph, 28, 87; $4-FC_6H_4CH_2$, 26, 91; $4-ClC_6H_4CH_2$, 28, 87; $PhCH_2CH_2$, 25, 84; $PhCH_2(CH_2)_2$, 30, 79.

For some amino acids, porcine pancreatic lipase (PPL) was found to be superior to the microbial lipases in both the reaction time and enantioselectivity. The results of experiments on the hydrolysis of the 2,2,2-trifluoroethyl esters are summarized in Table 1. Table 1 suggests a correlation between the side chain length of an amino acid and the optical purity of the resulting acid.

The preferential hydrolysis of the L-enantiomers by all the enzymes studied here was revealed by comparison with authentic samples, if available, or suggested from the regularity of the elution order on HPLC.

The reactions of the 2,2,2-trifluoroethyl esters of Z-amino acids with methanol in such an organic solvent as isopropyl ether proceeded smoothly in the presence

918

Table 1   *PPL-catalyzed hydrolysis of Z-NHCH(R)CO$_2$CH$_2$CF$_3$*

R	% Convn.	% e.e.[a]	R	% Convn.	% e.e.[a]
CH$_3$	39	21	Ph	44	97
C$_2$H$_5$	40	97	PhCH$_2$	40	>99
n-C$_3$H$_7$	39	98	3-FC$_6$H$_4$CH$_2$	40	95
n-C$_4$H$_9$	38	92	4-FC$_6$H$_4$CH$_2$	40	90
i-C$_4$H$_9$	40	98	4-ClC$_6$H$_4$CH$_2$	28	94
n-C$_5$H$_{11}$	40	87	PhCH$_2$CH$_2$	30	71
n-C$_6$H$_{13}$	33	61	PhCH$_2$(CH$_2$)$_2$	17	36
allyl	38	93	4-thiazolylmethyl	44	89

[a]Enantiomeric excess of the resulting acid.

of some microbial lipases. Among the enzymes tested, lipases from *Pseudomonas fluorescens* and *Rhizopus javanicus* afforded methyl esters of the L-configuration with fairly high enantioselectivities (Table 2).

Table 2   *Lipase-catalyzed transesterification between Z-NHCH(R)CO$_2$CH$_2$CF$_3$ and methanol in isopropyl ether*

	Lipase			
	P. fluorescens		*R. javanicus*	
R	% Convn.	% e.e.[a]	% Convn.	% e.e.[a]
n-C$_3$H$_7$	38	91	38	88
n-C$_4$H$_9$	38	89	32	90
n-C$_5$H$_{11}$	35	87	44	83
n-C$_6$H$_{13}$	36	90	37	78
PhCH$_2$	42	73	38	84

[a]Enantiomeric excess of the methyl ester.

In conclusion, the procedures employing lipases were found to complement conventional enzymatic methods for the resolution of amino acids.

References

1.   Miyazawa, T., Takitani, T., Ueji, S., Yamada, T. and Kuwata, S., J. Chem. Soc., Chem. Commun., (1988) 1214.

Cysteine *N,S*-blocking through thiazolidine formation for amide ligation by means of thiol capture

D.S. Kemp and Robert I. Carey

Department of Chemistry, Massachusetts Institute of Technology, Cambridge, MA 02139, U.S.A.

Introduction

Peptide synthesis by prior thiol capture [1] (Scheme 1) is a methodology for the chemical ligation of medium-sized polypeptides that uses a rationally designed spacing element or template [2] to effect the efficient intramolecular *O,N*-acyl transfer across a cysteine disulfide function, as shown in Scheme 1. In this strategy, the cysteine at the amino terminus of a peptide must be activated for unsymmetrical disulfide formation by conversion to its methoxycarbonyldithia (Scm) derivate. The temporary blocking group that is used to mask the cysteine *N* and *S* functions during the synthesis of the peptide must therefore not only be compatible with all the steps of the thiol capture method but must also allow for efficient conversion to the Cys(Scm)-activated peptide.

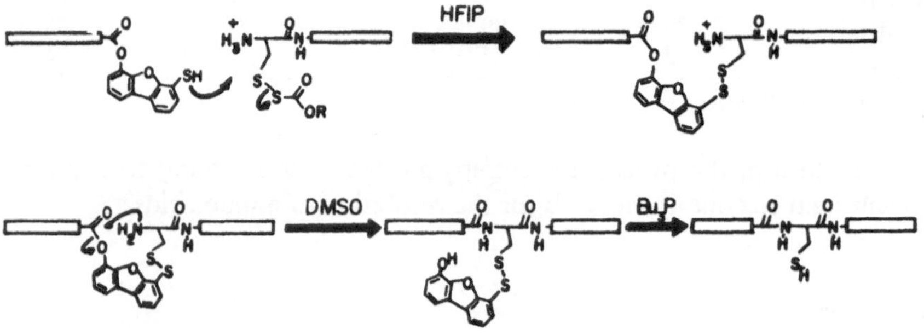

Scheme 1. Principle of the thiol capture methodology.

Results and Discussion

N-tert-butyloxycarbonyl-2,2-dimethyl-L-thiazolidine-4-carboxylic acid, Boc-Dmt-OH (**1**), is a versatile form of cysteine protection for this purpose (Fig. 1). Rate studies of coupling and racemization confirmed that **1** has a high coupling efficiency ($k_{coup} = 3.2 \times 10^{-2}$ M^{-1} s^{-1} with H-Val-OMe in THF at 22°C) under the conditions required for solution and SPPS. The particular stability shown by the pentafluorophenyl ester of **1** to racemization (no change in optical rotation after 144 h, 7×excess triethylamine, 22°C, THF) is consistent with previous

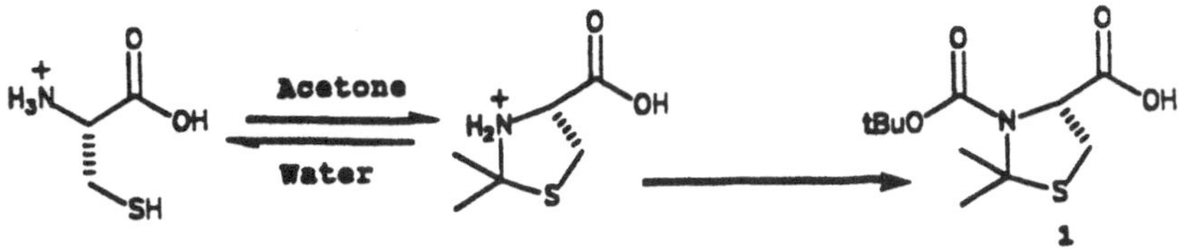

Fig. 1. Formation of Boc-Dmt-OH.

literature reports that showed thiazolidines to resist racemization [3]. Boc-Dmt-peptides can be converted to Cys(Scm)-peptides with or without prior Boc cleavage by any of the three different pathways shown in Scheme 2.

Boc-Dmt-OH has been used as the temporary protection for each cysteine in the 4+4 acyl transfer and thiol capture preparation of the Scm-activated C-terminal octapeptide fragment of BPTI (51–58), H-Cys(Scm)-Met-Arg-Thr-Cys(Dnp)-Gly-Gly-Ala-OH (Dnp = 2,4-dinitrophenyl), fully demonstrating the ability of 1 to be compatible with each of the steps in the thiol capture cycle. Our synthesis of the Scm-activated ColEl ROP fragment (52–63), H-Cys(Scm)-Leu-Ala-Arg-Phe-Gly-Asp-Asp-Gly-Glu-Asn-Leu-OH, also used 1 as temporary cysteine protection.

Reversible thiazolidine formation at amino-terminal cysteines offers a potential route to a thiol capture strategy of semisynthesis. For peptide or protein fragments containing multiple cysteine residues, the amino terminal cysteine can be reversibly blocked as a thiazolidine. Masking of the internal thiols followed by hydrolysis of the thiazolidine, and conversion to the Scm-activated peptide provides a semisynthetic fragment that can be coupled through the thiol capture method. Work toward realizing this goal is under investigation.

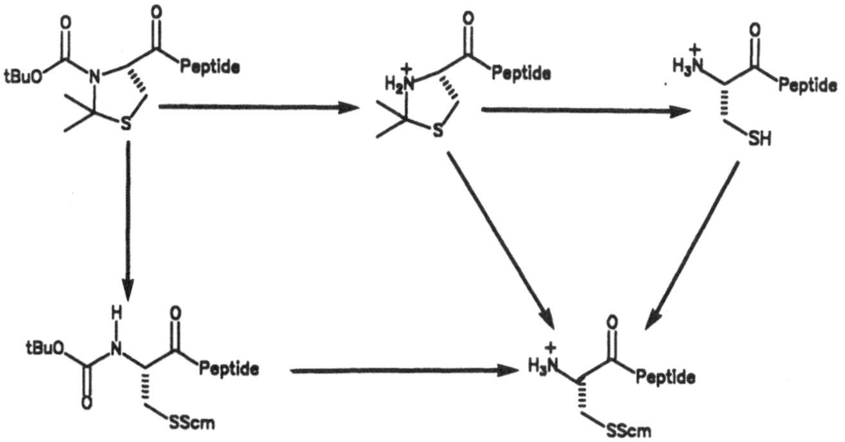

Scheme 2. Conversion of the Boc-Dmt-peptides into Cys(Scm)-peptides.

921

Acknowledgements

Generous support from the National Institutes of Health (GM-13453) is acknowledged. RIC is the recipient of an Applied Biosystems Travel Grant to attend this meeting.

References

1. Fotouhi, N., Galakatos, N. and Kemp, D., J. Org. Chem., 54 (1989) 2803.
2. Kemp, D., Carey, R., Dewan, J., Galakatos, N., Kerkman, D. and Leung, S.-L., J. Org. Chem., 54 (1989) 1589.
3. Barber, M. and Jones, J., In A. Loffet (Ed.) Peptides 1976 (Proceedings of the 14th European Peptide Symposium, Wepion, Belgium) Editions de l'Université de Bruxelles, Brussels, 1976, p. 109.

Fluorenylmethyl-based side-chain protecting groups: Towards a new strategy of peptide synthesis

F. Albericio, J. Rizo, E. Nicolás, M. Ruiz-Gayo, F. Cárdenas, C. Carreño, D. Andreu, E. Pedroso and E. Giralt

Department of Organic Chemistry, University of Barcelona, E-08028 Barcelona, Spain

Introduction

Synthesis of peptides requires a protection scheme with three different classes of protecting groups: temporary protecting groups for the α-amino function and two types of permanent protecting groups, one for the side chains and the other for the carboxyl group of the C-terminal residue, which in SPPS links the growing peptide chain to the support. Thus, the most efficient protection scheme should contain the maximum degree of chemoselectivity between these three classes of protecting groups. In this paper, we describe a new orthogonal strategy [1] of peptide synthesis based on fluorenylmethyl (Fm)-type groups for side-chain protection and the Boc group for the α-amino function. These two protecting groups, and with the photocleavable o-nitrobenzamidobenzyl (Nbb)-resin [2], exhibit three independent dimensions of orthogonality.

Results and Discussion

(1) Synthesis and stability of protected amino acids

H-Lys(Fmoc)-OH was prepared by reaction of Fmoc-azide with the Cu(II) complex of lysine, followed by destruction of this complex with EDTA. However, this method fails for other amino acids due to the poor solubility of the complexes. For aspartic and glutamic acids, we have investigated the method described by Nefkens and Zwanenburg [3], for dual protection, and thus, the corresponding β,β-boroxazolidones have been prepared from the amino acids and triethyl borane in refluxing THF. The resulting complexes were not isolated and were allowed to react with fluorenylmethanol (FmOH) in presence of catalytic amounts of DMAP. The ω-fluorenylmethyl esters of Asp and Glu were isolated as the hydrochlorides after bubbling hydrogen chloride through the solutions of the crude reactions. H-Cys(Fm)-OH was obtained by reaction of FmO-tosylate with cysteine in presence of DIEA. The introduction of the Boc group was carried out, by using 2-*tert*-butyloxycarbonyloxy-amino-2-phenylacetonitrile (Boc-ON). These Fm-based protecting groups are stable to the conditions used for the elongation of the peptide chain in SPPS (TFA-DCM, 3:7; DIEA-DCM, 1:19). The cysteine derivative, however, is completely stable to anhydrous HF. Likewise, all these groups can be removed by piperidine solutions, although the cysteine derivative requires higher concentrations of piperidine in DMF (1 : 1), and longer reaction time (2 h). Tetrabutylammonium fluoride can also remove, in a few minutes, the Fmoc group of the α-amino and the Fm carboxylic protecting groups, but not the corresponding of the ε-amino of lysine nor of the thiol of cysteine. Cleavage of Fm of cysteine in the presence of thiols (β-mercaptoethanol or DTT) leads to free cysteine, whereas, in their absence, the cystine is formed directly. Finally, Cys(Fm) is stable to I_2-DMF (1 : 19) and β-mercaptoethanol-DMF (1 : 19), implying that Fm protection for cysteine is orthogonal with Acm and S-tBu, thus, in principle, allowing selective formation of disulfide bridges.

(2) Orthogonal solid-phase peptide synthesis

Boc-Val-Lys(Fmoc)-Asp(OFm)-Gly-OCH$_2$-Nbb-resin was chosen as a model to demonstrate the feasibility of this strategy. Assembling of amino acids was carried out by a standard procedure [2]. Photolysis of the peptide-resin gave Boc-Val-Lys(Fmoc)-Asp(OFm)-Gly-OH (I), which was >90% pure by HPLC. To obtain the completely free peptide, two routes were tested. Thus, I was successively treated with TFA-DCM (1 : 1) for 10 min and piperidine-DMF (1 : 1) for 1 h. After extractions with CHCl$_3$, H-Val-Lys-Asp-Gly-OH was obtained with an excellent degree of purity as judged by HPLC. Alternatively, this peptide was also obtained with a comparable degree of purity by interchanging the order of treatments: first piperidine, and then TFA. All these peptides were characterized by AAA and NMR at 200 MHz.

(3) Convergent solid-phase peptide synthesis

One of the main problems associated with this strategy is the purification of the protected peptides due to their poor solubility. An alternative can be the selective deprotection of the side chains of Lys or Asp/Glu in order to enhance solubility, followed by the purification and reprotection before coupling to the growing peptide-resin. Thus, Boc-Val-Ala-Leu-Leu-Lys(Fmoc)-Ala-Leu-Tyr(Dcb)-Gly-OH, synthesized on a Nbb-resin, was treated with piperidine-DMF (1 : 1), and the resulting peptide (>90% pure by HPLC) was easily purified using a reversed-phase preparative column eluting with H$_2$O–MeOH mixtures. Finally, reaction of the purified peptide with ethyl trifluoroacetate led to the corresponding trifluoroacetyl peptide.

Acknowledgements

This work was supported by funds from CICYT (BT 86-18) and SKF.

References

1. Barany, G. and Albericio, F., J. Am. Chem. Soc., 107 (1985) 4936.
2. Albericio, F., Pons, M., Pedroso, E. and Giralt, E., J. Org. Chem., 54 (1989) 360.
3. Nefkens, G.H. and Zwanenburg, B., Tetrahedron, 39 (1983) 2995.

Use of Fmoc amino acid chlorides in solid-phase synthesis of medium-size peptides

M. Beyermann[a], M. Bienert[a], H. Niedrich[a], H.-G. Chao[b] and L.A. Carpino[b]

[a]AdW der DDR, Institut für Wirkstofforschung, Alfred-Kowalke-Str. 4,
DDR-1136 Berlin, DDR
[b]Department of Chemistry, University of Massachusetts, Amherst, MA 01003, U.S.A.

Introduction

Fmoc amino acid chlorides (FAAC) have been applied in the case of rapid, racemization-free coupling in the two-phase system $CHCl_3/Na_2CO_3$ [1,2]. Initial studies involving the use of amino acid chlorides in SPPS demonstrated that, with equimolar amounts of tertiary amines, the coupling reactions proceed relatively slowly, and racemization occurs when an excess of amine is used [3]. However, very fast and racemization-free couplings were achieved in the presence of pivalic acid/triethylamine mixtures [3]. In this communication we report the effect of related salts on solid-phase coupling reactions.

Results and Discussion

Table 1A demonstrates the remarkable effect of binary salts on such coupling reactions, with the combination HOBt/DIEA being most appropriate as both

Table 1   *Speed of coupling reactions of Fmoc-AA-chlorides (A) and Fmoc-AA-OPfp esters (B) in the presence of various t-amines and N-hydroxy compounds*

	Base (equiv.)	N-hydroxy compound	Coupling time (min)	Extent of coupling (%)	Racemization (%)
A	Pyridine (4)	–	10	62[a,c]	<0.1
	Pyridine (40)	–	10	100[c]	1.4
	DIEA (4)	–	10	90[a]	–
	TEA	HOBt	5	100[c]	<0.1
	DIEA	HONSu	3	15	–
	DIEA	N-OH-Phthalimide	3	20[a]	–
	DIEA	HOOBt	3	100[a]	–
	DIEA	HOBt	2	100[a,c]	<0.1
B	–	HOBt	3	37[b]	–
	DIEA	HOBt	3	100[b]	0.1

[a] Fmoc-Val-Cl + H-Leu-Pepsyn-KA, 4 equiv. of acylating agent (0.1 M) and 4 equiv. of binary salt or t-amine alone in DMF.
[b] Fmoc-Ile-OPfp + H-Ile-PepSyn-KA.
[c] Fmoc-Phe-Cl + H-Phe-Gly-OCH$_2$CO-BHA, 3 equiv. of acylating agent (0.18 M/DCM).

catalyst and HCl acceptor. Other N-hydroxy compounds, such as HONSu and N-hydroxyphthalimide, are less effective in combination with DIEA, whereas HOOBt is of nearly the same activity. The bases investigated so far (TEA, NMM and DIEA) are equally effective, although DIEA is superior in preventing possible premature liberation of the Fmoc group (no deblocking of Fmoc-Ile-resin in the presence of HOBt/DIEA within 4 h). All couplings carried out in the presence of the salts listed in Table 1A took place without any racemization, as shown by gas chromatography or HPLC. Also, anchoring to hydroxymethyl resins via FAAC results in racemization-free coupling in the presence of these additives (so far investigated for Ala, Phe, Leu). HOBt/DIEA also remarkably catalyzes the aminolysis of Fmoc amino acid pentafluorophenyl esters as demonstrated in Table 1B. However, Fmoc-Asn-OPfp and Fmoc-Gln-OPfp rapidly decompose in solutions of HOBt/DIEA in DMF. (For the more than ten-fold enhancement of the reaction rate of 2,4,5-tri-chlorophenylesters by the potassium salt of HOBt, see [4]).

Table 2 *Efficiency of Fmoc-Val-Cl couplings in comparison with other methods*[a]

Fmoc-Val-X (0.1 mmol) X	Coupling reagent (mmol)	HOBt (mmol)	DIEA (mmol)	Yield (%)
OH	BOP/0.1	–	0.25	83
OH	BOP/0.1	0.1	0.25	78
OH	TBTU/0.1	–	0.25	79
OPfp	–	0.1	0.1	81
OBt[b]	BOP/0.1	–	0.25	88
Cl	–	0.1	0.1	96

[a] Addition of coupling reagent or activated Fmoc-Val-X to a solution (2 ml DCM/DMF 1:1) of all other components using 100 mg MBHA-resin, 3 min (p-methylbenzhydrylamine, PS/1 % DVB, 0.55 mmol/g), reaction at ambient temperature for 3 min, washing with DMF and DCM with UV determination (301 nm) of Fmoc-incorporation (treatment with 20% Pip/DMF, 15 min).

[b] This experiment was performed by treatment of Fmoc-Val-OH with BOP/DIEA for 15 min at ambient temperature.

In Table 2, the coupling rates of Fmoc amino acid chlorides and pentafluorophenyl esters are compared with BOP- and TBTU-mediated couplings. The acid-chloride method gave the highest Fmoc-Val incorporation, demonstrating the high reactivity of acid chlorides with amino groups and/or with HOBt. To prove the utility of this novel method, we synthesized several short- and medium-sized peptides, either exclusively using FAAC, or in combination with pentafluorophenyl esters using both polystyrene and polyacrylamide resins Leu-enkephalin, Substance P analogs, ACP(65–74), Prothrombin(1–9). In most cases the crude peptides were of good purity. The synthesis of Leu-enkephalin on Pepsyn KA-resin was carried out using the Pepsynthesizer (Milligen 9050) with a cycle time of 20 min, thus demonstrating the usefulness of Fmoc amino acid chlorides as highly reactive, inexpensive reagents in automated SPPS.

Acknowledgements

M. Bienert acknowledges, with thanks, an ABI travel award.

References

1. Carpino, L.A., Cohen, B.J., Stephens, Jr., K.E., Sadat-Aalaee, D., Tien, J.H. and Landridge, D.C., J. Org. Chem., 51 (1986) 3732.
2. Beyermann, M., Bienert, M., Repke, H. and Carpino, L.A., In Theodoropoulos, D. (Ed.) Peptides 1986, Walter de Gruyter, Berlin, 1987, p. 107.
3. Beyermann, M., Granitza, D., Bienert, M., Haussner, M. and Carpino, L.A., In Jung, G. and Beyer, E. (Eds.) Peptides 1988, Walter de Gruyter, Berlin, 1989, p. 28.
4. Horiki, K. and Murakami, A., Heterocycles, 28 (1989) 615.

New approaches to prevention of side reactions in Fmoc solid-phase peptide synthesis

Cynthia G. Fields[a] and Gregg B. Fields[b]

[a]*Applied Biosystems Inc., Foster City, CA 94404, U.S.A.*
[b]*Department of Pharmaceutical Chemistry, University of California, San Francisco, CA 94118, U.S.A.*

Introduction

For the majority of Fmoc amino acids, several different loading protocols utilizing PSA have been found effective (reviewed in [1]). Certain amino acids are more problematic to load, such as Fmoc-Pro, Fmoc-Arg(Mtr/Pmc), Fmoc-Asn, Fmoc-Gln and Fmoc-His. Fmoc-Pro PSA loads slowly and Fmoc-Arg(Mtr/Pmc) PSA is not completely stable [2,3]; the formation of PSA results in extensive racemization of Fmoc-His and side-chain dehydration of Fmoc-Asn or Fmoc-Gln (reviewed in [1]). Fmoc-Pro, Fmoc-Arg(Mtr) and Tmob side-chain protected Fmoc-Asn and Fmoc-Gln are efficiently loaded in situ, using a mixed solvent system of DCM/DMF/NMP. Fmoc-Gln can also be loaded as an HOBt ester. DMF was used to keep the amino acid soluble and low temperature was used to reduce DMAP-induced racemization [4]. FABMS showed dehydration of Gln and dipeptide formation to be negligible. Alternatively, Fmoc-Asp(OH)OtBu or Fmoc-Glu(OH)OtBu HOBt esters may be loaded to an amide linker [5]; TFA cleavage converts the initial side chain carboxylic acid to an amide (R.L. Noble, unpublished results). Loading of τ-nitrogen protected Fmoc-His (i.e., Boc, Trt) as a pentafluorophenyl (Pfp) ester is sluggish, although in situ loading is extremely efficient. π-nitrogen protected Fmoc-His (i.e., Bum) may also be loaded efficiently in situ using an excess of DMF. Loading results are summarized in Tables 1 and 2.

A specific concern of Fmoc chemistry is the modification of Trp by arene sulfonyl groups during TFA deprotection of Arg residues [2,6]. Peptides prepared to study Trp modification were:

Peptide 1: Fmoc-Cys(Trt)-Pro-Asp(OtBu)-Phe-Gly-His(Trt)-Ile-Ala-Met-Glu-(OtBu)-Leu- Ser(tBu)-Val-Arg(Pmc)-Thr(tBu)-Trp-Lys(Boc)-Tyr(tBu)

Peptide 2: X-Arg(Pmc)-X-X-X-Arg(Pmc)-X-Trp-X-X-X-X-Lys(Boc)-NH$_2$

Peptide 3: Fmoc-X-X-X-Trp-X-Trp-X-Trp-Trp-X-Trp-Arg(Mtr/Pmc)-Arg(Mtr/Pmc)

Standard cleavage included a 1 h treatment with 82.5% TFA/5% phenol/5% H$_2$O/5% thioanisole/2.5% ethanedithiol (Reagent K) (D. King, personal

928

Table 1 *Efficiences of loading Fmoc amino acids*

Amino acid	Resin	Method	% Loading[a]
Fmoc-Pro	HMP	1	95
Fmoc-Pro	HMP	2	37
Fmoc-Arg(Pmc)	HMP	1	83
Fmoc-Arg(Pmc)	HMP	2	52
Fmoc-Asn(Tmob)	HMP	1	95
Fmoc-Gln(Tmob)	HMP	1	95
Fmoc-Gln	HMP	3	88
Fmoc-Gln	HMP	4	22
Fmoc-Asp(OH)OtBu	(DPAM)PM[b]	4	100
Fmoc-Glu(OH)OtBu	(DPAM)PM[b]	4	100
Fmoc-His(Trt)	HMP	1	90
Fmoc-His(Boc)OPfp	HMP	5	59
Fmoc-His(Bum)	HMP	4	26
Fmoc-His(Bum)	HMP	6	59
Fmoc-His(Bum)	HMP	7	82

[a] Quantitated by spectrophotometric analysis of the fulvene–piperidine adduct at 301 nm using $\epsilon = 7800$ $M^{-1}\cdot cm^{-1}$.
[b] (DPAM)PM = 4-(2',4'-dimethoxyphenylaminomethyl)phenoxymethyl.

Table 2 *Methodologies for Fmoc amino acid loadings*

No.	Solvent		A.A. (equiv.)	DCC/DIC (equiv.)	HOBt (equiv.)	DMAP (equiv.)	Time (h)
	ml/mmol	composition					
1	24.8	DCM/NMP/DMF (50: 44:6)	4	4	–	0.16	1
2	38.5	DCM	3	3	–	3	4
3	40.6	DCM/DMF (3:2)	3	3	3	3	4
4	17.0	DCM/DMF (25:75)	2	2	2	0.16	1
5	38.5	DCM	3	–	–	3	6
6	17.0	DCM/DMF (25:75)	4	2	–	0.16	1
7	17.0	DCM/DMF (25:75)	2	2	–	0.16	1

All reactions proceeded at room temperature, except nos. 2, 3 and 5, which were carried out at 4°C for the first hour and then at room temperature. Loading by protocol no. 4 to (DPAM)PM required no DMAP.

communication). Products were characterized by HPLC analysis [2], FABMS and/or Edman sequencing.

Cleavage of peptides **1**, **2**, or **3** by Reagent K yields one primary product of the proper composition. Elimination of phenol or replacement of thioanisole by 3% EMS has little effect on the homology of the cleaved products. Replacement of H_2O with 3% anisole results in a lowered yield of the desired products and an increased yield of: (i) peptide **1** containing Arg(Pmc), (ii) peptide **2** containing Lys modified by Pmc, and (iii) peptide **3** extensively Trp modified by either

929

Pmc or Mtr. Replacement of EDT with either 1 mg/ml indole or 1% dimethyl-aminophenol during cleavage of peptide 1 results in heterogeneous product profiles. The use of 2 mg/ml indole instead of EDT, or TFA cleavage mixtures containing 2.5% EDT/10 mg/ml kynurenine or 1% EDT/4% H_2O does not prevent Lys modification by Pmc in peptide 2. For peptide 3 cleavage, replacement of EDT by either 2 mg/ml kynurenine or 2-3 mg/ml indole greatly increases the amount of Trp modification by Pmc, but reduces Trp modification by Mtr. Trp modification by Mtr does occur during Reagent K cleavage of peptide 3 when the reaction time is extended to 2 h. Poor results were obtained when peptides 2 and 3 were cleaved for 4 or 24 h with TFA containing 5-7% pentamethylbenzene. These studies have shown that H_2O is an essential scavenger for reducing Pmc/Mtr modification of susceptible amino acid residues during TFA cleavage, and that anisole is not as effective a scavenger as H_2O.

References

1. Fields, G.B. and Noble, R.L., Int. J. Pept. Prot. Res., in press.
2. Harrison, J.L., Petrie, G.M., Noble, R.L., Beilan, H.S., McCurdy, S.N. and Culwell, A.R., In Hugli, T.E. (Ed.) Techniques in Protein Chemistry, Academic Press, New York, 1989, p. 506.
3. Wu, C.-R., Wade, J.D. and Tregear, G.W., Int. J. Pept. Prot. Res., 31 (1988) 47.
4. van Nispen, J.W., Polderdijk, J.P. and Greven, H.M., Recl. Trav. Chim. Pays-Bas., 104 (1985) 99.
5. Rink, H., Tetrahedron Lett., 28 (1987) 3787.
6. Sieber, P., Tetrahedron Lett., 28 (1987) 1637.

Efficient solid-phase peptide synthesis using an α-amino deprotecting procedure with methanesulfonic acid

Yoshiaki Kiso, Yoichi Fujiwara, Tooru Kimura, Makoto Yoshida and Kenichi Akaji

Department of Medicinal Chemistry, Kyoto Pharmaceutical University, Yamashina-ku, Kyoto 607, Japan

Introduction

We reported previously that the methanesulfonic acid (MSA) system (0.5 M MSA/CH$_2$Cl$_2$-dioxane (9:1)) was suitable for the removal of N$^\alpha$-Boc groups in SPPS, and superior to TFA systems in terms of stability of semipermanent side chain-protecting groups and undesired pyroglutamyl formation form N-terminal glutamine in peptide-resin [1]. Using this MSA deprotection system, we have developed an efficient peptide synthesis on an acid-stable phenacyl ester linkage (Pac)-resin [2] cleavable with fluoride ion [3–5].

Results and Discussion

First, we synthesized BK potentiating peptide 5a (BPP5a) from Boc-Pro-Pac-resin. In combination with the acid-labile Boc group for N$^\alpha$-protection, amino acid derivatives bearing protecting groups removable with fluoride ion were employed, i.e., Boc-Trp(Ppt)-OH [3] (Ppt = diphenylphosphinothioyl) and Boc-Lys(Fmoc)-OH [3,4]. To reduce the termination in the amino acid-Pac-resin [6], we omitted the base-wash with DIEA and employed BOP [7] or the new 2-(benzotriazol-1-ly)oxy-1,3-dimethyl-imidazolidinium hexafluorophosphate (BOI) [8] as a coupling reagent. In this method, an appropriate amount of DIEA and BOP or BOI reagent-activated N$^\alpha$-protected amino acid derivatives was added to the N$^\alpha$-deprotected peptide-resin, which was masked as its MSA salt. This in situ neutralization method reduced side reactions such as cyclization, because the terminal amino group was less exposed as a nucleophile by MSA masking and the coupling reaction was rapid when using BOP or BOI.

SPPS of BPP5a was achieved at 25°C, and the homogeneous peptide was obtained in excellent yield (67–68%). Similarly, we successfully synthesized Leu-enkephalin (79%) and neuromedin-N (59%). In these syntheses, we employed a new amino acid derivative bearing a protecting group removable with fluoride ion, Boc-Tyr(Dpp)-OH, (Dpp = diphenylphosphinyl).

Next, we applied our method to the automated SPPS of porcine brain natriuretic peptide (pBNP). The fully protected pBNP-resin was assembled from Boc-Tyr(BrZ)-Pam resin (Fig. 1) and treated with HF in the presence of *m*-cresol and dimethylsulfide at 0°C for 60 min to deprotect the protecting groups except

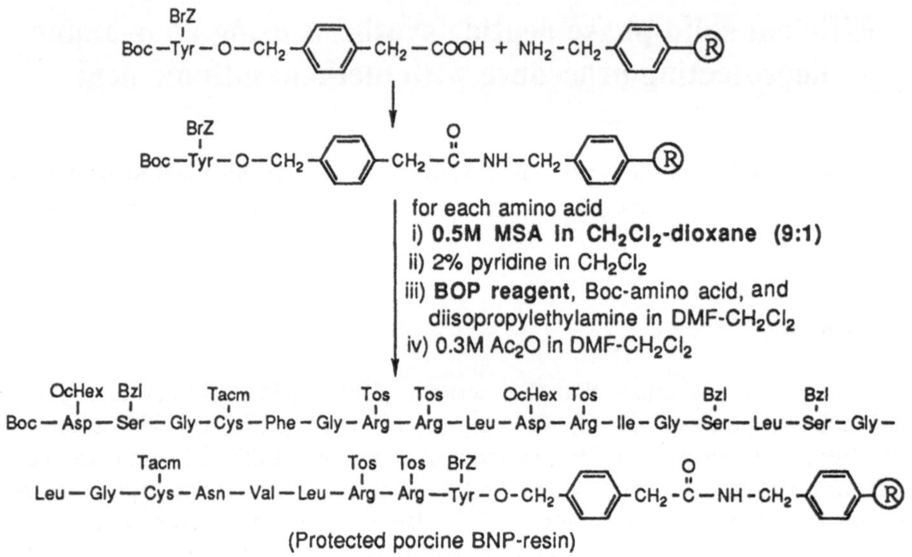

$$\text{BrZ} \atop \text{Boc}-\text{Tyr}-\text{O}-\text{CH}_2-\underset{}{\bigcirc}-\text{CH}_2-\text{COOH} + \text{NH}_2-\text{CH}_2-\underset{}{\bigcirc}-\text{®}$$

$$\text{BrZ} \atop \text{Boc}-\text{Tyr}-\text{O}-\text{CH}_2-\underset{}{\bigcirc}-\text{CH}_2-\overset{\text{O}}{\overset{\|}{\text{C}}}-\text{NH}-\text{CH}_2-\underset{}{\bigcirc}-\text{®}$$

for each amino acid
i) **0.5M MSA In CH$_2$Cl$_2$-dioxane (9:1)**
ii) 2% pyridine in CH$_2$Cl$_2$
iii) **BOP reagent**, Boc-amino acid, and
   diisopropylethylamine in DMF-CH$_2$Cl$_2$
iv) 0.3M Ac$_2$O in DMF-CH$_2$Cl$_2$

$$\text{OcHex} \quad \text{Bzl} \qquad \text{Tacm} \qquad\qquad \text{Tos} \quad \text{Tos} \qquad \text{OcHex} \; \text{Tos} \qquad\qquad \text{Bzl} \qquad \text{Bzl}$$
Boc — Asp — Ser — Gly — Cys — Phe — Gly — Arg — Arg — Leu — Asp — Arg — Ile — Gly — Ser — Leu — Ser — Gly —

$$\text{Tacm} \qquad\qquad \text{Tos} \quad \text{Tos} \; \text{BrZ}$$
Leu — Gly — Cys — Asn — Val — Leu — Arg — Arg — Tyr — O — CH$_2$ —$\bigcirc$— CH$_2$ —$\overset{\text{O}}{\overset{\|}{\text{C}}}$— NH — CH$_2$ —$\bigcirc$— ®

(Protected porcine BNP-resin)

Fig. 1. SPPS of porcine BNP.

for the S-trimethylacetamidomethyl(Tacm) [9] group. The crude [Cys(Tacm)[4,20]]-pBNP was purified and then treated with iodine to form a disulfide bond as described previously [9]. The synthetic pBNP was obtained in 10% yield.

References

1. Kiso, Y., Nishitani, A., Shimokura, M., Fujiwara, Y. and Kimura, T., In Shiba, T. and Sakakibara, S. (Eds.) Peptide Chemistry 1987, Protein Research Foundation, Minohshi, Osaka, Japan, 1988, p. 291.
2a. Weygand, F. and Obermeier, R., Z. Naturforsch., Teil B, 23 (1968) 1390.
 b. Mizoguchi, T., Shigezane, K. and Takamura, N., Chem. Pharm. Bull., 18 (1970) 1456.
3. Kiso, Y., Kimura, T., Shimokura, M. and Narukami, M., J. Chem. Soc., Chem. Commun., (1988) 287.
4. Kiso, Y., Kimura, T., Fujiwara, Y., Shimokura, M. and Nishitani, A., Chem. Pharm. Bull., 36 (1988) 5024.
5. Kiso, Y., Fujiwara, Y, Kimura, T., Shimokura, M. and Nishitani, A., In Ueki, M. (Ed.) Peptide Chemistry 1988, Protein Research Foundation, Minohshi, Osaka, Japan, 1989, p. 129.
6a. Birr, C., Wengert-Mueller, M. and Buku, A., Peptides: Proceedings of the 5th American Peptide Symposium, (1977) 510.
 b. Tam, J.P., Cunningham-Rundles, W.R., Erickson, B.W. and Merrifield, R.B., Tetrahedron Lett., (1977) 4001.
7. Castro, B., Dormoy, J.R., Evin, G. and Selve, C., Tetrahedron Lett., 14 (1975) 1219.
8. Kiso, Y., Fujiwara, Y., Kimura, T. and Yoshida, M., In Ueki, M. (Ed.) Peptide Chemistry 1988, Protein Research Foundation, Minohshi, Osaka, Japan, 1989, p. 123.
9. Kiso, Y., Yoshida, M., Kimura, T., Fujiwara, Y. and Shimokura, M., Tetrahedron Lett., 30 (1989) 1979.

Facile synthesis of peptides having C-terminal asparagine and glutamine by Fmoc-based solid-phase peptide synthesis

Gerhard Breipohl[a], Jochen Knolle[a] and Werner Stüber[b]

[a]Hoechst AG, Postfach 800320, D-6230 Frankfurt am Main 80, F.R.G.
[b]Behringwerke, Postfach 1140, D-3550 Marburg 1, F.R.G.

Introduction

SPPS of peptides having C-terminal Asn and Gln has always been troublesome, especially with Fmoc-based chemistry, because the carboxamide function of the lateral chain gives rise to side reactions. Among these, dehydration to the cyano group or formation of imides are predominant [1].

Side reactions of this type can be avoided if a protecting group (e.g., Mbh) is used. A major problem, however, is the coupling [2] of the protected Fmoc-Asn(Mbh)-OH or Fmoc-Gln(Mbh)-OH to the p-alkoxybenzylalcohol type of resin normally used in Fmoc-based SPPS. The sterical hindrance of the side-chain protecting group requires large excess of amino acid and special reaction conditions.

Results and Discussion

Based on our work in the development of new linkers [3,4] for the preparation of peptide amides by Fmoc-based SPPS, we found a very simple solution to these problems by using our acid-sensitive handle for peptide amides as the protecting group for the amide function of the side chain in asparagine and glutamine.

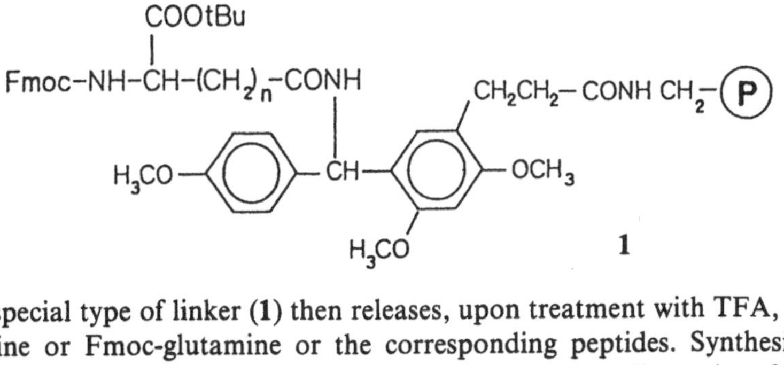

1

This special type of linker (1) then releases, upon treatment with TFA, Fmoc-asparagine or Fmoc-glutamine or the corresponding peptides. Synthesis of a resin with these properties is easily achieved by anchoring Fmoc-Asp-OtBu or Fmoc-Glu-OtBu with their side chain to a resin functionalized with our TMBPA

933

amide linker by standard Fmoc-protocol, e.g., by DIC/HOBt. Thus, peptides were synthesized according to our usual protocol with in situ preactivated amino acids or preformed HOBt esters.

References

1. Barany, G. and Merrifield, R.B., In Gross, E. and Meienhofer, J. (Eds.) The Peptides, Vol. 2, Academic Press, New York, 1980, p. 1. (Note discussion and references cited on pp. 200–201.)
2. Wu, C., Wade, J.D. and Tregear, G.W., Int. J. Pept. Prot. Res., 31 (1988) 47.
3. Breipohl, G., Knolle, J. and Stüber, W., Tetrahedron Lett., 28 (1987) 5651.
4. Breipohl, G., Knolle, J. and Stüber, W., Int. J. Pept. Prot. Res., 34 (1989) 262.

Synthesis of conformationally constrained phenylalanines and their incorporation into tachykinins

G. Chassaing, H. Josien and S. Lavielle

Laboratoire de Chimie Organique Biologique, CNRS UA 493, Université Paris VI,
4 place Jussieu, F-75005 Paris, France

Introduction

Structure–activity relationship and NMR analysis of SP and SP analogs (cyclic peptides, *N*-methylated amino acids or proline-containing peptides) led us to propose a model for the bioactive conformation of NK-1 agonists [1]. In order to determine the side-chain orientation (tg$^+$, gg or tg$^-$) of the Phe^7 and Phe^8 in the hydrophobic pocket of NK-1 type receptor, five conformationally constrained phenylalanines were introduced in the sequence of SP.

Results and Discussion

The syntheses of fluorenylglycine (Flg), diphenylalanine (Dpa) and indenylglycine (Ing) were based on the alkylation of the Schiff base of the ethyl glycinate [2] [$(C_6H_5)_2$-C$=$N-CH$_2$-COOCH$_2$CH$_3$] by 9-bromofluorene (yield 80%), bromodiphenylmethane (yield 85%) and 1-bromoindane (yield 60%), respectively. The two diastereoisomers of Ing were separated by silica gel chromatography. The corresponding Boc racemic amino acids have been synthesized by classical methods.

Table 1 *Biological potencies of SP analogs on guinea-pig ileum*
Arg-Pro-Lys-Pro-Gln-Gln-Phe^7-Phe^8-Gly-Leu-Met-NH$_2$

Substitution in position 7	EC$_{50}$ (nM)	R.A. (%)	Substitution in position 8	EC$_{50}$ (nM)	R.A. (%)
[L-Dpa^7]SP	70	1.1	[L-Dpa^8]SP	5.8	13
[D-Dpa^7]SP	1 800	0.04	[D-Dpa^8]SP	44	1.7
[L-Flg^7]SP	n.d.	–	[L-Flg^8]SP	1.9	40
[D-Flg^7]SP	n.d.	–	[D-Flg^8]SP	70	1.1
[L-Tiq^7]SP	80	0.9	[L-Tiq^8]SP	25	3
[Ac-Dic^7]SP(7–11)	5 000	0.01	SP	0.75	100

The L-2,3-dihydroindole-2-carboxylic acid (Dic) and the L-1,2,3,4-tetrahydroisoquinoleic-3-carboxylic acid (Tiq) have been prepared according to the literature [3,4].

Peptide syntheses were carried out manually, starting from MBHAR. The crude peptides were purified by low-pressure RPC. The D-diastereoisomer analogs

935

of SP have been tentatively assigned on the basis of their HPLC retention times and their biological activities.

The hydrophobic pocket of NK-1 receptor recognizes both tg^- and tg^+ rotamers of the Phe^8, whereas it accepts either tg^- or tg^+ rotamer of Phe^7.

References

1. Lavielle, S., Chassaing, G., Ploux, O., Loeuillet, D., Besseyre, J., Julien, S., Marquet, A., Convert, O., Beaujouan, J.C., Torrens, Y., Bergström, L., Saffroy, M. and Glowinski, J., Biochem. Pharmacol., 37 (1988) 41.
2. O'Donnell, M.J., Boniece, J.M. and Earp, S.E., Tetrahedron Lett., 30 (1978) 2641.
3. Vincent, M., Remond, G., Portevin, B., Serkiz, B. and Laubie, M., Tetrahedron Lett., 16 (1982) 1677.
4. Julian, P.L., Karpel, W.J., Magnani, A. and Meyer, E.W., J. Am. Chem. Soc., 70 (1948) 180.

Asymmetric synthesis of unusual amino acids: Synthesis of Tyr and Phe analogs

Ramalinga Dharanipragada, K.C. Russell, Ernesto Nicolas, Geza Toth and
Victor J. Hruby

Department of Chemistry, University of Arizona, Tucson, AZ 85721, U.S.A.

Introduction

Substitution of the diastereotopic β-hydrogens of many α-amino acids provides, in principle, an approach to topographic control of peptide structure. Asymmetric synthesis of the desired amino acids is needed to facilitate these studies. We have succeeded in extending the chiral imide enolate bromination methodology [1] to the successful synthesis of all the four individual isomers of β-methyl-phenylalanine in very high enantiomeric purities.

Results and Discussion

Commercially available racemic 3-phenylbutyric acid was resolved [2] into its optical isomers via fractional crystallization of the diastereomeric salts formed

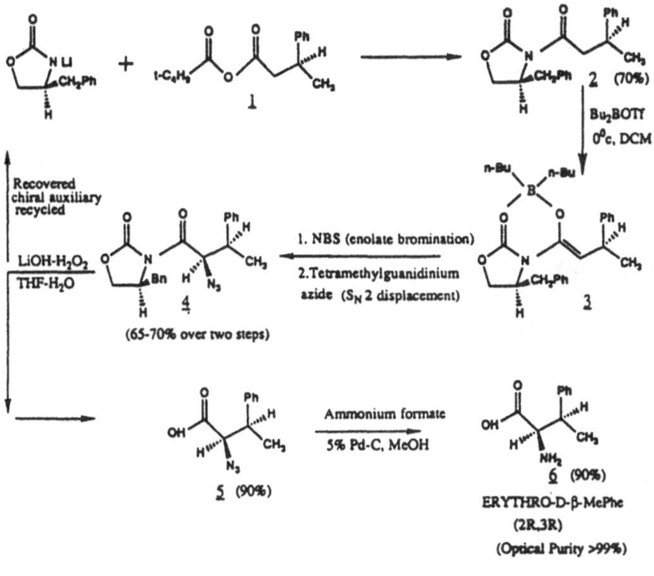

Scheme 1. *Asymmetric synthesis of Phe analogs.*

with (S)-(–)-methylbenzylamine. The (S)-(+)-acid, thus obtained, was treated with pivaloyl chloride to give the mixed anhydride **1** (Scheme 1). Acylation of the anhydride **1** with lithiated chiral auxiliary derived from L-Phe gave the *N*-acyl oxazolidinone **2**. Treatment of **2** with a 1 M solution of dibutylborontriflate in DCM at 0°C gave boron enolate **3**. Diastereoselective bromination of **3** was followed by S_N2 displacement to give the α-azidocarboximide **4**. Diastereoisomeric purity of **4** was > 99% as judged by ^{1}H NMR. Removal of chiral auxiliary and reduction of the resulting azido acid **5** by transfer hydrogenation with ammonium formate gave (2R,3R) β-methylphenylalanine. Starting with (R)-(–)-3-phenylbu-tyric acid, and utilizing the same chiral auxiliary, gave (2R,3S) β-methylphenyl-alanine. The other two isomers, i.e., (2S,3S) and (2S,3R) β-methylphenylalanine, were obtained utilizing the chiral auxiliary derived from D-Phe. Similar strategies were utilized for the synthesis of β-methyltyrosine analogs (Scheme 2).

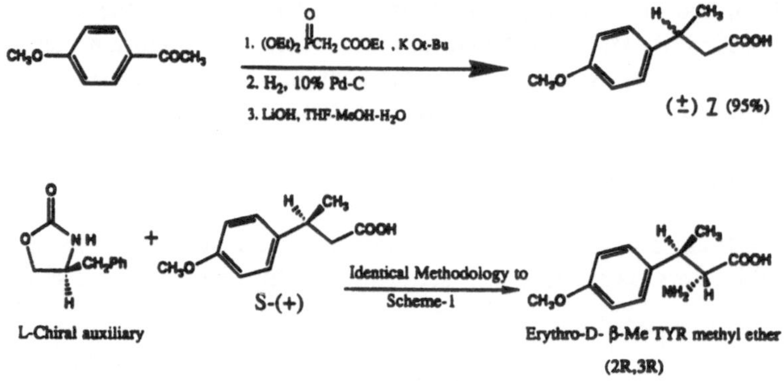

Scheme 2. *Asymmetric synthesis of Tyr analogs.*

In conclusion, we have found that extension of the chiral imide enolate bromination methodology [1] allows convenient access to optically pure isomers of β-methylphenylalanine and β-methyltyrosine.

Acknowledgements

Supported by grants from U.S. Public Health Service and the National Science Foundation.

References

1. Evans, D.A., Ellman, J. and Dorow, R., Tetrahedron Lett., 28 (1987) 1123.
2. Weidler, A. and Bergson, G., Acta Chim. Scand., 18 (1963) 1484.

938

Preparation of carboalkoxyalkylphenylalanine derivatives from tyrosine

Jefferson W. Tilley, Ramakanth Sarabu, Rolf Wagner and Kathleen Mulkerins
Roche Research Center, Hoffmann LaRoche Inc., Nutley, NJ 07110, U.S.A.

Introduction

Tyrosine phosphorylation by tyrosine kinases represents an important control point for cell growth and differentiation [1]. In addition, a number of neu-rohormones and secretory peptides such as gastrin [2], cholecystokinin [3], fibronectin [4], and leucosulfakinin [5] contain a sulfated tyrosine that is necessary for expression of their biological activity. In view of the ionic character and instability of tyrosine phosphates and sulfates, it would be of interest to have access to analogs incorporating less polar and more stable mimics of these groups for SAR. As part of an effort to address the stability problem, recent reports have described methods for the preparation of phosphono- [6], phosphonomethyl- [6] and sulphomethylphenylalanine [7] derivatives. In our own work, we have considered the possibility that tyrosine phosphates and sulfates might be productively substituted by carboxylalkylphenylalanines. Thus, we have devel-oped procedures for the conversion of suitably protected tyrosine derivatives to the corresponding carboalkoxyalkylphenylalanines.

Results and Discussion

For the preparation of carboalkoxymethylphenylalanines ($5, n = 1$), the tyrosine triflate derivative **1** is coupled with allyltributyl tin in the presence of $(Ph_3P)_2PdCl_2$ and lithium chloride to give an allylphenylalanine **2**. The allyl moiety is then subjected to a two-stage oxidation using ruthenium tetroxide/sodium periodate followed by sodium chlorite in phosphate buffer to give the acid **3**. Appropriate esterification of the newly formed carboxylic acid and selective deesterification then completes the synthesis of **5**, $n = 1$.

For the synthesis of carboalkoxyethylphenylalanines (**5**, $n = 2$), palladium-catalyzed coupling of **1** is effected with an appropriate acrylate ester or preferably a 2-trialkylstannylacrylate, to give the phenylalanineacrylate **4**. Hydrogenation of this material over palladium on carbon serves to simultaneously reduce the double bond and cleave the benzyl ester to afford **5**, $n = 2$. To ascertain whether racemization occurred during these transformations, the end-products were hydrolyzed, esterified with isopropanol and acylated with perfluoropropanoic anhydride prior to gas chromatographic analysis on a Chirasil-Val chiral column [8]. Employing conditions under which a racemic mixture gave two peaks of

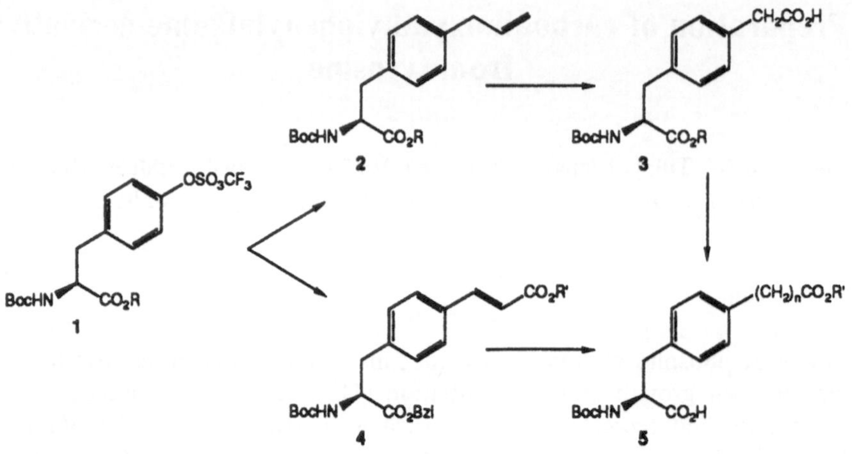

Scheme 1.

equal area, compounds **3** and **5** were determined to be $\geq 96.5\%$ enantiomerically pure. Since 1–2% racemization typically occurs during the hydrolysis procedure, it is apparent that the synthetic steps proceed with minimal racemization.

The chemical approaches outlined above provide a general entry to protected phenylalanine derivatives for use in the SPPS bearing carboxyalkylphenylalanines in the place of tyrosine phosphates and sulfates. Minor modifications to these procedures should permit the synthesis of intermediates with suitable carboxyl-protecting groups to accommodate various peptide synthetic strategies.

References

1a. Krebs, E.G., Biochem. Soc. Trans., 13 (1985) 813.
 b. Hunter, T. and Cooper, J.A., Annu. Rev. Biochem., 54 (1985) 897.
 c. Weinberg, R.A., Science, 230 (1985) 770.
 d. Goustin, A.S., Leof, E.B., Shipley, G.D. and Moses, H.L., Cancer Res., 46 (1986) 1015.
2. Gregory, H., Hardy, P.M., Jones, D.S., Kenner, G.W. and Sheppard, R.C., Nature, 204 (1964) 931.
3. Mutt, V. and Jorpes, J.E., Eur. J. Biochem., 6 (1968) 156.
4. Liu, M.-C. and Lipmann, F., Proc. Natl. Acad. Sci. U.S.A., 82 (1985) 34.
5a. Nachman, R.J., Holman, G.M., Haddon, W.F. and Ling, N., Science, 234 (1986) 71.
 b. Nachman, R.J., Holman, G.M., Haddon, W.F. and Hayes, T.K., Peptide Res., 2 (1989) 171.
6. Petrakis, K.S. and Nagabhushan, T.L., J. Am. Chem. Soc., 109 (1987) 2831.
7. Marseigne, I. and Roques, B.P., J. Org. Chem., 53 (1988) 3621.
8a. Frank, H., Nicholson, J.T. and Bayer, E., J. Chromatogr., 167 (1978) 187.
 b. Frank, H., Woiwode, W., Nicholson, G. and Bayer, E., Liebigs Ann. Chem., (1981) 354.

Stereoselective synthesis of (2*S*, 3*R*)-*N*-(Boc)- [3-²H] [4-¹³C]valine via regioselective opening of a chiral epoxide

Joseph P. Meara and Daniel H. Rich

School of Pharmacy, University of Wisconsin at Madison, 425 N. Charter St., Madison, WI 53706, U.S.A.

Introduction

Isotope-edited 2D NMR has recently been used to probe the conformation of an enzyme-inhibitor complex [1]. Isotope editing of NOESY spectra allows only NOEs arising from the protons attached to the labeled heteronucleus (e.g. ¹³C or ¹⁵N) to be observed, thus greatly simplifying spectral interpretation. Deuterium labeling is used to help confirm NOE assignments by their presence or absence in the isotope-edited spectra. In order to apply this technique to inhibitors bound to aspartic proteinases, we have developed a stereoselective synthesis for a series of chirally-labeled ¹³C and ¹³C/D valine derivatives for incorporation into pepstatin analogs. Here, we present a route for (2*S*,3*R*)-*N*-(Boc)-[4-¹³C][3-²H]valine employing the Sharpless epoxidation to set the stereochemistry at what will become the α and β carbons of the labeled amino acid [2].

Results and Discussion

The synthesis (Scheme 1) commences with the LAH reduction of 2-butyne-1-ol (**1**), followed by D₂O quenching to give the C-3 and *O*-deuterated *trans*-crotyl alcohol. The deteroxy deuteron was exchanged with hydrogen by passing it through a short silica gel flash column (yield **2**, 68%), because otherwise the alcohol gave poor results in the subsequent two reactions. Catalytic Sharpless epoxidation [3] of **2** was followed by in situ derivatization of the resulting epoxy alcohol by triphenylmethyl chloride to afford deuterated **3** in 52% yield [4]. The unlabeled material was obtained in 58–60% yield. The triphenylmethyl group directs the subsequent ring opening to C-3, greatly simplifies isolation of the epoxy alcohol, and allows for enantiomeric enrichment to >95% e.e. via simple recrystallization.

Unlabeled **3** was converted to **4** in 73–77% yield using two equivalents each of (CH₃)₂CuLi (from halide-free methyl lithium) and BF₃·Et₂O at –78°C. However, the use of methyl lithium containing a full equivalent of LiI resulted in a substantial amount (35%) of the iodohydrin, but only 40% of the desired product, **4**. Hence, [¹³C]-methyl lithium will be made from [¹³C]-chloromethane [5] rather than [¹³C]-iodomethane, to avoid this problem. (This work is currently in progress.)

Scheme 1. X = 1 and Y = 12 in the unlabeled series. X = 2 and Y = 13 in the labeled series. Yields of labeled series given in parentheses.

The monoprotected diol **4** was converted to the mesylate (2 equiv. MsCl, 2 equiv. TEA) and deprotected (*p*-TsOH (cat.), MeOH) in 81% yield to give alcohol **5**. The azide **6** was obtained in 69% from **5** by refluxing overnight with 4 equiv. tetramethyl-guanidinium azide in a 1:1 solution of dichloromethane/cyclohexane. Catalytic hydrogenation over 10% Pd-C in the presence of TFA, and subsequent *N*-protection (Boc-ON, TEA, 1:1 dioxane/water) afforded Boc-valinol (**7**) from **6** in 93% yield. RuO$_4$ oxidation under Sharpless conditions [6] gave *N*-Boc-L-valine in 70% yield.

Acknowledgements

Financial support from NIH (AR-32007) is greatly appreciated.

References

1. Fesik, S.W., Luly, J.R., Erickson, J.W. and Abad-Zapatero, C., Biochemistry, 27 (1988) 8297.
2a. Tung, R. and Rich, D.H., Tetrahedron Lett., 28 (1987) 3419
 b. A similar strategy has recently been published: Caron, M., Carlier, P.R. and Sharpless, K.B., J. Org. Chem., 53 (1988) 5185.
3. Gao, Y., Hanson, R.M., Klunder, J.M., Ko, S.Y., Masamune, H. and Sharpless, K.B., J. Am. Chem. Soc., 109 (1987) 5765.
4. A similar procedure for unlabeled 3 appeared while this work was in progress: Hoagland, S., Morita, Y., Lu Bai, D., Marki, H., Kees, K., Brown, L. and Heathcock, C.H., J. Org. Chem., 53 (1988) 4730.
5. Lusch, M.J., Phillips, W.V. and Sieloff, R.F., In Sememelhack, M. F. (Ed.) Organic Syntheses, 62 (1984) 101.
6. Carlson, P.H.J., Katsuki, T., Martin, V.S. and Sharpless, K.B., J. Org. Chem., 46 (1981) 3936.

Synthesis of substituted homoprolines

Robert T. Shuman, Paul L. Ornstein, Jonathan W. Paschal and
Paul D. Gesellchen

Lilly Research Laboratories, Eli Lilly & Company, Indianapolis, IN 46285, U.S.A.

Introduction

New and general routes for the synthesis of unnatural amino acids are of
considerable importance. Syntheses of homoprolines (sometimes called pipeco-
linic acids, or 2-carboxy-piperidines) have been reported [1–6], but these procedures generally suffer from low overall yields. Many of these procedures involve
a cyclization reaction, cumbersome reaction schemes or chromatographic purifications.

Methods

We now report an improved, general synthesis of substituted homoprolines
(Scheme 1). The key to this synthesis is the selective 2 cyanation of pyridines
and quinolines via the corresponding *N*-oxide. By using a modification of the
Reissert–Henze reaction, reported by Fife [7], high yields (nearly quantitative)
of 2-cyano-pyridines were obtained. The resulting nitriles were hydrolyzed to
the corresponding pyridine-2-carboxylic acids using 3 N hydrochloric acid and
refluxing for 24 h. Subsequent reduction of the aromatic ring with platinum
oxide in ethanol/water at 60°C and 60 psi of hydrogen in a Parr shaker apparatus
gave the homoprolines in yields [8] of 80 – 100%. In this fashion 3, 4 and 6-
substituted homoprolines were generated with good elemental, field desorption
mass spectral and NMR analyses.

Results and Discussion

Several of the homoprolines generated by this procedure were converted to
their corresponding *N-tert*-butyloxycarbonyl (Boc) derivatives in good yields.
When attempts were made to protect the amino group of **5b**, the yields were
unacceptably poor (i.e., 1% of Boc-**5b** was isolated and judged pure by ^{1}H NMR),
presumably due to steric hindrance by the methyl substituent at C_6.

The stereochemistry of the disubstituted amino acids **5b, 5c,** and **5d** was
examined by ^{1}H NMR decoupling and NOE experiments at 500 MHz. Taken
collectively, the data indicate that a *cis* configuration is present in all three
compounds.

The amino acids generated by this synthetic protocol are racemic (DL-*cis*)
due to the non-stereoselective nature of the final hydrogenation step [9]. However,

944

Scheme 1.

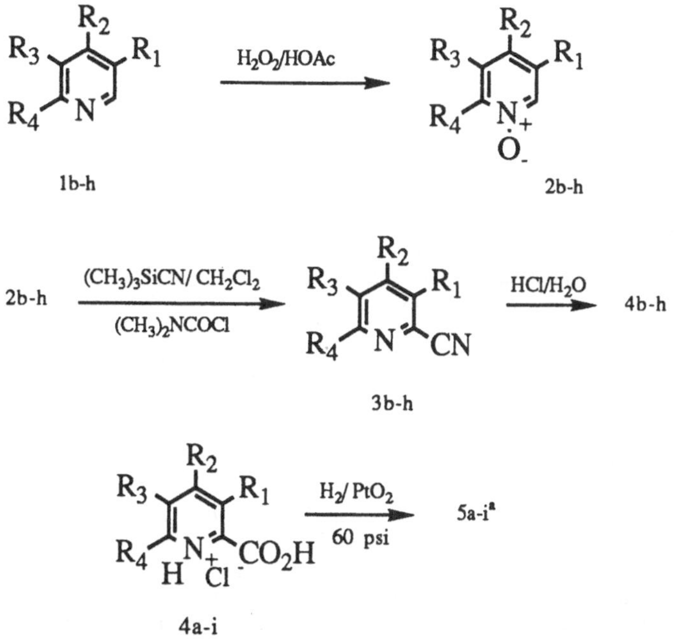

Compound	R_1	R_2	R_3	R_4
4-5a	H	H	H	H
1-5b	H	H	H	Me
1-5c	Me	H	H	H
1-5d	H	Me	H	H
1-5e	H	Et	H	H
1-5f	H	MeO	H	H
1-5g	H	t-Bu	H	H
1-5h	H	Me	b	b
4-5i	H	H	b	b

[a] All compounds isolated as HCl salt except **5a** and **5i** which were prepared from the pyridine and quinoline carboxylic acids.
[b] Fused ring (i.e., from quinoline precursors).

procedures have been published for the chemical [10] or enzymatic [11] resolution of DL-α-amino acids. Furthermore, the use of racemic amino acids during peptide synthesis results in a pair of diastereomeric peptides that often are separable by RPHPLC techniques [12].

In summary, the present method for the synthesis of substituted homoprolines allows the ready access to this class of unnatural amino acids. This procedure uses inexpensive reagents, generally gives high yields and eliminates the need for chromatographic purifications.

References

1. King, F.E., King, T.J. and Warwick, A.J., J. Chem. Soc., (1950) 3590.
2. Bonnett, R., Clark, V.M., Giddey, A. and Todd, A., J. Chem. Soc., (1959) 2087.
3. Fujii, T. and Miyoshi, M., Bull. Chem. Soc. Jpn., 48 (1975) 1341.
4. Asher, V., Becu, C., Anteunis, M.J.O. and Callens, R., Tetrahedron Lett., 21 (1981) 141.
5. Kisfaludy, L., Korenczki, F. and Katho, A., Synthesis, (1982) 163.
6. Murahashi, S. and Shiota, T., Tetrahedron Lett., 28 (1987) 6469.
7. Fife, W.K., J. Org. Chem., 48 (1983) 1375.
8. While this work was in progress a synthesis of the series of 3- and 4-(phosphonoalkyl)pyridine- and (phosphonoalkyl)piperidine-2-carboxylic acids was reported using the same general experimental protocol: Ornstein, P.L., Schaus, J.M., Chambers, J.W., Huser, D.L., Leander, J.D., Wong, D.T., Paschal, J.W., Jones, N.D. and Deeter, J.B., J. Med. Chem., 32 (1989) 827.
9. To obtain all four possible isomers (i.e. DL-*cis* and DL-*trans*), the reduction step can be carried out with nickel-aluminum alloy in the presence of 1 M KOH (See Lunn, G., J. Org. Chem., 52 (1987) 1043). In this fashion, 2-carboxy-4-methyl-pyridine hydrochloride, **4d**, was reduced to give a 60:40 mixture (estimate from integration of the 4-methyl resonances in the ^{1}H NMR) of the DL-*cis*, **5d**, and the DL-*trans* isomers.
10. Okamoto, S. and Hijikato, A., Biochem. Biophys. Res. Comm., 101 (1981) 440.
11. Greenstein, J.P. and Winitz, M., Chemistry of the Amino Acids, John Wiley and Sons, New York, 1961, p. 891.
12. Gesellchen, P.D., Tafur, S. and Shields, J.E., In Gross, E. and Meienhofer, J. (Eds.) Peptides – Structure and Biological Function (Proceedings of the 6th American Peptide Symposium), Pierce Chemical Comp., Rockford, IL, 1979, p. 117.

Solution and solid-phase synthesis of phosphopeptides relevant to phosphorylase kinase α (PKα) from rabbit muscle

Fidy Andriamanampisoa, Jean-Michel Lacombe and André A. Pavia*

Université d'Avignon, Laboratoire de Chimie Bioorganique, 33 rue Louis Pasteur, F-84000 Avignon, France

Introduction

Most of the few chemical syntheses of phosphopeptides reported to date reflect phosphorylation of the appropriate Ser, Thr or Tyr residue of a preformed peptide

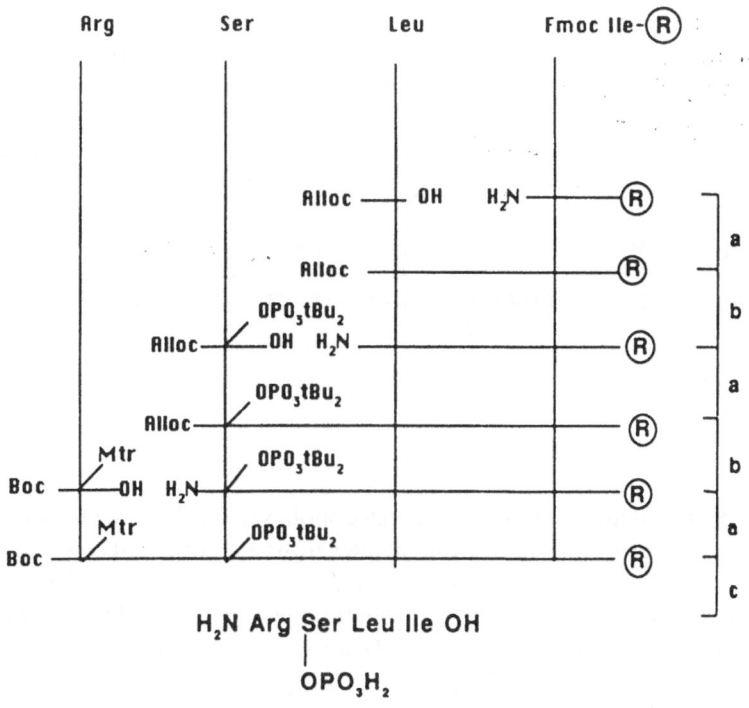

Scheme 1. SPPS of compound 8. a=BOP/DIEA peptide coupling; b=Alloc cleavage (Bu₃SnH, Pd(PPh₃)₄, H⁺); c=(1) TFA/CH₂CL₂, 1:1, (1 h): peptide resin cleavage as well as O-tert-Bu and Boc removal; (2) TFA/thioanisole 95%: 5%, 2 h, complete removal of peptide protecting groups.

---

*To whom correspondence should be addressed.

obtained by the standard procedure [1,2]. In one example of SPPS, phosphorylation was performed after completion of the peptide and before its removal
from the resin [3]. In this communication, we wish to report preliminary results
obtained in our search for a suitable, efficient route to synthesizing phosphopeptides in wich phosphorylserine and/or phosphorylthreonine are sequentially
introduced into glycopeptides just as are glycosylserine and glycosylthreonine
residues [4,5].

Results and Discussion

Our first efforts aimed at selecting protecting groups compatible with both
the requirements of peptide synthesis and the lability of Ser/Thr phosphate moiety
due to facile β-elimination. In that respect, Fmoc protection must be precluded.
Treatment with such mild bases as piperidine or even morpholine resulted in
the removal of phosphate and conversion of serine into dehydroalanine residue.

In the first approach, diphenylphosphorochloridate [Cl-PO(OPh)$_2$] was used
as the phosphorylating reagent. Hydrogenolysis of Z or Bzl protecting groups
had no effect on phenyl protection. Several phosphorylserine (**1**) and phosphorylthreonine (**2**) derivatives were prepared as well as two *O*-phosphorylserine-
containing tripeptides **3** and **4** relevant to isocitrate dehydrogenase (ICDH) from
E. coli and of PKα from rabbit skeletal muscle, respectively. Phenyl and/or
benzyl groups from precursors of compounds **1**, **2**, **3** and **4** on either peptide
or phosphate moiety, upon hydrogenolysis, were removed without problem.

Ac-NH-CH-COO-CH$_3$ Ac-NH-CH-COO-CH$_3$ Arg-Ser-Leu-OMe
       | | |
  CH$_2$-OPO$_3$H$_2$ CH$_3$CH-OPO$_3$H$_2$ CH$_2$-OPO$_3$H$_2$
        1   **2**   **3**

Arg-Leu-Ser-OMe   Boc-Arg(Z$_2$)-Leu-Ser[OPO(OPh)$_2$]-Ile-Ser-(OBzl)-Thr(OBzl)-OBzl
     |
  CH$_2$OPO$_3$H$_2$
   4   **5**

In contrast, deprotection of the phosphohexapeptide **5** from PKα upon
hydrogenolysis proved unsuccessful. Whatever the experimental conditions were,
deprotection was incomplete, leading to a mixture of partly protected compounds.

In the second approach, phosphorylation was achieved by reacting allyl-
oxycarbonylserine-p-nitrobenzyl ester (Alloc-Ser-*O*pNO$_2$Bzl) with *N,N*-diethyl
di*tert*butylphosphoramidite [NEt$_2$P(OtBu)$_2$] followed by oxidation from treatment with *m*-chloroperbenzoic acid to give compound **6** with orthogonal
protecting groups R^1, R^2, R^3.

R^2-NH-CH-COOR^3 **6**: R^1, *t*-Bu; R^2, Alloc; R^3, *p*-NO$_2$Bzl
     |
  CH$_2$OPO(OR^1)$_2$ **7**: R^1, *t*-Bu; R^2, Alloc; R^3, H

Treatment of compound **6** with Zn/AcOH afforded the key compound **7**, which was used in SPPS of compounds **8, 9** and **10** according to the general scheme reported below for compound **8**. Peptide coupling was achieved with the BOP reagent.

8 Arg-Ser*-Leu-Ile
9 Arg-Leu-Ser*-Ile
10 Arg-Leu-Ser*-Asn   (* = -OPO$_3$H$_2$)

Structures described in this communication were assigned by ^{1}H and ^{13}C NMR spectroscopy. The presence of a phosphate group was shown by chemical shifts of both protons (downfield chemical shift for H$_\alpha$- and H$_\beta$-Ser) and carbons (downfield chemical shift for C$_\beta$-Ser and upfield chemical shift for C$_\alpha$-Ser). In addition H-P and C-P couplings were observed (J$_{H-P}$, 5–6.5 Hz; J$_{C-P}$, 5–9.5 Hz) in all phosphorylated compounds. To our knowledge, this work constitutes the first SPPS of phosphopeptides involving the sequential incorporation of an adequately protected phosphoryl serine residue.

References

1. Grehn, L., Fransson, B. and Ragnarsson, U., J. Chem. Soc., Perkin Trans. I, (1987) 529.
2. Schlesinger, D.H., Buku, A., Wyssbrod, H. and Hay, D.Y., Int. J. Pept. Prot. Res., 30 (1987) 257.
3. Perich, J.W. and Johns, R.B., Tetrahedron Lett., 29 (1988) 2369.
4. Lacombe, J.M. and Pavia, A.A., J. Org. Chem., 48 (1983) 2557.
5. Ferrari, B. and Pavia, A.A., Int. J. Pept. Prot. Res., 22 (1983) 549.

Deprotection of peptides containing Arg(Pmc) and tryptophan or tyrosine: Elucidation of byproducts

Bernhard Riniker and Albert Hartmann

Pharmaceuticals Division, CIBA-GEIGY Ltd., CH-4002 Basle, Switzerland

Introduction

The Fmoc group for N^α-protection in SPPS allows the side-chain substituents and C-terminal bond with the resin to be labile by mild acidolysis. For arginine, Ramage et al. have recently developed the Pmc group [1,2]. Its lability in TFA is in good agreement with that of *t.*-butyl protections and with trityl for Cys, His [3], and the ω-trityl derivatives of Asn and Gln [4]. We have used Arg(Pmc) with good success for the synthesis of many peptides. The observation of byproducts when these contained Trp or Tyr prompted us to examine the cleavage of Pmc by TFA more closely.

The final product of the cleavage is pentamethylchroman [1], but in anhydrous solution, a mixed anhydride of TFA and the sulfonic acid **1** must be primarily formed. Similar compounds have been shown to form sulfones with suitably activated aromatic hydrocarbons [5,6]. By reaction of the mixed anhydride with nucleophiles present in the cleavage solution, various byproducts can be formed, some of which are described in the following.

Results and Discussion

Cleavage of Fmoc-Arg(Pmc)-OH with TFA

Fmoc-Arg(Pmc)-OH (2 g) in 20 ml TFA was evaporated after 8 min at 22°C and the mixture separated on a SiO_2 column. Besides Fmoc-Arg-OH (200 mg), **1** as Na-salt and **2** (120 mg) were obtained. The free acid **1** is unstable in the solid state àt room temperature. After 3 h, instead of 8 min, 400 mg **3** were obtained as an isomeric mixture with SO_3H in positions 2 or 3. With 90% TFA, only traces of **1** and **2** (120 mg), no **3**, were formed (Fig. 1).

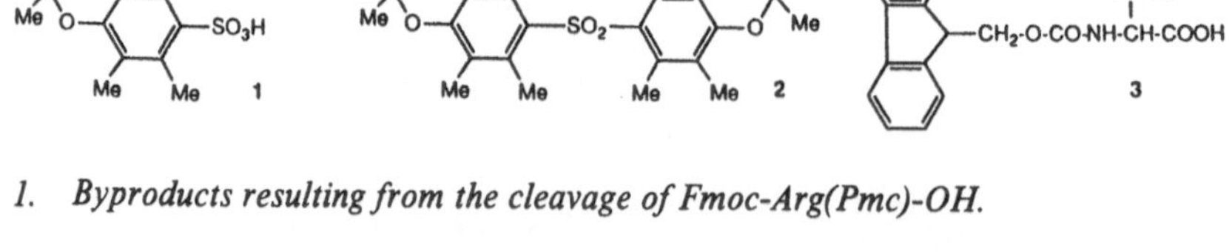

Fig. 1. Byproducts resulting from the cleavage of Fmoc-Arg(Pmc)-OH.

950

Formation of byproducts with Trp and Tyr

After 1 h at 30°C Fmoc-Arg(Pmc)-OH (1 g) and Ac-Trp-OH (0.5 g) in 20 ml 100% TFA gave 350 mg **4** and 80 mg of an isomeric mixture with Pmc in pos. 5 or 6 of indole. In 90% TFA, the yield of these compounds was 2–3 times lower. From Fmoc-Arg(Pmc)-OH (1 g) and Ac-Tyr-OH (0.5 g) in 20 ml 100% TFA, 1 h 30°C, 90 mg **5** and 100 mg **6** were obtained. In 90% TFA, no **5** and only traces of **6** were found (Fig. 2).

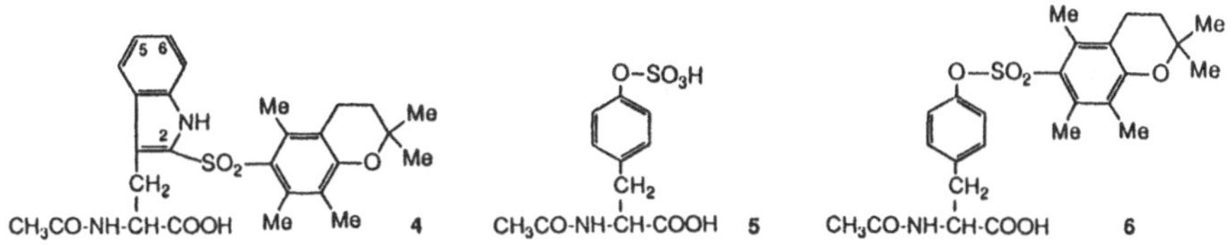

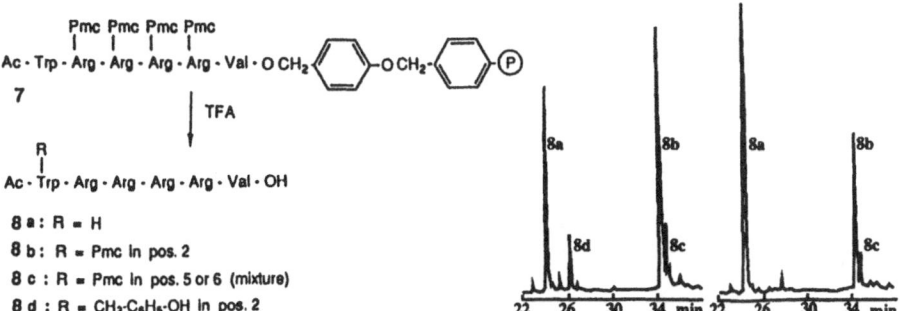

Fig. 2. *Byproducts resulting from the cleavage of Fmoc-Arg(Pmc)-OH in the presence of Trp or Tyr.*

Synthesis of a model peptide

As a hard test for the utility of the Pmc group, peptide **8a** was synthesized on a polystyrene resin with the Wang linker. Fmoc-AA-OTcp's, HOBt, and DIEA were used in a 3-fold excess DMA for 1 h. (Fmoc-Arg(Pmc)-OTcp was obtained crystalline from THF-hexane, m.p. 146°C). The protected, resin-bound **7** was cleaved for 5 min with 100% TFA (a), 95% TFA (b), or TFA-H$_2$O-ethanedithiol 76:4:20 (c). The peptides were precipitated and treated with the same solutions for 1 h at 30°C (a) and (b)/or for 2 h (c), again precipitated and analyzed by LC. (a) 100% TFA produced no **8a**, a major peak of **8b**, and several unidentified smaller peaks (LC diagram not shown). (b) The crude product with 95% TFA contained **8b** as the main component, 35% of the desired peptide **8a**, and some **8c** and **8d** (LC, left panel). **8d** originated from substitution of Trp by the Wang linker [7]. (c) With this mixture, **8a** was the main component (62%), and again **8b** (28%) and **8c** (6%) were present (LC, right panel) (Fig. 3).

Fig. 3. *Structures of model peptide and its byproducts upon TFA cleavage.*

The results demonstrate that, in anhydrous TFA, the Pmc group may be transferred to the side chains of Trp or Tyr, or give rise to the formation of sulfonated compounds. Most of these byproducts can be avoided by addition of water to the cleavage solution. In the presence of Trp, large amounts of scavengers are needed to obtain maximal suppression of substitution products by the Pmc residue.

Acknowledgements

We wish to thank Mr. F. Raschdorf (MS), Dr. H. Fuhrer (NMR), and Mr. S. Moss (IR) for the analytical work and Mrs. V. von Arx and Mr. R. Wille for expert technical assistance.

References

1. Ramage, R. and Green, J., Tetrahedron Lett., 28 (1987) 2287.
2. Green, J., Ogunjobi, O.M., Ramage, R., Stewart, A.S.J., McCurdy, S. and Noble, R., Tetrahedron Lett., 29 (1988) 4341.
3. Sieber, P. and Riniker, B., Tetrahedron Lett., 28 (1987) 6031.
4. Sieber, P. and Riniker, B., Helv. Chim. Acta, in preparation.
5. Bourne, E.J., Stacey, M., Tatlow, J.C. and Tedder, J.M., J. Chem. Soc., (1951) 718.
6. Tyobeka, T.E., Hancock, R.A. and Weigel, H., Tetrahedron, 44 (1988) 1971.
7. Riniker, B. and Kamber, B., In Jung, G. and Bayer, E. (Eds.) Peptides 1988, De Gruyter, Berlin, 1988, p. 115.

A facile method for decomposition of amino acid complexes using dithiocarbamates

Bjarne Due Larsen, Ole Buchardt and Arne Holm

Center for Medical Biotechnology, Chemical Laboratory II, The H.C. Ørsted Institute, University of Copenhagen, Universitetsparken 5, DK-2100 Copenhagen, Denmark

Introduction

Alteration of the side chain functional group in amino acids such as lysine, ornithine or tyrosine is usually done after masking the alfa-amino- and the carboxyl group with cupric ion which forms a stable chelate.

In order to recover the free amino acid derivative, it is necessary to decompose the copper complex. Present methods include the use of EDTA [1], KCN [2], HCl [1], H_2S [3–5], thioacetamide [6] or N,N-dialkyl-N'-(benzoyl)thiourea [7], but they suffer from certain drawbacks, such as toxicity of the reagents, unsatisfactory yields or long reaction times. Furthermore, they may require heating in strongly acidic solutions and the CuS formed may be in a colloidal state.

Results and Discussion

We report a new method for the decomposition of amino acid metal complexes using N,N-diethyl dithiocarbamate (DEDTC), producing an easily removed copper dithiocarbamate complex within 30 min at room temperature and affording side-chain substituted amino acids, such as ornithine or lysine in high yields and purity. The reagent is not destructive in a reductive way, such as H_2S, thereby allowing introduction of the trinitrophenyl (TNP) group without complications.

The copper complexes of H-L-Lys-OH, H-L-Orn-OH, H-L-Lys(Z)-OH, H-L-Orn(Z)-OH, H-L-Orn(TNP)-OH, H-L-Lys(TNP)-OH, and of H-L-Orn(Biotinyl)-OH were prepared according to known methods [3,8,9] and decomposed using the following general procedure: DEDTC (20 mmol) was added to a solution of the amino acid copper complex (2 mmol) in water (200 ml). The reaction mixture was stirred for 30 min at room temperature, filtered through celite which was washed with water (the solubility of the product/complex may be increased by adjusting the pH in the range ≤ 7 with HCl) and the combined filtrate evaporated to dryness. The crude material was washed with 5×25 ml of chloroform and the residue was dried over night. In the case of H-L-Lys(Z)-OH and H-L-Orn(Z)-OH, the crude material was redissolved in 2 N HCl, and the product precipitated by adjusting the pH to 7 with ammonia.

Table 1   *Yield and physical constants of amino acid derivatives*

Amino acid	Yield (%)	Melting point	$[\alpha]_{20}^D$
1. H-L-Lys(Z)-OH	90	250–252; (248–250) [7]	15.7; (15.1) [7]
2. H-L-Lys(TNP)-OH,HCl	90	203–204(dec.); (198–199) [9]	18.8[a]
3. H-L-Orn(Z)-OH	86	248–250; (248–250) [7]	17.0; (17.4) [7]
4. H-L-Orn(TNP)-OH,HCl	91	234–236 (dec.)	26.8[b]

[a] c = 0.5 in acetone/4 N HCl (1 : 1).
[b] c = 1.0 in acetone/2 N HCl (1 : 1).

DEDTC may be prepared as follows: diethylamine (20.6 ml; 0.2 mol) was dissolved in pentane (50 ml) and cooled to 0°C. Carbon disulfide (6.1 ml; 0.1 mol) was added very carefully to the stirred solution (heat evolution). The resulting precipitated DEDTC was collected, washed with pentane (50 ml), and dried in vacuo. Yield, 19.3 g (87%). M.p. 80–80.5°C (reported: 81–82°C).

Acknowledgements

This work was supported by Lundbeckfonden, The Danish Biotechnology Programme and The Danish National Science Foundation.

References

1. Kuwata, S. and Watanabe, H., Bull. Chem. Soc. (Japan), 38 (1965) 676; Ledger, R. and Stewart, F.H.C., Aust. J. Chem., 18 (1965) 933; Zaoral, M., Coll. Czech. Chem. Commun., 30 (1965) 1853; Wünsch, E., Fries, G. and Zwick, A., Chem. Ber., 91 (1958) 542; Morley, J.S., J. Chem. Soc. (C), (1967) 2410; Caldwell, J.B., Holt, L.A. and Milligan, B., Aust. J. Chem., 24 (1971) 435.
2. Zahn, H., Zuber, H., Ditscher, W., Wegerle, D. and Meienhofer, J., Chem. Ber., 89 (1956) 407.
3. Synge, R.L.M., Biochem. J., 42 (1948) 99.
4. Neuberger, A. and Sanger, F., Biochem. J., 37 (1943) 515.
5. Flaschka, H., Chemist-Anal., 44 (1955) 2.
6. Flaschka, H. and Jakoblejevich, H., Anal. Chim. Acta, 4 (1950) 482; Bodanszky, M., Tolle, J.C., Deshmane, S.S. and Bodanszky, A., Int. J. Pept. Prot. Res., 12 (1978) 57; Fölsch, G. and Serck-Hansen, K., Acta Chem. Scand., 13 (1959) 1243; Taylor, U.F., Dyckes, D.F. and Cox, J.R., Int. J. Pept. Prot. Res., 19 (1982) 158.
7. König, K.H., Kaul, L., Kuge, M. and Schuster, M. Liebig's Ann. Chem., (1987) 1115.
8. Bayer, E.A. and Wilchek, M., Methods Enzymol., 34 (1974) 265.
9. Okuyama, T. and Satake, K., J. Biochem., 47 (1960) 454.

Improved procedures for Na/liquid NH_3 deprotection and ion-exchange chromatography in large scale synthesis of oxytocin, vasopressin and DDAVP

Luiz Juliano, Regina S.H. Carvalho and Paul Boschcov

Department of Biophysics, Escola Paulista de Medicina, rua Tres de Maio 100,
04044 São Paulo-SP, Brazil

Introduction

The removal of benzyl-type and tosyl-protecting groups by Na/liquid NH_3 reduction is particularly useful for large scale procedures in peptide synthesis. However, its applications are limited by side reactions that are related to the reaction stoichiometry, nature of protecting groups and amino acid composition. The classical blue end-point is not necessary for the complete removal of protecting groups and the undesired side reactions are due to sodium excess [1,2]. In the present work the optimal conditions for Na/liquid NH_3 reductions, with regard to the amount of Na and the usefulness of additives (urea or sodium amide), were determined for deprotection of oxytocin, [Lys8]vasopressin and DDAVP.

Oxytocin, vasopressin and their analogs are usually purified by counter current distribution, partition and filtration chromatography on Sephadex [3]. A continuous free-flow electrophoresis procedure was described for large scale purification of DDAVP [4]. In order to avoid the concentration and lyophilization of large volumes of solution during the purification of oxytocin, [Lys8]vasopressin and DDAVP in batches larger than 20 g, ion-exchange chromatography conditions were developed that allowed concentration, desalting and purification of the peptides during the same chromatography step.

Results and Discussion

Oxytocin, [Lys8]vasopressin and DDAVP were synthesized by solution methods in scales of 10–100 g using benzyloxycarbonyl and benzyl protection for amino and thiol groups, respectively, and tosyl for Lys and Arg side chains. After deprotection, disulfide-bond formation was accomplished by potassium ferricyanide [5]. Reductions in Na/liquid NH_3 were done in 10–20 g scale using a convenient all-glass burette, which was designed to introduce gram quantities of Na, previously dissolved in NH_3, into the reaction vessel, in a controlled and quantitative way. The products of the Na/liquid NH_3 reductions were monitored by HPLC, and the best conditions for the reductions are as follows: (a) oxytocin: stoichiometry 2:1 (Na: protecting group) without blue color endpoint, urea (5×excess) reduces the amount of side product (Cys-Pro bond cleavage) with significantly increased yield ($>88\%$; **(b)** [Lys^8]vasopressin: stoi-chiometry 2:1 with persistent blue color, and urea as for oxytocin; **(c)** DDAVP: stoichiometry 1.5:1, with persistent blue color with urea having no effect.

Sodium amide in 10-fold molar excess, which is 5 times less than previously prescribed [6], presented highly deleterious effects, as yields were drastically decreased – ($<30\%$), and at least one unidentified peak in HPLC was generated.

Bio-Rex 70, analytical grade, a weakly acidic cation exchange resin with carboxylic group on a macroreticular acrylic polymer lattice, was used to purify oxytocin and DDAVP. Carboxymethylcellulose (cellex-CM, Bio-Rad) was used for the purification of [Lys^8]vasopressin. The following conditions were used: (a) oxytocin and DDAVP: column size 3.8×11 cm, loaded with 5 l of crude peptide solution (~1 mM), and washed with 2 l 0.01 M ammonium acetate (desalting), 2 l 0.1% acetic acid (elution of side product of Na/liquid NH$_3$ reduction), 4 l acetic acid (0.3% for oxytocin and 1% for DDAVP) (elution of 80–90% of monomer, 1 l 50% acetic acid (elution of dimers); (b) Lys-vasopressin; column size 3.8×13 cm, washings and elution were the same as for oxytocin. The flow rate of approximately 3.0 ml/min for all chromatographies was controlled by gravity. Crude peptide solutions and resins were equilibrated at pH 5. The purity for the three peptides was higher than 96% by HPLC. Using a set of 8 columns in a parallel configuration, up to 40 l of 1 mM oxytocin were purified in only one run.

References

1. Schön, I., Szirtes, T., Überhardt, T., Rill, A., Csehi, A. and Hegedüs, B., Int. J. Pept. Prot. Res., 22 (1983) 92.
2. Schön, I., Chem. Rev., 84 (1984) 287.
3. Manning, M., Wuu, T.-C. and Baxter, J.W.M., J. Chromatogr., 38 (1968) 396.
4. Zaoral, M., Flegel, M., Barth, T. and Machova, A., Coll. Czech. Chem. Commun., 43 (1978) 511.
5. Walti, M. and Hope, D.B., Experientia, 29 (1973) 389.
6. Katsoyannis, P.G., Zalut, C., Tometsko, A., Tilak, M., Johnson, S. and Trakatellis, A.C., J. Am. Chem. Soc., 93 (1971) 5871.

A new side reaction in solid-phase peptide synthesis: Solid support-dependent alkylation of tryptophan

Paul D. Gesellchen, Robert B. Rothenberger, Douglas E. Dorman,
Jonathan W. Paschal, Thomas K. Elzey and Charles S. Campbell

The Lilly Research Laboratories, Eli Lilly and Company, Indianapolis, IN 46285, U.S.A.

Introduction

The increasing use of the base-labile Fmoc-protecting group in SPPS has required the development of new acid-labile polymeric supports (resins). Recently the use of a 2,4-dimethoxybenzhydrylamine (2,4-DMBHA) resin, **1** (Fig. 1), has been advocated for the production of peptide amides [1,2]. We used this support for the Fmoc-aminopentafluorophenylester-mediated synthesis of tetragastrin (H-Trp-Met-Asp-Phe-NH$_2$). After removal of the final Fmoc group, the peptide was cleaved from the support by treatment with TFA in the presence of 4.5% phenol and 0.5% ethanedithiol for 16 h (as recommended by the supplier of the resin*. Subsequent RPHPLC analysis indicated that a major (5–10%) side reaction had occurred during this treatment. Results from ^{1}H NMR decoupling, NOE, 2D ^{1}H NMR and MS experiments showed that the tryptophan residue of the tetrapeptide amide had been alkylated at the C-2 position of the indole ring to give the impurity, **2** (Fig. 1).

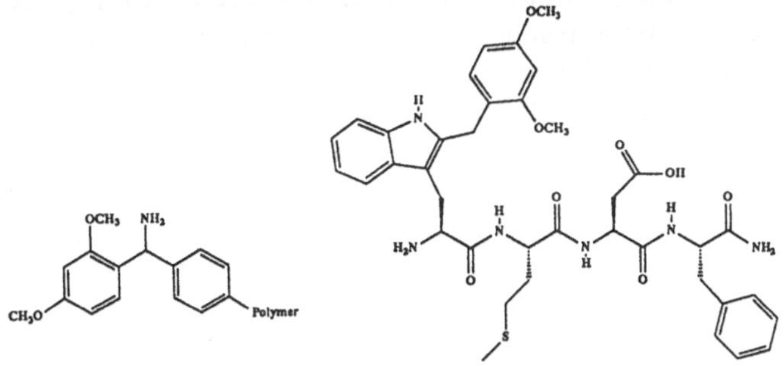

(1) (2)

Fig. 1. Structural formulas of 2,4-DMBHA (1) and compound 2.

Results and Discussion

There is ample precedent for the alkylation of tryptophan residues during

---

*Rapid Amide™ Resin. Available commercially from E.I. du Pont de Nemours and Co. (Inc.), Biotechnology Systems Division.

prolonged treatment with acid; however, these by-products usually arise from cationic species that are generated from side-chain protecting groups such as Boc [3–5] or Z(OMe) [6]. In the current study, the solid support (2,4-DMBHA resin) served as the source of the cationic species that produced the by-product **2**, Trp[2'(2,4-DMB)]-Met-Asp-Phe-NH$_2$. The intermolecular nature of this alkylation was shown by isolation of impurity **2**, following incubation of purified tetragastrin with a sample of Fmoc-Phe-(2,4-DMB resin) and TFA. Substitution of the tryptophan residue by a tyrosine or a histidine residue completely eliminated the side reaction, and thus verified the specificity for the indole moiety. Masking the amino group of tryptophan (Fmoc) or moving the tryptophan residue to the interior of the peptide (e.g., Met-Trp-Asp-Phe-NH$_2$) resulted in a significantly higher percentage of the 2,4-DMB impurity peptide. Furthermore, if the amino group was removed completely, the alkylated peptide became the major product. These data suggest that the alkylation reaction was mediated by a cationic species.

The alkylation reaction could be prevented by protection of the indole nitrogen of tryptophan with a formyl group. This result suggests that the initial reaction may be the kinetically controlled alkylation of the indole nitrogen group followed by rearrangement to the thermodynamically more stable C-2' position.

We cleaved the peptide-resin and generated high levels of pure tetragastrin with little or none of the 2,4-DMB impurity using standard liquid HF conditions. The absence of the impurity following the HF reaction may result from the instability of the N^{in}-2,4-DMB moiety [5].

Over 50 alternative TFA-based cleavage conditions were examined for the effect of scavenger on the production of the impurity. One of these procedures (TFA/thioanisole, 4:1) has been advocated for rapid (<3 h) removal of C-terminal phenylalanine peptides from the 2,4-DMBHA resin [2]. However, we find that these conditions result in incomplete peptide cleavage and also give high levels of the 2,4-DMB impurity peptide, **2**, along with a second major impurity (not characterized). We recommend the cleavage mixture of TFA/DMS/EDT/*m*-cresol (65:25:5:5) for routine use. This recommendation is based upon the 2,4-DMB (and other) impurity levels, the color of the final peptide and the compatibility of the scavenger mixture with a polystyrene-based resin.

Acknowledgements

The authors wish to thank Mr. Henry R. Wolfe for his helpful discussions concerning the 2,4-DMBHA resin.

References

1. Pietta, P.G. and Marshall,G.R., Chem. Commun., (1970) 650.
2. Penke, B. and Rivier, J., J. Org. Chem., 52 (1987) 1197.
3. Low, M., Kisfaludy, L. and Sohar, P., Hoppe-Seyler's Z. Physiol. Chem., 359 (1978) 1643.

4. Jaeger, E., Thamm, P., Knof, S. and Wunsch, E., Hoppe-Seyler's Z. Physiol. Chem., 359 (1978) 1629.
5. Chino, N., Masui, Y. and Sakakibara, S., In Shiba, T. (Ed.) Peptide Chemistry, 1977. Protein Research Foundation, Osaka, Japan, 1978, p. 27.
6. Ogawa, H., Sasaki, T. and Yajima, H. In Izumiya, M. (Ed.) Peptide Chemistry, 1978. Protein Research Foundation, Osaka, Japan, 1979, p. 21.

Side reactions in post-HF workup of peptides having the unusual Tyr-Trp-Cys sequence: Low-high acidolysis revisited

Berta Ponsati, Ernest Giralt and David Andreu

Department of Organic Chemistry, University of Barcelona, E-08028 Barcelona, Spain

Introduction

Low-high HF is now widely recognized as a reliable and effective method of peptide-resin acidolysis and has been adopted by many laboratories for routine cleavage and deprotection of complex synthetic peptides. The original work [1] pointed out some complications in the workup of crude Trp-containing peptides.

We wish to describe some recent results which suggest that Trp-Cys is a particularly troublesome sequence in this respect.

Results and Discussion

The following peptides, related to one of the multigenic families of African swine fever virus [2] and having in common the unusual Tyr-Trp-Cys sequence at the C-terminus, were synthesized by solid-phase method on MBHAR:

AGGHLRSTDNPPQEELGYWC-NH$_2$	[U124-20]
LQGFSTDNLLEEELRYWC-NH$_2$	[U104-18]
LVGQLRPTEDPPEEELEYWC-NH$_2$	[V118-20]

As shown in Fig. 1A, low-high HF of V118-20 resin gave a crude of satisfactory purity (>90% by HPLC) which, after lyophylization, became a complex mixture from which the original peptide was practically absent. The other two sequences suffered similar degradation, to an extent unparalleled by any other Trp-containing peptides in our experience. Purification (RPLC) and analysis (FABMS and ^{1}H NMR) of these modified peptides allowed their identification as methyl ($M+14$), benzyl ($M+90$), bromobenzyl ($M+169$; $M+171$) and p-tolylthiocarbonyl ($M+150$) substitution products at position 2 of tryptophan. Their formation was interpreted as resulting from nucleophilic attack by Trp on the various sulfonium salts derived from Me$_2$S and the different side-chain protecting groups.

Specific involvement of the Cys residue in the side reaction was demonstrated by the synthesis of the model peptides Ac-Tyr-Trp-Cys-NH$_2$ and Ac-Tyr-Trp-Ala-NH$_2$. Whereas the former behaved similarly to the previous ones, the latter, lacking the Cys residue, showed no alteration of its low-high HF crude after

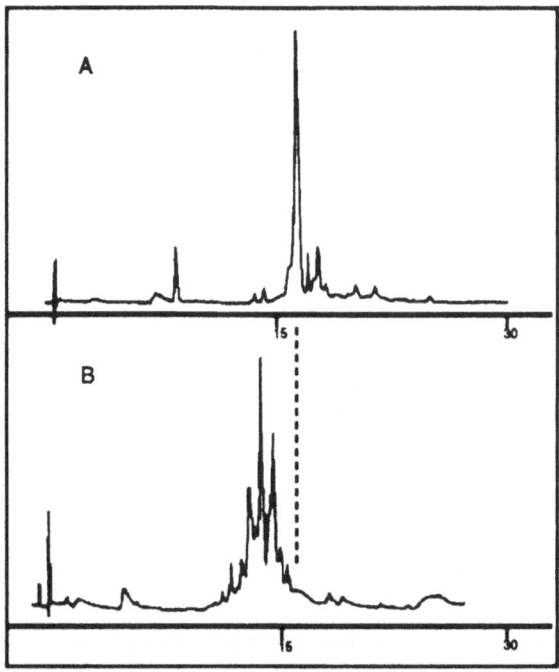

Fig. 1. A: HPLC analysis of crude V118-20 after low-high HF (experimental conditions as described in Ref. 1). Column: Vydac 218TP54. Elution: linear 5–65% gradient of MeCN (0.035% TFA) into water (0.045% TFA) over 30 min. B: Same crude after lyophilization from 10% HOAc.

lyophilization. More specifically, these findings suggested the possibility of anchimeric assistance from the thiol group of Cys in the Trp substitution reaction, possibly through one of the mechanisms outlined in Fig. 2. This hypothesis is further supported by molecular mechanics and AM1 calculations for Ac-Tyr-Trp-Cys, which gave an energy minimum conformation with a hydrogen bond between the carbonyl oxygen of Tyr and the amide proton of Cys in which the distance between the thiol and the C-2 of the indole ring is less than 2.6 Å.

Finally, we have found two work-up procedures effective in circumventing this side reaction. In one of them, lyophilization is simply avoided and the crude peptide (usually in 10% HOAc solution) is loaded onto an RPLC column and promptly purified. Alternatively, tryptophan (ca. 10 equiv.) is added to the crude to act as a scavenger during lyophilization.

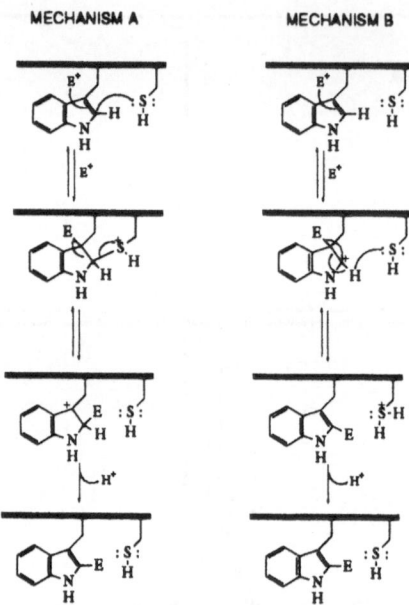

Fig. 2. Proposed mechanisms for tryptophan substitution with anchimeric assistance from cysteine. $E^+ = Me_2S^+R$, where, e.g., R = benzyl, bromobenzyl.

References

1. Tam, J.P., Heath, W.F. and Merrifield, R.B., J. Am. Chem. Soc., 105 (1983) 6442.
2. Viñuela, E., In Becker, Y. (Ed.) African Swine Fever, Martinus Nijhoff, Boston, MA, 1987, p. 31.

Selective cysteine alkylation in unprotected peptides: A useful tool for peptide side-chain modification

Yat Sun Or*, Richard F. Clark and Jay R. Luly

Abbott Laboratories, Pharmaceutical Products Division, Abbott Park, Il 60064, U.S.A.

Introduction

One of the first steps toward the optimization of biologically active peptides is the systematic replacement of individual residues with natural or unnatural amino acids. The range of substitutions that allows the determination of the importance of charge, steric bulk, hydrophobicity, aromaticity and chirality at each position is greatly extended by the use of unnatural amino acids. However, these studies, which involve the synthesis of unnatural amino acids, are sometimes rather time-consuming and tedious. We have recently developed an expedient method that allows modification of the amino acid residue side chain through selective cysteine alkylation of unprotected peptides. Alkylation of the thiol group in cysteine has been extensively used in protein modifications [1] and affinity labeling [2]. Selective alkylation of thiol is possible because of its high reactivity relative to other protein side chains. Although much of the thiol chemistry applied to protein derivatization should also be applicable to cysteine in small peptides, derivatization of cysteine in unprotected peptides has not found widespread application.

Results and Discussion

In order to test the general applicability of the cysteine alkylation methodology, an unprotected cysteine-containing peptide possessing several potential competing nucleophilic functionalities such as amino, hydroxy and carboxyl groups was chosen as a model peptide. The model tetrapeptide, H-Ser-Lys-Cys-Phe-OH·2TFA, **1**, was synthesized by standard SPPS using Merrifield resin, cleaved with HF and purified by RPHPLC. All solvents used for the extraction and purification were saturated with helium to eliminate oxygen. After lyophilization, the purified tetrapeptide was obtained as white amorphous powder and could be stored under nitrogen in the freezer for more than six months without any sign of disulfide formation by RPHPLC analysis.

For a typical alkylation, reaction of unprotected peptide **1** with 1.3 equivalents of various alkylating agents in saturated ammonia in methanol at 0°C proceeded cleanly to yield single alkylation products within 1 h, as shown by RPHPLC analysis. For the less reactive alkylating agent ethyleneimine, a large excess of

---

*To whom correspondence should be addressed.

reagent was used in saturated ammonia in DMF at room temperature. In most cases, acidic, basic, and neutral side chains were introduced by this method in over 80% yield after isolation of products.

	RX	R	Yield, %
a.	$ClCH_2CO_2H$	CH_2CO_2H	84
b.	$CH_2=CHCO_2Me$	$CH_2CH_2CO_2Me$	80
c.	$BrCH_2Ph$	CH_2Ph	84
d.	CH_2-CH_2 / NH	$CH_2CH_2NH_2$	81

The methodology is also applicable to the synthesis of cyclic peptides by intramolecular cysteine alkylation. Iodoacetyl-Ser-Lys-Cys-Phe-OH·TFA, **3**, prepared and purified as described before, was cyclized cleanly to **4** in refluxing liquid ammonia in 87% yield.

All the alkylation products were characterized by NMR, FABMS, AAA and elemental analysis. The indication of selective thiol alkylation was provided by NMR analysis and further verified by AAA. The methodology of peptide derivatization through the alkylation of unprotected peptides provides a facile way of synthesizing a large number of analogs. It is also applicable to the synthesis of cyclic peptides through intramolecular cysteine alkylation, and provides a useful alternative to cyclizations that occur through disulfide- or amide bond-forming reactions.

References

1. Means, G.E. and Feeney, R.E., Chemical Modifications of Proteins, Holden-Day, San Francisco, CA, 1971.
2. Jacoby, W.B. and Wilchek, M., (Eds.) Methods in Enzymology, Vol.46, Academic Press, New York/London, 1977.

Formation of chirally pure *N*-acyl-*N,N'*-dicyclohexylurea in DCC-mediated peptide synthesis

Aleksander M. Kolodziejczyk, Marek Slebioda and Zbigniew Wodecki

Department of Organic Chemistry, Technical University of Gdansk, 80-952 Gdansk, Poland

Introduction

Because of our interest in the mechanism of activation of amino acids by DCC as it relates to optical integrity, we have studied the formation of chirally pure *N*-acyl-*N,N'*-dicyclohexylurea in DCC-mediated peptide synthesis.

Results and Discussion

Utilizing HPLC on chiral stationary phases, we found that the *N*-acylureas **4** preserved their chiral integrity, although the peptides formed simultaneously

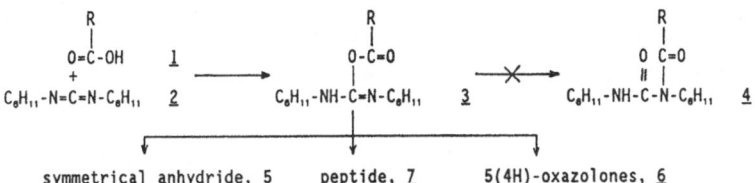

Scheme 1.   *Reaction products of DCC-activated amino acids.*

were racemized to a high extent (Table 1). Since the probability of chiral stability of *O*-acylisourea is rather low [1,2], the high chiral purity of *N*-acylurea **4** indicates that they are not formed from *O*-acylisourea **3**.

Table 1   *Chiral purity of peptides and* N-*acylureas formed in the reaction:*

X-L-AA + L-*LeuOMe* $\xrightarrow[25°C,\ 24\ h]{DCC}$ *Peptide (LL + DL)* + N-*acylurea*

		%, Contamination of	
X-L-AA	Solvent	DL-Peptide	D-N-acylurea
Bz-Phe	THF	56.7	<0.1
Ac-Leu	dioxane	28.2	<0.1

A new, efficient method (Table 2) for the synthesis of chirally pure *N*-acylureas was found. Its basis is a dropwise addition of *N*-protected amino acid or peptide solution to a boiling solution of DCC. Despite the nature of the *N*-protecting group (acyl- or urethane type), and the drastic reaction condition, all *N*-acylureas were obtained in high yield and were chirally pure (more than 99.9 %).

$$\text{X-AA} + \text{DCC} \xrightarrow[\text{THF or dioxane}]{\text{boiling}} \textit{N}\text{-acylurea}$$

Why is the chiral integrity of acylamino acid residues in *N*-acylureas preserved, while the same residues in peptide moieties are racemized to a high extent?

Table 2   *Reaction yields and physicochemical data of some N-acylureas*

N-acylurea	Yield (%)	M.p. (°C)	$[\alpha]_D^{25}$ (1,EtOH)
Ac-Phe-DCU	78	60– 61	+67.0
Ac-Leu-DCU	92	130–132	+53.0
Boc-Ala-DCU	76	135–137	+32.4
Z-Val-DCU	93	127–128	+49.0
Z-Gly-Phe-DCU	85	148–152	+ 2.7

In our opinion it is very likely that the carboxyl group activation by DCC begins as a 4-center addition, and the formed transient intermediate **8** is converted to *N*-acylurea **4** or *O*-acylisourea **3**. *N*-Acylurea is stable, whereas *O*-acylisourea undergoes racemization and could be converted to other, less active intermediates (**5, 6**), or directly aminolyzed to a peptide **7**.

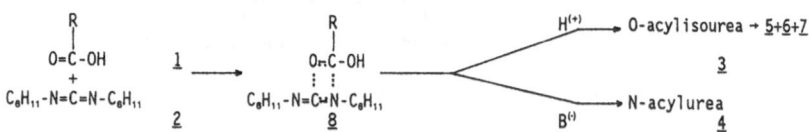

Scheme 2.   *Proposed mechanism of N-acylurea formation.*

References

1. Bodanszky, M., Principles of Peptide Synthesis, Springer-Verlag, Berlin, 1984.
2. Kolodziejczyk, A.M. and Slebioda, M., Int. J. Pept. Prot. Res., 28 (1986) 444.

Studies on racemization associated with the coupling of activated hydroxyamino acids

Rene Steinauer, Francis M.F. Chen and N. Leo Benoiton

Department of Biochemistry, University of Ottawa, Ottawa, Ontario, Canada K1H 8M5

Introduction

Segment couplings are traditionally done at glycine or proline, which are residues not subject to chiral inversion. Activation at other residues implies the possibility of racemization, the extent depending, among others, on the nature of the side chain. Knowing which residues racemize least would allow one to plan the points of chain assembly that are likely to generate products with the lowest content of diastereomers. Limited information on the tendency of different residues to racemize is available. (See Ref. 1 for a review.) The present study issued from the work of Kitada and Fugino [2], who reported unusually low levels of epimerization (1–3% epimer) in the DCC-mediated couplings of Boc-Asp(OBzl)-Thr-OH with four dipeptide esters (solvent unspecified). Prompted also by the findings that the DMAP-catalyzed reaction of Bpoc-Thr(Bzl)-OH with $HOCH_2$-anchor-resin showed no epimerization [3], we decided to examine the chiral lability during coupling of side-chain protected and unprotected hydroxyamino acids.

Results and Discussion

The stereomutation associated with the couplings of Z-Asp(OBzl)-Thr-OH and Z-Gly-Xxx(R)-OH with H-Val-OBzl.HCl.NMM, where Xxx = Ser, Thr and Tyr, and R = H, Bzl and tBu, was examined. Epimeric products were determined by RPHPLC after deprotection [4]. In contrast with the results of Kitada and Fugino [2], we found extremely high levels of stereomutation (27–40% epimer using DCC in DCM; 17–40% epimer using BOP [5] in DMF) (Tables 1 and 2). Moderately lower levels can be expected with BOP if the base is DIEA, instead of NMM [6,7]. BOP gave cleaner HPLC profiles than DCC. In summary, we conclude that *N*-Protected-Asp(OBzl)-Thr-OH is not exceptionally resistant to epimerization during coupling, and tyrosine and the two hydroxyamino acid residues, whether side-chain protected or not, undergo as much stereomutation when coupled as other residues for which more data have been collected.

R. Steinauer, F.M.F. Chen and N.L. Benoiton

Table 1   *Stereomutation of Z-Gly-Xxx(R)-OH during coupling[a]*

Xxx(R)	DCC/DCM	BOP-NMM/DMF	DCC-HOBt/DCM
Ser	40.3[b]	26.9	–
Thr	33.1	17.3[c]	<0.3
Ser(tBu)	27.1	23.0	<0.15
Ser(Bzl)	34.3	26.6	<0.2
Thr(Bzl)	40.4	40.7	0.8[d]
Tyr	36.8	30.8	1.4[d]

[a] With H-Val-OBzl.HCl.NMM. %Epimer formed.
[b] Coupling with H-Leu-OBzl.HCl.NMM.
[c] BOP-DIEA/DCM, 2.9%.
[d] Could be due to optically impure starting material.

Table 2   *Epimerization of Z-Asp(OBzl)-Thr-OH[a] during coupling[b]*

DCC/DCM	26.9%
EDAC/THF	35.1%
DCC-HOBt/DCM	<1.5%
BOP-NMM/DMF	21.8%

[a] Oil, characterized as dicyclohexylammonium salt, m.p. 137–138°C, $[\alpha]_D^{21}$ +9.4° (1, MeOH); C,H,N.
[b] With H-Val-OBzl.HCl.NMM. %epimer formed. HPLC on μBondapak C_{18} column (4), eluted with 0.01 M KH_2PO_4, pH 3.2 (96%), MeCN (4%) $t_{(LLL)}$ 8.08 min, $t_{(LDalloL)}$ 11.75 min.

Acknowledgement

This work was financially supported by the Medical Research Council of Canada.

References

1. Benoiton, N.L., In Gross, E. and Meienhofer, J. (Eds.) The Peptides – Analysis, Synthesis, Biology, Vol. 5, Academic Press, New York, NY, 1983, p. 217.
2. Kitada, C. and Fugino, M., Chem. Pharm. Bull., 26 (1978) 585.
3. Wang, S.S., Tam, J.P., Wang, B.S.H. and Merrifield, R.B., Int., J. Pept. Prot. Res., 18 (1981) 459.
4. Steinauer, R., Chen, F.M.F. and Benoiton, N.L., J. Chromatogr., 325 (1985) 111.
5. Castro, B., Dormoy, J.R., Evin, G. and Selve, C., Tetrahedron Lett., (1975) 1219.
6. Le Nguyen, D., Seyer, R., Heitz, A., and Castro, B., J. Chem. Soc. Perkin Trans I, (1985) 1025.
7. Steinauer, R., Chen, F.M.F. and Benoiton, N.L., Int. J. Pept. Prot. Res., 34 (1989) 295.

Synthesis of a saralasin derivative completely modified at every amide bond with a methyleneamino isostere

James S. Kaltenbronn, Dana E. DeJohn, James P. Hudspeth,
Christine C. Humblet, Elizabeth A. Lunney, Ernest D. Nicolaides,
Joseph T. Repine, W. Howard Roark and Francis J. Tinney

*Parke-Davis Pharmaceutical Research Division, Warner-Lambert Company, Ann Arbor,
MI 48105, U.S.A.*

Introduction

The reductive amination [1] of amino acid-derived aldehydes to prepare the methyleneamino isostere, $\psi[CH_2NH]$, has been used in both stepwise and fragment condensations to prepare a saralasin derivative completely modified at every amide bond.

Methods (Scheme 1)

The requisite aldehydes were prepared by both reductive [2] and oxidation [3] procedures. Partial racemization occurred in the preparation of Boc-Tyr-(OBzl)[CHO]. This aldehyde was elaborated to **1** in which the diastereomers present could be separated. Partial racemization also occurred in the preparation of Z-His(Trt)[CHO]. This was converted to **3** in which the diastereomers present were separated. A reductive amination between all combinations of **2a** and **2b** with **4a** and **4b** gave four diastereomers of **6**. Stepwise deprotection gave four diastereomers of **7**, designated as **7a** (fast-fast), **7b** (fast-slow), **7c** (slow-fast), and **7d** (slow-slow).

Results and Discussion

Earlier results from these laboratories [4] showed that angiotensin II (Ang II) analogs with one or two modifications at the amino-terminus maintained high potency when tested for their ability to inhibit [125I]Ang II binding to rat adrenal homogenates, whereas three modifications caused a 34-fold drop in potency. The completely modified derivatives described here were essentially inactive. Although molecular modeling showed that **7** could reasonably adopt conformations compatible with our current Ang II receptor-bound template, the energy required to overcome the loss of hydrogen bond stabilization and the dramatic increase in overall flexibility prevented an appropriate association with the receptor.

969

Scheme 1

Z-Sar[CHO] $\xrightarrow{a,b}$ Z-Sarψ[CH$_2$NZ]Arg(NO$_2$)-OCH$_3$

$\xrightarrow{c,d,e}$ Z-Sarψ[CH$_2$NZ]Arg(NO$_2$)[CHO]

$\xrightarrow{f,b,g}$ Z-Sarψ[CH$_2$NZ]Arg(NO$_2$)ψ[CH$_2$NZ]Val[CHO]

$\xrightarrow{h,b}$ Z-Sarψ[CH$_2$NZ]Arg(NO$_2$)ψ[CH$_2$NZ]Valψ[CH$_2$NZ]Tyr(OBzl)ψ[CH$_2$NZ]Ile-N(CH$_3$)OCH$_3$

   1a fast eluting diastereomer
   1b slow eluting diastereomer

$\xrightarrow{e}$ Z-Sarψ[CH$_2$NZ]Arg(NO$_2$)ψ[CH$_2$NZ]Valψ[CH$_2$NZ]Tyr(OBzl)ψ[CH$_2$NZ]Ile[CHO]
   2a,b

   Z-His(Trt)-N(CH$_3$)OCH$_3$ $\xrightarrow{e}$ Z-His(Trt)[CHO]

   $\xrightarrow{i}$ Z-His(Trt)ψ[CH$_2$N]Proψ[CH$_2$NBoc]Ala-O-t-Bu
   3a fast eluting diastereomer
   3b slow eluting diastereomer

   $\xrightarrow{j}$ His(Trt)ψ[CH$_2$N]Proψ[CH$_2$NBoc]Ala-O-t-Bu
   4a,b

2+4 $\xrightarrow{k}$ Z-Sarψ[CH$_2$NZ]Arg(NO$_2$)ψ[CH$_2$NZ]Valψ[CH$_2$NZ]Tyr(OBzl)ψ[CH$_2$NZ]-Ileψ[CH$_2$NH]His(Trt)ψ[CH$_2$N]Proψ[CH$_2$NBoc]Ala-O-t-Bu
   6

$\xrightarrow{l,m,j}$ Sarψ[CH$_2$NH]Argψ[CH$_2$NH]Valψ[CH$_2$NH]Tyrψ[CH$_2$NH]Ileψ[CH$_2$NH]His-ψ[CH$_2$N]Proψ[CH$_2$NH]Ala
   7

---

[a] (a) Arg(NO$_2$)-OCH$_3$, NaCNBH$_3$, 3A molecular sieves. (b) ZCl. (c) NaOH. (d) CH$_3$NHOCH$_3$, DCC, HOBT. (e) LAH. (f) Valinol, NaCNBH$_3$, 3A molecular sieves. (g) ClCOCOCl, DMSO. (h) Tyr(OBzl)ψ[CH$_2$NZ]Ile-N(CH$_3$)OCH$_3$, NaCNBH$_3$, 3A molecular sieves. (i) Proψ[CH$_2$NBoc]Ala-O-t-Bu, NaCNBH$_3$, 3A molecular sieves. (j) H$_2$, Pd/C. (k) NaCNBH$_3$, 3A molecular sieves. (l) HOAc, H$_2$O. (m) HCl gas in dioxane.

Acknowledgements

We thank Mr. A.D. Essenburg for the biological data.

References

1. Borch, R.F., Bernstein, M.D. and Durst, H.D., J. Am. Chem. Soc., 93 (1971) 2897.
2. Fehrentz, J.A. and Castro, B., Synthesis, (1983) 676.
3. Mancuso, A.J., Huang, S.L. and Swern, D., J. Org. Chem., 43 (1978) 2480.
4. Roark, W.H., Tinney, F.J. and Nicolaides, E.D., In Marshall, G.R. (Ed.) Peptides: Chemistry and Biology (Proceedings of the 10th American Peptide Symposium), ESCOM Leiden, 1988, p. 134.

A novel non-hydrolyzable isostere of the peptide transition state and a new synthesis of the hydroxy-ethylene isostere

D. Michael Jones[a], Brenda J. Leckie[b], Lennart Svensson[c] and Michael Szelke[a]

[a]*Ferring Research Institute, Southampton University Research Centre, Chilworth, Southampton SO1 7NP, U.K.*
[b]*MRC Blood Pressure Unit, Western Infirmary, Glasgow G11 6NT, U.K.*
[c]*AB Hässle, S-43183 Mölndal, Sweden*

Introduction

In recent years, much attention has been focused on the synthesis of renin inhibitors as potential antihypertensives [1]. A major strategy in their design has been to take a partial sequence of angiotensinogen and replace the scissile bond in the P_1-P'_1 dipeptide with structures (e.g. the hydroxyethylene isostere **2**) that mimic the transition state **1**, thereby giving tighter binding to the enzyme (Fig. 1).

Fig. 1.   Structure of isosteric replacements.

Results and Discussion

The active site of renin carries a net negative charge on two aspartic acids. We reasoned that replacing the hydroxy group of **2** by an amino function should lead to tighter interaction with renin by virtue of electrostatic attraction (e.g., see our work on amino statine [7]). We now report the synthesis of the aminoethylene isostere **3** (Scheme 1). Reductive amination of the keto-isostere **4** [5], followed by protection and separation of the diastereomers gave the required isostere **5** (proof of stereochemistry will be provided in a separate publication).

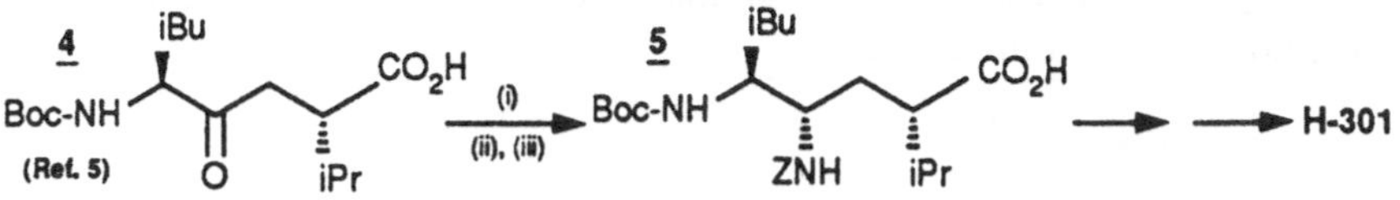

Reaction conditions: (i) NH$_4$OAc, MeOH, NaCNBH$_3$; (ii) ZONSu, CH$_2$Cl$_2$, Et$_3$N; (iii) separate diastereomers, Partisil(III)ODS, 40% MeCN-H$_2$O-0.1% TFA.

Scheme 1.

Table 1 *Structures and biopotencies of analogs of angiotensinogen fragments*

Compound No.	Structure (numbering as in human angiotensinogen) 7 8 9 10 11 12 13	IC_{50}(nM) vs. Renin⁺	Pepsin⁺	Ace⁺	Potency ratio R/P	R/ACE
H-269	Boc-Phe-His-Leu$\underline{OH^*}$Val-Ile-His-OH	2.4	42 700	15 000	1.8×10^4	6.3×10^3
H-301	Boc-Phe-His-Leu$\underline{A^*}$Val-Ile-His-OH	6.5	$>10^7$	10^6	$>1.5\times10^6$	1.5×10^5

*OH=-CH(OH)-CH_2- in place of -CONH-; A=-CH(NH_2)-CH_2- in place of -CONH-; + = human enzyme.

Incorporation of **3** into the human angiotensinogen (8–13) hexapeptide gave an inhibitor of human renin H-301 with potency comparable to that of the corresponding hydroxyethylene analog H-269 (Table 1). However, H-301 shows a hundred times greater selectivity vis-à-vis pepsin and ACE than H-269. This greater selectivity may be exploited to advantage in the design of novel inhibitors intended for clinical use.

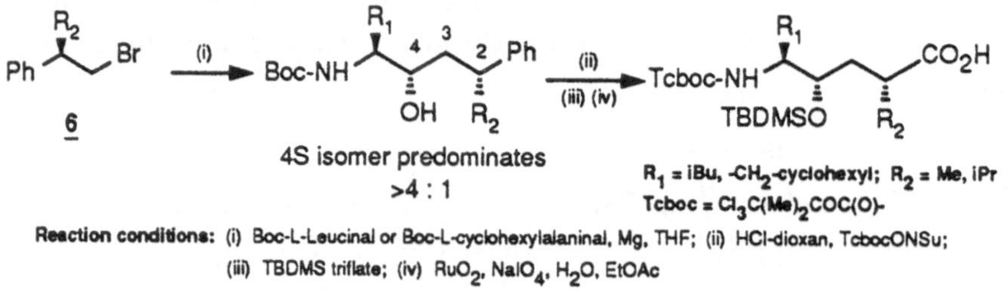
R2
Ph
Br
6
(i)
R1
Boc-NH
4
3
2
Ph
OH
R2
4S isomer predominates
>4 : 1
(ii)
(iii) (iv)
Tcboc-NH
TBDMSO
R2
CO2H
R1 = iBu, -CH2-cyclohexyl; R2 = Me, iPr
Tcboc = Cl3C(Me)2COC(O)-
Reaction conditions: (i) Boc-L-Leucinal or Boc-L-cyclohexylalaninal, Mg, THF; (ii) HCl-dioxan, TcbocONSu;
(iii) TBDMS triflate; (iv) RuO2, NaIO4, H2O, EtOAc

Scheme 2.

We also report an improved synthesis of the hydroxyethylene isostere which we introduced in 1981 [2]. Our original synthesis [3] was long and lacked adequate steric control. Other, more recent syntheses [4] have serious disadvantages, among them lack of versatility and incompatibility of the protecting groups used for subsequent incorporation into peptides. Our new synthesis (Scheme 2) begins with chiral bromides **6** (available by easy resolution from substituted phenylpropionic acids [6]). After Grignard reaction with protected α-amino aldehydes, followed by protection and separation of the major diastereomer, the phenyl substituent is degraded to a carboxyl group with RuO_4. Overall yields are high (30–40% from the protected aldehydes) and the route is amenable to scale-up. Protecting groups were chosen to be suitable for easy incorporation of the isosteres into peptidic enzyme inhibitors.

Acknowledgements

We acknowledge financial support by Ferring AB, Ferring Pharmaceuticals Ltd., AB Hässle and the Medical Research Council.

References

1. Greenlee, W.J., Pharm. Res., 4 (1987) 364.
2. Szelke, M., Jones, D.M. and Hallett, A., European Patent 45665 (1981).
3. Szelke, M., Jones, D.M., Atrash, B., Hallett, A., Leckie, B.J., In Hruby, V.J. and Rich, D.H. (Eds.) Peptides: Structure and Function, Pierce Chemical Co., Rockford, IL, 1983, p. 579.
4. Chakravarty, P.K., de Laszlo, S.E., Sarnella, C.S., Sringer, J.P. and Schuda, P.F., Tetrahedron Lett., 30 (1989) 415 (and refs. contained therein).
5. Szelke, M., In Kostka, V. (Ed.) Aspartic Proteinases and Their Inhibitors, De Gruyter, Berlin, New York, 1985, p. 421.
6. Aaron, C., Dull, D., Schmiegel, J.L., Jaeger, D., Ohashi, Y. and Mosher, H.S., J. Org. Chem., 32 (1967) 2797.
7. Jones, D.M., Sueiras-Diaz, J., Szelke, M., Leckie, B. and Beattie, S., In Deber, C.M., Hruby, V.J. and Kopple, K.D. (Eds.) Peptides: Structure and Function, Pierce Chemical Co., Rockford, IL, 1985, p. 759.

Electron density deformation and electrostatic potential in peptides

Mohamed Souhassou[a], Claude Lecomte[a], Virginie Pichon-Pesme[a],
Nour-Eddine Ghermani[a], André Aubry[a], Roland Wiest[b], Marie M. Rohmer[b],
Marc Benard[b] and Michel Marraud[c]

[a]UA-CNRS-809, University of Nancy I, BP 239, F-54506 Vandoeuvre Cedex, France
[b]Laboratoire de Chimie Quantique, Strasbourg, France
[c]UA-CNRS-494, ENSIC-INPL, BP 451, F-54001 Nancy Cedex, France

Introduction

Chemical modifications of the peptide backbone can induce conformational and electronic perturbations. In order to estimate the influence of α,β-dehydrogenation on the electronic distribution in peptides, we have studied Ac-Trp-NHMe (Trp) and Ac-Δ-Phe-NHMe (Δ-Phe) by low temperature (100 K) X-ray diffraction. Several multipole fits to the electron density were applied in order to get reliable structure factor phases [1–3], and thermal motion was calculated in terms of rigid group libration [4]. On the basis of the experimental electronic distribution, calculation of the electrostatic potential is in progress, giving some interesting preliminary results. Ab initio calculations have been performed using HF-SCF molecular wave functions [5], and theoretical results have been compared to the experimental ones.

Results and Discussion

Theoretical and experimental electron density maps are generally in good agreement (Fig. 1). No difference was found in theoretical maps between the two peptide groups for both Trp and ΔPhe. The same is true for Trp experimental maps. Some small but significant differences appear between the two Δ-Phe peptide groups for the experimental maps drawn from [3]. The electron density and the electrostatic potential increase for the C-terminal carbonyl predict a greater nucleophilicity due to electronic conjugation with the $C^\alpha = C^\beta$ double bond. We also observe a shift of the density deformation towards the N-atom for the N-terminal amide bond. This effect could contribute to explain the increasing resistance against enzymatic biodegradation for some α,β-dehydropeptides [6].

Analysis of thermal motion in terms of rigid group libration [4] is explicit in Table 1, in which the magnitude and axis of the libration mode together with the rigid groups involved are specified.

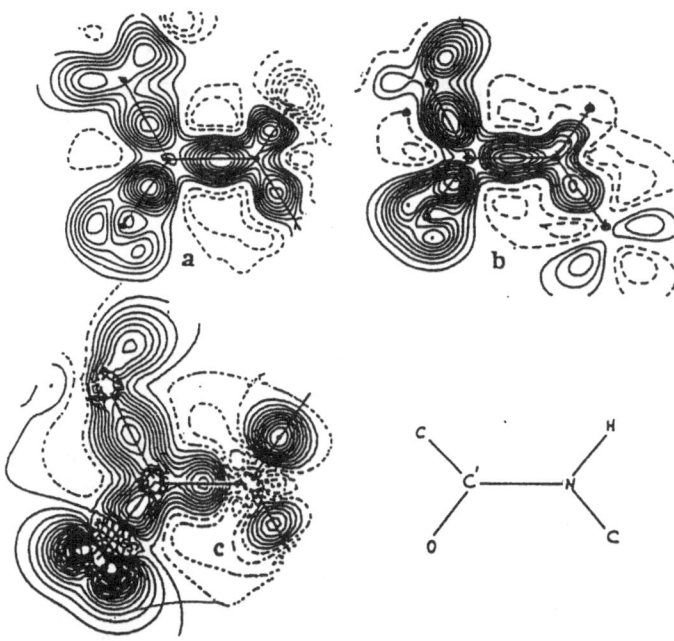

Fig. 1. Dynamic model deformation density maps from [3] for Δ-Phe: (a) is the N-terminal peptide link, (b) the C-terminal and (c) is the theoretical map for the C-terminal link.

Table 1 *Thermal motion analysis in terms of librational rigid groups*

Rigid groups	Libration axis	ΔPhe	Trp
N-terminal amide group	$N - C^\alpha$	4.4°	6.0°
C-terminal amide group	$C^\alpha - C'$	3.5°	4.7°
Phenyl or indole ring	$C^\alpha - C^\beta$	1.7°	1.7°
Phenyl ring	$C^\alpha - C^\beta - C^\gamma$ through C^β	1.5°	

Acknowledgements

This work was supported by C.E.C. (Grant ST2J-0184).

References

1. Stewart, R.F., Acta Crystallogr., A32 (1976) 565.
2. Hirshfeld, F.L., Isr. J. Chem., 16 (1977) 226.
3. Hansen, N.K. and Coppens, P., Acta Crystallogr., A34 (1978) 909.
4. Trueblood, K.N., THMA11 Program, Department of Chemistry and Biochemistry, University of California, Los Angeles, 1987.
5. Huzinaga, S., Approximate Atomic Functions, Technical Report, University of Alberta, 1971.
6. English, M.T. and Stammer, C.H., Biochem. Biophys. Res. Commun., 83 (1978) 1464.

975

Subtle amide bond surrogates: The effect of backbone thioamides on the physical properties, conformation, and biological activities of peptides

Leszek Lankiewicz, Douglas B. Sherman and Arno F. Spatola

Department of Chemistry, University of Louisville, Louisville, KY 40292, U.S.A.

Introduction

We and others have been interested in the effects of the thioamide substitution within the peptide backbone. Thioamides are useful probes for studying the role of the amide carbonyl because of the decreased electronegativity of sulfur and its larger covalent radius compared to oxygen. Following the synthesis of the first examples of cyclic thiopeptides [1], we have extended our efforts to the area of new linear peptide analogs and enzyme inhibitors. We now report the synthesis and biological activities of thyrotropin releasing hormone (TRH) analogs (Fig. 1) that contain a thionated pGlu (or Top) [2] residue, and we compare these results with those reported by Kruszynski et al. [3] for pGlu-His-Proψ[CSNH]H.

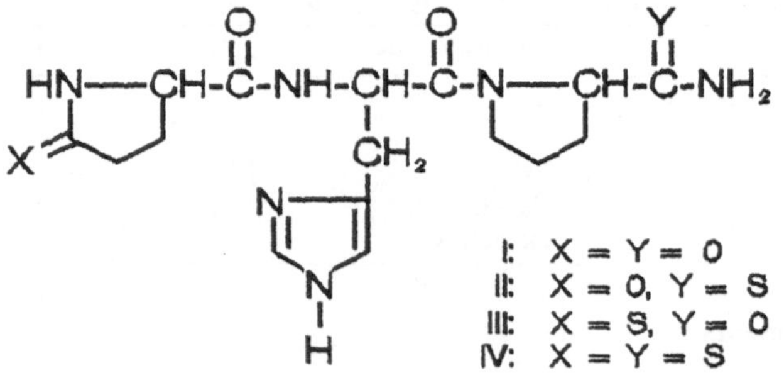

I: X = Y = O
II: X = O, Y = S
III: X = S, Y = O
IV: X = Y = S

Fig. 1. Structures of TRH and thionated analogs.

Results and Discussion

The thionated analogs shown in Fig. 1 were prepared by solution-phase methodologies. Top precursors were prepared by two different routes. In the first, pGlu-O^{t}Bu was thionated using Lawesson's Reagent (LR) in THF at 25°C for 60 min (94% yield). Following workup, the ester was hydrolyzed by HCl/dioxane (89% yield), and Top-OH was coupled to either His(Dnp)-Pro-NH$_2$ or His(Dnp)-Proψ[CSNH]H with EDAC/HOBT in 70–80% yield. In the second

976

route, pGlu-OTcp was prepared and thionated under identical conditions as above (74% yield). Top-OTcp was then coupled to the dipeptides listed above in 85–87% yields. Dnp was removed using mercaptoethanol, and the final peptides were purified by gel chromatography and RPHPLC. Each compound displayed the correct molecular ion peak on FABMS, gave correct AAA, and displayed consistent NMR spectral data. Overall, the yields were ≈40% and 41% by the free acid and active ester routes, respectively.

The physical properties shown in Table 1 confirm the expected lipophilic nature and characteristic chemical shift values for thioamide analogs. Although preliminary, the biological activity trends suggest the non-essentiality of the two carbonyl oxygens [3]. Conformational studies in progress should allow for further analysis of these subtle differences.

Table 1   *Properties of thionated TRH analogs*

Compound	Retention time (min)[a]	Thiocarbonyl ^{13}C (ppm)[b]	Preliminary biol. activity[c]
TRH	5.17	–	100%
[Pro³ψ[CSNH]H]-TRH[d]	9.99	206.25	≈100%
[Top¹]-TRH	9.62	204.99	>100%
[Top¹,Pro³ψ[CSNH]H]-TRH	14.59	205.02, 206.42	<100%

[a]5–35% CH_3CN/0.05% TFA, Vydac 4.6×250 mm ODS, flow 1.0 ml/min.
[b]In DMSO-d_6 at 30°C.
[c]TSH–releasing activity in vitro.
[d]Prepared according to the procedure in Ref. 3.

Acknowledgements

We thank Dr. C.Y. Bowers (Tulane) for the TRH bioassay results and Abbott Laboratories for the sample of TRH. This work was supported by NIH GM-33376 and AR-39573. We thank Applied Biosystems Inc. for a Travel Award.

References

1.   Sherman, D.B. and Spatola, A.F., J. Am. Chem. Soc., (1990) in press.
2.   Andersen, T.P., Rasmussen, P.B., Thomsen, I., Lawesson, S.-O., Jorgensen, P. and Lindhardt, P., Liebigs Ann. Chem., (1986) 269.
3.   Kruszynski, M., Kupryszewski, G., Ragnarsson, U., Alexandrova, M., Strbak, V., Tonon, M.C. and Vaudry, H., Experientia, 41 (1986) 1576.

Toward development of peptidomimetics: Diketopiperazine templates for the Trp-Met segment of CCK-4

Kazumi Shiosaki, Richard Craig, Chun Wel Lin, Ronald W. Barrett, Tom Miller, David Witte, Caroline A. W. Wolfram and Alex M. Nadzan

Abbott Laboratories, Neuroscience Research, Pharmaceutical Discovery Division, Abbott Park, IL 60064, U.S.A.

Introduction

CCK elicits a number of actions in mammalian peripheral tissues that are mediated via the type-A (e.g., pancreas) receptor. In the CNS, the majority of CCK-binding sites is of the type-B (e.g., cortex) class, which can be differentiated from type-A by their relative affinities for CCK fragments [1]. CCK has been hypothesized to modulate dopaminergic mechanisms based on their co-localization in the CNS. Attenuation of striatal dopamine release by CCK at the type-B receptor suggests that type-B selective compounds may ameliorate hyperactive dopaminergic states believed to underlie symptoms of schizophrenia [2]. The development of such a therapeutic agent will require its penetration into the CNS, but it is unlikely that peripherally administered CCK-4, a selective type-B ligand, would achieve significant levels in the brain. We have initiated efforts to design peptidomimetics of CCK-4 to improve the likelihood of CNS permeability. Restrained analogs of CCK-4 containing a diketopiperazine (DKP) nucleus were prepared to provide rigid templates for the eventual design of peptidomimetics.

Results and Discussion

The compounds were prepared by alkylating an amino acid benzyl ester **1** with methyl bromoacetate followed by coupling the *N*-alkylated product **2** with a Z-amino acid. Warming the dipeptide **3** in the presence of palladium and hydrogen gas effected the deprotection and cyclization to form directly the DKP **4**, which was coupled with the C-terminal dipeptide fragment to produce the final compound. The compounds were evaluated by radioligand binding studies in the guinea pig pancreas and cortex [3] and calcium mobilization studies in small cell lung cancer lines [4].

DKP analogs **5a** and **6a** lacking an R_1 group interacted poorly with type-A and type-B CCK receptors (Table 1). Introducing an alkyl appendage to the diastereomer (**5b**) containing L-Trp produced no improvement. However, using D-Trp to invert the optical center on the DKP nucleus yielded a potent and selective type-B ligand **6b**. Extending the aliphatic side chain by an additional

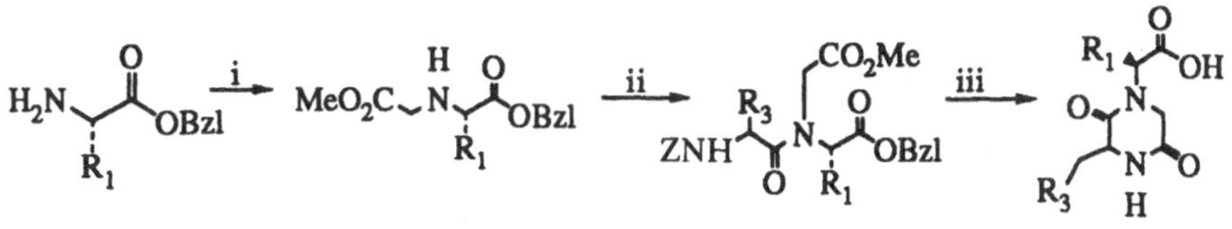

i: Methyl bromoacette, K₂CO₃, CH₃CN; ii: Z-amino acid, BOP, NMM;
iii: H₂, 10% Pd-C, MeOH, 50°C

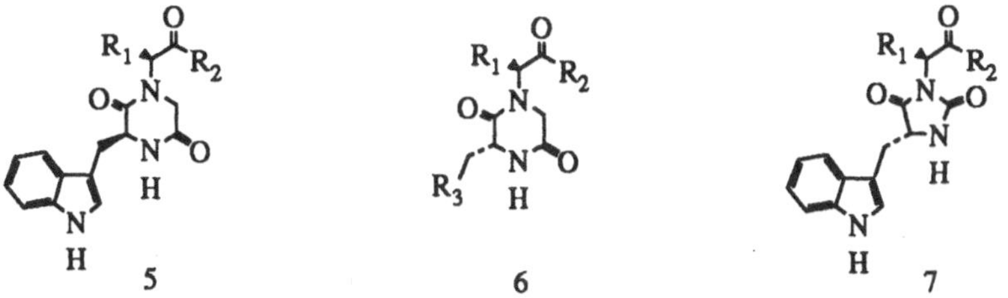

Fig. 1. Structural formulas of 1–7.

methyl group improved potency (**6c, 6d**). A critical placement of the aromatic moiety in space for favorable receptor recognition was suggested by the difference in potencies exhibited by the 1- versus 2-naphthyl derivatives (**6e, 6f**).

Additional restraints were incorporated into the Asp-Phe-NH₂ segment of the DKP-containing tetrapeptide. The dually constrained analogs (**6g, 6h, 6i**) containing various conformationally restrained Phe replacements were less potent

Table 1   *Binding affinities of DKP-based analogs*

	R_1	R_2	R_3	IC_{50} nM Type-B (cortex)	IC_{50} nM Type-A (pancreas)
5a	H	Asp-Phe-NH₂	---	> 10 000	> 10 000
5b		n-Propyl Asp-Phe-NH₂	---	> 10 000	> 10 000
6a	H	Asp-Phe-NH₂	3-Indolyl	> 10 000	7 200
6b	n-Propyl	Asp-Phe-NH₂	3-Indolyl	150	15 000
6c	n-Butyl	Asp-Phe-NH₂	3-Indolyl	36	13 800
6d	iso-Butyl	Asp-Phe-NH₂	3-Indolyl	38	14 500
6e	iso-Butyl	Asp-Phe-NH₂	1-Naphthyl	1 760	> 10 000
6f	iso-Butyl	Asp-Phe-NH₂	2-Naphthyl	167	> 10 000
6g	iso-Butyl	Asp-(N-Me)Phe-NH₂	3-Indolyl	204	14 000
6h	iso-Butyl	Asp-Tiq-NH₂[a]	3-Indolyl	> 10 000	> 10 000
6i	iso-Butyl	Asp-ΔPhe-NH₂[b]	3-Indolyl	565	5 000
7	iso-Butyl	Asp-Phe-NH₂	---	1 300	> 10 000

a Tiq = (S)-1,2,3,4-Tetrahydro-3-isoquinolinecarbonyl.
b ΔPhe = Dehydro-Phe.

979

than the parent compound. In addition, replacement of the DKP with a more conformationally restrained hydantoin nucleus produced analog **7** with significantly lower affinity for cortical receptors.

DKP analog **6d** was tested in an intracellular calcium mobilization assay of small cell lung cancer cell lines that express type-B CCK receptors. The DKP analog was shown to possess full intrinsic activity relative to CCK-8 and CCK-4 in stimulating calcium mobilization.

We are currently investigating the possible low-energy conformations adopted by the DKP-containing portion of our analogs. Identification of such conformations will provide a structural template from which the future design of peptidomimetics for the Trp-Met segment of CCK-4 will be possible.

References

1. Moran, T.H., Robinson, P.H., Goldrich, M.S. and McHugh, P.R., Brain Res., 362 (1986) 175.
2. Altar, C.A. and Boyar, W.C., Brain Res., 483 (1989) 321.
3. Lin, C.W. and Miller, T., J. Pharm. Exp. Ther., 232 (1985) 775.
4. Moody, T.W., Stanley, J. and Fiskum, G., J. Cell Biol., 107 (1989) 482A.

L-2-Thiolhistidine: A tool for the introduction of conformational constraints in peptides

Linda L. Maggiora, Clark W. Smith, Richard A. Hsi and Dennis E. Epps

The Upjohn Company, Kalamazoo, MI 49001, U.S.A.

Introduction

Conformational features, important for the interaction of peptides with their target molecule (receptor, enzyme, etc.), can often be probed through the use of conformationally constrained peptide analogs. The use of disulfide or thioether linkages between cysteine side chains is an often-used method to restrict conformational freedom [1]. A limitation of this method is that the side chain of the residue being replaced may be important to the biological activity of the peptide. To the repertoire of amino acids with derivatizable sulfur-containing side chains, we now offer the L-2-thiolhistidine residue, **1** (HisS). Because the pK of the imidazole nitrogen of HisS is greater than that of histidine [2], (pK 8.5 vs. pK 6.5), it is a possible surrogate, not only for a protonated histidine residue, but for a conformationally constrained lysine residue as well.

Results and Discussion

The protected amino acid, **2**, was prepared from commercially available **1** by alkylation of the sulfur in Na/NH_3 with α-bromo-p-xylene. The α-amino group was subsequently protected with the t-butyloxycarbonyl group. The Na/NH_3 reaction yielded exclusive derivatization of the sulfur, as determined by ^{13}C NMR. No alkylation of the α-amino nitrogen or imidazole nitrogen was detected. Incorporation of **2** into a peptide was accomplished using standard Boc SPPS with 'BOP reagent' [3] for coupling. Although treatment with HF to remove the peptide from the resin did not cleave the 4-methylbenzyl protecting group from the HisS residues, it was successfully removed by treatment with sodium in liquid ammonia. The re-alkylation of the HisS residues could then be accomplished in the same pot by adding the desired bromoalkane directly to the ammonia solution. If, on the other hand, it is desired to isolate the completely deprotected peptide, no precautions against oxidation are necessary because, by analogy with 2-thiolimidazoles, HisS exists predominately in the thioketo tautomeric state [4]. In fact, HisS does not give a positive Ellman's test [5], and air oxidation or 1,2-diiodoethane treatment fail to yield disulfide-bonded products.

Table 1   *Substituted 2-thiohistidine renin inhibitors*

Cmpd	X	Y	K_d molar [a]
3	-H	H	1.5×10^{-8}
4	-S-CH$_3$	-S-CH$_3$	5.9×10^{-7}
5	-S-CH$_2$CH$_2$CH$_3$	-S-CH$_2$CH$_2$CH$_3$	1.3×10^{-7}
6	-S-(CH$_2$)$_6$-S-		9.2×10^{-7}

[a]K_d determined by displacement of a fluorescent inhibitor from human renin as detailed in Ref. 7.

To explore the generality of the alkylation reaction of HisS residues under high dilution, and to obtain a series of HisS modified inhibitors of the aspartic acid proteinase, renin, we synthesized several analogs of peptide 3 (See Table 1, 4–6). All of the alkylation reactions were carried out with 0.05 mmol of peptide in 150 ml of ammonia. For the dimethyl, dipropyl and cyclic hexa-methylene peptides, 100, 10 and 1.5 equivalents of methylbromide, propylbromide and 1,6-dibromohexane were used, respectively. Mono-propyl peptide was produced in about equal quantity to peptide 5, but good yields of 4 and 6 (approximately 80 and 70% by HPLC, respectively) were obtained. The yield of 6, using only 1.5 equivalents of 1,6-dibromohexane, is evidence for the ease of intramolecular thioether bond formation in this system.

982

The dissociation constants from renin of the inhibitor peptides are listed in Table 1. The number of methylenes chosen for the bridge in **6** was determined for an extended conformation of **3**, with the His residues oriented on the same side of the peptide. This conformation was derived with computer-assisted molecular modeling techniques using dynamics simulations and energy minimization calculations, as described for docking renin inhibitory peptides to a model of human renin [6]. The magnitude of the decreased renin binding interactions shown by the 3–7-fold higher K_d values for the cyclic vs. the linear inhibitors (**6** vs. **4** or **5**) indicates that the hexamethylene bridge of the peptide imposes conformations that are allowed, but not preferred, for binding to renin, as compared to the linear peptides. In itself, the formal replacement of His by alkyl-HisS caused at least a 10-fold increase in the K_d.

We have demonstrated that L-2-thiolhistidine residues can be used to form constrained peptide analogs via thioether bridges: No precautions are necessary to prevent oxidative polymerization of HisS-containing peptides. Conversely, ring closure cannot be achieved through symmetrical or asymmetrical disulfide bond formation. The specificity for alkylation of sulfur vs. nitrogen that is achieved in ammonia as a solvent allows this strategy of cyclization to be applied to peptides containing amino acid residues with free amino groups.

References

1a. Schiller, P.W., Eggiman, B., DiMaio, J., Lemieux, C. and Nguyen, T.M.-D., Biochem. Biophys. Res. Commun., 101 (1981) 337.

 b. Mosberg, H.I. and Omnaas, J.R., J. Am. Chem. Soc., 107 (1985) 2986.

2. Sober, H.H. (Ed.) Handbook of Biochemistry, The Chemical Rubber Co., 1970, p. J-198.

3. Fournier, A., Wang, C.-T. and Felix, A.M., Int. J. Pept. Prot. Res., 31 (1988) 86.

4a. Bojarska-Olejnik, E., Stefaniak, L., Witanowski, M., Hamdi, B.T. and Webb, G.A., Mag. Res. Chem., 23 (1985) 166.

 b. Faure, R., Vincent, E.-J., Assef, G., Kister, J. and Metzger, J., J. Org. Mag. Res., 9 (1977) 688.

5. Stewart, J.M. and Young, J.D., Solid Phase Peptide Synthesis, Pierce Chemical Comp., Rockford, IL, 1984, p. 116.

6. Sawyer, T.K., Pals, D.T., Mao, B., Maggiora, L.L., Staples, D.J., deVaux, A.E., Schostarex, H.J., Kinner, J.H. and Smith, C.W., Tetrahedron, 44 (1987) 661.

7. Epps, D.E., Schostarez, H., Argoudelis, C., Poorman, R.A., Hinzmann, J. and Mandel, F., Anal. Biochem., 181 (1989) 172.

A new route to prepare conformationally restricted cyclic peptides as demonstrated by a potent NK-1 selective substance P analog

Gerardo Byk[a], Erez Gur[a], David Halle[b], Michael Chorev[c], Zvi Selinger[b] and Chaim Gilon[a,*]

[a]Department of Organic Chemistry, [b]Department of Biological Chemistry and [c]Department of Pharmaceutical Chemistry, The Hebrew University of Jerusalem, Jerusalem 91904, Israel

Introduction

Cyclization of linear peptides is generally used to confer metabolic stability and to restrict conformation and, thus, alter biological activity. Cyclization is usually achieved by end-to-end, side-chain-to-side-chain or end-to-side-chain connection. Recently, we have shown [1] that the tachykinin NK-1 receptor selective analogs, Ac-Arg-Septide {[Ac-Arg⁶, Pro⁹]SP⁶⁻¹¹} [2], have predominantly a type-I β-conformation in solution. Molecular models of the predominant conformation have indicated that further conformational restriction could be imposed on the peptide by cyclization. This cyclization in which Pro is replaced by N(γ-aminopropyl Gly) and the γ-amine is attached to the carboxyl end of succ-Arg locks the β-turn conformation. This cyclic SP analog was synthesized and characterized, and was found to be a highly selective and potent NK-1 agonist.

Results and Discussion

Peptide **1** was synthesized on a MBHA resin using Boc for N^α protection, Fmoc for side-chain protection of the N (amino propyl)-amino acid and Tos

Peptide 1

*To whom correspondence should be addressed.

for Arg. BOP was used for coupling and for cyclization on the resin after removal of the Fmoc protecting group. Succinylation was done by succinic anhydride and DMAP. The cyclic peptide was cleaved from the resin by HF and purified on semiprep RPHPLC. The pure peptide was characterized by NMR, FABMS and AAA. Evaluation of biological activity showed that peptide **1** is an NK-1 selective agonist with an IC_{50} of 1×10^{-8} M.

Conclusions

Cyclization of peptides via the backbone, as demonstrated in this article, opens new routes to impose conformational restriction on linear peptides.

References

1. Levian-Teitelbaum, D., Kolodny, N., Chorev, M., Selinger, Z. and Gilon, C., Biopolymers, 28 (1989) 51.
2. Papir-Kricheli, D., Frey, J., Laufer, R., Gilon, C., Chorev, M., Selinger, Z. and Devor, M., Pain, 31 (1987) 263.

Design and synthesis of peptide ligand for exploration of the thiol group in opioid receptors

Hiroaki Kodama[a], Yasuyuki Shimohigashi[b], Kaori Soejima[a], Teruo Yasunaga[a] and Michio Kondo[a]

[a]Department of Chemistry, Faculty of Science and Engineering, Saga University, Saga 840, Japan
[b]Laboratory of Biochemistry, Department of Chemistry, Faculty of Science, Kyushu University, Fukuoka 812, Japan

Introduction

Larsen et al. [1] suggested that, in the opioid receptors, there are at least two different types of thiol groups sensitive to N-ethylmaleimide. The first is the β-thiol of the cysteine residue of α_i-subunit in the GTP-binding protein G_i [2]. A second one has been suggested at or near the ligand binding site in the receptor protein [3], but no direct evidence for its presence has yet been reported. In the present study, to locate a thiol group(s) in the receptor protein, we have designed and synthesized enkephalin analogs containing leucinthiol as an affinity ligand for opioid receptors.

Results and Discussion

Leucinthiol [4] contains the mercaptomethyl group ($-CH_2SH$) instead of the carboxyl group of leucine, and its thiol group in the enkephalin molecule was activated by S-thiomethylation or S-3-nitro-2-pyridinesulfenylation (Fig. 1). Both, Enk-SSCH$_3$ and Enk-SNpys, have an asymmetric disulfide bond that can react with free receptor thiol(s) by the thiol-disulfide exchange reaction.

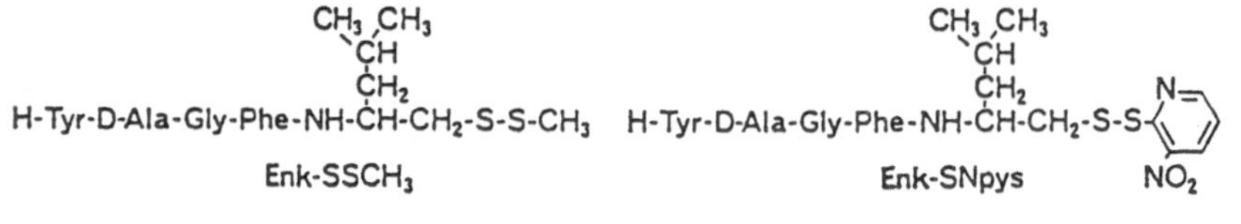

Fig. 1. Chemical structures of thiol-activated enkephalin analogs.

In the assay using the longitudinal muscle of GPI, in which the thiol of G_i is considered hidden, Enk-SSCH$_3$ or Enk-SNpys exhibited the activity profile shown in Fig. 2. Incubation (10 min) of 1 μM Enk-SSCH$_3$ or Enk-SNpys with GPI resulted in continuous stimulation of the μ-receptors even after many repeated washings. This sustained GPI activity was completely reversed with

986

the μ-antagonist naloxone (1 μM), and subsequent washings elicited again the full activity. This activity was completely eliminated by treatment with 1 mM DTT (30 min). These results suggest that S-activated enkephalin analogs interact with the μ-receptors and bind covalently to the receptor protein through the thiol-disulfide exchange reaction. It appears that such a receptor-thiol is present at the ligand binding site, a little apart from the naloxone binding site.

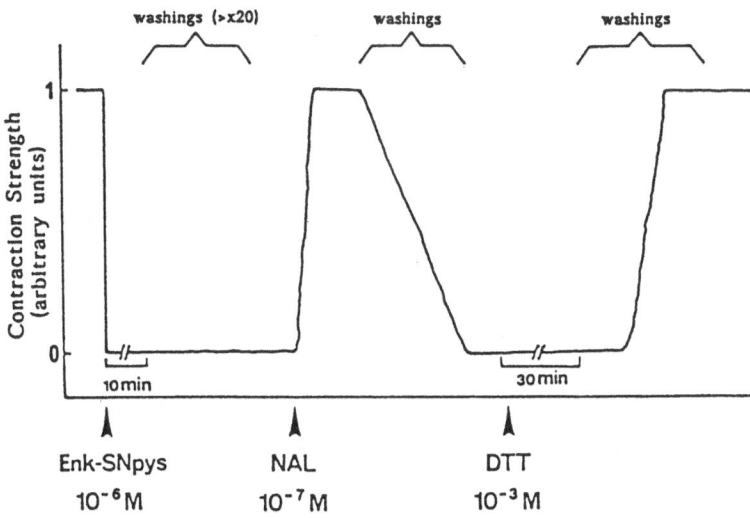

Fig. 2. Activity profiles of Enk-SNpys in the GPI assay.

In the MVD assay, the incubation with 1 μM Enk-SNpys retained considerable activity (51%) after washings. This activity varied exactly in the same manner as seen in GPI. When MVD was pre-treated with Enk-SNpys, DAGO (μ-selective ligand: H-Tyr-D-Ala-Gly-MePhe-Gly-ol) was 7–10 times less active than in normal MVD, whereas DADLE (δ-selective ligand: H-Tyr-D-Ala-Gly-Phe-D-Leu-OH) was equally active in both treated and untreated MVD. These results indicate that Enk-SNpys selectively binds to the μ-binding site in MVD.

References

1. Larsen, N.E., Mullikin-Kilpatrick, D. and Blume, A.J., Mol. Pharmacol., 20 (1981) 255.
2. Kurose, H., Katada, T., Amano, T. and Ui, M., J. Biol. Chem., 258 (1983) 4870.
3. Smith, J.R. and Simon, E.J., Proc. Natl. Acad. Sci. U.S.A., 77 (1980) 284.
4. Kondo, M., Uchida, H., Kodama, H., Kitajima, H. and Shimohigashi, Y., Chem. Lett., (1987) 997.

'Pseudo'-polyamino acids: New polypeptide mimetics

Qin-Xin Zhou, Gwendalyn C. Baumann and Joachim Kohn[a]

Department of Chemistry, Rutgers, The State University of New Jersey,
New Brunswick, NJ 08903, U.S.A.

Introduction

In analogy to the term 'pseudopeptide' [1], polyamino acids that contain non-amide backbone linkages may be defined as 'pseudo-polyamino acids'. Such polymers have been investigated as immunologically active adjuvants [2], and as implantable biomaterials for drug delivery and other medical applications [3,4].

Results and Discussion

Here we report on our studies of poly-L-serine ester (3), a structural isomer of conventional poly-L-serine (Fig. 1). Poly(Z-L-serine ester) (2) ($M_w = 50\,000$ Da) was synthesized by the ring-opening polymerization of N-Z-serine-β-lactone (1) [5], followed by removal of the Z group by hydrogenation.

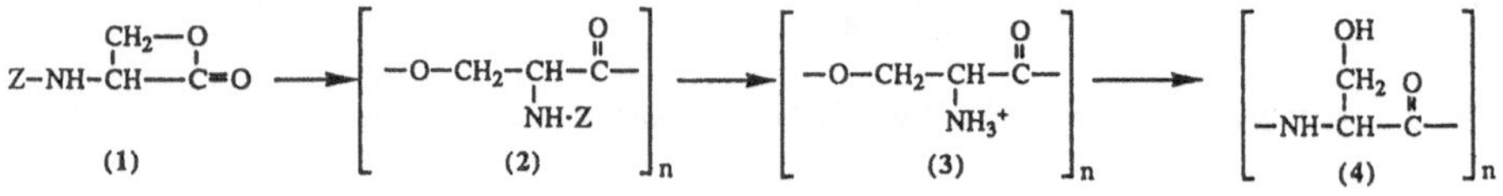

Fig. 1. Synthesis of poly-L-serine.

In order to ascertain the optical purity of (2), the polymer was hydrolyzed. No D-serine was detected in the hydrolysate by GC on a chiral column [6] (limit of detection: ~0.2%).

Spectroscopic analysis of aqueous solutions of (3) revealed the IR ester peak at 1759 cm^{-1} and a CD maximum at 208 nm (Fig. 2). The CD spectra of optically active polyesters, such as S(-)-poly(α-methyl-α- ethyl-β-propiolactone) [7], also exhibit a positive Cotton effect at about 210 nm. Since these polymers are believed to exist in a random coil conformation, we postulate, by analogy, that poly(L-serine ester) (3) also assumes a random coil conformation.

When the pH of aqueous solutions of (3) was adjusted to pH 8.5, the O → N shift of serine [8] resulted in the spontaneous rearrangement of (3) to poly-L-serine (4). The rearranged polymer was unambiguously identified as poly-L-serine, based on ^{13}C NMR, ^{1}H NMR, and FTIR spectral analysis, and by comparison with an authentic sample of poly-L-serine.

---

[a]To whom correspondence should be addressed.

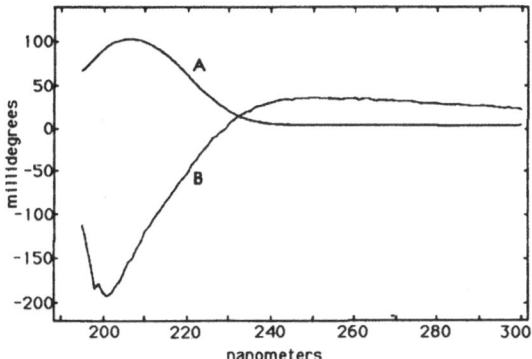

Fig. 2.   Curve A: CD Spectrum of unprotected poly(L-serine ester·HCl) (3) in water. Curve B: CD spectrum of the polymer after rearrangement to poly-L-serine (4). The spectra (average of 3 scans) were recorded on a model 60DS Aviv Spectropolarimeter (Aviv Associates, Lakewood, NJ), using a 1-mm cylindrical cell.

The CD spectrum of the rearranged polymer (Fig. 1) was deconvoluted into its component parts. The results were 67% β-pleated sheet, 3% β-turns, and 30% random coil. The predominantly β-pleated sheet structure of the rearranged polymer corresponds to observations by Fasman and Urry who had assigned a β-pleated sheet conformation to poly-L-serine [9,10].

While the chemical backbone rearrangement took place, changes in the CD spectrum of the solution were recorded at fixed wavelength (211 nm) (Fig. 3). A rapid change (half life approximately 10–15 min) from the positive Cotton effect of the ester to the negative Cotton effect of the amide was observed.

In all previous structural studies involving synthetic polyamino acids, per-

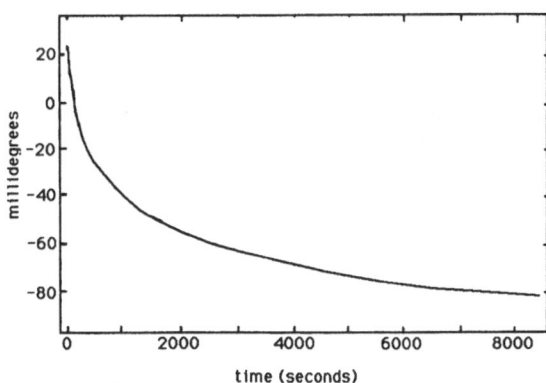

Fig. 3.   Change in ellipticity as a function of time during the backbone rearrangement of (3) to (4) at 211 nm. Sample concentrations were 3.81 mg/ml of poly(L-serine ester·HCl) in 4.75×10^{-2} M NaHCO$_3$.

989

turbations of the secondary structure were induced by physical changes. Our studies appear to be the first report of the spontaneous formation of secondary structure in a polyamino acid, induced by the chemical rearrangement of the polymer backbone. Furthermore, considering that high molecular weight poly-L-serine is not readily available via serine-*N*-carboxyanhydride, the formation of high-molecular-weight poly(L-serine ester) (**3**) followed by backbone rearrangement to (**4**) may be an attractive synthetic alternative.

Acknowledgements

The authors acknowledge the assistance of Dr. A. Felix (Hoffmann-La Roche, Nutley, NJ) who performed the GC analysis of the optical purity of poly(Z-L-serine ester), and the help of Prof. K. Breslauer, whose instrument was used to obtain the CD spectra. This work was supported by NIH grant GM39455 and in part by the Petroleum Research Fund, administered by the American Chemical Society. An ABI Travel Award to J.K. is gratefully acknowledged.

References

1. Spatola, A.F., In Weinstein, B. (Ed.) Chemistry and Biochemistry of Amino Acids, Peptides, and Proteins, Marcel Dekker, New York, 1983, p. 267.
2. Kohn, J., Niemi, S.M., Albert, E.C., Murphy, J.C., Langer, R. and Fox, J.G., J. Immunol. Methods, 95 (1986) 31.
3. Kohn, J. and Langer, R., J. Am. Chem. Soc., 109 (1987) 817.
4. Yu, H., Pseudopoly(Amino Acids): A Study of the Synthesis and Characterization of Polyesters Made from α-L Amino Acids, Ph. D. Thesis, MIT, 1988.
5. Zhou, Q. X. and Kohn, J., manuscript submitted; for preparation of N-Z-serine lactone, see: Arnold, L. D., Kalantar, T. H. and Vederas J. C., J. Am. Chem. Soc., 107 (1985) 7105.
6. Frank, H., Nicholson, G. J. and Bayer, E., J. Chromatol., 167 (1978) 187.
7. Grenier, D., Prud'Homme, R. E. and Spassky, N., J. Polymer Sci.(Polymer Chemistry Edition), 19 (1981) 1781.
8. Rabinowitz, J., Acta Chem. Scand., 13 (1959) 1463.
9. Quadrifoglio, F. and Urry, D. W., J. Am. Chem. Soc., 90 (1968) 2760.
10. Tooney, N. M. and Fasman, G. D., J. Mol. Biol., 36 (1968) 355.

Didemnins synthesis: The crucial cyclization step

Joël Poncet, Patrick Jouin and Bertrand Castro

Centre CNRS-INSERM, rue de la Cardonille, F-34094 Montpellier Cedex 2, France

Introduction

A symbiotic association between a Didemnid Ascidian and a unicellular alga [1] is responsible for the production of a highly potent antineoplastic group of natural cyclodepsipeptides [2]. Didemnin B, **1a**, the major component of the tunicate *Trididemnum cyanophorum* [3], is active against a variety of DNA and RNA viruses; it is also effective against B16 melanoma, P 388 and L 1210 leukemias [2]. The determination of the mechanism of action of didemnins awaits further investigation.

Results and Discussion

As part of an ongoing program devoted to the study of this last problem, we have developed an efficient synthesis of natural nordidemnin B, **1b**, [4] which exhibits analogous biological properties to didemnin B (S. Cros, personal communication). This synthetic strategy conveniently allows the production of analogs. The key step is the cyclization between the carboxylic function of the threonyl residue and the amino group of the γ-amino acid. When this cyclization was first performed in the heterogeneous conditions suggested by Shuman et al. [5], using DPPA as coupling reagent, the expected monomeric cyclic compound was isolated with an inadequate yield (Table 1, entry 7) that was dramatically improved by replacing DPPA with BOP reagent (54%) [4].

Table 1   *Cyclization of compound 1c*

Entry	Coupling conditions	Temp. (°C)	Time (h)	Conc. (M)	Yield (%)
1	KHCO$_3$/BOP	20	2	10^{-1}	44
2	KHCO$_3$/BOP	20	5	10^{-2}	56
3	KHCO$_3$/BOP	20	24	10^{-3}	41
4	NaHCO$_3$/BOP	20	5	10^{-2}	53
5	CS$_2$CO$_3$/BOP	20	2	10^{-2}	46
6	NMM/BOP	20	2	10^{-2}	42
7	KHCO$_3$/DPPA	0	72	10^{-2}	14
8	HOBt/DCC	0	72	10^{-2}	45

From these results, we further explored this process with regard to enhancing the yield of coupling. This study was carried out on the analog **1c**. All reactions were conducted in DMF. As shown in Table 1, BOP proved to be superior to either DPPA or DCC in terms of yield and reaction rate (entries 2, 7, 8). Moreover, BOP was more efficient in heterogeneous than in homogeneous conditions (entries 2, 6), and a possible template effect was evaluated using cesium cation. In this condition, the rate of the reaction was increased but with a lower yield (entry 5). Finally, for preparative purposes, the most efficacious concentration was determined to be 10 mM (entries 1, 2, 3).

Conclusion

Following this strategy, an average cyclization yield of 55 to 60% is routinely reached using BOP. Improvements are expected from new coupling reagents developed in our laboratory.

References

1.   Lafargue, F. and Duclaux, G., Ann. Inst. Oceanogr., Paris, 55 (1979) 163.
2.   Rinehart, Jr., K.L., In Marshall, G.R. (Ed.) Peptides: Chemistry and Biology, ESCOM, Leiden, 1988, p. 626 (and references cited therein).
3.   Banaigs, B., Jeanty, G., Francisco, C., Test, J., Jouin, P., Poncet, J., Heitz, A., Cavé, A., Promé, J.C., Walh, M. and Lafargue, F., Tetrahedron, 45 (1989) 181.
4.   Jouin, P., Poncet, J., Dufour, M.-N., Pantaloni, A. and Castro, B., J. Org. Chem., 54 (1989) 617.
5.   Shuman, R.T., Smithwick, E.L., Smiley, D.L., Brooke, G.S. and Gesellchen, P.D., In Hruby, V.J. and Rich, D.H. (Eds.) Peptides: Structure and Function, Pierce Chemical Co., Rockford, IL, 1983, p. 143.

Synthesis of cyclic (Asu^{1,7})-eel calcitonin by segment synthesis condensation on solid support

Pat Ho*, Dario Slavazza, Ding Chang, Kamaljit Bassi and Kang Chang

Peninsula Laboratories Inc., 611 Taylor Way, Belmont, CA 94002, U.S.A.

Introduction

Eel calcitonin, **1**, was isolated in 1974 by Otani et al. [1]. Calcitonins, which act to lower Ca^{+2} concentrations in the blood, have been used clinically in the treatment of osteoporosis. Eel calcitonin has one of the highest activities among natural calcitonins [1]. The [Asu^{1,7}] analog, **2**, in which L-α-aminosuberic acid is substituted for cysteine in sequence positions one and seven of eel calcitonin, has similar activities, but **2** is a more stable compound because it is more resistant to hydrolysis and degradation [2]. In 1976, Morikawa et al. reported the synthesis of [Asu^{1,7}]-eel calcitonin, **2**, by a conventional solution method [2]. However, solution procedures for peptide synthesis are tedious, extremely time consuming and expensive.

R$_1$————————R$_3$

R$_2$-Ser-Asn-Leu-Ser-Thr-R$_4$-Val-Leu-Gly-Lys-Leu-Ser-Gln-Glu-Leu-His-Lys-Leu-Gln-Thr-Tyr-Pro-Arg-Thr-Asp-Val-Gly-Ala-Gly-Thr-Pro-NH$_2$

1 Eel Calcitonin: R$_1$=R$_3$=Cys, R$_2$=R$_4$=S   **2** (Asu^{1,7})-Eel Calcitonin: R$_2$=NH, R$_4$=Asu, R$_3$=(CH$_2$)5, R$_4$=CO

Recently, side chain-to-side chain cyclization and side chain-to-amino terminal cyclization have been carried out on a solid support using either an active ester procedure or DCC and HOBT method [3]. The solid support cyclization generally required several days as well as repeated addition of fresh activating reagents and gave a rather low yield. In connection with our interest in the synthesis of cyclic peptides, we also extensively studied the solid-phase cyclization reaction and simple and efficient synthesis of [Asu^{1,7}]-eel calcitonin achieved under our conditions. Our synthetic approach is unique in that the synthesis, cyclization and segment condensation of our target peptide were all done on a solid support.

Results and Discussion

We have used two synthetic strategies. In the first approach, two fragments, **6** and **8**, were prepared separately (Scheme 1). **3** was synthesized on the resin by a combination of Fmoc protection of the amino function with benzyl protection

---

*To whom correspondence should be addressed.

of the side chain functions. This was followed by TFA deprotection to yield
4. The cyclization was performed on the resin using two equivalents of BOP,
HOBT, and DIEA in dry DMF for 48 h to give **5** [4]. This was then treated
with HF to yield the cyclic peptide **6**, which was purified by HPLC and confirmed
by AAA and FABMS. During this time fragment **8** was synthesized by standard
SPPS techniques. A small amount was cleaved by HF to be checked for purity
on HPLC and for composition by AAA. The final coupling was a segment
condensation using **8** still on the resin and one equivalent of **6**, with two equivalents
of BOP, HOBT, and DIEA in dry DMF for 48 h. The peptide was then cleaved
from the resin by HF and purified to give pure [Asu^{1,7}]-eel calcitonin, identical
in all respects to the authentic sample.

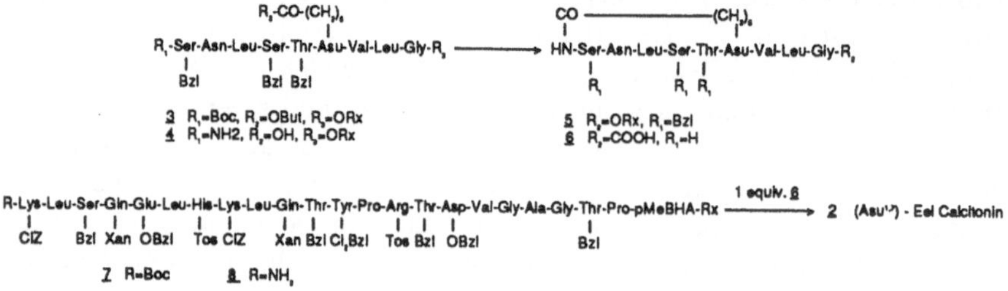

Scheme 1.

In the second approach, the entire sequence, H$_2$N-[2-Asu^{1,7}-31]-pMeBHA-Rx,
was synthesized on the resin along with the cyclization by a similar technique
to that used for **3**. The peptide was cyclized using six equivalents of BOP, HOBT,
and DIEA in dry DMF for 24 h. Finally, HF cleavage and purification with
preparative HPLC yielded better than 95% pure [Asu^{1,7}]-eel calcitonin. Comparison of the cyclized and non-cyclized crude products by HPLC showed that
the cyclization reaction proceeded to better than 90% completion in both
approaches.

Conclusion

Our approach has been successful for the preparation of eel calcitonin analogs
and is simple and straightforward. Using this methodology, a complete peptide
can be synthesized within a few days on a multimolar scale. The cyclization
does not require a large volume of solvent, and in the case of the segment
condensation approach, only a one-to-one equivalent of fragment **6** to fragment
8 is used. The second approach results in a complete peptide with good purity
for crude [Asu^{1,7}]-eel calcitonin. We believe that this method can be applied
to the synthesis of many other carbocyclic peptides.

References

1. Otani, M., Yamaguchi, H., Meguro, T., Kitasawa, S., Watanabe, S. and Orimo, H., J. Biochem., 79 (1976) 345.
2. Morikawa, T., Munekata, E., Sakakibara, S., Noda, T. and Otani, M., Experientia, 32 (1976) 1104.
3a. Bunker, A. and Schwartz, I.L., J. Prot. Chem., 4 (1985) 163.
  b. Lebl, M. and Hruby, V.J., Tetrahedron Lett., 104 (1984) 2067.
  c. Schiller, P.W., Nguyen, T.M. and Miller, J., Int. J. Pept. Prot., 25 (1985) 171.
4. Felix, Arthur, M., Wang, Ching-Tso, Heimer, Edgar P. and Fournier, A., Int. J. Pept. Res., 31 (1988) 231.

Cyclization of disulfide-containing peptides in solid-phase synthesis

Carlos García-Echeverría[a], M. Antonia Molins[a], Robert P. Hammer[b], Fernando Albericio[a,b], Miquel Pons[a], George Barany[b,]* and Ernest Giralt[a]

[a]*Department of Organic Chemistry of University of Barcelona, 08028-Barcelona, Spain*
[b]*Department of Organic Chemistry, University of Minnesota, Minneapolis, MN 55455, U.S.A.*

Introduction

Preparation of disulfide-containing peptides hinges on reliable chemistry to form disulfide bonds. Usually, a linear sequence is assembled by solid-phase or solution methods, and then protecting groups (as well as the anchoring linkage in the solid-phase case) are cleaved. There follows non-specific oxidation in dilute solution in order to minimize unwanted dimerization and oligomerization. The alternative of carrying out deprotection and oxidation of the cysteines, while the peptide chain remains anchored to the support, has been little studied [1, 2]. Such an approach takes advantage of the *pseudo-dilution* phenomenon which is expected to favor intramolecular processes [3]. In the present work, a tetrapeptide sequence known to prefer a β-turn conformation [2], and the nonapeptide amide oxytocin, are used as models to probe factors that influence disulfide bridge formation.

Results and Discussion

Linear sequences were assembled smoothly using Boc or Fmoc for N^α-amino protection. The β-thiols of cysteine or β,β-dimethylcysteine (penicillamine) have been protected using S-Acm and S-fluorenylmethyl (Fm). Anchoring to polystyrene supports was provided either with an MBHA linkage (HF-labile) in Boc-based syntheses, or via our new tris(alkoxy) or *o*-nitro benzylamide handles, which cleave under milder conditions and are used in conjunction with Fmoc [4]. Deprotection and/or oxidation of the peptide-resins, followed by cleavage, released crude peptide mixtures into solution that were assayed by HPLC. The amount of desired disulfide-containing peptide was calculated by comparison of HPLC areas with those of standard solutions of *pure* peptide of known concentration.

1. Type-II β-turn sequence: Ac-L-Cys-L-Pro-D-Val-L-Cys-NH$_2$
The best yields, 94% of the correct peptide [2], were obtained when the precursor

---

*To whom correspondence should be addressed.

bis(Acm)-tetrapeptide-resin was treated with Tl(tfa)$_3$ [5] (1.1 eq.) plus anisole (4.5% *v/v*) at 0°C, 1 h in TFA. The corresponding reaction in solution gave a 63% yield of the 14-membered cyclic disulfide. Returning to the resin-bound case, a range of solvents were explored. Yields were acceptable in DMF (72%), less so in propionic acid (45%) or DCM (37%). The principal by-product was unreacted bis(Acm)-tetrapeptide.

2. *Penicillamine analog: Ac-L-Pen-L-Pro-D-Val-L-Cys-NH$_2$*

Under comparable conditions, yields for deprotection/oxidation of the peptide-resin with Tl(tfa)$_3$ were best in TFA (75%), marginal in propionic-acid/acetic-acid (1:1, *v/v*) (48%), and worst in DMF (35%). The solution oxidation of the bis(Acm)-tetrapeptide in TFA occurred in 66% yield. Evidently, penicillamine slows the oxidation due to steric hindrance, but the desired bridges can still be created.

An alternative approach started with the bis(Fm)-tetrapeptide-resin. Removal of the Fm group with Piperidine (Pip)/DMF (1:1, *v/v*), 25°C, 3 h gave *directly* the disulfide (60% yield).

3. *Oxytocin: H-Cys-Tyr-Ile-Gln-Asn-Cys-Pro-Leu-Gly-NH$_2$*

Several routes have been explored. The bis(Fm)-nonapeptide-MBHA-resin, treated with Pip/DMF (1:1, *v/v*), 25°C, 3 h, gave *directly* a 47% yield of the 20-membered cyclic disulfide. Yields were improved somewhat by using a two-step deprotection/oxidation sequence. Thus, removal of the Fm group under argon with Pip/DMF/βME (10:10:0.7, *v/v*), 25°C, 3 h, gave the peptide-resin with cysteine residues as the free thiols. Subsequent oxidation with air or with DTNB (0.5 equiv.), while the resin was suspended in buffered DMF, gave fair yields of oxytocin (60–65%).

Direct oxidation of a bis(Acm)-nonapeptide-MBHA-resin under the best Tl(tfa)$_3$ conditions reported earlier gave moderate amounts (42%) of the desired disulfide. A larger excess of the oxidizing agent in fact lowered the yield. Non-optimized oxidations carried out on bis(Acm)-nonapeptide-resins assembled with Fmoc chemistry with either of our new handles [4] gave modest amounts of oxytocin (15–20%). It is encouraging that treatments with either Tl(tfa)$_3$ or I$_2$ did not adversely affect the yields of acidolytic or photolytic cleavage from the appropriate supports.

Acknowledgements

We are grateful to the National Institutes of Health (GM-28934 and 42722) and CICYT (BT 18/86) for financial support. C. G.-E. holds a doctoral fellowship from the Ministerio de Educacion y Ciencia (Spain), and R.P.H.'s support is from fellowships sponsored by the University of Minnesota Graduate School and by Amoco. F.A.'s travel and living expenses in Minnesota are covered by a NATO Collaborative Research Grant (0841/88), and he further thanks Applied Biosystems Inc. for a travel award to the Eleventh American Peptide Symposium.

References

1. Barany, G. and Merrifield, R.B., In Gross, E. and Meinenhofer, J. (Eds.) The Peptides, Vol.2, Academic Press, New York, 1979, pp. 1-284; esp. pp. 240-243.
2. García-Echeverría, C., Albericio, F., Pons, M., Barany, G., and Giralt, E., Tetrahedron Lett., 30 (1989) 2441 (and references cited therein).
3. Mazur, S. and Jayalekshmy, P., J. Am. Chem. Soc., 101 (1979) 677.
4. Albericio, F., Kneib-Cordonier, N., Gera, L., Hammer, R., Hudson, D., and Barany, G., In Marshall, G.R. (Ed.) Peptides: Chemistry and Biology (Proceedings of the Tenth American Peptide Symposium), ESCOM, Leiden, 1988, p. 159.
5. Yajima, H., Fujii, N., Funakoshi, S., Watanabe, T., Murayama, E. and Otaka, A., Tetrahedron, 44 (1988) 805.

Air oxidation of peptides and separation of their reduced and intra-chain disulfide-bridged forms by RPHPLC

Kok K. Lee[a], James A. Black[b] and Robert S. Hodges[a]

[a]Department of Biochemistry, The Medical Research Council (Canada) Group in Protein Structure and Function, Edmonton, Alberta, Canada T6G 2H7
[b]The Alberta Peptide Institute, University of Alberta, Edmonton, Alberta, Canada T6G 2H7

Introduction

The ability to obtain intrachain disulfide-bridged peptides in their purified forms is highly desirable, as many biologically active peptides and fragments of proteins are found to be disulfide-bonded in their native states. Whether the peptides are isolated as native fragments or prepared synthetically, a simple procedure is required to monitor easily the oxidation state of the peptide and to verify the purity of the disulfide-bridged form with respect to contamination by the reduced conformer.

Results and Discussion

In this study, the intra-chain disulfide bridge was formed by air oxidation of the peptides in volatile reagents. The retention behavior and rate of oxidation of the reduced and oxidized forms of 14 peptides were monitored using RP-HPLC. These peptides varied in length (9–29 residues), amino acid sequence, and in the number of amino acid residues within the disulfide loop (4–15 residues), as shown in Table 1. Reduced peptides were dissolved in 0.1 M ammonium bicarbonate (0.1 mg/ml) and air-oxidized by stirring at room temperature. Ammonium bicarbonate, unlike Tris or phosphate salts, can be removed by repeated freeze-drying. In this manner, a pure preparation of oxidized peptides can be obtained without further desalting of purification steps. The low peptide concentration reduces the possibility of interchain disulfide bridging that would lead to oligomer formation. It was observed that peptides have different rates of oxidation. For example, peptides 2, 5, 6 and 9 were completely oxidized after 1.5, 3, 6 and 20 h, respectively. The differences in rate of oxidation were not related to the size of the loop alone, as peptides 5, 6 and 9 have the same number of residues within the loop. Peptide conformation, an inherent property of the peptide sequence, has been suggested by White [1] to affect the rate of oxidation.

As shown in Table 1, most reduced and oxidized conformers can be separated by RPHPLC using a linear AB gradient (1% acetonitrile/min; flow of 1 ml/min; Aquapore C_8 220 × 4.6 mm I.D. column; pore size of 300Å), where A is 0.1% aqueous TFA and B is 0.1% TFA in acetonitrile. With the exception

Table 1 *Retention times of reduced and intra-chain disulfide-bridged peptides on RPHPLC*

Peptide No.	Amino acid sequence[a]	No. of residues within disulfide loop	Retention time (min)[b]		
			(-SH)	(-S-S-)	(ΔRt)
1	NH$_2$-GIVECSTSICSLY-NH$_2$	4	33.22	31.34	1.88
2	desamino-CYFQNCPRG-NH$_2$	4	26.82	26.82	0.00
3	NH$_2$-AGCKNFFWKTFTSC-OH	10	35.44	34.44	1.00
4	Ac-KCTSDQDEQFIPKGCSK-OH	12	23.23	21.24	1.99
5	Ac-ACAADQDEQFIPKGCSK-OH	12	25.84	24.09	1.75
6	Ac-ACAAAADEQFIPKGCSK-OH	12	27.56	25.19	2.37
7	Ac-ACAAAAAAQFIPKGCSK-OH	12	31.54	28.29	3.25
8	Ac-ITLTRTAADGLWKCTSDQDEQFIPKGCSK-OH	12	34.89	33.02	1.87
9	Ac-ACKSTQDPMFTPKGCDN-OH	12	24.67	24.67	0.00
10	Ac-CFGGRMDRIGAQSGLGC-NH$_2$	15	29.80	29.17	0.63
11	Ac-CFGGRMDRIGAQSGLGCNSFRY-OH	15	32.87	32.07	0.80
12	NH$_2$-SSCFGGGRIDRIGAQSGLGCNSFRY-OH	15	28.82	28.07	0.75
13	NH$_2$-SLRRSSCFGGRMDRIGAQSGLGCNSFRY-OH	15	29.36	28.73	0.63
14	NH$_2$-SLRRSSCFGGRMRRIGAQSGLGCNSFRY-OH	15	29.45	28.07	1.38

[a] The symbols, ▲ and o denote single amino acid sequence changes between peptides; ▽ denotes position of cysteine or cystine residues; Ac-, acetyl; -OH, C-terminal carboxyl group; NH$_2$, N-terminal amino group.
[b] ΔRt denotes the difference in retention time between reduced (-SH) and (-S-S-) oxidized peptides.

of peptides 2 and 9, the ΔRt for the reduced and oxidized conformers of the 12 peptides ranged from 0.63 to 3.25 min. In each instance where resolution occurred, the oxidized peptide was eluted before its corresponding reduced conformer. This result suggested that, on formation of the disulfide loop, some residues in the peptide were not as accessible to interact with the support as in the reduced peptide. In other words, the loop formation decreases the hydrophobicity of the peptide-binding domain.

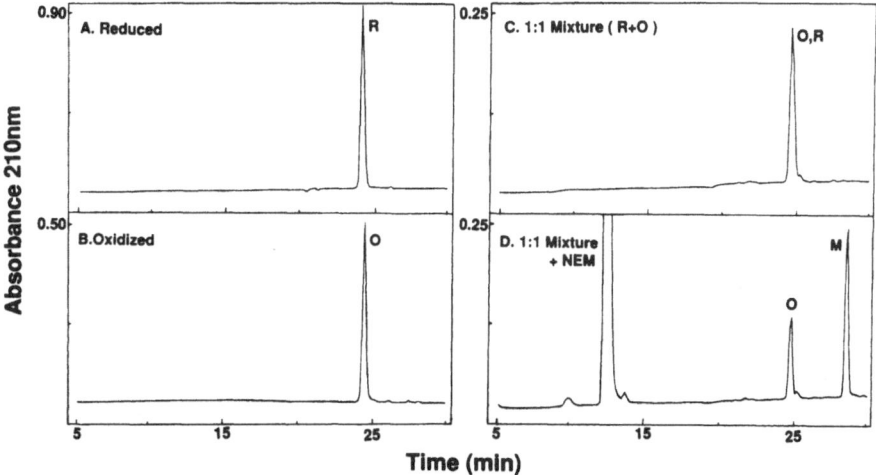

Fig. 1. The separation of reduced and oxidized forms of peptide 9 following modification of the reduced peptide with NEM. The NEM-modified peptide is denoted by M. The column used was an Aquapore RP-300 C-8 column of 220 × 4.6 mm I.D., 7 μm particle size and 300 Å pore size. The mobile phase consisted of a linear AB gradient of 1% B/min, where Eluent A is 0.05% aqueous TFA, and Eluent B is 0.05% TFA in acetonitrile (pH 2.0). Flow rate was 1 ml/min.

To verify peptide purity and the extent of oxidation, baseline resolution by RPHPLC of the reduced and oxidized conformers is required. This baseline resolution was obtained by modification of the reduced peptide with N-ethylmaleimide (NEM), a thiol-specific reagent [2]. The oxidation state of the peptide was followed by adding samples of the oxidation mixture (10 μg of peptides) at various time intervals to 10 μl of a solution (1 mg NEM/ml of 0.1% aqueous TFA) containing a 16 molar excess of NEM to peptide. This solution was incubated for 15 min at room temperature prior to RPHPLC. The NEM reagent increases the hydrophobicity of the reduced peptide (modified peptide denoted by M), resulting in an increase in retention and separation between the oxidized and modified peptide. As shown in Fig. 1, the change in retention time increased from zero to 3.8 min for peptide 9. The oxidized peptide did not react with NEM, and excess reagent was eluted off the column after 12 min. This simple procedure allows one to monitor by RPHPLC both reduced and oxidized peptide in an oxidation reaction, without the need for electrochemical detectors. It also has the advantage over titration methods such as Ellman's reagent [3] that only quantitates the amount of reduced peptide.

Acknowledgements

This work was funded by the Medical Research Council of Canada and the Alberta Heritage Foundation for Medical Research.

References

1.   White, F.H., J. Biol. Chem., 236 (1961) 1353.
2.   Lunte, S.M. and Kissinger, P.T., J. Liq. Chromatogr., 8 (1985) 691.
3.   Ellman, G.L., Arch. Biochem. Biophys., 82 (1959) 70.

Automated multiple peptide synthesis with BOP activation

H. Gausepohl, M. Kraft, Ch. Boulin and R. W. Frank

European Molecular Biology Laboratory, Meyerhofstr. 1, D-6900 Heidelberg, F.R.G.

Introduction

BOP (benzotriazole-1-yl-oxy-tris-(dimethylamino)-phosphoniumhexafluoro-phosphate) [1] has been demonstrated to be excellently suited for activation of Fmoc-amino acids [2,3] in SPPS. Employing our experience with BOP activation and with a robotic workstation [4], we constructed a fully automated instrument for the simultaneous synthesis of multiple peptides. The synthesis is performed either in microscale on polyethylene pins [5], or in mg quantities on conventional solid supports. A dedicated software was developed to compile peptide sequences from a protein sequence and to control all functions of the robot.

Results and Discussion

The design of the instrument is based on independent modules which were partly obtained from a commercial autosampler (Gilson M222). The modules include a robot arm with linear X-Y-Z movement, a microdispenser, a multi-needle assembly which is connected to a valve block, and a workbench carrying reagent containers and reactors (see Fig. 1). The system is controled by a

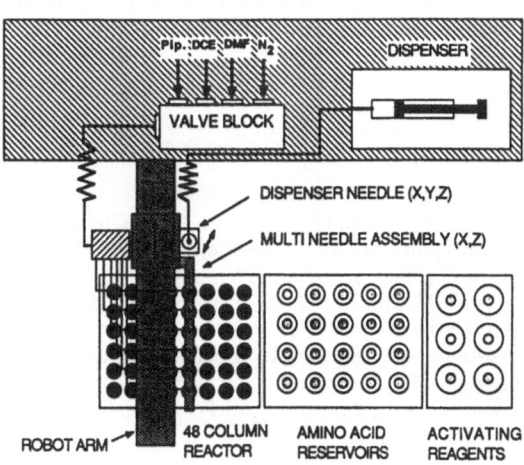

Fig. 1. Schematic layout of the robot workstation as seen from the top.

1003

Macintosh microcomputer, with software written using the Macintosh HyperCard package. The program generates peptide sequences of defined length and overlap from a given protein sequence, and calculates all steps neccessary to carry out the synthesis of the specified peptides. It also allows for very flexible specification of the chemistry to be performed, and calculates the amounts of reagents required.

Peptides are synthesized either covalently attached to polyethylene pins [5], that are immersed in a 96-well microtiterplate, or in mg amounts on conventional solid supports (1% crosslinked polystyrene, kieselguhr composite support), contained in a 48 column reactor. The synthesis cycles for the multi-column reactor are fully automated, and no operator intervention is required. Fmoc-amino acids are activated in situ by BOP and are distributed by the robot via the dispenser needle to the individual columns. DMF for washing, and 20% Pip/DMF for deprotection, are delivered by a multi-needle assembly also carried by the robot arm. Reagents and wash solutions are removed from the columns by a vacuum applied to the space below the supporting frit filters. For the final cleavage of the peptides from the resin, the waste manifold below the column array is simply replaced by an array of 48 sample tubes to collect the eluate. A 4-fold excess of amino acids is used, and the ratio of BOP to NMM is $1:2$. For a 25-μmol synthesis, the cycle time is 2 h and coupling is performed for 1.5 h.

Conclusions

A versatile robotic system for multiple peptide synthesis on either polyethylene pins or standard resins is presented. It employs in situ activation of Fmoc-amino acids by BOP in an open reaction system. In our laboratory, the workstation has shown excellent performance in the synthesis of sets of overlapping peptides for immunology (Pepscan method), as well as peptide analogs for systematic investigations, e.g., the characterization of the epitopes of neutralizing antibodies against human tumor necrosis factor at single amino-acid resolution (manuscript in preparation).

References

1. Castro, B., Dormoy, J.R., Evin, G. and Selve, C., J. Chem. Res., (1977) 2118.
2. Frank, R. and Gausepohl, H., In Tschesche, H. (Ed.) Modern Methods in Protein Chemistry, De Gruyter, Berlin, Vol. 3, 1988, p. 41.
3. Gausepohl, H., Kraft, M. and Frank, R.W., In Jung, G. and Bayer, E. (Eds.) Peptides 1988, De Gruyter, Berlin, 1989, p. 241.
4. Frank, R., Bosserhoff, A., Boulin, Ch., Epstein, A., Gausepohl, H. and Ashman, K., Biotechnology, 6 (1988) 1211.
5. Geysen, H.M., Rodda, S.J., Mason, T.J., Tribbick, G. and Schoofs, P.G., J. Immunol. Methods, 102 (1987) 259.

Z-DSP, Boc-DSP, Fmoc-DSP and pMZ-DSP: New agents for amino-protection in aqueous solution

Katsushige Kouge[a], Tatsuya Koizumi[a], Masahiro Tamura[b] and Hideo Okai[b]

[a]*Sanshin Chemical Ind. Co. Ltd., Hirao, Yamaguchi 742-11, Japan*
[b]*Department of Fermentation Technology, Faculty of Engineering, Hiroshima University,
Higashihiroshima, Hiroshima 724, Japan*

Introduction

In 1987, we reported that *p*-hydroxyphenyldimethylsulfonium methyl sulfate (HODSP) was found to be an excellent reagent for the peptide synthesis in aqueous solution [1]. We also pointed out that HODSP has the same high reactivity as that of conventional active ester reagents, such as *p*-nitrophenol or *N*-hydroxysuccinimide. By employing the HODSP active ester method, FMRF-amide (Phe-Met-Arg-Phe-NH$_2$), a molluscan neuropeptide from the ganglia of the clam *Macrocallisa nimbosa* [2], was synthesized [3,4].

Since we established the peptide synthesis using a water-soluble active ester, we then set out to develop a new type of water-soluble acylating reagent. We selected four amino protecting groups (Z-, Boc-, Fmoc- and pMz-) that are widely used in peptide synthesis and whose deprotecting conditions are all different. [*p*-(Benzyloxycarbonyloxy)(phenyl)]dimethylsulfonium methyl sulfate (Z-DSP), [*p*-(*t*-butoxycarbonyloxy)phenyl]dimethylsulfonium methyl sulfate (Boc-DSP), [*p*-(9-fluorenylmethyloxycarbonyloxy)phenyl]dimethylsulfonium methyl sulfate (Fmoc-DSP) and [*p*-(*p*-methoxybenzyloxycarbonyloxy)phenyl]dimethyl-sulfonium methyl sulfate (pMZ-DSP) were prepared and allowed to react with amino acids in aqueous media.

Results and Discussion

Preparation of water-soluble acylating reagents

Z-DSP, Boc-DSP, Fmoc-DSP and pMz-DSP were all easily synthesized from the reaction of HODSP with Z-Cl, (Boc)$_2$O, Fmoc-Cl and pMZ-Cl, respectively, in CH$_3$CN. These compounds were stable and did not undergo decomposition when stored below 10°C. They were soluble in water (solubility >30 wt%) and very soluble in DMF, alcohol, CH$_3$CN and CHCl$_3$.

Preparation of Z-, Boc-, Fmoc- and pMZ-amino acids

Reactions of water-soluble acylating reagents with amino acids were carried out in water using TEA or NaOH as a base. All amino acid derivatives were obtained in good yields.

Selective acylation of amino groups on the side chain of Lys and Orn

The other merit of the reaction in water is that we could presume selective acylation of amino groups of side chains of Lys and Orn. Z-DSP was allowed to react with Lys by controlling the pH of the reaction mixture automatically. The highest yield of H-Lys(Z)-OH was obtained at pH 11.5 (75%), and this condition kept the yield of Z-Lys(Z)-OH, which was obtained as a by-product and can be washed out with EtOAc easily, below 5.4%. Ornithine was also treated with Z-DSP in the same condition and converted to H-Orn(Z)-OH in an 80% yield. Presently, side-chain derivatives of Lys and Orn were prepared via the Cu(II) complex. This procedure needs three steps and decreases the yield of the product. By employing our water-soluble acylating reagent, those side-chain derivatives can be prepared in one step.

We dealt with four typical *N*-protecting groups in peptide synthesis. We also showed that the introduction of these groups into amino acids using the water-soluble active ester was carried out successfully. It is clear that the water-soluble active ester method can be applied to other urethane-type amino-protecting groups. Additionally, side-chain protected amino acids, such as H-Lys(Z)-OH, H-Orn(Z)-OH and others, can be prepared using the water-soluble active ester very easily. We believe that the water-soluble acylating reagent will also be a powerful tool for the chemical modification of proteins in the aqueous solution without causing conformational changes. We will soon show this elsewhere [5].

References

1. Kouge, K., Koizumi, T., Okai, H. and Kato, T., Bull. Chem. Soc. Jpn., 60 (1987) 2409.
2. Price, A.D. and Greenberg, J.M., Science, 197 (1977) 670.
3. Kouge, K., Soma, H., Katakai, Y., Okai, H., Takemoto, M. and Muneoka, Y., Bull. Chem. Soc. Jpn., 60 (1987) 4343.
4. Kouge, K., Katakai, Y., Soma, H., Kondo, M., Okai, H., Takemoto, M. and Muneoka, Y., Bull. Chem. Soc. Jpn., 60 (1987) 4351.
5. Azuse, I., Tamura, M., Kinomura, K., Kouge, K., Kawasaki, Y. and Okai, H., Chem. Express, in press.

Use of azides as intermediates in large-scale peptide synthesis

Robert B. Miller*, John C. Tolle and Kenneth W. Funk

Peptide Process Development Laboratory, Chemical and Agricultural Products Division, Abbott Laboratories, 1401 Sheridan Rd., North Chicago, IL 60064, U.S.A.

Introduction

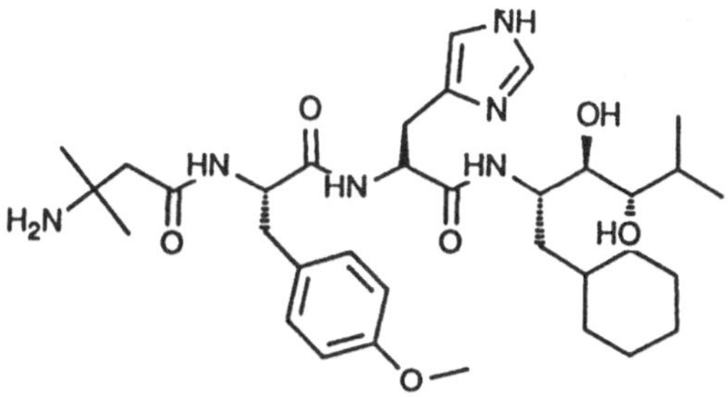

Fig. 1. Structure of A-64662.

A-64662 (Fig. 1) has been designated a clinical candidate for the renin inhibitor venture by Abbott Laboratories. A 17-step process was developed to produce kilogram quantities of A-64662. In the step to prepare the intermediate Boc-His-Amino Diol, (Fig. 2), the yield was always less than 45%. In addition, racemization of histidine during coupling made isolation difficult. A thorough study of the reaction was done to optimize this step.

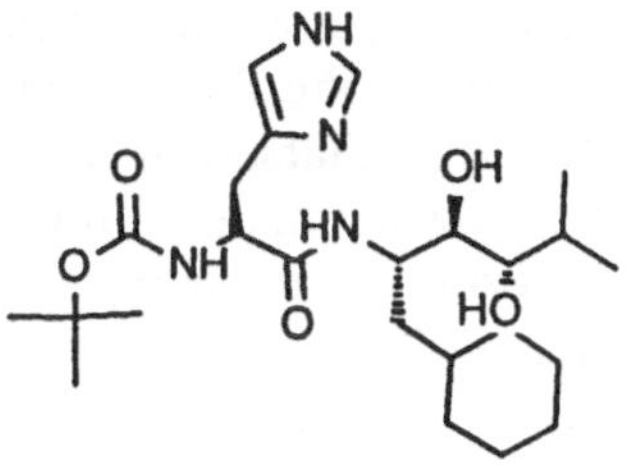

Fig. 2. Boc-His-Amino Diol.

---

*To whom correspondence should be addressed.

1007

Results and Discussion

The original coupling procedure used a WSCI coupling with the hydrochloride salt of Amino Diol [1] and Boc-L-histidine in the presence of NMM and HOBt. In an attempt to improve on the original procedure, various iterations of DCC coupling were investigated, including different stoichiometries, other racemization suppressants and the use of the unprotonated Amino Diol. The results of these experiments are shown in Table 1.

Table 1   *His-Amino Diol coupling results*

Coupling conditions[a]	Yield (%)	Optical purity (%)
(1) Boc-His, 1 equiv. HOBt	48	90
(2) Boc-His, 2 equiv. HOBt	45	91
(3) Boc-His, 2 equiv. HONb	42	92
(4) Boc-His, 1 equiv. HOBt, 1 equiv. DMAP	45	86
(5) Boc-His, 2 equiv. HONb, unprotonated Amino Diol	54	94
(6) Z-His, 2 equiv. HOOBt	54	96
(7) Z-His, 2 equiv. HOSu	59	91
(8) Z-His, 2 equiv. HONb	40	92
(9) Z-His, BOP-Cl	16	95
(10) Z-His, BOP-Cl, 2 equiv. HOBt	35	96

[a]Standard conditions: DIEA and the coupling agent added to stirred mixture of the amino acid, Amino Diol·HCl, and additive.

Few conditions were found to give adequate coupling conditions for the large-scale production of Boc- or Z-His-Amino Diol. The best of the trial reactions (coupling with 2 equiv. of HONb and the unprotonated Amino Diol) was selected for scale-up. Five kg Boc-His-Amino Diol have been prepared in this manner. However, the following problems were encountered in scale-up: the yields ranged from 65% to 50%; removal of DCU was difficult; the level of histidine racemization in the reaction mixture ranged from 8% to 10%; the material was isolated in about 91% purity, with 3–5% of the unwanted D-His diastereomer. Substitution of Z-histidine as starting material did not improve the reaction.

In spite of the various additives, the diimide-type coupling had reached its potential. The search for readily available and inexpensive histidine derivatives led to Z-His hydrazide [2] as starting material for coupling to Amino Diol via an azide reaction. Z-His hydrazide can easily be made in house, or be purchased at less cost than Boc-histidine. This reaction was examined to determine its suitability for preparation of the His-Amino Diol intermediate. The results are shown in Table 2.

The reaction of Z-His azide prepared in situ and Amino Diol consistently produced better than 65% yields. In addition, the Z-His-Amino Diol crystallized directly from the reaction mixture and could be isolated in 99+% purity, with no evidence of contaminating D-His diastereomer.

Table 2   *Z-His-Amino Diol azide couplings*

Coupling conditions[a]	Yield (%)	Optical purity (%)
(1) 1.2 equiv. hydrazide, 1.5 equiv. nitrite	41	100
(2) 1.6 equiv. hydrazide, 2.4 equiv. nitrite	48	100
(3) 2.0 equiv. hydrazide, 3.0 equiv. nitrite	34	100
(4) 2 + NMM	58	100
(5) 2 + HOBt	23	100
(6) 2 + HOOBt	20	100
(7) 2 + no DMF	78	100

[a]Standard conditions: see Ref. 2; reaction solvent is 1:1 DMF-EtOAc.

Acknowledgements

The assistance of Robert Oheim and Lorenzo Gutierrez of the Pharmaceutical Products Division, Analytical Department, and Melissa Knox of Chemical Development are greatly appreciated.

References

1.   Luly, J.R., Dellaria, J.F., Plattner, J.J., Soderquist, J.L. and Ba Maung, N.Y., J. Org. Chem., 52 (1987) 1487.
2.   Holley, R.W. and Sondheimer, E., J. Am. Chem. Soc., 76 (1954) 1326.

Total solid-phase synthesis of ubiquitin using BOP reagent: Biochemical and immunochemical properties of the purified synthetic product

J.P. Briand[a], A. Van Dorsselaer[b] and S. Muller[a]

[a]*IBMC, 15 rue René Descartes, F-67084 Strasbourg Cedex, France*
[b]*Institut de Neurochimie, 5 rue Blaise Pascal, F-67000 Strasbourg, France*

Introduction

The presence of auto-antibodies to ubiquitin and ubiquinated histones has been found in the serum of patients with systemic lupus erythematosus (SLE) [1,2]. During the course of this study, we found that the ubiquitin preparation was not pure and, in particular, that it was contaminated by ubiquitinated histone complexes (U-H2A, U-H2A – U-H2B) only detectable by specific antibodies in immunoblotting experiments. Exploration of the potential use of ubiquitin in diagnostic assays, as well as the investigation of the properties of ubiquitin in vivo or in vitro systems necessitates the availability of ubiquitin, and particularly ubiquitin analogs and derivatives under a pure form. This type of study precludes the use of natural ubiquitin, and thus, the chemical synthesis of the ubiquitin molecule (76 residues long) was carried out.

Results and Discussion

Assembly of the protected peptide chain was achieved by SPPS using BOP as coupling reagent [3]. Boc-Gly-PAM-resin (0.1 mmol) (200 mg) was placed in a 20-ml reaction vessel of a NPS 4000 synthesizer (NEOSYSTEM, France). Boc-amino acids (5 equiv.), BOP (5 equiv.) and HOBt (5 equiv.) were dissolved in 4 ml of 25% NMP/DCM (*v/v*) and the activation was performed in the reactor with 7.5 equiv. of DIEA. After 15 min, pure NMP (2.5 ml) was automatically added. The total coupling time was set at 30 min; monitoring with ninhydrin tests showed that all the couplings were complete within this time and no recoupling was necessary.

Removal of the protecting groups and cleavage of the peptide chain from the resin was performed by the low-high HF procedure. The crude product was purified by gel filtration, then by middle pressure liquid chromatography, and finally by ion exchange chromatography. These three purification steps yielded a ca. 95% pure peptide as determined by gradient elution on aquapore RP300 C_8 column. In isocratic elution, synthetic ubiquitin appeared as a single symmetrical peak. The yield calculated from the starting Gly resin was 1%.

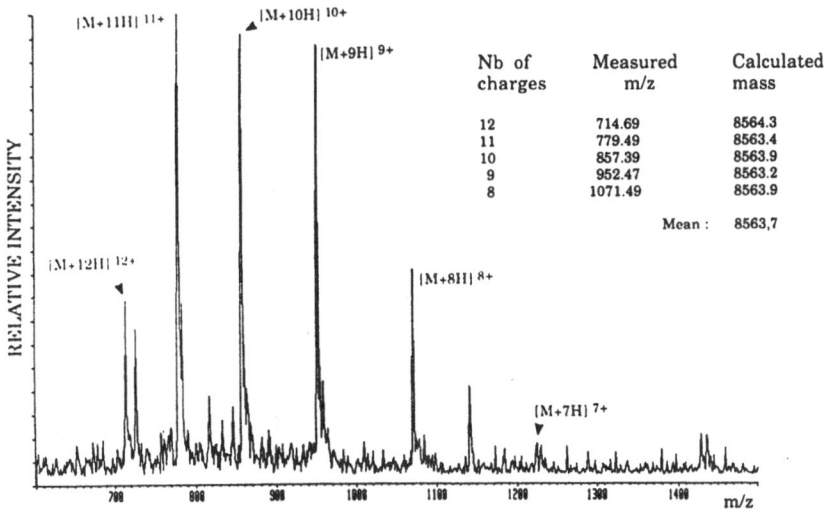

Fig. 1. Ion-spray spectra were obtained on a VG PMA-3000 quadrupole MS (VG Analytical Ltd., U.K.). Data were collected in the MCA mode. The final concentration of ubiquitin was about 1 µg/ml and 10 µl were injected.

Mass measurements were performed using two different techniques: LSIMS and ion spray. The spectrum obtained by ion spray is presented in Fig. 1. The average mean value calculated from the 5 major multiply charged ions generated is 8563.7 (expected value: 8564.9); this clearly shows that the major peak purified from the crude product has the correct sequence and does not correspond to an ubiquitin derivative containing modified amino acids. It is worthwhile to note that synthetic ubiquitin appears rather pure from the ion-spray spectrum.

Synthetic ubiquitin was found to be antigenically active in ELISA and immunoblotting experiments; it bound antibodies raised against natural ubiquitin and the synthetic peptide 22–45. Contrary to some recent claims, neither the synthetic ubiquitin nor a commercial ubiquitin preparation in our own experiments exhibited any proteolytic activity. The successful total synthesis of ubiquitin opens the way to the preparation of various stable analogs that should be useful for studying the intracellular metabolism of this molecule and its involvement in the protein degradation pathway.

References

1. Muller, S., Briand, J.P. and Van Regenmortel, M.H.V., Proc. Natl. Acad. Sci. U.S.A., 85 (1988) 8176.
2. Plaué, S., Muller, S. and Van Regenmortel, M.H.V., J. Exp. Med., 169 (1989) 1607.
3. Castro, B., Dormoy, J.R., Evin, G. and Selve, C., Tetrahedron Lett., 14 (1975) 1219.

An efficient facilitated method for solution phase peptide synthesis

David B. Head

Laboratory of Rational Drug Design, University Hospital E336, Boston University School of Medicine, Boston, MA 02118, U.S.A.

Introduction

Despite the popularity of solid phase methods, classical solution phase methods of peptide synthesis still offer advantages, e.g., ease of scale-up, better coupling efficiency, and fully characterizable intermediates. Classical methods [1] are often overlooked however, due to unpredictable solubilities and tedious purification of the intermediates. An improved method [2] facilitating the solution phase synthesis of peptides has been developed. The C-terminal residue is protected as an iodocholestane ester that provides a crystalline, homogenous 'handle'. This allows the growing peptide, after addition of each residue, to be purified by gel filtration on Sephadex LH-20.

Although Sephadex LH-20 is a popular method for purification of peptide sequences [3], as a standard tool it only becomes useful after several residues have been coupled. The bulky 'handle' reported here allows Sephadex LH-20 purification after the very first coupling. Also, use of a cholestane moiety as the 'handle' overcomes some of the problems of soluble heterogeneous carriers [4] and confers the advantage of crystallizability and resistance to diketopiperazine formation on the peptide.

Results and Discussion

To demonstrate this method, the Merrifield model tetrapeptide (LAGV) was synthesized using Fmoc protected pentafluorophenyl (OPfp)-activated amino acid esters. It is evident, though, that the method is not limited to this strategy. The C-terminal Fmoc-[^{3}H]Val residue (2 eq.) was protected by esterification to 2-cholestene (1 eq.) in quantitative yield, by addition of N-iodosuccinimide (NIS) [5] (1.3 eq.) to a solution of the above in chloroform. The product was the novel iodohydrin ester of the amino acid, 2-(Fmoc-[^{3}H]Valyl)-3-iodo-cholestane, or Fmoc-Val-OICh.

After a simple workup (wash with aq. sodium carbonate followed by aq. sodium thiosulfate), the N-terminus was selectively deprotected by removal of the Fmoc group (dimethylamine/DMF) followed by solvent reduction and purification on Sephadex LH-20 (1M toluene in DMF). Columns were loaded 300 mg/175 mL Sephadex (swelled resin), although as much as 2–3 g of

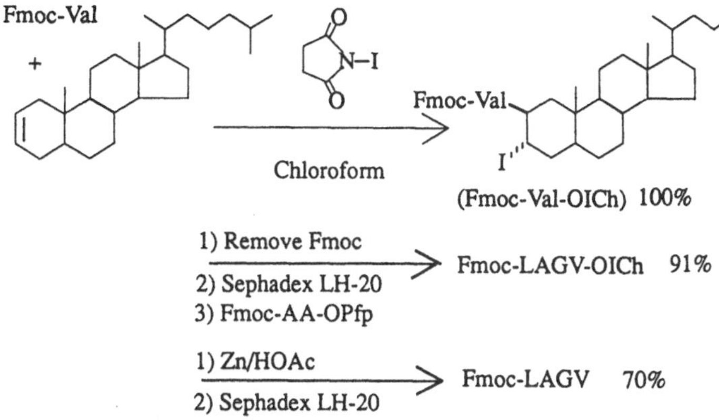

Fig. 1.

N-deprotected material can be purified by one loading on this size column. Elution was accelerated under air pressure increasing flow rate six-fold; elution (175 mL column) was accomplished in about 45 min.

Fractions containing the product, Val-OICh (determined by scintillation counting), were combined and coupled to Fmoc-Gly-OPfp by adding the activated ester directly to the combined fractions. Couplings were judged complete (TLC) within a few minutes. After coupling, *N*-deprotection was carried out as before, followed by Sephadex. The cycle was repeated for Ala and Leu residues. For each intermediate, after elution from Sephadex, a small portion of the eluted fractions was removed and fully characterized after solvent removal without any further purification (see Table 1).

Table 1

Compound	MW (calc)	MW (FABMS)	Elution vol (ml) Sephadex LH-20	Optical rotation	MP
Fmoc-Val-OICh	835	836	–	+21.0	76–78
Val-OICh	613	614	115	+49.5	–
Gly-Val-OICh	670	671	111	+45.0	78–80
Ala-Gly-Val-OICh	741	742	104	+40.5	72–74
Leu-Ala-Gly-Val-OICh	854	855	97	+22.5	–
Fmoc-Leu-Ala-Gly-Val-OICh	1 177	1 178	88	+22.5	123–126
Fmoc-Leu-Ala-Gly-Val	625	626	–	–	164 decomp

The fully protected tetrapeptide (Fmoc-LAGV-OICh) was isolated in 91% overall yield. The iodohydrin ester was cleaved selectively over the Fmoc protecting group treatment with excess zinc in acetic acid to afford the C-terminal residue deprotected Fmoc-LAGV in 70% yield (non-optimized).

Conclusion

A repetitive method simplifying solution phase peptide synthesis to a routine process that provides analytically pure intermediates as part of the standard purification cycle has been developed. The cholestene iodohydrin ester, as a C-terminal protecting group, is selectively removable under mildly reductive conditions and functions well as a bulky modifier of amino acids. A combination of features imparted by this group, i.e., simplification of purification protocols, selective removal, peptide solubility enhancement, and solution phase coupling efficiency may make its use ideal for the preparation of difficult peptide sequences for fragment condensations.

References

1. a) Bodanszky, M. and Du Vigneaud, V., J. Am. Chem. Soc., 81 (1959) 5688.
    b) Bodanszky, M., Funk, K.W. and Fink, M.L., J. Org. Chem., 3 (1973) 3565.
2. Burton, J., In Walter, R. and Meienhofer, J. (Eds.) Peptides – Chemistry, Structure, Biology (Proceedings of the 4th American Peptide Symposium), Ann Arbor Science, Ann Arbor, MI, 1975, p. 365.
3a. Kenner, G.W. and Galpin, I.J., In Gross, E. and Meienhofer, J. (Eds.) Peptides: Structure and Biological Function (Proceedings of the 6th American Peptide Symposium), Pierce Chemical Co., Rockford, IL, 1979, p. 431.
  b. Naithani, V.K. and Zahn, H., In Hearn, M.T.W. (Ed.) Peptide and Protein Reviews, 1984, p. 81.
4a. Andreatta, R.H. and Rink, H., Helv. Chim. Acta, 56 (1973) 1205.
  b. Mutter, M. and Bayer, E., In Gross, E. and Meienhofer, J. (Eds.) The Peptides: Analysis, Synthesis and Biology, Vol. 2, Academic Press, New York, NY, 1981, p. 285.
5a. Adinolfi, M., Parrilli, M., Barone, G., Laonigro, G. and Mangoni, L., Tetrahedron Lett., (1976) 3661.
  b. NIS is sensitive to air and only fresh reagent should be used. Transfer of NIS should be done under inert gas to ensure freshness of the unused portion.
  c. Yields of iodohydrin esters are somewhat lower after purification and range from fair to excellent for most of the other amino acids (unpublished reports).

New methods in peptide synthesis

A. Loffet[a], N. Galeotti[b], P. Jouin[b], B. Castro[b], F. Guibé[c], O. Dangles[c] and G. Balavoine[c]

[a]Propeptide, BP 12, F-91710 Vert-le-Petit, France
[b]CCIPE, rue de la Cardonille, F-34094 Montpellier Cedex, France
[c]Laboratoire de Chimie Organique des Eléments de Transition Bat 420, F-91405 Orsay Cedex, France

Introduction

Because of the recognized need for more efficient methods for the formation of *tert*-butyl esters and for new supports in SPPS, we have recently investigated the use of *tert*-butyl fluorocarbonate (Boc-F) for the formation of *tert*-butyl esters and independently, the use of an allylic anchor group.

Results and Discussion

tert-*Butyl esters*

A new access to the *tert*-butyl esters of *N*-protected amino acids by the use of Boc-F was developed. The catalytic ability of 4-dimethylamino-pyridine to produce esters from mixed carboxylic-carbonic anhydrides [1] of amino acids has been sparsely exploited for the preparation of *tert*-butyl esters. This general process involved the use of an alkyl chlorocarbonate that was not available in the case of the *tert*-butyl analog, so isopropenyl chlorocarbonate in the presence of *tert*-butanol had to be used [2].

A simple and convenient, high-yield synthesis of *tert*-butyl esters of *N*-protected amino acids resulted from the use of Boc-F in the presence of triethylamine and 4-dimethylamino-pyridine in dichloromethane. The esters formed were easily purified by usual workup. The use of *tert*-butanol in the solvent enhanced the yield of the method. We demonstrated that the reaction proceeded without observable racemization.

Further studies for selective *tert*-butylation of aspartic and glutamic acids using Boc-F are under active investigation in our laboratory.

Table 1 *Preparation of* tert-*butyl esters with Boc-F*

Amino ester	Yield (%)	Mp°C (lit.)	$[\alpha]^{20}_D$ (lit., c, solvent)
Z-Phe-OtBu	91	78–79 (81–82)	– 7 (–9.9, 2, MeOH)
Z-Pro-OtBu	88	42–43 (44–45)	–55 (–52, 2, EtOH)
Z-Cys-(Bzl)-OtBu	92	54–56 (oil)	–25 (–30, 2, MeOH)
Z-Met-OtBu	96	oil	–35 (–27, 5.7, EtOH)

1015

Allylic anchor group to solid phase supports

The use of an allylic anchor group and its palladium-catalyzed hydrostannolytic cleavage in the SPPS was developed.

In recent years, we and others have emphasized the potentialities of the allyloxycarbonyl groups that can be removed under very specific and mild conditions using homogeneous palladium catalysis [3]. A recent report relative to peptide synthesis on allyl-functionalized cellulose disks prompted us to disclose our own results [4].

We started with the inexpensive *cis*-2-butene-1,4 diol (1) for the elaboration of the allylic handle [4]. Volhard analysis on (5) gave a chloride content of 0.43 mequiv./g resin (80% functionalization based on starting amino substitution) that could not be improved by repeating the reaction.

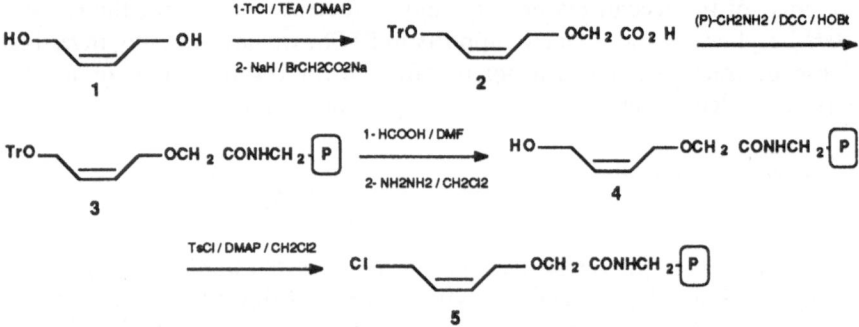

Fig. 1.   *Allylic handle for solid phase synthesis.*

Capping of the residual active sites was carried out with PhCOCl/EtN(iPr)$_2$. Boc Tyr(Dcb) was attached to the handle (5) by the Gisin method. The assembly of the model peptide Pyr-His-Trp-Ser(Bzl)-Tyr(Dcb) (6) was performed by the standard Boc methodology. Compound 6 was finally obtained after palladium-catalyzed hydrostannolytic cleavage from the resin.

References

1.   Kim, S., Kim, Y.C. and Lee, J.I., Tetrahedron Lett., 646 (1983) 119.
2.   Jouin, P., Castro, B., Zeggaf, C., Pantaloni, A., Senet, J.P., Lecolier, S. and Sennyey, G., Tetrahedron Lett., 28 (1987) 1661.
3.   Dangles, O., Guibé, F., Balavoine, G., Lavielle, S. and Marquet, A., J. Org. Chem., 52 (1987) 4984 (and references therein).
4.   Blankemeyer-Menge, B. and Frank, R., Tetrahedron Lett., 29 (1988) 5871.

A convenient preparation of C-terminal peptide alcohols by solid-phase synthesis

Joseph Swistok, Jefferson W. Tilley, Waleed Danho, Rolf Wagner* and
Kathleen Mulkerins

Roche Research Center, Hoffmann-La Roche Inc., Nutley, NJ 07110, U.S.A.

Introduction

Peptide alcohols occur in nature and are also of interest as analogs of bioactive peptides. These materials have generally been synthesized by hydride reduction of a peptide containing a C-terminal ester function [1,2], by a condensation strategy involving solution-phase coupling of a preformed peptide with an amino alcohol [3], or ammonolysis of the resin-bound peptide using the β-aminoalcohol [4]. The limitations of these methods prompted us to develop a more practical synthesis of these compounds.

Results and Discussion

We have found that N-protected β-aminoalcohols **1** react readily with succinic anhydride in the presence of 4-dimethylaminopyridine (DMAP) and pyridine in DMF to give the corresponding hemisuccinates **2**. These monoacids were coupled to a benzhydrylamine resin (BHA) using DIC/HOBT, to provide the resin bound intermediates **3**, that can be further elaborated

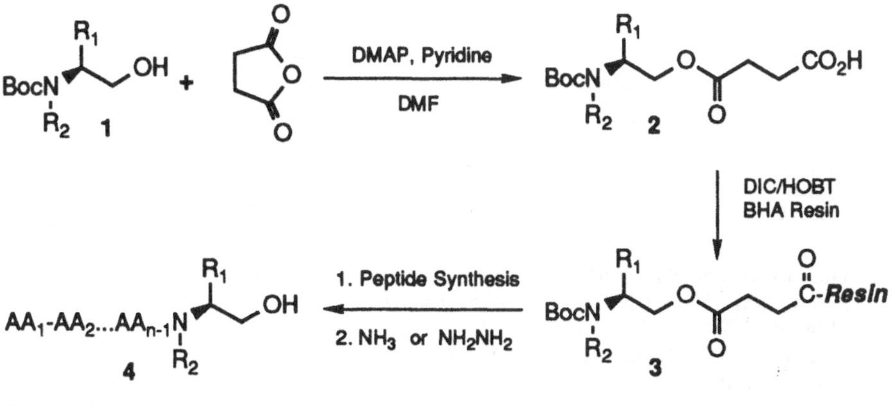

Scheme 1.

---

*Present address: Abbott Laboratories, Immunoscience Research, Abbott Park, IL 60064-2204, U.S.A.

by conventional SPPS. In order to illustrate this process, we have employed DIC/HOBT couplings of Boc-protected amino acids with TFA deprotections to construct a series of tetrapeptide derivatives. Liberation of the tetrapeptide alcohols **4a** through **4g** from the resin was accomplished by treating the intermediate succinic esters with ammonia/MeOH in a pressure bottle (72–96 h), or with an excess of hydrazine in DMF (24 h). As indicated in Table 1, both hydrolysis methodologies provide comparable yields. The benzyl-protecting group was removed from the tetrapeptide **4e** by simple Pd/C catalyzed hydrogenolysis in glacial acetic acid.

Table 1 *Peptide yields after ammonolytic cleavage*

		Hydrolysis (% yield) [5,6]	
		NH_3	N_2H_4
4a	Leu-Ala-Gly-Val-Glycinol	14	17
4b	Leu-Ala-Gly-Val-Isoleucinol	28	41
4c	Leu-Ala-Gly-Val-Cyclohexylalaninol	50	62
4d	Leu-Ala-Gly-Val-Phenylalaninol	78	78
4e	Leu-Ala-Gly-Val-L-Serinol(Bzl)	85	77
4f	Leu-Ala-Gly-Val-Methioninol	44	38
4g	Leu-Ala-Gly-Val-Prolinol	77	79

The method is compatible with Boc/OFm and Fmoc/OtBu strategies, provided that the timing of side-chain deprotection is appropriate. An exceptionally facile route to peptide alcohols has been described, and the application of this technology to the construction of biologically active peptide alcohols will be reported in due course.

Acknowledgements

The authors thank members of the Physical Chemistry Department, Hoffmann-La Roche Inc. for determination of the spectral and microanalytical data for the compounds reported herein.

References

1a. Giannis, A. and Sandhoff, K., Angew. Chem. Int. Ed. Engl., 28 (1989) 218
 b. Angrick, M., Monatsh. Chem., 116 (5) (1985) 645.
 c. Shigezane, K. and Mizoguchi, T., Chem. Pharm. Bull., 21 (1973) 972.
 d. Kawamura, K., Kondo, S., Maeda, K. and Umezawa, H., Chem. Pharm. Bull., 17 (1969) 1902.
 e. Yonemitsu, O., Hamada, H. and Kanaoka, Y., Chem. Pharm. Bull., 17 (1969) 2075.
 f. Yonemitsu, O., Hamada, H. and Kanaoka, Y., Tetrahedron Lett., 9 (1968) 3575.
2a. Stewart, J.M. and Morris, D.H., Pennwalt Corporation, U.S. Patent 4254023, (1981).
 b. Stewart, J.M., Morris, D. H. and Chipkin, R.E., Pennwalt Corporation, U.S. Patent 4254024, (1981).

3a. Matsueda, R., Yabe, Y., Kogen, H., Higashida, S., Koike, H., Iijima, Y., Kokubu, T., Hiwada, K., Murakami, E. and Imamura, Y., Chem. Lett., (7)(1985)1041.
  b. Pless, J., Sandrin, E. and Sandoz A.-G., U.S. Patent 4247543, (1981).
  c. Roemer, D., Buesher, H.H., Hill, R.C., Pless, J., Bauer, W., Cardinaux, F., Closse, A., Hauser, D. and Huguenin, R., Nature, 268(1977)547.
4a. Burton, J., Poulsen, K. and Haber, E., Biochemistry, 14(1975)3892.
  b. Krstenansky, J.L., Payne, M.H., Owen, T.J., Yates, M.T. and Mao, S.J.T., Thrombosis Res., 54(1989)319.
5. Yields based on **3**.
6. All final peptide products were purified by preparative RPHPLC and considered pure by analytical HPLC, AAA and FABMS.

Solid-phase synthesis of C-terminal peptide alcohols

W. Neugebauer, M.R. Lefebvre, R. Laprise and E. Escher

Department of Pharmacology, Faculty of Medicine, University of Sherbrooke,
Sherbrooke, Quebec, Canada J1H 5N4

Introduction

C-terminal hydroxymethyl peptides have earned rapid success as proteolysis-resistant and highly selective analogs of naturally occurring peptide hormones. The most successful are probably the enkephalin analog, DAGO [1], and its analogs and the somatostatin analog, Sandostatin [2], an octapeptide already in clinical use. Until now, all these peptides have been prepared by the classical solution method because the hydroxymethyl function does not permit use of the commonly applied anchoring procedures on solid support. For larger analog series necessary for SAR studies, this disadvantage may become quite important. Therefore, we sought and found a reaction scheme suitable for SPPS of peptide alcohols (see Scheme 1), and several analogs of enkephalin and somatostatin were prepared (see Table 1).

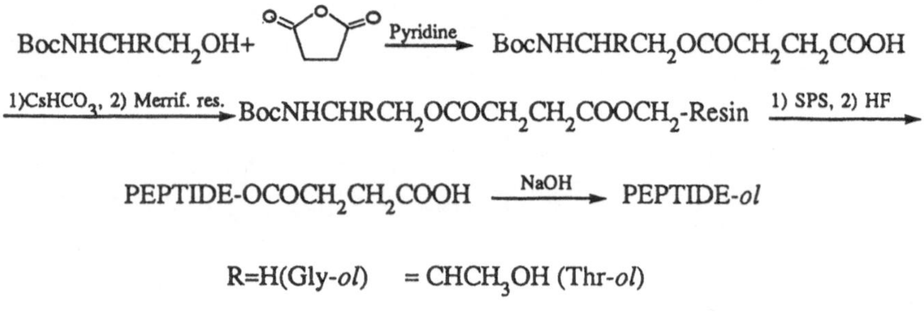

$$BocNHCHRCH_2OH + \text{(succinic anhydride)} \xrightarrow{\text{Pyridine}} BocNHCHRCH_2OCOCH_2CH_2COOH$$

$$\xrightarrow{\text{1)CsHCO}_3,\ \text{2) Merrif. res.}} BocNHCHRCH_2OCOCH_2CH_2COOCH_2\text{-Resin} \xrightarrow{\text{1) SPS, 2) HF}}$$

$$PEPTIDE\text{-}OCOCH_2CH_2COOH \xrightarrow{\text{NaOH}} PEPTIDE\text{-}ol$$

$$R=H(Gly\text{-}ol) \quad = CHCH_3OH\ (Thr\text{-}ol)$$

Scheme 1.

Results and Discussion

Compounds **1–4** are enkephalin analogs [1], specific for the μ-opiate receptor; peptides **3** and **4** are precursors of potential photolabels. Peptide **5** is the sandostatin analog of somatostatin [2], and peptides **6** and **7** are the precursors of potential photolabels of the somatostatin receptor.

The C-terminal amino alcohol was first *N*-protected with Boc-TCE and then esterified with an excess of succinic anhydride in pyridine. The resulting hemisuccinate was neutralized with $CsHCO_3$ and esterified [3] to chloromethyl-

Table 1 *Gly-ol = ethanolamine, Thr-ol = (2R,3R)-2 aminobutane-1,3-diol, NMePhe = N-methyl-L-phenylalanine*

1.	H-Tyr-D-Ala-Gly-NMePhe-Gly-*ol* (DAGO)
2.	H-Tyr-D-Ala-Gly-Phe-Gly-*ol*
3.	H-Tyr-D-Ala-Gly-NMePhe(4'-NO$_2$)-Gly-*ol*
4.	H-Tyr-D-Ala-Gly-Phe(4'-NO$_2$)-Gly-*ol*
5.	D-Phe-Cys-Phe-D-Trp-Lys-Thr-Cys-Thr-*ol*
6.	D-Phe(4'-NO$_2$)-Cys-Tyr-D-Trp-Lys-Thr-Cys-Thr-*ol*
7.	D-Tyr-Cys-Phe(4'-NO$_2$)-D-Trp-Lys-Thr-Cys-Thr-*ol*

ated polystyrene–1% divinylbenzene. The peptide sequences were built up using the Boc-TFA procedure; the completed peptides were cleaved from the solid support and the side-chain protecting groups by liquid HF (1h, 0°C anisol). Ammonolysis as an alternative was also considered but not used because of incompatibility with the cleavage of side chain protection, especially if Asp or Glu were present in the sequence. Disulfides were obtained immediately after cleavage (MeOBzl as thiol protection) by K$_3$Fe(CN)$_6$. The resulting hemisuccinate peptides were purified by gel filtration on LH-20-DMF, saponified with 0.3 N NaOH in MeOH and purified on preparative RPHPLC. Peptides were assessed for purity in analytical HPLC and TLC; quadrupole FABMS was used to identify the final products. Peptides **1**, **2** and **6** were also identified by co-elution with commercial reference material on HPLC. Purified peptides were obtained in yields of 6–7.5% (**5–7**) and of 15–26% (**1–4**).

Acknowledgements

This work has been supported by grants from the Medical Research Council of Canada and the Quebec Heart Foundation.

References

1. Roemer, D., Buescher, H.H., Hill, R.C., Pless, J., Bauer, W., Cardinaux, F., Closse, A., Hauser, D. and Hugenin, R., Nature, 268 (1977) 547.
2. Roemer, D. and Pless, J., Life Sci., 24 (1979) 621.
3. Gisin, B., Helv. Chim. Acta, 56 (1973) 1476.
4. Mackiewicz, Z., Belisle, S., Bellarbara, D., Gallo-Payet, N., Lehoux, J.-G. and Escher, E., Helv. Chim. Acta, 70 (1987) 423.

High-loading effect in solid phase peptide synthesis: Swelling and ESR approaches

Clovis R. Nakaie[a], Reinaldo Marchetto[a], Shirley Schreier[b] and Antonio C.M. Paiva[a]

[a]*Department of Biophysics, Escola Paulista de Medicina, C.P. 20388, CEP 04034, São Paulo, SP, Brazil*
[b]*Department of Biochemistry, Institute of Chemistry, University of São Paulo, C.P. 20780, CEP 01498, São Paulo, SP, Brazil*

Introduction

Owing to the advantages of peptide synthesis with high-titer resins, and following our studies of swelling properties of differently loaded benzhydrylamine resins (BHAR) and peptide-BHAR beads [1], we have investigated the influence of solvents other than DCM or DMF routinely employed in different steps of the synthetic cycle. In experimental protocols for syntheses involving heavily loaded resins, we have analyzed inter-chain aggregation by correlating swelling and ESR spectroscopy data.

Results and Discussion

The studies of swelling were done with the $(NANP)_4$ sequence anchored to a high-loaded BHAR (1.4 mmol/g), allowing a 68% (weight/weight) peptide-content resin. Solvation in DCM, DMF, ethanol, DMSO and TFE was examined by direct microscopic measurements of beads, and the swelling ratios found for BHAR and peptide-BHAR are shown in Table 1. For unprotonated forms, DCM is clearly the best solvent for BHAR. However, the polar peptide chain in $(NANP)_4$-BHAR caused it to be swollen by DMSO and TFE almost four and three times more, respectively, than by DCM and even DMF, which so far has been the solvent of choice for high-loaded peptide resins [2]. This outcome

Table 1. *Swelling ratio[a] of BHAR and $(NANP)_4$-BHAR beads (1.4 mmol/g) in unprotonated and protonated forms*

Compound	DMSO	TFE	DMF	DCM	EtOH
BHAR	4.3	3.9	4.9	9.6	1.9
$(NANP)_4$-BHAR	21.6	17.4	13.1	5.4	4.5
BHARH$^+$ (TFA$^-$)	13.5	5.7	6.3	2.8	5.5
$(NANP)_4$-BHARH$^+$ (TFA$^-$)	34.5	28.2	18.2	3.9	5.9

[a]Relative to dry styrene-1%-divinylbenzene taken as 1.

suggests the alternative use of DMSO and TFE for the unprotonated forms of these resins in the coupling and washing steps (after neutralization). In addition, the swelling behavior of protonated forms (Table 1) indicates again the lack of solvation in DCM, DMSO, TFE and DMF (in that order) are the more appropriate solvents for the washing step after deprotection in TFA.

The neutralization (with TEA) and deprotection (with TFA) steps were also evaluated comparatively in DCM and DMF by using the same 68%-peptide-content resin. For the TFA treatment, swelling in DCM was 6-fold that in DMF, whereas for the TEA treatment, swelling in DMF was 3-fold that in DCM, indicating the need for replacing DCM by DMF in the neutralization of high-loaded resins.

An ESR approach to monitor swelling [1] indicated extensive site–site interactions in high-titer BHAR and NANP-BHAR· beads spin-labeled with the nitroxyl Boc-amino acid 2,2,6,6-tetramethylpiperidine-N-oxyl-4-amino-4-carbo-xylic acid (Boc-TOAC) [3]. Site–site interactions, as evidenced by spectral line broadening, were proportional to the degree of labeling.

In a subsequent step, we examined the swelling properties of several low-labeled Boc-TOAC-BHAR (0.01 – 0.4 mmol/g) in DCM and DMF. A preliminary analysis of the spectral line broadening in terms of spin exchange, yielded minimal 'effective' concentration inside the bead and maximal site–site distances that still allow interactions. These values (9 mM and 60 Å, respectively) were obtained by using the known degree of labeling, the swelling data (volume of solvent inside the bead, number of beads per gram, number of sites per bead, etc.) of each resin and assuming a uniformly distributed cubic lattice of sites in the solid matrix. These results are in reasonable agreement with those obtained for Boc-TOAC free in solution and suggest that the coupled swelling-ESR strategy should provide a better understanding of the molecular events inside the beads during SPPS.

References

1. Nakaie, C.R., Marchetto, R., Schreier, S. and Paiva, A.C.M. In Marshall, G.R. (Ed.), Peptides: Chemistry and Biology (Proceedings of the 10th American Peptide Symposium), ESCOM, Leiden, 1988, p. 249.
2. Sarin, V.K., Kent, S.B.H. and Merrifield, R.B., J. Am. Chem. Soc., 102 (1980) 5463.
3. Nakaie, C.R., Schreier, S. and Paiva, A.C.M., Biochim. Biophys. Acta, 742 (1983) 63.

A novel Fmoc-based anchorage for the synthesis of protected peptides on solid phase

Ying-Zeng Liu[a], Shao-Hua Ding[b], Ji-Yu Chu[b] and Arthur M. Felix[c]

[a]*Shanghai Institute of Organic Chemistry, Academia Sinica, Shanghai, China*
[b]*East-China University of Chemical Technology, Shanghai, China*
[c]*Roche Research Center, Hoffmann-La Roche Inc., Nutley, NJ 07110, U.S.A.*

Introduction

Several reports have appeared in the literature describing base-labile anchor bonds for use in solid-phase peptide synthesis using the Boc/benzyl strategy. Recent interest on the use of the 9-fluorenylmethyl-group for carboxyl protection, and as a handle in SPPS, has prompted reports on 9-(hydroxymethyl)fluorene-4-carboxylic acid as the base-labile anchoring ligand for a polymer supported peptide synthesis [1,2]. Although the 9-(hydroxymethyl)fluorene-4-carboxylic acid system has been reported to be stable to all stages during peptide assemblage, including neutralization with DIEA, alternative bifunctional reagents are being developed which may be more stable to premature anchor bond cleavage. We have developed an homologous bifunctional reagent which has been evaluated for its application to the SPPS of protected peptide fragments using the Boc/benzyl strategy.

Results and Discussion

A new bifunctional compound, 9-(hydroxymethyl)-2-fluoreneacetic acid, **5**, was synthesized in 4 steps starting with 9-fluorenylmethanol, **1**, in overall yield of

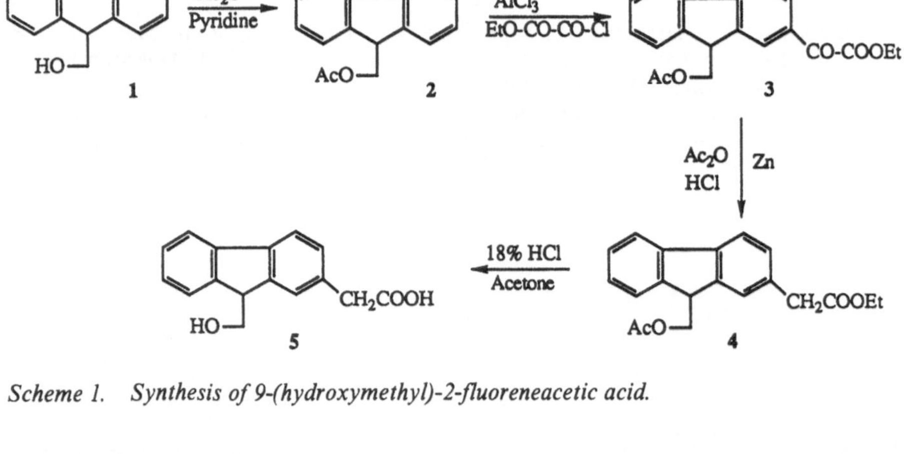

Scheme 1. Synthesis of 9-(hydroxymethyl)-2-fluoreneacetic acid.

40.9% (Scheme 1). The coupling of **5** to benzhydrylamine-resin to give **6** was mediated with DCC and subsequent conversion to **7** was achieved by DCC/DAMP coupling with a variety of Nα-Boc amino acid (Scheme 2). Conventional SPPS (Boc/benzyl strategy) was carried out using **7**. The target protected peptides with a free C-terminal COOH-function were released from the resin by the β-elimination reaction with 15% piperidine in DMF.

The stability of the anchor bond of **7** to the basicity of free amino groups was evaluated at 25°C in DCM or DMF using a variety of free NH$_2$-functions [Gly, Val, Arg(Tos), Phe, Ser(Bzl)] to determine the extent of premature cleavage.

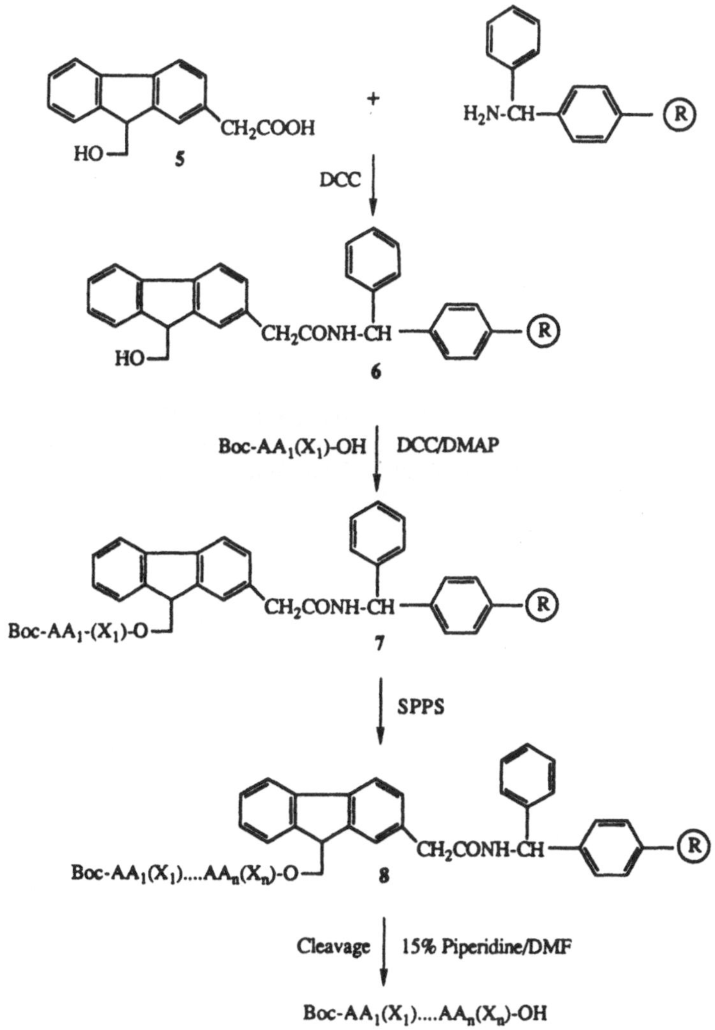

Scheme 2.   SPPS using 9-(hydroxymethyl)-2-fluoreneacetamido-BHA-resin.

Although the anchor-bond was stable in DCM, there was evidence in DMF for partial anchor-bond cleavage by the free NH_2-functions of Gly, Phe and Ser(Bzl). This premature cleavage was suppressed by the addition of HOBt in DMF which resulted in anchor-bond stability for all free NH_2-functions, with the exception of Phe.

Several protected peptides were synthesized with this new support in overall yields ranging from 38% to 75% after purification (Table 1).

Table 1   *Protected peptides synthesized on 9-(hydroxymethyl)-2-fluoreneacetamido-BHA-resin 6*

Structure		Yield (%)ᵃ	FABMS
TGF(1–7)	Boc-Val-Val-Ser(Bzl)-His(Tos)-Phe-Asn-Lys(Z)-OH	46	1153(M-Tos + 1)
GRF(1–6)	Boc-Tyr(Bzl)-Ala-Asp(Ochex)-Ala-Ile-Phe-OH	64.1	971(M + 1)
GRF(7–11)	Boc-Thr(Bzl)-Asn-Ser(Bzl)-Tyr(Bzl)-Arg(Tos)-OH	46.7	1164(M + 1)
GRF(12–17)	Boc-Lys(Z)-Val-Leu-Gly-Gln-Leu-OH	62	891(M + 1)
GRF(18–22)	Boc-Ser(Bzl)-Ala-Arg(Tos)-Lys(Z)-Leu-OH	75.3	1052(M + 1)
GRF(23–28)	Boc-Leu-Gln-Asp(OcHex)-Ile-Met-Ser(Bzl)-OH	38	978(M + 1)

ᵃCalculated from the loading value of first amino acid after purification.

References

1.   Mutter, M. and Bellof, D., Helv. Chim. Acta., 67 (1984) 2009.
2.   Tjeong, F.S., Zupec, M.E., Eubanks, S.R. and Adams, S.R., Deber, C.M. and Hruby, V.J., (Eds.) Proc. 9th American Peptide Symposium, 1985, p. 265.

Membranes as novel solid supports for peptide synthesis

Scott B. Daniels, Michael S. Bernatowicz, James M. Coull and Hubert Köster

MilliGen/Biosearch Division of Millipore, 186 Middlesex Turnpike, Burlington, MA 01803, U.S.A.

Introduction

The specific objectives of this work were the synthesis of high-quality peptide products with speed, simplicity, convenience and versatility, the potential for new applications, and the reduction of overall synthesis cost. It was decided that these goals, at least in part, could be achieved through the use of the Fmoc-protecting group strategy with a novel solid support and improvements in the chemistries used for anchoring Fmoc-protected amino acids to the solid support.

A polypropylene membrane coated with cross-linked polyhydroxypropyl-acrylate (Solvex™ manufactured by Millipore Corp.) was chosen for evaluation as a peptide synthesis support, because of its attractive physical and chemical properties, which include mechanical and chemical stability, good flow characteristics in a continuous-flow system, compatibility with aqueous solutions, and ease of handling. In contrast to conventional beaded supports, a sheet of porous contiguous polymer lends itself more easily to rapid simultaneous synthesis of large numbers of peptides, miniaturization of automated devices and novel reactor devices. Chemically, this membrane support allows for sufficient functionalization of the surface for synthesis and cleavage of the final peptide product, as well as the possibility of leaving the peptide covalently attached to the support to allow for affinity purification of biomolecules, new methods of epitope mapping, diagnostic testing and covalent sequence analysis.

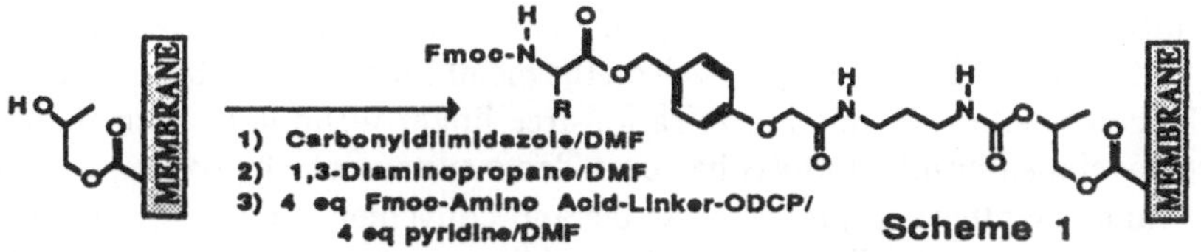

Scheme 1

When the Fmoc-protecting group strategy is used for peptide synthesis, the C-terminal amino acid is generally attached by chemical activation of the carboxyl group and its subsequent esterification to 4-alkoxybenzyl alcohol functionalized polymers. To achieve appropriate levels of polymer functionalization by such a process, it is necessary to use an acylation catalyst such as DMAP which can promote racemization and produce anchored dipeptide by-products. An attractive alternative method described here utilizes novel 2,4-dichlorophenyl-

Nα-Fmoc-aminoacyl-4-oxymethyl-phenoxyacetates (Scheme 1) for direct acylation of amine-functionalized polymers. These derivatives incorporate the acid
labile 4-oxymethylphenoxyacetyl linkage. A series of Fmoc-amino acid-linker
derivatives have been prepared and coupled to a variety of amino polymers
[1].

Results and Discussion

The usefulness and performance of this novel support were tested by the
assembly of four target peptides, prothrombin 1–9 (ANKGFLEEV-OH), acyl
carrier protein 64–75 (VQAAIDYING-OH), neurotensin (< ELYENKPRRPYIL-
OH) and Fos oncogene protein 147–172 (CVEQLSPEEEEKRRIRRERN-
KNAAA-OH) [2]. These peptides, varying in length (9–25 residues) and synthetic
difficulty, were assembled on the membrane and two beaded support, PepSyn
KA™ (polydimethylacrylamide on a kieselguhr matrix), and aminomethyl
polystyrene. Each of these beaded supports has been routinely used to prepare
many different peptides, and serves as a yardstick for comparing products
assembled on membranes.

The racemization-free acid-cleavable linker was introduced by reaction of the
appropriate Fmoc-amino acid-linker compound with the amino membrane (see
Scheme 1) or aminomethyl polystyrene. PepSyn KA™ is available with the C-
terminal amino acid connected to the resin by the same linkage. The amount
of Fmoc amino acid loaded on PepSyn KA™ and the membrane was 0.1 mmol
per g, and on polystyrene was 0.5 mmol/g. The PepSyn KA™ and polystyrene
resins were packed in columns, while the membrane was positioned and sealed
into a prototype disposable polypropylene cartridge. The syntheses were performed using a continuous-flow MilliGen/Biosearch 9050 PepSynthesizer with
the same protocol for each support examined and at a standard 0.2-mmol scale.

After the synthesis was complete, the peptide was cleaved from the support
with TFA containing 5% scavengers. All four target peptides were obtained
in crude yields >85% regardless of the solid support used. Amino-acid analysis
and MS data confirmed the identity of the products. HPLC data showed that
the peptides assembled equally as well on the membrane as on the beaded supports.

The combination of this racemization-free linker technology with the use of
these novel membrane supports has provided a system which gives reproducibly,
high-quality synthetic peptides, is simple and convenient to use, opens the door
to a variety of new applications, and should lower the overall cost of synthesis.

References

1. Bernatowicz, M.S., Kearney, T., Neves, R.S. and Köster, H., Tetrahedron Lett.,
   30 (1989) 4341.
2. Daniels, S.B., Bernatowicz, M.S., Coull, J.M. and Köster, H., Tetrahedron Lett.,
   30 (1989) 4345.

Fully automatic SMPS in micromolar scale: Quick and economic evaluation of SAR

Gerd Schnorrenberg[a] and Rudolf E. Lang[b]

[a]*Department of Medicinal Chemistry, Boehringer Ingelheim KG, D-6507 Ingelheim, F.R.G.*
[b]*Department of Pharmacology, University of Heidelberg, D-6900 Heidelberg, F.R.G.*

Introduction

The evaluation of SAR on peptides and proteins requires series of different peptides. In this respect, the SMPS turned out to be very useful [1]. We want here to report on a novel method of fully automatic simultaneous multiple peptide synthesis in micromolar scale that allows the quick and economic evaluation of structure–activity data by screening the crude peptides in biological or immunological assays.

Results and Discussion

Peptides are synthesized in 5 μM scale in the wells of a microtiterplate applying Fmoc strategy. All reagent and solvent handling is fully automatically performed by a robotic sample processor (Fig. 1). Special software has been developed allowing free choice of sequences and chain length [2].

The sample processor has two arms, one being responsible for dosage of reagents (Fmoc amino acid, HOBt, DIC for coupling reactions and piperidine/DMF solution for Fmoc deprotection), the other one performs all washing steps. Sucking off all solubles from the wells and dosage of DMF for washing of the resins are performed by a cannula connected to a motor driven syringe transferring all solutions to the waste. To avoid loss of resin, the tip of the cannula is protected by a narrow stainless steel net. Resin sticking to the net is rinsed back to the wells by dosage of DMF through a small attached cannula, thus initiating the next washing step. After all these operations, the tips of arm one and two are intensively washed in a wash station before proceeding to the next well to avoid cross-contamination.

To demonstrate the usefulness of the new method, 31 overlapping linear hexa- and heptapeptide segments of endothelin [3] have been synthesized, replacing all cysteines by alanine. 170 double couplings have been performed using 10 equivalents of coupling reagents each. Despite this, high excess syntheses are extremely economical due to the small scale. In total, for all 31 endothelin segments, 6.7 g of Fmoc amino acids were used. The yield was between 3 and 5 mg of crude peptides, which were analyzed by HPLC and FABMS, giving evidence for high purity. Only the C-terminal heptapeptide fragment exhibited a slight affinity in the endothelin radio-receptor assay.

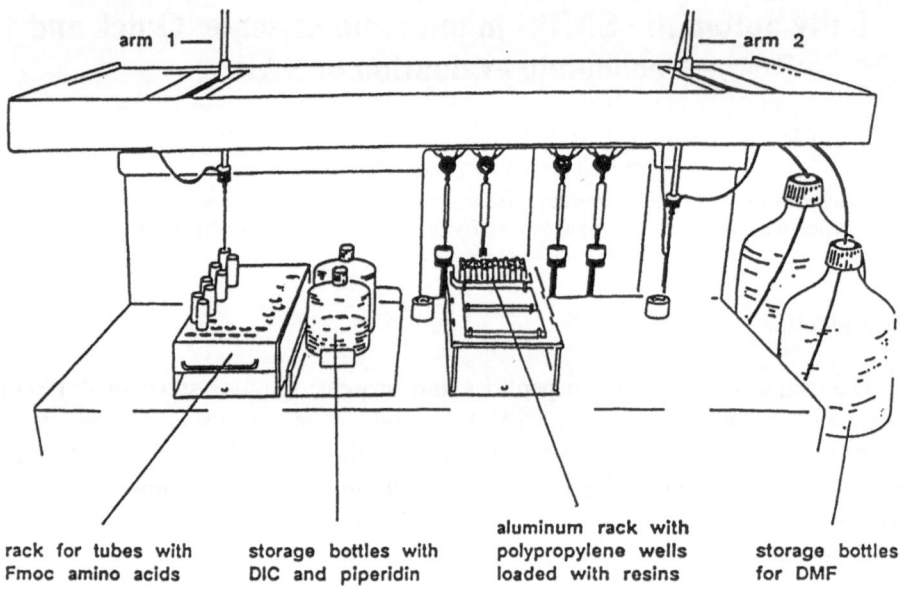

Fig. 1. Schematic representation of the automatic SMP synthesizer.

References

1. Houghten, R.A., Proc. Natl. Acad. Sci. U.S.A., 82(1985) 5131.
2. Schnorrenberg, G. and Knapp, W., German patent application P 3828576.
3. Yanagisawa, M., Kurihara, H., Kimura, S., Tomobe, Y, Kobayashi, M.M., Mitsui, Y., Yazaki, Y., Goto, K. and Masaki, T., Nature, 332(1988)411.

Advanced technologies in the synthesis of pharmaceutical peptides

Chr. Birr[a], G. Becker[a], Th. Nebe[a], H. Nguyen-Trong[a], Th. Müller[a],
M. Schramm[a], H. Kunz[b], B. Dombo[b] and W. Kosch[b]

[a]ORPEGEN, Med.-Molekularbiologische Forschungs-Gesellschaft mbH, Czernyring 22,
D-6900 Heidelberg, F.R.G.
[b]Institut für Organische Chemie, Universität Mainz, J.-J. Becher-Weg 18–20,
D-6500 Mainz, F.R.G.

Introduction

It is recognized that the chemistry for building peptide chains is less prone
to routine when compared to that used in digonucleotide synthesis. We have
utilized CAM (computer-assisted manufacturing) with the aid of the PSS-80™
from Applied Protein Technologies, Boston, to synthesize pharmaceutical peptides [1].

Results and Discussion

When coupling efficiencies reach the range of better than 99%, the precision
of automatic Dmt monitoring has been checked reproducibly with spectroscopic
transformation control measurements based on ninhydrin reaction (Table 1).

Table 1 *Precision of CAM in the peptide synthesis of HIV p24 PIVQNIQGQMVH-
QAISPRTLNAWVKV with chemical feedback control as cross-checked with ninhydrin
reaction spectroscopy*

Coupling onto BHA resin	Chemical feedback data Dmt measurement[a]	Ninhydrin reaction
Fmoc-Val	99.01 % [1)] 99,58 % [2)]	98.70 [1)] 99.80 [2)]
Fmoc-Lys(Boc)	99.60	99.80
Fmoc-Val	99.70	99.77
Fmoc-Trp	99.70	99.40
Fmoc-Ala	99.60	99.65

[a]Preset efficiency control level 99.5 %.

For the non-destructive release of pharma peptides from polymer phase (Fig.
1) we have introduced HYCRAM™ [2]. The catalyst is (triphenyl phosphine)₄
Pd[(0)] complex applicable in organic/aqueous solvents [3]. There are no hazardous
operations required, as in HF, TFMSA or HBr/TFA cleavages. For this
environmental reason, industrial scale-up of solid-phase production technologies
of pharma peptides is no longer inhibited by legal concerns (Table 2).

$$\text{PEPTIDE-}\overset{\overset{\text{O}}{\|}}{\text{C}}\text{-O-CH}_2\text{-CH=CH-}\overset{\overset{\text{O}}{\|}}{\text{C}}\text{-NH-CH}_2\text{-...polymer}$$

$$\text{Pd}^{(0)}$$

$$\text{PEPTIDE-}\overset{\overset{\text{O}}{\|}}{\text{C}}\text{-O-CH}_2\text{-CH-CH-}\overset{\overset{\text{O}}{\|}}{\text{C}}\text{-NH-CH}_2\text{-...polymer}$$

Acceptor

$$\text{Peptide-}\overset{\overset{\text{O}}{\|}}{\text{C}}\text{-O + Acc-CH}_2\text{-CH=CH-}\overset{\overset{\text{O}}{\|}}{\text{C}}\text{-NH-CH}_2\text{-...polymer + Pd}^{(0)}$$

Fig. 1. Peptide ester with hydroxycrotonoylamidomethyl (HYCRAM™) linkage to solid phase becomes activated by Pd$^{(0)}$ catalyst followed by allyl acceptor-mediated peptide acid release at neutral pH.

Thymosin-α_1-related peptides have been synthesized via CAM and HYCRAM and tested in vivo in bone marrow, thymus, lymphnodes, spleen, and peripheral blood. Data, as summarized in Table 3, are indicative of new therapeutic developments towards polyarthritis and allergy.

Table 2 *Computer-assisted manufacturing (CAM) of pharmaceutical peptides in polymer phase via HYCRAM release technology*

Pharmapeptide[a]	Yield-%[a] on HYCRAM resin	Peptide released from HYCRAM	
		Raw (%)[b]	Final (%)[b,c]
GnRH	a) 90	77	15
Eledoisin	b) 92	72	12
Somatostatin	a) 85	82	12
Thymopentin	b) 93	85	28
ANP	b) 85	78	8
Thymosin-α_1			
(13-28)	b) 80	68	24
Calcitonin	b) 78	70	3

[a] Synthesis strategy: a) Ddz-/*tert.* butyl-, b) Fmoc-/*tert.* butyl-.
[b] Yields based on initial load on polymer.
[c] Deprotected.

Table 3 *In vivo activity of Thymosin-α_1 (13–28) om the immune system, as determined by flow cytometry on 5-FU suppressed mice*

Sample	A(%)	B(%)	C(%)	D(%)	E(%)
Blank	1.5±1	1±0	40±0	12±2	7.5±2
Supp. + placebo	8.0±5	5±1	18±10	45±5	9.5±2
Supp. + T$_{13}$	4.0±2*	7±1*	5±2*	56±6*	12.5±2*

A: theta-T cells, bone marrow; B: pre-thymocytes, thymus;
C: immat. T. cells, Thymus; D: T-helper cells, thymus;
E: T-supp. cells, periph. blood.
* Significance $P>0.05$.

References

1. Horn, M., Novak, C. and Birr, C., In Jung, G. and Bayer, E. (Eds.) Peptides 1988, (Proceedings of the 20th European Peptide Symposium), De Gruyter, Berlin, 1989, p. 235.
2. Kunz, H. and Dombo, B., Angew. Chem. Int. Ed. Engl., 27 (1988) 711.
3. Trost, B.M., Tetrahedron, 33 (1977) 2615.

Multiple peptide synthesis using a single support (MPS3)

Foe S. Tjoeng, Derek S. Towery, Joseph W. Bulock, Deborah E. Whipple,
Kam F. Fok, Mark H. Williams, Mark E. Zupec and Steven P. Adams

*Department of Biological Sciences, Monsanto Company, 700 Chesterfield Village Parkway,
Chesterfield, MO 63198, U.S.A.*

Introduction

Increasing importance of peptides in many areas of biological and biochemical research places new demands on peptide chemistry. A large number of peptides must be synthesized in a short time, and this requirement can only be met by automated synthesis processes. Several research groups have developed a variety of manual solid-phase techniques [1,2] to synthesize a large number of peptides simultaneously. We have recently developed a procedure that allows for the simultaneous synthesis, cleavage and purification of several peptides in a single run. The technique is based on the synthesis of multiple peptides on a single solid-phase support and is easily adapted to manual or to automated methods.

Results and Discussion

This method has been employed in the synthesis of a series of magainin 2 and angiotensinogen analogs. In the coupling of the Boc-amino acid mixtures to the peptide resin, DCC/HOBt proved to be superior to DCC alone in achieving nearly uniform product ratios; it appears to be the coupling of choice. As demonstrated in the synthesis of angiotensinogen analogs (Fig. 1), the preferred conditions are easily and efficiently adaptable to automated operation; the method utilizes the standard program for the Applied Biosystems model 430A peptide synthesizer. The outstanding resolution of RPHPLC makes it possible to separate completely mixtures of closely related products, which is an essential element of the MPS3 method. Moreover, recent advances in FABMS provides a rapid, accurate assignment of the peptides in the synthetic mixture that dramatically facilitates the purification process. The MPS3 technique offers not only a reduction in the synthesis time, but more importantly, it also reduces the time required for the cleavage and purification, where the bottleneck normally occurs.

1034

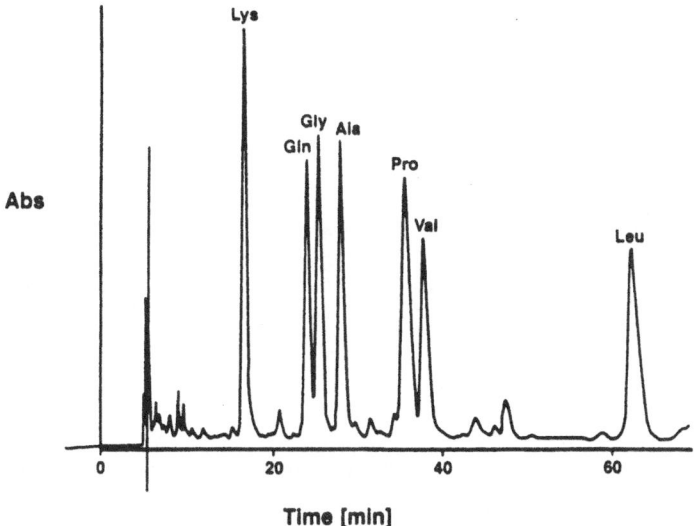

Fig. 1. *HPLC of the crude porcine [Lys^{11},Gln^{11}, Gly^{11},Ala^{11}, Pro^{11}, Val^{11},Leu^{11}]-angioten-*
*sinogen 1–14 peptides, Asp-Arg-Val-Tyr-Ile-His-Pro-Phe-His-Leu-**Leu**-Val-Tyr-Ser on a*
Vydac C-18 column (5 μm; 10 mm×25 cm). Solvent A: 0.05% TFA in acetonitrile;
solvent B: 0.05% TFA in water. Gradient: 22–24% solvent A in 60 min then 24–
35% solvent A in 40 min. Flow rate 3 ml/min. Detection at 215 nm.

References

1. Houghten, R.A., Proc. Natl. Acad. Sci. U.S.A., 82 (1985) 5131.
2. Geysen, H.M., Meloen, R.H. and Barteling, S.J., Proc. Natl. Acad. Sci. U.S.A.,
    81 (1984) 3998.

A simple approach to rapid parallel synthesis of multiple peptide analogs

Rolf H. Berg[a],*, Kristoffer Almdal[b], Walther Batsberg Pedersen[b], Arne Holm[a], James P. Tam[c] and R.B. Merrifield[c]

[a]*Department of General and Organic Chemistry, University of Copenhagen, DK-2100 Copenhagen, Denmark*
[b]*Chemistry Department, Risø National Laboratory, DK-4000 Roskilde, Denmark*
[c]*The Rockefeller University, 1230 York Avenue, New York, NY 10021, U.S.A.*

Introduction

Solid supports based on long-chain polystyrene-grafted polyethylene film matrices [1] can be handled as individually labeled and readily separable sheets, thus enabling a new, practical way of carrying out simultaneous synthesis of multiple-peptide analogs. We report here a rapid parallel synthesis of 13 melittin(7–21) analogs on a film support.

Results and Discussion

7   12 14
Boc-Lys(ClZ)-Val-Leu-Thr(Bzl)-Thr(Bzl)-Gly-Leu-Pro-Ala-Leu-Ile-Ser(Bzl)-
     21
Trp(CHO)-Ile-Lys(ClZ)-PAM-film

and twelve analogs derived from substitutions in position numbers 12 and 14 were assembled stepwise on 13 discrete sheets (each sheet: 1.5×3 cm, ca. 50 μm thickness, ca. 40 mg, substitution = ca. 0.6 mmol amino groups/g support) of 285 wt % long-chain, polystyrene-grafted polyethylene film (ca. 75% of the film is polystyrene). The common steps of deprotection, neutralization, washings, and coupling of identical amino acids were performed in a single reaction vessel, while the coupling of different amino acids was carried out in separate vessels, i.e., following a scheme similar to that outlined for Houghten's 'tea-bags' [2]. A standard solid-phase procedure employed double DCC coupling (3.5 equiv., 0.05 M, in 30% DMF/DCM) of all residues except Boc-Gln and Boc-Asn, and residues coupled after Gln, where the usual precautions were taken. Final deprotection and release of the peptides from the sheets were accomplished by the low/high-HF method [3]. From 1 cm² sheet (23.2 mg) of fully protected, film-bound melittin(7–21), the free 15-residue peptide was obtained in an overall synthetic yield of ca. 70%, and a homogeneity of unpurified product [see Fig. 1(A)]

---

*To whom correspondence should be addressed.

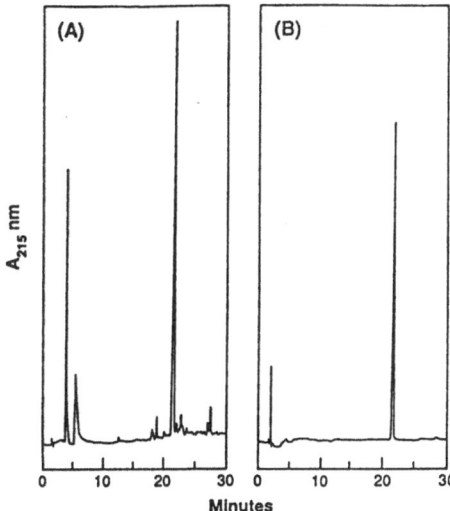

Fig. 1. Analytical HPLC chromatograms of (A) crude melittin(7–21) after low/high HF-cleavage and lyophilization, and (B) purified melittin(7–21). Buffer A, 5% CH₃CN/95% H₂O/0.0445% TFA; buffer B, 60% CH₃CN/ 40% H₂O/0.0390% TFA; linear gradient, 5– 95% of buffer B in 30 min; flow rate, 1.5 ml/min; column, Vydac C₁₈ (0.46 × 25 cm).

comparable to that obtained from other solid supports. Preparative RPHPLC afforded 3.2 mg of pure peptide [Fig. 1(B)] with the correct composition and molecular weight. The other twelve analogs were obtained in similar yield and purity.

Acknowledgements

We thank the Danish Research Academy and the Danish Natural Science Research Council for the partial support of this research.

References

1. Berg, R.H., Almdal, K., Batsberg Pedersen, W., Holm, A., Tam, J.P. and Merrifield, R.B., In Jung, G. and Bayer, E. (Eds.) Peptides 1988, De Gruyter, Berlin, 1989, p. 196.
2. Houghten, R.A., Proc. Natl. Acad. Sci. U.S.A., 82(1985)5131.
3. Tam, J.P., Heath, W.F. and Merrifield, R.B., J. Am. Chem. Soc., 105(1983)6442.

Use of carboxypeptidase Y for the introduction of probes into proteins via their carboxy terminus

Meir Wilchek[a], Alexander Schwarz[b], Christian Wandrey[b] and Edward A. Bayer[a]

[a]Department of Biophysics, The Weizmann Institute of Science, Rehovot 76100, Israel
[b]Institute of Biotechnology 2, Nuclear Research Center – Jülich, P.O. Box 1913,
D-5170 Jülich, F.R.G.

Introduction

Most labeling, or cross-linking, procedures result in a random group-specific modification of amino acid side chains (e.g., lysines, cysteines, etc.). In some cases, vectorial cross-linking can be achieved using heterobifunctional reagents, although the sites of modification of the two protein species generally occurs randomly. A more selective method for cross-linking two proteins involves their conjugation via the C-terminal group of one protein to the N-terminal residue of the other by genetic engineering, resulting in fused proteins. However, a simple chemical method for the selective incorporation of reporter groups at a given site in proteins is still lacking.

The pioneering work of Morihara, et al. [1,2] demonstrated the modification of porcine insulin to human insulin in which the C-terminal threonine residue was replaced by alanine. By using a similar enzymatic approach, we reasoned that a variety of potentially useful derivatized amino acids would also be subject to incorporation by this procedure. The present communication describes the labeling of proteins at their C-terminus with biocytin amide or N^ϵ-maleimido-propionyl-L-lysine amide using the nonspecific protease carboxypeptidase Y (CPD-Y). The procedure may also be used to label proteins with other probes (based on other trifunctional amino acid derivatives), provided that the probe-containing amino acid amide has a free α-amino group.

Methods

CPD-Y-mediated modification of proteins. Biocytin amide [3] or N^ϵ-maleimi-dopropionyl-L-lysine amide (100 mg/ml) was dissolved in a mixture of water/dimethylformamide/ethanol (2:1:1), and the pH of the solution was adjusted to 8.5 with 5 M NaOH. To this solution, the target protein and CPD-Y in a molar ratio of 100:1 were added, and the reaction was carried out overnight with shaking at 4°C. Free reagent was separated from the modified protein by subjecting the reaction mixture to gel filtration on Sephadex G-25 eluted with 1 M acetic acid. The solution was dialyzed against water or buffer and concentrated by lyophilization or ultrafiltration.

Analysis of biotinylated proteins. The incorporation of biotin into proteins

1038

was demonstrated by dot blots of the biotinylated protein which were stained with avidin-complexed alkaline phosphatase and a suitable, precipitable, colored substrate [4]. The C-terminal-biotinylated protein was further characterized by SDS-PAGE and blot transfer onto nitrocellulose membrane filters, followed by a similar staining procedure.

The percentage of biotinylated protein was determined by applying the preparation to an avidin-Sepharose affinity column. The fraction of protein which bound to the column was indicative of that which underwent biotinylation.

Analysis of maleimido-proteins. The modified protein (and unmodified controls) were subjected to treatment with cysteine, and the residual free cysteine was assayed colorimetrically using 5,5'-dithio-bis(2-nitrobenzoic) acid.

The percentage of maleimido-protein was also estimated by interaction of the preparation with SH-Eupergit. The immobilized protein was washed extensively with mercaptoethanol, and the percentage of free protein was determined.

Results and Discussion

The following proteins have been successfully labeled with the biotin moiety at the C-terminus: insulin; lysozyme; ribonuclease; trypsin; myoglobin; and cytochrome-c. By adsorption on an avidin-containing column, the fraction of biotinylated protein molecules could be determined. About 70% of the given protein sample was biotinylated in each case. Experimental verification of the incorporation of the maleimido derivative was accomplished by reaction of the resultant protein with cysteine or by adsorption of the modified protein onto thiol-containing polymers. The average yield of the incorporation of the maleimido group was about 60%, but was dependent on the protein species, varying between 75% for cytochrome-c and 40% for insulin.

The biotinylated proteins can be further cross-linked through their C-terminal residue using avidin, which has four binding sites for biotin. The maleimido-protein can be cross-linked to available protein SH groups (either native or introduced through chemical means). The cross-linking can also be performed through a C-terminal cysteine introduced into a second protein species by the enzymatic approach described here.

In the CPD-Y-mediated approach, the use of a good nucleophile results in a more efficient coupling to the target protein. Thus, an amino acid amide is superior to the corresponding underivatized amino acid. Amino-acid esters should not be employed in combination with CPD-Y due to multiple coupling and resulting oligomerization [5].

The reaction appears to proceed via transpeptidation, since we were unable to couple biocytin amide to the C-terminal proline of ovalbumin. Consequently, one limitation of the above-described approach is that proteins bearing a C-terminal proline are not susceptible to such modification. Nevertheless, examples of such proteins are relatively rare. It should also be noted that the reaction is usually incomplete; modified and native protein species should, therefore, be separated by suitable means, e.g., HPLC, ion-exchange chromatography, gel filtration, or affinity chromatography using a column that selectively adsorbs the incorporated probe. This approach is currently being applied for the incorporation of fluorescent groups onto the C-terminus of proteins.

References

1.   Morihara, K., Oka, T. and Tsuzuki, H., Nature, 280 (1979) 412.
2.   Morihara, K., Ueno, Y. and Sakina, K., Biochem. J., 240 (1986) 803.
3.   Hofmann, K., Finn, F.M. and Kiso, Y., J. Am. Chem. Soc., 100 (1978) 3585.
4.   Wilchek, M. and Bayer, E.A., Anal. Biochem., 171 (1988) 1.
5.   Breddam, K., Widmer, F. and Johansen, J.T., Carlsberg Res. Commun., 46 (1981) 361.

Chemical synthesis of the common-region peptide (PAS-57) of the human calcitonin precursor and identification in humans

H. Rink[a], W. Born[b] and J.A. Fischer[b]

[a]Pharmaceuticals Division, Biotechnology, Ciba-Geigy Ltd., CH-4002 Basel, Switzerland
[b]Research Laboratory for Calcium Metabolism, Department of Orthopedic Surgery and Department of Medicine, University of Zürich, CH-8008 Zürich, Switzerland

Introduction

The N-terminal peptide (1–57) (PAS-57) of procalcitonin has been isolated [1] from a human medullary thyroid carcinoma (MTC) and its structure found to be consistent with the cDNA-sequence [2]:

APFRSALESSPADPATLSEDEARLLLAALVQDYVQMKASELEQEQE-REGSSLDSPRS

We report a stepwise SPPS, a fragment-condensation synthesis approach on the resin, and monitoring of immunoextracted PAS-57 in normal humans and in MTC patients.

Results and Discussion

Syntheses: Stepwise Fmoc-SPPS with OBut, But and Pmc side-chain protection was performed on Wang-resin [double couplings (99.3% average yield) with DPCDI preformed HOBt esters (3 equiv.) in NMP at 45° for 40 min, Ac_2O capping, 2×60 min TFA/water/ethanedithiol 93/5/2 (vol), RT, for cleavage and deprotection]. After reduction of the Met-sulfoxide peptide and purification (gel filtration, CCD, RPHPLC) the overall yield was 6%. FABMS of the product and of protease digests was compatible with the structure. The apparently homogeneous peptide [HPLC (Fig. 1b), isoelectric focussing ($p_I = 4.2$), SDS-electrophoresis, TLC] still contained by-products (ca. 15%) according to capillary electrophoresis (CE) (Fig. 1c).

Protected fragments 1 (1–14), 2 (15–27) and 3 (28–49) were synthesized on an extremely acid-labile resin in high yields as described [3]. Fragments 3, 2 and 1 (2 equiv.) were coupled to resin-bound fragment (50–57) (BOP/DIEA in DMSO/NMP 1/1, 45°, 20–65 h) with 65%, 43% and 5% yield (Fmoc-determination).

Identification of PAS-57 in human plasma: Affigel-10 coated with antibodies to PAS-57 was used for immunoextraction. The binding capacity of the antibody-coated gel was higher than 10 ng synthetic PAS-57 per 10 µl of gel. Extracts

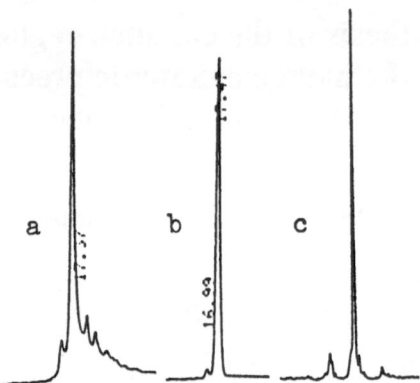

Fig. 1.   HPLC of (a) crude peptide, (b) purified peptide and (c) CE of purified peptide.

were analyzed by HPLC as previously described [4]. PAS-57 was measured by a homologous and a specific RIA in plasma extracts and in HPLC fractions. Immunoreactive components showed retention times on HPLC like those of synthetic PAS-57 and its sulfoxide.

Conclusion

The high yields described here clearly demonstrate the usefulness of stepwise Fmoc-SPPS for the efficient synthesis of large peptides. The detection of by-products wit CE in the apparently HPLC-homogeneous product exemplifies once more the need for great care in characterizing synthetic products. Coupling of large fragments (14–22 aa) on the resin was severely hampered; whether shorter segments can be coupled more efficiently on the resin has not been shown. This approach seems not to be generally applicable.

PAS-57 represents a major calcitonin gene-derived product in the circulation of normal humans and of MTC patients. Much like calcitonin, circulating PAS-57 is increased in MTC patients and is stimulated by Ca and pentagastrin given intravenously to normal subjects (3-fold) and to MTC patients (3.5-fold). In normal subjects, significantly larger amounts are found in men (194 ± 27 pgequiv./ml) than in women (102 ± 10 pgequiv./ml). The biological functions of PAS-57 remain to be elucidated.

References

1.   Conlon, J.M., Grimelius, L. and Thim, L., Biochem. J., 256(1988)245.
2.   Jonas, V., Lin, C.R., Kawashima, E., Semon, D., Swanson, L.W., Mermod, J.J., Evans, R.M. and Rosenfeld, M.G., Proc. Natl. Acad. Sci. U.S.A., 82(1985)1994.
3.   Rink, H, In Jung, G. and Bayer, E. (Eds.) Peptides 1988 (Proceedings of the 20th European Peptide Sympcsium), De Gruyter, Berlin, 1989, p. 139.
4.   Tobler, P.H., Jöhl, A., Born, W., Maier, R. and Fischer, J.A., Biochem. Biophys. Acta, 707(1982)59.

Activating peptide fragments for conformationally catalyzed resynthesis of proteins

Carmichael J.A. Wallace and Lori A. Campbell

Department of Biochemistry, Dalhousie University, Halifax, NS, Canada B3H 3N4

Introduction

Many examples exist of noncovalent complexes· of two or three protein fragments that adopt the conformation of the parent protein. If the match is exact, the novel termini are correctly oriented in close proximity, essential prerequisites for the catalysis of religation; but synthesis of peptide bonds from -COO⁻ and NH₃⁺ is thermodynamically unfavorable. However, when fragments are generated by CNBr cleavage, the terminal carboxylate is esterified as the homoserine lactone. Thus, [Hse[65]]cytochrome-c can be formed from a mixture of fragments 1–65 and 66–104 of the protein on standing in neutral aqueous solution [1]. We have attempted to mimic this process artificially and make a general method of it, because it has considerable advantages over traditional chemical methods of fragment condensation.

Conformationally catalyzed resynthesis is absolutely specific, so there are no side reactions, and no protection and deprotection are required. Furthermore, the synthesis proceeds to high yield under the mildest of conditions. The thermodynamic barrier to synthesis is overcome by esterification, but homoserine lactone is generally insufficiently activating. We have determined that dichloro-phenyl (Dcp) esters are ideal [1,2]. They can be introduced using the proteolytic enzyme used for fragment generation, under special conditions in which it acts in reverse to catalyze the condensation of peptides and free amino-acyl dichloro-phenyl esters. The approach has proved successful using trypsin [1–3]. In the present study, we have examined the potential of chymotrypsin (CT), since this is also a cheap and versatile enzyme.

Results and Discussion

The generation of contiguous two-fragment complexes with a site of CT specificity at the break point has been achieved in the roundabout ways shown in the flowchart (Fig. 1).

Activation of the 1–36 and 1–59 fragments and product purification followed the methods previously established with the substitution of CT for trypsin [1,2]. We found that better yields were obtained at room temperature, suggesting CT is sensitive to denaturation by the cosolvent, 1,4-butanediol, at higher temperatures. pH was varied in the range 5.6–7.1, and an optimum of 6.8 determined.

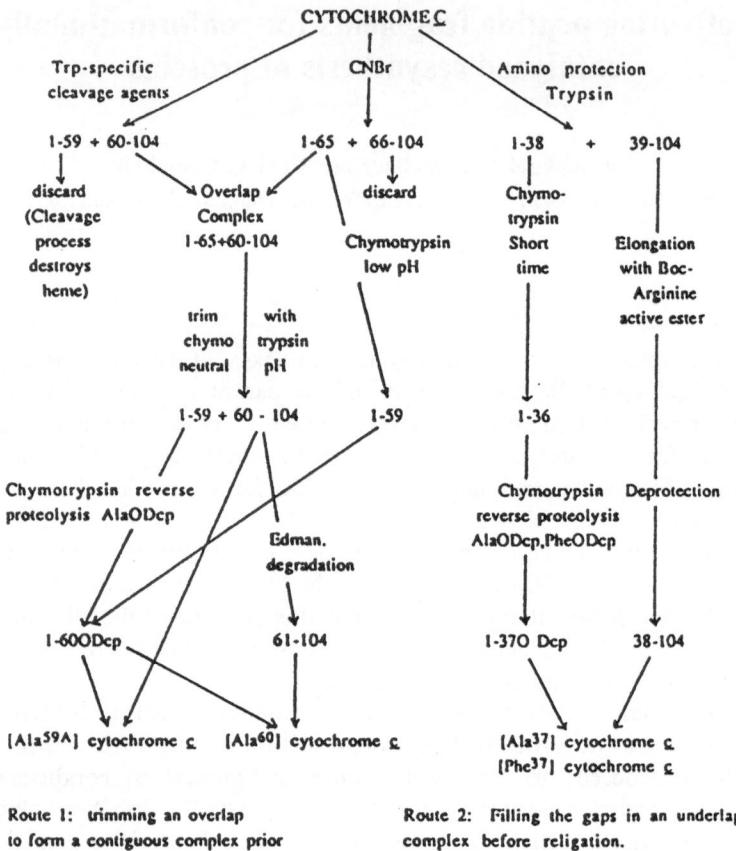

Fig. 1.   Reaction schemes as described in text.

This contrasts with that of 5.5 found for trypsin under analogous circumstances. There were marked differences between 1–36 (C-terminal Phe) and 1–59 (Trp) in the yield for extension with Ala-ODcp, and between Ala-ODcp and Phe-ODcp for the extension of 1–36, suggesting a strong sequence dependence for the cleavage/resynthesis equilibrium. Yields of extended-fragment Dcp esters were generally 20–50%, not as high as with trypsin, but nonetheless useful.

Couplings to complementary fragments were performed under conditions that with the 1–39/40–104 system gives up to 60% religation [1]. In neither of the present cases was the yield of resynthesized product greater than 5%. This result was anticipated in the case of the 1–60/61–104 system, since this is a 'weak' complex. It can be explained for the strong complex 1–37/38–104 by the putative role of Gly37, the residue we are attempting to substitute. It is the obligatory $i+2$ residue of a type-II 3_{10} bend [4]. Presumably with Ala or Phe, the folding required for conformationally catalyzed coupling is difficult to achieve.

Despite the limited success in coupling with these two complex systems, it is clear from our results to date that chymotrypsin can be as useful for activation in this approach to protein engineering as trypsin [1–3].

Acknowledgements

Supported by the Natural Sciences and Engineering Research Council of Canada. The development of this approach owes much to Drs. A.E.I. Proudfoot and K. Rose.

References

1. Proudfoot, A.E.I., Rose, K. and Wallace, C.J.A., J. Biol. Chem., 264 (1989) 8764.
2. Rose, K., Herrero, C., Proudfoot, A.E.I., Offord, R.E. and Wallace, C.J.A., Biochem. J., 249 (1988) 83.
3. Wallace, C.J.A., J. Biol. Chem., 262 (1987) 16767.
4. Takano, T., Trus, B.L., Mandel, N., Mandel, G., Kallai, O., Swanson, R. and Dickerson, R.E., J. Biol. Chem., 252 (1977) 776.

Practical peptide synthesis by fragment assembly on a polymer support

Kiyoshi Nokihara, Heribert Hellstern and Gerhard Höfle

*GBF-Gesellschaft für Biotechnologische Forschung, Mascheroder Weg 1,
D-3300 Braunschweig, F.R.G.*

Introduction

Toward the total synthesis of 126-amino acid residue pro-atrial hormone, human cardiodilatin (hCDD)/τANP, CDD-related peptides, αhANP/hCDD(99–126) and hCDD(1–38) were synthesized by fragment assembly to suppress the formation of deletion peptides and the loss of the resin-bound peptides [1].

Results and Discussion

The fragment assembly in SPPS is summarized as follows: (1) Preferred C-terminus of the fragment was a Gly residue. Gln, Ile, Pro, Val in the N-terminus and Ile, Val in the C-terminus were avoided. (2) N^αFmoc, Bzl-type side-chain protection was employed. β-Carboxyl function was protected by the 2-adamantyl ester (O2Ada) [2] to suppress the β-rearrangement. Arg was protected by the mesitylene-2-sulfonyl (Mts) group. Both O2Ada and Mts contributed to the superior solubility of fragments. (3) Benzyloxybenzyl alcohol resin was used for polymer support to allow cleavage with TFA. PAM-resin can also be used for the C-terminal fragment. Preparation of fragments with super acid labile resin [3] was attempted, but the resin was not economical, less stable during the synthesis, and side-chain protection was not suitable for the preparative HPLC employed. (4) First, amino acid was incorporated onto the Bzl-alcohol resin by the Mitsunobu reaction [4] to avoid racemization or Gly-Gly formation. (5) Assembly was carried out using protected fragment (2–3 fold excess), BOP, HOBT, and NMM (mol. ratio 1 : 1 : 1 : 1.5) in DMF with monitoring by ninhydrin test [5]. The reaction was generally performed in 1–2 h. Capping with 1 M acetyl imidazole was carried out. Incorporation of the fragment was almost quantitative [6]. (6) After removal of the N^αFmoc group, the peptidyl resin was cleaved with HF. And (7), HPLC, AAA, FABMS with isotope distribution, and sequencing were performed to characterize the peptides. D/L-analysis of acid hydrolysate was carried out using GC (CAT, Tübingen).

Synthesis of αhANP/hCDD(99–126). All fragments for αhANP (Fig.1) were easily prepared by stepwise SPPS. After HF-cleavage, the resulting Bis-Acm-αhANP (Fig. 2a) was treated with iodine and purified by HPLC to give biological active αhANP. No significant racemization was observed. As previously

(1– 3) :	Fmoc-N-P-M;
(4– 8) :	Fmoc-Y(BrZ)-N-A-V(Bzl);
(9– 13) :	Fmoc-N-A-D(O2Ada)-L-M;
(14– 18) :	Fmoc-D(O2Ada)-F-K(ClZ)-N-L;
(19– 23) :	Fmoc-L-D(O2Ada)-H-L-E(OBzl);
(24– 28) :	Fmoc-E(OBzl)-K(ClZ)-M-P-L;
(29– 38) :	Fmoc-E(OBzl)-D(O2Ada)-E(OBzl)-V-V-P-P-Q-V-L-Bzl.oxy-bzl-resin;
(99–100) :	Fmoc-S(Bzl)-L (solution synthesis);
(101–108) :	Fmoc-R(Mts)-R(Mts)-R(Mts)-S(Bzl)-C(Acm)-F-G-G;
(109–114) :	Fmoc-R(Mts)-M-D(O2Ada)-R(Mts)-I-G;
(115–118) :	Fmoc-A-Q-S(Bzl)-G;
(119–126) :	Fmoc-L-G-C(Acm)-N-S(Bzl)-F-R(Mts-Y(BrZ)-PAM-resin.
	[hCDD99–126] = α hANP].

Fig. 1.   Protected hCDD-fragments used for assembly.

described [1], the conventional SPPS with Boc-Bzl strategy for αhANP required triple or quadruple coupling to obtain the highest purity.

Synthesis of hCDD(1–38). Seven fragments (Fig. 1) were prepared stepwise, purified by preparative HPLC (yield 60–70%), and used for the assembly. After cleavage with HF, the crude peptide (Fig. 2b) was easily purifed by HPLC. hCDD(1–38) was assembled stepwise with the standard protocol of BOP/HOBT on TentaGel-Harz AC. To suppress nitrile formation, the side-chain amides were protected with trimethoxybenzyl group. The hCDD(1–38)-resin was cleaved with TFA, and the crude peptide (Fig. 2c) was purified on HPLC. The amount of D-Met in hCDD(1–38) prepared by the fragment method was 2–3-fold higher than that by stepwise synthesis.

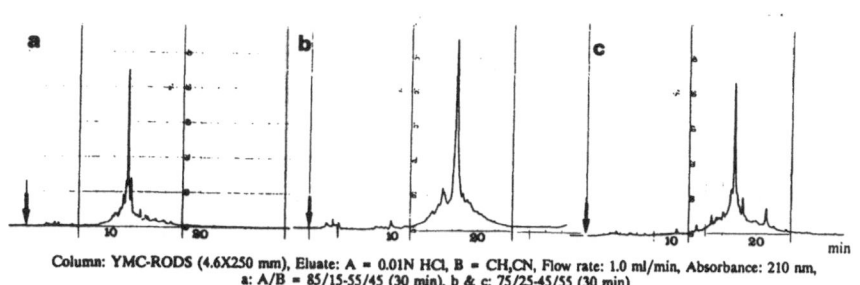

Column: YMC-RODS (4.6X250 mm), Eluate: A = 0.01N HCl, B = CH₃CN, Flow rate: 1.0 ml/min, Absorbance: 210 nm,
a: A/B = 85/15-55/45 (30 min), b & c: 75/25-45/55 (30 min)

Fig. 2.   HPLC profiles of HF-peptides, (a) Bis-Acm-αhANP/hCDD(99–126); (b) hCDD(1–38) by fragment assembly; (c) hCDD (1–38) by stepwise assembly.

In conclusion, the advantages of the present method are: the excess acyl components as protected fragments can be easily recovered, purified and re-used and the method generates various useful intermediate fragments. There were no significant differences for 30–40 amino acid residue peptides between the methods in purity or cost. Further elongation and production of CDD (stepwise with 100 coupling vs. fragment assembly with 20 coupling) is in progress.

Acknowledgements

We thank H. Dirks, S. Ruehe and R. Getzlaff for their assistance, and Dr. G. Chappuis, Novabiochem AG, for the use of a HF-reactor.

References

1a. Nokihara, K. and Semba, T., J. Am. Chem. Soc., 110 (1988) 7847
  b. Nokihara, K. and Semba, T., In Shiba, T. and Sakakibara, S. (Eds.) Peptide Chemistry 1987, Protein Research Foundation, Osaka, 1988, p. 371.
2. Okada, Y. and Iguchi, S., J. Chem. Soc. Perkin Trans. I (1988) 2129.
3. Rink, H., Tetrahedron Lett., 28 (1987) 3787.
4. Barlos, K., Gatos, D., Kallitsis, J., Papaioannou, D., Sotiriou, P. and Schäfer, W., Liebig's Ann. Chem., (1987) 1031.
5. Kaiser, E., Colescott, R.L., Bossinger, C.D. and Cook, P.I., Anal. Biochem., 34 (1970) 595.
6. Meienhofer, J., Waki, M., Heimer, E.P., Lambros, T.J., Makofske, R. and Chang, C.-D., Int. J. Pept. Prot. Res., 13 (1979) 35.

Building blocks for the covalent semisynthesis of apocytochrome c: SPPS and characterization of the N-terminal (1–66) sequence

C. Di Bello[a], C. Vita[b], L. Gozzini[c] and A. Hong[d]

[a]Institute of Industrial Chemistry, University of Padova, Via Marzolo 9,
I-35131 Padova, Italy
[b]Department of Organic Chemistry, University of Padova, Via Marzolo 1,
I-35131 Padova, Italy
[c]Farmitalia-Carlo Erba, R&D Biotechnology, Milano, Italy
[d]Applied Biosystems, Foster City, CA 94404, U.S.A.

Introduction

A possible strategy for the total synthesis of horse heart cytochrome c would require: (i) the synthesis of apocytochrome c; (ii) the stereochemically specific covalent attachment of the heme to Cys^{14} and Cys^{17} residues. Although the latter problem could be solved by using cytochrome c synthetase [1], the chemical synthesis of a peptide chain of over 100 residues remains a difficult task.

Interestingly, as recently observed [2], the two CNBr fragments [Hse > 65] (1–65) and (66–104) of horse heart cytochrome c bind non-covalently to the ferric heme segment (1–25) H to form a non productive three-fragment complex. Moreover, when the heme remains reduced at pH 5.6 for 48 h at 25°C, the peptide bond between the lactone activated Hse^{65} C-terminal residue of fragment (1–65) and the Glu^{66} N-terminal residue of the (66–104) fragment is restored to form [Hse^{65}] apocytochrome c with 20–40% yield. The [Hse^{65}] apocytochrome c thus obtained forms a complex with the ferri (1–25)H segment, which is indistinguishable from the analogous complex between the ferri (1–25)H segment and native apocytochrome c based on the intensity of the 695 nm absorption band, the rate of reduction by lactate dehydrogenase and CD spectra. Therefore, the present system, allowing the conformationally driven covalent semisynthesis of apocytochrome c, represents a useful tool for the preparation not only of the entire sequence of apocytochrome c, but also of analogs selectively modified both in the C-terminal and in the N-terminal regions of this important molecule.

The SPPS, purification and characterization of the C-terminal (66–104) fragment has been presented elsewhere [3]. Here we wish to report new data relative to the hexahexaconta peptide corresponding to the (1–66) N-terminal sequence.

Results and Discussion

The peptide corresponding to the (1–66) native sequence of horse heart cytochrome c has been synthesized by standard SPPS on a fully automated peptide synthesizer on PAM resin [4]. Double coupling with preformed symmetric anhydrides of Boc amino acids was used for each residue. In general, the first coupling was in DMF, the second in DMF/DCM. Asn, Gln and Arg were added as HOBt preformed active esters. Gly was activated in situ with DCC. Incompletely acylated peptides were capped with 20% acetic anhydride in pyridine. The following side-chain blocking groups were used: Asp(OBzl); Glu(OBzl); Thr(Bzl); Arg(Tos); Met(O); Tyr(Br-Z); Lys(Cl-Z); Cys(Acm). Fifty per cent TFA in DCM was added for removal of the Boc groups, and a 10% DIEA solution in DCM for neutralization. Couplings were monitored by a quantitative ninhydrin test. After HF cleavage from the resin and partial deblocking, the crude peptide showed a complex pattern in RPHPLC and was, therefore, submitted to a purification scheme involving: (i) preparative HPLC on a Waters SP-5PW cation exchange column eluted with a linear gradient of KCl in sodium acetate buffer, pH 5.0; (ii) preparative RPHPLC on μBondapak C_{18} eluted with a gradient of acetonitrile in 20 mM TEAP buffer, pH 2.2; (iii) preparative PRHPLC on μBondapak C_{18} column eluted with a linear gradient of acetonitrile in 0.1% TFA. The purified peptide eluted as a single peak in analytical HPLC. After acid hydrolysis, the amino acid composition was as expected. The presence of Trp[59] has been used to assess and monitor the homogeneity of this purified material, which showed a typical absorbance maximum at 280 nm and a fluorescence emission maximum at 350 nm upon excitation at 280 nm. To characterize the purified synthetic product further, tryptic and chymotryptic peptide mapping were also performed. Analysis of the peptides obtained by RPHPLC separation was in agreement with the expected fragmentation scheme. Better results [4] were obtained by removing the Acm protecting group immediately before the CNBr cleavage, which transforms the Met[65] residue in the activated Hse[65] lactone derivative.

In conclusion, the present approach based on the SPPS of the (1–66) sequence of apocytochrome c, followed by ion-exchange and RPHPLC purification, has allowed us to obtain milligram quantities of highly purified material. This peptide is at present used for the conformationally driven covalent semisynthesis of Hse[65] apocytochrome c in combination with native (1–25)H and the synthetic (66–104) fragment.

References

1. Basile, G., Di Bello, C. and Taniuchi, H., J. Biol. Chem., 255 (1980) 7181.
2. Gozzini, L., Taniuchi, H. and Di Bello, C., Fed. Proc., 45 (1986) 1617 and manuscript in preparation.
3. Di Bello, C., Tonellato, M., Lucchiari, A., Buso, O. and Gozzini, L., In Shiba, T. and Sakakibara, S. (Eds.) Peptide Chemistry 1987, Protein Research Foundation, Osaka, p. 409 and manuscript submitted to Int. J. Pept. Prot. Res.
4. Di Bello, C., Vita, C., Gozzini, L. and Hong, A., In Jung, G. and Bayer, E. (Eds.) Peptides 1988, De Gruyter, Berlin, 1989, p. 169.

Semisynthesis of carboxy-terminal fragments of thermolysin

Vincenzo De Filippis and Angelo Fontana

Department of Organic Chemistry, Biopolymer Research Centre of CNR, University of Padua, Via Marzolo 1, I-35131 Padua, Italy

Introduction

In recent years, great interest was devoted to protease-catalyzed 'semisynthesis' of protein molecules, defined as the rebuilding of a polypeptide chain from two components, one of which is obtained by chemical synthesis [1,2]. The advantages of this procedure lie in the possibility of obtaining large polypeptides with minimum use of protecting groups and chemical manipulations, and without the undesirable side-products usually generated in the chemical (solid-phase) synthesis of long peptides. Moreover, the procedure allows the introduction of unnatural or labeled amino acids at desired locations in the polypeptide chain, thus complementing current methods of protein engineering using genetic methods [3]. In this report we describe a high-yield procedure for the semisynthesis of long polypeptides encompassing the COOH-terminal region of thermolysin [4].

Results and Discussion

The semisynthesis procedure takes advantage of the availability of the COOH-terminal fragment 205–316 of thermolysin, obtained in high yields by specific autolysis of the 316-residue chain of the metalloprotease in the presence of EDTA [5]. This fragment, which autonomously folds into a stable and native-like globular structure, is a single polypeptide chain of 112 amino-acid residues, lacking disulfide or thiol groups, and contains a single glutamic acid residue in position 302 of the chain. This peculiar structural feature was exploited for specifically cleaving fragment 205–316 at Glu^{302} with *Staphylococcus aureus* V8 protease [6], and for preparing in high yields and homogeneous form fragment 205–302. This last fragment was coupled using V8-protease in the presence of organic solvents [2] to the peptide 303–316 (Val-Ala-Ser-Val-Lys-Gln-Ala-Phe-Asp-Ala-Val-Gly-Val-Lys) prepared by SPPS and purified to homogeneity by HPLC.

The protease-mediated synthesis of the peptide bond Glu^{302}-Val^{303} between fragment 205–302 and peptide 303–316 (molar ratio 1:5) was carried out in 0.1 M ammonium acetate buffer, pH 6.0, in the presence of 50% glycerol. The time-course of the resynthesis was followed by reverse-phase HPLC analysis of aliquots taken from the reaction mixture (Fig. 1). In the presence of V8-protease, a new peptide component was generated, eluting from the HPLC column after

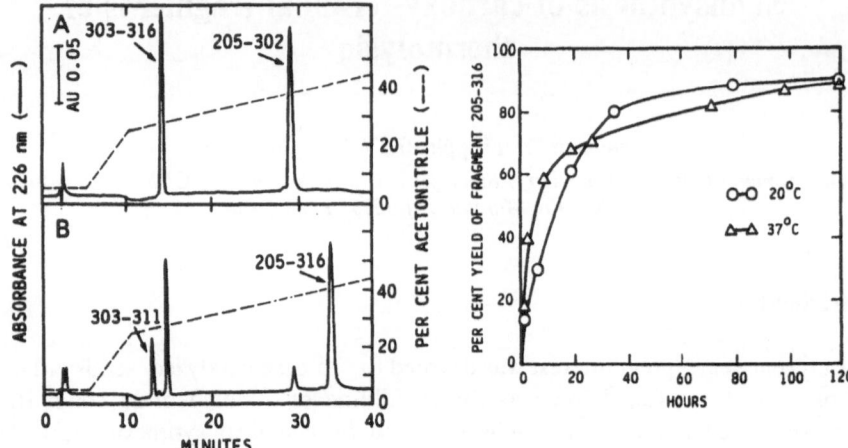

Fig. 1. Left. RPHPLC analysis of the V8-protease-catalyzed semisynthesis of the COOH-terminal thermolysin fragment 205–316. The analysis was carried out utilizing an Aquapore RP-300 C-8 column (4.6×100 mm) eluted at room temperature at a flow rate of 0.8 ml/min with a gradient of acetonitrile in 0.05% (v/v) aqueous TFA, as shown by the dashed line. The semisynthesis reaction was carried out dissolving fragment 205–302 (0.55 mg; 52 nmol) and the synthetic peptide 303–316 (0.4 mg; 281 nmol) in 100 μl of 0.1 M ammonium acetate buffer, pH 6.0, containing 50% glycerol and adding 3 μl of a solution of V8-protease (2 ng/ml in acetate buffer). (A) without V8-protease; (B) with V8-protease added to the reaction mixture and after 90 h incubation at 20°C. Numbers near the chromatographic peaks refer to the identities of the peptide material eluted from the column. Right. Time-course of the yields of V8-protease-mediated semisynthesis of fragment 205–316. The analysis was carried out by HPLC on aliquots taken from the reaction mixture incubated at 20 or 37°C. The experimental conditions of semisynthesis were those described above.

fragment 205–302, and at the same position as authentic fragment 205–316. Moreover, amino-acid composition and tryptic peptide mapping of the new component were as expected for fragment 205–316. These data led to the conclusion that, under the given experimental conditions, very efficient semisynthesis occurred, with 90% yields of coupling after 120 h of incubation at 20 or 37°C. An additional peak is seen in the RPHPLC chromatogram of the reaction mixture after 90 h reaction, preceeding the peak of peptide 303–316 (see Fig. 1, left). The peptide material of this peak gave an AAA consistent with sequence 303–311, indicating that, under the given experimental conditions of the semisynthesis reaction, V8-protease can cleave the synthetic peptide 303–316 at the level of the Asp[311]-Ala[312] peptide bond [6].

The semisynthetic procedure was also employed to obtain fragments 205–315 and 205–311 by V8-protease-mediated coupling of fragment 205–302 with 5 equivalents of peptide 303–315 or 303–311. These last peptides were prepared by proteolytic digestion of the synthetic peptide 303–316 (see above) with carboxypeptidase B and V8-protease, respectively, and isolated to homogeneity by semi-preparative HPLC. The yields of fragments 205–315 and 205–311 determined by HPLC analysis were ~90 and ~48% (based on fragment 205–302), respectively.

The novel procedure of semisynthesis described here provides new opportunities for additional studies on the structure-folding-stability relationships of COOH-terminal fragments of thermolysin, previously prepared in our laboratory by chemical and enzymatic fragmentation of the protein and shown to possess 'protein domain' properties [7–9]. Mutants of fragment 205–316 with amino-acid exchanges expected to alter the stability of the globular fragment, as well as mutants containing a tryptophan residue in position 310 of the polypeptide chain to be used as a fluorescent marker, are being synthesized and characterized. The results of these studies will be reported in subsequent publications.

References

1. Chaiken, I.M., CRC Crit. Rev. Biochem., 11 (1981) 255.
2. Seetharam, R. and Acharya, A.S., J. Cell. Biochem., 30 (1986) 87.
3. Offord, R.E., Prot. Eng., 1 (1987) 351.
4. Colman, P.M., Jansonius, J.N. and Matthews, B.W., J. Mol. Biol., 70 (1972) 701.
5. Fassina, G., Vita, C., Dalzoppo, D., Zamai, M., Zambonin, M. and Fontana, A., Eur. J. Biochem., 156 (1986) 221.
6. Drapeau, G.R., Methods Enzymol., 47 (1977) 189.
7. Vita C., Dalzoppo, D. and Fontana, A., Biochemistry, 23 (1984) 5512.
8. Dalzoppo, D., Vita, C. and Fontana, A., Biopolymers, 24 (1985) 767.
9. Dalzoppo, D., Vita, C. and Fontana, A., J. Mol. Biol., 182 (1985) 331.

Examples of the combination of solution-phase and solid-phase synthesis

H.U. Immer[a],*, I. Eberle[a], E. Moser[a], E. Bernath[b] and R. Pipkorn[b]

[a]Novabiochem AG, CH-4448 Läufelfingen, Switzerland
[b]Novabiochem GmbH, D-6902 Sandhausen, F.R.G.

Introduction

The assembling of large peptides from protected fragments prepared by SPPS is an attractive approach. We describe examples of this tactic using Fmoc-chemistry in conjunction with the hyper-acid-labile resin **1** developed by Rink [1] for the synthesis of appropriately protected fragments. BOP was used as coupling reagent for the stepwise elongations, as well as for the fragment couplings on the resin and in solution.

Glucagon fragments **2, 3, 4,** and **5** served as models to optimize the solid phase procedures.

Methods

```
Boc-His-Ser-Gln-Gly-Thr-Phe-Thr-Ser-Asp-R  2  Fmoc-Tyr-Ser-Lys-Tyr-Leu-Asp-R 3
     Trt  tBu           tBu   tBu tBu OtBu          tBu tBu Boc tBu      OtBu

Fmoc-Ser-Arg-Arg-Ala-Gln-Asp-R             4  Fmoc-Phe-Val-Gln-Trp-Leu-R      5
     tBu Pmc Pmc         OtBu
```

1 was synthesized and substituted with Fmoc-Asp(OtBu)-OH, Fmoc-Leu-OH, Fmoc-Tyr(tBu) and Fmoc-Ile-OH as described [1]. The substitution was, in all cases, ≈ 0.5 mM/g (determined by UV measurement at 300 nM of a cleaved aliquot). SPPS was done manually in a Milligen 504 shaker. Fmoc-cleavage: 20% Pip in DMA (20 ml/g resin; 20 min); icewashings after deprotection and couplings: 3×DMA, 3×isopropanol, 3×DMA. For couplings, the reagents were

---

*To whom correspondence should be addressed.

added in the following order: 3 eq. DIEA (0.5 m solution in DMA), 0.5 eq. HOBt (0.5 m solution in DMA), 1.5 eq. Fmoc-AA (0.5 – 0.2 m solution in DMA), 1.5 eq. BOP (0.5 m solution in DMA). Coupling times of 15 minutes were found to be sufficient in all couplings studied (monitoring by the Kaiser-test). In preliminary experiments, BOP was substituted successfully by PyBOP (benzotriazolyl-*N*-oxytris-[pyrrolidino]-phosphonium hexafluorophosphate) [2]. Cleavage from the resin: 10% AcOH in DCM (20 ml/g resin; 4 h).

Results and Discussion

Successive couplings in solution of **5** and **4** to **6** (prepared in solution), led to **7** (coupling conditions: 1.1 eq. BOP; 2 eq. DIEA, 0°; Fmoc-cleavage 10% diethylamine in DMF). Deprotection (TFA/ethanedithiol 4 : 1) afforded glucagon 16–29 in good yield and high purity after Mplc purification. No racemization was observed at the crucial Asp^{21} and Leu^{26}.

```
H-Met-Asn-Thr-OtBu      6
          |
          tBu
```

```
H-Ser-Arg-Arg-Ala-Gln-Asp-Phe-Val-Gln-Trp-Leu-Met-Asn-Thr-OtBu      7
 |    |   |           |                                |
 tBu  Pmc Pmc         OtBu                             tBu
```

*Synthesis of sarafotoxin S6b (**16**) [3]*

Fragments **8** (free acid), **9** (resinbound) and **10** (free acid) were synthesized as outlined for glucagon fragments. **10** was coupled with H-Trp-OtBu to afford **11** after Fmoc cleavage. Racemization to an extent > 10% was observed at Ile. The two diastereomers were separated by silica gel chromatography.

```
Boc-Cys-Ser-Cys-Lys-Asp-R      8    Fmoc-Phe-Cys-His-Gln-Asp-Val-Ile-R      10
    |   |   |   |                         |   |       |
    Acm tBu Trt Boc                       Trt Trt     OtBu
```

```
H-Met-Thr-Asp-Lys-Glu-Cys-Leu-Tyr-R      9    H-Phe-Cys-His-Gln-Asp-Val-Ile-Trp-OtBu      11
      |   |   |   |   |         |                    |   |       |
      tBu OtBuBoc OtBuTrt       tBu                   Trt Trt     OtBu
```

Coupling of **8** with **9** on the resin was done in the same way as described for single Fmoc AA (reaction time was extended to 30 min.). After cleavage from the resin and crystallization from DCM/MeOH the linear fragment 1–13 (**12**) was obtained in 60% yield (based on **9**). Cyclic **13** was obtained by I$_2$ oxidation in DCM/TFE as described for endothelin [4].

The linear peptide **14** was obtained without evidence of racemization (BOP at 0°C; 2 eq. DIEA). The oxidative ring closure with I$_2$ in MeOH to obtain **15** is analogous to that used in our endothelin synthesis [4]. The protecting groups were removed by TFA treatment with 2-methylindole as scavenger.

```
Boc-Cys-Ser-Cys-Lys-Asp-Met-Thr-Asp-Lys-Glu-Cys-Leu-Tyr-OH          12
    Acm tBu Trt Boc OtBu    tBu OtBuBoc OtBuTrt    tBu


Boc-Cys-Ser-Cys-Lys-Asp-Met-Thr-Asp-Lys-Glu-Cys-Leu-Tyr-OH          13
    Acm tBu     Boc OtBu    tBu OtBuBoc OtBu        tBu


Boc-Cys-Ser-Cys-Lys-Asp-Met-Thr-Asp-Lys-Glu-Cys-Leu-Tyr-Phe-Cys-His-Gln-Asp-Val-Ile-Trp-OtBu  14
    Acm tBu     Boc OtBu    tBu OtBuBoc OtBu        tBu   Trt Trt       OtBu


Boc-Cys-Ser-Cys-Lys-Asp-Met-Thr-Asp-Lys-Glu-Cys-Leu-Tyr-Phe-Cys-His-Gln-Asp-Val-Ile-Trp-OtBu  15
      tBu     Boc OtBu    tBu OtBuBoc OtBu        tBu           Trt       OtBu


H-Cys-Ser-Cys-Lys-Asp-Met-Thr-Asp-Lys-Glu-Cys-Leu-Tyr-Phe-Cys-His-Gln-Asp-Val-Ile-Trp-OH       16
```

```
Purification: Hplc on Vydac C18, 20 - 30 microns, 300 Å in 0.1 % TFA with a
              CH3CN gradient.
AAA: all AA in the expected ratio.
Racemization at crucial AA: Asp 0.2 % ; Tyr 0.8 % ; Ile 3.1 %
FAB-Ms: M+ at 2566
```

References

1. Rink, H., Tetrahedron Lett., 28 (1987) 3787.
2. Costa, J., Le-Nguyen, D. and Castro, B., Tetrahedron Lett., submitted.
3. Lee, C.Y. and Chiappinelli, V.A., Nature, 332 (1988) 343; Takasaki, C., Yanagisawa, M., Kimura, S., Goto, K. and Masaki, T., Nature, 332 (1988) 343.
4. Immer, H., Eberle, I., Fischer, W. and Moser, E., In Jung, G. and Bayer, E. (Eds.) Peptides 1988 (Proceedings of the 20th European Peptide Symposium), De Gruyter, Berlin, 1989, p. 94.

Comparison of solution with solid-phase synthesis of thymosin β4

Wolfgang Voelter, Hartmut Echner, Hubert Kalbacher, Afroditi Kapurniotu and
Peter Link

*Abteilung für Physikalische Biochemie, Physiologisch-chem. Institut, Universität Tübingen,
Hoppe-Seyler-Str. 4, D-7400 Tübingen, F.R.G.*

Introduction

Thymosin β4 (Ac-SDKPDMAEIEKFDKSKLKKTETQEKNPLPSKETIE-QEKQAGES-OH) is most widely distributed in the vertebrate kingdom and present in quantities of up to 100–200 μg/g tissue of different origin (thymus, lung, kidney, and so on [1]). Thymosin β4 has been reported to induce terminal deoxynucleotidyl transferase activity in bone-marrow cells [2], or to stimulate the hypothalamic secretion of LH/FSH-RH [3]; however, its real physiological function still remains uncertain. So far, only one solution [4] and one SPPS [5] of natural thymosin β4 were described in the literature. Herewith, we report an improved solution and 2 different SPPS of the polypeptide.

Solution synthesis: the solution synthesis is performed by stepwise condensation of the following 5 fragments I–V [4] starting with the two C-terminal segments IV, V: Ac-Ser(Bu^t)-Asp(OBu^t)-Lys(Boc)-Pro-Asp(OBu^t)-Met-Ala-Glu-(OBu^t)-Ile-Glu(OBu^t)-OH (I), Z-Lys(Boc)-Phe-Asp(OBu^t)-Lys-(Boc)-Ser(Bu^t)-Lys(Boc)-Leu-Lys(Boc)-Lys(Boc)-OH (II), Z-Thr(Bu^t)-Glu(OBu^t)-Thr(Bu^t)-Gln-Glu(OBu^t)-Lys (Boc)-OH (III), Z-Asn-Pro-Leu-Pro-Ser(Bu^t)-OH (IV) and Z-Lys(Boc)-Glu(OBu^t)-Thr(Bu^t)-Ile-Glu(OBu^t)-Gln-Glu(OBu^t)-Lys(Boc)-Gln-Ala-Gly-Glu-(OBu^t)-Ser(Bu^t)-OBu^t (V). For the condensations of these fragments, the BOP/HOBt coupling method proved to be the most efficient. The fully protected thymosin β4 was deblocked by TFA/ethyl methyl sulfide/ethanedithiol (90/5/5). The crude synthetic product containing the sulfoxide of thymosin β4 as well as thymosin β4 was further purified by preparative HPLC [Marcherey-Nagel Nucleosil C$_{18}$ column (250×10 mm, 7 μm particle size); gradient from 5% to 90% B in 30 min (A: 0.05% TFA in H$_2$O, B: CH$_3$CN/H$_2$O/TFA (600/400/0.5); flow rate: 3.1 ml/min; detection: UV at λ = 214 nm].

SPPS I, continuous-flow method (MilliGen 9050 PepSynthesizer, polymer: Fmoc-Ser(Bu^t)-PepSyn KA resin [6]; 2 g, 0.1 mmol/g). For couplings, penta-fluorophenyl esters with tert-butyl, side-chain protecting groups were used, except for Ser and Thr (HODhbt esters) in DMF. Fmoc groups were removed with 20% piperidine in DMF. All Fmoc amino acids, except the C-terminal one, were added automatically and delivered to the column as 0.33 M solutions in HOBt/DMF. Four-fold excess of acylating reagent was used in all cases. The N-terminal serine was acetylated with acetic anhydride/pyridine for 45 min, and the peptide resin treated with a mixture of TFA/thioanisole/ethanedithiol (10/2/1) for 2 h, then precipitated with ether and purified by HPLC (see above).

SPPS II, batch method (Biotronik Synostat P, polymer: Fmoc-Ser(Bu^t)-OH linked to p-benzyloxybenzyl alcohol resin [7]; 0.5 g, 0.3 mmol/g). The side chains of the Fmoc amino acids were protected as in SPPS I; however, for Asn and Gln, Mhb protection was used. The Fmoc amino acids HOBt and BOP were automatically dissolved in DMF in a four-fold excess and transferred to the reaction vessel. After addition of 8 equivalents of DIEA, coupling was carried out twice for 30 min. Deprotection was performed first with 50% piperidine/DMF (5 min) and second with 25% piperidine/DMF (15 min). The N-terminal serine was acetylated with acetic anhydride/pyridine (ten-fold excess) in DCM (60 min). The acetylated peptide resin was treated with a mixture of TFA/thioanisole/ethanedithiol (16/4/1) for 3 h; following precipitation with ether, the peptide was washed with cold ether and DCM. The crude peptide mixture was purified on a TSK-HW 40 S column (1.6×70 cm) using 5% acetic acid as a solvent and then the fractions containing thymosin β4 were lyophilized. All three synthetic and purified thymosin β4 samples are identical with that isolated from bovine thymus tissue, according to HPLC comparison, AAA and gel electrophoresis.

References

1. Horecker, B.L. and Morgan, J., Lymphokines, 9 (1981) 15.
2. Low, T.L.K., Hu, S.-K. and Goldstein, A.L., Proc. Natl. Acad. Sci. U.S.A., 78 (1981) 1162.
3. Rebar, R.W., Miyake, A., Low, T.L.K. and Goldstein, A.L., Science, 214 (1981) 669.
4. Kapurniotu, A., Link, P. and Voelter, W., In Jung, G. and Bayer, E. (Eds.) Peptides 1988, De Gruyter, Berlin, 1989, p. 97.
5. Low, T.L.K., Wang, S.-S. and Goldstein, A.L, Biochemistry, 22 (1983) 733.
6. Dryland, A. and Sheppard, R.C., J. Chem. Soc., Perkin Trans. I, (1986) 125.
7. Sieber, P., Tetrahedron Lett., 28 (1987) 6147.

Continuous solid-phase synthesis

Michal Lebl[a], Vladimír Gut[a] and Jutta Eichler[b]

[a]*Institute of Organic Chemistry and Biochemistry, Flemingovo 2, 16610 Prague 6, Czechoslovakia*
[b]*AdW der DDR, Institut für Wirkstofforschung, Alfred-Kowalke-Str. 4, 1136-Berlin, G.D.R.*

Introduction

In SPPS, material containing the starting functional group, usually the first amino acid, is placed into the reactor, and all steps of the synthesis are performed without isolation of intermediates. However, at the end of synthesis, the product must be taken out and worked up separately. Obviously, a system allowing for continuous input of starting material and reagents and continually producing the desired peptide would be very useful.

The idea of continuous SPPS may have come to the mind of peptide chemists after the introduction of SPPS by Merrifield [1], since it represents the next logical step in the development of this technology. However, its verification was possible only after continuous carriers became available. We have proven the soundness of continuous peptide synthesis with the use of cotton strips. However, the use of carriers modified only at the surface layer, allowing for a very high coupling rate, would be optimum.

Results and Discussion

The key component of the continuous synthesizer is the continuous carrier, which allows simultaneous performance of all synthetic steps at different locations of the carrier. During all operations, the carrier moves from one compartment to another, and the time of exposure to a particular bath is determined by the path length through this bath and the velocity of carrier movement. The carrier is led through the system of stirred or shaken baths in which washing is performed. To increase the effectiveness of washing, the carrier is, before reaching the next solution, compressed between two cylinders along with porous material (paper, textile), which removes the liquid from the previous wash. Use of the Fmoc protecting group requires less washing solution than using Boc protection. After deprotection in a bath containing the cleavage solution (in our case 20% piperidine in dimethylformamide) and thorough washing, the carrier is introduced to the solution of activated amino acid, and coupling proceeds. The exposure in this compartment must be determined experimentally to assure complete coupling. The most convenient method is continuous monitoring by bromophenol blue [2] which consists of observing decolorization of the blue carrier during coupling. After the coupling, the carrier again undergoes washing and it is ready to enter the next segment in which another amino acid is coupled to the peptide chain. In the last compartment of the synthesizer, the carrier is introduced to the cleavage solution, which releases the peptide from the carrier.

The search for a suitable carrier led us to use cotton, which had shown the most promising mechanical and chemical properties. Our experimental set-up on which we tested continuous peptide synthesis consisted of only one set of rollers and shaken baths; therefore, all the transfers had to be made manually. We used cotton strips 3 cm wide onto which Boc-glycine or Boc-alanine is coupled by the action of dicyclohexylcarbodiimide in the presence of dimethylaminopyridine. It is necessary to pretreat the cotton with trifluoroacetic acid and diisopropylethylamine before the coupling. Without this pretreatment, the coupling of protected amino acid onto the cotton is much less efficient. This carrier has a capacity of 3.1 μmol/cm^2 or 0.1 mmol/g. After cleavage of the Boc protecting group by TFA, we continue the synthesis with Fmoc-protected amino acids or attach onto the first amino acid the acidolytically cleavable handle (oxymethylphenoxypropionic acid) and perform the synthesis on this handle. The time required for complete coupling is determined by bromophenol blue monitoring, i.e., until the cotton strip looses its blue color. After the last step, the carrier is treated either with the 50% solution of trifluoroacetic acid and 5% dimethylsulfide in dichloromethane (for syntheses performed on the handle) or with a 1 M solution of sodium hydroxide. The products are purified by RPHPLC, and characterized by FABMS, then elemental and AAA. We have prepared the following peptides: Tyr-Gly-Phe-Met, Tyr-Gly-Gly-Phe-Met, Tyr-Gly-Gly-Phe-Met-Gly, Pro-Leu-Gly-Ala, Leu-Pro-Gly-Ala, Leu-Phe-Pro-Val-Ala, Leu-Phe-Pro-Val-Gly-Ala, Cys(Acm)-Tyr-Ile-Gln-Asn-Cys(Acm)-Pro-Leu-Gly, Cys(Acm)-Tyr-Ile-Met-Asn-Cys(Acm)-Pro-Leu-Gly.

References

1. Merrifield, R.B., J. Am. Chem. Soc., 85 (1964) 2149.
2. Krchňák, V., Vágner, J., Šafář, P. and Lebl, M., Coll. Czech. Chem. Commun., 53 (1988) 2542.

Chemical synthesis of the A-chain of human insulin

J.P. Mayer, G.S. Brooke and R.D. DiMarchi
Lilly Research Laboratories, Indianapolis, IN 46285, U.S.A.

Introduction

The chemical synthesis of the insulin A- and B-chains has been reported by classical [1] and solid-phase methods [2]. Previously, we reported a sizable synthetic loss in the automated synthesis of EGF that was directly attributable to the presence of multiple cysteine residues [3]. Therefore, the insulin A-chain, by virtue of four cysteine residues among its total of twenty-one (Gly-Ile-Val-Glu-Gln-Cys-Cys-Thr-Ser-Ile-Cys-Ser-Leu-Tyr-Gln-Leu-Glu-Asn-Tyr-Cys-Asn), was envisioned to represent an appreciable synthetic challenge.

Results and Discussion

The synthetic protocol for peptide chain assembly was as prescribed by Applied Biosystems, for use with their 430A instrument. Double couplings were followed by acetic anhydride capping of any remaining unreacted amine.

The total length of A-chain is of a size that the last residues in the synthesis extend to a region where amino acid couplings are generally less efficient. To compound the synthetic difficulty, an isoleucyl-valine sequence exists within this region. The presence of a single glycine as the first amino acid provides a diagnostic signal that can be used to monitor synthetic efficiency when compared to the C-terminal residues. AAA at various points in the synthesis of A-chain are shown in Table 1.

The resin analyses for synthesis with and without acetic anhydride capping are similar, with a single, notable exception. The glycine incorporation in the uncapped synthesis exceeded that in the capped by an absolute amount of 15.2%. This difference presumably represents peptide chains that had been purposefully terminated in the capped synthesis by acetylation, following incomplete couplings. The 65.5% incorporation of glycine in the capped synthesis indicates that approximately two-thirds of the initial asparagine had completed the synthesis. This equates to a 97.9% average efficiency at each step over the course of the twenty cycles.

The total synthetic yield was approximately 9%, with the major loss being attributed to adverse reactions. This specific step yield of 29% was calculated by analysis of the A-chain content immediately following cleavage and solubilization as the *S*-sulfonate. Losses prior to this analysis point were accounted for by amino acid content of the peptide resin before and after HF-cleavage. The adverse reactions represent the sum total of loss due to peptide modification

Table 1 *AAA^a of synthetic peptides*

Residue	Std.	Peptide resin		HPLC purified	
		Capped	Uncapped	Capped	Uncapped
Asp	100.0	100.0	100.0	100.0	100.0
Thr	82.3	72.0	74.9	86.4	82.2
Ser	62.4	64.3	65.1	76.7	64.7
Glu	101.0	83.3	88.0	99.6	100.0
Gly	99.0	65.5	80.7	97.4	99.0
Cys	92.5	79.8	84.4	92.4	92.2
Val	97.8	53.6	60.4	89.7	97.5
Ile	82.0	64.0	69.6	76.7	79.6
Leu	103.0	91.1	97.8	101.0	101.0
Tyr	97.0	95.3	100.2	95.1	97.6

[a]Expressed as a percentage relative to aspartic acid which was set at 100%.

in automated synthesis, HF-cleavage, and conversion to the S-sulfonate derivative. To determine the relative loss due to the specific presence of the four cysteine residues, an additional synthesis was completed in which each cysteine was substituted with glutamic acid. The amino acid content of this peptide resin was nearly identical to that of the previous synthesis, with the exception of the expected changes in cysteine and glutamic acid. In the analog synthesis, the yield of A-chain-related peptide was dramatically observed to be increased by more than two-fold relative to that of the previous syntheses. Slightly more than 10% of this increase was a result of increased recovery of peptide in the HF-cleavage step. Presumably, this resulted from improved solubilization of the product, and not from increased cleavage from the support.

Conclusion

As previously observed in the synthesis of EGF [3], the presence of multiple cysteine residues sharply diminished the synthetic yield of A-chain S-sulfonate. The inefficiency in chain assembly, attributed to a diminished yield but a high level of purity was, nonetheless, attained through the application of preparative RPLC. Small differences were detected in the peptides prepared by the uncapped and acetic anhydride capped syntheses.

References

1. Katsoyannis, P.G., Tometsko, A. and Fukuda, K., J. Am. Chem. Soc., 85 (1963) 2863.
2. Marglin, A. and Merrifield, R.B., J. Am. Chem. Soc., 88 (1966) 5051.
3. DiMarchi, R., Osborne, E., Roberts, E. and Slieker, L., In Marshall, G.R. (Ed.) Peptides: Chemistry, Structure and Biology (Proceedings of the 10th American Peptide Symposium), ESCOM Leiden, 1988, p. 202.

Progress on the process-scale synthesis of pentigetide

G.R. Nagarajan*, M.S. Verlander** and H. Jayakumar

Immunetech Pharmaceuticals, 11045 Roselle Street, San Diego, CA 92121, U.S.A.

Introduction

Pentigetide (DSDPR), derived from the ε heavy chain of human IgE (Immunoglobulin E), was shown to inhibit hypersensitivity in the Prausnitz-Küstner reaction [1]. Subsequent clinical studies of the peptide demonstrated its therapeutic potential in humans for the treatment of allergic rhinitis, with none of the side effects commonly associated with anti-allergy drugs [2,3].

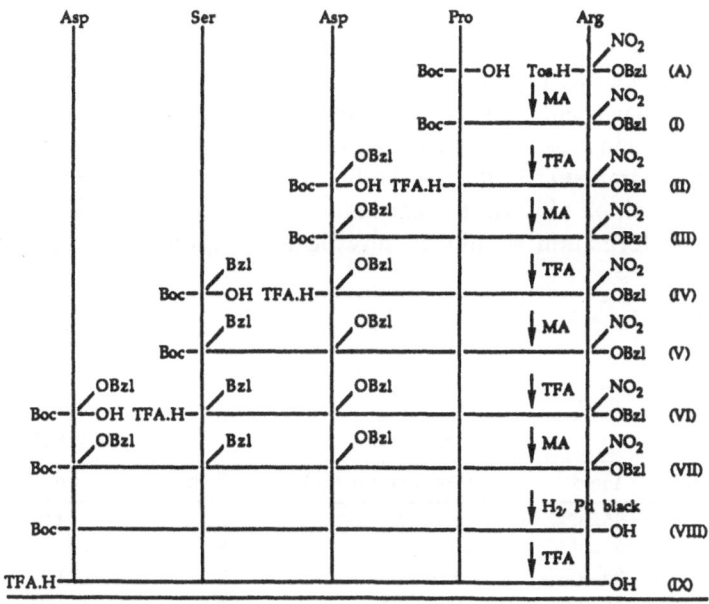

Scheme 1. Synthesis of pentigetide.

MA = mixed anhydride coupling method using isobutyl chloroformate/N-methylmorpholine. A 5–10% molar excess of carboxyl component was used in each case.

---

*To whom correspondence should be addressed.
**Present address: Bachem Inc., 3132 Kashiwa Street, Torrance, CA 90505, U.S.A.

G.R. Nagarajan, M.S. Verlander and H. Jayakumar

Results and Discussion

In order to develop a commercial-scale (multikilogram) synthesis of this potential drug, the classical, global-protection strategy has been investigated in conjunction with the mixed-anhydride method of coupling and only minimal purification of intermediates at each step (precipitation and/or trituration; Scheme 1). While yields by this procedure were good (75–95% for individual coupling steps), a number of persistent contaminants, attributed to the 'wrong-opening' of the mixed anhydride during the coupling reaction, were noted in the intermediates and the final protected pentapeptide. The side-product formation was maximum in the second coupling (Asp to Pro-Arg), and this coupling reaction was thus chosen as a model for optimization studies. For comparison, all four possible 'wrong-opening' products (N-isobutyloxycarbonyl-terminated peptides) were synthesized by direct acylation of the appropriate intermediates (A, II, IV, VI in Scheme 1) with isobutyl-N-succinimidyl carbonate, formed in situ.

Other coupling methods, such as the pivaloyl mixed anhydride, appeared to eliminate the side reaction, but the reaction rates were much slower and the yields disappointing. Addition of 1-hydroxy-benzotriazole (HOBt) as an additive [4,5] in the coupling step essentially eliminated the 'wrong-opening' product, and the tripeptide intermediate was obtained in excellent yields and high purity (Table 1). Similar results were obtained by this modified procedure in the third and fourth couplings as well. Surprisingly, however, the side-product formation was increased dramatically during the first coupling (Pro to Arg). Our results suggest that addition of HOBt to mixed-anhydride coupling reactions to suppress the urethane formation should be investigated carefully for each individual coupling.

Table 1 *Effect of addition of 1-hydroxybenzotriazole on purity and yields of intermediates*[a]

	Normal conditions		With HOBt	
Intermediate[b]	Yield	Iboc-peptide (%)	Yield	Iboc-peptide (%)
(I)	85–92%	1 – 2	90%	5–15[c]
(III)	83–90%	10 –16	88%	<0.8
(V)	85–95%	1.0– 1.5	92%	<0.3
(VII)	85–92%	0.3– 0.5	94%	<0.05

[a] 1-Hydroxybenzotriazole (1 equiv.) and N-methylmorpholine (1 equiv.) were added to the preformed mixed anhydride (–15°C) 15 min prior to addition of the amine component. After 20 min reaction, the pH was adjusted to approximately 6.5 with N-methylmorpholine. Reaction mixtures were analyzed by RPHPLC using a gradient of 30–60% acetonitrile in 0.1 M sodium phosphate, pH 4.5, comparing with synthetic standards.
[b] Refer to Scheme 1 for structures of intermediates.
[c] Various conditions were studied, including the literature conditions [5] in which the reaction mixture was adjusted to pH 8.0.

Deprotection of the protected pentapeptide was accomplished in two steps, first by catalytic hydrogenation (palladium on carbon or palladium black at atmospheric pressure in methanol : acetic acid : water 8 : 1 : 1), followed by acidolysis of the N-terminal t-Boc group with TFA in CH_2Cl_2 (room temperature, 30 min). Purification of the final product was achieved by ion-exchange chromatography on DEAE Sephadex.

The optimized process described above has routinely resulted in Pentigetide of high purity (Fig. 1) and good yield (about 40% overall) from kilogram-scale batches.

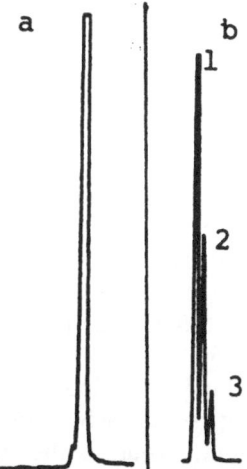

Fig. 1. Analytical HPLC of (a) purified pentigetide; (b) mixture of 1 (DSDPR), 2(β-DSDPR) and 3 [D(D-Ser)DPR]. Conditions: Ultrasphere ODS (4.6 mm × 25 cm); linear gradient of 0–12% acetonitrile in 0.1 M sodium phosphate, pH 4.5 over 30 min then isocratic for 15 min. Flow rate = 1 ml/min (see [6]).

References

1. Hamburger, R.N., Science, 189 (1975) 389.
2. Cohen, G.A., O'Connor, R.D. and Hamburger, R.N., Ann. Allergy, 52 (1984) 83.
3. Hahn, G.S., Nature, 324 (1986) 283.
4. Birr, C., Aspects of the Merrifield Peptide Synthesis, Springer, New York, 1978, p. 54.
5. Prasad, K.U., Iqbal, M.A. and Urry, D.W., Int. J. Pept. Prot. Res., 25 (1985) 408.
6. Nagarajan, G.R., Boone, J.S., Stolzer, T.J. and Richieri, S.P., In Hugli, T.E. (Ed.) Techniques in Protein Chemistry, Academic Press, San Diego, 1989, p. 357.

SPPS of the α and β domains of human liver metallothionein 2 and the metallothionein of *Neurospora crassa*

Thomas L. Ciardelli[a], Kristen G. Dillon[b], Timothy E. Elgren[b], F. Jon Kull[b], Michael F. Reed[b] and Dean E. Wilcox[b]

[a]*Department of Pharmacology and Toxicology, Dartmouth Medical School, Hanover, NH 03756, U.S.A.*
[b]*Department of Chemistry, Dartmouth College, Hanover, NH 03755, U.S.A.*

Introduction

Metallothionein (MT) [1] is a ubiquitous, small (61 residue) Cys-rich protein which binds both essential [Cu(I), Zn(II)] and toxic [Cd(II), Hg(II)] metal ions in vivo. Its biosynthesis is induced by several metals, and MT is thought to be involved in metal-ion homeostasis. All the cysteines are involved in metal-thiolate bonding and, as shown by ^{113}Cd NMR [2], proteolysis [3], X-ray crystallography [4] and 2D ^{1}H NMR [5], the protein consists of two domains: the N-terminal or β domain binds three dipositive metals with its 9 cysteines, and the C-terminal or α domain binds four dipositive metals with its 11 cysteines. MT from the fungus *Neurospora crassa* (NcMT) [6] contains only 25 residues; however, its 7 cysteines are homologous to the first 7 of the β domain of mammalian MT.

Results and Discussion

Using SPPS, we have synthesized the α and β domains of human liver MT-2 and NcMT; we have also prepared a modified form of the β domain with Cys^{26} → Ser and Cys^{29} → Ser substitutions, resulting in a β domain with the same number, as well as homology, of cysteines as in NcMT. Figure 1 shows the sequences of these peptides. Standard t-Boc chemistry, and low-high HF cleavage

α domain: -K S C C S C C P V S C A K C A Q G C I C K G A S D K C S C C A
β domain: Ac-M D P N C S C A A G D S C T C A G S C K C K E C K C T S C (K)-
NcMT: G D C G C S G A S S C N C G S G C S C S N C G S K
β domain: Ac-M D P N C S C A A G D S C T C A G S C K C K E C K S T S S K-
(modified)

Fig. 1. Amino acid sequences of the synthesized peptides.

from the resin and amino-acid deprotection have been employed. RPHPLC purification led to peptide products with the yield and purity, as determined by analytical RPHPLC designated in Table 1. AAA indicated that the correct residue content was achieved for each synthetic peptide.

Table 1 *Properties of the synthetic peptides*

Peptide	Yield	Purity (%)	Metal binding (mol equiv.)
α domain	4% (50 mg)	90	Cd(II): 4 Cu(I): ~6
β domain	3% (30 mg)	90	Cd(II): 2.5 Ag(I): 5
N. crassa MT	6% (39 mg)	95	Cd(II): 2 Cu(I): 6
modified β domain	4% (18 mg)	91	Cd(II): 2.5

To demonstrate that these synthetic peptides are identical to the natural domains and NcMT, we have determined their metal binding properties and compared our results with those reported for the proteolytically derived domains from rat and rabbit liver MT-2 and the natural NcMT. The stoichiometries of Cd(II), Ag(I) and Cu(I) binding by the peptides (see Table 1) were determined by AAA and match those found for proteolytically derived domains [7] and NcMT [8]. UV absorption and CD spectroscopy were also used to monitor metal binding by the peptides. Electronic features in the 220–300 nm region were observed upon binding metal ions to MT and have been assigned [9] as thiolate-to-metal charge-transfer transitions. Titrations of the synthetic peptides with Cd(II) or Cu(I) led to the appearance of these bands at similar energies as reported for the native proteins. These absorption and CD features showed systematic changes in intensity until the maximum metal-to-protein stoichiometry was achieved, thus supporting the atomic absorption analysis of metal ion binding. Finally, ^{113}Cd NMR spectral data for Cd(II) binding to the synthetic α domain were obtained and compared to those reported for the α domain from rat liver MT-2 [10] and the α domain Cd(II) ions of whole human liver MT-2 [11]. These results show that, in the synthetic α domain, three of the Cd(II) ions are in environments identical to those in the whole protein and in the proteolytically derived α domain; the chemical shift of the fourth Cd(II), which is located near the junction with the β domain, indicates that this metal ion is in a somewhat different environment than in the native protein.

The modified β domain was prepared to study differences between NcMT and the β domain, particularly the role of the non-Cys residues in metal ion binding. If the Cys residues predominantly determine metal ion stoichiometry and spectral properties, then this modification should result in a β domain with NcMT metal binding properties. Although the Cd(II) binding stoichiometry does not appear to be altered, the CD spectroscopic features of this modified β domain now resemble more closely those of NcMT. This suggests that the intervening residues play some role in determining the metalloprotein structure.

Conclusion

In this study, we have shown that it is possible to use SPPS to prepare the individual domains of metallothionein and the entire *Neurospora crassa* protein. This demonstrates the successful use of SPPS for peptides with high Cys content, and now provides a methodology for the facile-sequence modification of MT. Synthesis of the modified β domain indicates the potential for using this methodology to elucidate the biophysical properties of these Cys-rich metal-binding proteins.

References

1. Kagi, J.H.R. and Kojima, Y. (Eds.) Metallothionein II, Birkhauser Verlag, Basel, 1987.
2. Lerch, K., Nature, 284 (1980) 368.
3. Otvos, J.D. and Armitage, I.M., Proc. Natl. Acad. Sci. U.S.A., 77 (1980) 7094.
4. Winge, D.R. and Miklossy, K.-A., J. Biol. Chem., 257 (1982) 3471.
5. Furey, W.F., Robbins, A.H., Clancy, L.L., Winge, D.R., Wang, B.C. and Stout, C.D., Science, 231 (1986) 704.
6. Wagner, G., Neuhaus, D., Worgotter, E., Vasak, M., Kagi, J.H.R. and Wüthrich, K., Eur. J. Biochem., 157 (1986) 275.
7. Nielson, K.B. and Winge, D.R., J. Biol. Chem., 260 (1985) 8698.
8. Beltramini, M., Lerch, K. and Vasak, M., Biochemistry, 23 (1984) 3422.
9. Kagi, J.H.R. and Vallee, B.L., J. Biol. Chem., 236 (1961) 2436.
10. Boulanger, Y., Armitage, I.M., Miklossy, K.-A. and Winge, D.R., J. Biol. Chem., 257 (1982) 13717.
11. Boulanger, Y. and Armitage, I.M., J. Inorg. Biochem., 17 (1982) 147.

Synthetic studies on the K$^+$-channel antagonist charybdotoxin

Elizabeth E. Sugg[a], Maria L. Garcia[b], Bruce A. Johnson[c],
Gregory J. Kaczorowski[b], Arthur A. Patchett[a] and John P. Reuben[b]

[a]Department of Exploratory Chemistry, [b]Department of Membrane Biochemistry and Biophysics and [c]Department of Biophysical Chemistry, Merck, Sharp & Dohme Research Laboratories, P.O. Box 2000, Rahway, NJ 07065, U.S.A.

Introduction

The isolation and characterization of charybdotoxin (ChTX) from the venom of the scorpion, *Leiurus quinquestriatus*, was recently reported [1] as shown below. ChTX is a potent and selective inhibitor of the high conductance Ca^{2+}-activated K$^+$-channel in primary bovine aortic smooth muscle cells [1]. A second peptide, iberiatoxin (IbTX), was recently isolated from the venom of the scorpion *Buthus tamulus*. IbTX, with $>60\%$ sequence homology to ChTX (see sequence below), interacts noncompetitively with ChTX-binding sites.

ChTX	pQFTNVSCTTSKECWSVCQRLHNTSRGKCMNKKCRCYS
IbTX	pQFTDVDCSVSKECWSVCKRLFGVDRGKCMGKKCRCYQ

We now report the synthesis of ChTX and IbTX and of two hybrid analogs, ChTX$^{1-19}$IbTX$^{20-37}$ and IbTX$^{1-19}$ChTX$^{20-37}$. Assignment of the disulfide bonds of ChTX was accomplished by AAA of the fragments obtained from selective hydrolysis of native and synthetic ChTX with chymotrypsin and trypsin.

Results and Discussion

Analogs were prepared by SPPS on a Milligen 9050 using Fmoc-pentafluorophenyl ester methodology. The peptides were cleaved from the resin using TFA, cyclized by air oxidation at pH 8, and purified by gel filtration (Sephadex G-25) and RPHPLC. All four analogs were analytically pure by RPHPLC.

Synthetic ChTX was equivalent with native ChTX in the displacement of [^{125}I]-ChTX from bovine aortic sarcolemmal membrane vesicles (IC$_{50}$ = 0.02 nM). Synthetic IbTX, and the hybrid analogs ChTX$^{1-19}$-IbTX$^{20-37}$ and IbTX$^{1-19}$-ChTX$^{20-37}$, were about 30-fold less potent in this binding assay. Native and synthetic IbTX were non-competitive in the displacement of labeled ChTX-binding sites. The hybrid IbTX$^{1-19}$-ChTX$^{20-37}$ was a competitive inhibitor of ChTX binding, while the hybrid ChTX$^{1-19}$-IbTX$^{20-37}$ was noncompetitive. The IbTX$^{1-19}$-ChTX$^{20-37}$ hybrid had an electrophysiological profile very similar to ChTX on cultured bovine aortic smooth muscle membrane patches [2], while the profile for ChTX[1-19]-IbTX[20-37] resembled that of IbTX. This suggests that the C-terminal 17 amino acids are important for ChTX and IbTX-receptor selectivity.

Native and synthetic ChTX was treated with chymotrypsin and trypsin, and the resulting fragments were isolated by RPHPLC and treated with performic acid prior to AAA. Since tyrosine and tryptophan are destroyed by the performic acid treatment, the presence of Tyr and Trp was inferred from the UV intensity of the RPHPLC peaks. Table 1 lists results from AAA, and the assigned disulfide bonding pattern.

Table 1   *Fragments obtained from hydrolysis of ChTX with trypsin and chymotrypsin*

Peak[a]	Amino acid composition	Sequence assignment
15	His(0.84), Arg(1.2), Asx(1.3), Thr(1.2), Ser(1.4), Leu(1)	L[20]-H-N-T-S-R[25]
20	Lys(2.6), Asx(1.9), Thr(3), Ser(2.6), Cys(2.1), Val(1), Met(O)(0.86)	T-N-V-S-C[7]-T-T-S-K C[28]-M-N-K
21	Arg(1.01), Lys(0.94), Val(0.91), Glx(1.2), Ser(2.7), Thr(1.1), Asx(0.94),Cys(2.1)	S-V-C[17]-Q-R C[35]-Y
31	Arg(1.1), Glx(1.3), Cys(2.0)	E-C[13]-W C[33]-R

[a]Elution time (min) on Vydac C-18 (4.6 mm×25 cm), 0–20% acetonitrile in 0.1% TFA aq., 0.5%/min.

References

1.   Gimenez-Gallego, G., Navia, M.A., Reuben, J.P., Katz, G.M., Kaczorowski, G.J. and Garcia, M.L., Proc. Natl. Acad. Sci. U.S.A., 85 (1988) 3329.
2.   Hamil, O.P., Marty, A., Neher, E., Sakmann, B. and Sigworth, F.J., Pflüger's Arch., 391 (1981) 85.

Synthetic calcium binding EF hands: Testing the Acid Pair Hypothesis

Ronald E. Reid

Faculty of Pharmacy, University of Manitoba, Winnipeg, Manitoba, Canada R3T 2N2

Introduction

The Acid Pair Hypothesis was formulated to explain the differences in calcium affinity of the highly sequence-homologous EF hands found in such proteins as calmodulin (CaM), troponin C (STnC), S-100, calbindin, calcyclin and parvalbumin. The hypothesis assumes that the calcium affinity of EF hands can be described by the chemical nature of amino acids in positions 1, 3, 5, 7, 9 and 12 in the 12-residue loop region of the EF hands (Fig. 1). It evolved from the cobalt complex studies of Cotton and Bergman [1,2] who found that a polyatomic ligand, in which two chemically equivalent atoms are held much closer together in the cation complex than if the atoms were independent of each other, has a tendency to interact with the cation through the chemically equivalent atoms in such a way that the mean positions of the pairs of atoms lie roughly at the vertices of one of the usual coordination polyhedra.

The EF hands fulfill both requirements of polyatomic ligands with chemically equivalent atoms, and therefore, Cotton and Bergman's hypothesis was extended to the interaction of EF hands with calcium [3–5]. To test the hypothesis as it pertains to individual EF hands, a model peptide system was prepared using the sequence of bovine brain calmodulin calcium binding site III [6].

Results and Discussion

To determine the importance of the 6 amino acids in positions 1, 3, 5, 7, 9 and 12 of the loop region, a hybrid peptide of rabbit skeletal TnC site III (A^{98}STnC) and bovine brain calmodulin site III (L^{109}CaM) was synthesized (Fig. 1). The hybrid peptide (NDDL/4XZ CaM) consists of the natural sequence of calmodulin with the exception of the residues in positions 1, 3, 5, 7, 9 and 12 of the loop region that are identical to those of TnC site III (note: residue 109 was also changed from Met to Leu for synthetic reasons). The A^{98}STnC sequence has four acid residues paired on the x and z axes (Fig. 1) and has high affinity for calcium ($K_d = 28$ μM). The L^{109}CaM sequence has low affinity (735 μM), while the hybrid (NDDL/4XZ CaM) has a K_d for calcium of 19 μM, which is to be expected according to the Acid Pair Hypothesis.

The importance of the above 6 residues in calcium binding is obvious from the hybrid study. However, a more direct test of the hypothesis would come

K_D (µM)

LOOP
1 2 3 4 5 6 7 8 9 10 11 12

N-TERMINAL HELIX | C-TERMINAL HELIX

90 98 123	
28.2 ± 1.7 A⁹⁸STHC	K S E E E L A E A F R I F D R N A D G Y I D A E E L A E I F R A S G
81 95 97 101 109 113	
735 ± 61 L¹⁰⁹CAM	S E E E I R E A F R V F D K D G N G Y I S A A E L R H V L T N L G
19.1 ± 0.1 NDDL/4XZ CAM	S E E E I R E A F R V F D K N G D G Y I D A A E L R H.V L T N L G
524 ± 16 NDL/3X CAM	S E E E I R E A F R V F D K N G N G Y I D A A E L R H V L T N L G
58.8 ± 0.1 NDL/3Z CAM	S E E E I R E A F R V F D K N G D G Y I S A A E L R H V L T N L G

Fig. 1. Amino acid sequences and calcium dissociation constants of the synthetic analogs of the model helix-loop-helix calcium binding unit (EF hand).

from peptide analogs of CaM site III with 3 acidic residues and pairing of 2 of the 3 acid residues on axes of the coordinating tetrahedron. The first analog (NDL/3X CaM) has 3 acidic residues in the loop region at the x, $-x$ and $-z$ coordinating positions and is found to have a K_d for calcium of 524 µM. This affinity for calcium is slightly better than that of the natural sequence (L¹⁰⁹ CaM), but the increase in affinity is not considered large. The second analog (NDL/3Z CaM) has 3 acidic residues in the loop region at the x, z and $-z$ coordinates. The affinity of this peptide for calcium ($K_d = 54$ µM) is 14 times better than that of the natural CaM fragment (L¹⁰⁹ CaM) and 9 times better than that of the fragment with the acid pair on the x axis (NDL/3X CaM).

These results support the Acid Pair Hypothesis as an explanation for a molecular distinction between high and low calcium affinity in the EF hand calcium binding units. Additionally, we found that the acid pair on the z axis has greater effect on the calcium affinity than the acid pair on the x axis when the site contains 3 acidic residues in chelating positions in the loop region.

Acknowledgements

The author is indebted to Ms. Karen Kabaluk for her excellent technical assistance. This work is supported by Grant MA-8758 from the Medical Research Council of Canada.

References

1. Cotton, F.A. and Bergman, J.G., J. Am. Chem. Soc., 86 (1964) 2941.
2. Bergman, J.G. and Cotton, F.A., J. Inorg. Chem., 5 (1966) 1208.
3. Reid, R.E. and Hodges, R.S., J. Theoret. Biol., 84 (1980) 401.
4. Reid, R.E., J. Theoret. Biol., 105 (1983) 63.
5. Reid, R.E., J. Theoret. Biol., 114 (1985) 353.
6. Reid, R.E., Biochemistry, 26 (1987) 6070.

Synthesis of a 32-residue peptide inhibiting trypsin and carboxypeptidase A

Dung Le-Nguyen and Bertrand Castro

CCIPE, Rue de la Cardonille, F-34094 Montpellier Cedex 2, France

Introduction

We recently described the synthesis [1] and NMR study [2] of EETI II, a 28-residue peptide isolated from the Cucurbitaceae *Ecballium elaterium* [3]. This highly potent trypsin inhibitor ($Kd = 10^{-12}$) contains 6 Cys residues, the three Cys of the N-terminal half being connected with those of the C-terminal half in identical order. The molecule exhibits a compact structure in which the third disulfide bridge crosses a macrocycle limited by the two first ones. These topological characteristics are quite analogous to those described for CPI, a carboxypeptidase A (CPA) inhibitor found in potatoes [4]. The comparison between the 3D structures of EETI II and CPI is presented in Fig. 1.

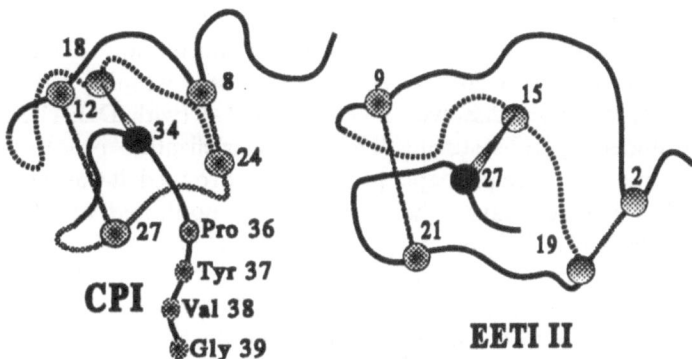

Fig. 1.   *3D-structure comparison between CPI and EETI II.*

A sequence alignment based on Fig. 1 shows that the CPI chain was elongated on both ends (Fig. 2), with the C-terminal tetrapeptide Pro-Tyr-Val-Gly being responsible for the inhibitory activity. These findings prompted us to synthesize a chimeric compound consisting of the EETI II molecule extended at its C-terminus by Pro-Tyr-Val-Gly in order to obtain a double-headed inhibitor named trypsin-carboxypeptidase A inhibitor (TCPI) acting on both trypsin and CPA.

D. Le-Nguyen and B. Castro

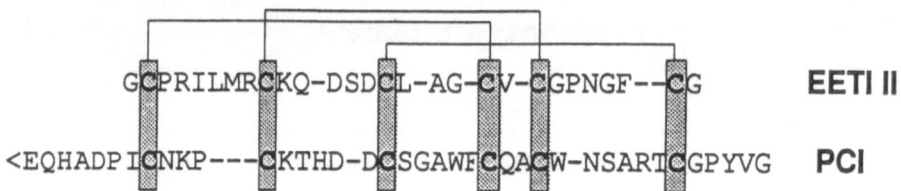

Fig. 2.   *CPI and EETI II: Sequence alignment.*

Results and Discussion

The synthesis was performed by the SPPS using 1% crosslinked chlorome-thylated (CM) resin and BOP for coupling following the procedure described previously [5]. Starting from Boc-Gly-CM resin (0.5 mmol Gly/g), the fully protected peptide was assembled manually. The reactive side chains of the amino acids were protected as follows: Tos for Arg; 2-ClZ for Lys; Mob for Cys; Xan for Asn and Gln; OcHx for Asp; Bzl for Ser; and Dcb for Tyr; Met was used unprotected. The couplings, monitored with ninhydrin tests, were all achieved within 30 min, the first five couplings requiring less than 10 min; no recoupling was needed.

After the synthesis was accomplished, the peptide resin was treated for 60 min with liquid HF at 0°C in the presence of anisole and DMS. The crude peptide was recovered in 20% AcOH in H_2O after washing with ether, and lyophilized. It was then dissolved in H_2O at pH 8 (with DIEA) and allowed to oxidize under vigorous stirring. During the cyclization process, the pH as monitored with moistened pH paper progressively dropped; it was then readjusted with DIEA when necessary. The reaction was monitored by RPHPLC and Ellman's reagent [6]. For this purpose, samples (1 drop) of the mixture were collected every hour and allowed to react with 1 drop of a solution containing dithiobis (2-nitrobenzoic acid) in 1M K_2HPO_4 buffer (pH 8). The cyclization was completed after 36 h stirring. Semi-preparative RPHPLC was utilized to purify the synthetic compound. The reaction mixture was filtered, pumped onto a Whatman M20 ODS-3 column (2.2 × 50 cm, 10 μm) and then eluted with a 0.1% TFA/CH_3CN gradient. One unique fraction was collected and lyophilized to yield the expected TCPI (yield for the cyclization-purification step: 9%). AAA were in agreement with the expected structure.

The antitrypsin activity was determined, as previously described [3], using *N*-benzoyl-L-arginine ethyl ester as the substrate. It was assayed after 5 min preincubation of the enzyme with TCPI. For anti-CPA assays, sodium *N*-hippuryl-β-phenyllactate was utilized. TCPI was shown separately to inhibit trypsin (Kd = 1.8×10^{-9}) and CPA (Kd = 3×10^{-9}). On another hand, it was shown that the enzymes could be inhibited concomitantly following a specially designed protocol. These results will be published elsewhere.

Acknowledgements

This work was supported by the Centre National de la Recherche Scientifique, France. We are grateful to H. Mattras for AAA and to Dr. M.A. Coletti-Previero for enzymatic assays.

References

1. Le-Nguyen, D., Nalis, D., Heitz, A. and Castro, B., In Jung, G. and Bayer, E. (Eds.) Peptides 1988 (Proceedings of the 20th European Peptide Symposium), De Gruyter, Berlin, 1989, p. 384.
2. Heitz, A., Chiche, L., Le-Nguyen, D. and Castro, B., Biochemistry, 28 (1989) 2392.
3. Favel, A., Mattras, H., Coletti-Previero, M.A., Zwilling, R., Robinson E.A. and Castro, B., Int. J. Pept. Prot. Res., 33 (1989) 202.
4. Rees, D.C. and Lipscom, W.N., J. Mol. Biol. 160 (1982) 475.
5. Le-Nguyen, D., Heitz, A. and Castro, B., J. Chem. Soc., Perkin Trans. I, (1987) 1915.
6. Ellman, G.L., Arch. Biochem. Biophys., 82 (1959) 70.

Light-harvesting peptides

Brian M. Peek, Sondra E. Vitols, Thomas J. Meyer and Bruce W. Erickson
Department of Chemistry, University of North Carolina, Chapel Hill, NC 27599-3290,
U.S.A.

Introduction

We are engineering a new class of peptides that convert light energy into stored chemical or electrical energy. These light-harvesting peptides will contain three new classes of amino acids: metal-ligand chromophores, electron donors, and electron acceptors. A metal-ligand complex, such as $Ru^{II}(bpy)_3^{2+}$, serves as the chromophore. Light absorbed by a metal-ligand complex promotes an electron from a d orbital of the metal Ru^{II} to a π^* orbital of the 2,2'-bipyridine (bpy) ligand field. The resulting metal-ligand charge-transfer complex $\{Ru^{III}[(bpy)_3\text{-}]\}^{2+}$ is both a powerful reductant and a powerful oxidant. The oxidized metal Ru^{III} can accept an electron from an electron donor, such as phenotiazine (PTZ), and the reduced ligand field $(bpy)_3\text{-}$ can donate an electron to an electron acceptor, such as paraquat (PQ^{2+}), which regenerates the initial chromophore. The net reaction is the light-induced transfer of an electron from PTZ to PQ^{2+} to create a redox-separated state ($PTZ^{\bullet+}$, $PQ^{\bullet+}$) having stored chemical energy.

Meyer et al. [1] covalently attached PTZ, PQ^{2+}, and $Ru^{II}(bpy)_3^{2+}$ to polystyrene and transiently observed the absorbance spectra of the free radicals $PTZ^{\bullet+}$ and $PQ^{\bullet+}$ produced by light harvesting. But electron transfer was not observed for a ternary mixture of PTZ-polystyrene, PQ^{2+}-polystyrene, and $RU^{II}(bpy)_3^{2+}$-polystyrene unless 9-methylanthracene was present to serve as an energy-plus-electron shuttle between the separate polymer strands [2]. Nonproductive back-electron transfer was slower for both polymeric systems than for a mixture of the three monomeric species. But proximity of the redox sites, due to their random spacing on the polystyrene strands as well as the flexibility of the polymer backbone, promoted rapid quenching of the redox-separated species. Segregation of the three types of redox sites on separate polymer strands further reduced the quantum efficiency of light harvesting.

Boc-Lys(Ptz)-OH   Boc-Lys(BRu)-OH   Boc-Gln(Pqt)-OH

Results and Discussion

We are exploring the design and synthesis of peptide structures containing these three redox sites. We have synthesized amino acids containing an electron donor or an electron acceptor by modifying the side chain of lysine or glutamic acid. We made Boc-L-Lys(Ptz)-OH, where Ptz is 3-(10-phenothiazinyl)propanoyl, and used it for SPPS of electrochemically active peptides, such as Lys(Ptz)-Lys(Ptz)-Ala. We have also synthesized Boc-L-Gln(Pqt)-OBzl, where Pqt is 3-(1'-methyl-4,4'bipyridinium-1-yl)propyl, and used it to prepare Boc-Lys(Ptz)-Gln(Pqt)-OBzl in solution. In this dipeptide, the electron donor Lys(Ptz) had a half-wave potential versus the sodium-saturated calomel electrode (0.70 V) close to that of the parent PTZ (0.74 V), and the electron acceptor Gln(Pqt) (−0.39 V) was close to that of the parent PQ^{2+} (−0.40 V). A synthesis of the amino acid Boc-Lys(BRu)-OH bearing a metal-ligand chromophore is in progress.

Once these three redox amino acids are available and characterized, novel molecular devices will be assembled by SPPS. Construction of α-helical peptides will allow control of the distances between redox centers. Peptides containing these redox modules will be assembled on an electrode surface bearing amino groups. Proper spacing of the chromophore C, acceptor A, and donor D should allow conversion of light energy into electrical energy (Fig. 1). For example, light absorbed by the chromophore Lys(BRu) of the peptide-modified electrode will generate the redox-separated state $(Lys(Ptz^{•+}), Gln(Pq^{•+}))$. The high-energy electron of $Pqt^{•+}$ could be transferred sequentially to a soluble electron acceptor A', to another electrode, through an external circuit, and back to the peptide-modified electrode to be accepted by $Ptz^{•+}$. This molecular device would be a peptide-based solar cell.

Fig. 1. Proposed scheme for electron transfer.

Acknowledgements

This work was supported in part by grants from NIH (GM-42031).

References

1. Meyer, T.J., Strouse, G., Younathan, J.N. and Danielson, E., unpublished results.
2. Olmsted, J., McClanahan, S.F., Danielson, E., Younathan, J.N. and Meyer, T.J., J. Am. Chem. Soc., 109 (1987) 3297.

A novel oligopeptide delivery system for poorly absorbed peptides and drugs

Istvan Toth, Richard A. Hughes, Michael R. Munday, Paolo Mascagni and
William A. Gibbons

*School of Pharmacy, University of London, 29–39 Brunswick Square,
London WC1N 1AX, U.K.*

Introduction

The development of biologically active peptides and other poorly absorbed compounds continues to be hampered by the absence of an effective means of delivering them. In an attempt to overcome this, a delivery system has been developed based on α-amino acids with long alkyl side chains, the so-called fatty amino acids (1) (see Fig. 1). The fatty amino acids can be condensed to form biodegradable and biocompatible oligopeptides (2) that possess structural similarities to the lipid bilayer of cell membranes.

Covalent conjugates of fatty amino acids or their oligopeptides with poorly absorbed peptides and drugs can be formed by linking the desired compounds to the N- or C-terminus of the fatty peptide via an amide, a simple ester or other easily cleavable connection. The conjugates should show increased membrane solubility and translocation characteristics over the parents, thus improving oral absorption and enhancing the activity of drugs that act at membranes.

Results and Discussion

The benzoquinolizine acid **3a** has antiinflammatory activity but is poorly absorbed. A series of fatty peptide derivatives of this compound **3b** were prepared by condensing the free amine of C-protected fatty peptides with the carboxylic acid function of the alkaloid. The oral absorption of C_2-^{3}H labeled benzoquinolizine conjugates **3b** in mice was found to be at least five-fold greater than that of the parent compound **3a**.

The berbane alkaloid **4a** is a highly selective α_2-adrenoceptor antagonist, but is of limited clinical use due to poor oral absorption. A series of esters of this compound with N-protected fatty peptides have been synthesized (**4b**) and tested for α_2-receptor antagonism.

A variety of compounds with GABA agonist properties have been investigated as a means of overcoming the limited passage of GABA across the blood–brain barrier. Examples **5** and **6** illustrate the use of two prodrug-type linkages between the active compound and the fatty oligopeptide delivery system. These conjugates are being evaluated for their ability to deliver and release GABA into the CNS.

CH₃ →

$$\text{CH}_3$$
$$(\text{CH}_2)_n$$
$$\text{H}_2\text{N-CH-COOH}$$

1

$$\left[\begin{array}{c}\text{CH}_3\\(\text{CH}_2)_n\\\text{H}-\text{-NH-CH-CO-}\end{array}\right]_m\text{-OH}$$ n=7-17   m>1

2

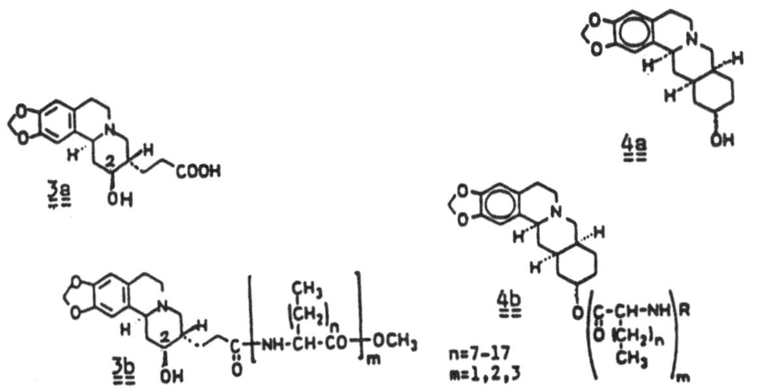

3a   COOH   OH

4a   OH

3b

$$\left[\begin{array}{c}\text{CH}_3\\(\text{CH}_2)_n\\\text{C-NH-CH-CO-}\end{array}\right]_m\text{OCH}_3$$

4b   n=7-17   m=1,2,3

$$\left(\text{O}\left[\begin{array}{c}\text{C-CH-NH}\\(\text{CH}_2)_n\\\text{CH}_3\end{array}\right]R\right)_m$$

H₂N(CH₂)₃COOCHCO-*ₘ-OCH₃
  (CH₂)ₙ
  CH₃

5

--EHWSYGLRPG-NH₂

EHWSYGLRPG-*-*

EHWSY-*-LRPG-NH₂

7

H₂N(CH₂)₃COO(CH₂)ₚO-*ₘ-H

6

*=fatty amino acid

n=7-17, m=1-4, p=1-3

Fig. 1. Illustration of structures (for details see U.K. Patent appl. No. GB 2217706A; International Patent appl. No. PCT/AU89/00166).

Analogs (7) of the peptide GnRH, containing fatty amino acids at the N- or C-terminus and within the sequence have been synthesized by SPPS. The absorption of radiolabeled analogs following oral administration is currently under investigation. The study will be extended to other peptides and vaccines of biological interest.

Author index

Abbadi, A.   809
Abbott, L.   691
Abrahamson, M.   375
Adams, S.P.   90, 269, 1034
Aebi, J.D.   49
Affholter, J.A.   46
Ahmad, M.   726
Ahmed, A.H.   439
Aiken, J.W.   205
Akaji, K.   931
Al-Obeidi, F.A.   527
Albericio, F.   895, 923, 996
Albert, J.   841
Albert, R.   437
Albrecht, E.   718
Albrightson, C.   287
Aldwin, L.   898
Ali, F.E.   94, 287
Alila, H.W.   214
Allaudeen, H.S.   815
Almdal, K.   1036
Almquist, R.G.   505
Amblard, M.   108
Ambrosius, D.   496
Amburn, E.   524
Ananthanarayanan, V.S.   527
Anantharamaiah, G.M.   672
Anderson, G.J.   649
Anderson, P.S.   72
Andreu, D.   923, 960
Andrews, J.   236
Andrews, P.C.   433
Andriamanampisoa, F.   947
Anolik, J.H.   664
Appel, J.   774
Arbegast, P.T.   43
Archer, R.   489
Arimura, A.   518
Arlinghaus, R.B.   714
Arnold, W.A.   582
Arrhenius, T.   870
Atrash, B.   78
Aubry, A.   535, 809, 974
Aurell, C.-J.   293
Aymard, M.   769

Babler, M.   396

Bahraoui, E.   847
Bairaktari, E.   578, 697
Baird, A.   524
Baizman, E.   877
Bajusz, S.   190
Baker, W.R.   393, 402
Balaji, V.N.   639
Balaram, P.   544, 630
Balasubramaniam, A.   314
Balavoine, G.   1015
Balbes, L.M.   590
Balboni, G.   321
Baldwin, R.L.   635
Bali, J.-P.   390
Balishin, N.L.   56
Ball, H.   435
Bannow, C.A.   205
Baran, J.S.   396
Barany, G.   895, 996
Barclay, P.L.   247
Bargiotta, E.   111
Barnish, I.T.   247
Barrett, R.W.   978
Barth, T.   297
Bartlett, P.A.   371
Bartley, T.D.   449
Barton, J.K.   785
Bassi, K.   993
Battersby, J.   430
Baumann, G.C.   988
Bayer, E.A.   1038
Bean, J.W.   637
Bean, M.A.   443
Becker, E.L.   630
Becker, G.   1031
Becker, J.M.   103, 111, 619
Belagaje, R.   99
Benard, M.   974
Benayad, A.   798
Benedetti, E.   118, 667
Bengtsson, B.   293
Benoit, R.   736
Benoiton, N.L.   889, 967
Berg, R.H.   1036
Bergeron, F.   106
Bernardi, A.   742
Bernath, E.   852, 1054

Subject index